LINEAR CONTROL SYSTEMS

with MATLAB Applications

B.S. Manke
Formerly, Prof. of Electrical Engineering,
Maulana Azad National Institute of Technology
Bhopal (M.P.)

KHANNA PUBLISHERS®
Investing in Learning®
Operational Office :
4575/15, Onkar House, Opp. Happy School,
Ground Floor, Daryaganj, New Delhi 110 002
Phones : 011-45033819 • *Mob.* 09811541460
email : contactus@khannapublishers.in

Published by :
Romesh Chander Khanna & Vineet Khanna
for **KHANNA PUBLISHERS**
2-B, Nath Market, Nai Sarak
Delhi-110 006 (India)

Website : www.khannapublishers.in

Hologram & Description

To all readers of our books, to prevent yourself from being defrauded by piraters, please make sure that there is an Hologram on the cover of our books with the below specifications. If you find any book without Hologram and Description, please mail us at **contactus@khannapublishers.in**

Thanking you

ISBN No. 978-81-7409-310-3

First Edition : 1986
Twelfth Edition : 2016
Sixth Reprint : 2024

***Price :* ₹ 639.00**

To my grandsons
Aadhit, Abhyudit
and granddaughters
Ira, Nitya

Preface to the First Edition

This textbook has been written to explain the basic principles of Linear Control Systems and an effort is made to present the subject in a simple and sequential manner to enable the students to acquire a good grasp of fundamentals of the subject.

The text presented covers the course content of the subject Linear Control Systems of Indian Universities and is meant for pre-final/final year students of electrical, electronics and mechanical engineering.

The material given in this book has been thoroughly class tested by the author while teaching the subject of control systems at undergraduate level for the past several years.

This book is divided in 9 chapters. The first four chapters give the basic concepts of the subject from the view point of control system representation. Chapter 5 presents the modelling of control systems and the respective mathematical models derived therein. The time response and steady state analysis is given in Chapter 6. Necessary derivations have been derived from the first principles. The stability analysis is described in Chapter 7. The methods of ascertaining stability: Routh-Hurwitz criterion, Nyquist criterion, Bode plot and root locus plot have been explained step by step in a simplified manner to make the explanation easily understandable. The compensation methods and introduction to state space analysis is described in chapters 8 and 9 respectively.

Suitable illustrative examples as well as solved examples have been incorporated in the text to make the subject clear and interesting. A list of references is given at the end.

Selective unsolved problems have been included at the end of each chapter to help the student to judge himself whether he has gained sufficient workable knowledge of basic principles involved. Answers to odd numbered problems being given in Appendix I.

The salient feature of this book is the inclusion of objective type multiple choice questions given in Appendix II covering the entire text which would be of great help for the students preparing for competitive examinations.

The author hopes that this book will serve the purpose of introducing basic principles of Linear Control Systems to undergraduate students for whom it is written.

The author would welcome any comments and suggestions to further improve the usefulness of this book.

The author acknowledges his indebtedness to Miss Saroj Rangnekar, Asst. Prof. in Elect. Engg., Maulana Azad College of Technology, Bhopal who thoroughly checked the manuscript and made useful suggestions.

Bhopal
October, 1987

B.S. Manke

Preface to the 12th Edition

The text written in the book deals with the concepts of feed-back control theory. The first five chapters stress on the fundamental concepts regarding representation and modelling of a control system. The subsequent chapters deal with the time response analysis, stability analysis, compensation method, state variable approach, and sampled data/discrete data systems.

Each chapter contains solved examples to support the theory developed. Unsolved problems have been included as an exercise.

The addition of analytical approach to derive the equation in *s*-plane for the root locus plot, can be used to calibrate the complex part of the root locus.

The root locus approach towards the design of compensators have been included in the text.

The answers to graphical solutions may slightly deviate due to graphical errors.

The chapter on computer solutions to control problems gives the use of MATLAB* software. The examples on various topics in the text have been solved using MATLAB software. This verifies the answers obtained using analytical solution.

Appendices given at the end of the book include :

Appendix I : Answers to Selected Problems

Appendix II : A Set of Objective Questions

Appendix III : Short Answer Type Questions

Appendix IV : List of Key Formulae, Charts and Calculation Tables

Any constructive suggestion for further improvement of the book will be appreciated.

The author acknowledges help given by Dr. D.M. Deshpande, ex. Prof. M.A.N.I.T. Bhopal for discussions with him towards the revision of this edition.

The author is thankful to Shri Kratu Khanna and staff of the Khanna Publishers, New Delhi for their role in bringing out this revised edition of this book.

B.S. Manke

B-92, Ishaan 1,
Off. Prahlad Nagar Road
Satellite
Ahmedabad (Gujarat)

*MATLAB is registered trade mark of Mathwork, Inc.

Contents

1.	**INTRODUCTION**		**1—15**
1.1	An Example of Control Action	...	2
1.2	Open-Loop Control System	...	2
1.3	Closed-Loop Control System	...	2
1.4	Use of Laplace Transformation in Control Systems	...	3
1.5	Laplace Transform	...	3
	1.5.1 Derivation of Laplace Transform	...	4
	1.5.2 Basic Laplace Transform Theorems	...	6
1.6	Solved Examples	...	7
	Problems	...	15
2.	**TRANSFER FUNCTION**		**16—24**
2.1	Poles and Zeros of a Transfer Function	...	17
2.2	Transfer Function and its Relationship with Impulse Response	...	19
2.3	Procedure for Determining the Transfer Function of a Control System	...	19
2.4	Solved Examples	...	20
	Problems	...	24
3.	**BLOCK DIAGRAMS**		**25—61**
3.1	Block Diagram Reduction	...	26
3.2	Solved Examples	...	32
	Problems	...	58
4.	**SIGNAL FLOW GRAPHS**		**62—78**
4.1	Rules for Drawing Signal Flow Graphs	...	63
4.2	Mason's Gain Formula	...	67
4.3	Drawing Signal Flow Graph from a Given Block Diagram	...	67
4.4	Solved Examples	...	69
	Problem	...	78
5.	**MODELLING A CONTROL SYSTEM**		**79—126**
5.1	Electrical Networks	...	79
5.2	Mechanical Systems	...	81
	5.2.1 Translation Mechanical Systems	...	81
	5.2.2 Rotational Mechanical System	...	84

Chapter				Pages
5.3	Hydraulic System		...	86
5.4	Pneumatic System		...	88
5.5	Thermal System		...	89
5.6	Servo Motors		...	90
5.7	Generators		...	97
5.8	Error Detectors		...	102
5.9	Solved Examples		...	106
	Problems		...	124
6.	**TIME RESPONSE ANALYSIS OF CONTROL SYSTEMS**			**127–215**
6.1	Transient and Steady State Response		...	128
6.2	Input Test Signals		...	128
6.3	Time Response of a First Order Control System		...	130
	6.3.1	Time Response of a First Order Control System Subjected to Unit Step Input Function	...	130
	6.3.2	Demarcation between the Transient Part and Steady State Part of the Time Response in terms of Time Constant	...	131
	6.3.3	Time Response of a First Order Control System Subjected to Unit Ramp Input Function	...	132
	6.3.4	Time Response of a First Order Control System Subjected to Unit Impulse Input Function	...	133
6.4	Time Response of a Second Order Control System		...	134
	6.4.1	Time Response of a Second Order Control System Subjected to Unit Step Input Function	...	134
	6.4.2	Critical Damping	...	139
	6.4.3	Characteristic Equation	...	140
	6.4.4	Transient Response Specifications of Second Order Control System	...	141
	6.4.5	Time Response of a Second Order Control System Subjected to Unit Ramp Input Function	...	144
	6.4.6	Time Response of a Second Order Control System Subjected to Impulse Input Function	...	146
6.5	Time Response of a Third Order Control System		...	147
	6.5.1	Effect of First Order Term Time Constant on Time Response of Third Order Control System	...	151
6.6	Time Response of Higher Order Control System		...	152
	6.6.1	An Example of Third Order Unstable Control System	...	152
6.7	Steady State Error		...	153
	6.7.1	Static Error Coefficients	...	155
	6.7.2	Type of Transfer Functions and Steady State Error	...	156
	6.7.3	Generalized Error Coefficients	...	159
	6.7.4	Performance Indices	...	162

Chapter		Pages
6.8 Sensitivity	...	168
6.8.1 Effect of Transfer Function Parameter variations in an Open-loop Control System	...	168
6.8.2 Effect of Forward Path Transfer Function Parameter variations in a Closed-loop Control System	...	168
6.8.3 Sensitivity of Overall Transfer Function *M*(*s*) with respect to Forward Path Transfer Function *G*(*s*)	...	169
6.8.4 Sensitivity of Overall Transfer Function *M*(*s*) with respect to Feedback Path Transfer Function	...	170
6.8.5 Effect of Feedback on Time Constant of a Control System	...	172
6.9 Control Actions	...	173
6.9.1 Proportional Control	...	173
6.9.2 Derivative Control	...	173
6.9.3 Integral Control	...	178
6.9.4 Proportional Plus Derivative Plus Integral Control (PID Control)	...	181
6.9.5 Derivative Feedback Control	...	181
6.10 Solved Examples	...	185
Problems	...	212
7. STABILITY ANALYSIS OF CONTROL SYSTEMS		**216—362**
7.1 Stability in Terms of Characteristic Equation of a Control System	...	218
7.1.1 Definition of Stability	...	218
7.1.2 Absolute and Relative Stability	...	218
7.1.3 Location of the Roots of Characteristic Equation in *s*-plane as related to Time Response and Prediction of Absolute Stability therefrom	...	218
7.2 To Determine the Number of Roots having Positive Real Parts for a Polynomial	...	219
7.3 Hurwitz Determinants of a Polynomial	...	220
7.4 Routh-Hurwitz Criterion	...	221
7.5 Solved Examples : Routh-Hurwitz Criterion	...	222
7.6 Nyquist Criterion	...	229
7.6.1 Procedure for Mapping from *s*-plane to *G*(*s*) *H*(*s*)-plane	...	229
7.6.2 Determination from the Nyquist Plot the Number of Zeros of *G*(*s*) *H*(*s*) which are located inside a Specified Region in *s*-plane	...	231
7.6.3 Application of Nyquist Criterion to Determine Stability of a Closed-loop Control System	...	231
7.7 Gain Margin and Phase Margin	...	242
7.8 Relative Stability from Nyquist Plot	...	244
7.9 Gain Phase Plot	...	245
7.10 Closed-loop Frequency Response of a Unity Feedback Control System from Nyquist Plot	...	246

Chapter			Pages
7.11	Constant Magnitude Loci : *M*-Circles	...	247
7.12	Constant Phase Angles Loci : *N*-Circles	...	249
7.13	Closed-Loop Frequency Response of Control System from *M* and *N*-Circles	...	250
7.14	Gain Adjustment using *M*-Circle	...	251
7.15	Nichols Chart	...	254
7.16	Cutoff Frequency and Bandwidth	...	257
7.17	Solved Examples : Nyquist Criterion	...	259
	7.17.1 Inverse Nyquist Plot	...	284
	7.17.2 1/*M*-Circles (Inverse *M*-Circles) and Constant Phase Angle Loci	...	287
	7.17.3 Gain Adjustment Using Inverse Polar Plot	...	288
7.18	Bode Plot	...	289
	7.18.1 Bode Plot (Logarithmic Plot) for Transfer Functions	...	289
	7.18.2 Graphs for the Gain Term *K*	...	290
	7.18.3 Graphs for the Term $\frac{1}{(j\omega)^N}$	...	290
	7.18.4 Graphs for the Term $(1 + j\omega T)$	...	291
	7.18.5 Graphs for the Term $\frac{1}{(1 + j\omega T)}$	...	292
	7.18.6 Initial Slope of Bode Plot	...	295
	7.18.7 Determination of Static Error Coefficients from Initial Slope of Bode Plot	...	297
	7.18.8 Graphs for Quadratic Term $\frac{\omega_n^2}{[(\omega_n^2\ \omega^2) + j2\zeta\ \omega_n\ \omega]}$	...	298
	7.18.9 Procedure for Drawing Bode Plot and Determination of Gain Margin, Phase Margin and Stability	...	301
	7.18.10 Bode Plot for Time Delay Element : e^{-sT}	...	304
	7.18.11 Minimum Phase, Non-minimum Phase and all Pass Transfer Function	...	306
7.19	Solved Examples : Bode Plot	...	308
7.20	Correlation between Transient Response and Frequency Response	...	315
7.21	Root Locus	...	319
7.22	Salient Features of Root Locus Plot	...	323
7.23	The Procedure for Plotting Root Locus	...	323
7.24	Solved Examples : Root Locus	...	324
7.25	Root Contours	...	352
	Problems	...	358
8.	**COMPENSATION OF CONTROL SYSTEMS**		**363—416**
8.1	Phase-Lead Compensation	...	364
8.2	Phase-Lag Compensation	...	366
8.3	Phase-Lag-Lead Compensation	...	368

Chapter			Pages
8.4	Concluding Remarks	...	369
8.5	Feedback Compensation	...	369
8.6	Solved Examples	...	372
8.7	Compensation Network Design: Root Locus Approach	...	380
	8.7.1 Phase Lead Compensator Design	...	381
	8.7.2 Phase Lag Compensator Design	...	394
	8.7.3 Lag-Lead Compensator Design	...	404
	Problems	...	414

9. INTRODUCTION TO STATE SPACE ANALYSIS OF CONTROL SYSTEMS 417—498

9.1	State Space Representation	...	418
9.2	The Concept of State	...	419
9.3	State Space Representation of Systems	...	420
9.4	Block Diagram for State Equation	...	427
9.5	Transfer Function Decomposition	...	428
	9.5.1 Direct Decomposition	...	429
	9.5.2 Cascade Decomposition	...	431
	9.5.3 Parallel Decomposition	...	432
	9.5.4 Parallel Decomposition of a Transfer Function having Repeated Roots of its Characteristic Equation	...	433
	9.5.5 Transfer Function Decomposition Using State Signal Flow Graph	...	434
9.6	Solution of State Equation	...	438
	9.6.1 Determination of State Transition Matrix	...	440
	9.6.2 Properties of State Transition Matrix	...	450
9.7	Transfer Matrix	...	450
9.8	Controllability	...	451
9.9	Observability	...	454
9.10	Solved Examples	...	456
9.11	State Variable Feedback	...	490
	Problems	...	494

10. SAMPLED DATA CONTROL SYSTEMS 499—545

10.1	Sampler	...	501
10.2	Sampling Process	...	502
10.3	Laplace Transform of Sampled Function	...	503
10.4	z-Transform	...	504
10.5	z-transform of Some Useful Functions	...	505
10.6	Inverse z-transform	...	508
10.7	Hold Circuit	...	515
10.8	Reconstruction of Signal : Minimum Sampling Frequency	...	518
10.9	Pulse Transfer Function (z-transfer Function)	...	519

Chapter			Pages
10.10	Stability Analysis of Sampled-data Control Systems	...	525
10.11	Solved Examples	...	532
	Problems	...	544

11. SOLUTION OF PROBLEMS USING COMPUTER 546–600

11.1	Introduction to MATLAB	...	546
11.2	Running MATLAB	...	549
	Solved Examples	...	549

12. CLASSIFIED SOLVED EXAMPLES 601–689

Contains highly graded solved problems selected from question papers of various Technical Universities, Competitive Examinations and self developed variety of questions for clearly and deeper understanding of matter/topics covered in text.

Appendix-I	:	**Answers to Selected Problems of Text**	**690–694**
Appendix-II	:	**Objective Type of Questions with Answers**	**695–736**
Appendix-III	:	**Short Answer Questions**	**737–759**
Appendix-IV	:	**Key Formulae, Tables and Charts**	**760–768**

References **769**

Index **770–772**

CHAPTER 1

Introduction

This chapter deals with basic ideas about the open-loop and closed-loop control systems. The differential equations describe the dynamic operation of control systems. The Laplace transform transforms the differential equation into an algebraic equation, the solution is obtained in the transform domain. The time domain solution is determined by taking the inverse Laplace transform.

CONTENTS

- *An Example of Control Action*
- *Open-Loop Control System*
- *Closed-Loop Control System*
- *Use of Laplace Transformation in Control Systems*
- *Laplace Transform*
- *Solved Examples*

CONTROL SYSTEM

A control system is a combination of elements arranged in a planned manner wherein each element causes an effect to produce a desired output. This cause and effect relationship is governed by a mathematical relation.

If the aforesaid mathematical relation is linear the control system is termed as linear control system. For a linear system the cause (independent variable or input) and the effect (dependent variable or output) are proportionally related and principle of superposition is applicable throughout the operating range of a system.

In a control system the cause acts through a control process which in turn results into an effect.

There may be variety of systems based on the principle mentioned above but all the systems have many features in common and as such common approach for the study and analysis of control systems is possible.

Control systems are used in many applications for example, systems for the control of position, velocity, acceleration, temperature, pressure, voltage and current etc.

1.1 AN EXAMPLE OF CONTROL ACTION

Control of a room temperature is achieved by switching ON and switching OFF of a power supply to a heating appliance. Thus power supply to an appliance is switched ON, when the room temperature is felt low and switched OFF, when the desired temperature is reached.

The above system can be modified, if the duration of application of power is predetermined to achieve the room temperature within desired limits.

However, a further refinement can be made by measuring the difference between the actual room temperature and the desired room temperature and this difference being the error is used to control the element which in turn controls the output, *i.e.* room temperature.

The above description indicates that in the former case the output (room temperature) has no control on the input and the control action is purely based on a sort of predetermined calibration only, where as in the latter case the control action is affected by a feedback received from the output to the input.

1.2 OPEN-LOOP CONTROL SYSTEM

Having explained the concept of control action, a control system can be described by a block diagram as shown in Fig. 1.2.1.

Fig. 1.2.1. Open-loop control system.

The input r controls the output c through a control action process. In the block diagram shown in Fig. 1.2.1, it is observed that the output has no effect on the control action. Such a system is termed as open-loop control system.

In an open-loop control system the output is neither measured nor fedback for comparison with the input. Faithfulness of an open-loop control system depends on the accuracy of input calibration.

1.3 CLOSED-LOOP CONTROL SYSTEM

In a closed-loop control system the output has an effect on control action through a feedback as shown in Fig. 1.3.1 and hence closed-loop control systems are also termed as feedback control systems. The control action is actuated by an error signal e which is the difference between the input signal r and the output signal c. This process of comparison between the output and input maintains the output at a desired level through control action process.

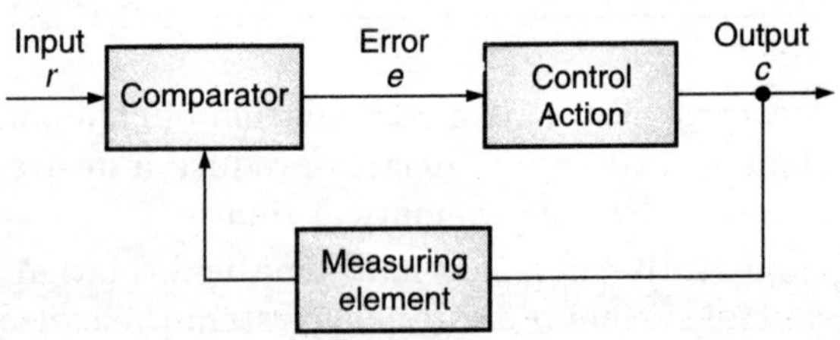

Fig. 1.3.1. Closed-loop control system.

The control systems without involving human intervention for normal operation are called automatic control systems.

A closed-loop (feedback) control system using a power amplifying device prior to controller and the output of such a system being mechanical *i.e.* position, velocity, acceleration is called servomechanism.

Comparison of Open-Loop and Closed-Loop Control System

Open-Loop	*Closed-Loop*
1. The accuracy of an open-loop system depends on the calibration of the input. Any departure from pre-determined calibration affects the output.	1. As the error between the reference input and the output is continuously measured through feedback, the closed-loop system works more accurately.
2. The open-loop system is simple to construct and cheap.	2. The closed-loop system is complicated to construct and costly.
3. The open-loop systems are generally stable.	3. The closed-loop systems can become unstable under certain conditions.
4. The operation of open-loop system is affected due to the presence of non-linearities in its elements.	4. In terms of the performance the closed-loop system adjusts to the effects of non-linearities present in its elements.

1.4 USE OF LAPLACE TRANSFORMATION IN CONTROL SYSTEMS

The control action for a dynamic control system whether electrical, mechanical, thermal, hydraulic etc. can be represented by a differential equation and the output response of such a dynamic system to a specified input can be obtained by solving the said differential equation. The system differential equation is derived according to physical laws governing a system in question.

In order to facilitate the solution of a differential equation describing a control system, the equation is transformed into an algebraic form. The differential equation wherein time being the independent variable is transformed into a corresponding algebraic equation by using Laplace transformation technique and the differential equation thus transformed is known as the equation in frequency domain. Hence, Laplace transform technique transforms a time domain differential equation into a frequency domain algebraic equation.

1.5 LAPLACE TRANSFORM

In order to transform a given function of time $f(t)$ into its corresponding Laplace transform first multiply $f(t)$ by e^{-st}, s being a complex number ($s = \sigma + j\omega$). Integrate this product w.r.t. time with limits as zero and infinity. This integration results in Laplace transform of $f(t)$, which is denoted by $F(s)$ or $\mathcal{L}f(t)$.

The mathematical expression for Laplace transform is,

$$\mathcal{L}f(t) = F(s) \qquad t \geq 0$$

or

$$F(s) = \int_0^{\infty} f(t) \cdot e^{-st}\, dt \qquad \text{...(1.1)}$$

The term "Laplace transform of $f(t)$" is used for the letter $\mathcal{L}f(t)$.

The time function $f(t)$ is obtained back from the Laplace transform by a process called inverse Laplace transformation and denoted as $\mathcal{L}^{-1}$ thus

$$\mathcal{L}^{-1}[\mathcal{L}f(t)] = \mathcal{L}^{-1}[F(s)] = f(t)$$

The time function $f(t)$ and its Laplace transform $F(s)$ are a transform pair.

Table 1.5 gives transform pairs of some commonly used functions and Laplace transform pairs for some functions are derived here under.

1.5.1 Derivation of Laplace transform

1. Laplace transform of e^{at}

$$\mathcal{L}e^{at} = \int_0^{\infty} e^{at} \cdot e^{-st}\, dt = \int_0^{\infty} e^{(a-s)t}\, dt = \frac{1}{(s-a)}$$

$$\therefore \quad \mathcal{L}e^{at} = \frac{1}{(s-a)} \qquad ...(1.2)$$

As the inverse Laplace transform is denoted by the letter $\mathcal{L}^{-1}$ and, therefore, the inverse Laplace transform of $\frac{1}{(s-a)}$ is e^{at} and expressed as below,

$$\mathcal{L}^{-1}\left[\frac{1}{(s-a)}\right] = e^{at} \qquad ...(1.3)$$

2. In the function $f(t) = e^{at}$ put $a = 0$

$$\therefore \quad e^{at} = e^{0t} = 1. \quad \text{Hence, } f(t) = 1$$

Therefore, using Eq. (1.2) $\mathcal{L}[1] = \frac{1}{(s-0)}$

or $$\mathcal{L}[1] = \frac{1}{s} \qquad ...(1.4)$$

and $$\mathcal{L}^{-1}\left[\frac{1}{s}\right] = 1 \qquad ...(1.5)$$

3. In the function $f(t) = e^{at}$ put $a = j\omega$

$$\therefore \quad e^{at} = e^{j\omega t}. \quad \text{Hence, } f(t) = e^{j\omega t}$$

Therefore, using Eq. (1.2) $\mathcal{L}e^{j\omega t} = \frac{1}{(s-j\omega)}$

$$\because \quad e^{j\omega t} = (\cos \omega t + j \sin \omega t)$$

$$\therefore \quad \mathcal{L}(\cos \omega t + j \sin \omega t) = \frac{1}{(s-j\omega)} = \frac{s+j\omega}{(s^2+\omega^2)}$$

Separating into real and imaginary parts,

$$\mathcal{L}\cos \omega t = \frac{s}{(s^2+\omega^2)} \qquad ...(1.6)$$

$$\mathcal{L}\sin \omega t = \frac{\omega}{(s^2+\omega^2)} \qquad ...(1.7)$$

and $$\mathcal{L}^{-1}\left[\frac{s}{(s^2+\omega^2)}\right] = \cos \omega t \qquad ...(1.8)$$

$$\mathcal{L}^{-1}\left[\frac{\omega}{(s^2+\omega^2)}\right] = \sin \omega t \qquad ...(1.9)$$

4. In the function $f(t) = e^{at}$ put $a = (-\alpha + j\omega)$

$$\therefore \quad e^{at} = e^{(-\alpha + j\omega)t}$$

Hence, $$f(t) = e^{(-\alpha + j\omega)t}$$

Therefore, using Eq. (1.2)

$$\mathcal{L}e^{(-\alpha+j\omega)t} = \frac{1}{s-(-\alpha+j\omega)} = \frac{1}{(s+\alpha)-j\omega}$$

$\because$ $$e^{(-\alpha+j\omega)t} = e^{-\alpha t}(\cos\omega t + j\sin\omega t)$$

$\therefore$ $$\mathcal{L}e^{-\alpha t}(\cos\omega t + j\sin\omega t) = \frac{1}{(s+\alpha)-j\omega} = \frac{(s+\alpha)+j\omega}{(s+\alpha)^2+\omega^2}$$

Separating into real and imaginary parts,

$$\mathcal{L}e^{-\alpha t}\cdot\cos\omega t = \frac{(s+\alpha)}{(s+\alpha)^2+\omega^2} \quad ...(1.10)$$

$$\mathcal{L}e^{-\alpha t}\cdot\sin\omega t = \frac{\omega}{(s+\alpha)^2+\omega^2} \quad ...(1.11)$$

and $$\mathcal{L}^{-1}\left[\frac{(s+\alpha)}{(s+\alpha)^2+\omega^2}\right] = e^{-\alpha t}\cdot\cos\omega t \quad ...(1.12)$$

$$\mathcal{L}^{-1}\left[\frac{\omega}{(s+\alpha)^2+\omega^2}\right] = e^{-\alpha t}\cdot\sin\omega t \quad ...(1.13)$$

5. In the function $f(t) = e^{at}$ put $a = 1$

$\therefore$ $$e^{at} = e^{1\cdot t} = e^t. \text{ Hence, } f(t) = e^t$$

Therefore, using Eq. (1.2) $\mathcal{L}e^t = \dfrac{1}{(s-1)}$

$\because$ $$e^t = 1 + t + \frac{t^2}{\angle 2} + \frac{t^3}{\angle 3} + \ldots$$

and $$\frac{1}{(s-1)} = \frac{1}{s} + \frac{1}{s^2} + \frac{1}{s^3} + \frac{1}{s^4} + \ldots$$

Table 1.5. Table of Laplace transform pairs

S.No.	$f(t)$	$F(s) = \mathcal{L}[f(t)]$
1	$\delta(t)$ unit impulse at $t = 0$	1
2	$u(t)$ unit step at $t = 0$	$\frac{1}{s}$
3	$u(t-T)$ unit step at $t = T$	$\frac{1}{s}e^{-sT}$
4	t	$\frac{1}{s^2}$
5	$\frac{t^2}{2}$	$\frac{1}{s^3}$
6	t^n	$\frac{\angle n}{s^{n+1}}$
7	e^{-at}	$\frac{1}{s+a}$
8	e^{at}	$\frac{1}{s-a}$

9	te^{-at}	$\frac{1}{(s+a)^2}$
10	te^{at}	$\frac{1}{(s-a)^2}$
11	$t^n e^{-at}$	$\frac{\angle n}{(s+a)^{n+1}}$
12	$\sin \omega t$	$\frac{\omega}{s^2+\omega^2}$
13	$\cos \omega t$	$\frac{s}{s^2+\omega^2}$
14	$e^{-\alpha t} \sin \omega t$	$\frac{\omega}{(s+\alpha)^2+\omega^2}$
15	$e^{-\alpha t} \cos \omega t$	$\frac{(s+\alpha)}{(s+\alpha)^2+\omega^2}$
16	$\sinh \alpha t$	$\frac{\alpha}{s^2-\alpha^2}$
17	$\cosh \alpha t$	$\frac{s}{s^2-\alpha^2}$

∴ Comparing the terms

$$\mathcal{L}[1] = \frac{1}{s}, \quad \mathcal{L}[t] = \frac{1}{s^2}$$

$$\mathcal{L}\left[\frac{t^2}{\angle 2}\right] = \frac{1}{s^2}$$

$$\mathcal{L}\left[\frac{t^n}{\angle n}\right] = \frac{1}{s^{n+1}} \quad \text{or} \quad \mathcal{L}[t^n] = \frac{\angle n}{s^{n+1}} \qquad ...(1.14)$$

and
$$\mathcal{L}^{-1}\left[\frac{\angle n}{s^{n+1}}\right] = t^n \qquad ...(1.15)$$

1.5.2 Basic Laplace Transform Theorems

Basic theorems of Laplace transform are given below :

1. Laplace transform of linear combination

$$\mathcal{L}[a\, f_1(t) + bf_2(t)] = aF_1(s) + bF_2(s) \qquad ...(1.16)$$

where $f_1(t)$, $f_2(t)$ are functions of time and a, b are constants.

2. If the Laplace transform of $f(t)$ is $F(s)$, then

(i) $\mathcal{L}\,\frac{df(t)}{dt} = [s\,F(s) - f(0+)]$

(ii) $\mathcal{L}\,\frac{d^2 f(t)}{dt^2} = [s^2 F(s) - s\,f(0+) - f'(0+)]$

(iii) $\mathcal{L}\,\frac{d^3 f(t)}{dt^3} = [s^3 F(s) - s^2 f(0+) - sf'(0+) - f''(0+)]$ ...(1.17) where $f(0+), f'(0+), f''(0+) \ldots$ are the values of $f(t), \dfrac{df(t)}{dt} \dfrac{d^2 f(t)}{dt^2} \ldots$ at $t = (0+)$

3. If the Laplace transform of $f(t)$ is $F(s)$, then

$$(i)\ \mathcal{L} \int f(t) = \left[\frac{F(s)}{s} + \frac{f^{-1}(0+)}{s}\right]$$

$$(ii)\ \mathcal{L} \iint f(t) = \left[\frac{F(s)}{s^2} + \frac{f^{-1}(0+)}{s^2} + \frac{f^{-2}(0+)}{s}\right]$$

$$(iii)\ \mathcal{L} \iiint f(t) = \left[\frac{F(s)}{s^3} + \frac{f^{-1}(0+)}{s^3} + \frac{f^{-2}(0+)}{s^2} + \frac{f^{-3}(0+)}{s}\right] \quad ..(1.18)$$

where $f^{-1}(0+), f^{-2}(0+), f^{-3}(0+) \ldots$ are the values of $\int f(t), \iint f(t), \iiint f(t) \ldots$ at $t = (0+)$.

4. If the Laplace transform of $f(t)$ is $F(s)$, then

$$\mathcal{L} e^{-at} f(t) = F(s+a)$$

5. If the Laplace transform of $f(t)$ is $F(s)$, then

$$\mathcal{L} t\, f(t) = -\frac{d}{ds} F(s)$$

6. Initial value theorem

$$\lim_{t \to 0} f(t) = \lim_{s \to \infty} s\, \mathcal{L} f(t) \quad ...(1.19\ a)$$

or

$$\lim_{t \to 0} f(t) = \lim_{s \to \infty} s\, F(s) \quad ...(1.19)$$

7. Final value theorem

$$\lim_{t \to \infty} f(t) = \lim_{s \to 0} s\, \mathcal{L} f(t) \quad ...(1.20\ a)$$

or

$$\lim_{t \to \infty} f(t) = \lim_{s \to 0} s\, F(s) \quad ...(1.20)$$

The final value theorem gives the final value ($t \to \infty$) of a time function using its Laplace transform and as such very useful in the analysis of control systems. However, if the denominator of $s\,F(s)$ has any root having real part as zero or positive, then the final value theorem is not valid.

1.6 SOLVED EXAMPLES

Example 1.6.1. *Find the inverse Laplace transform of the following functions :*

(i) $F(s) = \dfrac{1}{s(s+1)}$ (ii) $F(s) = \dfrac{s+6}{s(s^2+4s+3)}$

(iii) $F(s) = \dfrac{1}{s^2+4s+8}$ (iv) $F(s) = \dfrac{s+2}{s^2+4s+6}$

(v) $F(s) = \dfrac{5}{s(s^2+4s+5)}$ (vi) $F(s) = \dfrac{s^2+2s+3}{s^3+6s^2+12s+8}$.

Solution. (*i*) $F(s) = \dfrac{1}{s(s+1)}$

Using partial fraction expansion

$$F(s) = \frac{k_1}{s} + \frac{k_2}{s+1}$$

The coefficients can be determined as $k_1 = 1$ and $k_2 = -1$

$$\therefore \qquad F(s) = \frac{1}{s} - \frac{1}{s+1}$$

Taking inverse Laplace transform on both sides

$$\mathcal{L}^{-1} F(s) = \mathcal{L}^{-1}\left[\frac{1}{s} - \frac{1}{s+1}\right]$$

or

$$\mathcal{L}^{-1} F(s) = \mathcal{L}^{-1}\frac{1}{s} - \mathcal{L}^{-1}\frac{1}{s+1}$$

$$\therefore \qquad f(t) = (1 - e^{-t}) \quad \textbf{Ans.}$$

(*ii*)

$$F(s) = \frac{s+6}{s(s^2+4s+3)}$$

The term $(s^2 + 4s + 3)$ can be factorised as $(s + 1)(s + 3)$

$$\therefore \qquad F(s) = \frac{s+6}{s(s+1)(s+3)}$$

Using partial fraction expansion

$$F(s) = \frac{k_1}{s} + \frac{k_2}{s+1} + \frac{k_3}{s+3}$$

The coefficients can be determined as $k_1 = 2$, $k_2 = -2.5$ and $k_3 = \frac{1}{2}$.

$$\therefore \qquad F(s) = 2\cdot\frac{1}{s} - 2.5\cdot\frac{1}{s+1} + \frac{1}{2}\cdot\frac{1}{s+3}$$

Taking inverse Laplace transform on both sides

$$\mathcal{L}^{-1} F(s) = \mathcal{L}^{-1}\left[2\cdot\frac{1}{s} - 2.5\cdot\frac{1}{s+1} + \frac{1}{2}\cdot\frac{1}{s+3}\right]$$

or

$$\mathcal{L}^{-1} F(s) = \mathcal{L}^{-1}2\cdot\frac{1}{s} - \mathcal{L}^{-1}2.5\cdot\frac{1}{s+1} + \mathcal{L}^{-1}\frac{1}{2}\cdot\frac{1}{s+3}$$

$$\therefore \qquad f(t) = (2 - 2.5\, e^{-t} + \tfrac{1}{2}e^{-3t}) \quad \textbf{Ans.}$$

(*iii*)

$$F(s) = \frac{1}{s^2+4s+8}.$$

on completing the square, the term $(s^2 + 4s + 8)$ can be expressed as $[(s + 2)^2 + (2)^2]$

$$\therefore \qquad F(s) = \frac{1}{[(s+2)^2 + (2)^2]}$$

Taking inverse Laplace transform on both sides

$$\mathcal{L}^{-1} F(s) = \mathcal{L}^{-1}\frac{1}{[(s+2)^2 + (2)^2]}$$

or

$$\mathcal{L}^{-1} F(s) = \mathcal{L}^{-1}\frac{1}{2}\cdot\frac{2}{(s+2)^2 + (2)^2}$$

$$\therefore \quad f(t) = \frac{1}{2} e^{-2t} \sin(2t) \quad \textbf{Ans.}$$

(*iv*)
$$F(s) = \frac{s+2}{s^2+4s+6}$$

on completing the square the term $(s^2 + 4s + 6)$ can be expressed as $[(s+2)^2 + (\sqrt{2})^2]$

$$\therefore \quad F(s) = \frac{s+2}{(s+2)^2 + (\sqrt{2})^2}$$

Taking inverse Laplace transform on both sides

$$\mathcal{L}^{-1} F(s) = \mathcal{L}^{-1} \frac{s+2}{(s+2)^2 + (\sqrt{2})^2}$$

$$\therefore \quad f(t) = e^{-2t} \cos \sqrt{2}t \quad \textbf{Ans.}$$

(*v*)
$$F(s) = \frac{5}{s(s^2+4s+5)}$$

Using partial fraction expansion

$$\frac{5}{s(s^2+4s+5)} = \frac{k_1}{s} + \frac{k_2 s + k_3}{s^2+4s+5}$$

The coefficients are determined as $k_1 = 1$, $k_2 = -1$ and $k_3 = -4$

$$\therefore \quad F(s) = \frac{1}{s} - \frac{s+4}{s^2+4s+5}$$

on completing the square, the term $(s^2 + 4s + 5)$ can be expressed as $[(s + 2)^2 + (1)^2]$

$$\therefore \quad F(s) = \frac{1}{s} - \frac{s+4}{[(s+2)^2 + (1)^2]}$$

$$= \frac{1}{s} - \frac{s+2}{[(s+2)^2 + (1)^2]} - 2\frac{1}{[(s+2)^2 + (1)^2]}$$

Taking inverse Laplace transform on both sides

$$\mathcal{L}^{-1} F(s) = \mathcal{L}^{-1}\left[\frac{1}{s} - \frac{s+2}{(s+2)^2 + (1)^2} - 2\frac{1}{(s+2)^2 + (1)^2}\right]$$

or
$$\mathcal{L}^{-1} F(s) = \mathcal{L}^{-1}\frac{1}{s} - \mathcal{L}^{-1}\frac{s+2}{[(s+2)^2 + (1)^2]} - \mathcal{L}^{-1} 2\frac{1}{[(s+2)^2 + (1)^2]}$$

$$\therefore \quad f(t) = (1 - e^{-2t} \cos t - 2e^{-2t} \sin t) \quad \textbf{Ans.}$$

(*vi*)
$$F(s) = \frac{s^2+2s+3}{s^3+6s^2+12s+8}$$

The denominator $(s^2 + 6s^2 + 12s + 8)$ can be expressed as $(s \div 2)^3$

$$\therefore \quad F(s) = \frac{s^2+2s+3}{(s+2)^3}$$

Using partial fraction expansion

$$F(s) = \frac{k_1}{(s+2)} + \frac{k_2}{(s+2)^2} + \frac{k_3}{(s+2)^3}$$

The coefficients are determined as $k_1 = 1$, $k_2 = -2$ and $k_3 = 3$

$$\therefore \quad F(s) = \frac{1}{s+2} - \frac{2}{(s+2)^2} + \frac{3}{(s+2)^3}$$

Taking inverse Laplace transform on both sides

$$\mathcal{L}^{-1} F(s) = \mathcal{L}^{-1}\left[\frac{1}{s+2} - \frac{2}{(s+2)^2} + \frac{3}{(s+2)^3}\right]$$

or
$$\mathcal{L}^{-1} F(s) = \mathcal{L}^{-1}\frac{1}{s+2} - \mathcal{L}^{-1}\frac{2}{(s+2)^2} + \mathcal{L}^{-1}\frac{3}{(s+2)^3}$$

$$\therefore \quad f(t) = e^{-2t} - 2t\, e^{-2t} + \tfrac{3}{2}t^2 e^{-2t}$$

or
$$f(t) = e^{-2t}\,[1 - t(2 - \tfrac{3}{2}t)] \quad \textbf{Ans.}$$

Example 1.6.2. *Obtain the solution of the differential equation given below*

$$2\frac{dx}{dt} + 8x = 10; \text{ given } x(0+) = 2.$$

Solution. Taking Laplace transform on both sides the following equation is obtained :

$$\mathcal{L}\left[2\frac{dx}{dt} + 8x\right] = \mathcal{L}10 \quad \text{or} \quad \mathcal{L}\left[2\frac{dx}{dt}\right] + \mathcal{L}[8x] = \mathcal{L}10$$

$$\therefore \quad 2[s(X(s) - x(0+)] + 8[X(s)] = \frac{10}{s}$$

Substituting $x(0+) = 2$

$$2[s\,X(s) - 2] + 8[X(s)] = \frac{10}{s}$$

Simplifying, $X(s) = \dfrac{2s+5}{s(s+4)}$

Using partial fraction expansion

$$X(s) = \frac{k_1}{s} + \frac{k_2}{s+4}$$

The coefficients can be determined as $k_1 = 1.25$ and $k_2 = 0.75$

$$\therefore \quad X(s) = \frac{1.25}{s} + \frac{0.75}{s+4}$$

Taking inverse Laplace transform on both sides

$$x(t) = (1.25 + 0.75\, e^{-4t}). \quad \textbf{Ans.}$$

Example 1.6.3. *Obtain the solution of the differential equation given below*

$$\frac{d^2x}{dt^2} + 2\frac{dx}{dt} + 2x = 0, \text{ given } x(0+) = 0 \text{ and } x'(0+) = 1.$$

Solution. Taking Laplace transform on both sides the following equation is obtained :

$$\mathcal{L}\left[\frac{d^2x}{dt^2}\right] + \mathcal{L}\left[2\frac{dx}{dt}\right] + \mathcal{L}[2x] = 0$$

or
$$[s^2X(s) - sx(0+) - x'(0+)] + [2s\,X(s) - x(0+)] + 2\,[X(s)] = 0$$

Substituting $x(0+) = 0$ and $x'(0) = 1$

or $$[s^2X(s) - s \cdot 0 - 1] + 2[s\,X(s) - 0] + 2[X(s)] = 0$$

Simplifying, $$X(s) = \frac{1}{(s^2 + 2s + 2)}$$

On completing the square, the term $(s^2 + 2s + 2)$ can be expressed as $[(s+1)^2 + (1)^2]$

$$\therefore \quad X(s) = \frac{1}{[(s+1)^2 + (1)^2]}$$

Taking inverse Laplace transform on both sides

$$\mathcal{L}^{-1}X(s) = \mathcal{L}^{-1}\left[\frac{1}{(s+1)^2 + (1)^2}\right]$$

$$\therefore \quad x(t) = e^{-t} \sin t \quad \textbf{Ans.}$$

Example 1.6.4. *Find the Laplace transform of the differential equation given below and hence evaluate the time solution of the same given that $y(0+) = 0$ and $y'(0+) = 6$.*

$$\frac{d^2y}{dt^2} + 5\frac{dy}{dt} + 6y = 12\,e^t.$$

Solution. Taking Laplace transform on both sides, the following equation is obtained :

$$\mathcal{L}\left[\frac{d^2y}{dt^2}\right] + \mathcal{L}\left[5\frac{dy}{dt}\right] + \mathcal{L}[6y] = \mathcal{L}[12e^t]$$

$$\therefore \quad [s^2Y(s) - sy(0+) - y'(0+)] + 5[sY(s) - y(0+)] + 6[y(s)]$$

$$= \frac{12}{(s-1)}$$

Substituting $y(0+) = 0$ and $y'(0+) = 6$

$$[s^2Y(s) - s \cdot 0 - 6] + 5[sY(s) - 0] + 6[Y(s)] = \frac{12}{(s-1)}$$

or $$[s^2Y(s) - 6] + 5[sY(s)] + 6[Y(s)] = \frac{12}{(s-1)}$$

Simplifying, $$Y(s) = \frac{6s+6}{(s-1)(s^2+5s+6)}$$

The term $(s^2 + 5s + 6)$ can be expressed as $(s + 2) . (s + 3)$

$$\therefore \quad Y(s) = \frac{6s+6}{(s-1)(s+2)(s+3)}$$

Using partial fraction expansion

$$Y(s) = \frac{k_1}{s-1} + \frac{k_2}{s+2} + \frac{k_3}{s+3}$$

The coefficients can be determined as $k_1 = 1$, $k_2 = 2$ and $k_3 = -3$

$$\therefore \quad Y(s) = \frac{1}{s-1} + \frac{2}{s+2} - \frac{3}{s+3}$$

Taking inverse Laplace transform on both sides

$$\mathcal{L}^{-1}Y(s) = \mathcal{L}^{-1}\left[\frac{1}{s-1} + \frac{2}{s+2} - \frac{3}{s+3}\right]$$

or $$\mathcal{L}^{-1}Y(s) = \mathcal{L}^{-1}\frac{1}{s-1} + \mathcal{L}^{-1}\frac{2}{s+2} - \mathcal{L}^{-1}\frac{3}{s+3}$$

$\therefore$ $$y(t) = (e^{+t} + 2e^{-2t} - 3e^{-3t})$$ **Ans.**

Example 1.6.5. *The Laplace transformed equation for the charging current of a capacitor arranged in series with a resistance is given by,*

$$I(s) = \frac{Cs}{RCs+1}\cdot E(s)$$

The circuit is connected to a supply voltage of E. If E = 100 V, R = 2 MΩ, C = 1 μF; calculate the initial value of the charging current.

Solution. Since $E = 100 \quad \therefore \quad E(s) = \frac{100}{s}$

Substituting the given values,

$$I(s) = \frac{1\times 10^{-6}s}{[(2\times 10^{6}\times 1\times 10^{-6}s)+1]}\cdot\frac{100}{s}$$

Applying the initial value theorem,

$$i(0+) = \lim_{t\to 0} i(t) = \lim_{s\to\infty} sI(s)$$

$$= \lim_{s\to\infty} s\cdot\frac{1\times 10^{-6}s}{[(2\times 10^{6}\times 1\times 10^{-6}s)+1]}\cdot\frac{100}{s}$$

$$= \lim_{s\to\infty}\frac{1}{\left[2\times 10^{6}+\frac{1}{1\times 10^{-6}s}\right]}\cdot 100 = \frac{100}{2\times 10^{6}} = 50\ \mu\text{A}.$$ **Ans.**

Example 1.6.6. *A series circuit consisting of resistance R and an inductance of L is connected to a d.c. supply voltage of E. Derive an expression for the steady state value of the current flowing in the circuit.*

Solution. The differential equation relating the current $i(t)$ flowing in the circuit and the input voltage E is given by

$$E = Ri(t) + L\frac{di(t)}{dt}.$$

Taking Laplace transform of this equation yields,

$$E(s) = RI(s) + L[sI(s) - i(0+)]$$

because $$i(0+) = 0$$

$\therefore$ $$E(s) = RI(s) + LsI(s)$$

$\because$ E is constant (d.c. voltage)

$\therefore$ $$E(s) = \frac{E}{s} \quad \therefore \quad \frac{E}{s} = RI(s) + Ls\,I(s)$$

or $$I(s) = \frac{1}{R+Ls}\cdot\frac{E}{s}$$

Applying the final value theorem,

$$i_{ss} = \lim_{t\to\infty} i(t) = \lim_{s\to 0} sI(s)$$

$$= \lim_{s \to 0} s \cdot \frac{1}{R + Ls} \cdot \frac{E}{s} = \lim_{s \to 0} \frac{E}{R + Ls} = \frac{E}{R}$$

$$\therefore \quad i_{ss} = \frac{E}{R} \quad \textbf{Ans.}$$

Example 1.6.7. *A system is represented by a relation given below :*

$$X(s) = R(s)\frac{100}{s^2 + 2s + 50}$$

if r(t) = 1.0 unit, find the value of x(t) when t → ∞.

Solution. Since $r(t) = 1$

$$\therefore \quad R(s) = \frac{1}{s} \qquad \therefore \quad X(s) = \frac{1}{s} \cdot \frac{100}{(s^2 + 2s + 50)}$$

According to the final value theorem,

$$\lim_{t \to \infty} x(t) = \lim_{s \to 0} sX(s)$$

$$= \lim_{s \to 0} s\frac{1}{s}\frac{100}{(s^2 + 2s + 50)} = \frac{100}{50} = 2.0 \text{ units.} \quad \textbf{Ans.}$$

Example 1.6.8. *The Laplace transform of the error signal in a control system is expressed by a relation*

$$E(s) = R(s)\frac{s}{s^2 + 6s + 25}$$

Calculate the steady state value of the error if r(t) = t.

Solution. Since $r(t) = t$

$$\therefore \quad R(s) = \frac{1}{s^2} \qquad \therefore \quad E(s) = \frac{1}{s^2} \cdot \frac{s}{(s^2 + 6s + 25)}$$

According to final value theorem

$$e_{ss} = \lim_{s \to 0} s\,E(s) = \lim_{s \to 0} s\frac{1}{s^2}\frac{s}{(s^2 + 6s + 25)}$$

$$= \lim_{s \to 0} \frac{1}{s^2 + 6s + 25} = \frac{1}{25} = 0.4 \text{ unit.} \quad \textbf{Ans.}$$

Example 1.6.9. *Find f (0 +), f′(0 +) and f″(0 +) for the function whose Laplace transform is given below :*

$$F(s) = \frac{4s + 1}{s(s^2 + 2)}.$$

Solution. (1) $f(0+)$ is determined as follows :

$$f(0+) = \lim_{t \to 0} f(t) = \lim_{s \to \infty} s\,F(s) \text{ (Initial value theorem)}$$

$$= \lim_{s \to \infty} s \cdot \frac{4s + 1}{s(s^2 + 2)} = \lim_{s \to \infty} \frac{4s + 1}{(s^2 + 2)} = \lim_{s \to \infty} \frac{4 + \frac{1}{s}}{s + \frac{2}{s}} = 0$$

$$\therefore \quad f(0+) = 0. \quad \textbf{Ans.}$$

(2) $f'(0+)$ is determined as follows :

$$\mathcal{L}f'(t) = s\mathcal{L}f(t) - f(0+) = s \cdot \frac{4s+1}{s(s^2+2)} - 0 = \frac{4s+1}{(s^2+2)}$$

$$f'(0+) = \lim_{t\to 0} f'(t) = \lim_{s\to\infty} s\mathcal{L}f'(t)$$

$$= \lim_{s\to\infty} s\frac{4s+1}{s^2+2} = \lim_{s\to\infty} \frac{4s+1}{\left(s+\frac{2}{s}\right)} = \lim_{s\to\infty} \frac{4+\frac{1}{s}}{1+\frac{2}{s^2}} = 4$$

$\therefore$ $f'(0+) = 4$. **Ans.**

(3) $f''(0+)$ is determined as follows :

$$\mathcal{L}f''(t) = s^2\,\mathcal{L}f(t) - s\,f(0) - f'(0+)$$

$$= s^2 \cdot \frac{4s+1}{s(s^2+2)} - s\cdot 0 - 4 = s \cdot \frac{4s+1}{(s^2+2)} - 4 = \frac{s-8}{(s^2+2)}$$

$$f''(0+) = \lim_{t\to 0} f''(t) = \lim_{s\to\infty} s\mathcal{L}f''(t)$$

$$= \lim_{s\to\infty} s\,.\frac{(s-8)}{(s^2+2)} = \lim_{s\to\infty} \frac{(s-8)}{\left(s+\frac{2}{s}\right)} = \lim_{s\to\infty} \frac{\left(1-\frac{8}{s}\right)}{\left(1+\frac{2}{s^2}\right)} = 1$$

$f''(0+) = 1$. **Ans.**

Example 1.6.10. *The Laplace transform of $f(t)$ is given by*

$$F(s) = \frac{4}{s(s+2)}$$

Find the final value using the final value theorem and verify the result by determining $f(t)$ using inverse Laplace transform.

Solution. (1) The final value is the value of $f(t)$ as $t \to \infty$.

As per final value theorem,

$$\lim_{t\to\infty} f(t) = \lim_{s\to 0} s\,F(s)$$

$$= \lim_{s\to 0} s\frac{4}{[s(s+2)]} = \lim_{s\to 0} \frac{4}{(s+2)} = 2$$

$\therefore$ $\lim_{t\to\infty} f(t) = 2$. **Ans.**

(2) The expression for $f(t)$ is determined as follows :

$$F(s) = \frac{4}{s(s+2)}$$

Using partial fraction expansion

$$F(s) = \frac{k_1}{s} + \frac{k_2}{s+2}$$

The coefficients can be determined as $k_1 = 2$ and $k_2 = -2$.

$\therefore$ $$F(s) = \frac{2}{s} - \frac{2}{s+2}.$$

Taking inverse Laplace transform on both sides

$$\mathcal{L}^{-1} F(s) = \mathcal{L}^{-1}\left[\frac{2}{s} - \frac{2}{s+2}\right]$$

or
$$\mathcal{L}^{-1} F(s) = \mathcal{L}^{-1}\left[\frac{2}{s} - \frac{2}{s+2}\right]$$

$\therefore$
$$f(t) = 2 - 2e^{-2t}$$

and
$$\underset{t \to \infty}{f(t)} = 2 - 0 = 2. \text{ Verified.}$$

PROBLEMS

1.1. Find the inverse Laplace transform of the functions given below :

(i) $F(s) = \dfrac{1}{(s+1)(s+4)}$ (ii) $F(s) = \dfrac{s}{s^2 + 6s + 13}$

(iii) $F(s) = \dfrac{3s}{(s^2+1)(s^2+4)}$ (iv) $F(s) = \dfrac{s + 0.25}{(s+0.5)^2}$

(v) $F(s) = \dfrac{1}{s} \cdot \dfrac{2}{(s+1)(s+2)}$.

1.2. Express the following differential equations in Laplace transform form :

(i) $\dfrac{d^2x}{dt^2} + 6\dfrac{dx}{dt} + 8x = 0$, given $x(0+) = 0$ and $x'(0+) = 2$

(ii) $\dfrac{d^2x}{dt^2} + 4\dfrac{dx}{dt} + 5x = 5$, given $x(0+) = 0$ and $x'(0+) = 0$

(iii) $\dfrac{d\theta}{dt} + 6\theta + 5\int \theta\, dt = 0$, given $\theta(0+) = 0$, $\theta'(0+) = 0$ and $\theta^{-1}(0+) = 0.1$

(iv) $2\dfrac{dx}{dt} + 4x = te^{-3t}$, $x(0+) = 0$.

1.3. By using Laplace transform determine the ratio $\dfrac{C(s)}{R(s)}$ for the differential equation given below

$$\frac{d^2c(t)}{dt^2} + 4\frac{dc(t)}{dt} + c(t) = r(t).$$

Assume all initial conditions as zero. If $r(t) = 1$ evaluate the time solution for $c(t)$.

1.4. Find the initial value of the functions having following Laplace transform

(i) $F(s) = \dfrac{(s+\alpha)}{[(s+\alpha)^2 + \omega^2]}$ (ii) $F(s) = \dfrac{2(2s+1)}{s(s^2 + 0.5s + 4)}$.

1.5. Find the final value of the functions having following Laplace transform :

(i) $F(s) = \dfrac{(s+\alpha)}{(s+\alpha)^2 + \omega^2}$ where $\alpha > 0$ (ii) $F(s) = \dfrac{2(s+1)}{s(s^2 + 4s + 5)}$.

CHAPTER 2

Transfer Function

In this chapter, the transfer function derivation of a control system is explained. The concepts of poles-zeros of the transfer function are discussed. Having given a physical system the procedure for obtaining the transfer function is described.

CONTENTS

- *Poles and Zeros of a Transfer Function*
- *Transfer Function and its Relationship with Impulse Response*
- *Procedure for Determining the Transfer Function of a Control System*
- *Solved Examples*

TRANSFER FUNCTION

For any control system there exists an input termed as excitation or cause (denoted as r) which operates through a transfer operation termed as transfer function (denoted as g) and produces an effect resulting in output or response termed as controlled variable (denoted as c). Thus the cause and effect relationship between the output and input is related to each other through a transfer function. The relationship between the output and input is represented by a diagram known as block diagram as shown in Fig. 2.0.1.

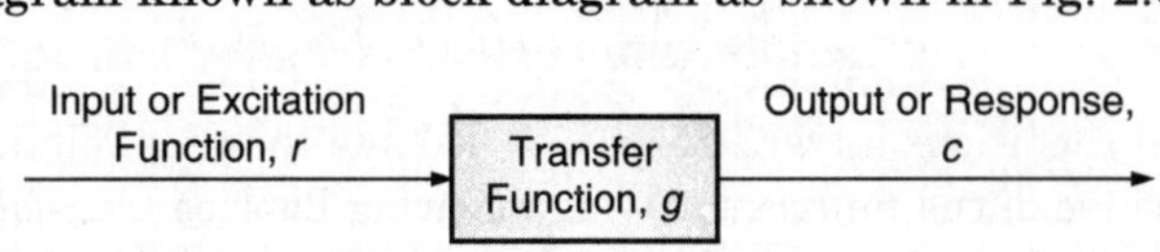

Fig. 2.0.1. Block diagram.

The arrow in the diagram indicates the direction of control action thus the block diagram has a unilateral property in the direction of arrow.

The transfer function is expressed as the ratio of output quantity to input quantity. Therefore,

$$g = \frac{c}{r} \tag{2.1}$$

As the cause and effect approach to control systems is general, it is not necessary that the output be of the same category as that of input. For example in the case of an electric motor the input is an electrical quantity and the output being a mechanical one.

With reference to control system wherein all mathematical functions are expressed by their corresponding Laplace transforms, therefore, the transfer function is also expressed as a ratio of Laplace transform of output to Laplace transform of input.

Thus if, $R(s)$ = Laplace transform of the input function

$C(s)$ = Laplace transform of the output function

then the block diagram for a control system is drawn as per Fig. 2.0.2.

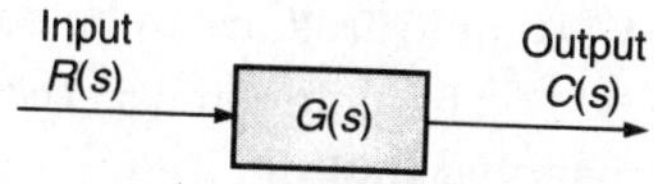

Fig. 2.0.2. Block diagram of a control system.

The transfer function is expressed as,

$$G(s) = \frac{C(s)}{R(s)} \qquad \text{...(2.2)}$$

While taking Laplace transformation for determining the transfer function of a control system it is assumed that all the initial conditions concerning the differential equation are zero.

Hence the transfer function is,

$$G(s) = \frac{\mathcal{L}\,c(t)}{\mathcal{L}\,r(t)} \qquad \text{...(2.3)}$$

"The transfer function of a control system is defined as the ratio of the Laplace transform of the output variable to the Laplace transform of the input variable assuming all initial conditions as zero."

2.1 POLES AND ZEROS OF A TRANSFER FUNCTION

The transfer function of a linear control system can be expressed in the form of a quotient of polynomials in the following form :

$$G(s) = \frac{A(s)}{B(s)} = \frac{a_0 s^n + a_1 s^{n-1} + a_2 s^{n-2} + ... + a_n}{b_0 s^m + b_1 s^{m-1} + b_2 s^{m-2} + ... + b_m} \qquad \text{...(2.4)}$$

The numerator and the denominator can be factored into n and m terms respectively, with such a factorisation the above expression for the transfer function can be expressed as,

$$G(s) = \frac{A(s)}{B(s)} = \frac{K(s - s_1)(s - s_2)...(s - s_n)}{(s - s_a)(s - s_b)...(s - s_m)} \qquad \text{...(2.5)}$$

where $K = \frac{a_0}{b_0}$ is known as the gain factor of the transfer function.

In the transfer function expression (2.5), if the numerator is equated to zero, then n roots of the equation are $s_1, s_2, s_3, ..., s_n$ and where as equating the denominator to zero, the m roots can be determined as $s_a, s_b, s_c, ..., s_m$.

Poles of the transfer function : In the transfer function expression (2.5), if s is put equal to $s_a, s_b, ..., s_m$ it is noted that the value of the transfer function is infinite, hence $s_a, s_b, ..., s_m$ are called the *Poles* of the transfer function.

Zeros of the transfer function : In the transfer function expression, if s is put equal to $s_1, s_2, ..., s_n$ it is noted that the value of the transfer function is zero, hence $s_1, s_2, ..., s_n$ are called the Zeros of the transfer function.

The poles s_a, s_b, ..., s_m or the zeros s_1, s_2, ..., s_n are either real or complex and the complex poles or zeros always appear in conjugate pairs.

It is possible that either poles or zeros may coincide. Such poles or zeros are called multiple poles or multiple zeros, otherwise non-coinciding poles or zeros are called simple poles or simple zeros.

The multiple poles/zeros occur due to the presence of repetitive factor in the denominator/numerator of a transfer function expression.

The graphical symbol for a pole is X and for a zero O. The said symbols are used when poles and zeros are to be shown on a real and imaginary axes (s-plane).

The poles and zeros mentioned above are the finite ones. From equation (2.5) it is observed that :

1. If $n > m$, then the value of the transfer function is found to be infinity for $s = \infty$. Hence it is concluded that there exists a pole of the transfer function at infinity $(s = \infty)$ and the multiplicity (order) of such a pole being $(n - m)$.
2. If $n < m$, then the value of the transfer function is found to be zero for $s = \infty$. Hence it is concluded that there exists a zero of the transfer function at infinity $(s = \infty)$ and the multiplicity (order) of such a zero is $(m - n)$.

Therefore, in addition to finite poles and zeros of a transfer function, if poles and zeros located at zero and infinity are considered, then for a rational function the total number of zeros is found to be equal to total number of poles.

For example, consider the following transfer function :

$$G(s) = \frac{(s+2)(s+4)}{s(s+3)(s+5)(s+2-j4)(s+2+j4)}$$

For the above transfer function, the poles are at (1) $s_a = 0$, (2) $s_b = -3$, (3) $s_c = (-2 + j4)$, (4) $s_d = (-2 - j4)$ and (5) $s_c = -5$. The zeros are at (1) $s_1 = -2$, (2) $s_2 = -4$.

As the number of zeros should be equal to number of poles, the remaining three zeros are located at $s = \infty$.

The pole-zero locations are plotted s-plane as shown in Fig. 2.1.1.

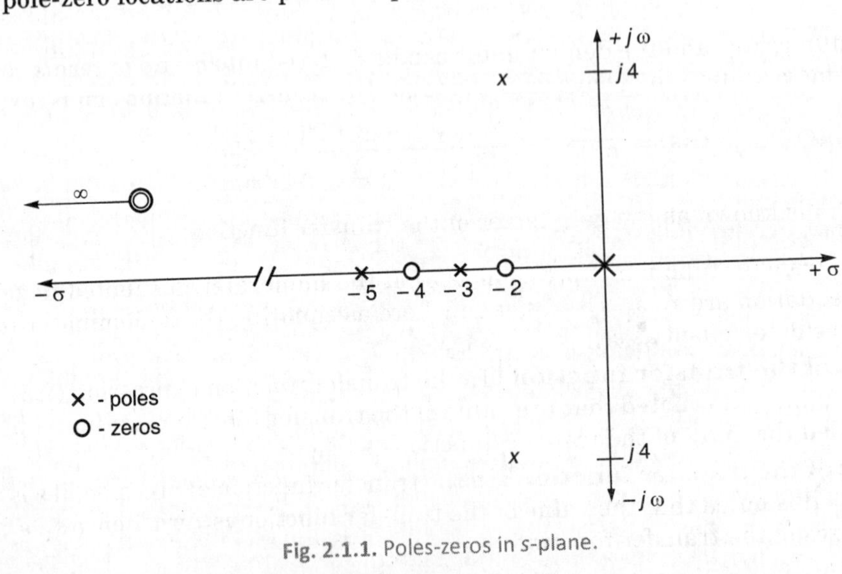

Fig. 2.1.1. Poles-zeros in s-plane.

In the transfer function expression of a control system the highest power of s in the numerator $A(s)$ is either equal or less than that of the denominator $B(s)$. *The transfer function of a system is completely specified in terms of its poles and zeros and the gain factor.

2.2 TRANSFER FUNCTION AND ITS RELATIONSHIP WITH IMPULSE RESPONSE

In a control system the output is related to the input by a transfer function as defined earlier, *i.e.*

$$\frac{C(s)}{R(s)} = G(s)$$

or

$$C(s) = R(s)\, G(s) \quad ...(2.6)$$

The output time response can be determined by taking inverse Laplace transform of relation (2.6). If the input is specified as unit impulse at $t = 0$, then $R(s) = 1$ and the transformed expression for the system output, is,

$$C(s) = 1 \,.\, G(s)$$

or

$$C(s) = G(s) \quad ...(2.7)$$

Thus the output time response is,

$$\mathcal{L}^{-1} C(s) = \mathcal{L}^{-1} G(s)$$

or

$$c(t) = g(t) \quad ...(2.8)$$

The inverse Laplace transform of $G(s)$ is, therefore, called the impulse response of a system, or the transfer function of a system is the Laplace transform of its impulse response. The output impulse response of a system in time domain is thus solely determined by its transfer function. The transfer function of a system depends on its elements, assuming initial conditions as zero and is independent of the input function. It will be shown in the later chapters that the transient performance of a control system is influenced only by the poles and zeros of its transfer function.

2.3 PROCEDURE FOR DETERMINING THE TRANSFER FUNCTION OF A CONTROL SYSTEM

The following steps give a procedure for determining the transfer function of a control system :

1. Formulate the equations for the system.
2. Take the Laplace transform of the system equations, assuming initial conditions as zero.
3. Specify the system output and the input.
4. Take the ratio of the Laplace transform of the output and the Laplace transform of the input.

The ratio as obtained in step (4) is the required transfer function.

*For the purpose of control system analysis, only finite poles/zeros and poles and zeros at the origin are considered.

2.4 SOLVED EXAMPLES

Example 2.4.1. *For the transfer function*

$$G(s) = \frac{1}{2} \cdot \frac{(s^2 + 4)(1 + 2.5s)}{(s^2 + 2)(1 + 0.5s)}$$

Plot the poles and zeros in s-plane and determine the value of the transfer function at s = 2.

Solution. $G(s) = \frac{1}{2} \frac{(s^2 + 4)(1 + 2.5s)}{(s^2 + 2)(1 + 0.5s)}$

The poles are determined from the equation

$$(s^2 + 2)(1 + 0.5s) = 0$$

Therefore, the poles are,

$$s = \pm j\sqrt{2},\ s = -\frac{1}{0.5} = -2$$

The zeros are determined from the equation.

$$(s^2 + 4)(1 + 2.5s) = 0$$

Therefore, the zeros are,

$$s = \pm j\sqrt{4} = \pm j2,$$

$$s = -\frac{1}{2.5} = -0.4$$

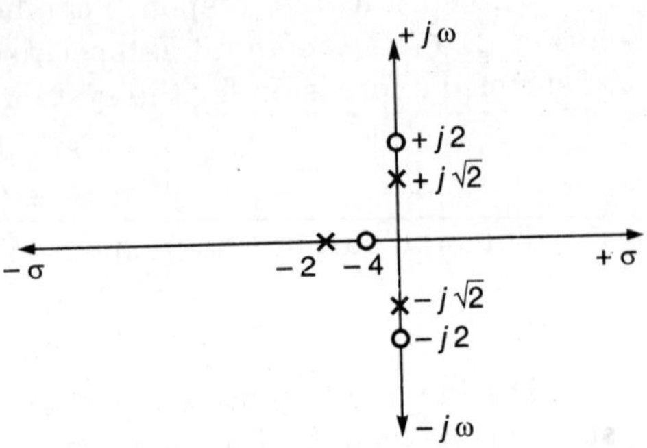

Fig. 2.4.1. Pole zero configuration for example 2.4.1.

The pole zero configuration is plotted in s-plane as shown in Fig. 2.4.1.

The value of $G(s)$ at $s = 2$ is calculated below,

$$G(s) = \frac{1}{2} \cdot \frac{(2^2 + 4)(1 + 2.5 \times 2)}{(2^2 + 2)(1 + 0.5 \times 2)} = \frac{1}{2} \cdot \frac{8 \times 6}{6 \times 2} = 2. \textbf{ Ans.}$$

Example 2.4.2. *The pole-zero configuration of a transfer function is given below (Fig. 2.4.2).*

The value of the transfer function at s = 1 is found to be 3.2. Determine the transfer function and gain factor K.

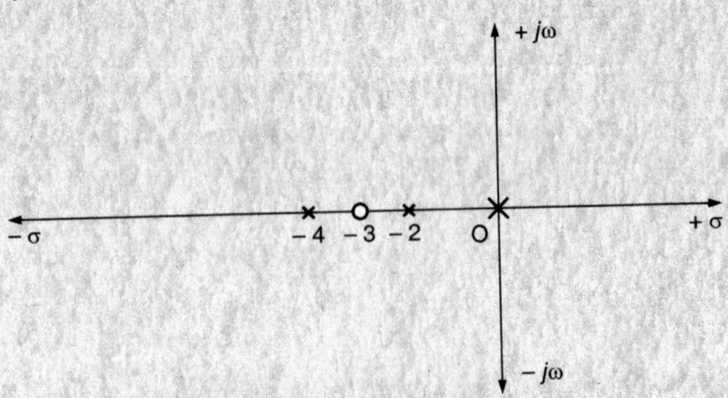

Fig. 2.4.2. Pole-zero configuration for example 2.4.2.

Solution. The transfer function has three poles and one zero, therefore, the transfer function consists of one term in the numerator and three terms in the denominator.

The poles are located at $s = 0, s = -2$ and $s = -4$

The zero is located at $s = -3$

The transfer function is thus

$$G(s) = \frac{K(s+3)}{s(s+2)(s+4)}$$

It is given that at $s = 1$, the value of $G(s)$ is 3.2

$$\therefore \qquad G(1) = 3.2 = \frac{K(1+3)}{1(1+2)(1+4)} = \frac{K \times 4}{1 \times 3 \times 5}$$

$$\therefore \qquad K = \frac{3.2 \times 1 \times 3 \times 5}{4} = 12$$

$$\therefore \qquad G(s) = \frac{12(s+3)}{s(s+2)(s+4)} \quad \textbf{Ans.}$$

Example 2.4.3. *Find the transfer function of the network given below (Fig. 2.4.3).*

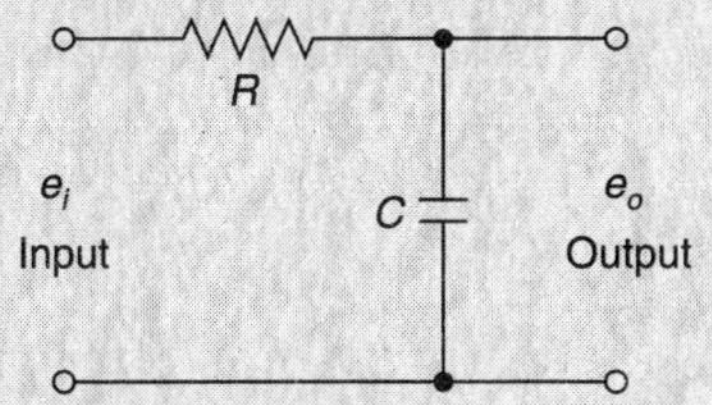

Fig. 2.4.3. Network for example 2.4.3.

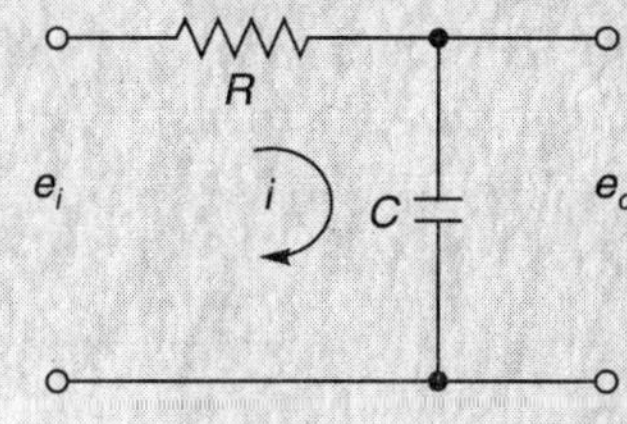

Fig. 2.4.4. R.C. network.

Solution. The network equations are

$$e_i - Ri + \frac{1}{C}\int i\,dt \qquad \text{...(1)}$$

and
$$e_o = \frac{1}{C}\int i\,dt \qquad \text{...(2)}$$

Assuming initial conditions as zero and taking Laplace transform on both sides of equations (1) and (2), the following equations are obtained :

$$E_i(s) = RI(s) + \frac{1}{C} \cdot \frac{I(s)}{s} \qquad \text{...(3)}$$

and
$$E_o(s) = \frac{1}{C} \cdot \frac{I(s)}{s} \qquad \text{...(4)}$$

or
$$E_i(s) = \left[R + \frac{1}{Cs}\right] I(s) \qquad \text{...(5)}$$

and
$$E_o(s) = \frac{1}{Cs} I(s) \qquad \text{...(6)}$$

From equations (5) and (6) the transfer function is obtained below,

$$\frac{E_o(s)}{E_i(s)} = \frac{\frac{1}{Cs}.I(s)}{\left[R + \frac{1}{Cs}\right]I(s)} \quad \text{or} \quad \frac{E_o(s)}{E_i(s)} = \frac{1}{RCs + 1} \quad \textbf{Ans.}$$

Eample 2.4.4. *A torque T Nm is applied to a shaft having a moment of inertia J and coefficient of viscous friction of f produces an angular of θ radians. Obtain the transfer function relating θ and T.*

Solution. The equation for the system is given by

$$T = J\frac{d^2\theta}{dt^2} + f\frac{d\theta}{dt} \qquad ...(1)$$

Assuming initial conditions as zero and taking the Laplace transform on both sides of equation (1), the following equation is obtained,

$$T(s) = Js^2\,\theta(s) + fs\,\theta(s)$$

or
$$T(s) = s(Js + f)\,\theta(s) \qquad ...(2)$$

From equation (2) the required transfer function is obtained below,

$$\frac{\theta(s)}{T(s)} = \frac{1}{s(Js+f)} \quad \textbf{Ans.}$$

Example 2.4.5. *A torque T_1 is applied to a load through a gear having a gear ratio N_2/N_1 (Fig. 2.4.5).*

If the moment of inertia and coefficient of viscous friction of the load is J_2 and f_2 respectively, determine the transfer function relating the applied torque T_1 and the angular shift θ_1 at the primary shaft side.

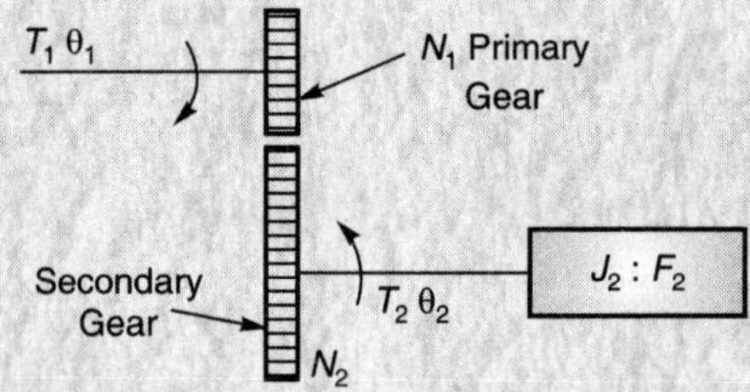

Fig. 2.4.5. Gear system for example 2.4.5.

Solution. The effect of the applied torque T_1 through the gear results in application of torque T_2 to the load and the torques T_1 and T_2 are related as follows :

$$\frac{T_1}{T_2} = \frac{N_1}{N_2}$$

The equation relating the torque T_2 and the angular θ_2 is formulated as,

$$T_2 = J_2\frac{d^2\theta_2}{dt^2} + f_2\frac{d\theta_2}{dt} \qquad ...(1)$$

$$\because \quad \frac{T_1}{T_2} = \frac{N_1}{N_2}$$

$$\therefore \quad T_2 = \frac{N_2}{N_1}T_1 \qquad ...(2)$$

Substituting for T_2 in equation (1) from equation (2) the following equation is obtained :

$$\frac{N_2}{N_1}T_1 = J_2\frac{d^2\theta_2}{dt^2} + f_2\frac{d\theta_2}{dt} \qquad ...(3)$$

Again, since $N_1\,\theta_1 = N_2\,\theta_2$

or $$\theta_2 = \frac{N_1}{N_2}\theta_1 \qquad ...(4)$$

Differentiating equation (4) w.r.t. time

$$\frac{d\theta_2}{dt} = \frac{N_1}{N_2}\cdot\frac{d\theta_1}{dt} \qquad ...(5)$$

Further, differentiating equation (5) w.r.t. time

$$\frac{d^2\theta_2}{dt^2} = \frac{N_1}{N_2}\cdot\frac{d^2\theta_1}{dt^2} \qquad ...(6)$$

Substituting for $\frac{d\theta_2}{dt}$ and $\frac{d^2\theta_2}{dt^2}$ in equation (3) from equations (5) and (6) the following equation is obtained.

$$\frac{N_2}{N_1}T_1 = J_2\frac{N_1}{N_2}\frac{d^2\theta_1}{dt^2} + f_2\frac{N_1}{N_2}\frac{d\theta_1}{dt} \qquad ...(7)$$

Simplifying equation (7)

$$T_1 = J_2\left(\frac{N_1}{N_2}\right)^2\frac{d^2\theta_1}{dt^2} + f_2\left(\frac{N_1}{N_2}\right)^2\frac{d\theta_1}{dt} \qquad ...(8)$$

put $$J_{1eq} = J_2\left(\frac{N_1}{N_2}\right)^2 \qquad ...(9)$$

and $$f_{1eq} = f_2\left(\frac{N_1}{N_2}\right)^2 \qquad ...(10)$$

The terms J_{1eq} and f_{1eq} are known as the moment of inertia and coefficient of viscous friction of the secondary side referred to the primary side of the gear. Substituting for $J_2\left(\frac{N_1}{N_2}\right)^2$ and $f_2\left(\frac{N_1}{N_2}\right)^2$ from expressions (9) and (10) in equation (8) the following equation relating T_1 and θ_1 is obtained.

$$T_1 = J_{1eq}\frac{d^2\theta_1}{dt^2} + f_{1eq}\frac{d\theta_1}{dt} \qquad ...(11)$$

Assuming initial conditions as zero and taking Laplace transform on both sides of equation (11)

$$T_1\,(s) = J_{1eq}\,s^2\,\theta_1(s) + f_{1eq}\,s\theta_1(s) \qquad ...(12)$$

or $$T_1\,(s) = s(J_{1eq}\,s + f_{1eq}\,)\,\theta_1(s) \qquad ...(13)$$

From equation (13) the required transfer function is obtained below

$$\frac{\theta_1(s)}{T_1(s)} = \frac{1}{s(J_{eq}\,s + f_{1eq})} \quad \textbf{Ans.}$$

PROBLEMS

2.1. The transfer function of a system is given below :

$$G(s) = \frac{8(s+3)(s+4)}{s(s+2)^2(s^2+2s+5)}$$

Determine the poles and zeros and show the pole-zero configuration in s-plane.

2.2. The transfer function of a system is given by

$$G(s) = \frac{K(s+8)}{s(s+4)(s+5)(s^2+13s+42)}$$

Determine the poles and zeros of the transfer function. Plot the pole-zero configuration in s-plane.

2.3. The impulse response of a system is given by

$$g(t) = e^{-t}(1 - \cos 2t)$$

Determine the transfer function of the system.

2.4. A force F when applied to a mass of M results in a displacement of X, obtain the transfer function relating the force applied and the mass displacement.

2.5. A series circuit consisting of a resistance R and the inductance of L is connected to a d.c. supply of E volts. Determine the transfer function relating the current flowing in the circuit and the input voltage.

2.6. Determine the transfer function of the network shown in Fig. 2P-1 relating $E_o(s)$ and $E_i(s)$.

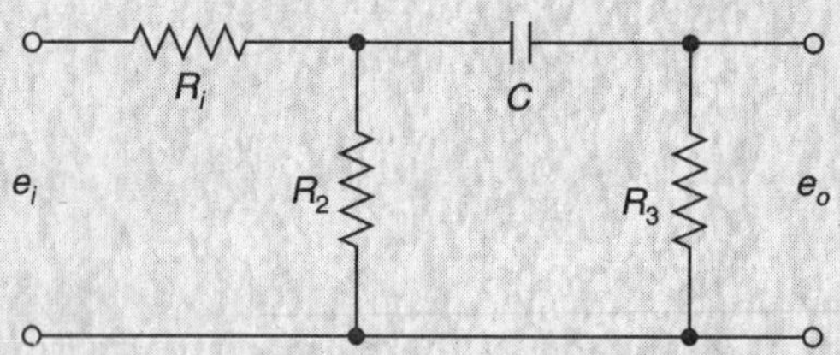

Fig. 2P-1. Figure for example 2.6.

2.7. Derive the transfer function of the system shown in Fig. 2P-2, K being the amplifier gain.

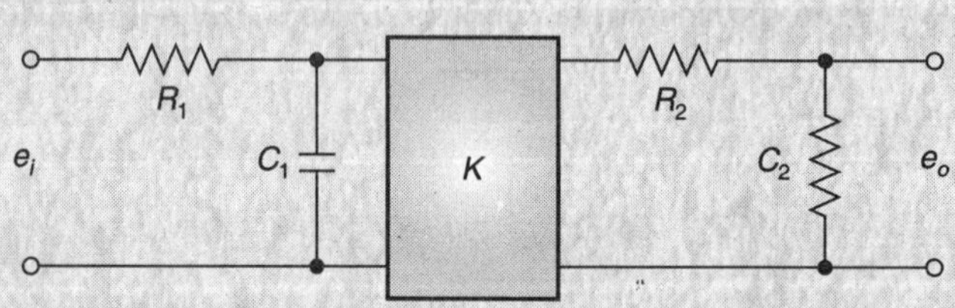

Fig. 2P-2. Figure for example 2.7.

CHAPTER 3

Block Diagrams

The block diagram gives the representation of a control system. The rules for block diagram reduction with reasoning are explained in this chapter. In one of the solved examples the procedure for shifting a take-off point after the summing point to a position before the said summing is described through steps.

CONTENTS

- *Block Diagram Reduction*
- *Solved Examples*

REPRESENTATION OF A CONTROL SYSTEM BY BLOCK DIAGRAM

It is not convenient to derive a complete transfer function for a complex control system, therefore, the transfer function of each element of a control system is represented by a block diagram and the concerned symbol mentioned in the block represents the transfer function of the element. This symbolic representation in a short form gives a pictorial representation relating the output and the input of a control system based on cause and effect approach.

A complete system can be arranged by interconnecting individual blocks as per requirements of the system.

A block diagram is shown in Fig. 3.0.1 wherein $G_1(s)$ and $G_2(s)$ represent the transfer function of individual elements of a control system. The system is a feedback type or closed-loop type. As the output signal $C(s)$ is fedback and compared with the input $R(s)$, the difference $E(s) = [R(s) - C(s)]$ is the actuating signal or error signal.

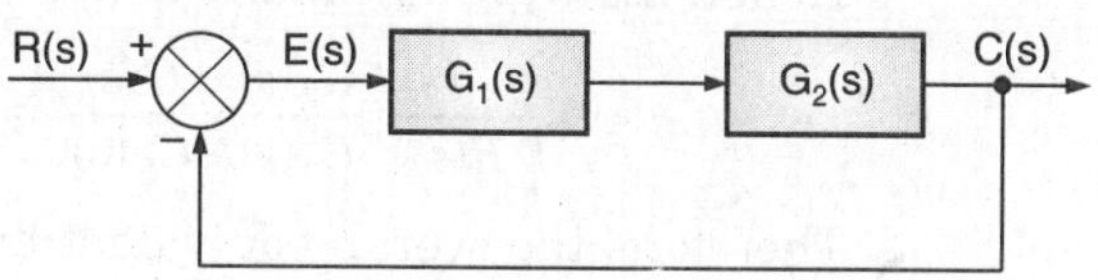

Fig. 3.0.1. Block diagram representation of a control.

3.1 BLOCK DIAGRAM REDUCTION

As described earlier several blocks are inter-connected together for arranging a complete control system. In order to obtain the overall transfer function (the ratio of final output and input in their Laplace transform), a procedure called block diagram reduction technique is followed. Some of the important rules for block diagram reduction are given below.

1. The block diagram shown in Fig. 3.1.1 relates the output and input as per the transfer function relation given below :

$$G(s) = \frac{C(s)}{R(s)} \quad \text{...(3.1)}$$

or $$C(s) = R(s) \cdot G(s) \quad \text{...(3.2)}$$

$G(s)$ is known as the transfer function of the system.

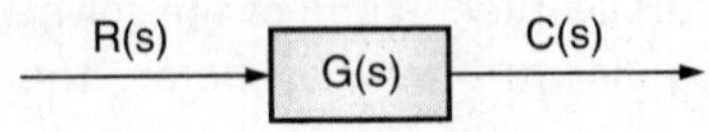

Fig. 3.1.1. Block diagram.

2. **Take off point.** Application of one input source to two or more systems is represented by a take off point as shown at point A in Fig. 3.1.2. It should be noted that such taking off from the input signal does not alter the input signal.

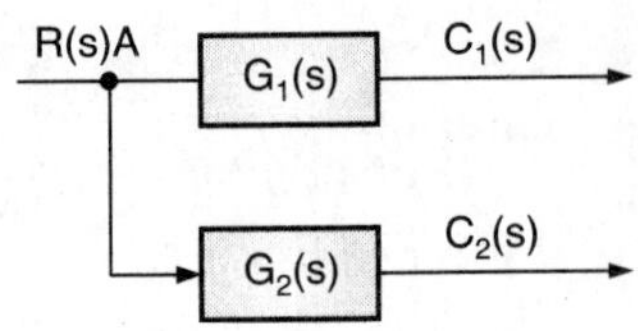

Fig. 3.1.2. Take off point.

3. **Blocks in cascade.** When several blocks are connected in cascade [Fig. 3.1.3 (*a*)] the overall equivalent transfer function is determined below :

$$\frac{C_1(s)}{R(s)} = G_1(s) \quad \text{...(3.3)}$$

$$\frac{C_2(s)}{C_1(s)} = G_2(s) \quad \text{...(3.4)}$$

$$\frac{C(s)}{C_2(s)} = G_3(s) \quad \text{...(3.5)}$$

while connecting the blocks in cascade it should be noted that the forward block does not alter the output signal of the preceding block.

Hence, multiplying equations (3.3), (3.4) and (3.5)

$$\frac{C_1(s)}{R(s)} \cdot \frac{C_2(s)}{C_1(s)} \cdot \frac{C(s)}{C_2(s)} = G_1(s) \cdot G_2(s) \cdot G_3(s) \quad \text{...(3.6)}$$

Therefore, the overall equivalent transfer function is,

$$\frac{C(s)}{R(s)} = G_1(s) \cdot G_2(s) \cdot G_3(s) \quad \text{...(3.7)}$$

The equivalence of Fig. 3.1.3 (*a*) (blocks is cascade) is shown by Fig. 3.1.3 (*b*).

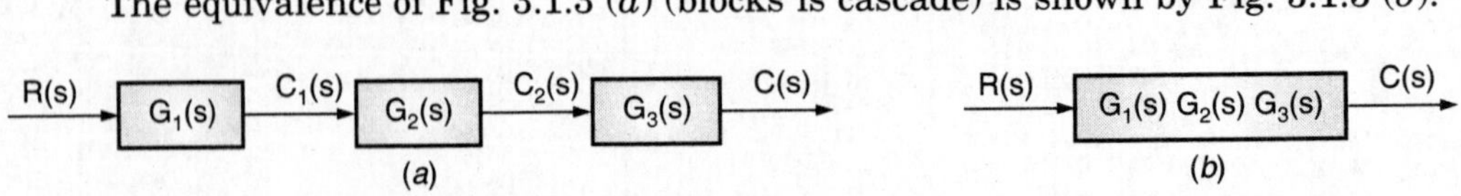

Fig. 3.1.3. Blocks in cascade and equivalence.

4. **Summing point.** Summing point represents summation of two or more input signals entering in a system. The output of a summing point being the sum of the entering inputs. Summing point is represented in Fig. 3.1.4.

   It is necessary to indicate the sign specifying the input signal entering a summing point.

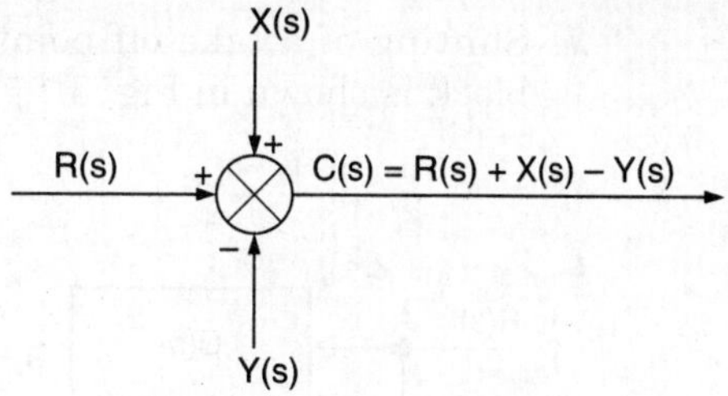

Fig. 3.1.4. Summing point.

5. Consecutive summing points can be interchanged, as this interchange does not alter the output signal as shown in Fig. 3.1.5 (*a*) and (*b*).

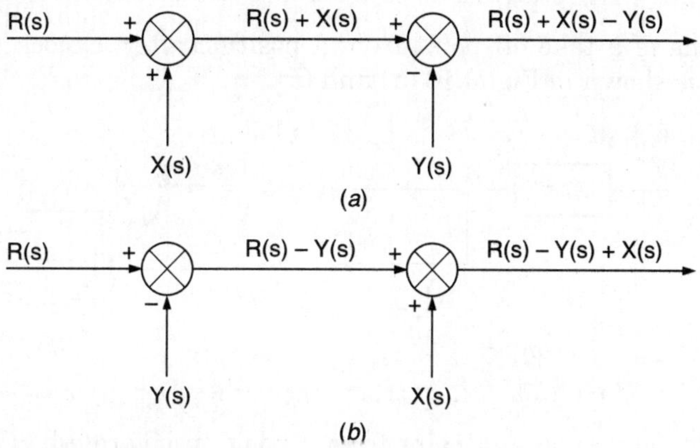

Fig. 3.1.5. Interchange of consecutive summing points.

6. **Blocks in parallel.** When one or more blocks are connected in parallel as shown in Fig. 3.1.6 (*a*), the overall equivalent transfer function is determined below,

$$C(s) = R(s)\,G_1(s) + R(s)\,G_2(s) + R(s)\,G_3(s) \quad \text{...(3.8)}$$

or

$$C(s) = R(s)\,[G_1(s) + G_2(s) + G_3(s)] \quad \text{...(3.9)}$$

Therefore, the overall equivalent transfer function is,

$$\frac{C(s)}{R(s)} = [G_1(s) + G_2(s) + G_3(s)] \quad \text{...(3.10)}$$

The equivalence of Fig. 3.1.6 (*a*) is shown by Fig. 3.1.6 (*b*).

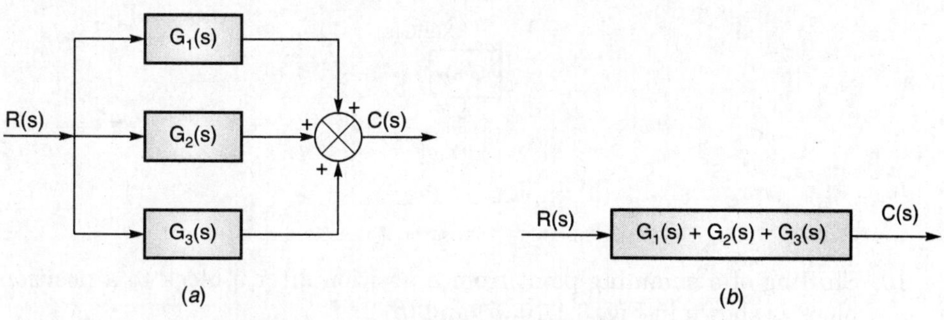

Fig. 3.1.6. Equivalence of blocks in parallel.

7. Shifting of a take off point from a position before a block to a position after the block is shown in Fig. 3.1.7 (*a*) and (*b*).

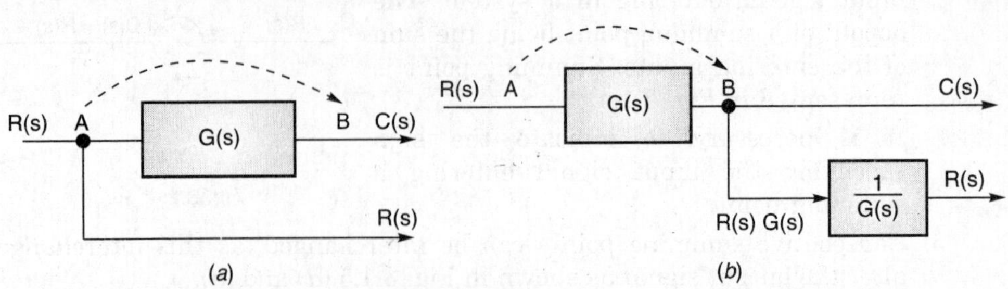

Fig. 3.1.7. Take off point at position *A* shifted to position *B*.

8. Shifting of a take off point from a position after a block to a position before the block is shown in Fig. 3.18 (*a*) and (*b*).

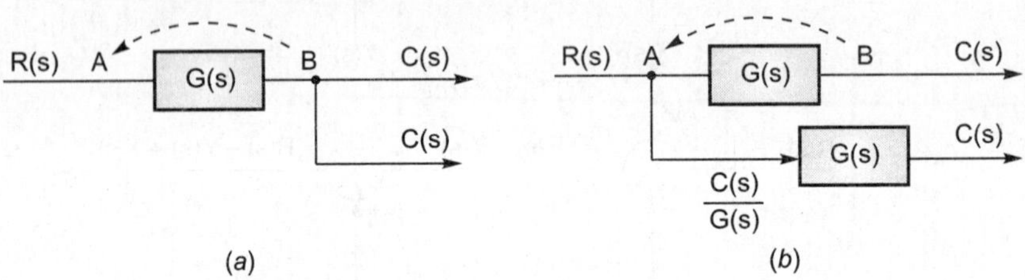

Fig. 3.1.8. Take off point at position *B* shifted to position *A*.

9. Shifting of a summing point from a position before a block to a position after the block is shown in Fig. 3.1.9 (*a*) and (*b*).

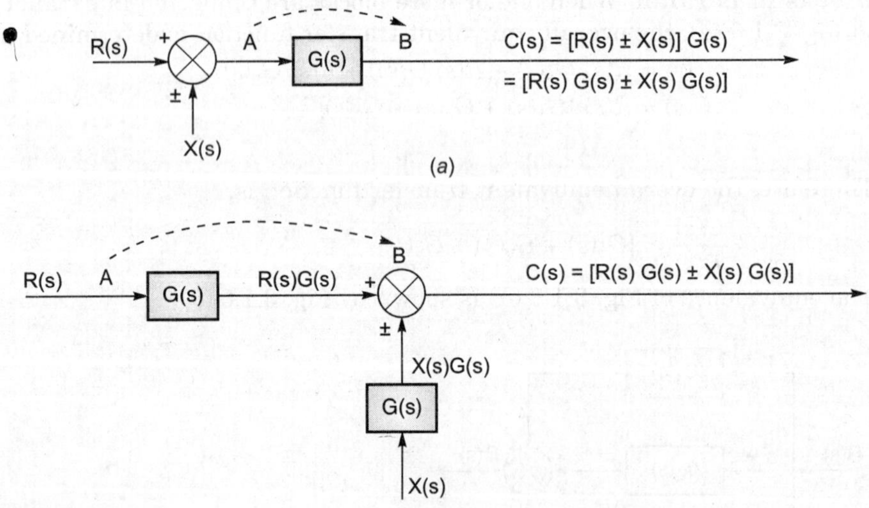

Fig. 3.1.9. Summing at position *A* is shifted to position *B*.

10. Shifting of a summing point from a position after a block to a position before the block is shown in Fig. 3.1.10 (*a*) and (*b*).

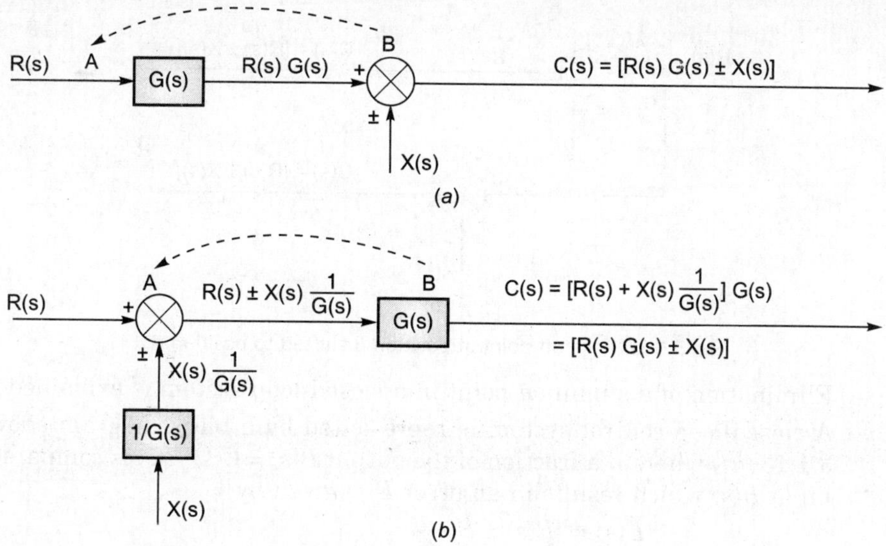

Fig. 3.1.10. Summing point at position *B* shifted to position *A*.

11. **Shifting of a take off point from a position before a summing point to a position after the summing point is shown in Fig. 3.1.11 (*a*) and (*b*).**

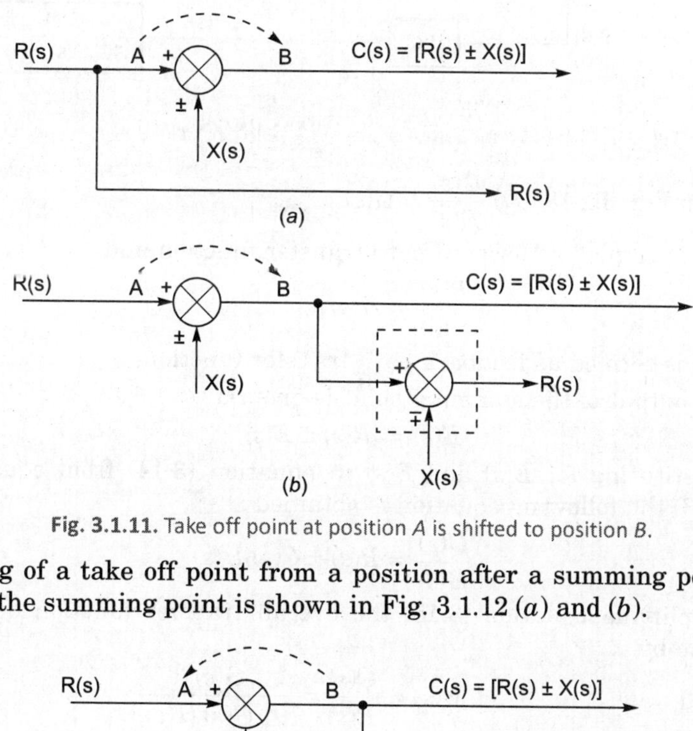

Fig. 3.1.11. Take off point at position *A* is shifted to position *B*.

12. **Shifting of a take off point from a position after a summing point to a position before the summing point is shown in Fig. 3.1.12 (*a*) and (*b*).**

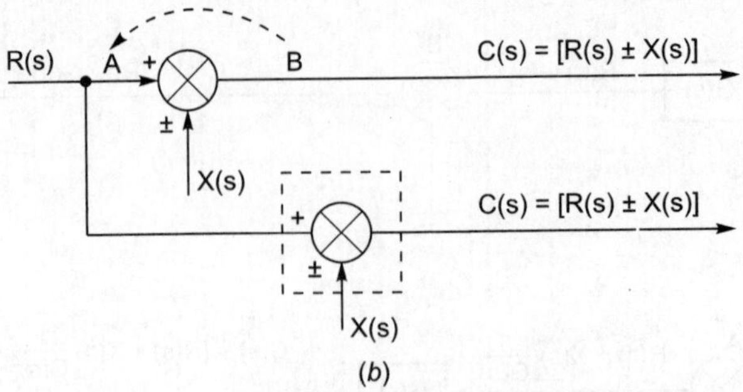

(b)

Fig. 3.1.12. Take off point at position *B* shifted to position *A*.

13. (*a*) Elimination of a summing point in a closed-loop system is explained below :

A closed-loop control system is represented by a block diagram shown in Fig. 3.1.13 (a_1) wherein a fraction of the output $B(s) = C(s)\,H(s)$ is compared with the input $R(s)$ which results in an error $E(s)$ given by

$$E(s) = [R(s) \pm B(s)] \qquad ...(3.11)$$

The equivalence of Fig. 3.1.13 (a_1) with Fig. 3.1.13. (a_2), is obtained as follows :

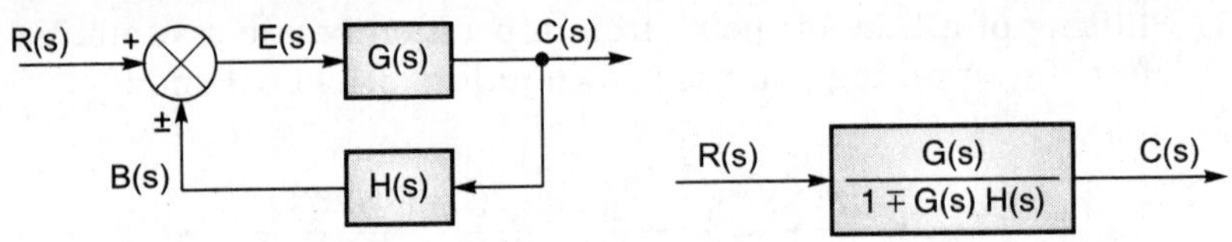

(a_1) (a_2)

Fig. 3.1.13 (*a*) Elimination of a summing point in a closed-loop (feedback) system.

From Fig. 3.1.13 (a_1) $\dfrac{C(s)}{E(s)} = G(s)$ ...(3.12)

$G(s)$ is termed as forward path transfer function and

$$\frac{B(s)}{C(s)} = H(s) \qquad ...(3.13)$$

$H(s)$ is termed as feedback path transfer function.

The output of the summing point is given by

$$E(s) = [R(s) \pm B(s)] \qquad ...(3.14)$$

Substituting for $E(s)$ and $B(s)$ in equation (3.14) from equations (3.12) and (3.13) the following equation is obtained :

$$\frac{C(s)}{G(s)} = R(s) \pm C(s)\,H(s) \qquad ...(3.15)$$

Simplifying equation (3.15) the overall transfer function denoted as $M(s)$ is given by

$$M(s) = \frac{C(s)}{R(s)} = \frac{G(s)}{1 \mp G(s)H(s)} \qquad ...(3.16)$$

The equivalent transfer function as obtained by equation (3.16) is represented by a single block diagram shown in Fig. 3.1.13 (a_2).

13. (*b*) A unity feedback control system is shown in Fig. 3.1.13 (b_1), wherein feedback path transfer function is unity and the equivalence of a unity feedback control system to a single block representation is shown in Fig. 3.1.13 (b_2).

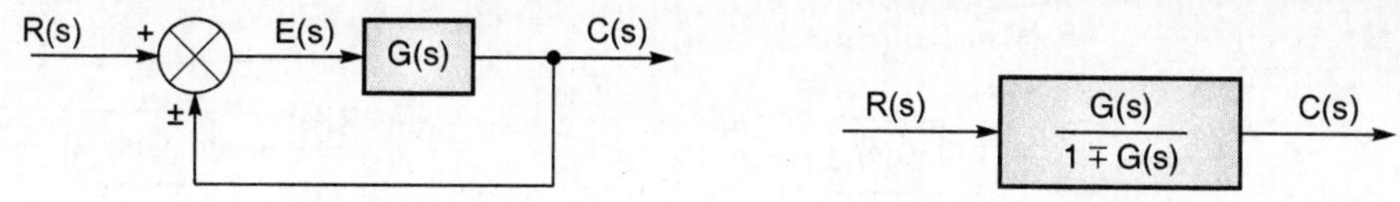

(b_1) (b_2)

Fig. 3.1.13 (*b*) Elimination of a summing point in a unity feedback control system.

Since $H(s) = 1$, the transfer function of a unity feedback control system relating the output and input is thus given by

$$\frac{C(s)}{R(s)} = \frac{G(s)}{1 \mp G(s) \cdot 1} \qquad ...(3.17)$$

or

$$\frac{C(s)}{R(s)} = \frac{G(s)}{1 \mp G(s)} \qquad ...(3.18)$$

13. (*c*) The transfer function relating $E(s)$ and $R(s)$ for a closed-loop control system shown in Fig. 3.1.13 (*c*) is determined below :

As $\quad E(s) = [R(s) \pm B(s)] \qquad ...(3.19)$

and $\quad B(s) = C(s)\, H(s) \qquad ...(3.20)$

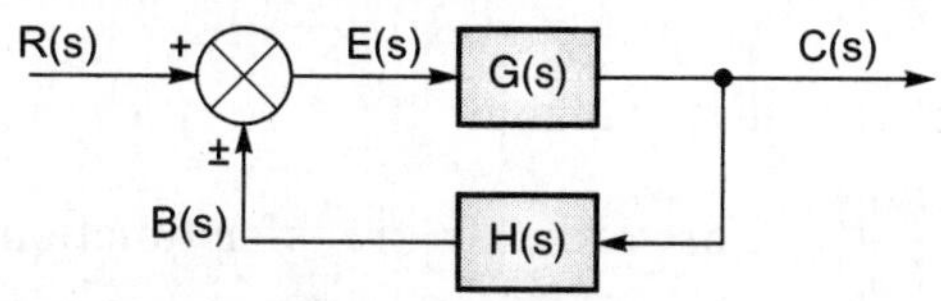

Fig. 3.1.13 (*c*) Block diagram of a closed-loop control system.

Therefore, substituting for $B(s)$ in (3.19) from (3.20) the following relation is obtained :

$$E(s) = [R(s) \pm C(s)\, H(s)] \qquad ...(3.21)$$

It is also seen that $\quad C(s) = E(s)\, G(s) \qquad ...(3.22)$

Eliminate $C(s)$ from (3.21) by substituting for $C(s)$ from (3.22).

Therefore, $\quad E(s) = [R(s) \pm E(s)\, G(s)\, H(s)] \qquad ...(3.23)$

Simplifying equation (3.23), the transfer function relating $E(s)$ and $R(s)$ is obtained below :

$$\frac{E(s)}{R(s)} = \frac{1}{1 \mp G(s) H(s)} \qquad ...(3.24)$$

14. When two or more inputs act on a system the total output is obtained by adding the effect of each individual input separately.

A case of two inputs acting on a system is shown in Fig. 3.1.14 (*a*).

In order to calculate the output due to $R_1(s)$, $R_2(s)$ is considered as zero as shown in Fig. 3.1.14 (*b*).

The output is given by $C_1(s) = R_1(s)\, G(s) \qquad ...(3.25)$

Similarly, while determining the output due to $R_2(s)$, $R_1(s)$ is considered as zero as shown in Fig. 3.1.14 (*c*).

The output for this case is given by

$$C_2(s) = R_2(s)\, G(s) \qquad ...(3.26)$$

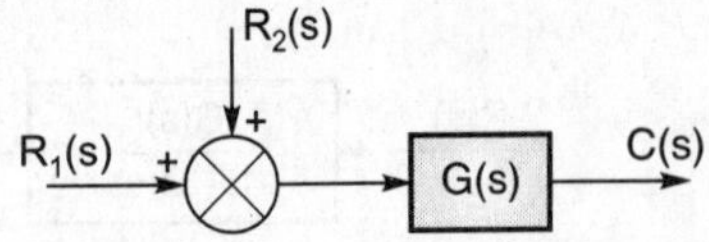

Fig. 3.1.14 (*a*) Two inputs acting on a system.

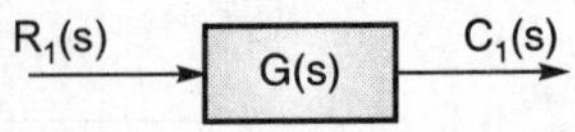

Fig. 3.1.14 (*b*) Considering only $R_1(s)$ input.

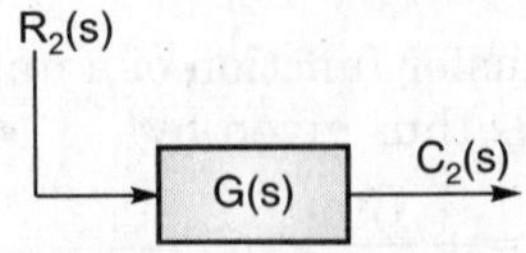

Fig. 3.1.14 (*c*) Considering only $R_2(s)$ input.

The total output is given by,

$$C(s) = C_1(s) + C_2(s) \quad ...(3.27)$$

or

$$C(s) = R_1(s)\, G(s) + R_2(s)\, G(s) \quad ...(3.28)$$

With reference to block diagram of a Control System the terms used are :

$R(s)$: Reference input signal

$C(s)$: Output signal

$E(s)$: Error signal

$G(s) = \dfrac{C(s)}{E(s)}$: Forward path transfer function

$B(s)$: Feedback signal

$H(s)$: Feedback path transfer function

$G(s)\, H(s)$: Open-loop transfer function

$M(s) = \dfrac{C(s)}{R(s)}$: Overall (closed-loop) transfer function or control ratio

$\dfrac{E(s)}{R(s)}$: Error ratio

$\dfrac{B(s)}{R(s)}$: Primary feedback ratio

The application of block diagram reduction rules is illustrated hereunder for determination of the overall transfer function of a system.

3.2 SOLVED EXAMPLES

Note. For convenience $G(s)$, $R(s)$, $C(s)$ etc. are denoted as G, R, C respectively.

Example 3.2.1. *Determine the transfer function C/R from the block diagram shown in Fig. 3.2.1.*

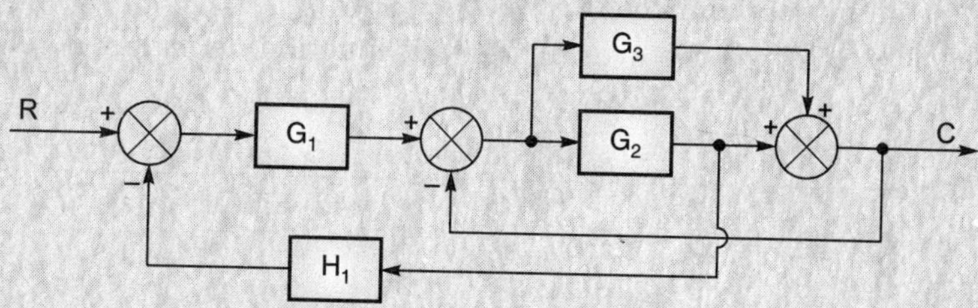

Fig. 3.2.1. Block diagram for example 3.2.1.

Solution. Shift the take off point after block G_2 to a position before block G_2.

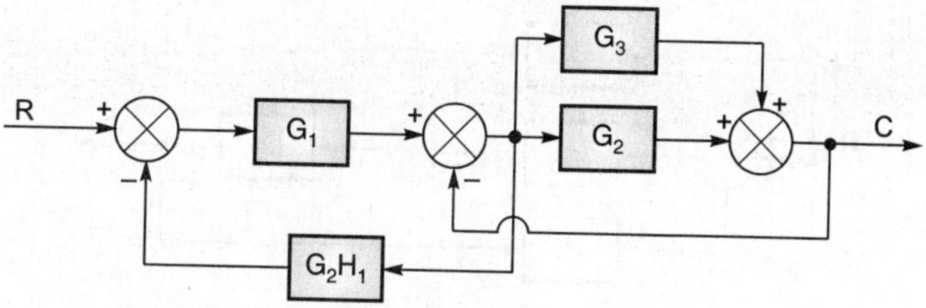

Fig. 3.2.1 (*a*) Block diagram reduction of system in Fig. 3.2.1.

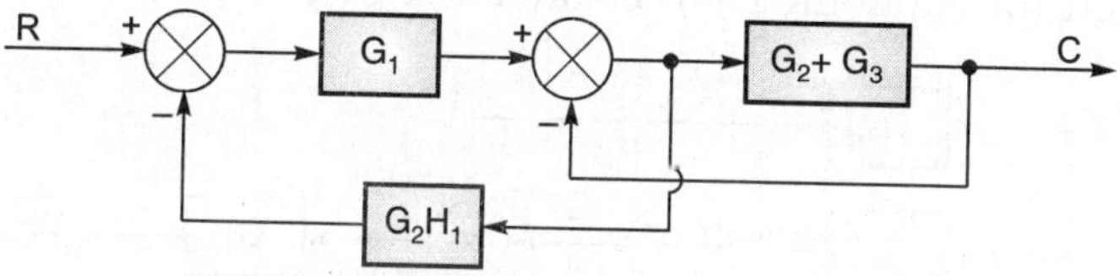

Fig. 3.2.1 (*b*) Block diagram reduction of system in Fig. 3.2.1 (*a*).

Eliminate the summing point after block G_2.

Shift the take off point before block $(G_2 + G_3)$ to a position after block $(G_2 + G_3)$.

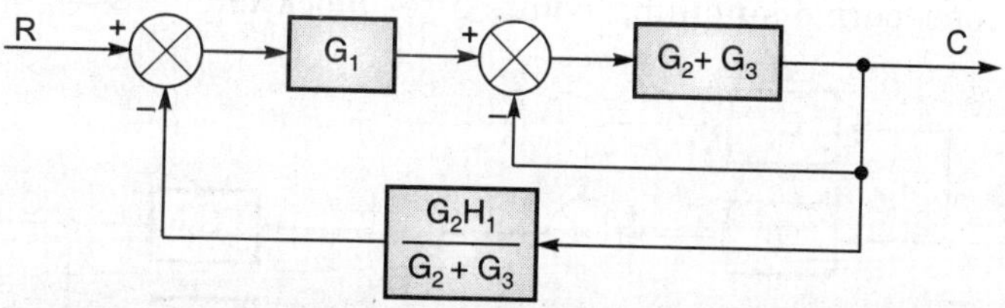

Fig. 3.2.1 (*c*) Block diagram reduction of system in Fig. 3.2.1 (*b*).

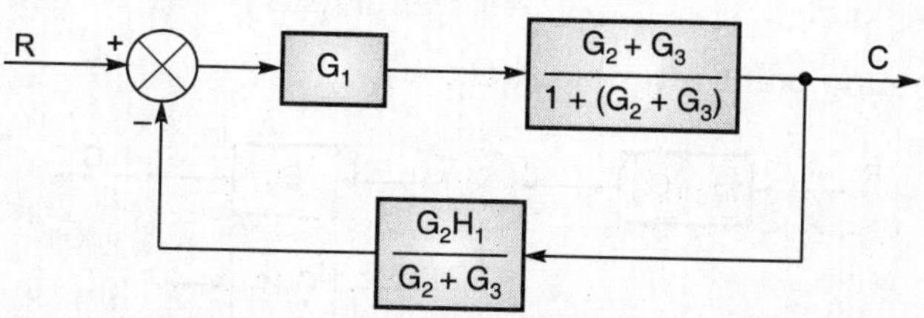

Fig. 3.2.1 (*d*) Block diagram reduction of system in Fig. 3.2.1 (*c*).

Eliminate the summing point before block $(G_2 + G_3)$

According to Fig. 3.2.1 (*d*) the overall transfer function determined below :

$$\frac{C}{R} = \frac{\dfrac{G_1(G_2+G_3)}{[1+(G_2+G_3)]}}{1+\dfrac{G_1(G_2+G_3)}{[1+(G_2+G_3)]}\cdot\dfrac{G_2H_1}{(G_2+G_3)}} = \frac{G_1G_2+G_1G_3}{1+G_2+G_3+G_1G_2H_1} \quad \textbf{Ans.}$$

Example 3.2.2. *Reduce the block diagram shown in Fig. 3.2.2 into*

(i) a form having one block in the forward path and one in feedback path

(ii) single block representation form.

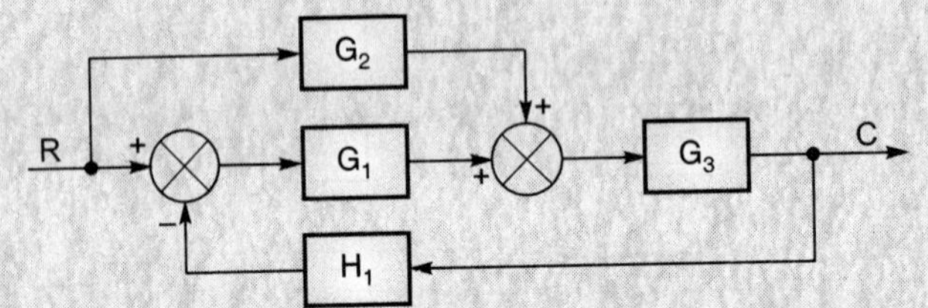

Fig. 3.2.2. Block diagram for example 3.2.2.

Solution. Shift the summing point before block G_1 to a position after block G_1.

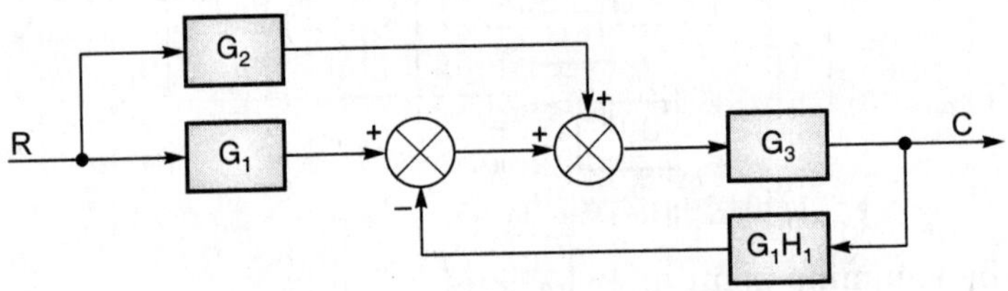

Fig. 3.2.2 (*a*) Block diagram reduction of system in Fig. 3.2.2.

Interchange consecutive summing points after block G_1.

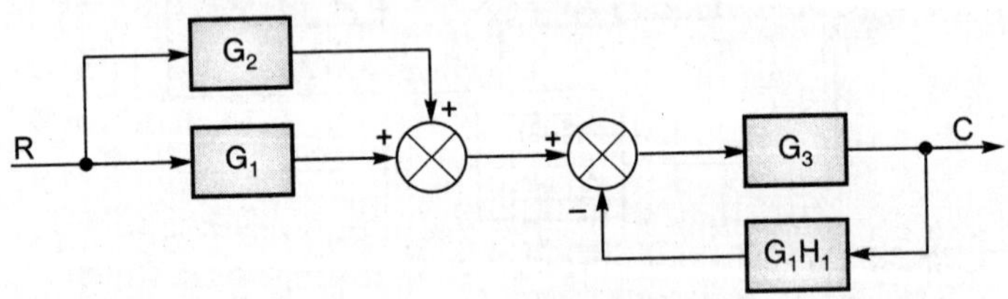

Fig. 3.2.2 (*b*) Block diagram reduction of system in Fig. 3.2.2 (*a*).

Eliminate summing point after block G_1.

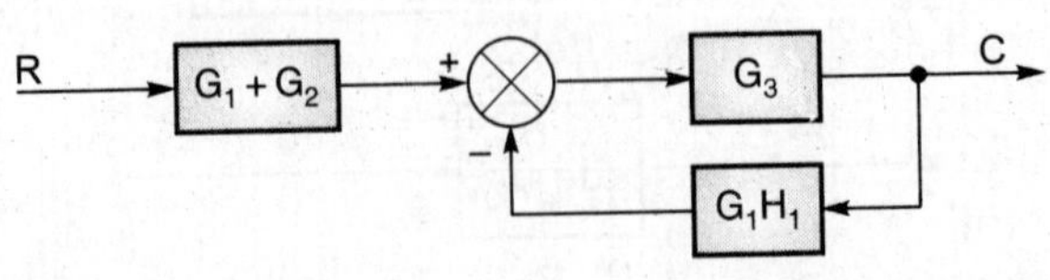

Fig. 3.2.2 (*c*) Block diagram reduction of system in Fig. 3.2.2 (*b*).

Shift the summing point after block $(G_1 + G_2)$ to a position before block $(G_1 + G_2)$ and combine blocks $(G_1 + G_2)$ and G_3.

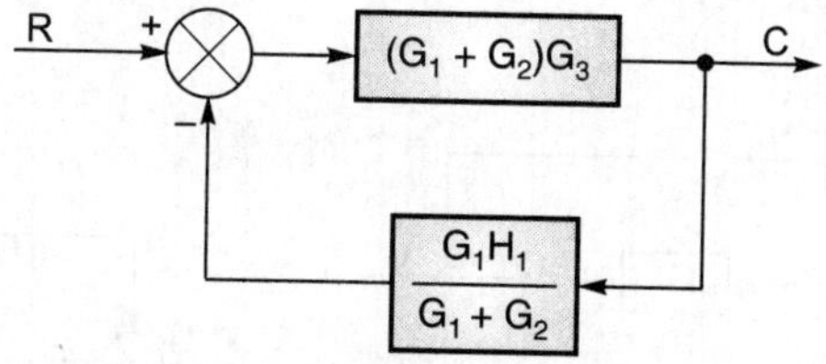

Ans. *(i)*

Fig. 3.2.2 *(d)* Block diagram reduction of example 3.2.2 into a form having one block in forward path and one block in feedback path.

The overall transfer function is determined below

$$\frac{C}{R} = \frac{(G_1 + G_2)G_3}{1 + (G_1 + G_2)G_3 \cdot \dfrac{G_1H_1}{(G_1 + G_2)}}$$

$$= \frac{G_1G_3 + G_2G_3}{1 + G_1G_3H_1} \qquad \text{...(1)}$$

According to equation (1) the single block diagram representation is shown in Fig. 3.2.2 *(e)*.

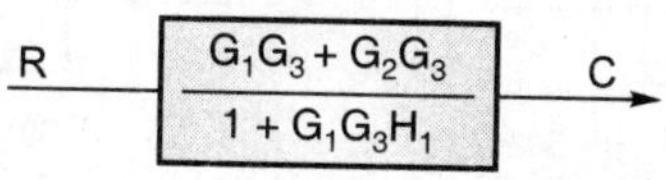

Ans. *(ii)*

Fig. 3.2.2 *(e)* Single block diagram representation of example 3.2.2.

Example 3.2.3. *Determine the overall transfer function for the block diagram shown below.*

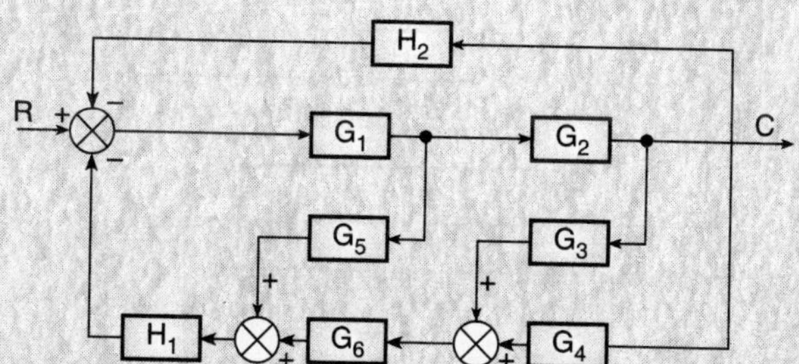

Fig. 3.2.3. Block diagram for example 3.2.3.

Solution. *(i)* Shift the take off point placed before block G_2 towards right side of block G_2

(ii) The blocks G_3 and G_4 are in parallel, the equivalence is $(G_3 + G_4)$, Fig. 3.2.3 *(a)*. Therefore :

(i) The blocks G_1 and G_2 are in cascade *(ii)*

The blocks G_5/G_2 and $G_6(G_3 + G_4)$ are in parallel, the equivalence is $\left\{\dfrac{G_5}{G_2} + G_6(G_3 + G_4)\right\}$, Fig. 3.2.3 (*b*).

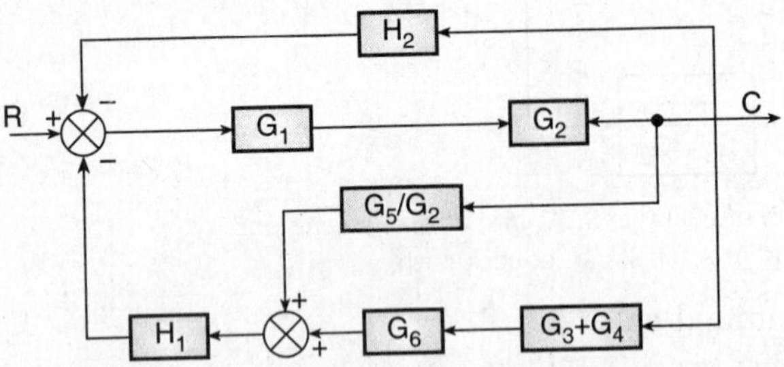

Fig. 3.2.3 (*a*) Block diagram reduction of Fig. 3.2.3.

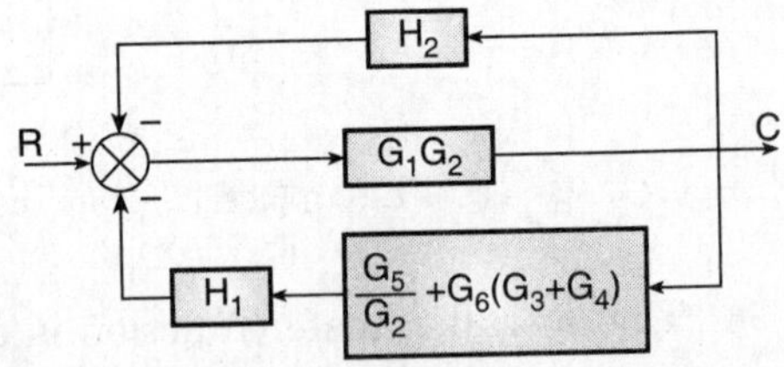

Fig. 3.2.3 (*b*) Block diagram reduction of Fig. 3.2.3 (*a*).

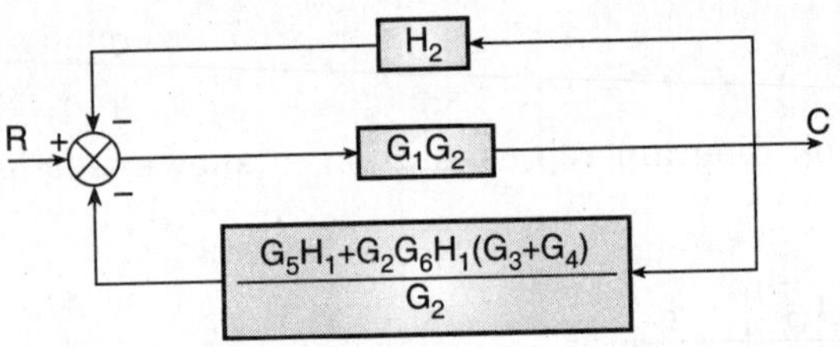

Fig. 3.2.3 (*c*) Block diagram reduction of Fig. 3.2.3 (*b*).

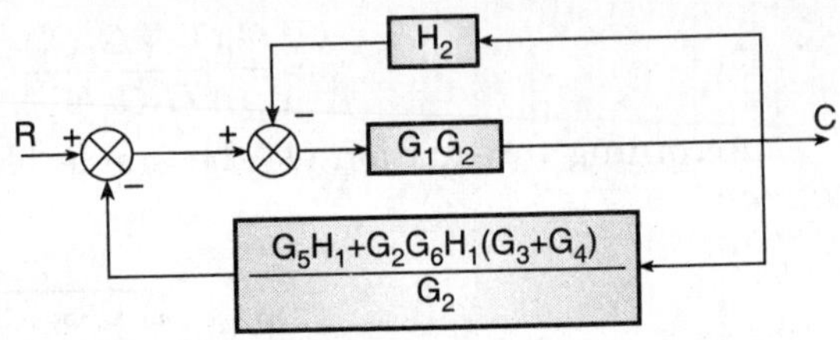

Fig. 3.2.3 (*d*) Block diagram reduction of Fig. 3.2.3 (*c*).

The blocks H_1 and $\dfrac{G_5}{G_2} + G_6(G_3 + G_4)$ are in cascade. Therefore :

Split the summing point to get : Fig. 3.2.3 (*d*).

Eliminate the summing point before block G_1G_2. Therefore :

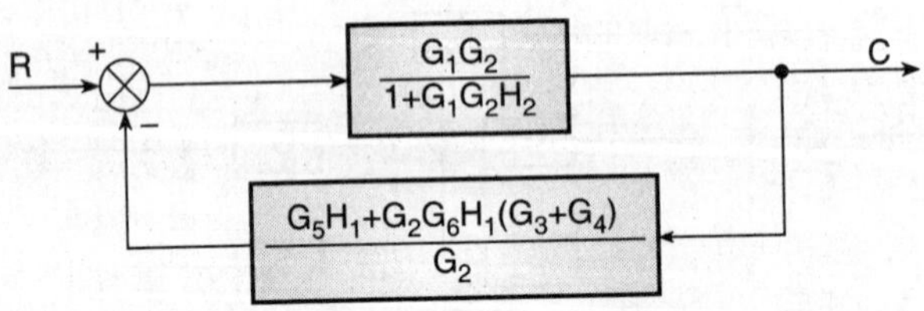

Fig. 3.2.3 (*e*) Block diagram reduction of Fig. 3.2.3 (*e*).

On eliminating the summing point in Fig. 3.2.3 (*d*)

$$\frac{C}{R} = \frac{\dfrac{G_1G_2}{1+G_1G_2H_2}}{1+\dfrac{G_1G_2}{1+G_1G_2H_2}\times\dfrac{G_5H_1+G_2G_6H_1(G_3+G_4)}{G_2}}$$

Simplifying, $$\frac{C}{R} = \frac{G_1G_2}{1+G_1G_2H_2+G_1G_5H_1+G_1G_2G_3G_6H_1+G_1G_2G_4G_6H_1}$$ **Ans.**

Example 3.2.4. *The block diagram of a control system is shown below. Determine the overall transfer function.*

Solution. Shift the take off point before block G_1 to a position after block G_1. The block diagram takes the form, Fig. 3.2.4 (*a*).

The blocks G_3/G_1 and G_2 are in parallel, the equivalence is $\dfrac{G_3}{G_1} + G_2 = \dfrac{G_1G_2 + G_3}{G_1}$.

Therefore : Fig. 3.2.4 (*b*).

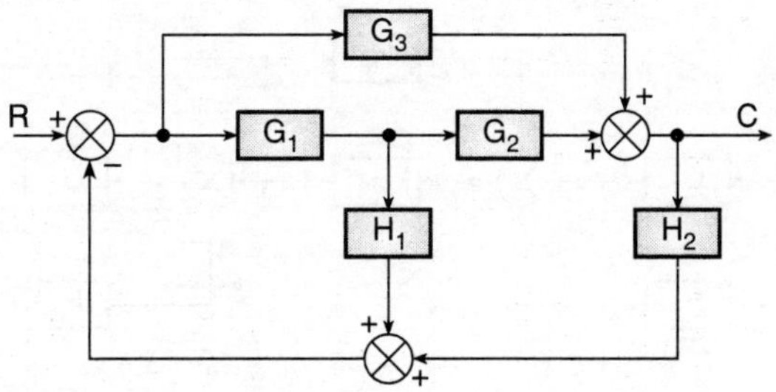

Fig. 3.2.4. Block diagram for example 3.2.4.

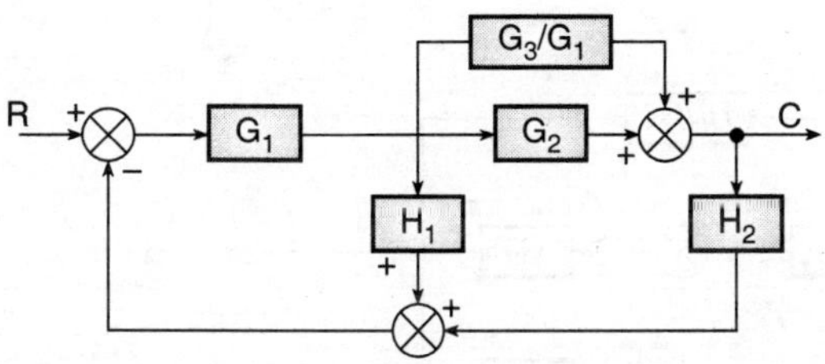

Fig. 3.2.4 (*a*) Block diagram reduction Fig. 3.2.4.

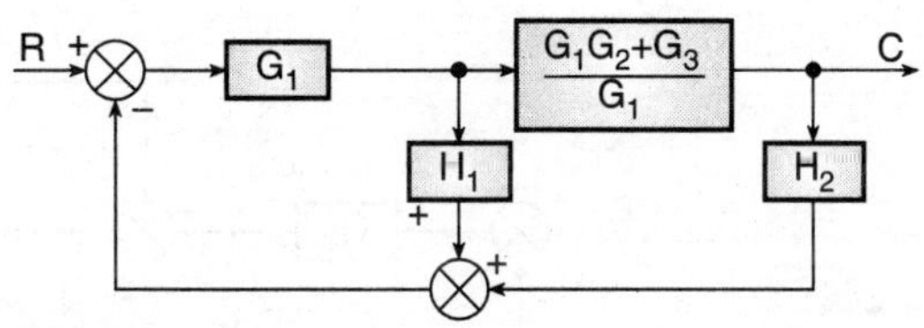

Fig. 3.2.4 (*b*) Block diagram reduction of Fig. 3.2.4 (*a*).

Shift the take off point located before block $(G_1G_2 + G_3)/G_1$ towards right side of the same block and redraw the block diagram [Fig. 3.2.4 (*c*)].

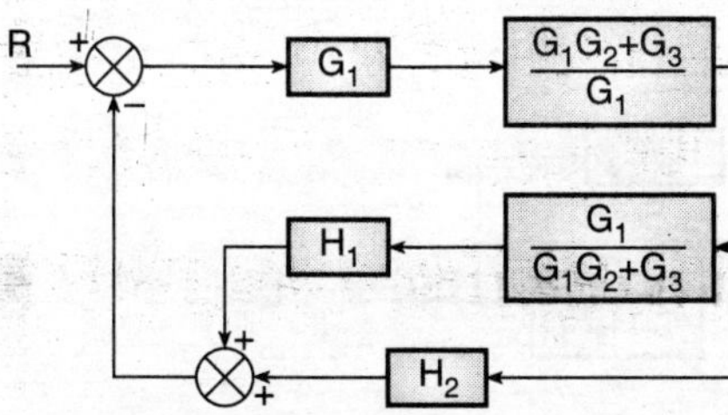

Fig. 3.2.4 (*c*) Block diagram reduction of Fig. 3.2.4 (*b*).

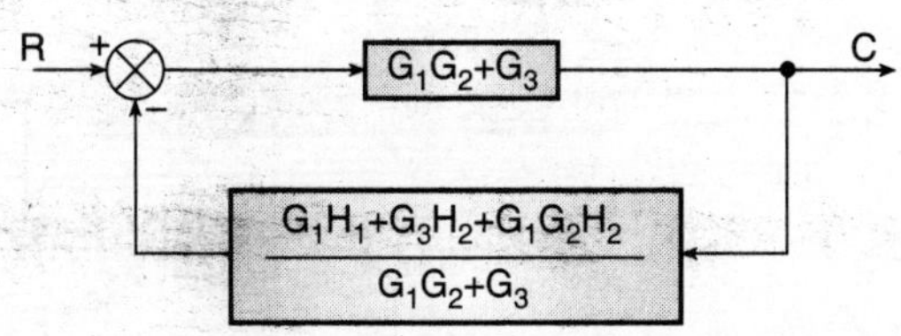

Fig. 3.2.4 (*d*) Block diagram reduction of Fig. 3.2.4 (*c*).

(*i*) The blocks G_1 and $(G_1G_2 + G_3)/G_1$ are in cascade, the equivalence is $G_1 \times (G_1G_2 + G_3)/G_1 = G_1G_2 + G_3$.

(*ii*) The blocks H_2 and $H_1\,G_1/(G_1G_2 + G_3)$ are in parallel, the equivalence is

$$H_2 + \frac{H_1G_1}{G_1G_2 + G_3} = \frac{G_1H_1 + G_3H_2 + G_1G_2H_2}{G_1G_2G_3}$$

The block diagram takes the form [Fig. 3.2.4 (*d*)]

Eliminating the summing point

$$\frac{C}{R} = \frac{G_1G_2 + G_3}{1 + (G_1G_2 + G_3) \times \dfrac{(G_1H_1 + G_3H_2 + G_1G_2H_2)}{G_1G_2 + G_3}}$$

Simplifying, $$\frac{C}{R} = \frac{G_1G_2 + G_3}{1 + G_1H_1 + G_3H_2 + G_1G_2H_2} \quad \textbf{Ans.}$$

Example 3.2.5. *Determine C/R for the block diagram given below (Fig. 3.2.5).*

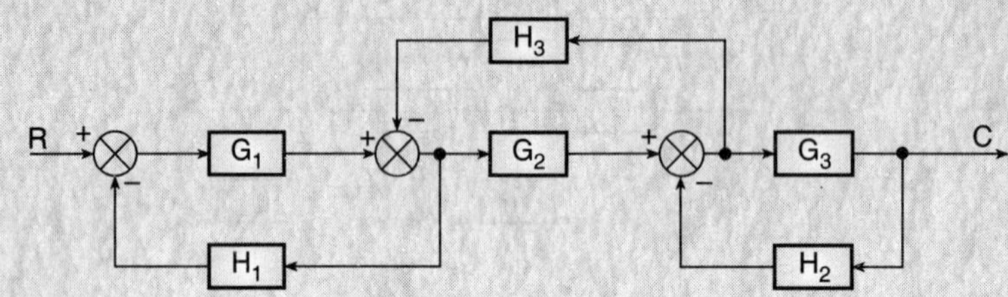

Fig. 3.2.5. Block diagram for Example 3.2.5.

Solution. Shift the take off point placed before block G_3 to a position after block G_3 to obtain Fig. 3.2.5 (*a*).

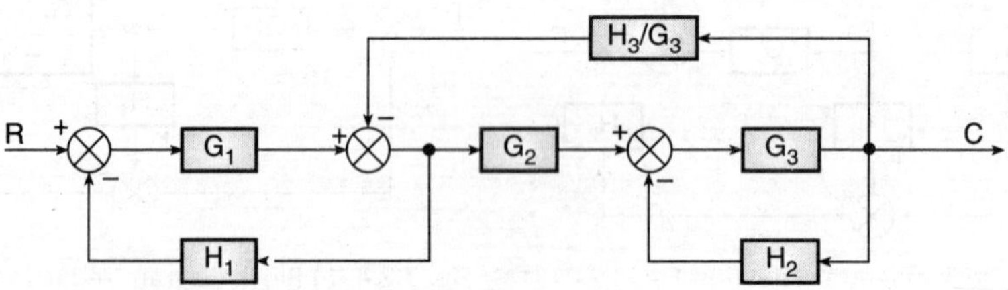

Fig. 3.2.5 (*a*) Block diagram reduction of Fig. 3.2.5.

Eliminate the summing point placed before block G_3 to get Fig. 3.2.5 (*b*).

The blocks G_2 and $G_3/(1 + G_3H_2)$ are in cascade, therefore, the block diagram takes the form [Fig. 3.2.5 (*c*)].

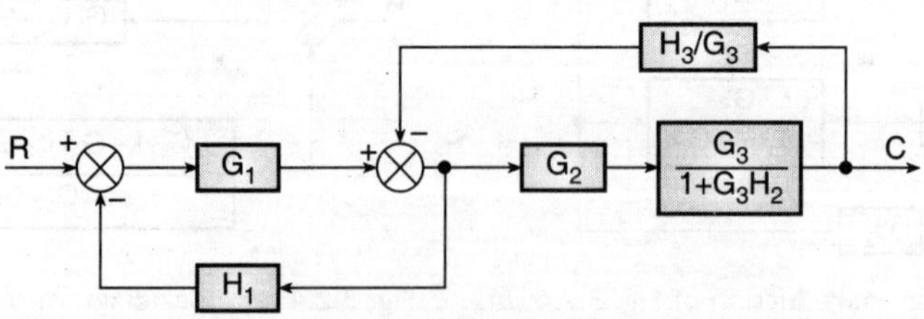

Fig. 3.2.5 (*b*) Block diagram reduction of Fig. 3.2.5 (*a*).

Shift the summing point placed after block G_1 to a location before block G_1. Therefore [Fig. 3.2.5 (*d*)].

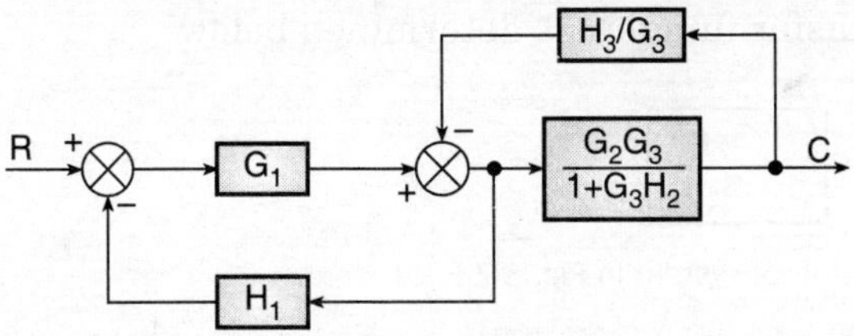

Fig. 3.2.5 (*c*) Block diagram reduction of Fig. 3.2.5 (*b*).

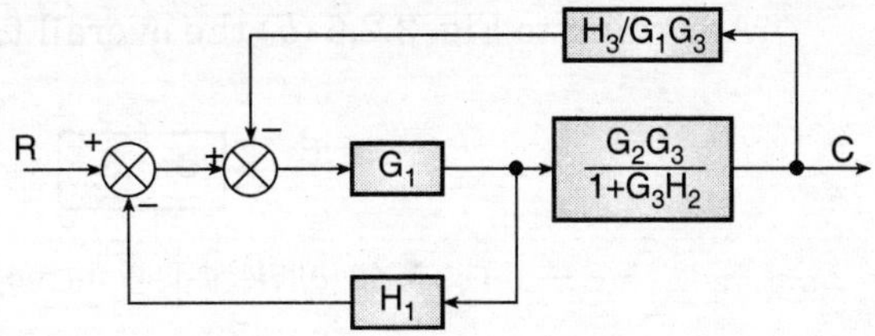

Fig. 3.2.5 (*d*) Block diagram reduction of Fig. 3.2.5 (*c*).

Interchange consecutive summing points and then eliminate the summing point before block G_1 to obtain the block diagram as below [Fig. 3.2.5 (*e*)].

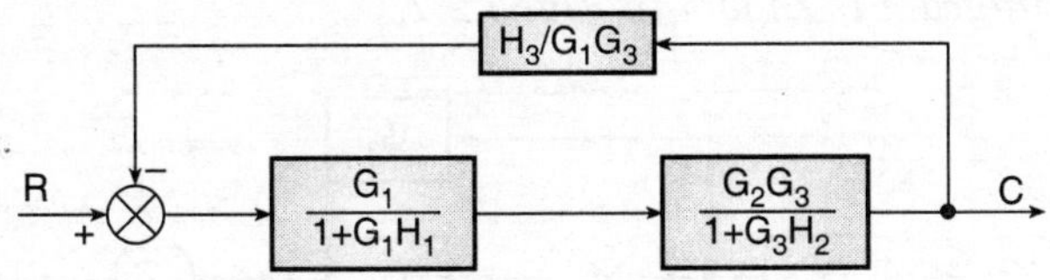

Fig. 3.2.5 (*e*) Block diagram reduction of Fig. 3.2.5 (*d*).

The forward path transfer function is :

$$\frac{G_1G_2G_3}{(1+G_1H_1)(1+G_3H_2)}$$

The feedback path transfer function is : H_3/G_1G_3

Therefore,
$$\frac{C}{R} = \frac{\dfrac{G_1G_2G_3}{(1+G_1H_1)(1+G_3H_2)}}{1+\dfrac{G_1G_2G_3}{(1+G_1H_1)(1+G_3H_2)}} \times \frac{H_1}{G_1G_3}$$

Simplifying,
$$\frac{C}{R} = \frac{G_1G_2G_3}{1+G_1H_1+G_2H_3+G_3H_2+G_1G_3H_1H_2} \quad \textbf{Ans.}$$

Example 3.2.6. *For the block diagram shown in Fig. 3.2.6. Determine the overall transfer function.*

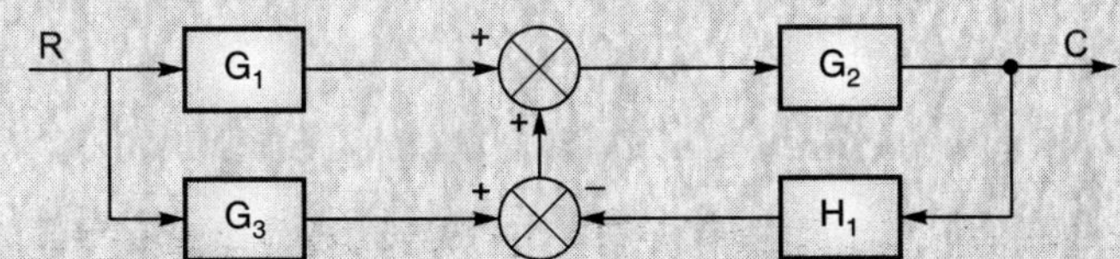

Fig. 3.2.6. Block diagram for example 3.2.6.

Solution. The block diagram of Fig. 3.2.6 can be redrawn as shown in Fig. 3.2.6 (*a*). Eliminate two summing points.

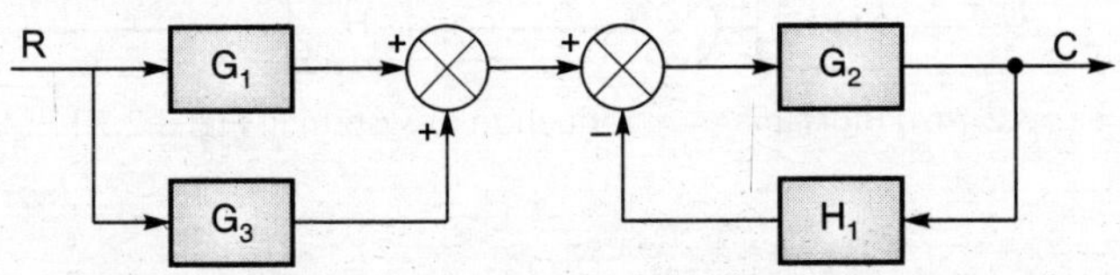

Fig. 3.2.6 (*a*) Block diagram of Fig. 3.2.6 redrawn.

According to Fig. 3.2.6 (*b*) the overall transfer function is determined below

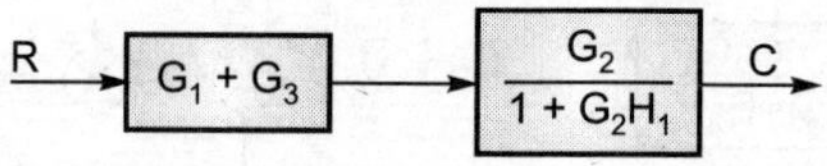

Fig. 3.2.6 (*b*) Block diagram reduction of system in Fig. 3.2.6 (*a*).

$$\frac{C}{R}=\frac{(G_1+G_3)G_2}{1+G_2H_1}=\frac{G_1G_2+G_2G_3}{1+G_2H_1} \quad \textbf{Ans.}$$

Example 3.2.7. *Determine the overall transfer function relating C and R for the system whose block diagram is shown in Fig. 3.2.7.*

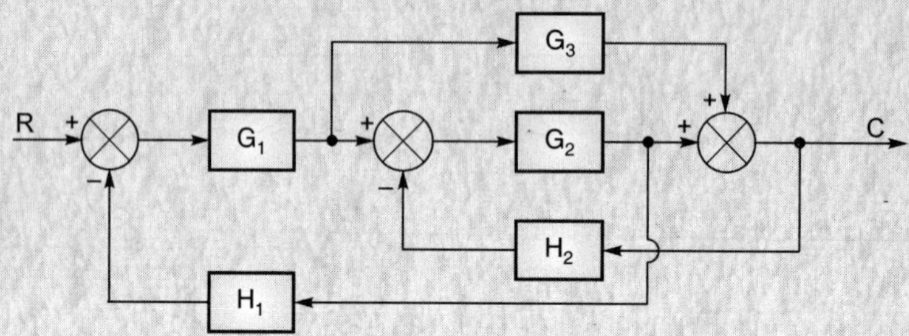

Fig. 3.2.7. Block diagram for example 3.2.7.

Solution. Shift the take off point after block G_2 to a position before block G_2.

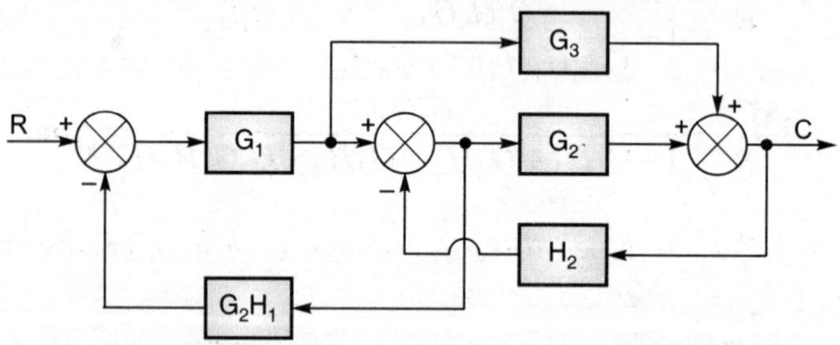

Fig. 3.2.7 (*a*) Block diagram reduction of system in Fig. 3.2.7.

Shift the take off point before block G_2 to a position after block G_1 (Rule 12).

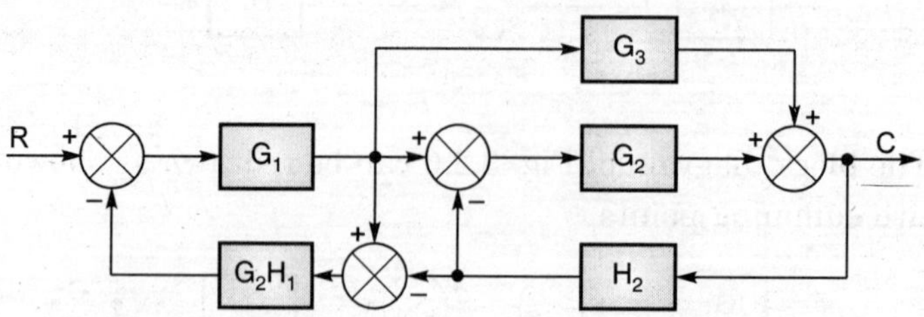

Fig. 3.2.7 (*b*) Block diagram reduction of system in Fig. 3.2.7 (*a*).

Shift summing point before block G_2 to a position after block G_2.

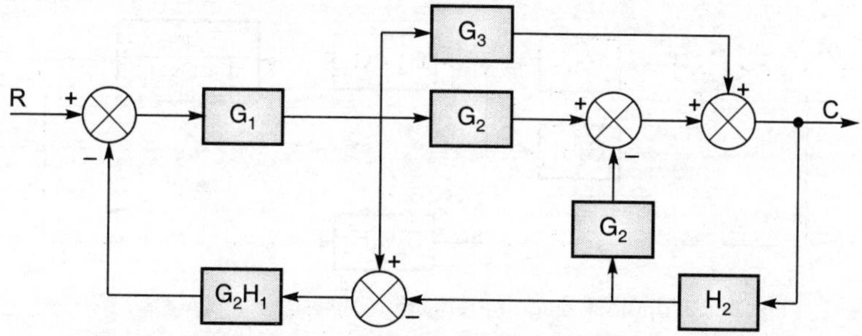

Fig. 3.2.7 (*c*) Block diagram reduction of system in Fig. 3.2.7 (*b*).

Interchange consecutive summing points after block G_2.

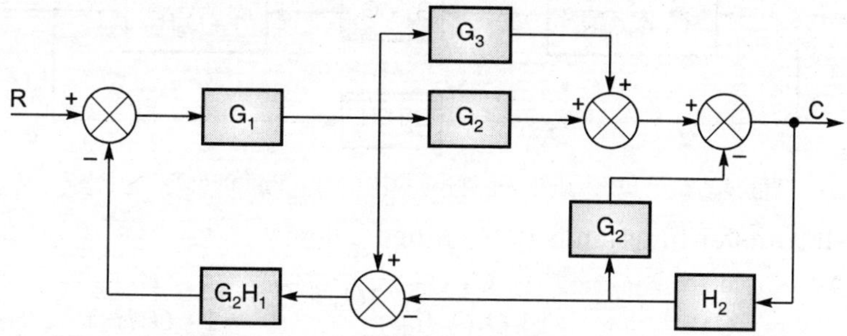

Fig. 3.2.7 (*d*) Block diagram reduction of system in Fig. 3.2.7 (*c*).

Shift the take off point after block H_2 to a position before block H_2.

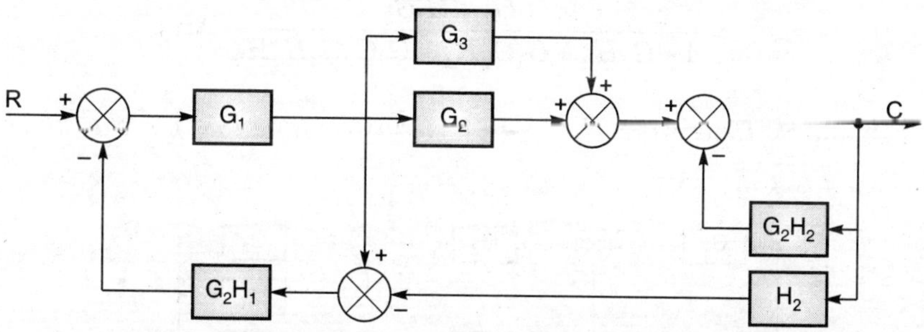

Fig. 3.2.7 (*e*) Block diagram reduction of system in Fig. 3.2.7 (*d*).

The block diagram shown in Fig. 3.2.7 (*e*) can be redrawn as shown in Fig. 3.2.7 (*f*).

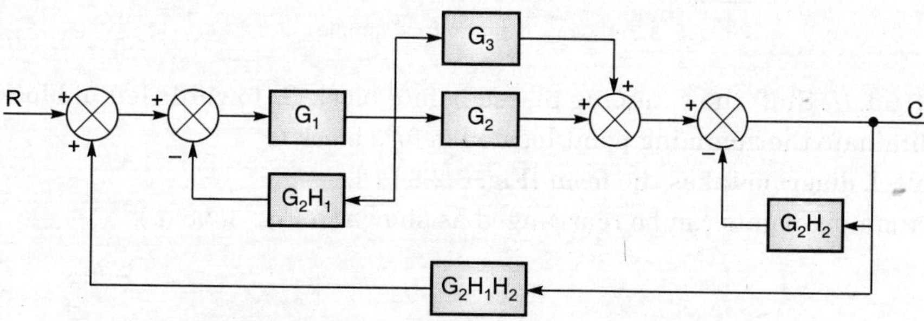

Fig. 3.2.7 (*f*) Block diagram of Fig. 3.2.7 (*e*) redrawn.

Eliminate two consecutive summing points after block G_2.

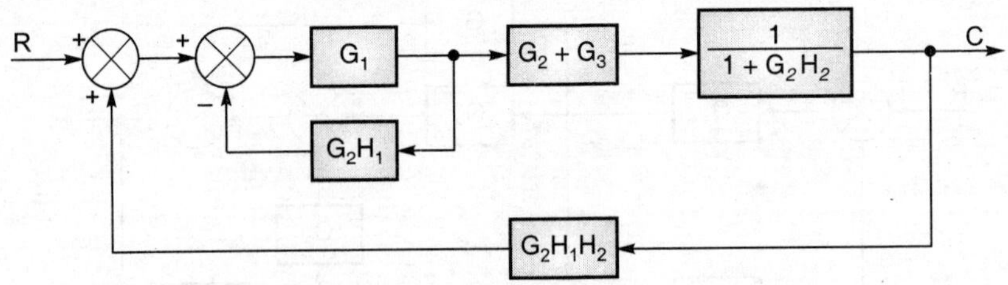

Fig. 3.2.7 (*g*) Block diagram reduction of system in Fig. 3.2.7 (*f*).

Eliminate the summing point before block G_1.

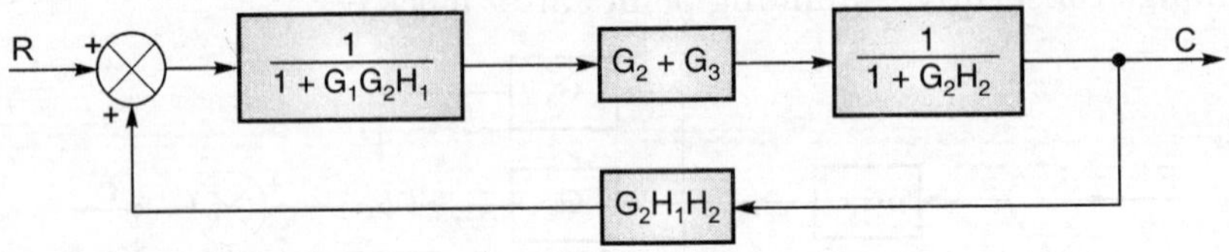

Fig. 3.2.7 (*h*) Block diagram reduction of system in Fig. 3.2.7 (*g*).

The overall transfer function is determined below :

$$\frac{C}{R} = \frac{\dfrac{G_1}{1+G_1G_2H_1} \cdot (G_2+G_3) \cdot \dfrac{1}{(1+G_2H_2)}}{1 - \dfrac{G_1}{(1+G_1G_2H_1)} \cdot (G_2+G_3) \cdot \dfrac{1}{(1+G_2H_2)} \cdot (G_2H_1H_2)}$$

$$= \frac{G_1G_2 + G_1G_3}{1 + G_2H_2 + G_1G_2H_1 - G_1G_2G_3H_1H_2} \quad \textbf{Ans.}$$

Example 3.2.8. *Determine the overall transfer function for the block diagram (Fig. 3.2.8).*

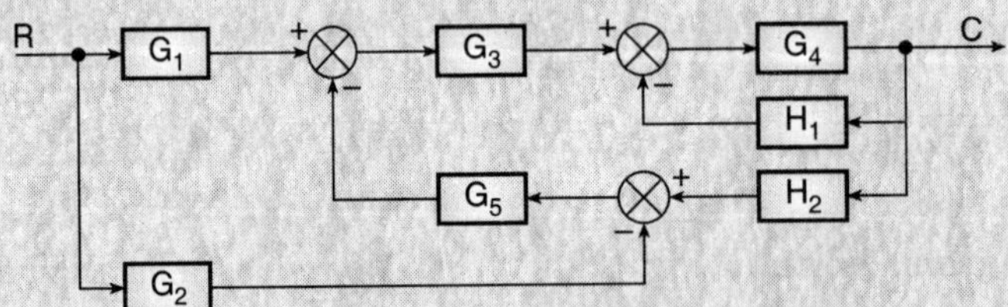

Fig. 3.2.8. Block diagram for example 3.2.8.

Solution. (*i*) Shift the summing placed before block G_5 towards left of block G_5.
(*ii*) Eliminate the summing point located before block G_4.
The block diagram takes the form [Fig. 3.2.8 (*a*)].
The summing points can be rearranged as shown in Fig. 3.2.8 (*b*).

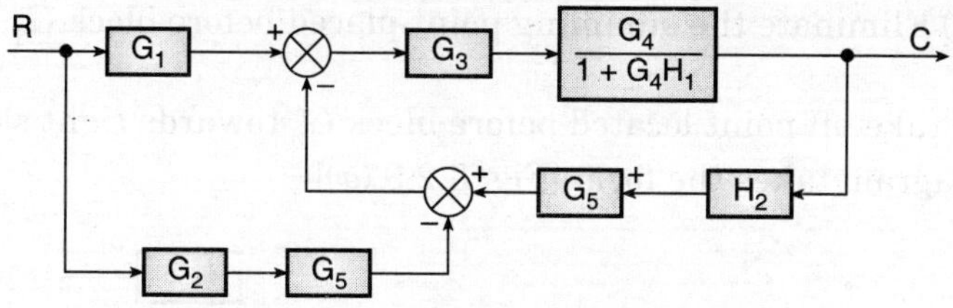

Fig. 3.2.8 (*a*) Block diagram reduction for Fig. 3.2.8.

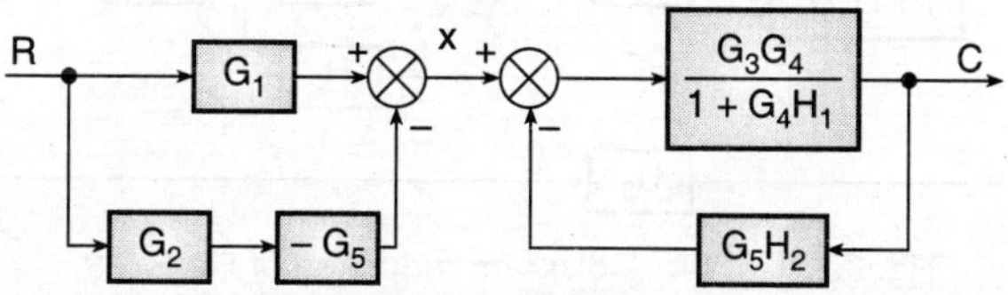

Fig. 3.2.8 (*b*) Block diagram reduction of Fig. 3.2.8 (*a*).

(*i*) **The blocks G_2 and $G_2(-G_5)$ are in parallel, the equivalence is**

$$G_1 - G_2(-G_5) = G_1 + G_2G_5$$

(*ii*) **Eliminating the summing point located before block $\dfrac{G_3G_4}{1+G_4H_1}$:**

$$\frac{C}{x} = \frac{\dfrac{G_3G_4}{1+G_4H_1}}{1+\dfrac{G_3G_4}{1+G_4H_1}\times G_5H_2} = \frac{G_3G_4}{1+G_4H_1+G_3G_4G_5H_2}$$

The block diagram takes the form as shown in Fig. 3.2.8 (*c*).

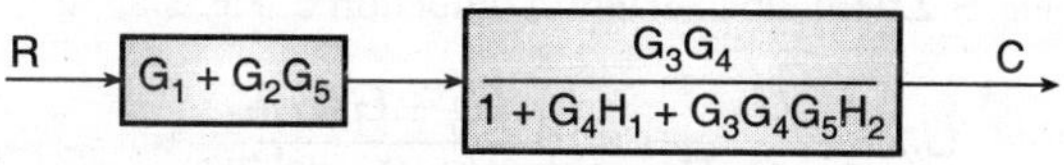

Fig. 3.2.8. (*c*) Block diagram reduction of Fig. 3.2.8 (*b*).

The two blocks are in cascade

$$\therefore \quad \frac{C}{R} = (G_1 + G_2G_5) \times \frac{G_3G_4}{1+G_4H_1+G_3G_4G_5H_1}$$

or $$\frac{C}{R} = \frac{G_1G_3G_4 + G_2G_3G_4G_5}{1+G_4H_1+G_3G_4G_5H_1} \quad \textbf{Ans.}$$

Example 3.2.9. *Determine the transfer function C/R for the system block diagram given in Fig. 3.2.9.*

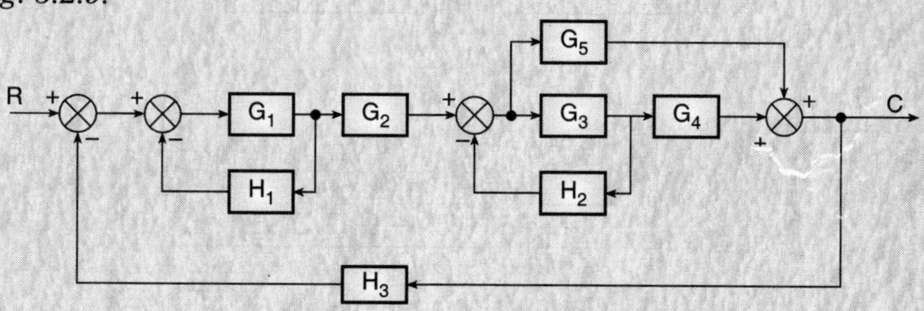

Fig. 3.2.9. Block diagram for example 3.2.9.

Solution. (*i*) Eliminate the summing point placed before block G_1, the equivalence is $G_1/(1 + G_1H_1)$.

(*ii*) Shift the take-off point located before block G_3 towards right side of block G_3. The block diagram takes the form [Fig. 3.2.9 (*a*)]

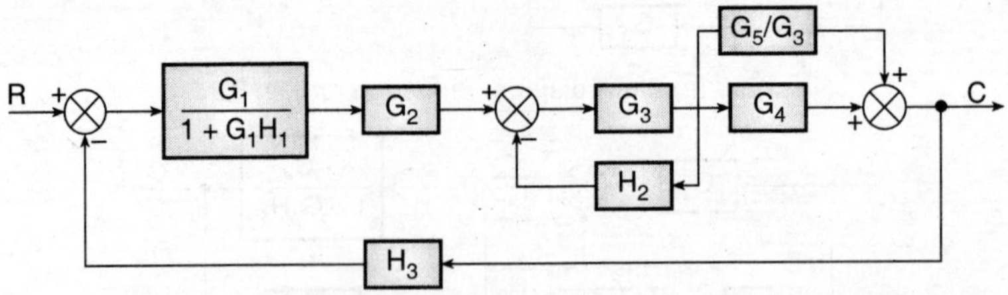

Fig. 3.2.9 (*a*) Block diagram reduction of Fig. 3.2.9.

(*i*) The blocks G_5/G_3 and G_4 are in parallel, the equivalence is $\dfrac{G_5}{G_3} + G_4 = \dfrac{G_3G_4 + G_5}{G_3}$.

(*ii*) Eliminate the summing point placed before block G_3, the equivalence is $G_3/(1 + G_3H_2)$.

The block diagram is arranged in Fig. 3.2.9 (*b*).

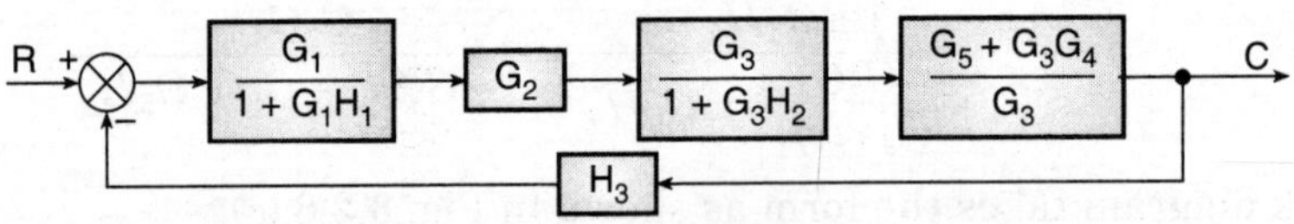

Fig. 3.2.9 (*b*) Block diagram reduction of Fig. 3.2.9 (*a*).

The blocks $\dfrac{G_1}{1+G_1H_1}, G_2, \dfrac{G_3}{1+G_3H_2}$ and $\dfrac{G_5 + G_3G_4}{G_3}$

are in cascade. The equivalence is :

$$\frac{G_1}{1+G_1H_1} \times G_2 \times \frac{G_3}{1+G_3H_2} \times \frac{(G_5 + G_3G_4)}{G_3} = \frac{G_1G_2G_5 + G_1G_2G_3G_4}{1+G_1H_1 + G_3H_2 + G_1G_3H_1H_2}$$

The block diagram is shown in Fig. 3.2.9 (*c*)

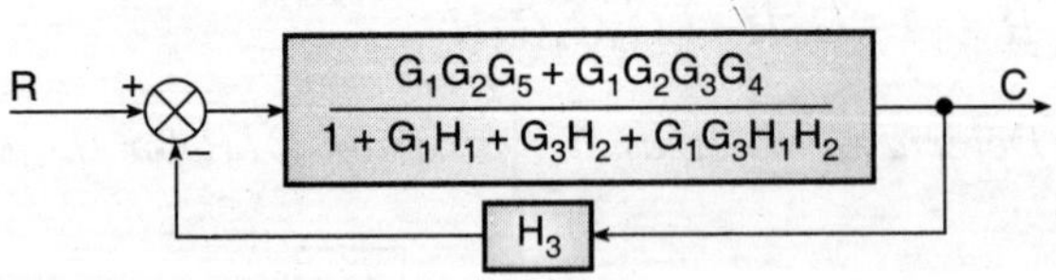

Fig. 3.2.9 (*c*) Block diagram reduction of Fig. 3.2.9 (*b*).

Eliminating the summing point

$$\frac{C}{R} = \frac{\dfrac{G_1G_2G_5 + G_1G_2G_3G_4}{1+G_1H_1 + G_3H_2 + G_1G_3H_1H_2}}{1 + \dfrac{G_1G_2G_5 + G_1G_2G_3G_4}{1+G_1H_1 + G_3H_2 + G_1G_3H_1H_2} \times H_3}$$

Simplifying,

$$\frac{C}{R} = \frac{G_1G_2G_5 + G_1G_2G_3G_4}{1 + G_1H_1 + G_3H_2 + G_1G_2G_5H_3 + G_1G_2G_3G_4H_3 + G_1G_2G_3H_1H_2} \quad \textbf{Ans.}$$

Example 3.2.10. *Obtain the overall transfer function for a control system having the block diagram (Fig. 3.2.10).*

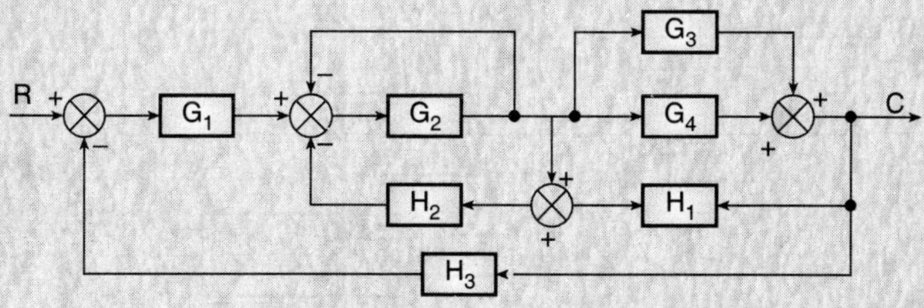

Fig. 3.2.10. Block diagram for example 3.2.10.

Solution. (*i*) The blocks G_3 and G_4 are in parallel, the equivalence is $(G_3 + G_4)$

(*ii*) Split the summing point located before block G_2

The block diagram takes the form [Fig. 3.2.10 (*a*)]

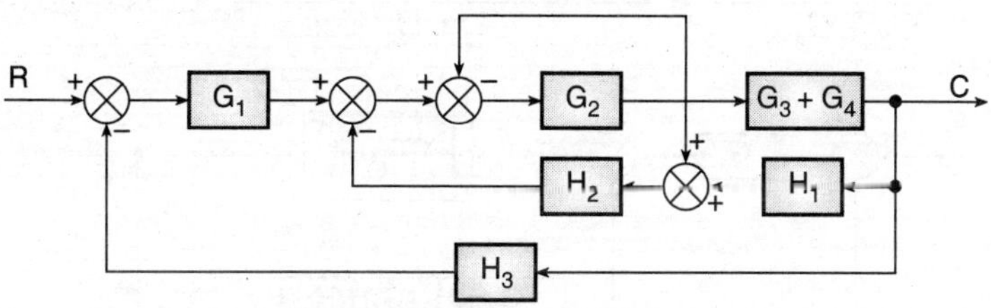

Fig. 3.2.10 (*a*) Block diagram reduction of Fig. 3.2.10.

(*i*) Shift the take-off point placed before block $(G_3 + G_4)$ towards right side of block $(G_3 + G_4)$.

(*ii*) Eliminate the summing point located before block G_2, the equivalence is $G_2/(1 + G_2)$.

Draw the block diagram as below [Fig. 3.2.10 (*b*)]

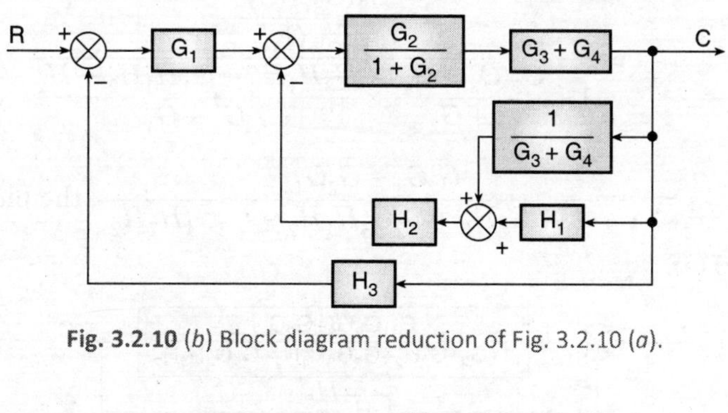

Fig. 3.2.10 (*b*) Block diagram reduction of Fig. 3.2.10 (*a*).

(*i*) The blocks H_1 and $1/(G_3 + G_4)$ are in parallel, the equivalence is

$$H_1 + \frac{1}{G_3 + G_4} = \frac{G_3H_1 + G_4H_1 + 1}{G_3 + G_4}$$

(*ii*) The blocks $G_2/(1 + G_2)$ and $(G_3 + G_4)$ are in cascade, the equivalence is

$$\frac{G_2}{1+G_2} \times (G_3 + G_4) = \frac{G_2(G_3 + G_4)}{1+G_2}$$

Therefore [Fig. 3.2.10 (*c*)]

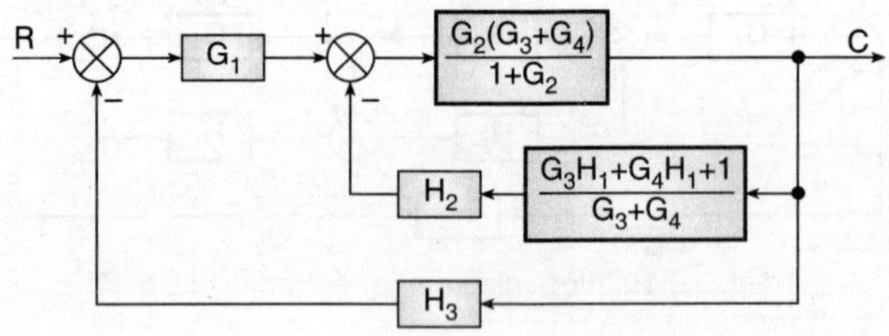

Fig. 3.2.10 (*c*) Block diagram reduction of Fig. 3.2.10 (*c*).

The blocks H_2 and $(G_3H_1 + G_4H_1 + 1)/(G_3 + G_4)$ are in cascade, the equivalence is

$$H_2 \times \frac{G_3H_1 + G_4H_1 + 1}{G_3 + G_4} = \frac{G_3H_1H_2 + G_4H_1H_2 + H_2}{G_3 + G_4}$$, thus [Fig. 3.2.10 (*d*)]

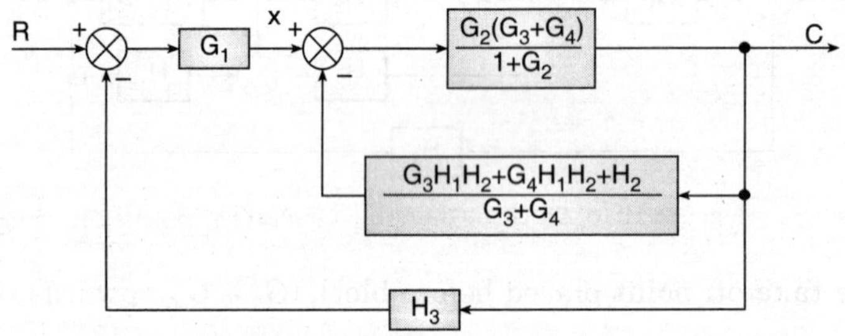

Fig. 3.2.10 (*d*) Block diagram reduction of Fig. 3.2.10 (*c*).

Eliminate the summing point located after block G_1.

$$\therefore \qquad \frac{C}{x} = \frac{\dfrac{G_2(G_3 + G_4)}{1+G_2}}{1 + \dfrac{G_2(G_3 + G_4)}{1+G_2} \times \dfrac{G_3H_1H_2 + G_4H_1H_2 + H_2}{G_3 + G_4}}$$

Simplifying, $\dfrac{C}{x} = \dfrac{G_2G_3 + G_2G_4}{1 + G_2 + G_2H_2 + G_2G_3H_1H_2 + G_2G_4H_1H_2}$, the block diagram takes the form Fig. 3.2.10 (*e*).

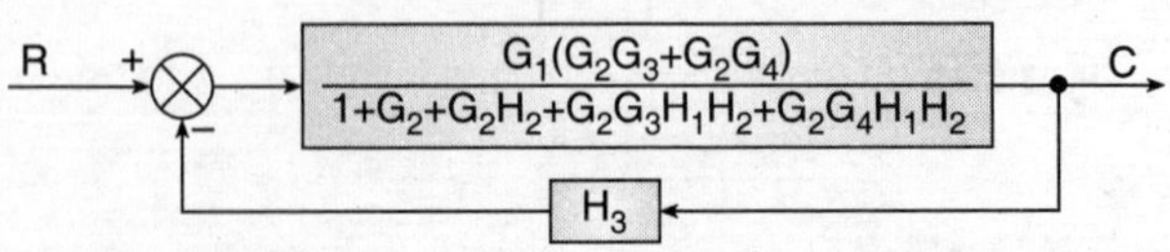

Fig. 3.2.10 (*e*) Block diagram reduction of Fig. 3.2.10 (*d*).

Eliminating the summing point,

$$\frac{C}{R} = \frac{\dfrac{G_1(G_2G_3 + G_2G_4)}{1 + G_2 + G_2H_2 + G_2G_3H_1H_2 + G_2G_4H_1H_2}}{1 + \dfrac{G_1(G_2G_3 + G_2G_4)}{1 + G_2 + G_2H_2 + G_2G_3H_1H_2 + G_2G_4H_1H_2} \times H_3}$$

Simplifying, $$\frac{C}{R} = \frac{G_1G_2G_3 + G_1G_2G_4}{1 + G_2 + G_2G_3H_1H_2 + G_2G_4H_1H_2 + G_1G_2G_3H_3 + G_1G_2G_4H_3}$$ **Ans.**

Example 3.2.11. *Reduce the system shown below to a single block representation.*

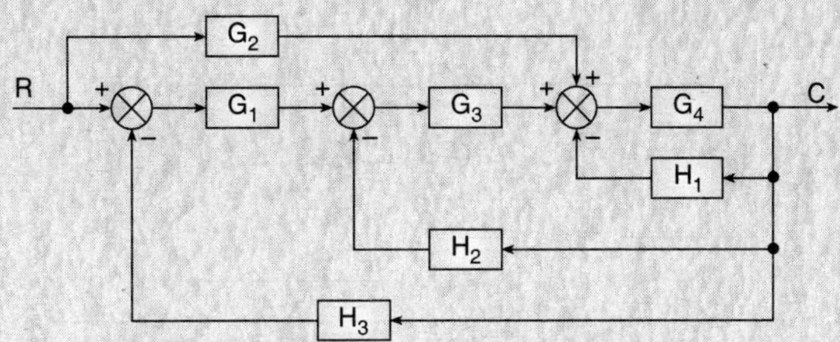

Fig. 3.2.11. Block diagram for example 3.2.11.

Solution. Split the summing point before block G_4 [Fig. 3.2.11 (*a*)]

(*i*) Eliminate summing point before block G_4, the equivalence is $G_4/(1 + G_4H_1)$.

(*ii*) Shift the summing point placed after block G_3 towards left side of block G_3. The block diagram takes the form [Fig. 3.2.11 (*b*)]

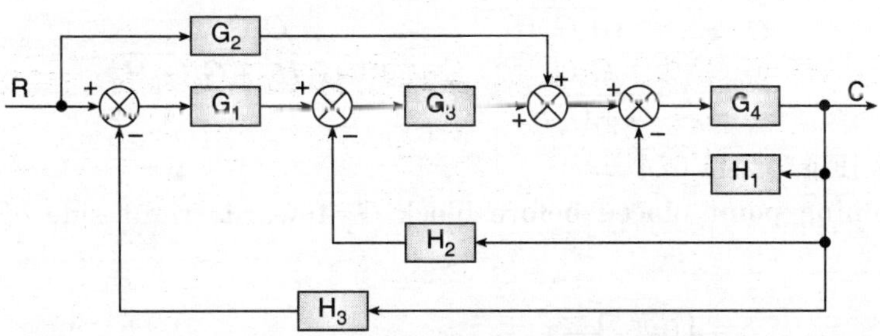

Fig. 3.2.11 (*a*) Block diagram reduction of Fig. 3.2.11.

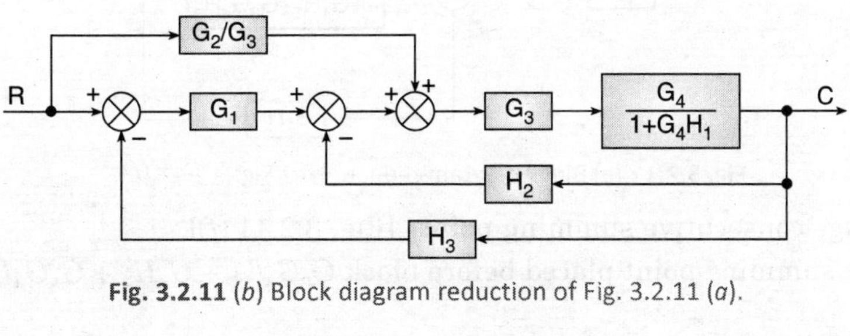

Fig. 3.2.11 (*b*) Block diagram reduction of Fig. 3.2.11 (*a*).

(*i*) Interchange consecutive summing points after block G_1.

(*ii*) Combine block G_3 and $\frac{G_4}{1+G_4H_1}$ as they are in cascade, the equivalence is $G_3G_4/(1 + G_4H_1)$.

The block diagram takes the form [Fig. 3.2.11 (*c*)]

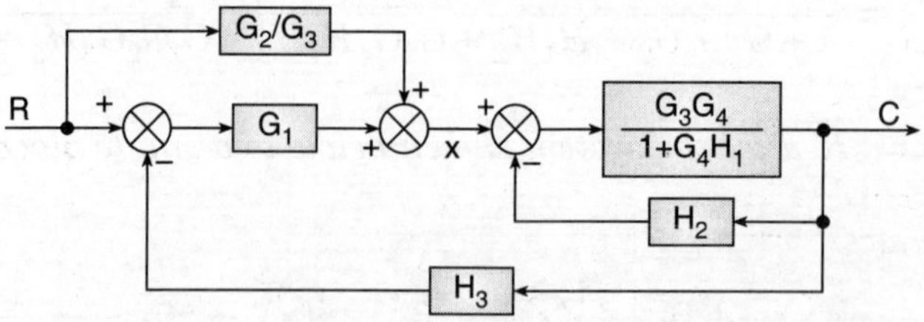

Fig. 3.2.11 (*c*) Block diagram reduction of Fig. 3.2.11 (*b*).

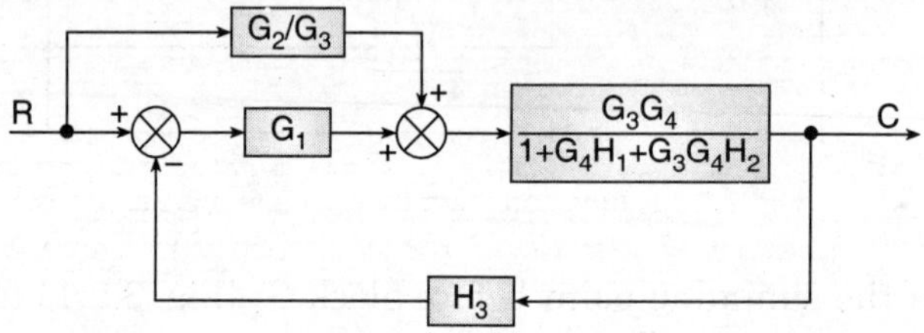

Fig. 3.2.11 (*d*) Block diagram reduction of Fig. 3.2.11 (*e*).

Eliminate the summing point placed before block $G_3G_4/1 + G_4H_1$

$$\therefore \qquad \frac{C}{x} = \frac{\dfrac{G_3G_4}{1+G_4H_1}}{1+\dfrac{G_3G_4}{1+G_4H_1}\times H_2} = \frac{G_3G_4}{1+G_4H_1+G_3G_4H_2}$$

Therefore, [Fig. 3.2.11 (*d*)]

Shift summing point placed before block G_1 towards right side of block G_1 [Fig. 3.2.11 (*e*)]

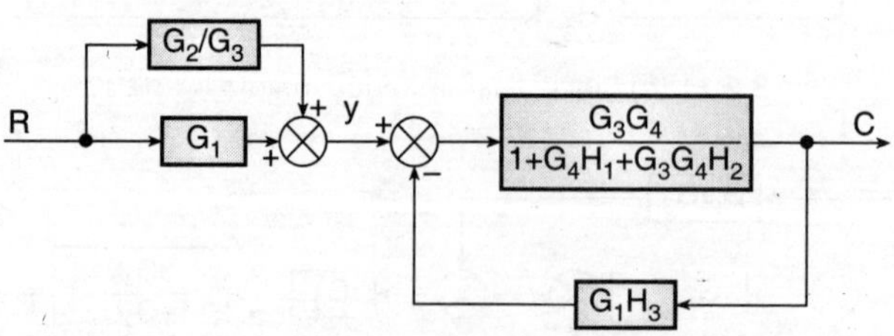

Fig. 3.2.11 (*e*) Block diagram reduction of Fig. 3.2.11 (*d*).

Interchange consecutive summing points [Fig. 3.2.11 (*f*)].

Eliminate summing point placed before block $G_3G_4/(1 + G_4H_1 + G_3G_4H_2)$.

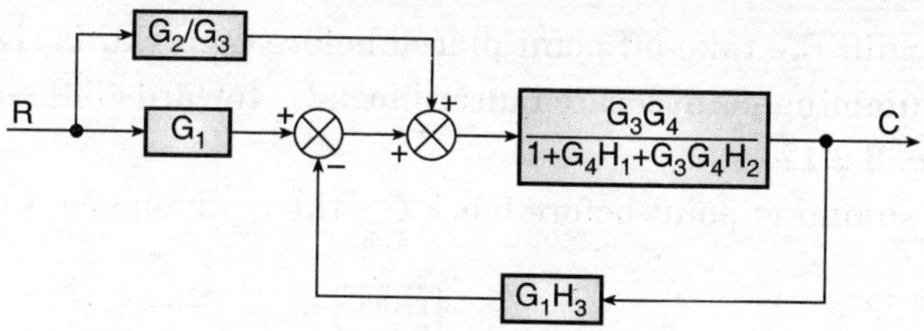

Fig. 3.2.11 (*f*) Block diagram reduction of Fig. 3.2.11 (*e*).

$$\therefore \quad \frac{C}{y} = \frac{\dfrac{G_3G_4}{1+G_4H_1+G_3G_4H_2}}{1+\dfrac{G_3G_4}{1+G_4H_1+G_3G_4H_2}\times G_1H_3}$$

or
$$\frac{C}{y} = \frac{G_3G_4}{1+G_4H_1+G_3G_4H_2+G_1G_3G_4H_3}$$

Thus [Fig. 3.2.11 (*g*)]

The blocks G_1 and G_2/G_3 are in parallel, the equivalence is $G_1 + \dfrac{G_2}{G_3} = \dfrac{G_1G_3+G_2}{G_3}$

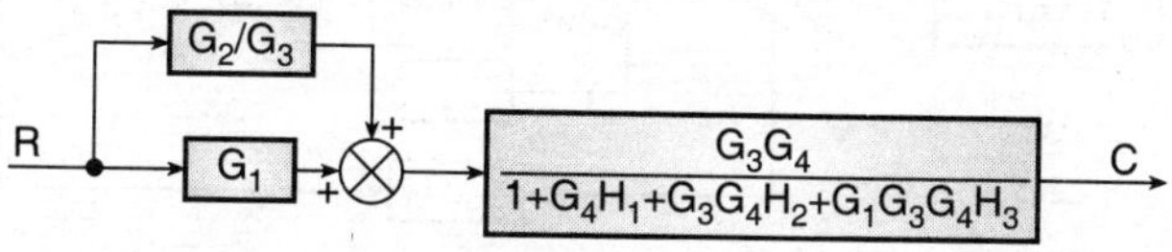

Fig. 3.2.11 (*g*) Block diagram reduction of Fig. 3.2.11 (*f*).

The block diagram takes the form [Fig. 3.2.11 (*h*)]

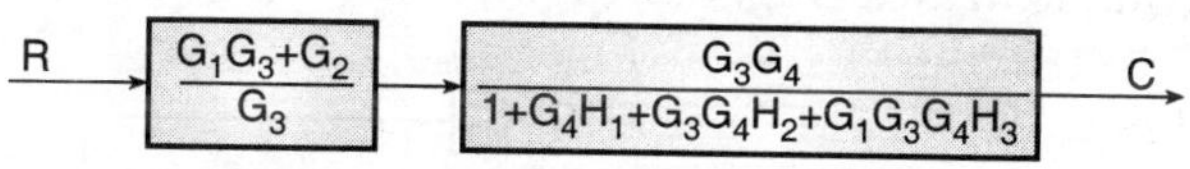

Fig. 3.2.11 (*h*) Block diagram reduction of Fig. 3.2.11 (*g*).

and
$$\frac{C}{R} = \frac{G_1G_3+G_2}{G_3}\times\frac{G_3G_4}{1+G_4H_1+G_3G_4H_2+G_1G_3G_4H_3}$$

or
$$\frac{C}{R} = \frac{G_1G_3G_4+G_2G_4}{1+G_4H_1+G_3G_4H_2+G_1G_3G_4H_3} \quad \textbf{Ans.}$$

Example 3.2.12. *Obtain the overall transfer function for the block diagram (Fig. 3.2.12).*

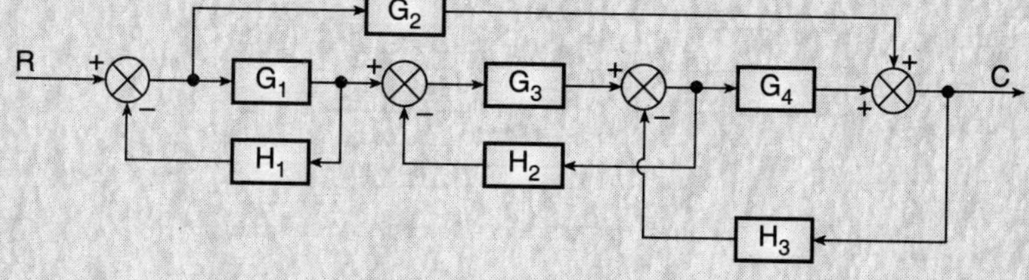

Fig. 3.2.12. Block diagram for example Fig. 3.2.12.

Solution. (*i*) Shift the take-off point placed before G_1 towards right side of block G_1.

(*ii*) Shift the summing point located after block G_3 towards left side.

Therefore [Fig. 3.2.12 (*a*)]

(*i*) Eliminate summing point before block G_1, the equivalence is $G_1/(1 + G_1H_1)$.

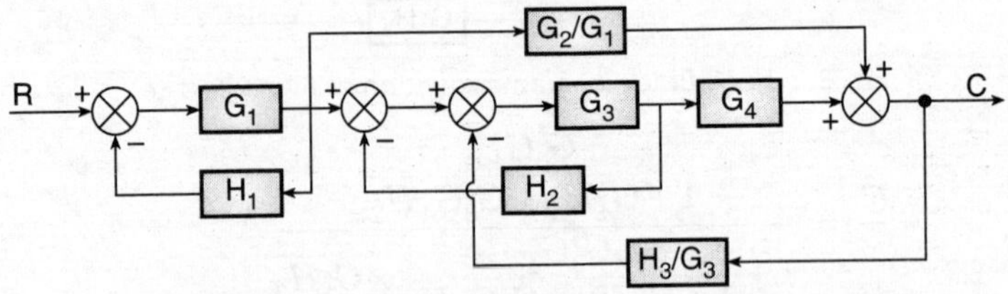

Fig. 3.2.12 (*a*) Block diagram reduction for Fig. 3.2.12.

(*ii*) Interchange consecutive summing points placed before block G_3.
The block is drawn [Fig. 3.2.12 (*b*)]

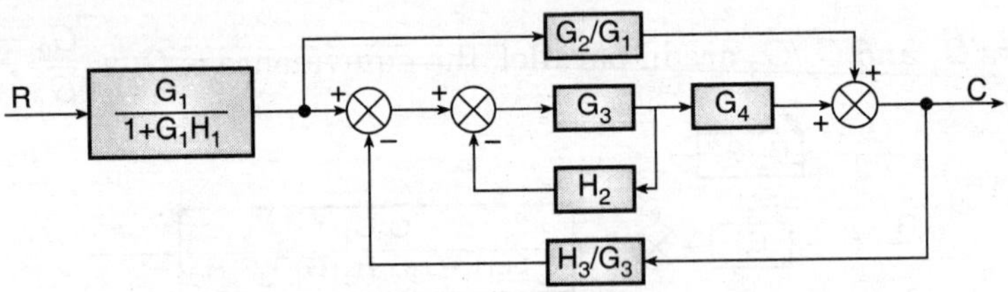

Fig. 3.2.12 (*b*) Block diagram reduction for Fig. 3.2.12 (*a*).

Eliminate the summing point located before block G_3, the equivalence is $G_3/(1 + G_3H_2)$. The block diagram is drawn [Fig. 3.2.12 (*c*)]

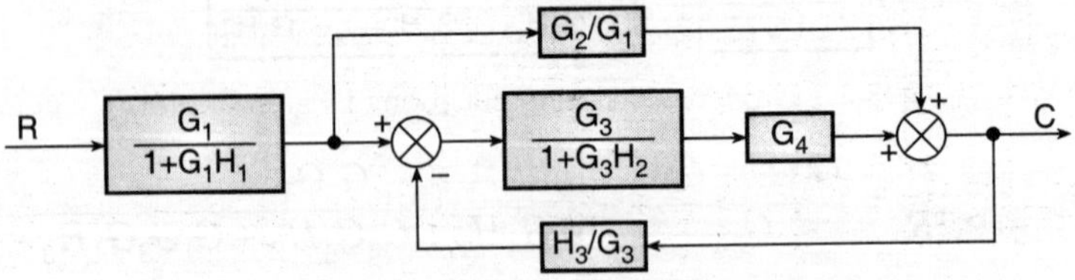

Fig. 3.2.12 (*c*) Block diagram reduction for Fig. 3.2.12 (*b*).

The blocks G_4 and $G_3/(1 + G_3H_2)$ are in cascade, the equivalence is $G_3G_4/(1 + G_3H_2)$. Therefore [Fig. 3.2.12 (*d*)]

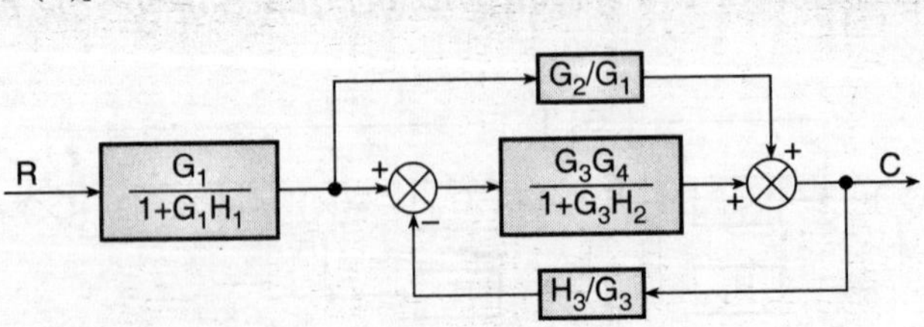

Fig. 3.2.12 (*d*) Block diagram reduction for Fig. 3.2.12 (*c*).

Shift the summing point located after block $G_3G_4/(1 + G_3H_2)$ towards left side to get [Fig. 3.2.12 (*e*)]

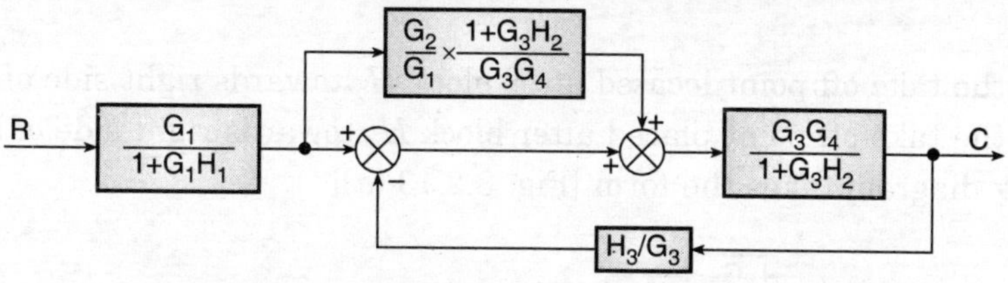

Fig. 3.2.12 (*e*) Block diagram reduction for Fig. 3.2.12 (*d*).

Interchange consecutive summing points to get [Fig. 3.2.12 (*f*)]

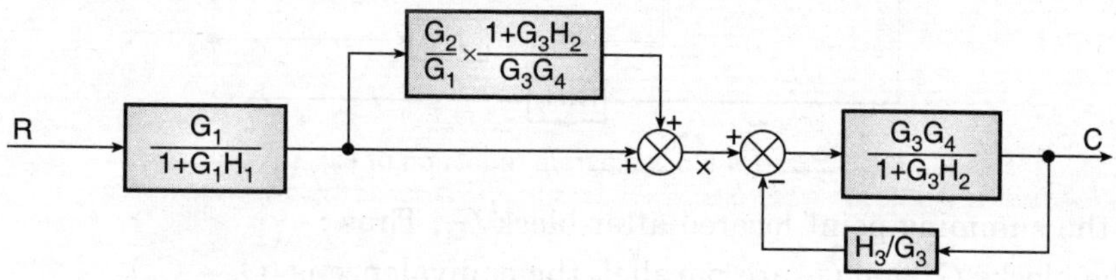

Fig. 3.2.12 (*f*) Block diagram reduction for Fig. 3.2.12 (*e*).

Eliminate summing points. Therefore :

$$(i)\ 1 + \frac{G_2}{G_1} \cdot \frac{1 + G_3H_2}{G_3G_4} = \frac{G_1G_3G_4 + G_2(1 + G_3H_2)}{G_1G_3G_4}$$

$$(ii)\ \frac{C}{x} = \frac{\dfrac{G_3G_4}{1 + G_3H_2}}{1 + \dfrac{G_3G_4}{1 + G_3H_2} \times \dfrac{H_3}{G_3}} = \frac{G_3G_4}{1 + G_3H_2 + G_4H_3}$$

Hence, the block diagram is arranged as [Fig. 3.2.12 (*g*)]

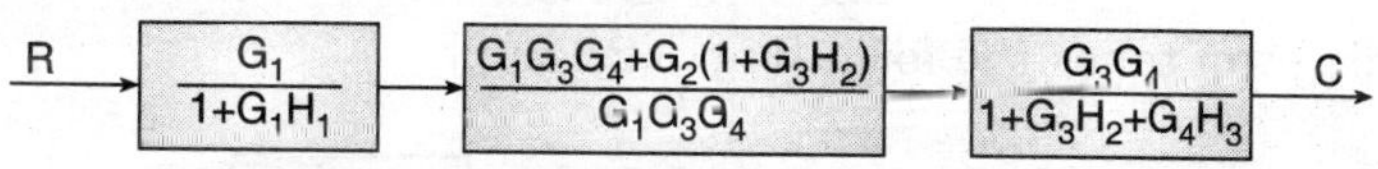

Fig. 3.2.12 (*g*) Block diagram reduction for Fig. 3.2.12 (*f*).

and
$$\frac{C}{R} = \frac{G_1}{1 + G_1H_1} \times \frac{G_1G_3G_4 + G_2(1 + G_3H_2)}{G_1G_3G_4} \times \frac{G_3G_4}{1 + G_3H_2 + G_4H_3}$$

Simplifying, $\dfrac{C}{R} = \dfrac{G_1G_3G_4 + G_2(1 + G_3H_2)}{1 + G_1H_1 + G_3H_2 + G_4H_3 + G_1G_3H_1H_2 + G_1G_4H_1H_3}$ **Ans.**

Example 3.2.13. *Obtain the overall transfer function for a system represented by the block diagram given (Fig. 3.2.13).*

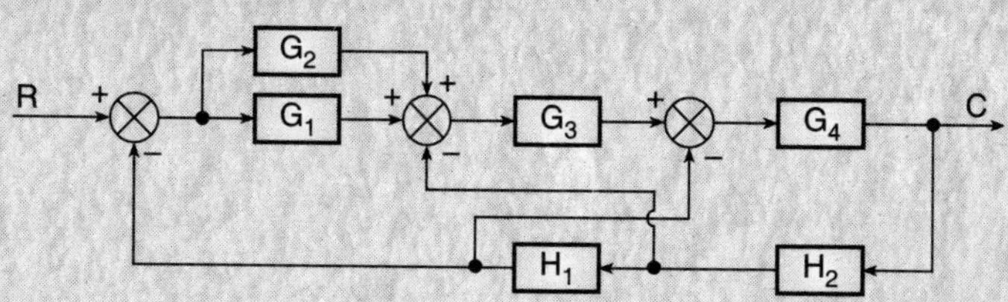

Fig. 3.2.13. Block diagram for example 3.2.13.

Solution.

(*i*) Shift the take-off point located after block H_2 towards right side of H_2.

(*ii*) Shift the take-off point placed after block H_1 towards right side of block H_2.

The block diagram takes the form [Fig. 3.2.13 (*a*)]

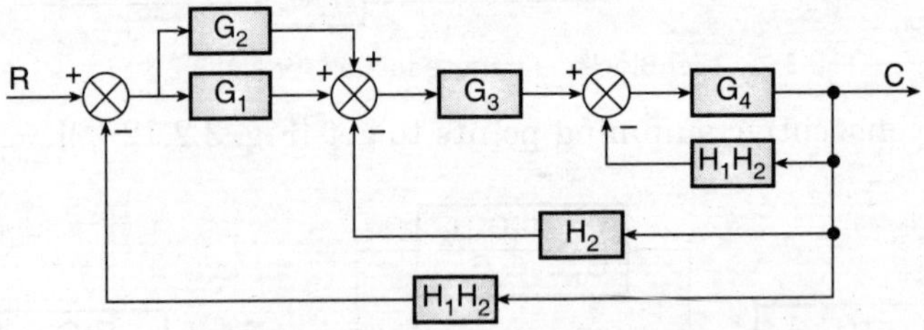

Fig. 3.2.13 (*a*) Block diagram reduction of Fig. 3.2.13.

Split the summing point located after block G_1. Thus :

(*i*) The blocks G_1 and G_2 are parallel, the equivalence is $(G_1 + G_2)$.

(*ii*) Eliminate the summing point placed before block G_4, the equivalence is $G_4/(1 + G_4H_1H_2)$.

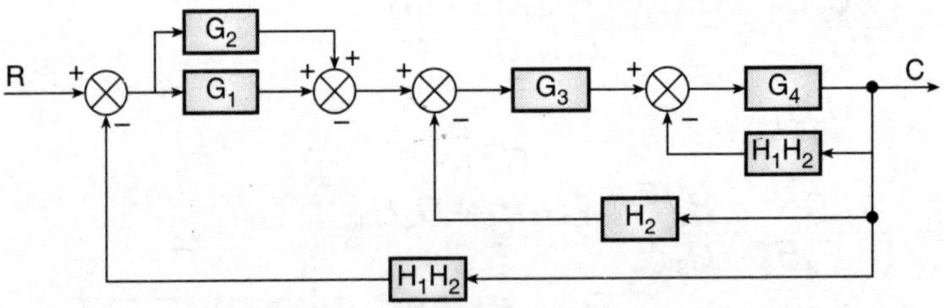

Fig. 3.2.13 (*b*) Block diagram reduction of Fig. 3.2.13 (*a*).

The block diagram takes the form [Fig. 3.2.13 (*c*)]

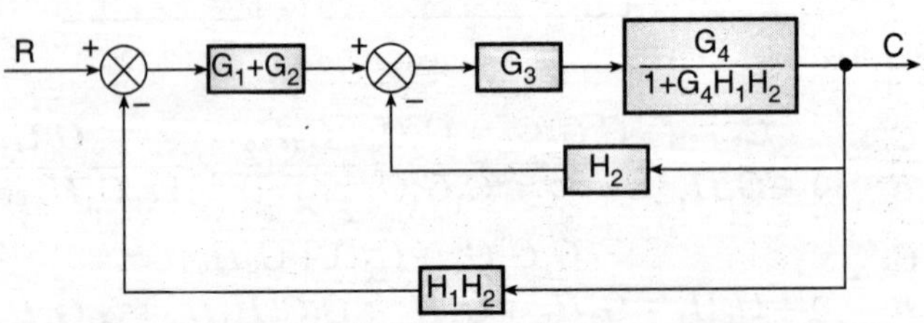

Fig. 3.2.13 (*c*) Block diagram reduction for Fig. 3.2.13 (*b*).

The blocks G_3 and $\dfrac{G_4}{1+G_4H_1H_2}$ are in cascade, the equivalence is $G_3G_4/(1 + G_4H_1H_2)$.

Therefore [Fig. 3.2.13 (*d*)]

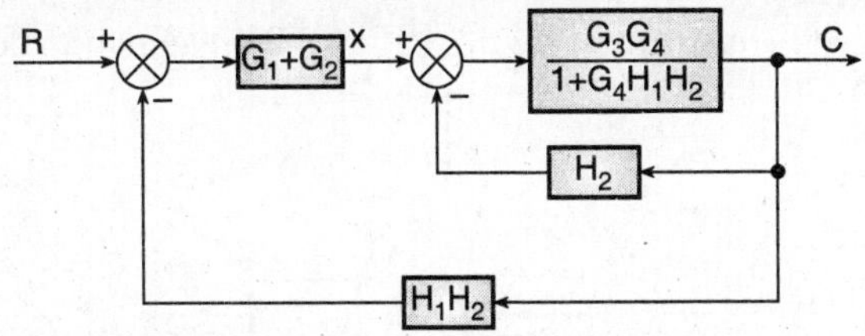

Fig. 3.2.13 (*d*) Block diagram reduction of Fig. 3.2.13 (*c*).

Eliminate summing point placed after block $(G_1 + G_2)$

$$\therefore \qquad \frac{C}{x} = \frac{\dfrac{G_3G_4}{1+G_4H_1H_2}}{1+\dfrac{G_3G_4}{1+G_4H_1H_2}\times H_2}$$

Simplifying, $\dfrac{C}{x} = \dfrac{G_3G_4}{1+G_4H_1H_2+G_3G_4H_2}$. Therefore [Fig. 3.2.13 (*e*)]

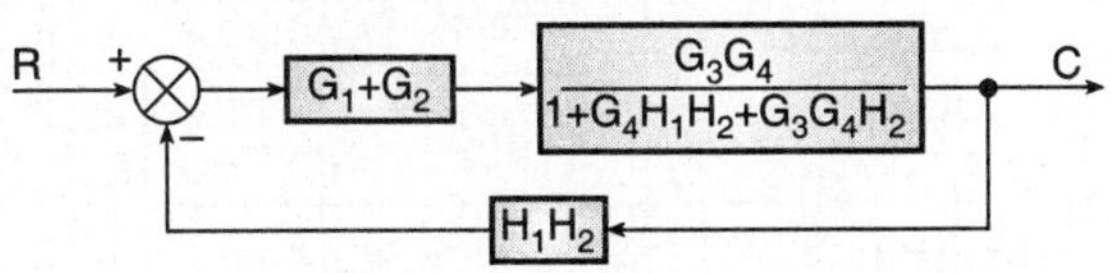

Fig. 3.2.13 (*e*) Block diagram reduction of Fig. 3.2.13 (*d*).

(*i*) The forward path transfer function is : $\dfrac{(G_1+G_2)G_3G_4}{1+G_4H_1H_2+G_3G_4H_2}$

(*ii*) The feedback path transfer function is : H_1H_2

$$\therefore \qquad \frac{C}{R} = \frac{\dfrac{(G_1+G_2)G_3G_4}{1+G_4H_1H_2+G_3G_4H_2}}{1+\dfrac{(G_1+C_2)G_3G_4}{1+G_4H_1H_2+G_3G_4H_2}\times H_1H_2}$$

Simplifying, $\dfrac{C}{R} = \dfrac{G_1G_3G_4+G_2G_3G_4}{1+G_3G_4H_2+G_4H_1H_2+G_1G_3G_4H_1H_2+G_2G_3G_4H_1H_2}$ **Ans.**

Example 3.2.14. *Determine the ratio C/R, C/D and the total output for the system whose block diagram is shown in Fig. 3.2.14.*

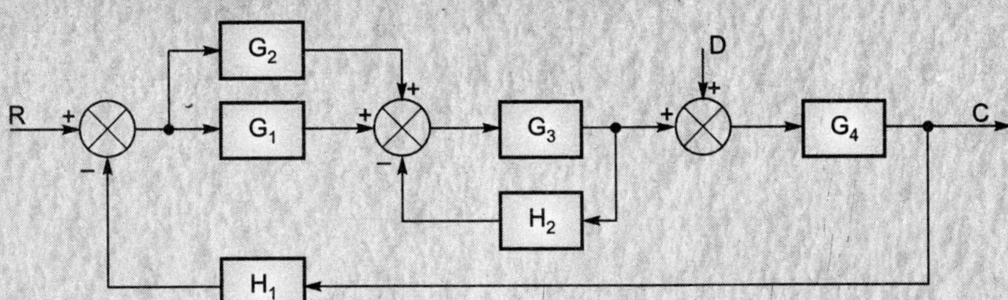

Fig. 3.2.14. Block diagram for example 3.2.14.

Solution. The block diagram shown in Fig. 3.2.14 can be redrawn as shown in Fig. 3.2.14 (*a*).

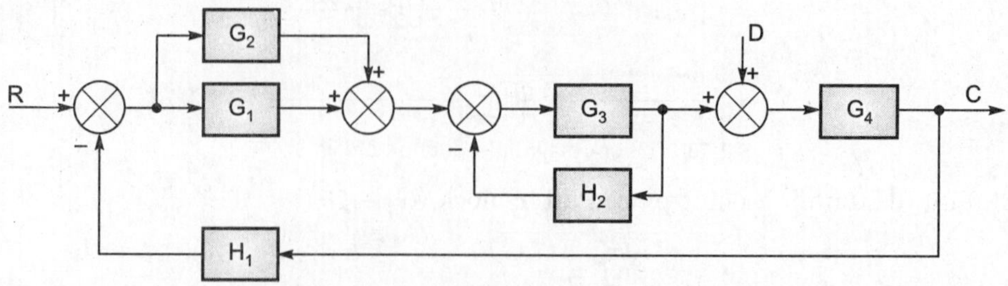

Fig. 3.2.14 (*a*) Block diagram in Fig. 3.2.14 redrawn.

Eliminate summing points after block G_1 and before block G_3.

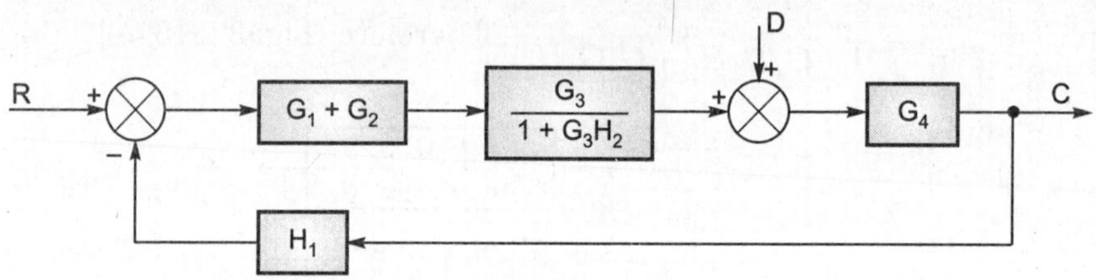

Fig. 3.2.14 (*b*) Block diagram reduction of system in Fig. 3.2.14 (*a*).

Combine blocks $(G_1 + G_2)$ and $\dfrac{G_3}{1+G_3H_2}$

(*i*) In order to determine the ratio *C*/*R* consider $D = 0$, hence the block diagram shown in Fig. 3.2.14 (*c*) takes the form as shown in Fig. 3.2.14 (*d*).

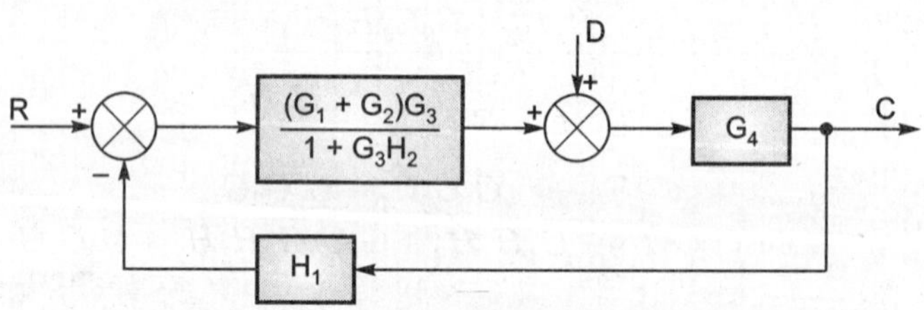

Fig. 3.2.14 (*c*) Block diagram reduction of system shown in Fig. 3.2.14 (*b*).

Consider $D = 0$, the corresponding output is denoted as C_r.

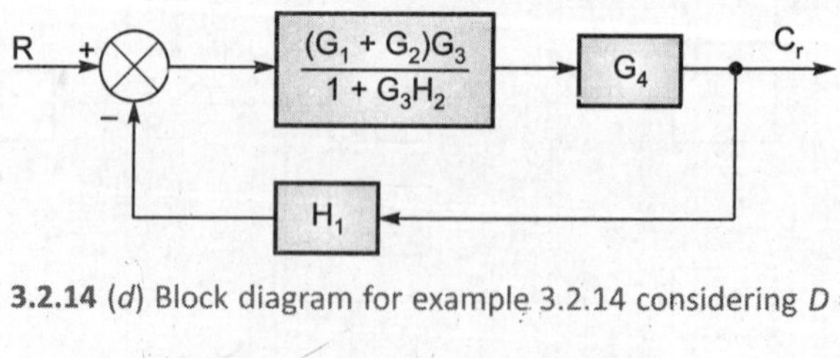

Fig. 3.2.14 (*d*) Block diagram for example 3.2.14 considering $D = 0$.

According to block diagram shown in Fig. 3.2.14 (*d*) the ratio C_r/R is determined below :

$$\frac{C_r}{R} = \frac{\frac{(G_1+G_2)G_3}{(1+G_3H_2)}\cdot G_4}{1+\frac{(G_1+G_2)G_3}{(1+G_3H_2)}\cdot G_4\cdot H_1}$$

$$= \frac{G_1G_3G_4 + G_2G_3G_4}{1+G_3H_2+G_1G_3G_4H_1+G_2G_3G_4H_1} \quad \textbf{Ans.}$$

(*ii*) Consider $R = 0$, the corresponding output is denoted as C_d, hence the block diagram shown in Fig. 3.2.14 (c) takes the form as shown in Fig. 3.2.14 (*e*).

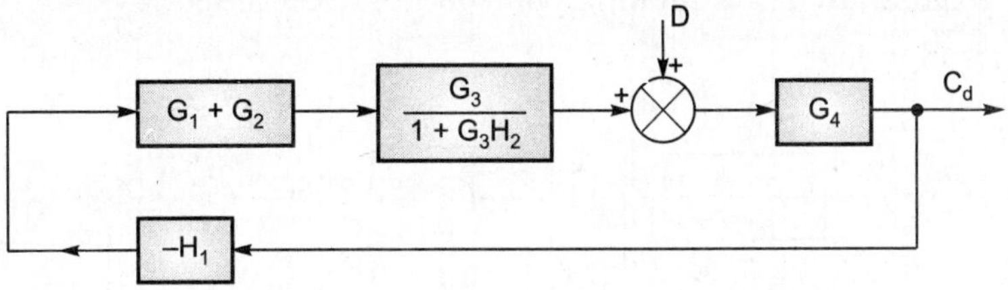

Fig. 3.2.14 (*e*) Block diagram for example 3.2.14 considering $R = 0$.

The block diagram shown in Fig. 3.2.14 (*e*) is redrawn as shown in Fig. 3.2.14 (*f*).

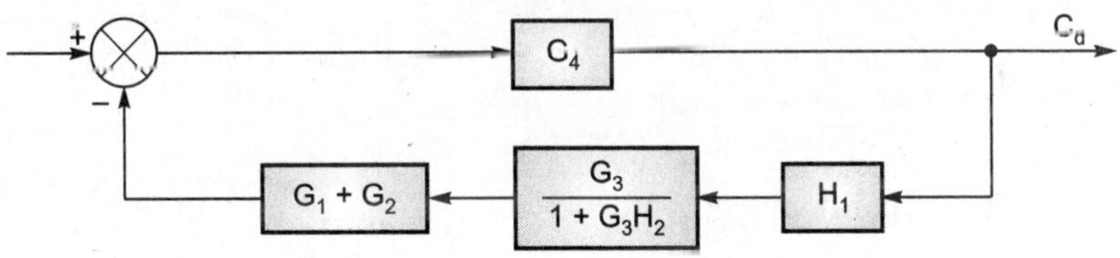

Fig. 3.2.14 (*f*) Block diagram of Fig. 3.2.14 (*e*).

According to block diagram shown in Fig. 3.2.14 (*f*) the ratio C_d/D is determined below.

$$\frac{C_d}{D} = \frac{G_4}{1+G_4\cdot\frac{(G_1+G_2)G_3H_1}{(1+G_3H_2)}} = \frac{G_4(1+G_3H_2)}{1+G_3H_2+G_1G_3G_4H_1+G_2G_3G_4H_1} \quad \textbf{Ans.}$$

The total output is given by,

$$C = C_rR + C_dD = \frac{G_1G_3G_4+G_2G_3G_4}{1+G_3H_2+G_1G_3G_4H_1+G_2G_3G_4H_1}\cdot R$$

$$+\frac{G_4(1+G_3H_2)}{1+G_3H_2+G_1G_3G_4H_1+G_2G_3G_4H_1}\cdot D \quad \textbf{Ans.}$$

Example 3.2.15. *Determine the expression for the output for the system block diagram (Fig. 3.2.15).*

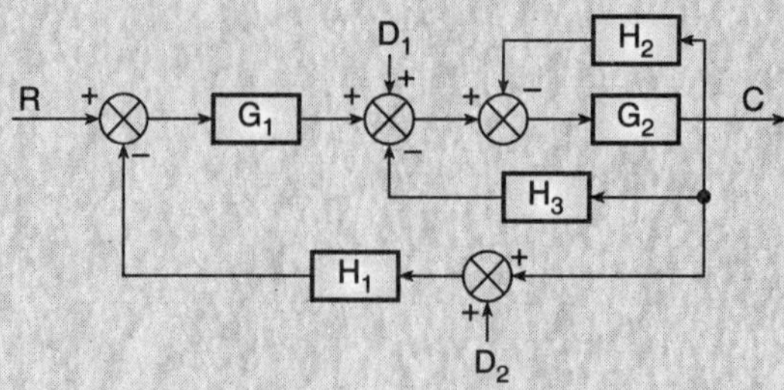

Fig.3.2.15. Block diagram for Example 3.2.15.

Solution. Eliminate the summing point located before block G_2 and draw the block diagram [Fig. 3.2.15 (*a*)]

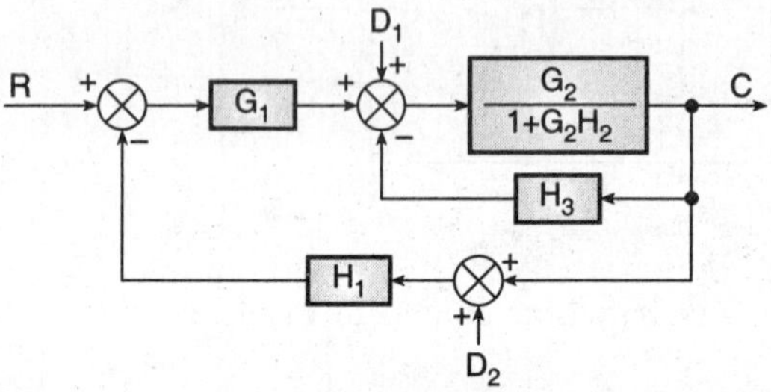

Fig. 3.2.15 (*a*) Block diagram reduction of Fig. 3.2.15.

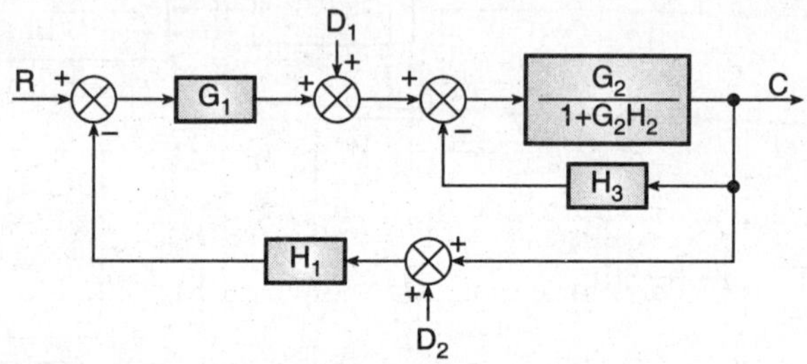

Fig. 3.2.15 (*b*) Block diagram reduction of Fig. 3.2.15 (*a*).

Split the summing point placed before block $\dfrac{G_2}{1+G_2H_2}$. Thus [Fig. 3.2.15 (*b*)]

Eliminate the summing point located before block $G_2/(1 + G_2H_2)$ to get [Fig. 3.2.15 (*c*)].

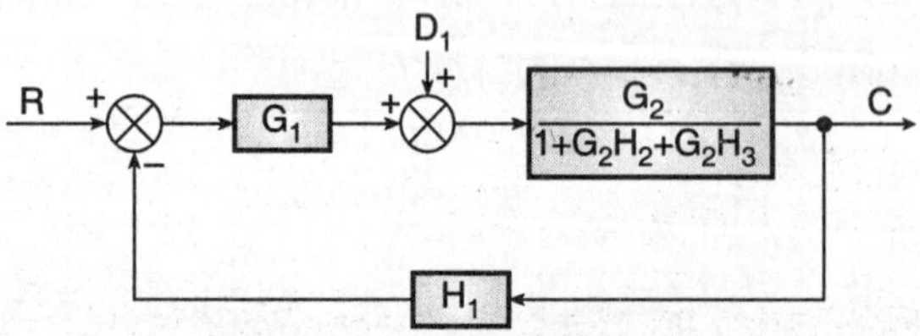

Fig. 3.2.15 (*c*) Block diagram reduction of Fig. 3.2.15 (*b*).

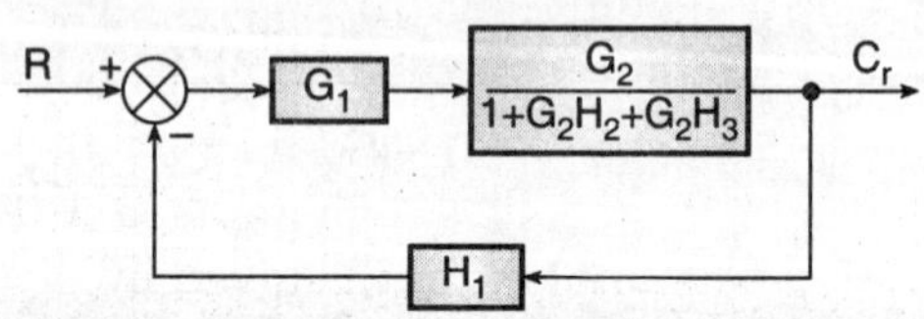

Fig. 3.2.15 (*d*) Block diagram reduction of Fig. 3.2.15 (*c*).

Consider $D_1 = 0$, $D_2 = 0$ and the output as C_r for the input R acting alone. Therefore, the block diagram takes the form [Fig. 3.2.15 (*d*)].

Eliminate summing point before block G_1

$$\therefore \qquad \frac{C_r}{R} = \frac{\dfrac{G_1G_2}{1+G_2H_2+G_2H_3}}{1+\dfrac{G_1G_2}{1+G_2H_2+G_2H_3}\times H_1}$$

Simplifying, $$C_r = \left(\frac{G_1G_2}{1+G_2H_2+G_2H_3+G_1G_2H_1}\right).R$$

Consider $R = 0$, $D_2 = 0$ and the output as C_{d1} for the input D_1 acting alone. Therefore, the block diagram takes the form [Fig. 3.2.15 (*e*)]

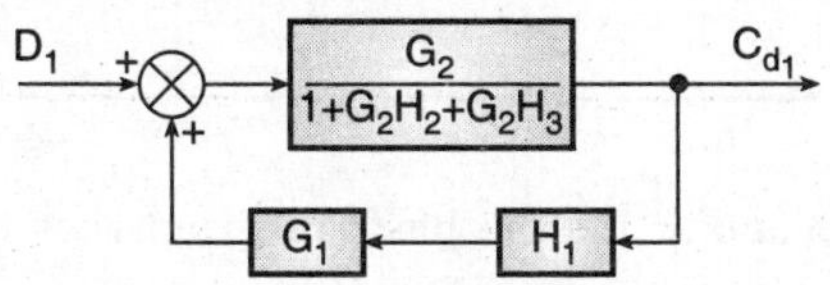

Fig. 3.2.15 (*e*) Block diagram reduction of Fig. 3.2.15 (*d*).

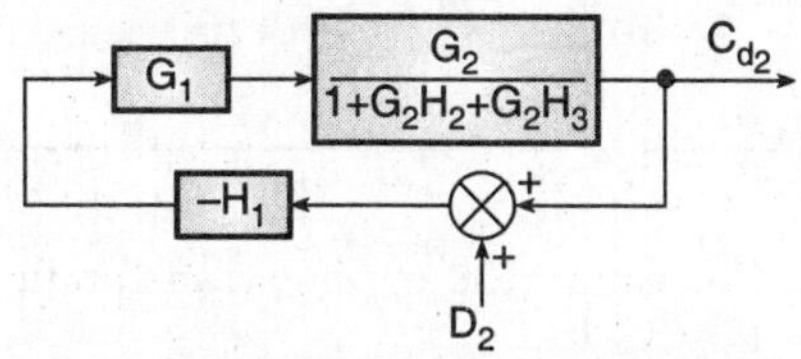

Fig. 3.2.15 (*f*) Block diagram reduction of Fig. 3.2.15 (*e*).

Eliminating the summing point

$$\frac{C_{d1}}{D_1} = \frac{\dfrac{G_2}{1+G_2H_2+G_2H_3}}{1-\dfrac{G_2}{1+G_2H_2+G_2H_3}.G_1(-H_1)}$$

Simplifying, $$C_{d1} = \left(\frac{G_2}{1+G_2H_2+G_2H_3+G_1G_2H_1}\right)D_1$$

Consider $R = 0$, $D_1 = 0$ and the output as C_{d_2} for the input D_2 acting alone. Therefore, the block diagram takes the form [Fig. 3.2.15 (*f*)]

On rearranging [Fig. 3.2.15 (*g*)]

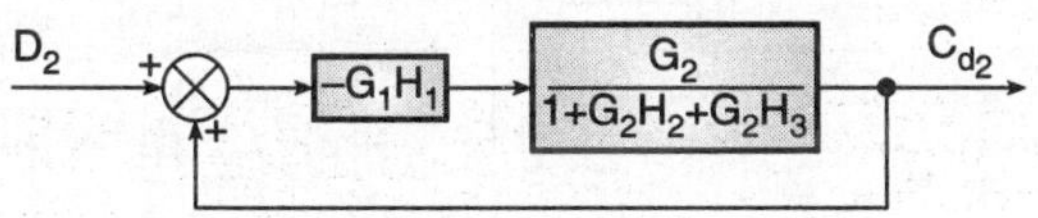

Fig. 3.2.15 (*g*) Block diagram rearrangement of Fig. 3.2.15 (*f*).

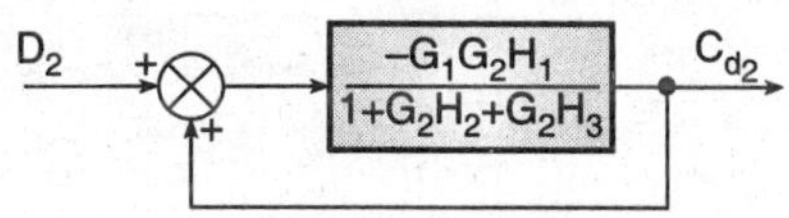

Fig. 3.2.15 (*h*) Block diagram reduction of Fig. 3.2.15 (*g*).

The blocks $(-G_1H_1)$ and $G_2/(1+G_2H_2+G_2H_3)$ are in cascade, the equivalence is $\dfrac{-G_1G_2H_1}{1+G_2H_2+G_2H_3}$. Therefore :

Eliminating the summing point

$$\frac{C_{d_2}}{D_2} = \frac{\dfrac{-G_1G_2H_1}{1+G_2H_2+G_2H_3}}{1-\dfrac{-G_1G_2H_1}{1+G_2H_2+G_2H_3}\times 1}$$

Simplifying, $$C_{d2} = \left(-\frac{G_1G_2H_1}{1+G_2H_2+G_2H_3+G_1G_2H_1}\right).D_2$$

The output $C = C_rR + C_{d1}D_1 + C_{d2}D_2$

or
$$C = \frac{G_1G_2}{1+G_2H_2+G_2H_3+G_1G_2H_1}\,.\,R + \frac{G_2}{1+G_2H_2+G_2H_3+G_1G_2H_1}\,.\,D_1$$
$$+\frac{-G_1G_2H_1}{1+G_2H_2+G_2H_3+G_1G_2H_1}\,.\,D_2 \quad \textbf{Ans.}$$

PROBLEMS

3.1. Determine the overall transfer function relating C and R for the block diagram shown in Fig. 3P-1.

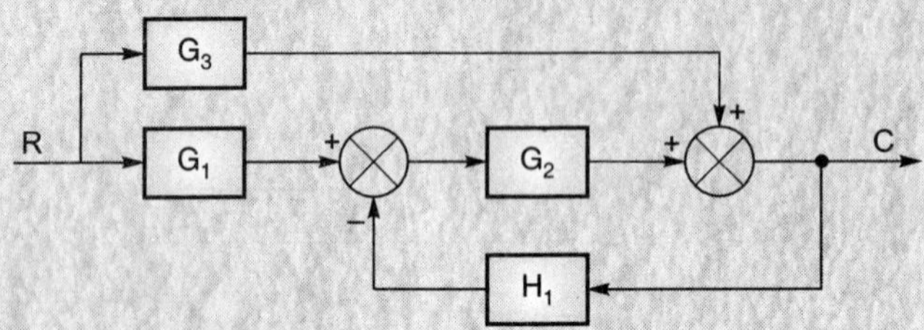

Fig. 3P-1. Block diagram for problem 3.1.

3.2. Reduce the block diagram shown in Fig. 3P-2 to a single block representation.

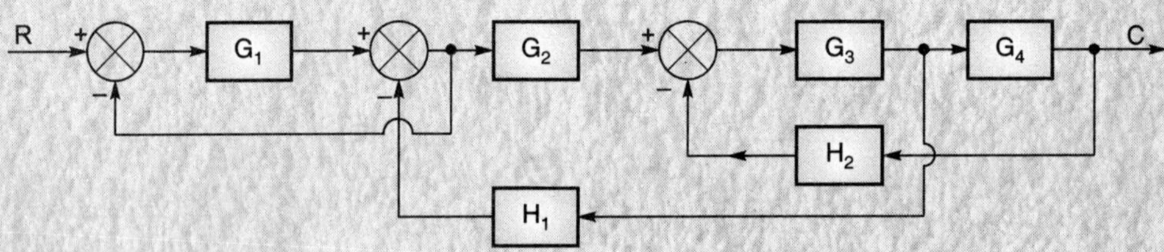

Fig. 3P-2. Block diagram for problem 3.2.

3.3. Determine the overall transfer function relating the output and input for a system represented by Fig. 3P-3.

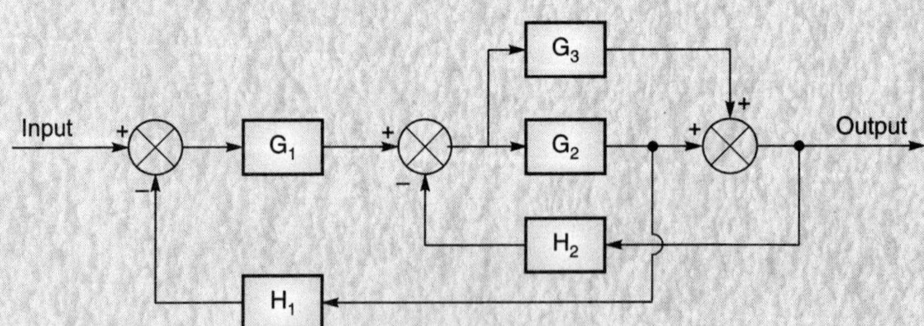

Fig. 3P-3. System representation for problem 3.3.

3.4. Determine the transfer function C_1/R_1, C_2/R_2, C_1/R_2 and C_2/R_1 from the block diagram in Fig. 3P-4.

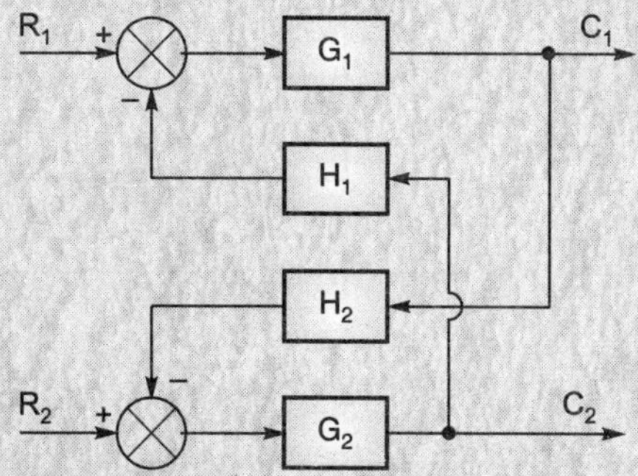

Fig. 3P-4. Block diagram for problem 3.4.

3.5. Reduce the block diagram shown in Fig. 3P-5 to a form having one block in forward path and one in feedback path.

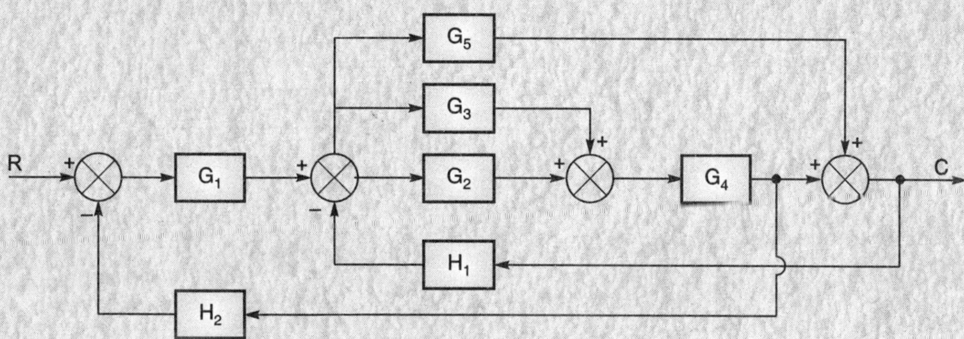

Fig. 3P-5. Block diagram for problem 3.5.

3.6. Reduce the block diagram shown in Fig. 3P-6 to a single block form.

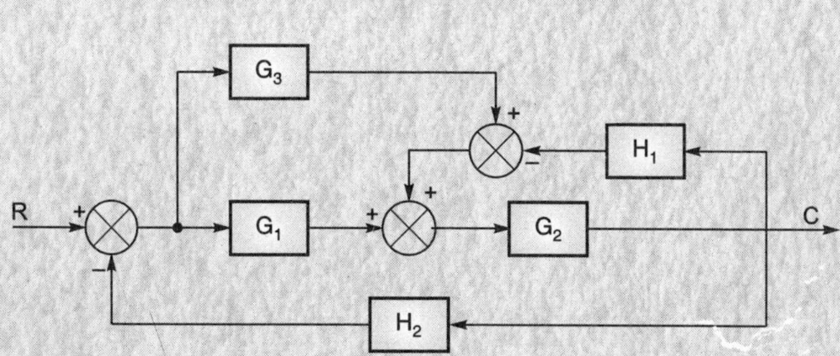

Fig. 3P-6. Block diagram for problem 3.6.

3.7. Determine the overall transfer function for the system represented by the block diagram shown in Fig. 3P-7.

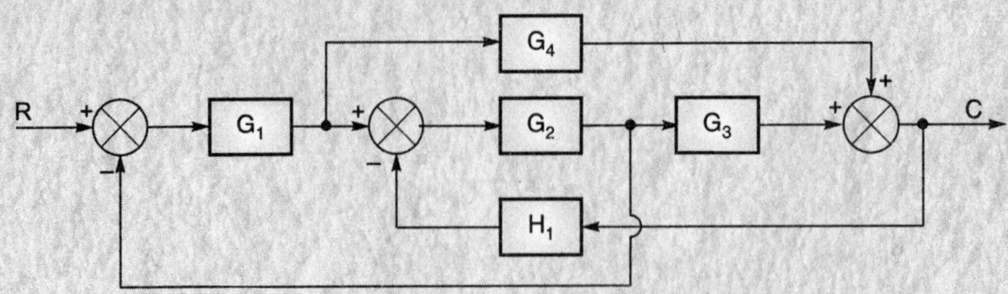

Fig. 3P-7. Block diagram for problem 3.7.

3.8. Determine *C/R* and *C/D* for the block diagram shown in Fig. 3P-8.

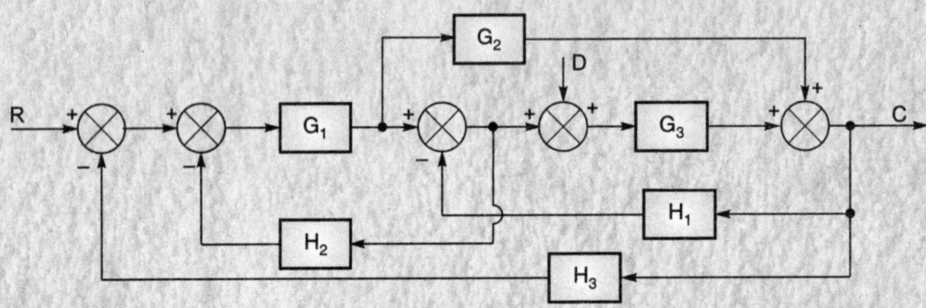

Fig. 3P-8. Block diagram for problem 3.8.

3.9. Determine the transfer functions C/R_1 and C/R_2 from the block diagram shown in Fig. 3P-9.

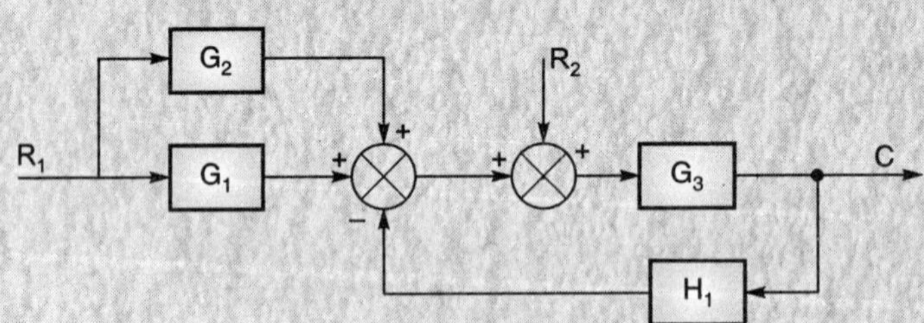

Fig. 3P-9. Block diagram for problem 3.9.

3.10. Derive an expression for the total output for the system represented by the block diagram in Fig. 3P-10.

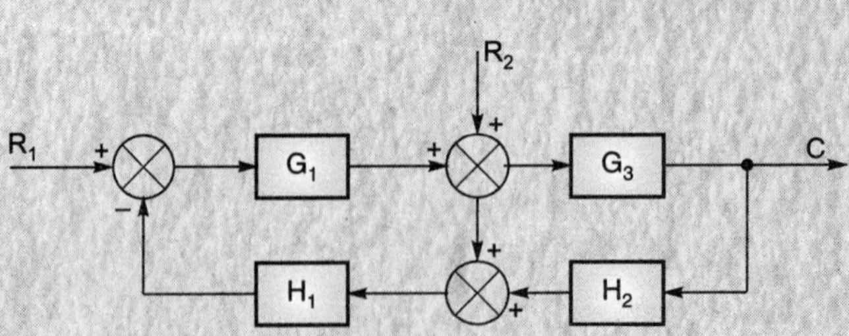

Fig. 3P-10. System representation for problem 3.10.

3.11. Determine the output for the block diagram (Fig. 3P-11)

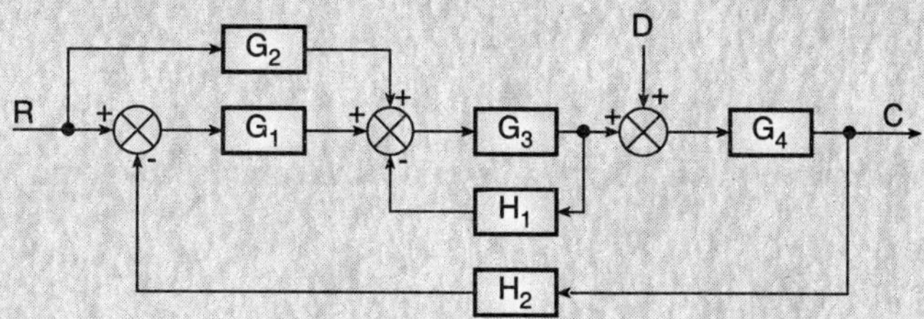

Fig. 3P-11

3.12. Determine the overall transfer function having given the block diagram (Fig. 3P-12).

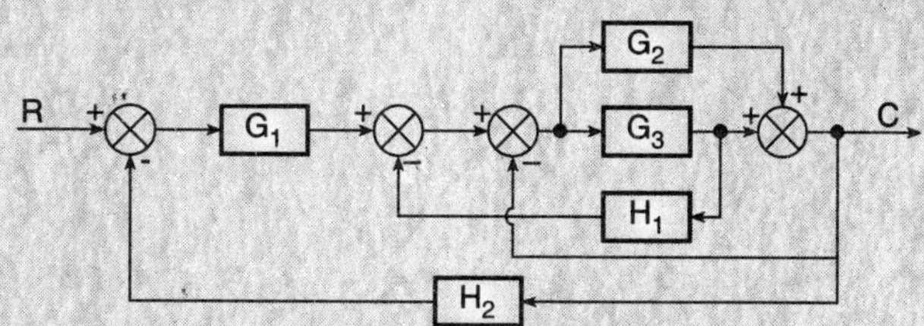

Fig. 3P-12

3.13. Determine the relation *C/R* for the block diagram shown in Fig. 3P-13.

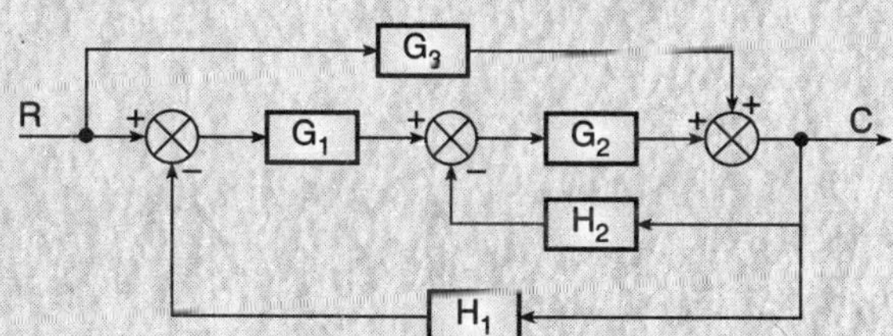

Fig. 3P-13

CHAPTER 4

Signal Flow Graphs

The modified form of a block diagram is a signal flow graph. The development of signal flow graph representation is explained through examples in a sequence and Mason's gain formula derived therefrom. The specific case of developing signal flow graph for the case of a take off point before the summing point is given separately.

CONTENTS

- *Rules for Drawing Signal Flow Graphs*
- *Mason's Gain Formula*
- *Drawing Signal Flow Graph from a Given Block Diagram*
- *Solved Examples*

REPRESENTATION OF A CONTROL SYSTEM BY SIGNAL FLOW GRAPH

Block diagram gives a pictorial representation of a control system by way of short handing the transfer function. Signal flow graph further shortens the representation of a control system by way of eliminating summing symbol, take off point and block. This elimination is achieved by representing the variables by points called "nodes" and the transfer function is termed as "transmittance" which is represented by a line called "branch". A signal on the branch travels in the direction of an arrow indicated on the branch.

The equation $x_2 = ax_1$ reveals that the variable x_1 when multiplied by the term a results in a new variable x_2.

If the aforesaid equation is represented by a block diagram, the configuration as shown in Fig. 4.0.1 is obtained.

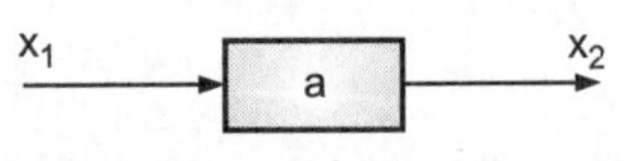

Fig. 4.0.1. Block diagram representation of equation $x_2 = ax_1$.

The same equation can be represented by a signal flow graph as shown in Fig. 4.0.2.

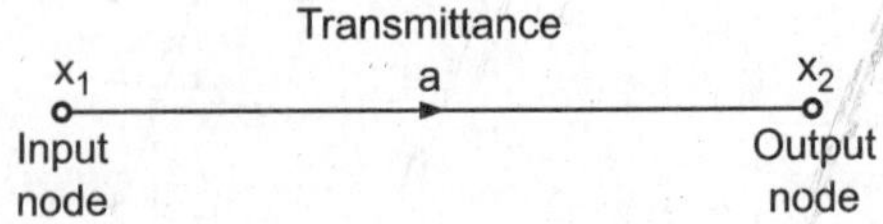

Fig. 4.0.2. Signal flow graph representation of equation $x_2 = ax_1$.

x_1 is the input variable node and x_2 is the output variable node, the line joining these two nodes with an arrow indicated on it gives the direction of signal flow from x_1 to x_2 and the transmittance being a.

Similarly, the equation given below is represented by a block diagram shown in Fig. 4.0.3.

$$x_1 = ax_0 + bx_1 \qquad ...(4.1)$$

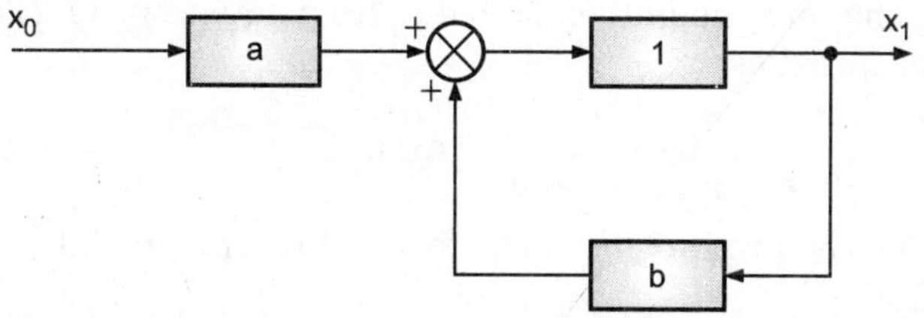

Fig. 4.0.3. Block diagram representation of equation $x_1 = ax_0 + bx_1$.

The signal flow graph representation for the above equation is shown in Fig. 4.0.4.

The variables are represented by input and output nodes respectively, a is the transmittance between x_1 and x_0 called forward path transmittance, b is the transmittance between x_1 and x_1 called loop transmittance.

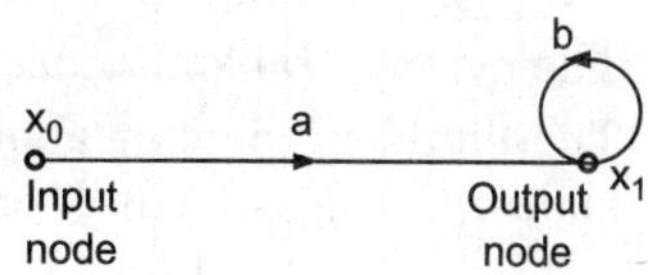

Fig. 4.0.4. Signal flow graph representation of equation $x_1 = ax_0 + bx_1$.

Above two examples form the basis of framing rules for drawing signal flow graphs representing an equation.

4.1 RULES FOR DRAWING SIGNAL FLOW GRAPHS

1. The signal travels along a branch in the direction of an arrow.
2. The input signal is multiplied by the transmittance to obtain the output signal.
3. Input signal at a node is the sum of all signals entering at that node.
4. A node transmits signals in all branches leaving that node.

Signal flow graphs shown in the following examples illustrate the application of the above rules to obtain overall transmittance (also called gain) expressed as the ratio of the output (effect) to input (cause).

Illustrative Example 4.1.1. *The signal flow graph shown in Fig. 4.1.1 has one forward path and single isolated loop. Determine overall transmittance relating x_3 and x_1.*

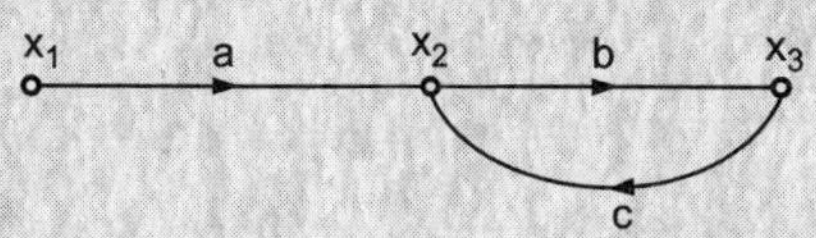

Fig. 4.1.1. Signal flow graph with one forward path and single isolated loop.

Solution.

x_1 is the input node (independent variable node)

x_2 is the intermediate node (intermediate variable node)

x_3 is the output node (dependent variable node)

ab is the forward path transmittance

bc is the loop transmittance.

The equations represented by the signal flow graph are expressed below

(*i*) at node x_2, $x_2 = ax_1 + cx_3$ ...(4.2)

and (*ii*) at node x_3, $x_3 = bx_2$ ...(4.3)

By eliminating the intermediate variable x_2 from equations (4.2) and (4.3), the overall transmittance is determined as

$$\frac{x_3}{x_1} = \frac{ab}{1-bc}. \quad \textbf{Ans.} \qquad ...(4.4)$$

In equation (4.4), the product *ab* is the forward path transmittance and denoted by a symbol *P*.

As *bc* forms a closed-loop on itself, the product *bc* is termed as loop transmittance and denoted by a symbol *L*. While traversing the closed-loop each node is encountered only once.

The overall transmittance is denoted by a symbol *T*.

Therefore, in view of above notations, the overall transmittance for the signal flow graph shown in Fig. 4.1.1 is expressed as

$$T = \frac{P}{1-L} \qquad ...(4.5)$$

Illustrative Example 4.1.2. *The signal flow graph shown in Fig. 4.1.2 has one forward path and two isolated (non-touching) loops. Determine the overall transmittance relating x_6 and x_1.*

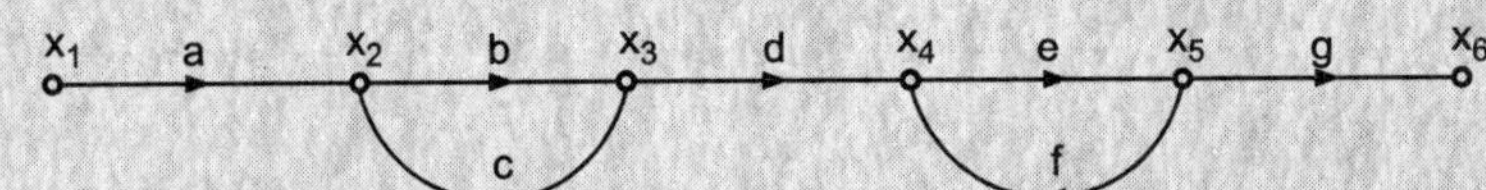

Fig. 4.1.2. Signal flow graph with one forward path and two isolated (non-touching) loops.

Solution. The equations represented are expressed below :

$$x_2 = ax_1 + cx_3 \qquad ...(4.6)$$

$$x_3 = bx_2 \qquad ...(4.7)$$

$$x_4 = dx_3 + fx_5 \qquad ...(4.8)^*$$

$$x_5 = ex_4 \qquad ...(4.9)$$

$$x_6 = gx_5 \qquad ...(4.10)$$

By eliminating the intermediate variables, *i.e.* x_2, x_3, x_4 and x_5 from above equations, the overall transmittance is determined as

$$\frac{x_6}{x_1} = \frac{abdeg}{1-(bc+ef)+bcef} \qquad ...(4.11)$$

or

$$T = \frac{P}{1-(L_1+L_2)+L_1L_2}. \quad \textbf{Ans.} \qquad ...(4.12)$$

where T = the overall transmittance between x_6 and x_1

---

*If the transmittance *d* between the nodes x_4 and x_3 is 1, then it be understand that the nodes x_3 and x_4 are not coincident. For such a case the signal cannot flow from the node x_4 to x_3. This situation arises in a block diagram representation where in a takeoff point preceds a summing symbol.

$P = abdeg$, forward path transmittance between x_6 and x_1
$L_1 = bc$, Loop transmittance of loop bc
$L_2 = ef$, Loop transmittance of loop ef.

Illustrative Example 4.1.3. *The signal flow graph shown in Fig. 4.1.3 has one forward path and two touching loops. Determine the overall transmittance relating x_6 and x_1.*

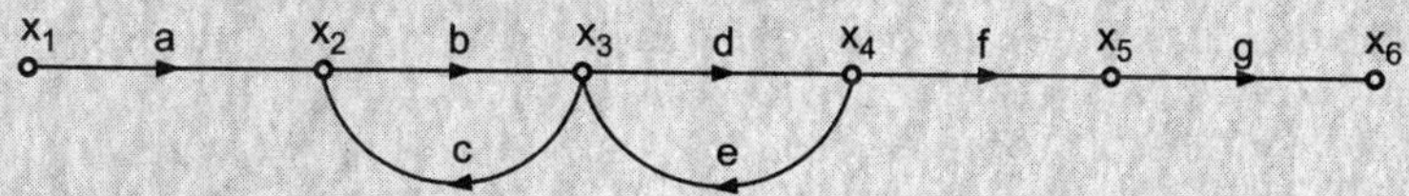

Fig. 4.1.3. Signal flow graph with one forward path and two touching loops.

Solution. The related equations represented are expressed as,

$$x_2 = ax_1 + cx_3 \quad ...(4.13)$$
$$x_3 = bx_2 + ex_4 \quad ...(4.14)$$
$$x_4 = dx_3 \quad ...(4.15)$$
$$x_5 = fx_4 \quad ...(4.16)$$
$$x_6 = gx_5 \quad ...(4.17)$$

By eliminating the intermediate variables, *i.e.* x_2, x_3, x_4 and x_5, the overall transmittance is determined as

$$\frac{x_6}{x_1} = \frac{abdfg}{1-(bc+de)}. \quad \textbf{Ans.} \quad ...(4.18)$$

or

$$T = \frac{P}{1-(L_1+L_2)} \quad ...(4.19)$$

where, T = the overall transmittance between x_6 and x_1
$P = abdfg$, the forward path transmittance between x_6 and x_1
$L_1 = bc$, the loop transmittance of loop bc
$L_2 = de$, the loop transmittance of loop de.

Illustrative Example 4.1.4. *The signal flow graph shown in Fig. 4.1.4 has two parallel paths in forward path. Determine the overall transmittance relating x_4 and x_1.*

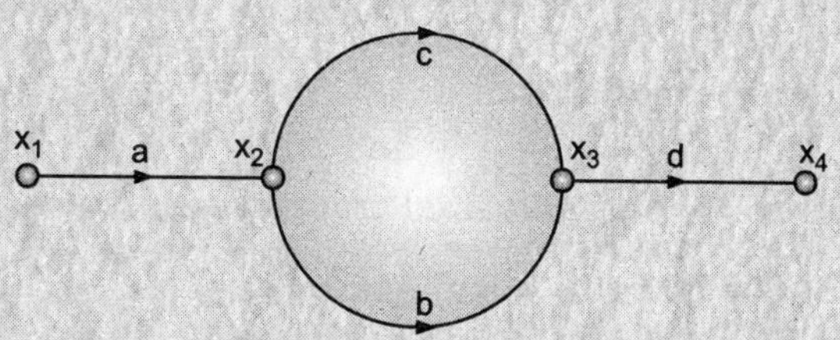

Fig. 4.1.4. Signal flow graph with two forward paths.

Solution. The related equations represented are expressed as

$$x_2 = ax_1 \quad \text{...(4.20)}$$

$$x_3 = bx_2 + cx_2 \quad \text{...(4.21)}$$

$$x_4 = dx_3 \quad \text{...(4.22)}$$

By eliminating the intermediate variables *i.e.*, x_2 and x_3 from the above equations, the overall transmittance is determined as

$$\frac{x_4}{x_1} = (abd + acd). \quad \textbf{Ans.} \quad \text{...(4.23)}$$

or
$$T = P_1 + P_2 \quad \text{...(4.24)}$$

where, T = the overall transmittance between x_4 and x_1

$P_1 = abd$, the forward path transmittance of the path 1

$P_2 = acd$, the forward path transmittance of the path 2.

In this case signal from the input node travels along two parallel paths and the equivalent transmittance is given by the sum of the individual forward path transmittances.

Illustrative Example 4.1.5. *The signal flow graph shown in Fig. 4.1.5 has two parallel paths and two isolated (non-touching) loops. Determine the overall transmittance relating x_2 and x_1.*

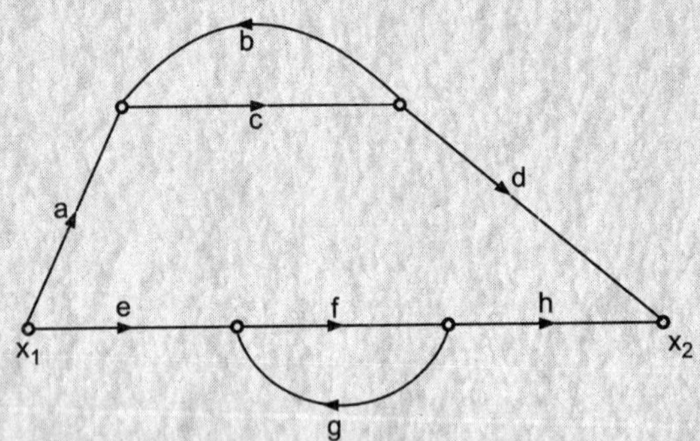

Fig. 4.1.5. Signal flow graph with two forward paths and two isolated (non-touching) loops.

Solution. The overall transmittance relating x_2 and x_1 is determined as follows :

There are two forward paths as mentioned below :

$$P_1 = acd \quad \text{and} \quad P_2 = efh$$

The two loop transmittances are :

$$L_1 = bc \quad \text{and} \quad L_2 = gf$$

Therefore, the corresponding individual transmittances are :

$$T_1 = \frac{P_1}{1 - L_1} \quad \text{...(4.25)}$$

and
$$T_2 = \frac{P_2}{1 - L_2} \quad \text{...(4.26)}$$

The overall transmittance is obtained by adding T_1 and T_2.

Hence,
$$\frac{x_2}{x_1} = T_1 + T_2 \quad \text{...(4.27)}$$

Substituting for T_1 and T_2 from (4.25) and (4.26) respectively in (4.27) following relation for the overall transmittance is obtained.

$$\frac{x_2}{x_1} = \frac{P_1}{1-L_1} + \frac{P_2}{1-L_2} \quad \text{...(4.28)}$$

$$= \frac{P_1(1-L_2)+P_2(1-L_1)}{1-(L_1+L_2)+L_1L_2} \quad \text{...(4.29)}$$

Substituting the values of P_1, P_2, L_1 and L_2 in (4.29), the overall transmittance is determined below :

$$\frac{x_2}{x_1} = \frac{acd\,(1-gf)+efh\,(1-bc)}{1-(bc+gf)+bcgf}. \quad \textbf{Ans.} \quad \text{...(4.30)}$$

4.2 MASON'S GAIN FORMULA

The overall transmittance (gain) can be determined by Mason's gain formula given below

$$T = \sum_{k=1}^{k} \frac{P_k \Delta_k}{\Delta} \quad \text{...(4.31)}$$

The terms in the Mason's gain formula are explained below :

P_k is the forward path transmittance of k_{th} path from a specified input node to an output node. In asserting P_k no node should be encountered more than once.

Δ is the graph determinant which involves closed-loop transmittances and mutual interactions between non-touching loops.

Δ = 1 – [Sum of all individual loop transmittances]
+ [Sum of loop transmittance products of all possible pairs of NON-TOUCHING loops]
– [Sum of loop transmittance products of all possible triplets of NON-TOUCHING loops]
+ [.........] – [.........]

Δ_k is the path factor associated with the concerned path and involves all closed loops in the graph which are isolated from the forward path under consideration.

The path factor Δ_k for the k_{th} path is equal to the value of the graph determinant of a signal flow graph which exists after ERASING the k_{th} path from the graph.

4.3 DRAWING SIGNAL FLOW GRAPH FROM A GIVEN BLOCK DIAGRAM

While drawing a signal flow graph from a given block diagram the adjacent summing points and take off points (but not a take off point preceding a summing point in the direction of signal flow) are represented by a node and the block transfer function is represented by a line joining the respective nodes. The direction of signal flow is indicated by an arrow on the line.

However, in the direction of signal flow if, a take off point precedes a summing point then such points are represented by two separate nodes with a transmittance of unity between them.

The illustrative examples given below give a procedure for obtaining signal flow graph from a given block diagram.

Illustrative Example 4.3.1. *Represent the block diagram given in Fig. 4.3.1 by signal flow graph and determine the overall transmittance relating C and R.*

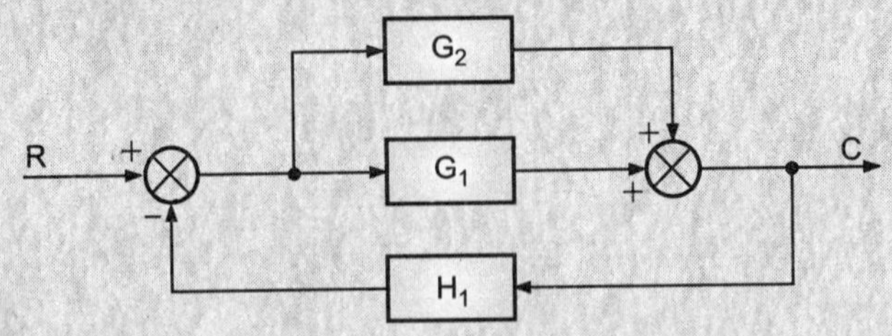

Fig. 4.3.1. Block diagram for example 4.3.1.

Solution. Represent adjacent summing points, take off points before and after the block G_1 respectively by nodes (1) and (2). Join nodes (1) and (2) by lines indicating respective transmittance as per block diagram. The resulting signal flow graph is shown in Fig. 4.3.1 (*a*).

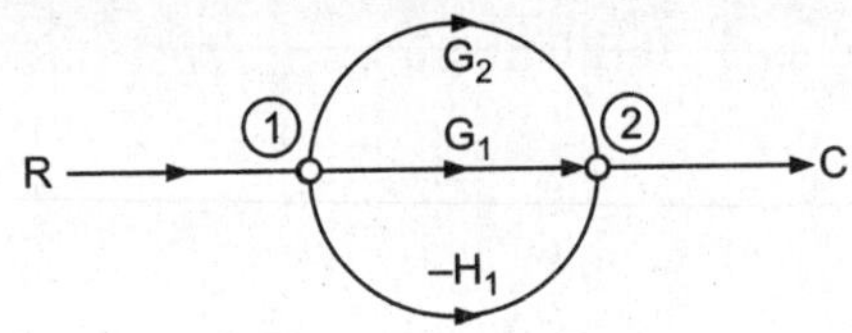

Fig. 4.3.1 (*a*) Signal flow graph representation of block diagram in Fig. 4.3.1.

The two forward paths are

$$P_1 = G_1 \text{ and } P_2 = G_2$$

The two closed loops are

$$L_1 = -G_1H_1 \text{ and } L_2 = -G_2H_1$$

As both the loops L_1 and L_2 are touching both forward paths the two path factors are

$$\Delta_1 = 1 \text{ and } \Delta_2 = 1$$

The graph determinants is

$$\Delta = 1 - (L_1 + L_2) = 1 - (-G_1H_1 - G_2H_1) = 1 + G_1H_1 + G_2H_1$$

Applying Mason's gain formula the overall transmittance relating C and R is determined below

$$\frac{C}{R} = \frac{P_1\Delta_1 + P_2\Delta_2}{\Delta} = \frac{G_1 .1 + G_2 .1}{1 + G_1H_1 + G_2H_1} = \frac{G_1 + G_2}{1 + G_1H_1 + G_2H_1} .$$ **Ans.**

Illustrative Example 4.3.2. *Draw a signal flow graph for the system whose block diagram is shown in Fig. 4.3.2. Determine the overall transmittance.*

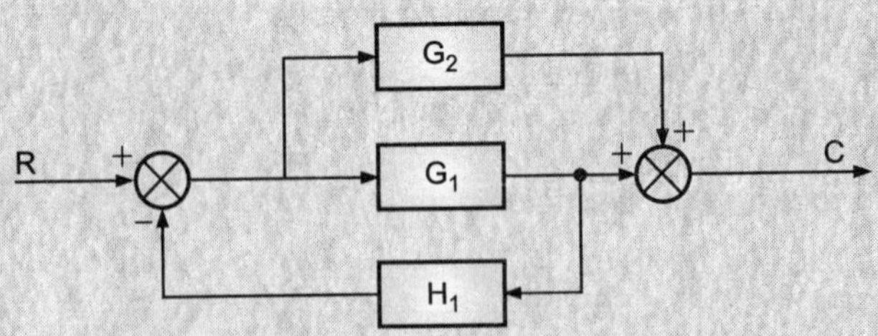

Fig. 4.3.2. Block diagram for example 4.3.2.

Solution. The adjacent summing point and take off point before the block G_1 is represented by node (1). As the take off point after the block G_1 is preceding a summing point, this is represented by two separate nodes (2) and (3) and the transmittance between these two nodes is unity. The signal flow graph for the system is shown in Fig. 4.3.2 (*a*).

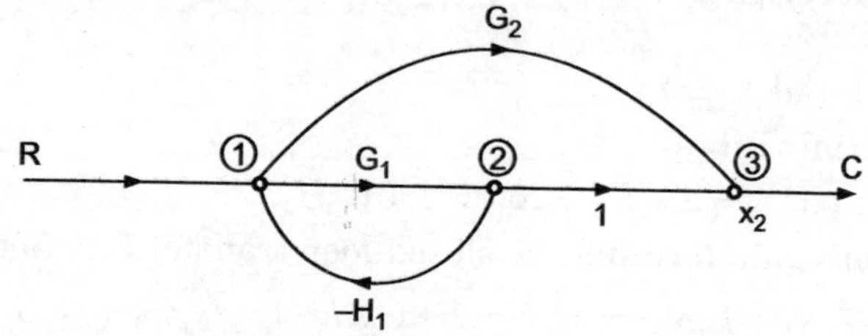

Fig. 4.3.2 (*a*) Signal flow graph representation of block diagram in Fig. 4.3.2.

The two forward paths are

$$P_1 = G_1 \text{ and } P_2 = G_2$$

The only loop is $L_1 = -G_1H_1$

As the loop L_1 touches both forward paths the two path factors are

$$\Delta_1 = 1 \text{ and } \Delta_2 = 1$$

The graph determinant is

$$\Delta = 1 - (L_1) = 1 - (-G_1H_1) = 1 + G_1H_1$$

Applying Mason's gain formula the overall transmittance relating C and R is determined below :

$$\frac{C}{R} = \frac{P_1\Delta_1 + P_1\Delta_2}{\Delta} = \frac{G_1 .1 + G_2 .1}{1 + G_1H_1} = \frac{G_1 + G_2}{1 + G_1H_1} \text{ . Ans.}$$

4.4 SOLVED EXAMPLES

Example 4.4.1. *From the block diagram shown in Fig. 4.4.1, draw the corresponding signal flow graph and evaluate close-loop transfer function relating the output and input.*

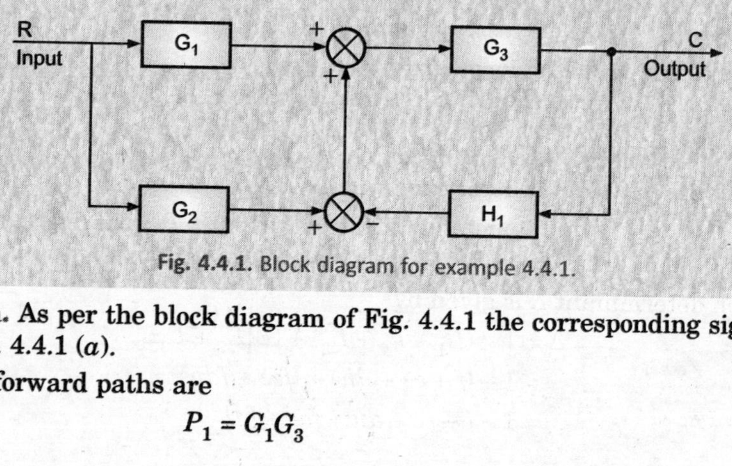

Fig. 4.4.1. Block diagram for example 4.4.1.

Solution. As per the block diagram of Fig. 4.4.1 the corresponding signal flow graph is drawn in Fig. 4.4.1 (*a*).

The two forward paths are

$$P_1 = G_1G_3$$

and $P_2 = G_2G_3$

The only individual loop is

$$L_1 = -G_3H_1$$

The closed-loop touches both the forward paths, therefore, path factors are

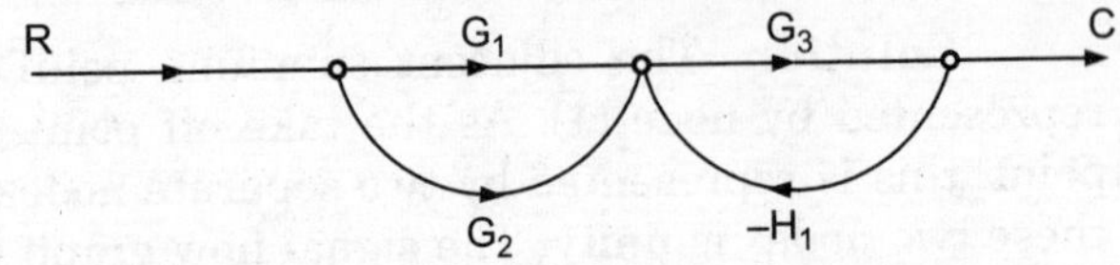

Fig. 4.4.1 (*a*) Signal flow graph of Fig. 4.4.1.

$$\Delta_1 = 1 \text{ and } \Delta_2 = 1$$

The graph determinant is

$$\Delta = 1 - (L_1) = 1 - (-G_3H_1) = 1 + G_3H_1.$$

Applying Mason's gain formula the closed-loop transfer function is determined below

$$\frac{C}{R} = \frac{P_1\Delta_1 + P_2\Delta_2}{\Delta} = \frac{G_1G_3 \cdot 1 + G_2G_3 \cdot 1}{1 + G_3H_1} = \frac{G_1G_3 + G_2G_3}{1 + G_3H_1}. \quad \textbf{Ans.}$$

Example 4.4.2. *Represent the following set of equations by a signal flow graph and determine the overall gain relating x_5 and x_1.*

$$x_2 = ax_1 + fx_2\,; \qquad x_3 = bx_2 + ex_4$$
$$x_4 = cx_3 + hx_5\,; \qquad x_5 = dx_4 + gx_2.$$

Solution. The variables x_1, x_2, x_3, x_4 and x_5 are represented by nodes and respective transmittances are joined by lines to draw signal flow graph as shown in Fig. 4.4.2.

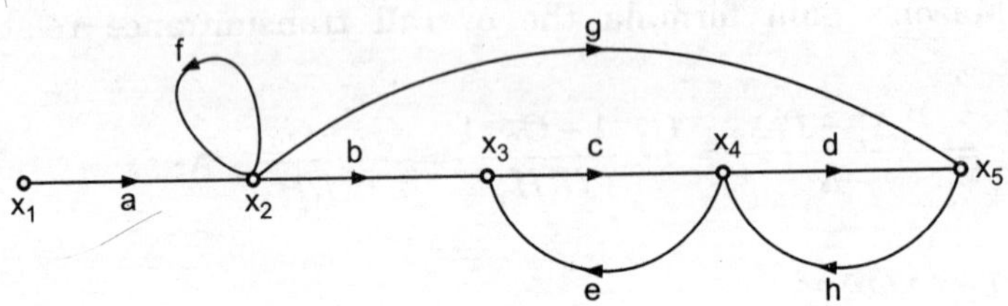

Fig. 4.4.2. Signal flow graph of example 4.4.2.

The two forward paths are

$$P_1 = abcd \text{ and } P_2 = ag$$

The individual loops are

$$L_1 = f,\; L_2 = ce \text{ and } L_3 = dh$$

Two pairs of non-touching loops are

$$L_1L_2 = fce \text{ and } L_1L_3 = fdh$$

All the loops touch the forward path P_1

Therefore, the path factor associated with P_1 is $\Delta_1 = 1$.

The loop $L_2 = ce$ is not touching the forward path P_2, therefore, the path factor associated with P_2 is $\Delta_2 = 1 - ce$.

The graph determinant Δ is given by

$$\Delta = 1 - (L_1 + L_2 + L_3) + (L_1L_2 + L_1L_3)$$
$$= 1 - (f + ce + dh) + (fce + fdh)$$
$$= 1 - f - ce - dh + fce + fdh$$

Applying Mason's gain formula the overall gain relating x_5 and x_1 is determined below

$$\frac{x_5}{x_1} = \frac{P_1\Delta_1 + P_2\Delta_1}{\Delta} = \frac{abcd\,.1 + ag\,.(1 - ce)}{1 - f - ce - dh + fce + fdh}$$

$$= \frac{abcd + ag\,.(1 - ce)}{1 - f - ce - dh + fce + fdh}. \quad \textbf{Ans.}$$

Example 4.4.3. *Obtain signal flow graph representation for a system whose block diagram is given in Fig. 4.4.3. Specify (a) forward path, (b) individual loops, (c) path factors, (d) non-touching loops, (e) determine the graph determinant.*

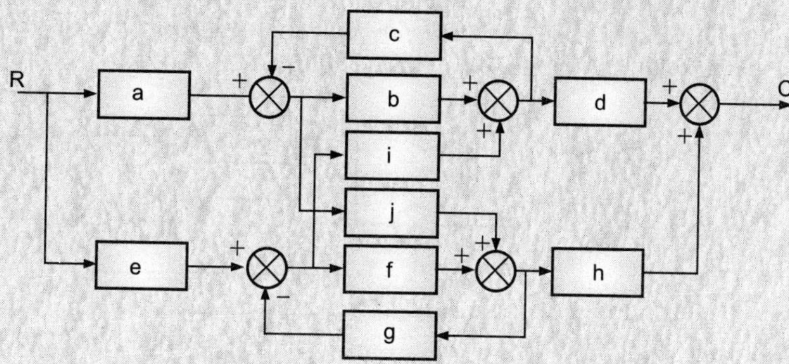

Fig. 4.4.3. Block diagram for example 4.4.3.

Solution. The signal flow graph is drawn as shown in Fig. 4.4.3 (*a*).

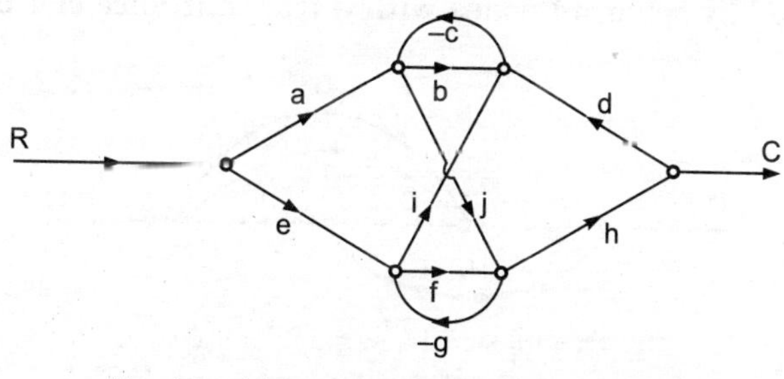

Fig. 4.4.3 (*a*) Signal flow graph of Fig. 4.4.3.

(*a*) There are six forward paths

$P_1 = abd$, $P_2 = efh$, $P_3 = ajh$, $P_4 = eid$,

$P_5 = -\,eicjh$ and $P_6 = -\,ajgid$. **Ans.**

(*b*) The individual loops are

$L_1 = -\,bc$, $L_2 = -\,fg$, $L_3 = icjg$. **Ans.**

(*c*) All the loops are touching the forward paths P_3, P_4, P_5 and P_6, therefore, the respective path factors are

$\Delta_3 = 1$, $\Delta_4 = 1$, $\Delta_5 = 1$ and $\Delta_6 = 1$. **Ans.**

The loop $L_2 = -\,fg$ is not touching the forward path P_1, therefore, the path factor associated with forward path P_1 is

$\Delta_1 = 1 - (-\,fg) = 1 + fg$. **Ans.**

Similarly, the loop $L_1 = -bc$ is not touching the forward path P_2, therefore, the path factor associated with forward path P_2 is

$$\Delta_2 = 1 - (-bc) = 1 + bc. \quad \textbf{Ans.}$$

(*d*) The two non-touching loops are

$$L_1 = -bc \text{ and } L_2 = -fg. \quad \textbf{Ans.}$$

(*e*) The graph determinant is determined below

$$\Delta = 1 - (L_1 + L_2 + L_3) + (L_1L_2)$$
$$= 1 - (-bc - fg + icjg) + (-bc)(-fg)$$
$$= 1 + bc + fg - icjg + bcfg. \quad \textbf{Ans.}$$

Example 4.4.4. *Determine the transfer function relating C and R for the block diagram given in Fig. 4.4.4. Use Mason's gain formula.*

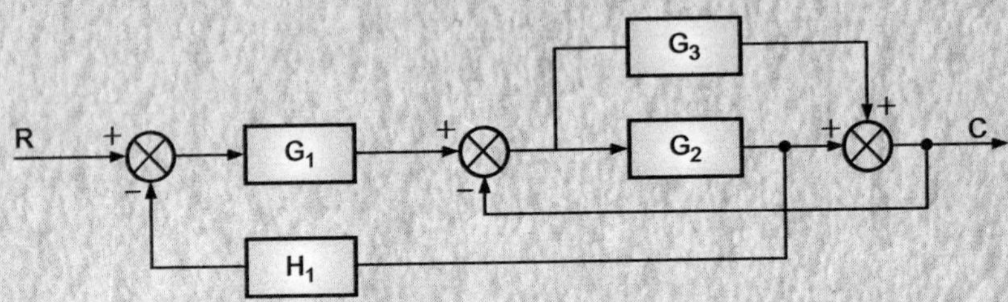

Fig. 4.4.4. Block diagram for example 4.4.4.

Solution. The block diagram given in Fig. 4.4.4 is represented by a signal flow graph drawn in Fig. 4.4.4 (*a*). As the take off point after the block G_2 precedes a summing point, the same being represented by separate nodes with a transmittance of 1 between them.

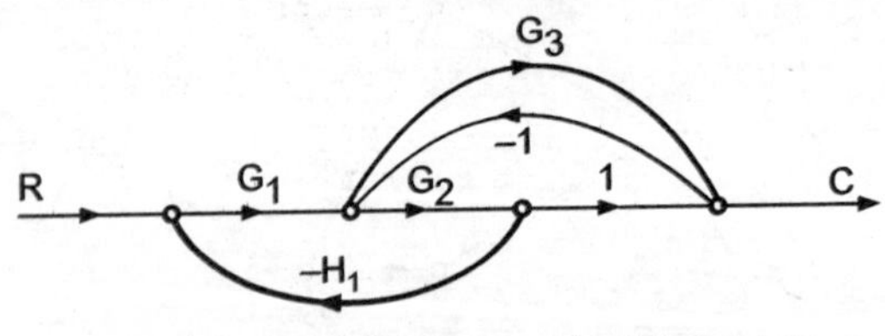

Fig. 4.4.4 (*a*) Signal flow graph of Fig. 4.4.4.

It is observed that

$P_1 = G_1G_2 \,.\, 1,$ $\quad\Delta_1 = 1$

$P_2 = G_1G_3,$ $\quad\Delta_2 = 1$

$L_1 = G_2(-1) = -G_2,$ $\quad L_2 = G_3(-1) = -G_3$

$L_3 = -G_1G_2H_1.$ $\quad$ There are no non-touching loops.

The graph determinants is

$$\Delta = 1 - (L_1 + L_2 + L_3) = 1 - (-G_2 - G_3 - G_1G_2H_1)$$
$$= 1 + G_2 + G_3 + G_1G_2H_1.$$

Using Mason's gain formula the transfer function is determined below

$$\frac{C}{R} = \frac{P_1\Delta_1 + P_2\Delta_2}{\Delta} = \frac{G_1G_2\,.1 + G_1G_3\,.1}{1 - (-G_2 - G_3 - G_1G_2H_1)}$$

$$= \frac{G_1G_2 + G_1G_3}{1 + G_2 + G_3 + G_1G_2H_1}. \quad \textbf{Ans.}$$

Example 4.4.5. *Draw signal flow graph for a system whose block diagram is shown in Fig. 4.4.5. Identify the forward paths, individual loops and determine the relation C/R.*

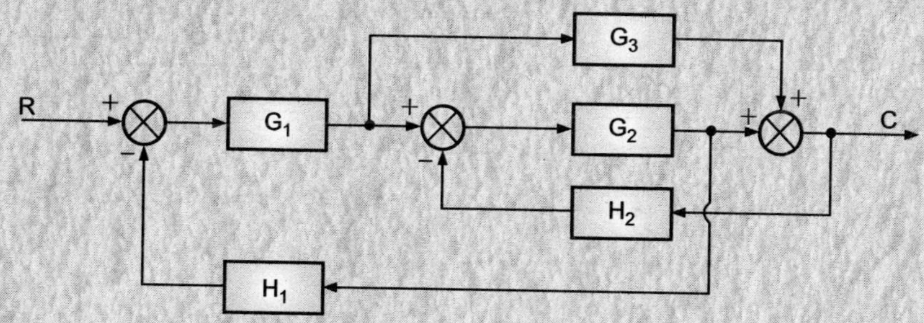

Fig. 4.4.5. Block diagram for example 4.4.5.

Solution. The take off points after the blocks G_1 and G_2 respectively are preceding a summing point, hence are represented by separate nodes and a transmittance of unity is incorporated between them. The signal flow graph is drawn in Fig. 4.4.5 (*a*).

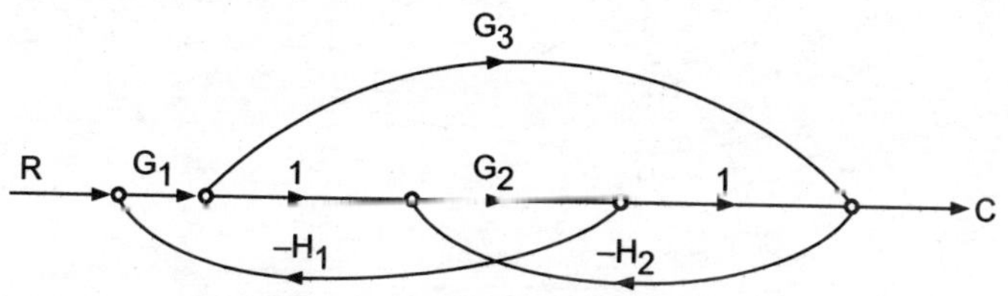

Fig. 4.4.5 (*a*) Signal flow graph of Fig. 4.4.5.

The two forward paths and associated path factors are

$$\left.\begin{aligned} P_1 &= G_1G_2, \quad \Delta_1 = 1 \\ \text{and} \qquad P_2 &= G_1G_3, \quad \Delta_2 = 1 \end{aligned}\right\}. \quad \textbf{Ans.}$$

The individual loops are

$$L_1 = -G_2H_2, \quad L_2 = -G_1G_2H_1$$

and $$L_3 = G_1G_3(-H_2)G_2(-H_1) = G_1G_2G_3H_1H_2. \quad \textbf{Ans.}$$

As all the loops are touching both forward paths, the associated path factors are $\Delta_1 = 1$ and $\Delta_2 = 1$.

The graph determinant is

$$\begin{aligned} \Delta &= 1 - (L_1 + L_2 + L_3) \\ &= 1 - (-G_2H_2 - G_1G_2H_1 + G_1G_2G_3H_1H_2) \\ &= 1 + G_2H_2 + G_1G_2H_1 - G_1G_2G_3H_1H_2 \end{aligned}$$

Applying Mason's gain formula the relation *C/R* is determined below

$$\frac{C}{R} = \frac{G_1G_2 + G_1G_3}{1 + G_2H_2 + G_1G_2H_1 - G_1G_2G_3H_1H_2}. \quad \textbf{Ans.}$$

Example 4.4.6. *Using Mason's gain formula determine the ratio C/R for the system represented by the block diagram of Fig. 4.4.6.*

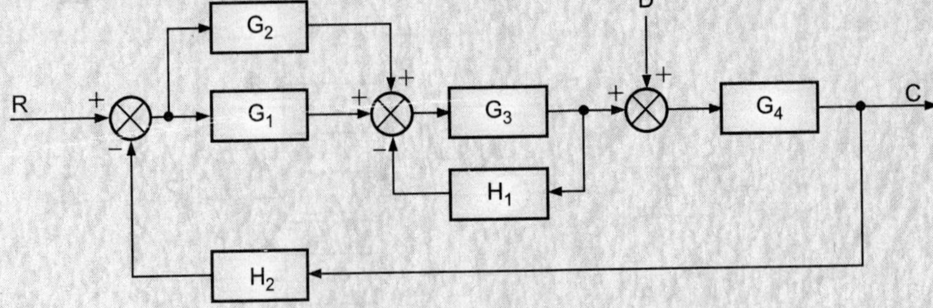

Fig. 4.4.6. Block diagram for example 4.4.6.

Solution. The block diagram of Fig. 4.4.6 is represented by a signal flow graph shown in Fig. 4.4.6 (*a*).

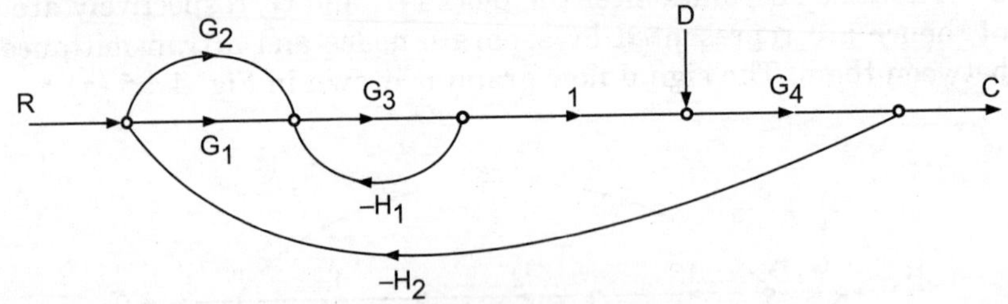

Fig. 4.4.6 (*a*) Signal flow graph of Fig. 4.4.6.

In order to determine the ratio C/R consider $D = 0$. Therefore, the signal flow graph of Fig. 4.4.6 (*a*) is redrawn as shown in Fig. 4.4.6 (*b*).

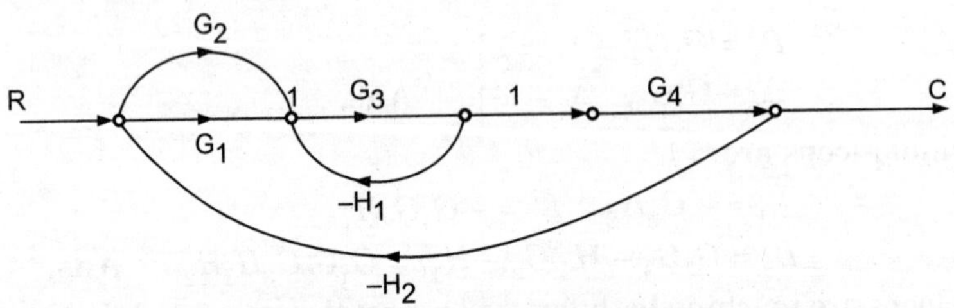

Fig. 4.4.6 (*b*) Signal flow graph of Fig. 4.4.6 (*a*) redrawn ($D = 0$).

From Fig. 4.4.6 (*b*) the elements of the signal flow graph are identified below

$$P_1 = G_1G_3G_4, \qquad \Delta_1 = 1$$
$$P_2 = G_2G_3G_4, \qquad \Delta_2 = 1$$
$$L_1 = -G_3H_1, \qquad L_2 = -G_1G_3G_4H_2$$
$$L_3 = -G_2G_3G_4H_2.$$

The graph determinant is

$$\Delta = 1 - (L_1 + L_2 + L_3)$$
$$= 1 - (G_3H_1 - G_1G_3G_4H_2 - G_2G_3G_4H_2)$$

$$= 1 + G_3H_1 + G_1G_3G_4H_2 + G_2G_3G_4H_2$$

The ratio C/R is

$$\frac{C}{R} = \frac{P_1\Delta_1 + P_2\Delta_2}{\Delta} = \frac{G_1G_3G_4 \,.1 + G_2G_3G_4 \,.1}{1 + G_3H_1 + G_1G_3G_4H_2 + G_2G_3G_4H_2}. \quad \textbf{Ans.}$$

Example 4.4.7. *Draw the signal flow graph and determine the overall transfer function. The block diagram is given below:*

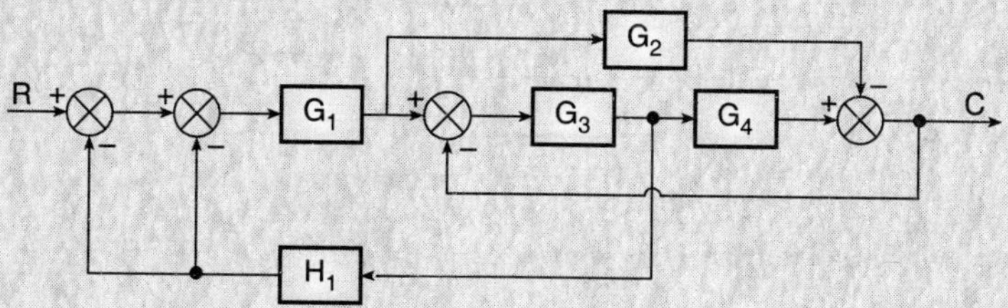

Fig. 4.4.7. Block diagram for example 4.4.7.

Solution. The signal flow graph for the given block diagram is shown in Fig. 4.4.7 (*a*).

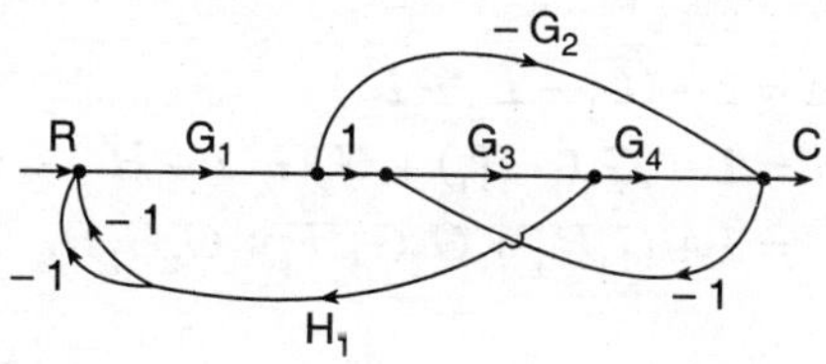

Fig. 4.4.7 (*a*) Signal flow graph for Fig. 4.4.7.

As per the signal flow graph the loop gains are :

$$L_1 = G_3G_4(-1),\ L_2 - G_1G_3H_1(-1),\ L_3 = G_1G_3H_1(-1)$$
$$L_4 = G_1(-G_2)(-1)\ G_3H_1(-1) \text{ and } L_5 = G_1(-G_2)(-1)\ G_3H_1(-1)$$

The graph determinant :

$$\Delta = 1 - (L_1 + L_2 + L_3 + L_4 + L_5)$$
$$= 1 - \{G_3G_4(-1) + G_1G_3H_1(-1) + G_1G_3H_1(-1)$$
$$+ G_1(-G_2)\,(-1)\,G_3H_1(-1) + G_1(-G_2)\,(-1)\,G_3H_1\,(-1)\}$$
$$= 1 + G_3G_4 + 2G_1G_3H_1 + 2G_1G_2G_3H_1$$

The forward paths are :

$$P_1 = G_1\,G_3\,G_4$$

All loops are touching path P_1, therefore,

The path factor : $\Delta_1 = 1$

$$P_2 = G_1(-G_2)$$

All loops are touching forward path P_2, therefore, the path factor : $\Delta_2 = 1$

$$\frac{C}{R} = \frac{P_1\Delta_1 + P_2\Delta_2}{\Delta} = \frac{G_1G_3G_4 \,.1 + G_1(-G_2)\,.1}{1 + G_3G_4 + 2G_1G_3H_1 + 2G_1G_2G_3H_1}. \quad \textbf{Ans.}$$

Example 4.4.8. *The block diagram representation of a control system is below (Fig. 4.4.8).*

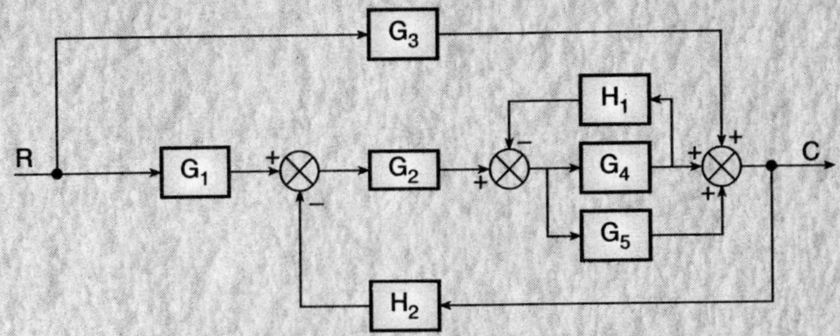

Fig. 4.4.8. Block diagram for example 4.4.8.

Draw the signal flow graph to determine the overall transfer function.

Solution. The signal flow graph for the given block diagram is shown in Fig. 4.4.8 (*a*).

As per signal flow graph, the loop gains are :

$$L_1 = G_4(-H_1), L_2 = G_2G_4 \,.\, 1 \,.\, (-H_2) \text{ and } L_3 = G_2G_5(-H_2).$$

The graph determinant :

$$\Delta = 1 - (L_1 + L_2 + L_3)$$
$$= 1 - \{G_4(-H_1) + G_2G_4\, 1(-H_2) + G_2G_5(-H_2)\}$$
$$= 1 + G_4H_1 + G_2G_4H_2 + G_2G_5H_2$$

The forward paths are :

$$P_1 = G_1G_2G_4 \,.\, 1$$

All loops are touching the path P_1, therefore,

Path factor : $\Delta_1 = 1$

$$P_2 = G_1G_2G_5 \,.\, 1$$

All loops are touching the path P_2, therefore,

The path factor : $\Delta_2 = 1$

$$P_3 = G_3$$

The loop $L_1 = -G_4H_1$ is not touching the path P_3, therefore,

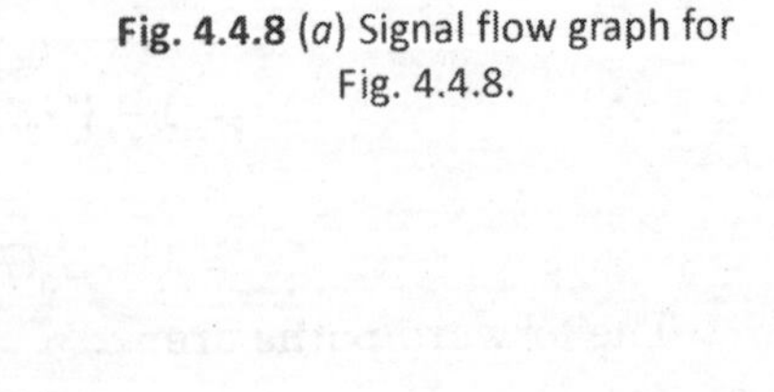

Fig. 4.4.8 (*a*) Signal flow graph for Fig. 4.4.8.

The path factor :

$$\Delta_3 = (1 + L_1) = \{1 + (-G_4H_1)\} = 1 + G_4H_1$$

$$\frac{C}{R} = \frac{P_1\Delta_1 + P_2\Delta_2 + P_3\Delta_3}{\Delta}$$

$$= \frac{G_1G_2G_4 \,.1 + G_1G_2G_5 \,.1 + G_3(1 + G_4H_1)}{1 + G_4H_1 + G_2G_4H_2 + G_2G_5H_2}$$

or

$$\frac{C}{R} = \frac{G_1G_2G_4 + G_1G_2G_5 + G_3(1 + G_4H_1)}{1 + G_4H_1 + G_2G_4H_2 + G_2G_5H_2}.$$ **Ans.**

Example 4.4.9. *The signal flow graph (Fig. 4.4.9) represents a feedfack control system. Draw the corresponding block diagram and determine the overall transfer function using block diagram reduction technique.*

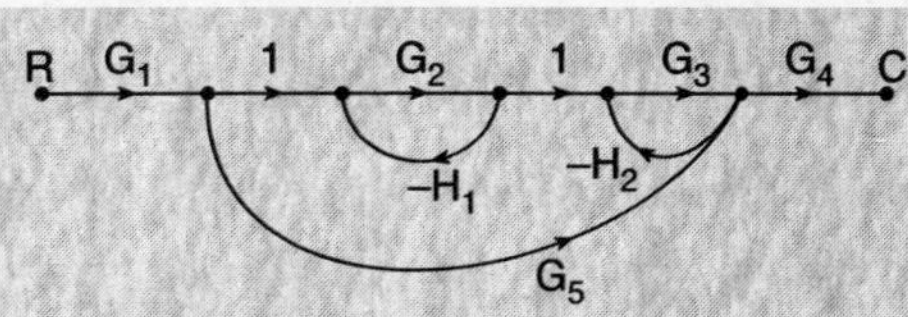

Fig. 4.4.9. Signal flow graph for example 4.4.9.

Solution. The block diagram is drawn as shown in Fig. 4.4.9 (*a*).

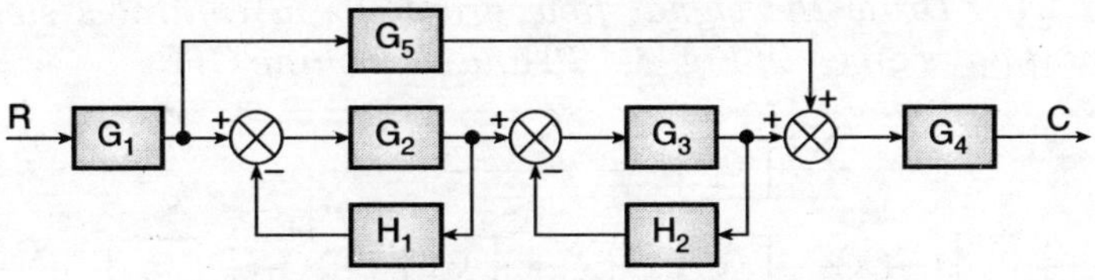

Fig. 4.4.9 (*a*) Block diagram for Fig. 4.4.9.

(*i*) Eliminate the summing point placed before block G_2.

(*ii*) Shift the summing point located after block G_3 towards left of block G_3

Therefore,

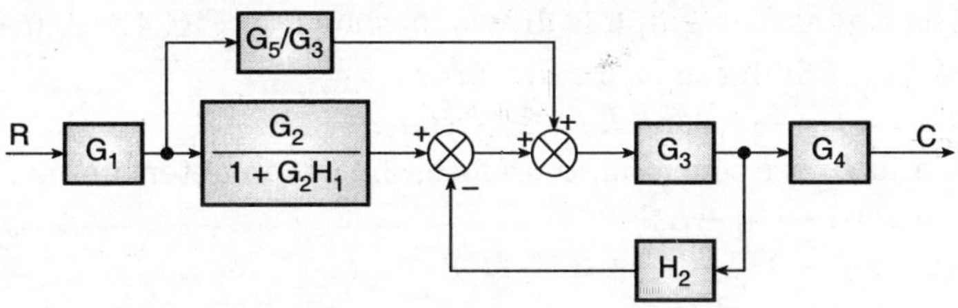

Fig. 4.4.9 (*b*) Block diagram reduction of Fig. 4.4.9 (*a*).

Interchange consecutive summing points placed before block G_3 to get :

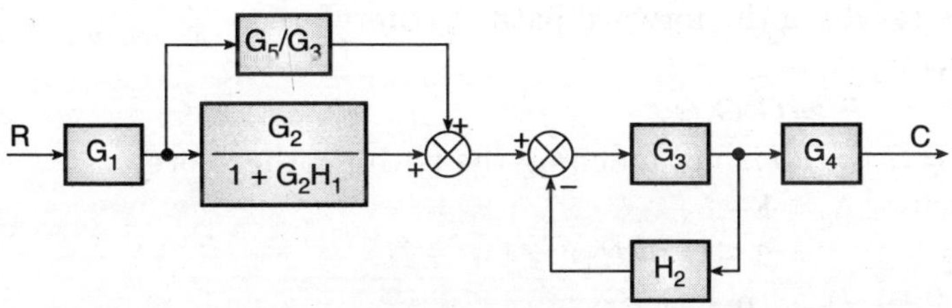

Fig. 4.4.9 (*c*) Block diagram reduction of Fig. 4.4.9 (*b*).

(*i*) The blocks G_5/G_3 and $G_2/(1 + G_2H_1)$ are in parallel, the equivalence is

$$\frac{G_5}{G_3} + \frac{G_2}{1+G_2H_1} = \frac{G_2G_3 + G_5(1+G_2H_1)}{G_3(1+G_2H_1)}$$

(*ii*) Eliminate the summing point located before block G_3 to get equivalence as $G_3/(1 + G_3H_2)$

The block diagram takes the following form :

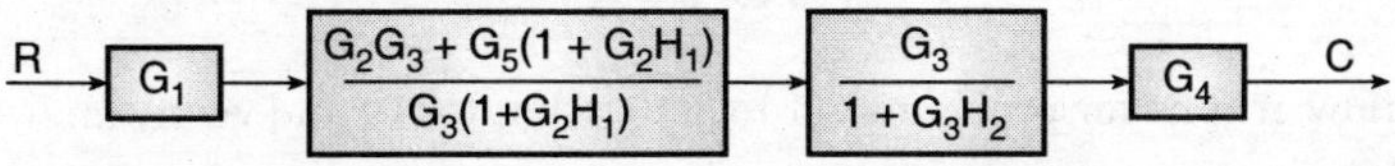

Fig. 4.4.9 (*d*) Block diagram reduction of Fig. 4.4.9 (*c*).

The blocks are placed in cascade. Therefore :

$$\frac{C}{R} = G_1 \times \frac{G_2G_3 + G_5(1+G_2H_1)}{G_3(1+G_2H_1)} \times \frac{G_3}{1+G_3H_2} \times G_4$$

or
$$\frac{C}{R} = \frac{G_1G_2G_3G_4 + G_1G_4G_5(1+G_2H_1)}{1+G_2H_1 + G_3H_2 + G_2G_3H_1H_2}. \quad \textbf{Ans.}$$

Example 4.4.10. *Draw the signal flow graph for a control system whose block diagram representation is given in Fig. 4.4.10 and determine C/R.*

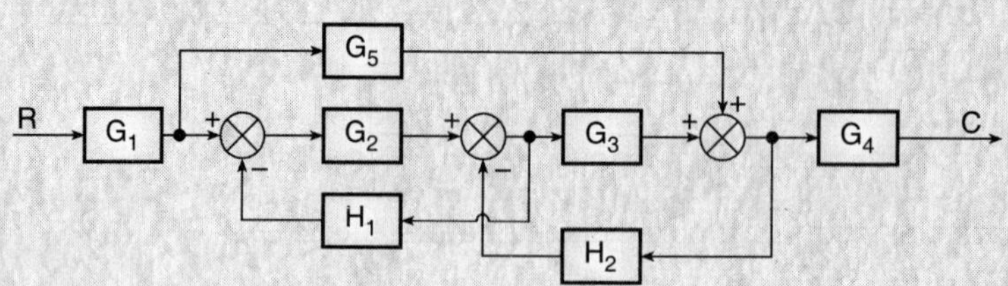

Fig. 4.4.10. Block diagram for example 4.4.10.

Solution. The signal flow graph is drawn as shown in Fig. 4.4.10 (*a*).

As per signal flow graph the loop gains are :

$$L_1 = G_2(-H_1) \text{ and } L_2 = G_3(-H_2)$$

The loops L_1 and L_2 are touching, therefore, the graph determinant is

$$\Delta = 1 - (L_1 + L_2)$$
$$= 1 - \{G_2(-H_1) + G_3(-H_2)\}$$
$$= 1 + G_2H_1 + G_3H_2$$

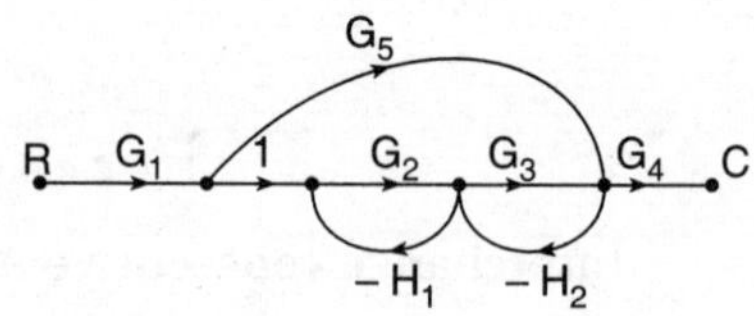

Fig. 4.4.10 (*a*) Signal flow graph for Fig. 4.4.10.

The forward paths are :

$$P_1 = G \,.\, 1 \,.\, G_2G_3G_4$$

All loops are touching the forward path P_1, therefore,

The path factor : $\Delta_1 = 1$

$$P_2 = G_1G_5G_4$$

The loop $L_1 = G_2(-H_1)$ is not touching the path P_2, therefore,

The path factor : $\Delta_2 = 1 - L_1$

$$= \{1 - G_2(-H_1)\} = 1 + G_2H_1$$

$$\frac{C}{R} = \frac{P_1\Delta_1 + P_2\Delta_2}{\Delta}$$

$$= \frac{G\,.1.\,G_2G_3G_4\,.1 + G_1G_5G_4(1+G_2H_1)}{1+G_2H_1+G_3H_2}$$

or
$$\frac{C}{R} = \frac{G_1G_2G_3G_4 + G_1G_5G_4(1+G_2H_1)}{1+G_2H_1+G_3H_2}. \quad \textbf{Ans.}$$

PROBLEM

4.1. Draw signal flow graphs for problems 3.1 to 3.13 (Chapter 3) and verify answers using Mason's gain formula.

CHAPTER 5

Modelling a Control System

A control system has to be mathematically modeled as per the physical laws governing its elements. In this chapter, transformed domain mathematical models have been developed. The approach being the model determination through transformed equation and development of corresponding block diagram. The rules of block diagram reduction are applied to obtain the model of the control system. The solved examples are given for the application of concepts explained in the text.

CONTENTS

- *Electrical Networks*
- *Mechanical Systems*
- *Hydraulic System*
- *Pneumatic System*
- *Thermal System*
- *Servo Motors*
- *Generators*
- *Error Detectors*
- *Solved Examples*

Introduction. As explained earlier, a control system is an interconnection of physical elements arranged in a planned manner. Now, hereunder the procedure for modelling a control system based on transfer function approach is described. Transfer functions of some elements commonly used in control systems have been derived and examples have been given to use the derived results for obtaining the mathematical model of a control system.

5.1 ELECTRICAL NETWORKS

The transfer function of electrical network consisting of resistance and capacitance configurations is determined by taking the ratio of Laplace of the output to input.

1. The network shown in Fig. 5.1.1 is a phase lag network.

The input, the output and the capacitance are expressed by their respective Laplace transform.

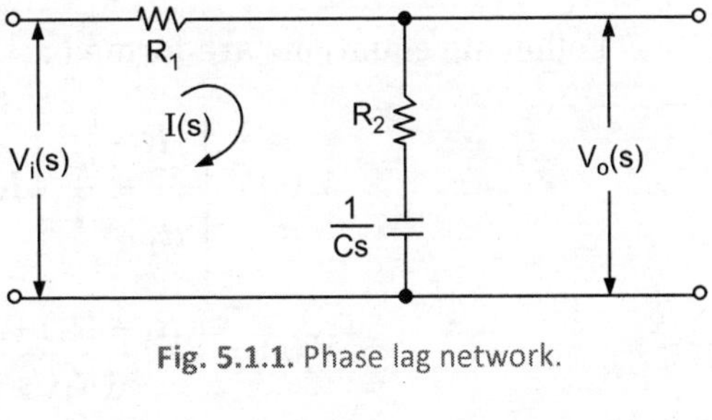

Fig. 5.1.1. Phase lag network.

Following equations are formed :

$$V_i(s) = R_1 I(s) + R_2 I(s) + \frac{1}{Cs} I(s)$$

or $$V_i(s) = \left[\frac{(R_1 + R_2)Cs + 1}{Cs}\right] I(s) \qquad ...(5.1)$$

and $$V_o(s) = R_2 I(s) + \frac{1}{Cs} I(s)$$

or $$V_o(s) = \left[\frac{(R_2 Cs + 1)}{Cs}\right] I(s) \qquad ...(5.2)$$

Therefore, the transfer function is given by

$$\frac{V_o(s)}{V_i(s)} = \frac{(R_2 Cs + 1)}{[(R_1 + R_2)Cs + 1]}$$

or $$\frac{V_o(s)}{V_i(s)} = \frac{(1 + sT_1)}{(1 + sT_2)} \qquad ...(5.3)$$

where $T_1 = R_2 C$ and $T_2 = (R_1 + R_2)C$.

The block diagram representation is given in Fig. 5.1.2.

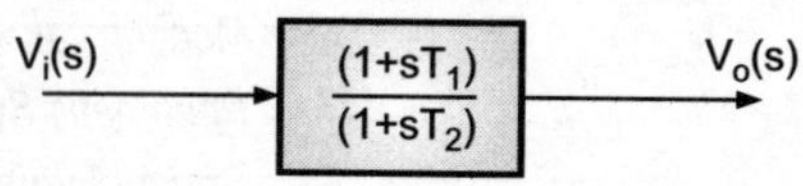

Fig. 5.1.2. Block diagram of phase lag network.

2. Phase lead network is shown in Fig. 5.1.3.

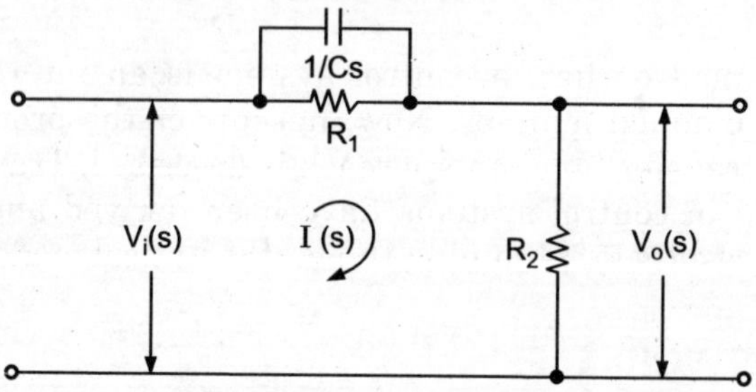

Fig. 5.1.3. Phase lead network.

Following equations are formed :

$$V_i(s) = \left[\frac{R_1 \frac{1}{Cs}}{R_1 + \frac{1}{Cs}}\right] I(s) + R_2 I(s)$$

or $$V_i(s) = \frac{[(R_1 + R_2) + R_1 R_2 Cs]}{(R_1 Cs + 1)} I(s) \qquad ...(5.4)$$

and $$V_o(s) = R_2 I(s) \qquad ...(5.5)$$

Therefore, the transfer function is given by

$$\frac{V_o(s)}{V_i(s)} = \frac{R_2}{(R+R_2)} \frac{(R_1 Cs+1)}{\left(\frac{R_1 R_2}{R_1+R_2} Cs+1\right)}$$

or
$$\frac{V_o(s)}{V_i(s)} = \frac{\alpha(1+sT_1)}{(1+sT_2)} \qquad ...(5.6)$$

where
$$\alpha = \frac{R_2}{(R_1+R_2)}$$

$$T_1 = R_1 C$$

and
$$T_2 = \frac{R_1 R_2}{(R_1+R_2)} C.$$

The block diagram representation is given in Fig. 5.1.3 (*a*).

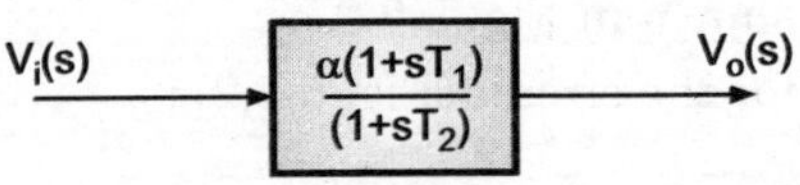

Fig. 5.1.3 (*a*) Block diagram of phase lead network.

5.2 MECHANICAL SYSTEMS

5.2.1. Translation Mechanical Systems

A translational spring-mass-damper system is shown in Fig. 5.2.1.

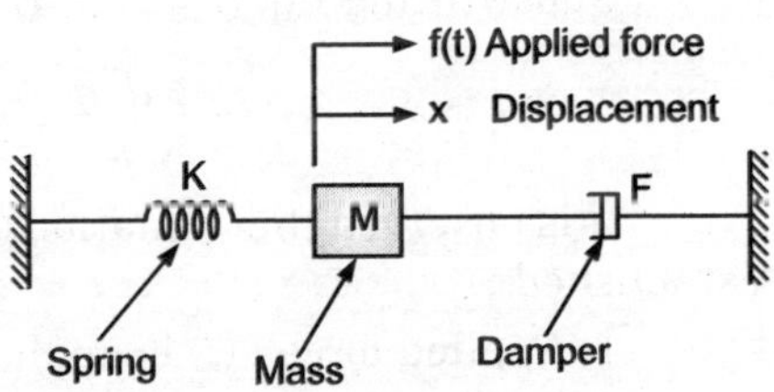

Fig. 5.2.1. Translational spring-mass damper system.

The mass of the system is M (unit of M is kg). The displacement of the mass due to applied force $f(t)$ [unit of force $f(t)$ is Newton] results in inertia force. This inertia force is the product of the mass and its acceleration. The spring deflection constant is K (unit of K is, Newton/meter). The restoring force f_k for the spring is proportional to its displacement. The viscous damping in the system is offered by a dashpot damper. The damping force varies in direct proportion to the velocity. The coefficient of viscous damping is f (unit of f is Newton/rad/sec).

Application of force $f(t)$ to the mass results in a displacement x (meters). The equation of motion for the system is obtained by applying Newton's second law of translational motion.

∴ Inertia force = Forces acting on the mass

or
$$M\frac{d^2x}{dt^2} = -f\frac{dx}{dt} - Kx + f(t) \qquad ...(5.7)$$

where $-f\frac{dx}{dt}$ ↓ Damping force, $-Kx$ ↓ Spring restoring force, $f(t)$ ↓ Applied force.

Rearranging Eq. (5.7), the following equation is obtained

$$M\frac{d^2x}{dt^2} + f\frac{dx}{dt} + Kx = f(t) \quad \text{...(5.8)}$$

Assuming initial conditions as zero, taking Laplace transform on both sides of equation (5.8) following equation is obtained :

$$Ms^2 X(s) + fs\,X(s) + KX(s) = F(s) \quad \text{...(5.9)}$$

If $X(s)$ is specified as the output and $F(s)$ as input then the transfer function of the system is given by the relation below :

$$\frac{X(s)}{F(s)} = \frac{1}{Ms^2 + fs + K} \quad \text{...(5.10)}$$

Force-voltage analogy

The analogy of equation (5.8) with the voltage equation of an electrical circuit can be established by considering electrical circuit shown in Fig. 5.2.2.

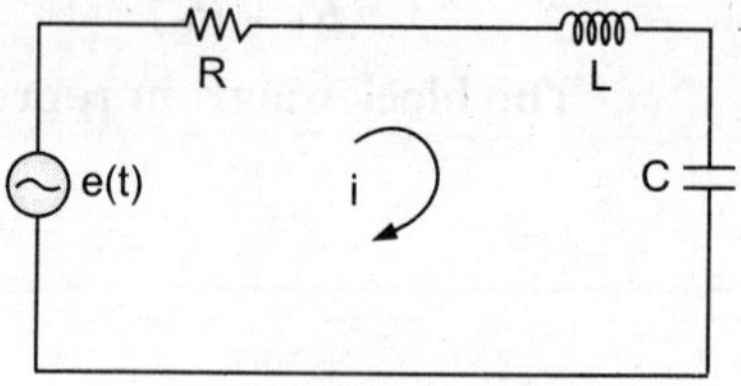

Fig. 5.2.2. Electrical circuit for force voltage analogy.

The voltage equations for the circuit (Fig. 5.2.2) is as follows :

$$L\frac{di}{dt} + Ri + \frac{1}{C}\int i\,dt = e(t) \quad \text{...(5.11)}$$

As the current is the rate of flow of electric charge.

$$\therefore \qquad i = \frac{dq}{dt} \quad \text{...(5.12)}$$

Substituting Eq. (5.12) in Eq. (5.11) following equation is obtained :

$$L\frac{d^2q}{dt^2} + R\frac{dq}{dt} + \frac{1}{C}q = e(t) \quad \text{...(5.13)}$$

It is observed that equation (5.8) is analogous to Eq. (5.13) and following analogies are established :

1. Applied force $f(t)$ is analogous to applied voltage $e(t)$.
2. Mass M is analogous to inductance L.
3. Coefficient of various friction f is analogous to resistance R.
4. Spring deflection constant K is analogous to reciprocal of capacitance $1/C$.
5. Displacement x is analogous to electric charge q.

As the quantities $L\frac{d^2q}{dt^2}, R\frac{dq}{dt}, \frac{1}{C}q, e(t)$ are voltages and $M\frac{d^2x}{dt^2}, f\frac{dx}{dt}, Kx, f(t)$ forces, therefore, aforesaid analogy is called force-voltage analogy.

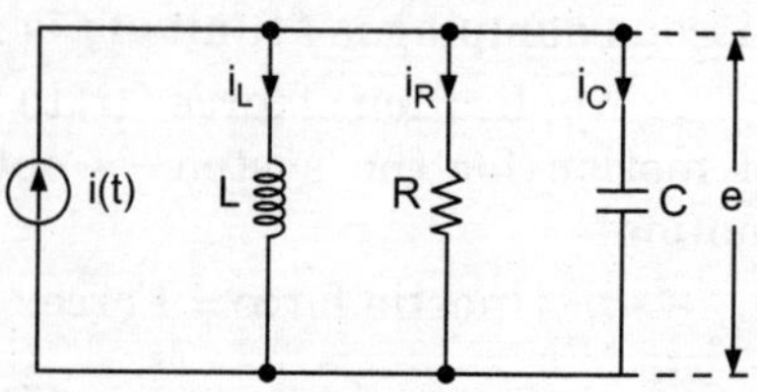

Fig. 5.2.3. Electric circuit for force current analogy.

Force-current analogy

The nodal equation as obtained for the circuit shown in Fig. 5.2.3 is analogous to spring-mass damper system described by the equation (5.8). The said analogy is established below.

According to nodal analysis

$$i_L + i_C + i_R = i(t) \quad ...(5.14)$$

The common voltage across the three parallel elements of the circuit (Fig. 5.2.3) is denoted as e. The following relations are obtained :

$$i_L = \frac{1}{L}\int edt \quad ...(5.15)$$

$$i_R = \frac{1}{R}e \quad ...(5.16)$$

$$i_C = C\frac{de}{dt} \quad ...(5.17)$$

Substituting relations (5.15), (5.16) and (5.17) in equation (5.14). The equation given below is obtained :

$$\frac{1}{L}\int e\,dt + \frac{1}{R}e + C\frac{de}{dt} = i(t) \quad ...(5.18)$$

The voltage e is related to flux linkages associated with inductance L as follows :

$$e = \frac{d\psi}{dt} \quad ...(5.19)$$

In view of the relation (5.19), the equation (5.18) is rewritten as follows :

$$\frac{1}{L}\psi + \frac{1}{R}\frac{d\psi}{dt} + C\frac{d^2\psi}{dt^2} = i(t) \quad ...(5.20)$$

Rearrange equation (5.20) as follows :

$$C\frac{d^2\psi}{dt^2} + \frac{1}{R}\frac{d\psi}{dt} + \frac{1}{L}\psi = i(t) \quad ...(5.21)$$

It is observed that equation (5.8) is analogous to equation (5.21) and following analogies are established :

1. Applied force $f(t)$ is analogous to injected current $i(t)$.
2. Mass M is analogous to capacitance C.
3. Coefficient of viscous friction f is analogous to reciprocal of resistance $1/R$.
4. Spring deflection constant K is analogous to reciprocal of inductance $1/L$.
5. Displacement x is analogous to flux linkages.

As the quantities $C\frac{d^2\psi}{dt^2}, \frac{1}{R}\frac{d\psi}{dt}, \frac{1}{L}\psi, i(t)$ are currents and $M\frac{d^2x}{dt}, f\frac{dx}{dt}, Kx, f(t)$ forces, therefore, aforesaid analogy is called force current analogy.

On the basis of force current analogy the equation for translational mechanical system can be derived by developing a mechanical circuit diagram. The development of mechanical circuit diagram simplifies the procedure for writing system equations.

Development of mechanical circuit diagram

A mechanical system having mass M attached to a spring and a dashpot damper is shown in Fig. 5.2.4. It is required to set up equation for the system on application of force $f(t)$ to the mass. The resulting displacement of the mass being x.

The fixed ends of the spring and deshpot are considered as reference points while the lower ends connected to the mass are considered as nodes. These lower ends and the mass have same deflection, *i.e.* x.

From the view point of developing a mechanical circuit diagram for the system the lower end of the mass is also considered as a fixed point although the same is not fixed.

The construction of the mechanical circuit is done considering the points having same displacement as nodes and fixed points as reference. The mechanical circuit diagram of Fig. 5.2.4 is shown in Fig. 5.2.5.

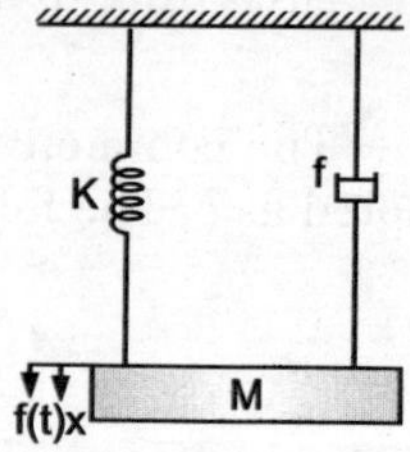

Fig. 5.2.4. Mechanical spring-mass damper system.

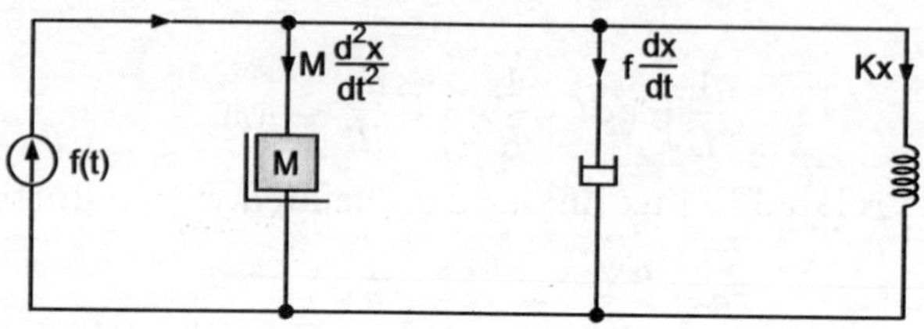

Fig. 5.2.5. Mechanical circuit diagram of Fig. 5.2.4.

From the mechanical circuit diagram of Fig. 5.2.5 it is noted that the forces developed in the system are as follows :

Inertia force $= M\dfrac{d^2x}{dt^2}$

Viscous friction force $= f\dfrac{dx}{dt}$

Spring restoring force $= Kx$

According to nodal analysis the sum of the forces mentioned above is equal to applied force $f(t)$. Thus the equation for the system relating displacement x and the applied force $f(t)$ is given below :

$$M\frac{d^2x}{dt^2} + f\frac{dx}{dt} + Kx = f(t) \qquad ...(5.22)$$

Assuming initial conditions as zero, taking Laplace transformation on both sides of equation (5.22) following transformed equation is obtained :

$$Ms^2X(s) + fsX(s) + KX(s) = F(s) \qquad ...(5.23)$$

If $X(s)$ is specified as the output and $F(s)$ as input, then the transfer function of the system is given by the relation below :

$$\frac{X(s)}{F(s)} = \frac{1}{Ms^2 + fs + K} \qquad ...(5.24)$$

5.2.2. Rotational Mechanical System

A rotational inertia-damper system is shown in Fig. 5.2.6.

The moment of inertia of the system is J (unit of J is kg-m^2). The inertia torque developed in the system is $J\alpha$, where α is the angular acceleration produced due to applied

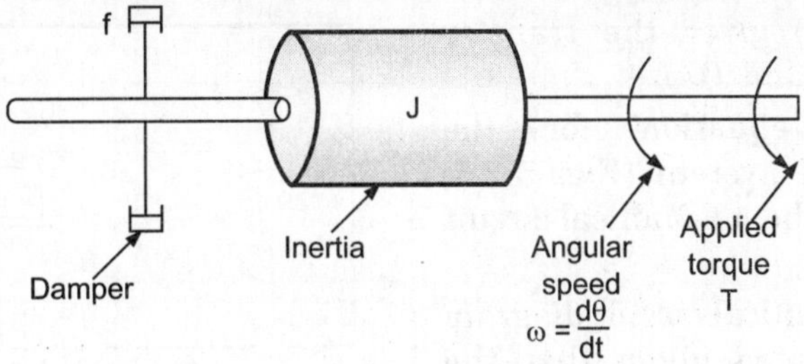

Fig. 5.2.6. Rotational inertia damper system.

torque T (unit of T is Nm). The viscous damping is offered by the damper. The damping torque varies in direct proportion to the angular velocity of the shaft. The coefficient of viscous damping is f (unit of f is Nm/rad/sec).

The equation for the system is obtained by applying Newton's second law of rotational motion, therefore,

Inertia torque = Torques acting on the system

or $$J\alpha = -f\omega + T \qquad ...(5.25)$$

If θ is the angular shaft shift in radian, then

$$\omega = \frac{d\theta}{dt} \qquad ...(5.26)$$

and $$\alpha = \frac{d^2\theta}{dt^2} \qquad ...(5.27)$$

Substituting relations (5.26) and (5.27), in equation (5.25) the system equation given below is obtained

$$J\frac{d^2\theta}{dt^2} = -f\frac{d\theta}{dt} + T$$

On rearranging terms above equation is rewritten as follows :

$$J\frac{d^2\theta}{dt^2} + f\frac{d\theta}{dt} = T \qquad ...(5.28)$$

Assuming initial conditions as zero, taking Laplace transform on both sides of the equation (5.28). The equation given below is obtained.

$$Js^2\,\theta(s) + fs\,\theta(s) = T(s)$$

If $\theta(s)$ is specified as the output and $T(s)$ as input the transfer function relating $\theta(s)$ and $T(s)$ is given below :

$$\frac{\theta(s)}{T(s)} = \frac{1}{Js^2 + fs} \qquad ...(5.29)$$

In order to determine the system transfer function relating shaft speed and torque $T(s)$ multiply equation (5.29) by s and get the following equation :

$$\frac{s\theta(s)}{T(s)} = \frac{s}{Js^2 + fs} \qquad ...(5.30)$$

As $\omega(s) = s\theta(s)$, equation (5.30) can be rewritten as below :

$$\frac{\omega(s)}{T(s)} = \frac{1}{Js + f} \qquad ...(5.31)$$

Equation (5.31) gives the transfer function relating ω(s) and T(s).

The system equation for the rotational mechanical system (Fig. 5.2.6) can be obtained from the mechanical circuit diagram in Fig. 5.2.7.

From the mechanical circuit diagram of Fig. 5.2.7 it is noted given that the torques developed in the system are as follows:

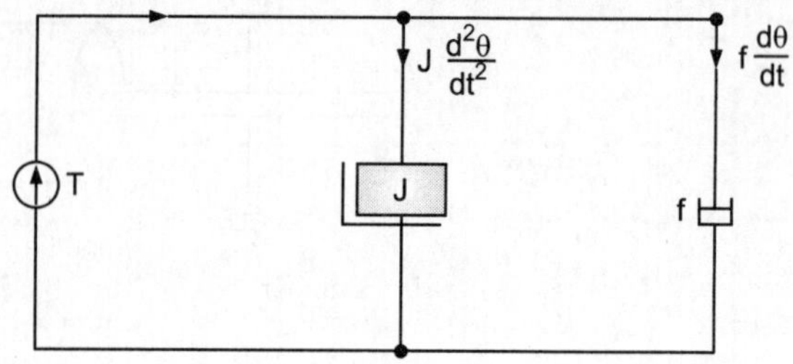

Fig. 5.2.7. Mechanical circuit diagram of Fig. 5.2.6.

Inertia torque $= J\dfrac{d^2\theta}{dt^2}$

Viscous friction torque $= f\dfrac{d\theta}{dt}$

According to nodal analysis the sum of two torques mentioned above is equal to applied torque T. Thus the equation for the system relating angular shaft shift θ and applied torque T is given below

$$T = J\frac{d^2\theta}{dt^2} + f\frac{d\theta}{dt} \qquad ...(5.32)$$

5.3 HYDRAULIC SYSTEM

Hydraulic actuator. Hydraulic actuator shown in Fig. 5.3.1 controls large displacement power at the shaft of main piston. The input power requirement is very small and provided by the displacement of the valve piston. The linear motion of the valve piston controls the flow of oil to either side of the main piston.

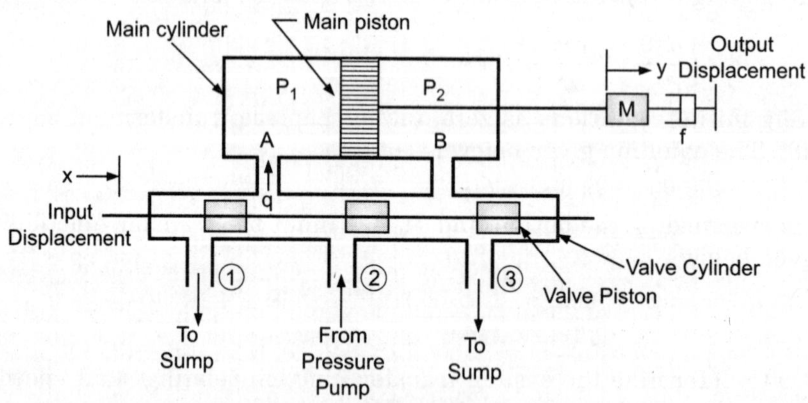

Fig. 5.3.1. Hydraulic actuator.

When the input force through the displacement x is applied to the valve piston, the valve piston moves to the right. Piston valve ports (2) and (3) are thus uncovered and oil at higher pressure from port (2) enters to the left side of the main piston through main port A. At the same time the main port B is connected to the sump through port (3). The pressure P_1

on the left side of the main piston is higher than the pressure P_2 on the right side of the piston. The pressure difference $(P_1 - P_2)$ causes the main piston to move from left to right and the resulting displacement of the mass attached to main piston is y.

The mathematical relation between the input displacement x and the output displacement y gives the model of the system.

The linearized performance characteristics of hydraulic actuator relating the oil flow q (m³/sec) and the pressure difference $(P_1 - P_2) = P_L$ (N/m²) acting on the main piston as a function of valve displacement x is shown in Fig. 5.3.2.

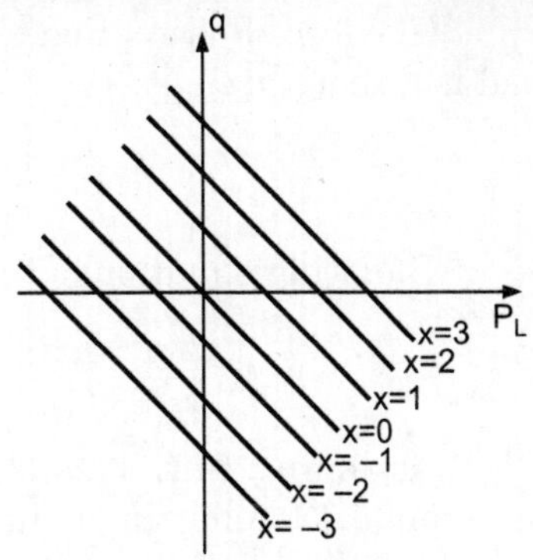

Fig. 5.3.2. Linearized performance characteristics of hydraulic actuator.

From Fig. 5.3.2, it is observed that the oil flow q is a function of valve displacement x and pressure difference P_L and can be expressed as

$$q = f(x, P_L) \quad \text{...(5.33)}$$

Performing partial differentiation on relation (5.33) equation (5.34) given below is obtained

$$dq = \frac{\partial q}{\partial x} dx + \frac{\partial q}{\partial P_L} dP_L \quad \text{...(5.34)}$$

Measuring x, P_L and q from initial zero values and assuming partial derivatives as constant, then on integration of equation (5.34), the equation given below is obtained

$$q = \left(\frac{\partial q}{\partial x}\right) x + \left(\frac{\partial q}{\partial P_L}\right) P_L \quad \text{...(5.35)}$$

In equation (5.35)

put $K_1 = \left(\frac{\partial q}{\partial x}\right)$ and $K_2 = \left(-\frac{\partial q}{\partial P_L}\right)$

Therefore, $q = K_1 x - K_2 P_L$ ...(5.36)

If main piston displacement is dy in time dt and assuming oil to be incompressible and neglecting leakage then the rate of flow of oil into main cylinder is given by

$$q = A\frac{dy}{dt} \quad \text{...(5.37)}$$

where A is the area of the piston. Substituting equation (5.37) in equation (5.36) following equation is obtained :

$$A\frac{dy}{dt} = K_1 x_1 - K_2 P_L \quad \text{...(5.38)}$$

From equation (5.38) following equation is obtained

$$P_L = \frac{1}{K_2}\left(K_1 x - A\frac{dy}{dt}\right) \quad \text{...(5.39)}$$

The force developed by the piston is AP_L, therefore, multiplying equation (5.39) by A

$$F = \frac{A}{K_2}\left(K_1 x - A\frac{dy}{dt}\right) \quad \text{...(5.40)}$$

The force F is applied to load having a mass M and coefficient of viscous friction at the load is f, therefore,

$$F = M\frac{d^2y}{dt^2} + f\frac{dy}{dt} \quad ...(5.41)$$

Equating equations (5.41) and (5.40) following equation is obtained :

$$M\frac{d^2y}{dt^2} + f\frac{dy}{dt} = \frac{A}{K_2}\left(K_1x - A\frac{dy}{dt}\right) \quad ...(5.42)$$

Assuming initial conditions as zero, taking Laplace transform on both sides of equation (5.42) following equation is obtained :

$$Ms^2Y(s) + fs\,Y(s) = \frac{A}{K_2}[K_1X(s) - As\,Y(s)] \quad ...(5.43)$$

From equation (5.43), the transfer function of hydraulic actuator relating $Y(s)$ and $X(s)$ is obtained below

$$\frac{Y(s)}{X(s)} = \frac{AK_1/K_2}{s[Ms + (f + A^2)]} \quad ...(5.44)$$

or

$$\frac{Y(s)}{X(s)} = \frac{K}{s(1 + sT)} \quad ...(5.45)$$

where $K = \dfrac{AK_1}{K_2(f + A^2)}$ and $T = \dfrac{M}{(f + A^2)}$

5.4 PNEUMATIC SYSTEM

Pneumatic actuating valve. Pneumatic actuating valve is used to obtain linear displacement of a plunger with pressurised air as input. The diagram of a pneumatic actuating valve is shown in Fig. 5.4.1.

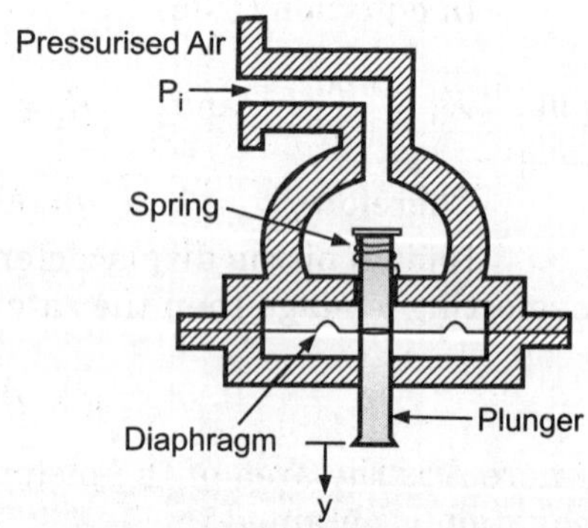

Fig. 5.4.1. Pneumatic actuating valve.

The air at pressure P_i is injected through the input manifold. As the plunger is attached to diaphragm, the pressurised air pushes the diaphragm as well as plunger and thus the plunger is displaced through a distance dy. The plunger arrangement has a mass M. The coefficient of viscous friction and the spring deflection constants are f and K respectively.

If the area of the diaphragm is A, then the force exerted on the system is AP_i. This force is opposed by inertia force, viscous damping force and spring restoring force of the system, therefore,

$$M\frac{d^2y}{dt^2} + f\frac{dy}{dt} + Ky = AP_i \quad ...(5.49)$$

Assuming initial conditions as zero, taking Laplace transform on both sides of equation (5.49) following equation is obtained :

$$Ms^2Y(s) + fsY(s) + KY(s) = AP_i(s) \qquad ...(5.50)$$

From equation (5.50) the transfer function relating the output $Y(s)$ and input $P_i(s)$ is obtained as given below :

$$\frac{Y(s)}{P_i(s)} = \frac{A}{Ms^2 + fs + K} \qquad ...(5.51)$$

5.5 THERMAL SYSTEM

Heat transfer system. A thermal system used for heating flow of water is shown in Fig. 5.5.1. An electric heating element is provided in the storage tank to heat the flow of water. The inlet water temperature is θ_i and the outlet water temperature is θ_o. The storage tank is insulated from surrounding atmosphere to reduce heat loss. The thermal resistance of the insulation is R (°C/joule/sec).

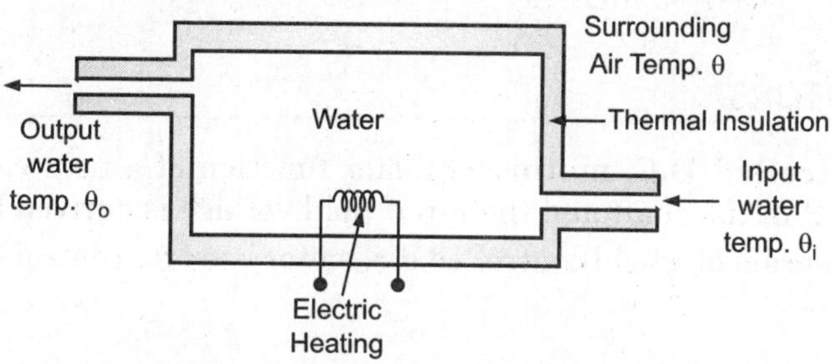

Fig. 5.5.1. Thermal water heating system.

The energy input to the system is q (joule/sec) and obtained from the electric heater. As the specific heat of the thermal insulation is very low there is negligible heat storage in the insulation. However, there is a heat loss from the water to surrounding atmosphere.

It is assumed that the water in the tank is thoroughly mixed during the heating process so that uniformity of the temperature is maintained.

If the thermal capacity is C (joule/°C) then the rate of heat flow for the water in the tank is given by

$$q_i = C\frac{d\theta}{dt} \qquad ...(5.52)$$

The rate of heat flow from the water to the surrounding atmosphere is

$$q_t = \frac{\theta_o - \theta}{R} \qquad ...(5.53)$$

According to heat transfer principle following relation is obtained :

$$q = q_i + q_t \qquad ...(5.54)$$

Substituting for q_t and q_i in equation (5.54) from Eqs. (5.53) and (5.52) following equation is obtained :

$$q = C\frac{d\theta_o}{dt} + \frac{\theta_o - \theta}{R}$$

or

$$q = C\frac{d\theta_o}{dt} + \frac{\theta_o}{R} - \frac{\theta}{R} \qquad ...(5.55)$$

The variation of the water temperature θ_o is over and above the ambient temperature, *i.e.* θ, therefore, the differential equation for the system is rewritten below neglecting the term θ/R from the equation (5.55)

$$q = C\frac{d\theta_o}{dt} + \frac{\theta_o}{R} \qquad ...(5.56)$$

In order to find the transfer function relating input heat energy q and the water temperature θ_o as output, take Laplace transform the both sides of the equation (5.56) assuming initial condition as zero, therefore,

$$Q(s) = Cs\theta_o(s) + \frac{1}{R}\theta_o(s)$$

From the above equation, the transfer function of the system relating $\theta_o(s)$ and $Q(s)$ is given by

$$\frac{\theta_o(s)}{Q(s)} = \frac{R}{(RCs+1)} \qquad ...(5.57)$$

5.6 SERVO MOTORS

1. **Field controlled D.C. motor.** Transfer function of a field controlled d.c. motor relating angular shift in the shaft and the input field voltage is derived hereunder.

The circuit diagram of a field controlled d.c. motor used for control systems is shown in Fig. 5.6.1.

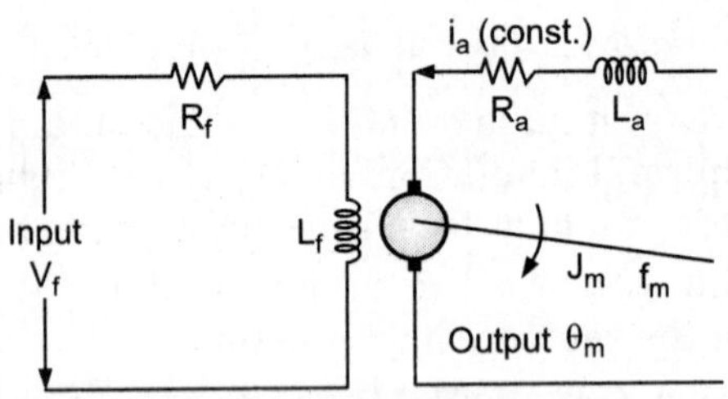

Fig. 5.6.1. Field controlled d.c. motor.

The input voltage V_f is applied to the field winding which has a resistance and inductance of R_f and L_f respectively. The armature current i_a supplied to the armature is kept constant and thus the motor shaft is controlled by the input voltage V_f. The field current i_f produces a flux in the machine which in turn produces a torque at the motor shaft. The moment of inertia and the coefficient of viscous friction at the motor shaft are J_m and f_m respectively. The angular shift resulting in the motor shaft being θ_m and the corresponding angular velocity being ω_m.

Since the armature current is kept constant, the relationship between the developed motor torque T_M and the field current i_f is given by,

$$T_M \propto i_f \qquad ...(5.58)$$

or

$$T_M = K_f i_f$$

where K_f is motor torque constant expressed in Nm/A.

The related dynamic equations are	*The corresponding Laplace transformed equations are*	
$V_f = R_f i_f + L_f \frac{di_f}{dt}$	$V_f(s) = R_f I_f(s) + sL_f I_f(s)$	...(5.59)
$T_M = K_f I_f$	$T_M(s) = K_f I_f(s)$	...(5.60)
$T_M = J_m \frac{d^2\theta_m}{dt^2} + f_m \frac{d\theta_m}{dt}$	$T_M(s) = s^2 J_m \theta_m(s) + sf_m \theta_m(s)$	...(5.61)
$T_M = J_m \frac{d\omega_m}{dt} + f_m \omega_m$	$T_M(s) = sJ_m \omega_m(s) + f_m \omega_m(s)$	...(5.62)

From equations (5.59) to (5.61), block diagram relating the output $\theta_m(s)$ and the input $V_f(s)$ is drawn and shown in Fig. 5.6.2 (*a*) which considers $\theta_m(s)$ as the output and simplified as shown in Fig. 5.6.2 (*b*).

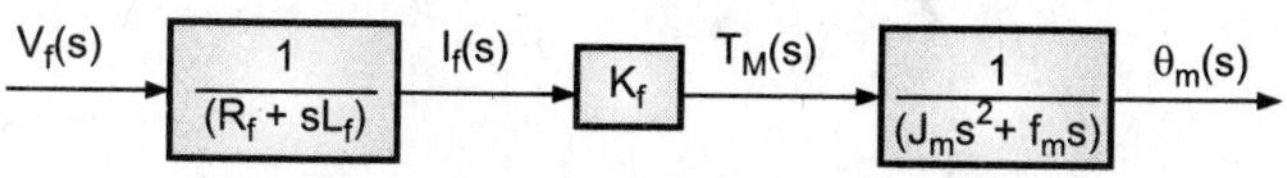

Fig. 5.6.2 (*a*) Block diagram of a field controlled d.c. motor relating $\theta_m(s)$ and $V_f(s)$.

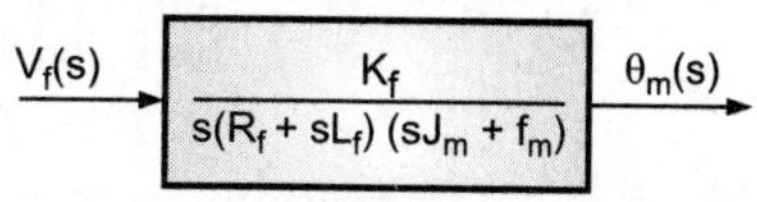

Fig. 5.6.2 (*b*) Block diagram reduction of Fig. 5.6.2 (*a*).

Hence, the transfer function relating the output $\theta_m(s)$ and the input $V_f(s)$ is given by

$$\frac{\theta_m(s)}{V_f(s)} = \frac{K_f}{s(R_f + sL_f)(sJ_m + f_m)} \qquad ...(5.63)$$

From equations (5.59), (5.60) and (5.62), the block diagram as shown in Fig. 5.6.3 (*a*) is obtained which considered $\omega_m(s)$ as the output.

The block diagram given in Fig. 5.6.3 (*a*) is reduced and shown in Fig. 5.6.3 (*b*).

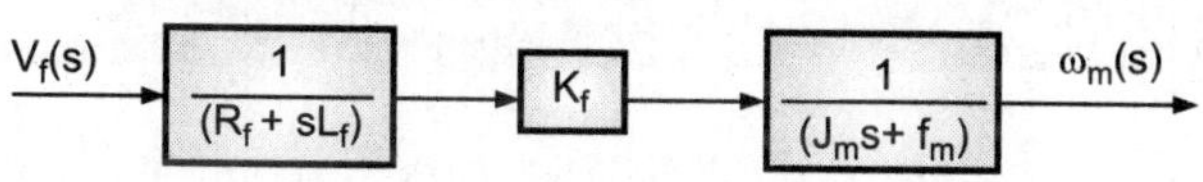

Fig. 5.6.3 (*a*) Block diagram of a field controlled d.c. motor relating $\omega_m(s)$ and $V_f(s)$.

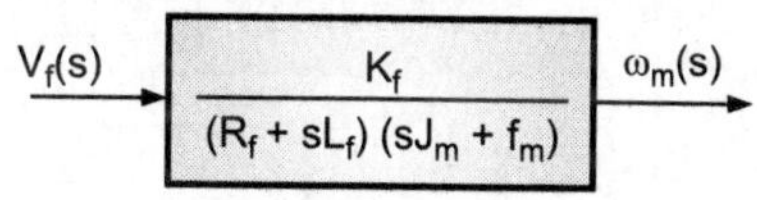

Fig. 5.6.3 (*b*) Block diagram reduction of Fig. 5.6.3 (*a*).

Hence, the transfer function relating the output $\omega_m(s)$ and the input $V_f(s)$ is given by

$$\frac{\omega_m(s)}{V_f(s)} = \frac{K_f}{(R_f + sL_f)(sJ_m + f_m)} \qquad ...(5.63\ a)$$

2. **Armature controlled d.c. motor.** The transfer function of an armature controlled d.c. motor relating angular shift in the shaft and the input armature voltage is derived here under.

The circuit diagram of an armature controlled d.c. motor used for control systems is shown in Fig. 5.6.4.

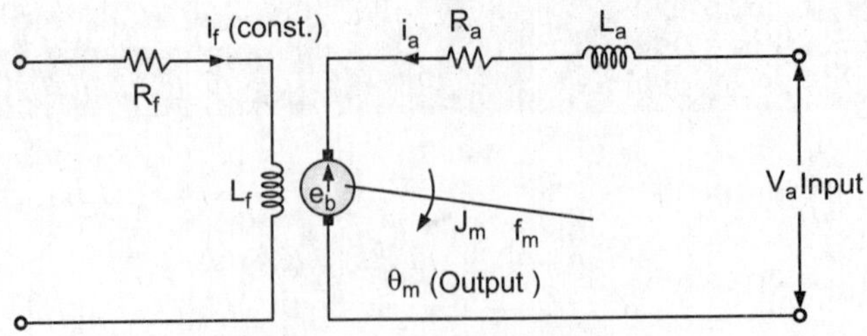

Fig. 5.6.4. Armature controlled d.c. motor.

The input voltage V_a is applied to the armature which has a resistance of R_a and inductance of L_a. The field current i_f supplied to the field winding is kept constant and thus the armature input voltage controls the motor shaft output. The moment of inertia and the coefficient of viscous friction at the motor shaft being J_m and f_m respectively. The angular shift in the motor shaft being θ_m and the corresponding angular velocity being ω_m.

As the field current is kept constant the relationship between the developed motor torque T_M and the armature current is given by,

$$T_M \propto i_a$$

or $$T_M = K_T i_a \qquad ...(5.64)$$

where K_T is motor torque constant expressed in Nm/A.

The applied input voltage V_a is opposed by the back e.m.f. e_b developed in the armature. The relationship between e_b and the motor speed ω_m is given by,

$$e_b \propto \omega_m \quad \because \quad \omega_m = \frac{d\theta_m}{dt}$$

$$\therefore \qquad e_b = K_b \frac{d\theta_m}{dt} \qquad ...(5.65)$$

where K_b is the back e.m.f. constant expressed in V/(rad/sec).

The related dynamic equations are	*The corresponding Laplace transformed equations are*	
$V_a - e_b = R_a i_a + L_a \dfrac{di_a}{dt}$	$V_a(s) - E_b(s) = R_a I_a(s) + sL_a I_a(s)$	...(5.66)
$e_b = K_b \dfrac{d\theta_m}{dt}$	$E_b(s) = sK_b\theta_m(s)$	...(5.67)
or $e_b = K_b\omega_m$	$E_b(s) = K_b\omega_m(s)$	...(5.67*a*)
$T_M = K_T i_a$	$T_M(s) = K_T I_a(s)$	...(5.68)
$T_M = J_m \dfrac{d^2\theta_m}{dt^2} + f_m \dfrac{d\theta_m}{dt}$	$T_M(s) = s^2 J_m \theta_m(s) + sf_m\theta_m(s)$	...(5.69)
$T_M = J_m \dfrac{d\omega_m}{dt} + f_m\omega_m$	$T_M(s) = sJ_m\omega_m(s) + f_m\omega_m(s)$	...(5.70)

From equations (5.66) to (5.69) the block diagram relating the output $\theta_m(s)$ and the input $V_a(s)$ is drawn and shown in Fig. 5.6.4 (*a*).

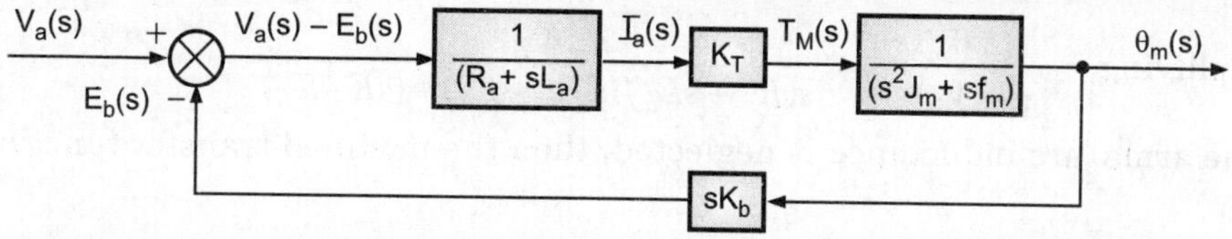

Fig. 5.6.4 (*a*) Block diagram of armature controlled d.c. motor relating $\theta_m(s)$ and $V_a(s)$.

The block diagram given in Fig. 5.6.4 (*a*) is reduced and shown in Fig. 5.6.4 (*b*).

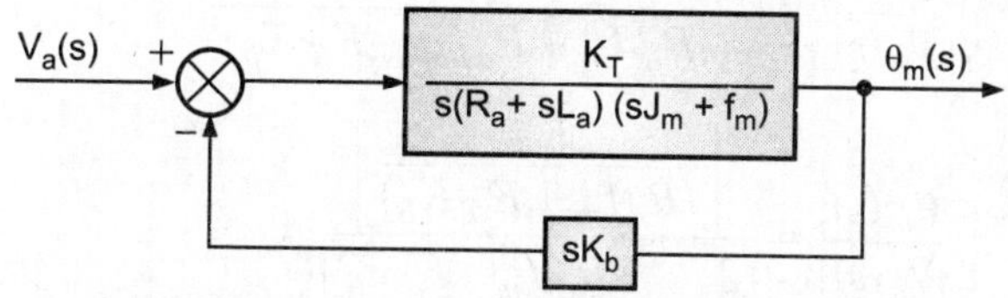

Fig. 5.6.4 (*b*) Block diagram reduction of Fig. 5.6.4 (*a*).

From equations (5.66), (5.67), (5.68) and (5.70) the block diagram as shown in Fig. 5.6.5 (*a*) is obtained which considers $\omega_m(s)$ as the output.

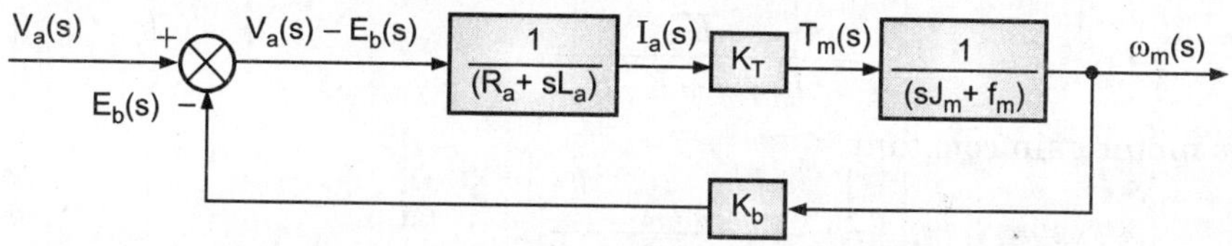

Fig. 5.6.5 (*a*) Block diagram of armature controlled d c. motor relating $\omega_m(s)$ and $V_a(s)$.

The block diagram given in Fig 5.6.5 (*a*) is reduced and shown in Fig. 5.6.5 (*b*).

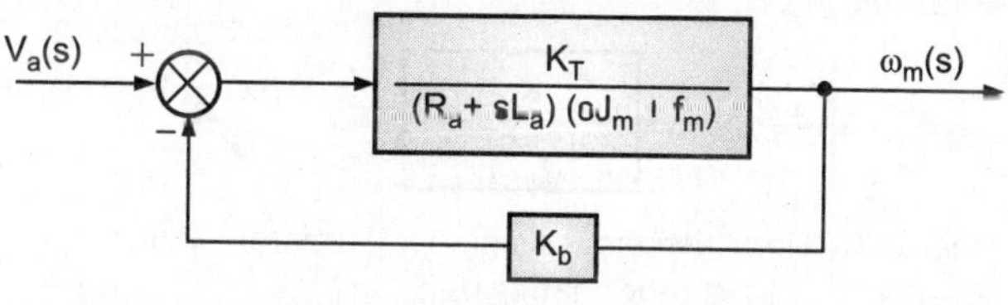

Fig. 5.6.5 (*b*) Block diagram reduction of Fig. 5.6.5 (*a*).

In the block diagram shown in Fig. 5.6.4 (*b*), the forward path transfer function is

$$G(s) = \frac{K_T}{s(R_a + sL_a)(sJ_m + f_m)}$$

and the feedback path transfer function is $H(s) = sK_b$

Hence, the overall transfer function relating the output $\theta_m(s)$ and the input $V_a(s)$ is given by the relation

$$\frac{\theta_m(s)}{V_a(s)} = \frac{G(s)}{1 + G(s)H(s)}$$

or

$$\frac{\theta_m(s)}{V_a(s)} = \frac{\dfrac{K_T}{s(R_a + sL_a)(sJ_m + f_m)}}{1 + \dfrac{K_T}{s(R_a + sL_a)(sJ_m + f_m)} . sK_b}$$

Simplifying, $$\frac{\theta_m(s)}{V_a(s)} = \frac{K_T}{s(R_a + sL_a)(sJ_m + f_m) + (sK_T K_b)}$$

If the armature inductance is neglected, then the modified transfer function is given by,

$$\frac{\theta_m(s)}{V_a(s)} = \frac{K_T}{sR_a(sJ_m + f_m) + sK_T K_b} \quad ...(5.71)$$

or $$\frac{\theta_m(s)}{V_a(s)} = \frac{K_T}{s(sR_a J_m + R_a f_m + K_T K_b)}$$

or $$\frac{\theta_m(s)}{V_a(s)} = \frac{\left[\dfrac{K_T}{(R_a f_m + K_T K_b)}\right]}{s\left[\dfrac{sR_a J_m}{R_a f_m + K_T K_b} + 1\right]}$$

or $$\frac{\theta_m(s)}{V_a(s)} = \frac{K_m}{s(1 + sT_m)} \quad ...(5.72)$$

where $$K_m = \frac{K_T}{(R_a f_m + K_T K_b)}$$

is termed as motor gain constant,

and $$T_m = \frac{R_a J_m}{(R_a f_m + K_T K_b)}$$

is termed as motor time constant.

The block diagram representing the above transfer function shown in Fig. 5.6.6.

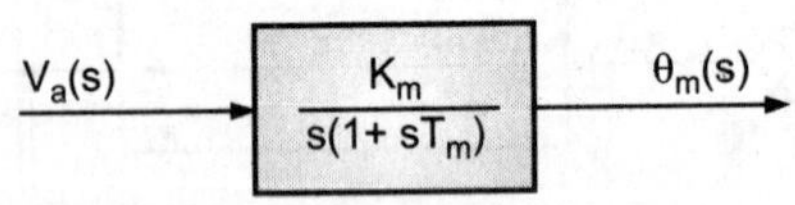

Fig. 5.6.6. Simplified block diagram of armature controlled d.c. motor relating $\theta_m(s)$ and $V_a(s)$.

Since $\omega_m(s) = s\theta_m(s)$

The block diagram relating $\omega_m(s)$ and $V_a(s)$ is shown in Fig. 5.6.7 (*a*) and simplified as shown in Fig. 5.6.7 (*b*).

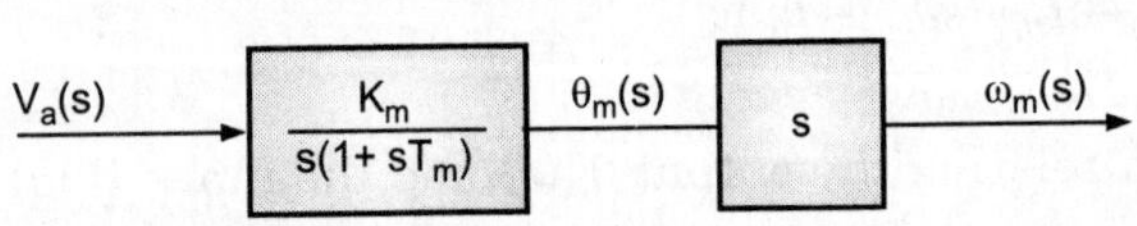

Fig. 5.6.7 (*a*) Block diagram of armature controlled d.c. motor relating $\omega_m(s)$ and $V_a(s)$.

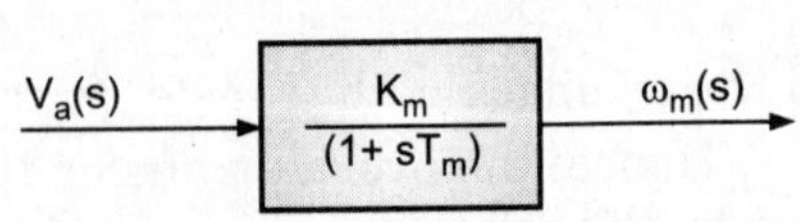

Fig. 5.6.7 (*b*) Simplified block diagram of armature controlled d.c. motor relating $\omega_m(s)$ and $V_a(s)$.

According to Fig. 5.6.7 (*b*) the transfer function relating $\omega_m(s)$ and $V_a(s)$ is thus,

$$\frac{\omega_m(s)}{V_a(s)} = \frac{K_m}{(1 + sT_m)} \quad ...(5.73)$$

Relation between torque constant K_T and back e.m.f. constant K_b

The mechanical power output of the motor is $T_M\omega_m$ which is equal to armature input $e_b i_a$ (neglecting armature losses and rotational losses). Therefore, in MKS units

$$T_M \omega_m = e_b i_a$$

Since $T_M = K_T i_a$ and $e_b = K_b \omega_m$.

Substituting in above equation

$$K_T i_a \omega_m = K_b \omega_m i_a. \qquad \text{Hence, } K_T = K_b.$$

3. **Two phase a.c. servomotor.** Transfer function of a two phase a.c. servomotor relating the angular shift in the shaft and the input control voltage is derived hereunder.

The circuit diagram of a two phase a.c. servomotor is shown in Fig. 5.6.8.

A two phase a.c. servomotor is a two phase induction motor having drag cup type rotor construction. The control voltage $V_c(t)$ is applied to the control winding and a fixed voltage having a phase difference of 90° w.r.t. control winding voltage is applied to the reference winding. The control voltage results in the development of the motor torque T_M. The torque speed characteristics of the motor shown in Fig. 5.6.9.

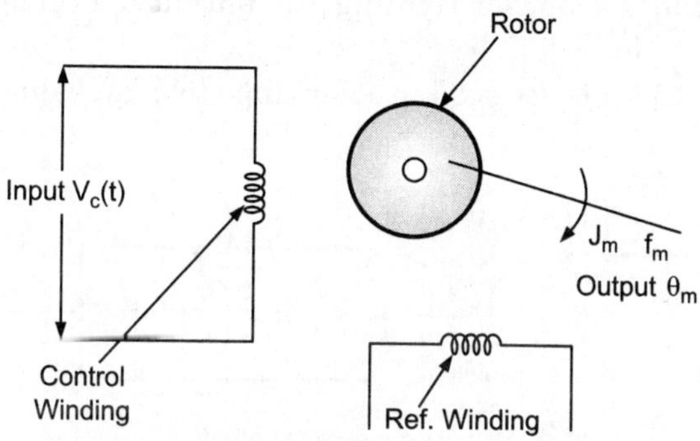

Fig. 5.6.8. Two phase a.c. servomotor.

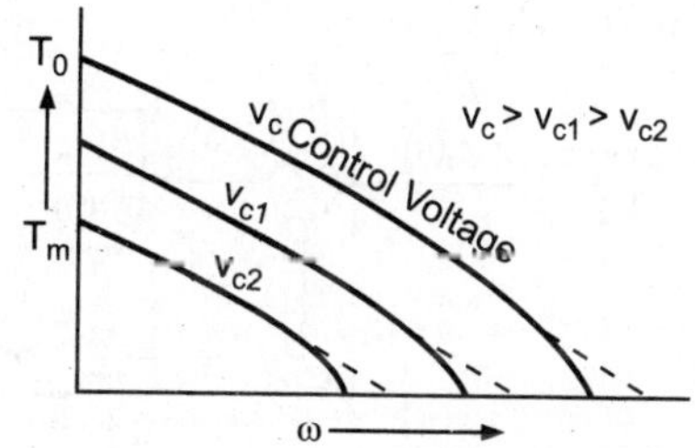

Fig. 5.6.9. Torque speed characteristics of a two phase a.c. servomotor.

The moment of inertia and the coefficient of viscous friction at the motor shaft are respectively J_m and f_m. The angular shift resulting in the motor shaft being θ_m and the corresponding angular velocity being ω_m.

From the torque-speed characteristics the dynamic equation relating the motor torque and the speed is formed below,

$$T_M = m\omega_m + KV_c \qquad \text{...(5.74)}$$

If T_0 is the stalling torque and ω_0 is the no load speed. The parameters m and K are determined in terms of T_0 and ω_0 as follows:

(*i*) When the speed is zero, the torque is T_0 (stalling torque) and this stalling torque is proportional to the control voltage V_c.

Therefore, $T_0 = KV_c$

or $$K = \frac{T_0}{V_c}, \text{ expressed in Nm/V}$$

(*ii*) The slope of the torque speed characteristics is

$$m = -\frac{T_0}{\omega_0}, \text{ expressed in Nm/(rad/sec) is a negative number.}$$

Since, $\omega_m = \dfrac{d\theta_m}{dt}$,

the torque equation can be expressed as,

$$T_M = m\frac{d\theta_m}{dt} + KV_c \qquad ...(5.75)$$

The related dynamic equations are	*The corresponding Laplace transformed equations are*
$T_M = m\dfrac{d\theta_m}{dt} + KV_c$	$T_M(s) = sm\theta_m(s) + KV_c(s)$ ...(5.76)
$T_M = J_m\dfrac{d^2\theta_m}{dt^2} + f_m\dfrac{d\theta_m}{dt}$	$T_M(s) = s^2J_m\theta_m(s) + sf_m\theta_m(s)$ ...(5.77)

In equations (5.76) and (5.77) the block diagram relating the output $\theta_m(s)$ and the input $V_c(s)$ is drawn and shown in Fig. 5.6.10 (*a*).

The block diagram given in Fig. 5.6.10 (*a*) is reduced in following steps and shown in Fig. 5.6.10 (*d*).

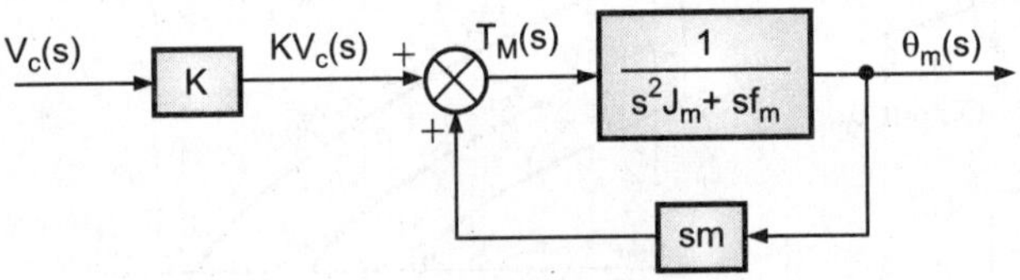

Fig. 5.6.10 (*a*) Block diagram of two phase a.c. servomotor relating $\theta_m(s)$ and $V_0(s)$.

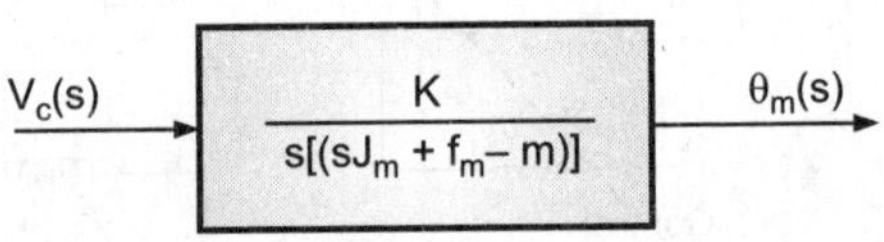

Fig. 5.6.10 (*b*) Block diagram reduction of Fig. 5.6.10 (*a*).

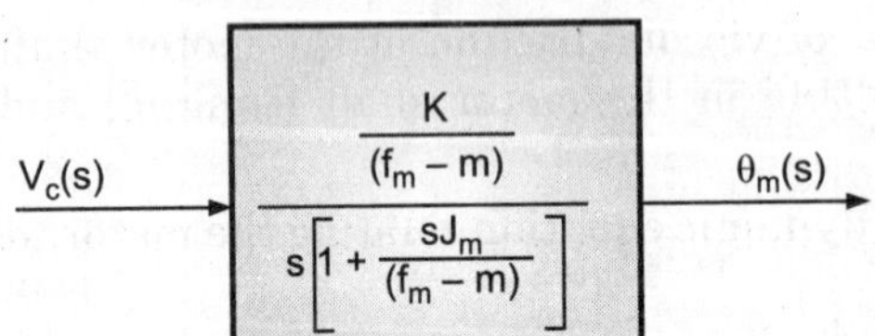

Fig. 5.6.10 (*c*) Block diagram reduction of Fig. 5.6.10 (*b*).

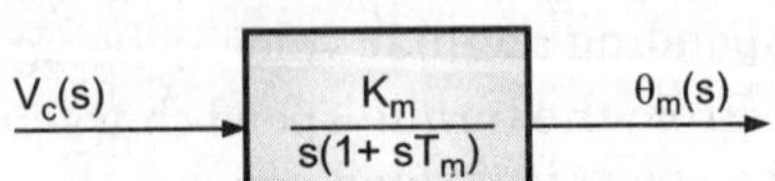

Fig. 5.6.10 (*d*) Simplified block diagram of two phase a.c. servomotor relating $\theta_m(s)$ and $V_0(s)$.

According to Fig. 5.6.10 (*d*) the transfer function of a two phase servomotor relating the output $\theta_m(s)$ and the input $V_c(s)$ is given by the relation

$$\frac{\theta_m(s)}{V_c(s)} = \frac{K_m}{s(1+sT_m)} \qquad ...(5.78)$$

where $K_m = \dfrac{K}{(f_m - m)}$

is termed as motor gain constant.

and $$T_m = \frac{J_m}{(f_m - m)}$$ is termed as motor time constant.

Since, $\omega_m(s) = s\theta_m(s)$ the block diagram relating $\omega_m(s)$ and $V_c(s)$ is shown in Fig. 5.6.11 (*a*) and simplified as shown in Fig. 5.6.11 (*b*).

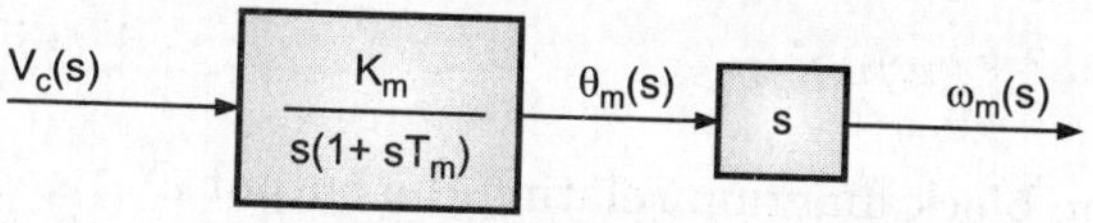

Fig. 5.6.11 (*a*) Block diagram of two phase a.c. servomotor relating $\omega_m(s)$ and $V_c(s)$.

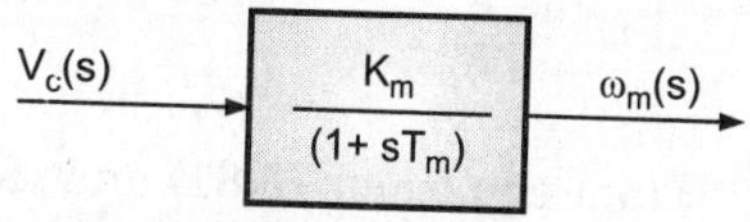

Fig. 5.6.11 (*b*) Simplified block diagram of an a.c. two phase motor relating $\omega_m(s)$ and $V_c(s)$.

The transfer function relating the output $\omega_m(s)$ and the input $V_c(s)$ is thus

$$\frac{\omega_m(s)}{V_c(s)} = \frac{K_m}{(1+sT_m)} \quad ...(5.79)$$

5.7 GENERATORS

1. (*a*) Separately excited d.c. generator. The transfer function of a separately excited d.c. generator relating the input field winding voltage V_f and the induced voltage developed across the armature is derived hereunder.

The circuit diagram is shown in Fig. 5.7.1. The resistance and the inductance of the field winding are respectively R_f and L_f. The generator being driven at a constant speed and is operating in the linear portion of the magnetising characteristics. The current i_f flowing in the field winding produces a flux in the generator which in turn induces a proportional voltage at the armature terminals. The relationship between the induced voltage e and the field current i_f is given by the magnetisation characteristics shown in Fig. 5.7.2.

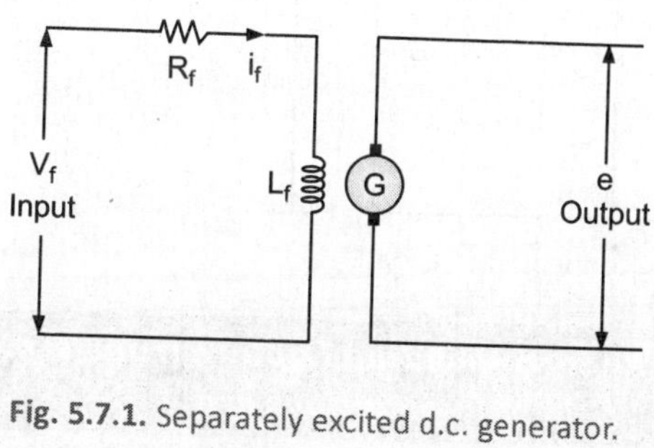

Fig. 5.7.1. Separately excited d.c. generator.

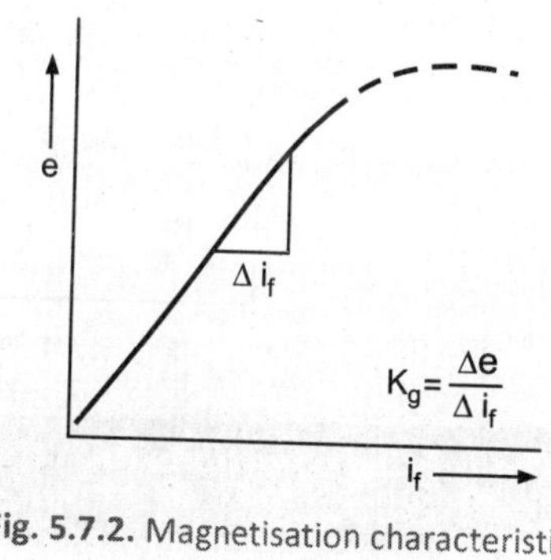

Fig. 5.7.2. Magnetisation characteristics.

The induced voltage e is related to field current i_f as follows:

$$e \propto i_f$$

or $$e = K_g i_f \quad ...(5.80)$$

where K_g is generator constant expressed in V/A.

The related equations are	*The corresponding Laplace transformed equations are*	
$V_f = R_f i_f + L_f \dfrac{di_f}{dt}$	$V_f(s) = R_f I_f(s) + sL_f I_f(s)$	...(5.81)
$e = K_g i_f$	$E(s) = K_g I_f(s)$	...(5.82)

From equations (5.81) and (5.82) the block diagram relating the output $E(s)$ and the input $V_f(s)$ is drawn and shown in Fig. 5.7.3 (*a*) and simplified as shown in Fig. 5.7.3 (*b*).

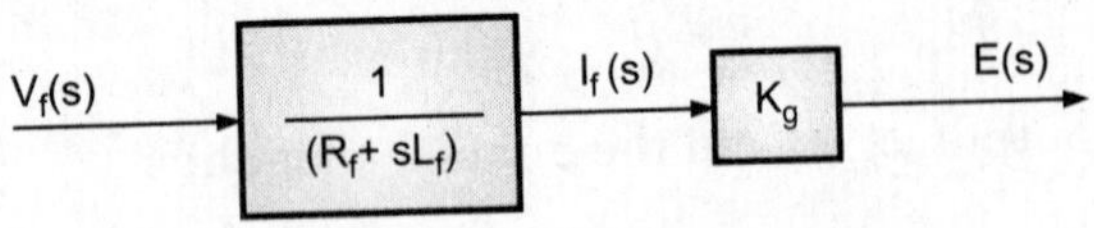

Fig. 5.7.3 (*a*) Block diagram of separately excited d.c. generator.

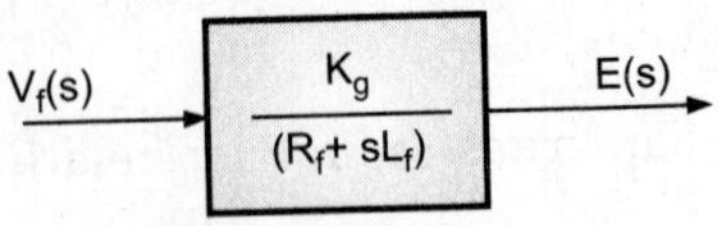

Fig. 5.7.3 (*b*) Simplified block diagram of separately excited d.c. generator.

The transfer function relating $E(s)$ and $V_f(s)$ is thus,

$$\frac{E(s)}{V_f(s)} = \frac{K_g}{(R_f + sL_f)} \qquad ...(5.83)$$

1. (*b*) Separately excited d.c. generator connected to load resistance R_L. Transfer function of a separately excited d.c. generator relating the input field winding voltage and the terminal voltage is derived hereunder.

In this case the output of the d.c. generator is connected to load having a resistance R_L. The resistance of the armature being R_a and the armature inductance is considered negligible. The circuit is shown in Fig. 5.7.4.

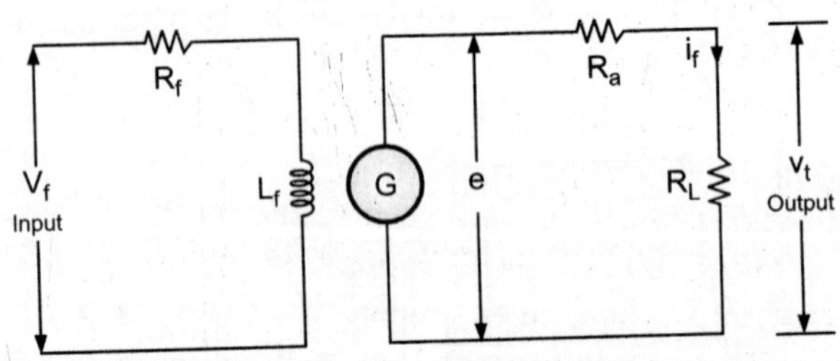

Fig. 5.7.4. D.C. generator connected to a load.

The related equations are	*The corresponding Laplace transformed equations are*	
$V_i = e - R_a i_L$	$V_t(s) = E(s) - R_a I_L(s)$	...(5.84)
$I_L = \dfrac{V_t}{R_L}$	$I_L(s) = \dfrac{V_t(s)}{R_L}$	...(5.85)

From equations (5.84) and (5.85) the block diagram relating $V_t(s)$ and $E(s)$ is drawn and shown in Fig. 5.7.5.

In order to obtain the block diagram relating the output terminal voltage $V_t(s)$ and the input field winding voltage $V_f(s)$, the two block diagrams shown in Figs. 5.7.3 (*b*) and 5.7.5 are joined as shown in Fig. 5.7.6 which is reduced in steps as shown in Fig. 5.7.7 (*a*) to (*c*).

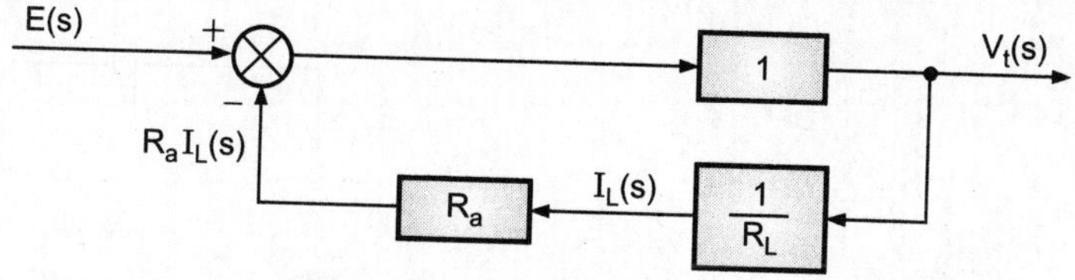

Fig. 5.7.5. Block diagram of d.c. generator connected to a load.

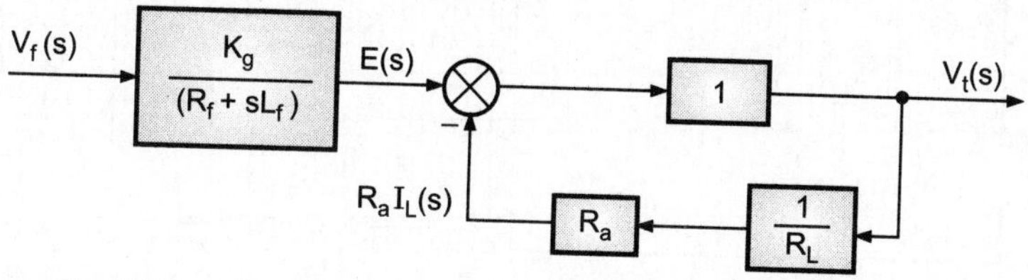

Fig. 5.7.6. Block diagram of Fig. 5.7.3 (*b*) and Fig. 5.7.5 connected.

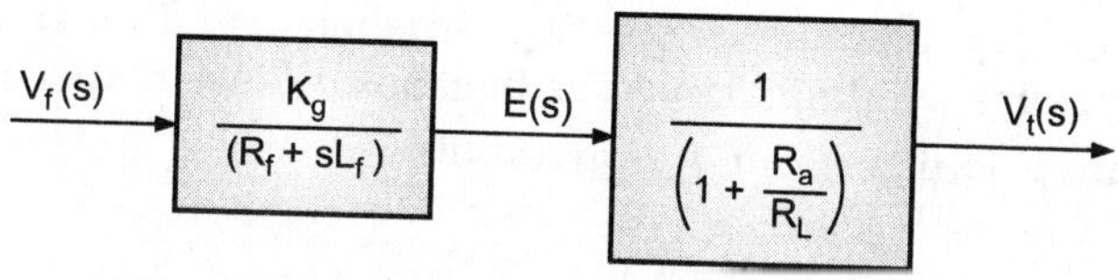

Fig. 5.7.7 (*a*) Block diagram reduction of Fig. 5.7.6.

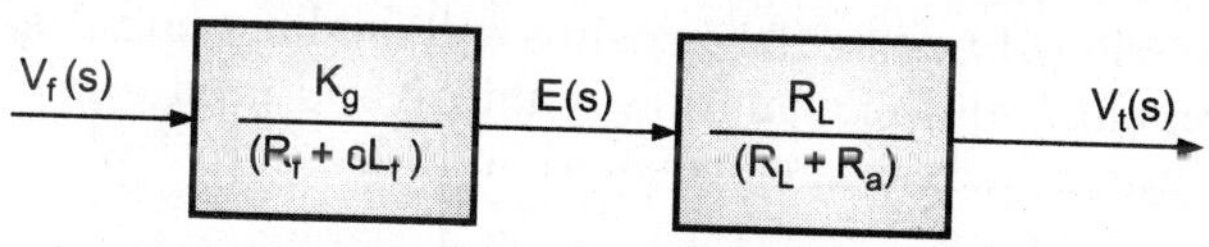

Fig. 5.7.7 (*b*) Block diagram reduction of Fig. 5.7.7 (*a*).

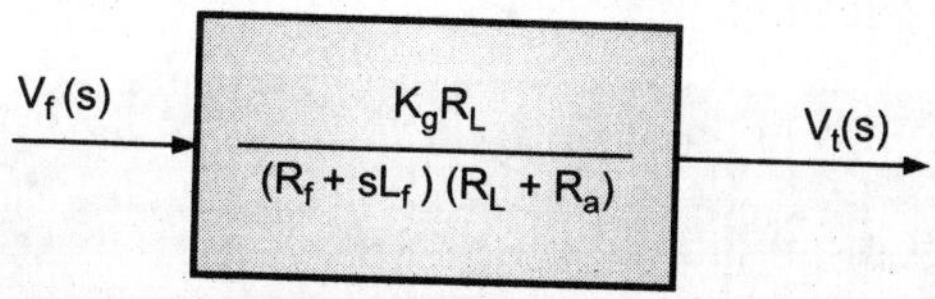

Fig. 5.7.7 (*c*) Block diagram of separately excited d.c. generator relating $V_t(s)$ and $V_f(s)$.

According to Fig. 5.7.7 (*c*) the transfer function relating the output $V_t(s)$ and the input $V_f(s)$ is given by,

$$\frac{V_t(s)}{V_f(s)} = \frac{K_g R_L}{(R_f + sL_f)(R_L + R_a)} \qquad \text{...(5.86)}$$

2. Amplidyne Generator. Transfer function of an amplidyne generator relating the output voltage and the input field winding voltage is derived hereunder.

An amplidyne is a two stage generator (rotating amplifier). The circuit is shown in Fig. 5.7.8. The input voltage e_f applied to the main field winding having a resistance and an inductance of R_f and L_f respectively and the output voltage e is developed across the brushes c and d.

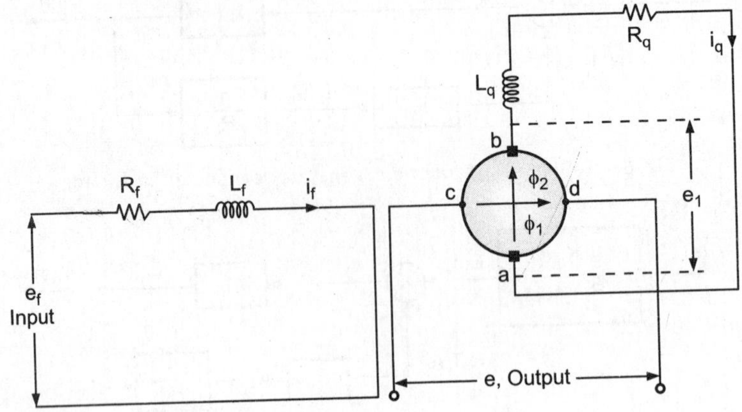

Fig. 5.7.8. Amplidyne generator.

In the first stage the current i_f flowing in the main field winding produces a flux ϕ_1 which in turn developes an e.m.f. e_1 across the brushes a and b. The resistance and the inductance of the armature winding across the brushes a-b are respectively R_q and L_q.

The relationship between e_1 and i_f is given by,

$$e_1 \propto i_f$$

or

$$e_1 = K_1 i_f \quad \text{...(5.87)}$$

where K_1 is a constant expressed in V/A.

The e.m.f. e_1 developed across the brushes a-b is short circuited on itself. The short circuit circulates a current i_q in the armature winding and produces a flux ϕ_2 which in turn develops an output voltage e across the brushes c and d. Therefore,

$$e \propto i_q$$

or

$$e = K_2 i_q \quad \text{...(5.88)}$$

where K_2 is a constant expressed in V/A.

The related equations are	*The corresponding Laplace transformed equations are*	
$e_f = R_f i_f + L_f \dfrac{di_f}{dt}$	$E_f(s) = R_f I_f(s) + sL_f I_f(s)$	...(5.89)
$e_1 = K_1 i_f$	$E_1(s) = K_1 I_f(s)$	...(5.90)
$e_1 = R_q i_q + L_q \dfrac{di_q}{dt}$	$E_1(s) = E_q I_q(s) + sL_q I_q(s)$	...(5.91)
$e = K_2 i_q$	$E(s) = K_2 I_q(s)$	...(5.92)

From equations (5.89) to (5.92) the block diagram relating the output $E(s)$ and the input $E_f(s)$ is drawn and shown in Fig. 5.7.9 (*a*) and reduced as shown in Fig. 5.7.9 (*b*).

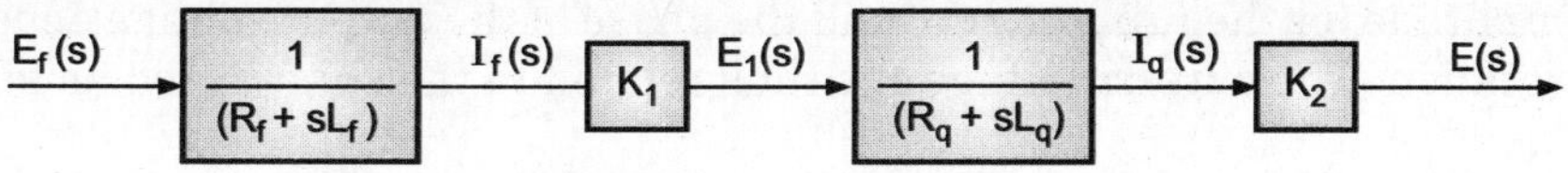

Fig. 5.7.9 (*a*) Block diagram of amplidyne generator.

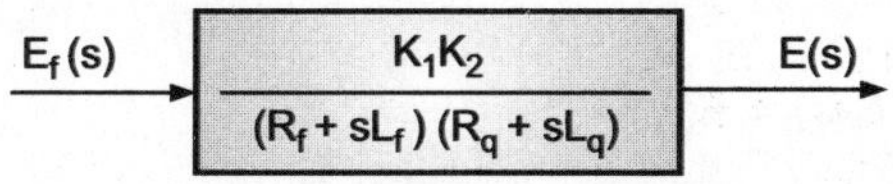

Fig. 5.7.9 (*b*) Simplified block diagram of amplidyne generator.

Hence, the transfer function relating the output $E(s)$ and the input $E_f(s)$ is given by,

$$\frac{E(s)}{E_f(s)} = \frac{K_1K_2}{(R_f + sL_f)(R_q + sL_q)} \quad ...(5.93)$$

3. D.C. Tachogenerator. **A D.C. tachogenerator (Fig. 5.7.10) is used to convert a rotational speed into a proportional d.c. voltage. It uses a permanent magnet for producing magnetic field. The tachogenerator is coupled to a shaft, of which speed is to be measured. As the magnetic flux is constant due to use of a permanent magnets, the induced voltage at the terminals of the armature is proportional to shaft speed and the polarity depends on the direction of rotation. Therefore,**

$$e \propto \omega$$

or

$$e = K_{tg}\,\omega \quad ...(5.94)$$

where K_{tg} is a tachogenerator constant expressed in V/(rad/sec).

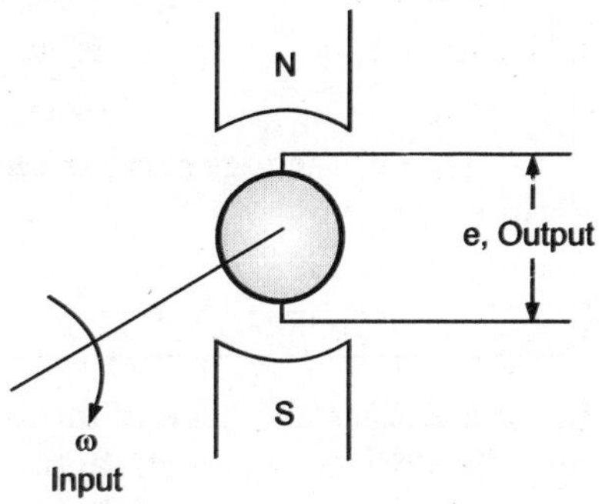

Fig. 5.7.10. D.C. Tachogenerator.

The Laplace transform of the above equation is given by,

$$E(s) = K_{tg}\omega(s) \quad ...(5.95)$$

The block diagram relating the output $E(s)$ and the input $\omega(s)$ is drawn and shown in Fig. 5.7.11. and the transfer function given by

$$\frac{E(s)}{\omega(s)} = K_{tg} \quad ...(5.96)$$

Fig. 5.7.11. Block diagram of d.c. tachogenerator.

A d.c. tachogenerator can also be used for a.c. system by using a modulator to convert d.c. voltage signal into an a.c. voltage signal.

4. A.C. Tachogenerator. **An a.c. tachogenerator (Fig. 5.7.12) is used to convert a rotational speed into a proportional a.c. voltage and works on the principal of induction generator.**

The tachogenerator is coupled to a shaft, of which speed is to be measured. The reference winding is supplied by a reference voltage and the output voltage *e* is induced across the output winding.

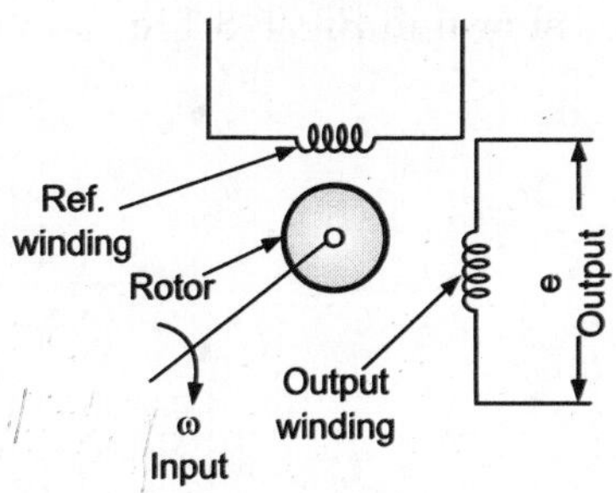

Fig. 5.7.12. A.C. Tachogenerator.

The amplitude (or the r.m.s. value) and the phase of the output voltage depends on the direction of rotation. The output voltage e is thus related to the shaft speed as follows :

$$e \propto \omega$$

or $$e = K_{tg}\,\omega \qquad ...(5.97)$$

where K_{tg} is a constant expressed in V/(rad/sec). The Laplace transform of the above equation is,

$$E(s) = K_{tg}\,\omega(s) \qquad ...(5.98)$$

Fig. 5.7.13. Block diagram of a.c. tachogenerator.

The block diagram relating the output $E(s)$ and the input $\omega(s)$ is drawn and shown in Fig. 5.7.13. and the transfer function is given by,

$$\frac{E(s)}{\omega(s)} = K_{tg} \qquad ...(5.99)$$

An a.c. tachogenerator can also be used on d.c. systems by using a phase sensitive rectifier.

5.8 ERROR DETECTORS

Error detectors are used to measure (sense) the error signal in control systems. The error is the difference between the actual magnitude of the output and the magnitude of the desired output. The magnitude of the desired output is held constant as a reference and also termed as input. Thus any deviation from the desired output (*i.e.* input or reference) results in an error signal.

In a way error detectors also perform the function of a transducer, *i.e.* to convert error signal obtained in one form to another form which is needed for the control action. For example in the control for the position of a shaft the error is the angular difference in the positions of the input (reference) shaft and the output shaft and is expressed in radian or degree. A potentiometer error detector measures this error and converts the same into a proportional voltage.

1. **Potentiometer error detector.** The potentiometer error detector consists of two identical potentiometers electrically connected in parallel and supplied by a voltage source as shown in Fig. 5.8.1 (*a*).

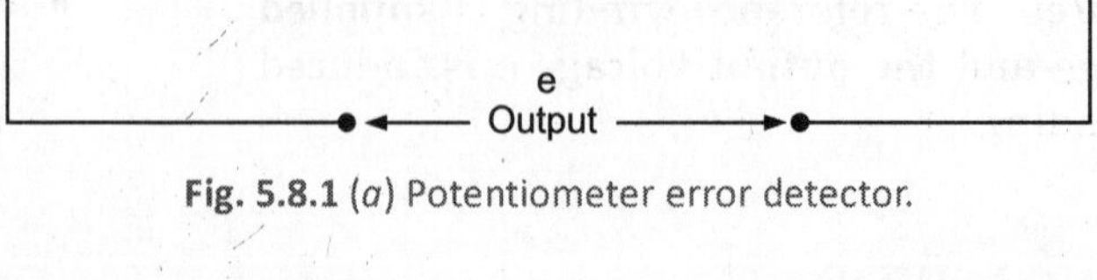

Fig. 5.8.1 (*a*) Potentiometer error detector.

The symbolic circuit representation of a potentiometer error detector is shown in Fig. 5.8.1 (*b*).

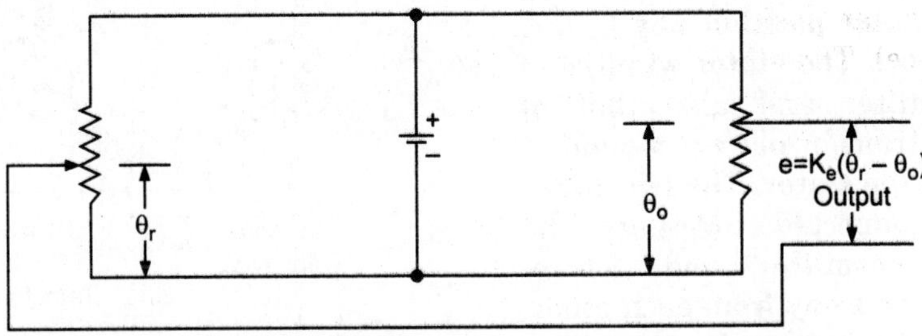

Fig. 5.8.1 (*b*) Circuit diagram of potentiometer error detector.

The input shaft is coupled to the potentiometer marked *A* and is held fixed at a desired angular position say θ_r while the output shaft is coupled to the potentiometer marked *B* and the position is indicated as θ_o. The potential difference between the variable points of potentiometers *A* and *B* is proportional to the angular difference $(\theta_r - \theta_o)$. Therefore, the error is given by,

$$e \propto (\theta_r - \theta_o)$$

or $$e = K_e(\theta_r - \theta_o) \qquad ...(5.100)$$

where K_e known as the gain of the error detector, and expressed is *V*/rad.

or $$e = K_e\theta_e \qquad ...(5.101)$$

where $\theta_e = (\theta_r - \theta_o)$.

The Laplace transform of the above equations is,

$$E(s) = K_e[\theta_r(s) - \theta_o(s)] \qquad ...(5.102)$$

or $$E(s) = K_e\theta_e(s) \qquad ...(5.103)$$

The block diagram relating the output *E*(*s*) and the input $\theta_r(s)$ drawn and shown in Fig. 5.8.2.

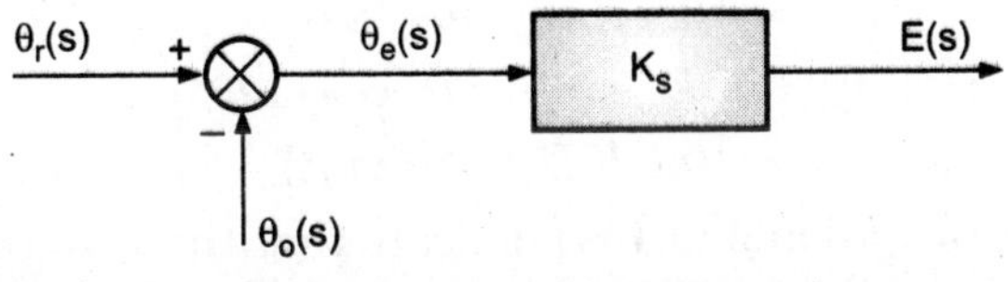

Fig. 5.8.2. Block diagram of potentiometer error detector.

and the transfer function is given by,

$$\frac{E(s)}{\theta_e(s)} = K_e \qquad ...(5.104)$$

or $$\frac{E(s)}{\theta_r(s) - \theta_o(s)} = K_e \qquad (5.105)$$

2. Synchro (selsyn) error detector. Fig. 5.8.3 shows synchro transmitter and synchro control transmitter arrangement for converting angular position difference into a proportional a.c. voltage.

The winding on the rotor of a synchro transmitter is connected to an a.c. supply voltage and this rotor is held fixed at a desired angular position any θ_r (*i.e.* input or reference). The stator winding of synchro transmitter and also that of synchro control transformer are wound at 120° in space on the stator. The two stator windings are connected together. The locations of transmitter and control transformer can be away from each other. The rotor of synchro transmitter is salient pole type and that of synchro control transformer is cylindrical type.

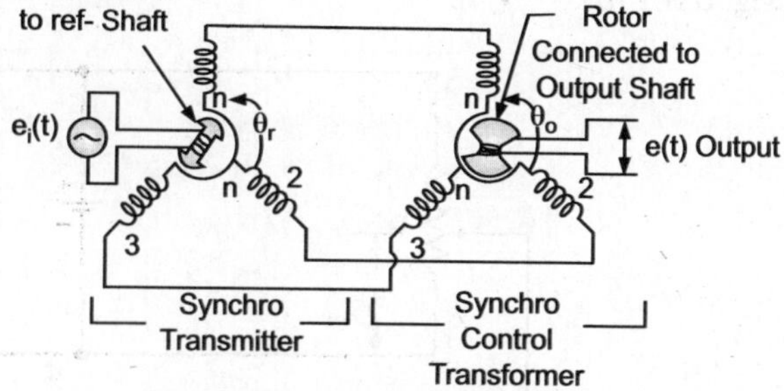

Fig. 5.8.3. Synchro error detector.

The rotor of the control transformer is coupled to the output shaft of the control system. If the position of the output shaft is indicated as θ_o, this results in an angular error $\theta_e = (\theta_r - \theta_o)$ between the positions of the input (reference) and the output shafts.

The process of conversion of the angular difference into a proportional voltage is explained below :

A single phase a.c. voltage,

$$e_i(t) = E_m \sin(2\pi ft) \qquad ...(5.106)$$

is applied to the rotor winding of the synchro transmitter, then if $\theta_r = 0$ the corresponding voltage induced by transformer section across the stator winding $1n$ is given by,

$$e_{1n} = K E_m \sin(2\pi ft) \qquad ...(5.107)$$

where K is a constant of proportionality. As the stator windings $2n$ and $3n$ are 240° and 120° apart (angle measured as positive in anti-clockwise direction) w.r.t. the winding $1n$, the voltages induced across them are,

$$e_{2n} = K E_m \sin(2\pi ft) \cos(240°)$$

$$= -0.5 K E_m \sin(2\pi ft) \qquad ...(5.108)$$

$$e_{3n} = K E_m \sin(2\pi ft) \cos(120°)$$

$$= -0.5 K E_m \sin(2\pi ft) \qquad ...(5.109)$$

Equations (5.107), (5.108) and (5.109) indicate that three voltages e_{1n}, e_{2n} and e_{3n} are single phase voltages of same frequency and have same phase.

Now, if the rotor of the synchro transmitter shifts in anti-clockwise direction through an angle θ, the voltages in the stator coils are,

$$e_{1n} = K E_m \sin(2\pi ft) \cos\theta \qquad ...(5.110)$$

$$e_{2n} = K E_m \sin(2\pi ft) \cos(240° - \theta) \qquad ...(5.111)$$

$$e_{3n} = K E_m \sin(2\pi ft) \cos(120° - \theta) \qquad ...(5.112)$$

Equations (5.110), (5.111) and (5.112) reveal that magnitudes of the voltages e_{1n}, e_{2n} and e_{3n} vary sinusoidally w.r.t. θ as shown in Fig. 5.8.4.

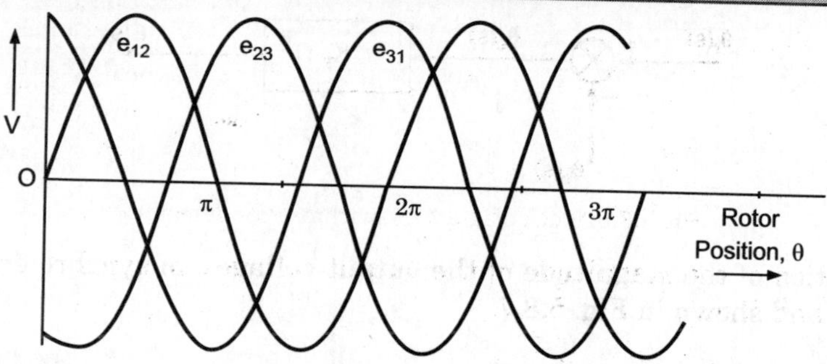

Fig. 5.8.4. Terminal voltage across stator of synchro transmitter w.r.t. rotor position (radian).

The three voltages e_{1n}, e_{2n} and e_{3n} are connected consecutively to three stator windings of the control transformer and produce a resultant flux in the air gap of the same, which in turn induces a voltage across the rotor winding of the control transformer. The magnitude of this induced voltage depends on the difference $(\theta_r - \theta_o)$. If the difference $(\theta_r - \theta_o)$ is zero, the induced voltage across the rotor winding terminals of the control transformer is zero, maximum for $(\theta_r - \theta_o) = 90°$, and again zero when $(\theta_r - \theta_o) = 180°$. After 180° the phase of the induced voltage reverses. The magnitude being again maximum with a reversed phase for $(\theta_r - \theta_o) = 270°$ and finally zero for $(\theta_r - \theta_o) = 360°$. The variation of the amplitude of induced voltage e across the rotor of the control transformer w.r.t. $(\theta_r - \theta_o)$ is shown in Fig. 5.8.5.

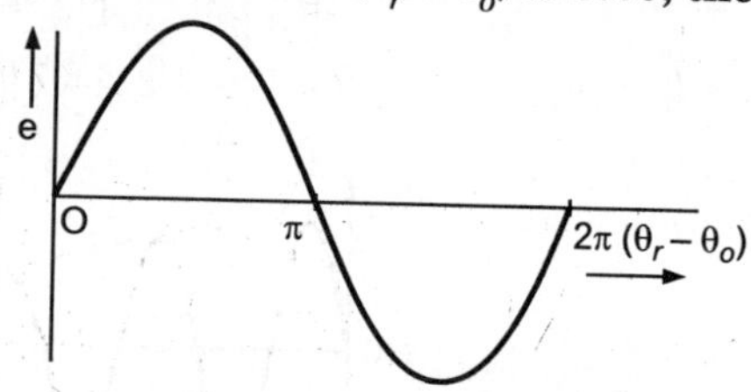

Fig. 5.8.5. Synchro error detector output.

Therefore, the magnitude of the output induced voltage e developed across the rotor of control transformer is given by the following relation :

$$e = K_s \sin(\theta_r - \theta_o) \quad ...(5.113)$$

where K_s is a constant of proportionality.

It is to be noted that the frequency of the voltage e is the same as that of supply and the magnitude is proportional to $\sin(\theta_r - \theta_o)$.

In control systems the angular error, *i.e.* $(\theta_r - \theta_o)$ is usually small and since $(\theta_r - \theta_o)$ is expressed in radian, therefore,

$$\sin(\theta_r - \theta_o) \cong (\theta_r - \theta_o)$$

Hence, $$e = K_s(\theta_r - \theta_o)$$

$(\theta_r - \theta_o)$ is expressed as θ_e (angular error)

$\therefore$ $$e = K_s\,\theta_e \quad ...(5.113)$$

K_s is expressed in V/rad.

The above equation indicates that angular error is converted into a proportional voltage. Laplace transform of the above equation is,

$$E(s) = K_s\,\theta_e(s) \quad ...(5.114)$$

The block diagram representation of synchro error detector is shown in Fig. 5.8.6. According to Fig. 5.8.6 the transfer function is given by,

$$\frac{E(s)}{\theta_e(s)} = K_s, \quad \text{or} \quad \frac{E(s)}{[\theta_r(s) - \theta_o(s)]} = K_s \quad ...(5.115)$$

K_s is known as the sensitivity or the gain of synchro-error detector.

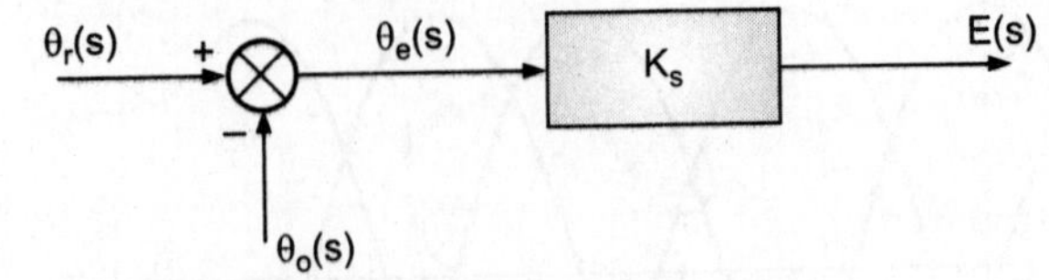

Fig. 5.8.6. Block diagram of synchro-error detector.

The variation of the magnitude of the output voltage e of synchro-error detector is a function of time and shown in Fig. 5.8.7.

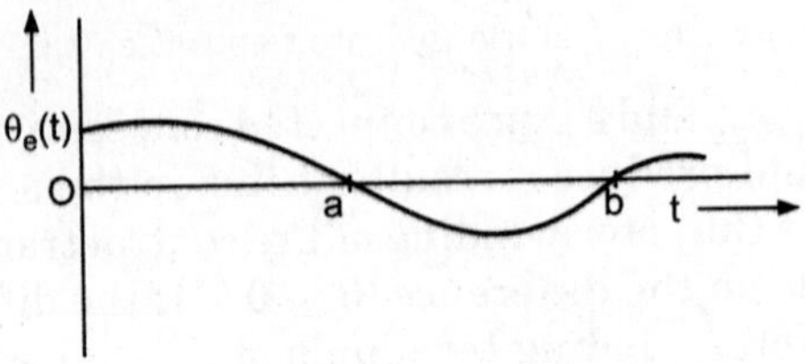

Fig. 5.8.7 (*a*) Variation of error $\theta_e(t)$ w.r.t. time.

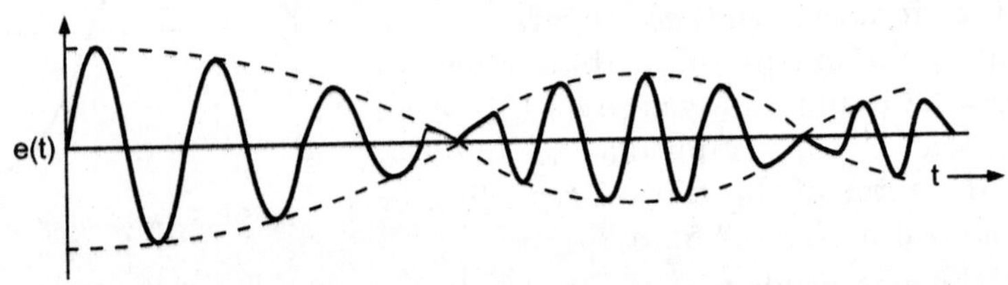

Fig. 5.8.7 (*b*) Variation of error $e(t)$ w.r.t. time (waveform).

It is to be noted that the phase of e (the output voltage) reverses at points a and b as the error $\theta_e(t)$ changes its sign.

5.9 SOLVED EXAMPLES

Example 5.9.1. *Find the transfer function relating displacements y and x for the mechanical system of Fig. 5.9.1.*

Solution. Let force $f(t)$ be applied at point A. The mechanical circuit diagram for the system is drawn in Fig. 5.9.1 (*a*).

From the mechanical circuit diagram the equations for the system are

$$f_2 \frac{dy}{dt} + Ky + f_1 \frac{d}{dt}(y - x) = 0 \quad \text{...(1)}$$

$$f_1 \frac{d}{dt}(x - y) = f(t) \quad \text{...(2)}$$

Assuming initial conditions as zero and taking Laplace transform on both sides of equation (1) following equation is obtained :

$$f_2 sY(s) + KY(s) + f_1 s[Y(s) - X(s)] = 0 \quad \text{...(3)}$$

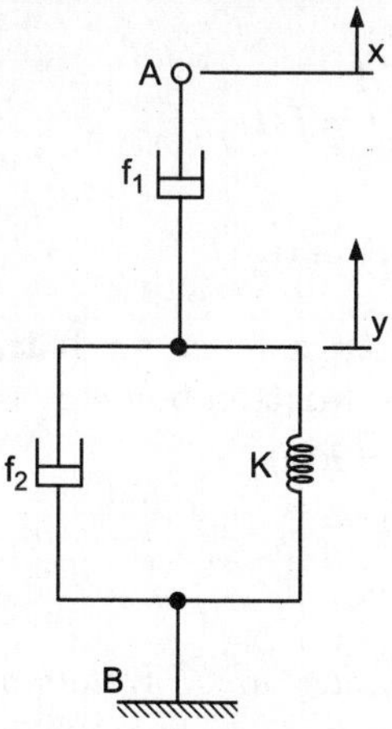

Fig. 5.9.1. Fig. for example 5.9.1.

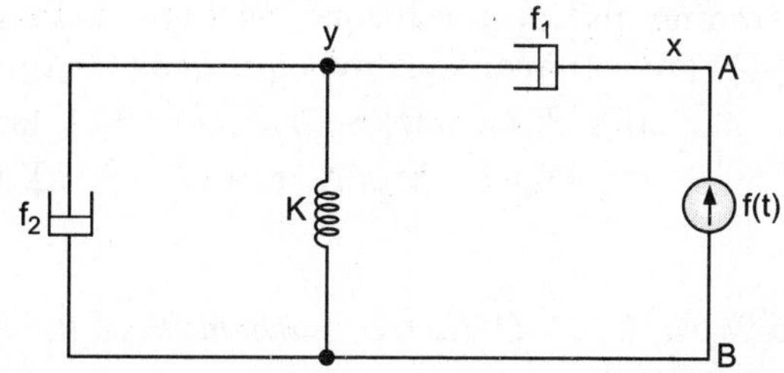

Fig. 5.9.1 (*a*) Mechanical circuit diagram for example 5.9.1.

From equation (3), the required transfer function is

$$\frac{Y(s)}{X(s)} = \frac{f_1 s}{[(f_1 + f_2)s + K]} = \frac{T_1 s}{(T_1 s + 1)} \quad \textbf{Ans.}$$

where $T_1 = \dfrac{f_1}{K}$ and $T_2 = \dfrac{(f_1 + f_2)}{K}$.

Example 5.9.2. *Draw the mechanical circuit diagram for the system shown in Fig. 5.9.2 and write system equations.*

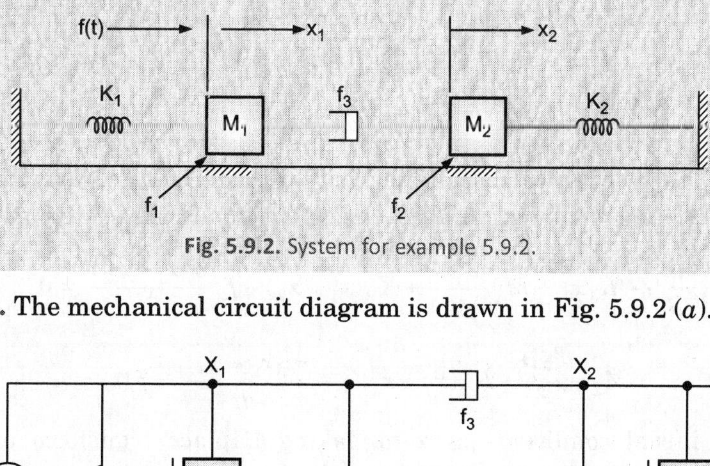

Fig. 5.9.2. System for example 5.9.2.

Solution. The mechanical circuit diagram is drawn in Fig. 5.9.2 (*a*).

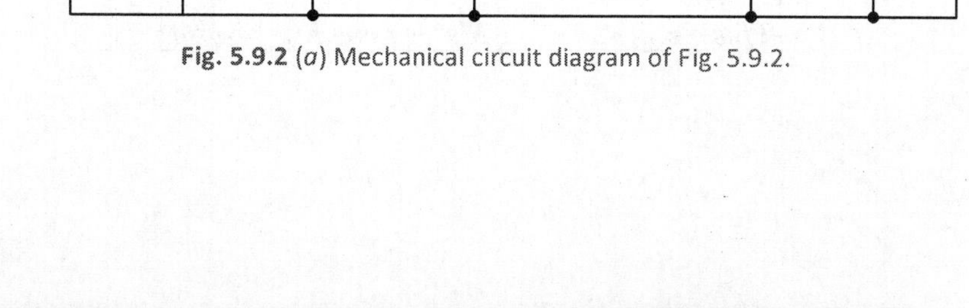

Fig. 5.9.2 (*a*) Mechanical circuit diagram of Fig. 5.9.2.

The equations for the system are written below

At node (1) $$K_1x_1 + M_1\frac{d^2x_1}{dt^2} + f_1\frac{dx_1}{dt} + f_3\frac{d}{dt}(x_1 - x_2) = f(t) \quad ...(1)$$

At node (2) $$K_2x_2 + M_2\frac{d^2x_2}{dt^2} + f_2\frac{dx_2}{dt} + f_3\frac{d}{dt}(x_2 - x_1) = 0 \quad ...(2)$$

Assuming initial conditions as zero, taking Laplace transform on both sides of equations (1) and (2), transformed equations for the system are written below :

$$M_1s^2 X_1(s) + (f_1 + f_3)\, sX_1(s) + K_1X_1(s) - f_3sX_2(s) = F(s) \quad ...(3)$$
$$- f_3sX_1(s) + M_2s^2X_2(s) + (f_2 + f_3)\, sX_2(s) + K_2X_2(s) = 0 \quad ...(4)$$

Ans.

Example 5.9.3. *Obtain a mathematical model for the mechanical system shown in Fig. 5.9.3.*

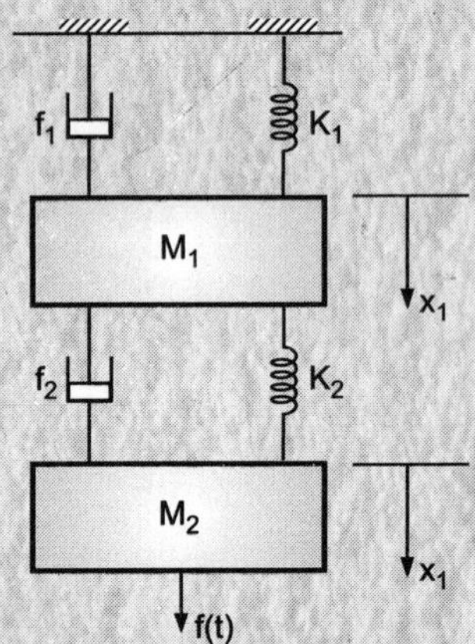

Fig. 5.9.3. System for example 5.9.3.

Solution. The mechanical circuit diagram for the circuit is drawn in Fig. 5.9.3 (*a*).

On the basis of force-current analogy the equations for the system are written below :

$$f_1\frac{dx_1}{dt} + K_1x_1 + M_1\frac{d^2x_1}{dt^2} + K_2(x_1 - x_2) + f_2\frac{d(x_1 - x_2)}{dt} = 0 \quad ...(1)$$

$$M_2\frac{d^2x_2}{dt^2} + K_2(x_2 - x_1) + f_2\frac{d(x_2 - x_1)}{dt} = f(t) \quad ...(2)$$

Assuming initial conditions as zero, taking Laplace transform on both sides of equations (1) and (2) and rearranging terms, following equations representing the system are obtained :

$$\begin{bmatrix} M^2s + (f_1 + f_2)s + (K_1 + K_2) & -(f_2s + K_2) \\ -(f_2s + K_2) & Ms^2 + f_2s + K_2 \end{bmatrix}\begin{bmatrix} X_1(s) \\ X_2(s) \end{bmatrix} = \begin{bmatrix} F(s) \\ 0 \end{bmatrix}$$ **Ans.**

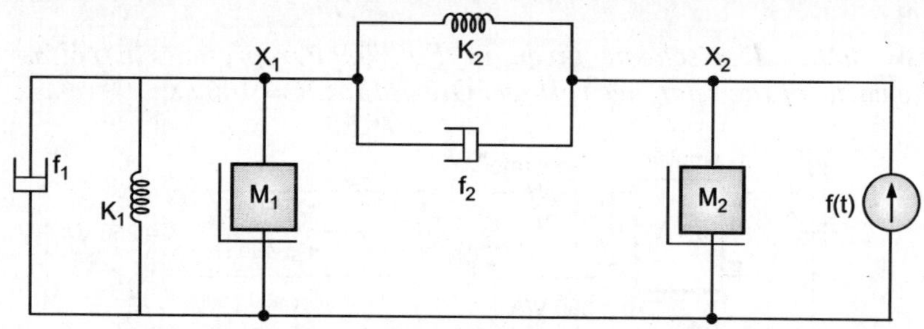

Fig. 5.9.3 (*a*) Mechanical circuit diagram for example 5.9.3.

Example 5.9.4. *The diagram given in Fig. 5.9.4 represents a closed-loop control system for regulating the speed of a field controlled d.c. motor. Determine the value of the reference voltage if the speed is to be maintained at 100 r.p.m.*

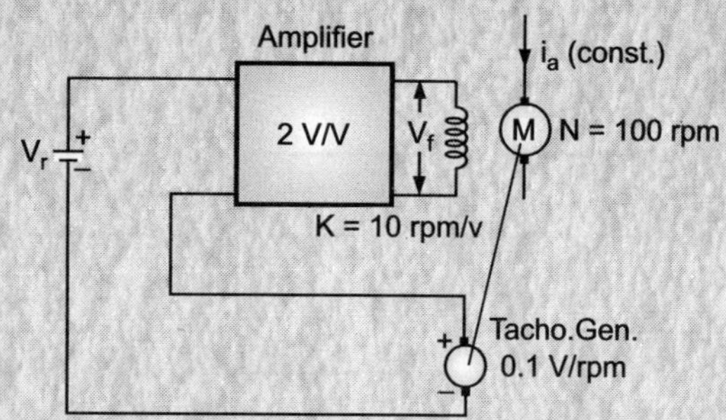

Fig. 5.9.4. Closed-loop speed regulating system for example 5.9.4.

Solution. As *K* the motor field constant is given in rpm/V; the resistance and inductance of the motor field winding are not needed for calculations.

The block diagram under steady state conditions is developed as shown in Fig. 5.9.4 (*a*).

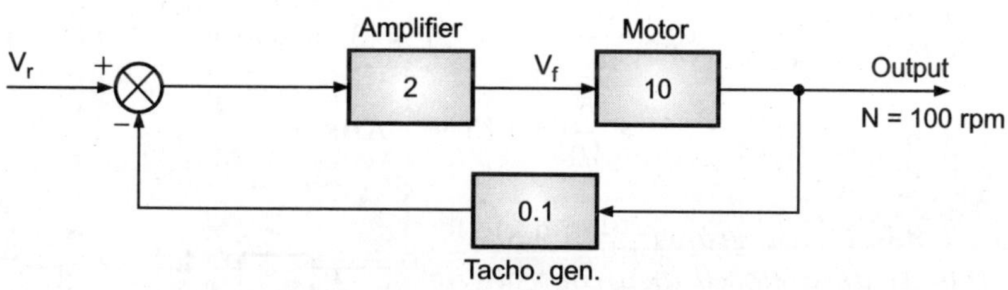

Fig. 5.9.4 (*a*) Block diagram of Fig. 5.9.4.

The overall transfer function for steady state condition is given by

$$\frac{N}{V_r} = \frac{2 \times 10}{1 + 2 \times 10 \times 0.1} = \frac{20}{3}$$

$$\therefore \quad V_r = N \times \frac{3}{20} \qquad [\because \; N = 100 \text{ rpm.}]$$

$$V_r = 100 \times \frac{3}{20} = 15 \text{ V.} \quad \textbf{Ans.}$$

Example 5.9.5. *The scheme given in Fig. 5.9.5 represents a voltage regulator, determine the value of the reference voltage. Given that $R_f = 100\ \Omega$.*

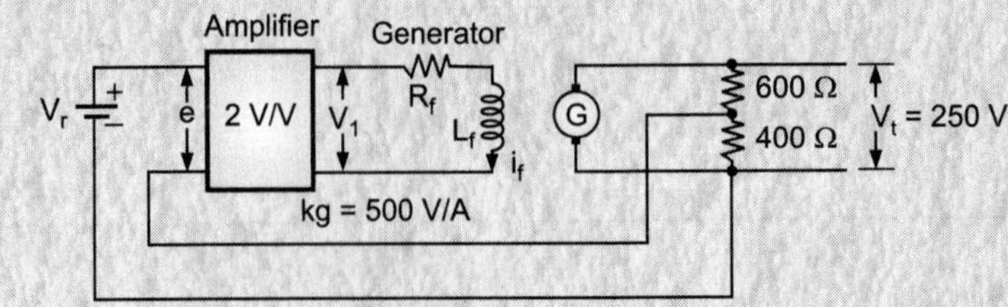

Fig. 5.9.5. Voltage regulator scheme for example 5.9.5.

Solution. The fraction of the output voltage compared with the reference voltage is 400/(400 + 600) = 0.4. The generator field inductance need not be considered as the system block diagram for the steady state condition is to be drawn. The block diagram considering steady state condition is developed in Fig. 5.9.5 (*a*).

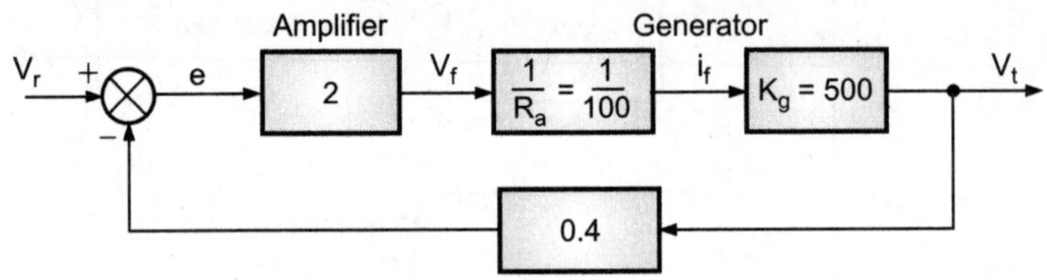

Fig. 5.9.5 (*a*) Block diagram of Fig. 5.9.5.

The overall transfer function for steady state condition is given by

$$\frac{V_t}{V_r} = \frac{2 \times \frac{1}{100} \times 500}{1 + 2 \times \frac{1}{100} \times 500 \times 0.4} = \frac{10}{5}$$

$$\therefore \quad V_r = V_t \times \frac{5}{10} \qquad [\because \ V_t = 250 \text{ V}]$$

$$V_r = 250 \times \frac{5}{10} = 125 \text{ V.} \quad \textbf{Ans.}$$

Example 5.9.6. *The diagram (Fig. 5.9.6) given below represents a closed-loop control system for regulating the terminal voltage of a d.c. generator. The generator field winding is excited by the amplified difference between the reference battery voltage E and 10% of the terminal voltage. The generator terminal voltage falls from 450 V to 400 V on switching the load, field current being 1.5 A.*

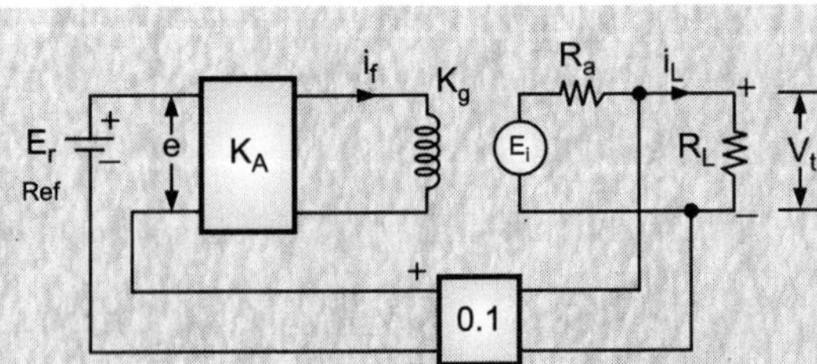

Fig. 5.9.6. Voltage regulator of example 5.9.6.

If the generator terminal voltage is to be kept constant within 2% of the full load value of 400 V, calculate the gain of the amplifier.

Solution. The steady state equations developed are as under:

$$E_i = V_t + i_L R_a \quad ...(1)$$

$$V_t = i_L R_L \quad ...(2)$$

From equations (1) and (2) following relation is derived

$$\frac{V_t}{E_i} = \frac{i_L R_L}{V_t + i_L R_L} = \frac{i_L R_L}{i_L R_L + i_L R_a}$$

or
$$\frac{V_t}{E_i} = \frac{R_L}{R_L + R_a} \quad ...(3)$$

The generator e.m.f. constant is given by

$$K_g = \frac{E_i}{i_f} = \frac{450}{1.5} = 300 \text{ V/A}$$

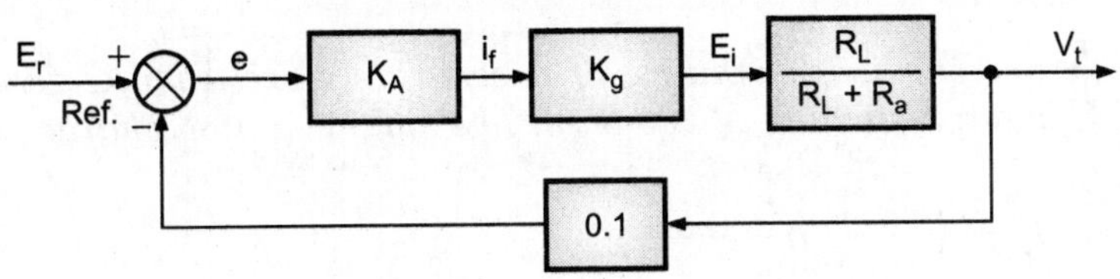

Fig. 5.9.6 (*a*) Block diagram of Fig. 5.9.6.

The block diagram under steady state condition for the system is developed as shown in Fig. 5.9.6 (*a*).

$$\left.\begin{aligned} V_t &= 400 \\ E_i &= 450 \end{aligned}\right\} \text{ given}$$

Therefore, from equation (3)

$$\frac{R_L}{R_L + R_a} = \frac{V_t}{E_i} = \frac{400}{450} = 0.88$$

Substituting the above values in Fig. 5.9.6 (*a*), the block diagram shown in Fig. 5.9.6 (*b*) is obtained.

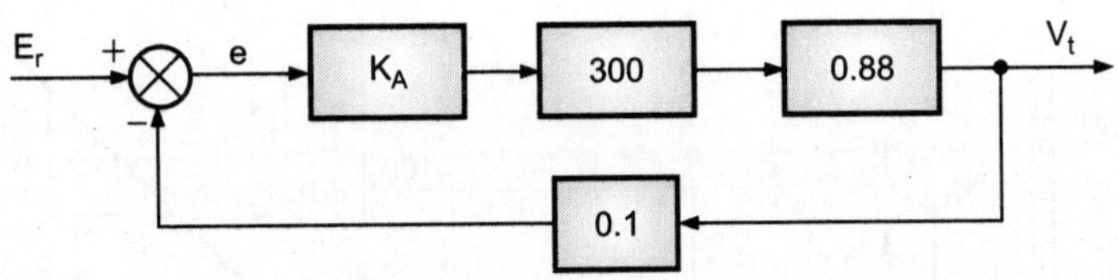

Fig. 5.9.6 (*b*) Block diagram of Fig. 5.9.6 (*a*) redrawn with numerical values.

The transfer function relating the error e and the reference voltage E_r is given by

$$\frac{e}{E_r} = \frac{1}{1 + K_A \times 300 \times 0.88 \times 0.1} \qquad \left[\text{Ref. } \frac{e}{R} = \frac{1}{1+GH}\right]$$

or
$$\frac{e}{E_r} = \frac{1}{1 + 26.4\, K_A} \quad ...(4)$$

The allowable error is 2%, therefore, the error voltage is

$$e = (400 \times 0.1) \times \frac{2}{100}$$

$$= 0.8 \text{ volt} \quad ...(5)$$

The reference voltage is thus,

$$E_r = (400 \times 0.1) + 0.8$$

$$= 40.8 \text{ volts} \quad ...(6)$$

Substituting equations (5) and (6) in equation (4) following relation is obtained.

$$\frac{0.8}{40.8} = \frac{1}{1 + 26.4\, K_A} \quad ...(7)$$

From relation (7) the gain of the amplifier is calculated as

$$K_A = 1.89 \text{ A/V}.$$ **Ans.**

Example 5.9.7. *The field winding of a separately excited d.c. generator has an inductance of 2 H and a resistance of 200 ohm. The magnetisation characteristics is given below :*

Field current i_f in A	*0*	*2*	*4*	*6*	*8*	*10*
Induced voltage e in V	*4*	*50*	*100*	*150*	*200*	*225*

The generator being driven at a constant speed of 1000 r.p.m.

Determine the transfer function of the generator relating the open circuit terminal voltage and the input field voltage.

Solution. The circuit diagram and the magnetisation characteristics of the generator are shown in Fig. 5.9.7 (*a*) and (*b*) respectively.

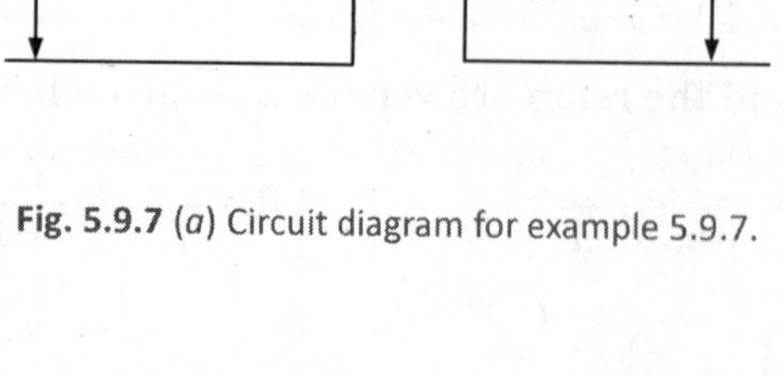

Fig. 5.9.7 (*a*) Circuit diagram for example 5.9.7.

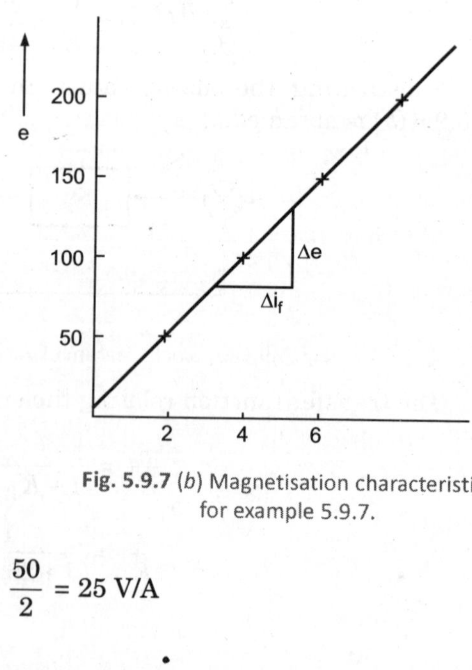

Fig. 5.9.7 (*b*) Magnetisation characteristics for example 5.9.7.

The generator field constant is $K_g = \frac{\Delta e}{\Delta i_f} = \frac{50}{2} = 25$ V/A

The system equations are $V_f(s) = R_f I_f(s) + sL_f I_f(s)$ ...(1)

$$E(s) = K_g I_f(s) \quad ...(2)$$

Eliminating $I_f(s)$ from equations (1) and (2) the transfer function relating $E(s)$ and $V_f(s)$ is obtained below

$$\frac{E(s)}{V_f(s)} = \frac{K_g}{R_f + sL_f} \quad ...(3)$$

Substituting the values of K_g, R_f and L_f in equation (3)

$$\frac{E(s)}{V_f(s)} = \frac{25}{200 + s \times 2} = \frac{0.125}{1 + 0.01\,s} \quad \textbf{Ans.}$$

Example 5.9.8. *From the diagram of an open-loop control system represented below (Fig. 5.9.8).*

(a) Obtain a block diagram representation.

(b) Determine the transfer function relating angular velocity of the load and the input voltage V_f.

(c) If the system is to be made a closed-loop one, suggest a method and incorporate the same in the block diagram.

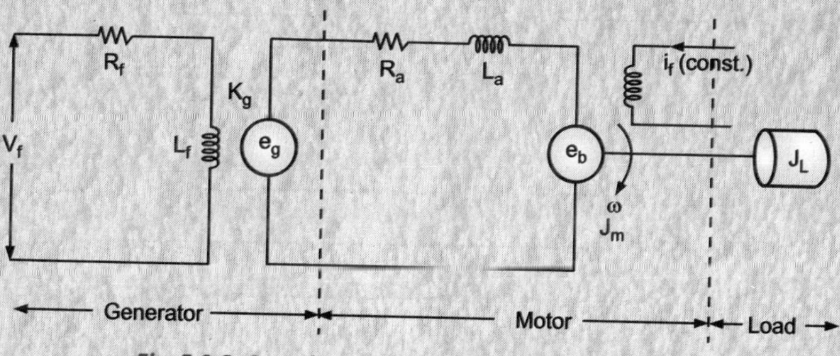

Fig. 5.9.8. Open-loop control system of example 5.9.8.

The system parameters are given below

$R_f = 1000\ \Omega$,	$L_f = 100\ H$,	$K_g = 1000\ V/A$
$R_a = 20$ ohm,	$L_a = 0.1\ H$,	$K_T = 1.2\ Nm/A$
$K_b = 1.2\ V/(rad/sec)$,		$J_L = 0.3 \times 10^{-4}$ kg-m^2,
$J_m = 0.2 \times 10^{-4}$ kg-m^2.		

Neglect friction.

Solution. (*a*) The generator, armature controlled motor and load are connected in cascade. The block diagrams for these elements are drawn as per procedure explained earlier in this chapter and the block diagram relating the output $\omega(s)$ and the input $V_f(s)$ is developed as shown in Fig. 5.9.8 (*a*).

Substituting the values of system parameters the block diagram shown in Fig. 5.9.8 (*b*) is obtained.

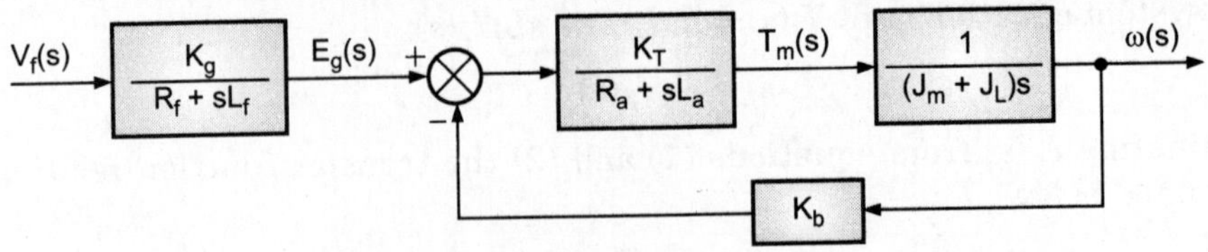

Fig. 5.9.8 (*a*) Block diagram of Fig. 5.9.8.

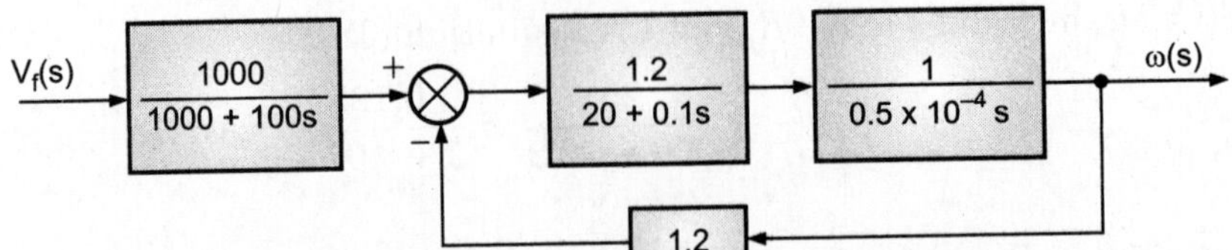

Fig. 5.9.8 (*b*) Values of system parameters substituted in Fig. 5.9.8 (*a*).

Using block diagram reduction technique the block diagram of Fig. 5.9.8 (*b*) is reduced in steps and shown in Fig. 5.9.8 (*c*) and (*d*).

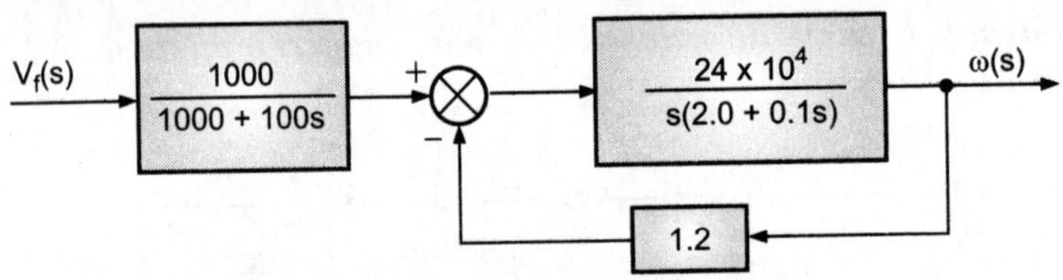

Fig. 5.9.8 (*c*) Block diagram reduction of Fig. 5.9.8 (*b*).

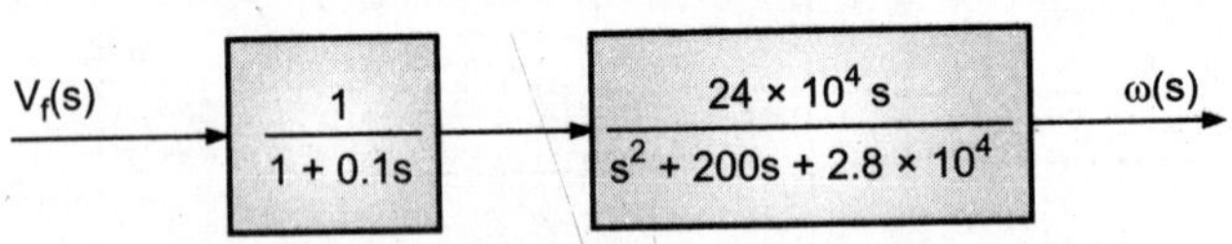

Fig. 5.9.8 (*d*) Block diagram reduction of Fig. 5.9.8 (*c*).

(*b*) From the block diagram of Fig. 5.9.8 (*d*) the transfer function relating ω(*s*) and $V_f(s)$ is determined below

$$\frac{\omega(s)}{V_f(s)} = \frac{1}{(1+0.1s)} \cdot \frac{24 \times 10^4}{(s^2 + 200s + 28.8 \times 10^4)}$$

$$= \frac{24 \times 10^4}{(1+0.1s)(s^2 + 200s + 28.8 \times 10^4)}. \quad \textbf{Ans.}$$

(*c*) Under steady state condition

$$\frac{\omega_{ss}}{V_f} = \frac{24 \times 10^4}{(1+0.1 \times 0)(0 + 200 \times 0 + 28.8 \times 10^4)}$$

$$\frac{\omega_{ss}}{V_f} = \frac{24}{28.8} = \frac{1}{1.2}$$

$$\therefore \quad V_f = 1.2\, \omega_{ss}.$$

In order to convert the system into a closed-loop one, a tachogenerator feedback is used. The transfer function of the feedback block will have a gain of 1.2 V/(rad/sec). The block diagram incorporating tachogenerator feedback is shown in Fig. 5.9.8 (*e*).

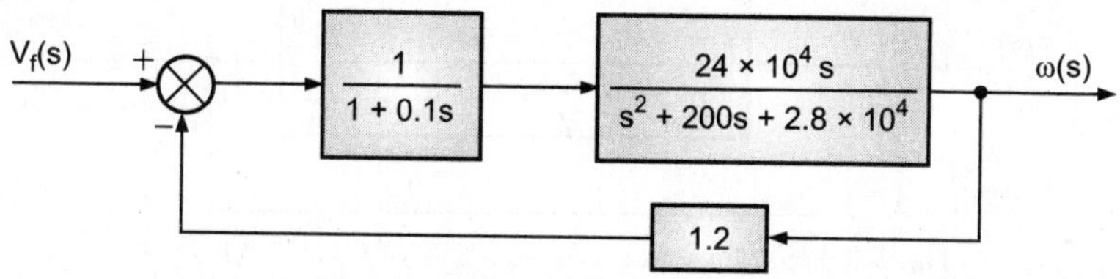

Fig. 5.9.8 (*e*) Open-loop system of Fig. 5.9.8 (*d*) converted into closed-loop system.

Example 5.9.9. *A torque controller uses a split field motor for position control. The error detector gain is 10 V per radian error and the amplifier transconductance is 100 mA / V. The motor torque constant is 5 × 10⁻⁴ Nm / mA. The moment of inertia and the coefficient of viscous friction at the motor shaft are respectively 1.25 × 10⁻⁵ kg-m² and 5 × 10⁻⁴ Nm / (rad / sec). The motor is coupled to the load through a gear having a ratio 20 : 1.*

Draw the block diagram of the system showing the transfer function of each block and determine the overall transfer function relating the output and input.

Solution. The system described in this example is a closed-loop control system for position control. The positional output is compared with the reference input by an error detector. The error thus obtained is amplified in the form of appropriate signal and applied to the motor for producing controlling torque. The output of the motor is coupled to the system output through a gear.

The block diagram of the system based on the aforesaid sequence is developed and drawn in Fig. 5.9.9.

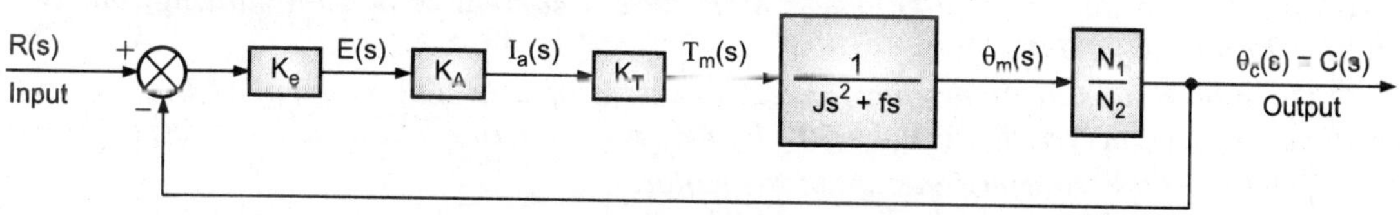

Fig. 5.9.9. Block diagram of the system of example 5.9.9.

Given values of the system parameters are follows:

$$K_e = 10 \text{ V/rad}, \; K_A = 100 \text{ mA/V}, \; K_T = 5 \times 10^{-4} \text{ Nm/mA}$$
$$J = 1.25 \times 10^{-5} \text{ kg-m}^2, \; f = 5 \times 10^{-4} \text{ Nm/(rad/sec)},$$
$$\frac{N_1}{N_2} = \frac{1}{20}$$

Substituting the above values in Fig. 5.9.9 the block diagram given in Fig. 5.9.9 (*a*) is obtained.

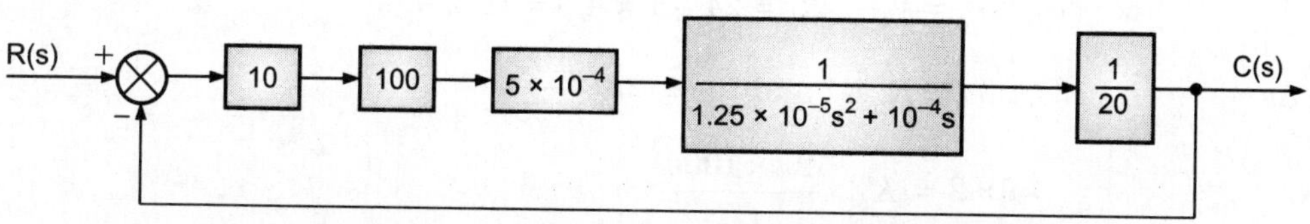

Fig. 5.9.9 (*a*) Values of system parameters substituted in Fig. 5.9.9.

The block diagram of Fig. 5.9.9 (*a*) is reduced as shown in Fig. 5.9.9 (*b*) and further simplified as shown in Fig. 5.9.9 (*c*).

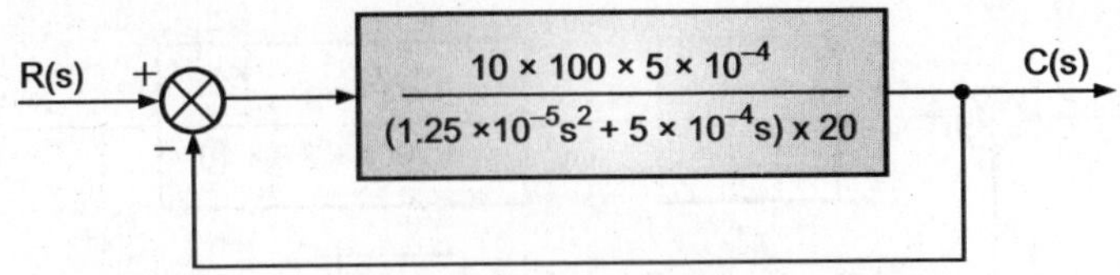

Fig. 5.9.9 (*b*) Block diagram reduction of Fig. 5.9.9 (*a*).

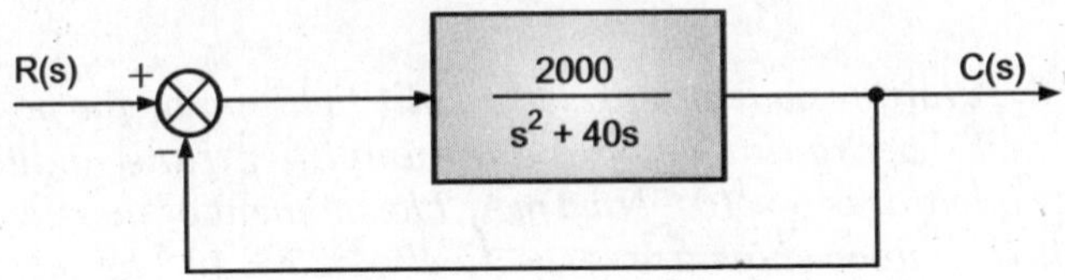

Fig. 5.9.9 (*c*) Block diagram of Fig. 5.9.9 (*b*) simplified.

From the block diagram (Fig. 5.9.9 *c*), the transfer function of the system is determined below :

$$\frac{C(s)}{R(s)} = \frac{\dfrac{2000}{(s^2+40s)}}{1+\left(\dfrac{2000}{s^2+40s}\right)\times 1} = \frac{2000}{(s^2+40s+2000)}. \quad \textbf{Ans.}$$

Example 5.9.10. *An armature controlled d.c. motor is supplied in series with a resistance from a 24 d.c. supply. The motor takes a current of 5 A on stalling and the stalling torque being 0.915 N-m.*

The motor runs at 1000 rpm taking a current of 1 A. The moment of inertia and coefficient of viscous friction are 4 × 10⁻³ kg-m² and 1.5 × 10⁻³ Nm/(rad/sec) respectively.

Determine the transfer function of the motor.

Solution. $T_{stall} = K_T \, i_{stall}$

$\therefore$ $0.915 = K_T \times 5$

$\therefore$ $K_T = 0.183$ Nm/A

$$i_{stall} = \frac{V}{R_a}, \quad \therefore \quad 5 = \frac{24}{R_a}$$

$\therefore$ $R_a = \dfrac{24}{5} = 4.8$ ohm

$e_b = V - i_a R_a = 24 - 1 \times 4.8 = 19.2$ V

As $e_b = K_b \omega$ and $\omega = 2\pi \times \dfrac{1000}{60}$

$\therefore$ $19.2 = K_b \times \dfrac{2\pi \times 1000}{60}$

or $K_b = 0.183$ V/(rad/sec)

If the armature inductance is neglected, the block diagram of an armature controlled d.c. motor [Refer Fig. 5.6.4 (*b*)] is shown in Fig. 5.9.10.

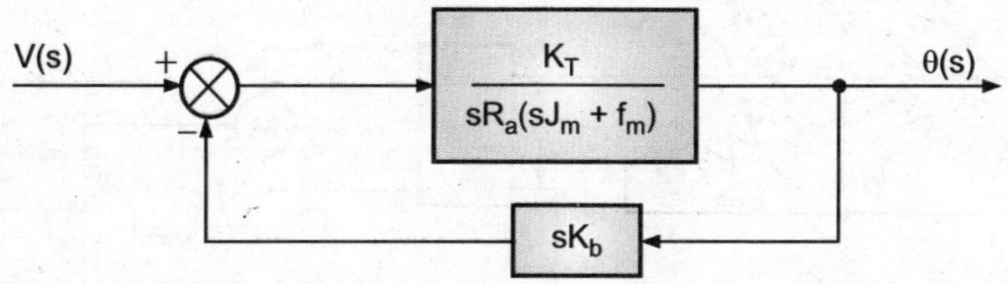

Fig. 5.9.10. Block diagram of an armature controlled d.c. motor neglecting armature inductance.

From Fig. 5.9.10 the transfer function of the motor is determined below :

$$\frac{\theta(s)}{V(s)} = \frac{\dfrac{K_T}{sR_a(sJ_m + f_m)}}{1 + \dfrac{K_T}{sR_a(sJ_m + f_m)} . sK_b}$$

or

$$\frac{\theta(s)}{V(s)} = \frac{K_T}{s(sR_aJ_m + R_af_m + K_TK_b)}$$

Substituting the values of K_T, K_b, R_a, J_m, and f_m in the above equation

$$\frac{\theta(s)}{V(s)} = \frac{0.183}{s(s4.8 \times 4 \times 10^{-3} + 4.8 \times 1.5 \times 10^{-3} + 0.183 \times 0.183)}$$

$$= \frac{0.183}{s(0.0192s + 0.034)} = \frac{5.38}{s(0.564s + 1)}. \quad \textbf{Ans.}$$

Example 5.9.11. *Fig. 5.9.11 shows a positional control system for controlling the position of a shaft. The potentiometer error detector measures the deviation of the output shaft w.r.t. reference position* θ_r. *The error is amplified by means of an amplifier, the output of which is fed to an armature controlled d.c. motor. The motor shaft is coupled to the load shaft through a gear.*

The system parameters are as follows:

Error detector gain $K_e = 2\ V/rad$

Amplifier gain $K_A = 10\ V/V$, $R_a = 0.2\ ohm$; L_a = *negligible*

Motor torque constant $K_T = 10 \times 10^{-5}\ Nm/A$

Motor back e.m.f. constant $K_b = 10 \times 10^{-5}\ V/(rad/sec)$

Equivalent moment of inertia referred to motor side

$$J_m = 10 \times 10^{-5}\ kg\text{-}m^2$$

Equivalent coefficient of viscous friction referred to motor side

$$f_m = 10 \times 10^{-5}\ Nm/(rad/sec)$$

Draw the block diagram and determine the overall transfer function relating the output and input.

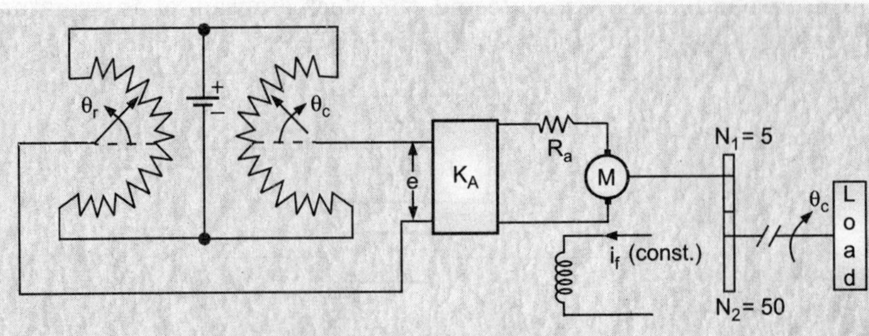

Fig. 5.9.11. Control system for example 5.9.11.

Solution. The error detector, amplifier, motor and gear are arranged in the block diagram shown in Fig. 5.9.11 (*a*).

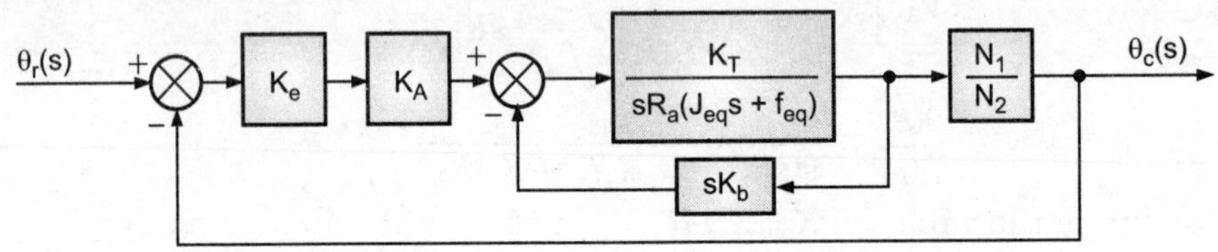

Fig. 5.9.11 (*a*) Block diagram of control system for example 5.9.11.

Substituting system parameter values the block diagram of Fig. 5.9.11 (*a*) is redrawn as shown in Fig. 5.9.11 (*b*).

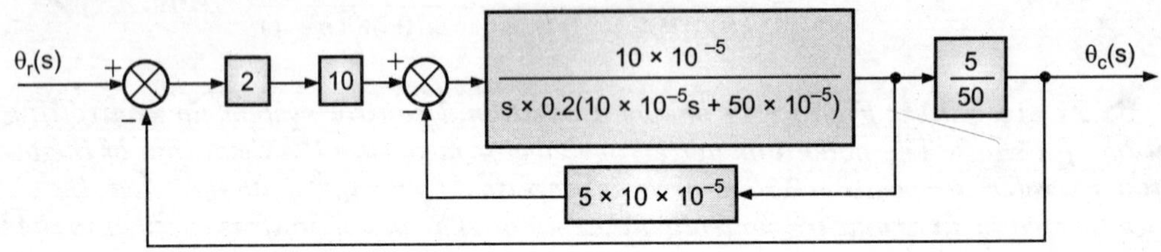

Fig. 5.9.11 (*b*) System parameter values substituted in Fig. 5.9.11 (*a*).

The block diagram of Fig. 5.9.11 (*b*) is reduced in following steps and shown in Fig. 5.9.11 (*c*), (*d*) and (*e*)

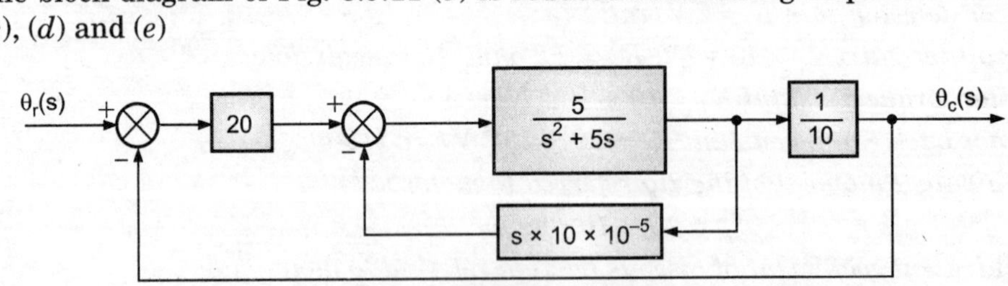

Fig. 5.9.11 (*c*) Block diagram reduction of Fig. 5.9.11 (*b*).

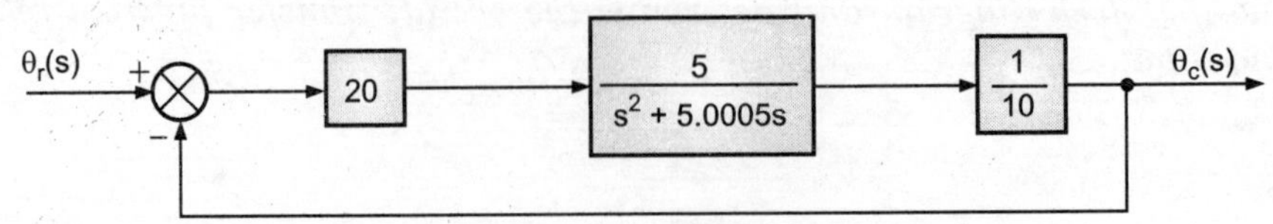

Fig. 5.9.11 (*d*) Block diagram reduction of Fig. 5.9.11 (*c*).

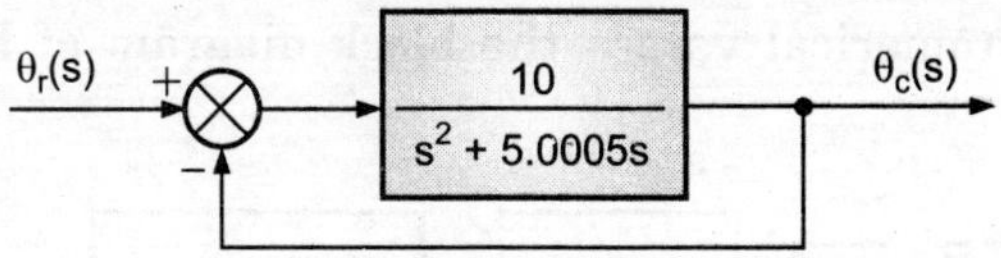

Fig. 5.9.11 (*e*) Block diagram reduction of Fig. 5.9.11 (*d*).

From Fig. 5.9.11 (*e*) the overall transfer function is determined below

$$\frac{\theta_c(s)}{\theta_r(s)} = \frac{\dfrac{10}{s^2+5.0005s}}{1+\dfrac{10}{s^2+5.0005s}\times 1} = \frac{10}{(s^2+5.0005s+10)} \quad \textbf{Ans.}$$

Example 5.9.12. *For the closed-loop control system shown in Fig. 5.9.12. Draw the block diagram and determine the overall transfer function relating θ_c and θ_r.*

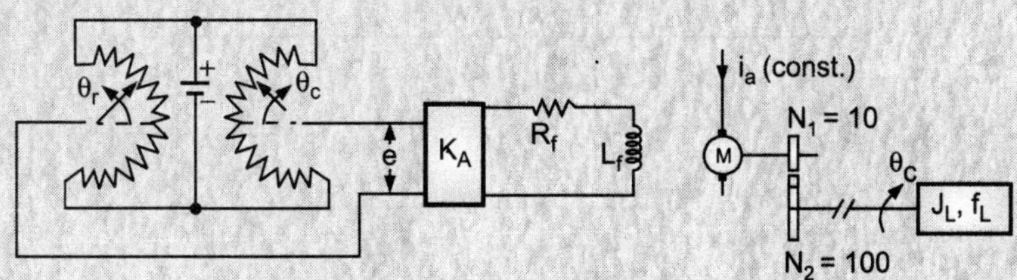

Fig. 5.9.12. Closed-loop control system for example 5.9.12.

The system has the following details:

Error detector gain $K_e = 8$ V / rad, amplifier gain $K_A = 10$ V / V, $R_f = 5$ ohm, $L_f = 0.25$ H, Motor torque constant $K_f = 0.05$ Nm / A. Moment of inertia for motor $J_m = 0.02$ kg-m^2.

Coefficient of viscous friction for motor $f_m = 0.03$ Nm / (rad / sec).

Moment of inertia for load $J_L = 3.0$ kg-m^2.

Coefficient of viscous friction for load $f_L = 4.5$ Nm / (rad / sec).

Solution. **The equivalent moment of inertia referred to motor side is**

$$J_{eq} = J_m + J_L\left(\frac{N_1}{N_2}\right)^2 = 0.02 + 3.0\left(\frac{10}{100}\right)^2 = 0.05 \text{ kg-m}^2$$

The equivalent coefficient of viscous friction referred to motor side is

$$f_{eq} = f_m + f_L\left(\frac{N_1}{N_2}\right)^2 = 0.03 + 4.5\left(\frac{10}{100}\right)^2 = 0.075 \text{ Nm/(rad/sec)}$$

The error detector, amplifier, motor gear and load are connected in cascade. The block diagrams for these elements are drawn as per the results obtained in earlier sections and the block diagram for the system is shown in Fig. 5.9.12 (*a*).

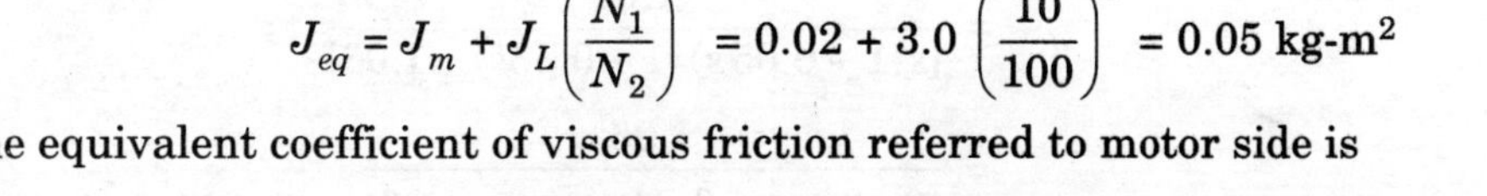

Fig. 5.9.12 (*a*) Block diagram for example 5.9.12.

Substituting the numerical values the block diagram of Fig. 5.9.12 is redrawn as shown in Fig. 5.9.12 (*b*).

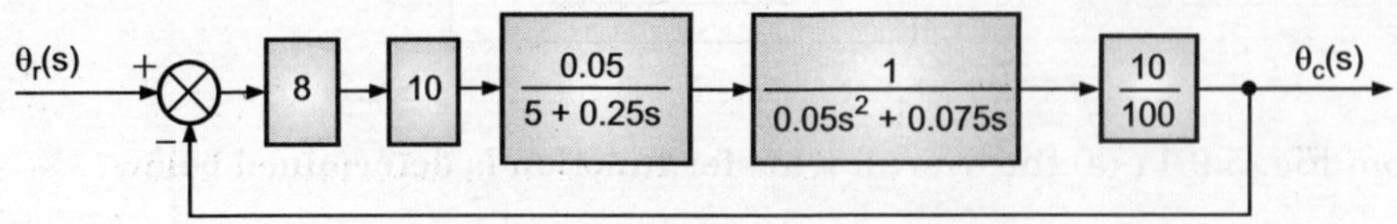

Fig. 5.9.12 (*b*) Block diagram of Fig. 5.9.12 (*a*) with numerical values substituted.

The block diagram in Fig. 5.9.12 (*b*) is reduced as shown in Fig. 5.9.12 (*c*).

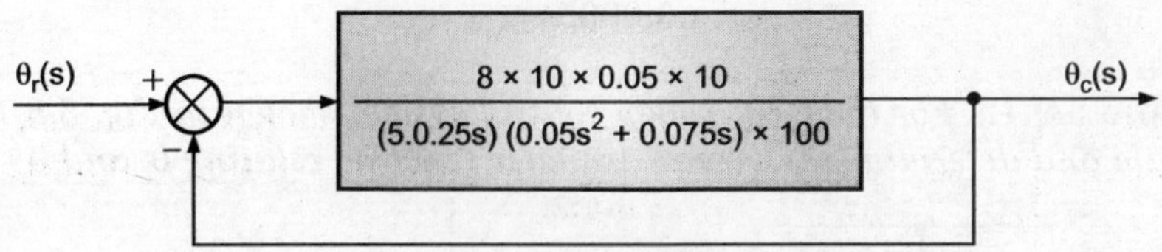

Fig. 5.9.12 (*c*) Block diagram reduction of Fig. 5.9.12 (*b*).

The block diagram of Fig. 5.9.12 (*c*) is simplified and shown in Fig. 5.9.12 (*d*).

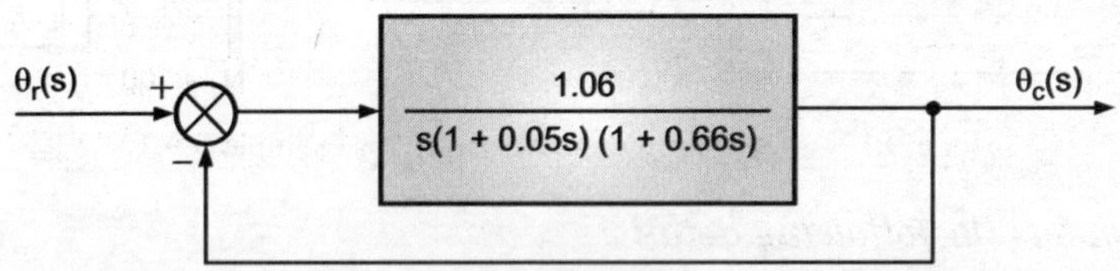

Fig. 5.9.12 (*d*) Block diagram of Fig. 5.9.12 (*c*) simplified.

The overall transfer function relating $\theta_c(s)$ and $\theta_r(s)$ is given by

$$\frac{\theta_c(s)}{\theta_r(s)} = \frac{\dfrac{1.06}{s(1+0.05s)(1+0.66s)}}{1+\dfrac{1.06}{s(1+0.05s)(1+0.66s)}\times 1}$$

$$= \frac{1.06}{[s(1+0.05s)(1+0.66s)+1.06]}$$

$$= \frac{1.06}{(0.033s^2+0.71s^2+s+1.06)}$$

or $$\frac{\theta_c(s)}{\theta_r(s)} = \frac{32.12}{(s^3+21.5s^2+30.3s+32.12)}.$$ **Ans.**

Example 5.9.13. *A closed-loop speed control system uses a two-phase a.c. motor. The error signal is amplified by an amplifier the output of which is connected to the motor. The motor shaft is connected to a load through a gear having a ratio 10 : 1.*

The amplifier gain is 20 V/V and tachogenerator used in the feedback path has a gain constant K_{tg} = 0.2 V/(rad/sec). The moment of inertia and coefficient of viscous friction referred to motor shaft side are respectively 1.5 × 10^{-4} kg-m^2 and 1 × 10^{-4} Nm/(rad/sec).

The motor has a stall torque of 0.12 Nm at 100 V input and the torque decreases by 50% when the speed increases to 2000 r.p.m.

Determine the transfer function of the system.

Solution. The characteristics of the two phase motor is drawn in Fig. 5.9.13 (*a*).

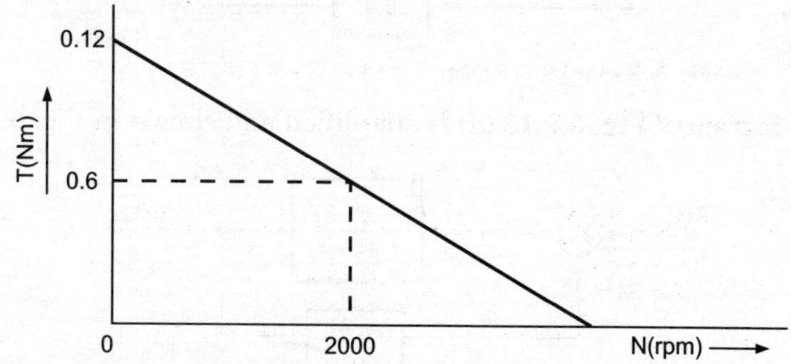

Fig. 5.9.13 (*a*) Torque-speed characteristics of two-phase motor of example 5.9.13.

From the torque speed characteristics [Fig. 5.9.13 (*a*)] the constants *m* and *K* of the motor are determined below

$$m = \frac{(0.06 - 0.12)}{2 \times \dfrac{2000\pi}{60}} = -3 \times 10^{-4} \text{ Nm/(rad/sec)}$$

$$K = \frac{0.12}{100} = 12 \times 10^{-4} \text{ Nm/V}$$

The block diagram for the system is developed and shown in Fig. 5.9.13 (*b*).

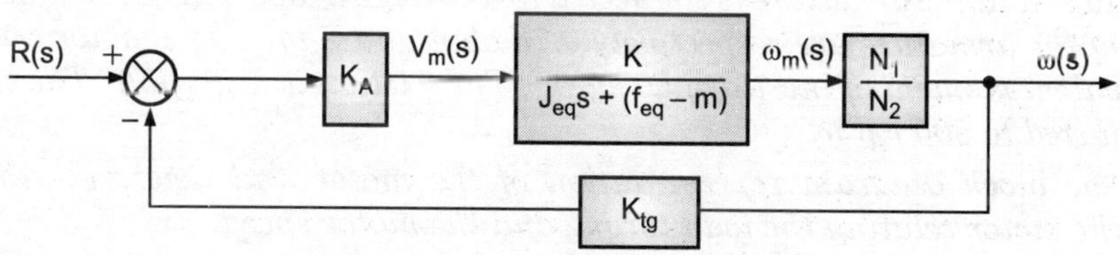

Fig. 5.9.13 (*b*) Block diagram for example 5.9.13.

Substituting numerical values, the block diagram of Fig. 5.9.13 (*b*) is shown in Fig. 5.9.13 (*c*).

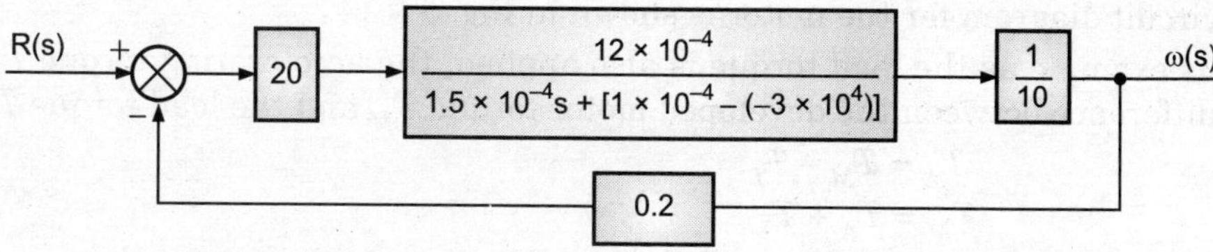

Fig. 5.9.13 (*c*) Block diagram of 5.9.13 (*b*) with numerical values.

The block diagram of Fig. 5.9.13 (*c*) is reduced and shown in Fig. 5.9.13 (*d*).

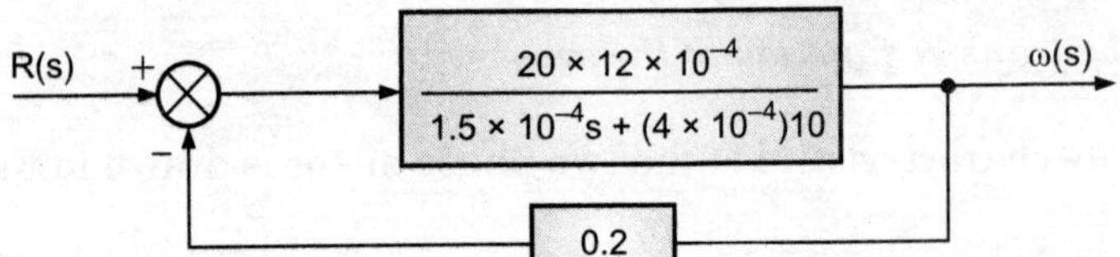

Fig. 5.9.13 (*d*) Block diagram reduction of Fig. 5.9.13 (*c*).

The block diagram of Fig. 5.9.13 (*d*) is simplified and shown in Fig. 5.9.13 (*e*).

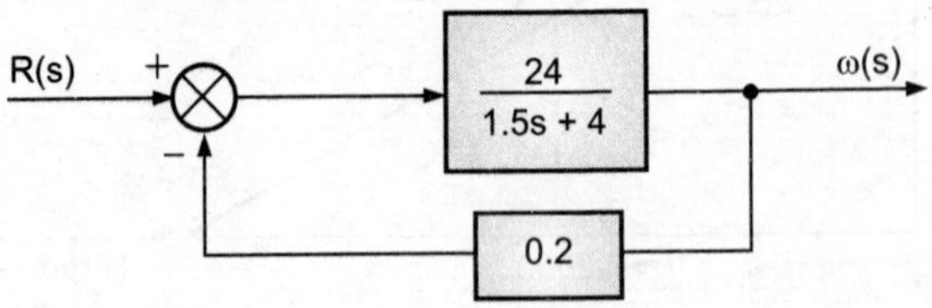

Fig. 5.9.13 (*e*) Block diagram simplification of Fig. 5.9.13 (*d*).

From the block diagram shown in Fig. 5.9.13 (*e*) the transfer function of the system is determined below

$$\frac{\omega(s)}{R(s)} = \frac{\dfrac{24}{(1.5s+5)}}{1+\dfrac{24}{(1.5s+4)}\times 0.2} = \frac{16}{(s+5.86)}. \quad \textbf{Ans.}$$

Example 5.9.14. *A separately excited d.c. motor is supplied at its field winding by a constant voltage and the armature is connected to a voltage of 500 V. The resistance and the inductance of the armature are respectively 0.75 ohm and 5 mH. The motor and the load has an equivalent moment of inertia of 50 kg-m*^{2}*. The friction is negligible. The speed of the motor is adjusted to 500 r.p.m.*

Draw the block diagram representation of the motor and determine the transfer function of the motor relating the load torque and the motor speed.

Solution. The motor speed in rad/sec is

$$\omega_m = \frac{2\pi N}{60} = \frac{2\pi \times 500}{60} = 52.35 \text{ rad/sec}$$

The circuit diagram for the motor is shown in Fig. 5.9.14.

In this example as the load torque is also applied, the accelerating torque T_A for the load is the difference between the developed motor torque T_M and the load torque T_L, hence

$$T_A = T_M - T_L$$

or

$$T_M = T_A + T_L \qquad ...(1)$$

As the motor is a armature controlled, the system equations are

$$V_a(s) - E_b(s) = R_a I_a(s) + sL_a I_a(s) \qquad ...(2)$$

$$\therefore \qquad I_a(s) = \frac{V_a(s) - E_b(s)}{R_a + sL_a} \qquad ...(3)$$

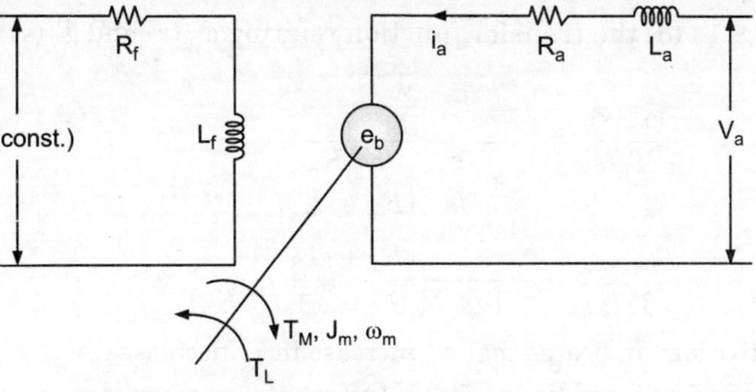

Fig. 5.9.14. Circuit diagram for example 5.9.14.

$$\because \quad T_M(s) = K_T I_a(s) \qquad ...(4)$$

Substituting for $I_a(s)$ in equation (4) from equation (3) following relation is obtained

$$T_M(s) = [V_s(s) - E_b(s)] \times \frac{K_T}{R_a + sL_a} \qquad ...(5)$$

Since friction is negligible,

$$T_M(s) = Js\ \omega_m(s)$$

Therefore, from equation (1)

$$T_M(s) = Js\ \omega_m(s) + T_L(s)$$

or

$$\omega_m(s) = \frac{1}{Js}[T_M(s) - T_L(s)] \qquad ...(6)$$

$$E_b(s) = K_b\ \omega_m(s) \qquad ...(7)$$

Using equations (5), (6) and (7) the block diagram as shown in Fig. 5.9.14 (*a*) is obtained.

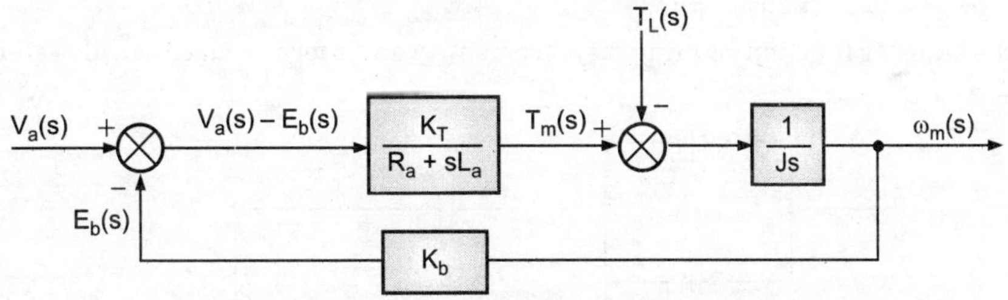

Fig. 5.9.14 (*a*) Block diagram for example 5.9.14.

There are two inputs $V_a(s)$ and $T_L(s)$. If $V_a(s)$ is held at a fixed value and the only the effect of $T_L(s)$ is considered, then in such a case the effect of $V_a(s)$ is taken as zero and the block diagram given in Fig. 5.9.14 (*a*) takes the form as shown in Fig. 5.9.14 (*b*).

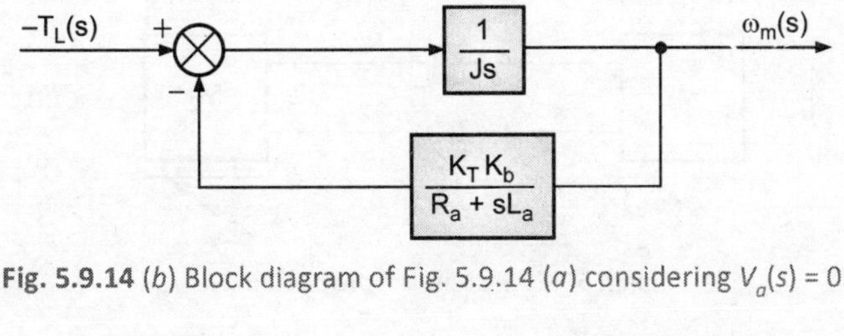

Fig. 5.9.14 (*b*) Block diagram of Fig. 5.9.14 (*a*) considering $V_a(s) = 0$.

From Fig. 5.9.14 (*b*) the transfer function relating $\omega_m(s)$ and $T_L(s)$ is determined below:

$$\frac{\omega_m(s)}{-T_L(s)} = \frac{\dfrac{1}{Js}}{1+\dfrac{1}{Js}\cdot\dfrac{K_T K_b}{(R_a+sL_a)}}$$

or
$$\frac{\omega_m(s)}{T_L(s)} = -\frac{(R_a+sL_a)}{[Js(R_a+sL_a)+K_T K_b]} \quad ...(8)$$

(**Note.** Negative sign indicates that an increase in T_L decreases ω_m.)

As the speed is 52.35 rad/sec at 500 V (effect of armature drop is neglected), therefore,

$$K_b = \frac{500}{52.35} = 9.55 \text{ V/(rad/sec)}$$

In MKS system of units

$$K_T = K_b \qquad \therefore \qquad K_T = 9.55 \text{ Nm/A}$$

Substituting numerical values in (8)

$$\frac{\omega_m(s)}{T_L(s)} = \frac{(0.75+5\times10^{-3}s)}{[50s(0.75+5\times10^{-3}s)+9.55\times9.55]}$$

$$= -\frac{(0.75+5\times10^{-3}s)}{(0.25s^2+37.5s+91.2)} = -\frac{(3+0.02s)}{(s^2+150s+364.8)}$$

$$= -\frac{3(1+0.066s)}{(s^2+150s+364.8)}. \quad \textbf{Ans.}$$

PROBLEMS

5.1. Find the transfer function relating displacements *y* and *x* for the mechanical system shown in Fig. 5P-1.

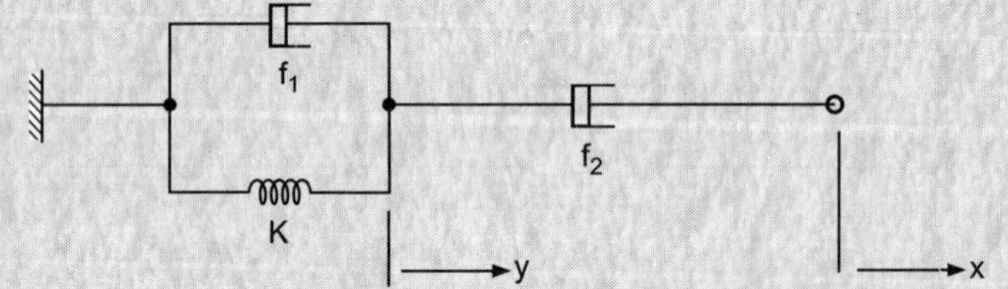

Fig. 5P-1. Mechanical system for problem 5.1.

5.2. On the basis of force-current analogy write the equations for the system given in Fig. 5P-2.

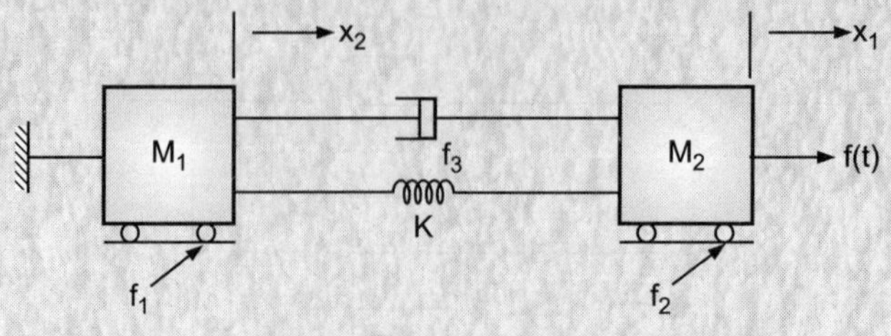

Fig. 5P-2. System for problem 5.2.

5.3. Obtain mathematical model for the mechanical system shown in Fig. 5P-3.

5.4. A full load current of 50 A at 150 V is supplied to a resistive load by a separately excited d.c. generator having an armature resistance of 0.5 ohm. The field current being 0.5 A supplied by a battery.

The battery voltage is compared with the load voltage and the error thus obtained is amplified through an amplifier and the output of the amplifier in turn is connected to the field winding of the generator forming a closed-loop control system.

Determine the gain of the amplifier for an improvement in the output voltage regulation by 10 times.

5.5. The diagram given below (Fig. 5P-4) represents a closed-loop control system for speed control using a field controlled d.c. motor. Motor rpm to field current ratio is 100 rpm/A. The tachogenerator constant is $K_{tg} = 0.25$ V/rpm. If the speed is to be maintained within 1% of 100 rpm, calculate the amplifier gain.

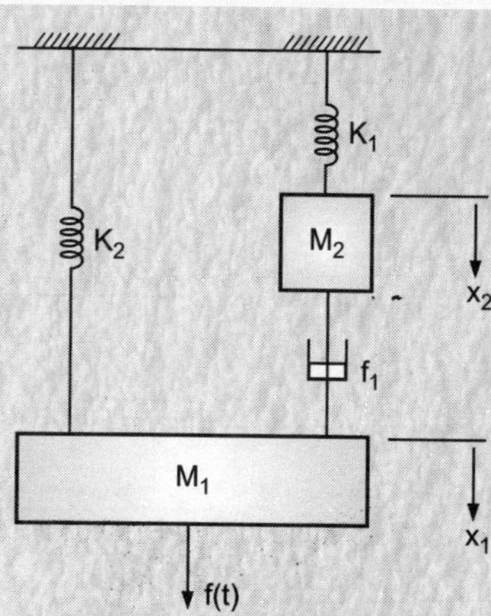

Fig. 5P-3. Mechanical system for problem 5.3.

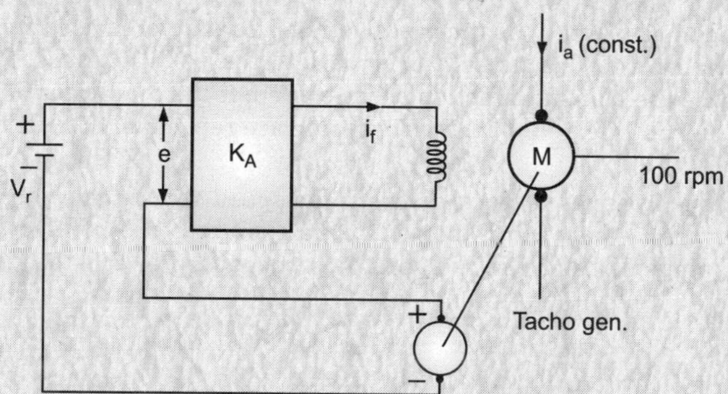

Fig. 5P-4. Diagram for problem 5.5.

5.6. A closed-loop position control system has a controller gain constant K_T Nm/radian error, the moment of inertia and the coefficient of viscous friction are respectively J kg-m^2 and f N-m/(rad/sec). Determine the transfer function,

(*a*) Relating the output and input
(*b*) Relating the output and error
(*c*) Relating the error and input.

5.7. The moment of inertia J_m and the coefficient of viscous friction f_m for a field controlled d.c. servo-motor are respectively 5×10^{-4} kg-m^2 and 12.5×10^{-4} Nm/(rad/sec). The motor torque constant K_f being 2.5 Nm/A. Draw the block diagram and determine the transfer function relating the angular speed of the shaft and the field current.

5.8. An armature controlled d.c. motor has an armature resistance of 0.37 ohm. The moment of inertia is 2.5×10^{-6} kg-m^2. A back e.m.f. of 2.09 V is generated per 100 rpm of the motor speed. The torque constant of the motor is 0.2 Nm/A. Determine the transfer function of the motor relating the motor shaft shift and the input voltage.

5.9. A control system is shown in Fig. 5P-5. The armature controlled d.c. motor is connected to a d.c. generator. Determine the transfer function of the system relating the output angular velocity and the input field voltage V_f. The parameters of the system are given bleow:

Generator e.m.f. constant $K_g = 1000$ V/A

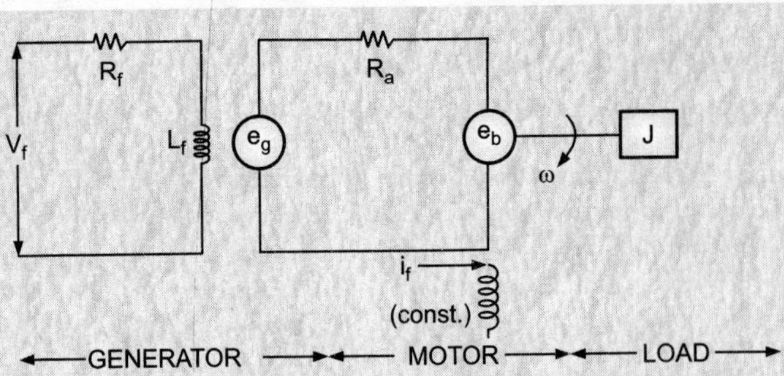

Fig. 5P-5. Control system for problem 5.9.

Motor back e.m.f. constant $K_b = 1$ V/(rad/sec)
Motor torque constant $K_T = 1$ Nm/A
Combined inertia of motor and load $J = 0.5 \times 10^{-4}$ kg-m^2
Friction is negligible, $R_f = 1000\ \Omega$, $L_f = 20$ H . $R_a = 50$ ohm
L_a is negligible.

5.10. The torque of a motor is given by the following equation :

$$T_M = (8\ V_m - 0.3\ N) \times 10^{-4}\ \text{Nm}$$

where N is speed in rpm and V_m is the input voltage. The motor is used through a gear ratio of 10 : 1 for position control. The gain of the error detector is 0.1 V/degree. The amplifier gain is 25 V/V. Determine the transfer function of the system if the coefficient of viscous friction and the moment of inertia referred to motor shaft side are repectively 50 × 10^{-4} Nm/(rad/sec) and 80 × 10^{-4} kg-m^2.

5.11. Fig. 5P-6 given below gives a block diagram for speed regulator system. The accelerating torque is the difference between the torque developed by the controller and load torque T_L. The controller gain is 0.00102 Nm/rad and the tachogenerator constant is 0.191 V/(rad/sec).

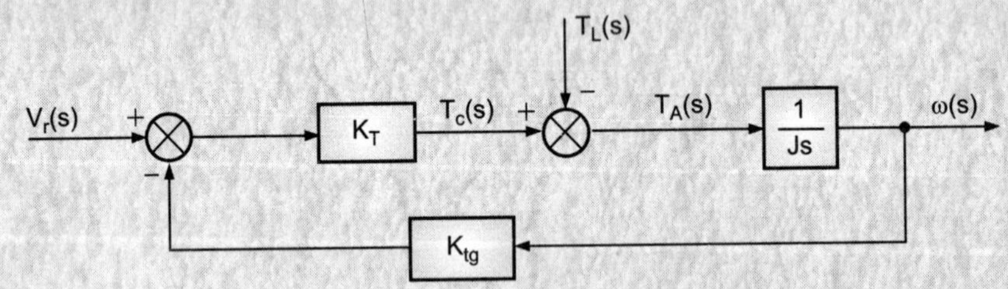

Fig. 5P-6. Block diagram for problem 5.11.

The no load speed is adjusted to 1000 rpm. The moment of inertia is J and the friction is negligible.
Calculate : (*a*) the reference voltage
(*b*) the speed if a constant load torque of 0.001 Nm is suddenly applied.

5.12. A separately excited d.c. motor is directly coupled to a load. The combined moment of inertia of the load and the motor is J and friction is negligible. The armature resistance is R_a and armature inductance is negligible.
Derive the transfer function relating the motor speed and the applied armature voltage and load torque as independent variables. Also obtain an expression for steady state speed.

CHAPTER 6

Time Response Analysis of Control Systems

This chapter deals with the time response analysis of feedback control systems. The test signals are specified as unit step, unit ramp, unit parabolic and unit impulse. The time response is divided in two parts, namely transient and steady state.

The transient response of a second order control system can be underdamped, critically damped, over-damped depending on the value of damping ratio. The reduction in damping ratio tends towards underdamped response and in the absence of damping sustained oscillations are noted.

The expressions for the time response for a specified input have been derived explaining each step.

The transient response specifications for a second order control system subjected to a unit step function are maximum overshoot, under shoot, natural and damped frequency of oscillations, rise time, peak time, settling time.

The error coefficient K_p, K_v and K_a assess the steady state error.

The higher order control system has the degree of s in the characteristic equation as more than 2. However, the higher order control system can be approximated to second order as per the s-plane location of the roots (poles of closed-loop transfer function) of the characteristic equation.

At the end of the transient response of higher order control systems, the concept of dominant poles is explained.

The improvement in the time response is achieved by applying control actions as described in the text.

CONTENTS

- *Transient and Steady State Response*
- *Input Test Signals*
- *Time Response of a First Order Control System*
- *Time Response of a Second Order Control System*
- *Time Response of a Third Order Control System*
- *Time Response of Higher Order Control System*
- *Steady State Error*
- *Performance Indices*
- *Sensitivity*

Introduction. In the earlier chapter the process of modelling a control system has been described, now hereunder time response analysis of control systems is dealt with. As the input to a control system cannot be assessed before hand, therefore, an input test signal is specified for ascertaining comparative time performance of a control system.

Specified input is one to which a control system is frequently subjected to. Once a control system responds satisfactorily to a test input signal, its time response to actual input signal is invariably upto the mark.

The time response of a control system means as to how a system behaves in accordance with time when a specified input test signal is applied.

6.1 TRANSIENT AND STEADY STATE RESPONSE

In the initial part of time response of a control system transients appear and during the post transient part steady state is achieved. Theoretically steady state means a state of the output of a control system as the time approaches infinity after initiation of the input. In actual practice the steady state period and transient period is identified in terms of time constants of a control system. Thus, the time response of a control system is divided in two parts namely :

(*i*) Transient response, (*ii*) Steady state response.

Typical time response of a control system for a specified input test signal is shown in Fig. 6.1.1.

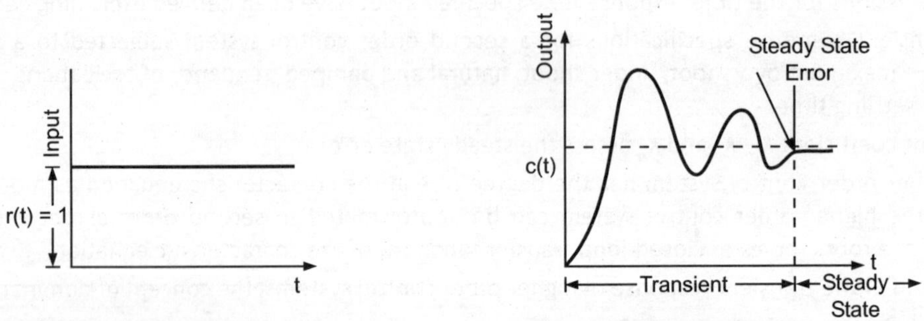

(*a*) Input test signal (*b*) Output response

Fig. 6.1.1. Typical time response of a control system.

The transient part of time the response reveals the nature of response (*i.e.* oscillatory or overdamped) and also gives an indication about its speed.

The steady state part of time response reveals the accuracy of a control system. Steady state error is observed if the actual output does not exactly match with the input.

6.2 INPUT TEST SIGNALS

Specified input test signals applied for time response analysis of a control system are described below :

(1) Step function. Step function is described as sudden application of input signal as illustrated in Fig. 6.2.1.

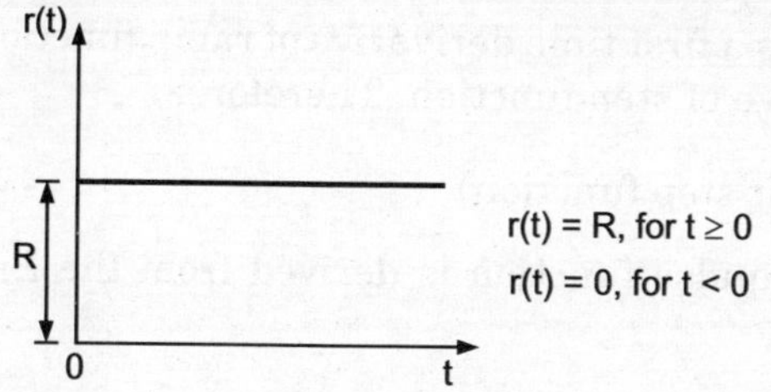

Fig. 6.2.1. Step function/displacement function.

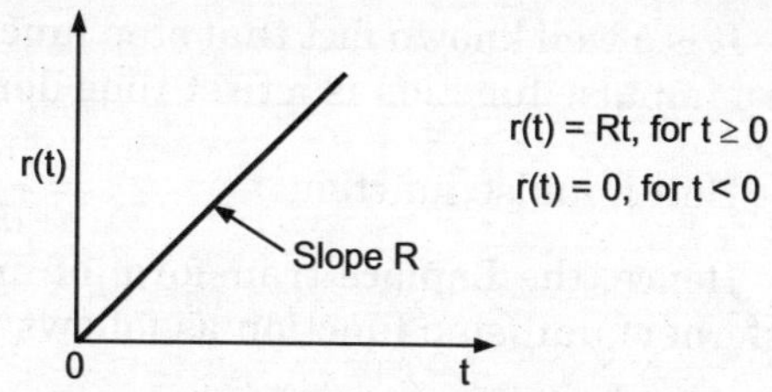

Fig. 6.2.2. Ramp function/velocity function.

If $R = 1$ unit, the step function is called unit step function and the corresponding Laplace transform is

$$\mathcal{L}1 = \frac{1}{s}.$$

Step function is also called displacement function.

(2) Ramp function. Ramp function is described as gradual application of input signal as illustrated in Fig. 6.2.2.

If $R = 1$, then $r(t) = t$ and the ramp function is called unit ramp function and the corresponding Laplace transform is

$$\mathcal{L}\,t = \frac{1}{s^2}.$$

Ramp function is also called velocity function.

(3) Parabolic function. Parabolic function is described as more gradual application of input in comparison with ramp function as illustrated in Fig. 6.2.3.

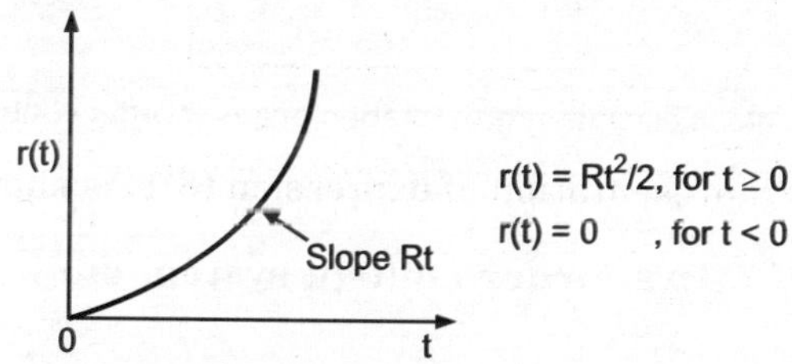

Fig. 6.2.3. Parabolic function/acceleration function.

If $R = 1$, then $r(t) = \frac{t^2}{2}$ and the parabolic function is called unit parabolic function and the corresponding Laplace transform is

$$\mathcal{L}\left(\frac{t^2}{2}\right) = \frac{1}{s^3}.$$

Parabolic function is also called acceleration function.

(4) Impulse function. In the case of impulse function; input is suddenly applied as a shock for a very short duration of time as illustrated in Fig. 6.2.4.

If the magnitude of impulse function is unity, the function is called unit impulse function.

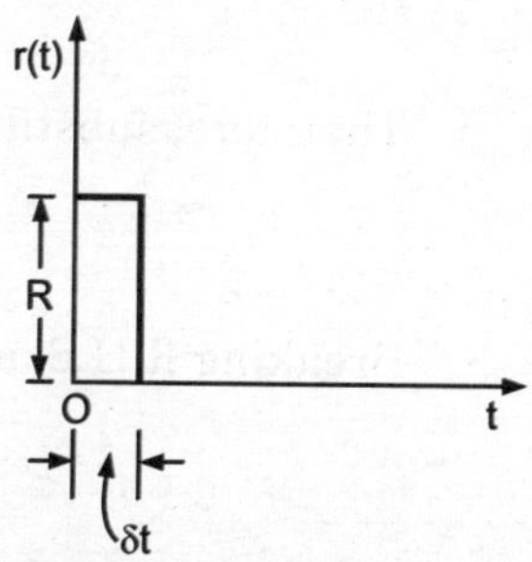

Fig. 6.2.4. Impulse function.

It is a well known fact that step function is a first time derivative of ramp function and in turn impulse function is a first time derivative of step function. Therefore,

$$\text{Unit impulse function} = \frac{d}{dt}(\text{unit step function})$$

Hence, the Laplace transform of unit impulse function is derived from the Laplace transform of unit step function as follows

$$\mathcal{L}(\text{unit impulse function}) = \mathcal{L}\frac{d}{dt}(\text{unit step function})$$

or $$\mathcal{L}(\text{unit impulse function}) = s\mathcal{L}(\text{unit step function})$$

$$\therefore \quad \mathcal{L}(\text{unit impulse function}) = s.\frac{1}{s} = 1.$$

6.3 TIME RESPONSE OF A FIRST ORDER CONTROL SYSTEM

A first order control system is one wherein highest power of s in the denominator of its transfer function equals 1. Thus a first order control system is expressed by a transfer function given below :

$$\frac{C(s)}{R(s)} = \frac{1}{sT+1} \qquad ...(6.1)$$

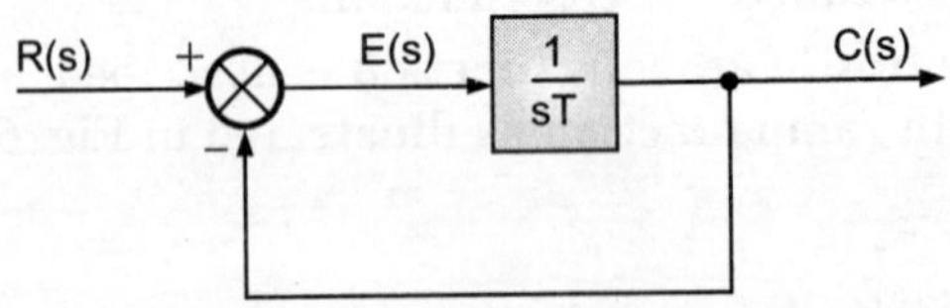

Fig. 6.3.1. Block diagram representation of a first order control system.

The block diagram representation of expression (6.1) is shown in Fig. 6.3.1.

6.3.1. Time response of a first order control system subjected to unit step input function

The output for the system is expressed as

$$C(s) = R(s)\frac{1}{sT+1} \qquad ...(6.2)$$

As the input is a unit step function

$$r(t) = 1 \text{ and } R(s) = \frac{1}{s}$$

Therefore, substituting in Eq. (6.2)

$$C(s) = \frac{1}{s}.\frac{1}{sT+1} \qquad ...(6.3)$$

Breaking R.H.S. of Eq. (6.3) into partial fractions

$$C(s) = \frac{1}{s} - \frac{1}{sT+1}$$

or $$C(s) = \frac{1}{s} - \frac{1}{s + \frac{1}{T}} \quad ...(6.4)$$

Taking inverse Laplace transform on both sides of Eq. (6.4)

$$\mathcal{L}^{-1} C(s) = \mathcal{L}^{-1}\left[\frac{1}{s} - \frac{1}{s + \frac{1}{T}}\right]$$

$$c(t) = 1 - e^{-t/T} \quad ...(6.5)$$

The error is given by $$e(t) = r(t) - c(t) = 1 - (1 - e^{-t/T})$$

$$= e^{-t/T} \quad ...(6.6)$$

The steady state error is $$e_{ss} = \lim_{t \to \infty} e^{-t/T} = 0$$

The time response in relation to Eq. (6.5) is shown in Fig. 6.3.2.

The graphical representation of the time response shown in Fig. 6.3.2 indicates that the response is exponential type and the steady state value is 1 unit. As the time increases the disparity between the output and input approaches to nil, hence, the steady state error is zero.

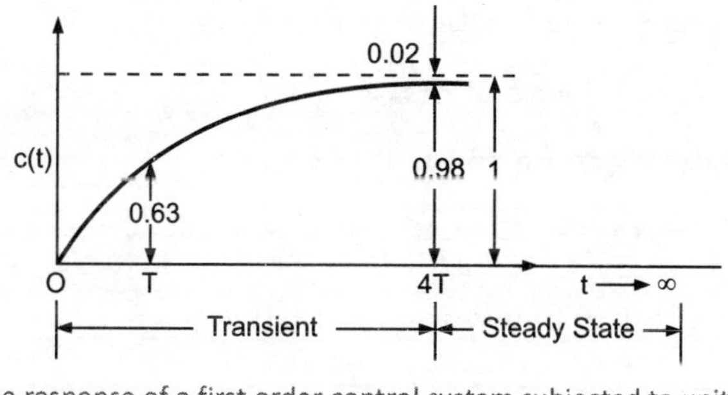

Fig. 6.3.2. Time response of a first order control system subjected to unit step function.

6.3.2. Demarcation between the transient part and steady state part of the time response in terms of time constant

The magnitude of the output time response equals 0.632 unit (63.2%) in time T (one time constant) and approximately 0.98 unit (98%) in $4T$ (four times the time constant). The final steady state value of the output in 1 unit.

The time $4T$ is the basis of demarcation between the transient part and the steady state part of the time response of a control system. When the actual output reaches within 2% of the desired value it is said that steady state has reached. The time $4T$ is termed as settling time (t_s) of a control system.

The measure of time from the initiation of the input signal upto the settling time of a control system is the transient period and beyond settling time steady state appears.

Effect of time constant on the speed of the time response of a control system: A decrease in the value of time constant of a control system indicates that the steady state reaches earlier. Therefore, lower the time constant faster is the time response of a control system as illustrated in Fig. 6.3.3.

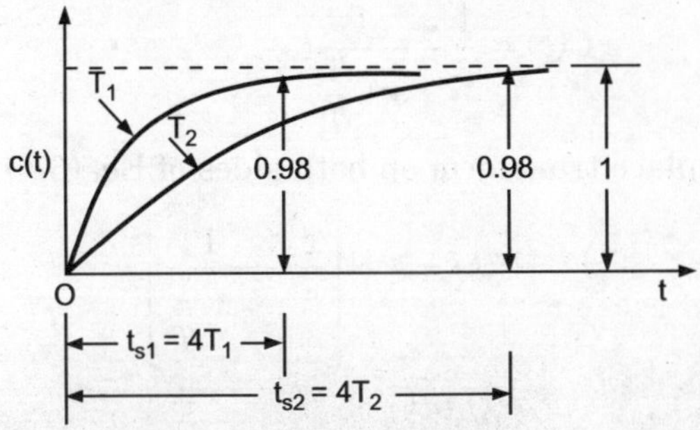

Fig. 6.3.3. Comparison of time response of two control systems having time constants T_1 and T_2, $T_1 < T_2$.

6.3.3. Time response of a first order control system subjected to unit ramp input function

The output for the system is expressed as

$$C(s) = R(s)\frac{1}{sT+1} \qquad \text{...(6.7)}$$

As the input is a unit ramp function

$$r(t) = t \text{ and } R(s) = \frac{1}{s^2}$$

Therefore, substituting in Eq. (6.7)

$$C(s) = \frac{1}{s^2} \cdot \frac{1}{sT+1} \qquad \text{...(6.8)}$$

Breaking R.H.S. of Eq. (6.8) into partial fractions

$$C(s) = \frac{1-sT}{s^2} + \frac{T^2}{sT+1}$$

$$C(s) = \frac{1}{s^2} - T\frac{1}{s} + T\frac{1}{s+\frac{1}{T}} \qquad \text{...(6.9)}$$

Taking inverse Laplace transform on both sides of Eq. (6.9)

$$\mathcal{L}^{-1}C(s) = \mathcal{L}^{-1}\left(\frac{1}{s^2} - T\frac{1}{s} + T\frac{1}{s+\frac{1}{T}}\right)$$

$$\therefore \qquad c(t) = (t - T + Te^{-t/T}) \qquad \text{...(6.10)}$$

The error is given as $e(t) = r(t) - c(t) = t - (t - T + Te^{-t/T})$

$$= (T - Te^{-t/T}) \qquad \text{...(6.11)}$$

The steady state error is $e_{ss} = \lim\limits_{t \to \infty} (T - Te^{-t/T}) = T$

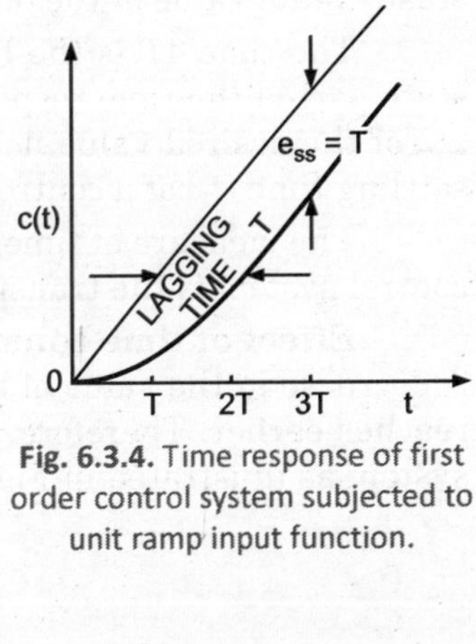

Fig. 6.3.4. Time response of first order control system subjected to unit ramp input function.

The time response in relation to Eq. (6.10) is shown in Fig. 6.3.4.

The time response shown in Fig. 6.3.4 indicates that during steady state the output velocity matches with the input velocity but lags behind the input by time T and a positional error of T unit exists in the system. It is also observed that lower the time constant lesser is the positional error and also lesser time lag.

6.3.4. Time response of a first order control system subjected to unit impulse input function

The output for the system is expressed as

$$C(s) = R(s)\,\frac{1}{sT+1} \qquad \text{...(6.12)}$$

As the input to the system is a unit impulse function, its Laplace transform is 1, *i.e.* $R(s) = 1$, therefore, substituting in Eq. (6.12)

$$C(s) = 1\frac{1}{sT+1} \qquad \text{...(6.12 } a\text{)}$$

Taking inverse Laplace transform on both sides of Eq. (6.12 *a*)

$$\mathcal{L}^{-1}C(s) = \mathcal{L}^{-1}\frac{1}{sT+1}$$

or

$$\mathcal{L}^{-1}C(s) = \mathcal{L}^{-1}\frac{1}{T}\cdot\frac{1}{s+\frac{1}{T}}$$

$$\therefore \qquad c(t) = \frac{1}{T}e^{-t/T} \qquad \text{...(6.13)}$$

The time response in relation to Eq. (6.13) is shown in Fig. 6.3.5.

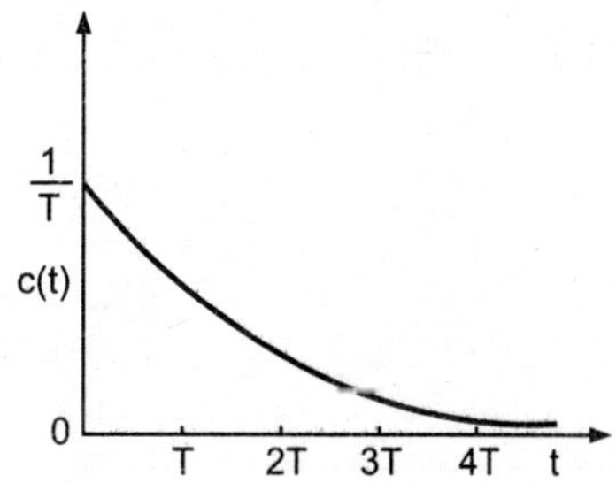

Fig. 6.3.5. Time response of a first order control system subjected to unit impulse input function.

Laplace transform of unit input functions and corresponding time response of a first order control system subjected to these input functions is given in table below :

Input function	*Time response expression*	*Remarks*
Unit ramp $\frac{1}{s^2}$	$c(t) = t - T + Te^{-t/T}$	Differentiate ↓
Unit step $\frac{1}{s}$	$c(t) = 1 - e^{-t/T}$	
Unit impulse 1	$c(t) = \frac{1}{T}e^{-t/T}$	Integrate ↑

It is observed that step function is first derivative of a ramp function and impulse function is first derivative of a step function. From the derived time response expression it is concluded that the output time response also follows the same sequence as that of input functions.

6.4 TIME RESPONSE OF A SECOND ORDER CONTROL SYSTEM

A second order control system is one wherein the highest power of s in the denominator of its transfer function equals 2.

A general expression for the transfer function of a second order control system is given by

$$\frac{C(s)}{R(s)} = \frac{\omega_n^2}{s^2 + 2\zeta\omega_n s + \omega_n^2} \quad ...(6.14)$$

The parameters ζ and ω_n will be explained later. The block diagram representation of the transfer function given by Eq. (6.14) is shown in Fig. 6.4.1.

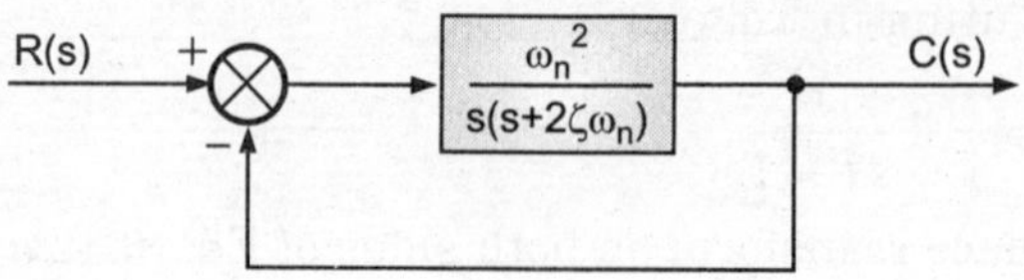

Fig. 6.4.1. Block diagram of a second order control system.

6.4.1. Time response of a second order control system subjected to unit step input function

The output for the system is given by

$$C(s) = R(s)\frac{\omega_n^2}{s^2 + 2\zeta\omega_n s + \omega_n^2} \quad ...(6.15)$$

As the input is a unit step function

$$r(t) = 1 \quad \text{and} \quad R(s) = \frac{1}{s}$$

Therefore, substituting in Eq. (6.15)

$$C(s) = \frac{1}{s}\cdot\frac{\omega_n^2}{s^2 + 2\zeta\omega_n s + \omega_n^2} \quad ...(6.16)$$

In Eq. (6.16) rewriting the term $(s^2 + 2\zeta\omega_n s + \omega_n^2)$ as $[(s + \zeta\omega_n)^2 + \omega_n^2(1 - \zeta^2)]$ and breaking R.H.S. into partial fractions

$$C(s) = \frac{1}{s} - \frac{s + 2\zeta\omega_n}{[(s + \zeta\omega_n)^2 + \omega_n^2(1 - \zeta^2)]} \quad ...(6.17)$$

put $\quad \omega_d = \omega_n\sqrt{1-\zeta^2}$

$$C(s) = \frac{1}{s} - \frac{s + 2\zeta\omega_n}{(s + \zeta\omega_n)^2 + \omega_d^2}. \quad ...(6.18)$$

Eq. (6.18) can be rewritten as follows

$$C(s) = \frac{1}{s} - \frac{s + \zeta\omega_n}{(s + \zeta\omega_n)^2 + \omega_d^2} - \frac{\zeta\omega_n}{\omega_d}\cdot\frac{\omega_d}{(s + \zeta\omega_n)^2 + \omega_d^2} \quad ...(6.18\ a)$$

Taking inverse Laplace transform on both sides of Eq. (6.18 *a*)

$$\mathcal{L}^{-1}C(s) = \mathcal{L}^{-1}\left[\frac{1}{s} - \frac{s+\zeta\omega_n}{(s+\zeta\omega_n)^2+\omega_d^2} - \frac{\zeta\omega_n}{\omega_d}\cdot\frac{\omega_d}{(s+\zeta\omega_n)^2+\omega_d^2}\right] \quad ...(6.19)$$

$$\therefore \quad c(t) = 1 - e^{-\zeta\omega_n t}\,.\cos\omega_d t - \frac{\zeta\omega_n}{\omega_d}\,.\,e^{-\zeta\omega_n t}\,.\sin\omega_d t \quad ...(6.20)$$

Since $\omega_d = \omega_n\sqrt{1-\zeta^2}$ Eq. (6.20) is rewritten as below

$$c(t) = 1 - \frac{e^{-\zeta\omega_n t}}{\sqrt{1-\zeta^2}}\,[(\sqrt{1-\zeta^2})\cos\omega_d t + \zeta\sin\omega_d t] \quad ...(6.21)$$

put $\quad \sin\phi = \sqrt{1-\zeta^2}, \quad \therefore \quad \cos\phi = \zeta$

$$\therefore \quad c(t) = 1 - \frac{e^{-\zeta\omega_n t}}{\sqrt{1-\zeta^2}}(\sin\phi\cos\omega_d t + \cos\phi\sin\omega_d t)$$

or

$$c(t) = 1 - \frac{e^{-\zeta\omega_n t}}{\sqrt{1-\zeta^2}}\sin(\omega_d t + \phi) \quad ...(6.22)$$

Since $\omega_d = \omega_n\sqrt{1-\zeta^2}$ and $\phi = \tan^{-1}\left(\frac{\sqrt{1-\zeta^2}}{\zeta}\right)$

Eq. (6.22) is rewritten as below

$$c(t) = 1 - \frac{e^{-\zeta\omega_n t}}{\sqrt{1-\zeta^2}}\sin\left[(\omega_n\sqrt{1-\zeta^2 t}) + \tan^{-1}\left(\frac{\sqrt{1-\zeta^2}}{\zeta}\right)\right] \quad ...(6.23)$$

The error is given as $e(t) = r(t) - c(t)$

and $\quad r(t) = 1$

$$\therefore \quad e(t) = 1 - \left[1 - \frac{e^{-\zeta\omega_n t}}{\sqrt{1-\zeta^2}}\sin\left(\omega_n\sqrt{1-\zeta^2}\,t + \tan^{-1}\frac{\sqrt{1-\zeta^2}}{\zeta}\right)\right]$$

or

$$e(t) = \frac{e^{-\zeta\omega_n t}}{\sqrt{1-\zeta^2}}\,.\sin\left[\omega_n\sqrt{1-\zeta^2}\,t + \tan^{-1}\left(\frac{\sqrt{1-\zeta^2}}{\zeta}\right)\right] \quad ...(6.24)$$

The steady state error $e_{ss} = \lim\limits_{t\to\infty}\frac{e^{-\zeta\omega_n t}}{\sqrt{1-\zeta^2}}\,.\sin\left[\omega_n\sqrt{1-\zeta^2}\,t + \tan^{-1}\left(\frac{\sqrt{1-\zeta^2}}{\zeta}\right)\right]$

The time response expression given by Eq. (6.23) indicates that for values of $\zeta < 1$ the response presents exponentially decaying oscillations having a frequency $\omega_n\sqrt{(1-\zeta^2)}$ and the time constant of exponential decay is $1/\zeta\,\omega_n$.

The term ω_n is called natural frequency of oscillations. The term $\omega_d = \omega_n\sqrt{1-\zeta^2}$ is called damped frequency of oscillations. The term ζ affects the damping and called damping ratio and the term $\zeta\omega_n$ is called damping factor or actual damping or damping coefficient.

The time response of a second order control system given by Eq. (6.23) is influenced by its damping ratio (ζ). The cases for the values of damping ratio as (*a*) $\zeta < 1$ (*b*) $\zeta = 0$ (*c*) $\zeta = 1$ (*d*) $\zeta > 1$ are considered below :

(*a*) The time response and the error in relation to Eqs. (6.23) and (6.24) for $\zeta < 1$ are plotted in Fig. 6.4.2 (*a*) and (*b*).

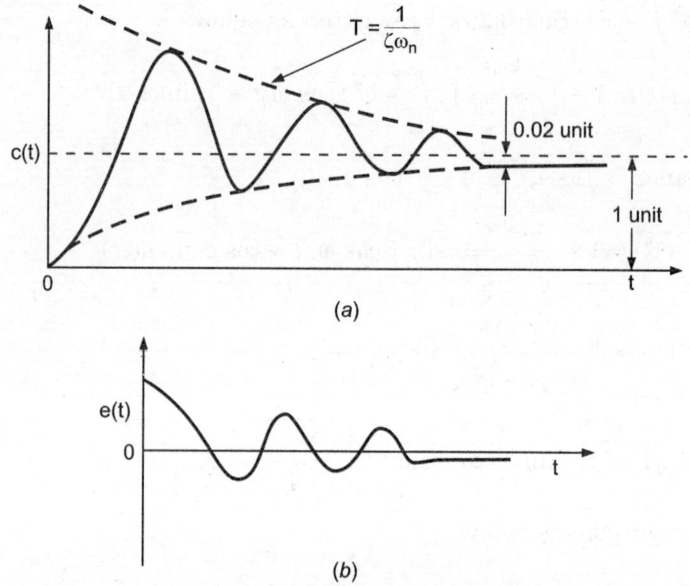

Fig. 6.4.2. Time response and error of a second order control system ($\zeta < 1$, under damped case) subjected to unit step input function.

As stated above, if $\zeta < 1$ the time response presents damped oscillations and such a response is called underdamped response.

The response settles within 2% of the desired value (1 unit) after damping out the oscillations in a time $4T$, where $T = 1/\zeta\omega_n$.

(*b*) An expression for the time response of a second order system having $\zeta = 0$ when subjected to unit step input function is derived hereunder :

Substituting $\zeta = 0$ in Eq. (6.23) following equation is obtained :

$$c(t) = 1 - \frac{e^{-o\omega_n t}}{\sqrt{1-o^2}} \sin\left[\omega_n \sqrt{1-o^2}\, t + \tan^{-1}\left(\frac{\sqrt{1-o^2}}{o}\right)\right]$$

or $$c(t) = 1 - \sin(\omega_n t + \tan^{-1}\infty)$$

or $$c(t) = 1 - \sin\left(\omega_n t + \frac{\pi}{2}\right)$$

or $$c(t) = (1 - \cos\omega_n t) \qquad ...(6.25)$$

The time response in relation to Eq. (6.25) is plotted in Fig. 6.4.3 which indicates sustained (undamped) oscillations.

(*c*) An expression for the time response of a second order control system having $\zeta = 1$ when subjected to unit step input function is derived hereunder :

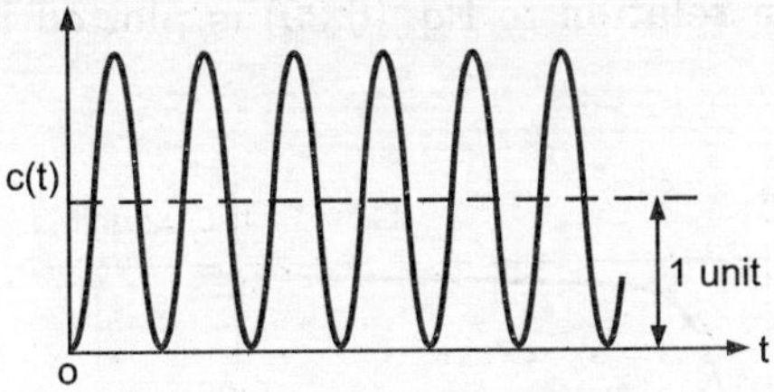

Fig. 6.4.3. Time response of a second order control system ($\zeta = 0$, undamped case) subjected to unit step input functions.

Taking limits of Eq. (6.23) as the value of ζ approaches 1, the equation derived below is obtained

$$c(t) = \lim_{\zeta \to 1}\left[1 - \frac{e^{-\zeta\omega_n t}}{\sqrt{1-\zeta^2}} \sin\left(\omega_n\sqrt{1-\zeta^2}\, t + \tan^{-1}\frac{\sqrt{1-\zeta^2}}{\zeta}\right)\right]$$

In the above equation put $\tan^{-1}\left(\frac{\sqrt{1-\zeta^2}}{\zeta}\right) = \phi$

$$\therefore \qquad c(t) = \lim_{\zeta \to 1}\left[1 - \frac{e^{-\zeta\omega_n t}}{\sqrt{1-\zeta^2}} \sin(\omega_n\sqrt{1-\zeta^2}\, t + \phi)\right]$$

or
$$c(t) = \lim_{\zeta \to 1}\left\{1 - \frac{e^{-\zeta\omega_n t}}{\sqrt{1-\zeta^2}} [\sin(\omega_n\sqrt{1-\zeta^2}\, t)\cos\phi + \cos(\omega_n\sqrt{1-\zeta^2}\, t)\sin\phi]\right\}$$

Since, $\tan\phi = \frac{\sqrt{1-\zeta^2}}{\zeta}$, $\cos\phi = \zeta$ and $\sin\phi = \sqrt{1-\zeta^2}$ hence, substituting for $\cos\phi$ and $\sin\phi$ in the above equation

$$c(t) = \lim_{\zeta \to 1}\left\{1 - \frac{e^{-\zeta\omega_n t}}{\sqrt{1-\zeta^2}} [\sin(\omega_n\sqrt{1-\zeta^2}\, t) . \zeta + \cos(\omega_n\sqrt{1-\zeta^2}\, t) . \sqrt{1-\zeta^2}\right\} \qquad ...(6.26)$$

Now $\lim_{\zeta \to 1} \sin(\omega_n\sqrt{1-\zeta^2}\, t) . \zeta \to \omega_n\sqrt{1-\zeta^2}\, t$ ...(6.27)

and $\lim_{\zeta \to 1} \cos(\omega_n\sqrt{1-\zeta^2}\, t) \to 1$ ...(6.28)

Substituting Eqs. (6.27) and (6.28) in Eq. (6.26),

$$c(t) = \lim_{\zeta \to 1}\left[1 - \frac{e^{-\zeta\omega_n t}}{\sqrt{1-\zeta^2}} (\omega_n\sqrt{1-\zeta^2}\, t + 1 . \sqrt{1-\zeta^2})\right]$$

$$= \lim_{\zeta \to 1}\left[1 - \frac{e^{-\zeta\omega_n t}}{\sqrt{1-\zeta^2}} \sqrt{1-\zeta^2}\,(\omega_n t + 1)\right]$$

or
$$c(t) = [1 - e^{-\omega_n t}(1 + \omega_n t)] \qquad ...(6.29)$$

The time response in relation to Eq. (6.29) is plotted in Fig. 6.4.4. The response is called critically damped.

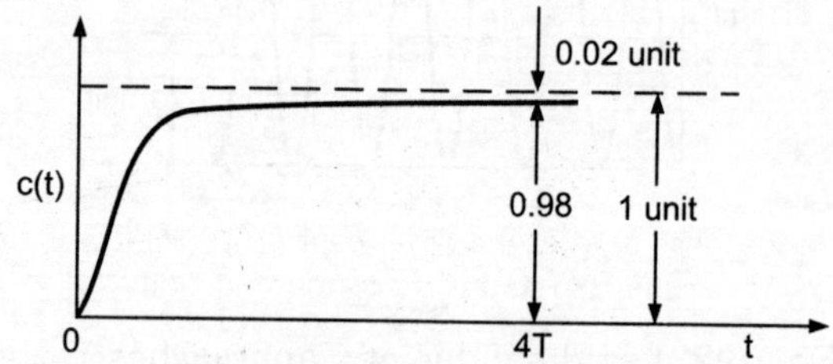

Fig. 6.4.4. Time response of a second order control system ($\zeta = 1$), critically damped case) subjected to unit step function.

(*d*) An expression for the time response of a second order control system having $\zeta > 1$ when subjected to unit step input function is derived hereunder:

The output for the system is given by

$$C(s) = R(s)\frac{\omega_n^2}{s^2 + 2\zeta\omega_n s + \omega_n^2} \quad ...(6.30)$$

As the input is a unit step function $r(t) = 1$ and $R(s) = 1/s$, therefore, substituting in Eq. (6.30)

$$C(s) = \frac{1}{s}\cdot\frac{\omega_n^2}{s^2 + 2\zeta\omega_n s + \omega_n^2} \quad ...(6.31)$$

Eq. (6.31) can also be written as

$$C(s) = \frac{1}{s}\cdot\frac{\omega_n^2}{(s+\zeta\omega_n)^2 - \omega_n^2(\zeta^2 - 1)}$$

or

$$C(s) = \frac{1}{s}\cdot\frac{\omega_n^2}{[s + (\zeta + \sqrt{\zeta^2 - 1})\,\omega_n]\,[s + (\zeta - \sqrt{\zeta^2 - 1})\,\omega_n]} \quad ...(6.32)$$

Expanding R.H.S. of Eq. (6.32) into partial fractions,

$$C(s) = \frac{1}{s} - \frac{1}{2\sqrt{\zeta^2 - 1}\,(\zeta - \sqrt{\zeta^2 - 1})\,[s + (\zeta - \sqrt{\zeta^2 - 1})\omega_n]}$$

$$+ \frac{1}{2\sqrt{\zeta^2 - 1}\,(\zeta + \sqrt{\zeta^2 - 1})\,[s + (\zeta + \sqrt{\zeta^2 - 1})\omega_n]} \quad ...(6.33)$$

Taking inverse Laplace transform on both sides of equation (6.33)

$$c(t) = 1 - \frac{e^{-(\zeta - \sqrt{\zeta^2 - 1})\omega_n t}}{2\sqrt{\zeta^2 - 1}\,(\zeta - \sqrt{\zeta^2 - 1})} + \frac{e^{-(\zeta + \sqrt{\zeta^2 - 1})\omega_n t}}{2\sqrt{\zeta^2 - 1}\,(\zeta + \sqrt{\zeta^2 - 1})} \quad ...(6.34)$$

The time response in relation to Eq. (6.34) is plotted in Fig. 6.4.5. The response is called overdamped.

Eq. (6.34) indicates that the time response has two time constants one slower $[T_1 = 1/(\zeta - \sqrt{\zeta^2 - 1})\,\omega_n]$ and the other faster $[T_2 = 1/(\zeta + \sqrt{\zeta^2 - 1})\,\omega_n]$. For values of ζ comparatively greater than 1, the effect of the faster time constant on the time response can be neglected and the time response expression takes the following form :

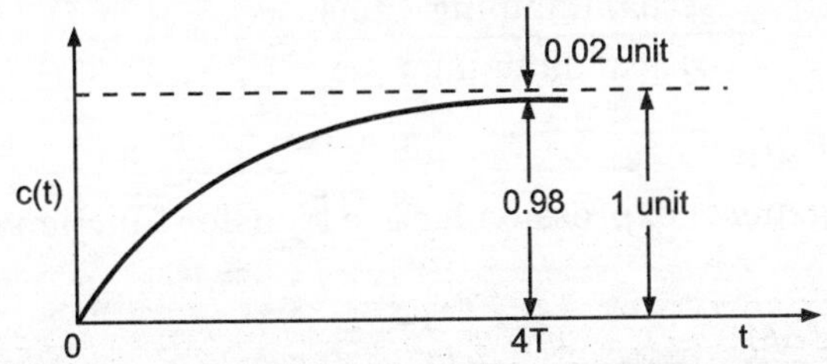

Fig. 6.4.5. Time response of a second order control system ($\zeta > 1$, overdamped case) subjected to unit step input function.

$$c(t) = 1 - \frac{e^{-(\zeta - \sqrt{\zeta^2 - 1})\omega_n t}}{2\sqrt{\zeta^2 - 1}\,(\zeta - \sqrt{\zeta^2 - 1})} \qquad ...(6.35)$$

The time constant of the response is

$$T = \frac{1}{(\zeta - \sqrt{\zeta^2 - 1})\omega_n}.$$

In relation to Eqs. (6.23), (6.25) and Eq. (6.34), the time response of a second order control system subjected to unit step input function is plotted in Fig. 6.4.6.

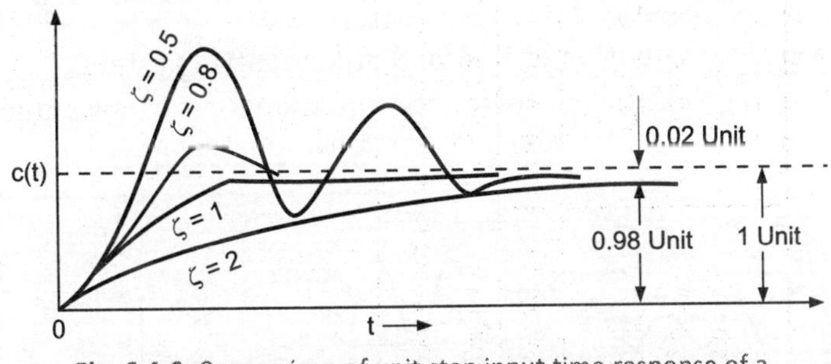

Fig. 6.4.6. Comparison of unit step input time response of a second order control system for different values of ζ.

6.4.2. Critical Damping

The time response expression

$$c(t) = 1 - \frac{e^{-\zeta\omega_n t}}{\sqrt{1-\zeta^2}} \sin\left[(\omega_n\sqrt{1-\zeta^2})t + \tan^{-1}\frac{\sqrt{1-\zeta^2}}{\zeta}\right]$$

reveals that the term $\zeta\omega_n$ is responsible for offering the damping in the system, when $\zeta = 0$ oscillations are sustained. For values of $\zeta < 1$ oscillations decay exponentially having a time constant $1/\zeta\omega_n$, whereas for $\zeta > 1$, the response does not exhibit any oscillations and the response is overdamped. Thus, the term $\zeta\omega_n$ represents actual damping offered in the time response of the system. When $\zeta = 1$ the actual damping equals ω_n. This actual damping when $\zeta = 1$ is known as critical damping. At the value of critical damping the oscillations in the time response just disappear.

The ratio of the actual damping to critical damping is known as the damping ratio as given by the following relation:

$$\frac{\text{Actual damping}}{\text{Critical damping}} = \frac{\zeta\omega_n}{\omega_n} = \zeta$$

6.4.3. Characteristic Equation

As mentioned earlier a general expression for the transfer function of a second order control system is given by

$$\frac{C(s)}{R(s)} = \frac{\omega_n^2}{s^2 + 2\zeta\omega_n s + \omega_n^2} \qquad ...(6.35\ a)$$

The denominator of the above expression is equated to zero and following equation is formed

$$s^2 + 2\zeta\omega_n s + \omega_n^2 = 0 \qquad ...(6.36)$$

The above equation is quadratic in s and two roots are

$$s_1, s_2 = -\zeta\omega_n \pm j\omega_n\sqrt{1-\zeta^2}$$

The roots s_1, s_2 are also the poles of transfer function expressed by relation (6.35 a). A study of roots s_1, s_2 gives a prediction about the nature of time response. The real part of the roots represents the damping and the imaginary part represents damped frequency of oscillations. Therefore, equation

$$s^2 + 2\zeta\omega_n s + \omega_n^2 = 0$$

is called the characteristic equation of a second order control system.

The location of roots of the characteristic equation for various values of ζ (keeping ω_n fixed) and the corresponding time response for a second order control system is shown in Fig. 6.4.7.

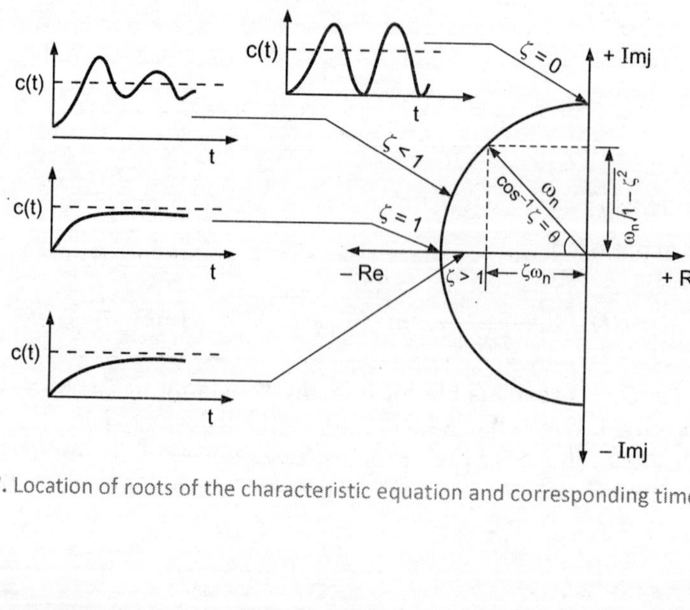

Fig. 6.4.7. Location of roots of the characteristic equation and corresponding time response.

From Fig. 6.4.7 it is inferred that the change over from underdamped to overdamped response takes place at $\zeta = 1$. The value of ζ from the location of roots is calculated as $\zeta = \cos\theta$.

6.4.4. Transient Response Specifications of Second Order Control System

The time response of an underdamped control system exhibits damped oscillations prior to reaching steady state. The specifications pertaining to time response during transient part are shown in Fig. 6.4.8 and explained below.

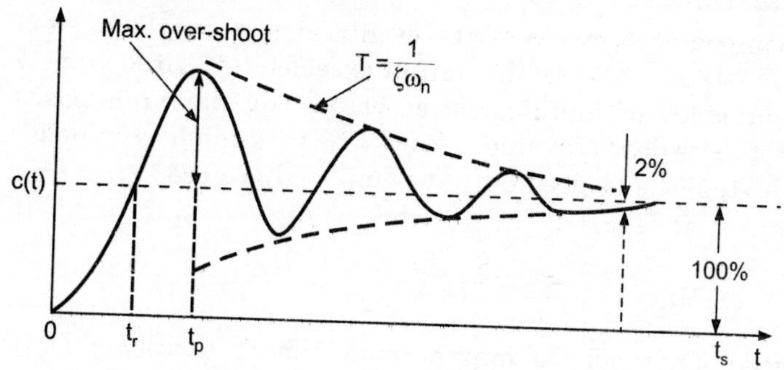

Fig. 6.4.8. Transient response specifications.

(1) The rise time : t_r

The rise time is the time needed for the response to reach from 10% to 90% or 0 to 100% of the desired value of the output at the very first instant. Usually 0-100% basis is used for underdamped systems and 10% to 90% for overdamped system.

The rise time (0 to 100%) for underdamped second order control system is calculated below. As expression for the time response of a second order control system is given by

$$c(t) = 1 - \frac{e^{-\zeta\omega_n t}}{\sqrt{1-\zeta^2}} \sin [(\omega_n \sqrt{1-\zeta^2})t + \phi]$$

At the first instant when time response reaches 100% of the desired value, *i.e.* $c(t) = 1$, the time is t_r, therefore, substituting $c(t) = 1$ is in above equation

$$1 = 1 - \frac{e^{-\zeta\omega_n t_r}}{\sqrt{1-\zeta^2}} \sin [(\omega_n \sqrt{1-\zeta^2})t_r + \phi]$$

or

$$\frac{e^{-\zeta\omega_n t_r}}{\sqrt{1-\zeta^2}} \sin [(\omega_n \sqrt{1-\zeta^2})t_r + \phi] = 0$$

As

$$\frac{e^{-\zeta\omega_n t_r}}{\sqrt{1-\zeta^2}} \text{ is finite}$$

$\therefore$

$$\sin [(\omega_n \sqrt{1-\zeta^2})t_r + \phi] = 0$$

The acceptable solution of the above equation is

$$[(\omega_n \sqrt{1-\zeta^2})t_r + \phi] = \pi$$

$$\therefore \quad t_r = \frac{\pi - \phi}{\omega_n \sqrt{1-\zeta^2}} \qquad \text{...(6.37)}$$

where $$\phi = \tan^{-1}\left(\frac{\sqrt{1-\zeta^2}}{\zeta}\right)$$

(2) Maximum overshoot : M_p and peak time : t_p

As underdamped system presents oscillations during transient part of its time response. At some instances of time the output exceeds its desired value while at some other instances the output is lower than its desired value. The maximum positive deviation of the output with respect to its desired value is known as maximum overshoot and denoted as M_p. If the input is unit step, the desired output is unity, therefore,

$$M_p = c(t)_{\max} - 1$$

$$\% M_p = \frac{c(t)_{\max} - 1}{1} \times 100$$

The time needed to reach the maximum overshoot is called peak time and denoted as t_p. It is observed that M_p occurs when the slope of the time response curve after the initiation of the input signal is zero. Therefore, if the time response expression is differentiated and equated to zero, the peak time t_p can be determined and by substituting the value of t_p in the time response expression for $c(t)_{\max}$ and M_p can be determined.

The expression for $c(t)$ is

$$c(t) = 1 - \frac{e^{-\zeta\omega_n t}}{\sqrt{1-\zeta^2}} \sin[(\omega_n\sqrt{1-\zeta^2})t + \phi]$$

$$\therefore \quad \frac{dc(t)}{dt} = -\frac{e^{-\zeta\omega_n t}}{\sqrt{1-\zeta^2}} \cdot \omega_n\sqrt{1-\zeta^2} \cdot \cos[(\omega_n\sqrt{1-\zeta^2})t + \phi] - \frac{(-\zeta\omega_n)e^{-\zeta\omega_n t}}{\sqrt{1-\zeta^2}} \sin[(\omega_n\sqrt{1-\zeta^2})t + \phi]$$

Put $$\frac{dc(t)}{dt} = 0$$

$$\therefore \quad \frac{e^{-\zeta\omega_n t}}{\sqrt{1-\zeta^2}}\{-\omega_n\sqrt{1-\zeta^2} \cdot \cos[(\omega_n\sqrt{1-\zeta^2})t + \phi] + \zeta\omega_n \sin[(\omega_n\sqrt{1-\zeta^2})t + \phi]\} = 0$$

In the above equation, since $\dfrac{e^{-\zeta\omega_n t}}{\sqrt{1-\zeta^2}}$ is finite

$$\therefore \quad \omega_n\sqrt{1-\zeta^2} \cdot \cos[(\omega_n\sqrt{1-\zeta^2})t + \phi] = \zeta\omega_n \sin[(\omega_n\sqrt{1-\zeta^2})t + \phi]$$

$$\therefore \quad \tan[(\omega_n\sqrt{1-\zeta^2})t + \phi] = \frac{\sqrt{1-\zeta^2}}{\zeta}$$

since, $\phi = \tan^{-1} \dfrac{\sqrt{1-\zeta^2}}{\zeta}$, substituting in above equation

$$\tan\left[(\omega_n\sqrt{1-\zeta^2})t + \tan^{-1}\frac{\sqrt{1-\zeta^2}}{\zeta}\right] = \frac{\sqrt{1-\zeta^2}}{\zeta} \quad ...(6.38)$$

The general solution of Eq. (6.38) is

$$(\omega_n\sqrt{1-\zeta^2})\,t = n\pi \quad ...(6.39)$$

where $n = 0, 1, 2...$

The instant of occurring maximum overshoot is identified by putting $n = 1$ in Eq. (6.39), hence,

$$t_p = \frac{\pi}{\omega_n\sqrt{1-\zeta^2}} \quad ...(6.40)$$

$c(t)_{\max}$ is determined by putting $t = t_p$ in the time response expression, therefore,

$$c(t)_{max} = 1 - \frac{e^{-\zeta\omega_n t_p}}{\sqrt{1-\zeta^2}} \,.\, \sin\,[(\omega_n\sqrt{1-\zeta^2})t_p + \phi]$$

$$= 1 - \frac{e^{-\zeta\omega_n\frac{\pi}{\omega_n\sqrt{1-\zeta^2}}}}{\sqrt{1-\zeta^2}} \,.\, \sin\left[(\omega_n\sqrt{1-\zeta^2})\frac{\pi}{\omega_n\sqrt{1-\zeta^2}} + \phi\right]$$

$$= 1 - \frac{e^{\frac{-\zeta\pi}{\sqrt{1-\zeta^2}}}}{\sqrt{1-\zeta^2}} \,.\, \sin\,(\pi + \phi) = 1 - \frac{e^{\frac{-\zeta\pi}{\sqrt{1-\zeta^2}}}}{\sqrt{1-\zeta^2}} \,.\, (-\sin\phi)$$

$\because \quad \phi = \tan^{-1}\dfrac{\sqrt{1-\zeta^2}}{\zeta}$, therefore, $\sin\phi = \sqrt{1-\zeta^2}$ hence,

$$c(t)_{\max} = 1 + \frac{e^{\frac{-\zeta\pi}{\sqrt{1-\zeta^2}}}}{\sqrt{1-\zeta^2}} \,.\, \sqrt{1-\zeta^2}$$

or

$$c(t)_{\max} = 1 + e^{-\frac{\zeta\pi}{\sqrt{1-\zeta^2}}} \quad ...(6.41)$$

$$M_p = c(t)_{\max} - 1$$

$$\therefore \quad M_p = \left(1 + e^{-\frac{\zeta\pi}{\sqrt{1-\zeta^2}}}\right) - 1$$

or

$$M_p = e^{\frac{-\zeta\pi}{\sqrt{1-\zeta^2}}} \quad ...(6.42)$$

$$\%M_p = e^{-\frac{\zeta\pi}{\sqrt{1-\zeta^2}}} \times 100 \quad ...(6.43)$$

Note: The instant at which first undershoot occurs can be determined by putting $n = 2$ in Eq. 6.39.

A graph relating M_p and ζ is plotted in Fig. 6.4.9.

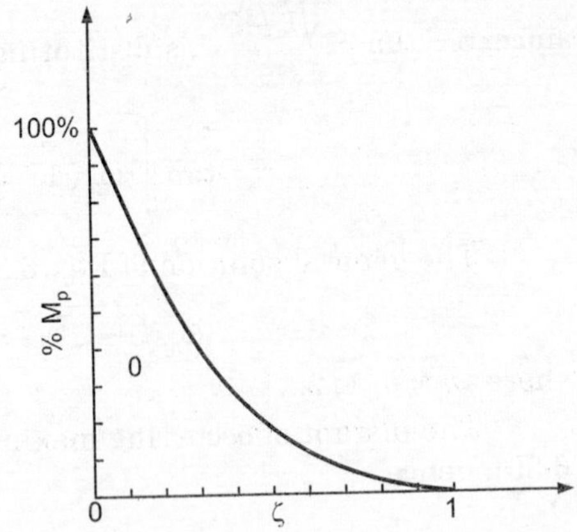

Fig. 6.4.9. Graph between M_p and ζ.

(3) **The settling time t_s.**

For an underdamped system the magnitude of the oscillations present in the output time response decay exponentially with a time constant $1/\zeta\omega_n$. The time needed to settle down aforesaid oscillations within 2% of desired value of the output is known as settling time and denoted as t_s. The settling time for a second order control system is approximately four times the time constant of the system, hence,

$$t_s = 4\frac{1}{\zeta\omega_n} \qquad \text{...(6.44)}$$

Note: On 5% basis the settling time for a second order control system is approximately three times the time constant, *i.e.*

$$t_s = 3\frac{1}{\zeta\omega_n}$$

6.4.5. Time response of a second order control system subjected to unit ramp input function

As the input is a unit ramp function

$$r(t) = t \text{ and } R(s) = \frac{1}{s^2}$$

Therefore, the output of a second order system is given by

$$C(s) = \frac{1}{s^2}\cdot\frac{\omega_n^2}{s^2 + 2\zeta\omega_n s + \omega_n^2} \qquad \text{...(6.45)}$$

In Eq. (6.45), rewriting the term $(s^2 + 2\zeta\omega_n s + \omega_n^2)$ as $(s + \zeta\omega_n)^2 + \omega_n^2(1 - \zeta^2)$ and breaking R.H.S. into partial fractions

$$C(s) = \frac{1}{s^2} - \frac{s + 2\zeta\omega_n}{s[(s+\zeta\omega_n)^2 + \omega_n^2(1-\zeta^2)]}$$

put $\omega_d = \omega_n\sqrt{1-\zeta^2}$

$$\therefore \quad C(s) = \frac{1}{s^2} - \frac{s + 2\zeta\omega_n}{s[(s+\zeta\omega_n)^2 + \omega_d^2]}$$

or

$$C(s) = \frac{1}{s^2} - \frac{2\zeta\omega_n}{s[(s+\zeta\omega_n)^2 + \omega_d^2]} - \frac{1}{[(s+\zeta\omega_n)^2 + \omega_d^2]}$$

or $$C(s) = \frac{1}{s^2} - \frac{2\zeta\omega_n}{\omega_n^2}\,\frac{\omega_n^2}{s[(s+\zeta\omega_n)^2+\omega_d^2]} - \frac{1}{[(s+\zeta\omega_n)^2+\omega_d^2]}$$

or $$C(s) = \frac{1}{s^2} - \frac{2\zeta}{\omega_n}\left[\frac{1}{s} - \frac{s+2\zeta\omega_n}{(s+\zeta\omega_n)^2+\omega_d^2}\right] - \frac{1}{(s+\zeta\omega_n)^2+\omega_d^2}$$

or $$C(s) = \frac{1}{s^2} - \frac{2\zeta}{\omega_n s} - \frac{2\zeta}{\omega_n}\cdot\frac{(s+\zeta\omega_n)}{(s+\zeta\omega_n)^2+\omega_d^2} + \frac{(2\zeta^2-1)}{[(s+\zeta\omega_n)^2+\omega_d^2]}$$

or $$C(s) = \frac{1}{s^2} - \frac{2\zeta}{\omega_n}\cdot\frac{1}{s} + \frac{2\zeta}{\omega_n}\cdot\frac{(s+\zeta\omega_n)}{[(s+\zeta\omega_n)^2+\omega_d^2]} + \frac{(2\zeta^2-1)}{\omega_d}\,\frac{\omega_d}{[(s+\zeta\omega_n)^2+\omega_d^2]} \qquad ...(6.46)$$

Taking inverse Laplace transform on both sides of Eq. (6.46)

$$c(t) = t - \frac{2\zeta}{\omega_n} + \frac{2\zeta}{\omega_n}e^{-\zeta\omega_n t}\,.\cos\omega_d t + \frac{(2\zeta^2-1)}{\omega_d}e^{-\zeta\omega_n t}\,.\sin\omega_d t$$

or $$c(t) = t - \frac{2\zeta}{\omega_n} + \frac{2\zeta}{\omega_n}e^{-\zeta\omega_n t}\,[\cos\sqrt{1-\zeta^2})t] + \frac{(2\zeta^2-1)}{\omega_n\sqrt{1-\zeta^2}}e^{-\zeta\omega_n t}\,.\sin[(\omega_n\sqrt{1-\zeta^2})t]$$

or $$c(t) = t - \frac{2\zeta}{\omega_n} + \frac{e^{-\zeta\omega_n t}}{\omega_n\sqrt{1-\zeta^2}}[2\zeta\sqrt{1-\zeta^2}\cos\omega_n\sqrt{1-\zeta^2}\,t$$
$$+ (2\zeta^2-1)\sin(\omega_n\sqrt{1-\zeta^2}\,t)] \qquad ...(6.47)$$

put $2\zeta\sqrt{1-\zeta^2} = \sin\phi$,

∴ $(2\zeta^2-1) = \cos\phi$, hence

$$c(t) = t - \frac{2\zeta}{\omega_n} + \frac{e^{-\zeta\omega_n t}}{\omega_n\sqrt{1-\zeta^2}}[\sin\phi\cos(\omega_n\sqrt{1-\zeta^2})\,t + \cos\phi\sin(\omega_n\sqrt{1-\zeta^2})\,t]$$

or $$c(t) = t - \frac{2\zeta}{\omega_n} + \frac{e^{-\zeta\omega_n t}}{\omega_n\sqrt{1-\zeta^2}}\sin[(\omega_n\sqrt{1-\zeta^2})t+\phi] \qquad ...(6.48)$$

where $\phi = \tan^{-1}\left(\frac{2\zeta\sqrt{1-\zeta^2}}{2\zeta^2-1}\right)$

It can also be shown that $\phi = 2\tan^{-1}\left(\frac{\sqrt{1-\zeta^2}}{\zeta}\right)$

The error signal is given as

$e(t) = r(t) - c(t)$ and $r(t) = t$

∴ $$e(t) = t - \left\{t - \frac{2\zeta}{\omega_n} + \frac{e^{-\zeta\omega_n t}}{\omega_n\sqrt{1-\zeta^2}}\sin[(\omega_n\sqrt{1-\zeta^2})t+\phi]\right\}$$

$$= \frac{2\zeta}{\omega_n} - \frac{e^{-\zeta\omega_n t}}{\omega_n\sqrt{1-\zeta^2}} \sin[(\omega_n\sqrt{1-\zeta^2})t + \phi] \quad ...(6.49)$$

As $t \to \infty$ the second term in Eq. (6.49) approaches zero, therefore, the steady state error is given by

$$e_{ss} = \frac{2\zeta}{\omega_n} \quad ...(6.50)$$

The time response in relation to Eq. (6.48) is plotted in Fig. 6.4.10.

As in the case of a first order system, second order control system also exhibits a positional error when subjected to unit ramp input function. The steady state error being $2\,\zeta/\omega_n$ unit and the output response lags by a time of $2\,\zeta/\omega_n$.

It is observed that the steady state error can be decreased either by decreasing ζ or increasing ω_n. After the disappearance of transient, the magnitude of the output is same as that of the input.

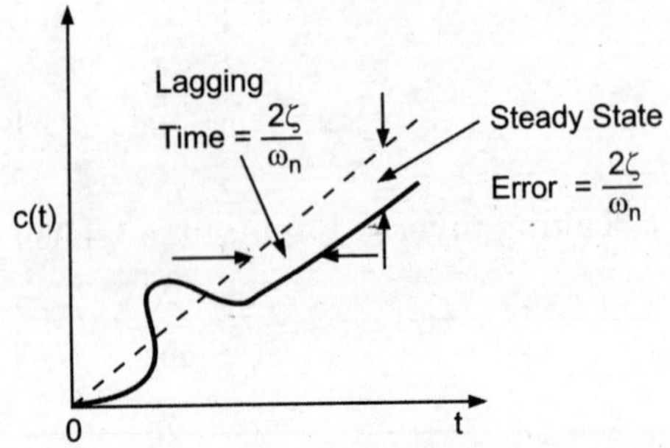

Fig. 6.4.10. The response of a second order control system subjected to unit ramp input function.

6.4.6. Time response of a second order control system subjected to impulse input function

As the input is a unit impulse function, therefore, $R(s) = 1$ and the output of a second order system is given by

$$C(s) = \frac{\omega_n^2}{s^2 + 2\zeta\omega_n s + \omega_n^2} \quad ...(6.51)$$

In Eq. (6.51) rewriting the term $(s^2 + 2\zeta\omega_n s + \omega_n^2)$ as $[(s + \zeta\omega_n)^2 + \omega_n^2(1 - \zeta^2)]$

$$C(s) = \frac{\omega_n^2}{[(s + \zeta\omega_n)^2 + \omega_n^2(1-\zeta^2)]}$$

or

$$C(s) = \frac{\omega_n^2}{\omega_n\sqrt{1-\zeta^2}} \cdot \frac{\omega_n\sqrt{1-\zeta^2}}{[(s+\zeta\omega_n)^2 + (\omega_n\sqrt{1-\zeta^2})^2]} \quad ...(6.52)$$

Taking inverse Laplace transform on both sides of Eq. (6.52)

$$c(t) = \frac{\omega_n}{\sqrt{1-\zeta^2}} \cdot e^{-\zeta\omega_n t} \sin[(\omega_n\sqrt{1-\zeta^2})\,t] \quad ...(6.53)$$

From Eq. (6.53) the time response for the following three values of ζ is determined and plotted in Fig. 6.4.11.

(a) $\zeta < 1$

$$c(t) = \frac{\omega_n}{\sqrt{1-\zeta^2}} . e^{-\zeta\omega_n t} \sin[(\omega_n\sqrt{1-\zeta^2})t] \quad ...(6.54)$$

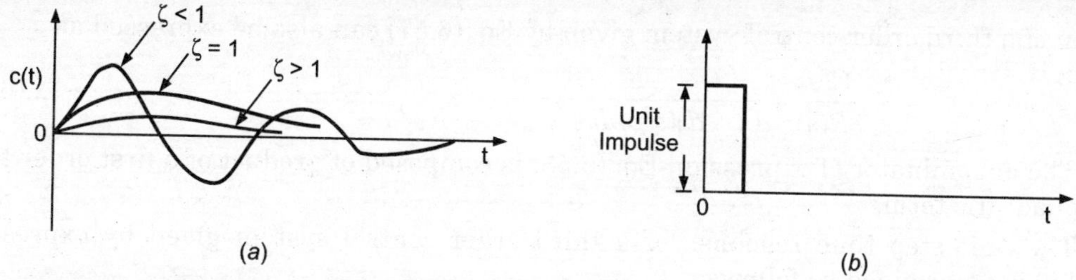

Fig. 6.4.11. Time response of a second control system subjected to unit impulse input function.

The time response given by above equation presents decaying exponential oscillations and the output at times goes negative also.

(*b*) ζ = 1

$$c(t) = \lim_{\zeta \to 1} \frac{\omega_n}{\sqrt{1-\zeta^2}} e^{-\zeta\omega_n t} \sin[(\omega_n\sqrt{1-\zeta^2})\,t]$$

$$= \frac{\omega_n}{\sqrt{1-\zeta^2}} e^{-\zeta\omega_n t} (\omega_n\sqrt{1-\zeta^2})\,t$$

$$\therefore \qquad c(t) = t\,\omega_n^2\, e^{-\omega_n t} \qquad \text{...(6.55)}$$

(*c*) ζ > 1

$$c(t) = \frac{\omega_n}{j\sqrt{\zeta^2-1}} e^{-\zeta\omega_n t} \sin[(j\omega_n\sqrt{\zeta^2-1})\,t]$$

$$= \frac{\omega_n}{j\sqrt{\zeta^2-1}} e^{-\zeta\omega_n t}\, j \sinh[(\omega_n\sqrt{\zeta^2-1})\,t]$$

$$= \frac{\omega_n}{\sqrt{\zeta^2-1}} e^{-\zeta\omega_n t} \left[\frac{e^{+\omega_n\sqrt{(\zeta^2-1)}t} - e^{-(\omega_n\sqrt{(\zeta^2-1)}t}}{2}\right]$$

or

$$c(t) = \frac{\omega_n}{2\sqrt{\zeta^2-1}} [e^{-(\zeta-\sqrt{(\zeta^2-1)})\omega_n t} - e^{-(\zeta+\sqrt{(\zeta^2-1)})\omega_n t}] \qquad (6.56)$$

6.5 TIME RESPONSE OF A THIRD ORDER CONTROL SYSTEM

A third order control system is one wherein the highest power of s in the denominator of its transfer function equals 3. Thus a third order control system is expressed by a transfer function given below

$$\frac{C(s)}{R(s)} = \frac{K}{s^2 + as^2 + bs + c} \qquad \text{...(6.57)}$$

In the above transfer function a, b and c are constants.

The denominator of Eq. (6.57) can be factorised as follows :

$$(s^3 + as^2 + bs + c) = (s + d)\,(s^2 + b_1 s + c_1)$$

The constant d, b_1 and c_1 can be determined in terms of a, b and c. Thus the transfer function of a third order control system given by Eq. (6.57) can also be expressed as

$$\frac{C(s)}{R(s)} = \frac{K}{(s+d)(s^2+b_1 s+c_1)} \qquad ...(6.58)$$

The denominator of expression Eq. (6.58) is composed of product of a first order term and a quadratic term.

The unit step time response of a third order control system given by expression Eq. (6.58) is determined as follows :

As input is $r(t) = 1$, therefore, $R(s) = 1/s$, hence

$$C(s) = \frac{1}{s} \cdot \frac{K}{(s+d)(s^2+b_1 s+c_1)} \qquad ...(6.59)$$

The R.H.S. of equation can be expanded using partial fraction technique and the corresponding time response is obtained by taking inverse Laplace transform. Therefore,

$$C(s) = \frac{K_1}{s} + \frac{K_2}{(s+d)} + \frac{K_3 s + K_4}{(s^2+b_1 s+c_1)}$$

and
$$c(t) = \mathcal{L}^{-1}\left[\frac{K_1}{s} + \frac{K_2}{(s+d)} + \frac{K_2 + s + K_4}{(s^2+b_1 s+c_1)}\right] \qquad ...(6.60)$$

Aforesaid procedure for obtaining the time response is adopted for the illustrative examples solved below :

Illustrative example : (*a*) Obtain an expression for unit step time response of a control system whose transfer function is given by

$$\frac{C(s)}{R(s)} = \frac{1.68}{s^3+12.3s^2+3.74s+1.68}$$

The denominator $f(s) = (s^3 + 12.3s^2 + 3.74s + 1.68)$ equals zero at $s = -12$, therefore, $(s + 12)$ is its first order factor and its quadratic factor is obtained by dividing $f(s)$ by $(s + 12)$, thus the quadratic factor is $(s^2 + 0.3s + 0.14)$, therefore,

$$\frac{C(s)}{R(s)} = \frac{1.68}{(s+12)(s^2+0.3s+0.14)}$$

As the input is $R(s) = 1/s$, therefore,

$$C(s) = \frac{1}{s} \cdot \frac{1.68}{(s+12)(s^2+0.3s+0.14)}$$

Expanding R.H.S. of the above expression into partial fraction

$$C(s) = \frac{K_1}{s} + \frac{K_2}{(s+12)} + \frac{K_3 s + K_4}{(s^2+0.3s+0.14)}$$

The coefficients can be determined as

$$K_1 = 1,\ K_2 = -9.47 \times 10^{-4},\ K_3 = -0.99 \text{ and } K_4 = -0.31$$

$$\therefore \qquad C(s) = \frac{1}{s} - \frac{9.47\times 10^{-4}}{(s+12)} - \frac{(0.99s+0.31)}{(s^2+0.3s+0.14)}$$

The term $(s^2 + 0.3s + 0.14)$ can be expressed as $[(s + 0.15)^2 + (0.34)^2]$ hence,

$$C(s) = \frac{1}{s} - \frac{9.47\times 10^{-4}}{(s+12)} - \frac{(0.99s+0.31)}{[(s+0.15)^2+(0.34)^2]}$$

$$= \frac{1}{s} - \frac{9.47\times10^{-4}}{(s+12)} - 0.99\frac{s+0.313}{[(s+0.15)^2+(0.34)^2]}$$

$$= \frac{1}{s} - \frac{9.47\times10^{-4}}{(s+12)} - 0.99\frac{(s+0.15+0.163)}{[(s+0.15)^2+(0.34)^2]}$$

$$= \frac{1}{s} - \frac{9.47\times10^{-4}}{(s+12)} - 0.99\frac{(s+0.15)}{[(s+0.15)^2+(0.34)^2]} - \frac{0.99\times0.163}{0.34}\cdot\frac{0.34}{[(s+0.15)^2+(0.34)^2]}$$

$$= \frac{1}{s} - \frac{9.47\times10^{-4}}{(s+12)} - 0.99\frac{(s+0.15)}{[(s+0.15)^2+(0.34)^2]} - 0.47\frac{0.34}{[(s+0.15)^2+(0.34)^2]}$$

Taking inverse Laplace transform on both sides of above equation

$$c(t) = [1 - 9.47 \times 10^{-4}\, e^{-12t} - 0.99\, e^{-0.15t}\cos(0.34\, t) - 0.47\, e^{-0.15t}\sin(0.34\, t)] \quad \textbf{Ans.} \quad ...(6.61)$$

Illustrative Example : (*b*) Obtain an expression for a unit step time response of a control system whose transfer function is given by

$$\frac{C(s)}{R(s)} = \frac{0.168}{s^3+1.5s^2+0.5s+0.168}$$

The denominator $f(s) = (s^3 + 1.5s^2 + 0.5s + 0.168)$ equals zero at $s = 1.2$, therefore, $(s + 1.2)$ is its first order factor and its quadratic factor is obtained by dividing $f(s)$ by $(s + 1.2)$, thus the quadratic factor is $(s^2 + 0.3s + 0.14)$, therefore,

$$\frac{C(s)}{R(s)} = \frac{0.168}{(s+1.2)(s^2+0.3s+0.14)}$$

As the input is $R(s) = 1/s$,

$$C(s) = \frac{1}{s}\cdot\frac{0.168}{(s+1.2)(s^2+0.3s+0.14)}$$

Expanding R.H.S. of above expression into partial fractions

$$C(s) = \frac{K_1}{s} + \frac{K_2}{(s+1.2)} + \frac{K_3 s + K_4}{(s^2+0.3s+0.14)}$$

The coefficient can be determined as

$$K_1 = 1, K_2 = -0.137, K_3 = -0.92 \text{ and } K_4 = -0.35$$

$$\therefore \quad C(s) = \frac{1}{s} - \frac{0.137}{(s+1.2)} - \frac{0.92s+0.35}{(s^2+0.3s+0.14)}$$

The term $(s^2 + 0.3s + 0.14)$ can be expressed as $[(s + 0.15)^2 + (0.34)^2]$ hence,

$$C(s) = \frac{1}{s} - \frac{0.137}{(s+1.2)} - \frac{0.92s+0.35}{[(s+0.15)^2+(0.34)^2]}$$

$$= \frac{1}{s} - \frac{0.137}{(s+1.2)} - 0.92\cdot\frac{s+0.38}{[(s+0.15)^2+(0.34)^2]}$$

$$= \frac{1}{s} - \frac{0.137}{(s+1.2)} - 0.92\cdot\frac{(s+0.15+0.23)}{[(s+0.15)^2+(0.34)^2]}$$

$$= \frac{1}{s} - \frac{0.137}{(s+1.2)} - 0.92 \cdot \frac{(s+0.15)}{[(s+0.15)^2+(0.34)^2]} - \frac{0.92 \times 0.23}{0.34} \cdot \frac{0.34}{[(s+0.15)^2+(0.34)^2]}$$

$$= \frac{1}{s} - \frac{0.137}{(s+1.2)} - 0.92 \frac{(s+0.15)}{[(s+0.15)^2+(0.34)^2]} - 0.62 \frac{0.34}{[(s+0.15)^2+(0.34)^2]}$$

Taking inverse Laplace transform on both sides of above equation

$$c(t) = [1 - 0.137\, e^{-1.2t} - 0.92\, e^{-0.15t} \cos(0.34t) - 0.62\, e^{-0.15t} \sin(0.34t)]$$ **Ans.** ...(6.62)

Illustrative Example : (*c*) Obtain an expression for unit step time response of a control system whose transfer function is given by

$$\frac{C(s)}{R(s)} = \frac{0.0168}{s^3 + 0.42s^2 + 0.17s + 0.0168}$$

The denominator $f(s) = (s^3 + 0.42s^2 + 0.17s + 0.0168)$ equals zero at $s = -0.12$, therefore, $(s + 0.12)$ is its first order factor and its quadratic factor is obtained by dividing $f(s)$ by $(s + 0.12)$, thus the quadratic factor is $(s^2 + 0.3s + 0.14)$,

$$\therefore \quad \frac{C(s)}{R(s)} = \frac{0.0168}{(s+0.12)(s^2+0.3s+0.14)}$$

As the input is $R(s) = 1/s$.

$$\therefore \quad C(s) = \frac{1}{s} \cdot \frac{0.0168}{(s+0.12)(s^2+0.3s+0.14)}$$

Expanding R.H.S. of above expression into partial fractions

$$C(s) = \frac{K_1}{s} + \frac{K_2}{(s+0.12)} + \frac{K_3 s + K_4}{(s^2+0.3s+0.14)}$$

the coefficients can be determined as

$$K_1 = 1, K_2 = -1.18, K_3 = 0.17 \text{ and } K_4 = -0.1$$

$$\therefore \quad C(s) = \frac{1}{s} - \frac{1.18}{(s+0.12)} + \frac{0.17s - 0.1}{(s^2+0.3s+0.14)}$$

The term $(s^2 + 0.3s + 0.14)$ can be expressed as $[(s + 0.15)^2 + (0.34)^2]$, hence

$$C(s) = \frac{1}{s} - \frac{1.18}{(s+0.12)} + \frac{0.17s - 0.1}{[(s+0.15)^2+(0.34)^2]}$$

$$= \frac{1}{s} - \frac{1.18}{(s+0.12)} + 0.17 \frac{s - 0.6}{[(s+0.15)^2+(0.34)^2]}$$

$$= \frac{1}{s} - \frac{1.18}{(s+0.12)} + 0.17 \frac{(s+0.15) - 0.75}{[(s+0.15)^2+(0.34)^2]}$$

$$= \frac{1}{s} - \frac{1.18}{(s+0.12)} + 0.17 \frac{s+0.15}{[(s+0.15)^2+(0.34)^2]} - \frac{0.17 \times 0.75}{0.34} \cdot \frac{0.34}{[(s+0.15)^2+(0.34)^2]}$$

$$= \frac{1}{s} - \frac{1.18}{(s+0.12)} + 0.17 \times \frac{(s+0.15)}{[(s+0.15)^2+(0.34)^2]} - 0.37 \cdot \frac{0.34}{[(s+0.15)^2+(0.34)^2]}$$

Taking inverse Laplace transform on both sides of the above equation

$$c(t) = [(1 - 1.18\, e^{-0.12\,t} + 0.17\, e^{-0.15\,t} \cos(0.34t) - 0.37\, e^{-0.15\,t} \sin(0.34t)] \quad \textbf{Ans.} \qquad ...(6.63)$$

6.5.1. Effect of first order term time constant on time response of third order control system

The respective first order term time constants for time response expressions (6.61), (6.62) and (6.63) are

$$(a)\ T_a = \frac{1}{12}\text{ sec} \qquad (b)\ T_b = \frac{1}{1.2}\text{ sec} \qquad (c)\ T_c = \frac{1}{0.12}\text{ sec}$$

The corresponding poles of the transfer function in relation to illustrative examples (*a*), (*b*) and (*c*) are $s_a = -12$, $s_b = -1.2$ and $s_c = -0.12$. The time constant and the damped frequency of oscillations in the relation to quadratic term are 1/0.15 sec and 0.34 rad/sec respectively. The complex poles are $(-0.15 \pm j\,0.34)$.

Aforesaid poles are plotted in Fig. 6.5.1 and corresponding nature of time response is plotted in Fig. 6.5.2.

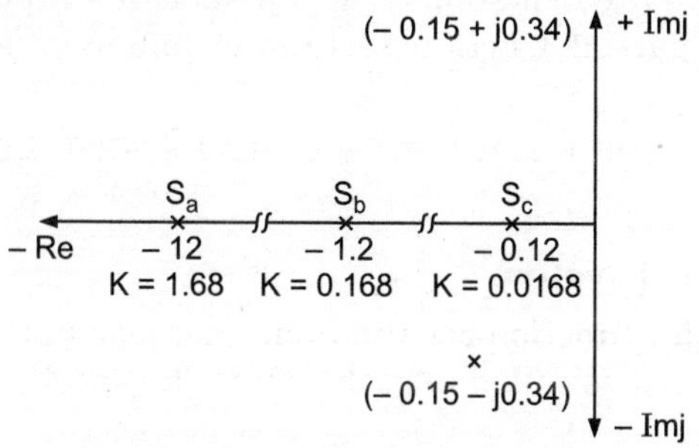

Fig. 6.5.1 Poles of illustrative examples (*a*), (*b*) and (*c*) plotted in s-plane.

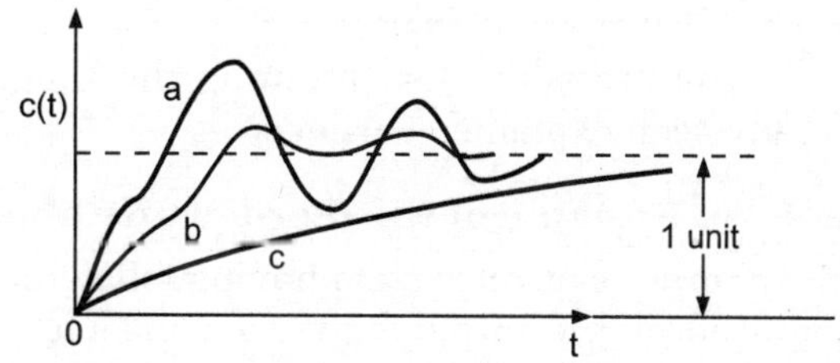

Fig. 6.5.2. Time response of illustrative examples (*a*), (*b*) and (*c*).

A comparative analysis of expressions (6.61), (6.62) and (6.63) and study of Fig. 6.5.2 reveal that the time response of a third order control system is governed by (*i*) first order term time constant (*ii*) quadratic term time constant.

The first order term time constant has a great influence on the time response of third order control system. More the value of the time constant in the first order term greater is its effect and *vice-versa*. The relative location (in *s*-plane) of the first order pole with respect to quadratic (second order) pole decides the nature of the time response of a third order control system.

If the first order pole is nearer to the imaginary axis in *s*-plane then the response is overdamped and if the quadratic (second order) complex poles are nearer to imaginary axis then the response is underdamped, *i.e.* **decaying oscillations.**

It is concluded that for a system having many poles, the nearness of the poles towards imaginary axis in *s*-plane dominates the nature of the time response, hence the poles (poles are also roots, of the characteristic equation) of a closed-loop transfer function nearer to imaginary axis are called dominant poles (or dominant roots of the characteristic equation).

6.6 TIME RESPONSE OF HIGHER ORDER CONTROL SYSTEM

A higher order control system is expressed by a transfer function

$$\frac{C(s)}{R(s)} = \frac{K}{s^n + a_1 s^{n-1} + a_2 s^{n-2} + ... + a_n} \quad ...(6.64)$$

In general any higher order polynomial can be factored as a product of first order terms and one quadratic term.

Therefore, relation (6.64) can be rewritten as

$$\frac{C(s)}{R(s)} = \frac{K}{(s+p_1)(s+p_2)(s+p_3)...(s^2 + 2\zeta\omega_n s + \omega_n^2)} \quad ...(6.65)$$

where p_1, p_2, p_3 ... and $(-\zeta\,\omega_n \pm j\,\omega_n\,\sqrt{1-\zeta^2})$ are the poles. In view of expression (6.65) the output is given by

$$C(s) = R(s)\,\frac{K}{(s+p_1)(s+p_2)...(s^2 + 2\zeta\omega_n s + \omega_n^2)} \quad ...(6.66)$$

In order to obtain the time response from (6.66) when subjected to a specified input function, the R.H.S. of (6.66) is expanded using partial fraction technique and inverse Laplace transform is taken.

The procedure for obtaining the time response of a third order control system has already been explained earlier.

6.6.1. An example of third order unstable control system

A third order control system having following transfer function results in an increase in the magnitude of the output w.r.t. time and such a response is called unstable one. The specified input being unit step input.

$$\frac{C(s)}{R(s)} = \frac{3}{s^3 + 1.5s^2 + 0.5s + 3}$$

The denominator $f(s) = s^3 + 1.5s^2 + 0.5s + 3$ equals zero at, $s = -2$, therefore, $(s + 2)$ is its first order factor and its quadratic factor is obtained by dividing $f(s)$ by $(s + 2)$, thus the quadratic factor is $(s^2 - 0.5s + 1.5)$, therefore,

$$\frac{C(s)}{R(s)} = \frac{3}{(s+2)(s^2 - 0.5s + 1.5)}$$

As the input is $R(s) = 1/s$, therefore,

$$C(s) = \frac{1}{s}\,\frac{3}{(s+2)(s^2 - 0.5s + 1.5)}$$

Expanding R.H.S. of above expression into partial fractions

$$C(s) = \frac{K_1}{s} + \frac{K_2}{(s+2)} + \frac{K_3 s + K_4}{(s^2 - 0.5s + 1.5)}$$

The coefficients can be determined as

$$K_1 = 1, K_2 = -0.23, K_3 = -0.83 \text{ and } K_4 = -0.36$$

$$\therefore \quad C(s) = \frac{1}{s} - \frac{0.23}{(s+2)} - \frac{0.83s + 0.36}{(s - 0.5s + 1.5)}$$

The term $(s^2 - 0.5s + 1.5)$ can be expressed as $[(s - 0.25)^2 + (1.2)^2]$, hence

$$C(s) = \frac{1}{s} - \frac{0.23}{(s+2)} - \frac{0.83s + 0.36}{[(s-0.25)^2 + (1.2)^2]}$$

$$= \frac{1}{s} - \frac{0.23}{(s+2)} - 0.83 \cdot \frac{s + 0.44}{[(s-0.25)^2 + (1.2)^2]}$$

$$= \frac{1}{s} - \frac{0.23}{(s+2)} - 0.83 \cdot \frac{(s - 0.25 + 0.69)}{[(s-0.25)^2 + (1.2)^2]}$$

$$= \frac{1}{s} - \frac{0.23}{(s+2)} - 0.83 \cdot \frac{(s-0.25)}{[(s-0.25)^2 + (1.2)^2]} - \frac{0.83 \times 0.69}{1.2} \cdot \frac{1.2}{[(s-0.25)^2 + (1.2)^2]}$$

$$= \frac{1}{s} - \frac{0.23}{(s+2)} - 0.83 \cdot \frac{(s-0.25)}{[(s-0.25)^2 + (1.2)^2]} - 0.477 \cdot \frac{1.2}{[(s-0.25)^2 + (1.2)^2]}$$

Taking inverse Laplace transform on both sides of above equation

$$c(t) = [1 - 0.23e^{-2t} - 0.83e^{+0.25t} \cos(1.2t) - (0.477)\, e^{+0.25t} \sin(1.2t)] \qquad ...(6.67)$$

The nature of the time response given by Eq. (6.67) and poles are plotted in Fig. 6.6.1.

From the graph (Fig. 6.6.1) it is observed that although the input is finite, the output tends to infinity as time approaches to infinity and, therefore, the system is called unstable. The system having order higher than 2 are sometimes liable to be unstable.

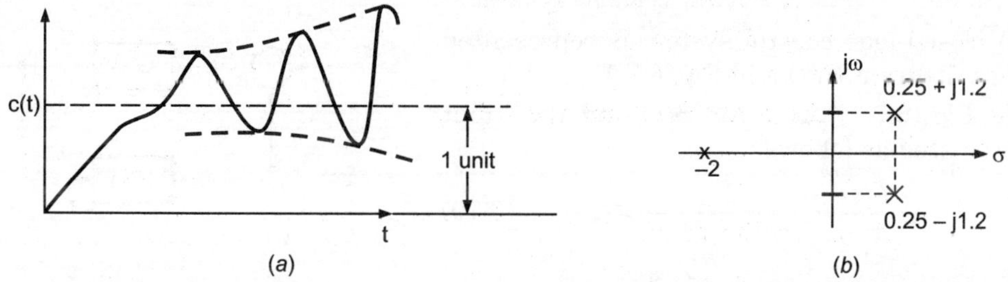

Fig. 6.6.1 Time response and pole locations of a third order unstable system.

6.7 STEADY STATE ERROR

A desirable feature of a control system is the faithful following of its input by the output. However, if the actual output of a control system during steady state deviates from the reference input (*i.e.* desired output), the system is said to possess a steady state error.

As the steady state error is an index of accuracy of a control system, therefore, the steady state error should be minimum as far as possible.

The steady state performance of a control system is assessed by the magnitude of the steady state error possessed by the system and the system input specified as either step or ramp or parabolic.

The magnitude of the steady state error in a closed-loop control system depends on its open-loop transfer function, *i.e.* $G(s)\ H(s)$ of the system. The classification of open-loop transfer function of a control system is explained below :

Classification of open-loop transfer function : The product of the forward path transfer function and the feedback path transfer function of a control system is known as open-loop transfer function. In general the open-loop transfer function of a system is given by

$$G(s)\,H(s) = \frac{K(1 + sT_a)(1 + sT_b)\,...}{s^N(1 + sT_1)(1 + sT_2)\,...} \qquad ...(6.68)$$

where K is the forward path gain

$$\left.\begin{aligned} -\frac{1}{T_a}, -\frac{1}{T_b}\ ... \text{ are the zeros} \\ -\frac{1}{T_1}, -\frac{1}{T_2}\ ... \text{ are the poles} \end{aligned}\right\} \qquad ...(6.69)$$

and N is the number of poles at the origin.

The number of poles at the origin possessed by an open-loop transfer function is the basis for classification of a control system

$N = 0$ for type '0' system

$N = 1$ for type '1' system

$N = 2$ for type '2' system

:

:

$N = N$ for type 'N' system.

A closed-loop control system is represented by a block diagram shown in Fig. 6.7.1.

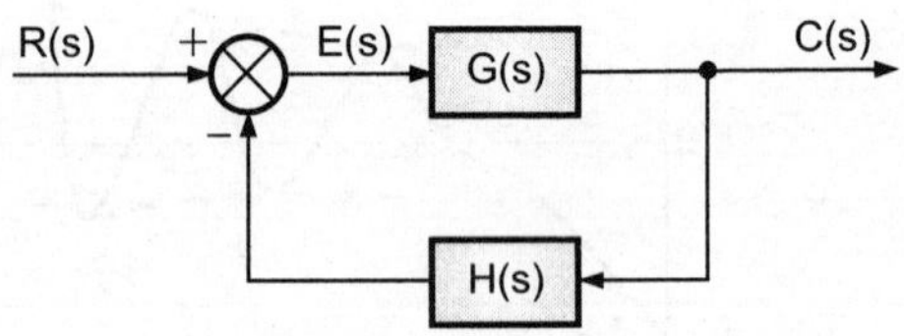

Fig. 6.7.1. Block diagram of a closed-loop control system.

In Fig. 6.7.1, the error $E(s)$ and the input $R(s)$ are related as follows :

$$\frac{E(s)}{R(s)} = \frac{1}{1 + G(s)H(s)} \qquad ...(6.70)$$

or

$$E(s) = R(s) \cdot \frac{1}{1 + G(s)H(s)}$$

Applying final value theorem, the steady state error can be determined as follows:

$$e_{ss} = \lim_{s \to 0} sE(s) = \lim_{s \to 0} sR(s) \cdot \frac{1}{1 + G(s)H(s)} \qquad ...(6.71)$$

From equation (6.71) it is noted that the steady state error possessed by a closed-loop control system depends on the input and the open-loop transfer function.

The actual output of a control system may be in any physical form, it is called as "*position*" or displacement. The first derivative of the actual output is called "*velocity*" and the second derivative is "acceleration".

The table given below explains the meaning of "*position*", "*velocity*" and "*acceleration*" with reference to physical quantities e (voltage), i (current) and θ (angular shift).

Position	e	i	θ
Velocity	$\frac{de}{dt}$	$\frac{di}{dt}$	$\frac{d\theta}{dt}$
Acceleration	$\frac{d^2e}{dt^2}$	$\frac{d^2i}{dt^2}$	$\frac{d^2\theta}{dt^2}$

6.7.1. Static error coefficients

As the steady state error for a control system is considered during the steady state period, the error is also called static error. For evaluating steady state error, the input function is specified as either unit step (displacement) or unit ramp (velocity) or unit parabolic (acceleration).

Static error coefficients associated with aforesaid inputs are determined below :

(i) Static positional error coefficient :

Static positional error coefficient K_p is associated with unit step input applied to a closed-loop control system and is determined below :

$$e_{ss} = \lim_{s \to 0} s\, R(s) \frac{1}{1 + G(s)H(s)} \qquad ...(6.72)$$

As the input is $R(s) = 1/s$

$$e_{ss} = \lim_{s \to 0} s \frac{1}{s} \frac{1}{1 + G(s)H(s)} = \frac{1}{1 + \lim_{s \to 0} G(s)H(s)}$$

put $K_p = \lim_{s \to 0} G(s)\, H(s)$, K_p is called static positional error coefficient, therefore

$$e_{ss} = \frac{1}{1 + K_p} \qquad ...(6.73)$$

(ii) Static velocity error coefficient :

Static velocity error coefficient is associated with unit ramp input applied to a closed-loop control system and is determined below :

$$e_{ss} = \lim_{s \to 0} R(s) \frac{1}{1 + G(s)H(s)}$$

As the input is $R(s) = 1/s^2$

$$e_{ss} = \lim_{s \to 0} s \frac{1}{s^2} \frac{1}{1 + G(s)H(s)}$$

$$= \lim_{s \to 0} \frac{1}{s + sG(s)H(s)} = \lim_{s \to 0} \frac{1}{sG(s)H(s)}$$

put $K_v = \lim_{s \to 0} s\, G(s)\, H(s)$, K_v is called static velocity error coefficient, therefore,

$$e_{ss} = 1/K_v \qquad ...(6.75)$$

(iii) Static acceleration error coefficient :

Static acceleration coefficient is associated with unit parabolic input applied to a closed-loop control system and is determined below :

$$e_{ss} = \lim_{s \to 0} sR(s) \cdot \frac{1}{1 + G(s)H(s)}$$

As the input is $R(s) = 1/s^3$

$$e_{ss} = \lim_{s \to 0} s \frac{1}{s^3} \cdot \frac{1}{1 + G(s)H(s)}$$

$$= \lim_{s \to 0} \frac{1}{s^2 + s^2 G(s)H(s)} = \lim_{s \to 0} \frac{1}{s^2 G(s)H(s)}$$

put $K_a = \lim_{s \to 0} s^2 G(s) H(s)$, K_a is called static acceleration error coefficient, therefore,

$$e_{ss} = 1/K_a \qquad \text{...(6.76)}$$

6.7.2. Type of transfer functions and steady state error

The steady state error depends on the type of open-loop transfer function of a closed-loop control system. An estimation of steady state error for a specified input in terms of forward path gain of an open-loop transfer function is given below :

1. *(a) Type '0' system with unit step input :*

For a type '0' system the open-loop transfer function is given by

$$G(s)\,H(s) = \frac{K(1 + sT_a)(1 + sT_b)...}{(1 + sT_1)(1 + sT_2)...}$$

$$K_p = \lim_{s \to 0} G(s)H(s)$$

$$= \lim_{s \to 0} \frac{K(1 + sT_a)(1 + sT_b)...}{(1 + sT_1)(1 + sT_2)...} \qquad \text{...(6.77)}$$

$\therefore$ $K_p = K$

Since $e_{ss} = 1/(1 + K_p)$

$\therefore$ $e_{ss} = 1/(1 + K)$ ...(6.78)

(b) Type '0' system with unit ramp input :

For a type '0' system the open-loop transfer function is given by

$$G(s)H(s) = \frac{K(1 + sT_a)(1 + sT_b)...}{(1 + sT_1)(1 + sT_2)...}$$

$$K_v = \lim_{s \to 0} s\,G(s)\,H(s) = \lim_{s \to 0} s \frac{K(1 + sT_a)(1 + sT_b)...}{(1 + sT_1)(1 + sT_2)...}$$

$\therefore$ $K_v = 0$

Since $e_{ss} = 1/K_v$

$\therefore$ $e_{ss} = 1/0 = \infty$ ...(6.79)

(c) Type '0' system with unit parabolic input :

For a type '0' system the open-loop transfer function is given by

$$G(s)\,H(s) = \frac{K(1 + sT_a)(1 + sT_b)...}{(1 + sT_1)(1 + sT_2)...}$$

$$K_a = \lim_{s \to 0} s^2 G(s)\,H(s)$$

$$= \lim_{s \to 0} s^2 \frac{K(1+sT_a)(1+sT_b)\ldots}{(1+sT_1)(1+sT_2)\ldots} \quad \ldots(6.80)$$

$\therefore$ $K_a = 0$

Since $e_{ss} = 1/K_a$ $\quad\therefore\quad$ $e_{ss} = 1/0 = \infty$

2. *(a) Type '1' system with unit step input :*

For a type '1' system the open-loop transfer function is given by

$$G(s)\,H(s) = \frac{K(1+sT_a)(1+sT_b)\ldots}{s(1+sT_1)(1+sT_2)\ldots}$$

$$K_p = \lim_{s \to 0} G(s)H(s)$$

$$= \lim_{s \to 0} \frac{K(1+sT_a)(1+sT_b)\ldots}{s\,(1+sT_1)(1+sT_2)\ldots} \quad \ldots(6.81)$$

$\therefore$ $K = \infty$. Since $e_{ss} = 1/(1 + K_p)$

$e_{ss} = 1/(1 + \infty)$

or $e_{ss} = 0$ $\quad \ldots(6.82)$

(b) Type '1' system with unit ramp input :

For a type '1' system the open-loop transfer function is given by

$$G(s)\,H(s) = \frac{K(1+sT_a)(1+sT_b)\ldots}{s(1+sT_1)(1+sT_2)\ldots}$$

$$K_v = \lim_{s \to 0} s\,G(s)\,H(s)$$

$$= \lim_{s \to 0} s \frac{K(1+sT_a)(1+sT_b)\ldots}{s\,(1+sT_1)(1+sT_2)\ldots} \quad \ldots(6.83)$$

$\therefore$ $K_v = K$. Since $e_{ss} = 1/K_v$

$\therefore$ $e_{ss} = 1/K$ $\quad \ldots(6.84)$

(c) Type '1' system with unit parabolic input :

For a type '1' system the open-loop transfer function is given by

$$G(s)\,H(s) = \frac{K(1+sT_a)(1+sT_b)\ldots}{s(1+sT_1)(1+sT_2)\ldots}$$

$$K_a = \lim_{s \to 0} s^2\,G(s)\,H(s)$$

$$= \lim_{s \to 0} s^2 \frac{K(1+sT_a)(1+sT_b)\ldots}{s(1+sT_1)(1+sT_2)\ldots} \quad \ldots(6.85)$$

$\therefore$ $K_a = 0$. Since $e_{ss} = 1/K_a$

$\therefore$ $e_{ss} = 1/0$

or $e_{ss} = \infty$ $\quad \ldots(6.86)$

3. *(a) Type '2' system with unit step input :*

For a type '2' system the open-loop transfer function is given by

$$G(s)\,H(s) = \frac{K(1+sT_a)(1+sT_b)\,...}{s^2(1+sT_1)(1+sT_2)\,...}$$

$$K_p = \lim_{s\to 0} G(s)\,H(s)$$

$$= \lim_{s\to 0} \frac{K(1+sT_a)(1+sT_b)\,...}{s^2(1+sT_1)(1+sT_2)\,...} \qquad ...(6.87)$$

$\therefore \quad K_p = \infty$. Since $e_{ss} = 1/(1+K_p)$

$\therefore \quad e_{ss} = 1/(1+\infty)$

or $\quad e_{ss} = 0 \qquad ...(6.88)$

(b) Type '2' system with unit ramp input :

For a type '2' system the open-loop transfer function is given by

$$G(s)\,H(s) = \frac{K(1+sT_a)(1+sT_b)\,...}{s^2(1+sT_1)(1+sT_2)\,...}$$

$$K_v = \lim_{s\to 0} s\,G(s)\,H(s)$$

$$= \lim_{s\to 0} s\frac{K(1+sT_a)(1+sT_b)\,...}{s^2(1+sT_1)(1+sT_2)\,...} \qquad ...(6.89)$$

$\therefore \quad K_v = \infty$. Since $e_{ss} = 1/K_v$

$\therefore \quad e_{ss} = 1/\infty$

or $\quad e_{ss} = 0 \qquad ...(6.90)$

(c) Type '2' system with unit parabolic input :

For a type '2' system the open-loop transfer function is given by

$$G(s)\,H(s) = \frac{K(1+sT_a)(1+sT_b)\,...}{s^2(1+sT_1)(1+sT_2)\,...}$$

$$K_a = \lim_{s\to 0} s^2\,G(s)\,H(s)$$

$$= \lim_{s\to 0} \frac{s^2 K(1+sT_a)(1+sT_b)\,...}{s^2(1+sT_1)(1+sT_2)\,...} \qquad ...(6.91)$$

$\therefore \quad K_a = K$. Since $e_{ss} = 1/K_a$

$\therefore \quad e_{ss} = 1/K \qquad ...(6.92)$

The steady state error estimated above as per the type of an open-loop transfer function with the specified input is shown in table below:

Input	$\mathcal{L}$(input)	Type '0'		Type '1'		Type '2'	
		Static error coeff.	e_{ss}	Static error coeff.	e_{ss}	Static error coeff.	e_{ss}
Unit step	$\frac{1}{s}$	$K_p = K$	$\frac{1}{1+K}$	$K_p = \infty$	0	$K_p = \infty$	0
Unit ramp	$\frac{1}{s^2}$	$K_v = 0$	∞	$K_v = K$	$\frac{1}{K}$	$K_v = \infty$	0
Unit parabolic	$\frac{1}{s^3}$	$K_a = 0$	∞	$K_a = 0$	∞	$K_v = K$	$\frac{1}{K}$

An observation of the above table gives the following information :

(*i*) For type '0' systems, the ramp and parabolic inputs are not acceptable.

(*ii*) For type '1' systems the parabolic input is not acceptable.

(*iii*) For type '2' systems all the three inputs the step, ramp and parabolic are acceptable.

(*iv*) Finite steady state error varies in inverse proportion to the forward path gain K.

(*v*) All the steady state errors are positional in nature.

6.7.3. Generalized error coefficients

The steady state error determined by using the static error coefficients does not give any indication as to how it varies with time and is calculated for step, ramp and parabolic inputs only. The generalized error coefficients to be described below give the steady state error performance of a feedback control system to any arbitrary input function. The error performance is indicated in terms of time and final steady state error is determined as $t \to \infty$.

The error transfer function of a feedback control system is given by

$$\frac{E(s)}{R(s)} = \frac{1}{1 + G(s)\,H(s)}$$

The R.H.S. of above equation can be expanded in a Maclaurin series in s

$$\frac{E(s)}{R(s)} = \frac{1}{1 + G(s)\,H(s)} = k_0 + k_1 s + k_2 s^2 + \ldots$$

or
$$E(s) = k_1\, R(s) + k_2 s\, R(s) + k_3 s^2\, R(s) + \ldots$$

Taking inverse Laplace Transform on both sides

$$e(t) = k_0\, r\,(t) + k_1\, r'\,(t) + k_2\, r''\,(t) + \ldots$$

In terms of static error coefficients following observations are made

k_0 is associated with step input : $k_0 = \dfrac{1}{1 + K_p}$

k_1 is associated with ramp input : $k_1 = \dfrac{1}{K_v}$

k_2 is associated with parabolic input : $k_2 = \dfrac{1}{K_a}$

Therefore the expression for $e(t)$ can be rewritten as

$$e(t) = \frac{1}{1+K_p} r(t) + \frac{1}{K_v} r''(t) + \frac{1}{K_a} r''(t) + \ldots \quad \ldots(6.94)$$

The error series given by (6.94) is valid for the case when all the poles of the transfer function lie in the L.H.S. of s-plane and sufficient time has elapsed after the initiation of the input *i.e.* transients generated by the poles of $1/1 + G(s)\,H(s)$ have decayed to reach very low magnitude.

The generalized error series for a feedback control system is derived below.

The error of a non-unity feedback control system is given by

$$E(s) = \frac{1}{1+G(s)\,H(s)} R(s)$$

Let $$F_e(s) = \frac{1}{1+G(s)\,H(s)} \quad \therefore \quad E(s) = F_e(s)\,.\,R(s)$$

Using convolution theorem the inverse Laplace transform of $E(s)$ is

$$e(t) = f_e(\tau) * r(t-\tau)$$

$$= \int_{-\infty}^{t} f_e(\tau)\,.\,r(t-\tau)\,d\tau \quad \ldots(6.95)$$

where, $f_e(\tau) = t\mathcal{L}^{-1} F_e(s)$; $r(\tau)$: input function and τ : dummy variable

The function $r(t - \tau)$ is valid for $t \geq 0$ hence the lower limit of integral is changed from $-\infty$ to 0.

The existence of n derivatives of $r(t)$ enables to use Taylor's series for expressing $r(t - \tau)$ as :

$$r(t-\tau) = r(t) - \tau\, r'(t) + \frac{\tau^2}{2!} r''(t) - \frac{\tau^3}{3!} r'''(t)$$

substituting for $r(t - \tau)$ from the above in (6.95) following expression for $e(t)$ is obtained :

$$e(t) = \int_0^t f_e(\tau) \left\{ r(t) - \tau r'(t) + \frac{\tau^2}{2!} r''(t) - \frac{\tau^3}{3!} r'''(t) + \ldots \right\} d\tau$$

Taking the limit of above expression as $t \to \infty$

$$\lim_{t\to\infty} e(t) = e_{ss} = \lim_{t\to\infty} \int_0^t f_e(\tau) \left\{ r(t) - \tau r'(t) + \frac{\tau^2}{2!} r''(t) - \frac{\tau^3}{3!} r'''(t) + \ldots \right\} d\tau$$

$$= \int_0^t r(t)\, f_e(\tau)\, d\tau - \int_0^\infty r'(t)\, \tau f_e(\tau)\, d\tau + \int_0^\infty r''(t) \frac{\tau^2}{2!} f_e(\tau)\, d\tau \ldots$$

$$= r(t) \int_0^\infty f_e(\tau)\, d\tau - r'(t) \int_0^\infty \tau f_e(\tau)\, d\tau + r''(t) \int_0^\infty \frac{\tau^2}{2!} f_e(\tau)\, d\tau \ldots$$

put $$C_0 = \int_0^\infty f_e(\tau)\, d\tau$$

$$C_1 = \int_0^\infty \tau f_e(\tau)\, d\tau$$

$$C_2 = \int_0^\infty \frac{\tau^2}{2!} f_e(\tau)\,d\tau$$

$$\vdots$$

$$C_n = \int_0^\infty \frac{\tau^n}{n!} f_e(\tau)\,d\tau$$

$$\therefore \qquad e_{ss} = C_0\, r(t) + C_1\, r'(t) + C_2 r''(t) + C_3 r'''(t) + \ldots \frac{C_n}{n!} r^n(t) \qquad \ldots(6.96)$$

The expression derived above is known as generalized error series for steady state error. The coefficients C_0, C_1, C_2, ... C_n are termed as generalized (or dynamic) error coefficients.

The generalized error coefficients mentioned above can be determined by substituting the values of $G(s)$ and $H(s)$ in the expression of $F_e(s)$ as follows :

As per the definition of Laplace Transform

$$F_e(s) = \int_0^\infty f_e(\tau)\, e^{-s\tau} d\tau$$

and

$$\lim_{s\to 0} F_e(s) = \lim_{s\to 0} \int_0^\infty f_e(\tau) e^{-s\tau} d\tau = \int_0^\infty f_e(\tau)\,d\tau = C_0$$

$$\therefore \qquad \boxed{C_0 = \lim_{s\to 0} F_e(s)}$$

Differentiating w.r.t. s

$$\frac{dF_e(s)}{ds} = (-1) \int_0^\infty \tau f_e(\tau) e^{-s\tau} d\tau$$

and

$$\lim_{s\to 0} \frac{dF_e(s)}{ds} = (-1) \int_0^\infty \tau f_e(\tau)\,d\tau = C_1$$

$$\therefore \qquad \boxed{C_1 = \lim_{s\to 0} \frac{dF_e(s)}{ds}}$$

Similarly

$$\boxed{C_2 = \lim_{s\to 0} \frac{d^2 F_e(s)}{ds^2}}$$

$$\vdots$$

and

$$\boxed{C_n = \lim_{s\to 0} \frac{d^n F_e(s)}{ds^n}}$$

Alternative method to determine generalized error coefficients.

The error transfer function $\dfrac{E(s)}{R(s)} = \dfrac{1}{1 + G(s)H(s)}$ can be arranged in the following form :

$$\frac{E(s)}{R(s)} = \frac{a_0 + a_1 s + a_2 s^2 + \ldots a_m s^m}{b_0 + b_1 s + b_2 s^2 + \ldots b_n s^n}; \quad m \le n$$

Dividing numerator of the above expression by denominator

$$\frac{E(s)}{R(s)} = k_0 + k_1 s + k_2 s^2 + \ldots$$

$$\therefore \quad E(s) = k_0 R(s) + k_1 s\, R(s) + k_2\, s^2 R(s) + \ldots$$

Taking inverse Laplace transform on both sides :

$$e(t) = k_0 r(t) + k_1 r'(t) + k_2 r''(t)$$

on comparing with (6.96) it is noted that :

$$k_0 = C_0\,;\; k_1 = C_1 \text{ and } k_2 = C_2 \text{ etc.}$$

6.7.4. Performance indices

Generally, it so happens that during the control system design process one or more parameters are to be selected to give best performance, for this purpose a measure called performance index can be expressed mathematically.

The system so designed with adjusted parameter is termed as optimal control system. The usual criteria for performance indices in terms of error function $e(t)$ are given below :

(1) Integral square-error criterion :

$$\text{ISE} = \int_0^\infty e^2(t)\,dt.$$

(2) Integral-of-time multiplied square-error criterion :

$$\text{ITSE} = \int_0^\infty t e^2(t)\,dt.$$

(3) Integral absolute-error criterion :

$$\text{IAE} = \int_0^\infty |\,e(t)|\,dt.$$

(4) Integral-of-time-multiplied absolute-error criterion :

$$\text{IATE} = \int_0^\infty t\,|\,e(t)\,|\,dt.$$

The error signal in a control system is given by

$$e(t) = r(t) - c(t)$$

where $r(t)$ and $c(t)$ being input and output respectively.

The error performance index approaches infinity if

$$\lim_{t\to\infty} e(t) = 0$$

However, for the case $\lim_{t\to\infty} e(t)$ does not approach zero then,

$$e(t) = c(\infty) - c(t)$$

As per above understanding, the performance index will be determined as finite number for a stable system.

Analytical evaluation of Integral square error (ISE) and Integral-of-time multiplied square-error (ITSE) performance index is given below :

Consider the following overall transfer function of the second order control system

$$\frac{C(s)}{R(s)} = \frac{\omega_n^2}{s^2 + 2\zeta\omega_n s + \omega_n^2} \qquad \ldots(6.97)$$

The best damping ratio for the underdamped case ($\zeta < 1$) in order to minimize ISE index is determined below :

Let $$R(s) = \frac{1}{s}, \text{ i.e. unit step input}$$

$$\therefore \quad C(s) = \frac{1}{s} \cdot \frac{\omega_n^2}{s^2 + 2\zeta\omega_n s + \omega_n^2}$$

$$E(s) = R(s) - C(s)$$

$$\therefore \quad E(s) = \frac{1}{s} - \frac{1}{s} \cdot \frac{\omega_n^2}{s^2 + 2\zeta\omega_n s + \omega_n^2}$$

or $$E(s) = \frac{s + 2\zeta\omega_n}{s^2 + 2\zeta\omega_n s + \omega_n^2} \qquad ...(6.98)$$

Taking inverse Laplace transform on both sides :

$$e(t) = \mathcal{L}^{-1} \frac{s + 2\zeta\omega_n}{s^2 + 2\zeta\omega_n s + \omega_n^2}$$

$$= \mathcal{L}^{-1} \frac{s + \zeta\omega_n}{(s + \zeta\omega_n)^2 + \omega_n^2 (1 - \zeta^2)} + \frac{\zeta\omega_n}{(s + \zeta\omega_n)^2 + \omega_n^2 (1 - \zeta^2)}$$

$$= \mathcal{L}^{-1} \frac{s + \zeta\omega_n}{(s + \zeta\omega_n)^2 + \omega_n^2 (1 - \zeta^2)} + \frac{\zeta}{\sqrt{1 - \zeta^2}} \cdot \frac{\omega_n \sqrt{1 - \zeta^2}}{(s + \zeta\omega_n)^2 + \omega_n^2 (1 - \zeta^2)}$$

$$\therefore \quad e(t) = e^{-\zeta\omega_n t} (\cos \omega_n \sqrt{1 - \zeta^2} t + \frac{\zeta}{\sqrt{1 - \zeta^2}} \sin \omega_n \sqrt{1 - \zeta^2} t) \qquad ...(6.99)$$

$$\therefore \quad e^2(t) = e^{-2\zeta\omega_n t} \left[\cos^2 \omega_n \sqrt{1 - \zeta^2} t + \frac{2\zeta}{\sqrt{1 - \zeta^2}} \cos \omega_n \sqrt{1 - \zeta^2} t \right.$$

$$\left. \sin \omega_n \sqrt{1 - \zeta^2} t + \frac{\zeta^2}{\sqrt{(1 - \zeta^2)}} \sin^2 \omega_n \sqrt{1 - \zeta^2} t \right]$$

$$= e^{-2\zeta\omega_n t} \left[\frac{1 + \cos 2\omega_n \sqrt{1 - \zeta^2} t}{2} + \frac{2\zeta}{\sqrt{1 - \zeta^2}} \cdot \frac{\sin 2\omega_n \sqrt{1 - \zeta^2} t}{2} + \frac{\zeta^2}{\sqrt{(1 - \zeta^2)}} \cdot \frac{1 - \cos 2\omega_n \sqrt{1 - \zeta^2} t}{2} \right]$$

$$= e^{-2\zeta\omega_n t} \left[\frac{1}{2(1 - \zeta^2)} + \frac{(1 - 2\zeta^2)}{2(1 - \zeta^2)} . \cos 2\omega_n \sqrt{1 - \zeta^2} t + \frac{\zeta}{\sqrt{(1 - \zeta^2)}} . \sin 2\omega_n \sqrt{1 - \zeta^2} t \right] \qquad ...(6.100)$$

Taking Laplace transform on both sides :

$$E^2(s) = \left[\frac{1}{2(1 - \zeta^2)(s + 2\zeta\omega_n)} + \frac{(1 - 2\zeta^2)}{2(1 - \zeta^2)} \cdot \frac{(s + 2\zeta\omega_n)}{(s + 2\zeta\omega_n)^2 + 4\omega_n^2 (1 - \zeta^2)} \right.$$

$$\left. + \frac{\zeta}{\sqrt{1 - \zeta^2}} \cdot \frac{2\omega_n \sqrt{1 - \zeta^2}}{(s + 2\zeta\omega_n)^2 + 4\zeta\omega_n (1 - \zeta^2)} \right] \qquad ...(6.101)$$

$$\text{ISE} = \int_0^\infty e^2(t)\,dt$$

ISE performance index given by the above integral can be determined using Laplace transform as follows :

According to real integration theorem,

$$\int_0^\infty f(t)\,dt = \lim_{t\to\infty} \int_0^t f(t)\,dt$$

Applying final value theorem,

$$\lim_{t\to\infty} \int_0^t f(t)\,dt = \lim_{s\to 0} s\,\mathcal{L}\left[\int_0^t f(t)\,dt\right]$$

$$= \lim_{s\to 0}\; s\,\frac{F(s)}{s} = \lim_{s\to 0} F(s) \qquad \text{...(6.102)}$$

Because $\qquad \text{ISE} = \int_0^\infty e^2(t)\,dt$

$\therefore$ In view of (6.102), ISE = $\lim_{s\to 0} E^2(s)$

$\because$ $E^2(s)$ is given by (6.101)

$$\therefore \qquad \text{ISE} = \lim_{s\to 0}\left[\frac{1}{2(1-\zeta^2)(s+2\zeta\omega_n)} + \frac{(1-2\zeta^2)}{2(1-\zeta^2)}\cdot\frac{s+2\zeta\omega_n}{(s+2\zeta\omega_n)^2+4\omega_n^2(1-\zeta^2)} + \frac{\zeta}{\sqrt{1-\zeta^2}}\cdot\frac{2\omega_n\sqrt{1-\zeta^2}}{(s+2\zeta\omega_n)^2+4\omega_n^2(1-\zeta^2)}\right]$$

$$= \frac{1}{2(1-\zeta^2)(2\zeta\omega_n)} + \frac{(1-2\zeta^2)}{2(1-\zeta^2)}\cdot\frac{2\zeta\omega_n}{4\omega_n^2} + \frac{\zeta}{\sqrt{1-\zeta^2}}\cdot\frac{2\omega_n\sqrt{1-\zeta^2}}{4\omega_n^2}$$

$$= \frac{1}{\omega_n}\left[\frac{1}{4\zeta(1-\zeta^2)} + \frac{\zeta(1-2\zeta^2)}{4(1-\zeta^2)} + \frac{\zeta}{2}\right]$$

$$= \frac{1}{\omega_n}\left[\frac{1+3\zeta^2-4\zeta^4}{4\zeta(1-\zeta^2)}\right] = \frac{1}{\omega_n}\left[\frac{(4\zeta^2+1)(1-\zeta^2)}{4\zeta(1-\zeta^2)}\right] = \frac{1}{\omega_n}\left(\frac{4\zeta^2+1}{4\zeta}\right) = \frac{1}{\omega_n}\left(\zeta+\frac{1}{4\zeta}\right)$$

$$\therefore \quad \omega_n(\text{ISE}) = \left(\zeta + \frac{1}{4\zeta}\right) \qquad \text{...(6.103)}$$

The plot of (ISE) for a fixed value of $\omega_n = 1$ as a function of damping ratio (ζ) is shown in Fig. 6.7.2.

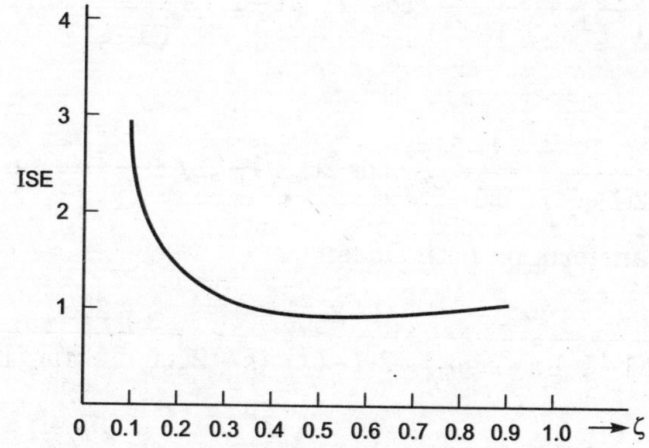

Fig. 6.7.2. The plot (ISE) as a function of ζ.

The value of ζ for minimization of ISE index is given by

$$\frac{d(ISE)}{d\zeta} = 0, \text{ with the condition } \frac{d^2(ISE)}{d\zeta^2} > 0$$

$$\therefore \quad \frac{d(ISE)}{d\zeta} = \frac{d}{d\zeta}\left[\frac{1}{\omega_n}\left(\zeta + \frac{1}{4\zeta}\right)\right] = \frac{1}{\omega_n}\left(1 - \frac{1}{4\zeta^2}\right)$$

$$\therefore \quad \frac{1}{\omega_n}\left(1 - \frac{1}{4\zeta^2}\right) = 0$$

$$\therefore \quad \zeta = 0.5 \text{ for minimum ISE index}$$

The minimum value of $\omega_n(ISE) = \zeta + \dfrac{1}{4\zeta} = 0.5 + \dfrac{1}{4 \times 0.5} = 1.0$

The best damping ratio for the underdamped ($\zeta < 1$) second order control system on the basis of minimizing Integral-of-time-multiplied square-error (ITSE) index is determined below :

The error function for the second order system subjected to unit step function is given by :

$$e(t) = e^{-\zeta\omega_n t}\left(\cos \omega_n \sqrt{1-\zeta^2}\, t + \frac{\zeta}{\sqrt{1-\zeta^2}} \sin \omega_n \sqrt{1-\zeta^2}\, t\right) \qquad \text{...(6.104)}$$

$$\text{ITSE} = \int_0^\infty t e^2(t)\, dt$$

The Laplace transform function $f(t)$ multiplied by t is given by

$$\mathcal{L}\, t f(t) = -\frac{d}{ds} F(s)$$

As per (6.100)

$$e^2(t) = e^{-2\zeta\omega_n t}\left[\frac{1}{2(1-\zeta^2)}+\frac{(1-2\zeta^2)}{2(1-\zeta^2)}.\cos 2\omega_n\sqrt{1-\zeta^2}\,t+\frac{\zeta}{\sqrt{1-\zeta^2}}\sin 2\omega_n\sqrt{1-\zeta^2}\,t\right] \quad ...(6.105)$$

Let $e^2(t) = f(t)$

$$\therefore \quad f(t) = e^{-2\zeta\omega_n t}\left[\frac{1}{2(1-\zeta^2)}+\frac{(1-2\zeta^2)}{2(1-\zeta^2)}.\cos 2\omega_n\sqrt{1-\zeta^2}\,t+\frac{\zeta}{\sqrt{1-\zeta^2}}\sin 2\omega_n\sqrt{1-\zeta^2}\,t\right]$$

Taking Laplace transform on both sides:

$$F(s) = \left[\frac{1}{2(1-\zeta^2)(s+2\zeta\omega_n)}+\frac{(1-2\zeta^2)}{2(1-\zeta^2)}.\frac{s+2\zeta\omega_n}{(s+2\zeta\omega_n)^2+4\omega_n^2(1-\zeta^2)}\right.$$

$$\left.+\frac{\zeta}{\sqrt{1-\zeta^2}}.\frac{2\omega_n\sqrt{1-\zeta^2}}{(s+2\zeta\omega_n)^2+4\omega_n^2(1-\zeta^2)}\right] \quad ...(6.106)$$

$$\therefore \quad \frac{dF(s)}{ds} = \frac{-1}{2(1-\zeta^2)(s+2\zeta\omega_n)^2}-\frac{(1-2\zeta^2)}{2(1-\zeta^2)}.\frac{(s+2\zeta\omega_n)^2-4\omega_n^2(1-\zeta^2)}{[(s+2\zeta\omega_n)^2+4\omega_n^2(1-\zeta^2)]^2}$$

$$-\frac{4\zeta(s+2\zeta\omega_n)}{[(s+2\zeta\omega_n)^2+4\omega_n^2(1-\zeta^2)]^2} \quad ...(6.107)$$

$$\therefore \quad -\frac{dF(s)}{ds} = \frac{1}{2(1-\zeta^2)(s+2\zeta\omega_n)^2}+\frac{(1-2\zeta^2)}{2(1-\zeta^2)}.\frac{(s+2\zeta\omega_n)^2-4\omega_n^2(1-\zeta^2)}{[(s+2\zeta\omega_n)^2+4\omega_n^2(1-\zeta^2)]^2}$$

$$+\frac{4\zeta(s+2\zeta\omega_n)}{[(s+2\zeta\omega_n)^2+4\omega_n^2(1-\zeta^2)]^2} \quad ...(6.108)$$

$$\text{ITSE} = \int_0^\infty t\,e^2(t)\,dt = \lim_{s\to 0} -\frac{dF(s)}{ds}$$

The term $-\dfrac{dF(s)}{ds}$ is given by (6.108)

$$\therefore \quad \text{ITSE} = \lim_{s\to 0}\frac{1}{2(1-\zeta^2)(s+2\zeta\omega_n)^2}+\frac{(1-2\zeta^2)}{2(1-\zeta^2)}.\frac{(s+2\zeta\omega_n)^2-4\omega_n^2(1-\zeta^2)}{[(s+2\zeta\omega_n)^2+4\omega_n^2(1-\zeta^2)]^2}$$

$$+\frac{4\zeta(s+2\zeta\omega_n)}{[(s+2\zeta\omega_n)^2+4\omega_n^2(1-\zeta^2)]^2}$$

$$= \frac{1}{4\zeta^2\omega_n^2(1-\zeta^2)}+\frac{(1-2\zeta^2)}{2(1-\zeta^2)}.\frac{(2\zeta^2-1)}{4\omega_n^2}+\frac{\zeta^2}{2\omega_n^2}$$

$$= \frac{1}{\omega_n^2}\left[\frac{1}{4\zeta^2(1-\zeta^2)}+\frac{(1-2\zeta^2)(2\zeta^2-1)}{8(1-\zeta^2)}+\frac{\zeta^2}{2}\right] = \frac{1}{\omega_n^2}\left(\zeta^2+\frac{1}{8\zeta^2}\right)$$

or

$$\text{ITSE} = \frac{1}{\omega_n^2}\left(\zeta^2+\frac{1}{8\zeta^2}\right) \quad ...(6.109)$$

The plot of (ITSE) for a fixed value of $\omega_n = 1$ as a function of damping ratio ζ is shown in Fig. 6.7.3.

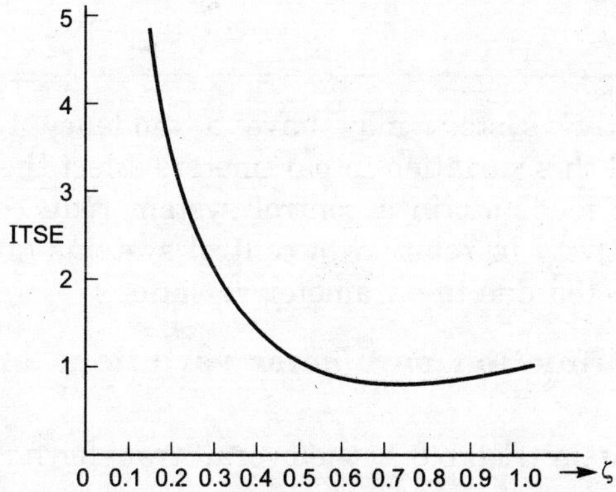

Fig. 6.7.3. The plot (ITSE) as a function of ζ.

The value of ζ for minimization of ITSE index is given by

$$\frac{d(ITSE)}{d\zeta} = 0, \text{ with the condition } \frac{d^2(ITSE)}{d\zeta^2} > 0$$

$$\therefore \quad \frac{d(ITSE)}{d\zeta} = \frac{d}{d\zeta}\left[\frac{1}{\omega_n^2}\left(\zeta^2 + \frac{1}{8\zeta^2}\right)\right] = \frac{1}{\omega_n^2}\left(2\zeta - \frac{1}{4\zeta^3}\right)$$

$$\therefore \quad \frac{1}{\omega_n^2}\left(2\zeta - \frac{1}{4\zeta^3}\right) = 0 \quad \text{or} \quad 2\zeta - \frac{1}{4\zeta^3} = 0 \text{ which gives } \zeta = 0.595$$

The minimum value of

$$\omega_n^2\,(ITSE) = \left(\zeta^2 + \frac{1}{8\zeta^2}\right) = \left(0.595^2 + \frac{1}{8 \times 0.595^2}\right) = 0.707$$

The comparison of error performance index graphs of a second order system is shown in Fig. 6.7.4.

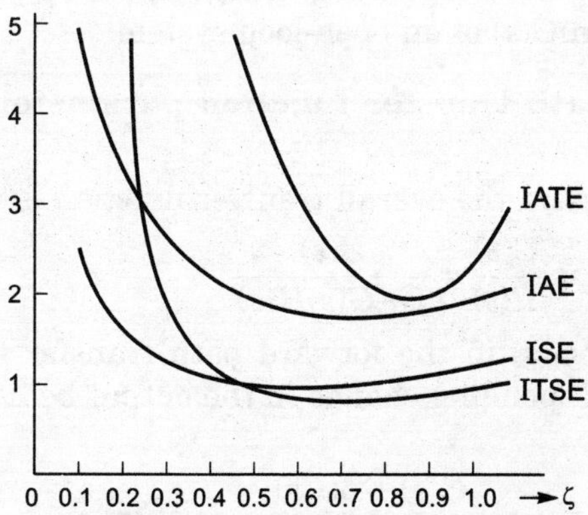

Fig. 6.7.4. Error performance indices: comparison.

It is noted that the ISE performance index does not have good selectivity. However, for a value of $\zeta = 0.7$ near optimal performance is achieved.

6.8 SENSITIVITY

The parameters of a control system may have a tendency to vary due to changing environment conditions and this variation in parameters affect the desired performance of a control system. The use of feedback in a control system reduces the effect of parameter variations. The term sensitivity in relation to control systems gives an assessment of the system performance as affected due to parameter variations.

6.8.1. Effect of transfer function parameter variations in an open-loop control system

In the case of open-loop system (Fig. 6.8.1) the overall transfer function is

$$M(s) = \frac{C(s)}{R(s)} = G(s)$$

A slight change say $\Delta G(s)$ in the transfer function $G(s)$ changes it to $[G(s) + \Delta G(s)]$ and let the corresponding change in the output be $\Delta C(s)$. The new output of the system is

$$C(s) + \Delta C(s) = [G(s) + \Delta G(s)]\, R(s)$$

or $$C(s) + \Delta C(s) = G(s)R(s) + \Delta G(s)R(s)$$

$\because$ $$C(s) = G(s)R(s)$$

$\therefore$ $$C(s) + \Delta C(s) = C(s) + [\Delta G(s)]\, R(s)$$

or $$\Delta C(s) = [\Delta G(s)]\, R(s) \qquad ...(6.111)$$

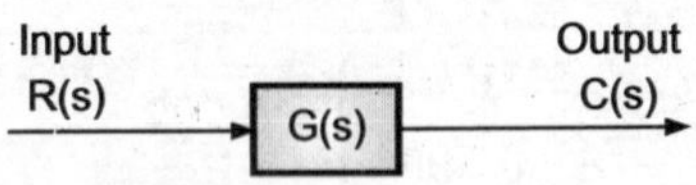

Fig. 6.8.1. Open-loop system.

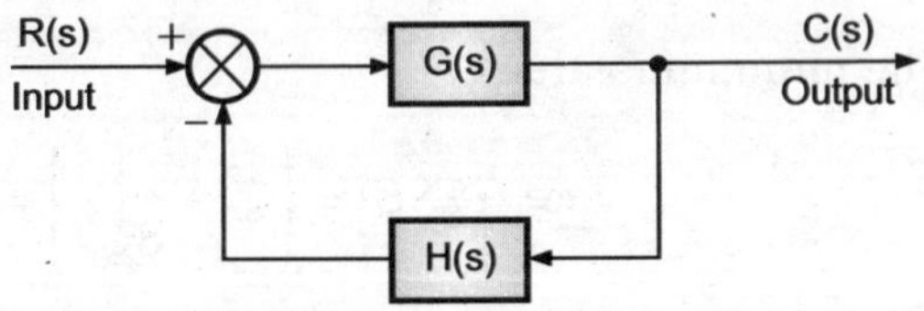

Fig. 6.8.2. Closed-loop control system.

The Eq. (6.111) gives the corresponding change in the output with reference to change $\Delta G(s)$ in the transfer function $G(s)$ of an open-loop system.

6.8.2. Effect of forward path transfer function parameter variations in a closed-loop control system

In the case of closed-loop system the overall transfer function is

$$M(s) = \frac{C(s)}{R(s)} = \frac{G(s)}{1 + G(s)H(s)} \qquad ...(6.112)$$

A slight change say $\Delta G(s)$ in the forward path transfer function $G(s)$ changes it to $G(s) + \Delta G(s)$ and let the corresponding change in the output be $\Delta C(s)$. The new output of the system is

$$C(s) + \Delta C(s) = \frac{G(s) + \Delta G(s)}{1 + [G(s) + \Delta G(s)]\, H(s)} \,.\, R(s)$$

or $$C(s) + \Delta C(s) = \frac{G(s) + \Delta G(s)}{1 + G(s)H(s) + [\Delta G(s)]\, H(s)} \,.\, R(s) \qquad ...(6.113)$$

As the change in $G(s)$ is considered slight, therefore $\Delta G(s) << G(s)$ and thus the term $[\Delta G(s)]H(s)$ can be neglected in comparison with $G(s)H(s)$.

$$\therefore \quad C(s) + \Delta C(s) \cong \frac{G(s) + \Delta G(s)}{1 + G(s)H(s)} R(s)$$

or
$$C(s) + \Delta C(s) \cong \frac{G(s)}{1 + G(s)H(s)} . R(s) + \frac{\Delta G(s)}{1 + G(s)H(s)} R(s)$$

$$\because \quad C(s) = \frac{G(s)}{1 + G(s)H(s)} . R(s)$$

$$\therefore \quad C(s) + \Delta C(s) \cong C(s) + \frac{\Delta G(s)}{1 + G(s)H(s)} . R(s)$$

or
$$\Delta C(s) \cong \frac{\Delta G(s)}{1 + G(s)H(s)} . H(s) \qquad ...(6.114)$$

The Eq. (6.114) gives the corresponding change in the output with reference to change $\Delta G(s)$ in the forward path transfer function $G(s)$ of a closed-loop control system.

As $| 1 + G(s)H(s) | >> 1$ therefore, comparing (6.111) and (6.114) it is concluded that due to feedback the variation in the output caused by the change in the forward path transfer function is reduced by a factor $[1 + G(s)H(s)]$, whereas in the open-loop system this reduction is not indicated. Thus, the output variation is more sensitive to variations in $G(s)$ in case of open-loop system as compared to closed-loop system.

The change in a variable due to variations in the parameter of a control system is expressed in terms of sensitivity as below :

Let the variable in a control system which changes its value be A such as output and this change is considered due to parameter variation of element K such as gain or feedback, then the control system sensitivity is expressed as :

$$\text{Sensitivity} = \frac{\%\text{ change in } A}{\%\text{ change in } K}$$

In mathematical terms sensitivity is written as :

$$S_K^A = \frac{\partial A / A}{\partial K / K} \qquad ...(6.115)$$

The notation S_K^A denotes sensitivity of variable A with respect to parameter K.

It is preferable that the sensitivity function S_K^A should be minimum.

6.8.3. Sensitivity of overall transfer function $M(s)$ with respect to forward path transfer function $G(s)$

The sensitivity function for the overall transfer function $M(s)$ with respect to variation in $G(s)$ is written as

$$S_G^M = \frac{\partial M(s) / M(s)}{\partial G(s) / G(s)}$$

or
$$S_G^M = \frac{G(s)}{M(s)} . \frac{\partial M(s)}{\partial G(s)} \qquad ...(6.116)$$

(1) **Open-loop control system.** For the open-loop control system shown in Fig. 6.8.1. The overall transfer function is

$$M(s) = \frac{C(s)}{R(s)} = G(s)$$

or
$$\frac{M(s)}{G(s)} = 1 \qquad \text{...(6.117)}$$

differentiating $M(s)$ w.r.t. $G(s)$

$$\frac{\partial M(s)}{\partial G(s)} = 1 \qquad \text{...(6.118)}$$

substituting (6.117) and (6.118) in (6.116), the sensitivity of $M(s)$ w.r.t. $G(s)$ for an open-loop control system is obtained as

$$S_G^M = \frac{G(s)}{M(s)} \cdot \frac{\partial M(s)}{\partial G(s)} = 1 \qquad \text{...(6.119)}$$

(2) **Closed-loop control system.** For the closed-loop control system shown in Fig. 6.8.2 the overall transfer function is

$$M(s) = \frac{G(s)}{1 + G(s)H(s)} \qquad \text{...(6.120)}$$

differentiating w.r.t. $G(s)$

$$\frac{\partial M(s)}{\partial G(s)} = \frac{[1 + G(s)H(s)] - [G(s)H(s)]}{[1 + G(s)H(s)]^2}$$

or
$$\frac{\partial M(s)}{\partial G(s)} = \frac{1}{[1 + G(s)H(s)]^2} \qquad \text{...(6.121)}$$

substituting (6.120) and (6.121) in (6.116), the sensitivity of $M(s)$ w.r.t. $G(s)$ for a closed-loop control system is obtained as

$$S_G^M = \frac{G(s)}{M(s)} \cdot \frac{\partial M(s)}{\partial G(s)} = \frac{G(s)}{\dfrac{G(s)}{1 + G(s)H(s)}} \cdot \frac{1}{[1 + G(s)H(s)]^2}$$

or
$$S_G^M = \frac{1}{1 + G(s)H(s)} \qquad \text{...(6.122)}$$

Comparing sensitivity functions (6.122) and (6.119) it is observed that the sensitivity of the overall transfer function w.r.t. forward path transfer function in the case of closed-loop control system is reduced by a factor $[1 + G(s)H(s)]$ as compared to open-loop system.

6.8.4. Sensitivity of overall transfer function *M*(*s*) with respect to feedback path transfer function

The sensitivity function for the overall transfer function $M(s)$ with respect to variations in the feedback path transfer function $H(s)$ is written as

$$S_H^M = \frac{\partial M(s) / M(s)}{\partial H(s) / H(s)}$$

or
$$S_H^M = \frac{H(s)}{M(s)} \cdot \frac{\partial M(s)}{\partial H(s)} \qquad \text{...(6.123)}$$

The overall transfer function $M(s)$ for the closed-loop control system (Fig. 6.8.2) is

$$M(s) = \frac{G(s)}{1+G(s)H(s)} \quad ...(6.124)$$

differentiating w.r.t. $H(s)$

$$\frac{\partial M(s)}{\partial H(s)} = \frac{[G(s)]^2}{[1+G(s)H(s)]^2} \quad ...(6.125)$$

Substituting (6.124) and (6.125) in (6.123) the sensitivity of $M(s)$ w.r.t. $H(s)$ for a closed-loop control system is obtained as

$$S_H^M = -\frac{G(s)H(s)}{1+G(s)H(s)} \quad ...(6.126)$$

Comparing sensitivity functions (6.122) and (6.126) it is concluded that a closed-loop control system is more sensitive to variations in feedback path parameters than the variations in forward path parameters therefore, the specifications of feedback elements in a closed-loop control system should be more rigid as compared to that of forward path elements.

Illustrative Example: *Determine the sensitivity of the overall (closed-loop) transfer function for the system shown in Fig. 6.8.3, at ω = 1 rad/sec w.r.t. (a) forward path transfer function (b) feedback path transfer function.*

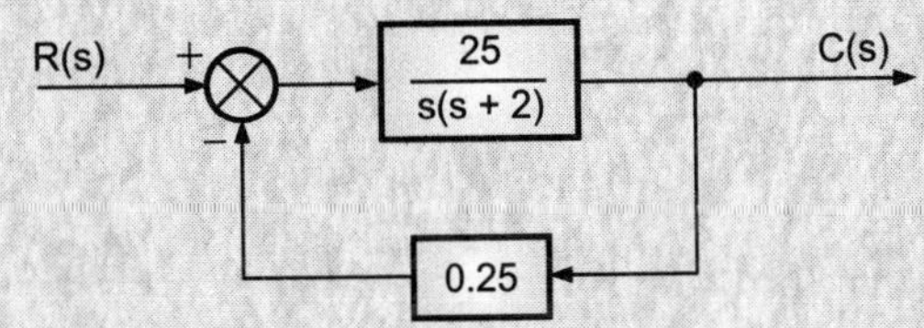

Fig. 6.8.3. Block diagram for illustrative example.

Solution. (*a*) The sensitivity of overall transfer function (M) w.r.t. forward path transfer function (G) is given by

$$S_G^M = \frac{1}{1+G(s)H(s)} = \frac{1}{1+\dfrac{25}{s(s+2)}\times 0.25} = \frac{s^2+2s}{s^2+2s+6.25}$$

put $\qquad s = j\omega$

$$\therefore \quad S_G^M = \frac{-\omega^2+j2\omega}{-\omega^2+j2\omega+6.25}$$

put $\qquad \omega = 1$

$$\therefore \quad S_G^M = \frac{-1^2+j2\times 1}{-1^2+j2\times 1+6.25} = \frac{|-1+j2|}{5.25+j2} = 0.398. \quad \textbf{Ans.}$$

(*b*) The sensitivity of overall transfer function (M) w.r.t. feedback path transfer function (H) is given by

$$S_H^M = -\frac{G(s)H(s)}{1+G(s)H(s)} = -\frac{\frac{25}{s(s+2)} \times 0.25}{1+\frac{25}{s(s+2)} \times 0.25} = \frac{6.25}{s^2+2s+6.25}$$

put $\quad s = j\omega$

$$\therefore \quad S_H^M = -\frac{6.25}{-\omega^2 + j2\omega + 6.25} \quad \text{put } \omega = 1$$

$$\therefore \quad S_H^M = -\frac{6.25}{-1^2 + j2\times 1\times 6.25} = -\frac{6.25}{5.25 + j2} = -1.11. \quad \textbf{Ans.}$$

6.8.5. Effect of feedback on time constant of a control system

In a control system the time response is improved by adjusting the feedback parameters. Consider an open-loop control system whose transfer function is

$$\frac{C(s)}{R(s)} = \frac{K}{1+sT} \qquad \text{...(6.127)}$$

where K is the gain and T the time constant.

The time response of the transfer function (6.127) subjected to unit step input is given by

$$c(t) = \frac{K}{T}(1 - e^{-t/T}) \qquad \text{...(6.128)}$$

If the system is made closed-loop with a feedback path transfer function $H(s) = b$. The closed-loop (overall) transfer function is

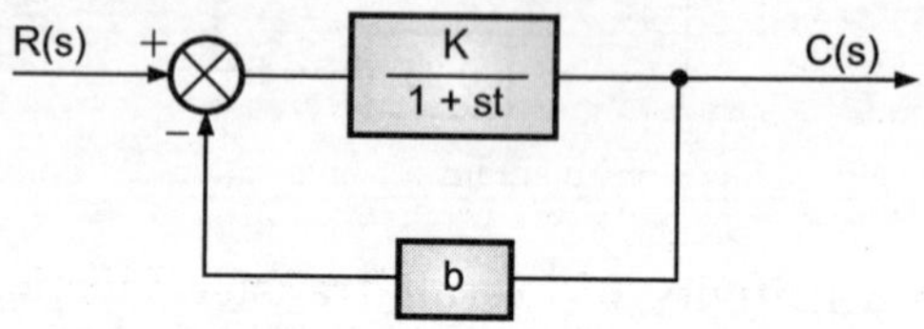

Fig. 6.8.4. Closed-loop system.

$$\frac{C(s)}{R(s)} = \frac{\frac{K}{1+sT}}{1+\frac{K}{1+sT}.b} \quad \text{or} \quad \frac{C(s)}{R(s)} = \frac{K}{1+sT+Kb}$$

or

$$\frac{C(s)}{R(s)} = \frac{K}{1+Kb}\left[\frac{1}{1+s\left(\frac{T}{1+Kb}\right)}\right] \qquad \text{...(6.129)}$$

The time response of the closed-loop transfer function subjected to unit step input is given by

$$c(t) = \frac{K}{1+Kb}\left[1 - e^{\frac{-\frac{t}{T}}{1+K_b}}\right] \qquad \text{...(6.130)}$$

The time constant in the case of open-loop system is T and that in closed-loop case is $T/(1 + Kb)$. Since $K > 1$ and b being positive $T/(1 + Kb) < T$, indicating improvement in time response. However, the overall gain is reduced.

6.9 CONTROL ACTIONS

An automatic control system is used to maintain its output within desirable limits by means of a control action. Any deviation of the output from the reference input is detected by an error detector. The error thus detected is used as an actuating signal for control action through a controller. The proportional control action and types of control actions used for improving transient and steady state response of a control system are described hereunder.

6.9.1. Proportional control

In proportional control the actuating signal for the control action in a control system is proportional to the error signal. The error signal being the difference between the reference input signal and the feedback signal obtained from the output.

For the system considered as shown in Fig. 6.9.1. The actuating signal is proportional to the error signal, therefore, the system is called proportional control system.

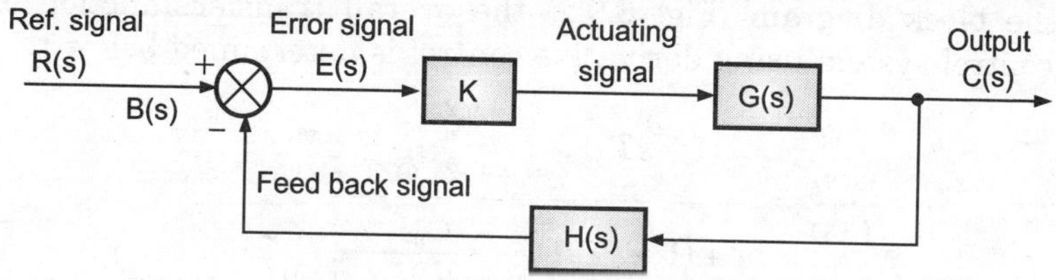

Fig. 6.9.1. Proportional control action.

It is desirable that the control system be underdamped for the point of view of quick response. An underdamped control system exhibits exponentially decaying oscillations in the output time response during the transient period.

The sluggish overdamped response of a control system can be made faster by increasing forward path gain of the system. The increase in forward path gain reduces the steady state error, but at the same time maximum overshoot is increased. For satisfactory performance of a control system a convenient adjustment has to be made between the maximum overshoot and steady state error.

Without sacrificing the steady state accuracy, the maximum overshoot can be reduced to same extent by modifying the actuating signal.

Aforesaid modification is carried out in a manner described below.

6.9.2. Derivative control

For the derivative control action the actuating signal consists of proportional error signal added with derivative of the error signal. Therefore, the actuating signal for derivative control action is given by

$$e_a(t) = e(t) + T_d \frac{de(t)}{dt} \quad ...(6.131)$$

where T_d is a constant.

The Laplace transform of the actuating signal incorporating derivative control is

$$E_a(s) = E(s) + s \,.\, T_d E(s) \qquad \text{...(6.132)}$$

For a second order unity feedback control system the block diagram representation using derivative control is shown in Fig. 6.9.2.

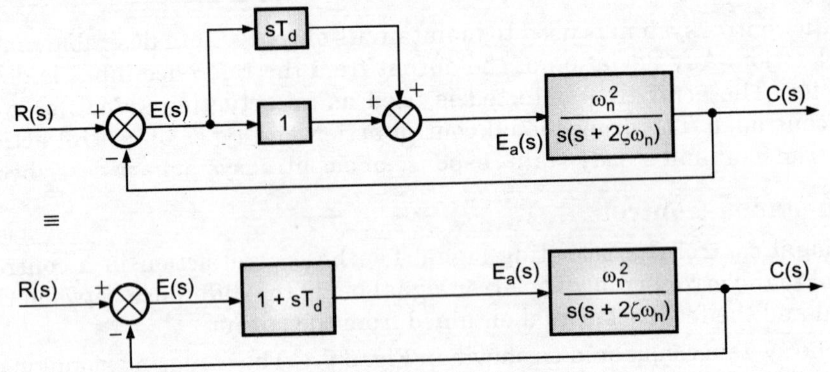

Fig. 6.9.2 Derivative control.

From the block diagram (Fig. 6.9.2) the overall transfer function of a closed-loop second order control system using derivative control is determined below :

$$\frac{C(s)}{R(s)} = \frac{(1+sT_d)\,.\,\dfrac{\omega_n^2}{s(s+2\zeta\omega_n)}}{1+(1+sT_d)\,.\,\dfrac{\omega_n^2}{s(s+2\zeta\omega_n)}}$$

or

$$\frac{C(s)}{R(s)} = \frac{(1+sT_d)\,\omega_n^2}{s^2+(2\zeta\,\omega_n+\omega_n^2 T_d)\,s+\omega_n^2} \qquad \text{...(6.133)}$$

The characteristic equation for the overall transfer function (6.133) is

$$s^2 + (2\zeta\,\omega_n + \omega_n^{\,2} T_d)\, s + \omega_n^{\,2} = 0 \qquad \text{...(6.134)}$$

As per characteristic Eq. (6.134) the damping ratio of the system using derivative control is

$$\zeta' = \frac{2\zeta\omega_n + \omega_n^2 T_d}{2\omega_n}$$

or

$$\zeta' = \zeta + \frac{\omega_n T_d}{2} \qquad \text{...(6.135)}$$

Thus, using derivative control the effective damping is increased and therefore, the maximum overshoot is reduced. The overall transfer function given by (6.133) can be written as

$$\frac{C(s)}{R(s)} = \omega_n^2 T_d \,.\, \frac{\left(s+\dfrac{1}{T_d}\right)}{s^2+2\zeta'\omega_n s+\omega_n^2} \qquad \text{...(6.136)}$$

The overall transfer function of a second order control system without using derivative control is given by

$$\frac{C(s)}{R(s)} = \frac{\omega_n^2}{s^2 + 2\zeta\omega_n s + \omega_n^2} \qquad ...(6.137)$$

In the block diagram (Fig. 6.9.2) the forward path transfer function is

$$G(s) = (1 + sT_d) \cdot \frac{\omega_n^2}{s(s + 2\zeta\omega_n)} \qquad ...(6.138)$$

and the feedback path transfer function is

$$H(s) = 1 \qquad ...(6.139)$$

The transfer function relating $E(s)$ and $R(s)$ is given by

$$\frac{E(s)}{R(s)} = \frac{1}{1 + G(s)\,H(s)} \qquad ...(6.140)$$

Substituting (6.138) and (6.139) in (6.140) following relation between the error and input signal for derivative control action is obtained

$$\frac{E(s)}{R(s)} = \frac{1}{1 + \dfrac{(1 + sT_d)\omega_n^2}{s(s + 2\zeta\omega_n)} \cdot 1}$$

or

$$\frac{E(s)}{R(s)} = \frac{s(s + 2\zeta\omega_n)}{[s^2 + (2\zeta\omega_n + \omega_n^2 T_d)\, s + \omega_n^2]} \qquad ...(6.141)$$

For a unit ramp input, $R(s) = 1/s^2$,

$$\therefore \quad E(s) = \frac{1}{s^2} \cdot \frac{s(s + 2\zeta\omega_n)}{[s^2 + (2\zeta\omega_n + \omega_n^2 T_d)\, s + \omega_n^2]} \qquad ...(6.142)$$

The steady state error for unit ramp input is determined below :

$$e_{ss} = \lim_{s \to 0} sE(s)$$

$$\therefore \quad e_{ss} = \lim_{s \to 0} s \cdot \frac{1}{s^2} \cdot \frac{s(s + 2\zeta\omega_n)}{[s^2 + (2\zeta\omega_n + \omega_n^2 T_d)\, s + \omega_n^2]}$$

or

$$e_{ss} = \frac{2\zeta}{\omega_n} \qquad ...(6.143)$$

Comparing (6.50) and (6.143) it is concluded that the steady state error is not affected by derivative control action. As the derivative of the error is used in actuating signal and as such if the error varies with time, then in that case the derivative control reduces the error.

On comparing transfer functions (6.136) and (6.137) it is noted that using derivative control the natural frequency ω_n is unchanged but a zero ($s = -1/T_d$) is added which results in different expression for time response wherein the rise time t_r is reduced. The determination of the time response of a second order system using derivative control is illustrated in the example below :

Illustrative Example: *A unity feedback control system is shown in Fig. 6.9.3. By using derivative control the damping ratio is to be made 0.8. Determine the value of T_d and compare the rise time, peak time and maximum overshoot for the system*

(a) Without derivative control.
(b) With derivative control.
The input to the system being unit step.

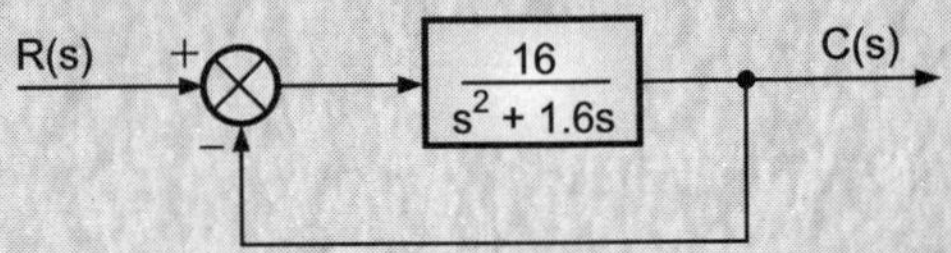

Fig. 6.9.3. Closed-loop control system without derivative control.

Solution. The overall transfer function for the block diagram shown in Fig. 6.9.3 is

$$\frac{C(s)}{R(s)} = \frac{\dfrac{16}{s^2+1.6s}}{1+\dfrac{16}{s^2+1.6s}\,.\,1} \quad \text{or} \quad \frac{C(s)}{R(s)} = \frac{16}{s^2+1.6s+16}$$

The characteristic equation is

$$s^2 + 1.6s + 16 = 0$$

Comparing above equation with second order system characteristic equation *i.e.*

$$s^2 + 2\zeta\omega_n\, s + \omega_n^2 = 0$$

$$2\zeta\omega_n = 1.6$$

and $$\omega_n = \sqrt{16} = 4 \text{ rad/sec}$$

Therefore, the damping ratio for the system without derivative control is

$$\zeta = \frac{1.6}{2\omega_n} = \frac{1.6}{2\times 4} = 0.2$$

The damping ratio with derivative control is given by

$$\zeta' = \zeta + \frac{\omega_n T_d}{2}$$

As the damping ratio is to be made 0.8 and ω_n = 4 rad/sec

$$\therefore \quad 0.8 = 0.2 + \frac{4T_d}{2} \quad \therefore \quad T_d = 0.3. \quad \textbf{Ans.}$$

The block diagram for the system with derivative control is shown in Fig. 6.9.4.

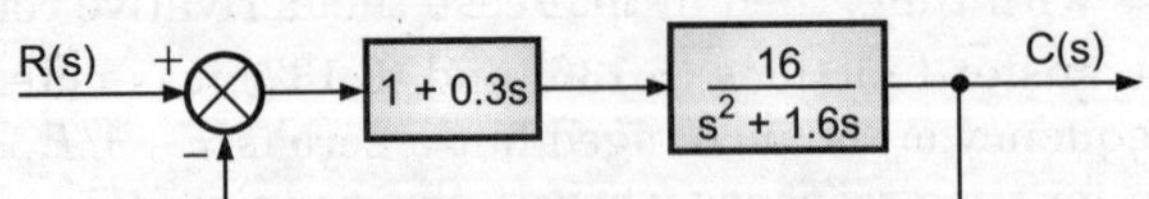

Fig. 6.9.4. Block diagram with derivative control.

(*a*) Without derivative control

Rise time, $$t_r = \frac{\pi - \tan^{-1}\dfrac{\sqrt{1-\zeta^2}}{\zeta}}{\omega_n\sqrt{1-\zeta^2}} = \frac{\pi - \tan^{-1}\dfrac{\sqrt{1-0.2^2}}{0.2}}{4\sqrt{1-0.2^2}}$$

$$= \frac{\pi - \tan^{-1} 4.89}{4 \times 0.98} = 0.45 \text{ sec.} \quad \textbf{Ans.}$$

Peak time, $$t_p = \frac{\pi}{\omega_n \sqrt{1-\zeta^2}} = \frac{\pi}{4\sqrt{1-0.2^2}} = \frac{\pi}{4 \times 0.98} = 0.8 \text{ sec.} \quad \textbf{Ans.}$$

% Maximum overshoot

$$M_p = e^{-\frac{\zeta\pi}{\sqrt{1-\zeta^2}}} \times 100 = e^{-\frac{0.2\pi}{\sqrt{1-0.2^2}}} \times 100$$

$$= e^{-0.64} \times 100 = 52.6\%. \quad \textbf{Ans.}$$

(*b*) With derivative control the overall transfer function is

$$\frac{C(s)}{R(s)} = \frac{(1+0.3s)16}{s^2 + 6.4s + 16}$$

Since the input is unit step

$$R(s) = 1/s$$

$$\therefore \quad C(s) = \frac{1}{s} \cdot \frac{(1+0.3s)16}{s^2+6.4s+16} = \frac{1}{s} \cdot \frac{16}{s^2+6.4s+16} + \frac{4.8}{s^2+6.4s+16}$$

$$= \frac{K_1}{s} + \frac{K_2 s + K_3}{s^2+6.4s+16} + \frac{4.8}{s^2+6.4s+16}$$

The coefficients can be determined as

$$K_1 = 1, K_2 = -1$$

and $$K_3 = -6.4$$

$$\therefore \quad C(s) = \frac{1}{s} - \frac{s+6.4}{s^3+6.4\,s+16} + \frac{4.8}{s^2+6.4\,s+16}$$

$$= \frac{1}{s} - \frac{s+6.4}{(s+3.2)^2+(16-3.2^2)} + \frac{4.8}{(s+3.2)^2+(16-3.2^2)}$$

$$= \frac{1}{s} - \frac{s+6.4}{(s+3.2)^2+(2.4)^2} + \frac{4.8}{(s+3.2)^2+(2.4)^2}$$

$$= \frac{1}{s} - \frac{s+3.2}{(s+3.2)^2+(2.4)^2} + \frac{1.6}{(s+3.2)^2+(2.4)^2}$$

$$= \frac{1}{s} - \frac{s+3.2}{(s+3.2)^2+(2.4)^2} + \frac{1.6}{2.4} \cdot \frac{2.4}{(s+3.2)^2+(2.4)^2}$$

$$= \frac{1}{s} - \frac{s+3.2}{(s+3.2)^2+(2.4)^2} + 0.66 \cdot \frac{2.4}{(s+2.3)^2+(2.4)^2}$$

Taking inverse Laplace transform on both sides

$$c(t) = 1 - e^{-3.2t}\cos 2.4t + 0.66e^{-3.2t}\sin 2.4t$$

The rise time t_r is determined by putting $c(t) = 1$ in the above time response expression

$$\therefore \quad 1 = 1 - e^{-3.2t_r}\cos 2.4t_r + 0.66e^{-3.2t_r}\sin 2.4t_r$$

or $$\cos 2.4t_r = 0.66 \sin 2.4t_r$$

or $$\frac{\sin 2.4t_r}{\cos 2.4t_r} = 1.51 \quad \text{or} \quad \tan 2.4t_r = 1.51$$

$$t_r = \frac{\tan^{-1} 1.51}{2.4} = 0.41. \quad \textbf{Ans.}$$

The peak time t_p is determined below:

$$c(t) = 1 - e^{-3.2t} \cos 2.4t + 0.66\, e^{-3.2t} \sin 2.4t$$

or $$c(t) = 1 - e^{-3.2t} (\cos 2.4t - 0.66 \sin 2.4t)$$

differentiating above expression w.r.t. time

$$\frac{dc\,(t)}{dt} = 0 - e^{-3.2t} (-\,2.4 \sin 2.4t - 0.66 \times 2.4 \cos 2.4t)$$
$$+ (\cos 2.4t - 0.66 \sin 2.4t)(-\,3.2\, e^{-3.2t})$$
$$= -\,e^{-3.2t} [(-2.4 \sin 2.4t - 1.58 \cos 2.4t) - (3.2 \cos 2.4t - 2.4 \sin 2.4t)]$$

or $$\frac{dc\,(t)}{dt} = -\,e^{3.2t} (-\,0.28 \sin 2.4t - 4.78 \cos 2.4t)$$

To obtain peak time t_p, put $\frac{dc\,(t)}{dt} = 0$

$\therefore$ $$-\,e^{-3.2t} (-\,0.28 \sin 2.4t - 4.78 \cos 2.4t) = 0$$

or $$(-\,0.28 \sin 2.4t - 4.78 \cos 2.4t) = 0$$

$\therefore$ $$\tan 2.4t = -\,17.07$$

and $$t_p = \frac{\tan^{-1}(-17.07)}{2.4} = \frac{\pi - \tan^{-1}(17.07)}{2.4} = 0.68 \text{ sec.} \quad \textbf{Ans.}$$

The maximum value $c(t)_{max}$ is determined by putting $t = t_p$ in the time response expression.

$\therefore$ $$c(t)_{max} = 1 - e^{-3.2 \times 0.68} [\cos (2.4 \times 0.68) - 0.66 \sin (2.48 \times 0.68)]$$
$$= 1 - 0.113\,(-\,0.061 - 0.658) = 1.0812$$

% Maximum overshoot

$$M_p = \frac{[c\,(t)_{max} - 1] \times 100}{1} = \frac{[1.0812 - 1] \times 100}{1} = 8.12\%. \quad \textbf{Ans.}$$

	Without derivative control	*With derivative control*
Damping ratio ζ	0.2	0.8
Rise time t_r	0.45 sec	0.41 sec
Peak time t_p	0.8 sec	0.68 sec
%Maximum overshoot M_p	52.6%	8.12%

6.9.3. Integral control

For integral control action the actuating signal consists of proportional-error signal added with integral of the error signal. Therefore, the actuating signal for integral control action is given by

$$e_a(t) = e(t) + K_i \int e(t)\, dt \qquad ...(6.144)$$

where K_i is a constant.

The Laplace transform of the actuating signal incorporating integral control is

$$E_a(s) = E(s) + K_i \frac{E(s)}{s} \qquad ...(6.145)$$

For a second order unity feedback control system the block diagram representation using integral control action is shown in Fig. 6.9.5.

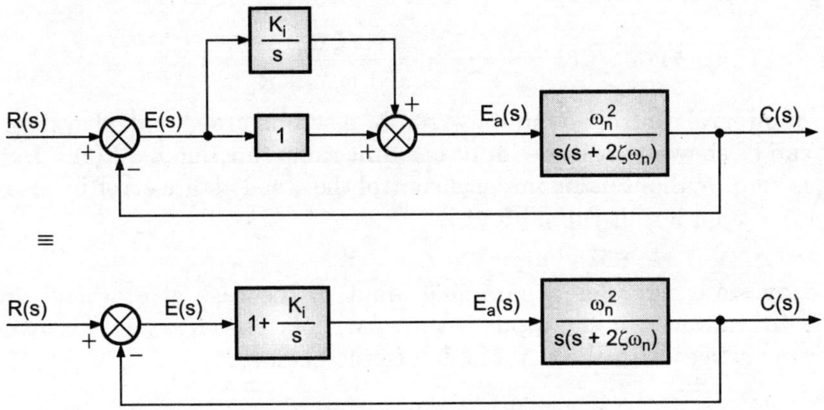

Fig. 6.9.5 Integral control.

From the block diagram (Fig. 6.9.5) the transfer function of a closed loop second order control system using integral control is determined below :

$$\frac{C(s)}{R(s)} = \frac{\left(1 + \frac{K_i}{s}\right)\left[\frac{\omega_n^2}{s(s + 2\zeta\omega_n)}\right]}{1 + \left(1 + \frac{K_i}{s}\right)\left[\frac{\omega_n^2}{s(s + 2\zeta\omega_n)}\right] \times 1}$$

$$\frac{C(s)}{R(s)} = \frac{(s + K_i)\,\omega_n^2}{s^3 + 2\zeta\omega_n s^2 + \omega_n^2 s + K_i \omega_n^2} \qquad ...(6.146)$$

The characteristic equation for the overall transfer function (6.146) is

$$s^3 + 2\zeta\omega_n s^2 + \omega_n^2 s + K_i \omega_n^2 = 0 \qquad ...(6.147)$$

The characteristic Eq. (6.147) is of third order and it will be shown later (Routh-Hurwitz Criterion) that, if $2\zeta\omega_n < K_i$, then out of three roots of the above characteristic equation two roots have positive real and as such system is unstable. However, if $2\zeta\omega_n > K_i$, then all the three roots have negative real parts indicating a stable system.

In the block diagram representation (Fig. 6.9.5) the forward path transfer function is

$$G(s) = \frac{(s + K_i)\,\omega_n^2}{s^2 (s + 2\zeta\omega_n)} \qquad ...(6.148)$$

and the feedback path transfer function is

$$H(s) = 1 \qquad ...(6.149)$$

The transfer function relating $E(s)$ and $R(s)$ is given by

$$\frac{E(s)}{R(s)} = \frac{1}{1+G(s)\,H(s)} \qquad \text{...(6.150)}$$

Substituting (6.148) and (6.149) in (6.150), the following relation between the error and input signal for integral control action is obtained.

$$\frac{E(s)}{R(s)} = \frac{1}{1+\dfrac{(s+K_i)\,\omega_n^2}{s^2\,(s+2\,\zeta\,\omega_n)}\cdot 1}$$

or

$$E(s) = R(s).\,\frac{s^2\,(s+2\,\zeta\,\omega_n)}{s^3+2\,\zeta\,\omega_n s^2+\omega_n^2 s+K_i\,\omega_n^2} \qquad \text{...(6.151)}$$

In case of integral control action the error for a second order control system is given by (6.151) and it can be shown that if the input is a unit ramp function *i.e.* $R(s) = 1/s^2$ the steady state error e_{ss} is zero. Without using integral control the steady state error in case of a second order control system for a unit ramp input is

$$e_{ss} = 2\,\zeta/\omega_n.$$

The steady state error for a parabolic input in the case of a second order control system is infinite. However, if the input is unit parabolic (acceleration) *i.e.* $R(s) = 1/s^3$ the steady state error using integral control is determined below :

$$E(s) = R(s).\,\frac{s^2\,(s+2\,\zeta\,\omega_n)}{s^3+2\,\zeta\,\omega_n s^2+\omega_n^2 s+K_i\,\omega_n^2}$$

$\because$ $$R(s) = 1/s^3$$

$\therefore$ $$E(s) = \frac{1}{s^3}\cdot\frac{s^2\,(s+2\,\zeta\,\omega_n)}{s^3+2\,\zeta\,\omega_n s^2+\omega_n^2 s+K_i\,\omega_n^2}$$

$$e_{ss} = \lim_{s\to 0} sE(s) = \lim_{s\to 0} s\cdot\frac{1}{s^3}\cdot\frac{s^2\,(s+2\,\zeta\,\omega_n)}{s^3+2\,\zeta\,\omega_n s^2+\omega_n^2 s+K_i\,\omega_n^2}$$

$$e_{ss} = \frac{2\,\zeta}{K_i\,\omega_n} \qquad \text{...(6.152)}$$

The table below gives the steady state error for a second order control system with and without using integral control action.

Input	*Steady state error*	
	Without Integral Control	*With Integral Control*
Unit ramp	$\dfrac{2\zeta}{\omega_n}$	0
Unit parabolic	∞	$\dfrac{2\zeta}{K_i\,\omega_n}$

6.9.4. Proportional Plus Derivative Plus Integral Control (PID Control)

For PID control the actuating signal consists of proportional error signal added with derivative and integral of the error signal. Therefore, the actuating signal for PID control is

$$e_a(t) = e(t) + T_d \frac{de(t)}{dt} + K_i \int e(t)\, dt \qquad \text{...(6.153)}$$

The Laplace transform of the actuating signal incorporating PID control is

$$E_a(s) = E(s) + sT_d E(s) + \frac{K_i}{s} E(s)$$

or

$$E_a(s) = E(s) \left[1 + sT_d + \frac{K_i}{s}\right] \qquad \text{...(6.154)}$$

The block diagram of a second order control system incorporating PID control is shown in Fig. 6.9.6.

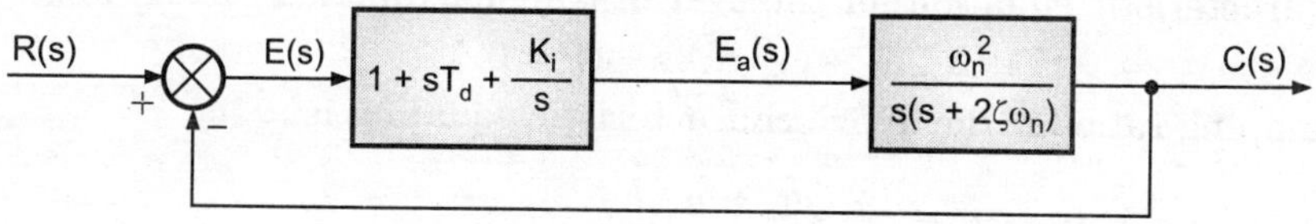

Fig. 6.9.6. PID control.

PID control combines the advantages of proportional, derivative and integral control actions.

6.9.5. Derivative Feedback Control

The derivative feedback control is also known as rate feedback or tachometer feedback control. For derivative feedback control the actuating signal is obtained as the difference between the proportional error signal and derivative (rate) of the output signal. Therefore, the actuating signal for derivative feedback control action is

$$e_a(t) = e(t) - K_t \frac{dc(t)}{dt} \qquad \text{...(6.155)}$$

where K_t is a constant.

The Laplace transform of the actuating signal incorporating derivative control

$$E_a(s) = E(s) - sK_t C(s) \qquad \text{...(6.156)}$$

For a second order unity feedback control system block diagram incorporating derivative feedback control is shown in Fig. 6.9.7.

From the block diagram (Fig. 6.9.7) the transfer function for a unity feedback second order control system using derivative feedback control is determined below :

$$\frac{C(s)}{R(s)} = \frac{\dfrac{\omega_n^2}{s^2 + (2\zeta\omega_n + \omega_n^2 K_t)s}}{1 + \dfrac{\omega_n^2}{s^2 + (2\zeta\omega_n + \omega_n^2 K_t)s} \cdot 1}$$

or

$$\frac{C(s)}{R(s)} = \frac{\omega_n^2}{s^2 + (2\zeta\omega_n + \omega_n^2 K_t)s + \omega_n^2} \qquad \text{...(6.157)}$$

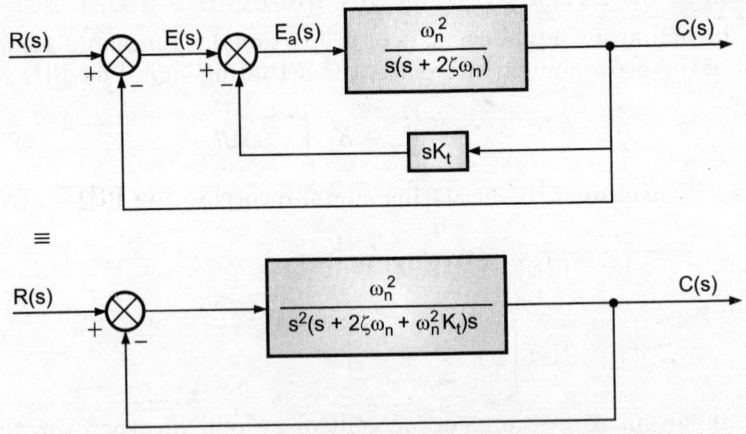

Fig. 6.9.7. Derivative feedback (tachometer feedback) control.

The characteristic equation for the overall transfer function (6.157) is

$$s^2 + (2\,\zeta\,\omega_n + \omega_n^2\,K_t)\,s + \omega_n^2 = 0 \qquad \text{...(6.158)}$$

The damping ratio for the above characteristic equation is

$$\zeta' = \frac{2\,\zeta\omega_n + \omega_n^{\,2} K_t}{2\,\omega_n}$$

or

$$\zeta' = \zeta + \frac{\omega_n K_t}{2}. \qquad \text{...(6.159)}$$

The damping ratio is increased by using derivative feedback control and therefore, the maximum overshoot is decreased. However, the rise time is increased.

In the block diagram Fig. 6.9.7 the forward path transfer function is

$$G(s) = \frac{\omega_n^{\,2}}{s^2 + (2\,\zeta\;\omega_n + \omega_n^{\,2} K_t)\,s} \qquad \text{...(6.160)}$$

and the feedback path transfer function is

$$H(s) = 1 \qquad \text{...(6.161)}$$

The transfer function relating E (s) and R (s) is given by

$$\frac{E(s)}{R(s)} = \frac{1}{1 + G(s)\,H(s)} \qquad \text{...(6.162)}$$

Substituting (6.160) and (6.161) in (6.162) the following relation between the error and input signal for the derivative feedback control action is obtained.

$$\frac{E(s)}{R(s)} = \frac{1}{1 + \dfrac{\omega_n^{\,2}}{s^2 + (2\,\zeta\omega_n + \omega_n^{\,2} K_t)\,s} \cdot 1} \qquad \text{...(6.163)}$$

or

$$\frac{E(s)}{R(s)} = \frac{s^2 + (2\zeta\,\omega_n + \omega_n^{\,2}\,K_t)\,s}{s^2 + (2\,\zeta\,\omega_n + \omega_n^{\,2}\,K_t)\,s + \omega_n^{\,2}} \qquad \text{...(6.164)}$$

For a unit ramp function,

$$R(s) = 1/s^2$$

$$\therefore \qquad E(s) = \frac{1}{s^2} \cdot \frac{s^2 + (2\zeta\omega_n + \omega_n^2 K_t)s}{s^2 + (2\zeta\omega_n + \omega_n^2 K_t)s + \omega_n^2} \qquad \text{...(6.165)}$$

The steady state error is determined below :

$$e_{ss} = \lim_{s \to 0} s\,E(s)$$

$$= \lim_{s \to 0} s \cdot \frac{1}{s^2} \cdot \frac{s^2 + (2\zeta\omega_n + \omega_n^2 K_t)s}{s^2 + (2\zeta\omega_n + \omega_n^2 K_t)s + \omega_n^2}$$

or

$$e_{ss} = \frac{2\zeta}{\omega_n} + K_t \qquad \text{...(6.166)}$$

From (6.166) it is noted that using derivative feedback control the steady state error is increased and as the damping ratio is increased the maximum overshoot is decreased.

In both types of control actions derivative error as well as derivative feedback, the damping ratio is increased which results in decreasing maximum overshoot but the time response expression is different in two cases. In case of derivative control the rise time is decreased and the steady state error remains unchanged whereas in derivative feedback control the rise time is decreased and the steady state error is increased as illustrated in the example below :

Illustrative Example : *The overall transfer function of a control system is given by*

$$\frac{C(s)}{R(s)} = \frac{16}{s^2 + 1.6s + 16}$$

It is desired that the damping ratio be 0.8. Determine the derivative rate feedback constant K_t and compare rise time, peak time, maximum overshoot and steady state error for unit ramp input without and with derivative feedback control.

Solution. The characteristic equation is $s^2 + 1.6s + 16 = 0$. Comparing with the second order characteristic equation *i.e.*

$$s^2 + 2\zeta\omega_n s + \omega_n^2 = 0$$

$$2\zeta\omega_n = 1.6$$

and

$$\omega_n = \sqrt{16} = 4 \text{ rad/sec}$$

therefore, the damping ratio for the system without derivative feedback control is

$$\zeta = \frac{1.6}{2\,\omega_n} = \frac{1.6}{2 \times 4} = 0.2$$

The damping ratio with derivative feedback control is given by

$$\zeta' = \zeta + \frac{\omega_n K_t}{2}$$

As the damping ratio is to be made 0.8 and $\omega_n = 4$ rad/sec

$$\therefore \qquad 0.8 = 0.2 + \frac{4K_t}{2}. \qquad K_t = 0.3 \quad \textbf{Ans.}$$

(*a*) Without derivative feedback control

Rise time
$$t_r = \frac{\pi - \tan^{-1} \cdot \dfrac{\sqrt{1-\zeta^2}}{\zeta}}{\omega_n \sqrt{1-\zeta^2}} = \frac{\pi - \tan^{-1} \dfrac{\sqrt{1-0.2^2}}{0.2}}{4\sqrt{1-0.2^2}}$$

$$= \frac{\pi - \tan^{-1} 4.89}{4 \times 0.98} = 0.45 \text{ sec.} \quad \textbf{Ans.}$$

Peak time
$$t_p = \frac{\pi}{\omega_n \sqrt{1-\zeta^2}} = \frac{\pi}{4\sqrt{1-0.2^2}} = \frac{\pi}{4 \times 0.98} = 0.8 \text{ sec.} \quad \textbf{Ans.}$$

% maximum overshoot

$$M_p = e^{-\frac{\zeta \pi}{\sqrt{1-\zeta^2}}} \times 100 = e^{-\frac{0.2\pi}{\sqrt{1-0.2^2}}} \times 100$$

$$= e^{-0.64} \times 100 = 52.6\%. \quad \textbf{Ans.}$$

Steady state error
$$e_{ss} = \frac{2\zeta}{\omega_n} = \frac{2 \times 0.2}{4} = 0.1. \quad \textbf{Ans.}$$

(*b*) With derivative feedback control

The overall transfer function of the system using derivative feedback control is given by

$$\frac{C(s)}{R(s)} = \frac{\omega_n^{\,2}}{s^2 + (2\,\zeta\,\omega_n + \omega_n^{\,2} K_t)\, s + \omega_n^{\,2}}$$

given that
$$\zeta = 0.2,\ \omega_n = 4 \text{ and } K_t = 0.3$$

$$\therefore \quad \frac{C(s)}{R(s)} = \frac{4^2}{s^2 + (2 \times 0.2 \times 4 + 4^2 \times 0.3)\, s + 4^2}$$

or
$$\frac{C(s)}{R(s)} = \frac{16}{s^2 + 6.4s + 16}$$

$$2\,\zeta'\,\omega_n = 6.4 \text{ and } \omega_n = 4$$

$$\therefore \quad \zeta' = \frac{6.4}{2\,\omega_n} = \frac{6.4}{2 \times 4} = 0.8$$

Rise time
$$t_r = \frac{\pi - \tan^{-1} \dfrac{\sqrt{1-\zeta'^2}}{\zeta'}}{\omega_n \sqrt{1-\zeta'^2}} = \frac{\pi - \tan^{-1} \dfrac{\sqrt{1-0.8^2}}{0.8}}{4\sqrt{1-0.8^2}}$$

$$= \frac{\pi - \tan^{-1} 0.75}{2.4} = \frac{\pi - 0.64}{2.4} = 1.04 \text{ sec.} \quad \textbf{Ans.}$$

Peak time
$$t_p = \frac{\pi}{\omega_n \sqrt{1-\zeta'^2}} = \frac{\pi}{4\sqrt{1-0.8^2}} = 1.3 \text{ sec.} \quad \textbf{Ans.}$$

% Maximum overshoot

$$M_p = e^{-\frac{\zeta'\pi}{\sqrt{1-\zeta'^2}}} \times 100 = e^{-\frac{0.8\pi}{\sqrt{1-0.8^2}}} \times 100 = e^{-4\pi/3} \times 100 = 1.52\%. \quad \textbf{Ans.}$$

$$\text{Steady state error } e_{ss} = \frac{2\zeta}{\omega_n} + K_t = \frac{2\times 0.2}{4} + 0.3 = 0.4. \quad \textbf{Ans.}$$

	Without derivative feedback control	*With derivative feedback control*
Damping ratio ζ	0.2	0.8
Rise time t_r	0.45 sec	1.04 sec
Peak time t_p	0.8 sec	1.3 sec
% maximum overshoot M_p	52.6%	1.52%
Steady state error e_{ss}	0.1	0.4

6.10 SOLVED EXAMPLES

Example 6.10.1. *A control system is shown in Fig. 6.10.1. Determine the transfer function and derive an expression relating the output time if the input is a step having a magnitude of 2 unit.*

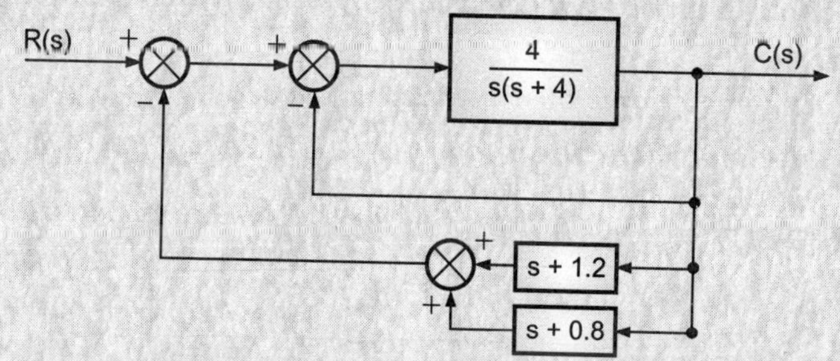

Fig. 6.10.1. Control system for example 6.10.1.

Solution. The block diagram of Fig. 6.10.1 is reduced in steps below.

From the block diagram the overall transfer function is determined below :

$$\frac{C(s)}{R(s)} = \frac{\dfrac{4}{[s(s+4)+4]}}{1+\dfrac{4}{[s(s+4)+4]}\cdot(s+2)}$$

or $$\frac{C(s)}{R(s)} = \frac{4}{(s+2)(s+6)} \quad \textbf{Ans.}$$

$$\therefore \quad C(s) = R(s)\,\frac{4}{(s+2)(s+6)}$$

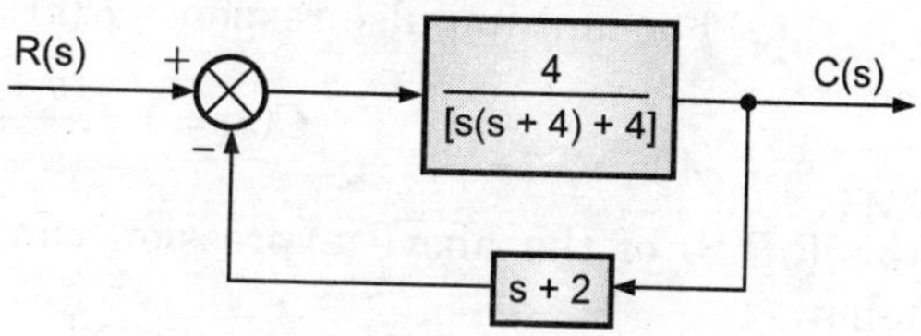

Fig. 6.10.1 (*a*) Block diagram reduction of Fig. 6.10.1.

As $r(t) = 2, R(s) = 2/s$

$$\therefore \quad C(s) = \frac{2}{s} \cdot \frac{4}{(s+2)(s+6)}$$

or
$$C(s) = \frac{8}{s(s+2)(s+6)}$$

Expanding R.H.S. into partial fractions

$$\frac{8}{s(s+2)(s+6)} = \frac{K_1}{s} + \frac{K_2}{(s+2)} + \frac{K_3}{(s+6)}$$

The values of K_1, K_2 and K_3 can be calculated as

$$K_1 = \frac{2}{3}, K_2 = -1 \text{ and } K_3 = \frac{1}{3}$$

$$\therefore \quad C(s) = \frac{2}{3} \cdot \frac{1}{s} - \frac{1}{(s+2)} + \frac{1}{3} \cdot \frac{1}{(s+6)}$$

Taking inverse Laplace transform on both sides

$$\mathcal{L}^{-1} C(s) = \mathcal{L}^{-1} \left[\frac{2}{3} \frac{1}{s} - \frac{1}{(s+2)} + \frac{1}{3} \cdot \frac{1}{(s+6)} \right]$$

$$\therefore \quad c(t) = \tfrac{2}{3} - e^{-2t} + \tfrac{1}{3} e^{-6t}.$$ **Ans.**

Example 6.10.2. *A unity feedback control system has its open-loop transfer function given by $G(s) = (4s + 1)/4s^2$.*

Determine an expression for the time response when the system is subjected to

(a) Unit impulse input function and (b) Unit step input function.

Solution. The overall transfer function for the system is determined as

$$\frac{C(s)}{R(s)} = \frac{\dfrac{4s+1}{4s^2}}{1 + \dfrac{(4s+1)}{4s^2} . 1} = \frac{(4s+1)}{(4s^2+4s+1)} = \frac{(s^2+0.25)}{(s^2+s+0.25)}$$

It is noted that the denominator of the transfer function is a perfect square, therefore,

$$\frac{C(s)}{R(s)} = \frac{(s+0.25)}{(s+0.5)^2}$$

(*a*) For unit impulse response $R(s) = 1$

$$\therefore \quad C(s) = 1 . \frac{(s+0.25)}{(s+0.5)^2}$$

the R.H.S. of the above expression can be expanded into partial fraction expansion as follows :

$$\frac{(s+0.25)}{(s+0.5)^2} = \frac{K_1}{(s+0.5)} + \frac{K_2}{(s+0.5)^2}$$

The coefficients K_1 and K_2 can be determined as $K_1 = 1$ and $K_2 = -0.25$.

$$\therefore \qquad C(s) = \frac{1}{(s+0.5)} - \frac{0.25}{(s+0.5)^2}$$

Taking inverse Laplace transform on both sides

$$c(t) = (e^{-0.5t} - 0.25t\, e^{-0.5t})$$

or $$c(t) = e^{-0.5t}(1 - 0.25t). \quad \textbf{Ans.}$$

(*b*) For unit step input

$$R(s) = \frac{1}{s} \quad \therefore \quad C(s) = \frac{1\,(s+0.25)}{s\,(s+0.5)^2}$$

The R.H.S. of the above expression can be expanded into partial fractions as follows:

$$\frac{1\,(s+0.25)}{s\,(s+0.5)^2} = \frac{K_1}{s} + \frac{K_2}{(s+0.5)} + \frac{K_3}{(s+0.5)^2}$$

The coefficients, K_1, K_2, and K_3 can be determined as $K_1 = 1$, $K_2 = -1$ and $K_3 = 0.5$.

$$\therefore \qquad C(s) = \frac{1}{s} - \frac{1}{(s+0.5)} + \frac{0.5}{(s+0.5)^2}$$

Taking inverse Laplace transform on both sides

$$c(t) = (1 - e^{-0.5t} + 0.5t\, e^{-0.5t})$$

or $$c(t) = 1 - e^{-0.5t}(1 - 0.5t). \quad \textbf{Ans.}$$

Example 6.10.3. *The forward path transfer function of a unity feedback control system is given by*

$$G(s) = \frac{2}{s(s+3)}$$

Obtain an expression for unit step response of the system.

Solution. The overall transfer function for the system is

$$\frac{C(s)}{R(s)} = \frac{\dfrac{2}{s(s+3)}}{1 + \dfrac{2}{s(s+3)}\cdot 1} \quad \text{or} \quad \frac{C(s)}{R(s)} = \frac{2}{(s^2+3s+2)}$$

It is noted that the denominator of the above expression can be factored as $[(s+1)(s+2)]$

$$\therefore \qquad C(s) = R(s)\,\frac{2}{[(s+1)(s+2)]}$$

As the input is a unit step

$$R(s) = 1/s$$

$$\therefore \qquad C(s) = \frac{1}{s}\cdot\frac{2}{[(s+1)(s+2)]}$$

The R.H.S. of the above expression can be expanded into partial fraction as follows :

$$\frac{1}{s}\cdot\frac{2}{(s+1)(s+2)} = \frac{K_1}{s} + \frac{K_2}{(s+1)} + \frac{K_3}{(s+2)}$$

The coefficients K_1, K_2 and K_3 can be determined as

$$K_1 = 1, K_2 = -2 \text{ and } K_3 = 1$$

$$\therefore \quad C(s) = \frac{1}{s} - \frac{2}{(s+1)} + \frac{1}{(s+2)}$$

Taking inverse Laplace transform on both sides $c(t) = (1 - 2e^{-t} + e^{-2t})$. **Ans.**

Example 6.10.4. *Fig. 6.10.2 represents a control system. Obtain an expression relating the output and time. The input being unit step.*

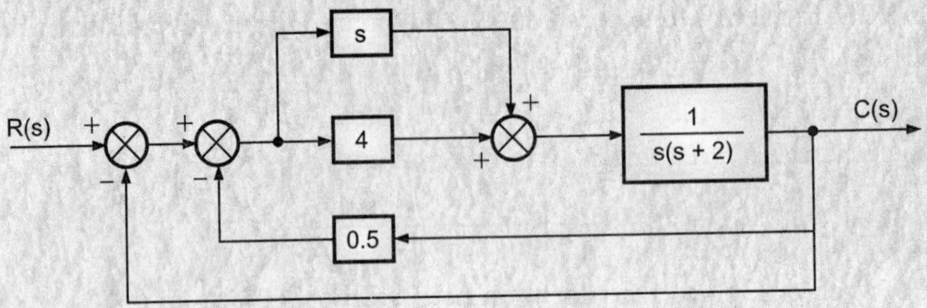

Fig. 6.10.2. Control system for example 6.10.2 (*a*).

Solution. The block diagram of Fig. 6.10.2 is reduced in steps below :

From the block diagram of Fig. 6.10.2 (*b*) the transfer function of the system is determined below :

$$\frac{C(s)}{R(s)} = \frac{\dfrac{(s+4)}{(s^2+2.5s+2)}}{1 + \dfrac{(s+4)}{(s^2+2.5s+2)}.1} \quad \text{or} \quad \frac{C(s)}{R(s)} = \frac{(s+4)}{(s^2+3.5s+6)}$$

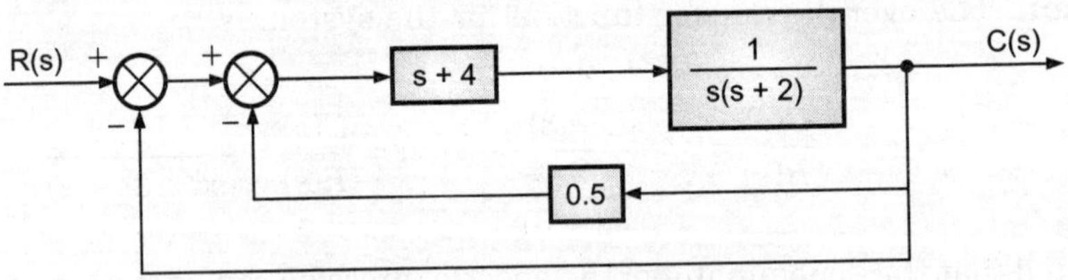

Fig. 6.10.2 (*a*) Block diagram reduction of Fig. 6.10.2.

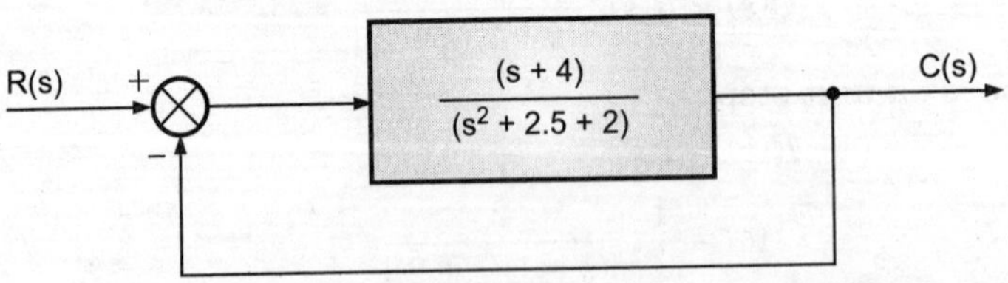

Fig. 6.10.2 (*b*) Block diagram reduction of Fig. 6.10.2 (*a*).

$$\therefore \quad C(s) = R(s)\frac{(s+4)}{(s^2+3.5s+6)}$$

As $\quad r(t) = 1, R(s) = \dfrac{1}{s}$

$$\therefore \qquad C(s) = \frac{1}{s} \cdot \frac{(s+4)}{(s^2+3.5s+6)}$$

The R.H.S. of the above equation can be expanded into partial fraction

$$\therefore \qquad \frac{1}{s} \cdot \frac{(s+4)}{(s^2+3.5s+6)} = \frac{K_1}{s} + \frac{K_2 s + K_3}{(s^2+3.5s+6)}$$

$$K_1 = 2/3,\ K_2 = -2/3 \text{ and } K_3 = -4/3$$

$$\therefore \qquad C(s) = \frac{2}{3} \cdot \frac{1}{s} - \frac{2}{3} \cdot \frac{s}{(s^2+3.5s+6)} - \frac{4}{3} \cdot \frac{1}{(s^2+3.5s+6)}$$

$$= \frac{2}{3} \cdot \frac{1}{s} - \frac{2}{3} \cdot \frac{s}{[(s^2+3.5s+(1.75)^2+6-(1.75)^2]} - \frac{4}{3} \cdot \frac{1}{[(s^2+3.5s+(1.75)^2+6-(1.75)^2]}$$

$$= \frac{2}{3} \cdot \frac{1}{s} - \frac{2}{3} \cdot \frac{s}{[(s+1.75)^2+(1.71)^2]} - \frac{4}{3} \cdot \frac{1}{[(s+1.75)^3+(1.71)^2]}$$

$$= \frac{2}{3} \cdot \frac{1}{s} - \frac{2}{3} \cdot \frac{[(s+1.75)-1.75]}{[(s+1.75)^2+(1.71)^2]} - \frac{4}{3} \cdot \frac{1}{[(s+1.75)^2+(1.71)^2]}$$

$$= \frac{2}{3} \cdot \frac{1}{s} - \frac{2}{3} \cdot \frac{(s+1.75)}{[(s+1.75)^2+(1.71)^2]} + \frac{2}{3} \cdot \frac{1.75}{[(s+1.75)^2+1.71^2]}$$

$$- \frac{4}{3} \cdot \frac{1}{[(s+1.75)^2+(1.71)^2]}$$

$$= \frac{2}{3} \cdot \frac{1}{s} - \frac{2}{3} \cdot \frac{(s+1.75)}{[(s+1.75)^2+1.71^2]} - \frac{0.5}{3} \cdot \frac{1}{[(s+1.75)^2+(1.71)^2]}$$

$$= \frac{2}{3} \cdot \frac{1}{s} - \frac{2}{3} \cdot \frac{(s+1.75)}{[(s+1.75)^2+(1.71)^2]} - \frac{0.5}{3 \times 1.71} \cdot \frac{1.71}{[(s+1.75)^2+(1.71)^2]}$$

Taking inverse Laplace transform

$$c(t) = \frac{2}{3} - \frac{2}{3} e^{-1.75t} \cdot \cos(1.71t) - \frac{0.29}{3} \cdot e^{-1.75t} \cdot \sin(1.71t). \qquad \textbf{Ans.}$$

Example 6.10.5. *The block diagram of a unity feedback control system is shown in Fig. 6.10.3.*

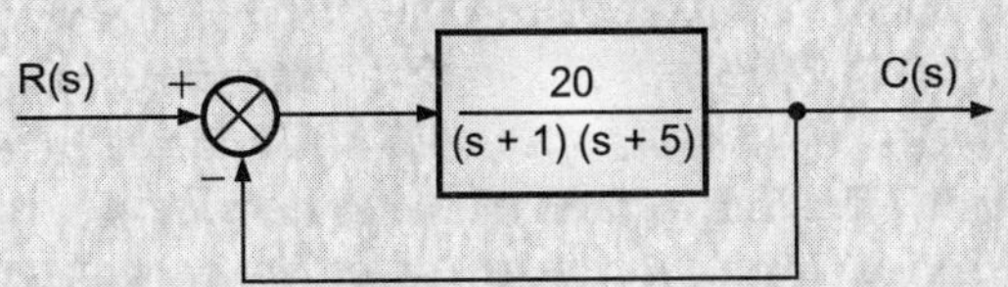

Fig. 6.10.3. Block diagram for example 6.10.5.

Determine the characteristic equation of the system, ω_n, ζ, ω_d, t_p, M_p, the time at which the first undershoot occurs, the time period of oscillations and the number of cycles completed before reaching the steady state.

Solution. The overall transfer function is given by

$$\frac{C(s)}{R(s)} = \frac{\dfrac{20}{[(s+1)(s+5)]}}{1+\dfrac{20}{[(s+1)(s+5)]}.1} = \frac{20}{(s^2+6s+25)} = \frac{20}{25}\cdot\frac{25}{(s^2+6s+25)}$$

The characteristic equation is given by

$(s^2 + 6s + 25) = 0.$ **Ans.**

From the characteristic equation following quantities are calculated :

$$\omega_n = \sqrt{25} = 5 \text{ rad/sec.} \quad \textbf{Ans.}$$

$$2\,\zeta\,\omega_n = 6$$

$$\therefore \qquad \zeta = \frac{6}{2\omega_n} = \frac{6}{2\times 5} = 0.6. \quad \textbf{Ans.}$$

$$\omega_d = \omega_n\sqrt{1-\zeta^2} = 5\sqrt{1-(0.6)^2} = 4 \text{ rad/sec.} \quad \textbf{Ans.}$$

The time at which the maximum overshoot occurs is

$$t_p = \frac{\pi}{\omega_n\left(\sqrt{1-\zeta^2}\right)} = \frac{\pi}{5\sqrt{1-0.6^2}} = 0.78 \text{ sec.} \quad \textbf{Ans.}$$

The maximum overshoot is given by

$$M_p = e^{-\frac{\zeta\pi}{\sqrt{1-\zeta^2}}}\times 100 = e^{-\frac{0.6\pi}{\sqrt{1-0.6^2}}}\times 100 = 9.4\%. \quad \textbf{Ans.}$$

The time at which the first undershoot occurs is

$$t = \frac{2\pi}{\omega_n\sqrt{1-\zeta^2}} = \frac{2\pi}{5\sqrt{1-(0.6)^2}} = 1.56 \text{ sec.} \quad \textbf{Ans.}$$

The period of oscillation $= \dfrac{2\pi}{\omega_d} = \dfrac{2\pi}{4} = 1.56$ sec. **Ans.**

As the steady state is reached in time $4/\zeta\omega_n$, the number of oscillations completed before reaching steady state is

$$\frac{\omega_d}{2\pi}\times\frac{4}{\zeta\omega_n} = \frac{4}{2\pi}\times\frac{4}{0.6\times 5} = 0.84 \text{ cycle.} \quad \textbf{Ans.}$$

Example 6.10.6. *A load is controlled by an electric motor. The position of the input shaft is adjusted to θ_r. The motor torque constant is 360 Nm per radian of error between the input and output position. The moment of inertia and the coefficient of viscous friction are 10 kg-m² and 60 Nm/(rad/sec) respectively. The system being initially at rest. Determine an expression relating time and angular position of the output following a sudden shift of 50° in the input shaft.*

Solution. The block diagram for the system is shown in Fig. 6.10.4.

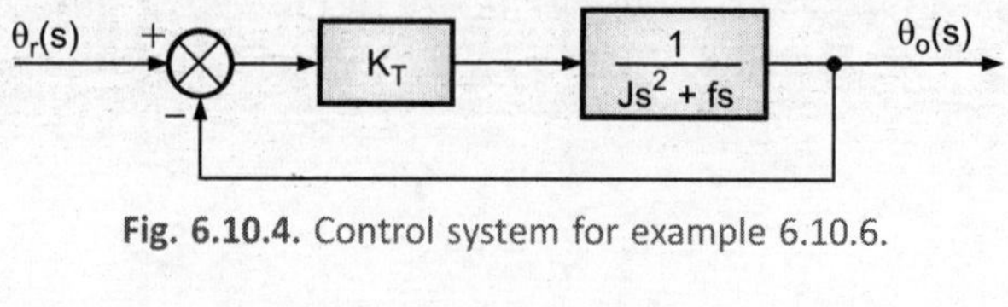

Fig. 6.10.4. Control system for example 6.10.6.

The overall transfer function is determined as

$$\frac{\theta_o(s)}{\theta_r(s)} = \frac{K_T \cdot \dfrac{1}{Js^2 + fs}}{1 + K_T \cdot \dfrac{1}{Js^2 + fs} \cdot 1} = \frac{K_T}{Js^2 + fs + K_T}$$

Substituting the values

$$\frac{\theta_o(s)}{\theta_r(s)} = \frac{360}{10s^2 + 60s + 360}$$

or

$$\frac{\theta_o(s)}{\theta_r(s)} = \frac{36}{s^2 + 6s + 36}$$

$$\therefore \quad \theta_o(s) = \theta_r(s) \frac{36}{s^2 + 6s + 36}$$

Since $\theta_r = 50 \times \frac{\pi}{180} = \frac{5\pi}{18}$ rad

$$\therefore \quad \theta_r(s) = \frac{5\pi}{18} \cdot \frac{1}{s}$$

$$\therefore \quad \theta_o(s) = \frac{5\pi}{18} \cdot \frac{1}{s} \cdot \frac{36}{(s^2 + 6s + 36)}$$

In the above relation

$$\omega_n = \sqrt{36} = 6 \text{ rad/sec}$$

$$2\zeta\omega_n = 6$$

$$\therefore \quad \zeta = \frac{6}{2\omega_n} = \frac{2}{2 \times 6} = 0.5$$

The time response expression is given by

$$\theta_o(t) = \frac{5\pi}{18}\left\{1 - \frac{e^{-\zeta\omega_n t}}{\sqrt{1-\zeta^2}} \sin\left[\left(\omega_n \sqrt{1-\zeta^2}\, t + \tan^{-1}\left(\frac{\sqrt{1-\zeta^2}}{\zeta}\right)\right)\right]\right\}$$

Substituting the values

$$\theta_o(t) = \frac{5\pi}{18}\left\{1 - \frac{e^{-0.5 \times 6t}}{\sqrt{1-(0.5)^2}} \sin\left[\left(6\sqrt{1-0.5^2}\, t + \tan^{-1}\frac{\sqrt{1-0.5^2}}{0.5}\right)\right]\right\}$$

or

$$\theta_o(t) = \frac{5\pi}{18}\left\{1 - \frac{e^{-3t}}{0.86} \sin\,[5.19t + \tan^{-1} 1.732]\right\}. \quad \textbf{Ans.}$$

Example 6.10.7. *The maximum overshoot for a unity feedback control system having its forward path transfer function as G(s) = K/s (sT + 1) is to be reduced from 60% to 20%. The system input is a unit step function.*

Determine the factor by which K should be reduced to achieve aforesaid reduction.

Solution. The overall transfer function is given by

$$\frac{C(s)}{R(s)} = \frac{\dfrac{K}{s(sT+1)}}{\left[1+\dfrac{K}{s(sT+1)}.1\right]} = \frac{K/T}{\left(s^2+\dfrac{1}{T}s+\dfrac{K}{T}\right)}$$

The characteristic equation is

$$s^2 + \frac{1}{T}s + \frac{K}{T} = 0$$

$$\therefore \quad \omega_n \sqrt{K/T} \quad \text{and} \quad 2\zeta\omega_n = 1/T$$

$$\therefore \quad \zeta = \frac{1}{T}\cdot\frac{1}{2\omega_n} = \frac{1}{T}\cdot\frac{1}{2\sqrt{K/T}} = \frac{1}{2\sqrt{KT}}$$

Let K_1 be the forward path gain when $M_{p1} = 60\%$ and the corresponding damping ratio be ζ_1.

Since $$M_{p1} = e^{-\frac{\zeta_1\pi}{\sqrt{1-\zeta_1^2}}} \times 100\% \quad \therefore \quad 60 = e^{-\frac{\zeta_1\pi}{\sqrt{1-\zeta_1^2}}} \times 100$$

or $$\log_e(0.6) = -\frac{\zeta_1\pi}{\sqrt{1-\zeta_1^2}}\log_e(e) \quad \text{or} \quad -0.51 = -\frac{\zeta_1\pi}{\sqrt{1-\zeta_1^2}}.1$$

$$\therefore \quad \zeta_1^2 = \frac{0.51^2}{\pi^2}(1-\zeta_1^2)$$

or $$\zeta_1^2 = 0.026\,(1-\zeta_1^2) \quad \text{or} \quad \zeta_1^2 = 0.025$$

$$\therefore \quad \zeta_1 = 0.158$$

Let K_2 be the forward path gain when $M_{p2} = 20\%$ and the corresponding damping ratio be ζ_2.

Since $$M_{p2} = e^{-\frac{\zeta_2\pi}{\sqrt{1-\zeta_2^2}}} \times 100\% \quad \therefore \quad 20 = e^{-\frac{\zeta_2\pi}{\sqrt{1-\zeta_2^2}}} \times 100$$

From the above relation the value of ζ_2 can be calculated as

$$\zeta_2 = 0.447$$

Assuming time constant T to be constant

$$\zeta_1 = \frac{1}{2}\cdot\frac{1}{\sqrt{K_1T}} \quad \text{and} \quad \zeta_2 = \frac{1}{2}\cdot\frac{1}{\sqrt{K_2T}}$$

$$\frac{\zeta_1}{\zeta_2} = \frac{1}{2}\cdot\frac{1}{\sqrt{K_1T}} \times \frac{2\sqrt{K_2T}}{1}$$

Hence, $$\frac{K_2}{K_1} = \left(\frac{\zeta_1}{\zeta_2}\right)^2 = \left(\frac{0.158}{0.447}\right)^2 = \frac{1}{8}. \quad \textbf{Ans.}$$

Example 6.10.8. *An automatic speed control system uses an error detector whose gain in 1 V/(rad/sec), the error signal is given to an electric motor having a torque constant 75 Nm/V. The combined moment of inertia and coefficient of viscous friction of the motor and load are 10 kg-m² and 5 Nm/(rad/sec) respectively. If the input setting is suddenly increased by 60 r.p.m. obtain an expression relating the load speed and time. Also calculate :*

(a) The steady state speed and (b) The steady state error.

Solution. The system equations are

$$\omega_e(s) = \omega_r(s) - \omega_o(s) \quad ...(1)$$

$$E(s) = 1 \,.\, \omega_e(s) \quad ...(2)$$

$$T_M(s) = 75 \,.\, E(s) \quad ...(3)$$

$$\omega_o(s) = T_M(s) \,.\, \frac{1}{(10s+5)} \quad ...(4)$$

From the above equations the block diagram shown in Fig. 6.10.5 is obtained.

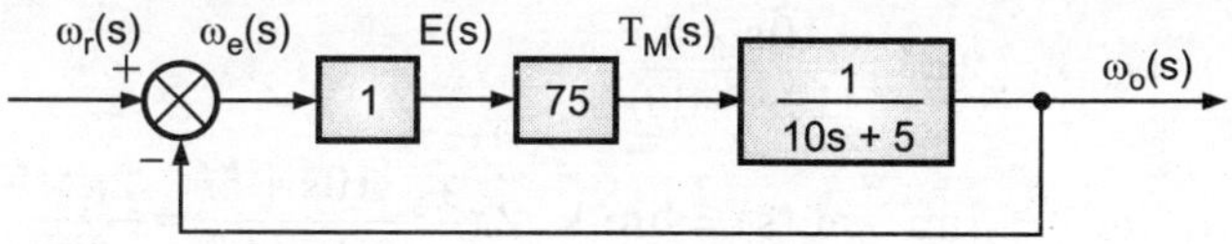

Fig. 6.10.5. Block diagram for example 6.10.8.

The transfer function relating $\omega_o(s)$ and $\omega_r(s)$ is

$$\frac{\omega_o(s)}{\omega_r(s)} = \frac{1\times 75\times \frac{1}{(10s+5)}}{1+1\times 75\times \frac{1}{(10s+5)}\,.\,1}$$

or $$\frac{\omega_o(s)}{\omega_r(s)} = \frac{7.5}{(s+8)} \qquad \therefore \quad \omega_o(s) = \omega_r(s)\frac{7.5}{(s+8)}$$

Since $$\omega_r = \frac{2\pi}{60}\times 60 = 2\pi \text{ rad/sec}$$

$$\therefore \quad \omega_r(s) = 2\pi\frac{1}{s}$$

$$\therefore \quad \omega_o(s) = 2\pi\frac{1}{s}\,.\,\frac{7.5}{(s+8)} = 2\pi\times 7.5\left[\frac{K_1}{s}+\frac{K_2}{(s+8)}\right]$$

The coefficients can be determined as

$$K_1 = 0.125 \quad \text{and} \quad K_2 = -\,0.125$$

$$\therefore \quad \omega_o(s) = 2\pi\times 7.5\left[\frac{0.125}{s} - \frac{0.125}{(s+8)}\right]$$

Taking inverse Laplace transform on both sides

$$\omega_o(t) = 2\pi \times 7.5(0.125 - 0.125\, e^{-8t})$$

or $\omega_o(t) = 1.875\pi\,(1 - e^{-8t})$. **Ans.**

(*a*) $$\omega_{oss} = \lim_{s\to 0} s\,\omega_o(s) = \lim_{s\to 0} s\, 2\pi \frac{1}{s} \cdot \frac{7.5}{(s+8)}$$

$$= \frac{2\pi \times 7.5}{8} = 1.875\,\pi \text{ rad/sec}$$

In r.p.m. $N_{oss} = 1.875\,\pi \times (60/2\pi) = 56.25$ r.p.m. **Ans.**

(*b*) $$\frac{\omega_e(s)}{\omega_r(s)} = \frac{1}{1 + 1 \times 75 \times \dfrac{1}{(10s+5)} . 1} = \frac{10s+5}{(10s+80)}$$

$\therefore$ $$\omega_e(s) = \omega_r(s)\,\frac{10s+5}{(10s+80)}$$

Since $$\omega_r(s) = 2\pi \frac{1}{s}$$

$\therefore$ $$\omega_e(s) = 2\pi \frac{1}{s} \cdot \frac{10s+5}{(10s+80)}$$

$$\omega_{ess} = \lim_{s\to 0} s\,\omega_e(s) = \lim_{s\to 0} s \,.\, 2\pi \frac{1}{s} \cdot \frac{10s+5}{10s+80} = \frac{2\pi \times 5}{80} = \frac{\pi}{8} \text{ rad/sec}$$

In r.p.m. $$N_{ess} = \frac{\pi}{8} \times \left(\frac{60}{2\pi}\right) = 3.75 \text{ r.p.m.}$$ **Ans.**

Example 6.10.9. *A second order control system is represented by a transfer function given below:*

$$\frac{\theta_o(s)}{T(s)} = \frac{1}{Js^2 + fs + K}$$

where θ_o is the proportional output and T is the input torque.

A step input of 10 Nm is applied to the system and test results are given below:

(a) M_p = 6%, (b) t_p = 1 sec and (c) The steady state value of the output is 0.5 radian. Determine the values of J, f and K.

Solution. $$\theta_o(s) = T(s)\,\frac{1}{Js^2 + fs + K}$$

Since $$\theta_{oss} = \lim_{s\to 0} s\,\theta_o(s)$$

$\therefore$ $$\theta_{oss} = \lim_{s\to 0} s \,.\, T(s) \,.\, \frac{1}{Js^2 + fs + K}$$

Since $$T(s) = 10\frac{1}{s}$$

$\therefore$ $$\theta_{oss} = \lim_{s\to 0} s \,.\, \frac{10}{s} \cdot \frac{1}{Js^2 + fs + K}$$

$\therefore \quad \theta_{oss} = 10/K$

As $\quad \theta_{oss} = 0.5$ rad (given)

$\therefore \quad 0.5 = 10/K,$ hence $K = 20$ N. **Ans.**

$$\%M_p = e^{-\frac{\zeta\pi}{\sqrt{1-\zeta^2}}} \times 100$$

As $\quad M_p = 6\%$ (given)

$$\therefore \quad 6 = e^{-\frac{\zeta\pi}{\sqrt{1-\zeta^2}}} \times 100 \quad \text{or} \quad \frac{1}{0.06} = e^{\frac{\zeta\pi}{\sqrt{1-\zeta^2}}}$$

$$\therefore \quad \log_e\left(\frac{1}{0.06}\right) = \frac{\zeta\pi}{\sqrt{1-\zeta^2}} \log_e(e)$$

or

$$2.8 = \frac{\zeta\pi}{\sqrt{1-\zeta^2}} \times 1$$

$$\therefore \quad \frac{(1-\zeta^2)}{\zeta^2} = \left(\frac{\pi}{2.8}\right)^2, \quad \therefore \quad \zeta = 0.66$$

$$t_p = \frac{\pi}{\omega_n\sqrt{1-\zeta^2}}$$

As $\quad t_p = 1$ sec (given) and $\zeta = 0.66$

$$\therefore \quad 1 = \frac{\pi}{\omega_n\sqrt{1-(0.66)^2}} \quad \therefore \quad \omega_n = 4.2 \text{ rad/sec.}$$

The characteristic equation is

$$s^2 + \frac{f}{J}s + \frac{K}{J} = 0$$

$$\therefore \quad \omega_n = \sqrt{K/J} \quad \therefore \quad 4.2 = \sqrt{20/J}$$

$\therefore \quad J = 1.13$ kg-m^2. **Ans.**

As $\quad 2\zeta\omega_n = f/J$

$\therefore \quad 2 \times 0.66 \times 4.2 = f/1.13$

$\therefore \quad f = 6.26$ Nm/(rad/sec). **Ans.**

Example 6.10.10. *A two phase a.c. servomotor having a torque constant of 0.045 Nm/V controls a position load through a gear ratio of 10 : 1. The effective moment of inertia and coefficient of viscous friction referred to load side are 0.25 kg-m^2 and 1.0 N-m (rad/sec). The synchro (megslip) error detector produces an error signal of 0.1 V per degree error in misalignment. Develop the block diagram representation of the control system and therefrom obtain the transfer function.*

(a) Calculate the natural frequency of oscillations and damping ratio.

(b) If the damping is to be made critical using a derivative error control, determine the time constant of the phase advance circuit.

Solution. The error detector gain is

$$K_e = 0.1 \text{ V/degree} = 0.1 \times \frac{180}{\pi} = 5.73 \text{ V/rad}$$

Referred to motor side

$$J = 0.25 \left(\frac{N_1}{N_2}\right)^2 = 0.25 \times \frac{1}{10^2} = 0.25 \times \frac{1}{100} \text{ kg-m}^2$$

$$f = 1 \times \left(\frac{N_1}{N_2}\right)^2 = 1 \times \frac{1}{10^2} = 1 \times \frac{1}{100} \text{ Nm/(rad/sec)}$$

The block diagram of the system is shown in Fig. 6.10.6.

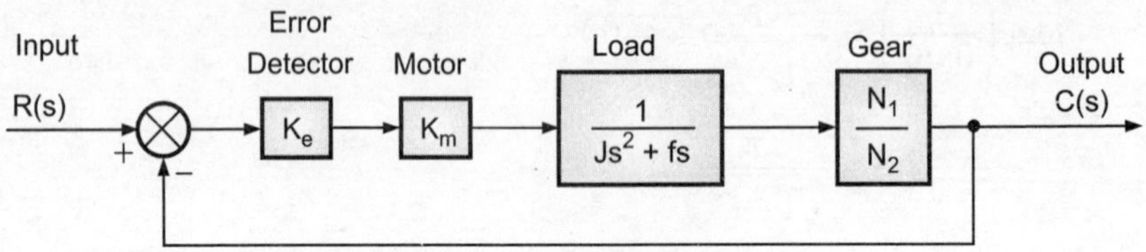

Fig. 6.10.6. Block diagram for example 6.10.10.

Substituting numerical values the block diagram shown in Fig. 6.10.6 (*a*) obtained.

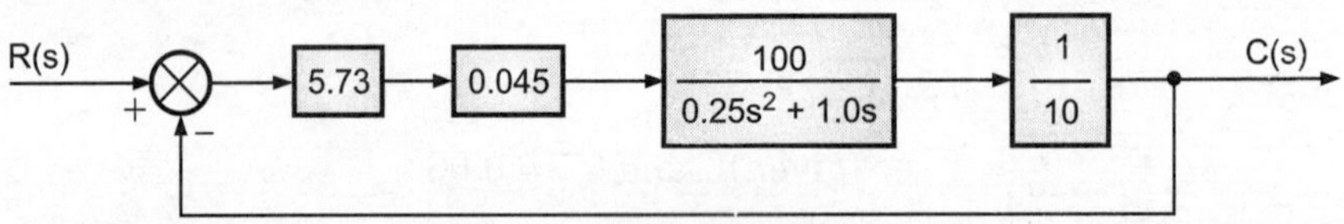

Fig. 6.10.6. (*a*) Block diagram of Fig. 6.10.6 with numerical values substituted.

The block diagram [Fig. 6.10.6 (*a*)] is simplified and shown in Fig. 6.10.6 (*b*).

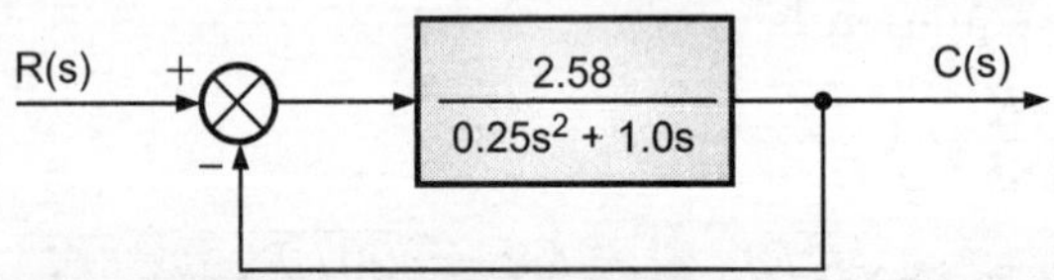

Fig. 6.10.6. (*b*) Block diagram of Fig. 6.10.6 (*a*) simplified.

The transfer function of the system is

$$\frac{C(s)}{R(s)} = \frac{\dfrac{2.58}{(0.25s^2 + s)}}{\left[1 + \dfrac{2.58}{0.25s^2 + s} . 1\right]}$$

or
$$\frac{C(s)}{R(s)} = \frac{10.32}{(s^2 + 4s + 10.32)}$$

The characteristic equation of the system is

$$s^2 + 4s + 10.32 = 0$$

(*a*)

$$\omega_n = \sqrt{10.32} = 3.21 \text{ rad/sec}. \quad \textbf{Ans.}$$

$$2\zeta\omega_n = 4$$

$$\therefore \quad \zeta = \frac{4}{2\omega_n} = \frac{4}{2 \times 3.21} = 0.623. \quad \textbf{Ans.}$$

(*b*) To make the damping critical (*i.e.* $\zeta = 1$) the derivative error control is incorporated using a phase advance circuit as shown in Fig. 6.10.6 (*c*).

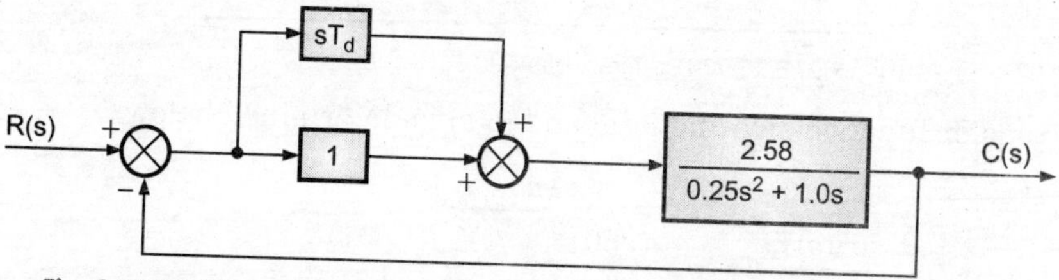

Fig. 6.10.6 (*c*) Control system of example 6.10.10 with derivative error control.

The block diagram shown in Fig. 6.10.6 (*c*) is simplified and shown in Fig. 6.10.6 (*d*).

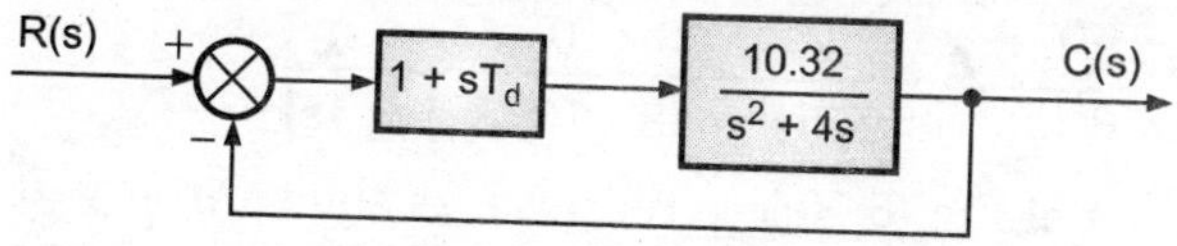

Fig. 6.10.6 (*d*) Block diagram of Fig. 6.10.6 (*c*) simplified.

The transfer function for the system incorporating derivative error control is thus

$$\frac{C(s)}{R(s)} = \frac{(1 + sT_d) \cdot \dfrac{10.32}{(s^2 + 4s)}}{\left[1 + (1 + sT_d) \cdot \dfrac{10.32}{(s^2 + 4s)} \cdot 1\right]}$$

or
$$\frac{C(s)}{R(s)} = \frac{10.32(1 + sT_d)}{[s^2 + (10.32T_d + 4)\, s + 10.32]}$$

The characteristic equation of the system incorporating derivative control is

$$[s^2 + (10.32\, T_d + 4)\, s + 10.32] = 0$$

$$\therefore \quad \omega_n = \sqrt{10.32} = 3.21 \text{ rad/sec}$$

and
$$2\zeta\omega_n = (10.32\, T_d + 4)$$

As the damping is to be critical, $\zeta = 1$

$$\therefore \quad 2 \times 1 \times 3.21 = (10.32\, T_d + 4) \qquad \therefore \quad 10.32\, T_d = 2.42$$

or
$$T_d = \frac{2.42}{10.32} = 0.234 \text{ sec.} \quad \textbf{Ans.}$$

Example 6.10.11. *The moment of inertia referred to the load side of a unity feedback proportional type control system is 250 kg-m² while the controller gain is 8×10^4 Nm per radian error of misalignment between the output and input shafts.*

The system is to be critically damped by incorporating output derivative (rate) feedback. Determine the characteristic equation and the steady state error if the input is 5 r.p.m.

Solution. Let the coefficient of viscous friction on the load side be f. The block diagram for the system is developed as shown in Fig. 6.10.7.

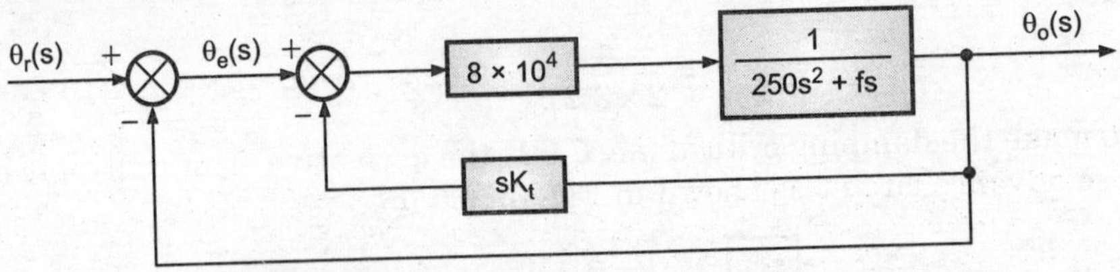

Fig. 6.10.7. The block diagram for example 6.10.11.

The transfer function relating $\theta_o(s)$ and $\theta_e(s)$ is determined below

$$\frac{\theta_o(s)}{\theta_e(s)} = \frac{\dfrac{8\times10^4}{(250s^2+fs)}}{\left[1+\dfrac{8\times10^4}{(250s^2+fs)}.sK_t\right]}$$

or

$$\frac{\theta_o(s)}{\theta_e(s)} = \frac{8\times10^4}{[(250s^2+(8\times10^4K_t+f)s]}$$

The block diagram of Fig. 6.10.7 is redrawn as shown in Fig. 6.10.7 (*a*).

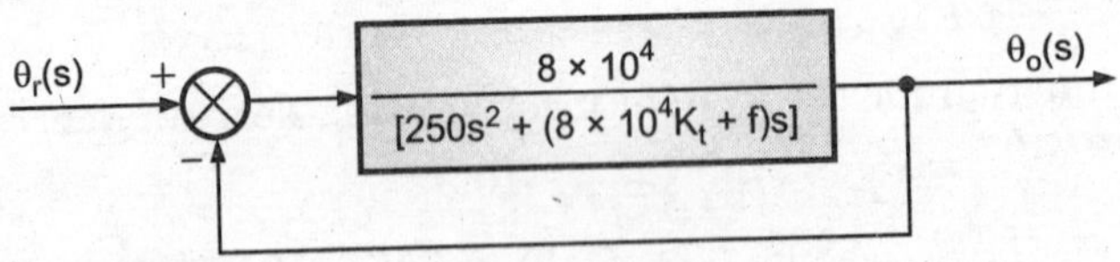

Fig. 6.10.7 (*a*) Block diagram of Fig. 6.10.7 redrawn.

The transfer function relating $\theta_o(s)$ and $\theta_r(s)$ is determined below :

$$\frac{\theta_o(s)}{\theta_r(s)} = \frac{\dfrac{8\times10^4}{[250s^2+(8\times10^4K_t+f)s]}}{1+\dfrac{8\times10^4}{[(250s^2+(8\times10^4K_t+f)s]}.1}$$

or

$$\frac{\theta_o(s)}{\theta_r(s)} = \frac{8\times10^4}{[250s^2+(8\times10^4K_t+f)s+8\times10^4]}$$

or

$$\frac{\theta_o(s)}{\theta_r(s)} = \frac{320}{s^2+\left[\dfrac{(8\times10^4K_t+f)}{250}\right]s+320}$$

The characteristic equation of the system of

$$s^2+\left[\frac{(8\times10^4K_t+f)}{250}\right]s+320=0 \quad \therefore \quad \omega_n=\sqrt{320}=17.88 \text{ rad/sec}$$

and

$$2\zeta\omega_n = \left[\frac{(8\times10^4K_t+f)}{250}\right]$$

For critical damping $\zeta = 1$

$\therefore$ $2\zeta\omega_n = 2 \times 1 \times 17.88 = 35.7$, hence the characteristic equation for critical damping is

$$s^2 + 35.7s + 320 = 0. \quad \textbf{Ans.}$$

The input is 5 r.p.m., *i.e.* $\dfrac{5 \times 2\pi}{60} = \dfrac{\pi}{6}$ rad/sec.

As the steady state error for a unit ramp input is $2\,\zeta/\omega_n$, therefore, when the input is $\pi/6$ rad/sec. The steady state error is

$$e_{ss} = \frac{2\zeta}{\omega_n} \cdot \frac{\pi}{6} = \frac{2 \times 1}{17.88} \cdot \frac{\pi}{6} = 0.058 \text{ rad.} \quad \textbf{Ans.}$$

Example 6.10.12. *The open-loop transfer function of a unity feedback control system is given by*

$$G(s) = \frac{25}{s(s+5)}$$

(*a*) *Calculate:*

(*i*) *The natural frequency of oscillations, damped frequency of oscillations, damping factor, damping ratio and the maximum overshoot of a unit step input.*

(*ii*) *The steady state error for a unit ramp input.*

(*b*) *If the damping ratio is to be made 0.75 using a tachometer feedback, calculate the tachometer constant and determine the maximum overshoot.*

Solution. (*a*) The block diagram is drawn in Fig. 6.10.8.

The overall transfer function is

$$\frac{C(s)}{R(s)} = \frac{\dfrac{25}{s(s+5)}}{1 + \dfrac{25}{[s\,(s+5)]} \cdot 1}$$

$$\frac{C(s)}{R(s)} = \frac{25}{s^2 + 5s + 25}$$

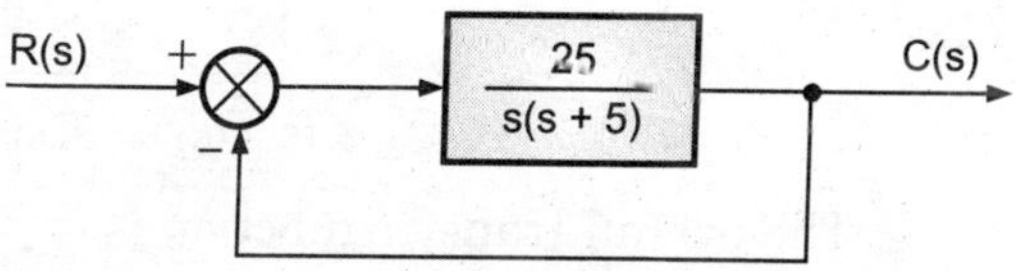

Fig. 6.10.8. Block diagram for example 6.10.12.

The characteristic equation is

$$(s^2 + 5s + 25) = 0$$

(*i*) $\omega_n = \sqrt{25} = 5$ rad/sec. **Ans.**

$$2\,\zeta\omega_n = 5$$

$\therefore$ Damping factor, $\zeta\omega_n = \dfrac{5}{2} = 2.5$. **Ans.**

Damping ratio $\zeta = \dfrac{5}{2\omega_n} = \dfrac{5}{2 \times 5} = 0.5$. **Ans.**

Damped frequency of oscillations

$$\omega_d = \omega_n\sqrt{1-\zeta^2} = 5\sqrt{1-(0.5)^2} = 4.3 \text{ rad/sec.} \quad \textbf{Ans.}$$

$$\%M_p = e^{-\frac{\zeta\pi}{\sqrt{1-\zeta^2}}} \times 100 = e^{-\frac{0.5\pi}{\sqrt{1-(0.5)^2}}} \times 100 = 16.3\%. \quad \textbf{Ans.}$$

(*ii*) The steady state error is given by

$$e_{ss} = \frac{2\zeta}{\omega_n} = \frac{2 \times 0.5}{5} = 0.2 \text{ rad.} \quad \textbf{Ans.}$$

(*b*) The damping ratio is to be increased from 0.5 to 0.75 by incorporating tachometer feedback. The block diagram incorporating tachometer feedback is drawn in Fig. 6.10.8 (*a*).

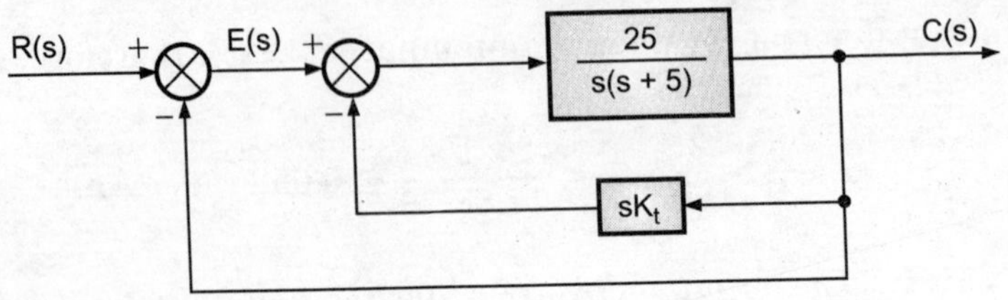

Fig. 6.10.8 (*a*) Block diagram for example 6.10.12 incorporating tachometer feedback.

The transfer function relating $C(s)$ and $E(s)$ is

$$\frac{C(s)}{E(s)} = \frac{\dfrac{25}{s(s+5)}}{1+\left[\dfrac{25}{s(s+5)}\right]sK_t}$$

$$\frac{C(s)}{E(s)} = \frac{25}{[s^2 + (25K_t + 5)s]}$$

The block diagram relating $C(s)$ and $R(s)$ is redrawn in Fig. 6.10.8 (*b*).

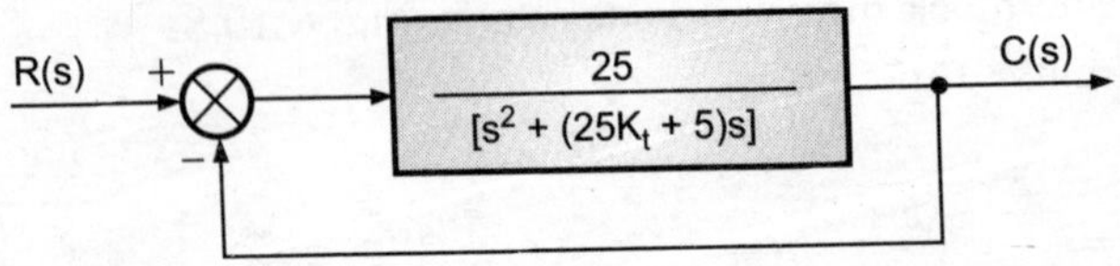

Fig. 6.10.8 (*b*) Block diagram of Fig. 6.10.8 (*a*) redrawn.

The overall transfer function is

$$\frac{C(s)}{R(s)} = \frac{25}{[(s^2 + (25K_t + 5)s + 25]}$$

The characteristic equation is

$$[s^2 + (25K_t + 5)\, s + 25] = 0$$

As the new value of ζ is 0.75

$$\therefore \quad 2\zeta\omega_n = (25K_t + 5) \text{ and since } \omega_n = \sqrt{25} = 5 \text{ rad/sec}$$

$$\therefore \quad 2 \times 0.75 \times 5 = (25K_t + 5)$$

$$\therefore \quad K_t = 0.1 \text{ (V/rad/sec).} \quad \textbf{Ans.}$$

$$\%M_p = e^{-\frac{\zeta\pi}{\sqrt{1-\zeta^2}}} \times 100 = e^{-\frac{0.75\pi}{\sqrt{1-(0.75)^2}}} \times 100 = 2.8\%. \quad \textbf{Ans.}$$

Example 6.10.13. *A closed-loop control system is shown in Fig. 6.10.9.*

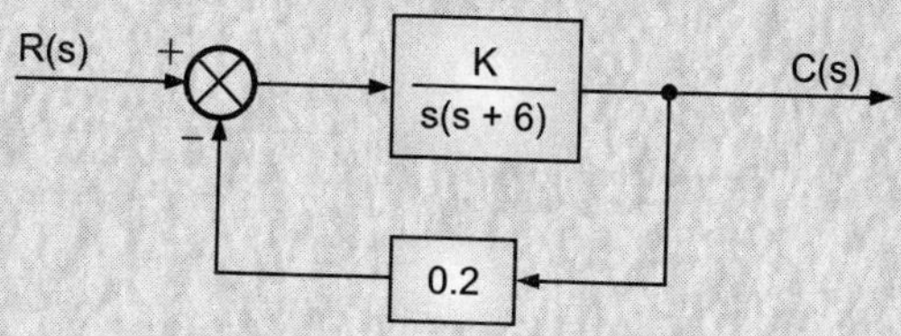

Fig. 6.10.9. Block diagram for example 6.10.13.

The system is to have a damping ratio of 0.7. Determine the value of K to satisfy this condition and calculate the settling time, peak time and maximum overshoot for the value of K thus determined.

Solution. The overall transfer function relating $C(s)$ and $R(s)$ is

$$\frac{C(s)}{R(s)} = \frac{\dfrac{K}{s(s+6)}}{1+\left[\dfrac{K}{s(s+6)}\right]0.2}$$

or
$$\frac{C(s)}{R(s)} = \frac{K}{(s^2+6s+0.2K)}$$

The characteristic equation is

$$s^2 + 6s + 0.2\,K - 0$$

$\therefore$
$$\omega_n = \sqrt{0.2K} \text{ and } 2\,\zeta\omega_n = 6$$

Since
$$\zeta = 0.7 \text{ (given)}$$

$\therefore$
$$2 \times 0.7 \times \sqrt{0.2K} = 6$$

or
$$K = \left(\frac{6}{2\times 0.7}\right)^2 \times \frac{1}{0.2} = 91.8$$

$\therefore$
$$\omega_n = \sqrt{0.2\times 91.8} = 4.28 \text{ rad/sec.}$$ **Ans.**

Settling time
$$t_s = \frac{4}{\zeta\omega_n} = \frac{4}{0.7\times 4.28} = 1.33 \text{ sec.}$$ **Ans.**

Peak time
$$t_p = \frac{\pi}{\omega_n\sqrt{1-\zeta^2}} = \frac{\pi}{4.28\sqrt{1-(0.7)^2}} = 1.03 \text{ sec.}$$ **Ans.**

Maximum overshoot

$$M_p = e^{-\frac{\zeta\pi}{\sqrt{1-\zeta^2}}} \times 100 = e^{-\frac{0.7\pi}{\sqrt{1-(0.7)^2}}} \times 100 = 4.59\%.$$ **Ans.**

Example 6.10.14. *For the control system shown in Fig. 6.10.10.*

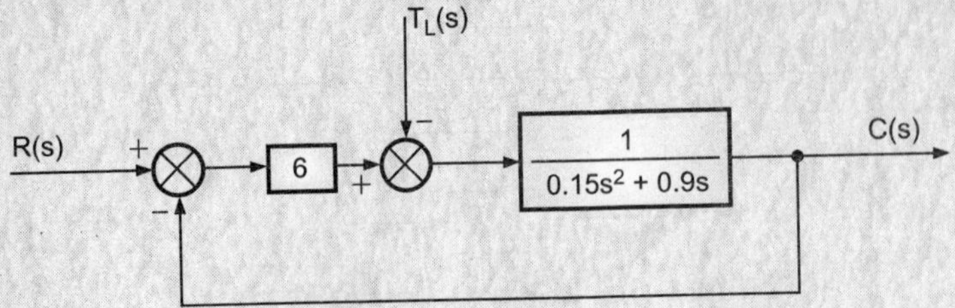

Fig. 6.10.10. Block diagram for example 6.10.14.

(a) Determine

(i) M_p for a unit step input

(ii) e_{ss} for unit ramp input.

(b) Calculate the steady state value of the output when the input shaft is held fixed and a sudden torque T_L = 1 Nm is applied.

Solution. There are two inputs

(1) Reference input $R(s)$, and

(2) Load torque $T_L(s)$.

(*a*) While considering the effect of $R(s)$ put $T_L(s) = 0$, therefore, the block diagram takes the form as shown in Fig. 6.10.10 (*a*).

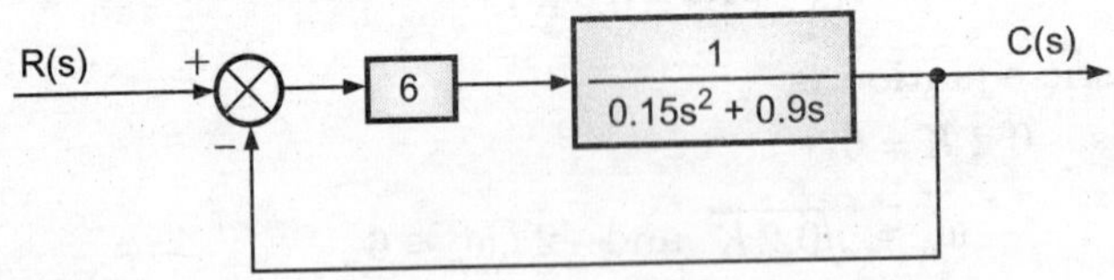

Fig. 6.10.10 (*a*) Block diagram for example 6.10.14 considering $T_L(s) = 0$.

The transfer function for the system is

$$\frac{C(s)}{R(s)} = \frac{\dfrac{6}{0.15s^2 + 0.9s}}{1 + \left[\dfrac{6}{0.15s^2 + 0.9s}\right] \times 1}$$

or
$$\frac{C(s)}{R(s)} = \frac{6}{(0.15s^2 + 0.9s + 6)} \quad \text{or} \quad \frac{C(s)}{R(s)} = \frac{40}{(s^2 + 6s + 40)}$$

The characteristic equation is

$$(s^2 + 6s + 40) = 0$$

$$\therefore \qquad \omega_n = \sqrt{40} = 6.32 \text{ rad/sec} \quad \text{and} \quad 2\zeta\,\omega_n = 6$$

$$\therefore \qquad \zeta = \frac{6}{2 \times 6.32} = 0.474$$

(*i*)
$$M_p = e^{-\frac{\zeta\pi}{\sqrt{1-\zeta^2}}} \times 100 = e^{-\frac{0.474\pi}{\sqrt{1-0.474^2}}} \times 100 = 18.39\%. \quad \textbf{Ans.}$$

(*ii*) $$e_{ss} = \frac{2\zeta}{\omega_n} = \frac{2\times 0.474}{6.32} = 0.15 \text{ rad.} \quad \textbf{Ans.}$$

(*b*) For considering the effect of $T_L(s)$, put $R(s) = 0$, therefore, the block diagram of Fig. 6.10.10 takes the form as shown in Fig. 6.10.10 (*b*).

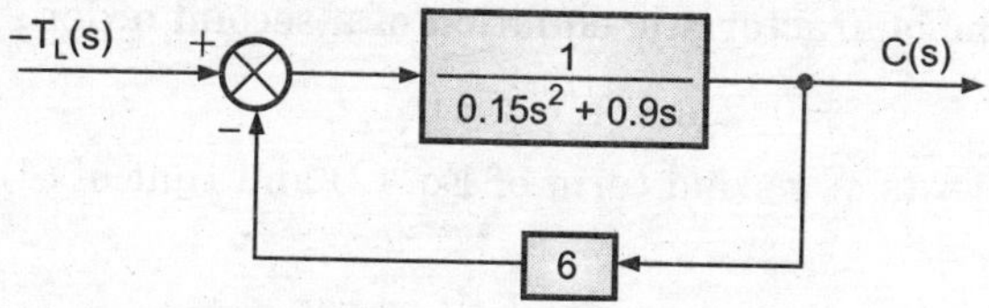

Fig. 6.10.10 (*b*) Block diagram for example 6.10.14 considering $R(s) = 0$.

The transfer function for the system is

$$\frac{C(s)}{-T_L(s)} = \frac{\dfrac{1}{0.15s^2 + 0.9s}}{1 + \left[\dfrac{1}{0.15s^2 + 0.9s}\right]6}$$

or $$\frac{C(s)}{-T_L(s)} = \frac{1}{(0.15s^2 + 0.9s + 6)}$$

or $$C(s) = -T_L(s)\frac{1}{(0.15s^2 + 0.9s + 6)}$$

Since $T_L = 1$ Nm, $T_L(s) = 1/s$

$\therefore$ $$C(s) = -\frac{1}{s}\cdot\frac{1}{(0.15s^2 + 0.9s + 6)}$$

$\therefore$ $$C_{ss} = \lim_{s\to 0} s\left[-\frac{1}{s}\cdot\frac{1}{(0.15s^2 + 0.9s + 6)}\right]$$

or $$C_{ss} = -\frac{1}{6} = -0.166 \text{ rad.} \quad \textbf{Ans.}$$

(**Note.** Negative sign indicates that the output position lags behind the input.)

Example 6.10.15. *The open-loop transfer function of a unity feedback control system is given by:*

$$G(s) = \frac{K(s+2)}{s^3 + \beta s^2 + 4s + 1}$$

Determine the value of K and β such that the closed-loop unit step response has $\omega_n = 3$ rad/sec and $\zeta = 0.2$.

Solution. The characteristics equation of the system is

$$1 + G(s) = 0$$

$\therefore$ $$1 + \frac{K(s+2)}{s^3 + \beta s^2 + 4s + 1} = 0$$

or $$s^3 + \beta s^2 + 4s + 1 + K(s + 2) = 0$$

or $$s^3 + \beta s^2 + (K + 4)s + (2K + 1) = 0 \quad ...(1)$$

Let $$(s + a)(s^2 + bs + c) = 0 \quad ...(2)$$

$\therefore$ $$s^3 + (a + b)s^2 + (ab + c)s + ac = 0 \quad ...(3)$$

It is known that the characteristic equation of a second order system is given by

$$s^2 + 2\zeta\omega_n s + \omega_n^2 = 0 \quad ...(4)$$

Comparing coefficients of second term of Eq. (2) and that of (4), the values of b and c are obtained below :

$$c = \omega_n^2 = (3)^2 = 9$$

$$b = 2\zeta\omega_n = 2 \times 0.2 \times 3 = 1.2$$

On comparing coefficients of Eq. (1) and (3)

$$(a + b) = \beta \quad ...(5)$$

$$(ab + c) = K + 4 \quad ...(6)$$

$$ac = 2K + 1 \quad ...(7)$$

From equation (7) : $a = \dfrac{2K + 1}{c} \quad \because \quad c = 9$

$\therefore$ $$a = (2K + 1)/9$$

Since, $$b = 1.2, \text{ and } c = 9$$

$$(ab + c) = K + 4$$

$\therefore$ $$(a \times 1.2 + 9) = K + 4$$

$\because$ $$a = (2K + 1)/9$$

$\therefore$ $$[(2K + 1)/9] \times 1.2 + 9 = K + 4 \quad \text{or} \quad 6.6K = 45$$

$$K = \frac{45}{6.6} = 6.81. \quad \textbf{Ans.}$$

$\therefore$ $$a = \frac{2K + 1}{9} = \frac{2 \times 6.81 + 1}{9} = 1.51$$

From equation (5) $\beta = (a + b) = (1.51 + 1.2) = 2.71.$ **Ans.**

Example 6.10.16. *Determine the damping ratio and undamped natural frequency of oscillatory roots and $\%M_p$ for a unit step input given that :*

$$\frac{C(s)}{E(s)} = \frac{1}{s(1 + 0.5\,s)(1 + 0.2\,s)}$$

and the system is unity feedback type.

Solution. $$\frac{C(s)}{E(s)} = \frac{1}{0.5 \times 0.2\,s(s + 2)(s + 5)} = \frac{10}{s(s + 2)(s + 5)} = \frac{10}{s^3 + 7s^2 + 10s}$$

$$\frac{C(s)}{R(s)} = \frac{G(s)}{1 + G(s) \times 1}$$

$\because$ $$G(s) = \frac{10}{s^3 + 7s^2 + 10s}$$

$$\frac{C(s)}{R(s)} = \frac{\frac{10}{s^3 + 7s^2 + 10s}}{1 + \frac{10}{s^3 + 7s^2 + 10s} \times 1} \quad \text{or} \quad \frac{C(s)}{R(s)} = \frac{10}{s^3 + 7s^2 + 10s + 10}$$

The characteristic equation of the system is given by

$$s^3 + 7s^2 + 10s + 10 = 0$$

To a fair approximation one of the three roots of the above characteristic equation can be determined as $s = -5.5$, thus the quadratic term of the above equation can be determined as

$$\frac{s^3 + 7s^2 + 10s + 10}{s + 5.5} = s^2 + 1.5s + 1.8$$

$$\therefore \quad \frac{C(s)}{R(s)} = \frac{10}{(s + 5.5)(s^2 + 1.5s + 1.8)}$$

The characteristic equation of the system can be rewritten as

$$(s + 5.5)(s^2 + 1.5s + 1.8) = 0$$

For oscillatory roots :

$$\omega_n^2 = 1.8$$

or $\omega_n = \sqrt{1.8} = 1.34$ rad/sec. **Ans.**

or $2\zeta\omega_n = 1.5$

$$\therefore \quad \zeta = \frac{1.5}{2\omega_n} = \frac{1.5}{2 \times 1.34} = 0.56. \quad \textbf{Ans.}$$

The time constant of quadratic term is $1/\zeta\omega_n = \frac{1}{0.56 \times 1.34} = 1.33$ sec

The time constant due to first order term is 1/5.5 = 0.18 sec, its effect is neglected as compared to quadratic term time constant 1.33 sec

$$\%M_p = e^{-\frac{\zeta\pi}{\sqrt{1-\zeta^2}}} \times 100 = e^{\frac{-0.56\pi}{\sqrt{1-0.56^2}}} \times 100$$

$$= e^{-\frac{1.76}{0.83}} \times 100 = e^{-2.12} \times 100 = 11.9\%. \quad \textbf{Ans.}$$

Example 6.10.17. *The open-loop transfer function of a control system is given below:*

$$G(s)\,H(s) = \frac{2(s^2 + 3s + 20)}{s(s + 2)(s^2 + 4s + 10)}$$

Determine the static error coefficients and steady state error for the input given as:

(a) 5 *(b)* $4t$ *(c)* $4t^2/2$.

Solution.

(a) $$K_p = \lim_{s \to 0} G(s)\,H(s) = \lim_{s \to 0} \frac{2(s^2 + 3s + 20)}{s(s + 2)(s^2 + 4s + 10)} = \infty$$

$$e_{ss} = \frac{1}{1+K_p} = 5\frac{1}{1+\infty} = 0. \quad \textbf{Ans.}$$

(b)
$$K_v = \lim_{s\to 0} sG(s)\,H(s) = \lim_{s\to 0} \frac{s\,.\,2(s^2+3s+20)}{s(s+2)(s^2+4s+10)} = 2$$

$$e_{ss} = \frac{1}{K_v}\times 4 = \frac{1}{2}\times 4 = 2. \quad \textbf{Ans.}$$

(c)
$$K_a = \lim_{s\to 0} s^2 G(s)\,H(s) = \lim_{s\to 0} \frac{s^2\,.\,2(s^2+3s+20)}{s(s+2)(s^2+4s+10)} = 0$$

$$e_{ss} = \frac{1}{K_a}\times 4 = \frac{1}{0}\times 4 = \infty. \quad \textbf{Ans.}$$

Example 6.10.18. *Determine the position, velocity and acceleration error coefficients for a system given by :*

(i) $G(s) = \dfrac{K}{s(s^2+2\zeta\omega_n s+\omega_n^2)}$ *(unity feedback)*

(ii) $G(s)H(s) = \dfrac{100(s+2)(s+40)}{s^3(s^2+4s+200)}$.

Solution.

(i)
$$K_p = \lim_{s\to 0} G(s) = \lim_{s\to 0} \frac{K}{s(s^2+2\zeta\omega_n s+\omega_n^2)} = \infty. \quad \textbf{Ans.}$$

$$K_v = \lim_{s\to 0} sG(s) = \lim_{s\to 0} \frac{sK}{s(s^2+2\zeta\omega_n s+\omega_n^2)} = \frac{K}{\omega_n^2}. \quad \textbf{Ans.}$$

$$K_a = \lim_{s\to 0} s^2 G(s) = \lim_{s\to 0} \frac{s^2 K}{s(s^2+2\zeta\omega_n s+\omega_n^2)} = 0. \quad \textbf{Ans.}$$

(ii)
$$K_p = \lim_{s\to 0} G(s)\,H(s) = \lim_{s\to 0} \frac{100(s+2)(s+40)}{s^3(s^2+4s+200)} = \infty. \quad \textbf{Ans.}$$

$$K_v = \lim_{s\to 0} sG(s)\,H(s) = \lim_{s\to 0} \frac{s\,.\,100(s+2)(s+40)}{s^3(s^2+4s+200)} = \infty. \quad \textbf{Ans.}$$

$$K_a = \lim_{s\to 0} s^2 G(s)\,H(s) = \lim_{s\to 0} \frac{s^2\,.\,100(s+2)(s+40)}{s^3(s^2+4s+200)} = \infty. \quad \textbf{Ans.}$$

Example 6.10.19. *The closed-loop transfer function of a unity feedback control system is given below :*

$$\frac{C(s)}{R(s)} = \frac{Ks+\beta}{s^2+\alpha s+\beta}$$

Determine the steady state error for unit ramp input.

Solution. For a unity feedback control system

$$\frac{C(s)}{R(s)} = \frac{G(s)}{1+G(s)} \quad \therefore \quad \frac{Ks+\beta}{s^2+\alpha s+\beta} = \frac{G(s)}{1+G(s)}$$

or $$(Ks + \beta)\,[1 + G(s)] = G(s)\,(s^2 + \alpha s + \beta)$$

or $$G(s) = \frac{Ks+\beta}{s^2 + s(\alpha - K)}$$

$$\frac{E(s)}{R(s)} = \frac{1}{1+G(s)}$$

$\therefore$ $$\frac{E(s)}{R(s)} = \frac{1}{1+\dfrac{Ks+\beta}{s^2+s(\alpha-K)}}$$

$\therefore$ $$E(s) = R(s)\,\frac{s(s+\alpha-K)}{s^2+s\alpha+\beta}$$

$\because$ $$R(s) = \frac{1}{s^2}$$

$$e_{ss} = \lim_{s\to 0} s\,E(s) = \lim_{s\to 0} s\,.\,\frac{1}{s^2}\,.\,\frac{s(s+\alpha-K)}{s^2+s\alpha+\beta}$$

or $$e_{ss} = \frac{\alpha - K}{\beta}.$$ **Ans.**

Example 6.10.20. *The open loop transfer function of a unity feedback control system is given by :*

$$G(s) = \frac{20}{s(s+2)}.$$ *The input function is* $r(t) = a_0 + a_1 t + \frac{a_2}{2}t^2$

Determine the generalized error coefficients and expression for steady state error.

Solution. $$r(t) = a_0 + a_1 t + \frac{a_2}{2}t^2$$

$$r'(t) = a_1 + a_2 t\ ;\ r''(t) = a_2$$

The generalized error coefficients C_0, C_1 and C_2 are to be determined because derivatives higher than $r''(t)$ are zero.

$$\frac{E(s)}{R(s)} = \frac{1}{1+G(s)} \quad \text{or} \quad \frac{E(s)}{R(s)} = \frac{1}{1+\dfrac{20}{s(s+2)}}$$

or $$\frac{E(s)}{R(s)} = \frac{s^2+2s}{s^2+2s+20} \quad \therefore \quad F_e(s) = \frac{s^2+2s}{s^2+2s+20}$$

$$C_0 = \lim_{s\to 0} F_e(s) = \lim_{s\to 0} \frac{s^2+2s}{s^2+2s+20}$$

$$= \lim_{s \to 0} \frac{s(s+2)}{s^2 + 2s + 20} = 0 \quad \therefore \quad C_0 = 0. \quad \textbf{Ans.}$$

$$C_1 = \lim_{s \to 0} \frac{d F_e(s)}{ds} = \lim_{s \to 0} \frac{d}{ds}\left(\frac{s^2 + 2s}{s^2 + 2s + 20}\right)$$

$$= \lim_{s \to 0} \frac{(s^2 + 2s + 20)(2s+2) - (s^2 + 2s)(2s+2)}{(s^2 + 2s + 20)^2}$$

$$= \lim_{s \to 0} \frac{40s + 40}{(s^2 + 2s + 20)} = 0.1 \quad \therefore \quad C_1 = 0.1. \quad \textbf{Ans.}$$

$$C_2 = \lim_{s \to 0} \frac{d^2 F_e(s)}{ds^2} = \lim_{s \to 0} \frac{d}{ds}\left\{\frac{40s + 40}{(s^2 + 2s + 20)^2}\right\}$$

$$= \lim_{s \to 0} \frac{(s^2 + 2s + 20)^2 (40) - (40s + 40)(2)(s^2 + 2s + 20)(2s+2)}{(s^2 + 2s + 20)^4}$$

$$= \frac{(20)^2 (40) - (40)(2)(20)(2)}{(20)^4} = 0.08 \quad \therefore \quad C_2 = 0.08. \quad \textbf{Ans.}$$

The expression for the steady state error is

$$e(t) = C_0 r(t) + C_1 r'(t) + \frac{C_2}{2!} r''(t)$$

$$= 0 \times \left(a_0 + a_1 t + \frac{a_2}{2} t^2\right) + 0.1\,(a_1 + a_2 t) + \frac{0.08}{2!} a_2$$

or

$$e(t) = 0.1(a_1 + a_2 t) + 0.04\, a_2. \quad \textbf{Ans.}$$

Alternative method :

$$\frac{E(s)}{R(s)} = \frac{s^2 + 2s}{s^2 + 2s + 20}$$

Rearranging, $\dfrac{E(s)}{R(s)} = \dfrac{2s + s^2}{20 + 2s + s^2}$

By long division

$$\frac{2s + s^2}{20 + 2s + s^2} = 0.1s + 0.04s^2 - 0.009s^3 \ldots$$

$$\therefore \quad E(s) = 0.1s\, R(s) + 0.04s^2 R(s) - 0.009s^3 R(s) \ldots$$

Taking inverse Laplace transform on both sides

$$e(t) = 0.1\, r'(t) + 0.04\, r''(t) - 0.009\, r'''(t)$$

$$r(t) = a_0 + a_1 t + \frac{a_2}{2} t^2$$

$$r'(t) = a_1 + a_2 t$$

$$r''(t) = a_2$$

$$r'''(t) = 0$$

$$\therefore \quad e(t) = 0.1(a_1 + a_2 t) + 0.04\, a_2. \quad \textbf{Ans.}$$

Example 6.10.21. *A unity feedback control system has* $G(s) = \dfrac{1}{s(s+2)}$

The input to the system is given by $r(t) = 2 + 3t + 2t^3$. *Determine the generalized error coefficients and steady state error.*

Solution. $r(t) = 2 + 3t + 2t^3$; $r'(t) = 3 + 6t^2$; $r''(t) = 12t$; $r'''(t) = 12$; $r''''(t) = 0$.

The coefficients C_0, C_1, C_2 and C_3 need to be determined.

$$\frac{E(s)}{R(s)} = \frac{1}{1+G(s)} = \frac{1}{1+\dfrac{1}{s(s+2)}} = \frac{s^2+2s}{s^2+2s+1} = \frac{s^2+2s}{(s+1)^2}$$

$$\therefore \quad E(s) = \frac{s^2+2s}{(s+1)^2} R(s) \quad \therefore \quad F_e(s) = \frac{s^2+2s}{(s+1)^2}$$

$$C_0 = \lim_{s\to 0} F_e(s) = \lim_{s\to 0} \frac{s^2+2s}{(s+1)^2} = 0 \quad \therefore \quad C_0 = 0. \quad \textbf{Ans.}$$

$$C_1 = \lim_{s\to 0} \frac{dF_e(s)}{ds} = \lim_{s\to 0} \frac{d}{ds}\left\{\frac{s^2+2s}{(s+1)^2}\right\}$$

$$= \lim_{s\to 0} \frac{(s+1)^2(2s+2) - (s^2+2s)\,(2)\,(s+1)(1)}{(s+1)^4}$$

$$= \lim_{s\to 0} \frac{2(s+1)^3 - 2s\,(s+2)(s+1)}{(s+1)^4}$$

$$= \lim_{s\to 0} \frac{2(s+1)\,\{(s+1)^2 - s(s+2)\}}{(s+1)^4}$$

$$= \lim_{s\to 0} \frac{2(s+1)\,(s^2+2s+1-s^2-2s)}{(s+1)^4} = \lim_{s\to 0} \frac{2}{(s+1)^3}$$

$$= 2 \quad \therefore \quad C_1 = 2. \quad \textbf{Ans.}$$

$$C_2 = \lim_{s\to 0} \frac{d^2F_e(s)}{ds^2} = \lim_{s\to 0} \frac{d}{ds}\left\{\frac{dF_e(s)}{ds}\right\} = \lim_{s\to 0} \frac{d}{ds}\left\{\frac{2}{(s+1)^3}\right\}$$

$$= \lim_{s\to 0} \frac{(s+1)^3(0) - 2(3)\,(s+1)^2\,(1)}{(s+1)^6} = \lim_{s\to 0} -\frac{6}{(s+1)^4}$$

$$= -6 \quad \therefore \quad C_2 = -6. \quad \textbf{Ans.}$$

$$C_3 = \lim_{s\to 0} \frac{d^3F_e(s)}{ds^3} = \lim_{s\to 0} \frac{d}{ds}\left\{\frac{d^2F_e(s)}{ds^2}\right\}$$

$$= \lim_{s \to 0} \frac{d}{ds} \left\{ -\frac{6}{(s+1)^4} \right\}$$

$$= \lim_{s \to 0} -\left\{ \frac{(s+1)^4 (0) - 6 \times 4 (s+1)^3 (1)}{(s+1)^8} \right\}$$

$$= \lim_{s \to 0} -\left\{ \frac{-24 (s+1)^3}{(s+1)^8} \right\} = \lim_{s \to 0} \frac{24}{(s+1)^5}$$

$$= 24 \quad \therefore \quad C_3 = 24. \textbf{ Ans.}$$

Example 6.10.22. *A closed-loop control system for regulating the generator terminal valtage E_o is shown in Fig. 6.10.11. The forward path gain $G = 200$ and feedback factor $H = 0.1$. Determine*

(a) The reference voltage E_r for maintaining $E_o = 250$.

(b) The % change in generator terminal voltage if the forward path gain is reduced by 10%.

Determine the results (a) and (b) if the system is made open-loop.

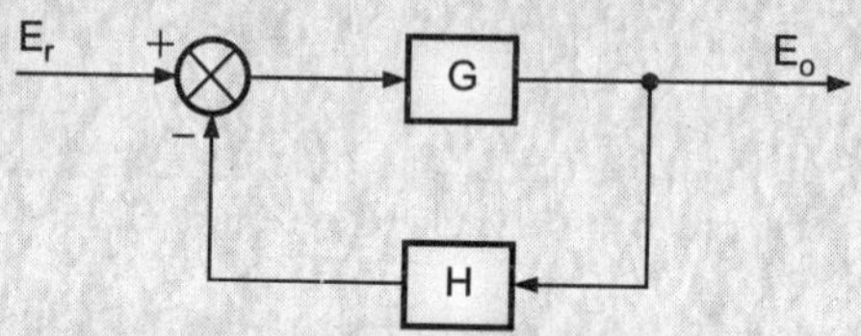

Fig. 6.10.11. Block diagram for example 6.10.22.

Solution. (*a*) For closed-loop system

$$\frac{E_o}{E_r} = \frac{G}{1+GH} = \frac{200}{1+200 \times 0.1} = \frac{200}{21}$$

$$\therefore \qquad E_r = \frac{21}{200} \cdot E_o = \frac{21}{200} \times 250 = 26.25 \text{ V. } \textbf{Ans.}$$

For open-loop system, $E_o/E_r = G$

$$\therefore \qquad E_r = \frac{E_o}{G} = \frac{250}{200} = 1.25 \text{ V. } \textbf{Ans.}$$

(*b*) *For closed-loop system.* As the forward path gain is reduced by 10%, the new value of gain is 180.

The output E_o is proportional to overall gain M and the sensitivity function of overall gain M w.r.t. forward path gain G is given by

$$S_G^M = \frac{1}{1+GH} = \frac{1}{1+180 \times 0.1} = \frac{1}{19}$$

$$\because \qquad S_G^M = \frac{\% \text{ change in } M}{\% \text{ change in G}} \quad \therefore \quad \frac{1}{19} = \frac{\% \text{ change in } M}{10}$$

$$\therefore \qquad \% \text{ change in } M = \frac{10}{19} = 0.526\%$$

$$\% \text{ change in } E_o = \% \text{ change in } M = 0.526\%. \textbf{ Ans.}$$

For open-loop system. Change in terminal voltage ΔE_o and change in gain ΔG are related as

$$\Delta E_o = \Delta G \,.\, E_r$$

$$\therefore \quad \Delta E_o = \frac{\Delta E_o}{E_o} \times 100 = \frac{25}{250} \times 10 = 10\%. \quad \textbf{Ans.}$$

Example 6.10.23. *Determine the sensitivity of the overall transfer function for the system shown in Fig. 6.10.12 with respect to change in parameter (a)* ***K****; (b)* ***H****.*

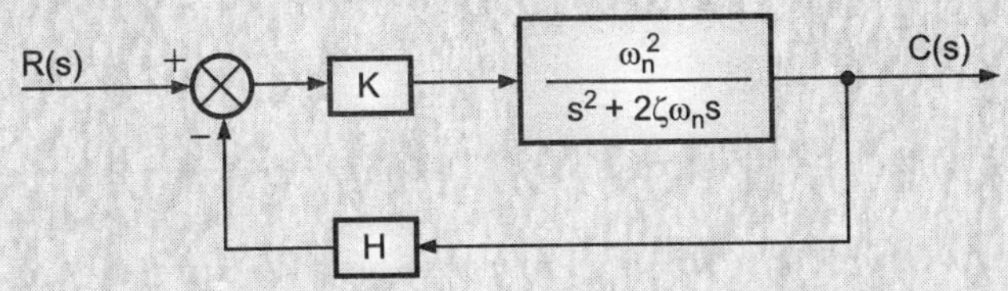

Fig. 6.10.12. Block diagram for example 6.10.23.

Solution. The overall transfer function is given by

$$M(s) = \frac{K.\dfrac{\omega_n^2}{s^2 + 2\zeta\omega_n s}}{1 + \dfrac{K\omega_n^2}{s^2 + 2\zeta\omega_n s}\,.\,H} = \frac{K\omega_n^2}{s^2 + 2\zeta\omega_n s + KH\omega_n^2}$$

(*a*) The sensitivity of $M(s)$ w.r.t. K is given by

$$S_K^M = \frac{\partial M}{\partial K} \cdot \frac{K}{M}$$

$$\because \quad M(s) = \frac{K\omega_n^2}{s^2 + 2\zeta\omega_n s + KH\omega_n^2}$$

$$\frac{\partial M}{\partial K} = \frac{(s^2 + 2\zeta\omega_n s + KH\omega_n^2)\,\omega_n^2 - K\omega_n^2 H \omega_n^2}{(s^2 + 2\zeta\omega_n s + KH\omega_n^2)^2} = \frac{\omega_n^2 (s^2 + 2\zeta\omega_n s)}{(s^2 + 2\zeta\omega_n s + KH\omega_n^2)^2}$$

$$\therefore \quad S_K^M = \frac{\omega_n^2 (s^2 + 2\zeta\omega_n s)}{(s^2 + 2\zeta\omega_n s + KH\omega_n^2)^2} \cdot \frac{K}{\dfrac{K\omega_n^2}{(s^2 + 2\zeta\omega_n s + KH\omega_n^2)}}$$

or $$S_K^M = \frac{s^2 + 2\zeta\omega s}{s^2 + 2\zeta\omega_n s + KH\omega_n^2} \quad \text{or} \quad S_K^M = 1 + \frac{\dfrac{1}{KH\omega_n^2}}{s^2 + 2\zeta\omega_n s}. \quad \textbf{Ans.}$$

(*b*) Sensitivity of $M(s)$ w.r.t. $H(s)$ is given by

$$S_H^M = \frac{\partial M}{\partial H} \cdot \frac{H}{M}$$

$$\because \quad M(s) = \frac{K\omega_n^2}{s^2 + 2\zeta\omega_n s + KH\omega_n^2}$$

$$\therefore \quad \frac{\partial M}{\partial H} = \frac{(s^2 + 2\zeta\omega_n s + KH\omega_n^2).0 - K\omega_n^2 (K\omega_n^2)}{(s^2 + 2\zeta\omega_n s + KH\omega_n^2)^2} = -\frac{K^2\omega_n^4}{(s^2 + 2\zeta\omega_n s + KH\omega_n^2)^2}$$

$$\therefore \quad S_H^M = -\frac{K^2\omega_n^4}{(s^2 + 2\zeta\omega_n s + KH\omega_n^2)^2} \cdot \frac{\frac{H}{K\omega_n^2}}{(s^2 + 2\zeta\omega_n s + KH\omega_n^2)}$$

$$= -\frac{KH\omega_n^2}{(s^2 + 2\zeta\omega_n s + KH\omega_n^2)}. \quad \textbf{Ans.}$$

PROBLEMS

6.1. Determine an expression for a unit step response for the control system shown in Fig. 6.P-1.

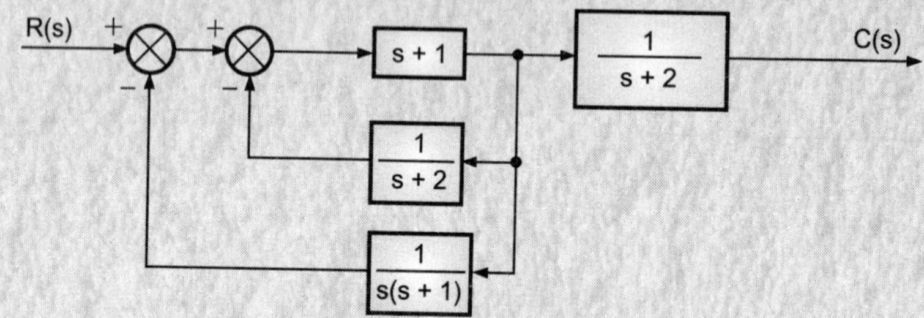

Fig. 6.P-1. Block diagram for problem 6.1.

6.2. Derive an expression relating the output $c(t)$ and the time for the control system shown in Fig. 6.P-2. The input to the system being a unit step function.

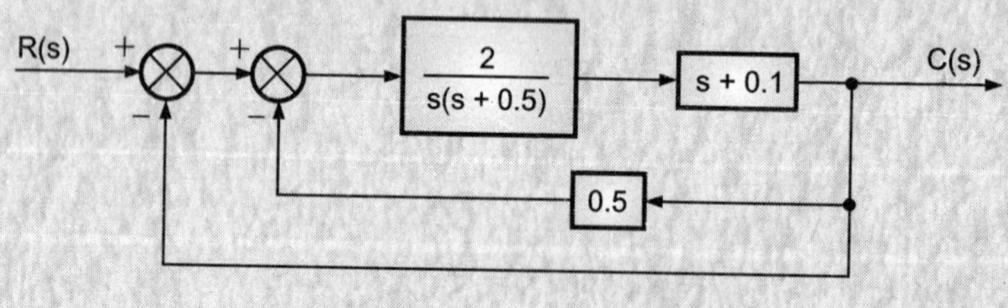

Fig 6.P-2. Block diagram for problem 6.2.

6.3. A control system is represented by a signal flow graph shown in Fig. 6.P-3. Determine the overall transmittance and expression for unit step time response of the system.

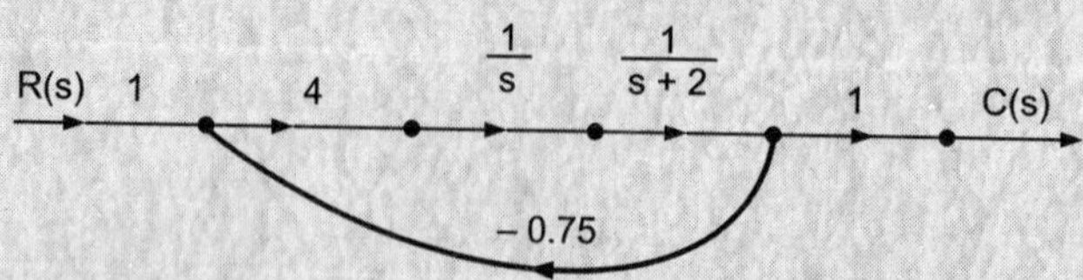

Fig. 6.P-3. Signal flowgraph for problem 6.3.

6.4. A positional control system has a damping ratio of 0.6. The damped frequency of oscillations is 8 Hz. Derive an expression relating the output and the time when the input is suddenly changed from 0 to θ_r and calculate the percentage maximum overshoot.

6.5. The maximum overshoot and settling time for a control system whose overall transfer function is given by

$$\frac{C(s)}{R(s)} = \frac{\omega_n^2}{s^2 + 2\zeta\omega_n s + \omega_n^2}$$

are 10% and 4 second respectively. The input being a unit step function. Determine the values of ζ and ω_n in order to satisfy these requirements.

6.6. A unity feedback control system has forward path transfer function given by $G(s) = \frac{s+2}{s(s+1)}$.

Determine: (*a*) The expression for unit step response; (*b*) The rise time and (*c*) The maximum overshoot.

6.7. A control system is used to drive a load. The moment of inertia and the coefficient viscous friction referred to the motor shaft size are 16×10^{-4} kg-m^2 and 64×10^{-4} Nm/(rad/sec). The motor produces a torque of 0.075 Nm per degree of error. The load is connected to the motor through a gear having a ratio of 100 : 1. Determine:

(*a*) The natural frequency of oscillations; (*b*) The damping factor; (*c*) The steady state error if the reference input is 10 rpm; (*d*) The maximum overshoot for step input and (*e*) The settling time.

6.8. The gain constant of a controller is 0.75 Nm/rad. The controller is used for a unity feedback control system controlling the position of a load having a moment of inertia 12×10^{-4} kg-m^2 and viscous friction 6×10^{-2} Nm/(rad/sec). Obtain an expression relating the time and error angle. The input shaft being fed with a velocity of 5 r.p.m.

6.9. A closed-loop control system when subjected to a unit step input has an expression for the time response given by

$$c(t) = 0.5 + 1.25\, e^{-3t} - 1.75\, e^{-12t}.$$

Determine the overall transfer function of the system.

6.10. The forward path transfer function of a unity feedback control system is given by:

$$G(s) = \frac{K(s+2)}{s^3 + \beta s^2 + 4s + 1}$$

Determine the values of K and β such that the system exhibits sustained oscillations having a frequency of 4.0 rad/sec.

6.11. Determine the unit step time response of a system having closed-loop transfer function given below:

$$\frac{C(s)}{R(s)} = \frac{1}{(s+1)(s^2+1)}$$

6.12. Following data refers to a unity feedback control system shown in Fig. 6.P-4.

(*a*) The maximum overshoot for a unit step function applied to input 1 is 18.39%.

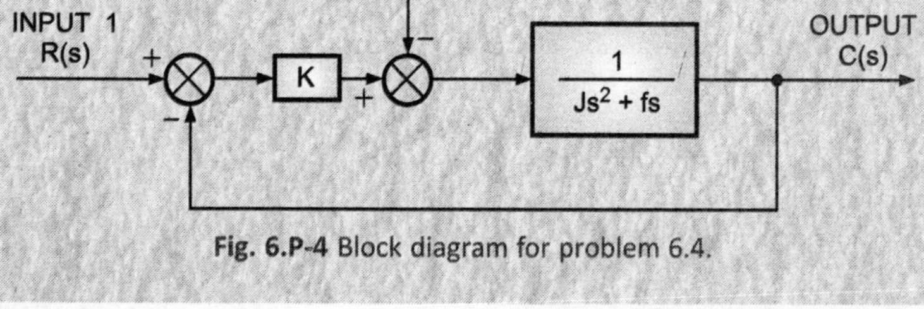

Fig. 6.P-4 Block diagram for problem 6.4.

(b) Steady state error for a ramp input function of magnitude 2 rad/sec applied to input 1 is 0.3 rad.

(c) Input 1 being held fixed and a torque of 1 Nm suddenly applied. The output shaft exhibits a shift of – 0.166 rad.

Calculate the values of controller gain K, the moment of inertia J and coefficient of viscous friction f.

6.13. Determine the position, velocity and acceleration error constants and the steady state error for (i) unit step input (ii) unit ramp input (iii) unit parabolic input. The open-loop transfer function for the unity feedback control system is given below:

(a) $G(s) = \dfrac{K}{s(s^2 + 3s + 100)}$ (b) $G(s) = \dfrac{K(1 + 0.5s)(1 + 2s)}{s^2(s^2 + 4s + 5)}$.

6.14. The forward path and feedback transfer functions of a control system are given below:

$$G(s) = \frac{K}{s(s+1)} \; ; H(s) = \frac{1}{(s+5)}.$$

Permissible error is 5° for a steady state velocity of 0.1 π rad/sec. Determine the value of K.

6.15. The open-loop transfer function of a unity feedback control system is given by

$$G(s) = \frac{K}{s(1 + sT)}$$

The input to the system is 1 rpm and the steady state error being 0.25°. Calculate the natural frequency of oscillations if the system is critically damped.

6.16. The overall transfer function of a unity feedback control system is given by

$$\frac{C(s)}{R(s)} = \frac{10}{s^2 + 6s + 10}$$

Find (a) K_p, K_v and K_a

(b) Determine the steady state error if the input is

$$r(t) = 1 + t + t^2.$$

6.17. Determine the steady state error for a system whose open-loop transfer function is given by

$$G(s)\,H(s) = \frac{5(s+4)}{s(s+1)(s+5)}$$

The input to the system being

$$r(t) = 0 \quad \text{for } t < 0$$
$$r(t) = 2 + 2t \quad \text{for } t \geq 0.$$

6.18. The open-loop transfer function of a unity feedback control system is given by

$$G(s) = \frac{5(s+1)}{s^2(s+4)}$$

Determine the steady state error if the input is

$$r(t) = 2 - t + 2t^2.$$

6.19. (a) Determine the transfer function of the d.c. generator shown in Fig. 6.P-5. The data given: amplifier gain K_A = 50 V/V, R_f = 20 ohm.

L_f = 1.0 H, generator constant K_g = 10 V/A.

(b) Determine the value of feedback factor and increase in e_i if the time constant is to be made 10 ms.

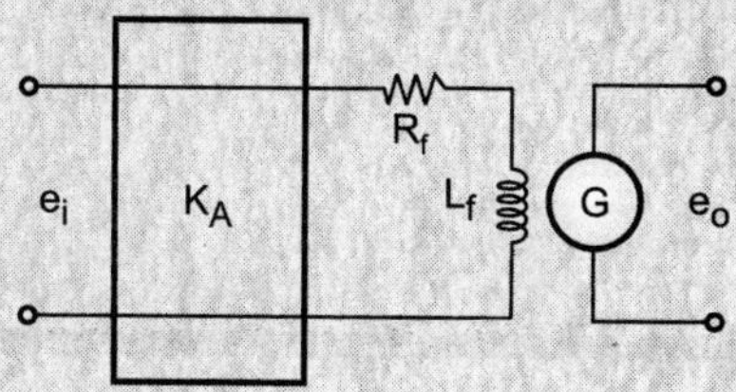

Fig. 6.P-5. Block diagram for problem 6.19

6.20. Calculate the reference voltage E_r for the output voltage $E_o = 250$ V, given that the forward path gain $K = 200$ for the open-loop system shown in Fig. 6.P-6.

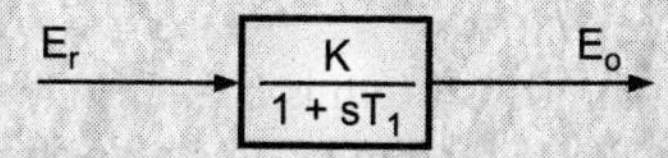

Fig. 6.P-6 Block diagram for problem 6.20.

(*a*) Calculate the % change in E_o if the gain K is reduced by 10%.

(*b*) Calculate the reference voltage E_r and the data in (*a*) if the system is made closed-loop with a feedback factor 0.1.

6.21. Determine the sensitivity of the overall gain for the system shown in Fig. 6.P-7 and calculate the value of K such that the sensitivity is 0.2 under steady state conditions.

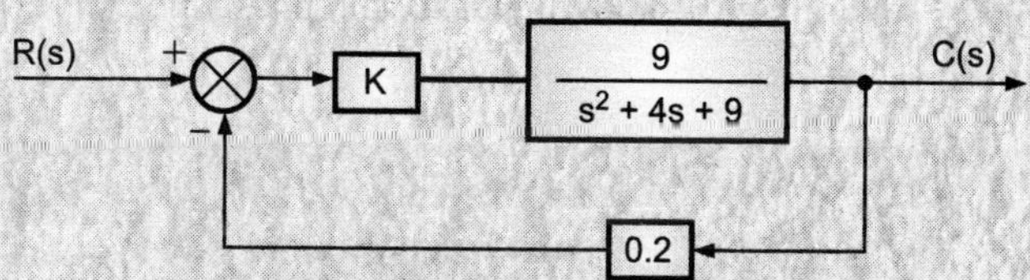

Fig. 6.P-7. Block diagram for problem 6.21.

6.22. The block diagram shown in Fig. 6.P-8 shows a speed control system. Compare the time response

(*a*) with tachometer feedback

(*b*) without tachometer feedback.

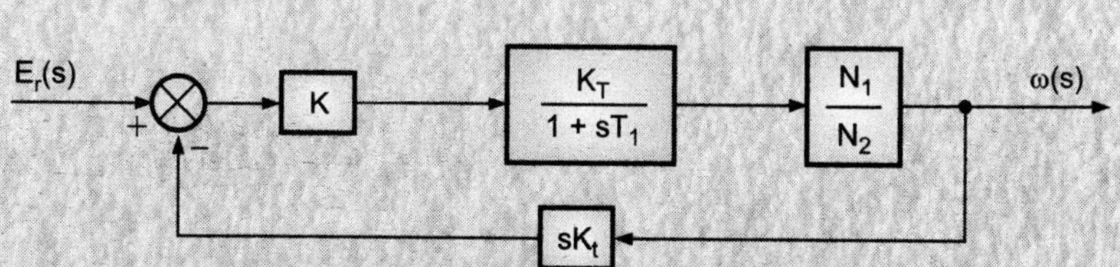

Fig. 6.P-8. Block diagram for problem 6.22.

CHAPTER 7

Stability Analysis of Control Systems

The stability analysis of a feedback control system is based on identifying the *s*-plane location of the roots of its characteristic equation. For a system to be stable the said roots should be located in the left hand side of *s*-plane. The roots need not be determined by solving the characteristic equation.

The method of assessing absolute stability is given by Routh-Hurwitz criterion. The relative stability is determined by using frequency response methods. Nyquist plot and Bode plot. The Root Locus plot approach gives the *s*-plane variations of the roots of a characteristic equation, as the parameter K or any other parameter is varied.

The frequency response specifications are : gain margin, phase margin, resonant peak and band width.

The Gain-phase plot, M-N circles are used to construct Nichol's chart. The closed-loop frequency response can be obtained by calibrating the superimposed gain phase plot on Nichol's chart.

CONTENTS

- *Stability in Terms of Characteristic Equation of a Control System*
- *To Determine the Number of Roots having Positive Real Parts for a Polynomial*
- *Hurwitz Determinants of a Polynomial*
- *Routh-Hurwitz Criterion*
- *Solved Examples : Routh-Hurwitz Criterion*
- *Nyquist Criterion*
- *Gain Margin and Phase Margin*
- *Relative Stability from Nyquist Plot*
- *Gain Phase Plot*
- *Closed-loop Frequency Response of a Unity Feedback Control System from Nyquist Plot*
- *Constant Magnitude Loci : M-Circles*
- *Constant Phase Angle Loci : N-Circles*
- *Closed-loop Frequency Response of Control System from M and N-Circles*
- *Gain Adjustment Using M-Circle*
- *Nichols Chart*
- *Cutoff Frequency and Bandwidth*
- *Solved Examples : Nyquist Criterion*
- *Inverse Nyquist Plot*

- *I/M Circles*
- *Gain Adjustment using 1/M Circle*
- *Bode Plot*
- *Solved Examples : Bode Plot*
- *Bode Plot for Time Delay Element e^{-sT}*
- *Correlation Between Transient Response and Frequency Response*
- *Root Locus*
- *Salient Features of Root Locus Plot*
- *The Procedure for Plotting Root Locus*
- *Solved Examples : Root Locus*
- *Root Contours*

Introduction. In the chapter on time response of control system, the time solution of system equations were obtained. The time solutions thus obtained give the variation of the system output with time as independent variable. The time solution can be classified into two categories, the one wherein the output tending to a finite value and the other wherein the output approaching towards infinite value with the advancement of time. The former case is termed as stable system while the latter is termed as unstable system.

An illustration to explain stability and instability is shown in Fig. 7.0.1.

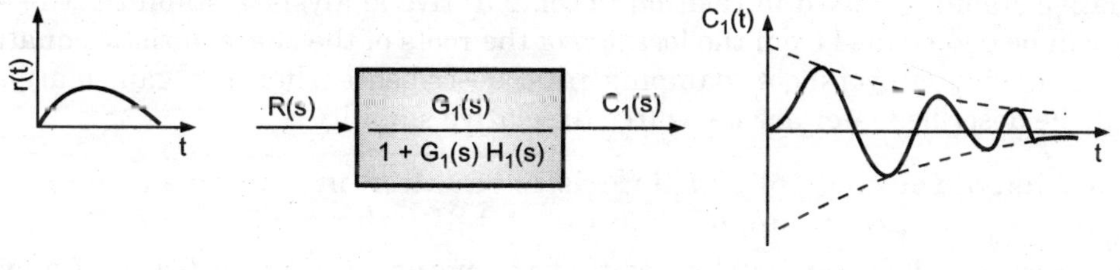

(*a*) Stable system

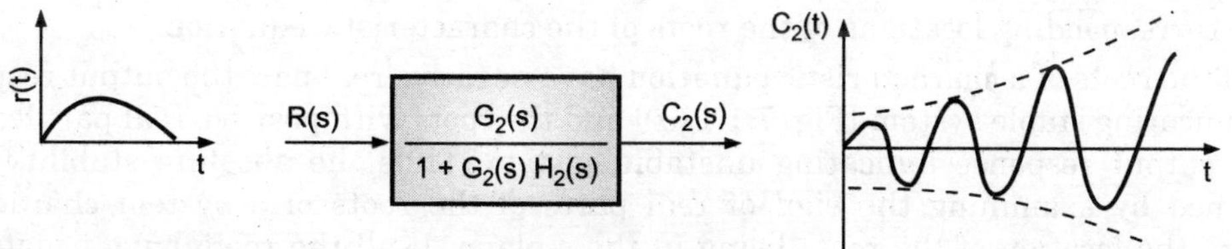

(*b*) Unstable system

Fig. 7.0.1

A bounded input is applied to systems shown in Fig. 7.0.1. For the time response shown in Fig. 7.0.1 (*a*) the output response is also bounded and the system is said to be stable whereas for the system shown in Fig. 7.0.1 (*b*) the output response is unbounded and the system is said to unstable.

7.1 STABILITY IN TERMS OF CHARACTERISTIC EQUATION OF A CONTROL SYSTEM

If the order of the characteristic equation of a system is upto 2, the time solution of the system is conveniently obtainable and from the time solution the stability or instability can be judged. However, if the order of the system is higher, it is not convenient to obtain the time solution and as such stability of such a system cannot be easily determined by direct solution.

At first instance stability analysis requires a qualitative assessment, that is if a system characteristic equation or transfer function of a system is given then without obtaining a direct time solution of characteristic equation how one can say whether the system is stable or unstable. In other words, to get a YES or NO answer pertaining to stability, means qualitative analysis.

7.1.1. Definition of stability

An adequate definition of stability with reference to linear control systems is given below :

"If any oscillations setup in a system in consequence to application of an input are damped out with respect to time, the system is said to be stable".

Conversely for unstable systems oscillations are increasing in magnitude. If the magnitude of the oscillations is sustained the system is marginally stable.

7.1.2. Absolute and relative stability

The term absolute stability is used in relation to qualitative analysis of stability and the term relative stability is used in relation to comparative analysis of stability. The absolute stability can be determined from the location of the roots of the characteristic equation in s-plane. The maximum overshoot, damping ratio described earlier and gain margin phase margin, to be described later are measures to relative stability.

7.1.3. Location of the roots of characteristic equation in s-plane as related to time response and prediction of absolute stability therefrom

It has been explained in earlier chapter that the nature of time response of a system is related to the location of the roots of a characteristic equation in s-plane. Fig. 7.1.1 (*a*), (*b*) and (*c*) illustrate the nature of the time response of a second order system for a step input and the corresponding locations of the roots of the characteristic equation.

If the roots of a characteristic equation have negative real part the output response is finite indicating stable system [Fig. 7.1.1 (*a*)] and the roots with positive real part lead to an infinite output response indicating unstable system. Thus the absolute stability can be determined by examining the sign of real parts of the roots of a system characteristic equation, the location of the roots being in the s-plane. If all the roots have negative real parts then the system is stable and even if one root is having a positive real part the system is unstable.

It is, therefore, inferred that the problem of ascertaining absolute stability of a control system is associated with determination of location of roots of system characteristic equation in s-plane.

In mathematical terms, for determining absolute stability of a control system it is to be ascertained whether the system characteristic equation $1 + G(s)\,H(s) = 0$ has any root with positive real part ? And answer to this question forms a basis of stability studies pertaining to control systems.

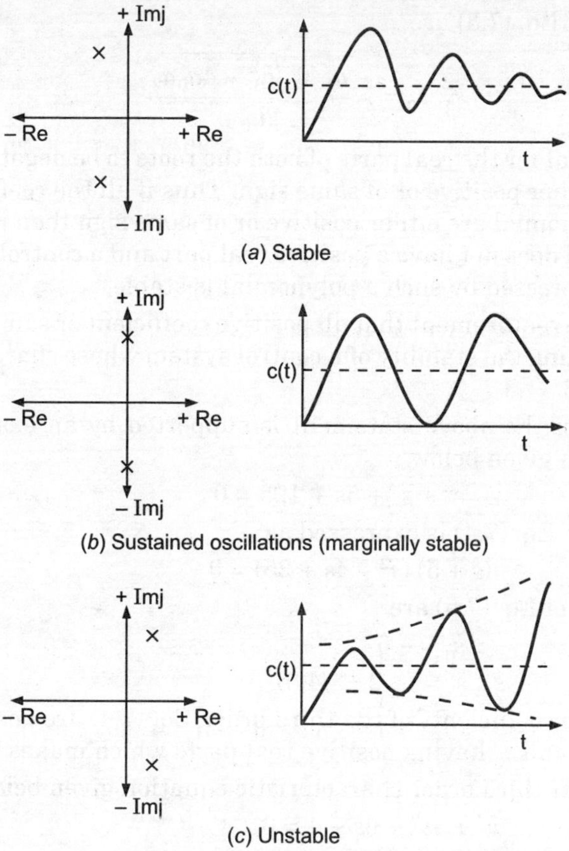

Fig. 7.1.1. Time response and location of roots of characteristic equation.

In general the characteristic equation $1 + G(s)\,H(s) = 0$ can be expressed in polynomial form as follows :

$$a_0 s^n + a_1 s^{n-1} + a_2 s^{n-2} + \ldots + a_n = 0 \qquad \ldots(7.1)$$

Only the sign of the real part of the roots of the polynomial (7.1) is to be ascertained which in turns indicates absolute stability or instability.

7.2 TO DETERMINE THE NUMBER OF ROOTS HAVING POSITIVE REAL PARTS FOR A POLYNOMIAL

(*a*) Consider a first order characteristic equation expressed by a polynomial given below :

$$a_0 s + a_1 = 0 \qquad \ldots(7.2)$$

The root of Eq. (7.2) is $s = -\dfrac{a_1}{a_0}$

(*b*) Consider a second order characteristic equations expressed by a polynomial given below :

$$a_0 s^2 + a_1 s + a_2 = 0 \qquad \ldots(7.3)$$

The two roots of Eq. (7.3) are

$$s_1, s_2 = \frac{-a_1 \pm \sqrt{a_1^2 - 4a_0a_2}}{2a_0}$$

It is observed that for the real parts of both the roots to be negative, the coefficients a_0, a_1 and a_2 should be either positive or of same sign. Thus if all the coefficients of a first order and second order polynomial are either positive or of same sign then it is concluded that the roots of the polynomial does not have a positive real part and a control system whose characteristic equation is expressed by such a polynomial is stable.

(*c*) However, the requirement that all positive coefficient or same sign coefficient is not sufficient for ascertaining the stability of a control system whose characteristic equation has an order higher than 2.

A clarification of the above statement is supported by an example of a third order characteristic equation given below :

$$s^3 + s^2 + 5s + 125 = 0 \quad ...(7.4)$$

On factorization Eq. (7.4) is expressed as

$$(s + 5)(s^2 - 4s + 25) = 0 \quad ...(7.5)$$

The three roots of Eq. (7.5) are

$$s_1 = -5$$

$$s_2, s_3 = (+2 \pm j4.5)$$

Although all the coefficients of the third order Eq. (7.4) are positive the equation has distinctly two roots s_2 and s_3 having positive real parts which makes the system unstable.

Consider another third order characteristic equation given below :

$$s^3 + 4s^2 + 6s + 4 = 0 \quad ...(7.6)$$

On factorization Eq. (7.6) is expressed as

$$(s + 2)(s^2 + 2s + 2) = 0 \quad ...(7.7)$$

The three roots of Eq. (7.7) are

$$s_1 = -2$$

$$s_2, s_3 = (-1 \pm j1).$$

It is observed that none of the roots of Eq. 7.6 have positive real parts, therefore, the system is stable.

Hence, for determining stability of higher order systems, in addition to aforesaid necessary condition, a sufficient condition is further needed and this sufficient condition is given by Hurwitz determinants as given below.

7.3 HURWITZ DETERMINANTS OF A POLYNOMIAL

$$a_0s^n + a_1s^{n-1} + a_2s^{n-2} + ... + a_n = 0 \quad ...(7.8)$$

For the roots of the nth order polynomial (7.8) to have only negative real parts (*i.e.*, all the roots to lie in left half of s-plane) it is required that the polynomial's Hurwitz determinants D_k, (k = 1, 2, 3, ..., n) be greater than 0. Hurwitz determinants are given by

$$D_1 = a_1 > 0$$

$$D_2 = \begin{vmatrix} a_1 & a_3 \\ a_0 & a_2 \end{vmatrix} > 0$$

$$D_n = \begin{vmatrix} a_1 & a_3 & & \ldots\ldots a_{2n-1} \\ a_0 & a_2 & & \ldots\ldots a_{2n-2} \\ 0 & a_1 & a_3 & \ldots\ldots a_{2n-3} \\ 0 & 0 & a_2 & \ldots\ldots a_{2n-4} \\ \vdots & \vdots & & \vdots \\ \vdots & \vdots & & \vdots \\ \vdots & \vdots & & \vdots \\ 0 & 0 & & a_n \end{vmatrix} > 0 \qquad \ldots(7.9)$$

For evaluating Hurwitz determinants mentioned above lot of time is involved. However, in relation to evaluation of Hurwitz determinants Routh has simplified the procedure for ascertaining the number of positive real part roots of a polynomial and the method is called Routh-Hurwitz criterion.

7.4 ROUTH-HURWITZ CRITERION

In order to determine the existence of a root having positive real part for a polynomial equation given by

$$a_0 s^n + a_1 s^{n-1} + a_2 s^{n-2} + \ldots + a_n = 0 \qquad \ldots(7.10)$$

Two rows consisting of alternate coefficients of polynomial (7.10) are formed as follows :

$$\begin{matrix} a_0 & a_2 & a_4 & a_6 \ldots \\ a_1 & a_3 & a_5 & \ldots\ldots \end{matrix}$$

and the following array is formed (illustrated for a 6th order polynomial).

$$\begin{matrix} s^6 & a_0 & a_2 & a_4 & a_6 \\ s^5 & a_1 & a_3 & a_5 & 0 \\ s^4 & b_1 & b_3 & b_5 & 0 \\ s^3 & c_1 & c_3 & 0 & 0 \\ s^2 & d_1 & d_3 & 0 & 0 \\ s^1 & e_1 & 0 & 0 & 0 \\ s^0 & f_1 & 0 & 0 & 0 \end{matrix} \qquad \ldots(7.11)$$

In the Routh array (7.11) the coefficients b_1, b_3, b_5, c_1 etc. are related to the coefficients of the polynomial (7.10) as under

$$b_1 = \frac{a_1 a_2 - a_0 a_3}{a_1}, \quad b_3 = \frac{a_1 a_4 - a_0 a_5}{a_1}$$

$$b_5 = \frac{a_1 a_6 - a_0 \times 0}{a_1} = a_6$$

$$c_1 = \frac{b_1 a_3 - a_1 b_3}{b_1}, \quad d_1 = \frac{c_1 b_3 - b_1 c_3}{c_1} \text{ etc.}$$

From the array (7.11) so formed, it is observed that the elements of the first column are related to Hurwitz determinants [for example $b_1 = (D_2/D_1)$]. The number of changes of sign in the first column elements of Routh array gives the number of positive real part roots of the polynomial. Therefore, for a stable system there should be no change of sign in the first column of Routh array formed from the coefficients of the characteristic equation expressed in polynomial form.

While applying Routh-Hurwitz criterion it is imperative that no powers of s in the characteristic equation be absent. Any absence of such powers indicate the presence of at least one positive real part root and confirms system instability by inspection.

However, if the characteristic equation contains either only odd or even powers of s, this indicates that the roots have no real parts and possess only imaginary parts and therefore the system has sustained oscillations in its output response.

Following difficulties are faced while applying Routh-Hurwitz criterion :

1. The element in one of the first columns of Routh array is zero.
2. All the elements in one of the rows of Routh array are zero.

Above cases are considered in examples 7.5.4 and 7.5.5, which follows :

7.5 SOLVED EXAMPLES : ROUTH-HURWITZ CRITERION

Example 7.5.1. *A closed-loop control system has the characteristic equation given by, $s^3 + 4.5s^2 + 3.5s + 1.5 = 0$. Investigate the stability using Routh-Hurwitz criterion.*

Solution. The Routh array is formed as follows :

s^3	1	3.5
s^2	4.5	1.5
s^1	$3.16 = \dfrac{4.5 \times 3.5 - 1 \times 1.5}{4.5}$	0
s^0	$1.5 = \dfrac{3.16 \times 1.5 - 4.5 \times 0}{3.16}$	0

In the first column of the Routh array there is no change in sign of the elements, hence there is no root of the characteristic equation with positive real part and the system is stable.

Example 7.5.2. *Apply Routh criterion to determine the stability of the system having the characteristic equation as*

$$s^3 + 4 \times 10^2 s^2 + 5 \times 10^4 s + 2 \times 10^6 = 0.$$

Solution. The calculations in the Routh table are simplified by substituting $s = \alpha p$, ($\alpha = 10^2$) in the characteristic equation, thus the characteristic equation takes the following form :

$$(10^2 p)^3 + 4 \times 10^2 (10^2 p)^2 + 5 \times 10^4 (10^2 p) + 2 \times 10^6 = 0$$

or $$p^3 + 4p^2 + 5p + 2 = 0$$

The Routh table is formed below :

p^3	1	5
p^2	4	2
p^1	4.5	0
p^0	2	0

All the first column elements are of the same sign, therefore, the system is stable.

Example 7.5.3. *Determine the stability of a system whose overall transfer function is given below*

$$\frac{C(s)}{R(s)} = \frac{2s + 5}{s^5 + 1.5s^4 + 2s^3 + 4s^2 + 5s + 10}$$

If the system is found unstable, how many roots it has with positive real part ?

Solution. The characteristics equation is

$$s^5 + 1.5s^4 + 2s^3 + 4s^2 + 5s + 10 = 0$$

The Routh array is formed below

s^5	1	2	5
s^4	1.5	4	10
Sign change			
s^3	– 0.66	– 1.66	0
Sign change			
s^2	0.227	10	0
s^1	27.4	0	0
s^0	10	0	0

As there are two changes in sign of the elements in the first column of Routh table, it is concluded that the system has two roots with positive real part and as such the system is unstable.

Example 7.5.4. *Determine the stability of a closed-loop control system whose characteristic equation is*

$$s^5 + s^4 + 2s^3 + 2s^2 + 11s + 10 = 0.$$

Solution. The Routh array is formed below :

s^5	1	2	11
s^4	1	2	10
s^3	$0 = \dfrac{1\times 2 - 1\times 2}{1}$,	$1 = \dfrac{1\times 11 - 1\times 10}{1}$	0
s^2			

while forming the Routh array as above the third element in the first column is zero and thus the Routh criterion fails at this stage. The difficulty is solved if zero in the third row of the first column is replaced by a symbol ε and Routh array is formed as follows :

s^5	1	2	11
s^4	1	2	10
s^3	ε	1	0
Sign change			
s^2	$\lim_{\varepsilon \to 0}\left(\frac{\varepsilon \times 2 - 1 \times 1}{\varepsilon}\right) = -\infty$	10	0
Sign change			
s^1	$\lim_{\varepsilon \to 0}\left(1 - \frac{10\,\varepsilon^2}{2\varepsilon - 1}\right) = 1$	0	0
s^0	10		

The limits of fourth and fifth element in the first column as $\varepsilon \to 0$ from positive side are $-\infty$ and $+1$ respectively, indicating two sign changes, therefore, the system is unstable and the number of roots with positive real part of the characteristic equation is 2. **Ans.**

The presence of a zero in the first column of the Routh's tabulation leads to following conclusions :

(*i*) equal real roots with opposite signs. As one of the roots is *+ve*, the system is unstable as indicated by the sign change in the first column.

(*ii*) pair of conjugate root on imaginary axis. This gives marginal stability provided there is no sign change in the first column.

In the both cases as per procedure 0 is replaced by ε.

The conclusions as above are verified below :

(*i*) $F(s) = (s - 1)(s + 1)^2 (s + 2)$: two equal roots with opposite sign $s = -1$, $s = +1$

or $$F(s) = s^4 + 3s^3 + s^2 - 3s + 2$$

The Routh's tabulation is formed below :

s^4	1	1	2	
s^3	3	– 3		
s^2	2	– 2		
s^1	0 (ε)			+ as $\varepsilon \to 0$
s^0	– 2			

There is one sign change in the first column ; the system being unstable.

(*ii*) $F(s) = (s + 1)(s + 2)(s^2 + 1)$: pair of conjugate root on imaginary axis $s = \pm j\,1$

or $$F(s) = s^4 + 3s^3 + 3s^2 + 2$$

The Routh's tabulation is formed below :

s^4	1	3	2	
s^3	3	3		
s^2	2	2		
s^1	0 (ε)			+ as $\varepsilon \to 0$
s^0	1			

There is no sign change in the first column and noting that the polynomial $F(s)$ has a pair of conjugate root on imaginary axis, the system is marginally stable.

Example 7.5.5. *Determine the stability of a system having following characteristic equation :*

$$s^6 + s^5 + 5s^4 + 3s^3 + 2s^2 - 4s - 8 = 0.$$

Solution. The Routh table is formed below :

s^6	1	5	2	– 8	
s^5	1	3	– 4	0	
s^4	2	6	– 8	0	*Auxiliary equation*
s^3	0	0	0	0	
s^2					

It is observed in this example that all the elements in the fourth row vanish and the application of Routh criterion fails. This situation occurs when the array has two consecutive rows having the same ratio of corresponding elements.

This difficulty faced is overcome by forming an auxiliary equation using elements of the last but one vanishing row. The derivative of this auxiliary equation is taken w.r.t. s and the coefficients of the differentiated equation are taken as the elements of the following row :

For the example under consideration the auxiliary equation is

$$A(s) = (2s^4 + 6s^2 - 8)$$

and
$$\frac{dA(s)}{ds} = (8s^3 + 12s - 0)$$

The coefficients of the fourth row are thus 8, 12 and 0. The modified Routh array is given below :

s^6	1	5	2	– 8	
s^5	1	3	– 4	0	
s^4	2	6	– 8	0	
*s^3	8	12	0	0	**coefficients of*
s^2	3	– 8	0	0	*differentiated auxiliary eqn.*
s^1	$\frac{100}{3}$	0	0	0	
Sign change					
s^0	– 8	0	0	0	

As there is one sign change in the first column, the system has one root with positive real part indicating that the system is unstable.

All the elements of a row in Routh's tabulation being zero indicate a pair of conjugate root on imaginary axis and on substitution of coefficients of $dA(s)/ds$ in the zero element row on sign change is noted in the first column of Routh's tabulation, then the system is marginally stable.

However, if any two rows of a Routh's tabulation are zero it can be concluded that a pair of conjugate roots on imaginary axis having multiplicity order of two is present. This shows unstable system even though no change of sign in the first column is observed on following usual procedure for completing Routh's tabulation.

The statements mentioned above are verified as per the examples given below :

$F(s) = (s^2 + 1)(s + 1)(s + 2)(s + 3)$: pair of conjugate root on imaginary axis $s = \pm j\,1$

$= s^5 + 6s^4 + 6s^3 + 12s^2 + 5s + 6$

s^5	1	6	5	
s^4	6	12	6	Aux. equation $A(s) = 6s^2 + 6$
s^3	4	4		$\frac{dA(s)}{ds} = 12s$
s^2	6	6		* coeff. of $\frac{dA(s)}{ds}$
s^1	0(12)*	0		
s^0	6			

There is no sign change in the first column, the system is marginally stable.

$F(s) = (s^2 + 2)^2 (s + 1)(s + 2)$: pair of conjugate root on imaginary axis of multiplicity order two *i.e.* $s = \pm\, j\sqrt{2}$, $s = \pm\, j\sqrt{2}$

$$F(s) = s^6 + 3s^5 + 6s^4 + 12s^3 + 12s^2 + 12s + 8$$

s^6	1	6	12	8	
s^5	3	12	12		(1) $A(s) = 2s^4 + 8s^2 + 8$
s^4	2	8	8		$\frac{dA(s)}{ds} = 8s^2 + 16s$ *coeff. of $\frac{dA(s)}{ds}$
s^3	0(8)*	0(16)*	0		(2) $A(s) = 4s^2 + 8$
s^2	4	8			$\frac{dA(s)}{ds} = 8$ ** coeff. of $\frac{dA(s)}{ds}$
s^1	0(8)**	0			
s^0	8				

There is no change of sign in the first column of Routh's tabulation but existence of two rows having zero elements make the system unstable.

Example 7.5.6. *Using Routh criterion investigate the stability of a unity feedback control system whose open-loop transfer function is given by*

$$G(s) = 2\,\frac{e^{-sT}}{s(s+2)}$$

Solution. The overall transfer function is

$$\frac{C(s)}{R(s)} = \frac{\dfrac{e^{-sT}}{s(s+2)}}{\left[1 + \dfrac{e^{-sT}}{s(s+2)}\cdot 1\right]}$$

$$= \frac{e^{-sT}}{(s^2 + 2s + e^{-sT})} \qquad ...(1)$$

The characteristic equation is

$$s^2 + 2s + e^{-sT} = 0 \qquad ...(2)$$

Since $$e^{-sT} = \left(1 - sT + \frac{s^2 T^2}{2!} + \cdots\right) \qquad ...(3)$$

Considering only first two terms of series (3), the characteristic equation can be written as

$$s^2 + 2s + (1 - sT) = 0$$

or $$s^2 + (2 - T)\, s + 1 = 0 \qquad ...(4)$$

The Rough array for equation (4) is formed below :

s^2	1	1
s^1	$(2 - T)$	0
s^0	1	0

For the system to be stable the sign of the first element in the Routh table should be positive, therefore, the condition for the stability of the system is given by

$$(2 - T) > 0$$

$$\therefore \qquad T < 2. \quad \textbf{Ans.}$$

Example 7.5.7. *(a) Using Routh-Hurwitz criterion determine the relation between K and T so that unity feedback control system whose open-loop transfer function given below is stable*

$$G(s) = \frac{K}{s[s(s+10)+T]}$$

Solution. (*a*) The characteristic equation for the system is

$$s\,[s(s + 10) + T] + K = 0$$

or $$s^3 + 10s^2 + Ts + K = 0$$

The Routh array is formed below

s^3	1	T
s^2	10	K
s^1	$\dfrac{10T - K}{10}$	0
s^0	K	0

For the given system to be stable

(*i*) $$\frac{(10\,T - K)}{10} > 0$$

or $$(10T - K) > 0$$

$$\therefore \qquad K < 10T \qquad ...(1)$$

(*ii*) $$K > 0 \qquad ...(2)$$

From (1) and (2) the relation for stability is given below

$$0 < K < 10T. \quad \textbf{Ans.}$$

Example 7.5.7. *(b) Determine the value of K such that the roots of the characteristic equation given below lie to the left of line s = –1*

$$s^3 + 10s^2 + 18s + K = 0.$$

Solution. Put $s = s_1 - 1$

$$\therefore \quad (s_1 - 1)^3 + 10\,(s_1 - 1)^2 + 18(s_1 - 1) + K = 0$$

or $$s_1^3 + 7s_1^2 + s_1 + (K - 9) = 0$$

The Routh's tabulation is given below :

s_1^3	1	1
s_1^2	7	$(K - 9)$
s_1^1	$\dfrac{7-(K-9)}{7}$	
s_1^0	$K - 9$	

$$K < 16 \text{ and } K > 9,$$

K to have values between 9 and 16. **Ans.**

Example 7.5.8. *The open-loop transfer function of a unity feedback control system is given by*

$$G(s) = \frac{K}{s(sT_1 + 1)(sT_2 + 1)}$$

Applying Routh-Hurwitz criterion determine the value of K in term of T_1 and T_2 for the system to be stable.

Solution. The characteristic equation is given by

$$s(sT_1 + 1)(sT_2 + 1) + K = 0$$

or $$T_1T_2s^3 + (T_1 + T_2)\,s^2 + s + K = 0$$

The Routh array is formed below

s^3	T_1T_2	1
s^2	$(T_1 + T_2)$	K
s^1	$\dfrac{[(T_1 + T_2) - KT_1T_2]}{(T_1 + T_2)}$	
s^0	$K,$	

For the system to be stable

(*i*) $$K > 0 \quad \text{...(1)}$$

(*ii*) $$\frac{[(T_1 + T_2) - KT_1T_2]}{(T_1 + T_2)} > 0$$

or $$[(T_1 + T_2) - KT_1T_2] > 0$$

or $$KT_1T_2 < (T_1 + T_2)$$

$$\therefore \quad K < \left(\frac{1}{T_1} + \frac{1}{T_2}\right) \quad \text{...(2)}$$

In view of relations (1) and (2) following condition for stability is obtained below :

$$0 < K < \left(\frac{1}{T_1} + \frac{1}{T_2}\right) \cdot \textbf{ Ans.}$$

7.6 NYQUIST CRITERION

Nyquist criterion is used to identify the presence of roots of a characteristic equation of a control system in a specified region of *s*-plane. From the stability view point the specified region being the entire right hand side beyond the imaginary axis of complex *s*-plane.

Although the purpose of using Nyquist criterion is similar to Routh-Hurwitz criterion but the approach differs in following respects :

(1) The open-loop transfer function $G(s)\,H(s)$ is considered instead of closed-loop characteristic equation $1 + G(s)\,H(s) = 0$.

(2) Inspection of graphical plot of $G(s)\,H(s)$ enables to get more than Yes or No answer of Routh-Hurwitz method pertaining to stability of control systems.

7.6.1. Procedure for mapping from *s*-plane to *G*(*s*) *H*(*s*)-plane

The first step to understand the application of Nyquist criterion in relation to determination of stability of control systems is mapping from *s*-plane to $G(s)\,H(s)$-plane. "*s*" being considered as independent complex variable and the corresponding value of $G(s)\,H(s)$ being the dependent variable plotted in another complex plane called $G(s)\,H(s)$-plane.

Thus for every point in *s*-plane there exists a corresponding point in $G(s)\,H(s)$-plane. During the process of mapping the independent variable "*s*" is varied along a specified path in "*s*" plane and the corresponding points in $G(s)\,H(s)$ plane are joined. This completes the process of mapping from *s*-plane to $G(s)\,H(s)$-plane.

Example 1. *Let* $G(s)\,H(s) = (s + 1.5)$ *and "s" being transversed in s-plane along a line through points* $s_a = (1 + j1)$ *and* $s_b = (2.8 + j0.5)$ *as shown in Fig. 7.6.1.*

Solution.

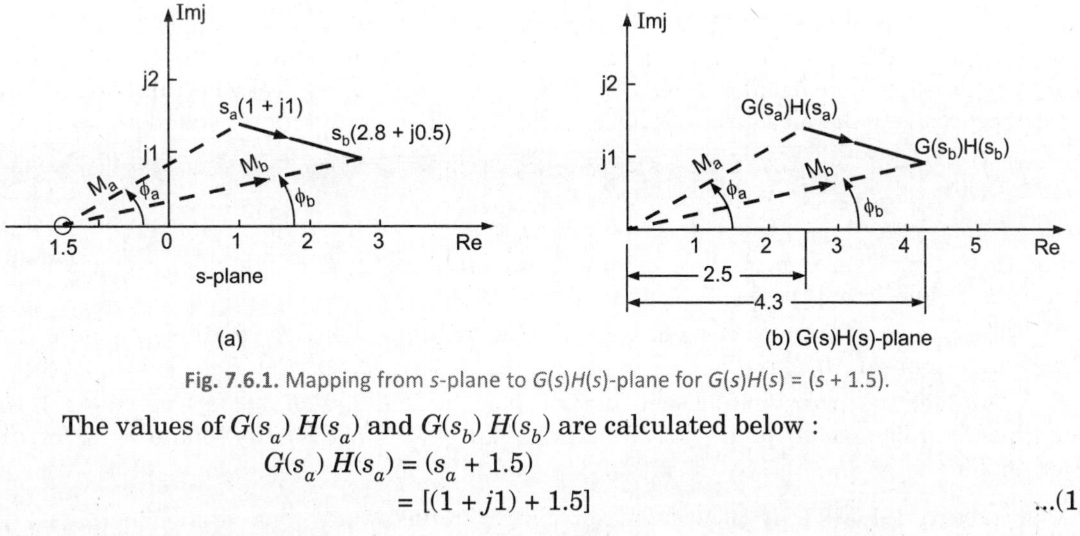

Fig. 7.6.1. Mapping from *s*-plane to *G*(*s*)*H*(*s*)-plane for *G*(*s*)*H*(*s*) = (*s* + 1.5).

The values of $G(s_a)\,H(s_a)$ and $G(s_b)\,H(s_b)$ are calculated below :

$$G(s_a)\,H(s_a) = (s_a + 1.5)$$
$$= [(1 + j1) + 1.5] \qquad \text{...(1)}$$

$$= (2.5 + j1)$$

$$G(s_b)\,H(s_b) = (s_b + 1.5) = [(2.8 + j0.5) + 1.5]$$

$$= (4.3 + j0.5) \qquad \text{...(2)}$$

In Fig. 7.6.1 (*a*) the zero of the transfer function is shown at location $s = -1.5$ in s-plane. A phasor M_a is drawn from the point $s = -1.5$ to the point s_a. The magnitude M_a and phase ϕ_a of this phasor gives the value of the $G(s)\,H(s)$ at s_a in polar form. Similarly M_b and gives the value of $G(s)\,H(s)$ at s_b. The values obtained in equations (1) and (2) are plotted in $G(s)\,H(s)$-plane.

From Fig. 7.6.1 (*b*) it is observed that the magnitude and the phase of the transfer function $G(s)\,H(s) = (s + 1.5)$ at a point in $G(s)\,H(s)$-plane is given by the magnitude and the phase of a phasor drawn from the origin of $G(s)\,H(s)$-plane.

The change in phase of $G(s)\,H(s)$ as s is varied from s_a to s_b is $(\phi_b - \phi_a)$.

Example 2. *Another example is given below wherein change in the phase of G(s) H(s) is observed when the specified path of traversing "s" in s-plane being a closed one. The said closed path is traversed in clockwise direction. The transfer function under consideration has two zeros, the one lies inside the closed path and the other outside the closed path.*

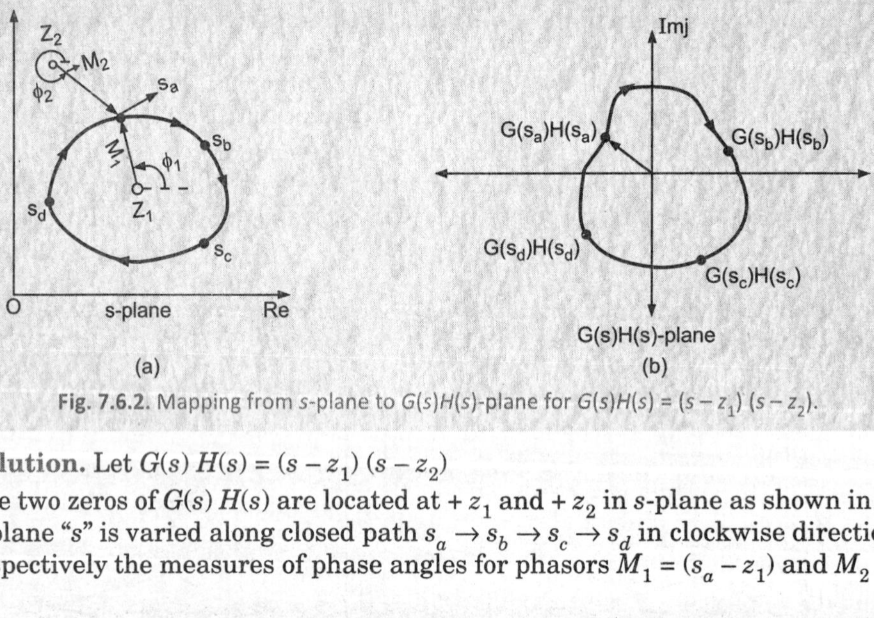

Fig. 7.6.2. Mapping from s-plane to $G(s)H(s)$-plane for $G(s)H(s) = (s - z_1)\,(s - z_2)$.

Solution. Let $G(s)\,H(s) = (s - z_1)\,(s - z_2)$

The two zeros of $G(s)\,H(s)$ are located at $+\,z_1$ and $+\,z_2$ in s-plane as shown in Fig. 7.6.2 (*a*). In s-plane "s" is varied along closed path $s_a \to s_b \to s_c \to s_d$ in clockwise direction. ϕ_1 and ϕ_2 are respectively the measures of phase angles for phasors $M_1 = (s_a - z_1)$ and $M_2 = (s_a - z_2)$ at s_a.

$$\therefore \qquad |G(s_a)\,H(s_a)| = M_1\,M_2 \qquad \text{...(7.12)}$$

and

$$\angle G(s_a)\,H(s_a) = (\phi_1 + \phi_2) \qquad \text{...(7.13)}$$

while traversing closed path starting from s_a and ending at s_a through s_b, s_c, s_d in clockwise direction it is observed that :

The phasor M_1 undergoes a change of $-\,2\pi$, *i.e.* one clockwise rotation and change in the phase of phasor M_2 is NIL.

Thus for the transfer function under consideration $G(s)\,H(s) = (s - z_1)\,(s - z_2)$, the corresponding change in phase* of $G(s)\,H(s)$ is also -2π in $G(s)\,H(s)$-plane as shown in Fig. 7.6.2 (*b*).

*Change of phase in $G(s)\,H(s)$-plane is measured w.r.t. origin.

Therefore, if, the number of zeros of $G(s)\,H(s)$ located in a specified region in s-plane is Z and the independent variable s is varied along a path closing the boundary of such a region in clockwise direction the corresponding change in the argument (phase) of $G(s)\,H(s)$ in $G(s)\,H(s)$-plane is $-2\pi Z$.

On similar reasoning, if the number of poles of $G(s)\,H(s)$ in a specified region in s-plane is P and the independent variable "s" is varied along a path closing the boundary of such a region in clockwise direction, the corresponding change in the argument (phase) of $G(s)\,H(s)$ in $G(s)\,H(s)$-plane is $+2\pi P$.

Considering Z zeros and P poles of $G(s)\,H(s)$ together as located inside a specified region in s-plane and "s" being varied as mentioned above, the mathematical expression for corresponding change in the argument of $G(s)\,H(s)$ in $G(s)\,H(s)$-plane is given below :

$$\Delta \text{Aug} = 2\pi\,(P - Z) \qquad \text{...(7.14)}$$

Conversely, if a region is specified in s-plane and $G(s)\,H(s)$ is plotted as dependent variable and as 's' is varied as independent variable along a closed path in the said specified region in s-plane and the number of poles of $G(s)\,H(s)$ in specified region being given as P, then the existence of number of zeros of $G(s)\,H(s)$ (*i.e.* Z) inside the specified region mentioned above is identified by inspecting the corresponding change in argument of $G(s)\,H(s)$ as obtained from $G(s)\,H(s)$ plot in $G(s)\,H(s)$-plane, the relation (7.14) being used for this purpose.

The plot of $G(s)\,H(s)$ in $G(s)\,H(s)$-plane plotted as per aforesaid procedure is known as Nyquist plot.

7.6.2. Determination from the Nyquist plot the number of zeros of $G(s)\,H(s)$ which are located inside a specified region in s-plane

The plots shown in Fig. 7.6.3 illustrate the application of the $\Delta\text{Aug} = 2\pi\,(P - Z)$ to determine the number of zeros of $G(s)\,H(s)$ located inside a specified region having given the number of poles located in that region.

7.6.3. Application of Nyquist criterion to determine stability of a closed-loop control system

The overall transfer function of a closed-loop control system is given by

$$\frac{C(s)}{R(s)} = \frac{G(s)}{1 + G(s)\,H(s)} \qquad \text{...(7.15)}$$

$G(s)\,H(s)$ is the open-loop transfer function* and $1 + G(s)\,H(s) = 0$ is the characteristic equation.

Let $$G(s)\,H(s) = \frac{K(s - s_0)\,(s - s_2)\cdots}{(s - s_1)\,(s - s_3)\cdots} \qquad \text{...(7.16)}$$

$$\therefore \quad 1 + G(s)\,H(s) = 1 + \frac{K(s - s_0)\,(s - s_2)\cdots}{(s - s_1)\,(s - s_3)}$$

$$= \frac{(s - s_1)\,(s - s_3)\cdots + K\,(s - s_0)\,(s - s_2)}{(s - s_1)\,(s - s_3)\cdots} \qquad \text{...(7.17)}$$

*The transfer function $G(s)\,H(s)$ considered has the number of poles greater than the number of zeros.

Variation of independent variable s in s-plane along closed path $s_a - s_b - s_c - s_d - s_a$	*Corresponding shape of G(s)H(s) plot in G(s)H(s)-plane*	*Conclusion*
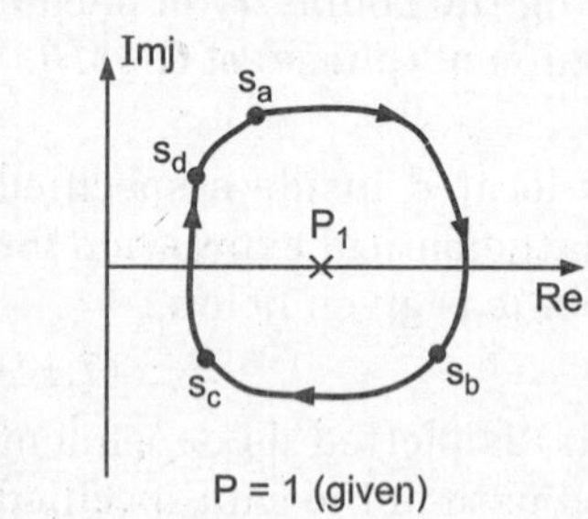 P = 1 (given)	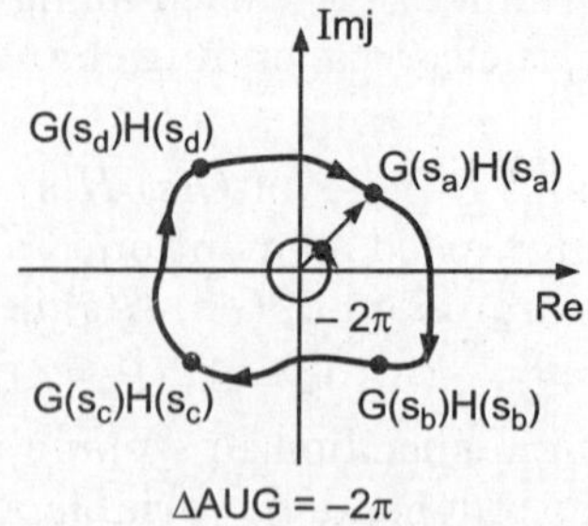 ΔAUG = –2π	Δ Aug = $2\pi (P - Z)$ $-2\pi = 2\pi (1 - Z)$ $\therefore \quad Z = 2$
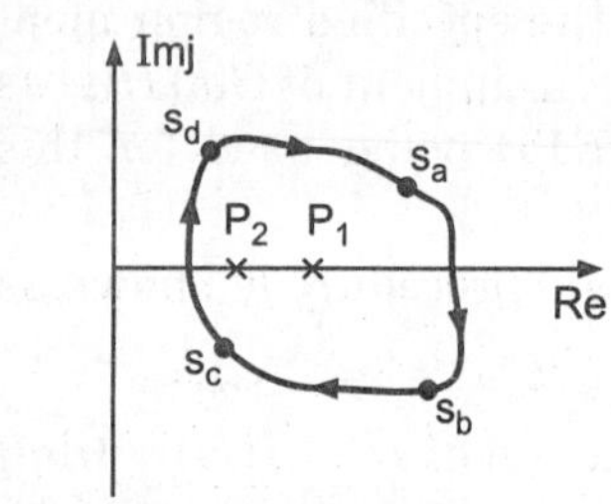 P = 2 (given)	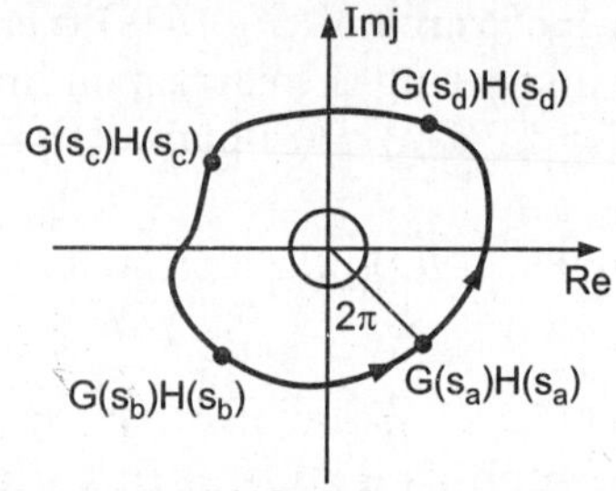 ΔAUG = +2π	Δ Aug = $2\pi (P - Z)$ $2\pi = 2\pi (2 - Z)$ $\therefore \quad Z = 1$
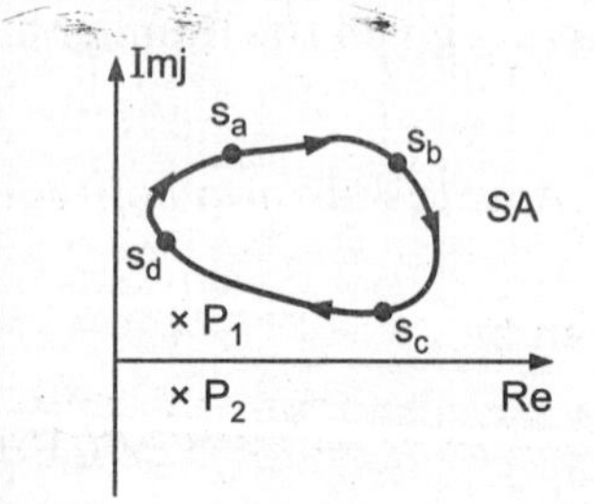 P = 0 (given)	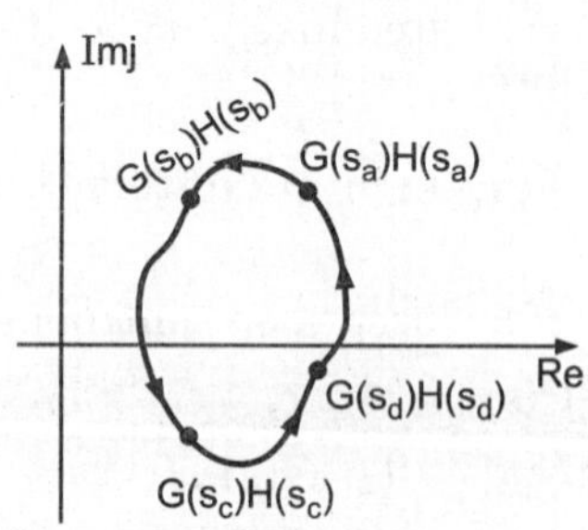 ΔAUG = 0	Δ Aug = $2\pi (P - Z)$ $0 = 2\pi (P - Z)$ $\therefore \quad Z = 0$
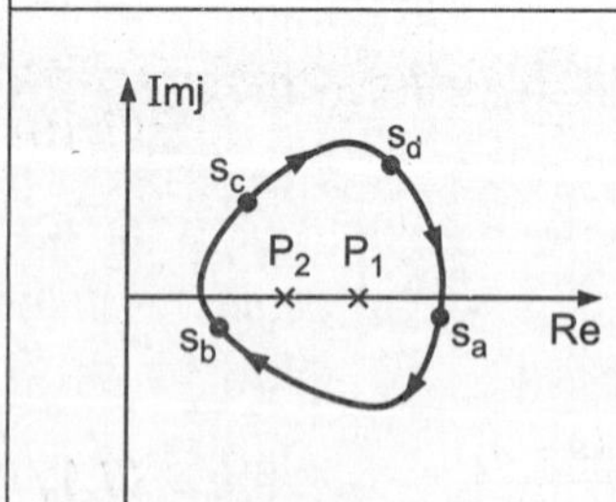 P = 2 (given)	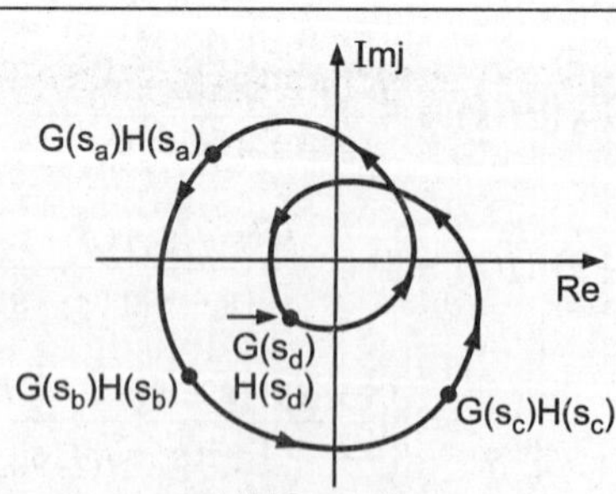 ΔAUG = + 4π	Δ Aug = $2\pi (P - Z)$ $4\pi = 2\pi (2 - Z)$ $\therefore \quad Z = 0$

Fig. 7.6.3. Application of the relation Δ Aug = $2\pi (P - Z)$ to determine the number of zeros of $G(s)H(s)$ having given the number of poles.

Let the numerator of equation be as

$$K'(s - s_a)(s - s_b) \ldots$$

$$\therefore \qquad 1 + G(s)\,H(s) = \frac{K'(s - s_a)(s - s_b)\ldots}{(s - s_1)(s - s_3)\ldots} \qquad \ldots(7.18)$$

It is observed that the two functions given by Eqs. (7.16) and (7.18) have same poles, *i.e.* s_1, s_3, s_5 etc. but different zeros.

The zeros of $1 + G(s) H(s) = 0(7.18)$ are s_a, s_b, s_c etc. As the function (7.18), when equated to zero is the characteristic equation of the closed-loop system. These zeros are also its roots, hence from the stability view point the zeros of the function (7.18) are to be identified. For the system to be stable these identified zeros should not have positive real part. Thus the basis of applying Nyquist criterion for ascertaining stability of a control system is that, the specified region for identifying the presence of zeros of $1 + G(s) H(s) = 0$ should be the entire right hand side beyond the imaginary axis of *s*-plane and the closed path along which the independent variable *s* is to be varied is shown in Fig. 7.6.4.

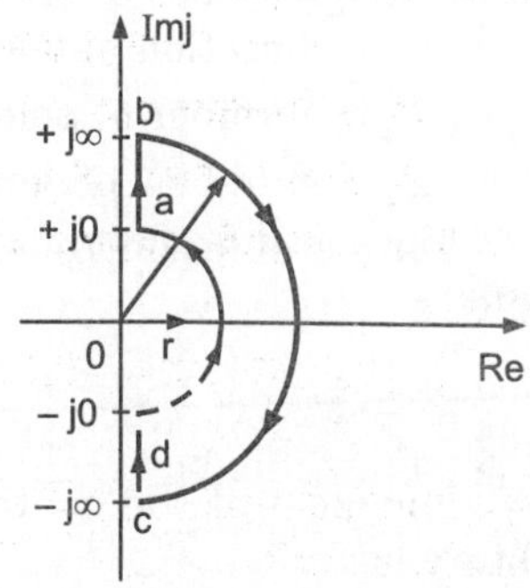

Fig. 7.6.4. Closed path for variation of *s* in *s*-plane.

The closed path is $ab \to bc \to cd \to da$ starting at $a(s = +j0)$ through $b(s = +j\infty)$, along a semi-circular arc *bc* of infinite radius, $c(s = -j\infty)$, $d(s = -j0)$ and finally from *d* to *a* along a semicircular arc of radius *r* in anti-clockwise direction in such a manner that $r \to 0$.

On traversing the closed path for independent variable *s* in *s*-plane as above, the corresponding point to point values of $G(s) H(s)$ are plotted in $G(s) H(s)$ plane and change in argument of $G(s) H(s)$ is noted.

A number indicating the zeros of $G(s) H(s)$ which are located in the right hand portion of *s*-plane is calculated by using the relation :

$$\Delta \text{ Aug } G(s)\,H(s) = 2\pi\,(P - Z)$$

where **P* is the number of poles of $G(s) H(s)$ having positive real part and is a known quantity.

Note that the change in argument of $G(s) H(s)$ is measured w.r.t. origin (0, 0) in $G(s) H(s)$ plane. The aforesaid procedure calculates the number of zeros of $G(s) H(s) = 0$ and not the zeros of $1 + G(s) H(s) = 0$.

However, from the plot of $G(s) H(s)$ as plotted above the number of zeros of $1 + G(s) H(s) = 0$ which have positive real part can also be determined if the origin of $G(s) H(s)$ plane is shifted to the point $(-1 + j0)$ and the change in argument of $G(s) H(s)$ plot is measured w.r.t. the point $(-1 + j0)$ in $G(s) H(s)$ plane.

The number of zeros thus calculated is also the number of roots of the characteristic equation $1 + G(s) H(s) = 0$.

As the *s*-plane region wherein only poles and zeros of $1 + G(s) H(s) = 0$ with positive real part is under investigation, the expression relating the number of poles and zeros present in the R.H.S. of *s*-plane is written as

*If $G(s)H(s)$ has a pole on imaginary axis then such poles are excluded from the path and the modified path around such a pole being a semi-circle in the RHS of *s*-plane. The radius of the semi-circle being approaching to zero.

$$\Delta \text{Aug } G(s) H(s) = 2\pi (P_+ - Z_+) \quad \text{...(7.19)}$$
(origin $-1 + j0$)

or
$$\frac{\Delta \text{ Aug } G(s) H(s)}{2\pi} = (P_+ - Z_+)$$

or
$$N = (P_+ - Z_+) \quad \text{...(7.20)}$$

where, N = Number of encirclements of the point $(-1 + j0)$ by $G(s) H(s)$ plot. The positive direction of encirclements being anti-clockwise.

P_+ = Number of poles of $G(s) H(s)$ with positive real part

Z_+ = Number* of zeros of $1 + G(s) H(s) = 0$ with positive real part.

For a stable control system $Z_+ = 0$, therefore, the condition for a control system to be stable is

$$N = (P_+ - 0)$$

or
$$N = P_+ \quad \text{...(7.21)}$$

For most of the control systems $P_+ = 0$, therefore, for such cases the conditions for stability is

$$N = 0 \quad \text{...(7.22)}$$

which means that the encirclement of the point $(-1 + j0)$ is nil, *i.e.*, there are no encirclements of the point $(-1 + j0)$ by $G(s) H(s)$ plot or the point $(-1 + j0)$ lies outside $G(s) H(s)$ plot in $G(s) H(s)$-plane.

Closing of Nyquist Plot from $s = -j0$ to $s = +j0$:

As mentioned earlier the origin (0, 0) of the s-plane is avoided from the specified closed path, because the poles at the origin of $G(s) H(s)$ are a deciding factor in closing the corresponding plot in $G(s) H(s)$-plane from $G(-j0) H(-j0)$ to $G(+j0) H(+j0)$.

The example given below illustrates the procedure for closing the Nyquist plot for the points mentioned above.

Let
$$G(s) H(s) = \frac{K}{s(sT + 1)} \quad \text{...(7.23)}$$

As s is varied from a to b (Fig. 7.6.4) on the imaginary axis, substitute $s = j\omega$ in Eq. (7.23)

$\therefore$
$$G(j\omega) H(j\omega) = \frac{K}{j\omega(j\omega T + 1)} \quad \text{...(7.24)}$$

For plotting $G(j\omega) H(j\omega)$, the independent variable "s" is varied on the entire imaginary axis from $\omega = +0$ to $\omega = +\infty$ in s-plane and the corresponding plot of $G(j\omega) H(j\omega)$ is plotted in $G(s) H(s)$-plane. The plot for variation from $\omega = -0$ to $\omega = -\infty$ is vertical image of the plot from $\omega = 0$ to $\omega = +\infty$.

As only the nature of the $G(j\omega) H(j\omega)$ plot is needed, all the values of ω need not be substituted to obtain the plot, only typical values say $\omega = +0$ and $\omega = +\infty$ be substituted. The corresponding plot is obtained as explained below :

(*a*) Put $\omega = +0$ in Eq. (7.24)

$\therefore$
$$G(j0) H(j0) = \frac{K}{j0\,(j0T + 1)}$$

Z_+ is also the number of roots of the characteristic equation $1 + G(s) H(s) = 0$ with positive real part.

$j0T$ is neglected in comparison with 1

$$\therefore \qquad G(j0)\,H(j0) = \frac{K}{j0} = -j\infty = |\infty| \angle -90^\circ$$

(*b*) Put $\omega = +\infty$ in Eq. (7.24)

$$\therefore \qquad G(j\infty)\,H(j\infty) = \frac{K}{j\infty\,(j\infty T + 1)}$$

1 is neglected in comparison with $j\infty$.

$$\therefore \qquad G(j\infty)\,H(j\infty)\ \frac{K}{j^2\,\infty T} = -0 = |0| \angle -180^\circ$$

It is thus observed that as ω increases from $\omega = +0$ to $\omega = +\infty$ the plase angle of $G(j\omega)$ $H(j\omega)$ changes from -90° to -180° and the respective magnitudes are $|\infty|$ and $|0|$.

The nature of the plot is shown in Fig. 7.6.5. The plot for values $\omega = -0$ to $\omega = -\infty$ is shown by dotted lines. The arrows marked on the plot indicate the direction of increasing ω.

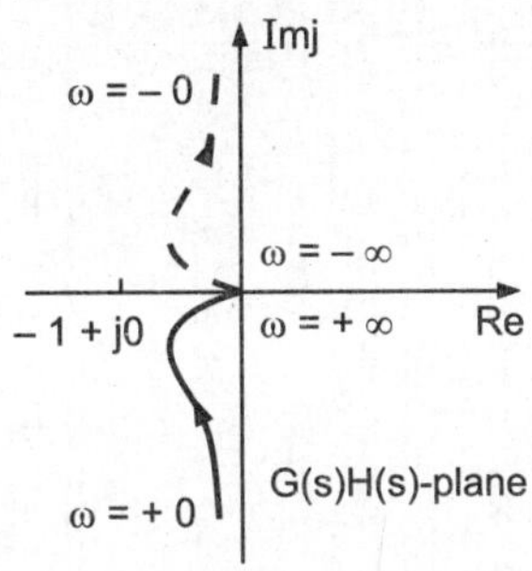

Fig. 7.6.5. Nyquist plot for $G(s)\,H(s) = \dfrac{K}{s(sT+1)}$ for variation $\omega = +0$ to $\omega = +\infty$ and $\omega = -\infty$ to $\omega = -0$.

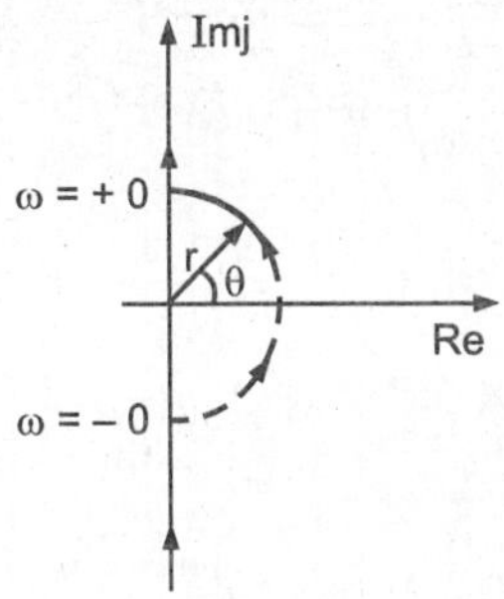

Fig. 7.6.6. Variation of s from $s = -0$ to $s = +0$ in s-plane.

In respect of variation of ω from $\omega = -0$ to $\omega = +0$ the plot is not complete. The completion of the plot depends on the number of poles of $G(s)\,H(s)$ located at the origin in s-plane.

Consider a general case where the number of poles at the origin is n, for such a case the transfer function is given by

$$G(s)\,H(vs) = \frac{K}{s^n\,(sT_1+1)\,(sT_2+1)\cdots} \qquad \text{...(7.25)}$$

As $s \to 0$, the terms sT_1, sT_2, ... etc., also approaches to zero value. Hence,

$$\underset{s\to 0}{G\,(s)\ H(s)} = \frac{K}{s^n} \qquad \text{...(7.26)}$$

As shown in Fig. 7.6.6 "s" is varied in s-plane from $s = -0$ to $s = +0$ in anti-clockwise direction along a semi-circular arc of radius r in such a manner that $r \to 0$, because if "s" is varied from $s = -0$ to $s = +0$ in the clockwise direction, then the region to the L.H.S. of imaginary axis is covered which is not permitted.

The equation of the phasor along the semi-circular arc shown in Fig. 7.6.6 is given by

$$(s - 0) = re^{j\theta}$$

$$s = re^{j\theta} \qquad ...(7.27)$$

substituting for s in Eq. (7.26) from (7.27) following relation is obtained

$$\underset{s \to 0}{G(s)\ H(s)} = \frac{K}{r^n\ e^{jn\theta}} = \frac{K}{j^n} e^{-jn\theta} \qquad ...(7.28)$$

In s-plane

At $s = -j0$, $\qquad \theta = -\pi/2$

and at $s = +j0$, $\qquad \theta = +\pi/2$

Therefore, change in θ as s is varied in s-plane from $s = -j0$ to $s = +j0$ is $(+\pi/2) - (-\pi/2) = +\pi$.

The corresponding change in phase of $G(s)\ H(s)$ in $G(s)\ H(s)$ plane is determined below :

Since $\qquad \underset{r \to 0}{G(s)\ H(s)} = \frac{K}{r^n} e^{-jn\theta}$

and at $\quad s = -j0,\ \theta = -\pi/2$

$$\therefore \qquad G(-j0)\ H(-j0) = \lim_{r \to 0} \frac{K}{r^n} e^{-jn\left(-\frac{\pi}{2}\right)}$$

$$= \lim_{r \to 0} \frac{K}{r^n} e^{\frac{jn\pi}{2}} \qquad ...(7.29)$$

Similarly at $s = +j0$, $\theta = \pi/2$

$$\therefore \qquad G(+j0)\ H(+j0) = \lim_{r \to 0} \frac{K}{r^n} e^{-jn\left(\frac{\pi}{2}\right)}$$

$$= \lim_{r \to 0} \frac{K}{r^n} e^{\frac{-jn\pi}{2}} \qquad ...(7.30)$$

From Eqs. (7.29) and (7.30) it is noted that the corresponding change in the phase of $G(s)\ H(s)$ in $G(s)\ H(s)$-plane as s is varied in s-plane from $s = -j0$ to $s = +j0$ as per direction (anticlockwise) shown in Fig. 7.6.6 is given by

$$\left(-\frac{n\pi}{2}\right) - \left(+\frac{n\pi}{2}\right) = -n\pi \text{ (clockwise direction)} \qquad ...(7.31)$$

It is, therefore, concluded that, if in the s-plane "s" changes from $s = -j0$ to $s = +j0$ by π radian (anti-clockwise) then the corresponding change in phase of $G(s)\ H(s)$ in $G(s)\ H(s)$-phase is $-n\pi$ radian (clockwise) and the magnitude of $G(s)\ H(s)$ during this phase change is infinite. "n" denotes the type of the transfer function or the number of poles of $G(s)\ H(s)$ at the origin.

The closing angle of Nyquist plot from $s = -j0$ to $s = j0$ is given in the following table according to type of $G(s)\ H(s)$:

n	Type of $G(s)\,H(s)$	Angle through which Nyquist plot is to be closed from $\omega = -0$ to $\omega = +0$	Magnitude of $G(j\omega)\,H(j\omega)$
0	0	0	The points $\omega = -0$ $\omega = +0$ are coincident
1	1	$-\pi$	∞
2	2	-2π	∞
3	3	-3π	∞
"	,	,	
"	,	,	
n	n	$-n\pi$	∞

Note. From Nyquist plot the stability of a closed-loop system as well as that of open-loop system can be determined. If the origin of $G(s)\,H(s)$-plane is considered as (0, 0) the open-loop stability is obtained and consideration of the point $(-1 + j0)$ as origin gives the stability of closed-loop system.

Examples given below illustrate the application of Nyquist criterion to determine the stability of closed-loop control systems.

Illustrative Example 1. *Examine the closed-loop stability of a control system whose open-loop transfer function is given below :*

$$G(s)\,H(s) = \frac{K}{s(sT + 1)}.$$

Solution.
$$G(s)\,H(s) = \frac{K}{s\,(sT + 1)}$$

Put $s = j\omega$

$$\therefore \quad G(j\omega)\,H(j\omega) = \frac{K}{j\omega\,(j\omega T + 1)} \qquad ...(1)$$

Rationalizing Eq. (1) and separating into real and imaginary parts following equation is obtained

$$G(j\omega)\,H(j\omega) = -\frac{KT}{(\omega^2 T^2 + 1)} - \frac{jK}{\omega\,(\omega^2 T^2 + 1)} \qquad ...(2)$$

From Eq. (2) it is observed that as ω increases from $\omega = +0$ to $\omega = +\infty$ both the real part and the imaginary part lie in the third quadrant of $G(s)\,H(s)$-plane.

At $\omega = +0$, $\angle G(+j0)\,H(+j0) = -90°$ and the magnitude approaches infinity.

At $\omega = +\infty$, $\angle G(+j\infty)\,H(+j\infty) = -180°$ and the magnitude approaches zero.

The Nyquist plot as ω varied from $\omega = +0$ to $\omega = +\infty$ is shown in Fig. 7.6.7. The plot for $\omega = -\infty$ to $\omega = -0$ is shown by dotted lines.

As the system is type 1 the plot is closed from $\omega = -0$ to $\omega = +0$ through an angle of $-\pi$ (clockwise) with an infinite radius. The arrowheads shown in Fig. 7.6.7 are in the direction of increasing ω.

As the point $(-1 + j0)$ is not encircled by the plot, therefore,

$$N = 0$$

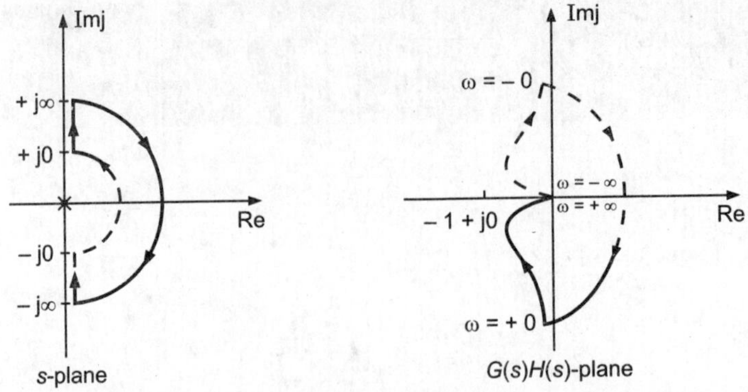

Fig. 7.6.7. Nyquist plot for $G(s)\,H(s) = \dfrac{K}{s(sT+1)}$.

The number of zeros (roots) of the characteristic equation with positive real part is determined by using relation*

$$N = (P_+ - Z_+)$$

because,
$$P_+ = 0 \qquad \therefore \quad Z_+ = 0$$

Hence, the number of zeros (roots of the characteristics equation) with positive real part is nil and the closed-loop system is stable.

Illustrative Example 2. *A closed-loop control system is described by the block diagram given below (Fig. 7.6.8), determine the stability using Nyquist criterion.*

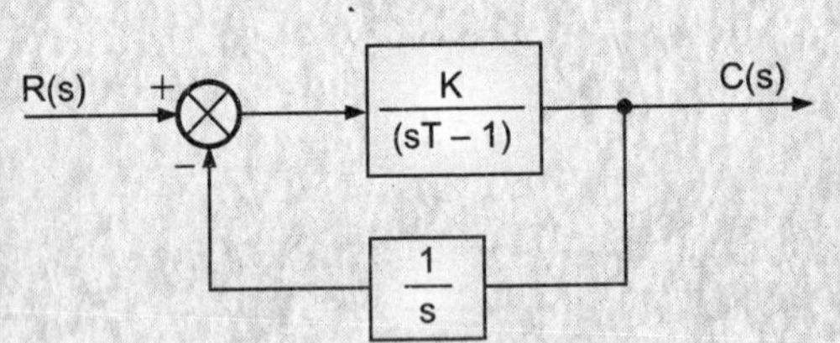

Fig. 7.6.8. Block diagram for illustrative example 2.

Solution. The open-loop transfer function is given by

$$G(s)\,H(s) = \frac{K}{(sT-1)}\cdot\frac{1}{s} = \frac{K}{s\,(sT-1)}$$

Put $\quad s = j\omega$

$$\therefore \qquad G(j\omega)\,H(j\omega) = \frac{K}{j\omega\,(j\omega T - 1)} \qquad ...(1)$$

Rationalizing Eq. (1) and separating into real and imaginary parts following equation is obtained :

---

*The value of N is determined by imagining a phasor whose one end rests on the point $(-1 + j0)$, the other end of this phasor is allowed to trace the Nyquist plot starting at $\omega = +0$ through $\omega = +\infty$, $\omega = -\infty$, $\omega = -0$ and finally ending at $\omega = +0$. The net counter clockwise rotations of this phasor gives the value of N. A clockwise rotation of this phasor is noted by a negative number of N.

$$G(j\omega)\,H(j\omega) = -\frac{KT}{(\omega^2T^2+1)} + \frac{jK}{\omega\,(\omega^2T^2+1)} \quad ...(2)$$

From Eq. (2) it is observed that as ω increases from $\omega = +0$ to $\omega = +\infty$ the real part and the imaginary part lie in the second quadrant of $G(s)\,H(s)$-plane.

At $\omega = +0$, $\angle G(+j0)\,H(+j0) = 90°$ and the magnitude approaches infinity.

At $\omega = +\infty$, $\angle G(j\infty)\,H(j\infty) = 180°$ and the magnitude approaches zero.

The Nyquist plot as ω is varied from $\omega = +0$ to $\omega = +\infty$ is shown in Fig. 7.6.9. The plot for the values of ω from $\omega = -\infty$ to $\omega = -0$ is shown by dotted lines.

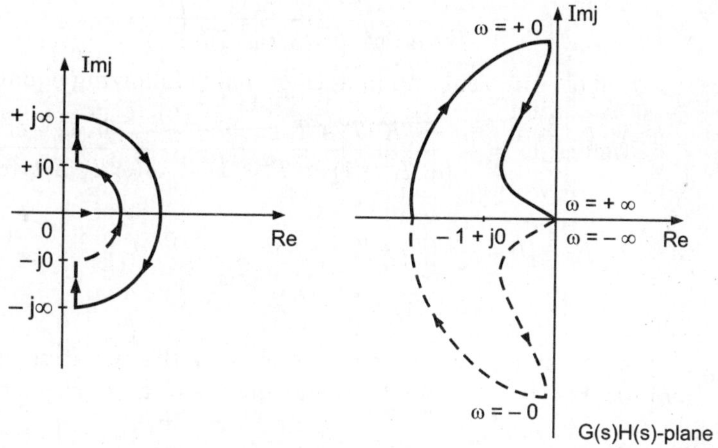

Fig. 7.6.9. Nyquist plot for $G(s)\,H(s) = K/s(sT+1)$.

As the system is type 1 the plot is closed from $\omega = -0$ to $\omega = +0$ through an angle of $-\pi$ (clockwise) with an infinite radius. The arrowheads shown in Fig. 7.6.9 are in the direction of increasing ω.

As the point $(-1 + j0)$ is encircled by the plot once in the clockwise direction, therefore, $N = -1$.

The number of zeros (roots of the characteristic equation) with positive real part is determined using relation

$$N = (P_+ - Z_+)$$

For the system under consideration it is observed that $P_+ = 1$, *i.e.* $s = \frac{1}{T}$ and $N = -1$, therefore,

$$-1 = (1 - Z_+) \qquad \therefore \qquad Z_+ = 2$$

Hence, the closed-loop system given in the example has two roots with positive real part and closed-loop system is unstable.

Illustrative Example 3. *Using Nyquist criterion investigate the stability of a closed-loop control system whose open-loop transfer function is given below :*

$$G(s)\,H(s) = \frac{K}{s\,(sT_1+1)\,(sT_2+1)}.$$

Solution. $$G(s)\,H(s) = \frac{K}{s\,(sT_1+1)\,(sT_2+1)}$$

Put $s = j\omega$

$$\therefore \quad G(j\omega)\,H(j\omega) = \frac{K}{j\omega\,(j\omega T_1+1)\,(j\omega T_2+1)}$$

$$= \frac{-jK\,[(-j\omega T_1+1)\,(-j\omega T_2+1)]}{\omega\,(\omega^2T_1^2+1)\,(\omega^2\,T_2^2+1)}$$

$$= \frac{K\,[-\omega\,(T_1+T_2) - j\,(1-\omega^2T_1\,T_2)]}{\omega\,(\omega^2T_1^2+1)\,(\omega^2\,T_2^2+1)} \quad ...(1)$$

Separating equation (1) into real and imaginary parts following equation is obtained :

$$G(j\omega)\,H(j\omega) = -\frac{K\,(T_1+T_2)}{(\omega^2T_1^2+1)\,(\omega^2T_2^2+1)} - \frac{jK\,(1-\omega^2T_1T_2)}{\omega\,(\omega^2T_1^2+1)\,(\omega^2\,T_2^2+1)} \quad ...(2)$$

At, $\omega = +0$

$$G(+j0)\,H(j+0) = -K(T_1+T_2) - j\infty$$

$$= |\infty| \angle -\frac{\pi}{2} \quad ...(3)$$

From Eq. (1) it is observed the Nyquist plot crosses the negative real axis and the value of ω at which this occurs is determined by equating the imaginary part of Eq. (2) to zero therefore,

$$\frac{K\,(1-\omega^2T_1T_2)}{\omega\,(\omega^2T_1^2+1)\,(\omega^2\,T_2^2+1)} = 0 \quad ...(4)$$

The solution of Eq. (4) is

$$\omega = \pm\frac{1}{\sqrt{T_1T_2}}$$

$$\therefore \quad \left|G\,(j\omega)\,H\,(j\omega)\right|_{\omega=\frac{1}{\sqrt{T_1T_2}}} = -\frac{K\,(T_1+T_2)}{\left[\left(\frac{1}{T_1T_2}T_1^2+1\right)\left(\frac{1}{T_1T_2}T_2^2+1\right)\right]}$$

$$= -\frac{KT_1T_2}{(T_1+T_2)} \quad ...(5)$$

At, $\omega = +\infty$

$$G(+j\infty)\,H(+j\infty) = -0 + j0 = |0| \angle\frac{\pi}{2} \quad ...(6)$$

In view Eqs. (3), (5) and (6) the Nyquist plot is drawn as shown in Fig. 7.6.10. The plot for values of ω from $\omega = -\infty$ to $\omega = -0$ is shown by dotted lines. As the system is type 1, the plot is closed from $\omega = -0$ to $\omega = +0$ in the clockwise direction through an angle of π.

The intersection of Nyquist plot with the real axis is given by Eq. (5) therefore, if

$$\frac{K\,T_1T_2}{(T_1+T_2)} < 1$$

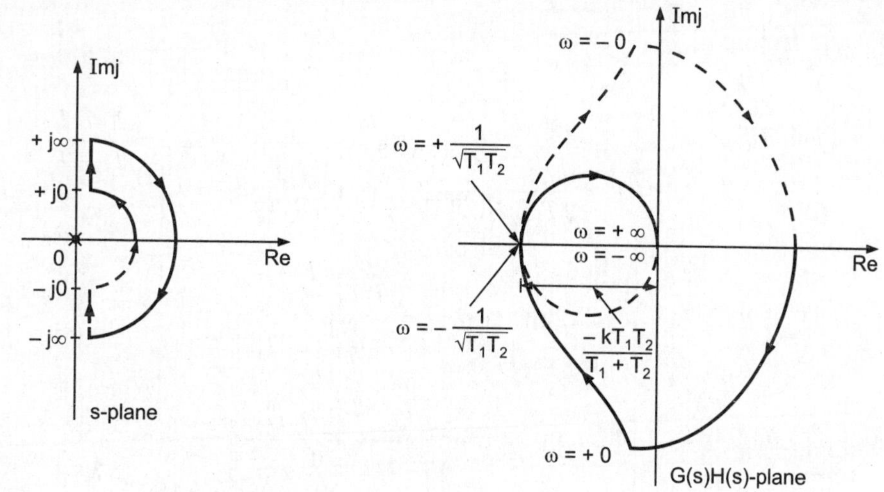

Fig. 7.6.10. Nyquist plot for $G(s)\,H(s) = K/s(sT_1 + 1)(sT_2 + 1)$.

the point $(-1 + j0)$ is not encircled by the Nyquist plot and the closed-loop system is found to be stable (Note that $P_+ = 0$).

However, if $$\frac{K\,T_1T_2}{(T_1 + T_2)} > 1$$

the point $(-1 + j0)$ is encircled twice in clockwise direction indicating that there are two roots of the characteristics equation with positive real part and the closed-loop system is unstable.

The stability condition is derived below :

For stability $$\frac{K\,T_1T_2}{(T_1 + T_2)} \le 1 \quad \text{or} \quad K \le \left(\frac{1}{T_1} + \frac{1}{T_2}\right)$$

Alternatively, general shape of the Nyquist plot (Fig. 7.6.10) in relation to open-loop transfer function

$$G(j\omega)\,H(j\omega) = \frac{K}{j\omega\,(j\omega T_1 + 1)\,(j\omega T_2 + 1)}$$

can be drawn using following method.

$$|G(j\omega)\,H(j\omega)| = \frac{K}{\omega\sqrt{\omega^2T_1^2 + 1^2}\,\sqrt{\omega^2T_2^2 + 1^1}}$$

and $$\angle G(j\omega)\,H(j\omega) = -90° - \tan^{-1}\omega T_1 - \tan^{-1}\omega T_2$$

$$= -90° - \tan^{-1}\frac{\omega(T_1 + T_2)}{1 - \omega^2T_1T_2} \quad \text{for } \omega < \frac{1}{\sqrt{T_1T_2}}$$

$$= -90° - \tan^{-1}\frac{\omega(T_1 + T_2)}{1 - \omega^2T_1T_2} - 180° \quad \text{for } \omega > \frac{1}{\sqrt{T_1T_2}}$$

(1) As $\omega \to 0$

$$|G(j0)\,H(j0)| \to \infty$$
$$|G(j0)\,H(j0)| \to -90°$$

(2) At $\omega = \dfrac{1}{\sqrt{T_1 T_2}}$

$$\left| G(j\omega)\, H(j\omega) \right|_{\omega = \frac{1}{\sqrt{T_1 T_2}}} = \frac{K}{\sqrt{\frac{1}{T_1 T_2}} \sqrt{\left(\frac{1}{T_1 T_2} T_1^2 + 1\right)\left(\frac{1}{T_1 T_2} T_2^2 + 1\right)}} = \frac{K\, T_1 T_2}{T_1 + T_2}$$

$$\angle G(j\omega) H(j\omega) \Big|_{\omega = \frac{1}{\sqrt{T_1 - T_2}}} = -90° - \tan^{-1} \frac{\frac{1}{\sqrt{T_1 T_2}}(T_1 + T_2)}{1 - \frac{1}{\sqrt{T_1 T_2}} \cdot T_1 T_2}$$

$$= -90° - \tan^{-1} \frac{(T_1 + T_2)/\sqrt{T_1 T_2}}{0}$$

$$= -90° - \tan^{-1} \infty = -90° - 90° = -180°$$

(3) As $\omega \to \infty$

$$| G(j\infty)\, H(j\infty) | \to 0$$

and
$$\angle G(j\infty)\, H(j\infty) \to (-90° - 0°) - 180° = -270°$$

The Nyquist plot crosses the negative real axis of $G(s)\, H(s)$-plane at

$$\omega = \frac{1}{\sqrt{T_1 T_2}} \text{ and } \left| G_j\left(\frac{1}{\sqrt{T_1 T_2}}\right) H\left(j \frac{1}{\sqrt{T_1 T_2}}\right) \right| = -\frac{K\, T_1 T_2}{T_1 + T_2}$$

As per above derivations general shape of the Nyquist plot is shown in Fig. 7.6.10 and closed-loop stability conditions mentioned above can be verified.

7.7 GAIN MARGIN AND PHASE MARGIN

A study of the above illustrative example reveals that as the gain K is increased (the values of K, T_1 and T_2 being positive) following possibilities regarding stability of the system arise :

(*a*) $K < \left(\dfrac{1}{T_1} + \dfrac{1}{T_2}\right)$ the system is stable

(*b*) $K = \left(\dfrac{1}{T_1} + \dfrac{1}{T_2}\right)$ the system exhibits sustained oscillations, therefore the system is on the verge of instability or marginally stable.

(*c*) $K > \left(\dfrac{1}{T_1} + \dfrac{1}{T_2}\right)$ the system is unstable.

Fig. 7.7.1 shows the Nyquist plot for the above mentioned cases and the corresponding step response. For clarity the Nyquist plot is shown for values* of ω from ω = + 0 to ω = + ∞.

Gain margin. From Fig. 7.7.1, it is observed that Nyquist plot gives the information about the system stability and also indicates that how far the system is stable and further as to how the performance of the system can be improved in view of stability.

* The plot of $G(j\omega)\, H(j\omega)$ for values of ω from ω = +0 to ω = + ∞ is also known as polar plot.

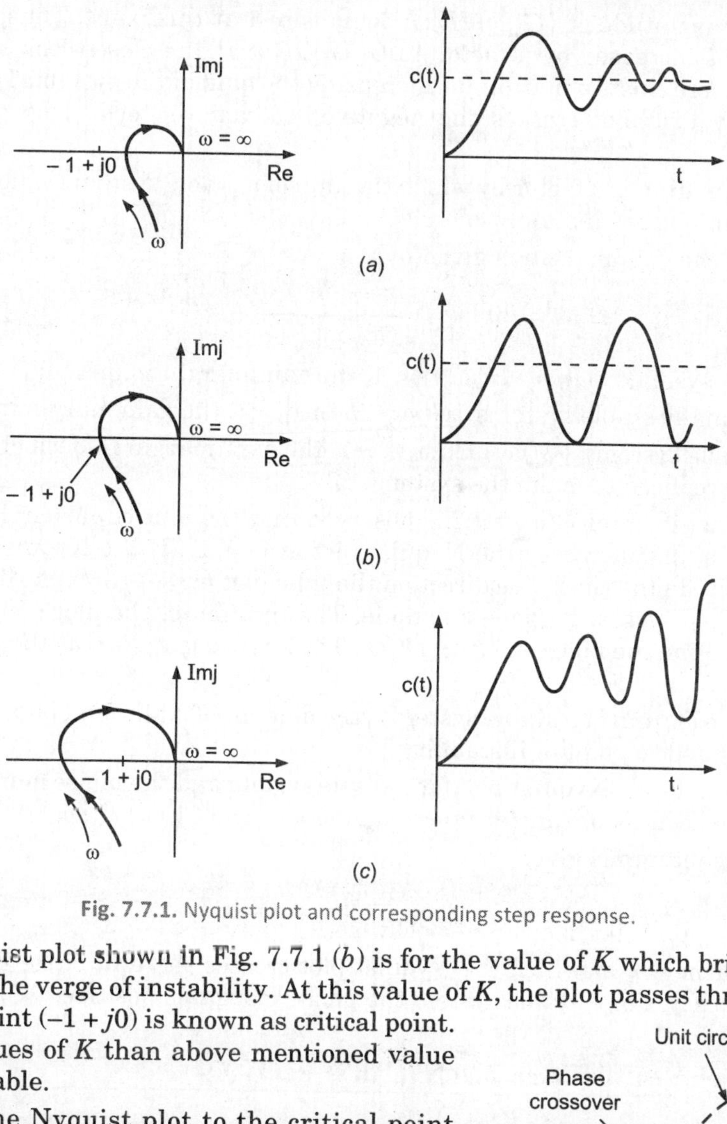

Fig. 7.7.1. Nyquist plot and corresponding step response.

The Nyquist plot shown in Fig. 7.7.1 (b) is for the value of K which brings the closed-loop system on the verge of instability. At this value of K, the plot passes through the point $(-1 + j0)$. The point $(-1 + j0)$ is known as critical point. For smaller values of K than above mentioned value the system is stable.

Nearer the Nyquist plot to the critical point $(-1 + j0)$ the system tends towards more oscillations in its output. This nearness of the Nyquist plot to the critical point $(-1 + j0)$ determines the relative stability of a control system expressed in terms of gain margin and phase margin.

The Nyquist plot for a stable system is shown in Fig. 7.7.2. The plot crosses the negative real axis at $\omega = \omega_2$ and the corresponding magnitude $|G(j\omega_2) H(j\omega_2)|$ is less than 1. If the gain K of the system is further increased so that the plot passes through the point $(-1 + j0)$, the system starts tending towards

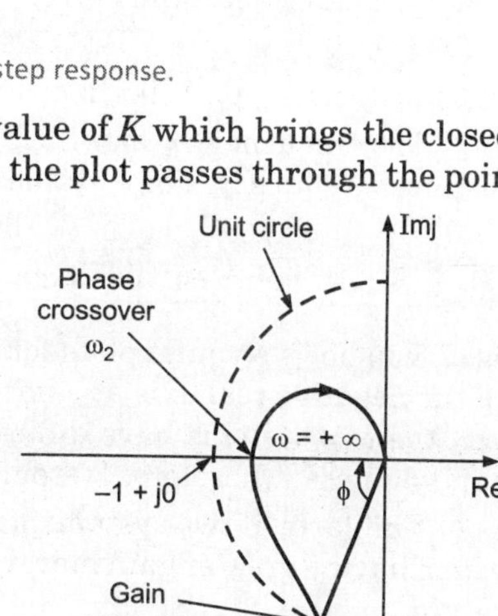

Fig. 7.7.2. Gain margin and phase margin from Nyquist plot.

instability. The magnitude of $|G(j\omega) H(j\omega)|$ equals to 1 at this point. Thus, it is concluded that if the gain is increased by a factor $|1/G(j\omega_2) H(j\omega_2)|$ the closed-loop system becomes unstable. Thus there exists a margin for increasing the gain of the system. The frequency ω_2 at which the Nyquist plot crosses the negative real axis is termed as phase crossover frequency.

"The gain margin is a factor by which the gain of a stable system is allowed to increase before the system reaches instability."

In decibel the gain margin is given by

$$\text{G.M.} = 20 \log_{10} \frac{1}{|G(j\omega_2) H(j\omega_2)|} \text{ db} \qquad ...(7.32)$$

For stable systems $|G(j\omega_2) H(j\omega_2)| < 1$, the gain margin in decibel is positive.

For marginally stable systems $|G(j\omega_2) H(j\omega_2)| = 1$, the gain margin in decibel is zero.

For unstable systems $|G(j\omega_2) H(j\omega_2)| > 1$, the gain margin in decibel is negative and the gain is to be reduced to make the system stable.

Phase Margin. From Fig. 7.7.2 is observed that the unit circle (circle with centre at origin and radius unity) crosses the Nyquist plot at $\omega = \omega_1$. If the Nyquist plot is rotated through an angle ϕ in the clockwise direction then the plot passes through the point $(-1 + j0)$ and the closed-loop system becomes unstable. The rotation of the plot mentioned above is achieved by changing the phase of $G(j\omega) H(j\omega)$. The angle ϕ is known as the phase margin of the system.

"The phase margin of a stable system is the amount of additional phase lag required to bring the system to the point of instability."

Fig. 7.7.2 gives the Nyquist plot for a stable system and the phase margin ϕ is positive. The phase of the open-loop transfer function at $\omega = \omega_1$ is $\angle G(j\omega_1) H(j\omega_1)$.

The phase margin is given by

$$\text{P.M.} = 180° + \angle G(j\omega_1) H(j\omega_1) \qquad ...(7.33)$$

$\angle G(j\omega_1) H(j\omega_1)$ being measured clockwise.

The frequency ω_1 at which the Nyquist plot crosses the unit circle is known as gain cross over frequency. For a stable system the phase margin is positive.

7.8 RELATIVE STABILITY FROM NYQUIST PLOT

Fig. 7.8.1 shows Nyquist plots for two systems which are stable. As both the plots are crossing the negative real axis at same point, the two systems have the same gain margin. However, the two systems have different phase margin. The system B has more phase margin than system A. Therefore, system B is relatively more stable than system A.

Similarly, if two systems have same phase margin but different gain margin then the system having greater gain margin is relatively more stable than the system with lesser gain margin.

Conditionally stable systems. In the Nyquist plot shown in Fig. 7.8.2 the location of the point $(-1 + j0)$ depends on the value of forward path gain K. For smaller range of K the point $(-1 + j0)$ lies between oa, any increase in the value of K beyond this range brings the point $(-1 + j0)$ between ab and if, K is further increased then the point $(-1 + j0)$ lies between bc.

It is given that the open-loop transfer function of the system has the number of poles with positive real part as nil, therefore, if the point $(-1+j0)$ lies between oa or bc the number of encirclements of the point $(-1+j0)$ by the Nyquist plot are -2 indicating that the closed-loop system is unstable.

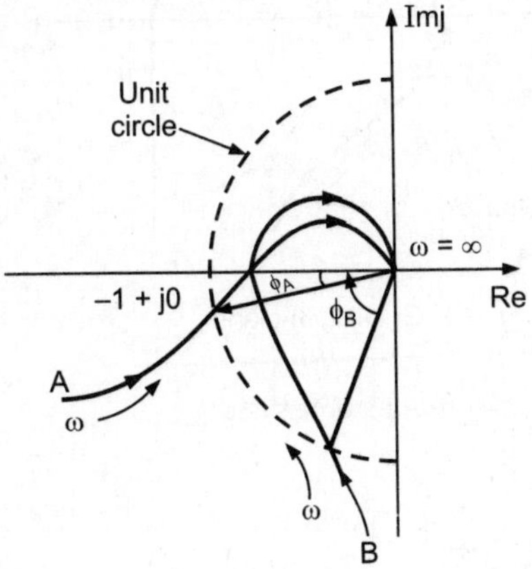

Fig. 7.8.1. Nyquist plot for two systems having same gain margin but different phase margin.

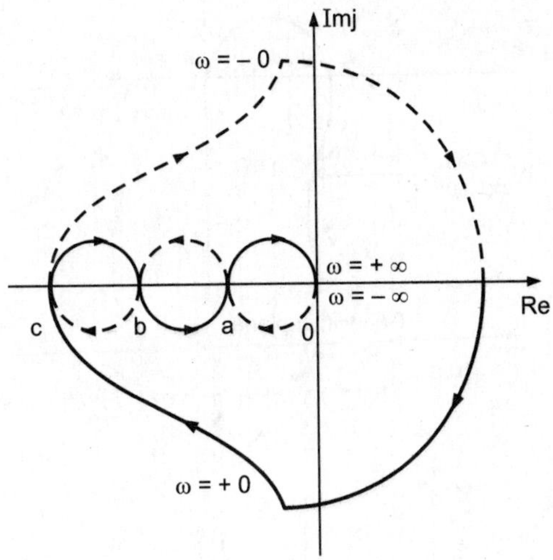

Fig. 7.8.2. Nyquist plot of conditionally stable system.

However, if the point $(-1+j0)$ lies between ab the point $(-1+j0)$ is encircled once in clockwise direction and then once again in the anti-clockwise direction resulting in net encirclements of the point $(-1+j0)$ as nil indicating that the system is stable.

The system discussed above is stable only for a particular range of K any decrease or increase in the value of K makes the system unstable. Such systems are called conditionally stable systems.

The concepts of gain margin and phase margin are not applicable to conditionally stable systems.

7.9 GAIN PHASE PLOT

Fig. 7.9.1 (*a*) and (*b*) give the polar plots of $G(j\omega)\,H(j\omega)$ for a stable and unstable system respectively.

In Fig. 7.9.1 (*a*), $|G(j\omega)\,H(j\omega)|$ at the phase crossover frequency ω_2 is less than 1, hence the gain margin is positive, the phase margin is also positive, therefore, the closed-loop system is stable.

In Fig. 7.9.1 (*b*), $|G(j\omega)\,H(j\omega)|$ at the phase crossover frequency ω_2 is more than 1, hence the gain margin is negative, the phase margin is also negative, therefore the closed-loop system is unstable. From Fig. 7.9.1 (*a*) and (*b*) the gain phase plot for the system can be drawn. The gain phase plot is a graph between the gain (magnitude) of $G(j\omega)\,H(j\omega)$ in decibel and the phase of $G(j\omega)\,H(j\omega)$. The gain phase plots as derived from polar plots of Fig. 7.9.1 (*a*) and (*b*) are plotted in Fig. 7.9.2 (*a*) and (*b*) respectively. The gain margin and phase margin is shown on the gain phase plot. It is observed that for a stable system gain crossover point occurs earlier than the phase crossover point, *i.e.* gain crossover frequency ω_1 is less than the phase crossover frequency ω_2 whereas for unstable system ω_2 is less than ω_1.

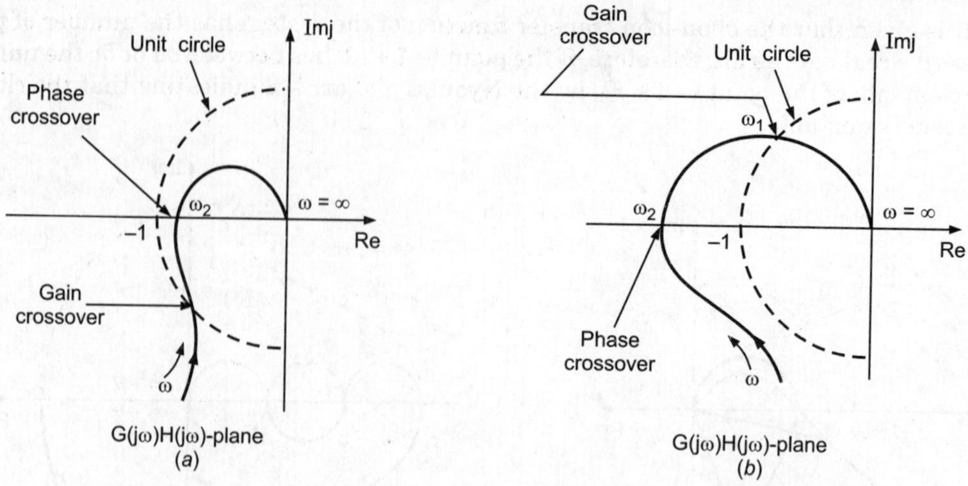

Fig. 7.9.1 (*a*) Polar plot for stable system. (*b*) Polar plot for unstable system.

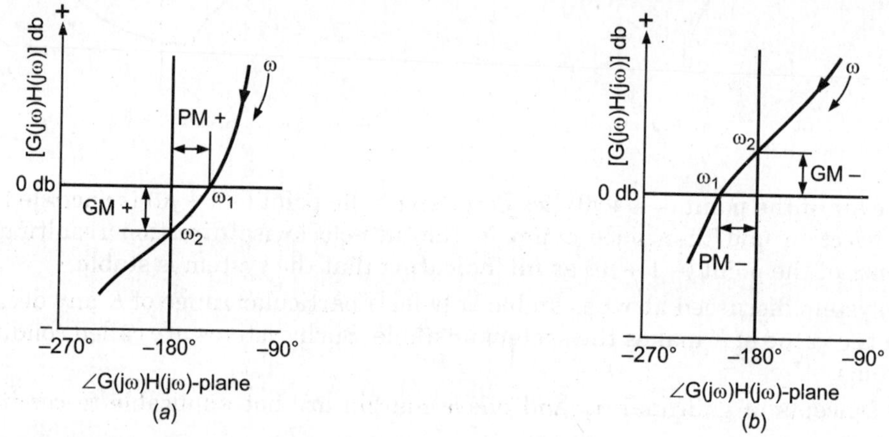

Fig. 7.9.2 (*a*) Gain phase plot for stable system. (*b*) Gain phase plot for unstable system.

7.10 CLOSED-LOOP FREQUENCY RESPONSE OF A UNITY FEEDBACK CONTROL SYSTEM FROM NYQUIST PLOT

A point on the Nyquist plot gives the magnitude of open-loop transfer function $|G(j\omega)\,H(j\omega)|$ as well as its phase $\angle G(j\omega)\,H(j\omega)$ at a particular value of frequency ω. Thus the Nyquist plot also represents frequency response of $G(j\omega)\,H(j\omega)$ in polar form.

From the open-loop frequency response as given by the Nyquist plot, the closed-loop frequency response of a unity feedback control system can be obtained. For a unity feedback system $H(j\omega) = 1$. The overall sinusoidal transfer function of a unity feedback control system is given by

$$\frac{C(j\omega)}{R(j\omega)} = \frac{G(j\omega)}{1 + G(j\omega)} \qquad \text{...(7.34)}$$

The frequency response (polar plot) for transfer function (7.34) can be obtained as shown in Fig. 7.10.1.

In Fig. 7.10.1 it is noted that

$$\vec{OB} = G(j\omega)$$

$$\vec{OA} = -1$$

$$\therefore \quad \vec{AB} = G(j\omega) - (-1) = 1 + G(j\omega)$$

$$\therefore \quad M(\omega) = \frac{C(j\omega)}{R(j\omega)} = \frac{G(j\omega)}{1+G(j\omega)} = \frac{|\vec{OB}|}{|\vec{AB}|} \quad \text{...(7.35)}$$

and $$\angle\frac{C(j\omega)}{R(j\omega)} = \frac{\angle\vec{OB}}{\angle\vec{AB}} = \frac{\angle\alpha}{\angle\theta} = \angle(\alpha - \theta) \quad \text{...(7.36)}$$

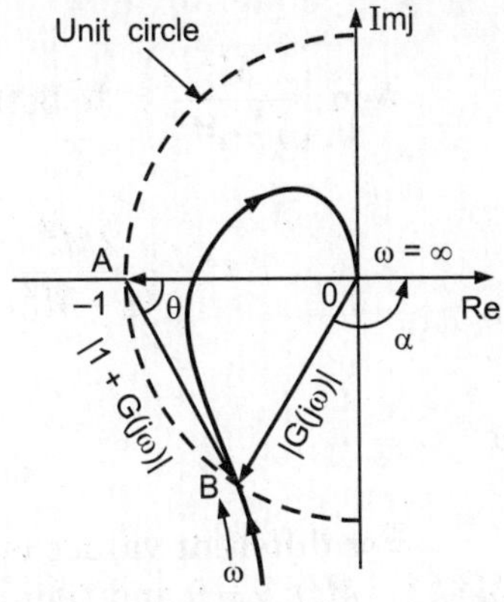

Fig. 7.10.1. Nyquist plot (polar plot) for transfer function 7.34.

As per Eqs. (7.35) and (7.36) the equation given below is obtained

$$\frac{C(j\omega)}{R(j\omega)} = M\,e^{j\phi} \quad \text{...(7.37)}$$

where $\phi = (\alpha - \theta)$

The magnitude of $C(j\omega)/R\,(j\omega)$ is given by M and the phase by ϕ. By taking measurements of phasors $\vec{OB}, \vec{AB}$ and angles θ, α at different values of ω on the Nyquist plot the closed-loop frequency response can be obtained. The closed-loop frequency response consists of two parts (*i*) a graph between $|C(j\omega)\,/\,R(j\omega)|$ and ω (*ii*) a graph between $\angle C(j\omega)\,/\,R(j\omega)$ and ω.

The loci of constant magnitude and constant phase angle of closed-loop transfer function $C(j\omega)\,/\,R(j\omega)$ can be drawn in a complex $G(j\omega)$-phase and from these loci the closed-loop frequency response can be obtained.

7.11 CONSTANT MAGNITUDE LOCI : M-CIRCLES

The open-loop transfer function $G(j\omega)$ of a unity feedback control system is a complex quantity and can be expressed as

$$G(j\omega)\,.\,1 = x + jy$$

$$\because \quad M = \frac{C(j\omega)}{R(j\omega)} = \frac{G(j\omega)}{1+G(j\omega)}$$

$$\therefore \quad M = \frac{x + jy}{1 + x + jy} \quad \text{...(7.38)}$$

$$\therefore \quad |M| = \frac{\sqrt{x^2 + y^2}}{\sqrt{(1+x)^2 + y^2}} \quad \text{...(7.39)}$$

On squaring both sides of Eq. (7.39) and simplifying, following equation is obtained :

$$(1 - M^2)\,x^2 - 2M^2\,x + (1 - M^2)\,y^2 = M^2 \quad \text{...(7.40)}$$

or $$x^2 - \frac{2M^2}{(1-M^2)}\,x + y^2 = \frac{M^2}{(1-M^2)} \quad \text{...(7.40 a)}$$

Add $\left(\frac{M^2}{1-M^2}\right)^2$ to both sides of Eq. (7.40 *a*)

$$\therefore \quad x^2 - \frac{2M^2}{(1-M^2)}x + \left(\frac{M^2}{(1-M^2)}\right)^2 + y^2 = \frac{M^2}{(1-M^2)} + \left(\frac{M^2}{1-M^2}\right)^2 \qquad ...(7.41)$$

or

$$\left(x - \frac{M^2}{1-M^2}\right)^2 + y^2 = \left(\frac{M}{1-M^2}\right)^2 \qquad ...(7.42)$$

For different values of M, Eq. (7.42) represents a family of circles with centre at $x = (M^2 / 1 - M^2)$, $y = 0$ and radius as $(M / 1 - M^2)$. On a particular circle the value of M (magnitude of closed-loop transfer function) is constant, therefore, these circles are called M-circles.

The centres and radii of M-circles for different values of M are given in the following table and M-circles are drawn in Fig. 7.11.1 :

M	Centre $x = \frac{M^2}{1-M^2}$, $y = 0$	Radius $r = \frac{M}{1-M^2}$
0.5	0.33	0.67
1.0	∞	∞*
1.2	– 3.27	2.73
1.6	– 1.64	1.03
2.0	– 1.33	0.67
3.0	– 1.13	0.38

*Intersection with real axis at $x = -0.5$.

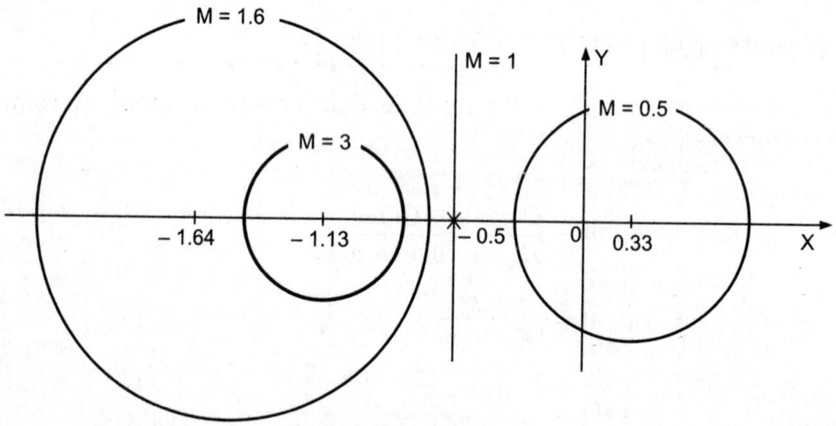

Fig. 7.11.1. *M*-circles.

In $G(j\omega)$ plane the Nyquist plot is superimposed on M-circle and the points of intersection give the magnitude of $C(j\omega) / R(j\omega)$ at different values of ω.

7.12 CONSTANT PHASE ANGLES LOCI : N-CIRCLES

In view of Eq. (7.38) the phase angle of the closed-loop transfer function of a unity feedback control system is given by

$$\angle \frac{C(j\omega)}{R(j\omega)} = \angle \frac{x + jy}{1 + x + jy} \qquad ...(7.43)$$

The phase angle is denoted by ϕ, therefore,

$$\phi = \tan^{-1}\left(\frac{y}{x}\right) - \tan^{-1}\left(\frac{y}{1+x}\right)$$

or

$$\tan\phi = \tan\left[\tan^{-1}\left(\frac{y}{x}\right) - \tan^{-1}\left(\frac{y}{1+x}\right)\right]$$

$$= \frac{\tan\left[\tan^{-1}\left(\frac{y}{x}\right)\right] - \tan\left[\tan^{-1}\left(\frac{y}{1+x}\right)\right]}{1 + \tan\left[\tan^{-1}\left(\frac{y}{x}\right)\right]\cdot\tan\left[\tan^{-1}\left(\frac{y}{1+x}\right)\right]} = \frac{\frac{y}{x} - \frac{y}{1+x}}{1 + \frac{y}{x}\cdot\frac{y}{(1+x)}}$$

or

$$\tan\phi = \frac{y}{x^2 + x + y^2} \qquad ...(7.44)$$

Substitute $\tan\phi = N$ in Eq. (7.44)

$$\therefore \qquad N = \frac{y}{x^2 + x + y^2}$$

or

$$x^2 + x + y^2 - \frac{y}{N} = 0 \qquad ...(7.45)$$

Add $\left(\frac{1}{4} + \frac{1}{4N^2}\right)$ on both sides of Eq. (7.45)

$$\therefore \qquad \left(x^2 + x + \frac{1}{4}\right) + \left(y^2 - \frac{y}{N} + \frac{1}{4N^2}\right) = \left(\frac{1}{4} + \frac{1}{4N^2}\right)$$

or

$$\left(x + \frac{1}{2}\right)^2 + \left(y - \frac{1}{2N}\right)^2 = \left(\frac{1}{4} + \frac{1}{4N^2}\right) \qquad ...(7.46)$$

For different values of N Eq. (7.46) represents a family of circles with centre at $x = -\frac{1}{2}$, $y = \frac{1}{2N}$ and radius as $\sqrt{\frac{1}{4} + \frac{1}{4N^2}}$.

On a particular circle, the value of N or the value of phase angle of the closed-loop transfer function is constant, therefore, these circles are called N-circles.

The centre and the radii for N-circles are given in the following table and N-circles are drawn in Fig. 7.12.1.

ϕ	$N = \tan \phi$	*Centre* $x = -\frac{1}{2}, y = \frac{1}{2N}$	*Radius* $R = \sqrt{\frac{1}{4} + \frac{1}{4N^2}}$
– 90°	∞	0	0.5
– 60°	– 1.732	– 0.289	0.577
– 50°	– 1.19	– 0.42	0.656
– 30°	– 0.577	0.866	1.0
– 10°	– 0.176	– 2.84	2.88
0°	0	∞	∞
+ 10°	0.176	2.84	2.88
+ 30°	0.577	0.866	1.0
+ 50°	0.19	0.42	0.656
+ 60°	1.732	0.289	0.577
+ 90°	∞	0	0.5

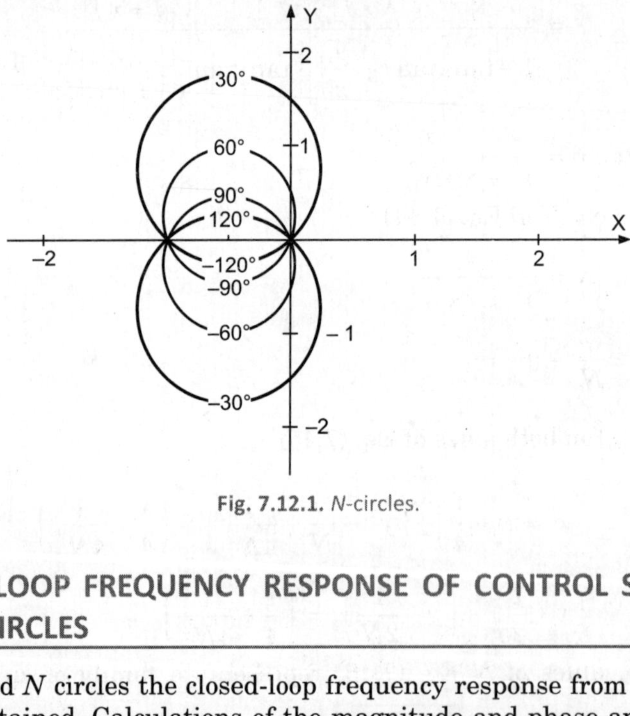

Fig. 7.12.1. *N*-circles.

7.13 CLOSED-LOOP FREQUENCY RESPONSE OF CONTROL SYSTEM FROM M AND N-CIRCLES

By the use of *M* and *N* circles the closed-loop frequency response from the given open-loop response can be obtained. Calculations of the magnitude and phase angle are not needed. The open-loop frequency response (the Nyquist plot) is superimposed on *M* and *N* circles and the magnitude and phase angle are read at the point of intersection whereas the frequency is read from the open-loop frequency response plot.

Fig. 7.13.1 (*a*) gives superimposed open-loop frequency response on *M* and *N* circles and the corresponding closed-loop frequency response is drawn in Fig. 7.13.1 (*b*).

The *M*-circle tangential to open-loop frequency response plot gives the maximum magnitude of the closed-loop transfer function and the resonant frequency at this point is denoted as ω_r. The maximum magnitude is denoted as M_r.

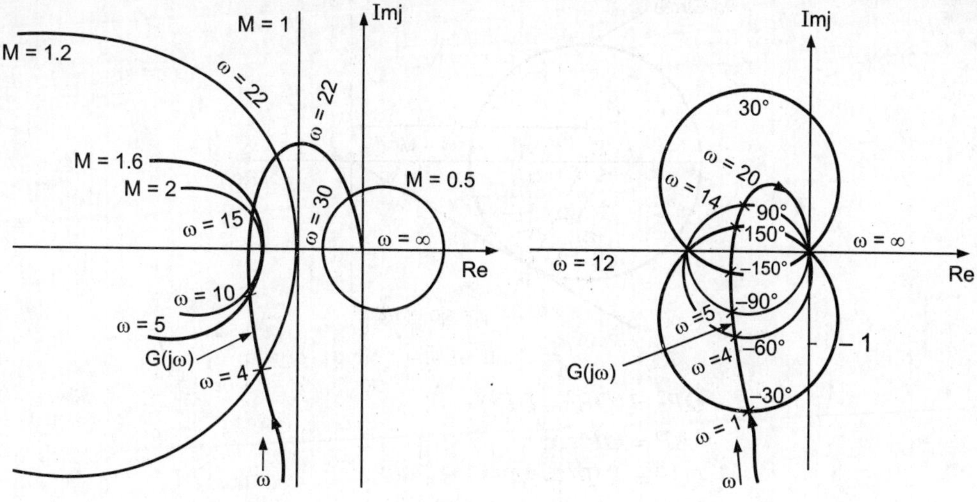

Fig. 7.13.1 (*a*)

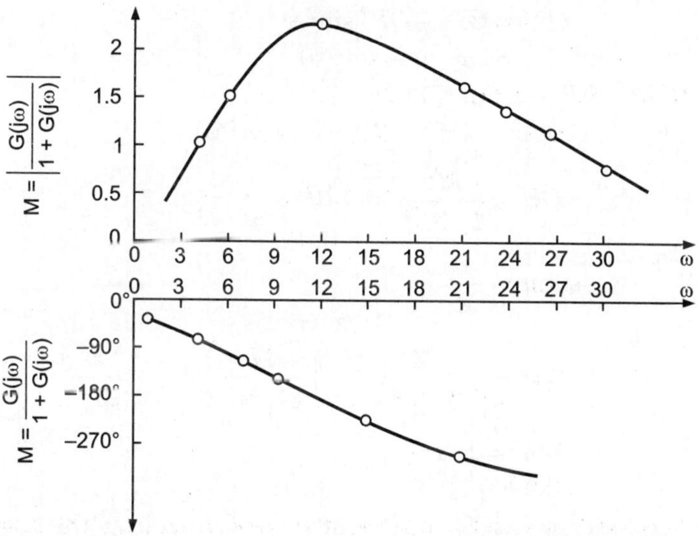

Fig. 7.13.1. (*b*) Closed-loop frequency response from *M* and *N* circles.

7.14 GAIN ADJUSTMENT USING M-CIRCLE

In Fig. 7.14.1 M-circle is drawn such that $M > 1$. A tangential line drawn from the origin cuts the M-circle at point P. The angle made by the tangential line with the negative real axis being ψ, then

$$\sin\psi = \left|\frac{\dfrac{M}{1-M^2}}{\dfrac{M^2}{1-M^2}}\right| = \frac{1}{M} \qquad ...(1)$$

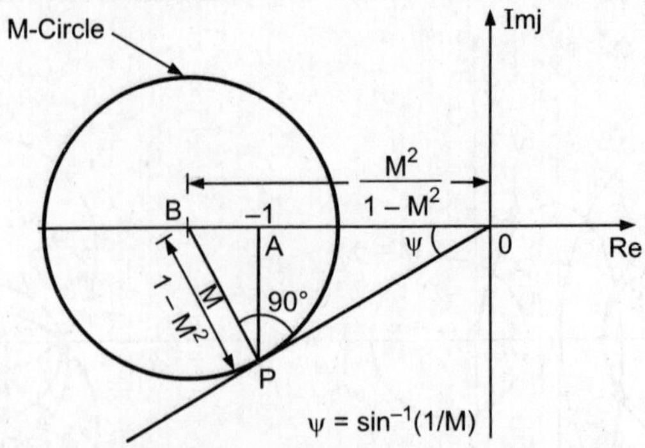

Fig. 7.14.1. *M*-circle and tangent from the origin.

$$OP^2 = (OB^2 - BP^2) \quad \text{...(2)}$$

$$AP = OP \sin \psi \quad \text{...(3)}$$

and

$$OA^2 = (OP^2 - AP^2) \quad \text{...(4)}$$

Eliminating AP^2 from Eq. (4)

$$OA^2 = OP^2 - OP^2 \sin^2 \psi$$

$$= OP^2 (1 - \sin^2 \psi) \quad \text{...(5)}$$

Put $OP^2 = (OB^2 - BP^2)$ in Eq. (5)

$$\therefore \quad OA^2 = (OB^2 - BP)^2 (1 - \sin^2 \psi)$$

Since

$$OB = \frac{M^2}{1-M^2} \text{ and } BP = \frac{M}{1-M^2}$$

$$\sin \psi = \frac{1}{M}$$

$$OA^2 = \left[\left(\frac{M^2}{1-M^2}\right)^2 - \left(\frac{M}{1-M^2}\right)^2\right]\left(1 - \frac{1}{M^2}\right)$$

$$\therefore \quad OA^2 = 1$$

or

$$OA = |1|$$

Thus the intersection point of the perpendicular drawn from the tangential point P on the negative real axis intersects at $(-1 + j0)$.

The analysis given above can be applied for the adjustment of the gain of a unity feedback control system to obtain desired value of resonant peak M_r (refer section 7.18.8). The procedure is illustrated in the example given below :

Let the open-loop transfer function of a unity feedback control system be given as below :

$$G(j\omega) = \frac{K}{j\omega\,(j0.1\,\omega + 1)\,(j0.2\,\omega + 1)}$$

It is desired to determine the value of gain K such that resonant peak $M_r = 2$. Consider $K = 1$, therefore,

$$G_1(j\omega) = \frac{1}{j\omega\,(0.1\,\omega + 1)\,(j0.2\,\omega + 1)}$$

where $G_1(j\omega) = G(j\omega)$ for $K = 1$

Rationalizing $G_1(j\omega)$ and separating into real and imaginary part

$$G_1\,(j\omega) = -\frac{0.3}{(0.01\,\omega^2 + 1)\,(0.04\,\omega^2 + 1)} - \frac{j\,(1 - 0.02\,\omega^2)}{\omega(0.01\,\omega^2 + 1)\,(0.04\,\omega^2 + 1)}$$

The polar plot of $G_1(j\omega)$ is plotted in Fig. 7.14.2 as per calculations given in the following table :

ω rad/sec	2	3	4	7.07	10	∞
$Re\ G_1\ (j\omega)$	– 0.248	– 0.2	– 0.158	– 0.1	– 0.03	– 0
$Imj\ G_1\ (j\omega)$	– *j*0.38	– *j*0.184	– *j*0.022	– *j*0	+ *j*0.001	+ *j*0

As the desired value of resonant peak M_r is 2.0, therefore

$$\psi = \sin^{-1}\left(\frac{1}{M_r}\right)$$

$$\psi = \sin^{-1}\left(\frac{1}{2}\right)$$

$$\psi = 30°$$

The line at an angle of $\psi = 30°$ is drawn as shown in Fig. 7.14.2 and a circle tangential to the polar plot as well as the line is drawn. The trangential point of the line and the circle is marked *P* and the perpendicular from the point *P* intersects the negative real axis at *A* = – 0.22.

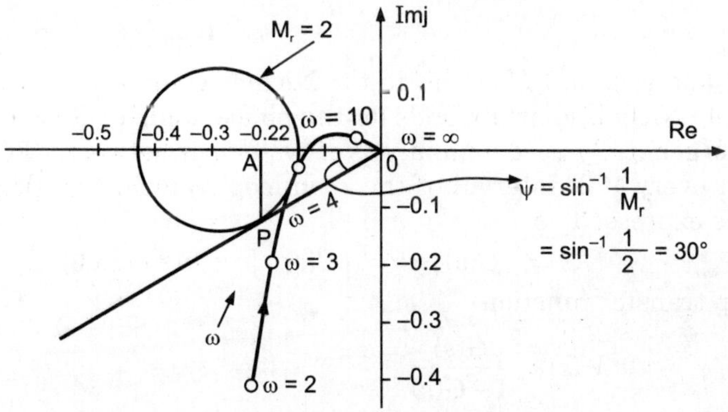

Fig. 7.14.2. Gain adjustment for $G(j\omega) = K/j\omega(j0.1\omega + 1)(j0.2\omega + 1)$.

If the circle drawn in Fig. 7.14.2 be the desired *M* circle corresponding to $M_r = 2$, then the point *A* should be $(-1 + j0)$. This is possible if the scale is changed by a factor $1/|OA|$ and hence the value of *K* to obtain the desired resonant peak $M_r = 2$ is :

$$K = \frac{1}{|OA|} = \frac{1}{0.22} = 4.54$$

7.15 NICHOLS CHART

The constant M and constant N circles in $G(j\omega)$ plane can be used for the analysis and design of control systems. However the constant M and constant N circles in gain phase plane *i.e.* graph having gain in decibel along the ordinate and phase angle along the abscissa are prepared for system design and analysis as these plots supply information with less manipulations. The M and N circles of $G(j\omega)$ in the gain phase plane are transformed into M and N contours in rectangular co-ordinates. A point on the constant M loci in $G(j\omega)$ plane is transferred to the gain phase plane by drawing the vector directed from the origin of $G(j\omega)$ plane to the particular point on the M circle and then measuring the length db, angle in and degree. These values of length and angles are the coordinates of the corresponding point in the gain phase plane as shown in Fig. 7.15.1 and 7.15.2. The critical point in $G(j\omega)$, plane corresponds to the point of zero decibel and – 180° in the gain-phase plane. Plot of M and N circles in gain phase plane is shown in Fig. 7.15.3 and known as 'Nichols chart'.

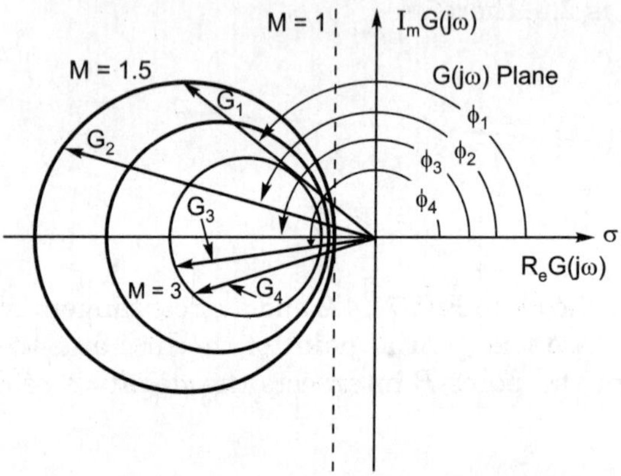

Fig. 7.15.1. *M*-circle transformation to Nichols chart.

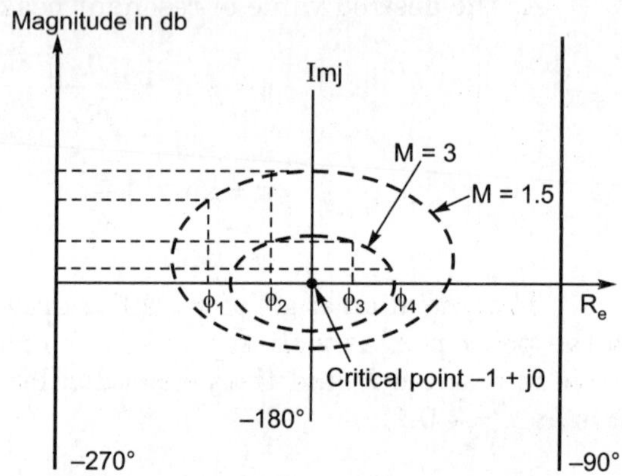

Fig. 7.15.2. *M*-circles in gain-phase plane.

Constant M and constant N circles in the Nichols chart are deformed into squashed circles. The complete Nichols chart extends for the phase angle of $G(j\omega)$ from 0 to – 360° but the region of $\angle Gj(\omega)$ generally used for analysis of systems is between – 90° and –270°. These curves repeat after every 180° interval. If the open-loop transfer function of the unity feedback system $G(s)$ is expressed as

$$G(s) = |G(s)|\, e^{j\theta} = |G(s)|\, [\cos\theta + j\sin\theta]$$

and the closed-loop transfer function

$$M(s) = \frac{G(s)}{1+G(s)}$$

Substituting ($s = j\omega$) in the above equations the frequency functions are

$$G(j\omega) = |G(j\omega)|\, [\cos\theta + j\sin\theta]$$

and $$M(j\omega) = Me^{j\phi} = \frac{G(j\omega)}{1+G(j\omega)}$$

Eliminating $G(j\omega)$ from above two equations

$$M = \frac{|G(j\omega)|}{\sqrt{|G(j\omega)|^2 + 2|G(j\omega)|\cos\theta + 1}}$$

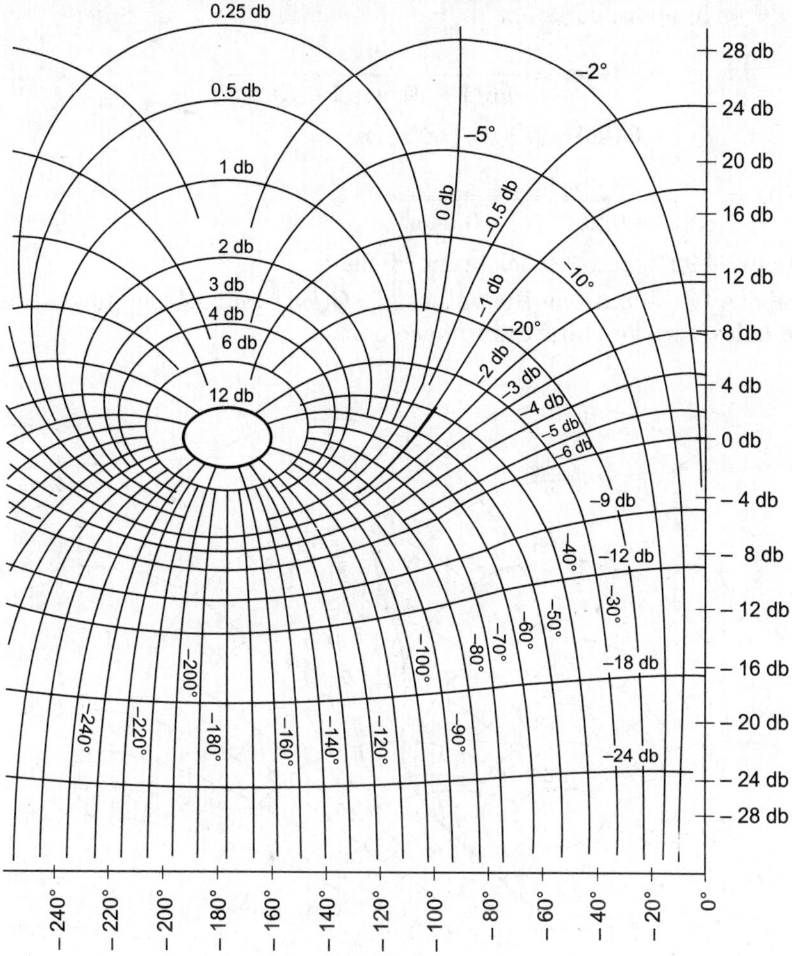

Fig. 7.15.3. Nichols chart.

and
$$\phi = \tan^{-1} \frac{\sin \theta}{|G(j\omega)| + \cos \theta}$$

These equations define the plots in Nichols chart shown in Fig. 7.15.3.

Illustrative Example. *Use Nichols chart to obtain the closed-loop frequency response, $C(j\omega)/R(j\omega)$ for a unity feedback control system having open-loop transfer function given below :*

$$G(s) = \frac{20}{s\,(s+2)\,(s+5)}.$$

Solution. For the given system

$$G(s) = \frac{20}{s\,(s+2)\,(s+5)} = \frac{2}{s\,(1+0.5s)\,(1+0.2s)}$$

substituting $s = j\omega$ in above equation

$$G(j\omega) = \frac{2}{j\omega(1 + j0.5\omega)(1 + j0.2\omega)}$$

or

$$G(j\omega) = |G(j\omega)| \angle G(j\omega)$$

where

$$|G(j\omega)| = \frac{2}{\omega\sqrt{(1+0.25\omega^2)(1+0.04\omega^2)}}$$

and

$$\angle G(j\omega) = -90° - \tan^{-1} 0.5\omega - \tan^{-1} 0.2\omega$$

From above two equations the values of $|G(j\omega)|$ and $\angle G(j\omega)$ for various frequencies are tabulated below for sketching gain-phase plot.

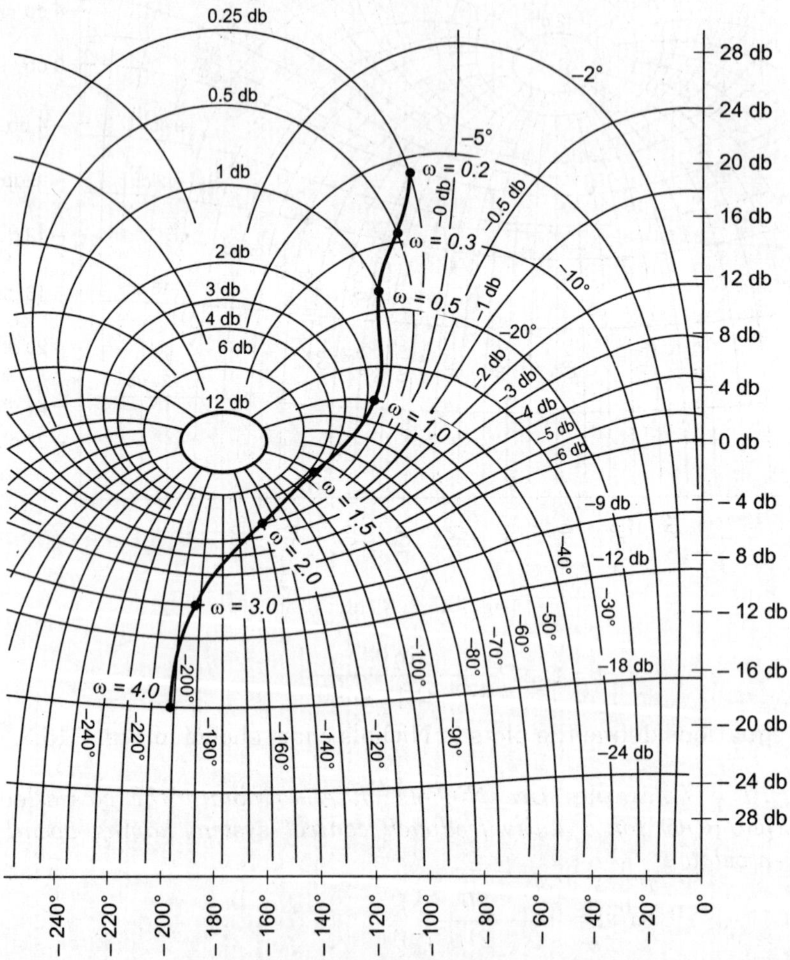

Fig. 7.15.4. Gain phase plot superimposed on Nichols chart for illustrative example.

ω rad/sec.	0.2	0.3	0.5	1.0	1.5	2.0	3.0	4.0
$\|G(j\omega)\|$-db	20	16	11.8	4.2	0.0	– 4.0	– 10.0	– 16.0
$\angle G(j\omega)°$	– 99.8	– 103	– 110	– 120	– 145	– 157	– 177	– 195

From the tabulated values of $|G(j\omega)|$ db and $\angle G(j\omega)$ as above, the gain-phase plot of the open-loop function $G(j\omega)$ is sketched on the Nichols chart *i.e.* Fig. 7.15.4.

From the intersection points of M-circle locus (db) with superimposed gain-phase plot on Nichols chart the closed-loop response for $C(j\omega) / R(j\omega)$-db versus ω is obtained.

The intersection of gain-phase plot with N-circle locus on Nichols chart give the closed-loop frequency response for $\angle C(j\omega) / R(j\omega)$ versus ω.

The closed-loop response as above is tabulated below :

ω rad/sec	0.2	0.3	0.5	1.0	1.5	2.0	3.0	4.0
$\lvert C(j\omega)/R(j\omega)\rvert$ db	0.25	0.5	1.0	– 2.0	4.0	0.0	– 10	– 18
$\angle C(j\omega)/R(j\omega)°$	– 6	– 10	– 16	– 40	– 100	– 145	– 180	– 200

From the data available in above table, the closed-loop frequency response of the system is plotted as shown in Fig. 7.15.5 (*a*) and (*b*). **Ans.**

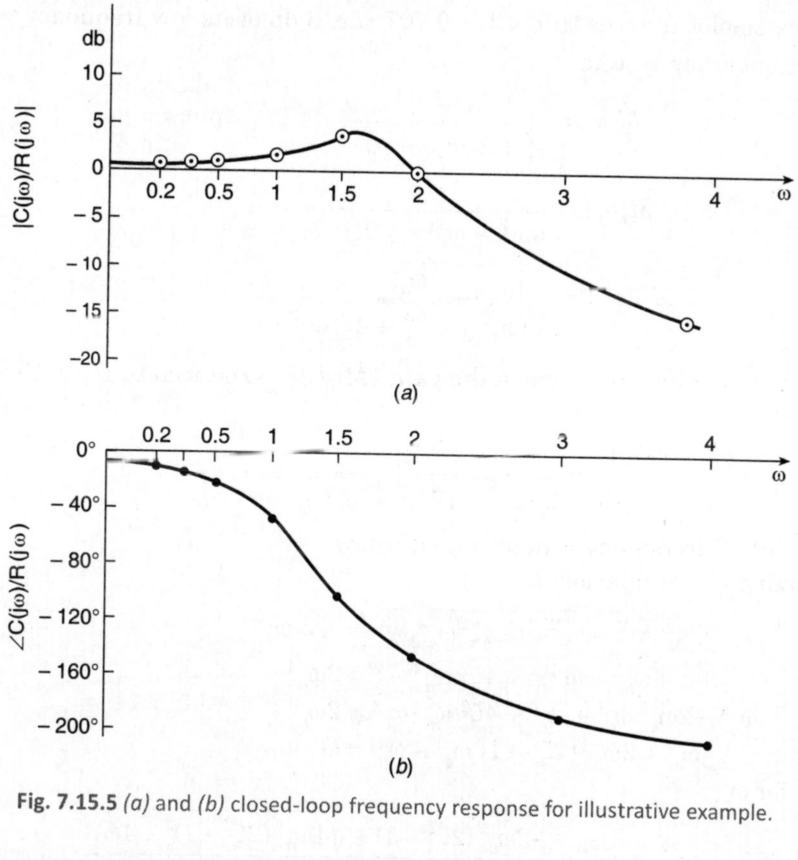

Fig. 7.15.5 *(a)* and *(b)* closed-loop frequency response for illustrative example.

7.16 CUTOFF FREQUENCY AND BANDWIDTH

The closed-loop frequency response of a system is shown in Fig. 7.16.1. The response falls by 3 db from its low frequency value to a frequency value ω_c. The frequency ω_c is called cut off frequency and the frequency range 0 to ω_c is called the bandwidth of the system. The response peak M_r occurs at resonance frequency ω_r.

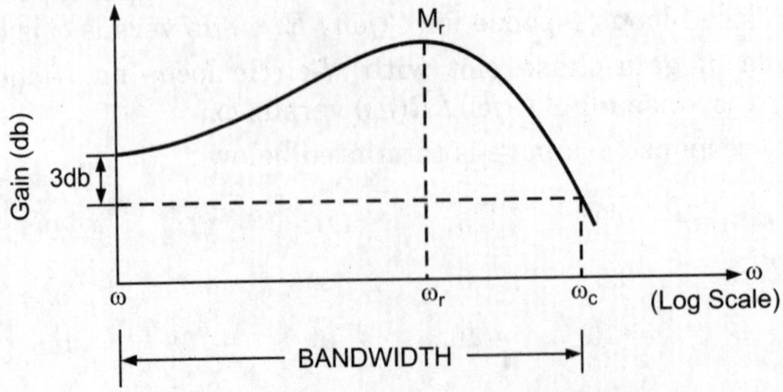

Fig. 7.16.1. Cut off frequency and bandwidth.

The bandwidth is defined as the frequency at which the magnitude gain of the frequency response plot reduces to $1/\sqrt{2} = 0.707$; *i.e.* 3 db of its low frequency value.

For a second order system

$$M(s) = \frac{\omega_n^2}{s^2 + 2\zeta\omega_n s + \omega_n^2}, \qquad \text{put } s = j\omega$$

$$\therefore \quad M(j\omega) = \frac{\omega_n^2}{(\omega_n^2 - \omega^2) + j\,2\zeta\omega_n\omega}$$

$$\therefore \quad |\,M(j\omega)\,| = \frac{\omega_n^2}{\sqrt{(\omega_n^2 - \omega^2)^2 + 4\zeta^2\omega_n^2\omega^2}}$$

Let, ω_c be the frequency at which the gain $|M(j\omega)|$ is reduced to $1/\sqrt{2}$ times of its low frequency value.

$$\therefore \quad \frac{1}{\sqrt{2}} = \frac{\omega_n^2}{\sqrt{(\omega_n^2 - \omega_c^2)^2 + 4\zeta^2\omega_n^2\omega_c^2}}$$

ω_c : The cut off frequency is determined below :
Rearranging above relation

$$\sqrt{(\omega_n^2 - \omega_c^2)^2 + 4\zeta^2\omega_n^2\omega_c^2} = \sqrt{2}\,\omega_n^2$$

or
$$(\omega_n^2 - \omega_c^2)^2 + 4\zeta^2\,\omega_n^2\,\omega_c^2 = 2\omega_n^4$$

or
$$\omega_n^4 - 2\omega_n^2\,\omega_c^2 + \omega_c^4 + 4\zeta^2\,\omega_n^2\,\omega_c^2 = 2\omega_n^4$$

or
$$\omega_c^4 + 2\omega_n^2\,(2\zeta^2 - 1)\,\omega_c^2 - \omega_n^4 = 0$$

Solving for ω_c^2 :

$$\omega_c^2 = \frac{-2\omega_n^2(2\zeta^2 - 1) + \sqrt{4\omega_n^4(2\zeta^2 - 1)^2 + 4\omega_n^4}}{2}$$

$$= \omega_n^2\,(1 - 2\zeta^2) + \omega_n^2\,\sqrt{(2\zeta^2 - 1)^2 + 1}$$

$$= \omega_n^2\left\{(1 - 2\zeta^2) + \sqrt{4\zeta^4 - 4\zeta^2 + 2}\right\}$$

$$\therefore \qquad \omega_c = \omega_n\,(1 - 2\zeta^2 + \sqrt{4\zeta^4 - 4\zeta^2 + 2})^{1/2}$$

The negative sign under quadratic solution as well as while taking square root is not taken into consideration because, ω_c is a real and positive quantity.

Hence, the bandwidth of a second order system having non-zero magnitude at $\omega = 0$ is given by

$$\text{B.W.} = \omega_n\,(1 - 2\zeta^2 + \sqrt{4\zeta^4 - 4\zeta^2 + 2})^{1/2}$$

The capacity of a feedback control system to reproduce input signal is indicated by its bandwidth. Frequencies greater than the cut off frequency are attenuated, where as, frequencies lower than the cut off frequency are transmitted.

A system with large bandwidth has lower rise time.

7.17 SOLVED EXAMPLES : NYQUIST CRITERION

Example 7.17.1. *Examine the closed-loop stability of a system whose open-loop transfer function is given by*

$$G(s)\,H(s) = \frac{50}{(s+1)(s+2)}.$$

Solution. $$G(s)\,H(s) = \frac{50}{(s+1)(s+2)}$$

Put $s = j\omega$

$$G(j\omega)\,H(j\omega) = \frac{50}{(j\omega+1)(j\omega+2)}$$

$$|G(j\omega)\,H(j\omega)| = \frac{50}{\sqrt{\omega^2+1^2}\,\sqrt{\omega^2+2^2}}$$

and
$$\angle G(j\omega)\,H(j\omega) = -\tan^{-1}\omega - \tan^{-1}\frac{\omega}{2}$$

$$= -\tan^{-1}\frac{3\omega}{2-\omega^2} \qquad \text{for } \omega < \sqrt{2}$$

$$= -\tan^{-1}\frac{3\omega}{2-\omega^2} - 180° \quad \text{for } \omega > \sqrt{2}$$

(1) As $\omega \to 0$

$$|G(j0)\,H(j0)| \to 25$$

and
$$\angle G(j0)\,H(j0) \to 0°$$

(2) At $\omega = \sqrt{2}$

$$|G(j\sqrt{2})\,H(j\sqrt{2})| = \frac{50}{\sqrt{\sqrt{2}^2+1^2}\,\sqrt{\sqrt{2}^2+2^2}} = 11.78$$

and
$$\angle G(j\sqrt{2})\,H(j\sqrt{2}) = -\tan^{-1}\frac{3\sqrt{2}}{2-2} = -\tan^{-1}\infty = -90°$$

(3) As $\omega \to \infty$

$$|G(j\infty)\, H(j\infty)| \to 0$$

and
$$\angle G(j\omega)\, H(j\omega) \to -180°$$

As per above calculations, general shape of the Nyquist plot is shown in Fig. 7.17.1.

As the given open-loop transfer function is type O, hence, the points $\omega = +0$ and $\omega = -0$ are coincident.

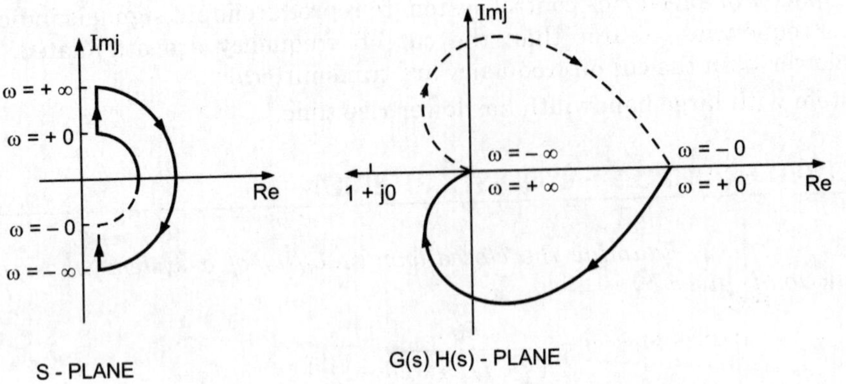

Fig. 7.17.1. Nyquist plot for example 7.17.1.

The point $(-1 + j0)$ is not encircled by the Nyquist plot, therefore, $N = 0$ and $P_+ = 0$ (given)

$\because$
$$N = P_+ - Z_+$$

or
$$0 = 0 - Z_+$$

$\therefore$
$$Z_+ = 0$$

As $Z_+ = 0$, there are no roots of closed-loop characteristic equation having positive real part, hence the closed-loop system is stable. **Ans.**

Example 7.17.2. *Using Nyquist criterion investigate the closed-loop stability of the system whose open-loop transfer function is given below :*

$$G(s)\, H(s) = \frac{K(s+1)}{(s+0.5)(s-2)}.$$

Consider (i) $K = 1.25$, (ii) $K = 2.5$.

Also determine the limiting value of K for stability.

Solution.
$$G(s)\, H(s) = \frac{K\,(s+1)}{(s+0.5)\,(s-2)}$$

Put $s = j\omega$

$$G(j\omega)\, H(j\omega) = \frac{K\,(j\omega+1)}{(j\omega+0.5)\,(j\omega-2)}$$

$$|G(j\omega)\, H(j\omega)| = \frac{K\,\sqrt{\omega^2+1^2}}{\sqrt{\omega^2+0.5^2}\,\sqrt{\omega^2+2^2}}$$

and $\quad \angle G(j\omega)\, H(j\omega) = \tan^{-1} \omega - \tan^{-1} \dfrac{\omega}{0.5} - \tan^{-1} \dfrac{\omega}{-2} - 180°$*

$$= \tan^{-1} \omega - \tan^{-1} 2\omega + \tan^{-1} 0.5\omega - 180°$$

$$= \tan^{-1} \frac{1.5\,\omega}{1 - 0.5\,\omega^2} - \tan^{-1} 2\omega - 180°$$

$$= \tan^{-1} \left\{ \frac{\dfrac{1.5\,\omega}{1 - 0.5\,\omega^2} - 2\omega}{1 + \dfrac{1.5\,\omega}{1 - 0.5\omega^2} \times 2\omega} \right\} - 180°$$

$$= \tan^{-1} \frac{\omega(\omega^2 - 0.5)}{1 + 2.5\omega^2} - 180°$$

(1) As $\omega \to 0$

$$|G(j0)\, H(j0)| \to K$$

and $\quad \angle G(j0)\, H(j0)| \to -180°$

(2) As $\omega = \sqrt{0.5}$

$$|G(j\sqrt{0.5})\, H(j\sqrt{0.5})| = \frac{K\sqrt{\sqrt{(0.5)^2 + 1^2}}}{\sqrt{\sqrt{(0.5)^2 + (0.5)^2}}\sqrt{\sqrt{0.5^2 + 2^2}}} = 0.66\, K$$

and $\quad \angle G(j\sqrt{0.5})\, H(j\sqrt{0.5}) \to -180°$

(3) As $\omega \to \infty$

$$|G(j\infty)\, H(j\infty)| \to 0$$

and $\quad \angle G(j\infty)\, H(j\infty) \to -90°$

As per above calculations the Nyquist plot (general shape) is drawn in Fig. 7.17.2.

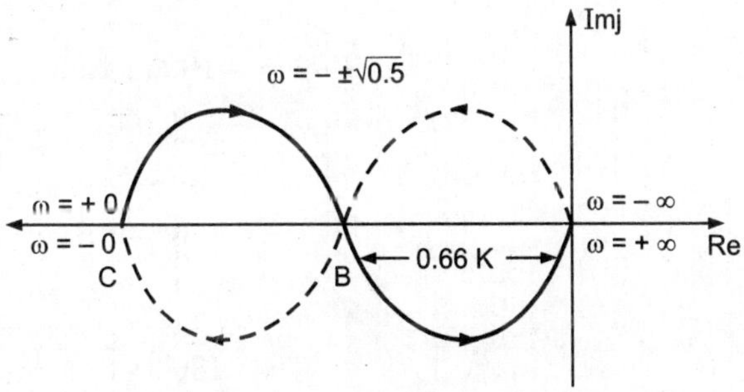

Fig. 7.17.2. Nyquist plot for example 7.17.2.

(*i*) For $K = 2.5$, $AB = 0.66 \times 2.5 = 1.65$ thus the point $(-1 + j0)$ lies between A and B and its number of encirclement N by the plot is + 1 (anti-clockwise). Since $P_+ = 1$, *i.e.* $s = 2$, therefore,

$$N = P_+ - Z_+$$

$$\therefore \quad +1 = 1 - Z_+ \qquad \therefore \quad Z_+ = 0$$

Thus the closed-loop characteristics equation has no roots with positive real part and the system is stable. Ans.

(*ii*) For $K = 1.25$, $AB = 0.66 \times 1.25 = 0.825$ thus the point $(-1 + j0)$ lines between B and C and its number of encirclement N by the plot is – 1 (clockwise).

---

* The real part of term $1/(j\omega - 2)$ is –ive. Hence the corresponding phase angle is $\left(-\tan^{-1}\dfrac{\omega}{2} - 180°\right)$

Since $P_+ = 1$, *i.e.* $s = 2$, therefore

$$N = P_+ - Z_+$$

$$\therefore \quad -1 = 1 - Z_+ \qquad \therefore \qquad Z_+ = 2$$

Thus the closed-loop characteristic equation has two roots with positive real part and the system is unstable. **Ans.**

The limiting value of K for the stability is given by

$$0.66K \geq 1 \quad \text{or} \quad K \geq 1.56. \quad \textbf{Ans.}$$

Example 7.17.3. *The open-loop transfer function of a unity feedback control system is given below :*

$$G(s) = \frac{(s + 0.25)}{s^2\,(s+1)\,(s+0.5)}$$

Determine the closed-loop stability by applying Nyquist criterion.

Solution.

$$G(s) = \frac{(s + 0.25)}{s^2\,(s+1)\,(s+0.5)}$$

Put $s = j\omega$

$$G(j\omega) = \frac{(j\omega + 0.25)}{(j\omega)^2\,(j\omega+1)\,(j\omega+0.5)}$$

$$|G(j\omega)| = \frac{\sqrt{\omega^2 + 0.25^2}}{\omega^2\,\sqrt{\omega^2+1^2}\,\sqrt{\omega^2+0.5^2}}$$

and

$$\angle G(j\omega) = -180° + \tan^{-1}\frac{\omega}{0.25} - \tan^{-1}\frac{\omega}{1} - \tan^{-1}\frac{\omega}{0.5}$$

$$= -180° + \tan^{-1}4\omega - \tan^{-1}\omega - \tan^{-1}2\omega$$

$$= -180° + \tan^{-1}4\omega - \tan^{-1}\frac{3\omega}{1-2\omega^2}$$

$$= -180° + \tan^{-1}\frac{4\omega - \dfrac{3\omega}{1-2\omega^2}}{1 + \dfrac{4\omega\cdot 3\omega}{1-2\omega^2}}$$

$$= -180° + \tan^{-1}\frac{\omega\,(1-8\omega^2)}{1+10\omega^2}$$

(1) As $\omega \to 0$

$$|G(j0)| \to \infty$$

and

$$\angle G(j0) \to -180°$$

(2) At $\omega = \sqrt{1/8} = \sqrt{0.125}$

$$|G(j\sqrt{0.125})| = \frac{\sqrt{\sqrt{0.125}^2 + 0.25^2}}{\sqrt{0.125}^2\,\sqrt{\sqrt{0.125}^2+1^2}\,\sqrt{\sqrt{0.125}^2+0.5^2}} = 5.3$$

and $\angle G(j\sqrt{0.125}) = -180°$

(3) As $\omega \to \infty$

$|G(j\infty)| \to 0$

and $\angle G(j\infty) \to -270°$

As per above calculations, general shape of the Nyquist plot is drawn in Fig. 7.17.3. As the given open-loop transfer function is of type 2, the plot is completed from $\omega = -0$ to $\omega = +0$ through an angle -2π with infinite radius.

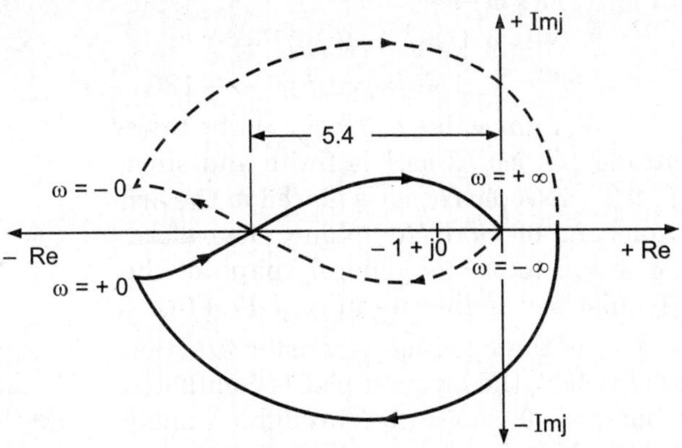

Fig. 7.17.3. Nyquist plot for example 7.17.3.

The number of encirclements N of the point $(-1 + j0)$ by the Nyquist plot is -2, (clockwise), and $P_+ = 0$, therefore,

$$N = P_+ - Z_+$$

$$\therefore \quad -2 = 0 - Z_+ \quad \therefore \quad Z_+ = 2.$$

Thus, the closed-loop characteristic equation has two roots with positive real part indicating that the system is unstable.

Example 7.17.4. *Examine the closed-loop stability*

(a) $T_1 > T_2$ *(b)* $T_1 < T_2$

Given that

$$G(s)\,H(s) = \frac{K\,(1 + sT_1)}{s^2\,(1 + sT_2)}$$

K, T_1 *and* T_2 *are positive constants.*

Solution. $$G(s)\,H(s) = \frac{K\,(1 + sT_1)}{s^2\,(1 + sT_2)}$$

Put $s = j\omega$

$$\therefore \quad G(j\omega)\,H(j\omega) = \frac{K\,(1 + j\omega T_2)}{(j\omega)^2\,(1 + j\omega T_2)}$$

$$|G(j\omega)\,H(j\omega)| = \frac{K\sqrt{1^2 + \omega^2 T_1^2}}{\omega^2\sqrt{1^2 + \omega^2 T_2^2}}$$

and $$\angle G(j\omega)\,H(j\omega) = -180° + \tan^{-1}\omega T_1 - \tan^{-1}\omega T_2$$

$$= -180° + \tan^{-1}\frac{\omega(T_1 - T_2)}{1 + \omega^2 T_1 T_2}$$

(a) $T_1 > T_2$

(1) As $\omega \to 0$

$$|G(j0)\,H(j0)| \to \infty$$

$$\angle G(j0)\,H(j0) \to -180° + 0°$$

(2) As $\omega \to \infty$

$$|G(j\infty)\, H(j\infty)| \to 0$$
$$\angle G(j\infty)\, H(j\infty) \to -180° + 0°$$

For any value of $0 < \omega < \infty$, the magnitude $|G(j\omega)\, H(j\omega)|$ is finite and since $T_1 > T_2$, $\angle G(j\omega)\, H(j\omega)$ is placed in the 3rd quadrant of $G(s)\, H(s)$-plane. Therefore, for $\omega = 0$ to $\omega = \infty$ general shape of the Nyquist plot is shown in Fig. 7.17.4 (*a*).

As the open-loop transfer function is of type 2, the Nyquist plot is completed from $\omega = -0$ to $\omega = +0$. Through an angle -2π ($-360°$) as shown in Fig. 7.17.4 (*a*).

It is noted that the number of encirclements of the critical point $(-1 + j0)$ the Nyquist plot is nil,

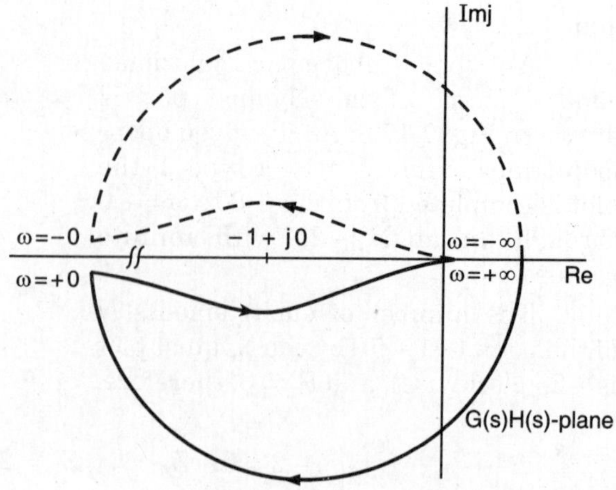

Fig. 7.17.4 (*a*) Nyquist plot $T_1 > T_2$.

$$\therefore \quad N = 0$$
$$P_+ = 0$$
$$N = P_+ - Z_+$$
$$\therefore \quad 0 = 0 - Z_+$$
$$\therefore \quad Z_+ = 0$$

The closed-loop system is stable.

(*b*) $T_1 < T_2$

$$|G(j\omega)\, H(j\omega)| = \frac{K\sqrt{1^2 + \omega^2 T_1^2}}{\omega^2 \sqrt{1^2 + \omega^2 T_2^2}}$$

and

$$\angle G(j\omega)\, H(j\omega) = -180° - \tan^{-1} \omega T_1 + \tan^{-1} \omega T_2$$
$$= -180° - \tan^{-1} \frac{\omega(T_2 - T_1)}{1 + \omega^2 T_1 T_2}$$

$$\because \quad T_1 < T_2$$

$$\angle G(j\omega)\, H(j\omega) = -180° - \tan^{-1} \frac{\omega(T_2 - T_1)}{1 + \omega^2 T_1 T_2}$$

(1) As $\omega \to 0$

$$|G(j0)\, H(j0)| \to \infty$$
$$\angle G(j0)\, H(j0) \to -180° - 0°$$

(2) As $\omega \to \infty$

$$|G(j\infty)\, H(j\infty)| \to 0$$
$$\angle G(j\infty)\, H(j\infty) \to -180° - 0°$$

Thus, for any value $0 < \omega < \infty$, the magnitude $|G(j\omega)\, H(j\omega)|$ is finite and since $T_1 < T_2$, $\angle G(j\omega)\, H(j\omega)$ is placed in the second quadrant of $G(s)\, H(s)$-plane.

General shape of the Nyquist plot is shown in Fig. 7.17.4 (*b*). As the open-loop transfer function is of type 2, the plot is completed from $\omega = -0$ to $\omega = +0$ through an angle -2π ($-360°$) as shown in Fig. 7.17.4 (*b*).

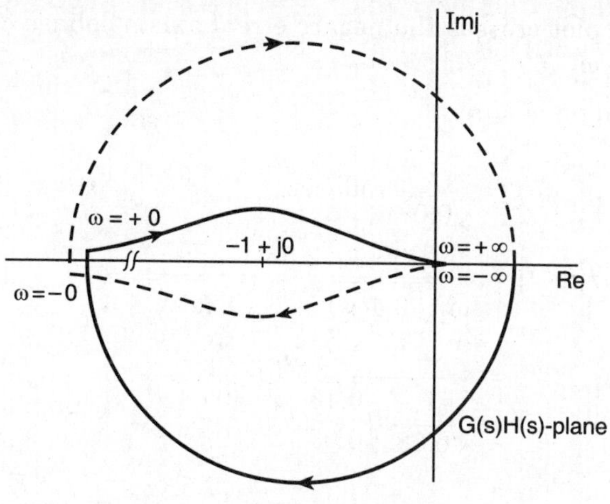

Fig. 7.17.4 (*b*) Nyquist plot $T_1 < T_2$.

The Nyquist plot encircles the critical point (–1 + j0) twice in clockwise direction,

$$\therefore \qquad N = -2$$
$$P_+ = 0$$
$$N = P_+ - Z_+$$
$$\therefore \qquad -2 = 0 - Z_+$$
$$\therefore \qquad Z_+ = 2$$

The closed-loop system is unstable.

Example 7.17.5. *Plot the Nyquist plot to determine*
(a) phase crossover frequency and gain margin for

$$G(s)\,H(s) = \frac{2.5K}{s\,(0.4s + 1)\,(0.2s + 1)}$$

(b) determine phase margin for K = 3.

Solution. (*a*) Put $s = j\omega$

$$\therefore \qquad G(j\omega)\,H(j\omega) = \frac{2.5\,K}{j\omega\,(j\,0.4\,\omega + 1)\,(j0.2\omega + 1)}$$

$$|\,G(j\omega)\,H(j\omega)\,| = \frac{2.5\,K}{\omega\sqrt{(0.4\omega)^2 + 1^2}\,\sqrt{(0.2\omega)^2 + 1^2}}$$

and
$$\angle\,G(j\omega)\,H(j\omega) = -\,90° - \tan^{-1} 0.4\omega - \tan^{-1} 0.2\omega$$
$$= -\,90° - \tan^{-1}\frac{0.4\omega + 0.2\omega}{1 - 0.4\omega \times 0.2\omega}$$
$$= -\,90° - \tan^{-1}\frac{0.6\omega}{1 - 0.08\omega^2}$$

(1) As $\omega \to 0$

$$|\,G(j0)\,H(j0)\,| \to \infty$$

and
$$\angle G(j0)\,H(j0) \to -\,90°$$

(2) The Nyquist plot crosses the negative real axis at phase crossover frequency ω_2. Therefore, $\angle G(j\omega_2)\, H(j\omega_2) = -180°$

$$\therefore \quad 1 - 0.08\,\omega_2^2 = 0$$

or
$$\omega_2 = \frac{1}{\sqrt{0.08}} \text{ rad./sec.}$$

$$|G(j\omega_2)\, H(j\omega_2)| = \frac{2.5K}{\omega_2\sqrt{(0.4\omega_2)^2+1}\sqrt{(0.2\omega_2)^2+1}}$$

$$= \frac{2.5K}{\frac{1}{\sqrt{0.08}}\sqrt{\frac{0.16}{0.08}+1}\sqrt{\frac{0.04}{0.08}+1}}$$

$$= \frac{2.5}{\frac{1}{\sqrt{0.08}}\sqrt{3}\sqrt{1.5}} = \frac{2.5K}{\sqrt{4.5/0.08}} = \frac{2.5K}{\sqrt{56.25}} = \frac{K}{3}$$

Thus, the Nyquist plot crosses the negative real axis of $G(s)\, H(s)$-plane at $-\dfrac{K}{3}$.

(3) As $\omega \to \infty$

$$|G(j\infty)\, H(j\infty)| \to 0$$

and
$$\angle G(j\infty)\, H(j\infty) \to -270°$$

The open-loop transfer function is of type 1, therefore, completion of the Nyquist plot for the variation from $\omega = -0$ to $\omega = +0$ is effected through an angle $-\pi$ ($-180°$) as shown in Fig. 7.17.5.

$$\text{G.M.}\,|_{db} = 20\log_{10}\frac{1}{(K/3)} = 20\log_{10}\frac{3}{K}$$

$$= 20\log_{10}3 - 20\log_{10}K. \textbf{ Ans.}$$

(*b*) $K = 3$

$$\text{G.M.} = 20\log_{10}3 - 20\log_{10}3 = 0 \text{ db.}$$

Since, G.M. = 0 db. Therefore P.M. = 0°. **Ans.**

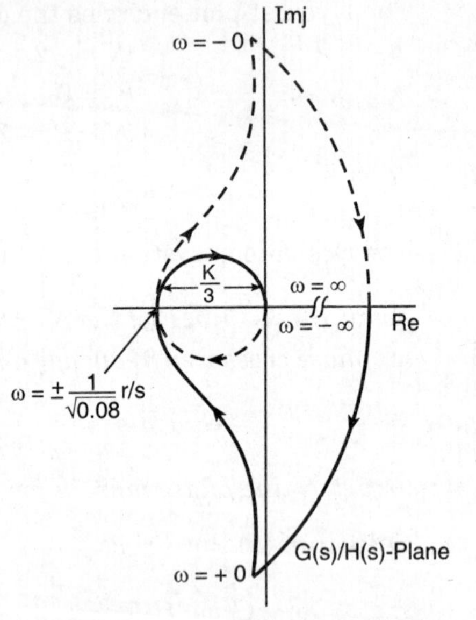

Fig. 7.17.5. Nyquist plot for Example 7.17.5.

Example 7.17.6. *Investigate the closed-loop stability of a system whose open-loop transfer function is given below :*

$$G(s)\,H(s) = \frac{15\,(1+4s)}{(s^2+100)\,(1+0.25s)}.$$

Solution. $\quad G(s)\,H(s) = \dfrac{15\,(1+4s)}{(s^2+100)\,(1+0.25\,s)}$

Put $\quad s = j\omega$

$$G(j\omega)\,H(j\omega) = \frac{15\,(1+j4\omega)}{(100-\omega^2)\,(1+j\omega 0.25)} \qquad ...(1)$$

Rationalizing Eq. (1) and separating into real and imaginary part following equation is obtained :

$$G(j\omega)\,H(j\omega) = \frac{15\,(1+\omega^2)}{(100-\omega^2)\,(1+0.625\omega^2)} + \frac{j\,56.25\,\omega}{(100-\omega^2)\,(1+0.0625\omega^2)} \quad ...(2)$$

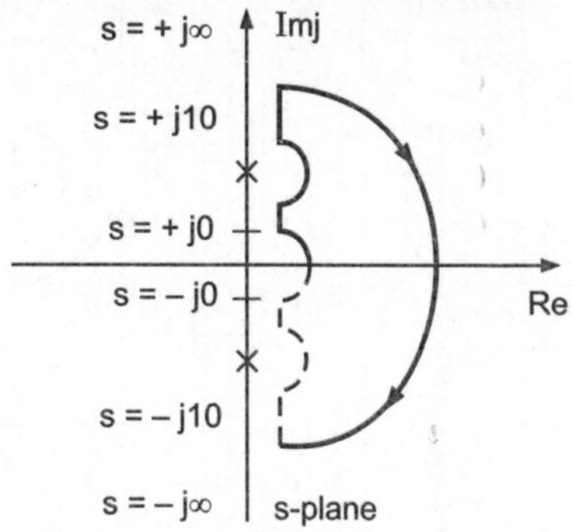

Fig. 7.17.6 (a) Variation s in s-plane.

The open-loop transfer function has poles on imaginary axis

$$(s = \sqrt{-100} = \pm j\,10),$$

these poles are avoided from the Nyquist path in s-plane as shown in Fig. 7.17.6 (*a*).

The real and imaginary parts of $G(j\omega)\,H(j\omega)$ for some values of ω are calculated as given in the following table :

ω	0	9	10	11	∞
$Re\ G(j\omega)\ H(j\omega)$	+ 0.15	+ 10.68	+ ∞	– 10.18	– 0
$Imj\ G(j\omega)\ H(j\omega)$	+ $j0$	+ $j4.39$	+ $j\infty$	– $j3.43$	– $j0$

The real part as well as imaginary part of $G(j\omega)\,H(j\omega)$ are both positive for values of ω just less than 10 and, therefore, at these frequencies the phase of $G(j\omega)\,H(j\omega)$ is given by

$$\underset{\omega<10}{\angle G(j\omega)\,H(j\omega)} = \tan^{-1}\left[\frac{+56.25\times\omega}{+15\,(1+\omega^2)}\right]$$

$$= \tan^{-1}\left[\frac{56.25\times 10}{15\,(1+10^2)}\right] = 20.4°$$

For values of ω just greater than 10 the real part and imaginary part of $G(j\omega)\,H(j\omega)$ are both negative and at these values of frequency the phase of $G(j\omega)\,H(j\omega)$ is given by

$$\underset{\omega\to>10}{\angle G(j\omega)\,H(j\omega)} = \tan^{-1}\left[\frac{-56.25\times\omega}{-15\,(1+\omega^3)}\right]$$

$$= \tan^{-1}\left[\frac{-56.25\times 10}{-15\,(1+10^2)}\right] = -159.6°$$

Thus at $\omega = 10$ the phase of $G(j\omega)\,H(j\omega)$ suddenly changes through – 159.6 – (+ 20.4) = – 180° and the magnitude being infinite. Therefore as per above analysis and calculated values of $G(j\omega)\,H(j\omega)$ at various of ω, the Nyquist plot takes the shape as shown in Fig. 7.7.6 (*b*). As the type of the system is 1 the points $\omega = -0$ and $\omega = +0$ are coincident.

The point $(-1 + j0)$ is not encircled by the Nyquist plot and $P_+ = 0$

$$\therefore \qquad N = P_+ - Z_+$$

$$\therefore \qquad 0 = 0 - Z_+ \quad \therefore \quad Z_+ = 0$$

As $Z_+ = 0$, the number of roots of the characteristics equation of the closed-loop system having positive real part is nil indicating that the system is stable.

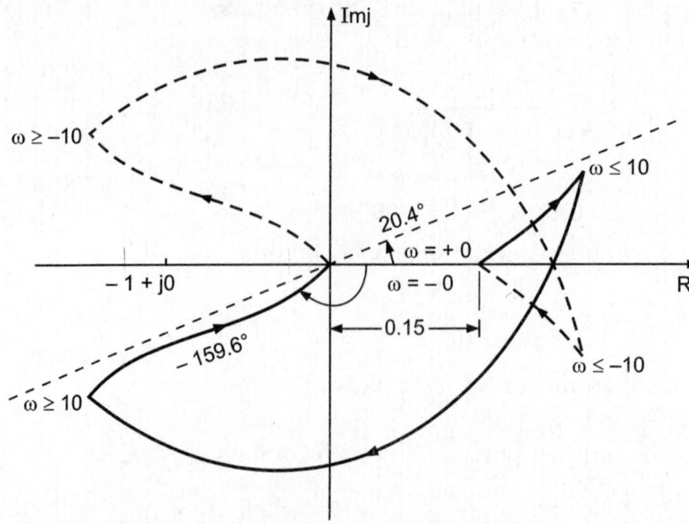

Fig. 7.17.6 (*b*) Nyquist plot for example 7.17.6.

Example 7.17.7. *Draw the Nyquist plot for the open-loop transfer function given below and comment on the closed-loop stability.*

$$G(s)\,H(s) = \frac{2.2}{s(s+1)(s^2+2s+2)}.$$

Solution. $G(s)\,H(s) = \dfrac{2.2}{s(s+1)(s^2+2s+2)}$

Put $s = j\omega$

$$G(j\omega)\,H(j\omega) = \frac{2.2}{j\omega\,(j\omega+1)\,[(2-\omega^2)+j2\omega]} \qquad ...(1)$$

Rationalizing Eq. (1) and separating into real and imaginary part following equation is obtained

$$G(j\omega)\,H(j\omega) = -\frac{2.2\,[\omega\,(4-\omega^2)]}{\omega\,(\omega^2+1)\,[(2-\omega^2)^2+4\omega^2]} - \frac{j2.2\,(2-3\omega^2)}{\omega\,(\omega^2+1)\,[(2-\omega^2)^2+4\omega^2]} \qquad ...(2)$$

From Eq. (2) it is observed that the Nyquist plot cuts the real axis at $\omega = \sqrt{(2/3)}$ $= \pm\,0.816$ and its intersection with the imaginary axis is at $\omega = \sqrt{4} = \pm\,2$.

The real and imaginary part of $G(j\omega)\,H(j\omega)$ for values of $\omega = 0, +\,0.816, +\,2$ and $+\,\infty$ etc. are calculated as given in the following table.

ω	0	0.816	1	$\sqrt{2}$	2	∞
$Re\ G(j\omega)\,H(j\omega)$	– 2.2	– 1	– 0.66	– 0.29	– 0	+ 0
$Imj\ G(j\omega)\,H(j\omega)$	$-j\infty$	$j0$	$+j0.22$	$+j0.32$	$+j0.022$	$+j0$

The shape of Nyquist plot is drawn in Fig. 7.17.7.

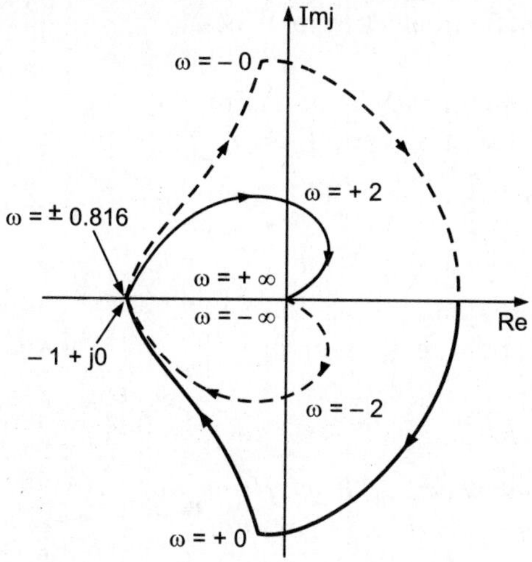

Fig. 7.17.7. Nyquist plot for example 7.17.7.

As the system is type 1 the Nyquist plot is closed from $\omega = -0$ to $\omega = +0$ through an angle of π in the clockwise direction with an infinite radius.

As the Nyquist plot passes through the point $(-1 + j0)$, it is concluded that the closed-loop system is marginally stable and the output presents sustained oscillations.

Example 7.17.8. *Using Nyquist criterion to determine the closed-loop stability of a system whose open-loop transfer function is given by*

$$G(s)\,H(s) = \frac{K}{s^3\,(sT + 1)}.$$

Solution. $$G(s)\,H(s) = \frac{K}{s^3(sT + 1)}$$

Put $s = j\omega$

$$G(j\omega)\,H(j\omega) = \frac{K}{(j\omega)^3(j\omega T + 1)} \qquad \text{...(1)}$$

Rationalizing Eq. (1) and separating into real and imaginary part

$$G(j\omega)\,H(j\omega) = \frac{jK(-j\omega T + 1)}{\omega^3(\omega^2 T^2 + 1)} = \frac{K\,\omega T}{\omega^3\,(\omega^2 T^2 + 1)} + \frac{jK}{\omega^3\,(\omega^2\,T^2 + 1)}$$

(*i*) As $\omega \to 0$, $|G(j\omega)\,H(j\omega)| \to \infty$

and $$\angle G(j\omega)\,H(j\omega) \to \frac{\pi}{2}$$

(*ii*) As $\omega \to \infty$, $|G(j\omega)\,H(j\omega)| \to 0$

and $$\angle G(j\omega)\,H(j\omega) \to 0.$$

The Nyquist plot is shown in Fig. 7.17.8. As the system is type 3 the plot from $\omega = -0$ to $\omega = +0$ is closed through an angle of 3π in clockwise direction.

The number of encirclements N of the point $(-1 + j0)$ are noted as -2 (clockwise) and $P_+ = 0$

$\therefore$ $$N = (P_+ - Z_+)$$

or $$-2 = (0 - Z_+)$$

$\therefore$ $$Z_+ = 2$$

Thus, the closed-loop system has 2 roots having positive real part and the system is unstable.

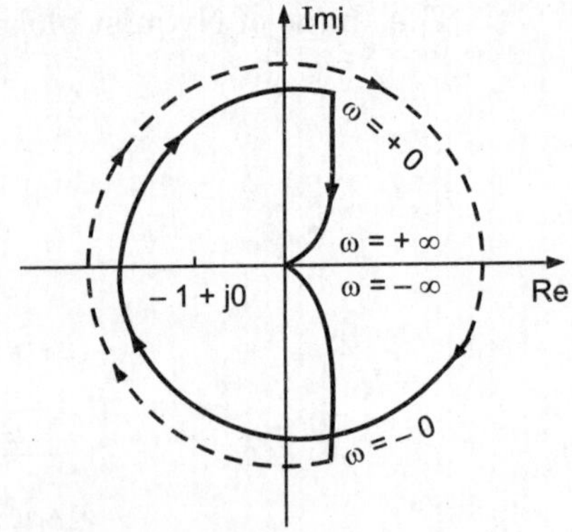

Fig. 7.17.8. Nyquist plot for example 7.17.8.

Example 7.17.9. *Determine* $|G(j\omega)\ H(j\omega)|$ *at* $\omega = \sqrt{2}$ *rad./sec. Calculate gain margin and gain crossover frequency. Given that :*

$$G(s)\ H(s) = \frac{32}{s\,(s + \sqrt{6})^3}$$

Comment on stability.

Solution. Put $s = j\omega$

$\therefore$ $$G(j\omega)\ H(j\omega) = \frac{32}{j\omega\,(j\omega + \sqrt{6})^3}$$

$$|G(j\omega)\ H(j\omega)| = \frac{32}{\omega\,(\sqrt{\omega^2 + 6})^3}$$

put $\omega = \sqrt{2}$

$\therefore$ $$|G(j\sqrt{2}\,)\ H(j\sqrt{2}\,)| = \frac{32}{\sqrt{2}(2+6)^3} = \frac{32}{\sqrt{2}(\sqrt{8})^3} = \frac{32}{\sqrt{2}\sqrt{8}\sqrt{8}\sqrt{8}} = 1$$

$\therefore$ gain crossover frequency $\omega_1 = \sqrt{2}$ rad./sec. **Ans.**

$$\angle\, G(j\omega)\ H(j\omega) = -\,90° - 3\tan^{-1}\frac{\omega}{\sqrt{6}}$$

At the phase crossover frequency ω_2,

$$\angle\, G(j\omega_2)\ H(j\omega_2) = -\,180°.$$

$\therefore$ $$-\,180° = -\,90 - \tan^{-1}\frac{\omega_2}{\sqrt{6}}$$

or $$\tan^{-1}\frac{\omega_2}{\sqrt{6}} = 30°$$

$\therefore$ $$\frac{\omega_2}{\sqrt{6}} = \tan 30°$$

$\therefore$ $$\omega_2 = (\sqrt{6}\,)\tan 30° = \sqrt{6} \times \frac{1}{\sqrt{3}} = \sqrt{2}\ \text{rad./sec.}$$

$\because \quad \omega_1 = \omega_2$

G.M. = 0 db. **Ans.**

and P.M. = 0°

Since, $\omega_1 = \omega_2$, the system is marginally stable.

Example 7.17.10. *Determine the value of K for a unity feedback control system having open-loop transfer function*

$$G(s)\,H(s) = \frac{K}{s\,(s+2)\,(s+4)}, \text{ such that}$$

(i) Gain margin = 20 db, and (ii) Phase margin = 60°.

Solution. (*i*) G.M. = 20 $\log_{10}(1/a)$

where '*a*' is the intersection point of the Nyquist plot on the negative real axis. Therefore,

$$20 = 20 \log_{10}(1/a) \quad \therefore \quad \log_{10}(1/a) = 1$$

$$\therefore \quad (1/a) = 10 \quad \text{or} \quad a = 0.1$$

$$G(s)\,H(s) = \frac{K}{s\,(s+2)\,(s+4)}$$

Put $s = j\omega$

$$\therefore \quad G(j\omega)\,H(j\omega) = \frac{K}{j\omega\,(j\omega+2)\,(j\omega+4)} = \frac{-jK(-j\omega+2)(-j\omega+4)}{(\omega)(j\omega+2)(-j\omega+2)(j\omega+4)(-j\omega+4)}$$

$$= -\frac{jK(-j\omega+2)(-j\omega+4)}{\omega(\omega^2+4)(\omega^2+16)} = -\frac{jK[(-\omega^2+8)-j\,6\omega]}{\omega(\omega^2+4)(\omega^2+16)}$$

$$= -\frac{6K\omega}{\omega(\omega^2+4)(\omega^2+16)} - \frac{jK(-\omega^2+8)}{\omega(\omega^2+4)(\omega^2+16)}$$

At the point of intersection of the Nyquist plot with the real axis the imaginary part of the above expression is zero. Therefore, the phase crossover frequency ω_2 is determined below :

$$-\frac{K(-\omega_2^2+8)}{\omega_2^2(\omega_2^2+4)(\omega_2^2+16)} = 0$$

$$\therefore \quad \omega_2 = \sqrt{8} \text{ rad/sec}$$

The intersection of the Nyquist plot with the real axis occurs at $\omega = \omega_2 = \sqrt{8}$

$$\therefore \quad G(j\sqrt{8})\,H(j\sqrt{8}) = \frac{6K\sqrt{8}}{\sqrt{8}(\sqrt{8}^2+4)(\sqrt{8}^2+16)} = -\frac{K}{48}$$

To have a gain margin of 20 db *i.e.* $a = 0.1$

$$-\frac{K}{48} = -0.1$$

$$\therefore \quad K = 4.8. \quad \textbf{Ans.}$$

(*ii*) $$G(j\omega)\,H(j\omega) = \frac{K}{j\omega\,(j\omega+2)\,(j\omega+4)}$$

$$\therefore \qquad \angle G(j\omega)\,H(j\omega) = -90° - \tan^{-1}\left(\frac{\omega}{2}\right) - \tan^{-1}\left(\frac{\omega}{4}\right)$$

If the gain crossover frequency is ω_1, the phase margin is :

Phase margin = $\angle G(j\omega_1)\,H(j\omega_1) + 180°$

$$\therefore \qquad 60° = -90° - \tan^{-1}\left(\frac{\omega_1}{2}\right) - \tan^{-1}\left(\frac{\omega_1}{4}\right) + 180°$$

or
$$30° = \tan^{-1}\left(\frac{\omega_1}{2}\right) + \tan^{-1}\left(\frac{\omega_1}{4}\right)$$

or
$$\tan 30° = \frac{\frac{\omega_1}{2} + \frac{\omega_1}{4}}{1 - \frac{\omega_1}{2}\cdot\frac{\omega_1}{4}} = \frac{\frac{6\omega_1}{8}}{1 - \frac{\omega_1^2}{8}} = \frac{6\omega_1}{8 - \omega_1^2}$$

or
$$\frac{1}{\sqrt{3}} = \frac{6\omega_1}{8 - \omega_1^2}$$

or
$$\omega_1^2 + (6\sqrt{3})\omega_1 - 8 = 0$$

$$\therefore \qquad \omega_1 = \frac{-6\sqrt{3} \pm \sqrt{(6\sqrt{3})^2 + 4 \times 8}}{2} = \frac{-10.39 \pm 11.83}{2}$$

Considering positive value of ω_1

$$\omega_1 = \frac{-10.39 + 11.83}{2} = 0.72 \text{ rad/sec}$$

At $\omega = \omega_1$

$$|G(j\omega_1)\,H(j\omega_1)| = 1$$

$$\therefore \qquad \left|\frac{K}{j\omega_1\,(j\omega_1 + 2)\,(j\omega_1 + 4)}\right| = 1$$

or
$$\frac{K}{\sqrt{(\omega_1^2)}\sqrt{(\omega_1^2 + 4)}\sqrt{(\omega_1^2 + 16)}} = 1$$

or
$$\frac{K}{\sqrt{0.72^2}\sqrt{(0.72^2 + 4)}\sqrt{(0.72^2 + 16)}} = 1$$

or
$$K = \sqrt{0.72^2}\sqrt{(0.72^2 + 4)}\sqrt{(0.72^2 + 16)}$$

or $K = 6.2$. **Ans.**

Example 7.17.11. *Sketch Nyquist plot for a unity feedback control system having given*

$$G(s) = \frac{Ks^3}{(s + 1)(s + 2)}$$

and determine the stability condition.

Solution. Put $s = j\omega$

$$\therefore \quad G(j\omega) = \frac{K(j\omega)^3}{(j\omega+1)(j\omega+2)} \quad \text{or} \quad G(j\omega) = \frac{-jK\omega^3}{(2-\omega^2)+j3\omega}$$

or
$$G(j\omega) = \frac{-jK\omega^3\,[(2-\omega^3)-j3\omega]}{(2-\omega^2)^2+(3\omega)^2}$$

or
$$G(j\omega) = \frac{-3K\omega^4}{(2-\omega^2)^2+(3\omega)^2} - \frac{jK\omega^3\,(2-\omega^2)}{(2-\omega^2)^2+(3\omega)^2}$$

(*a*) $\underset{\omega\to 0}{G(j\omega)}$: $Re\,(G(j0) \to -0 \quad \text{Imj}\,(Gj0) \to +j0$

(*b*) The intersection of $G(j\omega)$ plot with –ive real axis is obtained by equating imaginary part of $G(j\omega)$ to zero and solving for ω. Therefore,

$$-\frac{K\omega^2\,(2-\omega^2)}{(2-\omega^2)^2+(3\omega)^2} = 0$$

or
$$(2-\omega^2) = 0$$

The intersection with –ive real axis occurs at :

$$\omega = \pm\sqrt{2} \text{ rad/sec}$$

The intersection is obtained by substituting $\omega = \sqrt{2}$ in real part of $G(j\omega)$, *i.e.*

$$G(j\sqrt{2}) = \frac{-3K(\sqrt{2})^4}{(2-\sqrt{2^2})^2+(3\sqrt{2})^2} = \frac{-3K\times 2\times 2}{(2-2)^2+3\times 3\times 2}$$

or
$$G(j\sqrt{2}) = -\frac{2}{3}K$$

(*c*) $\underset{\omega\to 0}{G(j\omega)}$: $Re\,G(j\infty) \to -3K$

$$\text{Imj } G(j\infty) \to +j\infty$$

As per steps (*a*), (*b*) and (*c*), the general shape of the Nyquist plot is drawn as shown in Fig. 7.17.9.

$$\underset{\omega\to 0}{G(j\omega)} = \frac{K(j\omega)^3}{(j\omega+1)(j\omega+2)} = \frac{K(j\omega)^3}{(j\omega)(j\omega)} = Kj\omega$$

or
$$\underset{\omega\to 0}{G(j\omega)} = Kj\infty = |\infty|\angle 90°$$

The closing of Nyquist plot from $\omega = \infty$ to $\omega = -\infty$ is done as explained below :

In *s*-plane the RHS region is closed from $s = j\infty$ to $s - j\infty$ through a semicircle of infinite radius in the clockwise direction, hence the corresponding points in $G(s)$-plane *i.e.* $G(j\infty)$ to $G(-j\infty)$ are closed through a semicircle of infinite radius to account for effective numerator term *i.e.* $Kj\infty$ of $G(j\omega)$ as $\omega \to \infty$.

The completed Nyquist plot is shown in Fig. 7.17.9.

It is given that, the number of poles of $G(s)$ having positive real part is nil *i.e.* $P_+ = 0$. The encirclements of critical point $(-1 + j0)$ are determined below :

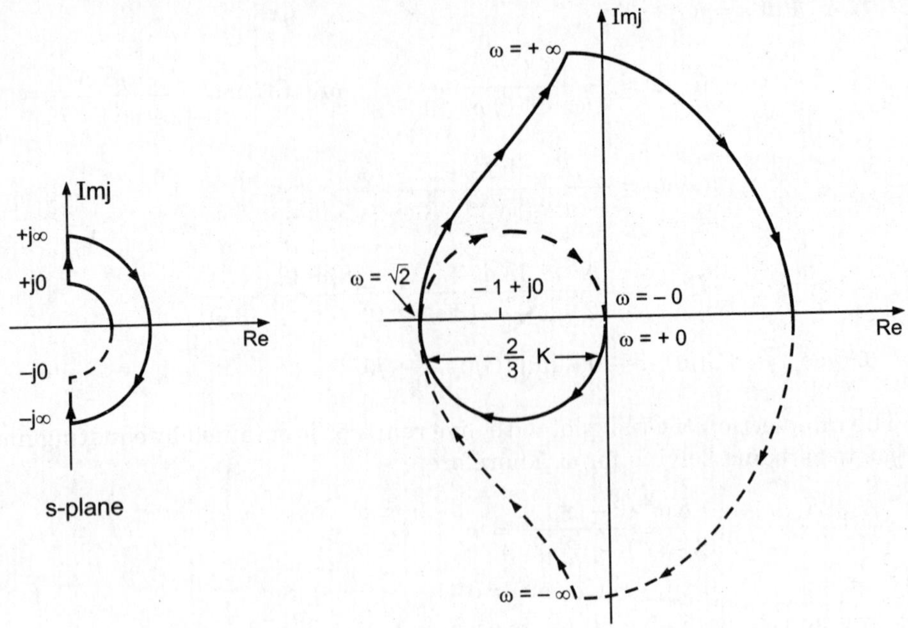

Fig. 7.17.9. Nyquist plot for example 7.17.11.

(1) If $K < \dfrac{3}{2}$

The critical point $(-1 + j0)$ lies outside the Nyquist plot, hence $N = 0$

$$\therefore \quad N = P_+ - Z_+$$

or

$$0 = 0 - Z_+$$

$$\therefore \quad Z_+ = \text{Nil}$$

The system is stable

(2) If $K > \dfrac{3}{2}$

The critical point $(-1 + j0)$ and encircled twice in the clockwise direction by the Nyquist plot, hence $N = -2$.

$$\therefore \quad N = P_+ - Z_+$$

or

$$-2 = 0 + Z_+ \quad \therefore \quad Z_+ = 2$$

The system is unstable. For stability $K \le \dfrac{3}{2}$. **Ans.**

Example 7.17.12. *Sketch the Nyquist plot and examine the closed-loop stability of a control system having open-loop transfer function given below :*

$$G(s)\,H(s) = \frac{Ks\,(1+2s)}{s^3 + 4s + 8}.$$

Solution. For the application of Nyquist plot it is required to know the number of poles (P_+) of $G(s)\,H(s)$ which have positive real part. For the transfer function :

$$G(s)\,H(s) = \frac{Ks\,(1+2s)}{s^3 + 4s + 8}$$

P_+ can be determined by applying Routh-Hurwitz criterion to following equation :

$$s^3 + 4s + 8 = 0$$

s^3	1	4
s^2	(0) ε	8
s^1	$\dfrac{4\varepsilon - 8}{\varepsilon}$	
s^0	8	

As $\varepsilon \to 0$ the sign of $\dfrac{4\varepsilon - 8}{\varepsilon} \to -$ hence, there are two sign changes in the first column of Routh's table and as such $P_+ = 2$.

Put $s = j\omega$ in the expression for $G(s)\,H(s)$, thus

$$G(j\omega)\,H(j\omega) = \frac{K\,j\omega\,(1+j2\omega)}{(j\omega)^3 + j4\omega + 8}$$

or

$$G(j\omega)\,H(j\omega) = \frac{K\,(j\omega - 2\omega^2)}{-j\omega^3 + 4j\omega + 8} = \frac{K\,(j\omega - 2\omega^2)}{8 + j\omega\,(4-\omega^2)}$$

or

$$G(j\omega)\,H(j\omega) = \frac{K\,(j\omega - 2\omega^2)\,[8 - j\omega\,(4-\omega^2)]}{[8 + j\omega\,(4-\omega^2)]\,[8 - j\omega\,(4-\omega^2)]}$$

$$= \frac{-K\omega^2\,(\omega^2 + 12) + j\omega K\,(8 + 10\omega^2 - 2\omega^4)}{64 + \omega^2\,(4-\omega^2)^2}$$

Separating into real and imaginary part :

$$G(j\omega)\,H(j\omega) = \frac{-K\omega^2\,(\omega^2+12)}{64 + \omega^2\,(4-\omega^2)^2} + \frac{j\omega K\,(8 + 10\omega^2 - 2\omega^4)}{64 + \omega^2\,(4-\omega^2)^2}$$

(*a*) $\underset{\omega \to 0}{G(j\omega)H(j\omega)}$: $Re\ G(j0) \to -0$

$$Imj\ G(j0) \to j0$$

(*b*) The intersection of the Nyquist plot with – ive real axis occurs at a value of ω given by solving

$$\frac{\omega K\,(8 + 10\omega^2 - 2\omega^4)}{64 + \omega^2\,(4-\omega^2)^2} = 0$$

or $$-2\omega^4 + 10\omega^2 + 8 = 0$$

or $$-(\omega^2)^2 + 5\,(\omega^2) + 4 = 0$$

The solution of the above equation is :

$$\omega^2 = 2.5 \mp 3.2$$

Considering +ive value of ω^2 :

$$\omega^2 = 5.7$$

$$\therefore \quad \omega = \pm\sqrt{5.7} = \pm j2.38 \text{ rad/sec}$$

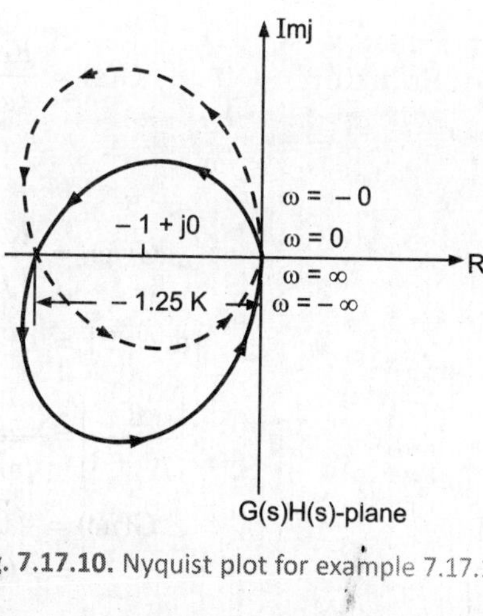

Fig. 7.17.10. Nyquist plot for example 7.17.12.

At $\omega = +j2.38$ rad/sec

$$G(j2.38) = \frac{-K(2.38)^2(2.38^2+12)}{64+(2.38)^2(4-2.38^2)^2} = \frac{-K(5.7)(5.7+12)}{64+(5.7)(4-5.7)^2}$$

$$= -\frac{100.9\,K}{80.4} = -1.25\,K$$

The intersection of the Nyquist plot with – ive real axis is thus, – 1.25 *K*.

(c) $G(j\omega)H(j\omega)_{\omega\to\infty}$: $Re\ G(j\infty) \to -0$

$$Imj\ G(j\infty) \to -j0$$

From the data determined above the Nyquist plot is drawn as shown in Fig. 7.17.10.

(1) It is noted that, for $1.25\,K > 1$, the critical point $(-1+j0)$ is encircled twice in the anticlockwise direction by the Nyquist plot, hence $N = 2$, $P_+ = 2$.

$$\therefore \quad N = P_+ - Z_+$$

or $\quad 2 = 2 - Z_+ \qquad \therefore\ Z_+ = \text{Nil}$

and the closed-loop system is stable. **Ans.**

(2) However, for $1.25\,K < 1$

$$N = 0$$

$$P_+ = 2 \qquad \therefore\ N = P_+ - Z_+$$

or $\quad 0 = 2 - Z_+ \qquad \therefore\ Z_+ = 2$

and the closed-loop system is unstable. **Ans.**

The condition for stability is, thus :

$$1.25\,K \geq 1 \quad \text{or} \quad K \geq 0.8. \quad \textbf{Ans.}$$

Example 7.17.13. *Determine the critical value of K for the stability of a unity feedback control system whose open-loop transfer function is given by*

$$G(s) = \frac{Ke^{-0.5s}}{(s+1)}.$$

Use Nyquist plot method.

Solution.

$$G(s) = \frac{Ke^{-0.5s}}{(s+1)}$$

Put $\quad s = j\omega$

$$\therefore \quad G(j\omega) = \frac{Ke^{-0.5j\omega}}{(j\omega+1)}$$

$$\because \quad |e^{-j0.5\omega}| = 1$$

$$\therefore \quad \left|\frac{K}{j\omega+1}\right| = \frac{K}{\sqrt{\omega^2+1^2}}$$

and

$$\angle G(j\omega) = -\tan^{-1}\omega - 0.5\,\omega$$

$$= -(\tan^{-1}\omega + 0.5\,\omega)$$

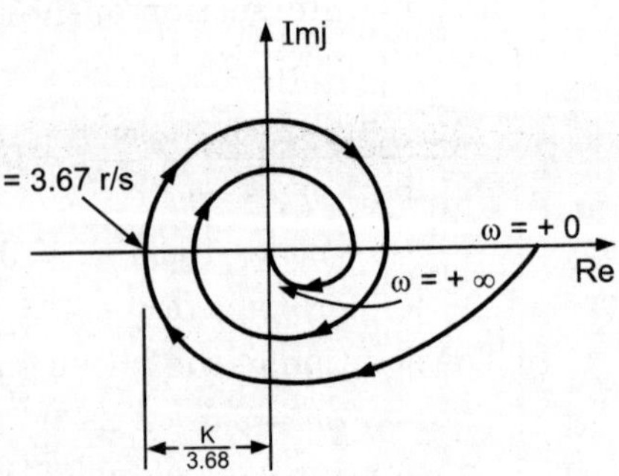

Fig. 7.17.11. Nyquist plot (from ω = + 0 to ω = + ∞) for example 7.17.13.

(1) As $\omega \to 0$

$$|G(j\omega)| \to K$$

and $$\angle G(j\omega) \to -0°$$

(2) As $\omega \to \infty$

$$|G(j\omega)| \to 0$$

and $$\angle G(j\omega) \to -\infty$$

As ω varies from $\omega = 0$ to $\omega = \infty$, $|G(j\omega)|$ decreases and $\angle G(j\omega)$ changes continuously from 0° to ∞. The Nyquist plot takes the shape of a spiral intersecting real and imaginary axis many times. The plot for the variation from $\omega = 0$ to $\omega = \infty$ is shown in Fig. 7.17.11.

(3) The intersection points of the plot with the negative real axis of $G(s)\,H(s)$-plane are determined using the relation given below :

$$-(\tan^{-1}\omega + 0.5\,\omega) = -180\,(2k+1)$$

$k = 0, 1, 2, ...$

For $\omega > 0$, at the first instant, the said intersection is given by considering $k = 0$. The frequency at this intersection point is the phase crossover frequency ω_2. Thus,

$$-(\tan^{-1}\omega_2 - 0.5\,\omega_2) = -180°$$

Using trial and error, this equation is satisfied at $\omega_2 = 3.67$ rad./sec.

For stability $|G(j\omega_2)| < 1$.

$$|G(j\omega_2)| = \left|\frac{Ke^{-j0.5\,\omega_2}}{j\omega_2 + 1}\right|$$

$$\because \quad |e^{-j0.5\,\omega_2}| = 1$$

$$\therefore \quad |G(j\omega_2)| = \frac{K}{\sqrt{\omega_2^2 + 1^2}}$$

$\omega_2 = 3.67$ rad./sec.

$$\therefore \quad |G(j\,3.67)| = \frac{K}{\sqrt{(0.367)^2 + 1^2}} = \frac{K}{3.68}$$

For stability, the critical point $(-1 + j0)$ should not be encircled by the Nyquist plot.

$$\therefore \quad \frac{K}{3.68} < 1$$

or $$K < 3.68.$$ **Ans.**

Example 7.17.14. *Determine the closed-loop stability given that :*

$$G(s)\,H(s) = \frac{Ke^{-0.2s}}{s\,(s+1)\,(s+2)}.$$

Solution. Put $s = j\omega$

$$\therefore \quad G(j\omega)\,H(j\omega) = \frac{Ke^{-j0.2\omega}}{j\omega\,(j\omega+1)\,(j\omega+2)}$$

$$\angle G(j\omega)\,H(j\omega) = -0.2\,\omega \times \frac{180}{\pi} - 90° - \tan^{-1}\frac{\omega}{1} - \tan^{-1}\frac{\omega}{2}$$

$$= -11.46\omega - 90^\circ - \tan^{-1}\frac{3\omega}{2-\omega^2}\,;\ (\omega < \sqrt{2}\,)$$

Intersection of the Nyquist plot with the negative real axis of $G(s)\,H(s)$-plane is determined using following relation

$$\angle G(j\omega)\,H(j\omega) = -180\,(2k+1)$$

$k = 0, 1, 2, \ldots$

At the first instant the Nyquist plot intersects the negative real axis of $G(s)\,H(s)$-plane for $k = 0$ and the frequency at the intersection point is ω_2. Therefore,

$$\angle G(j\omega_2)\,H(j\omega_2) = -180^\circ$$

$$\therefore \quad -11.46\,\omega_2 - 90^\circ - \tan^{-1}\frac{3\omega_2}{2-\omega_2^2} = -180^\circ$$

or

$$-11.46\,\omega_2 - \tan^{-1}\frac{3\omega_2}{2-\omega_2^2} = -90^\circ$$

The equation above is satisfied at $\omega_2 = 1.114$ rad./sec. (phase crossover frequency)

$$|G(j\omega_2)\,H(j\omega_2)| = \left|\frac{K}{j\omega_2(j\omega_2+1)\,(j\omega_2+2)}\right|$$

(**Note:** $|e^{-j0.2\omega}| = 1$)

$$= \frac{K}{\omega_2\sqrt{\omega_2^2+1^2}\,\sqrt{\omega_2^2+2^2}}$$

$$\omega_2 = 1.114 \text{ rad./sec.}$$

$$\therefore \quad |G(j1.114)\,H(j1.114)| = \frac{K}{1.114\sqrt{1.114^2+1^2}\,\sqrt{1.114^2+2^2}} = \frac{K}{3.86}$$

The general shape of the Nyquist plot for $\omega = 0$ to $\omega = -\infty$ is drawn as shown in Fig. 7.17.12.

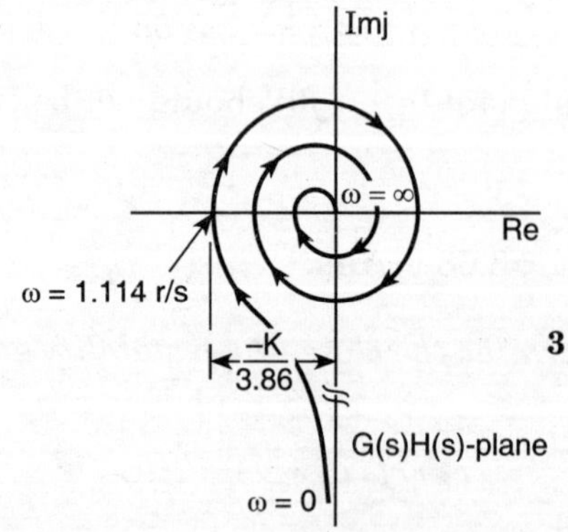

3

Fig. 7.17.12. Nyquist plot for example 7.17.14.

For stability the critical point $(-1 + j0)$ be placed outside the Nyquist plot.

$$\therefore \quad \frac{K}{3.86} < 1$$

or $\quad K < 3.86$. **Ans.**

Example 7.17.15. *Determine the value of T for stability limit for a unity feedback control system having open-loop transfer function as*

$$G(s) = e^{-sT} \cdot \frac{10}{s\,(0.1\,s+1)}.$$

Solution. In the absence of time delay element e^{-sT}, the open-loop transfer function is

$$G_1(s) = \frac{10}{s\,(0.1\,s+1)}$$

Put $s = j\omega$

$$\therefore \quad G_1(j\omega) = \frac{10}{j\omega\,(j0.1\omega+1)}$$

The Nyquist plot for $G_1(j\omega)$ is shown in Fig. 7.17.13.

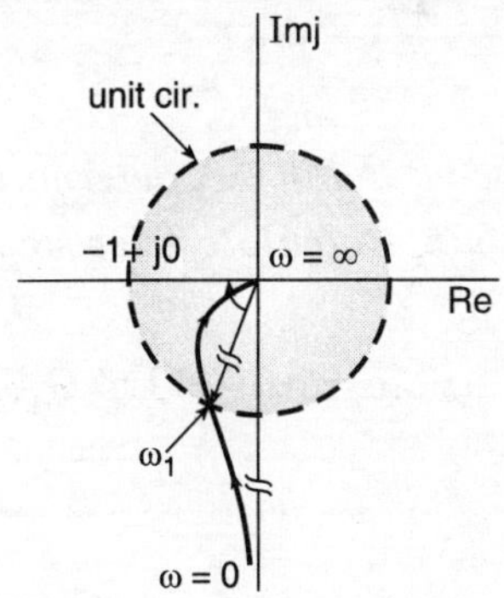

Fig. 7.17.13. Nyquist plot for $G(s) = 10/s(0.1s + 1)$.

The gain crossover frequency ω_1 is determined below :

At ω_1, $\quad |Gj(\omega_1)| = 1$

$$\therefore \quad |G(j\omega_1)| = \left|\frac{10}{j\omega_1\,(j0.1\omega+1)}\right| = 1$$

or

$$\frac{10}{\omega_1\sqrt{(0.1\omega_1)^2+1^2}} = 1$$

$$\therefore \quad 0.01\,\omega_1^4 + \omega_1^2 - 100 = 0$$

Solving

$$\omega_1^2 = \frac{-1 \pm \sqrt{1^2 - 4\times 0.01\,(-100)}}{2\times 0.01}$$

$$= (-0.5 \pm 1.115) \times 100$$

Considering +ive value

$$\omega_1^2 = 61.5$$

$$\therefore \quad \omega_1 = \sqrt{61.5} = 7.84 \text{ rad./sec.}$$

Incorporating the time delay element e^{-sT}, the magnitude of $G_1(j\omega)$ is not affected but the phase angle is modified as below :

$$\angle G_1(j\,\omega_1) = -90° - \tan^{-1} 0.1\,\omega_1$$

$$= -90° - \tan^{-1} 0.1 \times 7.84$$

$$= -90° - 38.09° = -128.09°$$

$$\text{P.M.} = 180° + \angle G_1(j\omega_1)$$

$$= 180° - 128.09° = 51.81°$$

If the phase angle $(-\omega_1 T \times 180/\pi)$ provided by the time delay element equals $-51.81°$, then the limiting value of T for stability is determined as below :

$$\frac{-\omega_1 T \times 180}{\pi} = -51.81°$$

or $$\frac{-7.84T \times 180}{\pi} = -51.81°$$

$\therefore$ $$T = 0.115 \text{ sec.} \quad \textbf{Ans.}$$

Example 7.17.16. *For a unity feedback control system*

$$G(s) = \frac{e^{-sT}}{s(0.5\,s+1)(0.2\,s+1)}$$

(a) For T = 0, determine the phase margin

(b) Calculate T for stability limit.

Solution. (*a*) Let $G_1(s) = \dfrac{1}{s\,(0.5\,s+1)\,(0.2\,s+1)}$

Put $s = j\omega$

$\therefore$ $$G_1(j\omega) = \frac{1}{j\,\omega(j\,0.5\,\omega+1)\,(j\,0.2\,\omega+1)}$$

To determine gain crossover frequency ω_1 put $|G(j\omega_1)| = 1$

$\therefore$ $$\left|\frac{1}{j\omega_1\,(j\,0.5\,\omega_1+1)\,(j\,0.2\,\omega_1+1)}\right| = 1$$

$\therefore$ $$\frac{1}{\omega_1\sqrt{(0.5\,\omega_1)^2+1^2}\,\sqrt{(0.2\,\omega_1)^2+1^2}} = 1$$

or $$\frac{1}{\omega_1^2\,(0.25\omega_1^2+1)\,(0.04\omega_1^2+1)} = 1$$

$\therefore$ $$\omega_1^2\,(0.25\omega_1^2+1)\,(0.04\omega_1^2+1) - 1 = 0$$

put $\omega_1^2 = x$

$\therefore$ $$x\,(0.25\,x+1)\;(0.04\,x+1) - 1 = 0$$

Using trial and error method above equation is nearly satisfied at x = 0.8062 and acceptable being +ive.

$\therefore$ $$\omega_1 = \sqrt{0.8062} = 0.897 \text{ rad./sec.}$$

Phase margin $= 180° - \angle G_1(j\omega_1)$

$= 180° - \tan^{-1} 0.5\omega_1 - \tan^{-1} 0.2\omega_1$

$= 180° - \tan^{-1} 0.5 \times 0.897 - \tan^{-1} 0.2 \times 0.897$

$= 180° - 24.156° - 10.17° = 55.6°$. **Ans.**

The closed-loop system is stable with a phase margin 55.6°.

(*b*) To make the system marginally stable, the phase angle to be provided by the time delay element is – 55.6°.

The phase contributed with the insertion of time delay element ($e^{-j\omega T}$) is $-\dfrac{\omega_1 T \times 180}{\pi}$

$$\therefore \qquad -\frac{\omega_1 T \times 180}{\pi} = -55.6$$

$$\therefore \qquad T = \frac{55.6 \times \pi}{\omega_1 \times 180}$$

$$\omega_1 = 0.897 \text{ rad./sec.}$$

$$\therefore \qquad T = \frac{55.6 \times \pi}{0.897 \times 180} = 1.08 \text{ sec.} \quad \textbf{Ans.}$$

Example 7.17.17. *The open-loop transfer function of a unity feedback control system is given below :*

$$G(s) = \frac{10Ke^{-sT}}{s(s+1)(s+7)}$$

(a) For K = 1, determine T such that the system is marginally stable.

(b) For T = 1 sec., determine the limiting value of K such that the system is marginally stable.

Solution. (*a*) For $T = 0$ and $K = 1$ determine the gain crossover frequency ω_1

$$\therefore \qquad G(j\omega) = \frac{10}{j\omega(j\omega+1)(j\omega+7)}$$

At the phase crossover frequency $\omega = \omega_1$; $|G(j\omega_1)| = 1$,

$$\left|\frac{10}{j\omega_1(j\omega_1+1)(j\omega_1+7)}\right| = 1$$

or

$$\frac{10}{\omega_1\sqrt{\omega_1^2+1^2}\sqrt{\omega_1^2+7^2}} = 1$$

or

$$\frac{10}{\omega_1\sqrt{\omega_1^2+1}\sqrt{\omega_1^2+49}} = 1$$

By inspection $\omega_1 = 1$ rad./sec.

$$\angle G(j\omega) = -90° - \tan^{-1}\frac{\omega}{1} - \tan^{-1}\frac{\omega}{7}$$

$$\therefore \qquad G(j\omega_1) = -90° - \tan^{-1}\omega_1 - \tan^{-1}\frac{\omega_1}{7}$$

$\omega_1 = 1$ rad./sec.

$$\therefore \qquad G(j\omega_1) = -90° - \tan^{-1}1 - \tan^{-1}\frac{1}{7}$$

$$= -90° - 45° - 8.13° = -141.13°$$

Incorporating time delay element $e^{-j\omega T}$, the condition for marginal stability is given below :

$$\angle G(j\omega_1) + \angle e^{-j\omega T} = -180°$$

$$\therefore \quad -141.13 - \frac{\omega_1 T \times 180}{\pi} = -180$$

$$\because \quad \omega_1 = 1 \text{ rad./sec.}$$

$$\therefore \quad -141.13 - \frac{1 \times T \times 180}{\pi} = 180$$

$$\therefore \quad T = \frac{(180 - 141.13)\,\pi}{1 \times 180} = 0.678 \text{ sec.} \quad \textbf{Ans.}$$

(*b*) $T = 1$ sec.

$$\therefore \quad G(s) = \frac{10Ke^{-s.1}}{s\,(s+1)\,(s+7)}$$

Put $s = j\omega$

$$\therefore \quad G(j\omega) = \frac{10Ke^{-j\omega}}{j\omega\,(j\omega+1)\,(j\omega+7)}$$

$$\angle G(j\omega) = -\frac{\omega \times 180}{\pi} - 90° - \tan^{-1}\frac{\omega}{1} - \tan^{-1}\frac{\omega}{7}$$

$$= -\frac{\omega \times 180}{\pi} - 90° - \tan^{-1}\frac{\omega + \dfrac{\omega}{7}}{1 - \omega \cdot \dfrac{\omega}{7}}$$

$$= -\frac{\omega \times 180}{\pi} - 90° - \tan^{-1}\frac{8\omega}{7 - \omega^2}; \qquad \omega < \sqrt{7}$$

At the phase crossover frequency ω_2, $\angle G(j\omega_2) = -180°$

$$\therefore \quad -\frac{\omega_2 \times 180}{\pi} - 90° - \tan^{-1}\frac{8\omega_2}{7 - \omega_2^2} = -180°$$

or

$$= -57.3\,\omega_2 - \tan^{-1}\frac{8\omega_2}{7 - \omega_2^2} = -90°$$

Solving by trial-error method : $\omega_2 = 0.79$ rad./sec.

$$|G(j\omega_2)| = \left|\frac{10K}{j\omega_2\,(j\omega_2+1)\,(j\omega_2+7)}\right|$$

$$= \frac{10K}{\omega_2\sqrt{\omega_2^2 + 1^2}\,\sqrt{\omega_2^2 + 7^2}}$$

$\omega_2 = 0.79$ rad./sec.

$$\therefore \quad |G(j0.79_2)| = \frac{10K}{0.79\,\sqrt{(0.79)^2 + 1^2}\,\sqrt{(0.79)^2 + 7^2}} = \frac{K}{0.708}$$

For stability $\quad \dfrac{K}{0.708} < 1;\ K < 0.708$

For marginally stability : $K = 0.708$. **Ans.**

Example 7.17.18. *The open-loop transfer function of a unity feedback control system is given by*

$$G(s) = \frac{K}{s\,(0.5\,s+1)\,(0.25\,s+1)}$$

assuming K = 1 sketch the gain phase plot. Determine the value of K such that
(a) G.M. = 8 db and (b) P.M. = 20°.

Solution. The gain (magnitude) in db is given by

$$20 \log_{10} |G(j\omega)| = 20 \log_{10} \left| \frac{K}{j\omega\,(j\,0.5\,\omega+1)\,(j\,0.25\,\omega+1)} \right|$$

$K = 1$ (given)

$$\therefore \qquad 20 \log_{10} |G(j\omega)| = 20 \log_{10} \left| \frac{1}{j\omega\,(j\,0.5\,\omega+1)\,(j\,0.25\,\omega+1)} \right|$$

$$= 20 \log_{10} \left[\frac{1}{\omega\,\sqrt{(0.5\,\omega)^2+1^2}\,\sqrt{(0.25\,\omega)^2+1^2}} \right]$$

The phase angle is given by

$$\angle G(j\omega) = -\,90° - \tan^{-1} 0.5\,\omega - \tan^{-1} 0.25\,\omega$$

$|G(j\omega)|$ db and $\angle G(j\omega)$ for ω are calculated and tabulated below :

ω rad/sec	0.1	0.2	0.5	1	2.5	5	7
$20 \log_{10} \lvert G(j\omega)\rvert$ db	20	14	5.7	– 1.2	– 13.5	– 2.6	– 34
$\angle G(j\omega)°$	– 94	– 98	– 111	– 130	– 173	– 209	– 224

The gain phase plot is drawn as shown in Fig. 7.17.14. It is noted from the gain phase plot that

(*a*) To have G.M. = 8 db the gain is to be increased by (15 db – 8 db) = 7 db

i.e. the gain *K* is to be increased by a factor of

$$\text{antilog}_{10} \left(\frac{7}{20} \right) = 2.23$$

∴ For G.M. = 8 db :

$$K = 1 \times 2.23 = 2.23. \quad \textbf{Ans.}$$

(*b*) To have P.M. = 20° the gain is to be increased by 8.6 db *i.e.* the gain *K* is to be increased by a factor of

$$\text{antilog}_{10} \left(\frac{8.6}{20} \right) = 2.69$$

∴ For P.M. = 20° :

$$K = 1 \times 2.69 = 2.69. \quad \textbf{Ans.}$$

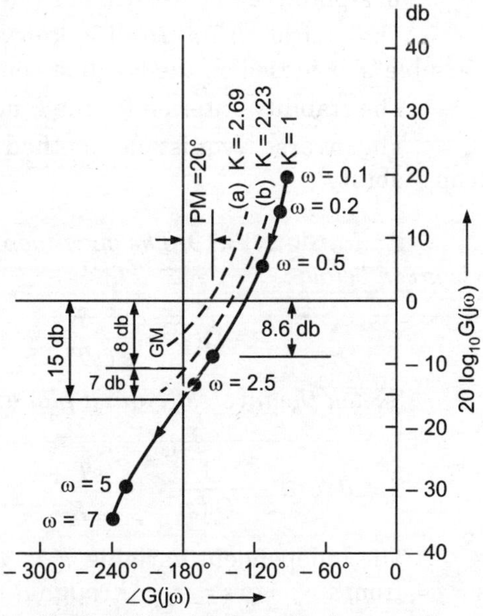

Fig. 7.17.14. Gain phase plot for example

7.17.1. Inverse Nyquist Plot

The open-loop transfer function of a feedback control system is given below :

$$G(s)\,H(s) = \frac{K\,(s+z_1)(s+z_2)\ldots(s+z_m)}{(s+p_1)(s+p_2)\ldots(s+p_n)} \qquad m \le n \qquad \ldots(1)$$

$$1 + G(s)\,H(s) = 1 + \frac{K\,(s+z_1)(s+z_2)\ldots(s+z_m)}{(s+p_1)(s+p_2)\ldots(s+p_n)}$$

$$1 + G(s)\,H(s) = \frac{(s+p_1)(s+p_2)\ldots(s+p_n) + K\,(s+z_1)(s+z_2)\ldots(s+z_m)}{(s+p_1)(s+p_2)\ldots(s+p_n)} \qquad \ldots(2)$$

The numerator of (2) can be factored as below :

$$\boxed{1 + G(s)\,H(s) = \frac{(s+z'_1)(s+z'_2)\ldots(s+z'_m)}{(s+p_1)(s+p_2)\ldots(s+p_n)}} \qquad \ldots(3)$$

Dividing (3) by (1)

$$1 + \frac{1}{G(s)\,H(s)} = \frac{(s+z'_1)(s+z'_2)\ldots(s+z'_m)}{(s+p_1)(s+p_2)\ldots(s+p_n)} \times \frac{(s+p_1)(s+p_2)\ldots(s+p_n)}{K\,(s+z_1)(s+z_2)\ldots(s+z_m)} \qquad \ldots(4)$$

or

$$\boxed{1 + \frac{1}{G(s)\,H(s)} = \frac{(s+z'_1)(s+z'_2)\ldots(s+z'_m)}{K\,(s+z_1)(s+z_2)\ldots(s+z_m)}}$$

On comparing (4) and (3), it is noted that the zeros (roots of the characteristic equation $z'_1, z'_2, \ldots z'_m$) of the function $1 + G(s)\,H(s)$ and $1 + \dfrac{1}{G(s)\,H(s)}$ are same.

Therefore, as per Nyquist's method, presence of the zeros of $1 + G(s)\,H(s)$ in a specified region of s-plane can be identified from the plot of $1/[G(s)\,H(s)]$.

The plot of $1/[G(s)\,H(s)]$ is known as inverse Nyquist plot wherein the independent variable 's' is varied on the Nyquist contour.

The stability criterion for the function $1/[G(s)\,H(s)]$ is the same as that for $[G(s)\,H(s)]$.

The inverse Nyquist plot method is applied to following examples to determine closed-loop stability.

Example 7.17.19. *The open-loop transfer function of a unity feedback control system is given below :*

$$G(s) = \frac{K}{s\,(sT+1)}$$

Sketch the inverse Nyquist plot and determine the closed-loop stability.

Solution. $$\frac{1}{G\,(s)} = \frac{s\,(sT+1)}{K}$$

The independent variable 's' is varied in s-plane as per Nyquist path from $s = j0$ to $s = j\infty$, from $s = j\infty$ to $s = -j\infty$ through $R \to \infty$ by a semicircle to cover the entire RH-side, from $s = -j\infty$ to $s = -j0$ and from $s = -j0$ to $s = +j0$ along a semicircle of radius $\delta \to 0$ such that any pole of $1/[G(s)\,H(s)]$ at the origin of s-plane is excluded.

Put $s = j\omega$

$$\therefore \qquad \frac{1}{G(j\omega)} = \frac{j\omega(j\omega T + 1)}{K}$$

$$\left|\frac{1}{G(j\omega)}\right| = \frac{\omega\sqrt{\omega^2 T^2 + 1^2}}{K}$$

and $$\angle 1/G(j\omega) = 90° + \tan^{-1}\omega T$$

(1) As $\omega \to 0$

$$\left|\frac{1}{G(j0)}\right| \to 0$$

$$\angle 1/G(j0) \to 90°$$

(2) As $\omega \to \infty$

$$\left|\frac{1}{G(j\infty)}\right| \to \infty$$

$$\angle 1/G(j\infty) \to 180°$$

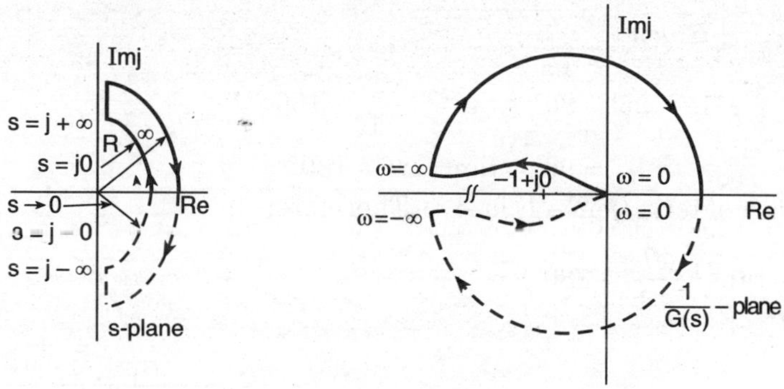

Fig. 7.17.15. Inverse Nyquist plot for example 7.17.19.

(3) Since, number of poles of $1/G(s)$ at the origin of s-plane is nil. Therefore, corresponding points $\omega = 0$ and $\omega = -0$ in $1/G(s)$-plane are coincident.

(4) $$\left.\frac{1}{G(j\omega)}\right|_{\omega\to\infty} \to \infty\, e^{j2\theta}$$

In the s-plane θ changes from $\frac{\pi}{2}$ to $-\frac{\pi}{2}$ rad. as ω changes from ∞ to $-\infty$. Thus, the change in θ is $-\frac{\pi}{2} - \left(\frac{\pi}{2}\right) = -\pi$ rad. The corresponding change in $\angle 1/G(j\omega)$ is $2(-\pi) = -2\pi$ rad.

Hence, the inverse Nyquist plot is completed by a circular path of radius $\to \infty$ through an angle -2π rad. ($-360°$) as shown in Fig. 7.17.15.

P_+ : Number of poles $1/G(s)$ having positive real part = Nil.

As per the Nyquist plot, the critical point $(-1 + j0)$ is not encircled.

$$\therefore \quad N = 0$$
$$N = P_+ - Z_+$$
$$\therefore \quad 0 = 0 - Z_+$$
$Z_+ = 0$. The closed-loop system is stable. **Ans.**

Example 7.17.20. *Apply Nyquist criterion to determine the stability of a unity feedback control system having given :*

$$G(s) = \frac{K}{s(sT-1)}.$$

Solution.
$$\frac{1}{G(s)} = \frac{s(sT-1)}{K}$$

Put $s = j\omega$

$$\therefore \quad \frac{1}{G(j\omega)} = \frac{j\omega(j\omega T - 1)}{K}$$

$$\left|\frac{1}{G(j\omega)}\right| = \frac{\omega\sqrt{\omega^2T^2 + 1^2}}{K}$$

and
$$\angle 1/G(j\omega) = 90° + \tan^{-1}\frac{\omega T}{-1} - 180°$$
$$= 90° - \tan^{-1}\omega T - 180°$$

(**Note :** The real part of term $(j\omega T - 1)$ is –ive. Therefore, the phase angle is $\tan^{-1}\frac{\omega T}{-1} - 180°$)

(1) As $\omega \to 0$

$$\left|\frac{1}{G(j0)}\right| \to 0$$
$$\angle 1/G(j0) \to -90°$$

(2) As $\omega \to \infty$

$$\left|\frac{1}{G(j\infty)}\right| \to \infty$$
$$\angle 1/G(j\infty) \to -180°$$

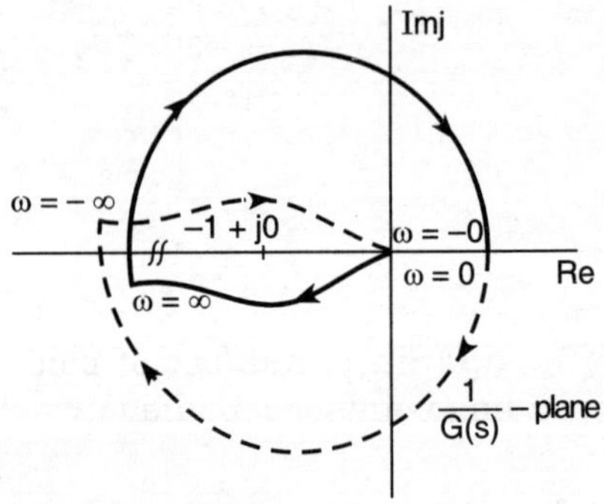

Fig. 7.17.16. Inverse Nyquist plot for example 7.17.20.

(3) Since, the number of poles of $1/G(s)$ at the origin of s-plane is nil. Therefore, points corresponding to $\omega = -0$ and $\omega = 0$ are coincident in $1/G(s)$-plane.

(4)
$$\left.\frac{1}{G(j\omega)}\right|_{\omega \to \infty} \to \infty\, e^{j2\theta}$$

In the s-plane θ changes from $\frac{\pi}{2}$ to $-\frac{\pi}{2}$ rad. as ω varied from ∞ to $-\infty$. Thus, the change in θ in s-plane is $-\frac{\pi}{2} - \left(\frac{\pi}{2}\right) = -\pi$ rad. $(-180°)$.

The corresponding change in $\angle 1/G(j\omega)$ is $2(-\pi) = -2\pi$ rad ($-360°$). Hence, the inverse Nyquist plot is completed by a circular path of radius $\to \infty$ through an angle -2π rad. ($-360°$) as shown in Fig. 7.17.16.

P_+ : Number of poles of $1/G(s)$ having positive real part = Nil

As per Nyquist plot, the critical point $(-1 + j0)$ is encircled twice in the clockwise direction.

$$\therefore \quad N = -2$$

$$N = P_+ - Z_+$$

$$\therefore \quad -2 = 0 - Z_+$$

$Z_+ = 2$. The closed-loop system is unstable.

7.17.2. 1/*M*-Circles (Inverse *M*-circles) and constant phase angle loci

The closed-loop transfer function of a unity feedback control system is given below :

$$\frac{C(s)}{R(s)} = \frac{G(s)}{1+G(s)}$$

In sinusoidal form above transfer function can be expressed as

$$\frac{C(j\omega)}{R(j\omega)} = \frac{G(j\omega)}{1+G(j\omega)}$$

Let

$$\frac{C(j\omega)}{R(j\omega)} = Me^{n\phi}$$

where, M : magnitude ϕ : phase angle

$$\therefore \quad \frac{R(j\omega)}{C(j\omega)} = \frac{1}{G(j\omega)} + 1 = \frac{1}{M} e^{-j\phi}$$

The function $1/G(j\omega)$ is (say) plotted in complex plane as shown in Fig. 7.17.17. Any point $(x + jy)$ on the plot satisfies

$$\frac{R(j\omega)}{C(j\omega)} = \frac{1}{G(j\omega)} + 1$$

$$\therefore \quad \frac{R(j\omega)}{C(j\omega)} = x + jy + 1$$

or

$$x + jy + 1 = \frac{1}{m} e^{-j\phi}$$

$$\therefore \quad (x+1)^2 + y^2 = \frac{1}{M}$$

For constant value of M, the equation above is of a circle having centre at $(-1 + j0)$ and radius $1/M$.

$$\angle \frac{R(j\omega)}{C(j\omega)} = \tan^{-1} \frac{y}{x+1}$$

or

$$\angle \frac{C(j\omega)}{R(j\omega)} = \tan^{-1} \frac{y}{x+1}$$

$$\therefore \qquad \phi = -\tan^{-1} \frac{y}{x+1}$$

The equation above is of constant phase angle.

The constant $\frac{1}{M}$ circles and constant phase angle loci are shown in Fig. 7.17.18.

The point of intersection of $1/G(j\omega)$ plot and $\frac{1}{M}$ circle/constant phase angle locus give the value of $\frac{1}{M}/\phi$ at the corresponding frequency ω. The $\frac{1}{M}$ circle tangential to $1/G(j\omega)$ plot gives the value of $\frac{1}{M_r}$ and the frequency at the tangential point is ω_r.

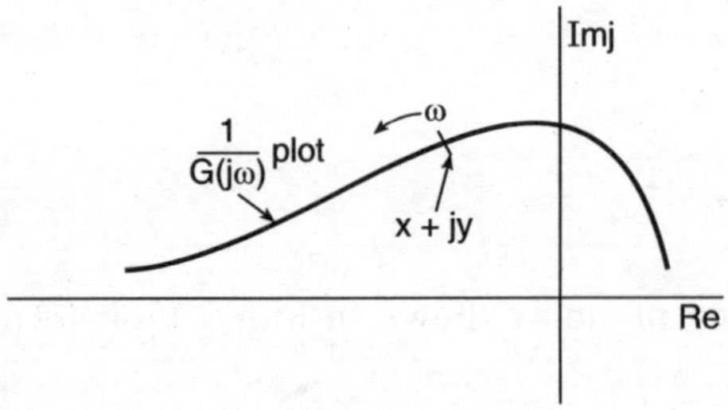

Fig. 7.17.17. $1/G(j\omega)$ plot.

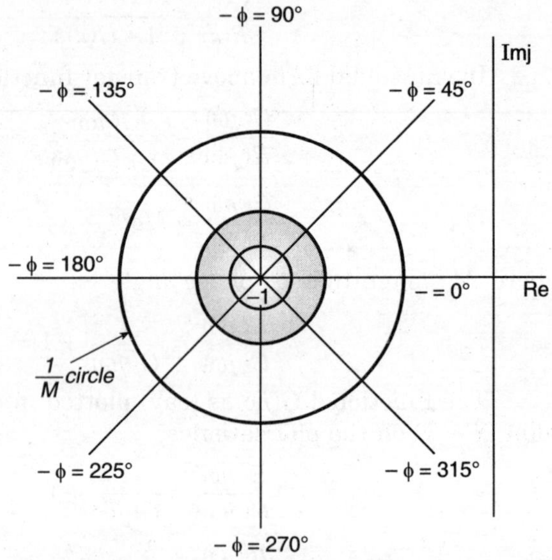

Fig. 7.17.18. $1/M$ circles and constant phase angle loci.

7.17.3. Gain adjustment using inverse polar plot

The procedure to determine the gain for a desired value of M_r using inverse polar plot is explained below.

The inverse transfer function $1/G(j\omega)$ is written below :

$$\frac{1}{G(j\omega)} = \frac{(j\omega + p_1)(j\omega + p_2)\ldots}{K\,(j\omega + z_1)(j\omega + z_2)\ldots}$$

(1) Put $K = 1$ and plot $1/G(j\omega)$ from $\omega = 0$ to $\omega = \infty$. Draw the inverse polar plot (Fig. 7.17.19).

(2) Draw the line OAB such that $\phi = \sin^{-1}\frac{1}{M_r}$.

(3) Draw a circle such that it is tangent to both $1/G(j\omega)$ plot and the line OAB. The centre of this circle is located at C.

(4) If the circle drawn is to be $\frac{1}{M_r}$ circle, then its centre must be $(-1 + j0)$. Therefore, distance $OC = 1$, for this is to be possible, the new gain $K' = oc$. Thus, the original gain K is to be changed by a factor $\frac{K'}{K}$.

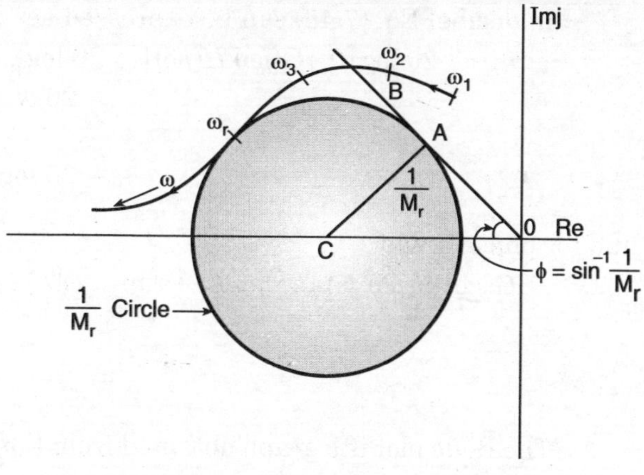

Fig. 7.17.19. Gain adjustment using $1/M$ circle.

7.18 BODE PLOT

The Bode plot method gives a graphical procedure for determining the stability of a control system based on sinusoidal frequency response. The transfer function of a system for sinusoidal input response can be obtained by substituting $j\omega$ in place of Laplace operator "s". Therefore, if the open-loop transfer function of a system is $G(s)\,H(s)$, the corresponding sinusoidal open-loop transfer function is $G(j\omega)\,H(j\omega)$ which can be expressed in the form of magnitude and phase angle.

The variation of the magnitude of sinusoidal transfer function expressed in decibel and the corresponding phase angle in degrees being plotted w.r.t. frequency on a logarithmic scale (*i.e.* $\log_{10}\omega$) in rectangular axes. The plot thus obtained is known as Bode Plot. The logarithmic Bode plot has an advantage of being approximated by asymptotic straight lines.

Relative stability of a closed-loop control system can be conveniently assessed by plotting its open-loop transfer function by Bode plot method. The gain margin and phase margin is determined directly from the Bode plot.

For complicated systems Bode plot can also be plotted from experimental results and therefrom the transfer function of the system under test can be determined.

7.18.1. Bode plot (Logarithmic plot) for transfer functions

The open-loop transfer function of a closed-loop control system can be expressed in a form given below :

$$G(s)\,H(s) = \frac{K\,[(1+sT_1)\,(1+sT_2)\,...]\,\omega_n^2}{s^N\,[(1+sT_a)\,(1+sT_b)\,...\,(s^2+2\zeta\omega_n s+\omega_n^2)]} \qquad ...(7.47)$$

The sinusoidal form a transfer function (7.47) is obtained by substituting $s = j\omega$, therefore

$$G(j\omega)\,H(j\omega) = \frac{K\,[(1+j\omega T_1)\,(1+j\omega T_2)\,...]\,\omega_n^2}{(j\omega)^N\,\{(1+j\omega T_a)\,(1+j\omega T_b)\,...\,[(\omega_n^2-\omega^2)+j2\zeta\omega_n\omega]\}} \qquad ...(7.48)$$

The procedure for drawing the Bode plot for transfer function (7.48) is explained below :

In decibel Eq. (7.48) can be expressed as

$$20 \log 10 \ |G(j\omega)\, H(j\omega)| = 20 \log_{10} K + 20 \log_{10} |\ 1 + j\omega\, T_1| + ...$$
$$... - 20\, N \log_{10} |j\omega| - 20 \log_{10} |\ 1 + j\omega\, T_a| + ...$$
$$... - 20 \log_{10} \left| \frac{(\omega_n^2 - \omega^2)}{\omega_n^2} + \frac{(j\, 2\zeta\omega_n \omega)}{\omega_n^2} \right| \quad ... (7.49)$$

and for phase angle

$$\angle G(j\omega)\, H(j\omega) = \tan^{-1} \omega\, T_1 + \tan^{-1} \omega\, T_2 + ... - N \angle 90° - \tan^{-1} \omega\, T_a - \tan^{-1} \omega\, T_b + ...$$
$$- \tan^{-1} \left[\frac{2\zeta\omega_n \omega}{(\omega_n^2 - \omega^2)} \right] \quad ...(7.50)$$

The Bode plot is a graph obtained from Eqs. (7.49) and (7.50) consisting of two parts as follows :

(*i*) Magnitude of $G(j\omega)\, H(j\omega)$ in decibel, [*i.e.* $20 \log_{10} |G(j\omega)\, H(j\omega)|$] versus $\log_{10} \omega$.

(*ii*) Phase angle $\angle G(j\omega)\, H(j\omega)$ versus $\log_{10} \omega$.

For plotting magnitude versus $\log_{10} \omega$ plot each term of Eq. (7.49) is considered separately and graphs are drawn. To obtain final plot the contributions due to each term are added as described below :

7.18.2. Graphs for the gain term K

The magnitude in decibel for the term of K is given by

$$K\,(\text{db}) = 20 \log_{10} (K) \quad ...(7.51)$$

Eq. (7.51) indicates that the magnitude is independent of $\log_{10} \omega$ and as K is considered positive real ; the phase angle is always zero whatever may be the value of ω. The graphs are shown in Fig. 7.18.1.

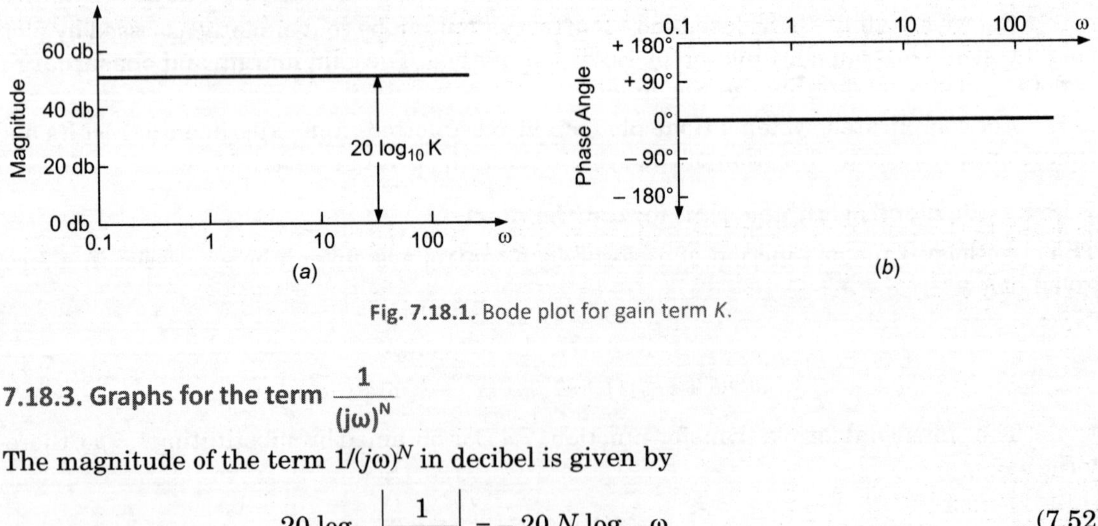

Fig. 7.18.1. Bode plot for gain term *K*.

7.18.3. Graphs for the term $\frac{1}{(j\omega)^N}$

The magnitude of the term $1/(j\omega)^N$ in decibel is given by

$$20 \log_{10} \left| \frac{1}{(j\omega)^N} \right| = -\, 20\, N \log_{10} \omega \quad ...(7.52)$$

The phase angle is given by $\angle \frac{1}{(j\omega)^N} = -\, 90°\, N$ ...(7.53)

In view of Eqs. (7.52) and (7.53) the graphs are shown in Fig. 7.18.2. The graph for the magnitude versus $\log_{10} \omega$ is a straight line having a slope of – 20 N db/decade.

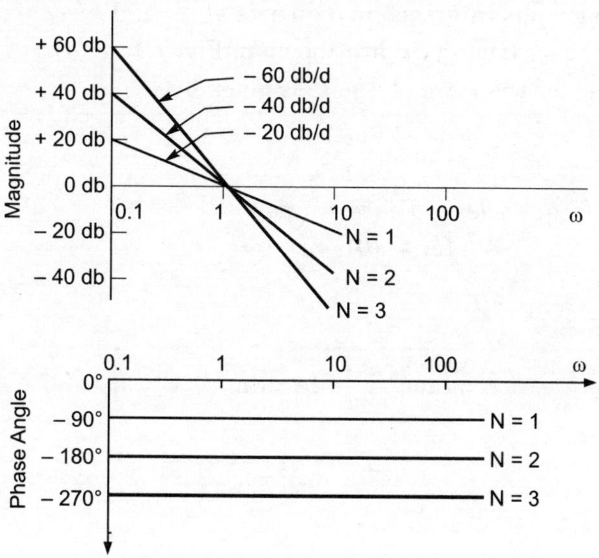

Fig. 7.18.2. Bode plot for the term $1/(j\omega)^N$.

As the term $1/(j\omega)^N$ has only imaginary term in the denominator the phase angle is – 90 N°.

7.18.4. Graphs for the Term $(1 + j\omega T)$

The magnitude in decibel for the term $(1 + j\omega T)$ is given by

$$20 \log_{10} |(1 + j\omega T)| = 20 \log_{10} \sqrt{1 + \omega^2 T^2} \quad ...(7.54)$$

Consider following two cases :

(*i*) $\omega T << 1$ (very low frequencies), ωT is negligible as compared to 1.

$$\therefore \quad 20 \log_{10} |(1 + j\omega T)| \simeq 20 \log_{10} 1 = 0 \text{ db} \quad ...(7.55)$$

(*ii*) $\omega T >> 1$ (very high frequencies), 1 is negligible as compared to ωT

$$\therefore \quad 20 \log_{10} |(1 + j\omega T)| \simeq 20 \log_{10} \sqrt{\omega^2 T^2} = 20 \log_{10} \omega T$$

$$= 20 \log_{10} \omega - 20 \log_{10} (1/T) \quad ...(7.56)$$

In view of Eqs. (7.55) and (7.56) the graph for case (*i*) lies on 0 db axis, whereas for case (*ii*) the graph has a slope of 20 db/decade*. These two graphs intersect on 0 db axis at a point determined by equating the R.H.S. of Eq. (7.56) to zero.

---

*On logarithmic scale a decade means a ratio of 10. Sometimes the term octave is used to represent a ratio of 2 on logarithmic scale. The relation between octave and decade is obtained below

$$\text{Octave} = \frac{\log_{10} (10)}{\log_{10} (2)} = \frac{1}{0.301} \text{ decade}$$

$$\therefore \quad \pm 20 \text{ N db/decade} = \pm 20 \text{ N} \times 0.30 = \pm 6 \text{ N db/octave.}$$

$$\therefore \qquad 0 = 20 \log_{10} \omega - 20 \log_{10} (1/T)$$

or $$20 \log_{10} \omega = 20 \log_{10} (1/T), \quad \therefore \quad \omega = (1/T)$$

Hence, the two graphs intersect on 0 db axis at $\omega = 1/T$.

The graphs for cases (*i*) and (*ii*) are shown in Fig. 7.18.3 (*a*).

The phase angle for the term $(1 + j\omega T)$ is given by

$$\phi = \tan^{-1}\left(\frac{\omega T}{1}\right)$$

(*i*) At very low frequencies ωT is very small

$$\therefore \qquad \phi \cong \tan^{-1}(0) \quad \text{or} \quad \phi = 0°$$

(*ii*) At $\omega = \dfrac{1}{T}$

$$\phi = \tan^{-1}\left(\frac{1}{T} \cdot 1\right) = \tan^{-1}(1)$$

or $$\phi = 45°$$

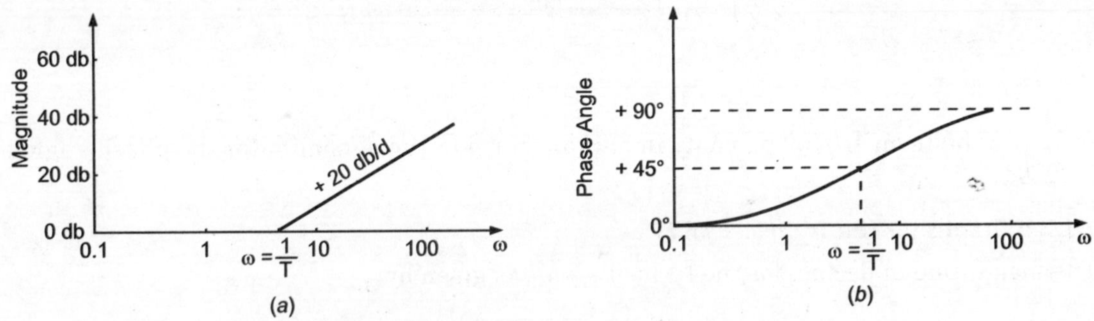

Fig. 7.18.3. Bode plot for the term $(1 + j\omega T)$

(*iii*) At very high frequencies ωT is very large

$$\therefore \qquad \phi \cong \tan^{-1}(\infty)$$

or $$\phi \cong 90°$$

The graph for phase angle is shown in Fig. 7.18.3 (*b*).

7.18.5. Graphs for the term 1/(1 + jωT)

The magnitude for the term $(1/(1 + j\omega T)$ in decibel is given by :

$$10 \log_{10} \left|\frac{1}{(1 + j\omega T)}\right| = 20 \log_{10} (1/\sqrt{1 + \omega^2 T^2})$$

$$= -20 \log_{10} \sqrt{1 + j\omega^2 T^2} \qquad \text{...(7.57)}$$

Consider following two cases :

(*i*) $\omega T << 1$ (very low frequencies), ωT is negligible as compared to 1.

$$\therefore \quad 20 \log_{10} \left|\frac{1}{(1 + j\omega T)}\right| \simeq -20 \log_{10} 1 = 0 \text{ db} \qquad \text{...(7.58)}$$

(*ii*) $\omega T >> 1$ (very high frequencies), 1 is negligible as compared to ωT.

$$\therefore \quad 20 \log_{10} \left| \frac{1}{(1 + j\omega T)} \right| \simeq -20 \log_{10} \sqrt{\omega^2 T^2} = -20 \log_{10} \omega T$$

$$= -20 \log_{10} \omega + 20 \log_{10} \left(\frac{1}{T}\right) \qquad ...(7.59)$$

In view of Eqs. (7.58) and (7.59) graph for case (*i*) lies on 0 db axis whereas for case (*ii*) the graph has a slope of – 20 db/decade. These two graphs intersect on 0 db axis at a point determined by equating R.H.S. of Eq. (7.59) to zero.

$$\therefore \quad 0 = -20 \log_{10} \omega + 20 \log_{10} \left(\frac{1}{T}\right)$$

or

$$20 \log_{10} \omega = 20 \log_{10} \left(\frac{1}{T}\right), \quad \therefore \quad \omega = \frac{1}{T}$$

Hence, the two graphs intersect on 0 db axis at $\omega = \frac{1}{T}$. The graphs for cases (*i*) and (*ii*) are shown in Fig. 7.18.4 (*a*).

The phase angle for the term $1/(1 + j\omega T)$ is given by

$$\phi = -\tan^{-1} \left(\frac{\omega T}{1}\right)$$

(*i*) At very low frequencies ωT is very small

$$\therefore \quad \phi \cong -\tan^{-1}(0)$$

or

$$\phi \cong 0°$$

(*ii*) At $\quad \omega = \frac{1}{T}$

$$\phi = -\tan^{-1} \left(\frac{1}{T} \cdot T\right) = -\tan^{-1}(1)$$

or

$$\phi = -45°$$

(*iii*) At very high frequencies ωT is very large

$$\therefore \quad \phi \cong -\tan^{-1}(\infty)$$

or

$$\phi \cong -90°$$

The graph for phase angle is shown in Fig. 7.18.4 (*b*).

The magnitude versus $\log_{10} \omega$ graphs shown in Figs. 7.18.3 and 7.18.4 for the terms $(1 + j\omega T)$ and $1/(1 + j\omega T)$ are asymptotic plots. A comparison of asymptotic plot and exact plot is shown in Fig. 7.18.5. The exact plot is shown dotted. The exact and asymptotic plots match each other at frequencies lower and higher than $\omega = 1/T$. The frequency $\omega = 1/T$ is called corner frequency. The maximum error between the exact and asymptotic plot occurs at corner frequency, this error is calculated below :

From asymptotic plot it is observed that the magnitude at $\omega = 1/T$ is 0 db.

From exact plot the magnitude in decibel for the term $(1 + j\omega T)$ at $\omega = 1/T$ is

$$20 \log_{10} \left| \left(1 + j\frac{1}{T} \cdot T\right) \right| = 20 \log_{10} \sqrt{1 + \frac{1}{T^2} \cdot T^2} = 20 \log_{10} \sqrt{2} = 3 \text{ db.}$$

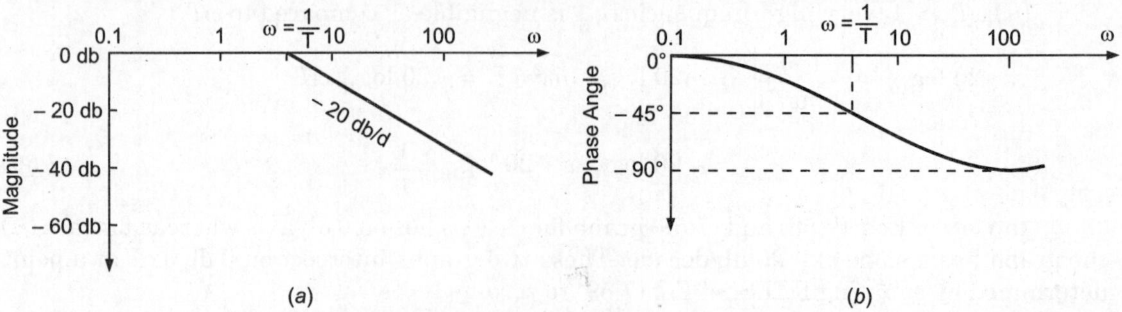

Fig. 7.18.4. Bode plot for the term $1/(1 + j\omega T)$.

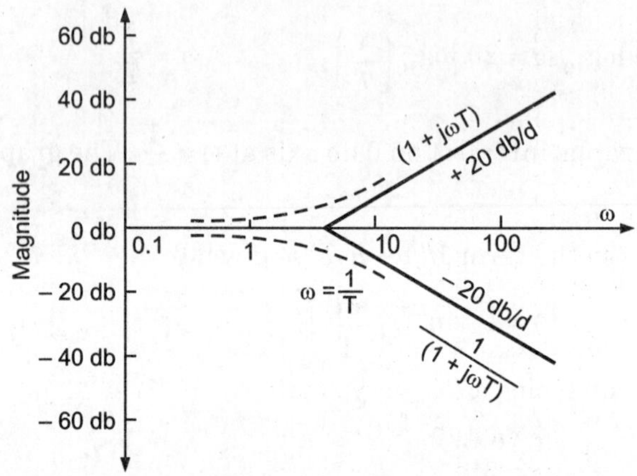

Fig. 7.18.5. Exact and asymptotic (approximate) bode plots for the terms $(1 + j\omega T)$ and $1/(1 + j\omega T)$.

Thus the maximum error due to asymptotic plot of the term $(1 + j\omega T)$ is 3 db. Similarly the maximum error for the term $1/(1 + j\omega T)$ is – 3 db. An error of ± 3 db is permissible.

For multiple first order terms such as $(1 + j\omega T)^{\pm N}$, the magnitude in decibel is given by

$$20 \log_{10} |(1 + j\omega T)^{\pm N}| = \pm 20 \text{ N} \log_{10} \sqrt{1 + \omega^2 T^2}$$

Consider the following two cases :

(*i*) $\omega T << 1$ very low frequencies, ωT is negligible as compared to 1.

$$\therefore \quad 20 \log_{10} |(1 + j\omega T)^{\pm N}| = \pm 20 \text{ N} \log_{10} \sqrt{1} = 0 \text{ db.}$$

(*ii*) $\omega T >> 1$ very high frequencies, 1 is negligible as compared to ωT.

$$\therefore \quad 20 \log_{10} |(1 + j\omega T)^{\pm N}| = \pm 20 \text{ N} \log_{10} \sqrt{\omega^2 T^2}$$

$$= \pm 20 \text{ N} \log_{10} \omega \mp 20 \text{ N} \log_{10} (1/T).$$

The intersection point of the plots for the cases (*i*) and (*ii*) is at $\omega = 1/T$.

The exact magnitude at $\omega = 1/T$ is given by

$$20 \log_{10} \left| \left(1 + j\frac{1}{T} \cdot T\right)^{\pm N} \right| = \pm 20 \text{ N} \log_{10} \sqrt{1 + \frac{1}{T^2} \cdot T^2}$$

$$= \pm 20 \text{ N} \log_{10} \sqrt{2} = \pm 3 \text{ N db.}$$

Thus for transfer function where multiple first order terms are involved the error between the asymptotic plot and exact plot is appreciable (± 3 N db) and the correction to the magnitude has to be applied at corner frequency occurring due to such multiple terms.

Further, if corner frequencies are very near to each other the error between the asymptotic plot and exact plot is appreciable and correction to the asymptotic plot have also to be applied in such cases.

The corrections are applied by referring to the error curve for the term $(1 + j\omega)^{\pm 1}$ shown in Fig. 7.18.6. The ± 3 db point is located at $\omega T = 1$ and ± 1 db points are located at $0.5\ \omega T$ and $2\omega T$ respectively, the points thus obtained are joined by a smooth curve.

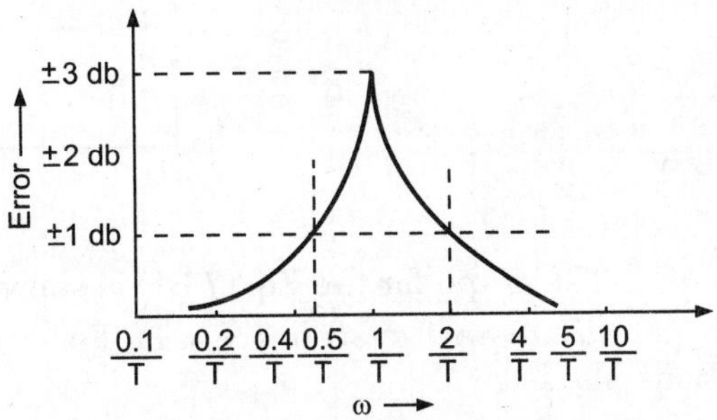

Fig. 7.18.6. Error curve for term $(1 + j\omega T)^{\pm 1}$.

In the case of multiple first order terms the phase angle is also multiplied by $\pm N$, therefore, at ω =/T

$$\angle(1 + j\omega T)^{\pm N} = \pm\ 45\ \text{N}°$$

and at high frequencies (*i.e.* $\omega T >> 1$)

$$\angle|(1 + j\omega T)^{\pm N}| = \pm\ 90\ \text{N}°$$

7.18.6. Initial slope of bode plot

The corner frequencies due to first order terms $(1 + j\omega T_1)$, $(1 + j\omega T_2)$..., $1/(1 + j\omega T_a)$ $1/(1 + j\omega T_b)$ ... etc. are given by

$$\omega = \frac{1}{T_1}, \frac{1}{T_2}, \ldots, \frac{1}{T_a}, \frac{1}{T_b} \ \ldots \text{etc.}$$

For the frequencies lower than the lowest corner frequency the contribution towards the gain (magnitude in decibel) of the transfer function is nil, therefore, the sinusoidal transfer function for frequencies lower than the lowest corner frequency can be expressed as follows :

$$G(j\omega)\ H(j\omega) \simeq \frac{K}{(j\omega)^N}$$

The magnitude of above expression in decibel can be expressed as

$$20 \log_{10}|G(j\omega)\ H(j\omega)| = 20 \log_{10}\left|\frac{K}{(j\omega)^N}\right|$$

$$= 20 \log_{10} K - 20\ \text{N} \log_{10} \omega \qquad \ldots(7.60)$$

As per above equation, the graph of magnitude (in decibel) versus $\log_{10} \omega$ has initial slope of – 20 N db/decade, as N denotes the type of the transfer function, therefore, the initial slope of Bode plot is decided by the type of a system.

(*i*) **Type 0 system (N = 0).** For a type 0 system the equation for the initial part of the Bode plot is

$$20 \log_{10}|G(j\omega)\ H(j\omega)| = 20 \log_{10}\left|\frac{K}{(j\omega)^{\circ}}\right| = 20 \log_{10} K - 20 \log_{10}(1)$$

$$= 20 \log_{10} K \qquad \ldots(7.61)$$

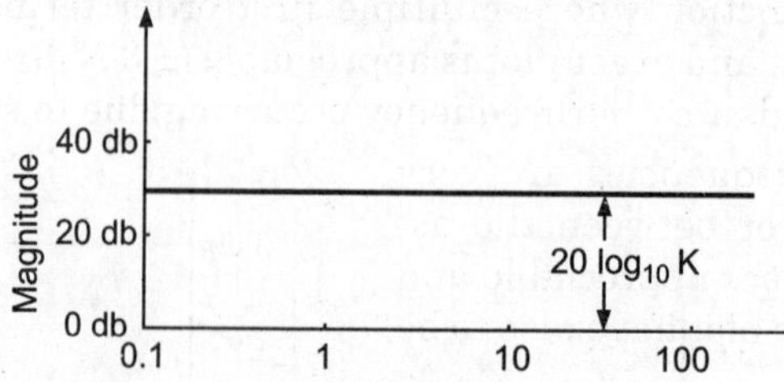

Fig. 7.18.7. Initial part of Bode plot for type 0 system.

The graph for the Eq. (7.61) is shown in Fig. 7.18.7.

(*ii*) **Type 1 system (N = 1).** For a type 1 system equation for the initial part of the Bode plot is

$$20 \log_{10} |G(j\omega)\, H(j\omega)| = 20 \log_{10} \left|\frac{K}{(j\omega)^1}\right|$$

$$= 20 \log_{10} K - 20 \log_{10} \omega \qquad \text{...(7.62)}$$

The graph for the Eq. (7.62) has a slope of – 20 db/decade and its intersection with 0 db axis is given by equating R.H.S. of Eq. (7.62) to zero.

$$\therefore \quad 0 = 20 \log_{10} K - 20 \log_{10} \omega$$

or $\quad 20 \log_{10} K = 20 \log_{10} \omega$

$$\therefore \quad \omega = K.$$

Hence, the graph intersects 0 db axis at $\omega = K$. The graph for Eq. (7.62) is shown in Fig. 7.18.8.

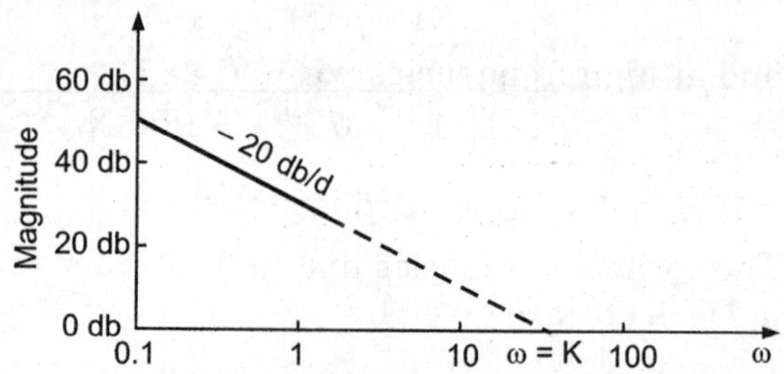

Fig. 7.18.8. Initial part of Bode plot for type 1 system.

(*iii*) **Type 2 system (N = 2).** For a type 2 system the equation for the initial part of the Bode plot is

$$20 \log_{10} |G(j\omega)\, H(j\omega)| = 20 \log_{10} \left|\frac{K}{(j\omega)^2}\right| = 20 \log_{10} K - 20 \log_{10} \omega^2$$

$$= 20 \log_{10} K - 40 \log_{10} \omega \qquad \text{...(7.63)}$$

The graph for the Eq. (7.63) has a slope of – 40 db/decade and its intersection with 0 db axis is given by equating R.H.S. of Eq. (7.63) to zero.

$$\therefore \quad 0 = 20 \log_{10} K - 40 \log_{10} \omega$$

or $\quad 20 \log_{10} K = 20 \log_{10} \omega^2$

$$\therefore \quad \omega^2 = K$$

or $\quad \omega = K^{1/2}$

Hence the graph intersects 0 db axis at $\omega = \sqrt{K}$. The graph for Eq. (7.63) is shown in Fig. 7.18.9.

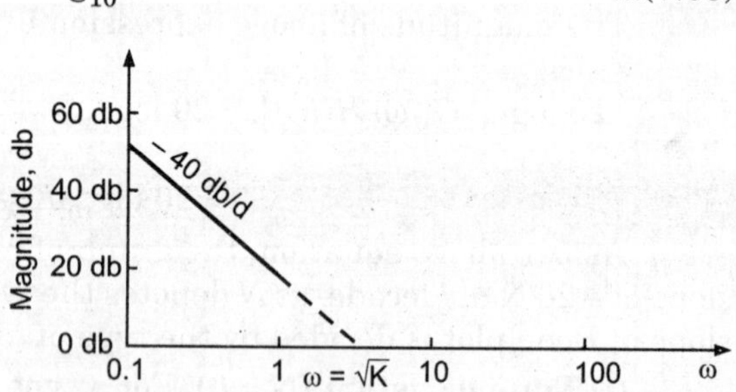

Fig. 7.18.9. Initial part of Bode plot for type 2 system.

Conclusions concerning initial slope of Bode plot derived from Eqs. (7.61), (7.62) and (7.63) are given in the table below :

Type of System N	*Initial slope*	*Intersection with 0 db axis at*
0	0 db/decade	Parallel to 0 db axis
1	– 20 db/decade	$\omega = K$
2	– 40 db/decade	$\omega = K^{1/2}$
3	– 60 db/decade	$\omega = K^{1/3}$
"	"	"
"	"	"
"	"	"
N	– 20 N db/decade	$\omega = K^{1/N}$

In the case of zeros of the open-loop transfer function $G(s)\,H(s)$ located at the origin of s-plane :

$$G(s)\,H(s) = \frac{K\,s^N\,(sT_1+1)\,(sT_2+1)\,...}{(sT_a+1)\,(sT_b+1)\,...}$$

N : the number of zeros of $G(s)\,H(s)$ located at the origin of s-plane.

For the frequencies lower than the lowest corner frequency the sinusoidal form of open-loop transfer function takes the form as below :

$$G(j\omega)\,H(j\omega) = \frac{K(j\omega)^N}{1}$$

It can be shown that the initial slope of the Bode plot is +20 Ndb/decade intersecting 0-db axis at $\omega = \dfrac{1}{(K)^{1/N}}$.

7.18.7. Determination of static error coefficients from initial slope of Bode plot

In sinusoidal form static error coefficients K_p, K_v and K_a described in section 6.7.1 can be written as

For type 0 system : $K_p = \lim\limits_{\omega \to 0} G(j\omega)\,H(j\omega) = K$ ...(7.64)

For type 1 system : $K_v = \lim\limits_{\omega \to 0} (j\omega)\,G(j\omega)\,H(j\omega) = K$ ...(7.65)

For type 2 system : $K_a = \lim\limits_{\omega \to 0} (j\omega)^2\,G(j\omega)\,H(j\omega) = K$ ...(7.66)

In view of Eqs. (7.64), (7.65), (7.66) and initial slope of Bode plot for type 0, type 1, and type 2 system respectively. K_p, K_v and K_a can be determined as follows :

K_p equals the value of K as determined from the initial part of log magnitude plot of a type 0 system.

K_v is given by the value of ω at which the initial slope (– 20 db/decade) of log magnitude plot of a type 1 system intersects with 0 db axis.

K_a is related to ω at which the initial slope (– 40 db/decade) of log magnitude plot of a type 2 system intersects with 0 db axis and the relation is $K_a = \omega^2$.

7.18.8. Graphs for quadratic term $\omega_n^2/[(\omega_n^2 - \omega^2) + j2\zeta\omega_n\omega]$

The magnitude of the term $\dfrac{\omega_n^2}{[(\omega_n^2 - \omega^2) + j2\zeta\,\omega_n\,\omega]}$ in decibel is given by

$$20 \log_{10} \left| \frac{\omega_n^2}{[(\omega_n^2 - \omega^2) + j2\zeta\,\omega_n\,\omega]} \right| = 20 \log_{10} \left| \frac{1}{\left(1 - \dfrac{\omega^2}{\omega_n^2}\right) + j2\,\zeta\left(\dfrac{\omega}{\omega_n}\right)} \right|$$

$$= 20 \log_{10} \left| \frac{1}{\sqrt{\left(1 - \dfrac{\omega^2}{\omega_n^2}\right)^2 + \left(2\zeta\,\dfrac{\omega}{\omega_n}\right)^2}} \right| = -\,20 \log_{10} \left[\sqrt{\left(1 - \dfrac{\omega^2}{\omega_n^2}\right)^2 + \left(2\zeta\,\dfrac{\omega}{\omega_n}\right)^2} \right] \quad ...(7.67)$$

Consider the following two cases :

(*i*) $\omega <<< \omega_n$ (very low frequencies than ω_n)

The terms $\dfrac{\omega^2}{\omega_n^2}$ and $\dfrac{\omega}{\omega_n}$ in Eq. (7.67) can be neglected, therefore,

$$20 \log_{10} \left| \frac{\omega_n^2}{[(\omega_n^2 - \omega^2) + j2\zeta\,\omega_n\,\omega]} \right| \simeq -\,20 \log_{10} \sqrt{1}$$

$$\simeq -\,20 \log_{10} (1) = 0 \text{ db} \quad ...(7.68)$$

(*ii*) $\omega >> \omega_n$ (very high frequencies than ω_n)

The terms 1 and $\left(2\zeta\,\dfrac{\omega}{\omega_n}\right)^2$ can be neglected in comparison with $\left(\dfrac{\omega^2}{\omega_n^2}\right)^2$ in Eq. (7.67),

therefore,

$$20 \log_{10} \left| \frac{\omega_n^2}{[(\omega_n^2 - \omega^2) + j2\zeta\,\omega_n\,\omega]} \right|$$

$$\simeq -\,20 \log_{10} \left| \sqrt{\left(-\frac{\omega^2}{\omega_n^2}\right)^2} \right| \simeq -\,20 \log_{10} \left(\frac{\omega^2}{\omega_n^2}\right)$$

$$\simeq -\,40 \log_{10} \omega + 40 \log_{10} \omega_n \quad ...(7.69)$$

From Eqs. (7.68) and (7.69) it is observed that the magnitude in decibel versus $\log_{10} \omega$ graph for case (*i*) lies on 0 db axis and for case (*ii*) graph has a slope of – 40 db/decade.

The intersection of the two graphs on 0 db axis is given by equating Eq. (7.69) to zero,

$$\therefore \qquad 0 = -40 \log_{10} \omega + 40 \log_{10} \omega_n$$

or $$40 \log_{10} \omega = 40 \log_{10} \omega_n \quad \therefore \quad \omega = \omega_n$$

Hence, the two graphs intersect on 0 db axis at $\omega = \omega_n$ and the corner frequency is ω_n (ω_n is also natural frequency of oscillation for the second order term).

The graphs mentioned above are approximated asymptotic plots. However, the magnitude near about frequency ω_n depends on the value of damping ratio ζ. The maximum magnitude is determined below :

The magnitude of the second order term is given by

$$\left| \frac{\omega_n^2}{[(\omega_n^2 - \omega^2) + j2\zeta\,\omega_n\,\omega]} \right| = \frac{1}{\sqrt{\left(1 - \dfrac{\omega^2}{\omega_n^2}\right)^2 + \left(2\zeta\,\dfrac{\omega}{\omega_n}\right)^2}} \qquad ...(7.70)$$

The frequency at which the maximum value of expression (7.70) is attained, is called resonant frequency and denoted by ω_r. The resonant frequency ω_r is determined by taking first derivative of Eq. (7.70) w.r.t. ω, equating to zero and solving for ω.

$$\therefore \quad \frac{d}{d\omega}\left[\frac{\omega_n^2}{[(\omega_n^2 - \omega^2) + j2\zeta\,\omega_n\,\omega]}\right] = \frac{d}{d\omega}\left[\frac{1}{\sqrt{(1 - \omega^2/\omega_n)^2 + [2\zeta\,(\omega/\omega_n)]^2}}\right]$$

$$= \frac{d}{d\omega}\left[\frac{1}{\sqrt{\left(1 - \dfrac{2\omega^2}{\omega_n^2} + \dfrac{\omega^4}{\omega_n^2} + 4\zeta^2\,\dfrac{\omega^2}{\omega_n^2}\right)}}\right] = \frac{d}{d\omega}\left[\sqrt{\left(1 - \frac{2\omega^2}{\omega_n^2} + \frac{\omega^4}{\omega_n^4} + 4\zeta^2\,\frac{\omega^2}{\omega_n^2}\right)}\right]^{-1}$$

$$= -\frac{1}{2}\left[\left(1 - \frac{2\omega^2}{\omega_n^2} + \frac{\omega^4}{\omega_n^4} + 4\,\frac{\zeta^2\,\omega^2}{\omega_n^2}\right)\right]^{-\frac{3}{2}}\left[0 - 4\,\frac{\omega}{\omega_n^2} + 4\,\frac{\omega^3}{\omega_n^4} + 8\,\frac{\zeta^2\,\omega}{\omega_n^2}\right]$$

$$= -\frac{1}{2}\,\frac{\left[-4\,\dfrac{\omega}{\omega_n^2} + 4\,\dfrac{\omega^3}{\omega_n^4} + 8\zeta^2\,\dfrac{\omega}{\omega_n^2}\right]}{\left[\left(1 - \dfrac{2\omega^2}{\omega_n^2} + \dfrac{\omega^4}{\omega_n^4} + 4\,\dfrac{\zeta^2\omega^2}{\omega_n^2}\right)\right]^{3/2}} \qquad ...(7.71)$$

Equating the numerator of the R.H.S. of Eq. (7.71) to zero, the following equation is obtained :

$$\left[-4\,\frac{\omega}{\omega_n^2} + 4\,\frac{\omega^3}{\omega_n^4} + 8\zeta^2\,\frac{\omega}{\omega_n^2}\right] = 0, \quad \text{or} \quad \frac{4\omega}{\omega_n^2}\left(-1 + \frac{\omega^2}{\omega_n^2} + 2\zeta^2\right) = 0$$

As $\left(\dfrac{4\omega}{\omega_n^2}\right) \neq 0$, $\qquad \therefore \quad \left(-1 + \dfrac{\omega^2}{\omega_n^2} + 2\zeta^2\right) = 0$

$$\therefore \qquad \omega = \omega_n\,\sqrt{1 - 2\zeta^2}$$

Hence, the resonant frequency is given by

$$\omega_r = \omega_n\,\sqrt{1 - 2\zeta^2} \qquad ...(7.72)$$

The expression (7.72) for resonant frequency ω_r holds good only for $\zeta \le 0.707$. At resonant frequency ω_r, the magnitude of the quadratic term $\omega_n^2/[(\omega_n^2 - \omega^2) + j2\zeta\omega_n\omega]$ attains maximum value and this maximum value is known as resonant peak denoted by M_r. The resonant peak M_r is calculated below in terms of damping ratio ζ.

$$M_r = \frac{1}{\sqrt{[(1 - \omega_r^2/\omega_n^2)^2 + (2\zeta\,\omega_r/\omega_n)^2]}}$$

Since $\qquad \omega_r = \omega_n\sqrt{1 - 2\zeta^2} \quad \text{for} \quad \zeta < 0.707$

$$\therefore \qquad M_r = \frac{1}{\sqrt{\left[1 - \dfrac{\omega_n^2(1 - 2\zeta^2)}{\omega_n^2}\right]^2 + \left[2\zeta\,\omega_n \dfrac{\sqrt{1 - 2\zeta^2}}{\omega_n}\right]^2}}$$

$$= \frac{1}{\sqrt{[1 - (1 - 2\zeta^2)]^2 + [2\zeta\sqrt{1 - 2\zeta^2}\,]^2}}$$

$$= \frac{1}{\sqrt{(4\zeta^2 - 4\zeta^4)}} = \frac{1}{\sqrt{4\zeta^2(1 - \zeta^2)}}$$

or

$$M_r = \frac{1}{2\zeta\sqrt{(1 - \zeta^2)}} \qquad \text{...(7.73)}$$

In decibel, $\qquad M_r\,(\text{db}) = 20\log_{10}\left[\dfrac{1}{2\zeta\sqrt{1 - \zeta^2}}\right]$

The phase angle is given by

$$\angle\frac{\omega_n^2}{[(\omega_n^2 - \omega^2) + j\,2\zeta\omega_n\,\omega]} = -\tan^{-1}\left[\frac{2\zeta\dfrac{\omega}{\omega_n}}{\left(1 - \dfrac{\omega^2}{\omega_n^2}\right)}\right]$$

At corner frequency *i.e.*, at $\omega = \omega_n$

$$\angle\frac{\omega_n^2}{[(\omega_n^2 - \omega^2) + j2\zeta\omega_n\,\omega]} = -\tan^{-1}\left[\frac{2\zeta\dfrac{\omega_n}{\omega_n}}{\left(1 - \dfrac{\omega_n^2}{\omega_n^2}\right)}\right]$$

$$= -\tan^{-1}(\infty) = -90° \qquad \text{...(7.74)}$$

At resonant frequency, *i.e.*, at $\omega = \omega_n\sqrt{1 - 2\zeta^2}$

$$\angle\frac{\omega_n^2}{[(\omega_n^2 - \omega^2) + j2\zeta\omega_n\,\omega]} = -\tan^{-1}\left[\frac{\dfrac{2\zeta\omega_n\sqrt{1 - 2\zeta^2}}{\omega_n}}{1 - \omega_n^2\left(\dfrac{1 - 2\zeta^2}{\omega_n^2}\right)}\right] = -\tan^{-1}\left(\frac{\sqrt{1 - 2\zeta^2}}{\zeta}\right)$$

$$= -90° + \sin^{-1}\left[\frac{\zeta}{\sqrt{1-\zeta^2}}\right] \quad ...(7.75)$$

In view of mathematical derivations obtained above the Bode plot for the quadratic term $\omega_n^2/[(\omega_n^2 - \omega^2) + j2\zeta\omega_n\omega]$ is drawn in Fig. 7.18.10.

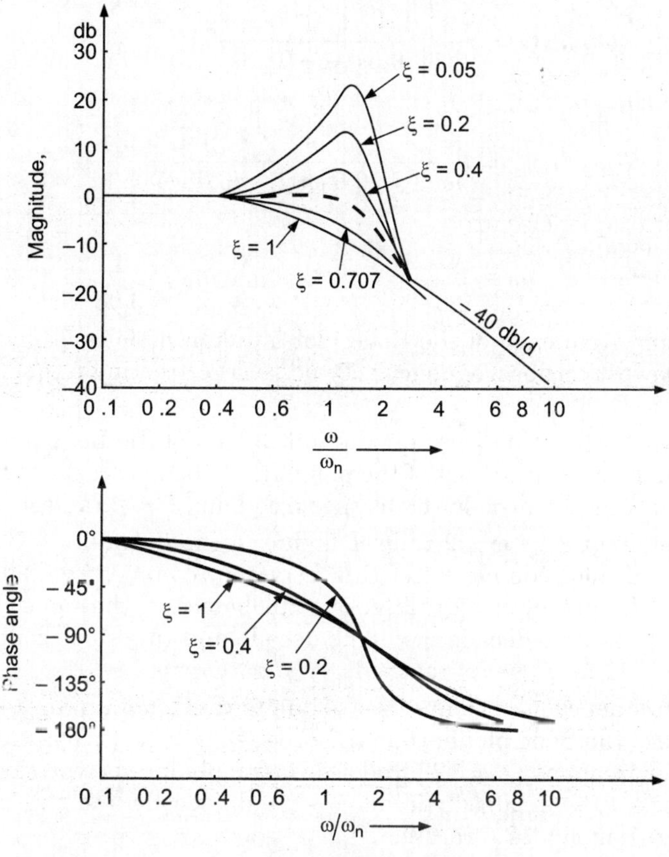

Fig. 7.18.10. Bode plot for quadratic term $\omega_n^2/[(\omega_n^2 - \omega^2) + j2\zeta\omega_n\omega]$.

7.18.9. Procedure for drawing Bode plot and determination of gain margin, phase margin and stability

In section 7.7 the stability in terms of gain margin and phase margin as obtained from the Nyquist plot has been discussed. The gain margin and phase margin can also be obtained from the Bode plot of the open-loop transfer function and the stability of closed-loop system can be determined.

The two illustrative examples given below describe the procedure for constructing the Bode plot and determination of stability in terms of gain margin and phase margin therefrom.

Illustrative Example 1. *Construct Bode plot for the system whose open-loop transfer function is given below and determine (a) the gain margin (b) the phase margin and (c) the closed-loop stability.*

$$G(s)\,H(s) = \frac{4}{s\,(1+0.5s)\,(1+0.08s)}.$$

Solution. $$G(s)\,H(s) = \frac{4}{s\,(1+0.5s)\,(1+0.08s)}$$

Put $s = j\omega$

$$G(j\omega)\,H(j\omega) = \frac{4}{j\omega\,(1+j0.5\,\omega)\,(1+j0.08\,\omega)}$$

1. The corner frequencies are

$$\omega = \frac{1}{0.5} = 2 \text{ rad/sec} \quad \text{and} \quad \omega = \frac{1}{0.08} = 12.5 \text{ rad/sec}$$

2. The starting frequency of the Bode plot is taken as lower than the lowest corner frequency. As the lowest corner frequency is 2 rad/sec the starting frequency is taken as say 1 rad/sec.

3. As the transfer function is type 1 the initial part of the Bode plot is – 20 db/decade and the intersection of the initial part of the plot with 0 db axis occurs at $\omega = K$, *i.e.*, at $\omega = 4$. The initial slope continues till the lowest corner frequency $\omega = 2$ rad/sec is reached.

4. The corner frequency $\omega = 2$ rad/sec is due to the denominator term $1/(1 + j0.5\omega)$, therefore, the slope of the Bode plot after this frequency changes by – 20 db/decade. As the slope prior to corner frequency $\omega = 2$ rad/sec is – 20 db/decade, the slope of the plot after $\omega =$ 2 rad/sec is [– 20 + (– 20)] db/decade = – 40 db/decade and this slope continues till the next corner frequency $\omega = 12.5$ rad/sec is reached.

The corner frequency $\omega = 12.5$ rad/sec is due to the denominator term $1/(1 + j0.08\,\omega)$, therefore, the slope of the Bode plot beyond this frequency changes by – 20 db/decade. As the slope prior to corner frequency $\omega = 12.5$ rad/sec is – 40 db/decade, the slope of the plot after $\omega = 12.5$ rad/sec is [– 40 + (– 20)] db/decade = – 60 db/decade and this slope continues for frequencies greater than $\omega = 12.5$ rad/sec.

5. $\angle G(j\omega)\,H(j\omega)$ for frequencies between $\omega = 1$ rad/sec to $\omega = 100$ rad/sec is calculated as below :

$$\angle G(j\omega)\,H(j\omega) = -\,90° - \tan^{-1}(0.5\omega) - \tan^{-1}(0.08\omega)$$

ω (rad/sec)	1	2	8	10	20	50
$\angle G(j\omega)\,H(j\omega)°$	– 121	– 144	– 198	– 207	– 234	– 252

The Bode plot $|G(j\omega)\,H(j\omega)|$ db and $\angle G(j\omega)\,H(j\omega)$ *versus* ω (log scale) is drawn and shown in Fig. 7.18.11.

(*a*) The gain in db at phase crossover frequency is the gain margin. If gain is negative the gain margin is positive. In Fig. 7.18.11 the phase crossover frequency is 5 rad/sec and $G\,(j5)\,H(j5) = -12$ db hence, the gain margin is

$$\text{G.M.} = 12 \text{ db}$$

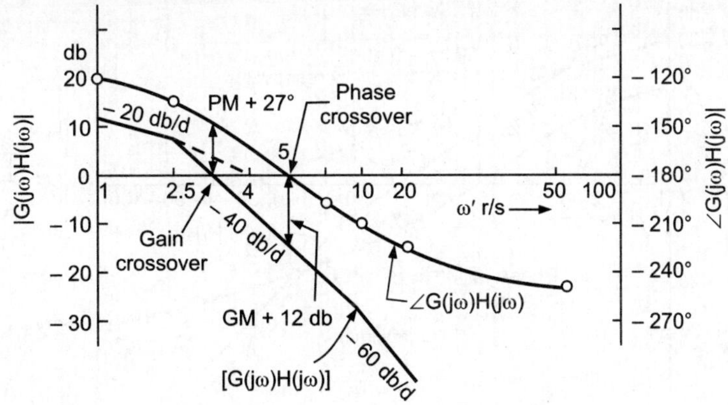

Fig. 7.18.11. Bode plot for $G(s)\ H(s) = \dfrac{4}{s(1+0.5s)(1+0.08s)}$.

(*b*) The phase margin is given by

$$180° + \angle G(j\omega)\ H(j\omega)| \text{ at gain crossover frequency}$$

In Fig. 7.18.11 the gain crossover frequency is 2.5 rad./sec and $G(j2.5)\ H(j2.5) = -153°$ hence the phase margin is

$$\text{P.M.} = 180° + (-153°) = 27°$$

(*c*) The gain margin and phase margin both are positive, therefore, the closed-loop system is stable.

Illustrative Example 2. *The open-loop transfer function of a system is given by*

$$G(s)\ H(s) = \frac{30}{s(1+0.5s)(1+0.08s)}$$

Determine (a) gain margin (b) phase margin and (c) closed-loop stability.

Solution. 1. The corner frequencies in this example are same as those in the previous example, *i.e.*

$$\omega = \frac{1}{0.5} = 2 \text{ rad/sec} \quad \text{and} \quad \omega = \frac{1}{0.08} = 12.5 \text{ rad/sec}$$

2. The transfer function is type 1, therefore, the initial slope of the plot is – 20 db/decade intersecting 0 db axis at $\omega = K$, *i.e.*, $\omega = 30$. The initial slope – 20 db/decade continues till the lowest corner frequency $\omega = 2$ rad/sec is reached. The slope of the plot between corner frequencies $\omega = 2$ rad/sec and $\omega = 12.5$ rad/sec is [– 20 + (– 20)] db/decade = – 40 db/decade and the slope of the plot beyond $\omega = 12.5$ rad/sec is [– 40 + (– 20)] db/decade = – 60 db/decade.

3. $\angle G(j\omega)\ H(j\omega)$ plot is same as that in the previous example.

The Bode plot $|G(j\omega)\ H(j\omega)|$ db and $\angle G(j\omega)\ H(j\omega)$ *versus* ω (log scale) is drawn and shown in Fig. 7.18.12.

In Fig. 7.18.12 the gain at phase crossover frequencies $\omega = 5$ rad/sec is $G(j5)\ H(j5) = 9$ db hence, the gain margin is

$$\text{G.M.} = -9 \text{ db}$$

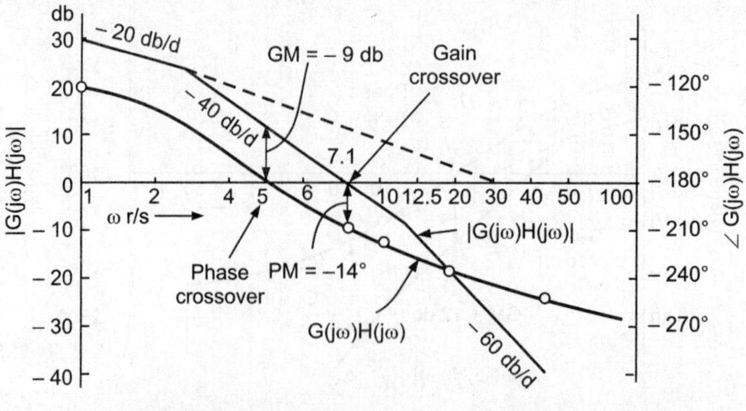

Fig. 7.18.12. Bode plot for $G(s)\,H(s) = 30/s(1 + 0.5s)(1 + 0.08s)$.

The phase margin is given by

$$\text{P.M.} = 180° + \angle G(j\omega)\,H(j\omega)\,|\text{ at gain crossover frequency}$$

In Fig. 7.18.12 the gain crossover frequency is $\omega = 7.1$ rad/sec and $\angle G(j7.1)\,H(j7.1) = -194°$ hence the phase margin is

$$\text{P.M.} = 180° + (-194°) = -14°$$

Since the gain margin and phase margin both are negative closed-loop system is unstable.

From the above two examples it is concluded that :

1. *For stable systems :* The gain crossover frequency < phase crossover frequency.
2. *For unstable systems :* The gain crossover frequency > phase crossover frequency.
3. *For marginally stable systems :* The gain crossover frequency = phase crossover frequency.

7.18.10. Bode plot for time delay element : e^{-sT}

The time delay element (transportation lag) has the transfer function as

$$G(s) = e^{-sT}$$

The sinusoidal form of above transfer function is

$$G(j\omega) = e^{-j\omega T}$$

or
$$G(j\omega) = 1\angle -\omega T$$
$$= 1\angle -57.3\,\omega T°$$

(1) As $|G(j\omega)| = 1$, the magnitude (db) versus ωT plot is a straight line on 0 db axis.

(2) $\angle G(j\omega) = -57.3\,\omega T°$.

Therefore, the phase angle varies in linear proportion to ωT.

The magnitude (db) and phase angle plots with reference to time delay elements ($e^{-j\omega T}$) are shown in Fig. 7.18.13.

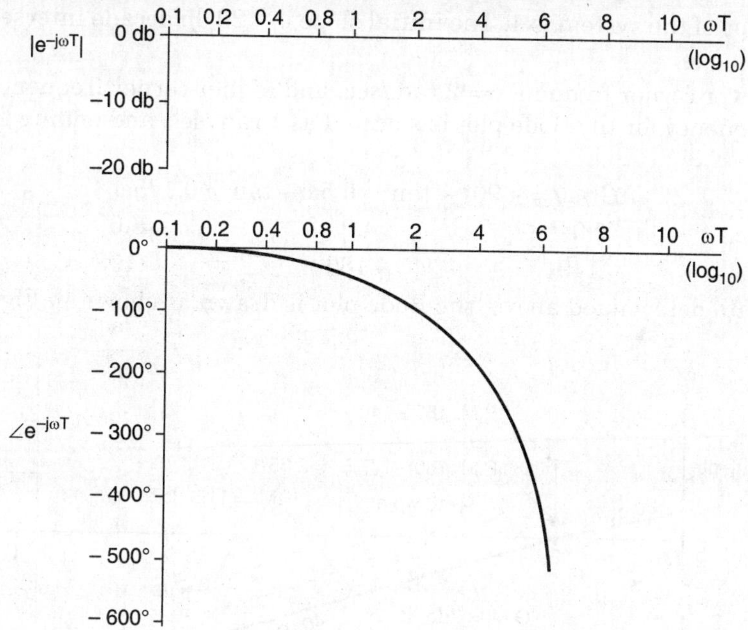

Fig. 7.18.13. $e^{-j\omega T}$ Mag. (db), phase-angle plot.

Illustrative Example 3. *The open-loop transfer function is given below :*

$$G(s) = \frac{2e^{-0.5s}}{s(0.5s+1)(0.125s+1)}$$

Plot the Bode plot

(a) without time delay element.

(b) with time delay element

Determine gain margin and phase margin.

Solution. Put $s = j\omega$

$$\therefore \qquad G(j\omega) = \frac{2e^{-j0.5\omega}}{j\omega(j0.5\,\omega+1)(j\,0.125\,\omega+1)}$$

(*a*) without time delay element,

$$G(j\omega) = \frac{2}{j\omega(j0.5\,\omega+1)(j\,0.125\,\omega+1)}$$

The corner frequencies are

$$\left.\begin{aligned} \omega &= \frac{1}{0.5} = 2 \text{ rad./sec.} \\ \omega &= \frac{1}{0.125} = 8 \text{ rad./sec.} \end{aligned}\right\} \text{ due to denominator term (zero)}$$

As the type of the system is 1. The initial slope is – 20 db/decade intersecting 0 db-axis at $\omega = K = 2$.

As the lower corner frequency = 2 rad./sec. and higher corner frequency = 8 rad./sec., the starting frequency for the Bode plot is selected as 1 rad./sec. and ending frequency as 10 rad./sec.

$$\angle G(j\omega) = -90° - \tan^{-1} 0.5\omega - \tan^{-1} 0.125\omega$$

ω rad./sec.	2.0	4.0	6.0
$\angle G(j\omega)°$	– 149	– 180	– 198

As per data determined above, the Bode plot is drawn as shown in Fig. 7.18.14.

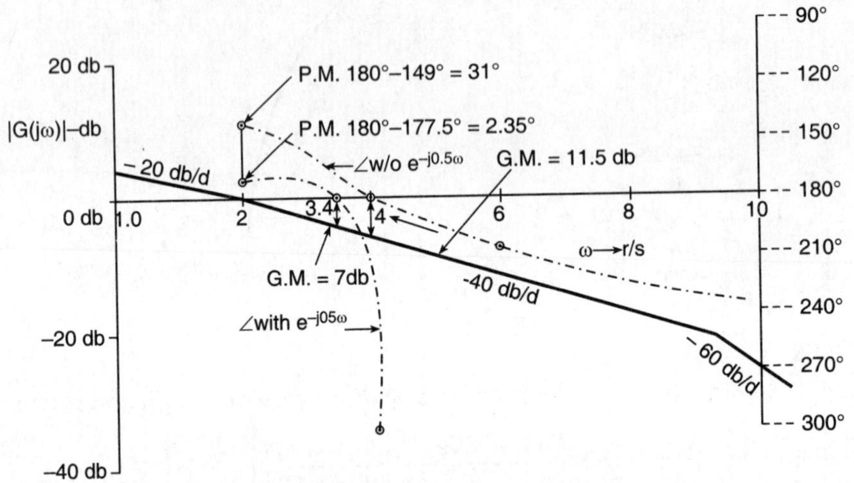

Fig. 7.18.14. Bode plot for example 3.

It is noted that, $\left.\begin{aligned} G.M. &= 11.5 \text{ db} \\ P.M. &= 31° \end{aligned}\right\}$ **Ans.**

(*b*) $\because \quad |e^{-j\,0.5\,\omega}| = 1$

The magnitude (db) plot is not affected.

The $\angle G(j\omega)$ with time delay element is given by

$$\angle G(j\omega) = \angle e^{-j\,0.5\,\omega} - 90° - \tan^{-1} 0.5\,\omega - \tan^{-1} 0.125\,\omega$$

$$= \frac{-0.5\,\omega \times 180}{\pi} - 90° - \tan^{-1} 0.5\,\omega - \tan^{-1} 0.125\,\omega$$

$$= -28.65\,\omega - 90° - \tan^{-1} 0.5\,\omega - \tan^{-1} 0.125\,\omega$$

ω rad./sec.	2.0	3.4	4.0
$\angle G(j\omega)°$	– 177.5	– 180°	– 294.6

It is noted that, $\left.\begin{aligned} G.M. &= 7.0 \text{ db} \\ P.M. &= 2.35° \end{aligned}\right\}$ **Ans.**

7.18.11. Minimum phase, non-minimum phase and all pass transfer function

Transfer functions considered in previous sections have no poles and zeros in the RHS of *s*-plane are called minimum phase transfer functions. Transfer functions having at least one pole or zero in the RHS of *s*-plane are called non-minimum phase transfer functions.

Let $$G_1(s) = \frac{1 + sT_1}{1 + sT_2} \quad \text{and} \quad G_2(s) = \frac{1 - sT_1}{1 + sT_2}$$

In sinusoidal form above transfer functions are written as

$$G_1(j\omega) = \frac{1 + j\omega T_1}{1 + j\omega T_2} \qquad \text{...(7.76)}$$

$$G_2(j\omega) = \frac{1 - j\omega T_1}{1 + j\omega T_2} \qquad \text{...(7.77)}$$

The transfer function given by Eq. (7.76) is a minimum phase transfer function and that given by Eq. (7.77) is non-minimum phase transfer function. The pole-zero configurations and Bode plots for such transfer functions are shown in Figs. 7.18.15 and 7.18.16 respectively.

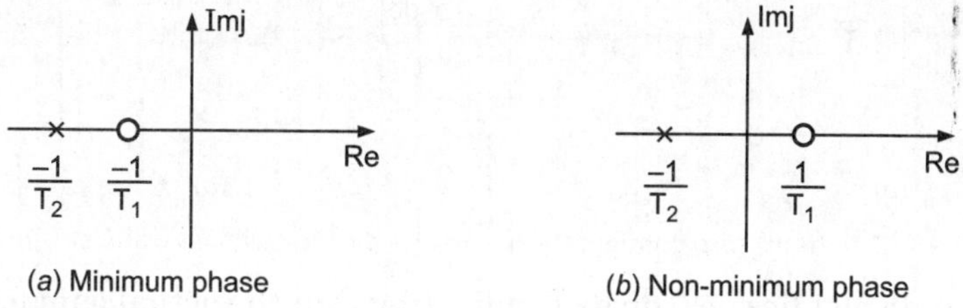

Fig. 7.18.15. Pole zero configuration for minimum phase and non-minimum phase transfer functions.

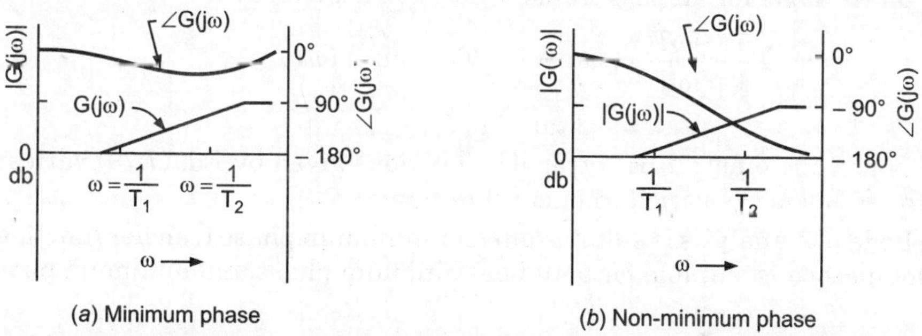

Fig. 7.18.16. Bode plot for minimum phase and non-minimum phase transfer functions.

From Bode plot Fig. 7.18.16 it is observed that for a minimum phase transfer function the slope of the magnitude plot is uniquely related to phase angle. However, this unique relationship is not valid for non-minimum phase transfer function.

The phase angle for a minimum phase transfer function at $\omega = \infty$ is $[-90\,(m - n)]°$, where n and m are the orders of the numerator and denominator polynomials of the transfer function respectively, whereas the phase angle for non-minimum phase transfer function is not $[-90\,(m - n)]°$. For both the minimum phase and non-minimum phase transfer functions the slope of the magnitude plot at $\omega = \infty$ is $[-20\,(m - n)]$ db/decade.

Thus from the observation of Bode plots the minimum phase and non-minimum phase transfer functions can be identified.

If the transfer function has symmetric pole and zero about the imaginary axis in s-plane then the transfer function is called all pass transfer function and given by

$$G(s) = \frac{1 - sT}{1 + sT}$$

In sinusoidal form above transfer function is written as

$$G(j\omega) = \frac{1 - j\omega T}{1 + j\omega T} \qquad ...(7.78)$$

The pole-zero configuration and Bode plot for all pass transfer function is shown in Fig. 7.18.17.

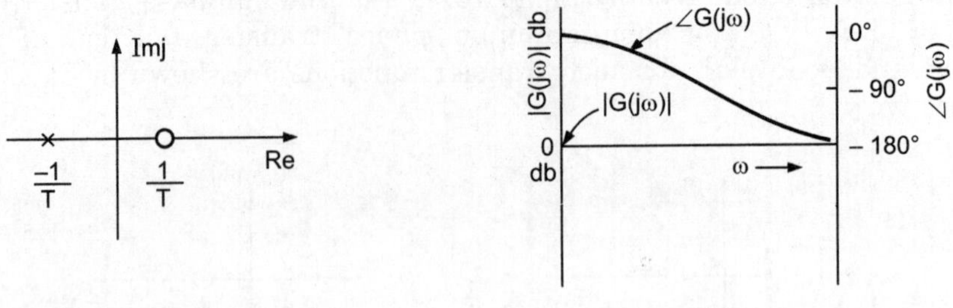

(a) (b)

Fig. 7.18.17 *(a)* Pole-zero configuration *(b)* Bode plot for all pass transfer function.

The magnitude plot lies on 0 db axis indicating that the actual gain for the frequencies is 1, thus the transfer function passes all frequencies.

The phase angle for all pass transfer function is given by

$$\angle\frac{1 - j\omega T}{1 + j\omega T} = \tan^{-1}(-\omega T) - \tan^{-1}(\omega T)$$

$$= -2\tan^{-1}(\omega T) \qquad ...(7.79)$$

The phase angle for all pass transfer functions given by Eq. (7.79) varies from 0° to − 180° as the frequency is varied from $\omega = 0$ to $\omega = \infty$.

The Bode plot analysis is suitable only for minimum phase transfer functions whereas Nyquist plot method is suitable for both non-minimum phase and minimum phase transfer functions.

7.19 SOLVED EXAMPLES : BODE PLOT

Example 7.19.1. *Sketch the Bode plot for the open-loop transfer function for the unity feedback system given below and assess stability*

$$G(s) = \frac{50}{(s+1)(s+2)}.$$

Solution. Rewrite the transfer function is Bode form,

$$\therefore \qquad G(s) = \frac{\frac{50}{2}}{(s+1)\left(\frac{s}{2}+1\right)} = \frac{25}{(s+1)(0.5s+1)}$$

Sinusoidal form of the transfer function is

$$G(j\omega) = \frac{25}{(j\omega + 1)(j0.5\,\omega + 1)}$$

As the system is type 0, the initial slope of the Bode plot is 0 db/decade and the initial ordinate is given by

$$20 \log_{10} 25 = 27.9 \text{ db}$$

The two corner frequencies due to denominator terms are

$$\omega = 1 \text{ rad/sec} \quad \text{and} \quad \omega = \frac{1}{0.5} = 2 \text{ rad/sec}$$

The slope of the magnitude plot after $\omega = 1$ rad/sec is $(0 - 20) = -20$ db/decade and after $\omega = 2$ rad/sec is $(-20 - 20) = -40$ db/decade.

The frequency range of the plot is considered between $\omega = 0.1$ rad/sec and $\omega = 10$ rad/sec. The Bode plot is drawn in Fig. 7.19.1.

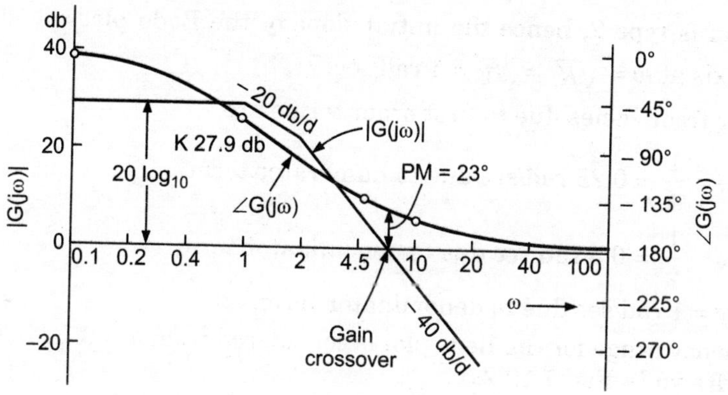

Fig. 7.19.1. Bode plot for example 7.19.1.

The phase angle is given by

$$\angle G(j\omega) = -\tan^{-1}(\omega) - \tan^{-1}(0.5\omega)$$

The phase angle for frequency range between $\omega = 0.1$ rad/sec and 10 rad/sec is calculated and given below :

ω rad/sec	0.1	1	5	10
∠ G(jω)°	– 8.5	– 71.5	– 146.8	– 163

The phase curve is asymptotic to – 180° line at high frequencies and the gain at high frequencies is negative as such the gain margin is + infinite. At the gain crossover frequency the phase angle is – 157°, hence the phase margin is

$$\text{P.M.} = 180° + -(157°) = 23°.$$

The gain margin as well as phase margin are both positive, therefore, the system is stable and further the gain margin is infinite hence, the system is inherently stable. **Ans.**

Example 7.19.2. *Sketch the asymptotic Bode plot for the transfer function given below*

$$G(s)\,H(s) = \frac{2\,(s + 0.25)}{s^2\,(s + 1)\,(s + 0.5)}$$

From the Bode plot determine :
(a) the phase crossover frequency, *(b) the gain crossover frequency*
(c) the gain margin and *(d) the phase margin.*
Is the system stable ?

Solution. The transfer function is rewritten as below :

$$G(s)\,H(s) = \frac{\dfrac{2\times 0.25}{0.5}\left(\dfrac{s}{0.25}+1\right)}{s^2\,(s+1)\left(\dfrac{s}{0.5}+1\right)} = \frac{1\,(4s+1)}{s^2\,(s+1)\,(2s+1)}$$

Sinusoidal form of the transfer function is

$$G(j\omega)\,H(j\omega) = \frac{1\,(j4\omega+1)}{(j\omega)^2\,(j\omega+1)\,(j2\omega+1)}$$

The system is type 2, hence the initial slope of the Bode plot = – 40 db/decade and intersects 0 db axis at $\omega = \sqrt{K} = \sqrt{1} = 1$ rad/sec.

The corner frequencies due to first order terms are

$\omega = \frac{1}{4} = 0.25$ rad/sec due to numerator term

$\omega = \frac{1}{2} = 0.5$ rad/sec due to denominator term

$\omega = 1$ rad/sec due to denominator term.

The frequency range for the Bode plot is considered from $\omega = 0.1$ rad/sec to $\omega = 10$ rad/sec. The plot is drawn in Fig. 7.19.2.

As the initial slope of the plot is – 40 db/decade and the corner frequency $\omega = 0.25$ rad/sec is due to numerator first order term, the slope of the plot after $\omega = 0.25$ rad/sec is (– 40 + 20) = – 20 db/decade. The corner frequencies $\omega = 0.5$ rad/sec and $\omega = 1$ rad/sec are due to denominator first order terms, the Bode plot has following slope :

After $\omega = 0.5$ rad/sec, (– 20 – 20) = – 40 db/decade.

After $\omega = 1$ rad/sec, (– 40 – 20) = – 60 db/decade.

The phase angle is given by

$$\angle G(j\omega)\,H(j\omega) = \tan^{-1}(4\omega) - 180° - \tan^{-1}(\omega) - \tan^{-1}(2\omega).$$

The phase angle for the frequency range of the Bode plot is calculated and given below:

ω rad/sec	0.1	0.25	0.5	1	5
$\angle G(j\omega)\,H(j\omega)°$	– 175.2	– 175.6	– 188	– 212.4	–225

(*a*) The gain plot crosses 0 db axis at $\omega = 1.24$ rad/sec therefore, the gain crossover frequency is

$\omega = 1.24$ rad/sec. **Ans.**

(*b*) The phase plot crosses – 180° line at $\omega = 0.4$ rad/sec, therefore, the phase crossover frequency is

$\omega = 0.4$ rad/sec. **Ans.**

(*c*) At phase crossover the gain is 20 db, therefore, the gain margin is

G.M. = – 20 db. **Ans.**

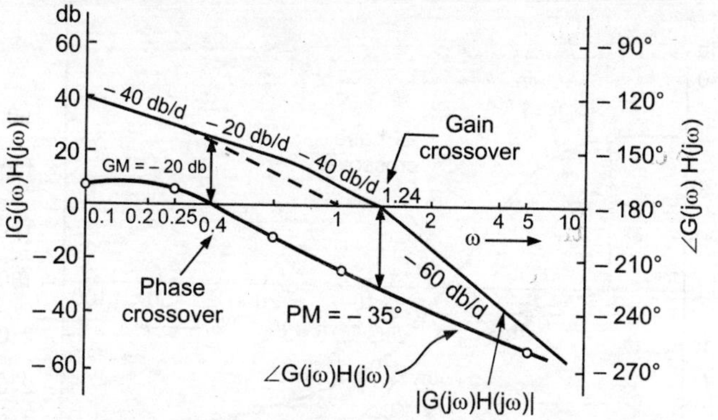

Fig. 7.19.2. Bode plot for example 7.19.2.

(*d*) At gain crossover the phase angle is – 215°, therefore, the phase margin as

$$\text{P.M.} = 180 + (-125) = -35°. \quad \textbf{Ans.}$$

As both the gain margin and phase margin are negative the system is unstable.

Example 7.19.3. *Draw the Bode plot for the transfer function given below :*

$$G(s)\,H(s) = \frac{48\,(s+10)}{s\,(s+20)\,(s^2+2.4s+16)}$$

Apply correction to the magnitude plot for the quadratic term and comment on the stability.

Solution. Rewrite the transfer function in normalized (Bode) form as follows .

$$G(s)\,H(s) = \frac{48 \times 10\left(\dfrac{s}{10}+1\right)\times 16}{20 \times 16 \times s\left(\dfrac{s}{20}+1\right)(s^2+2.4s+16)} = \frac{1.5\,(0.1s+1)\times 16}{s(0.05s+1)\,(s^2+2.4s+16)}$$

Put $s = j\omega$

$$G(j\omega)\,H(j\omega) = \frac{1.5\,(j0.1\,\omega+1)\times 16}{j\omega\,(j0.05\,\omega+1)\,[j2.4\,\omega+(16-\omega^2)]}$$

The type of the system is 1, therefore, the initial slope of the plot is – 20 db/decade intersecting 0 db axis at ω = 1.5 rad/sec (*K* = 1.5).

The Bode plot is plotted in the frequency range of ω = 1 rad/sec to ω = 100 rad/sec as shown in Fig. 7.19.3.

The three corner frequencies and the slopes of the magnitude plot are as follows :

$\omega = \omega_n = \sqrt{16} = 4$ rad/sec is due to denominator quadratic term, and the slope of the magnitude plot after ω = 4 rad/sec is (– 20 – 40) = – 60 db/decade.

$\omega = \dfrac{1}{0.1} = 10$ rad/sec is due to numerator first order term, the slope of the magnitude plot after ω = 10 rad/sec is (– 60 + 20) = – 40 db/decade.

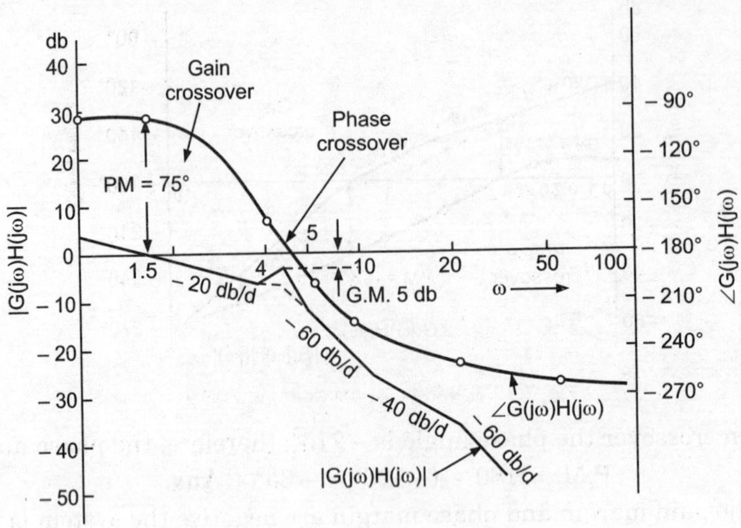

Fig. 7.19.3. Bode plot for example 7.19.3.

$\omega = \dfrac{1}{0.05} = 20$ rad/sec is due to denominator first order term, the slope of the magnitude plot after $\omega = 20$ rad/sec is $(-40-20) = -60$ db/decade.

The damping ratio is calculated below.

Since $\qquad 2\zeta\,\omega_n = 2.4 \quad$ and $\quad \omega_n = 4$

$$\therefore \qquad \zeta = \frac{2.4}{2\omega_n} = \frac{2.4}{2\times 4} = 0.3$$

The exact magnitude at $\omega = 4$ rad/sec is determined analytically as follows :

$$20\log_{10}|G(j4)\,H(j4)| = 20\log_{10}\left|\frac{1.5(j0.1\times 4+1)\times 16}{j4\,(j0.05\times 4+1)\,[j2.4\times 4+(16-4^2)]}\right|$$

$$= 20\log_{10}\frac{1.5[\sqrt{(0.1\times 4)^2+1^2}\,]\times 16}{[\sqrt{4}^2\;\sqrt{(0.05\times 4)^2+1^2}\;\sqrt{(2.4\times 4)^2}\,]} = -4.0 \text{ db.}$$

The phase angle is given by

$$\angle G(j\omega)\,H(j\omega) = \tan^{-1}(0.1\omega) - 90° - \tan^{-1}(0.05\omega) - \tan^{-1}\left(\frac{2.4\omega}{16-\omega^2}\right);\ \omega < 4$$

$$\angle G(j\omega)\,H(j\omega) = \tan^{-1}(0.1\omega) - 90° - \tan^{-1}(0.05\omega) - 180° - \tan^{-1}\frac{2.4\omega}{16-\omega^2};\ \omega > 4$$

The phase angle for the frequency range of the Bode plot is calculated and given below :

ω rad/sec	1	4	5	10	20	50
$\angle G(j\omega)\,H(j\omega)°$	– 96.25	– 169.5	– 204.3	– 235.6	– 244.6	– 256.8

The gain margin is, G.M. = 5 db

The phase margin is, P.M. = 180° + (– 105°) = + 75°

As both the gain margin and phase margin are positive the system is stable. **Ans.**

Example 7.19.4. *Determine the value of gain K for the open-loop transfer function given below so that (a) the gain margin is 15 db and (b) phase margin is 60°.*

$$G(j\omega)\,H(j\omega) = \frac{K}{j\omega(j0.1\,\omega + 1)\,(j\omega + 1)}$$

Solution. Bode plot is plotted considering $K = 1$.

As the system is type 1, the initial slope of the Bode plot is – 20 db/decade intersecting 0 db axis at $\omega = 1$ ($K = 1$).

The two corner frequencies due to denominator first order terms are

$$\omega = 1 \text{ rad/sec} \quad \text{and} \quad \omega = \frac{1}{0.1} = 10 \text{ rad/sec}$$

The frequency range of the plot is considered from $\omega = 0.1$ rad/sec to $\omega = 100$ rad/sec. The Bode plot is plotted in Fig. 7.19.4.

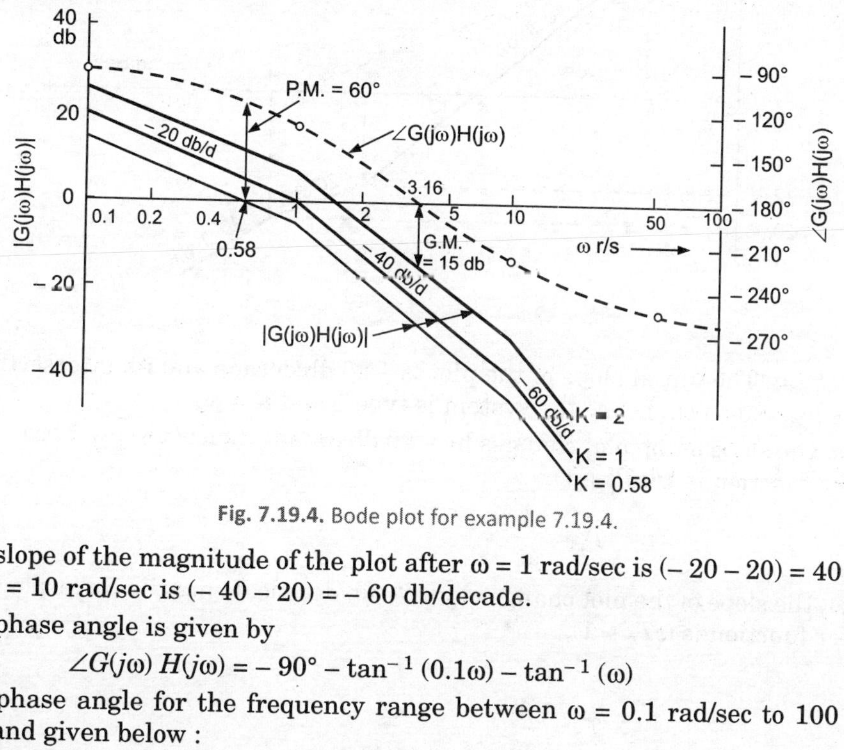

Fig. 7.19.4. Bode plot for example 7.19.4.

The slope of the magnitude of the plot after $\omega = 1$ rad/sec is (– 20 – 20) = 40 db/decade and after $\omega = 10$ rad/sec is (– 40 – 20) = – 60 db/decade.

The phase angle is given by

$$\angle G(j\omega)\,H(j\omega) = -\,90° - \tan^{-1}(0.1\omega) - \tan^{-1}(\omega)$$

The phase angle for the frequency range between $\omega = 0.1$ rad/sec to 100 rad/sec is calculated and given below :

ω rad/sec	0.1	1	5	10	50
$\angle G(j\omega)\,H(j\omega)°$	– 96.3	– 140.7	– 195.6	– 219.7	– 257.4

The phase crossover is at $\omega = 3.16$ rad/sec (Fig. 7.19.4), therefore, to have a gain margin of 15 db the gain at $\omega = 3.16$ rad/sec (phase crossover) should be – 15 db which is obtained by shifting the magnitude plot upwards such that the gain at $\omega = 3.16$ rad/sec is – 15 db. The initial slope of this shifted plot intersects the 0 db axis at $\omega = 2$ rad/sec, hence, the value of gain K of the system to have a gain margin of 15 db is

$$K = 2. \quad \textbf{Ans.}$$

The phase margin of 60° is marked on the Bode plot (Fig. 7.19.4) at $\omega = 0.58$ rad/sec. In order to have a phase margin of 60° the gain crossover should be at $\omega = 0.58$ rad/sec which is obtained by shifting the magnitude plot downwards such that the gain at $\omega = 0.58$ rad/sec (gain crossover) is 0 db. The initial slope of this shifted plot intersects the 0 db axis at $\omega =$ 0.58 rad/sec, hence the value of gain K of the system to have a phase margin of 60° is

$$K = 0.58. \quad \textbf{Ans.}$$

Example 7.19.5. *Determine the open-loop transfer function from the Bode plot shown in Fig. 7.19.5.*

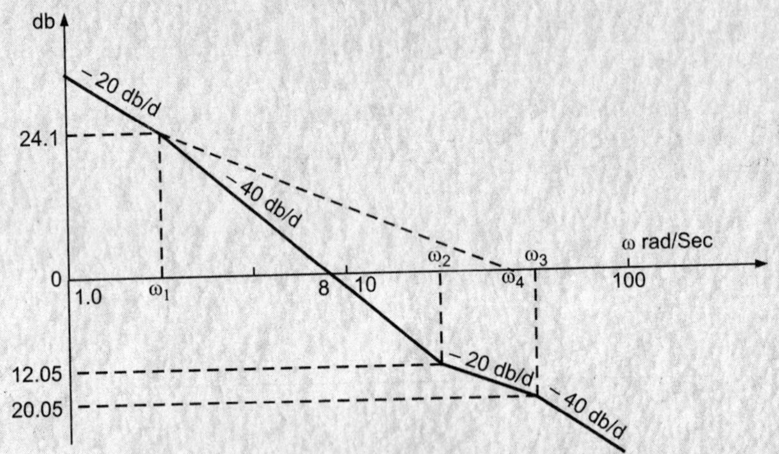

Fig. 7.19.5. Bode plot for example 7.19.5.

Solution. The initial slope of the plot is – 20 db/decade and its intersection with the 0 db axis is located as ω_4, hence the system is type 1 and $K = \omega_4$.

At ω_1 the slope of the plot changes by – 20 db/decade, hence the corresponding term of the transfer function is $1/(sT_1 + 1)$,

where $$T_1 = \frac{1}{\omega_1}.$$

At ω_2 the slope of the plot changes by + 20 db/decade, hence the corresponding term of the transfer function is $(sT_2 + 1)$,

where $$T_2 = \frac{1}{\omega_2}.$$

At ω_3 the slope of the plot changes by – 20 db/decade, hence the corresponding term of the transfer function is $1/(sT_3 + 1)$,

where $$T_3 = \frac{1}{\omega_3}.$$

The calculations are given below :

1. $\log_{10}\left(\dfrac{8}{\omega_1}\right) = \dfrac{24.1}{40}$

$$\therefore \quad \frac{8}{\omega_1} = \text{antilog}_{10}\left(\frac{24.1}{40}\right) = 4 \quad \therefore \quad \omega_1 = 2 \text{ rad/sec.}$$

and $$T_1 = \frac{1}{\omega_1} = \frac{1}{2} = 0.5 \text{ sec.}$$

2. $$\log_{10}\left(\frac{\omega_2}{\omega_1}\right) = \frac{12.05 + 24.1}{40} = \frac{36.15}{40}$$

$$\therefore \quad \frac{\omega_2}{\omega_1} = \text{antilog}_{10}\left(\frac{36.15}{40}\right) = 8 \quad \therefore \quad \omega_2 = 8\omega_1 = 8 \times 2 = 16 \text{ rad/sec.}$$

and $$T_2 = \frac{1}{\omega_2} = \frac{1}{16} = 0.0625 \text{ sec.}$$

3. $$\log_{10}\left(\frac{\omega_3}{\omega_2}\right) = \frac{20.05 - 12.05}{20} = \frac{8}{20}$$

$$\therefore \quad \frac{\omega_3}{\omega_2} = \text{antilog}_{10}\left(\frac{8}{20}\right) = 2.5 \quad \therefore \quad \omega_3 = 2.5\omega_2 = 2.5 \times 16 = 40 \text{ rad/sec.}$$

and $$T_3 = \frac{1}{\omega_3} = \frac{1}{4} = 0.025 \text{ sec.}$$

4. $$\log_{10}\left(\frac{\omega_4}{\omega_1}\right) = \frac{24.1}{20}$$

$$\therefore \quad \frac{\omega_4}{\omega_1} = \text{antilog}_{10}\left(\frac{24.1}{20}\right) = 16 \quad \therefore \quad \omega_4 = 16\omega_1 = 16 \times 2 = 32 \text{ rad/sec.}$$

and $K = 32.$

$$\therefore \quad G(s)\,H(s) = \frac{K\,(sT_2 + 1)}{s(sT_1 + 1)\,(sT_3 + 1)} = \frac{32\,(0.0625s + 1)}{s(0.5s + 1)\,(0.025s + 1)} \quad \textbf{Ans.}$$

7.20 CORRELATION BETWEEN TRANSIENT RESPONSE AND FREQUENCY RESPONSE

Transient and frequency response are correlated to each other and mathematical relationship between them can be established.

For a second order system the resonant peak occurs at resonant frequency

$$\omega_r = \omega_n\sqrt{1 - 2\zeta^2} \qquad \text{and given by}$$

$$M_r = \frac{1}{2\zeta\sqrt{1 - \zeta^2}} \quad \text{for} \quad 0 \le \zeta \le 0.707 \qquad \text{...(7.80)}$$

The maximum overshoot is given by

$$M_p = e^{-\frac{\pi\zeta}{\sqrt{1-\zeta^2}}} \qquad \text{...(7.81)}$$

The damped frequency of oscillations during the transient response is $\omega_d = \omega_n\sqrt{1 - \zeta^2}$.

It is noted from Eqs. (7.80) and (7.81) that M_r as well as M_p occur for values of ζ as $0 \le \zeta \le 0.707$.

The phase margin of a second order system can be determined in terms of ζ as follows :

The open-loop transfer function of a unity feedback second order control system is given by

$$G(s) = \frac{\omega_n^2}{s\,(s + 2\zeta\,\omega_n)} \qquad ...(7.82)$$

In sinusoidal form Eq. (7.82) can be written as

$$G(j\omega) = \frac{\omega^2}{j\omega\,(j\omega + 2\zeta\,\omega_n)}$$

$$= \frac{\omega_n^2}{-\,\omega^2 + j2\zeta\,\omega_n\,\omega} \qquad ...(7.83)$$

The magnitude of Eq. (7.83) is

$$|\,G(j\omega)\,| = \frac{\omega_n^2}{[(-\,\omega^2)^2 + (2\zeta\,\omega_n)^2]^{1/2}} \qquad ...(7.84)$$

At gain cross-over frequency ω_1 the magnitude $|\,G(j\omega)\,|$ is unity. Hence,

$$[(-\,\omega_1^2)^2 + (2\zeta\,\omega_n\omega_1)^2]^{1/2} = \omega_n^2 \qquad ...(7.85)$$

or

$$\omega_1^4 + 4\zeta^2\,\omega_n^2\omega_1^2 - \omega_n^2 = 0$$

$$\therefore \quad \omega_1^2 = -\,2\zeta^2\,\omega_n^2 \pm \omega_n^2\,\sqrt{4\zeta^4 + 1}$$

$$\therefore \quad \omega_1 = \omega_n\sqrt{(4\zeta^4 + 1)^{1/2} - 2\zeta^2} \qquad ...(7.86)$$

As per Eq. (7.83)

$$\angle G(j\omega_1) = \tan^{-1}\left(\frac{2\zeta\,\omega_n\omega_1}{\omega_1^2}\right) = \tan^{-1}\left(2\zeta\frac{\omega_n}{\omega_1}\right) \qquad ...(7.87)$$

Substituting for ω_1 in Eq. (7.87) from (7.86) following equation is obtained :

$$\angle G(j\omega_1) = \tan^{-1}\left[\frac{2\zeta\omega_n}{\omega_n\sqrt{(4\zeta^4 + 1)^{1/2} - 2\zeta^2}}\right] \qquad ...(7.88)$$

The phase margin is thus,

$$\phi_m = -\,180° + \tan^{-1}\left[\frac{2\zeta\omega_n}{\omega_n\sqrt{(4\zeta^4 + 1)^{1/2} - 2\zeta^2}}\right]$$

or

$$\phi_m = \tan^{-1}\left[\frac{2\zeta}{\sqrt{(4\zeta^4 + 1)^{1/2} - 2\zeta^2}}\right] \qquad ...(7.89a)$$

On solving Eqn. (7.89*a*)

$$\xi = \frac{1}{2}\sqrt{\sin\phi_m\,\tan\phi_m} \qquad ...(7.89b)$$

In view of equations derived for ω_d, ω_r, M_r, M_p and phase margin ϕ_m in terms of damping ratio ζ following correlations are established between the transient and frequency response for a second order control system.

1. For small values of ζ the values of damped frequency ω_d and resonant frequency ω_r are nearly same. Hence for small values of ζ, ω_r indicates the speed of transient response, greater the value of ω_r quicker is the transient response.

2. Fig. 7.20.1 shows resonant peak M_r and maximum overshoot M_p as a function of damping ratio ζ. As ζ decreases M_r and M_p both increase but maximum value of M_p is limited to 1 and that of M_r attains very large values for $\zeta < 0.4$.

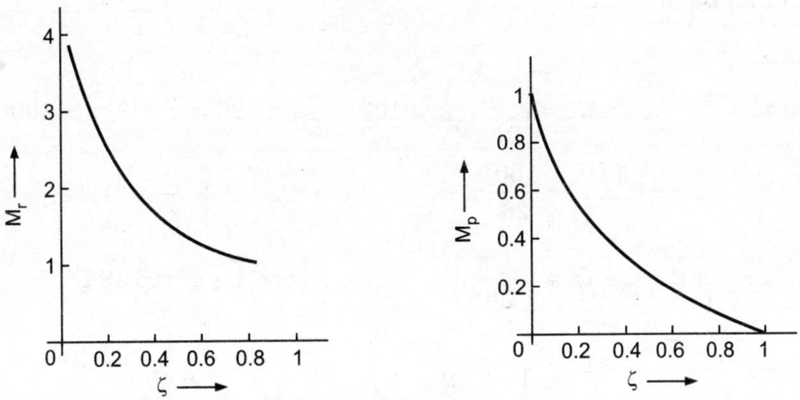

Fig. 7.20.1. M_r and M_p as function of ζ.

3. The graph between ζ and phase margin ϕ is shown in Fig. 7.20.2. An increase in phase margin is observed as the value of ζ increases.

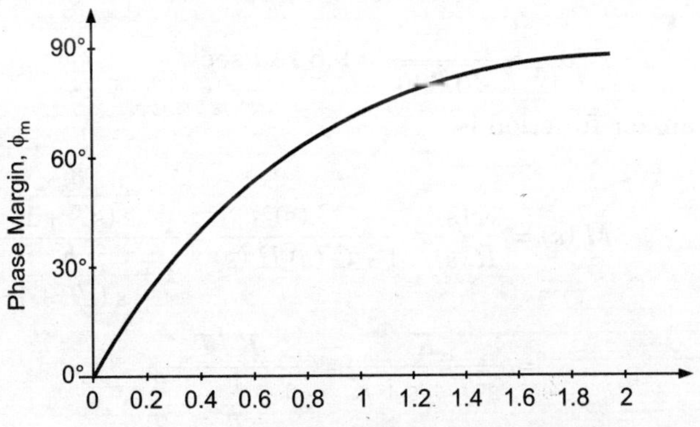

Fig. 7.20.2. Phase margin ϕ_m as a function of ζ.

Usually frequency response forms the basis of design of a control system. However, it is not convenient and simple to correlate transient and frequency response of higher order control systems as in the case of second order systems. The correlation between the transient and frequency response for higher order systems depends on the location of closed-loop poles (roots of the characteristic equation) in s-plane. The transient response and frequency response of a higher order system is dominated by the dominant poles (poles nearer to imaginary axis in s-plane) and the mathematical correlation established for second order system can be approximately extended to higher order systems considering dominant complex poles.

Illustrative Example. *A unit step input is applied to a unity feedback control system whose open-loop transfer function is given by*

$$G(s) = \frac{K}{s(sT+1)}$$

Determine the values of K and T given that maximum overshoot M_p = 26% and resonant frequency ω_r = 8 rad/sec. Calculate the resonance peak M_r, gain crossover frequency and phase margin.

Solution. $\%M_p = e^{-\frac{\zeta\pi}{\sqrt{1-\zeta^2}}} \times 100 \quad \therefore \quad 26 = e^{-\frac{\zeta\pi}{\sqrt{1-\zeta^2}}} \times 100$

or $e^{+\frac{\zeta\pi}{\sqrt{1-\zeta^2}}} = \frac{100}{26} \quad \therefore \quad \frac{\zeta\pi}{\sqrt{1-\zeta^2}} = 1.34$

or $1-\zeta^2 = \left(\frac{\zeta\pi}{1.34}\right)^2$ or $1-\zeta^2 = 5.49\,\zeta^2$

$\therefore \quad \zeta^2 = \frac{1}{1+5.49} \quad \therefore \quad \zeta = \frac{1}{6.49} = 0.39$

Resonant frequency :

$$\omega_r = \omega_n\sqrt{1-2\zeta^2}$$

$$\therefore \quad 8 = \omega_n\sqrt{1-2\zeta^2} = \omega_n\sqrt{1-2\times 0.39^2} = \omega_n\sqrt{0.699}$$

$$\therefore \quad \omega_n = \frac{8}{\sqrt{0.699}} = 9.6 \text{ rad/sec}$$

The overall transfer function is

$$M(s) = \frac{C(s)}{R(s)} = \frac{G(s)}{1+G(s)H(s)} = \frac{\dfrac{K}{s(sT+1)}}{1+\dfrac{K}{s(sT+1)}\cdot 1}$$

$$= \frac{K}{s^2T+s+K} = \frac{K/T}{s^2+\dfrac{1}{T}s+\dfrac{K}{T}}$$

The characteristic equation is

$$s^2 + \frac{1}{T}s + \frac{K}{T} = 0$$

On comparing above equation with $s^2 + 2\zeta\,\omega_n s + \omega_n^2 = 0$ (characteristic equation of a second order system)

$$2\zeta\,\omega_n = \frac{1}{T}$$

$$\therefore \quad T = \frac{1}{2\zeta\,\omega_n} = \frac{1}{2 \times 0.39 \times 9.6} = 0.13. \quad \textbf{Ans.}$$

and
$$\omega_n^2 = \frac{K}{T}$$

$$\therefore \quad K = \omega_n^2\, T = (9.6)^2\,(0.13) = 11.9. \quad \textbf{Ans.}$$

$$\text{Resonant peak} : M_r = \frac{1}{2\zeta\sqrt{1-\zeta^2}} = \frac{1}{2 \times 0.39\sqrt{1-0.39^2}} = 1.39. \quad \textbf{Ans.}$$

Gain crossover frequency

$$\omega_1 = \omega_n\sqrt{(4\zeta^4+1)^{1/2} - 2\zeta^2} = 8.69\sqrt{(4 \times 0.39^4+1)^{1/2} - 2\times 0.39^2}$$

$$= 8.69\sqrt{0.74} = 7.47 \text{ rad/sec.} \quad \textbf{Ans.}$$

Phase margin :

$$\phi_m = \tan^{-1}\left[\frac{2\zeta}{\sqrt{(4\zeta^4+1)^{1/2} - 2\zeta^2}}\right] = \tan^{-1}\left[\frac{2\times 0.39}{\sqrt{(4\times 0.39^4+1)^{1/2} - 2\times 0.39^2}}\right]$$

$$= \tan^{-1}(1.054) = 46.5°. \quad \textbf{Ans.}$$

7.21 ROOT LOCUS

The stability of a closed-loop control system is determined from the location of the roots* of the characteristic equation 1 + $G(s)\,H(s)$ = 0. For a system to be stable the roots of its characteristic equation should be located in the L.H.S. of s-plane.

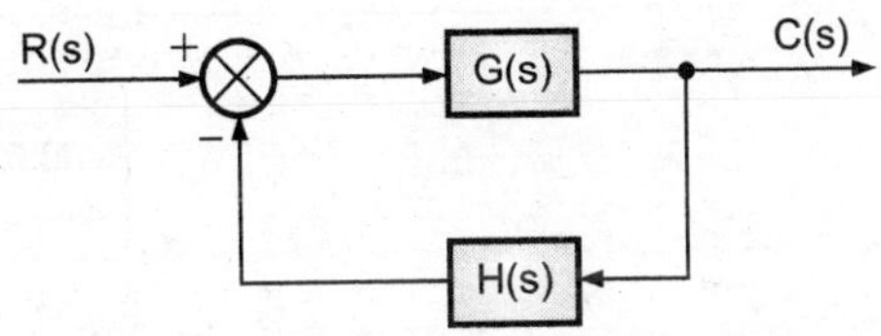

Fig. 7.21.1. Block diagram of closed-loop control system.

A closed-loop control system having forward path gain K is shown by the block diagram in Fig. 7.21.1.

The characteristic equation of the system is given by

$$1 + G(s)\,H(s) = 0 \qquad ...(7.90)$$

Any root of Eq. (7.90) satisfies following two conditions :

$$|\,G(s)\,H(s)\,| = 1 \qquad ...(7.91)$$

and
$$\angle G(s)\,H(s) = \pm\,(2k+1)\,180°$$

where k = 0, 1, 2, ...(7.92)

The root locus method of analysis is a process of determining the points in s-plane satisfying Eqs. (7.91) and (7.92). Usually the forward path gain factor K is considered as an independent variable and the roots of 1 + $G(s)\,H(s)$ = 0 as dependent variables, the roots are plotted in s-plane with K as variable parameter. From the location of the roots in s-plane the nature of time response and system stability can be ascertained.

The root locus for the three examples given below is plotted by determining the roots of the characteristic equation, the variable parameter being K. The information obtained from the root locus plotted as above is used to frame the procedure for plotting the root locus.

*The poles of closed-loop transfer function are also the roots of the characteristic equation.

Example 7.21.1. *Let* $G(s)\,H(s) = \dfrac{Ks}{(s+1)}$.

Solution. The characteristic equation of the system is

$$1 + G(s)\,H(s) = 0 \qquad \therefore \qquad 1 + \frac{Ks}{(s+1)} = 0$$

or $$\frac{(s+1)+Ks}{(s+1)} = 0 \qquad \text{or} \qquad s\,(1+K) + 1 = 0$$

The root of the characteristic equation is

$$s = -\frac{1}{1+K}$$

For various values (+ ve) of K the roots are

K	0	1	9	99	∞
s	– 1	– 0.5	– 0.1	– 0.99	0

The roots determined above are plotted in s-plane as shown in Fig. 7.21.2.

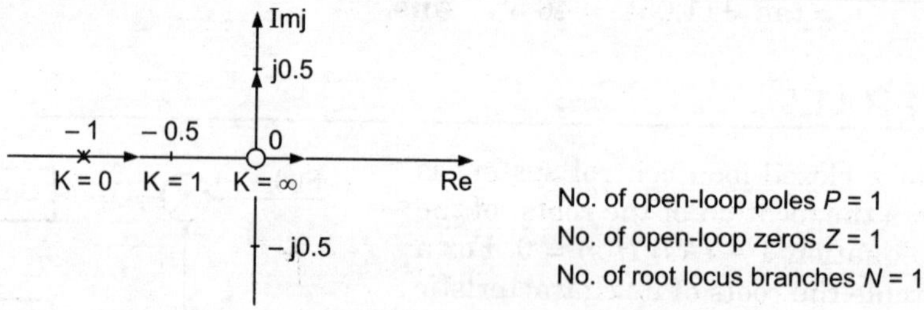

Fig. 7.21.2. Root locus for $G(s)\,H(s) = \dfrac{Ks}{(s+1)}$.

Example 7.21.2. *Let* $G(s) = \dfrac{K}{s\,(s+4)}$

and $H(s) = 1$.

Solution. The characteristic equation of the system is

$$1 + G(s)\,H(s) = 0$$

$$\therefore \qquad 1 + \frac{K}{s\,(s+4)}\cdot 1 = 0$$

or $$s^2 + 4s + K = 0$$

The two roots of above equation are

$$s_1, s_2 = -2 \pm \sqrt{4-K}$$

For various values of K the roots are

K	0	2	4	6	∞
s_1	0	– 1	– 2*	$-2 + j\sqrt{2}$	$-2 + j\infty$
s_2	– 4	– 3	– 2*	$-2 - j\sqrt{2}$	$-2 - j\infty$

*Coincident roots on real axis between two open-loop poles.

The roots determined above are plotted in s-plane as shown in Fig. 7.21.3.

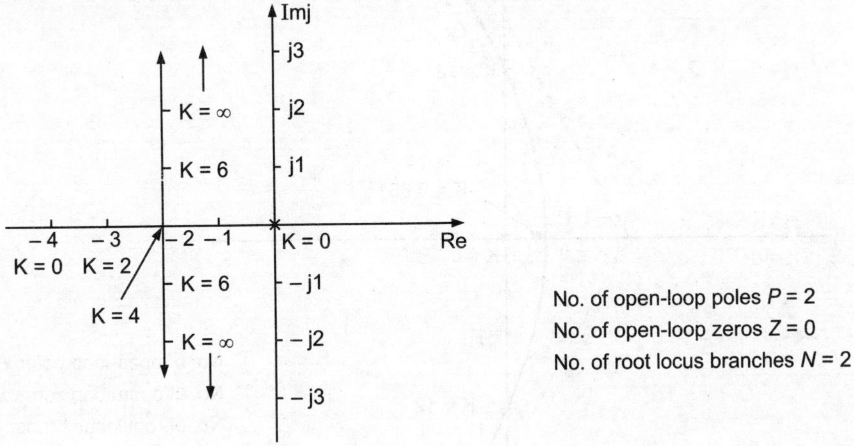

Fig. 7.21.3. Root locus for $G(s)\,H(s) = K/s(s+4)$.

Following conclusions are drawn from the study of root locus plotted in Fig. 7.21.3.

1. For the values of K between $K = 0$ and $K = 4$ the two roots are real negative indicating overdamped time response.

2. For $K > 4$ the two roots are complex with negative real part indicating that the time response is underdamped, *i.e.* exponentially decaying oscillations.

3. Whatever may be the value (+ ve) of K there is no chance for the real part of the roots to become positive, hence the system is inherently stable.

Example 7.21.3. *Let* $G(s) = \dfrac{K}{s\,(s+1)\,(s+3)}$.

Solution. The characteristic equation of the system is

$$1 + G(s)\,H(s) = 0$$

$$\therefore \quad 1 + \frac{K}{s\,(s+1)\,(s+3)} = 0$$

or $$s^3 + 4s^2 + 3s + K = 0$$

For various values of K the three roots of the characteristic equation are given below :

K	0	0.631	5	12	40
s_1	0	– 0.45*	– 0.27 + j0.98	+ j1.732	+ 0.5 + j2.74
s_2	– 1	– 0.45*	– 0.27 – j0.98	– j1.732	+ 0.5 – j2.64
s_3	– 3	– 3.1	– 3.55	– 4.1	– 5

The roots determined above are plotted in s-plane as shown in Fig. 7.21.4.

From the root locus plot shown in Fig. 7.21.4 it is observed that for the values of $K < 12$ all the three roots lie in the LHS of s-plane indicating that the system is stable. However, for the values of $K > 12$ the two roots have positive real part and as such the system is unstable.

*Coincident roots on real axis between two open-loop poles.

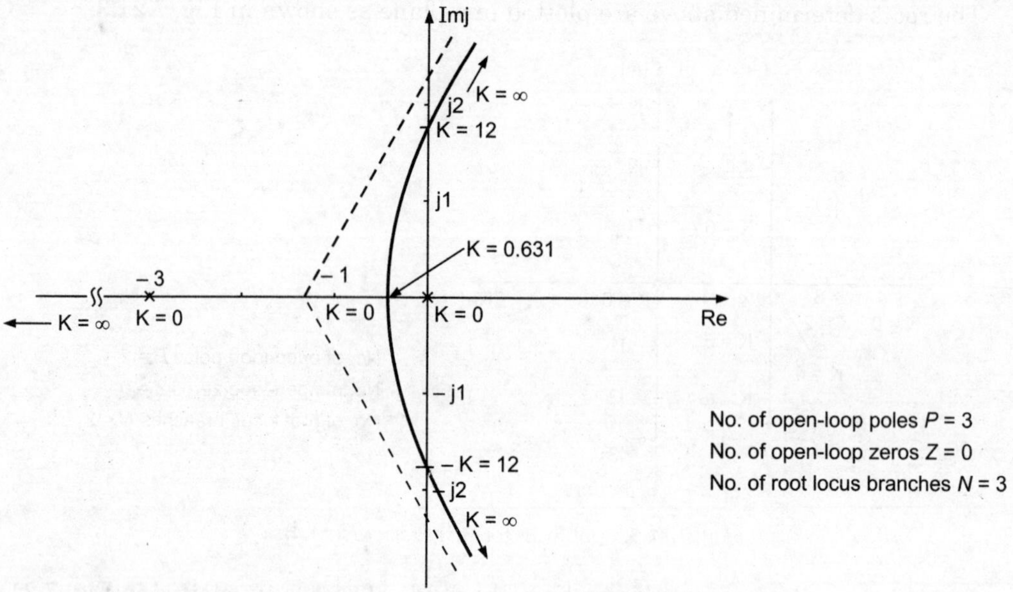

Fig. 7.21.4. Root locus for $G(s)\,H(s) = K/s(s+1)(s+3)$.

Example 7.21.4. *A typical root locus plot for a system having 5 open-loop poles and 1 open-loop zero is shown in Fig. 7.21.5.*

Solution.

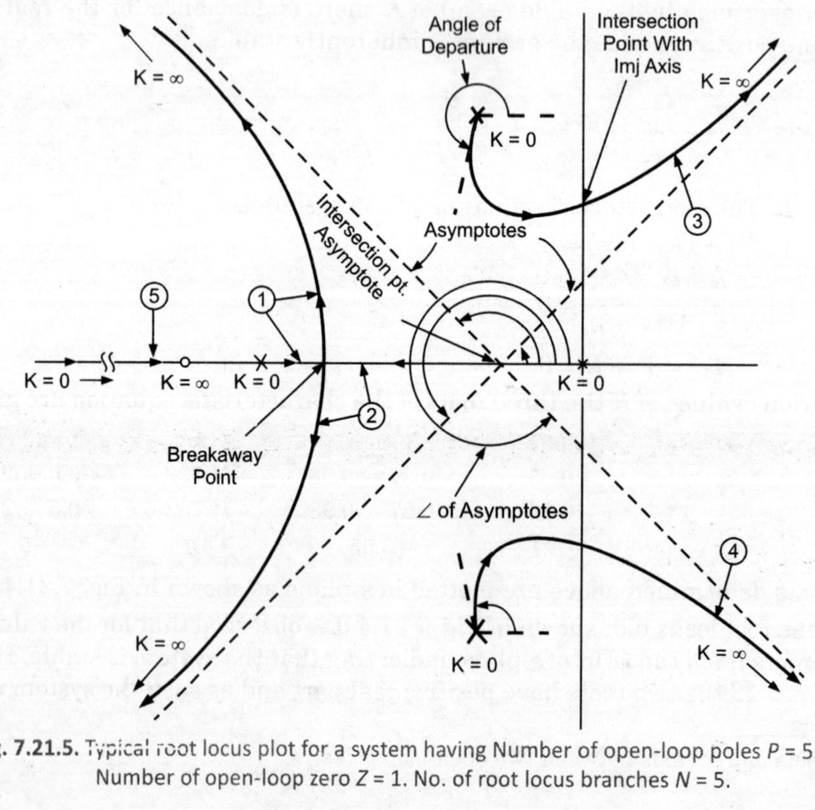

Fig. 7.21.5. Typical root locus plot for a system having Number of open-loop poles $P = 5$ and Number of open-loop zero $Z = 1$. No. of root locus branches $N = 5$.

7.22 SALIENT FEATURES OF ROOT LOCUS PLOT

The salient features as observed from the root locus plots for the examples given in section 7.21.1 are as follows :

1. The root locus starts ($K = 0$) from the open-loop poles and terminates ($K = \infty$) on either finite open-loop zeros or infinity.

2. The number of separate branches of the root locus equals either the number of open-loop poles or number of open-loop zeros whichever is greater.

3. A section of root locus lies on the real axis if the total number of open-loop poles and zeros to the right of the section is odd.

4. The root locus is symmetrical with respect to the real axis.

5. At some point on the root locus between two open-loop poles the root locus breaks away.

6. For higher values of K the root locus can be approximated by asymptotic lines and these asymptotic lines intersect at a point on the real axis.

7. The asymptotic lines make an angle with real axis at an intersection point (centre of gravity).

8. If the root locus intersects the imaginary axis then the points of intersection are conjugate.

9. From the open-loop complex pole the root locus departs making an angle with the horizontal line.

7.23 THE PROCEDURE FOR PLOTTING ROOT LOCUS

The stepwise procedure for plotting the root locus for a given open-loop transfer function based on salient features mentioned in section 7.22 is given below :

1. **Starting points.** The root locus starts ($K = 0$) from the open-loop poles.

2. **Ending points.** The root locus terminates ($K = \infty$) either on open-loop zero or infinity.

3. **Number of branches.** The number of branches of the root locus are

$$N = P, \quad \text{if } P > Z$$
$$= Z, \quad \text{if } Z > P$$

Usually, $P > Z$, therefore, $N = P$.

4. **Existence on real axis.** The existence of the root locus on a section of real axis is confirmed if the sum of the open-loop poles and zeros to the right of the section is odd.

5. **Breakaway points.** On the root locus between two open-loop poles the roots move towards each other as the gain factor K is increased till they are coincident. At the coincident point the value of K is maximum as far as the portion of the root locus between the two open-loop poles is concerned. Any further increase in the value of K, breaks the root locus in two parts. The breakaway point can be determined by rewriting the characteristic equation and therefrom solving for the value of s from the equation given below :

$$\frac{dK}{ds} = 0$$

6. **The angle of asymptotes.** For higher values of K the root locus branches are approximated by asymptotic lines making an angle with the real axis given by

$$\frac{(2k+1)\times 180^\circ}{P-Z} \quad \text{where } k = 0, 1, 2, \ldots \text{ upto } (P-Z)-1$$

7. **Intersection of asymptotes on real axis.** The asymptotes intersect at a point x on the real axis given by

$$x = \frac{\Sigma \text{ Poles} - \Sigma \text{ Zeros}}{P-Z}$$

8. **Intersection points with imaginary axis.** The value of K and the point at which the root locus branch crosses the imaginary axis is determined by applying Routh criterion to the characteristic equation. The roots at the intersection point are imaginary.

9. **The angle of departure from complex pole.** The angle of departure of the root locus from a complex pole is given by

$$\phi_d = 180^\circ - (\phi_p - \phi_z)$$

where ϕ_p is the sum of all the angles subtended by remaining poles.

and ϕ_z is the sum of all the angles subtended by zeros.

The angle of departure is tangent to the root locus at the complex pole.

10. **The angle of arrival at complex zeros.** The angle of arrival of the root locus at a complex zero is given by

$$\phi_a = 180^\circ - (\phi_z - \phi_p)$$

where ϕ_z is the sum of all the angles subtended by remaining zeros.

and ϕ_p is the sum of all the angles subtended by poles.

The angle of arrival is tangent to the root locus at the complex zero.

The steps given above are sufficient to plot the root locus for a given open-loop transfer function. The calculation of the value of K at a specified point on the root locus, determination of damping ratio ζ, assessment of time response and stability etc. are illustrated in the following solved examples.

7.24 SOLVED EXAMPLES : ROOT LOCUS

Example 7.24.1. *A unity feedback control system has an open-loop transfer function*

$$G(s) = \frac{K}{s\,(s+4)}$$

Draw the root locus and determine the value of K, if the damping ratio ζ is to be 0.707.

Solution. 1. The root loci start ($K = 0$) from $s = 0$ and $s = -4$.

2. As there is no open-loop zero root loci terminate ($K = \infty$) at infinity.

3. As the number of poles is 2 the number of root locus branches $N = 2$.

4. The root locus on the real axis exists between $s = 0$ and $s = -4$.

5. Breakaway points.

The characteristic equation is

$$s\,(s + 4) + K = 0 \qquad \text{or} \qquad K = -(s^2 + 4s)$$

$$\therefore \quad \frac{dK}{ds} = -(2s+4)$$

Put $$\frac{dK}{ds} = 0 \quad \therefore \quad (2s+4) = 0$$

$\therefore$ $s = -2$ is a breakaway point.

6. The angle of asymptotes is given by

$$\frac{(2k+1)\,180°}{P-Z} \text{ where } k = 0 \text{ and } 1$$

The two asymptotes are at an angle of

(i) $$\frac{(2\times 0+1)\,180°}{2-0} = 90°$$

(ii) $$\frac{(2\times 1+1)\,180°}{2-0} = 270°$$

7. The asymptotes intersect on real axis at a point given by

$$x = \frac{\Sigma \text{ Poles} - \Sigma \text{ Zeros}}{P-Z} = \frac{(0-4)-0}{2} = -2$$

With the data calculated in steps (1) to (7) the root locus is plotted in Fig. 7.24.1.

Since $\zeta = 0.707$

and $\cos\theta = \zeta$ (Refer Fig. 6.4.7)

$\therefore$ $\cos\theta = 0.707$

$\theta = 45°$

From the origin a line at an angle $\theta = 45°$ is drawn as shown in Fig. 7.24.1 which intersects the root locus plot at $s = (-2 + j2)$. As the point $s = -2 + j12$ lies on the root locus the following equation (magnitude condition) should be satisfied at $s = (-2 + j2)$

$$|G(s)\,H(s)| = 1$$

or $$\left|\frac{K}{s\,(s+4)}\right| = 1$$

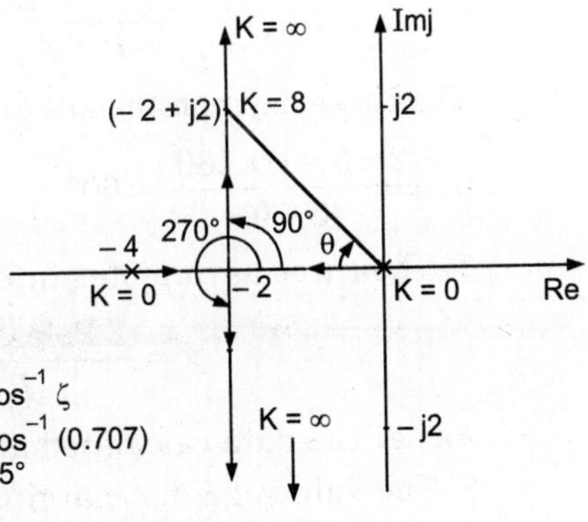

Fig. 7.24.1. Root locus for $G(s)\,H(s) = K/s(s+4)$.

$$\therefore \quad \left|\frac{K}{(-2+j2)\,[(-2+j2)+4]}\right| = 1$$

$$\left|\frac{K}{(-2+j2)\,(2+j2)}\right| = 1$$

$$\left|\frac{K}{\sqrt{(2^2+2^2)}\,\sqrt{(2^2+2^2)}}\right| = 1$$

$$\frac{K}{\sqrt{8}\,\sqrt{8}} = 1 \text{ or } K = 8. \quad \textbf{Ans.}$$

Example 7.24.2. *Sketch the root locus for the open-loop transfer function of a unity feedback control system given below and determine*

(i) The value of K for $\zeta = 0.5$ *(ii) the value of K for marginal stability (iii) the value of K at s = – 4 and obtain the closed-loop transfer function for K = 1.66.*

$$G(s) = \frac{K}{s(s+1)(s+3)}.$$

Solution. 1. The root loci start ($K = 0$) at $s = 0$, $s = -1$, $s = -3$.

2. As there is no open-loop zero the root loci terminate at infinity.

3. $P = 3$, $Z = 0$ ($P > Z$), therefore, number of branches of the root locus $N = 3$.

4. The root locus on the real axis exists between $s = 0$ and $s = -1$ and beyond $s = -3$.

5. Between open-loop poles $s = 0$ and $s = -1$, there exists a breakaway point as calculated below :

The characteristic equation is

$$s(s+1)(s+3) + K = 0$$

or
$$K = -(s^3 + 4s^2 + 3s)$$

$\therefore$
$$\frac{dK}{ds} = -(3s^2 + 8s + 3)$$

Put
$$\frac{dK}{ds} = 0$$

$\therefore$
$$3s^2 + 8s + 3 = 0$$

or
$$s = -1.33 \pm 0.88$$

$\therefore$
$$s_1 = -0.45 \text{ and } s_1 = -2.21$$

As the breakaway point has to lie between $s = 0$ and $s = -1$, the breakaway point is $s = -0.45$.

6. The angle of asymptotes is given by

$$\frac{(2k+1)\,180°}{P-Z} \quad \text{where } k = 0, 1 \text{ and } 2$$

The three asymptotes are at any angle of

$$(i)\ \frac{(2\times0+1)\times180°}{(3-0)} = 60° \quad (ii)\ \frac{(2\times1+1)\times180°}{(3-0)} = 180° \quad (iii)\ \frac{(2\times2+1)\times180°}{(3-0)} = 300°$$

7. The three asymptotes intersect on the real axis at

$$x = \frac{\Sigma\text{ Poles} - \Sigma\text{ Zeros}}{P-Z} = \frac{(0-1-3)-(0)}{3-0} = -1.33.$$

As per the data calculated above the root locus plot is sketched in Fig. 7.24.2.

8. The value of K for marginal stability is determined by applying Routh criterion to the characteristic equation.

$$s(s+1)(s+3) + K = 0 \quad \text{or} \quad s^3 + 4s^2 + 3s + K = 0$$

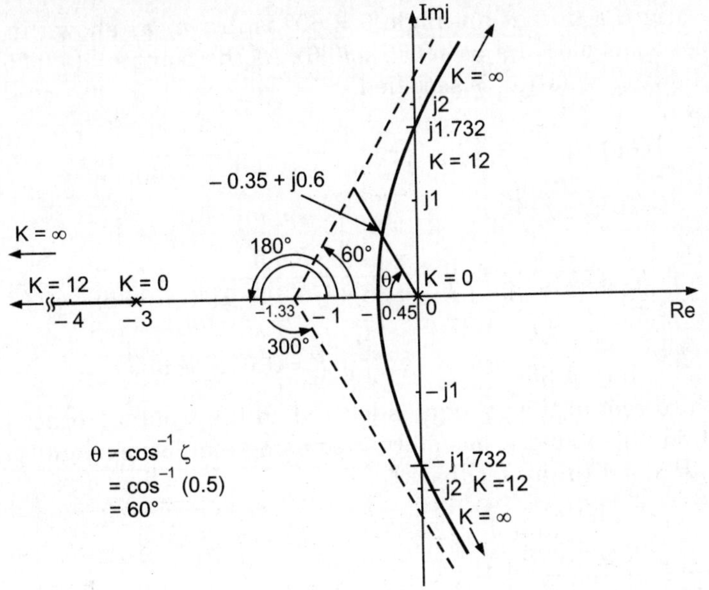

Fig. 7.24.2. Root locus for $G(s) = K/s(s + 1)(s + 3)$.

The Routh array is tabulated below

s^3	1	3
s^2	4	K
s^1	$\dfrac{(12 - K)}{4}$	
s^0	K	

The value of K at which the root locus plot crosses the imaginary axis is determined by equating the first term in s^1 row to zero, therefore,

$$\frac{(12 - K)}{4} = 0 \qquad \therefore \qquad K = 12$$

(*i*) For $K > 12$ the roots lie in the R.H.S. of s-plane, hence $K = 12$ is the value for marginal stability. **Ans.**

At $K = 12$ the root locus plot intersects the imaginary axis and the value of s at the intersection point is determined by solving auxiliary equation formed from the s^2 terms in Routh array, therefore,

$$4s^2 + K = 0 \quad \text{or} \quad 4s^2 + 12 = 0 \qquad \therefore \qquad s = \pm j1.73$$

At $K = 12$ the system exhibits sustained oscillations, the frequency of these oscillations is given by the intersection point of the root locus plot with the imaginary axis, therefore, at $K = 12$, the frequency of sustained oscillations is $\omega = 1.73$ rad/sec.

(*ii*) Since $\zeta = 0.5$

and $\cos\theta = \zeta$ (Refer Fig. 6.4.7.)

$\cos\theta = 0.5$

$\therefore \qquad \theta = 60°$

From the origin a line at an angle $\theta = 60°$ is drawn as shown in Fig. 7.24.2 which intersects the root locus plot at $s = (-0.35 + j0.6)$. As the point $s = (-0.35 + j0.6)$ lies on the root locus the following equation is satisfied.

$$|G(s)\,H(s)| = 1 \quad \text{or} \quad \left|\frac{K}{s\,(s+1)\,(s+3)}\right| = 1$$

Put $\quad s = (-0.35 + j0.6)$

$$\therefore \quad \left|\frac{K}{(-0.35 + j0.6)\,[(-0.35 + j0.6) + 1]\,[(-0.35 + j0.6) + 3]}\right| = 1$$

$$\therefore \quad \frac{K}{1.66} = 1 \quad \text{or} \quad K = 1.66. \quad \textbf{Ans.}$$

(*iii*) The third root of the ch. eqn. is located on the *s*-plane negative real axis beyond $s = -3$. For $K = 1.66$, this root can be determined using root locus magnitude condition $|G(s)\,H(s)| = 1$. Since, $H(s) = 1$ (unity feedback)

$$\therefore \quad |G(s)| = 1$$

$$\therefore \quad \left|\frac{K}{s\,(s+1)\,(s+3)}\right| = 1$$

Put $\quad K = 1.66$

$$\therefore \quad \left|\frac{1.66}{s\,(s+1)\,(s+3)}\right| = 1$$

The equation above is satisfied (to a fair approximation) at $s = -3.231$.

The overall transfer function is given by :

$$\frac{C\,(s)}{R\,(s)} = \frac{G\,(s)}{1 + G\,(s)} = \frac{K/[s\,(s+1)\,(s+3)]}{1 + \{K/[s\,(s+1)\,(s+3)]\} \times 1} = \frac{K}{s\,(s+1)\,(s+3) + K}$$

Put $\quad K = 1.66 \therefore \dfrac{C\,(s)}{R\,(s)} = \dfrac{1.66}{s\,(s+1)\,(s+3) + 1.66}$

The characteristic equation n is $s\,(s+1)\,(s+3) + 1.66 = 0$. As per root locus plot, the three roots are $s_1, s_2 = (-0.35 \pm j\,0.6)$, $s_3 = -3.231$.

$$\therefore \quad \frac{C\,(s)}{R\,(s)} = \frac{1.66}{(s + 3.231)\,(s + 0.35 + j\,0.6)\,(s + 0.35 - j\,0.6)}$$

or

$$\frac{C\,(s)}{R\,(s)} = \frac{1.66}{(s + 3.231)\,(s^2 + 0.7\,s + 0.4825)}. \quad \textbf{Ans.}$$

Example 7.24.3. *The open-loop transfer function of a control system is given by*

$$G(s)\,H(s) = \frac{K}{s\,(s+6)\,(s^2 + 4s + 13)}$$

Sketch the root locus and determine : (a) The breakaway points, (b) The angle of departure from complex poles and (c) The stability condition.

Solution. The four open-loop poles $P = 4$ are

$$s = 0,\ s = -6,\ s = (-2 + j3) \text{ and } s = (-2 - j3)$$

The number of open-loop zeros $Z = 0$.

The number of root locus branches $N = P = 4$. On the real axis the root locus lies between two poles $s = 0$ and $s = -6$.

The angle of asymptotes is given by

$$\frac{(2k+1)\times 180°}{P-Z} \qquad \text{where } k = 0, 1, 2, \text{ and } 3$$

The four angles of asymptotes are

(i) $\frac{(2\times 0+1)\times 180°}{4-0} = 45°$ (ii) $\frac{(2\times 1+1)\times 180°}{4-0} = 135°$

(iii) $\frac{(2\times 2+1)\times 180°}{4-0} = 225°$ (iv) $\frac{(2\times 3+1)\times 180°}{4-0} = 315°$

The asymptotes intersect on the real axis at

$$x = \frac{\Sigma \text{ Poles} - \Sigma \text{ Zeros}}{P-Z} = \frac{(0-6-2+j3-2-j3)-0}{4-0} = -\frac{10}{4} = -2.5$$

The characteristic equation is

$$s(s+6)(s^2+4s+13)+K = 0$$

or $$K = -(s^4+10s^3+37s^2+78s)$$

$\therefore$ $$\frac{dK}{ds} = -(4s^3+30s^2+74s+78)$$

Put $$\frac{dK}{ds} = 0 \quad \therefore \quad 4s^3+30s^2+74s+78 = 0$$

The breakaway point on the real axis is given by the real root of the above equation, the real root is determined approximately as follows :

Let $$f(s) = 4s^3+30s^2+74s+78$$

It is known that the breakaway point on the real axis lies somewhere between 0 and -6.

$$f(-5) = -42$$
$$f(-4.5) = -12$$
$$f(-4) = +6$$

Fig. 7.24.3 gives the graphical solution of

$$f(s) = 4s^3+30s^2+74s+78 = 0$$

the equation is satisfied at $s = -4.2$, hence, the breakaway point is $s = -4.2$. **Ans.**

The angle of departure from the pole $(-2+j3)$ is calculated below (Refer Fig. 7.24.4).

$$\phi_{(-2+j3)} = 180° - (\phi_{p1}+\phi_{p2}+\phi_{p3})$$

$$\phi_{p1} = 180° - \tan^{-1}\left(\frac{3}{2}\right) = 123.6°$$

$$\phi_{p2} = \tan^{-1}\left(\frac{3}{4}\right) = 36.86°$$

$$\phi_{p3} = 90°$$

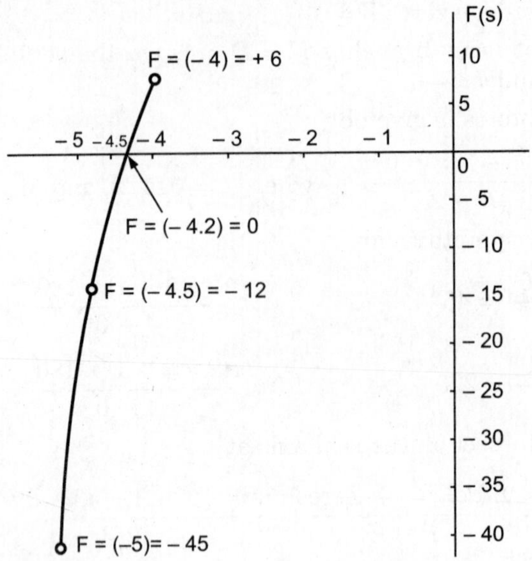

Fig. 6.24.3. Solution of $f(s) = 4s^3 + 30s^2 + 74s + 78 = 0$.

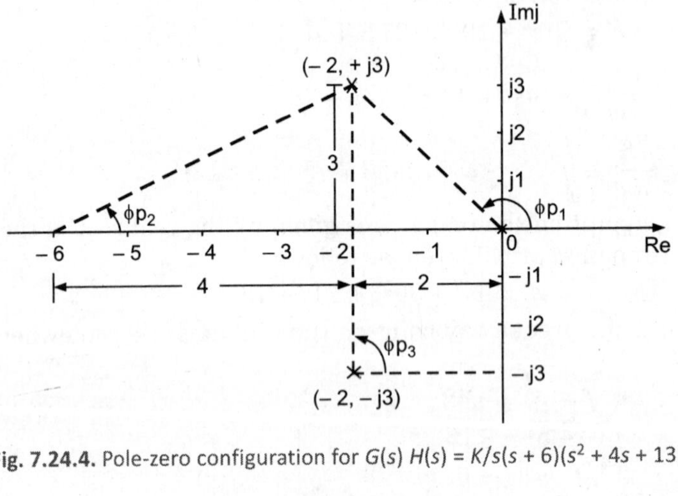

Fig. 7.24.4. Pole-zero configuration for $G(s)\,H(s) = K/s(s + 6)(s^2 + 4s + 13)$.

$\therefore$ $$\phi_{(-2+j3)} = 180° - (123.6° + 36.86° + 90°) = -70.46°. \quad \textbf{Ans.}$$

and $$\phi_{(-2-j3)} = +70.46°. \quad \textbf{Ans.}$$

The intersection of the root locus plot with the imaginary axis is determined by applying Routh criterion to the characteristic equation,

i.e. $$s^4 + 10s^3 + 37s^2 + 78s + K = 0$$

The Routh array is arranged below :

s^4	1	37	K
s^3	10	78	
s^2	29.2	K	
s^1	$(78 - 0.34\,K)$		
s^0	K		

The intersection of the root locus plot with the imaginary axis is given by the value of K obtained by solving following equation :

$$78 - 0.34\,K = 0$$

$$\therefore \quad K = 229.4$$

The intersection point is determined by solving auxiliary equation for s^2 term

$$\therefore \quad 29.2\,s^2 + K = 0$$

or $$29.2\,s^2 + 229.4 = 0$$

or $$s = \pm j2.98$$

The root locus plot is shown in Fig. 7.24.5.

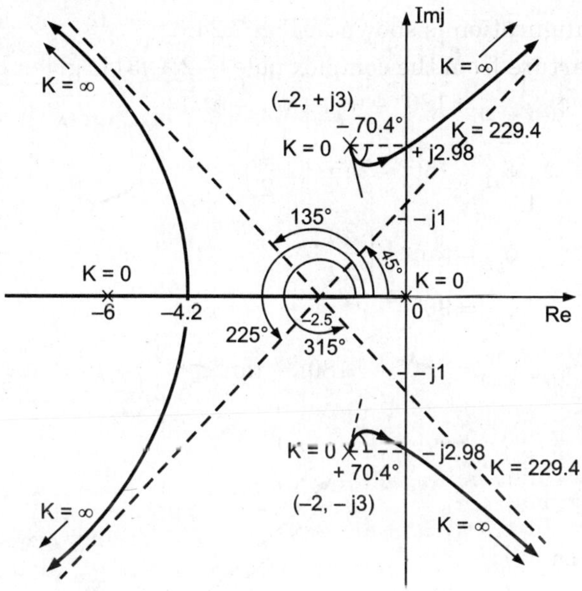

Fig. 7.24.5. Root locus plot for $G(s)\,H(s) = K/s(s + 6)(s^2 + 4s + 13)$.

From the root locus plot (Fig. 7.24.5) it is noted that for $K > 229.4$ the two roots have positive real part, therefore, the system is stable if,

$$K < 229.4. \quad \textbf{Ans.}$$

Example 7.24.4. *Sketch the root locus plot for the system having open-loop transfer function is given by*

$$G(s)\,H(s) = \frac{K}{s\,(s+4)\,(s^2+4s+13)}.$$

Solution. The number of open-loop poles $P = 4$, $s = 0$, $s = -4$, $s = (-2 + j3)$ and $s = (-2 - j3)$

The number of open-loop zeros $Z = 0$

Therefore, the number of branches of root locus $N = P = 4$

The four angles of asymptotes are given by $\dfrac{(2k+1)\times 180°}{P-Z}$, $k = 0, 1, 2,$ and 3

(i) $\dfrac{(2\times 0+1)+180°}{4-0} = 45°$ (ii) $\dfrac{(2\times 1+1)\times 180°}{4-0} = 135°$

(iii) $\dfrac{(2\times 2+1)\times 180°}{4-0} = 225°$ (iv) $\dfrac{(2\times 3+1)\times 180°}{4-0} = 315°$

The asymptotes intersect on the real axis at

$$x = \frac{\Sigma\,\text{Poles} - \Sigma\,\text{Zeros}}{P-Z} = \frac{(0-4-2+j3-2-j3)-(0)}{4-0} = \frac{-8}{4} = -2$$

The pole-zero configuration is shown in Fig. 7.24.6.

The angle of departure from the complex pole $(-2+j3)$ is given by (Refer Fig. 7.24.6)

$$\phi_{(-2+j3)} = 180° - (\phi_{p1} + \phi_{p2} + \phi_{p3})$$

$$\phi_{p1} = 180° - \tan^{-1}\left(\frac{3}{2}\right)$$

$$\phi_{p2} = \tan^{-1}\left(\frac{3}{2}\right)$$

$$\phi_{p3} = 90°$$

$$\therefore \quad \phi_{(-2+j3)} = 180° - \left[180° - \tan^{-1}\left(\frac{3}{2}\right) + \tan^{-1}\left(\frac{3}{2}\right) + 90°\right] = -90°$$

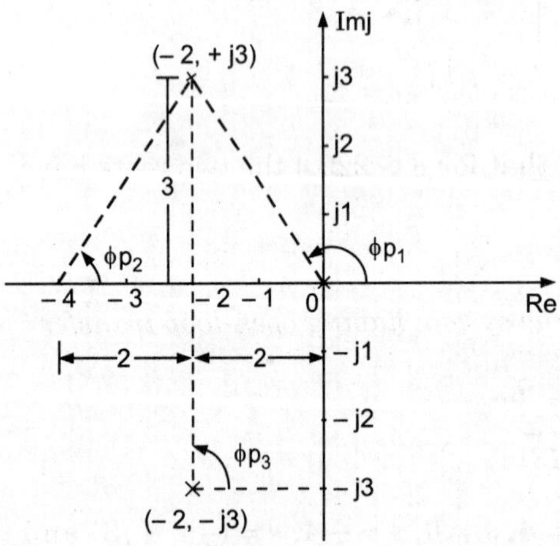

Fig. 7.24.6. Pole-zero configuration for $G(s)\,H(s) = K/s(s+4)(s^2+4s+13)$.

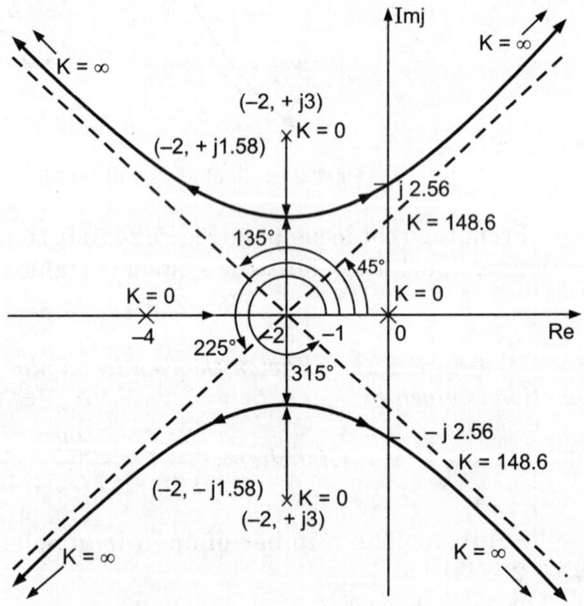

Fig. 7.24.7. Root locus plot for $G(s)\,H(s) = K/s(s+4)(s^2+4s+13)$.

and the angle of departure from the complex pole $(-2-j3)$ is

$$\phi_{(-2-j3)} = +90°.$$

The root locus plot takes the shape as shown in Fig. 7.24.7. There are three break-away points. One of the breakaway point is identified at $s = -2$, the other two breakaway points are complex and can be obtained as follows :

The characteristic equation is

$$s(s+4)(s^2+4s+13) + K = 0$$

or
$$K = -(s^4 + 8s^3 + 29s^2 + 52s)$$

$$\therefore \quad \frac{dK}{ds} = (-4s^3 + 24s^2 + 58s + 52)$$

Put $\frac{dK}{ds} = 0 \qquad \therefore \quad (4s^3 + 24s^2 + 58s + 52) = 0$

The breakaway points are given by the roots of the above equation, one of the root is identified as $s = -2$ (at $s = -2$ above equation is satisfied), therefore,

$$\frac{(4s^3 + 24s^2 + 58s + 52)}{(s+2)} = (4s^2 + 16s + 26)$$

and thus
$$(s+2)(4s^2 + 16s + 26) = 0$$

The two complex breakaway points are given by the roots of the equation $(4s^2 + 16s + 26) = 0$, therefore the complex breakaway points are

$$s = \frac{-16 \pm \sqrt{(16^2 - 4 \times 4 \times 26)}}{2 \times 4} = (-2 + j1.58)$$

The intersection of the root locus plot with the imaginary axis is determined by applying Routh criterion to the characteristic equation

$$s^4 + 8s^3 + 29s^2 + 52s + K = 0$$

The Routh array is arranged below

s^4	1	29	K
s^3	8	52	
s^2	22.5	K	
s^1	$(52 - 0.35\,K)$		
s^0	K		

For stability $(52 - 0.35\,K) \geq 0$

$$\therefore \quad K \leq 148.6$$

The value of s at the intersection point with the imaginary axis where $K = 148.6$ is given by solving the auxiliary equation for s^2 term.

$$\therefore \quad 22.5s^2 + K = 0 \quad \text{or} \quad 22.5s^2 + 148.6 = 0$$

$$\therefore \quad s = \pm j\sqrt{\frac{148.6}{22.5}} = \pm j2.56$$

The root locus plot is sketched in Fig. 7.24.7.

Example 7.24.5. *The open-loop transfer function of a control system is given by*

$$G(s)\,H(s) = \frac{K\,(s+1)^2}{(s+2)^2}\cdot$$ *Show that the root locus lies on a circle.*

Solution. $$G(s)\,H(s) = \frac{K\,(s+1)^2}{(s+2)^2}$$

Let any point in the s-plane $s = (\sigma + j\omega)$ lie on the root locus

$$\therefore \quad G(\sigma + j\omega)\,H(\sigma + j\omega) = \frac{K\,(\sigma + j\omega + 1)^2}{(\sigma + j\omega + 2)^2}$$

and $$\angle G(\sigma + j\omega)\,H(\sigma + j\omega) = 2\tan^{-1}\frac{\omega}{\sigma+1} - 2\tan^{-1}\frac{\omega}{\sigma+2}$$

For the point to lie on root locus, then as per the phase angle condition.

$$\angle G(\sigma + j\omega)\,H(\sigma + j\omega) = \pm\,(2k+1)\,180^\circ\ ;\ k = 0, 1, 2, \ldots$$

Consider $k = 0$

$$\therefore \quad \angle G(\sigma + j\omega)\,H(\sigma + j\omega) = 180^\circ$$

$$\therefore \quad 2\tan^{-1}\frac{\omega}{\sigma+1} - 2\tan^{-1}\frac{\omega}{\sigma+2} = -\,180^\circ$$

or $$2\left\{\tan^{-1}\frac{\left(\dfrac{\omega}{\sigma+1} - \dfrac{\omega}{\sigma+2}\right)}{1+\dfrac{\omega^2}{(\sigma+1)(\sigma+2)}}\right\} = -\,180^\circ$$

or $$\tan^{-1}\frac{\dfrac{\omega}{\sigma+1} - \dfrac{\omega}{\sigma+2}}{1+\dfrac{\omega^2}{(\sigma+1)(\sigma+2)}} = -\,90^\circ$$

Taking tangent on both sides,

$$\frac{\dfrac{\omega}{\sigma+1} - \dfrac{\omega}{(\sigma+1)(\sigma+2)}}{1+\dfrac{\omega^2}{(\sigma+1)(\sigma+2)}} = \tan(-\,90^\circ)$$

or $$\frac{\dfrac{\omega}{\sigma+1} - \dfrac{\omega}{\sigma+2}}{1+\dfrac{\omega^2}{(\sigma+1)(\sigma+2)}} = \infty$$

$$\therefore \quad 1+\frac{\omega^2}{(\sigma+1)(\sigma+2)} = 0$$

or $$(\sigma+1)(\sigma+2) + \omega^2 = 0$$

or $$\sigma^2 + 3\sigma + 2 + \omega^2 = 0$$

or $$\sigma^2 + 3\sigma + 2.25 - 2.25 + 2 + \omega^2 = 0$$

or $$(\sigma + 1.5)^2 + (\omega - 0)^2 = (0.5)^2$$

The equation derived above is of a circle in s-plane with centre $(-1.5, j0)$ and radius 0.5. **Ans.**

Example 7.24.6. *For the open-loop transfer function given below. Determine the value of K at s = – 2 and comment on the stability and time response of the system*

$$G(s)\,H(s) = \frac{K\,(s+1)}{(s^2 + 0.4s + 0.4)}.$$

Solution. The number of open-loop poles $P = 2$

The two open-loop poles are

$$s_1, s_2 = \frac{-0.4 \pm \sqrt{(0.4^2 - 4 \times 1 \times 0.4)}}{2 \times 1} = (-0.2 \pm j0.6)$$

The no. of open-loop zero $Z = 1$ at $(s = -1)$.

As $P > Z$, the number of root locus branches $N = P = 2$. On the real axis the root locus exists between $s = -1$ (zero) and $s = -\infty$.

The angle of asymptote is given by

$$\frac{(2k+1)\times 180^\circ}{P - Z} \qquad \text{where } k = 0$$

Therefore, the angle of asymptote is

$$\frac{(2\times 0+1)\times 180^\circ}{2-1} = 180^\circ$$

The characteristic equation is

$$(s^2 + 0.4s + 0.4) + K\,(s + 1) = 0$$

or $$K = -\frac{(s^2 + 0.4s + 0.4)}{(s+1)}$$

$\therefore$ $$\frac{dK}{ds} = -\left[\frac{(s+1)\,(2s+0.4) - (s^2 + 0.4s + 0.4)\times 1}{(s+1)^2}\right] = -\frac{(s^2+2s)}{(s+1)}$$

In order to determine breakaway point put $(dK/ds = 0)$ and solve for s, therefore,

$$\frac{(s^2+2s)}{(s+1)} = 0$$

or $$s(s + 2) = 0$$

$\therefore$ $$s = 0 \quad \text{and} \quad s = -2$$

As the point $s = 0$ does not lie on the root locus and such is not a breakaway point. The point $s = -2$ lies on the locus and is a breakaway point (between two zeros the root locus breaks away).

It is noted that the breakaway point $s = -2$ and the poles $s = (-0.2 \pm j0.6)$ are equidistant from the zero $s = -1$, hence the root locus plot is an arc of the circle with centre at $s = -1$ and radius 1. The root locus plot is drawn in Fig. 7.24.8.

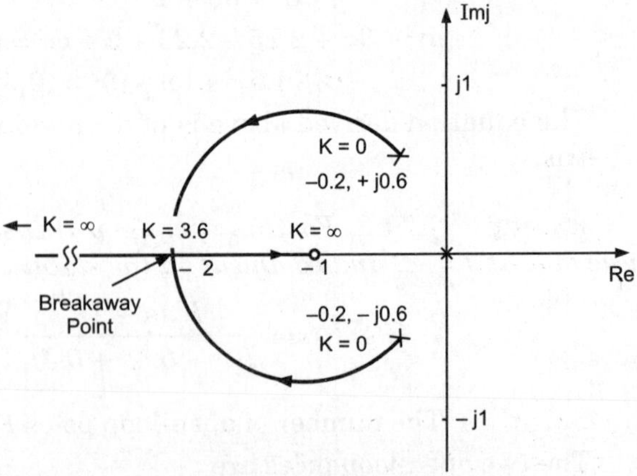

Fig. 7.24.9. Root locus plot for $G(s)\,H(s) = K(s + 1)/(s^2 + 0.4s + 0.4)$.

The value of K at $s = -2$ (breakaway point) is determined by substituting $s = -2$ in the following equation :

$$|G(s)\,H(s)| = 1$$

or
$$\left|\frac{K\,(s+1)}{(s^2+0.4s+0.4)}\right| = 1$$

or
$$\left|\frac{K\,(-2+1)}{[(-2)^2+(0.4)(-2)+0.4]}\right| = 1$$

$$\therefore \qquad K = 3.6.$$

For all values (+ ve) of K the two roots lie in the LHS of s-plane, hence the system is inherently stable.

For $K < 3.6$ the roots are complex with –ve real parts as such the time response is underdamped. At $K = 3.6$ the two roots are coincident and the response is critically damped.

For $K > 3.6$ the two roots are negative real, therefore, the response is overdamped.

Ans.

Example 7.24.7. *The transfer function of a unity feedback control system is given by:*

$$G(s) = \frac{K}{s\,(s+2)\,(s+4)}$$

Determine :

(a) The value of K to have 40% overshoot for unit step input.

(b) The value of K for sustained oscillations in output.

(c) The value of K_v corresponding to value of K as obtained in (a).

(d) The value of settling time t_s.

Solution.

1. The root loci start ($K = 0$) at $s = 0$, $s = -2$ and $s = -4$.
2. As there is no open-loop zero. The root loci terminate ($K = \infty$) at infinity.
3. $P = 3$, $Z = 0$ ($P > Z$). Therefore, number of branches of the root locus $N = 3$.
4. The root locus branches on the real axis exists between $s = 0$ and $s = -2$ and beyond $s = -4$.
5. Between open-loop poles $s = 0$ and $s = -2$, there exists a breakaway point as calculated below :

The characteristic equation is

$$s\,(s + 2)\,(s + 4) + K = 0$$

or
$$K = -(s^3 + 6s^2 + 8s) \quad \therefore \quad \frac{dK}{ds} = -(3s^2 + 12s + 8)$$

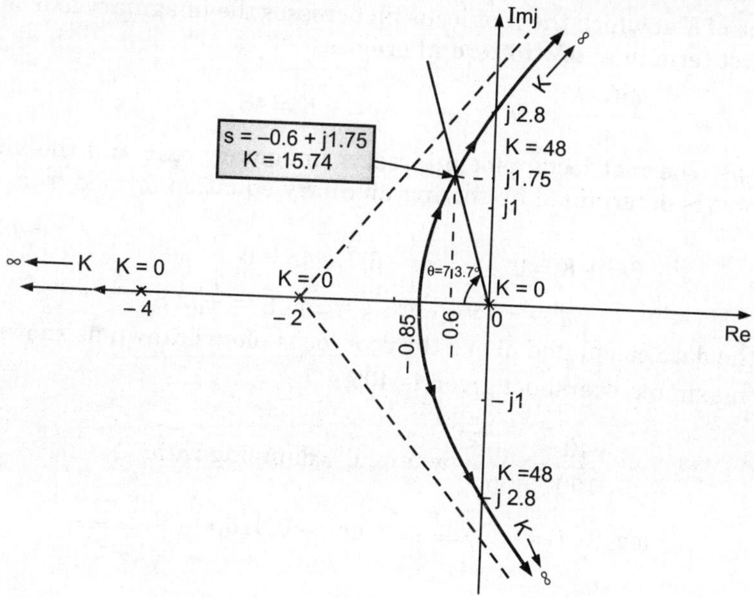

Fig. 7.24.9. Root locus plot for example 7.24.7.

Put $\dfrac{dK}{ds} = 0 \quad \therefore \quad 3s^2 + 12s + 8 = 0 \quad$ or $\quad s^2 + 4s + 2.66 = 0$

$$\therefore \qquad s = \frac{-4 \pm \sqrt{4^2 - 4 \times 1 \times 2.66}}{2} = \frac{-4 \pm 2.3}{2} = -3.15, -0.85.$$

As the breakaway point has to lie between $s = 0$ and $s = -2$, the breakaway point is $s = -0.85$.

6. The angle of asymptotes is given by $\dfrac{(2k+1)}{P-Z} \times 180°$ where, $k = 0, 1, 2$

The three asymptotes are at an angle of

(*i*) $\dfrac{(2 \times 0 + 1)}{3-0} \times 180° = 60°$, (*ii*) $\dfrac{(2 \times 1 + 1)}{3-0} \times 180° = 180°$ and (*iii*) $\dfrac{(2 \times 2 + 1)}{3-0} \times 180° = 300°$

7. The three asymptotes intersect on the real axis at

$$x = \frac{\Sigma \text{ Poles} - \Sigma \text{ Zeros}}{P-Z} = \frac{(0-2-4)-(0)}{3-0}$$

or $\qquad x = -2.$

8. The value of K for marginal stability is determined by applying Routh-Hurwitz criterion to the characteristic equation.

$$s(s+2)(s+4) + K = 0 \qquad \text{or} \qquad s^3 + 6s^2 + 8s + K = 0$$

The Routh array is tabulated below :

s^3	1	8
s^2	6	K
s^1	$\dfrac{48-K}{6}$	
s^0	K	

The value of K at which the root locus plot crosses the imaginary axis is determined by equating the first term in s^1 row to zero, therefore,

$$\frac{48-K}{6}=0 \qquad \therefore \quad K=48$$

At $K = 48$. The root locus plot intersects imaginary axis and the value of s at the intersection point is determined by solving auxiliary equation formed from the s^2 terms in Routh array.

$$\therefore \qquad 6s^2 + K = 0 \quad \text{or} \quad 6s^2 + 48 = 0$$

$$\text{or} \qquad s^2 = -48/6 \quad \text{or} \quad s = \pm j\sqrt{8} = \pm j2.8$$

As per the data calculated above the root locus plot is drawn as shown in Fig. 7.24.9.

(*a*) The maximum overshoot given is 40%

$$\therefore \qquad \frac{40}{100} = e^{\frac{-\zeta\pi}{\sqrt{1-\zeta^2}}} \quad \text{where, } \zeta = \text{damping ratio}$$

$$\text{or} \qquad \log_e(0.4) = -\frac{\zeta\pi}{\sqrt{1-\zeta^2}} \quad \text{or} \quad -0.916 = -\frac{\zeta\pi}{\sqrt{1-\zeta^2}}$$

$$\text{or} \qquad \frac{\zeta^2}{1-\zeta^2} = \left(\frac{+0.916}{\pi}\right)^2 = 0.0846$$

$$\text{or} \qquad \zeta^2 = 0.0846 - 0.0846\,\zeta^2 \qquad \text{or} \quad \zeta^2 = 0.078 \quad \therefore \quad \zeta = 0.28$$

$$\theta = \cos^{-1}\zeta = \cos^{-1} 0.28 = 73.7^\circ$$

A line passing through origin and making an angle of 73.7° with – ve real axis is drawn on the root locus plot. This line intersects root locus plot at $s = -0.6 + j1.75$.

The value of K at $s = (-0.6 + j1.75)$ is calculated below, which is the corresponding value for 40% overshoot.

The point $s_1 = -0.6 + j1.75$ lies on the root locus. Therefore,

$$|1| = \frac{K}{s_1(s_1+2)(s_1+4)}$$

$$\therefore \qquad |1| = \frac{K}{|(0.6+j1.75)||(-0.6+j1.75+2)||(-0.6+j1.75+4)|}$$

$$= \frac{K}{|(0.6+j1.75)||(1.4+j1.75)||(3.4+j1.75)|}$$

$$= \frac{K}{\sqrt{0.6^2+1.75^2}\,\sqrt{1.4^2+j1.75^2}\,\sqrt{3.4^2+1.75^2}} = \frac{K}{1.84\times 2.24\times 3.82}$$

$$\therefore \qquad K = 1.84 \times 2.24 \times 3.82 \quad \text{or} \quad K = 15.74. \quad \textbf{Ans.}$$

(*ii*) The root locus plot crosses the imaginary axis at $K = 48$, hence, for sustained oscillations $K = 48$. **Ans.**

(*c*)
$$K_v = \lim_{s\to 0} s\,G(s)\,H(s) = \lim_{s\to 0} s\,\frac{K}{s(s+2)(s+4)}$$

$\because$ $K = 15.74$ as determined in (*a*)

$$K_v = \lim_{s\to 0} \frac{15.74}{(s+2)(s+4)} = \frac{15.74}{2\times 4} = 9.96. \quad \textbf{Ans.}$$

(*d*) $s = -0.6 + j1.75$ $\therefore$ $\zeta\omega_n = 0.6$

Settling time $t_s = \dfrac{4}{\zeta\omega_n} = \dfrac{4}{0.6} = 6.66$ sec. **Ans.**

Example 7.24.8. *Draw the root locus plot for a system having open-loop transfer function as :* $G(s)\,H(s) = \dfrac{K}{s\,(s+1)\,(s+3)}$

Determine : (a) Gain margin and (b) Phase margin, for $K = 6$.

Solution. 1. The root locus branches start ($K = 0$) from open-loop poles $s = 0$, $s = -1$ and $s = -3$.

2. There are no finite zeros, as such root locus branches terminate ($K = \infty$) at infinity.

3. $P = 3$, $Z = 0$, hence number of branches $N = 3$.

4. On the real axis section, the root locus exists between $s = 0$ and $s = -1$ and beyond $s = -3$.

5. Angle of asymptotes is given by $\dfrac{(2k+1)}{P-Z} \times 180°$ where $k = 0, 1$, and 2.

$\therefore$ (*i*) $\dfrac{(2\times0+1)}{3-0} \times 180° = 60°$, (*ii*) $\dfrac{(2\times1+1)}{3-0} \times 180° = 180°$ (*iii*) $\dfrac{(2\times2+1)}{3-0} \times 180° = 300°$

6. The asymptotes intersect on real axis at :

$$x = \frac{\Sigma P - \Sigma Z}{P - Z} = \frac{(0-1-3)-(0)}{3} = -1.33$$

7. The breakaway points are determined below :

The characteristic equation is given by $1 + G(s)\,H(s) = 0$

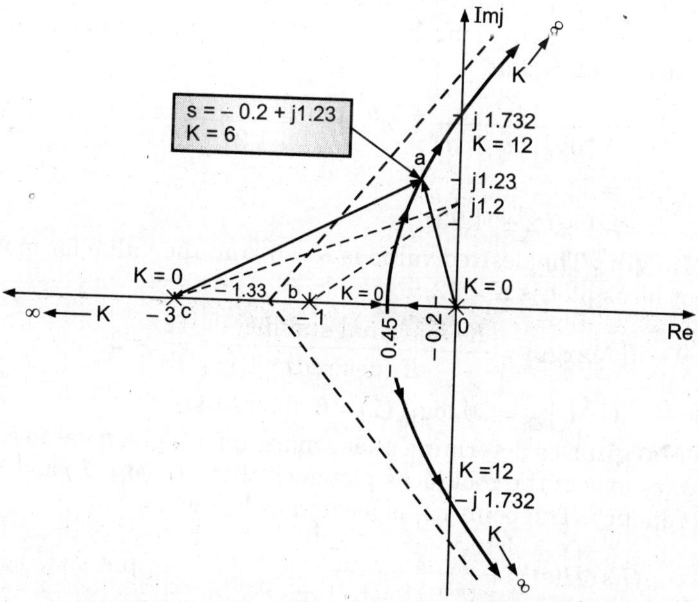

Fig. 7.24.10. Root locus plot for example 7.24.8.

$$\therefore \qquad 1+\frac{K}{s\,(s+1)\,(s+3)}=0$$

or $\qquad s\,(s + 1)\,(s + 3) + K = 0 \quad \text{or} \quad K = -\,(s^3 + 4s^2 + 3s)$

$\therefore \qquad dK/ds = -\,(3s^2 + 8s + 3)$

put $\quad dK/ds = 0 \qquad \therefore \quad 3s^2 + 8s + 3 = 0$

or $$s=\frac{-8\pm\sqrt{8^2-4\times3\times3}}{2\times3} \quad \text{or} \quad s = -\,1.33 \pm 0.88 = -\,0.45, -\,2.21$$

As the breakaway point has to lie between $s = 0$ and $s = -\,1$, the breakaway point is $s = -\,0.45$.

8. The value of K for marginal stability is determined by applying Routh criterion to the characteristic equation. Therefore,

$$s\,(s + 1)\,(s + 3) + K = 0 \qquad \text{or} \qquad s^3 + 4s^2 + 3s + K = 0$$

The Routh array is tabulated below :

s^3	1	3
s^2	4	K
s^1	$\frac{12-K}{4}$	
s^0	K	

For marginal stability

$$12 - K = 0 \qquad \text{or} \qquad K = 12$$

The intersection of root locus plot with imaginary axis is given by solving the equation,

$$4s^2 + K = 0$$

or $$4s^2 + 12 = 0 \qquad \text{or} \qquad s = \pm\, j1.732.$$

From the data determined above, the root locus plot is drawn as shown in Fig. 7.24.10.

On the root locus plot at $s = a$ the value $K = 6$ is marked (by trial and error).

This is obtained by drawing lengths oa, ba and ca from open-loop poles at $s = 0$, $s = -\,1$ and $s = -\,3$ respectively, such that,

$$|1|=\frac{K}{|oa|\,|ba|\,|ca|} \quad \text{or} \quad |1|=\frac{K}{1.21\times1.6\times3.1} \quad \text{or} \quad K = 1.21 \times 1.6 \times 3.1$$

$\therefore \qquad K = 6$

At "a", $\qquad s = (-\,0.2 \pm j1.23)$

(*a*) **Gain Margin :** The desired value is $K = 6$, and the value for marginal stability as obtained from root locus plot is $K = 12$.

$$\therefore \qquad \text{Gain Margin} = \frac{K\,(\text{marginal stability})}{K\,(\text{desired})}=\frac{12}{6}=2$$

or $$\text{G.M.}\,|_{db} = 20 \log_{10} (2) = 6 \text{ db.} \quad \textbf{Ans.}$$

(*b*) **Phase Margin :** To determine phase margin for a given value $K = 6$, locate a point $j\omega_1$ on the imaginary axis of the root locus plot such that $|G(j\omega_1)\,H(j\omega_1)| = 1$, thus, ω_1 is the gain crossover frequency. The point $j\omega_1$ is located as below

$$G(s)\,H(s)=\frac{K}{s\,(s+1)\,(s+3)} \qquad \text{put } s = j\omega$$

$$\therefore \qquad G(j\omega)\,H(j\omega) = \frac{K}{j\omega\,(j\omega+1)\,(j\omega+3)}$$

At the gain crossover frequency ω_1, $|G(j\omega_1)\,H(j\omega_1)| = 1$.

To determine the gain crossover frequency, put $|G(j\omega_1)\,H(j\omega_1)| = 1$ and the given value $K = 6$

$$\therefore \qquad 1 = \left|\frac{K}{j\omega_1\,(j\omega_1+1)\,(j\omega_1+3)}\right| \quad \text{or} \quad 1 = \frac{6}{\omega_1\sqrt{\omega_1^2+1}\,\sqrt{\omega_1^2+9}}$$

or $\qquad \omega_1^2\,(\omega_1^2+1)\,(\omega_1^2+9) = 36$

Above equation is nearly satisfied at $\omega_1 = 1.2$ rad/sec.

Hence, $\omega_1 = 1.2$ rad/sec. is the gain crossover frequency.

$$\text{Phase margin} = 180° + \angle G(j\omega_1)\,H(j\omega_1) = 180° + (-\theta_{p_1} - \theta_{p_2} - \phi_{p_3})$$

$$= 180° + \left(-90° - \tan^{-1}\frac{1.2}{1} - \tan^{-1}\frac{1.2}{3}\right)$$

$$= 180° - (90° + 50.2° + 21.8°) = 180° - (162°) = 18°$$

Hence, P.M. = 18°. **Ans.**

***Note :** $\theta_{p_1}, \theta_{p_2}$, and θ_{p_3} are taken as – ve angles as they are contributed by the poles of $G(s)\,H(s)$.

Example 7.24.9. *Sketch the root locus plot for the open-loop transfer function given below :*

$$G(s)\,H(s) = \frac{K\,(s^2+4)}{s\,(s+2)}$$

Calculate the value of K at (a) breakaway point (b) s = – 0.69 + j0.9.

Solution. 1. The root locus branches start ($K = 0$) from open-loop poles $s = 0$ and $s = -2$.

2. The root locus branches terminate ($K = \infty$) on open-loop zeros $s = +j2$ and $s = -j2$.

3. $P = Z = 2$, hence, the number of root locus branches $N = 2$.

4. The root locus branches on the real axis appear between $s = 0$ and $s = -2$.

5. The breakaway point is determined below :

The characteristic equation is

$$1 + G(s)\,H(s) = 0 \quad \therefore \quad 1 + \frac{K\,(s^2+4)}{s\,(s+2)} = 0$$

or $\quad s(s+2) + K\,(s^2+4) = 0$

$$\therefore \qquad K = -\frac{s(s+2)}{(s^2+4)} = -\frac{s^2+2s}{s^2+4}$$

$$\frac{dK}{ds} = -\frac{[(s^2+4)(2s+2) - (s^2+2s)(2s)]}{(s^2+4)^2} = -\frac{-2s^2+8s+8}{(s^2+4)^2} = \frac{2s^2-8s-8}{(s^2+4)}$$

Put $\dfrac{dK}{ds} = 0$

$\therefore \quad 2s^2 - 8s - 8 = 0 \quad \text{or} \quad s^2 - 4s - 4 = 0$

$$\therefore \quad s = \frac{+4 \pm \sqrt{4^2 - 4 \times 1 \times (-4)}}{2}$$

$$= +2 \pm 2\sqrt{2} = 4.82, -0.82.$$

As the point $s = -0.82$ lies on the root locus, therefore, $s = -0.82$ is a breakaway point.

6. The root locus branches for $K \to \infty$ does not terminate on zeros at infinity, therefore, asymptotes does not appear on the root locus plot.

7. As the root locus branches terminate ($K = \infty$) on zeros $s = \pm j2$ located on imaginary axis, it does not cross the same.

8. The angle of arrival of root locus branches at complex zeros ($s = \pm j2$) is calculated below :

$$\theta_{a(+j2)} = 180° - (\theta_z - \theta_p)$$

$$= 180° - [\theta_{z_2} - (\theta_{p_1} + \theta_{p_2})]$$

$$= 180° - \left\{90° - \left[90° + \tan^{-1}\left(\frac{2}{2}\right)\right]\right\}$$

$$= 180° - [90° - (90° + 45°)] = 225°$$

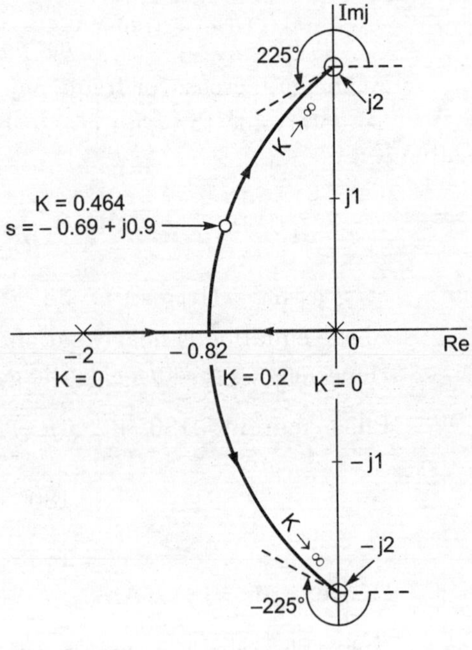

Fig. 7.24.11. Root locus plot for example 7.24.9.

$$\theta_{a(-j2)} = 180° - (\theta_z - \theta_p) = 180° - [\theta_{z_1} - (\theta_{p_1} + \theta_{p_2})]$$

$$= 180° - \left\{270° - \left[270° - \tan^{-1}\left(\frac{2}{3}\right)\right]\right\}$$

$$= 180° - [270° - (270° - 45°)] = 135° \quad \text{or} \quad -225°$$

With the data calculated in steps (1) to (8) the root locus plot is drawn as shown in Fig. 7.24.12.

(*a*) The value of K at breakaway point $s = -0.82$ is calculated below :

$$G(s)\,H(s) = \frac{K\,(s^2 + 4)}{s\,(s + 2)}$$

put $s = -0.82$ and $|G(s)\,H(s)| = 1$

$$\therefore \quad |1| = \frac{K\,|[(-0.82)^2 + 4]|}{|(-0.82)|\,|(-0.82 + 2)|}$$

or $$|1| = \frac{K\,(4.67)}{(0.82)\,(1.18)} \quad \text{or} \quad K = \frac{(0.82)(1.18)}{4.67} = 0.2.$$ **Ans.**

(*b*) The value of K at $s = (-0.69 + j0.9)$ is calculated below :

$$G(s)\,H(s) = \frac{K\,(s^2 + 4)}{s\,(s + 2)}$$ put $s = (-0.69 + j0.9)$ and $|G(s)\,H(s)| = 1$

$$\therefore \quad 1 = \frac{K\,|[(-0.69 + j0.9)^2 + 4]|}{|(-0.69 + j0.9)|\,|(-0.69 + j0.9 + 2)|}$$

$$= \frac{K\,|(3.66 + j1.242)|}{|(-0.69 + j0.9)|\,|(1.31 + j0.9)|} = \frac{K\sqrt{3.66^2 + 1.242^2}}{\sqrt{0.69^2 + 0.9^2} \times \sqrt{1.31^2 + 0.9^2}}$$

$$= \frac{K\sqrt{14.93}}{\sqrt{1.28} \times \sqrt{2.5}} = \frac{K \times 3.86}{1.13 \times 1.581}$$

$$\therefore \qquad K = \frac{1.13 \times 1.581}{3.86} = 0.464. \quad \textbf{Ans.}$$

Example 7.24.10. *(a) A control system is shown in Fig. 7.24.12.*

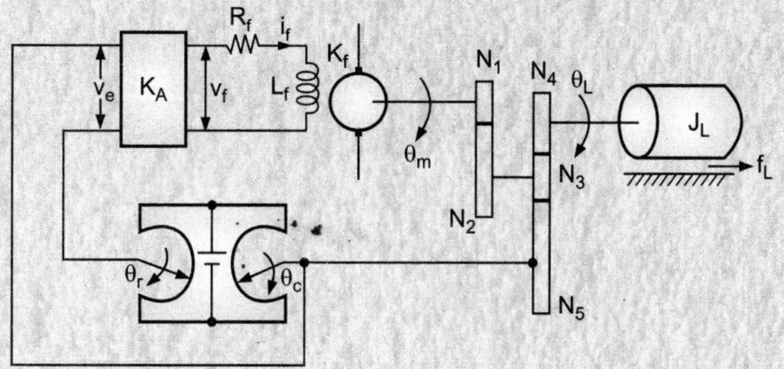

Fig. 7.24.12. Control system for example 7.24.10.

The parameters are given below :

Error detector gain :
$K_e = 0.05\ V/rad$

Amplifier gain : K_A

Motor torque constant :
$K_f = 2\ Nm/A$

Motor field winding resistance : $R_f = 10$ *ohm*

Motor field winding inductance : $L_f = 2.0\ H$

$$\frac{N_1}{N_2} = \frac{1}{10};\ \frac{N_3}{N_4} = \frac{1}{5} \text{ and } \frac{N_3}{N_5} = \frac{1}{10}$$

Moment of inertia of motor shaft :
$J_m = 0.1 \times 10^{-4}\ kg\text{-}m^2$

Viscous friction coeff. of motor shaft :
$f_m = 0.25 \times 10^{-4}\ Nm/rad/sec.$

Moment of inertia of load :
$J_L = 1000 \times 10^{-4}\ kg\text{-}m^2$

Viscous friction coeff. of load :
$f_L = 1875 \times 10^{-4}\ Nm/rad/sec.$

(a) Draw the block diagram representation of the system and determine open-loop transfer function.

(b) Sketch the root locus plot for the open-loop transfer function.

(c) Determine the value of K_A for (i) marginal stability (ii) for damping ratio $\zeta = 1$ and $\zeta = 0.5$.

Solution. The transfer function of a field controlled d.c. motor is given by :

$$\frac{\theta_m(s)}{V_f(s)} = \frac{K_f}{(R_f + sL_f)(J_{eq}s^2 + f_{eq}\,s)}$$

The equivalent moment of inertia ref. to motor shaft side is determined below :

$$J_{eq} = J_m + J_L\left(\frac{N_1}{N_2} \times \frac{N_3}{N_4}\right)^2$$

$$= 0.1 \times 10^{-4} + 1000 \times 10^{-4} \left(\frac{1}{10} \times \frac{1}{5}\right)^2$$

$$= 0.1 \times 10^{-4} + 0.4 \times 10^{-4} = 0.5 \times 10^{-4} \text{ kg-m}^2$$

The equivalent viscous friction coeff. ref. to motor shaft side is determined below :

$$f_{eq} = f_m + f_L \left(\frac{N_1}{N_2} \times \frac{N_3}{N_4}\right)^2$$

$$= 0.25 \times 10^{-4} + 1875 \times 10^{-4} \left(\frac{1}{10} \times \frac{1}{5}\right)^2$$

$$= 0.25 \times 10^{-4} + 0.75 \times 10^{-4} = 1 \times 10^{-4} \text{ Nm/rad/sec.}$$

(*a*) The block diagram representation is shown in Fig. 7.24.13 (*a*) and (*b*).

The take-off point before block N_3/N_4 in Fig. 7.24.13 (*a*) is shifted after the same block as shown in Fig. 7.24.13 (*b*).

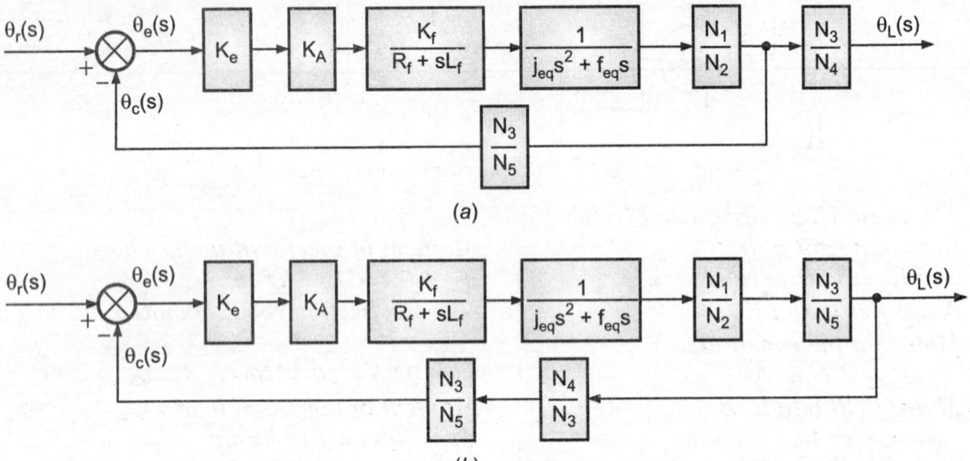

Fig. 7.24.13 (*a*) and (*b*) Block diagram for example 7.24.10.

On substituting given numerical values the block diagram of Fig. 7.24.13 (*b*) is re-drawn and shown in Fig. 7.24.14 (*a*) and (*b*). **Ans.**

(*b*) The forward path transfer function is determined below :

$$G(s) = \frac{0.05K_A \times 2 \times 1}{(10 + 2s)\,(0.5 \times 10^{-4}\, s^2 + 1 \times 10^{-4} s)} \times \frac{1}{50}$$

$$= \frac{0.1K_A}{50 \times 10^{-4}\, s\,(2s + 10)\,(0.5s + 1)}$$

$$= \frac{20K_A}{s\,(2s + 10)\,(0.5s + 1)} = \frac{20K_A}{2 \times 0.5\, s\,(s + 2)\,(s + 5)}$$

or
$$G(s) = \frac{20K_A}{s\,(s + 2)\,(s + 5)}, \quad \because \quad H(s) = 0.5$$

The open-loop transfer function is :

$$G(s)\,H(s) = \frac{20K_A \times 0.5}{s\,(s + 2)\,(s + 5)} \quad \text{or} \quad G(s)\,H(s) = \frac{10K_A}{s\,(s + 2)\,(s + 5)}$$

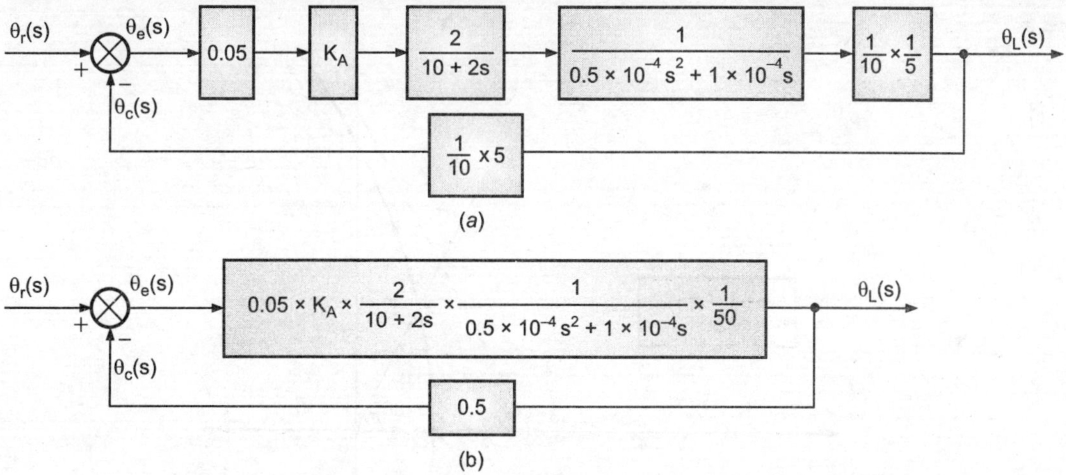

Fig. 7.24.14 (*a*) and (*b*) Block diagram with numerical values for example 7.24.10.

put $$10K_A = K$$

$$\therefore \quad G(s)\,H(s) = \frac{K}{s\,(s+2)\,(s+5)} \cdot \quad \textbf{Ans.}$$

The root locus plot for the open-loop transfer function given above is drawn in Fig. 7.24.15 as per the procedure given below :

1. The root locus branches start ($K = 0$) from the open-loop poles $s = 0$, $s = -2$ and $s = -5$.

2. As there are no finite open-loop zeros, the root locus branches terminate ($K = \infty$) at infinity.

3. $P = 3$, $Z = 0$, therefore, the number of root locus branches $N = 3$.

4. On real axis, the root locus appears between open-loop poles $s = 0$ and $s = -2$ and beyond $s = -5$.

5. Between open-loop poles $s = 0$ and $s = -2$, there exists a breakaway point as calculated below :

The characteristic equation is given by :

$$1 + G(s)\,H(s) = 0$$

$$\therefore \quad 1 + \frac{K}{s\,(s+2)\,(s+5)} = 0 \quad \text{or} \quad s^3 + 7s^2 + 10s + K = 0$$

$$\therefore \quad K = -(s^3 + 7s^2 + 10s)$$

$$\frac{dK}{ds} = -(3s^2 + 14s + 10) \quad \text{put} \quad \frac{dK}{ds} = 0$$

$$\therefore \quad 3s^2 + 14s + 10 = 0$$

or $$s^2 + 4.66s + 3.33 = 0$$

$$s = \frac{-4.66 \pm \sqrt{4.66^2 - 4 \times 1 \times 3.33}}{2}$$

$$\therefore \quad s = -3.78,\ -0.88$$

The point $s = -0.88$ lies on the root locus, hence it is a breakaway point.

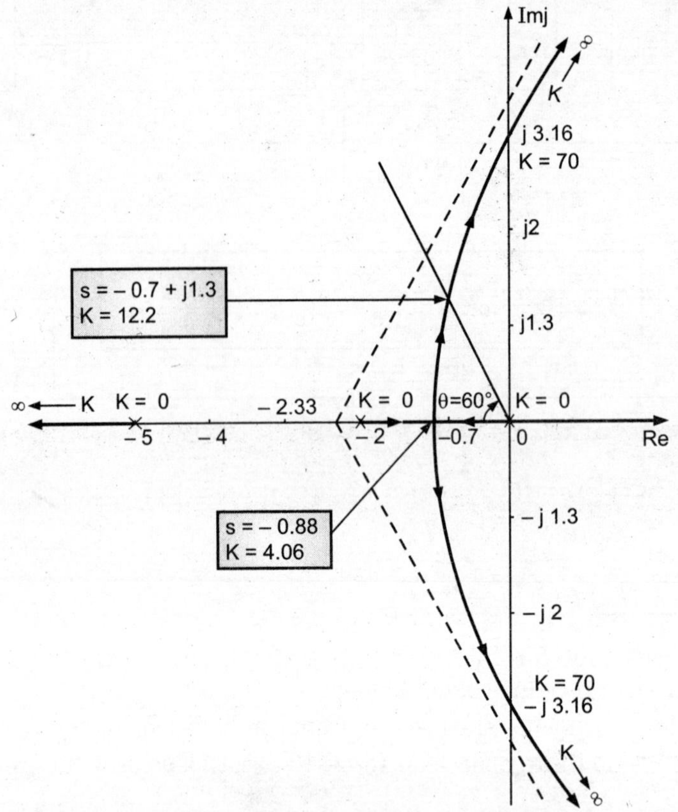

Fig. 7.24.15(a). Root locus plot for example 7.24.10.

6. The angle of asymptotes is given by $\dfrac{(2k+1)}{P-Z} \times 180°$, where $K = 0, 1, 2$

(i) $\dfrac{(2\times0+1)}{3-0} \times 180° = 60°$, (ii) $\dfrac{(2\times1+1)}{3-0} \times 180° = 180°$ and (iii) $\dfrac{(2\times2+1)}{3-0} \times 180° = 300°$.

7. The asymptotes intersect on real axis at :

$$x = \frac{\Sigma P - \Sigma Z}{P-Z} = \frac{(0-2-5)-(0)}{3-0} \quad \text{or} \quad x = -2.33$$

8. The value of K for marginal stability is determined by applying Routh criterion to the characteristic equation :

$$s\,(s+2)\,(s+5) + K = 0 \qquad \text{or} \quad s^3 + 7s^2 + 10s + K = 0$$

The Routh array is tabulated below :

s^3	1	10
s^2	7	K
s^1	$\dfrac{70-K}{7}$	
s^0	K	

The value of K at which root locus crosses imaginary axis is given by :

$$\frac{70-K}{7} = 0 \qquad \text{or} \qquad \text{at} \qquad K = 70$$

The value of s at the crossing point of root locus and imaginary axis is calculated below :

$$7s^2 + K = 0 \quad \text{or} \quad 7s^2 + 70 = 0 \quad \text{or} \quad s = \pm\, j\sqrt{10} = \pm j3.16$$

(c) (i) At the breakaway point $s = -0.88$, the damping ratio is $\zeta = 1$. The value of K at $s = -0.88$ is determined below :

$$|1| = \left|\frac{K}{s(s+2)(s+5)}\right| \qquad \text{put} \qquad s = -0.88$$

$$\therefore \qquad |1| = \left|\frac{K}{(-0.88)(-0.88+2)(-0.88+5)}\right| = \left|\frac{K}{(-0.88)(1.12)(4.12)}\right|$$

$$\therefore \qquad K = (0.88) \times (1.12) \times (4.12) \quad \text{or} \quad K = 4.06$$

$$\because \qquad 10K_A = K$$

$$\therefore \qquad K_A = \frac{K}{10} = \frac{4.06}{10} = 0.406. \quad \textbf{Ans.}$$

(ii) A line at an angle of $\theta = 60°$ with real axis corresponds to damping ratio $\zeta = 0.5$ ($\cos^{-1} \zeta = \cos^{-1} 0.5 = 60°$) and cuts the root locus at $s = (-0.7 + j1.3)$. The value of K at $s = (-0.7 \pm j1.3)$ is determined below :

$$|1| = \left|\frac{K}{s(s+2)(s+5)}\right| \qquad \text{put} \quad s = (-0.7 + j1.3)$$

$$\therefore \qquad |1| = \left|\frac{K}{(-0.7+j1.3)(-0.7+j1.3+2)(-0.7+j1.3+5)}\right|$$

$$= \left|\frac{K}{\sqrt{(-0.7+j1.3)(1.3+j1.3)(4.3+j1.3)}}\right|$$

$$= \frac{K}{\sqrt{(0.49+1.69)}\sqrt{(1.69+1.69)}\sqrt{(18.49+1.69)}}$$

$$= \frac{K}{\sqrt{2.18}\sqrt{3.38}\sqrt{20.18}} = \frac{K}{\sqrt{148.7}} = \frac{K}{12.2}$$

$$\therefore \qquad K = 12.2 \qquad \because \qquad K = 10\,K_A$$

$$\therefore \qquad K_A = \frac{K}{10} = \frac{12.2}{10} = 1.22. \quad \textbf{Ans.}$$

Example 7.24.10 (b). *The open-loop transfer function of a non unity feedback control system is given below:*

$$G(s)\,H(s) = \frac{K}{s(s+2)(s+4)}$$

The damping ratio $\xi = 0.5$. The variation of K is from 0 to ∞.

Determine intersection point of the damping ratio line with the complex part of the root locus by analytical method.

Determine the value of dominant roots of the characteristic equation and the value of K at the aforesaid intersection point, the value of the third real root.

Solution. Let $s = \sigma + j\omega$ be the intersection point of the damping ratio line and the complex part of the root locus as shown in Fig. 7.24.15 (*b*) wherein the general shape of the root locus plot for given $G(s)H(s)$ is drawn.

The point $s = \sigma + j\omega$ is located on the root locus thus the angle condition of the root locus is satisfied at $s = \sigma + j\omega$ as below,

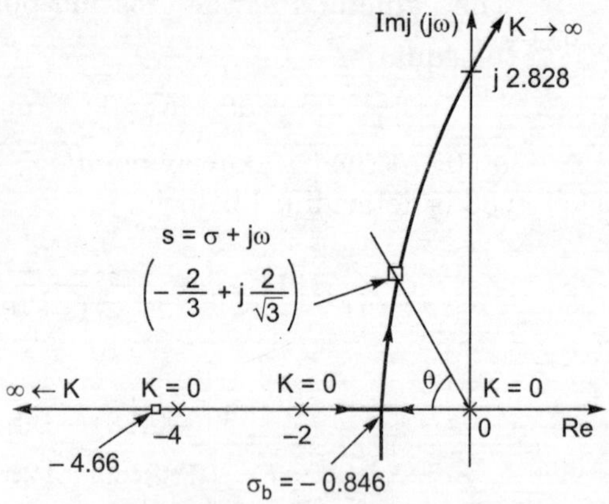

Fig. 7.24.15(*b*) Root locus analytical solution.

$$\left. \angle \frac{K}{s(s+2)(s+4)} \right|_{s=\sigma+j\omega} = -180^\circ$$

$$\therefore \quad \angle \frac{K}{(\sigma+j\omega)(\sigma+j\omega+2)(\sigma+j\omega+4)} = -180^\circ$$

Hence, $-\tan^{-1}\dfrac{\omega}{\sigma} - \tan^{-1}\dfrac{\omega}{\sigma+2} - \tan^{-1}\dfrac{\omega}{\sigma+4} = -180^\circ$

Rearranging, $\tan^{-1}\dfrac{\omega}{\sigma+2} + \tan^{-1}\dfrac{\omega}{\sigma+4} = 180^\circ - \tan^{-1}\dfrac{\omega}{\sigma}$

$$\therefore \quad \tan^{-1}\left[\frac{\dfrac{\omega}{\sigma+2}+\dfrac{\omega}{\sigma+4}}{1-\dfrac{\omega}{(\sigma+2)}\cdot\dfrac{\omega}{(\sigma+4)}}\right] = 180^\circ - \tan^{-1}\frac{\omega}{\sigma}$$

or
$$\frac{\dfrac{\omega}{\sigma+2}+\dfrac{\omega}{\sigma+4}}{1-\dfrac{\omega^2}{(\sigma+2)(\sigma+4)}} = -\frac{\omega}{\sigma}, \qquad (\tan^{-1} 180^\circ = 0)$$

Simplifying, $\omega^2 = 3\sigma^2 + 12\sigma + 8$

The equation for the complex part of the root locus is derived above.

The equation for the damping ratio line is determined below.

The damping ratio line makes an angle θ with the s-plane –ve real axis. Therefore, the equation for the damping ratio line is

$$\frac{\omega}{\sigma} = -\tan\theta \text{ (–ve slope)}$$

$\because \quad \xi = 0.5$ and $\theta = \cos^{-1}\xi = \cos^{-1} 0.5 = 60^\circ$

$$\therefore \quad \frac{\omega}{\sigma} = -\tan 60^\circ = -\sqrt{3} \quad \text{or} \quad \omega = -\sigma\sqrt{3}$$

The intersection is given by substituting $\omega = -\sigma\sqrt{3}$ in the equation for root locus

$$\therefore \quad (-\sigma\sqrt{3})^2 = 3\sigma^2 + 12\sigma + 8$$

$$\therefore \quad 12\sigma = -8 \quad \text{and} \quad \sigma = \frac{-8}{12} = \frac{-2}{3}$$

also,
$$\omega = -\sigma\sqrt{3} = -\left(-\frac{2}{3}\right)\sqrt{3} = \frac{2}{\sqrt{3}}$$

Hence, the intersection occurs at $s = \sigma + j\omega$ or $s = \dfrac{-2}{3} + j\dfrac{2}{\sqrt{3}}$. **Ans.**

The break away point of the root locus on the –ve real axis is obtained by putting $\omega = 0$ in the equation

$$\omega^2 = 3\sigma^2 + 12\sigma + 8 \qquad \therefore \quad 3\sigma^2 + 12\sigma + 8 = 0$$

Solving above equation $\sigma = -0.846, -3.154$

$\sigma = -3.154$ does not lie on the real axis part of the root locus. Therefore the accepted break away point is $\sigma_b = -0.846$.

The complex part intersection crossing the s-plane imaginary axis is obtained by substituting $s = 0$ in the equation $\omega^2 = 3\sigma^2 + 12\sigma + 8$

$$\therefore \quad \omega^2 = 3 \times 0 + 12 \times 0 + 8 \qquad \therefore \quad \omega = \pm 2\sqrt{2} = \pm 2.828$$

The value of gain K at $j\omega = j\,2\sqrt{2}$ is calculated applying root locus magnitude condition as below.

$$\therefore \quad |G(s)\,H(s)|_{s = j2\sqrt{2}} = 1 \qquad \because \quad G(s)\,H(s) = \frac{K}{s(s+2)(s+4)}$$

$$\therefore \quad \left|\frac{K}{s(s+2)(s+4)}\right|_{s = j2\sqrt{2}} = 1$$

or

$$K = |s(s+2)(s+4)|_{s = j2\sqrt{2}} = |(j(2\sqrt{2})(2 + j2\sqrt{2})(4 + j\,2\sqrt{2})|$$

$$= \sqrt{[(2\sqrt{2})^2][(2)^2 + (2\sqrt{2})^2][(4)^2 + (2\sqrt{2})^2]}$$

Simplifying, $K = 28$

The intersection point is $s = -\frac{2}{3} + j\frac{2}{\sqrt{3}}$, the complex conjute is $s = -\frac{2}{3} - j\frac{2}{\sqrt{3}}$. Hence, the two complex roots of the characteristic equation are $s = -\frac{2}{3} \pm j\frac{2}{\sqrt{3}}$.

The value of gain K at $s = -\frac{2}{3} + j\frac{2}{\sqrt{3}}$ is determined below,

The root $s = -\frac{2}{3} + j\frac{2}{\sqrt{3}}$ lies on the root locus satisfying root locus magnitude condition,

$$\therefore \quad \left|\frac{K}{s(s+2)(s+4)}\right|_{s = -\frac{2}{3} + j\frac{2}{\sqrt{3}}} = 1$$

or

$$K = |s(s+2)(s+4)|_{s = -\frac{2}{3} + j\frac{2}{\sqrt{3}}}$$

$$= \left|\left(-\frac{2}{3} + j\frac{2}{\sqrt{3}}\right)\left(-\frac{2}{3} + j\frac{2}{\sqrt{3}} + 2\right)\left(-\frac{2}{3} + j\frac{2}{\sqrt{3}} + 4\right)\right|$$

$$= \left|\left(-\frac{2}{3} + j\frac{2}{\sqrt{3}}\right)\left(\frac{4}{3} + j\frac{2}{\sqrt{3}}\right)\left(\frac{10}{3} + j\frac{2}{\sqrt{3}}\right)\right|$$

$$= \sqrt{\left(\frac{4}{9} + \frac{4}{3}\right)\left(\frac{16}{9} + \frac{4}{3}\right)\left(\frac{100}{9} + \frac{4}{3}\right)}$$

$$K = 8.29$$

$$\therefore \quad \text{Simplifying, } G(s)\,H(s) = \frac{K}{s(s+2)(s+4)} \qquad \therefore \quad G(s)\,H(s) = \frac{8.29}{s(s+2)(s+4)}$$

The third real root lies beyond $s = -4$ on the s-plane –ve real axis. The third real root satisfies the magnitude condition of root locus

$$\therefore \quad \left| \frac{8.29}{s(s+2)(s+4)} \right| = 1$$

The above equation is satisfied at $s = -0.466$ to a fair approximation. Hence the third real root is located at $s = -4.66$. **Ans.**

The closed-loop poles *i.e.* roots of the characteristic equation $1 + G(s) H(s) = 0$ obtained by the analytical solution are $s_1 = -4.666$, $s_2, s_3 = -\frac{2}{3} \pm j\frac{2}{\sqrt{3}}$.

Thus, the characteristic equation is $(s + 4.66)\left(s + \frac{2}{3} + j\frac{2}{\sqrt{3}}\right)\left(s + \frac{2}{3} - j\frac{2}{\sqrt{3}}\right) = 0$. **Ans.**

Example 7.24.11. *Sketch the root locus from the open-loop transfer function given below and determine the stability.* $G(s)\,H(s) = \dfrac{K\,(s + 0.1)}{s\,(s - 0.2)\,(s^2 + s + 0.6)}$.

Solution. The number of open-loop poles $P = 4$, *i.e.*, [$s = 0$, $s = 0.2$, $s = (-0.5 + j0.59)$ and $s = (-0.5 - j0.59)$].

The number of open-loop zeros $Z = 1$, *i.e.* ($s = -0.1$)

As $P > Z$, the number of root locus branches $N = 4$. On the real axis the root locus lies between (*i*) $s = -\infty$ and $s = -0.1$ (zero) (*ii*) $s = 0$ (pole) and $s = 0.2$ (pole).

The angle of asymptotes is given by

$$\frac{(2k+1)\times 180°}{P-Z}\ ; \qquad \text{where} \quad k = 0, 1, 2.$$

Therefore, the three angles of asymptotes are

$$(i)\ \frac{(2\times 0+1)\times 180°}{4-1} = 60°,\ (ii)\ \frac{(2\times 1+1)\times 180°}{4-1} = 180° \text{ and } (iii)\ \frac{(2\times 2+1)\times 180°}{4-1} = 300°$$

The asymptotes intersect at a point on real axis at

$$x = \frac{\Sigma \text{ poles} - \Sigma \text{ zeros}}{P-Z} = \frac{(0+0.2-0.5+j0.59-0.5-j0.59)-(-0.1)}{4-1} = \frac{-0.7}{3} = -0.23$$

The pole-zero configuration is shown in Fig. 7.24.16.

The angle of departure from the complex pole $(-0.5 + j0.59)$ is determined below (refer Fig. 7.24.16).

$$\phi_{p_1} = 180° - \tan^{-1}\left(\frac{0.59}{0.7}\right) = 139.8°, \qquad \phi_{p_2} = 180° - \tan^{-1}\left(\frac{0.59}{0.5}\right) = 130.2°$$

$$\phi_{p_3} = 90°, \qquad \phi_{z_1} = 180° - \tan^{-1}\left(\frac{0.59}{0.4}\right) = 124.1°$$

The angle of departure from the pole $(-0.5 + j\,0.59)$ is

$$\phi_{(-0.5+j0.59)} = 180° - (139.8° + 130.2° + 90° - 124.1°) = -55.9°$$

The angle of departure from the pole $(-0.5 - j0.59)$ is $\phi_{(-0.5-j0.59)} = +55.9°$

The root locus starts from complex poles $(-0.5 + j0.59)$ and $(-0.5 - j0.59)$ breaks on the negative real axis between $s = -\infty$ and $s = -0.1$. The breakaway points are determined as the real roots of the equation $dK/ds = 0$.

The characteristic equation is

$$[s(s-0.2)(s^2+s+0.6)+K(s+0.1)] = 0$$

or
$$K = -\frac{(s^4+s^3+0.4s^2-0.12s)}{(s+0.1)}$$

$$\therefore \quad \frac{dK}{ds} = -\frac{(3s^4+2s^3+0.64s^2+0.08s-0.12)}{(s+0.1)^2}$$

Put
$$\frac{dK}{ds} = 0$$

$$\therefore \quad (3s^4+2s^3+0.64s^2+0.08s-0.12) = 0$$

Above equation is nearly satisfied at $s = 0.08$ and $s = -0.37$, thus the breakaway points can be considered to be located respectively at $s = -0.37$ and $s = +0.08$ with a fair approximation.

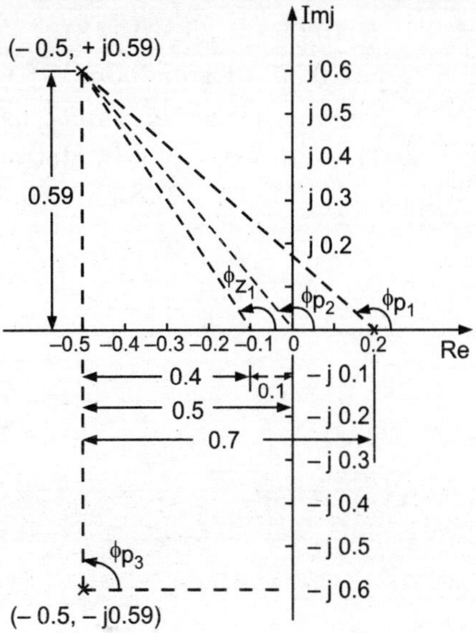

Fig. 7.24.16. Pole-zero configuration for $G(s)H(s) = \dfrac{K(s+0.1)}{s(s-0.2)(s^2+s+0.6)}$.

The intersection of the root locus plot with the imaginary axis is determined by applying Routh criterion to the characteristic equation, *i.e.*

$$[s(s-0.2)(s^2+s+0.6)+K(s+0.1)] = 0$$

or
$$[s^4+0.8s^3+0.4s^2+(K-0.12)s+0.1K] = 0$$

The Routh array is arranged below :

s^4	1	0.4	0.1 K
s^3	0.8	$(K-0.12)$	
s^2	$(0.55-1.25K)$	0.1 K	
s^1	$\dfrac{(-1.25K^2+0.63K-0.066)}{(0.55-1.25K)}$		
s^0	0.1 K		

The value of K at which the root locus plot crosses the imaginary axis is determined by equating the first term in s^1 row to zero, therefore,

$$(-1.25K^2+0.63K-0.066) = 0$$

$$\therefore \quad K = \frac{-0.63+\sqrt{0.63^2-4\times(-1.25)\times(-0.066)}}{2\times(-1.25)}$$

$$K = \frac{-0.372}{-2.5}, \frac{-0.888}{-2.5} = 0.148, 0.352.$$

The root locus crosses the imaginary axis at $K = 0.148$ and $K = 0.352$.

For finding the crossing points the auxiliary equation formed from s^2 terms is solved, by putting $K = 0.148$ and $K = 0.352$ respectively, therefore, (*i*) $(0.55 - 1.25 \times 0.148)\, s^2 + 0.1 \times 0.148 = 0$ $\quad\therefore\quad s = \pm j0.2$

(*ii*) $(0.55 - 1.25 \times 0.352)\, s^2 + 0.1 \times 0.352 = 0$ $\quad\therefore\quad s = \pm j0.56$

The root locus plot thus crosses the imaginary axis at $s = j\pm j0.2$ and $s = \pm\ j\,0.56$.

The root locus plot is plotted in Fig. 7.24.17 and it is inferred from the plot that the system is stable for $0.148 < K < 0.352$, thus the system is conditionally stable.

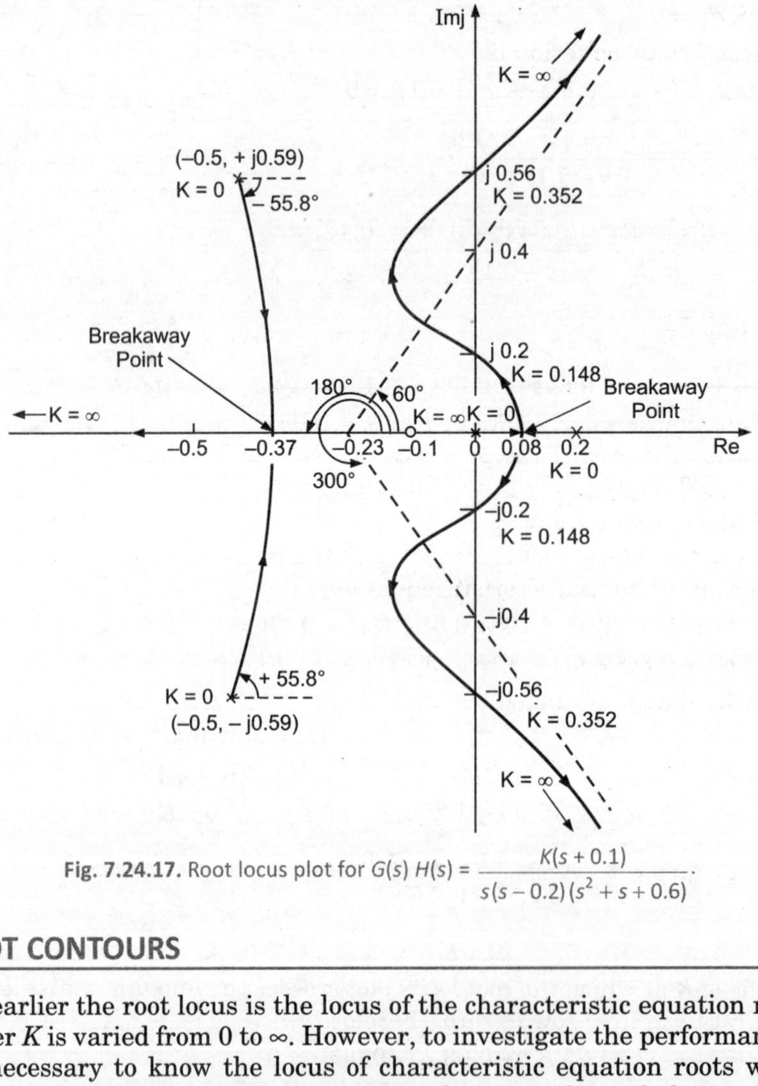

Fig. 7.24.17. Root locus plot for $G(s)\,H(s) = \dfrac{K(s+0.1)}{s(s-0.2)(s^2+s+0.6)}$.

7.25 ROOT CONTOURS

As described earlier the root locus is the locus of the characteristic equation roots when the gain parameter K is varied from 0 to ∞. However, to investigate the performance of a control system it is necessary to know the locus of characteristic equation roots when a system parameter other than K is varied. In an actual system loci of the roots of modified characteristic equation for different parameter variations are sketched by using a generalized root locus method. The loci thus drawn are known as root contours.

Generally, the variations of closed-loop poles which are same as the roots of characteristic equations are plotted in s-plane keeping the gain factor K to a fixed value while either the open-loop poles or zeros are varied in order to obtain a root contour. This generalized root locus called root contour follows all the rules as applicable to the root locus approach described earlier.

Root contour plots are essential for describing and investigating the properties of the compensation networks as they clearly indicate the selection of parameter values.

Consider a feedback control system having open-loop transfer function as :

$$G(s)\,H(s) = \frac{K\,P\,(s)}{Q\,(s)} \qquad ...(7.93)$$

where $P\,(s)$ and $Q\,(s)$ are polynomials of s in usual terms. The characteristic equation is given by :

$$1 + G(s)\,H(s) = 0 \quad \text{or} \quad 1 + \frac{K\,P\,(s)}{Q\,(s)} = 0 \quad \text{or} \quad Q\,(s) + K\,P\,(s) = 0$$

The equation given above contains a variable parameter say α other than K.

For plotting the root contour keeping K as constant and varying α from 0 to ∞. The aforesaid equation can be modified in the following form :

$$Q\,(s) + K\,P\,(s) = Q_1\,(s) + \alpha P_1\,(s) = 0 \qquad ...(7.94)$$

The locus of the roots of $Q_1\,(s) + \alpha\,P_1\,(s) = 0$ being plotted as the parameter α is varied from 0 to ∞, is the root contour for the system under consideration and obtained from the pole-zero configuration of $P_1\,(s)/Q_1\,(s)$ by applying procedure as followed for root locus.

The open-loop transfer function with variation of the parameter x given below determines the modified characteristic equation/open-loop transfer function in relation to the sketch of root contour plot.

$$G(s)\,H(s) = \frac{5\,(s+8)}{s\,(s+x)}$$

The root contour is to be plotted as the value of parameter x (open-loop pole) is varied from 0 to ∞.

The original characteristic equation is

$$1 + G(s)\,H(s) = 0 \qquad \text{or} \qquad 1 + \frac{5\,(s+8)}{s\,(s+x)} = 0$$

$$\therefore \quad s(s+x) + 5\,(s+8) = 0 \qquad \text{or} \qquad s^2 + 5s + 40 + sx = 0$$

which can be written as : $\quad 1 + \dfrac{sx}{s^2 + 5s + 40} = 0$

The transfer function for plotting the root contour is thus given by :

$$G_1\,(s)\,H_1\,(s) = \frac{sx}{s^2 + 5s + 40}.$$

Example 7.25.1. *A control system is shown in the block diagram given below :*

Sketch the root contour as the value of the parameter α is varied from 0 to ∞. Determine the value of α for the transient response to have critical damping.

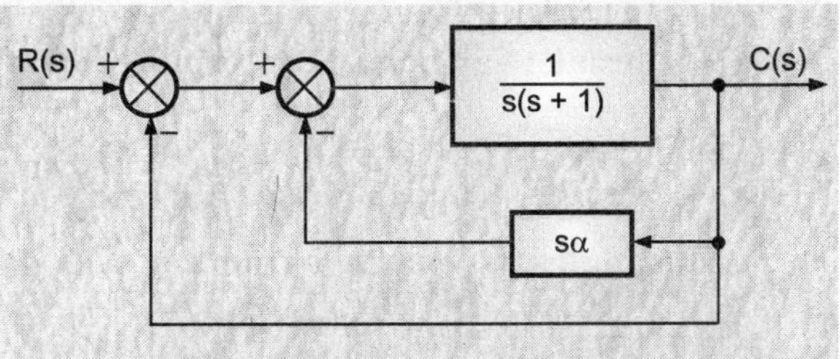

Fig. 7.25.1. Block diagram for example 7.25.1.

Solution. The overall transfer function is determined as :

$$\frac{C(s)}{R(s)} = \frac{\frac{1}{s(s+1)}}{1+\frac{1}{s(s+1)}\times(s\,\alpha+1)} \quad \text{or} \quad \frac{C(s)}{R(s)} = \frac{1}{s^2+(\alpha+1)s+1}$$

The characteristic equation is :

$$s^2 + (\alpha + 1)\, s + 1 = 0$$

which can be rearranged as :

$$s^2 + s + 1 + \alpha\, s = 0 \quad \text{or} \quad 1 + \frac{\alpha\, s}{s^2+s+1} = 0$$

Therefore, the open-loop transfer function for sketching the root contour is given by :

$$G_1(s)\, H_1(s) = \frac{\alpha\, s}{s^2+s+1}$$

open-loop zero : $s = 0$, open-loop poles : $s = \frac{-1 \pm \sqrt{1^2 - 4\times 1}}{2} = -0.5 \pm j\,0.866$

The number of root contour branches = 2

The starting point ($\alpha = 0$) of root contours is at $s = -0.5 \pm j\,0.866$

The terminating point ($\alpha \to \infty$) of root contours is at $s = 0$ and $\alpha \to \infty$

The angle of asymptotes :

$$\angle \text{Asymptotes} = \frac{(2k+1)}{P-Z} \times 180° \; ; \quad k = 0$$

$$= \frac{(2\times 0+1)}{2-1} \times 180° = 180°$$

The root contour is present on entire –ve real axis.

The characteristic equation is :

$$1+\frac{\alpha\, s}{s^2+s+1} = 0$$

or

$$\alpha = -\frac{s^2+s+1}{s}$$

$$\therefore \quad \frac{d\alpha}{ds} = -\left\{\frac{s\,(2\,s+1)-(s^2+s+1)\times 1}{s^2}\right\} = -\left\{\frac{s^2-1}{s^2}\right\}$$

The breakaway point is determined using $\frac{d\alpha}{ds} = 0$

$$\therefore \quad s^2 - 1 = 0 \quad \text{or} \quad s = \pm 1$$

The breakaway point is identified at $s = -1$ as it lies on root counter branch.

The angles of departure from complex poles $s = -0.5 \pm j\,0.866$ are determined below :

$$\underset{(-0.5+j\,0.0866)}{\theta_d} = 180° - \{(\theta_{p_2} - \phi_z)\}$$

$$= 180° - \left\{90° - \left(90° + \tan^{-1}\frac{0.5}{0.866}\right)\right\} = 180° - \{90° - (90° + 30°)\} = 210°$$

$$\underset{(-0.5 - j\,0.0866)}{\theta_d} = 180° - \{(\phi_{p_1} - \phi_z)\}$$

$$= 180° - \left\{-90° - \left(-90° - \tan^{-1}\frac{0.5}{0.866}\right)\right\}$$

$$= 180° - \{-90° - (-90° - 30°)\} = 150°$$

As per data calculated above the root contour plot is drawn and shown in Fig. 7.25.2.

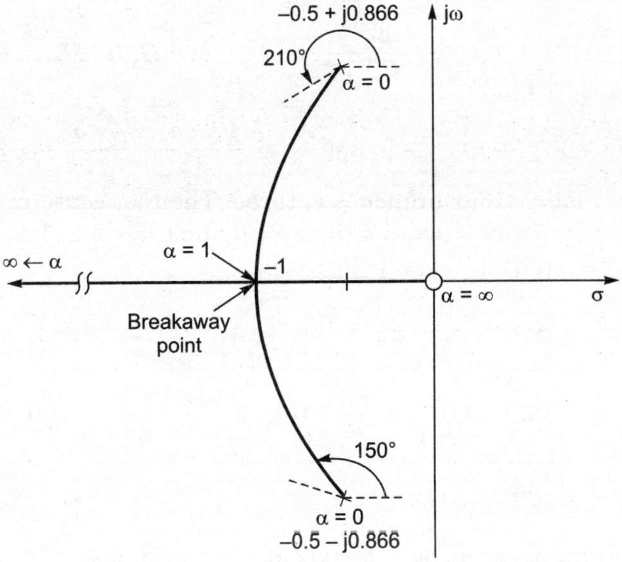

Fig. 7.25.2. Root contours for example 7.25.1.

The critical damping occurs at breakaway point on real axis *i.e.* $s = -1$. As the point $s = -1$ lies on the root contour.

Therefore, $|G_1(-1)\,H_1(-1)| = 1$

$$\therefore \qquad 1 = \left|\frac{\alpha(-1)}{(-1)^2 + (-1) + 1}\right|$$

$\therefore \qquad \alpha = 1.$ **Ans.**

Example 7.25.2. *The block diagram of a unity feedback control system is shown in figure given below :*

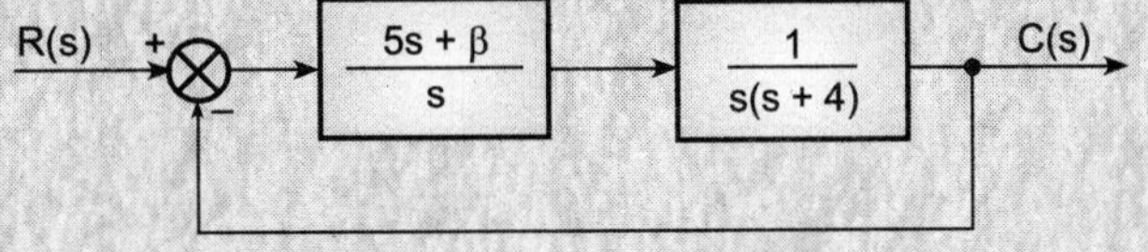

Fig. 7.25.3. Block diagram for example 7.25.2.

Draw the root contours as the parameter β is varied from 0 to ∞ and determine the value of β so that the damping ratio due to dominant close-loop pole is 0.6.

Solution. The overall transfer function is determined as :

$$\frac{C(s)}{R(s)} = \frac{\frac{5s+\beta}{s} \times \frac{1}{s(s+4)}}{1 + \frac{5s+\beta}{s^2(s+4)} \times 1} \quad \text{or} \quad \frac{C(s)}{R(s)} = \frac{5s+\beta}{s^3+4s^2+5s+\beta}$$

Therefore, the original characteristic equation is $s^3 + 4s^2 + 5s + \beta = 0$ and therefore, modified transfer function is

$$G_1(s)H_1(s) = \frac{\beta}{s^3+4s^2+5s} \quad \text{or} \quad G_1(s)H_1(s) = \frac{\beta}{s(s^2+4s+5)}$$

The poles of $G_1(s)H_1(s)$ are $s = 0$, and $s = \frac{-4 \pm \sqrt{4^2 - 4\times 5}}{2} = -2 \pm j1$.

The number of root contour branches is three. The root contour exists on entire – ve and real axis. The root contours start ($\beta = 0$) at $s = 0$ and $s = -2 \pm j1$.

The angle of asymptotes are determine below :

$$\angle\text{Asy.} = \frac{(2k+1)}{p-Z} \times 180° \; ; \quad k = 0, 1, 2$$

$$(i) \quad = \frac{(2\times 0+1)}{3-0} \times 180° = 60°, \qquad (ii) \quad = \frac{(2\times 1+1)}{3-0} \times 180° = 180°,$$

$$(iii) \quad = \frac{(2\times 2+1)}{3-0} \times 180° = 300°.$$

The asymptotes intersection on real axis at :

$$x = \frac{\Sigma P - \Sigma Z}{P - Z} = \frac{(0-2+j1-2-j1)-0}{3-0} = -1.33$$

The characteristic equation w.r.t. root contour is given by :

$$1 + G_1(s)H_1(s) = 0$$

or
$$1 + \frac{\beta}{s^3+4s^2+5s} = 0 \quad \text{or} \quad s^3 + 4s^2 + 5s + \beta = 0$$

The Routh's tabulation for above equation is formed below :

s^3	1	5
s^2	4	β
s^1	$\frac{20-\beta}{4}$	
s^0	β	

The root contour intersects imaginary axis for $\beta = 20$, the point of intersection can be calculated by solving

$$4s^2 + 20 = 0 \quad i.e. \quad s = \pm\, j\sqrt{5}$$

Breakaway point :

$$\because \quad 1 + \frac{\beta}{s(s^2+4s+5)} = 0$$

$$\therefore \qquad \beta = -s(s^2 + 4s + 5) = -(s^3 + 4s^2 + 5s)$$

and $$\frac{d\beta}{ds} = -(3s^2 + 8s + 5)$$

to determine breakaway point, put $\frac{d\beta}{ds} = 0$ *i.e.* $3s^2 + 8s + 5 = 0$

or $$s^2 + \frac{8}{3}s + \frac{5}{3} = 0 \qquad \text{or} \qquad s^2 + 2.66s + 1.33 = 0$$

$$\therefore \qquad s = \frac{-2.66 \pm \sqrt{2.66^2 - 4 \times 1.33}}{2} = -1.01, -1.66.$$

The two breakaway points $s = -1.01$ and $s = -1.66$ lie on the root contour.

Angle of departure from complex poles :

The angle of departure from the pole $(-2 + j\,1)$ is calculated below :

$$\phi_{(-2+j1)} = 180° - (\theta_{p_2} + \theta_{p_3}) = 180° - \left\{(+90°) + \left(90° + \tan^{-1}\frac{2}{1}\right)\right\} = \tan^{-1}\frac{2}{1} = -63.43°$$

The angle of departure from the pole $(-2 - j1)$ is calculated below :

$$\phi_{(-2-j1)} = 180° - (\theta_{p_1} + \theta_{p_3})$$

$$= 180° - \left\{(-90°) + \left(-90° - \tan^{-1}\frac{2}{1}\right)\right\} = 296.57° = (360° - 296.57°) = 63.43°.$$

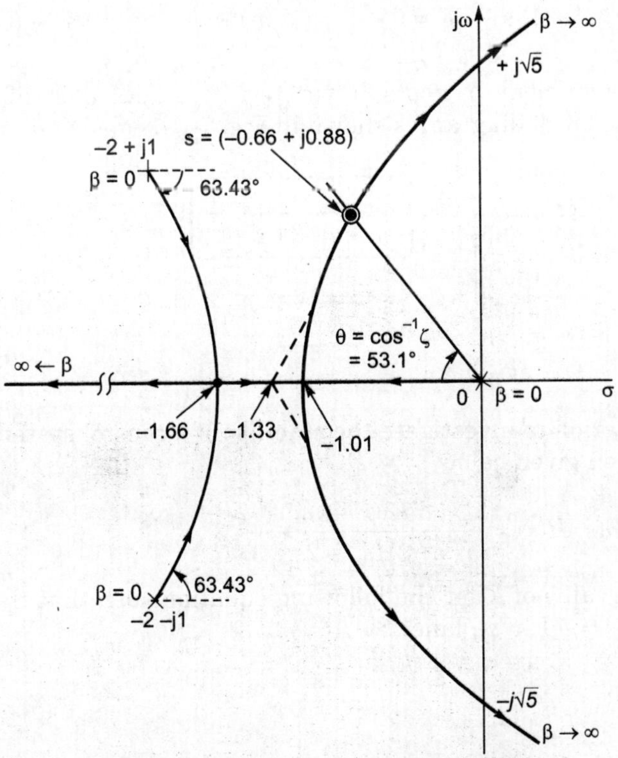

Fig. 7.25.4. Root contours for example 7.25.2.

As per the data calculated above the root contour is plotted as shown in Fig. 7.25.4. The damping ratio line drawn at an angle

$$\theta = \cos^{-1} \zeta = \cos^{-1} 0.6 = 53.1°$$

intersects root contour plot at $s = (-0.66 + j\,0.88)$. As the point $s = (-0.66 + j\,0.88)$ lies on the root contour.

Therefore, $$|G_1(s)\, H_1(s)|_{s = (-0.66 + j\,0.88)} = 1$$

$$\therefore \quad |1| = \frac{\beta}{|(-0.66 + j\,0.88)\{(-0.66 + j\,0.88)^2 + 4(-0.66 + j\,0.88) + 5\}|}$$

or $$1 = \frac{\beta}{1.1(1.21 + 4 \times 1.11 + 5)} \quad \therefore \quad \beta = 11.7. \quad \textbf{Ans.}$$

PROBLEMS

7.1. Determine the stability of the systems whose characteristic equations are given below :

(a) $s^4 + 10s^3 + 30s^2 + 100s + 25 = 0$ (b) $s^4 + 8s^3 + 36s^2 + 80s + 20 = 0$

(c) $s^5 + 13s^4 + 54s^3 + 82s^2 + 170s + 330 = 0$

(d) $s^5 + 4s^4 + s^3 + 4s^2 - s + 6 = 0$ (e) $s^4 + s^3 + 5s^2 + 4s + 4 = 0$.

7.2. For the characteristic equations given below determine the limiting value of K for stability

(a) $s^3 + 4Ks^2 + (K + 4)\,s + 5K = 0$ (b) $s^4 + 7s^3 + 10s^2 + K = 0$

(c) $s^4 + 4s^3 + 6s^2 + (K + 5)\,s + 2K = 0$.

7.3. Determine the number of roots having positive real part of the characteristic equation whose block diagram is shown in Fig. 7-P-1.

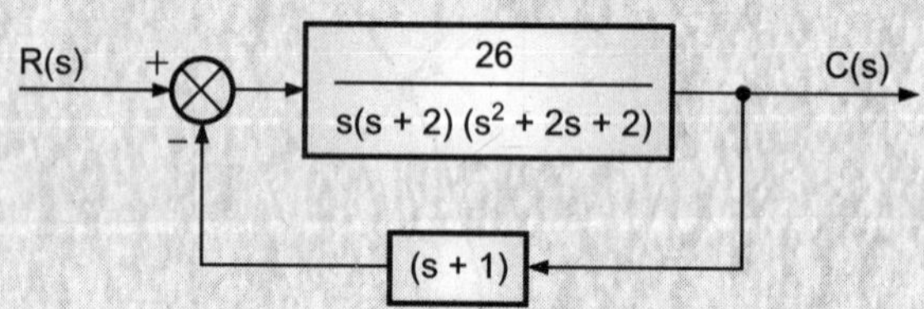

Fig. 7-P-1. Block diagram for problem 7.3.

7.4. Use Routh criterion to investigate the stability of a control system having open-loop transfer function given below

$$G(s)\,H(s) = \frac{K\,(s + 2)}{s\,(s + 3)\,(s^2 + 2s + 3)}.$$

7.5. Determine the value of K for the following equation such that the roots lie towards the left of line $s = -1$ in s-plane.

$$s^3 + 3s^2 + 3s + K = 0.$$

7.6. The output position of a control system is given by

$$\frac{d^3c(t)}{dt^3} + 5\frac{d^2c(t)}{dt^2} + (10-\alpha)\frac{dc(t)}{dt} + 10c(t) = 10\,u(t)$$

If $u(t) = 1$, determine the stability in terms of parameter α.

7.7. Plot the Nyquist plot and determine the closed-loop stability for the control systems whose open-loop transfer function is given below :

(*a*) $G(s)\,H(s) = \dfrac{1.5}{(s+1)(0.54s+1)(0.25s+1)}$ (*b*) $G(s)\,H(s) = \dfrac{5}{s^2(s+1)}$

(*c*) $G(s)\,H(s) = \dfrac{10}{s(s+1)(s^2+4)}$ (*d*) $G(s)\,H(s) = \dfrac{1}{s(s+1)(s^2+2s+2)}$; determine G.M.

(*e*) $G(s)\,H(s) = \dfrac{125}{s(s^2+5s+25)}$.

7.8. The block diagram of a control system is shown in Fig. 7-P-2. Apply Nyquist criterion to determine the value of K such that the system is stable.

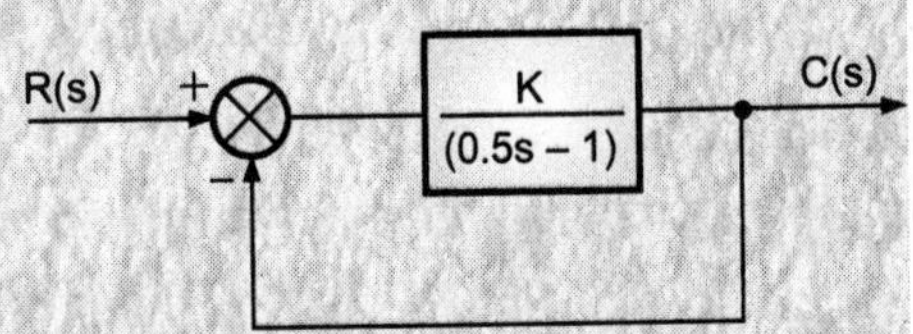

Fig. 7-P-2. Block diagram for problem 7.8.

7.9. (*a*) The open-loop transfer function of a control system is given by,

$$G(s)\,H(s) = \frac{K(s-1)}{(s+2)(s+3)}$$

Using Nyquist plot determine the stability in terms of K and check the result by Routh criterion.

(*b*) Sketch Nyquist plot having given, $G(s)\,H(s) = \dfrac{Ks^2}{s^3+4s+4}$ and examine closed-loop stability in terms of parameter K.

7.10. For the open-loop transfer function given below sketch the Nyquist plot when (*a*) $K = 1.5$ (*b*) $K = 2.5$ and comment on the closed-loop stability.

$$G(s)\,H(s) = \frac{K(s+4)}{s(s-2)}.$$

7.11. The characteristic equation of a unity feedback system is given by, $0.5\,Ts^3 + (0.5+T)s^2 + s + 10 = 0$

(*a*) Determine the open-loop transfer function

(*b*) Apply Nyquist criterion to the open-loop transfer function to determine the value of T such that the system is stable.

7.12. The open-loop transfer function of a feedback control system is given by, $G(s)\,H(s) = \dfrac{K}{s^2+s-2}$

Plot the Nyquist plot and show that the closed-loop system is stable if $K \geq 2$.
(**Hint :** Factor the denominator to find P_+).

7.13. Plot the gain-phase plot from the data given below :

ω rad/sec	0.2	0.5	1	2	2.5	5	10
$20 \log_{10} \lvert G(j\omega) H(j\omega) \rvert$	30 db	21 db	14.5 db	5.0 db	2.0 db	– 10 db	– 26 db
$\angle G(j\omega) H(j\omega)°$	– 98	– 110	– 130	– 161	– 173	– 209	– 236

Determine :

(*a*) G.M. and P.M.

(*b*) the factor by which the gain is to be multiplied to have G.M. = 10 db

(*c*) the factor by which the gain is to be multiplied to have P.M. = 30°.

7.14. The gain frequency measurements for a control system are given below. Sketch the gain phase plot and determine the gain margin, phase margin

ω rad/sec	1	2	4	8	10	20
$\lvert G(j\omega) \rvert$	4.8	2.3	0.9	0.26	0.16	0.026
$\angle G(j\omega)$	– 106°	– 123°	– 141°	– 186°	– 198°	– 229°

By what factor the gain should be multiplied in order to have

(*a*) a gain margin of 20 db and (*b*) a phase margin of 20°.

7.15. Plot approximate logarithmic plot for the open-loop transfer function given below and investigate closed-loop stability.

$$G(s)\,H(s) = \frac{5}{s^2(1+0.1s)\,(s^2+0.4s+1)}.$$

7.16. Using Bode plot method determine the gain crossover and phase crossover frequencies for the transfer function

$$G(s)\,H(s) = \frac{1}{s\,(0.025s+1)\,(0.25s+1)}.$$

7.17. Determine the value of K in the transfer function given below such that

(*a*) the gain margin is 20 db and (*b*) the phase margin is 30°

$$G(j\omega)\,H(j\omega) = \frac{K}{j\omega\,(j0.1\,\omega+1)\,(j0.05\omega+1)}.$$

7.18. (*a*) From the approximate logarithmic plots given below in Figs. 7-P-3 (*a*) (*i*) and (*ii*) determine the open-loop transfer function.

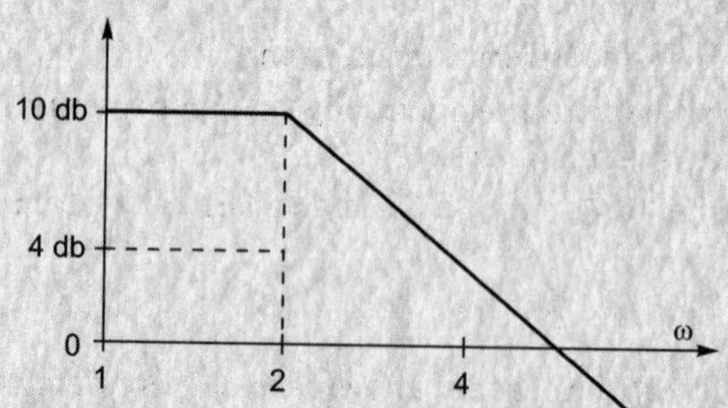

Fig. 7-P-3 (*a*) (*i*) Logarithmic plot for example 7.18 (*a*) (*i*).

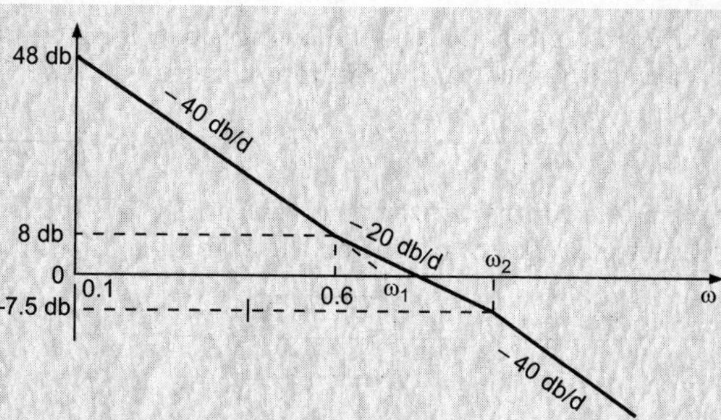

Fig. 7-P-3 (*a*) (*ii*) Logarithmic plot for example 7.18 (*a*) (*ii*).

(*b*) The magnitude (*db*)-frequency (rad/sec) plot given in Fig. 7-*P*-3*b*. Determine the open-loop transfer function and sketch the phase angle frequency plot. Obtain the gain margin and phase margin and determine from the Bode plot the value of gain such that the system is marginally stable.

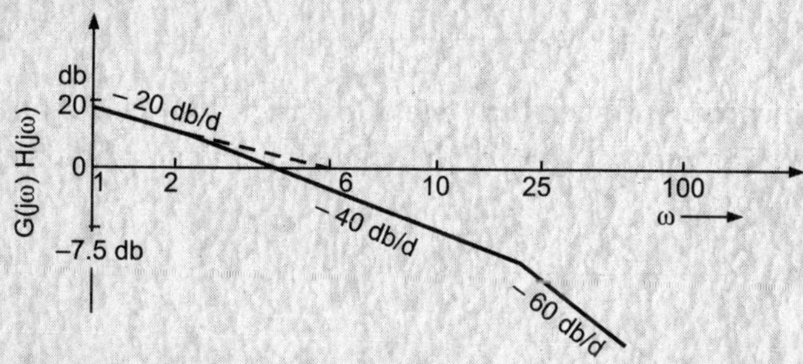

Fig. 7-P-3 (*b*) Magnitude (db)-frequency (rad/sec) plot for problem 7.18 (*b*).

7.19. Sketch the root locus plot for each of the following open-loop transfer function by determining :

(*i*) Number of root locus branches

(*ii*) Angle of asymptotes

(*iii*) Intersection points of asymptotes on real axis

(*a*) $G(s)\,H(s) = \dfrac{K}{s\,(s+2.5)}$ (*b*) $G(s)\,H(s) = \dfrac{K}{s^2\,(s+4)}$

(*c*) $G(s)\,H(s) = \dfrac{K\,(s+1)}{s^2\,(s+4)}$ (*d*) $G(s)\,H(s) = \dfrac{K\,(s+4)}{s^2\,(s+1)}$.

7.20. The transfer functions of a feedback control system are given below

(*a*) $G(s) = \dfrac{K}{s\,(s+1)\,(s+3)}$ and $H(s) = 1$ (*b*) $G(s)\,H(s) = \dfrac{K}{(s+1)\,(s+10)\,(s+30)}$

Sketch the root locus plot as K is varied from 0 to ∞ and determine :

(*i*) the value of K such the system becomes unstable.

(*ii*) the value of K such that the system is critically damped.

7.21. Sketch the root locus plots for the following open-loop transfer functions and determine the value of K such that the damping ratio is 0.707.

(*a*) $G(s)\,H(s) = \dfrac{K(s+20)}{s\,(s+1)\,(s+10)\,(s+30)}$ (*b*) $G(s)\,H(s) = \dfrac{K(s+30)}{(s+1)\,(s+10)\,(s+20)}$.

7.22. The block diagram of a control system is given in Fig. 7-P-4. Plot the root locus as a function of parameter K and comment on the stability.

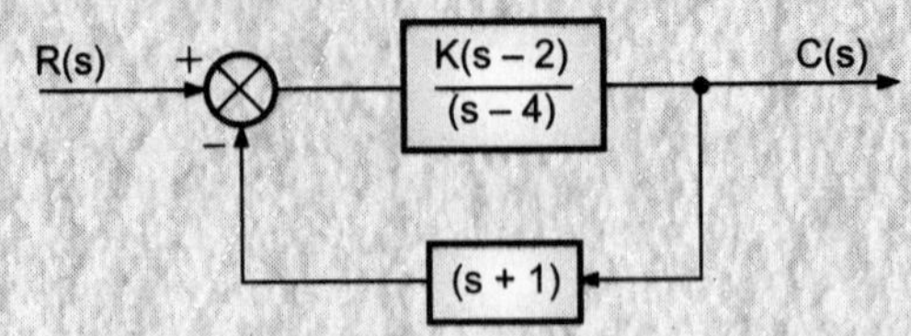

Fig. 7-P-4. Block diagram for problem 7.22.

7.23. Draw the root locus plot for a unity feedback system whose open-loop transfer function is given below

$$G(s) = \frac{K}{s\,(s^2+5s+12.5)}$$

Determine the value of K such that one of the root is $(-1 + j2.5)$.

7.24. Draw the root locus plot for each of the following transfer functions and investigate closed-loop stability :

(*a*) $G(s)\,H(s) = \dfrac{K}{s\,(s+3)\,(s^2+2s+3)}$ (*b*) $G(s)\,H(s) = \dfrac{K}{s^2\,(s+3)\,(s+6)}$

(*c*) $G(s)\,H(s) = \dfrac{50}{s\,(s+1)\,(s+2)\,(s^2+1.6s+1.64)}$ (*d*) $G(s)\,H(s) = \dfrac{K\,(s+4)}{s\,(s+6)\,(s+8)\,(s^2+3s+4)}$.

7.25. The characteristic equation of a control system is given by, $s^3 + 3s^2 + (K+2)\,s + 3K = 0$. Plot the root locus with K as variable parameter from 0 to ∞. Determine the value of K such that the system is

(*i*) marginally stable (*ii*) critically damped.

7.26. Show that for $0 < K < 4$, the root locus lies on a circle. Given that, $G(s)\,H(s) = \dfrac{K\,(s+2)}{(s+1)^2}$.

7.27. Draw the root locus for $G(s)\,H(s) = \dfrac{K}{s\,(s+1)\,(s+4)}$

Determine the gain margin and phase margin for the value of K such that damping ratio $\zeta = 0.707$.

7.28. A closed-loop control system is shown in Fig. 7-P-5 wherein the feedback path has a pole located at $s = -x$. Sketch the root contour plot when the parameter x is varied from 0 to ∞ and examine the closed-loop stability.

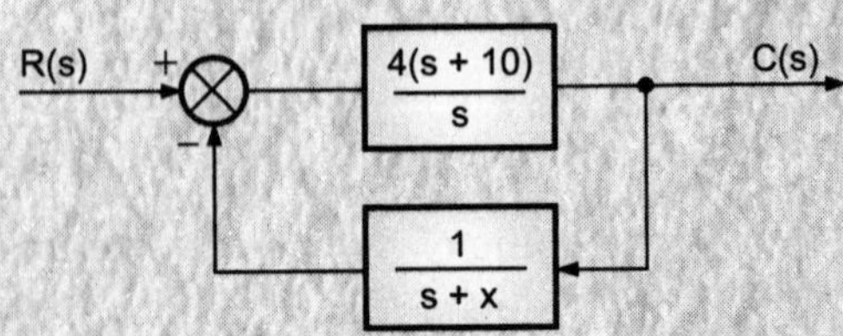

Fig. 7-P-5. Block diagram for problem 7.28.

CHAPTER 8

Compensation of Control Systems

A control system is to be compensated to obtain the desired performance. This chapter introduces the Bode plot and root locus approach to meet the desired specifications of a system. The design of lead, log and lag-lead compensation network to be connected in the forward path as cascade elements is discussed through examples.

The concept of feedback compensation is explained in the form of derivative feedback control.

CONTENTS

- *Phase-Lead Compensation*
- *Phase-Lag Compensation*
- *Phase-Lag-Lead Compensation*
- *Concluding Remarks*
- *Feedback Compensation*
- *Root Locus Approach*
- *Solved Examples*

Introduction. If the performances of a control system is not upto expectations as per desired specifications, then it is required that some change in the system is needed to obtain the desired performance. The change can be in the form of adjustment of forward-path gain or inserting a compensating device in control systems.

For example, the steady state error in a control system can be reduced by increasing forward-path gain but on the other hand this increase in forward-path gain results in making the system more oscillatory or sometimes unstable.

Thus the gain adjustment improves the steady state accuracy of the system at the cost of driving the system towards instability. However, improvement in steady state performance of a control system by gain adjustment is possible in cases where the system is found to be within desirable stability limit even after altering forward-path gain. The forward-path gain adjustment method is not suitable wherein the system becomes unstable due to such adjustment. In such cases a compensation network is introduced in the system.

The compensation network can be introduced in forward-path as shown in Fig. 8.0.1.

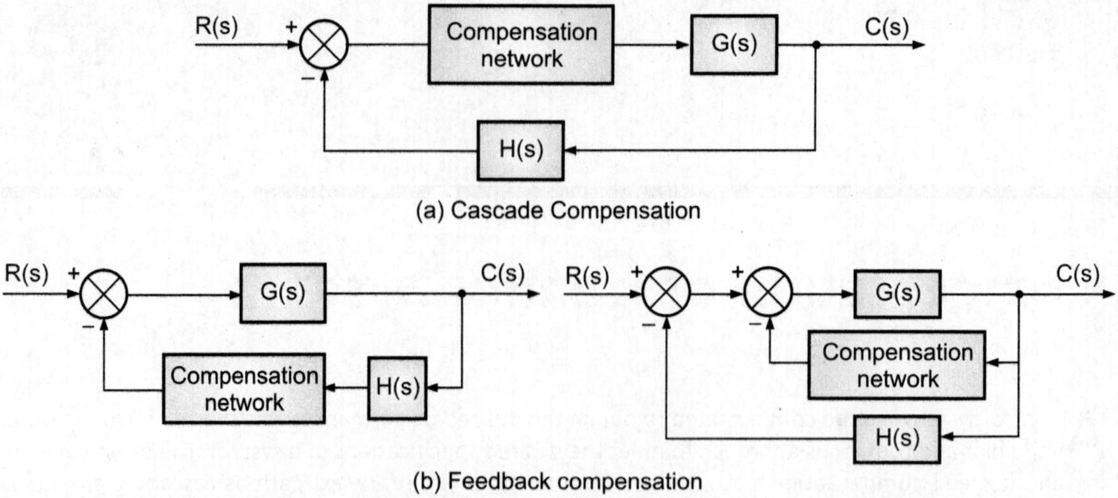

Fig. 8.0.1. Arrangements for introducing compensation network in a control system.

The compensation network introduces additional poles/zeros in the original system thereby altering its original transfer function. The effect of adding zero to an unstable system results in making the system stable as shown in root locus plots of Fig. 8.0.2.

Fig. 8.0.2 (*a*) shows the root locus plot of an unstable system for higher values of forward-path gain. Addition of a zero to the system transfer function changes the shape of root locus plot as shown in Fig. 8.0.2 (*b*) which indicates a stable system for all values of forward-path gain.

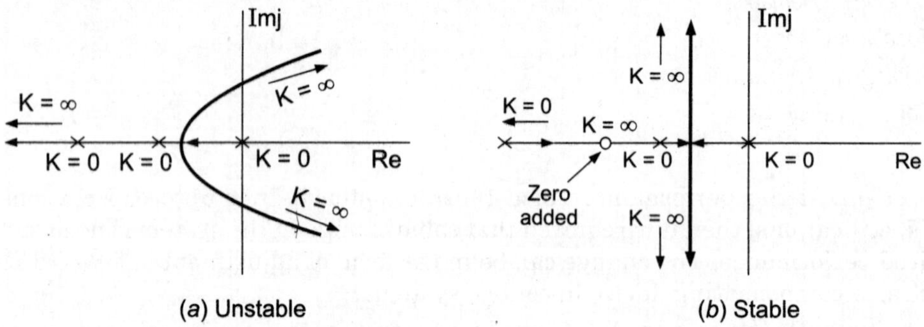

Fig. 8.0.2. Unstable system made stable by adding a zero.

Usually, the performance specifications of a control system are given in frequency domain, *i.e.* gain margin and phase margin, therefore, the use of Bode plot method is very convenient in designing compensation networks for control systems. The networks used for cascade compensation are described below :

8.1 PHASE-LEAD COMPENSATION

Fig. 8.1.1 shows a phase-lead network wherein the phase of output voltage leads the phase of input voltage for sinusoidal inputs.

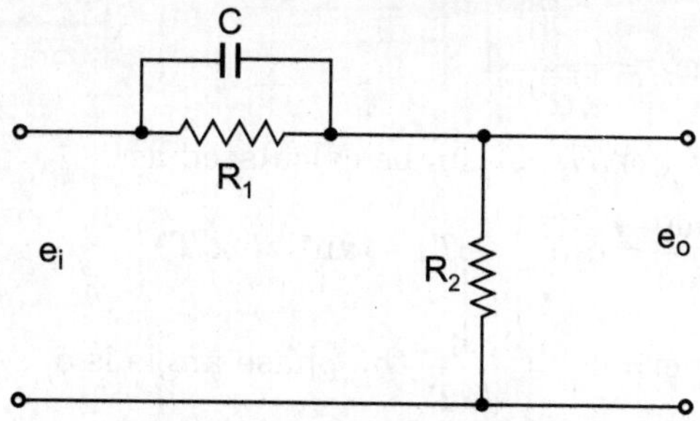

Fig. 8.1.1. Phase-lead network.

The transfer function of a phase-lead network shown in Fig. 8.1.1 is given by

$$\frac{E_o(s)}{E_i(s)} = \frac{\alpha(1+sT)}{(1+s\alpha T)} \quad ...(8.1)$$

where $\alpha < 1$

$$\alpha = \frac{R_2}{R_1 + R_2} \quad \text{and} \quad T = R_1 C$$

The transfer function given by Eq. (8.1) can be expressed in sinusoidal form as

$$\frac{E_o(j\omega)}{E_i(j\omega)} = \frac{\alpha(1+j\omega T)}{(1+j\omega\alpha T)} \quad ...(8.2)$$

The pole-zero configuration and Bode plot for transfer function (8.2) is shown in Fig. 8.1.2.

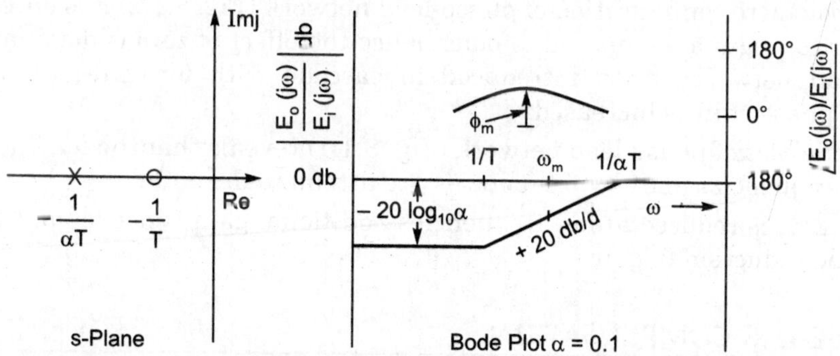

Fig. 8.1.2. Pole-zero configuration and Bode plot for phase-lead network.

The two corner frequencies are

$$\omega = \frac{1}{T} \text{ lower corner frequency}$$

$$\omega = \frac{1}{\alpha T} \text{ upper corner frequency.}$$

The maximum phase-lead ϕ_m occurs at mid-frequency ω_m between upper and lower corner frequencies.

$$\therefore \quad \log_{10}\omega_m = \frac{1}{2}\left[\log_{10}\left(\frac{1}{T}\right) + \log_{10}\left(\frac{1}{\alpha T}\right)\right]$$

$$\therefore \qquad \omega_m = \frac{1}{\sqrt{\alpha}T} \qquad ...(8.3)$$

The phase angle $\angle E_o(j\omega)/E_i(j\omega)$ can be calculated as

$$\angle \frac{E_o(j\omega)}{E_i(j\omega)} = \tan^{-1}(\omega T) - \tan^{-1}(\omega\alpha T)$$

At $\omega = \omega_m = \dfrac{1}{\sqrt{\alpha}T}$ the phase angle is ϕ_m,

$$\therefore \qquad \phi_m = \tan^{-1}\left(\frac{1}{\sqrt{\alpha}T}.T\right) - \tan^{-1}\left(\frac{1}{\sqrt{\alpha}T}\alpha T\right)$$

or

$$\phi_m = \tan^{-1}\left(\frac{1}{\sqrt{\alpha}}\right) - \tan^{-1}(\sqrt{\alpha})$$

$$\therefore \qquad \tan\phi_m = \frac{\dfrac{1}{\sqrt{\alpha}} - \sqrt{\alpha}}{1 + \dfrac{1}{\sqrt{\alpha}}.\sqrt{\alpha}}$$

or

$$\tan\phi_m = \frac{1-\alpha}{2\sqrt{\alpha}} \qquad ...(8.4)$$

and

$$\sin\phi_m = \frac{1-\alpha}{1+\alpha} \qquad ...(8.5)$$

From pole-zero configuration of phase-lead network (Fig. 8.1.2) it is observed that the zero is nearer to origin as compared to pole, hence the effect of zero is dominant, therefore, the phase-lead network, when introduced in cascade with forward-path of a transfer function, the phase shift is increased.

The Bode plot of phase-lead network (Fig. 8.1.1) reveals that the lead network allows to pass high frequencies and low frequencies are attenuated.

As the gain is reduced at low frequencies additional gain is needed in the system to account for the reduction in gain.

8.2 PHASE-LAG COMPENSATION

Fig. 8.2.1 shows a phase-lag network wherein the phase of output voltage lags the phase of input voltage for sinusoidal inputs.

The transfer function of phase-lag network shown in Fig. 8.2.1 is given by

$$\frac{E_o(s)}{E_i(s)} = \frac{1+sT}{1+s\beta T} \qquad ...(8.6)$$

where $\beta > 1$

$$\beta = \frac{R_1 + R_2}{R_2}$$

and $T = R_2 C$

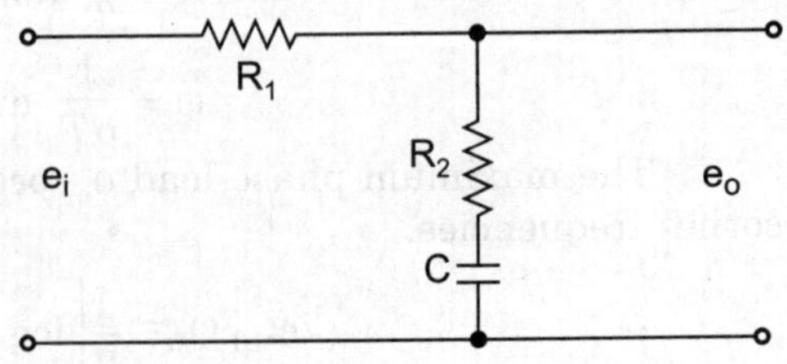

Fig. 8.2.1. Phase-lag network.

The transfer function given by Eq. (8.6) can be expressed in sinusoidal form as

$$\frac{E_o(j\omega)}{E_i(j\omega)} = \frac{1 + j\omega T}{1 + j\omega\beta T} \qquad ...(8.7)$$

The pole-zero configuration and Bode plot for transfer function (8.7) is shown in Fig. 8.2.2.

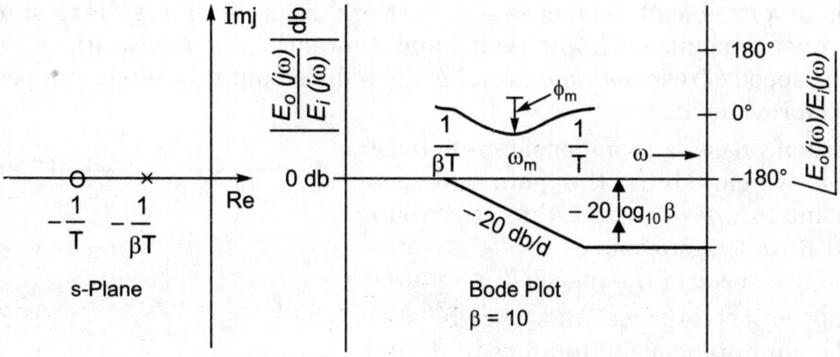

Fig. 8.2.2. Pole-zero configuration and Bode plot for phase-lag network.

The two corner frequencies are

$$\omega = \frac{1}{T} \text{ upper corner frequency}$$

$$\omega = \frac{1}{\beta T} \text{ lower corner frequency}$$

The maximum phase-lag ϕ_m occurs at mid frequency ω_m between upper and lower corner frequencies.

$$\therefore \qquad \log_{10} \omega_m = \frac{1}{2}\left[\log_{10}\left(\frac{1}{\beta T}\right) + \log_{10}\left(\frac{1}{T}\right)\right]$$

$$\therefore \qquad \omega_m = \frac{1}{\sqrt{\beta}T} \qquad ...(8.8)$$

The phase angle $\angle E_o(j\omega)/E_i(j\omega)$ can be calculated as

$$\angle\frac{E_o(j\omega)}{E_i(j\omega)} = \tan^{-1}(\omega T) - \tan^{-1}(\omega\beta T)$$

At $\qquad \omega = \omega_m = \dfrac{1}{\sqrt{\beta}T}$, the phase angle is ϕ_m

$$\therefore \qquad \tan\phi_m = \frac{1-\beta}{2\sqrt{\beta}} \qquad ...(8.9)$$

and $$\sin\phi_m = \frac{1-\beta}{1+\beta} \qquad ...(8.10)$$

Note. As $\beta > 1$, $(1-\beta)/(1+\beta)$ is negative indicating that ϕ_m is negative which means that the phase angle is lagging.

From the pole-zero configuration of phase-lag network (Fig. 8.2.2) it is observed that the pole is nearer to origin compared to zero, hence the effect pole is dominant, therefore, the phase-lag network when introduced in cascade with forward-path of a transfer function, the phase shift is reduced. The Bode plot of phase-lag network (Fig. 8.2.2) reveals that the lag network allows to pass low frequencies and high frequencies are attenuated.

8.3 PHASE-LAG-LEAD COMPENSATION

Introduction to phase-lead compensation network in a control system shifts the gain crossover frequency point to a higher value and, therefore, the bandwidth is increased thus improving the speed of response and overshoot is reduced but the steady state error does not show much improvement.

The use of phase-lag compensation network in a control system shifts the gain crossover frequency point to a lower value thus decreasing the bandwidth and improvement in steady state error is noted but speed of the response is reduced.

The speed of response and, steady state error can be simultaneously improved, if both phase-lag and phase-lead compensation networks are used. However, instead of using two separate lag and lead networks, a single network known as phase-lag-lead network shown in Fig. 8.3.1 is used which combines the effects of both lag and lead networks.

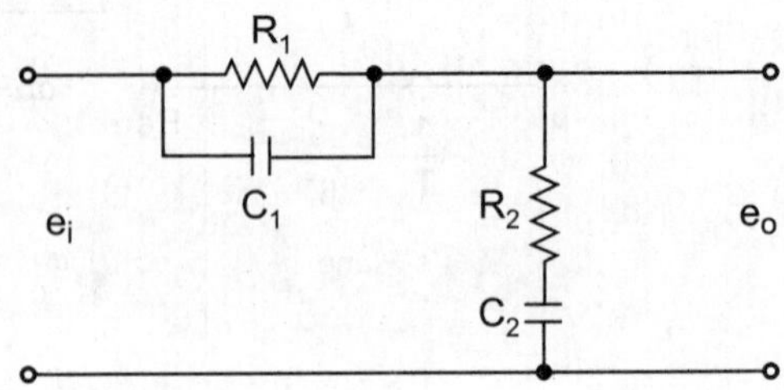

Fig. 8.3.1. Phase lag-lead network.

It can be shown that the transfer function of phase-lag-lead network is given by

$$\frac{E_o(s)}{E_i(s)} = \underbrace{\frac{(1+sT_1)}{(1+s\alpha T_1)}}_{\text{Lead}} \cdot \underbrace{\frac{(1+sT_2)}{(1+s\beta T_2)}}_{\text{Lag}} \quad \text{...(8.11)}$$

where $T_1 = R_1C_1$, $T_2 = R_2C_2$, $\alpha < 1$, $\beta > 1$ and $\alpha\beta = 1$.

Fig. 8.3.2. Pole zero configuration and Bode plot phase-lag-lead network.

The transfer function given by (8.11) can be expressed in sinusoidal form as

$$\frac{E_o(j\omega)}{E_i(j\omega)} = \frac{(1+j\omega T_1)(1+j\omega T_2)}{(1+j\omega\alpha T_1)(1+j\omega\beta T_2)} \qquad ...(8.12)$$

The pole zero configuration and Bode plot for the transfer function (8.12) is shown in Fig. 8.3.2.

8.4 CONCLUDING REMARKS

Introducing compensation network in a control system after the error signal results in reshaping the original transfer function and adjustment of the time constant of compensation network, enables to obtain the system performance as per desired specifications. Lead compensation provides phase-lead.

The gain margin and phase margin can be adjusted to a desired value by using a phase lead compensation network but the bandwidth is increased. The increase in bandwidth results in reduction in settling time and thus speed of the time response is improved.

Additional increase in gain is needed when lead compensation is used. In cases where the phase lead more than 60° is desired, excessive increase in gain is needed and two lead networks are used in cascade.

The inherent unstable systems are some times of type 2 and higher, lead compensation provides phase lead and thus these systems can be made stable by introducing lead compensation networks.

The steady state accuracy is improved by using lag network but as the bandwidth is reduced the speed of time response is deteriorated to a certain extent.

Lag-lead compensation is employed where fast time response as well as better steady state accuracy is desired.

8.5 FEEDBACK COMPENSATION

In feedback compensation method the compensation element is introduced in the feedback path of a control system. Usually, tachogenerator feedback is used wherein a signal proportional to derivative (rate) of the output is feedback to improve the performance of a control system.

The block diagram of an uncompensated system is shown in Fig. 8.5.1.

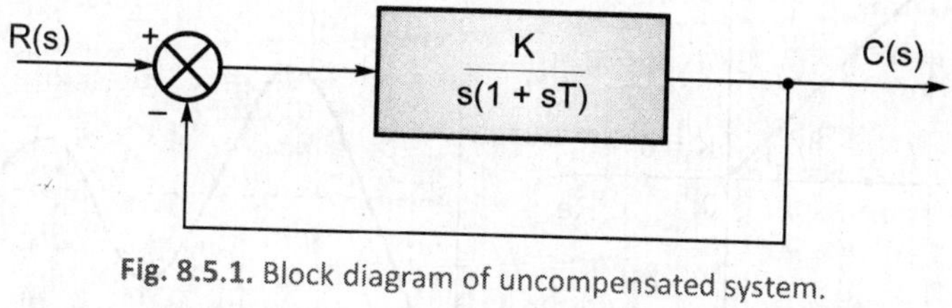

Fig. 8.5.1. Block diagram of uncompensated system.

The open-loop transfer function of the uncompensated system shown in Fig. 8.5.1 is

$$G(s)\,H(s) = \frac{K}{s(1+sT)}.1 = \frac{K}{s(1+sT)} \qquad ...(8.13)$$

Fig. 8.5.2 shows the block diagram of a system using tachometer feedback (output derivative/rate), K_t being tachogenerator constant.

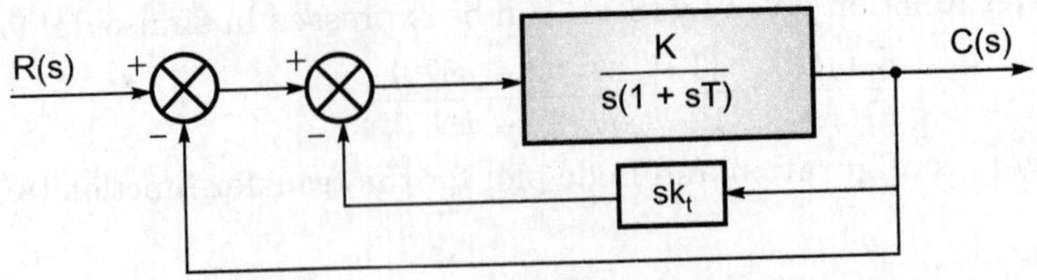

Fig. 8.5.2. Feedback compensation using tachogenerator.

The block diagram shown in Fig. 8.5.2 can be reduced in the form as shown in Fig. 8.5.3.

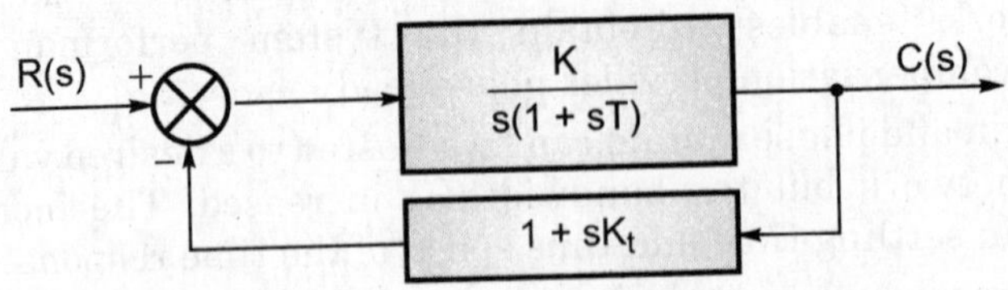

Fig. 8.5.3. Block diagram of Fig. 8.5.2 reduced.

The open-loop transfer function of the compensated system as obtained from Fig. 8.5.3 is given by

$$G(s)\,H(s) = \frac{K}{s(1+sT)}\,.\,(1+sK_t) = \frac{K(1+sK_t)}{s(1+sT)} \qquad \text{...(8.14)}$$

Comparison of transfer function given by Eqs. (8.16) and (8.17) indicate that a zero at $s = -1\ (1/K_t)$ is added to the system due to feedback compensation. The performance of the compensated system depends on the values of K_t and T. Considering the following cases :

(1) $\mathbf{K_t = 0}$

When $K_t = 0$ the feedback compensation is ineffective and the system is uncompensated. Fig. 8.5.4 shows the root locus and the corresponding closed-loop time response of transfer function (8.17) for $K_t = 0$. The value of K being sufficiently high so that the response is underdamped.

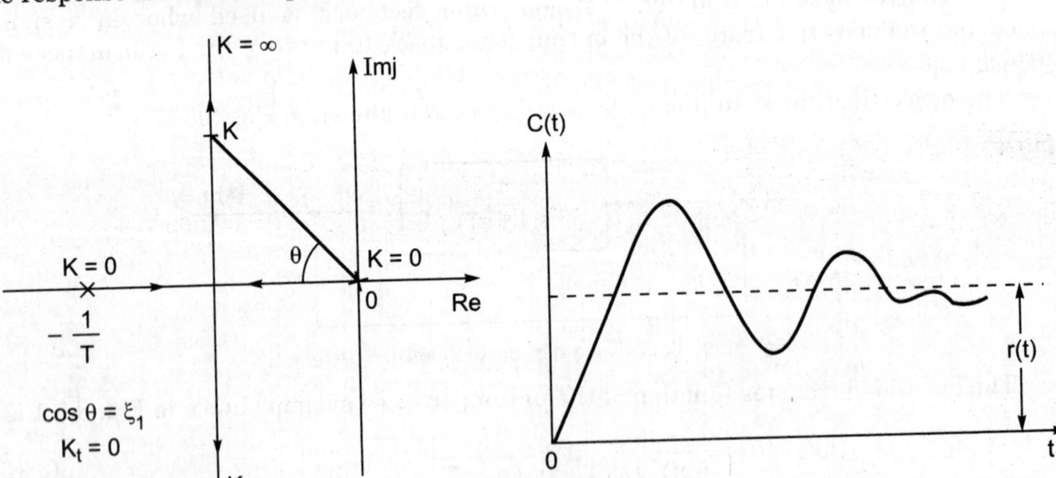

Fig. 8.5.4. Root locus plot and time response of uncompensated system ($K_t = 0$).

(2) $\mathbf{K_t < T}$

The poles of the transfer function (8.17) are at $s = 0$ and $s = (-1/T)$, the zero at $s = (-1/K_t)$ such that $K_t < T$.

Fig. 8.5.5 shows the root locus plot and the corresponding time response of transfer function (8.17) for $K_t < T$.

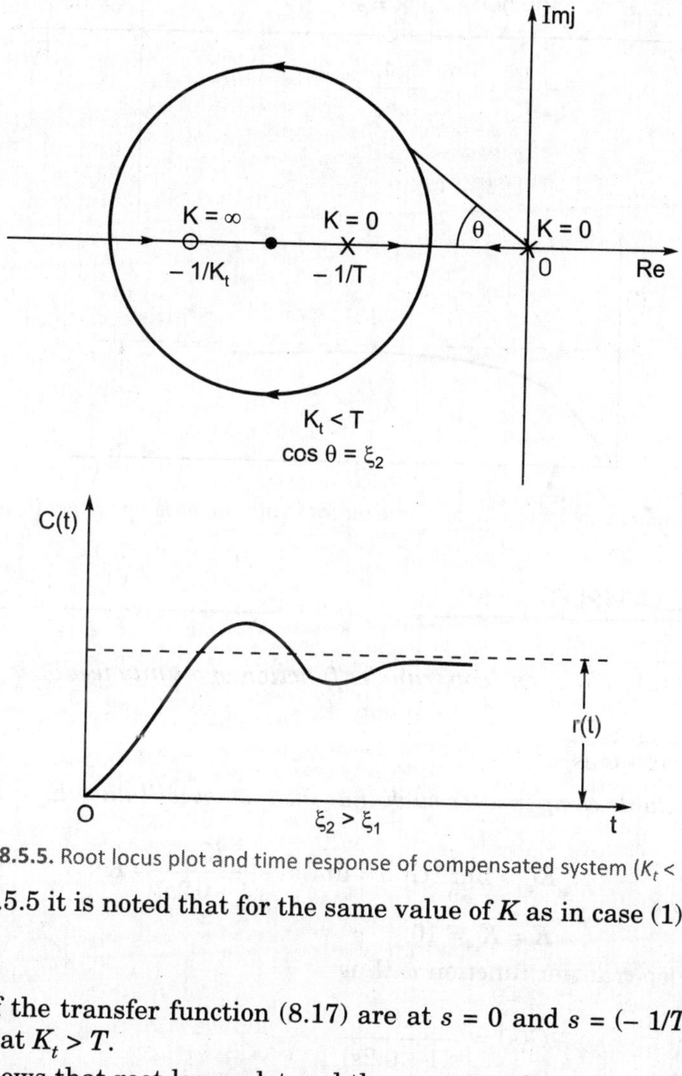

Fig. 8.5.5. Root locus plot and time response of compensated system ($K_t < T$).

From Fig. 8.5.5 it is noted that for the same value of K as in case (1). The damping is reduced.

(3) $\mathbf{K_t > T}$

The poles of the transfer function (8.17) are at $s = 0$ and $s = (-1/T)$. The zero is at $s = (-1/K_t)$ such that $K_t > T$.

Fig. 8.5.6 shows that root locus plot and the corresponding time response of transfer function (8.17) for $K_t > T$. As the roots have no imaginary part, the closed-loop time response is overdamped.

The analysis given above reveals that the time response of a control system can be adjusted by using tachogenerator feedback.

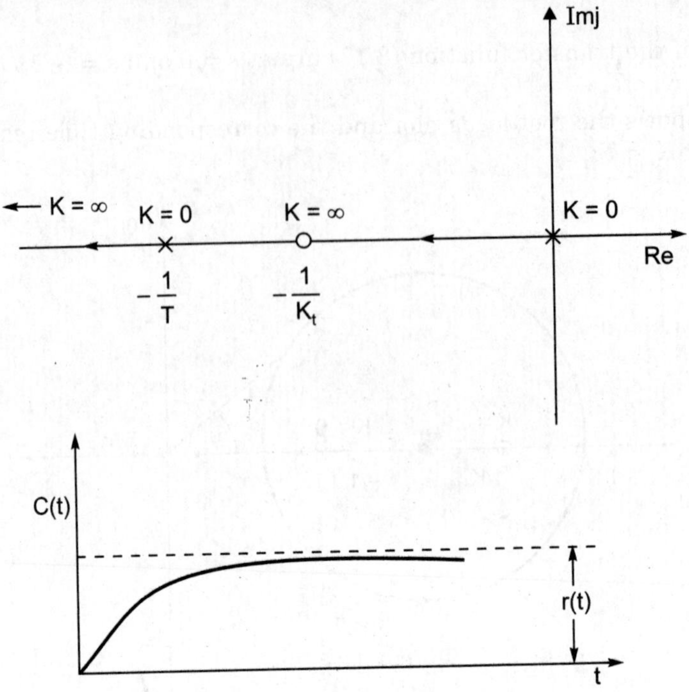

Fig. 8.5.6. Root locus plot and time response of compensated system ($K_t > T$).

8.6 SOLVED EXAMPLES

Example 8.6.1. *The open-loop transfer function of a unity feedback control system is given by,* $G(s) = \dfrac{K}{s(1+0.2s)}$

Design a suitable compensator such that the system will have $K_v = 10$ *and P.M.* $= 50°$.

Solution.

$$K_v = \lim_{s \to 0} sG(s) = \lim_{s \to 0} s \cdot \frac{K}{s(1+0.2s)} = K$$

$\therefore$
$$K = K_v = 10$$

The open-loop transfer function is thus

$$G(s) = \frac{10}{s(1+0.2s)}$$

The Bode plot for the above transfer function is plotted in Fig. 8.6.1. The frequency range selected is ω = 1 rad/sec to ω = 10 rad/sec. The phase angle is given by

$$\angle G(j\omega) = -90° - \tan^{-1}(0.2\omega)$$

Therefore,

ω (rad/sec)	1	5	10	50
$\angle G(j\omega)°$	− 101.3	− 135	− 153.3	− 177.3

From the Bode plot (Fig. 8.6.1) it is noted that uncompensated system is stable having a phase margin of 30° and the gain crossover occurs at $\omega = 6.75$ rad/sec.

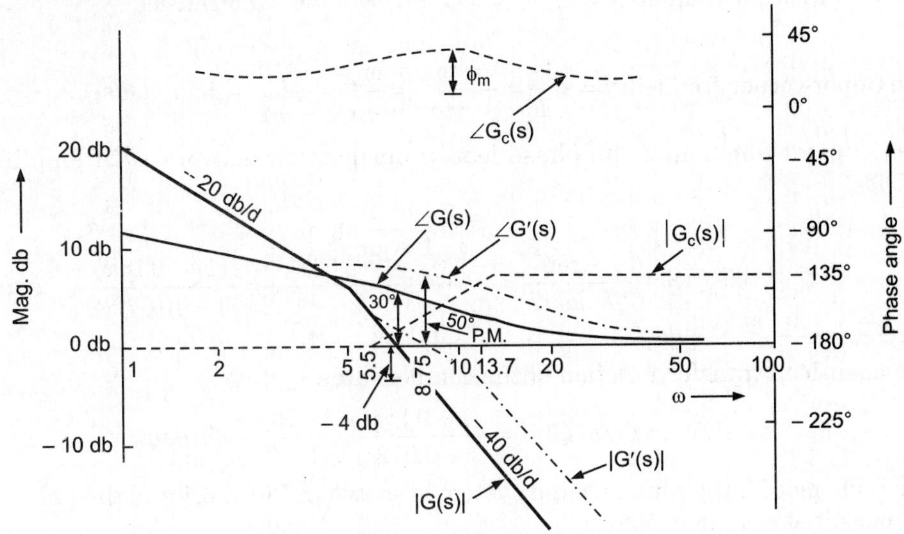

Fig. 8.6.1. Bode plot for example 8.6.1.

To increase the phase margin to 50°, the gain crossover point is to be shifted to a higher value of frequency and this is possible by introducing a phase lead compensation network. The required phase lead is determined below :

$$\phi_m = 50° - 30° + \text{Tolerance}$$

Considering a tolerance of say 5°

$$\phi_m = 50° - 30° + 5° = 25°$$

The value of α parameter of the phase-lead network is given by

$$\alpha = \frac{1 - \sin \phi_m}{1 + \sin \phi_m} = \frac{1 - \sin 25°}{1 + \sin 25°} = \frac{1 - 0.423}{1 + 0.423} = 0.4$$

The maximum phase lead $\phi_m = 25°$ is contributed at midfrequency ω_m of the phase lead compensation network. Allowing an amplification of $1/\alpha$ the magnitude of phase-lead compensation network transfer function at $\omega = \omega_m$ is determined below.

$$\because \qquad \omega_m = \frac{1}{\sqrt{\alpha T}}$$

$$\therefore \qquad |G_c(j\omega_m)| = \left| \frac{1 + j\omega_m T}{1 + j\omega_m \alpha T} \right| = \left| \frac{1 + j\dfrac{1}{\sqrt{\alpha T}} . T}{1 + j\dfrac{1}{\sqrt{\alpha T}} . \alpha T} \right| = \frac{1}{\sqrt{\alpha}}$$

In decibel $20 \log_{10} | G_c(j\omega_m) | = 20 \log_{10} \dfrac{1}{\sqrt{\alpha}} = 20 \log_{10} \dfrac{1}{0.4} = 4.0$ db

From the Bode plot (Fig. 8.6.1) it is noted that for uncompensated system at a gain of – 4.0 db the frequency is 8.7 rad/sec. The new gain crossover frequency for the compensated system is selected at $\omega = \omega_m = 8.7$ rad/sec.

The two corner frequencies for the phase-lead network are

$$\text{The lower corner frequency} = \frac{1}{T} = \sqrt{\alpha}\,\omega_m = \sqrt{0.4} \times 8.7 = 5.5 \text{ rad/sec}$$

$$\text{The upper corner frequency} = \frac{1}{\alpha T} = \frac{\sqrt{\alpha}\,\omega_m}{\alpha} = \frac{\omega_m}{\sqrt{\alpha}} = \frac{8.7}{\sqrt{0.4}} = 13.7 \text{ rad/sec.}$$

The transfer function of the phase lead compensation network with amplifier gain $(1/\alpha) = 2.5$ is

$$G_c(j\omega) = \frac{1 + j\omega T}{1 + j\omega\alpha T} = \frac{\left(1 + j\omega\frac{1}{5.5}\right)}{\left(1 + j\omega \times 0.4 \times \frac{1}{5.5}\right)} = \frac{(1 + j0.18\omega)}{(1 + j0.072\omega)}. \quad \textbf{Ans.}$$

The open-loop transfer function of the compensated system is

$$G'(s) = G_c(s)\, G(s) = \frac{(1 + 0.18s)}{(1 + 0.072s)} \cdot \frac{10}{s(1 + 0.2s)}. \quad \textbf{Ans.}$$

The Bode plot for the compensated system is shown in Fig. 8.6.1 and the phase margin of the compensated system is 50°.

Example 8.6.2 *Design a phase lead compensation network for a system having open-loop transfer function given below such that the system will have acceleration error coefficient $K_a = 100$ sec^{-2} and suitable phase margin for stable operation*

$$G(s)\,H(s) = \frac{K}{s^2(1 + 0.05s)}.$$

Solution.
$$K_a = \lim_{s \to 0} s^2 G(s) H(s) = \lim_{s \to 0} s^2 \cdot \frac{K}{s^2(1 + 0.05s)} = K$$

$\therefore \quad K = K_a = 100$

The open-loop transfer function is thus

$$G(s)H(s) = \frac{100}{s^2(1 + 0.05s)}$$

The Bode plot for the above transfer function is plotted in Fig. 8.6.2. The frequency range selected is ω = 1 rad/sec to ω = 100 rad/sec. The phase angle is given by

$$\angle G(j\omega)\, H(j\omega) = -180° - \tan^{-1}(0.05\omega)$$

Therefore,

ω (rad/sec)	1	10	20	50	100
$\angle G(j\omega)H(j\omega)°$	– 182.8	– 206.5	– 225	– 248.2	– 258.6

From the Bode plot (Fig. 8.6.2) it is noted that the P.M. is – 26° and the system is unstable. A suitable phase margin varying between 20° to 50° is assumed, thus

(*a*) Assuming P.M. = 20° the required phase lead is

$$\phi_m = 20° - (-26°) + \text{Tolerance}$$

As the system is type 2, the phase angle decreases rapidly hence a higher tolerance say 10° is considered

$$\therefore \quad \phi_m = 20° - (-26°) + 10° = 56°$$

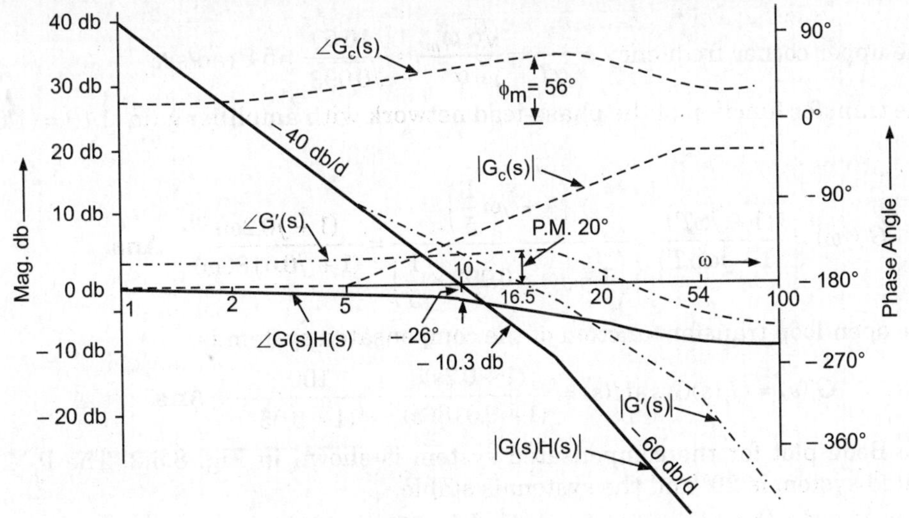

Fig. 8.6.2. Bode plot for example 8.6.2 ($\phi_m = 56°$).

(*b*) Assuming P.M. = 50° the required phase lead is

$$\phi_m = 50° - (-26°) + 10° = 86°$$

In case (*a*) the phase lead contributed is 56° whereas in case (*b*) the phase lead is 86°, whenever the phase lead to be contributed by compensation network exceeds 60° then it is preferred to use two networks in cascade each contributing phase lead of $\phi_m/2$.

The compensation networks for cases (*a*) and (*b*) are designed below :

Case (*a*) $\phi_m = 56°$

$$\alpha = \frac{1 - \sin 56°}{1 + \sin 56°} = \frac{1 - 0.829}{1 + 0.829} = \frac{0.17}{1.829} = 0.093.$$

The maximum phase-lead 56°, is contributed at mid-frequency ω_m of the phase-lead network. Allowing an amplification of $1/\alpha$ the magnitude of phase-lead compensation network transfer function at $\omega = \omega_m$ is determined below :

$$\because \quad \omega_m = \frac{1}{\sqrt{\alpha T}}$$

$$\therefore \quad |G_c(j\omega_m)| = \left|\frac{1 + j\omega_m T}{1 + j\omega_m \alpha T}\right| = \left|\frac{1 + \frac{1}{\sqrt{\alpha T}} . T}{1 + \frac{1}{\sqrt{\alpha T}} . \alpha T}\right| = \frac{1}{\sqrt{\alpha}}$$

In decibel, $20 \log_{10} |G_c(j\omega_m)| = 20 \log_{10} \frac{1}{\sqrt{\alpha}} = 20 \log_{10} \frac{1}{\sqrt{0.093}} = 10.3$ db.

From the Bode plot (Fig. 8.6.2) it is noted that for uncompensated system at a gain of – 10.3 db the frequency is 16.5 rad/sec. The new gain crossover frequency for the compensated system is selected as $\omega = \omega_m = 16.5$ rad/sec.

The two corner frequencies for the phase-lead network are

The lower corner frequency $= \frac{1}{T} = \sqrt{\alpha}\, \omega_m = \sqrt{0.093} \times 16.5 = 5$ rad/sec

The upper corner frequency $= \dfrac{1}{\alpha T} = \dfrac{\sqrt{\alpha}\,\omega_m}{\alpha} = \dfrac{16.5}{\sqrt{0.093}} = 54$ rad/sec

The transfer function of the phase-lead network with amplifier gain (1/α) = (1/0.093) = 15.75 is

$$G_c(j\omega) = \frac{(1 + j\omega T)}{(1 + j\omega\alpha T)} = \frac{\left(1 + j\omega\dfrac{1}{5}\right)}{\left(1 + j\omega \times 0.093 \times \dfrac{1}{5}\right)} = \frac{(1 + j0.2\omega)}{(1 + j0.0186\omega)} \quad \textbf{Ans.}$$

The open-loop transfer function of the compensated system is

$$G'(s) = G_c(s)G(s)\,H(s) = \frac{(1 + 0.2s)}{(1 + 0.0186s)} \cdot \frac{100}{s^2(1 + 0.05s)} \quad \textbf{Ans.}$$

The Bode plot for the compensated system is shown in Fig. 8.6.2. The P.M. of the compensated system is 20° and the system is stable.

Case (*b*). As the phase lead required is 86° (> 60°) the same is provided by two identical phase-lead networks each contributing a phase lead of 43°, therefore,

$$\alpha = \frac{1 - \sin 43^\circ}{1 + \sin 43^\circ} = \frac{1 - 0.681}{1 + 0.681} = 0.19$$

The phase lead ϕ_m = 43° due to each phase-lead network is contributed by each compensation network at mid-frequency ω_m of the phase lead network. Allowing an amplification of 1/α for each compensation network, the magnitude of each phase-lead compensation network is determined below :

$$\because \qquad \omega_m = \frac{1}{\sqrt{\alpha T}}$$

$$\therefore \qquad |G_c(j\omega_m)| = \left|\frac{1 + j\omega_m T}{1 + j\omega_m \alpha T}\right| = \left|\frac{1 + j\dfrac{1}{\sqrt{\alpha T}}.T}{1 + j\dfrac{1}{\sqrt{\alpha T}}.\alpha T}\right| = \frac{1}{\sqrt{\alpha}}$$

In decibel, $20 \log_{10} |G_c(j\omega_m)| = 20 \log_{10} \dfrac{1}{\sqrt{\alpha}} = 20 \log_{10} \dfrac{1}{\sqrt{0.19}} = 7.2$ db

The total magnitude due to two identical phase-lead networks is 7.2 + 7.2 = 14.4 db.

From Bode plot (Fig. 8.6.3) it is observed that for uncompensated system, at a gain of – 14.4 db the frequency is 21 rad/sec. The new gain crossover frequency for the compensated system is selected as $\omega = \omega_m$ = 21 rad/sec.

The two corner frequencies for the phase-lead networks are

The lower corner frequency $= \dfrac{1}{T} = \sqrt{\alpha}\,\omega_m = \sqrt{0.19} \times 21 = 9.15$ rad/sec

The upper corner frequency $= \dfrac{1}{\alpha T} - \dfrac{\sqrt{\alpha}\omega_m}{\alpha} = \dfrac{\omega_m}{\sqrt{\alpha}} = \dfrac{21}{0.19} = 48.2$ rad/sec

The transfer function of the phase-lead network with amplifier gain (1/α)(1/α) = (1/0.19)(1/0.19) = 27.7 is

$$G_c(j\omega) = \frac{(1 + j\omega T)(1 + j\omega T)}{(1 + j\omega\alpha T)(1 + j\omega\alpha T)} = \frac{(1 + j\omega T)^2}{(1 + j\omega\alpha T)^2}$$

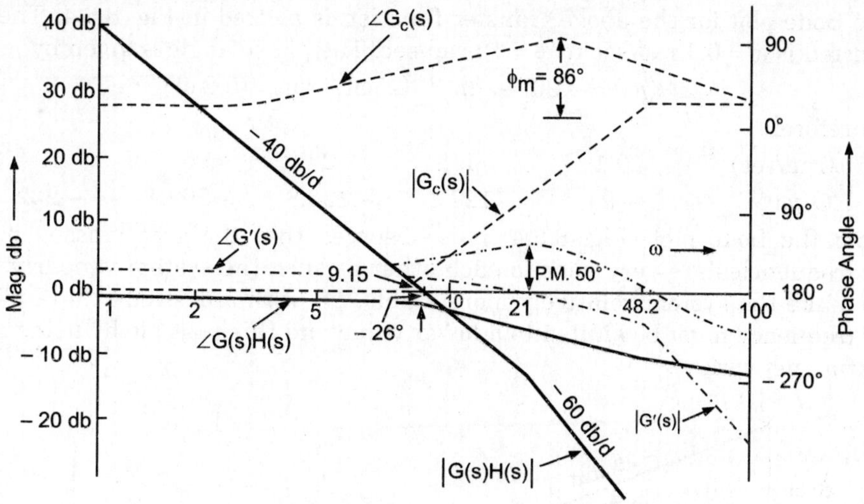

Fig. 8.6.3. Bode plot for example 8.6.2 (ϕ_m = 86°).

$$= \frac{\left(1 + j\omega \frac{1}{9.15}\right)^2}{\left(1 + j\omega \times 0.19 \times \frac{1}{9.15}\right)^2} = \frac{(1 + j0.11\omega)^2}{(1 + j0.02\omega)^2}. \quad \textbf{Ans.}$$

The open-loop transfer function of the compensated system is

$$G'(s) = G_c(s)G(s)H(s) = \frac{(1 + 0.11s)^2}{(1 + 0.02s)^2} \cdot \frac{100}{s^2(1 + 0.05s)}. \quad \textbf{Ans.}$$

The Bode plot for the compensated system is shown in Fig. 8.6.3. The phase margin of the compensated system is 50° and the system is stable.

Example 8.6.3. *The open-loop transfer function of a unity feedback control system is given by*

$$G(s) = \frac{K}{s(1 + 0.5s)(1 + 0.2s)}$$

It is desired that

(i) for a unit ramp input the steady state error of the output position be less than 0.125 degrees / (degree / second)

(ii) the P.M. ≥ 30° and *(iii) the G.M. ≥ 10 db*

Design a suitable compensation network.

Solution. $$K_v = \frac{\text{Desired output velocity}}{\text{Steady state error}} = \frac{1}{0.125} = 8 \text{ sec}^{-1}$$

As the system is type "1"

$$\therefore \quad K = K_v = 8$$

The open-loop transfer function for the system is

$$G(s) = \frac{8}{s(1 + 0.5s)(1 + 0.2s)}$$

The Bode plot for the above transfer function is plotted in Fig. 8.6.4. The frequency range selected is ω = 0.1 rad/sec to ω = 10 rad/sec. The phase angle is given by

$$\angle G(j\omega) = -90^\circ - \tan^{-1}(0.5\omega) - \tan^{-1}(0.2\omega)$$

Therefore,

ω(rad/sec)	0.1	1	2	5	10
$\angle G(j\omega)^\circ$	– 94	– 127.1	– 156.8	– 203.1	– 232

From the Bode plot (Fig. 8.6.4) it is observed that the gain crossover and phase crossover frequencies are very near to each other as such the system is nearly marginally stable, to make the system relatively more stable, *i.e.* with a desired P.M. ≥ 30°, the gain crossover frequency must be shifted to a lower value and this possible by using a phase-lag compensation network.

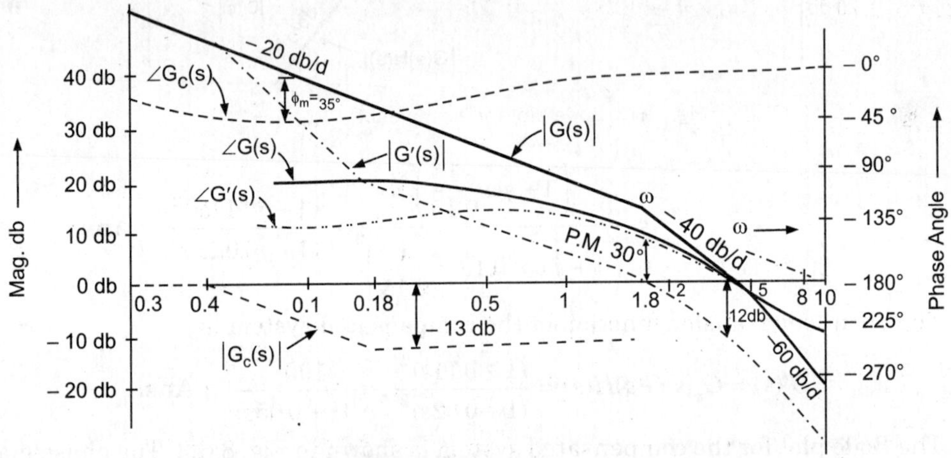

Fig. 8.6.4. Bode plot for example 8.6.3.

However, the use phase-lag network will reduce the phase angle to some extent at new gain crossover frequency, therefore, some tolerance say 5° is taken into consideration, thus the phase angle of the compensated system at new gain crossover frequency is 30° (desired P.M.) + 5° = 35°.

For the uncompensated system, at ω = 1.8 rad/sec the phase angle is 30°. The new gain crossover frequency is selected as ω = 1.8 rad/sec. At this new gain crossover frequency the magnitude of the uncompensated system is 13 db which is to be neutralized by phase-lag network, therefore, at new gain crossover frequency

$$-20\log_{10}\beta = -13 \qquad \therefore \qquad \beta = 4.46$$

The upper corner frequency of the phase lag compensation network is selected one decade earlier than the new gain crossover frequency, hence the upper corner frequency for the phase lag network is ω = (1.8/10) = 0.18 rad/sec.

$$\therefore \quad \text{The upper corner frequency} = \frac{1}{T} = 0.18 \text{ rad/sec}$$

$$\text{The lower corner frequency} \quad = \frac{1}{\beta T} = \frac{0.18}{4.46} = 0.04 \text{ rad/sec}$$

The transfer function of the phase-lag network is

$$G_c(j\omega) = \frac{(1+j\omega T)}{(1+j\omega\beta T)} = \frac{\left(1+j\omega\dfrac{1}{0.18}\right)}{\left(1+j\omega\times 4.46\times\dfrac{1}{0.18}\right)} = \frac{(1+j5.55\omega)}{(1+j24.7\omega)}$$

The open-loop transfer function of the compensated system is

$$G'(s) = G_c(s)\,G(s) = \frac{(1+5.55s)}{(1+24.7s)}\cdot\frac{8}{s(1+0.5s)(1+0.2s)}$$

The Bode plot for the compensated system is shown in Fig. 8.6.4 and P.M. = 30°, G.M. = 12 db is obtained.

Example 8.6.4. *The open-loop transfer function of a unity feedback control system is given by*

$$G(s) = \frac{K}{s(s+2)}$$

The system is to have 25% maximum overshoot and peak time 1.0 second. Determine the value of K and tachometer feedback constant K_t.

Solution. The block diagram of uncompensated system is drawn in Fig. 8.6.5.

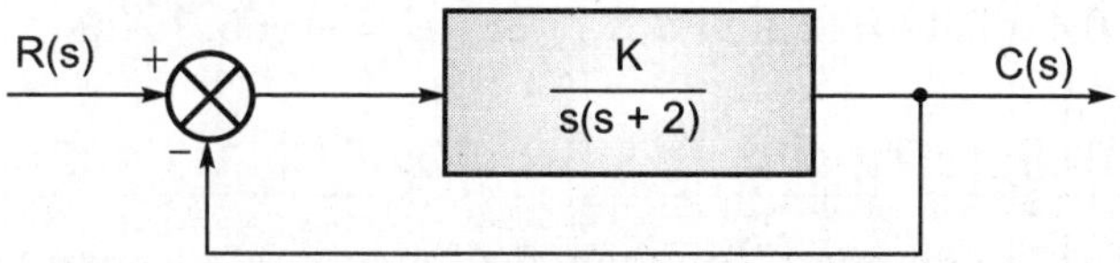

Fig. 8.6.5. Block diagram of uncompensated system for example 8.6.4.

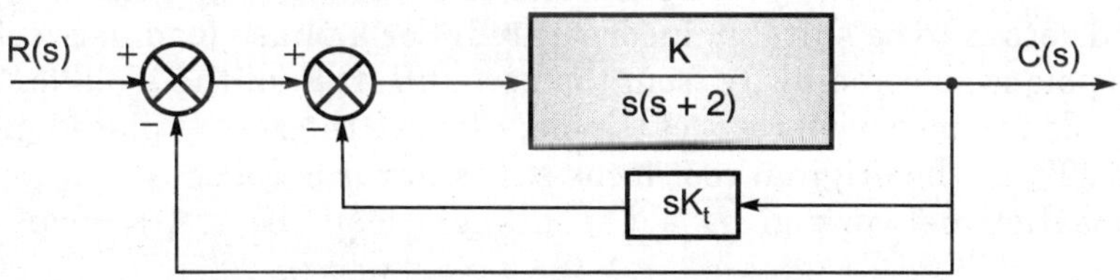

Fig. 8.6.6. Block diagram of compensated system for example 8.6.4.

With tachometer feedback (feedback compensation) the block diagram of Fig. 8.6.5 is redrawn as shown in Fig. 8.6.6.

The overall transfer function of the compensated system is

$$\frac{C(s)}{R(s)} = \frac{K}{s^2+(2+KK_t)s+K}$$

Since $\quad M_\phi = 25\%$

$$\therefore \quad 0.25 = e^{-\frac{\zeta\pi}{\sqrt{1-\zeta^2}}} \quad \therefore \quad \log_e 0.25 = -\frac{\zeta\pi}{\sqrt{1-\zeta^2}}\log_e(e)$$

$$\therefore \quad -1.38 = -\frac{\zeta\pi}{\sqrt{1-\zeta^2}}\times 1 \quad \therefore \quad \zeta = 0.4$$

$$t_p = 1 \text{ sec} \qquad \text{(given)}$$

$\because$
$$t_p = \frac{\pi}{\omega_n\sqrt{1-\zeta^2}}$$

$\therefore$
$$1 = \frac{\pi}{\omega_n\sqrt{1-\zeta^2}} \qquad \therefore \quad \omega_n\sqrt{1-\zeta^2} = \pi$$

or
$$\omega_n = \frac{\pi}{\sqrt{1-\zeta^2}} = \frac{\pi}{\sqrt{1-(0.4)^2}} = 3.42 \text{ rad/sec}$$

The characteristic equation for the compensated system is

$$s^2 + (2 + KK_t)\, s + K = 0$$

From the characteristic equation it is noted that

$$\omega_n = \sqrt{K} \quad \text{and} \quad 2\,\zeta\omega_n = (2 + KK_t)$$

$\therefore$
$$3.42 = \sqrt{K}$$

or
$$K = (3.42)^2 = 11.7. \quad \textbf{Ans.}$$

$$2\zeta\omega_n = (2 + KK_t)$$

$\therefore$
$$2 \times 0.4 \times 3.42 = (2 + 11.7\, K_t) \quad \text{or} \quad K_t = 0.065. \quad \textbf{Ans.}$$

8.7 COMPENSATOR NETWORK DESIGN: ROOT LOCUS APPROACH

The addition of a "zero" to an open-loop transfer function of a control system improves its time response. However, it is not possible to add an isolated pole/zero to a transfer function. Therefore, in order to realize a transfer function of a compensation network a pair of pole and zero is to be suitably incorporated. For a phase lead network, the suitable location of the pole is selected away from the zero in L.H.S. of the s-plane. This condition gives the zero a dominance with respect to pole, because the zero is placed near the s-plane imaginary axis. Thus, the original root locus shifts towards L.H.S. of the s-plane. The use of this compensation network gives a phase angle load. Hence, known as phase lead compensation.

The addition of a pole to an open-loop transfer function such that the said pole is placed near L.H.S. of the s-plane imaginary axis and the zero suitably away from this pole introduces a phase angle lag. This type of compensation is known as phase lag compensation. The use of lag compensation increases steady state error coefficient, thereby improves (reduces) steady state error. The angle contribution due to the phase lag compensator is very small as such the shape of the original root locus almost remains the same. Therefore, to a fair approximation the time response is not affected.

In general the transfer function of a compensator is given by

$$G_c(s) = \frac{(s+a)}{(s+b)} \qquad \text{...(8.15)}$$

Depending on the s-plane location of pole $(-b)$/zero $(-a)$ the transfer function $G_c(s)$ is decided as that of lead or lag type.

1. **Lead Compensation**

For the case $a < b$ the pole-zero location of the lead compensator is shown in Fig. 8.7.1.

The zero at $s = -a$ of the lead compensator is place near to the imaginary axis ($j\omega$) with respect to pole at $s = -b$. Thus, the effect of the compensator zero is more as compared to pole in modifying the performance of a given control system. The compensation thus introduced in cascade with the open-loop transfer function of a control system gives a phase lead.

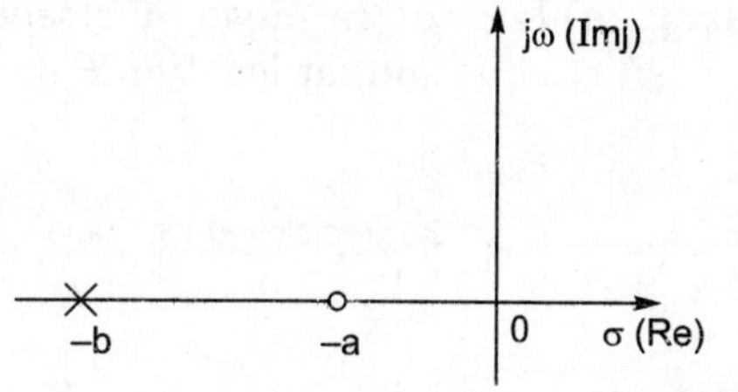

Fig. 8.7.1. Pole-Zero plot : Lead compensation.

As $a < b$, the gain less than unity is introduced. Thus, to adjust the gain to its original value, the transfer function of a phase lead compensator takes the form as below:

$$G_c(s) = \frac{K_c(s+a)}{(s+b)},\ a < b \qquad \text{...(8.16)}$$

The insertion of a phase lead compensator improves the time response.

2. **Lag Compensation**

For the case $a > b$ the s-plane pole-zero location is shown in Fig. 8.7.2.

The corresponding transfer function is given below:

$$G_c(s) = \frac{(s+a)}{(s+b)},\ a > b \qquad \text{...(8.17)}$$

The pole of the compensator at $s = -b$ is placed near to the s-plane imaginary axis ($j\omega$) with respect to the zero at $s = -a$. Thus, the effect of the compensator pole is more as compared to the zero. The compensation network connected in cascade with the open-loop transfer function gives a phase lag.

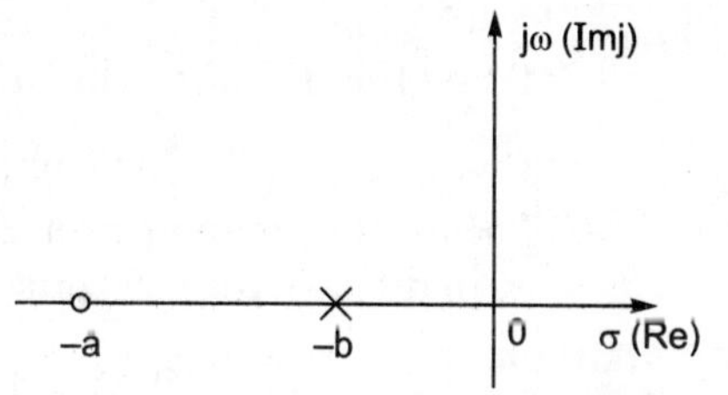

Fig. 8.7.2. Pole-Zero plot: Lag compensation.

As $a > b$ the steady-state error coefficient is increased which improves (reduces) the steady-state error. The gain of the lag compensator is considered as unity.

3. **Lag-Lead Compensation**

For the case wherein both time response as well as steady-state error are to be improved lag-lead compensation is used.

The transfer function of a lag-lead compensator is given below:

$$G_c(s) = \frac{K_c(s+a_1)(s+a_2)}{\underbrace{(s+b_1)}_{\text{lead}}\underbrace{(s+b_2)}_{\text{lag}}} \qquad \text{...(8.18)}$$

$$a_1 < b_1,\quad a_2 > b_2$$

8.7.1. Phase Lead Compensator Design

The design of a phase lead compensator is carried out as per the following methods:

Method no. 1: Graphical method

(*a*) Locate the desired closed-loop pole "s_d" (*i.e.* desired root of the characteristic equation) at location *P* on the *s*-plane (Fig. 8.7.3).

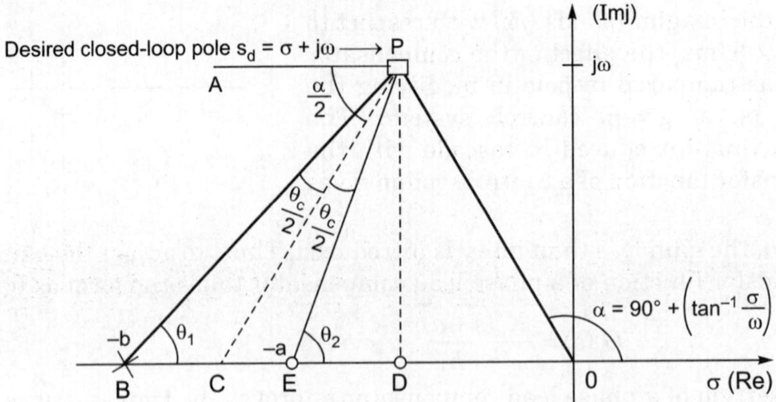

Fig. 8.7.3. Phase lead compensation (graphical method)

Draw the lines *OP* and *PA*. The line *PA* is parallel to *s*-plane real axis.

(*b*) Bisect ∠*OPA*, the bisector intersects the real axis at point *C*.

(*c*) Draw line *PE* such that $\angle EPC = \frac{\theta_c}{2}$ where θ_c is the phase angle lead required.

(*d*) Draw line *PB* such that $\angle BPC = \frac{\theta_c}{2}$.

(*e*) Draw perpendicular from the point *P* to intersect *s*-plane real axis at point *D*.

(*f*) Locate the compensator zero at $s = -a$ at point *E* and the compensator pole $s = -b$ at point *B* (with reference to Fig. 8.7.3).

$$\angle OPA = \alpha, \quad \therefore \quad \angle BPA = \frac{\alpha}{2} - \frac{\theta_c}{2}$$

$$\because \quad \angle BPA = \angle PBC = \theta_1$$

$$\therefore \quad \theta_1 = \frac{\alpha}{2} - \frac{\theta_c}{2} \quad \text{...(8.19)}$$

Also, $$\angle EPA = \angle PED = \theta_2$$

$$\therefore \quad \theta_2 = \frac{\alpha}{2} + \frac{\theta_c}{2} \quad \text{...(8.20)}$$

The required phase lead (*i.e.* the phase angle contributed by the phase lead compensator) is ∠*BPE*.

Subtracting Eqn. (8.19) from Eqn. (8.20)

$$\theta_c = \theta_2 - \theta_1 \quad \text{...(8.21)}$$

The lengths *DB* and *DE* are determined below:

$$\tan \theta_1 = \frac{\omega}{DB} \quad \text{and} \quad \tan \theta_2 = \frac{\omega}{DE}$$

Place the compensator zero "*a*" at location *E*

$$\therefore \quad -a = -(\sigma + DE) \quad \text{...(8.22)}$$

Place the compensator pole "b" at location B

$$\therefore \quad -b = -(\sigma + DB) \qquad ...(8.23)$$

The transfer function of a phase lead compensator is

$$G_c(s) = \frac{K_c(s+a)}{(s+b)}, \ a < b$$

The parameters a and b are determined as above. The open-loop transfer function of a unity feedback control system using phase lead compensator is given by:

$$G(s)\, G_c(s) = G(s) \frac{K_c(s+a)}{(s+b)}, \ a < b$$

Consider,
$$G(s) = \frac{K(s+z_1)\ldots(s+z_n)}{s(s+p_1)\ldots(s+p_m)}, \ n < m$$

$$\therefore \quad G(s)\, G_c(s) = \frac{KK_c(s+z_1)\ldots(s+z_n)(s+a)}{s(s+p_1)\ldots(s+p_m)(s+b)} \qquad ...(8.24)$$

Determination of gain KK_c and K

To determine the value of compensated gain KK_c (gain-factor) apply root locus magnitude condition at desired closed-loop pole $s = s_d$.

$$|\, G(s)\, G_c(s) = 1\, |_{s = s_d} \qquad ...(8.25)$$

The uncompensated (original) gain K is determined by applying root locus magnitude condition at the original operating point $s = s_1$ which lies on the uncompensated root locus plot *i.e.* prior to the addition of phase lead compensator.

Therefore $\quad |\, G(s)\, | = 1\, |_{s = s_1}$

The gain K_cK and K are determined as above and $K_c = \dfrac{KK_c}{K}$.

2. **Method no. 2**

Place the compensator zero $s = -a$ coinciding with one of the real open-loop pole of the given control system.

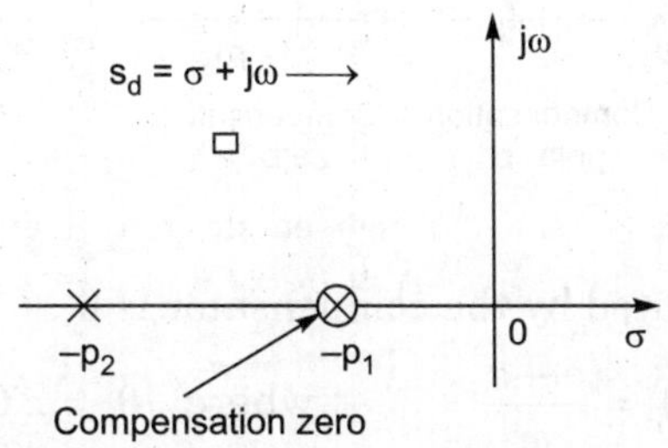

Fig. 8.7.4. Placement of compensator zero : method no. 2.

Thus, select $a = p_1$, where say p_1 is the said open-loop pole.

Consider,

$$G(s) = \frac{K}{s(s+p_1)(s+p_2)}$$

$$\therefore \quad G_c(s) = K_c \frac{(s+p_1)}{(s+b)} \quad p_1 < b \qquad ...(8.26)$$

The phase angle contribution by the phase-lead compensator

The phase angle θ_c contributed by the phase-lead compensator at the desired closed-loop pole $s_d = \sigma + j\omega$ can be determined by the root locus angle condition *i.e.*

$$\theta_c + \angle G(s)\big|_{s = s_d = \sigma + j\omega} = -180°$$

The values of σ and ω are given and therefrom the value of the phase lead angle can be obtained. The phase lead angle θ_c is equated with the expression for the angle contribution by the lead compensator.

$$\therefore \qquad \theta_c = \angle \frac{K_c(s+a)}{(s+b)}\Bigg|_{s = s_d = \sigma + j\omega}, \qquad a < b$$

As $a = p_1$, thus

$$\theta_c = \angle \frac{K_c(s+p_1)}{(s+b)}\Bigg|_{s = s_d = \sigma + j\omega} = \angle \frac{\sigma + j\omega + p_1}{\sigma + j\omega + b}$$

$$\therefore \qquad \theta_c = \tan^{-1}\frac{\omega}{\sigma + p_1} - \tan^{-1}\frac{\omega}{b + \sigma}$$

Having given, the values of θ_c, $s_d = \sigma + j\omega$ and p_1, the value of compensator pole "b" can be determined. As the open-loop pole is cancelled by the compensator zero placed at p_1 the order of the characteristic equation remains unchanged.

Method no. 3

Place the zero "a" of the compensator just below the desired closed-loop pole "s_d" as shown in Fig. 8.7.5.

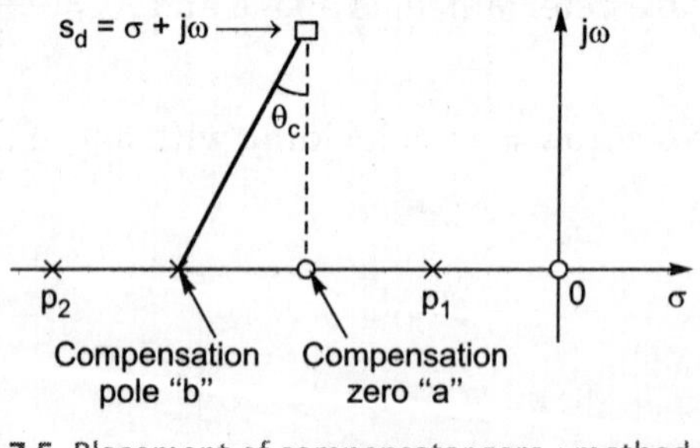

Fig. 8.7.5. Placement of compensator zero : method no. 3.

The angle lead contributed by the compensator is θ_c.

$$\tan\theta_c = \frac{b-a}{\omega} \qquad \text{where} \quad \theta_c = \angle G_c(s)\big|_{s = s_d = \sigma + j\omega}$$

The values of ω,"a" and θ_c are shown in Fig. 8.7.5. Also $a = \sigma$. Thus, the unknown compensator parameter "b" is determined.

Solved Examples : Phase Lead Compensation

Example 8.7.1. *The forward path transfer function of a unity feedback control system is given below:*

$$G(s) = \frac{K}{s(s+9)}$$

It is desired to have a maximum overshoot as 9.4% and natural frequency of oscillations ω_n = 12 rad./sec. The input is unit step.

Design a phase lead compensator.

Solution.

1. The desired closed-loop pole "s_d"

$$\%M_p = e^{\frac{-\xi\pi}{\sqrt{1-\xi^2}}}$$

$$\therefore \quad \frac{9.4}{100} = e^{\frac{-\xi\pi}{\sqrt{1-\xi^2}}}$$

From the above relation the value of the damping ratio is calculated as $\xi = 0.6$. The desired closed-loop is given by

$$s_d = -\xi\omega_n + j\omega_n\sqrt{1-\xi^2}$$

$$\because \quad \xi = 0.6 \quad \text{and} \quad \omega_n = 12 \text{ rad./sec.}$$

$$\therefore \quad s_d = -0.6 \times 12 + j\,12\sqrt{1-0.6^2}$$

or

$$s_d = -7.2 + j\,9.6$$

2. Uncompensated root locus plot

As per the root locus plot procedure the root locus for the open-loop transfer function

$$G(s) = \frac{K}{s(s+9)}$$

is shown in Fig. 8.7.6.

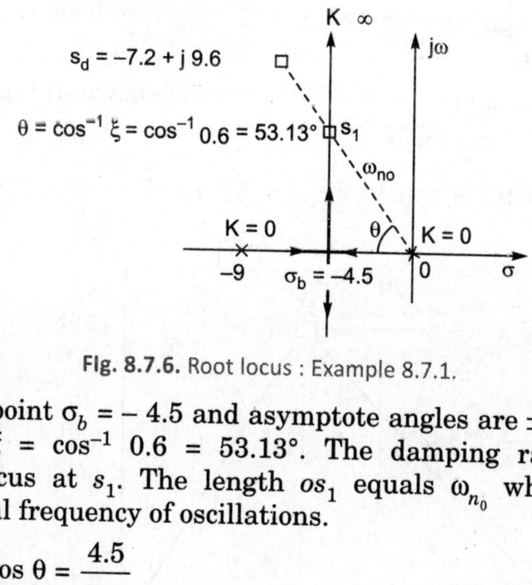

Fig. 8.7.6. Root locus : Example 8.7.1.

The break away point $\sigma_b = -4.5$ and asymptote angles are $\pm 90°$. The damping ratio line angle $\theta = \cos^{-1}\xi = \cos^{-1} 0.6 = 53.13°$. The damping ratio line intersects the uncompensated root locus at s_1. The length os_1 equals ω_{n_0} where ω_{n_0} is the original (uncompensated) natural frequency of oscillations.

$$\therefore \quad \cos\theta = \frac{4.5}{\omega_{n_0}}$$

or

$$\omega_{n_0} = \frac{4.5}{\cos\theta} = \frac{4.5}{0.6} = 7.5 \text{ rad./sec.}$$

The second order characteristic equation of the given system is

$$s^2 + 9s + K = 0$$

and the standard form of the second order characteristic equation is

$$s^2 + 2\xi\omega_{n_0} s + \omega_{n_0}^2 = 0$$

On comparing above euqations,

$$K = \omega_{n_0}^2$$

$$\because \quad \omega_{n_0} = 7.5 \text{ rad./sec.}$$

$$\therefore \quad K = 7.5^2 = 56.25$$

3. Compensated case

The desired closed-loop pole $s_d = -7.2 + j\,9.6$ does not lie on the uncompensated root locus (Fig. 8.7.6). Hence, determine the phase angle lead θ_c to be contributed ocmpensator.

The lead angle θ_c is calculated as below, applying root locus angle condition at $s_d = -7.2 + j\,9.6$:

$$\therefore \quad \theta_c + \angle G(s)\Big|_{s = s_d = -7.2 + j\,9.6} = -180°$$

or $$\theta_c + \angle \frac{K}{s(s+9)}\Bigg|_{s = s_d = -7.2 + j\,9.6} = -180°$$

or $$\theta_c + \angle \frac{K}{(-7.2 + j\,9.6)(-7.2 + j\,9.6 + 9)} = -180°$$

or $$\theta_c - \left(\tan^{-1}\frac{9.6}{-7.2} + \tan^{-1}\frac{9.6}{1.8}\right) = -180°$$

$$\therefore \quad \theta_c - [(180° - 53.13°) + 79.37°] = -180°$$

Simplifying, $$\theta_c = 26.24°$$

4. To determine the parameters "a" and "b" of phase-lead compensator.

Graphical method:

The desired closed-loop pole ($s_d = -7.2 + j\,9.6$) is shown in Fig. 8.5.7 at P.

$$\angle OPA = \angle POM$$

$$\therefore \quad \angle OPA = 90° + \tan^{-1}\left(\frac{OD}{PD}\right) = 90° + \tan^{-1}\frac{7.2}{9.6} = 126.86°$$

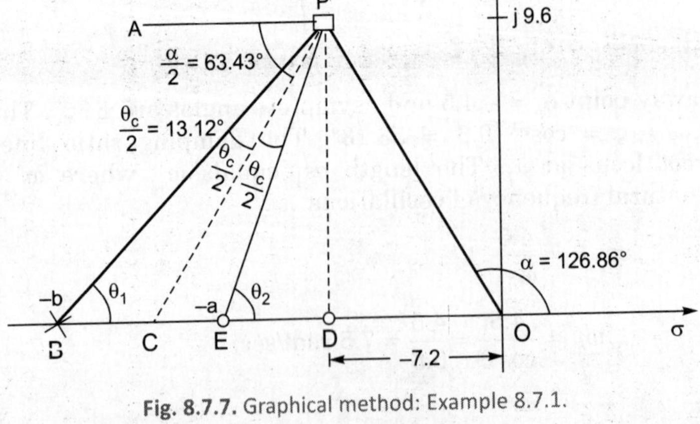

Fig. 8.7.7. Graphical method: Example 8.7.1.

The line *PC* bisects the $\angle OPA$.

$$\therefore \quad \angle APC = \frac{126.86}{2} = 63.43°$$

The required phase lead $\theta_c = 26.24°$ is contributed by the compensator. Draw lines *PB* and *PE* making angle $\theta_c/2$ with the line *PC* and cut the *s*-plane real axis at *B* and *E* respectively.

$$\angle APB = \angle PBC$$

$$\angle APB = 63.43° - \frac{\theta_c}{2} = 63.43° - \frac{26.24}{2} = 50.31°$$

$$\therefore \quad \angle\theta_1 = 50.31°$$

Also,

$$\angle APE = \angle PED$$

$$\angle APE = 63.43° + \frac{\theta_c}{2} = 63.43° + \frac{26.24}{2} = 76.55°$$

$$\therefore \quad \theta_c = 76.55°$$

$$\tan\theta_1 = \frac{PD}{BD} = \frac{9.6}{BD}, \quad \therefore \quad BD = \frac{9.6}{\tan\theta_1} = \frac{9.6}{\tan 50.31°} = 8$$

The parameter "*b*" of the lead compensator w.r.t. Fig. 8.7.6 is

$$b = (BD + DO) = 8 + 7.2 = 15.2$$

Also,

$$\tan\theta_2 = \frac{PD}{ED} = \frac{9.6}{ED}, \quad \therefore \quad ED = \frac{9.6}{\tan\theta_2} = \frac{9.6}{\tan 76.55°} = 2.3$$

The parameter "*a*" of the lead compensator w.r.t. Fig. 8.7.6 is

$$a = ED + DO = 2.3 + 7.2 = 9.5$$

$$\therefore \quad G_c(s) = \frac{K_c(s+9.5)}{(s+15.2)}$$

5. **To determine compensator gain K_c**

The open-loop transfer function using phase lead compensator is given below:

$$G(s)\,G_c(s) = \frac{KK_c(s+9.5)}{s(s+9)(s+15.2)}$$

The desired closed-loop pole $s_d = -7.2 + j\,9.6$ has to lie on the phase lead compensated root locus. Hence, the magnitude condition of the root locus is to satisfied at $s_d = -7.2 + j\,9.6$.

$$\therefore \quad \left| G(s)\,G_c(s) \right|_{s = s_d = -7.2 + j\,9.6} = 1$$

$$\therefore \quad \left| \frac{KK_c(s+9.5)}{s(s+9)(s+15.2)} \right|_{s = s_d = -7.2 + j\,9.6} = 1$$

Rearranging,

$$KK_c = \left| \frac{s(s+9)(s+15.2)}{s+9.5} \right|_{s = s_d = -7.2 + j\,9.6}$$

$$\therefore \quad KK_c = \left| \frac{(-7.2 + j\,9.6)(-7.2 + j\,9.6 + 9)(-7.2 + j\,9.6 + 15.2)}{(-7.2 + j\,9.6 + 9.5)} \right|$$

$$= \left| \frac{(-7.2 + j\,9.6)(1.8 + j\,9.6)(8 + j\,9.6)}{(2.3 + j\,9.6)} \right|$$

$$= \sqrt{\frac{(51.84 + 92.16)(3.24 + 92.16)(64 + 92.16)}{(5.29 + 92.16)}}$$

$$\therefore \quad KK_c = 148.33$$

$$K_c = \frac{148.33}{K}$$

$$\because \quad K = 56.25 \text{ (determined earlier)}$$

$$\therefore \quad K_c = \frac{148.33}{56.25} = 2.63$$

The compensated open-loop transfer function is given below

$$G(s)\,G_c(s) = \frac{56.25 \times 2.63\,(s + 9.5)}{s(s + 9)(s + 15.2)}. \quad \textbf{Ans.}$$

Example 8.7.2. *Determine the transfer function of a cascade compensator for a unity feedback control system having the following open-loop transfer function,*

$$G(s) = \frac{K}{(s + 2)(s + 4)}$$

Following specifications are to be satisfied:

(i) $\%M_p = 16.3\%$

(ii) Reduction in setting time 60%

(iii) The order of the characteristic equation to remain the same.

Solution. 1. Uncompensated case

As per the root locus plot procedure. The break away point is determined as $\sigma = -3$. The angle of asymptotes are $\pm 90°$. The shape of the root locus plot is shown in Fig. 8.7.8.

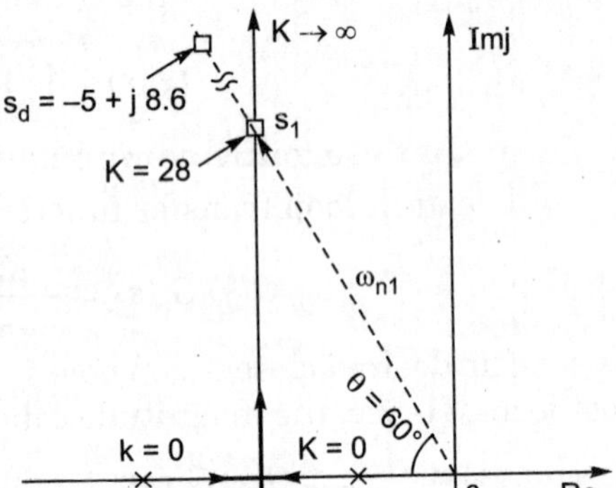

Fig. 8.7.8. Root locus for Example 8.7.2.

$$\%M_p = e^{-\frac{\xi\pi}{\sqrt{1-\xi^2}}}$$

$$\therefore \quad \frac{16.4}{100} = e^{-\frac{\xi\pi}{\sqrt{1-\xi^2}}}$$

Solving above equation, the damping ratio $\xi = 0.5$. The angle between the s-plane real axis and the damping ratio line is $\theta = \cos^{-1} \xi = \cos^{-1} 0.5 = 60°$.

Intersection of the damping ratio line with root locus

Damping ratio line equation is,

$$\frac{\omega}{\sigma} = -\tan\theta = -\tan 60° = -\sqrt{3}$$

$$\therefore \quad \omega = -\sigma\sqrt{3}$$

$$\because \quad \sigma = -3, \quad \therefore \quad \omega = 3\sqrt{3} \text{ rad./sec.}$$

The intersection point is $s_1 = \sigma + j\omega = -3 + j\,3\sqrt{3}$.

The value of gain K (uncompensated) is determined applying the root locus magnitude condition at $s = s_1 = (-3 + j\,3\sqrt{3})$

$$\therefore \quad |G(s)|_{s=s_1=-3+j3\sqrt{3}} = 1$$

$$\therefore \quad \left|\frac{K}{(s+2)(s+4)}\right|_{s=s_1=-3+j3\sqrt{3}} = 1$$

$$\therefore \quad K = |(s+2)(s+4)|_{s=s_1=-3+j3\sqrt{3}}$$

$$= |(-3+j\,3\sqrt{3}+2)(-3+j\,3\sqrt{3}+4)|$$

$$= 28$$

2. Compensated case

$$t_{s_1} = \frac{4}{\xi\omega_{n_1}}, \quad \because \ \xi\omega_n = 3 \text{ (root locus breaks away)}$$

$$\therefore \quad t_{s_1} = \frac{4}{3} \text{ sec.}$$

$$t_{s_2} = 0.6\, t_{s_1} \text{ (given)}$$

$$\therefore \quad t_{s_2} = 0.6 \times \frac{4}{3} = 0.8 \text{ sec.}$$

Also, $$t_{s_2} = \frac{4}{\xi\omega_{n_2}}, \quad \therefore \ \omega_{n_2} = \frac{4}{\xi t_{s_2}} = \frac{4}{0.5 \times 0.8} = 10 \text{ rad./sec.}$$

Hence, the desired closed-loop pole s_d for the compensated case is

$$s_d = -\xi\omega_{n_2} + j\omega_{n_2}\sqrt{1-\xi^2}$$

$$= -0.5 \times 10 + j\,10\sqrt{1-0.5^2}$$

Simplifying, $$s_d = -5 + j\,8.66$$

The desired closed-loop pole $s_d = -5 + j\,8.66$ does not lie on the root locus plot (Fig. 8.7.8). Hence, the phase nagle θ_c contributed by the phase lead compensator is determined below, by applying root locus angle condition at $s_d = -5 + j\,8.66$

i.e. $$\theta_c + \angle\left.\frac{K}{(s+2)(s+4)}\right|_{s=s_d=-5+j8.66} = -180^\circ$$

$$\therefore \quad \theta_c + \angle\frac{K}{(-5+j8.66+2)(-5+j\,8.66+4)} = -180^\circ$$

$$\therefore \quad \theta_c - \left[\left(180^\circ - \tan^{-1}\frac{8.66}{3}\right) + \left(180^\circ - \tan^{-1}\frac{8.66}{1}\right)\right] = -180^\circ$$

Simplifying, $$\theta_c - [(180^\circ - 70.89^\circ) + (180^\circ - 83.41^\circ)] = -180^\circ$$

or $$\theta_c = 25.7^\circ$$

The phase lead $\theta_c = 25.7^\circ$ is to be provided by the lead compensator.

The transfer function of a phase lead compensator is given below:

$$G_c(s) = \frac{K_c(s+a)}{(s+b)}, \quad a < b$$

As the order of the characteristic is to remain unchanged, the zero of the compensator $s = -a$ is placed at the open-loop pole $s = -2$.

The phase lead compensator takes the form

$$G_c(s) = \frac{K_c(s+2)}{(s+b)}$$

The angle θ_c contributed by the lead compensator at $s_d = -5 + j\,8.66$ is given by

$$\theta_c = \angle \left. \frac{K_c(s+2)}{(s+b)} \right|_{s = s_d = -5 + j\,8.66}$$

$\because$ $\theta_c = 25.7°$ (determined earlier)

$$25.7° = \angle \left. \frac{K_c(s+2)}{(s+b)} \right|_{s = s_d = -5 + j\,8.66}$$

$$\therefore \quad 25.7° = \angle \frac{(-5 + j8.66 + 2)}{(b-5) + j\,8.66} = \angle \frac{-3 + j8.66}{(b-5) + j\,8.66}$$

$$= \left(180° - \tan^{-1}\frac{8.66}{3}\right) - \tan^{-1}\frac{8.66}{b-5}$$

$$= (180° - 70.89°) - \tan^{-1}\frac{8.66}{b-5}$$

Simplifying,

$$\tan^{-1}\frac{8.66}{b-5} = 83.41°$$

$$\therefore \quad \frac{8.66}{b-5} = \tan 83.41° = 8.65$$

Simplifying, $b = 6$

Thus, the transfer function of phase lead compensator is

$$G_c(s) = \frac{K_c(s+2)}{(s+6)}$$

The open-loop transfer function using phase-lead compensator is given below:

$$G(s)\,G_c(s) = \frac{KK_c(s+2)}{(s+2)(s+4)(s+6)} = \frac{KK_c}{(s+4)(s+6)}$$

The desired closed-loop $s_d = -5 + j\,8.66$ lies on the compensated root locus. Therefore, the root locus magnitude condition is to be satisfied at $s_d = -5 + j\,8.66$.

$$\therefore \quad \left| G(s)\,G_c(s) \right|_{s = s_d = -5 + j\,8.66} = 1$$

Therefore,

$$\left| \frac{KK_c}{(s+4)(s+6)} \right|_{s = s_d = -5 + j\,8.66} = 1$$

or

$$KK_c = \left| (s+4)(s+6) \right|_{s = s_d = -5 + j\,8.66}$$

$$= \left| (-5 + j\,8.66 + 4)(-5 + j\,8.66 + 6) \right|$$

$$= |(-1 + j\,8.66)(1 + j\,8.66)|$$

$$= \sqrt{(1^2 + 8.66^2)(1^2 + 8.66^2)} = 76$$

$\because$ $K = 28$ (determined earlier)

$$\therefore \quad K_c = \frac{76}{28} = 2.71$$

$$\therefore \quad G(s)\,G_c(s) = \frac{28 \times 2.71}{(s+4)(s+6)}. \quad \textbf{Ans.}$$

Example 8.7.3. *The open-loop transfer function of a unity feedback control system operating with damping ratio $\xi = 0.5$ is given below:*

$$G(s) = \frac{K}{s(s+2)(s+5)}$$

The settling time is 5.63 sec.

Design a suitable compensator to reduce the settling time by a factor of 0.6.

Solution. As the settling time is to be reduced, a phase lead compensator is selected.

1. Uncompensated System

$$t_{s_1} = \frac{4}{\xi\omega_{n_1}}, \quad \because \quad t_{s_1} = 5.63 \text{ sec} \quad \text{and} \quad \xi = 0.5$$

$$\therefore \quad \omega_{n_1} = \frac{4}{t_{s_1}} = \frac{4}{5.63}. = 1.42 \text{ rad./sec.}$$

and $\xi\omega_{n_1} = 0.5 \times 1.42 = 0.71$

The general shape of the root locus is shown in Fig. 8.7.9. The break away point can be determined as $\sigma_b = -0.88$

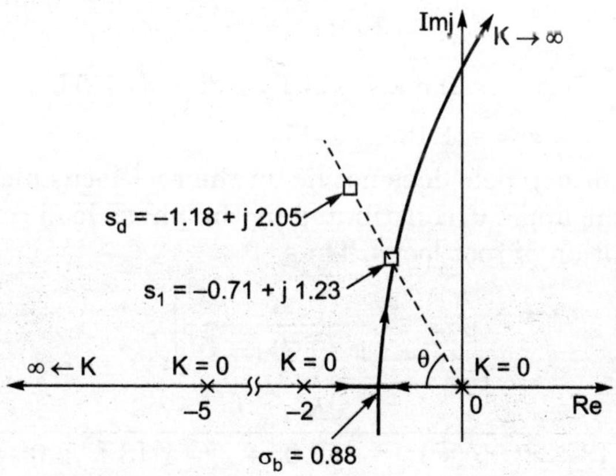

Fig. 8.7.9. Root locus for Example 8.7.3.

As the values of $\xi\omega_{n_1}$ and ξ are derived from the given settling time, the uncompensated closed-loop pole is given by

$$s_1 = -\xi\omega_{n_1} + j\omega_{n_1}\sqrt{1-\xi^2}$$

$$= -0.5 \times 1.42 + j\,1.42\,\sqrt{1-0.5^2}$$

$$\therefore \qquad s_1 = -0.71 + j\,1.23$$

The value of uncompensated gain K is determined by applying magnitude condition to original root lcous at $s_1 = 0.71 + j\,1.23$.

$$\therefore \qquad \left|\frac{K_c}{s(s+2)(s+5)}\right|_{s=s_1=0.71+j\,1.23} = 1$$

Rearranging, $\qquad K = |\,s(s+2)(s+5)\,|_{s=s_1=0.71+j\,1.23}$

$$= |(-0.71 + j1.23)(-0.71 + j1.23 + 2)(-0.71 + j1.23 + 5)|$$

$$= |(-0.71 + j1.23)(1.29 + j1.23)(4.29 + j1.23)|$$

$$= \sqrt{(0.71^2 + 1.23^2)(1.29^2 + 1.23^2)(4.29^2 + 1.23^2)}$$

$$\therefore \qquad K = 11.27$$

2. **Compensated case**

$$t_{s_2} = 0.6\,t_{s_1}$$

$$\because \qquad t_{s_1} = 5.63 \text{ sec.}$$

$$t_{s_2} = 0.6 \times 5.63 = 3.378 \text{ sec.}$$

As the value of damping ratio $\xi = 0.5$ remains same for the compensated/uncompensated case

$$t_{s_2} = \frac{4}{\xi\omega_{n_2}}$$

$$\therefore \qquad \omega_{n_2} = \frac{4}{\xi t_{s_2}} = \frac{4}{0.5 \times 3.378} = 2.368 \text{ rad./sec.}$$

The desired closed-loop pole is

$$s_d = -\xi\omega_{n_2} + j\omega_{n_2}\sqrt{1-\xi^2}$$

$$= -0.5 \times 2.368 + j\,2.368\,\sqrt{1-0.5^2}$$

Simplifying, $\qquad s_d = -1.18 + j\,2.05.$

The desired closed-loop pole does not lie on the root locus plot (Fig. 8.7.9). Therefore, at $s_d = -1.18 + j\,2.05$, the angle θ_c contributed by the phase lead compensator plus $\angle G(s)|_{s_d}$ satisfies the angle condition of root locus. Thus,

$$\therefore \qquad \theta_c + \angle\left.\frac{K}{s(s+2)(s+5)}\right|_{s=s_d=-1.18+j\,2.05} = -180^\circ$$

or

$$\theta_c + \angle\frac{K}{(-1.18 + j2.05)(-1.18 + j2.05 + 2)(-1.18 + j2.05 + 5)} = -180^\circ$$

$$\therefore \qquad \theta_c + \left[-\left(180^\circ - \tan^{-1}\frac{2.05}{1.18}\right) - \tan^{-1}\frac{2.05}{0.82} - \tan^{-1}\frac{2.05}{3.82}\right] = -180^\circ$$

or

$$\theta_c + [-(180^\circ - \tan^{-1} 1.73) - \tan^{-1} 2.5 - \tan^{-1} 0.536] = -180^\circ$$

$\therefore \quad \theta_c + [-(180° - 59.97°) - 68.19° - 28.19°] = -180°$

Simplifying, $\quad \theta_c = 36.41°$

2. Phase Lead Compensator

The transfer function of a phase lead compensator is given below:

$$G_c(s) = \frac{K_c(s+a)}{(s+b)}, \; a < b$$

The phase lead compensator provides a lead angle θ_c at $s_d = -1.18 + j2.05$.

$$\therefore \quad \theta_c = \angle \left. \frac{K_c(s+a)}{(s+b)} \right|_{s = s_d = -1.18 + j2.05}$$

$$\because \quad \theta_c = 36.41° \text{ (calculated earlier)}$$

$$\therefore \quad 36.41° = \angle \left. \frac{K_c(s+a)}{(s+b)} \right|_{s = s_d = -1.18 + j2.05}$$

Select $a = 2$ such that the open-loop pole at $s = -2$ is cancelled and the order of the characteristic equation remains unchanged.

$$\therefore \quad 36.41° = \angle \left. \frac{K_c(s+2)}{(s+b)} \right|_{s = s_d = -1.18 + j2.05}$$

$$= \angle \frac{(-1.18 + j2.05 + 2)}{(b - 1.18) + j2.05}$$

$$= \tan^{-1} \frac{2.05}{0.82} - \tan^{-1} \frac{2.05}{b - 1.18}$$

$$\therefore \quad 36.41° = 68.19° - \tan^{-1} \frac{2.05}{b - 1.18}$$

Rearranging,

$$\tan^{-1} \frac{2.05}{b - 1.18} = 31.78°$$

$$\therefore \quad \tan 31.78° = \frac{2.05}{b - 1.18}$$

$$\therefore \quad 0.619 = \frac{2.05}{b - 1.18}$$

Solving above equation,

$$b = 4.49$$

Compensated gain KK_c

The value of the compensated gain KK_c at $s_d = -1.18 + j2.05$ is determined using root locus magnitude condition to the compensated open-loop transfer function.

The compensated transfer function is

$$G(s)\,G_c(s) = \frac{KK_c(s+2)}{s(s+2)(s+5)(s+4.99)}$$

Simplifying, $$G(s)\,G_c(s) = \frac{KK_c}{s(s+5)(s+4.99)}$$

$$\therefore \quad \left| \frac{KK_c}{s(s+5)(s+4.99)} \right|_{s = s_d = -1.18 + j\,2.05} = 1$$

Rearranging, $KK_c = | s(s+5)(s+4.99) |_{s = s_d = -1.18 + j\,2.05}$

$$= |(-1.18 + j\,2.05)(-1.18 + j\,2.05 + 5)(1.18 + j\,2.05 + 4.99)|$$

$$= |(-1.18 + j\,2.05)(3.82 + j\,2.05)(3.81 + j\,2.05)|$$

$$= \sqrt{(1.18^2 + 2.05^2)(3.82^2 + 2.05^2)(3.81^2 + 2.05^2)}$$

$$\therefore \quad KK_c = 44.31$$

$$\because \quad K = 11.27 \text{ (determined earlier)}$$

$$\therefore \quad K_c = \frac{44.31}{11.27} = 3.93$$

Therefore, $G(s)\,G_c(s) = \dfrac{11.27 \times 3.93}{s(s+5)(s+4.49)}$. **Ans.**

8.7.2. Phase Lag Compensator Design

The lag compensator is used in cascade with the forward path (open-loop) transfer function of a feedback control system, wherein the steady state error is to be improved (*i.e.* reduced).

The lag compensator, thus introduced does not appreciably alter the original (uncompensated) root locus near the desired cloesd-loop pole. This is possible by selecting a lag compensator contributing lag angle in the range of 0° to – 3°.

The distances of poles and zeros of the lag compensator are measured small as compared with the distance of closed-loop pole. Thus, the lag compensator does not affect the root locus shape near the desired closed-loop pole much.

The complex part of the root locus slightly shifts towards the R.H.S., but this shift is insignificant.

Design steps : Lag Compensator

(*a*) Locate the desired closed-loop pole on the root locus of the uncompensated system.

(*b*) Determine the value of gain K (gain factor) for the uncompensated system at the desird closed-loop pole.

(*c*) Calculate the value of the new error coefficient say K_n to meet the steady state error requirement.

(*d*) The transfer function a lag compensator is given below,

$$G_c(s) = \frac{(s+a)}{(s+b)}, \quad a > b$$

The gain (gain factor) is considered as unity. The use of lag compensator increases the steady state error coefficient, thereby reduces steady state error.

Determination of parameters "*a*" (zero) and "*b*" (pole) of a phase lag compensator.

(*i*) **Type "o" transfer function**

The original steady state error e_{ss_o} for the uncompensated case in terms of positional error coefficient K_{p_o} is given below.

$$e_{ss0} = \frac{1}{1+K_{p_o}} \quad ...(8.27)$$

The steady state error e_{ss_n} due to the use of phase lag compensator in terms of positional error coefficient K_{p_n} is given below.

$$e_{ss_n} = \frac{1}{1+K_{p_n}} \quad ...(8.28)$$

$$\therefore \quad \frac{e_{ss0}}{e_{ss_n}} = \frac{1+K_{p_n}}{1+K_{p0}} \quad ...(8.29)$$

It can be shown that

$$\frac{e_{ss0}}{e_{ss_n}} = \frac{a}{b}$$

$$\therefore \quad \frac{a}{b} = \frac{1+K_{p_n}}{1+K_{p_o}} \quad ...(8.30)$$

Having given the values of K_{p_n}, K_{p0} and selecting the appropriate value of the parameter "*b*", the parameter "*a*" is determined.

(*ii*) **Type "1" transfer function**

The original steady state error e_{ss0} in terms of velocity error coefficient K_{v0} is given below:

$$e_{ss0} = \frac{1}{K_{v0}} \quad ...(8.31)$$

The relation between the specified stedy start error e_{ss_n} and the velocity error coefficient K_{v_n} due the use of phase lag compensator is given below,

$$e_{ss_n} = \frac{1}{K_{v_o}} \quad ...(8.32)$$

It can be shown that,

$$\frac{e_{ss0}}{e_{ss_n}} = \frac{a}{b}$$

$$\therefore \quad \frac{a}{b} = \frac{K_{v_n}}{K_{v0}} \quad ...(8.33)$$

Having given the values of K_{v_n} and K_{v0} and selecting appropriate value of the parameter "*b*", the parameter "*a*" is determined.

Solved Examples : Lag Compensator

Example 8.7.4. *Design a compensator for a unity feedback control system having open-loop transfer function as*

$$G(s) = \frac{K}{s(s+8)}$$

The required specifications are:

Damping ratio $\xi = 0.5$ *and the steady state error due to unit ramp input is reduced by a factor of 6.25.*

Solution. As the steady state error is to be reduced a lag compensator is needed.

1. Uncompensated case

As per procedure, the root locus is plotted as shown in Fig. 8.7.10.

The characteristic equation is

$$1 + G(s) = 0$$

$$\therefore \quad 1 + \frac{K}{s(s+8)} = 0$$

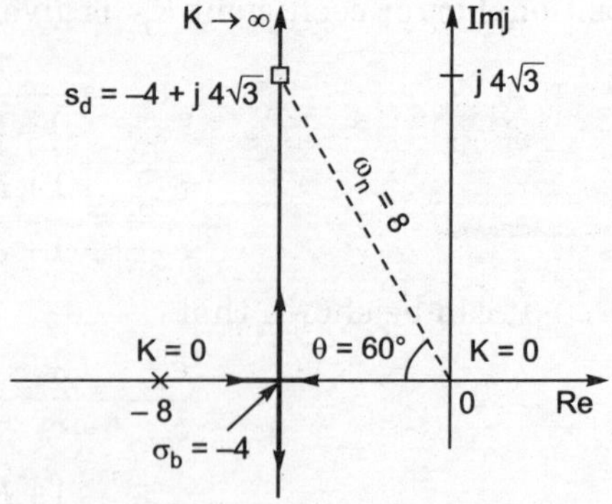

Fig. 8.7.10. Root locus plot for Example 8.7.4.

Simplifying,

$$K = -(s^2 + 8s)$$

The break away point is obtained by relation

$$\frac{dK}{ds} = 0$$

$$\therefore \quad \frac{dK}{ds} = -(2s+8)$$

or $$-(2s + 8) = 0$$

$$\therefore \quad s = -\frac{8}{2} = -4$$

Hence, the break away point on the root is located at

$$\sigma_b = -4$$

Angle of asymptotes are given by

$$\angle \text{Asym} = \frac{(2k+1)}{p-z} \times 180° \qquad k = 0.1$$

$$= \frac{(2 \times 0 + 1)}{2 - 0} \times 180° = 90°$$

and $$= \frac{(2 \times 1 + 1)}{2 - 0} \times 180° = 270° = -90°$$

2. Lag Compensator

$$-\sigma_b = -\xi\omega_n$$

$$\therefore \quad -\xi\omega_n = -4$$

$$\because \quad \xi = 0.5 \text{ (given)}$$

$$\omega_n = \frac{4}{0.5} = 8 \text{ rad./sec.}$$

The desired closed-loop pole is given by

$$s_d = -\xi\omega_n + j\omega_n\sqrt{1-\xi^2}$$

$$= -4 + j\,8\sqrt{1-0.5^2}$$

Simplifying, $$s_d = -4 + j\,4\sqrt{3}$$

As per given data the steady state error is to be reduced by a factor of 6.25.

$$\therefore \qquad \frac{K_{v_n}}{K_{v_o}} = 6.25$$

The transfer function of a lag compensator is given below,

$$G_c(s) = \frac{(s+a)}{(s+b)}, \quad a > b$$

Also
$$\frac{K_{v_n}}{K_{v_0}} = \frac{a}{b}$$

and
$$\frac{a}{b} = 6.25$$

select $b = 0.05$, $\therefore \quad a = 0.05 \times 6.25 = 0.3125$

$$\therefore \qquad G_c(s) = \frac{(s+0.3125)}{(s+0.05)}$$

The angle contributed to the root locus (uncompensated) due to use of lag compensator at $s = s_d = -4 + j\,4\sqrt{3}$ (desired closed-loop pole) is determined below,

$$\angle G_c(s) = \angle \left. \frac{s+0.3125}{s+0.05} \right|_{s = s_d = -4 + j\,4\sqrt{3}}$$

$$= \angle \frac{-4 + j4\sqrt{3} + 0.3125}{-4 + j4\sqrt{3} + 0.05} = \angle \frac{-3.68 + j4\sqrt{3}}{-3.95 + j4\sqrt{3}}$$

$$= \left(180° - \tan^{-1}\frac{4\sqrt{3}}{3.68}\right) - \left(180° - \tan^{-1}\frac{4\sqrt{3}}{3.95}\right)$$

$$= (180° - 62.02) - (180° - 60.25)$$

$$= -1.77°$$

As the angle $-1.77°$ falls within a range $0°$ to $-3°$. Hence, the lag compensator design is accepted.

The phase lag compensated open-loop transfer function is given below:

$$G(s)\,G_c(s) = \frac{K(s+0.3125)}{s(s+8)(s+0.05)}$$

The value of K (uncompensated gain at ($s = s_d = -4 + j4\sqrt{3}$) for a second order characteristic equation is $K = \omega_n^2 = 8 \times 8 = 64$

$$\therefore \qquad G(s)\,G_c(s) = \frac{64(s+0.3125)}{s(s+8)(s+0.05)}. \quad \textbf{Ans.}$$

Example 8.7.5. *The open-loop transfer function of a unity feedback control system is given below:*

$$G(s) = \frac{K}{(s+1)(s+6)}$$

The damping ratio = 0.6.

Design a suitable compensator such that the steady state error coeff is $K_p = 41.37$ for a unit step input.

Solution. As per the procedure, the root locus plot for $G(s) = K/(s + 1)(s + 6)$ is drawn in Fig. 8.7.11.

$$\theta = \cos^{-1} \xi = \cos^{-1} 0.6 = 53.15°$$

Let, the damping ratio line intersect the root locus say at $s_d = \sigma + j\omega$.

The equation for the damping ratio line is derived below.

The break away point can be calculated as $\sigma = -3.5$ as per root locus plot procedure.

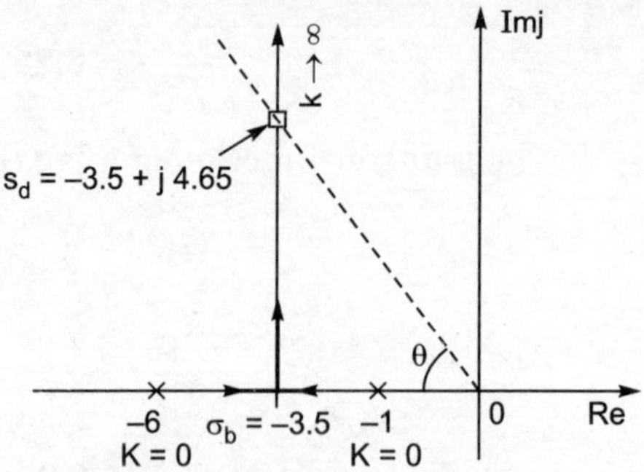

Fig. 8.7.11. Root locus for Example 8.7.5

$$\therefore \quad \frac{\omega}{\sigma} = -\tan\theta$$

(–ve slope) is the equation for the damping ratio line.

$$\because \quad \sigma = -3.5$$

$$\therefore \quad \frac{\omega}{-3.5} = -\tan 53.15° = -1.33$$

$$\therefore \quad \omega = (-3.5)(-1.33) = 4.65$$

Hence, the desired closed-loop pole is

$$s_d = -3.5 + j\,4.65$$

The value of gain K at $s = s_d = -3.5 + j\,4.65$ is determined using magnitude condition of root locus *i.e.*

$$|G(s)|_{s = s_d = -3.5 + j\,4.65} = 1$$

$$\therefore \quad \left|\frac{K}{(s+1)(s+6)}\right|_{s = s_d = -3.5 + j\,4.65} = 1$$

or

$$K = |(s+1)(s+6)|_{s = s_d = -3.5 + j\,4.65}$$

$$= |(-3.5 + j\,4.65 + 1)(-3.5 + j\,4.65 + 6)|$$

$$= |(-2.5 + j\,4.65)(2.5 + j\,4.65)|$$

$$= \sqrt{(2.5^2 + 4.65^2)(2.5^2 + 4.65^2)}$$

Simplifying, $K = 27.87$

$$\therefore \quad G(s) = \frac{27.87}{(s+1)(s+6)}$$

The open-loop transfer function as above is type "0".

$$\therefore \quad e_{ss_0} = \frac{1}{1 + K_{p_0}}$$

and

$$K_{p_0} = \lim_{s \to 0} G(s) = \lim_{s \to 0} \frac{27.87}{(s+1)(s+6)} = \frac{27.87}{6} = 4.64$$

$$\therefore \quad K_{p_0} = 4.64$$

and $K_{p_n} = 41.37$ (Given)

Hence, steady state error is to be improved and lag compensator is needed.

The transfer function of a lag compensator is given below,

$$G_c(s) = \frac{(s+a)}{(s+b)}, \quad a > b$$

The parameters "a" and "b" are determined as below,

$$\because \qquad \frac{a}{b} = \frac{e_{ss_0}}{e_{ss_n}} = \frac{1+K_{p_n}}{1+K_{p_0}} \text{ (mentioned earlier)}$$

$$\frac{a}{b} = \frac{1+K_{p_n}}{1+K_{p_0}} = \frac{1+41.37}{1+4.64} = 7.5$$

Thus, $\frac{a}{b} = 7.5$

Select the parameter $b = 0.01$

$$\therefore \qquad a = 0.01 \times 7.5 = 0.075$$

Hence, the transfer function of the lag compensator is

$$G_c(s) = \frac{s+0.075}{s+0.01}. \quad \textbf{Ans.}$$

The angle contributed by the lag compensator at $s = s_d = -3.5 + j\,4.65$ is determined below,

$$\angle G_c(s) = \angle \left. \frac{s+0.075}{s+0.01} \right|_{s = s_d = -3.5 + j\,4.65}$$

$$= \angle \frac{-3.5 + j\,4.65 + 0.075}{-3.5 + j\,4.65 + 0.01} = \angle \frac{-3.425 + j4.65}{-3.49 + j4.65}$$

$$= \left(180° - \tan^{-1}\frac{4.65}{3.425}\right) - \left(180° - \tan^{-1}\frac{4.65}{3.49}\right)$$

$$= (180° - 53.6°)\,(180° - 53.1°)$$

$$= -0.5°$$

The angle contribution is negligible. Hence, there is no appreciable change in the shape of the root locus and the design is accepted.

The compensated transfer function is

$$G(s)\,G_c(s) = \frac{27.87(s+0.075)}{(s+1)(s+6)(s+0.01)}. \quad \textbf{Ans.}$$

Example 8.7.6. *The closed-loop poles of a control system are located at $s = -2 \pm j\,2\sqrt{2.5}$. The open-loop transfer function with unity feedback is given as*

$$G(s) = \frac{16.25}{s(s+5)}$$

Design a lag compensator such that the vel. error coefficient $K_v = 19.8 \text{ sec}^{-1}$.

Solution. 1. Uncompensated case

$$K_{v_0} = \lim_{s \to 0} s\,G(s)$$

$$= \lim_{s \to 0} \frac{s \times 16.25}{s(s+5)} = 3.25 \text{ sec}^{-1}$$

2. Compensated case

$$K_{v_n} = 19.8 \text{ sec}^{-1} \text{ (Given)}$$

The transfer function of a lag compensator is given below

$$G_c(s) = \frac{(s+a)}{(s+b)}, \quad a > b$$

$$\because \quad \frac{a}{b} = \frac{K_{v_n}}{K_{v_0}} \text{ (mentioned earlier)}$$

$$\therefore \quad \frac{a}{b} = \frac{19.8}{3.25} = 6$$

Select parameter $b = 0.05$

$\therefore$ parameter $a = 6 \times 0.05 = 0.3$

and
$$G_c(s) = \frac{(s+0.3)}{(s+0.05)}$$

The angle contribution by the lag compensator at the given closed-loop pole $s = -2 + j2\sqrt{2.5}$ is determined below,

$$\angle G_c(s) = \angle \frac{(s+0.3)}{(s+0.05)}\bigg|_{s = s_d = -2 + j2\sqrt{2.5}}$$

$$= \angle \frac{-2 + j2\sqrt{2.5} + 0.3}{-2 + j2\sqrt{2.5} + 0.05} = \angle \frac{-1.7 + j3.16}{-1.95 + j3.16}$$

$$= \left(180° - \tan^{-1}\frac{3.16}{1.7}\right) - \left(180° - \tan^{-1}\frac{3.16}{1.95}\right)$$

$$= -3.4°$$

As the angle – 3.4° does not fit within the acceptable range *i.e.* 0° to –3°, the design with $b = 0.05$ and $a = 0.3$ is not accepted.

Further, changing the value of the parameter as $b = 0.025$ and $a = 6 \times 0.025 = 0.15$. The angle contributed by the lag compensator is determined as

$$\angle G_c(s) = \angle \frac{(s+0.15)}{(s+0.025)}\bigg|_{s = s_d = -2 + j2\sqrt{2.5}}$$

$$= \angle \frac{-2 + j2\sqrt{2.5} + 0.15}{-2 + j2\sqrt{2.5} + 0.025}$$

$$= \angle \frac{-1.85 + j3.16}{-1.975 + j3.16}$$

$$= \left(180° - \tan^{-1}\frac{3.16}{1.85}\right) - \left(180° - \tan^{-1}\frac{3.16}{1.975}\right)$$

$$= -1.66°$$

which is in the range of 0° to – 3°. Hence, the design is accepted. The compensated open-loop transfer function is,

$$G(s)\, G_c(s) = \frac{16.25(s+0.15)}{s(s+5)(s+0.025)}. \quad \textbf{Ans.}$$

Example 8.7.7. *The forward path transfer function of a unity feedback control system is given below:*

$$G(s) = \frac{K}{s(s+2)(s+8)}$$

Design a lag compensator such that

(a) Maximum over shoot 16.4% for a unit step input.

(b) The steady state error 0.16 rad. for unit ramp input.

Solution. The general shape of the root locus plot is for the open-loop transfer function

$$G(s) = \frac{K}{s(s+2)(s+8)}$$

is shown in Fig. 8.7.12.

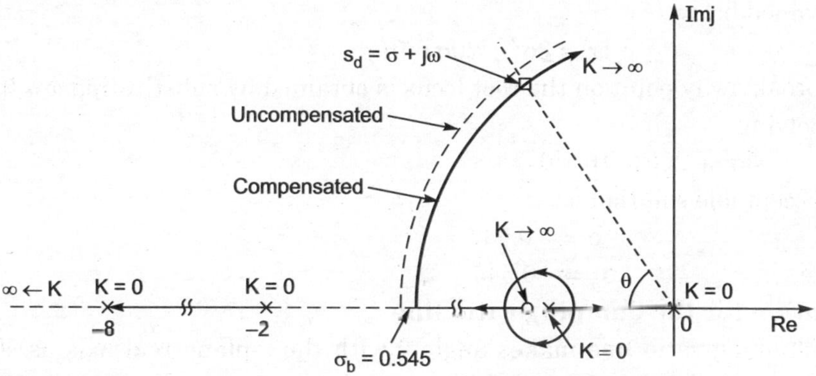

Fig. 8.7.12. Root locus for Example 8.7.7.

1. Uncompensated case

The uncompensated closed-loop pole is obtained by the intersection complex part of the root locus plot with the damping ratio line.

Equation for the complex part of root locus.

The equation for the complex part of the root locus is obtained below.

Consider a point $s = \sigma + j\omega$ on the complex part of the root locus,

$$\therefore \quad G(\sigma + j\omega) = \frac{K}{(\sigma + j\omega)(\sigma + 2 + j\omega)(\sigma + 8 + j\omega)}$$

For the point $s = \sigma + j\omega$, to lie on the root locus the phase angle condition is to be satisfied. Therefore,

$$\angle G(\sigma + j\omega) = -180°$$

$$\therefore \quad \angle \frac{K}{(\sigma + j\omega)(\sigma + 2 + j\omega)(\sigma + 8 + j\omega)} = -180°$$

Thus,

$$-\tan^{-1}\frac{\omega}{\sigma} - \tan^{-1}\frac{\omega}{\sigma+2} - \tan^{-1}\frac{\omega}{\sigma+8} = -180°$$

Rearranging, $$\tan^{-1}\frac{\omega}{\sigma+2}+\tan^{-1}\frac{\omega}{\sigma+8}=180^\circ-\tan^{-1}\frac{\omega}{\sigma}$$

Simplifying, $$\tan^{-1}\left(\frac{\dfrac{\omega}{\sigma+2}+\dfrac{\omega}{\sigma+8}}{1-\dfrac{\omega^2}{(\sigma+2)(\sigma+8)}}\right)=180^\circ-\tan^{-1}\frac{\omega}{\sigma}$$

or $$\tan^{-1}\left[\frac{\omega(2\sigma+10)}{\sigma^2+10\sigma+16-\omega^2}\right]=180^\circ-\tan^{-1}\frac{\omega}{\sigma}$$

$\because$ $$\tan^{-1}180^\circ=0$$

$\therefore$ $$\frac{\omega(2\sigma+10)}{\sigma^2+10\sigma+16-\omega^2}=-\frac{\omega}{\sigma}$$

Rearranging the equation for the complex part of the root locus relating ω and σ in s-plane is obtained below,

$$\omega^2 = 3\sigma^2 + 20\sigma + 16$$

The break away point on the root locus is obtained by substituting $\omega = 0$.

$\therefore$ Solving

$$3\sigma^2 + 20\sigma + 16 = 0$$

The acceptable solution is

$$\sigma = -0.545$$

Hence, $$\sigma_b = -0.545$$

Equation for the damping ratio line

The damping ratio line makes angle θ with the s-plane real axis as shown in Fig. 8.7.12, where $\theta = \cos^{-1}\xi$

The value of the damping ratio ξ is determined using relation below,

$$M_p = e^{\frac{-\xi\pi}{\sqrt{1-\xi^2}}}$$

$\because$ $$\%M_p = 16.4\%$$

$\therefore$ $$\frac{16.4}{100} = e^{\frac{-\xi\pi}{\sqrt{1-\xi^2}}}$$

On solving above equation,

$$\xi = 0.5$$

$\therefore$ $$\theta = \cos^{-1}0.5 = 60^\circ$$

and $$\tan\theta = \tan 60^\circ = \sqrt{3}$$

Thus, the damping ratio line has a slope as under,

$$\frac{\omega}{\sigma} = -\tan\theta \text{ (–ve slope)}$$

Therefore, the equation for the damping ratio line is given below,

$$\omega = -\sqrt{3}\,(\sigma)$$

As already derived the equation for the complex part of root locus is

$$\omega^2 = 3\sigma^2 + 20\sigma + 16 = 0$$

Substitute $\omega = -\sqrt{3}(\sigma)$ in the above equation

$$\therefore \qquad 3\sigma^2 = 3\sigma^2 + 20\sigma + 16 = 0$$

$$\therefore \qquad \sigma = -0.8 \quad \text{and} \quad \omega = -\sqrt{3}(\sigma) = -\sqrt{3}(-0.8) = 0.8\sqrt{3}$$

Therefore, the intersection point is at

$$s_d = -0.8 + j\,0.8\sqrt{3}$$

which is the closed-loop pole (uncompensated).

Determination of the value of K at $s_d = -0.8 + j0.8\sqrt{3}$

The value of K at $s = s_d$ is determined using root locus magnitude condition *i.e.*

$$\left|\frac{K}{s(s+2)(s+8)}\right|_{s = s_d = -0.8 + j\,0.8\sqrt{3}} = 1$$

$$\therefore \qquad \left|\frac{K}{s(s+2)(s+8)}\right|_{s = s_d = -0.8 + j\,0.8\sqrt{3}} = 1$$

or

$$K = |\,s(s+2)(s+8)\,|_{s = s_d = -0.8 + j\,0.8\sqrt{3}}$$

$$= |(-0.8 + j\,0.8\sqrt{3})(-0.8 + j\,0.8\sqrt{3} + 2)(-0.8 + j\,0.8\sqrt{3} + 8)|$$

$$= |(-0.8 + j\,0.8\sqrt{3})(1.2 + j\,0.8\sqrt{3})(7.2 + j\,0.8\sqrt{3})|$$

$$= \sqrt{(0.64 + 1.92)(1.44 + 1.92)(51.84 + 1.92)}$$

Simplifying, $\quad K = 21.5$

2. Lag Compensator

The original (uncompensated) velocity error coefficient is determined as below:

$$K_{v_0} = \lim_{s \to 0} s\,G(s)$$

$$= \lim_{s \to 0} \frac{sK}{s(s+2)(s+8)} = \frac{K}{16}$$

$$\because \qquad K = 21.5$$

$$K_{v_0} = \frac{21.5}{16} = 1.34 \text{ sec}^{-1}.$$

On lag compensation the new velocity coefficient is required as below

$$K_{v_n} = \frac{1}{e_{sp_n}} = \frac{1}{0.16} = 6.25 \text{ sec}^{-1}. \qquad (e_{ss_n} = 0.16 \text{ given})$$

The transfer function of a lag compensator is given by

$$G_c(s) = \frac{(s+a)}{(s+b)}, \quad a > b$$

$$\because \qquad \frac{a}{b} = \frac{K_{v_n}}{K_{v_o}}$$

$$\therefore \qquad \frac{a}{b} = \frac{6.25}{1.34} = 4.66$$

Select the parameter $b = 0.02$.

The parameter $a = 0.02 \times 4.66 = 0.0932$

$$\therefore \quad G_0(s) = \frac{(s+0.0932)}{(s+0.02)}$$

The phase angle contributed by the lag compensator at $s = s_d = -0.8 + 0.8\sqrt{3}$ is calculated below,

$$\angle G_c(s) = \angle \left. \frac{(s+0.0932)}{(s+0.02)} \right|_{s = s_d = -0.8 + j\,0.8\sqrt{3}}$$

$$= \angle \frac{(-0.8 + j\,0.8\sqrt{3} + 0.032)}{(-0.8 + j\,0.8\sqrt{3} + 0.02)}$$

$$= \angle \frac{(-0.7068 + j1.385)}{(-0.78 + j1.385)}$$

$$= \left(180° - \tan^{-1}\frac{1.385}{0.7068}\right) - \left(180° - \tan^{-1}\frac{1.385}{0.78}\right)$$

$$= (180° - 62.95°) - (180° - 60.6°)$$

$$= -2.35°$$

As the phase angle contributed by the lag compensator at $s_d = -0.8 + j\,0.8\sqrt{3}$, is within the range 0° to – 3°, the design is accepted.

The open-loop transfer function cascaded with lag compensator is given below,

$$G(s)\,G_c(s) = \frac{21.5(s+0.0932)}{s(s+2)(s+8)(s+0.02)}. \quad \textbf{Ans.}$$

Note. The distances of the lag compensator pole-zeros are measured to be very small from the *s*-plane origin as such the root locus shape around the aforesaid pole-zero is also very small as compared to the original root locus plot.

However, in Fig. 8.7.12 the root locus around lag compensator pole-zero is plotted to a large scale for clarity.

8.7.3. Lag-Lead Compensator Design

The lead compensation improves the time response of a feedback control system whereas the lag compensation improves (reduces) the steady state error. However, if both the aforesaid requirements are to be met, then the compensator consists of a cascade combination lag and lead compensation known as lag-lead compensation.

In this process lead compensator is designed to meet the desired time response specifications, then, the lag compensator parameters are determined as per the steady state error requirements.

The procedure of designing lead as well as lag compensator being already described earlier.

Solved Examples : Lag-Lead Compensator

Example 8.7.8. *The open-loop transfer function of a unity feedback control system is given below:*

$$G(s) = \frac{6}{s(s+4)}$$

Design a lag-lead compensator such that the closed-loop desired pole is placed as

$$s_d = -3 \pm j\,1.5\sqrt{2}$$

and the velocity error coefficient. $K_v = 15\ sec^{-1}$.

Solution. $$G(s) = \frac{6}{s(s+4)}$$

$$\underset{\text{uncompensated}}{K_{v_0}} = \lim_{s\to 0} s\,G(s)$$

$$= \lim_{s\to 0} \frac{s6}{s(s+4)} = 1.5\ \text{sec}^{-1}$$

Lead Compensator

The desired closed-loop pole $s_d = -3 + j1.5\sqrt{2}$ does not lie on the uncompensated root locus. The angle θ_c to be contributed by the lead compensator is determined below using angle condition of root locus at $s = s_d = -3 + j\,1.5\sqrt{2}$

$$\theta_c + \angle \frac{6}{s(s+4)}\bigg|_{s=s_d=-3+j\,1.5\sqrt{2}} = -180$$

$$\therefore \quad \theta_c + \angle \frac{6}{(-3+j1.5\sqrt{2})(-3+j1.5\sqrt{2}+4)} = -180°$$

or

$$\theta_c + \angle \frac{6}{(-3+j\,1.5\sqrt{2})(1+j\,1.5\sqrt{2})} = -180°$$

$$\therefore \quad \theta_c + \left[-\left(180° - \tan^{-1}\frac{1.5\sqrt{2}}{3}\right) - \tan^{-1}\frac{1.5\sqrt{2}}{1}\right] = -180°$$

$$\therefore \quad \theta_c - 180° + 35.26° - 64.75 = -180°$$

Simplifying, $\theta_c = 29.49°$

The transfer function of a lead compensator is given by,

$$\underset{\text{lead}}{G(s)} = \frac{K_c(s+a_1)}{(s+b_1)}, \qquad a_1 < b_1$$

Select, $a_1 = 4$ to cancel the open-loop pole $s = -4$

$$\therefore \quad \underset{\text{lead}}{G_c(s)} = \frac{K_c(s+4)}{(s+b_1)}$$

and

$$\theta_c = \angle \underset{\text{lead}}{G_c(s)}\bigg|_{s=s_d=-3+j\,1.5\sqrt{2}}$$

or

$$\theta_c = \angle \frac{K_c(-3+j\,1.5\sqrt{2}+4)}{(b_1-3)+j\,1.5\sqrt{2}}$$

$\because \quad \theta_c = 29.49°$ (calculated earlier)

$$\therefore \quad 29.49° = \tan^{-1}\frac{1.5\sqrt{2}}{1} - \tan^{-1}\frac{1.5\sqrt{2}}{(b_1 - 3)}$$

or
$$29.49° = 64.75 - \tan^{-1}\frac{1.5\sqrt{2}}{(b_1 - 3)}$$

Simplifying,

$$\tan^{-1}\frac{1.5\sqrt{2}}{(b_1 - 3)} = 35.26$$

$$\therefore \quad \frac{1.5\sqrt{2}}{b_1 - 3} = \tan 35.26 = 0.7069$$

$$\therefore \quad b_1 = 6$$

$$K = 6 \text{ (given)}$$

$$\therefore \quad G(s)\, \underset{\text{lead}}{G_c(s)} = \frac{6K_c(s+4)}{s(s+4)(s+6)} = \frac{6K_c}{s(s+6)}$$

The value of $6K_c$ is determined applying root locus magnitude condition at $s = s_d = (-3 + j\,1.5\sqrt{2})$

$$\therefore \quad \left| G(s)\, \underset{\text{lead}}{G_c(s)} \right|_{s = s_d = -3 + j\,1.5\sqrt{2}} = 1$$

$$\therefore \quad \left| \frac{6K_c}{s(s+6)} \right|_{s = s_d = -3 + j\,1.5\sqrt{2}} = 1$$

Rearranging,
$$6K_c = \left| s(s+6) \right|_{s = s_d = -3 + j\,1.5\sqrt{2}}$$

$$= \left| (-3 + j\,1.5\sqrt{2})(-3 + j1.5\sqrt{2} + 6) \right|$$

$$= \left| (-3 + j\,1.5\sqrt{2})(3 + j\,1.5\sqrt{2}) \right|$$

$$= 13.5, \quad K_c = \frac{13.5}{6} = 2.25$$

$$\therefore \quad G(s)\, \underset{\text{lead}}{G_c(s)} = \frac{13.5}{s(s+6)}$$

The output signal of $G(s)\underset{\text{lead}}{G_c(s)}$ is the input to lag compensator.

$$\therefore \quad K_{v_0} = \lim_{s \to 0} s\, G(s)\, G_c(s)$$

$$= \lim_{s \to 0} \frac{s \times 13.5}{s(s+6)} = 2.25 \text{ sec}^{-1}$$

$$\because \quad K_{v_n} = 15 \text{ sec}^{-1} \text{ (given)}$$

Lag Compensator

The lag compensator transfer function is given by

$$\underset{\text{lead}}{G_c(s)} = \left(\frac{s + a_2}{s + b_2}\right), \quad a_2 < b_2$$

$$\frac{a_2}{b_2} = \frac{K_{v_n}}{K_{v_0}}$$

$$\because \quad \frac{K_{v_n}}{K_{v_0}} = \frac{15}{2.25} = 6.66$$

$$\therefore \quad \frac{a_2}{b_2} = 6.66$$

Select $b_2 = 0.02$, therefore, $a_2 = 0.02 \times 6.66 = 0.133$

$$\therefore \quad \underset{\text{lag}}{G_c(s)} = \frac{(s+0.133)}{(s+0.02)}$$

The angle contributed by the lag compensator at $s = s_d = -3 + j\,1.5\,\sqrt{2}$ is calculated as

$$\angle \underset{\text{lag}}{G_c(s)} = \angle \frac{(s+0.133)}{(s+0.02)}\Bigg|_{s=s_d=-3+j\,1.5\sqrt{2}}$$

$$= \angle \frac{(-3+j1.5\sqrt{2}+0.133)}{(-3+j1.5\sqrt{2}+0.02)}$$

$$= \angle \frac{(-2.867+j1.5\sqrt{2})}{(-2.98+j1.5\sqrt{2})}$$

$$= \left(180° - \tan^{-1}\frac{1.5\sqrt{2}}{2.867}\right) - \left(180° - \tan^{-1}\frac{1.5\sqrt{2}}{2.98}\right)$$

$$= -1.05°$$

As the contributed angle is within the range 0° to – 3°, the design is accepted.

$$G(s)\underset{\text{lead}}{G_c(s)}\underset{\text{lag}}{G_c(s)} = \frac{6 \times 2.25(s+0.133)}{s(s+6)(s+0.02)}. \quad \textbf{Ans.}$$

Example 8.7.9. *The unity feedback control system has open-loop transfer function as below:*

$$G(s) = \frac{72.5}{s(s+3)(s+12)}$$

Design a compensator such that the closed-loop dominant (desired) pole is located at $s = -2.5 + j2.5\sqrt{3}$ and the velocity error coefficient. for a unit ramp input is $K_v = 18\ sec^{-1}$.

Solution. The general shape of the root locus is shown in Fig. 8.7.13. The break away point on the s-plane real axis is determined below.

The characteristic equation is given by

$$1 + G(s) = 0$$

or $$1 + \frac{K}{s(s+3)(s+12)} = 0, \text{ where } K = 72.5 \text{ (given)}$$

$$\therefore \quad K = -[s(s+3)(s+12)]$$

$$= -[s^3 + 15s^2 + 36s]$$

put $$\frac{dK}{ds} = 0, \quad \therefore \quad 3s^2 + 15s + 36 = 0$$

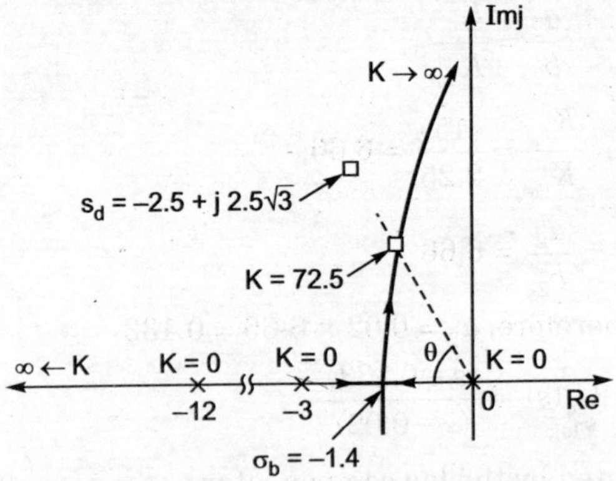

Fig. 8.7.13. Root locus for Example 8.7.9.

Solving above equation $s = -1.4$ and $s = -8.6$.

The break away point $\sigma_b = -1.4$ (lies on root locus), hence accepted.

It is noted that the desired closed-loop pole $s_d = -2.5 + j\,2.5\sqrt{2}$ does not lie on the uncompensated root-lcous and a lead compensator is to be incorporated.

The angle θ_c to be contributed by the lead compensator is determined below applying root locus angle condition at $s = s_d = -2.5 + j\,2.5\sqrt{3}$

$$\therefore \qquad \theta_c + \angle \left.\frac{72.5}{s(s+3)(s+12)}\right|_{s = s_d = -2.5 + j\,2.5\sqrt{3}} = -180^\circ$$

$$\text{or} \qquad \theta_c + \angle \frac{72.5}{(-2.5 + j\,2.5\sqrt{3})(-2.5 + j\,2.5\sqrt{3} + 3)(-2.5 + j\,2.5\sqrt{3} + 12)} = -180^\circ$$

$$\therefore \qquad \theta_c + \angle \frac{72.5}{(-2.5 + j\,2.5\sqrt{3})(0.5 + j\,2.5\sqrt{3})(9.5 + j\,2.5\sqrt{3})} = -180^\circ$$

$$\text{or} \qquad \theta_c + \left[-\left(180^\circ - \tan^{-1}\frac{2.5\sqrt{3}}{2.5}\right) - \left(\tan^{-1}\frac{2.5\sqrt{3}}{0.5}\right) - \left(\tan^{-1}\frac{2.5\sqrt{3}}{9.5}\right)\right] = -180^\circ$$

$$\therefore \qquad \theta_c + [-(180^\circ - \tan^{-1}\sqrt{3}) - (\tan^{-1} 5\sqrt{3}) - (\tan^{-1} 0.263\sqrt{3})] = -180^\circ$$

$$\therefore \qquad \theta_c + [-(180^\circ - 60^\circ) - 83.41^\circ - 24.46^\circ] = -180^\circ$$

Simplifying, $\qquad \theta_c = 47.87^\circ$

Lead Compensator

The transfer function of a lead compensator is given by,

$$G_{c}(s)_{\text{lead}} = \frac{K_c(s + a_1)}{(s + b_1)}, \qquad a_1 < b_1$$

Select, $a_1 = 3$, to cancel open loop pole at $s = -3$

$$\therefore \qquad G_{c}(s)_{\text{lead}} = \frac{K_c(s+3)}{(s+b_1)}$$

$$\because \qquad \theta_c = \angle \left.\frac{K_c(s+3)}{(s+b_1)}\right|_{s = s_d = -2.5 + j\,2.5\sqrt{3}}$$

$$= \angle \frac{K_c(-2.5 + j2.5\sqrt{3} + 3)}{(b_1 - 2.5) + (j2.5\sqrt{3})}$$

$\because$ $\theta_c = 47.87$ (calculated earlier)

$$\therefore \quad 47.87° = \angle \frac{K_c(0.5 + j2.5\sqrt{3})}{(b_1 - 2.5) + j2.5\sqrt{3}}$$

$$= 83.41° - \tan^{-1}\frac{2.5\sqrt{3}}{b_1 - 2.5}$$

$$\therefore \quad \tan^{-1}\frac{2.5\sqrt{3}}{b_1 - 2.5} = (83.41° - 47.87°) = 35.54°$$

$$\therefore \quad \frac{2.5\sqrt{3}}{b_1 - 2.5} = \tan 35.54°$$

or

$$\frac{2.5\sqrt{3}}{b_1 - 2.5} = 0.714$$

Simplifying, $b_1 = 8.56$

Thus, $\underset{\text{lead}}{G_c(s)} = \dfrac{K_c(s+3)}{(s+8.56)}$ and $G(s)\underset{\text{lead}}{G_c(s)} = \dfrac{72.5K_c(s+3)}{s(s+3)(s+2)(s+8.56)}$

or

$$G(s)\underset{\text{lead}}{G_c(s)} = \frac{72.5K_c}{s(s+12)(s+8.56)}$$

Since, $s_d = -2.5 + j\,2.5\sqrt{2}$ lies of lead compensated root locus, the magnitude condition of the root locus is applicable,

$$\therefore \quad \left| G(s)G_c(s) \right|_{s = s_d = -2.5 + j\,2.5\sqrt{3}} = 1$$

$$\therefore \quad \left| \frac{72.5K_c}{s(s+12)(s+8.56)} \right|_{s = s_d = -2.5 + j\,2.5\sqrt{3}} = 1$$

$$\therefore \quad 72.5\,K_c = \left| s(s+12)(s+8.56) \right|_{s = s_d = -2.5 + j\,2.5\sqrt{3}}$$

$$= |(-2.5 + j\,2.5\sqrt{3})(-2.5 + j\,2.5\sqrt{3} + 12)(-2.5 + j\,2.5\sqrt{3} + 8.56|$$

$$= |(-2.5 + j\,2.5\sqrt{3})(9.5 + j\,2.5\sqrt{3})(6.06 + j\,2.5\sqrt{3})|$$

$$= \sqrt{(6.25 + 18.75)(90.25 + 18.75)(36.72 + 18.75)}$$

Simplifying, $72.5\,K_c = 388.7$

$$\therefore \quad K_c = \frac{388.7}{72.5} = 5.36$$

and

$$G(s)\underset{\text{lead}}{G_c(s)} = \frac{72.5 \times 5.36}{s(s+12)(s+8.56)}$$

The velocity error coefficient K_{v_0} for the transfer function $G(s)\underset{\text{lead}}{G_c(s)}$ is calculated below,

$$K_{v_0} = \lim_{s \to 0} s\,G(s)\underset{\text{lead}}{G_c(s)}$$

$$= \lim_{s \to 0} s \frac{72.5 \times 5.36}{s(s+12)(s+8.56)}$$

$$\therefore \qquad K_{v_0} = 3.78$$

Lag Compensator

The transfer function of a lag compensator is given below:

$$\underset{\text{lag}}{G_c(s)} = \frac{(s+a_2)}{(s+b_2)}, \qquad a_2 > b_2$$

$$\frac{a_2}{b_2} = \frac{K_{v_n}}{K_{v_0}} = \frac{18}{3.78} = 4.76$$

Select $b_2 = 0.03$, $\therefore$ $a_2 = 0.03 \times 4.76 = 0.1428$

$$\therefore \qquad \underset{\text{lag}}{G_c(s)} = \frac{(s+0.1428)}{(s+0.03)}$$

The angle contributed by the lag compensator at $s = s_d = -2.5 + j\,2.5\sqrt{3}$ is determined below

$$\angle G_c(s) = \angle \left. \frac{(s+0.1428)}{(s+0.03)} \right|_{s = s_d = -2.5 + j\,2.5\sqrt{3}}$$

$$= \angle \frac{(-2.5 + j2.5\sqrt{3} + 0.1428)}{(-2.5 + j2.5\sqrt{3} + 0.03)}$$

$$= \angle \frac{(-2.357 + j4.33)}{(-2.47 + j4.33)}$$

$$= \left(180° - \tan^{-1}\frac{4.33}{2.357}\right) - \left(180° - \frac{4.33}{2.47}\right)$$

$$= (180° - 61.43°) - (180° - 60.29°)$$

$$= -1.14°$$

As the angle contributed by the lag compensator is $-1.14°$, lies in the range ($0°$ to $-3°$), the design is accepted

$$\therefore \qquad G(s) \underset{\text{lead}}{G_c(s)} \underset{\text{lag}}{G_c(s)} = \frac{72.5 \times 5.36(s+0.1428)}{s(s+12)(s+8.56)(s+0.03)}. \quad \textbf{Ans.}$$

Example 8.7.10. *Design a lag-lead compensator for a control system, the open-loop transfer function (unity feedback) is given as,*

$$G(s) = \frac{3.57}{s(s+1)(s+5)}$$

The damping ratio $\xi = 0.5$, the settling time $t_s = 5$ sec. and the velocity error coeff. $K_v = 5$ sec^{-1} for a unit ramp input.

Solution. The general shape of the root locus for open-loop transfer function

$$G(s) = \frac{K}{s(s+1)(s+5)}$$

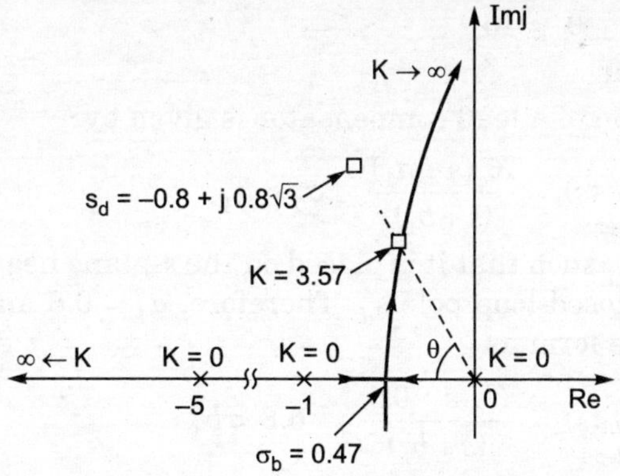

Fig. 8.7.14. Root locus for Example 8.7.10.

is shown in Fig. 8.7.14. The break away point can be determined as $\sigma_b = -0.478$. The damping ratio line makes an angle θ with the –ve real axis of *s*-plane.

where $\theta = \cos^{-1} \xi = \cos^{-1} 0.5 = 60°$

The damping ratio line intersects the root locus such that $K = 3.57$ (given)

The desired settling time $t_s = 5$ sec.

$$\because \quad t_s = \frac{4}{\xi\omega_n}, \quad \therefore \quad 5 = \frac{4}{\xi\omega_n}, \quad \text{thus } \xi\omega_n = \frac{4}{5} = 0.8$$

Further, $\xi\omega_n = 0.8$ and $\xi = 0.5$, $\therefore$ $\omega_n = \dfrac{0.8}{0.5} = 1.6$ rad./sec.

The desired closed-loop pole s_d is given by the following relation,

$$s_d = -\xi\omega_n + j\omega_n\sqrt{1-\xi^2}$$

$$\therefore \quad s_d = -0.8 + j\,1.6\sqrt{1-0.5^2}$$

Simplifying, $$s_d = -0.8 + j0.8\sqrt{3}$$

It is noted that, the desired closed loop pole $s_d = -0.8 + j\,0.8\sqrt{3}$ does not lie on the uncompensated root locus. Therefore, lead compensator is needed.

The closed-loop pole $s_d = -0.8 + j\,0.8\sqrt{3}$ has to lie on the lead compensated root locus, satisfying the root locus angle condition. The angle to be contributed by the lead compensator is θ_c.

Therefore, $$\theta_c + \angle \left.\frac{3.57}{s(s+1)(s+5)}\right|_{s = s_d = -0.8 + j\,0.8\sqrt{3}} = -180°$$

Thus, $$\theta_c + \angle \frac{3.57}{(-0.8 + j\,0.8\sqrt{3})(-0.8 + j\,0.8\sqrt{3} + 1)(-0.8 + j\,0.8\sqrt{3} + 5)} = -180°$$

$$\therefore \quad \theta_c + \left[-\left(180° - \frac{0.8\sqrt{3}}{0.8}\right) - \left(\tan^{-1}\frac{0.8\sqrt{3}}{0.2}\right) - \left(\tan^{-1}\frac{0.8\sqrt{3}}{4.2}\right)\right] = -180°$$

or $$\theta_c + [-(180° - 60°) - (81.78°) - (18.21)°] = -180°$$

Simplifying, $\theta_c \simeq 40°$

Lead Compensator

The transfer function of a lead compensator is given by,

$$\underset{\text{lead}}{G_c(s)} = \frac{K_c(s+a_1)}{(s+b_1)}, \quad a_1 < b_1$$

Select the value of a_1 such that it is placed on the s-plane negative real axis just below the real part of desired closed-loop pole s_d. Therefore, $a_1 = 0.8$ and the lead compensator transfer function takes the form as

$$\underset{\text{lead}}{G_c(s)} = \frac{K_c(s+0.8)}{(s+b_1)}, \quad 0.8 < b_1$$

The angle contributed by the lead compensator at the desired closed-loop pole $s_d = -0.8 + j\,0.8\sqrt{3}$ is determined using following relation,

$$\theta_c = \angle \left. \frac{K_c(s+0.8)}{(s+b_1)} \right|_{s = s_d = -0.8 + j\,0.8\sqrt{3}}$$

$$= \angle \frac{-0.8 + j\,0.8\sqrt{3} + 0.8}{(b_1 - 0.8) + j\,0.8\sqrt{3}} = \angle \frac{j\,0.8\sqrt{3}}{(b_1 - 0.8) + j\,0.8\sqrt{3}}$$

$$= 90° - \tan^{-1} \frac{0.8\sqrt{3}}{b_1 - 0.8}$$

$\because$ $\theta_c = 40°$ (determined earlier)

$$\therefore \quad 40° = 90° - \tan^{-1} \frac{0.8\sqrt{3}}{b_1 - 0.8}$$

Rearranging, $$50° = \tan^{-1} \frac{0.8\sqrt{3}}{b_1 - 0.8}$$

$$\therefore \quad \tan 50° = \frac{0.8\sqrt{3}}{b_1 - 0.8}$$

$$\therefore \quad 1.191 = \frac{0.8\sqrt{3}}{b_1 - 0.8}$$

Solving above equation,

$$b_1 = 1.96$$

The lead compensated open-loop transfer function is written below,

$$G(s)\underset{\text{lead}}{G_c(s)} = \frac{3.57K_c(s+0.8)}{(s+1)(s+5)(s+1.96)}$$

The desired closed-loop pole $s_d = -0.8 + j\,0.8\sqrt{3}$, lies on the lead compensator root locus. Hence, the value of gain (gain factor) K_c is determined applying root locus magnitude condition to above transfer function at $s = s_d = -0.8 + j\,0.8\sqrt{3}$.

$$\therefore \quad \left| \frac{3.57K_c(s+0.8)}{s(s+1)(s+5)(s+1.96)} \right|_{s = s_d = -0.8 + j\,0.8\sqrt{3}} = 1$$

Rearranging,

$$3.57\,K_c = \left|\frac{s(s+1)(s+5)(s+1.96)}{(s+0.8)}\right|_{s=s_d=-0.8+j\,0.8\sqrt{3}}$$

$$= \left|\frac{(-0.8+j0.8\sqrt{3})(-0.8+j0.8\sqrt{3}+1)(-0.8+j0.8\sqrt{3}+5)(-0.8+j0.8\sqrt{3}+1.96)}{(-0.8+j0.8\sqrt{3}+0.8)}\right|$$

$$= \left|\frac{(-0.8+j\,0.8\sqrt{3})(0.2+j\,0.8\sqrt{3})(4.2+j\,0.8\sqrt{3})(1.16+j\,0.8\sqrt{3})}{j\,0.8\sqrt{3}}\right|$$

$$= \sqrt{\frac{(0.64+1.92)(0.04+1.92)(17.64+1.92)(1.34+1.92)}{(0.8\sqrt{3})^2}}$$

$$\therefore \quad 3.57\,K_c = 12.84. \quad \text{Hence,} \quad K_c = \frac{12.84}{3.57} = 3.596$$

$$K_{v_0} = \lim_{s\to 0} s\,G(s)\,\underset{\text{lead}}{G_c(s)}$$

$$= \lim_{s\to 0} \frac{s12.84(s+0.8)}{s(s+1)(s+5)(s+1.96)}$$

$$= 1.047\ \text{sec}^{-1}.$$

Lag Compensator

The transfer function of a lag compensator is given by,

$$\underset{\text{lag}}{G_c(s)} = \frac{(s+a_2)}{(s+b_2)}, \qquad a_2 > b_2$$

also

$$\frac{K_{v_n}}{K_{v_0}} = \frac{a_2}{b_2}$$

$\because$ $K_{v_n} = 5\ \text{sec}^{-1}$ (given), $K_{v_0} = 1.047\ \text{sec}^{-1}$ (calculated earlier)

$$\therefore \quad \frac{a_2}{b_2} = \frac{5}{1.047} = 4.77$$

Select, $b_2 = 0.025, \quad \therefore \quad a_2 = 0.025 \times 4.77 = 0.119$

Thus, $$\underset{\text{lag}}{G_c(s)} = \frac{(s+0.119)}{(s+0.025)}$$

The angle θ contributed by the lag compensator at $s_d = -0.8 + j\,0.8\sqrt{3}$ is determined below,

$$\theta = \angle \left.\frac{(s+0.119)}{(s+0.025)}\right|_{s=s_d=-0.8+0.8\sqrt{3}}$$

$$= \angle \frac{(-0.8+j\,0.8\sqrt{3}+0.119)}{(-0.8+j\,0.8\sqrt{3}+0.025)}$$

$$= \angle \frac{(-0.681+j\,0.8\sqrt{3})}{(-0.775+j\,0.8\sqrt{3})}$$

$$= \left(180° - \tan^{-1}\frac{0.8\sqrt{3}}{0.681}\right) - \left(180° - \tan^{-1}\frac{0.8\sqrt{3}}{0.775}\right)$$

$$= -2.97°$$

Which lies within the range 0° to – 3°. Hence, the design is accepted.

The lag-lead open-loop transfer function is written below,

$$G(s)\, \underset{\text{lead}}{G_c(s)}\, \underset{\text{lag}}{G_c(s)} = \frac{12.84(s+0.8)(s+0.119)}{s(s+1)(s+5)(s+1.96)(s+0.025)}. \quad \textbf{Ans.}$$

PROBLEMS

8.1. For the open-loop transfer function, $G(s)\,H(s) = \dfrac{K}{s(s+0.5s)}$ determine the value of gain K such that the system will have a phase margin of at least 45°.

8.2. Design a lead compensator such that the system having following open-loop transfer function will have a phase margin of about 40°.

$$G(s)\,H(s) = \frac{5}{(1+s)(1+0.5s)^2}.$$

8.3. A unity feedback control system has following open-loop transfer function, $G(s) = \dfrac{1}{(1+0.5s)^3}$. Design a suitable compensator such that (*i*) $K_p = 3$ (*ii*) G.M. ≥ 15 db (*iii*) P.M. ≥ 30°.

8.4. Following specifications :

(*i*) $K_v \geq 4$ (*ii*) G.M. ≥ 6 db (*iii*) P.M. ≥ 20° refer to a system whose open-loop transfer function is $G(s)\,H(s) = \dfrac{K}{s(1+0.2s)^2}$

Design a phase-lead compensation network to satisfy the above specifications.

8.5. The system shown in Fig. 8P-1 is to be compensated by tachometer feedback such that the maximum overshoot is limited to 30%. Determine the value of tachometer constant K_t.

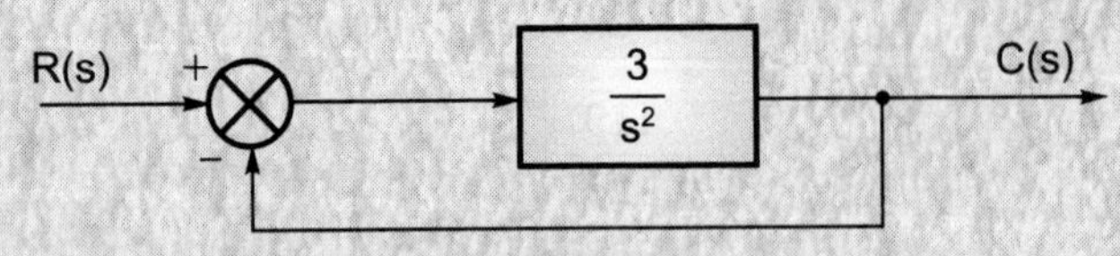

Fig. 8P-1. System for problem 8.5.

8.6. The damping ratio for the system shown in Fig. 8P-2 (*a*) is 0.166. A tachometer feedback as shown in Fig. 8P-2 (*b*) is in corporated in the system to increase the damping. Determine the value of K_t so that the damping ratio is 0.6. Further determine the maximum overshoot and the steady state error for unit ramp input both the cases.

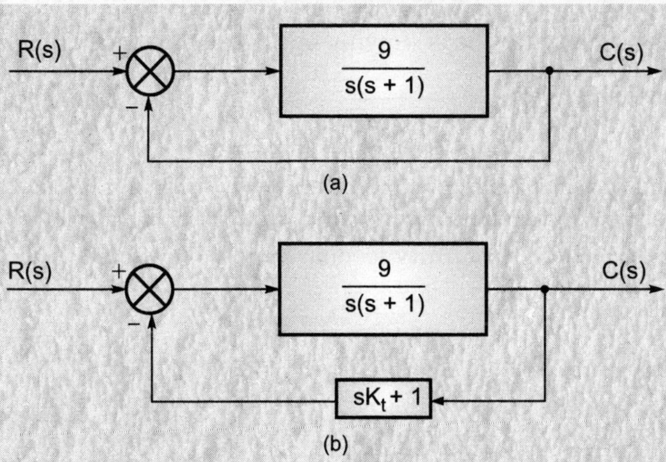

Fig. 8P-2. System for problem 8.6.

8.7. The open-loop transfer function of a unity feedback control system is given below,

$$G(s) = \frac{K}{s(s+10)}$$

The max. overshoot is 24.3% (unit step input), maintaining the same damping ratio, design a compensator such that the settling time is reduced by a factor of 2.

8.8. A unity feedback control system having open-loop transfer function as

$$G(s) = \frac{K}{s(s+6)}$$

operates at a daming ratio $\xi = 0.6$.

It is desired to have a settling time $t_s = 1.0$ sec. Design a suitable compensator.

8.9. Design a lead compensator such that a unity feedback control system having following transfer function,

$$G(s) = \frac{K}{s^2+1}$$

becomes stable with damping ratio $\xi = 0.5$ and settling time $t_s = 2.5$ sec. The uncompensated system operates at closed-loop poles $s = \pm j\sqrt{5}$.

8.10. The open-loop transfer function (unity feedback) is given below

$$G(s) = \frac{K}{s(s+4)}$$

The damping ratio is $\xi = 0.4$. It is specified that the velocity error coefficient for unit ramp input is $K_v = 25 \text{ sec}^{-1}$. Design a lag compensator.

8.11. The forward path transfer function of a unity feedback control system is given by,

$$G(s) = \frac{K}{s(s+4)(s+8)}$$

Design a lag compensator such that the maximum overshoot is $\leq 16.3°$ and steady state velocity error coefficient to a ramp input is $K_v = 12 \text{ sec}^{-1}$.

8.12. A control system (unity feedback) having open-loop transfer function,

$$G(s) = \frac{K}{s(s+1)(s+5)}$$

operates with damping ratio $\xi = 0.5$. The damping ratio line intersects the uncompensated root locus plot at $s = -0.416 + j\,0.72$. Design a lag compensator such that velocity error coefficient for unit ramp input is $K_v = 4.0 \text{ sec}^{-1}$.

8.13. The following data refers to a unity feedback control system:

$$G(s) = \frac{K}{s(s+2)(s+5)}$$

(*a*) The damping ratio $\xi = 0.6$, settling time $t_s = 2.5 \text{ sec}^{-1}$ and $K_v = 1.5 \text{ sec}^{-1}$.

(*b*) The steady state error to be improved by a factor of 5.0.

Design a lag-lead compensator.

8.14. The transfer function of a unity feedback control system is given by

$$G(s) = \frac{21.5}{s(s+2)(s+3)}$$

The damping ratio $\xi = 0.5$. The requirement is that the natural frequency $\omega_n = 4$ rad./sec. and steady state error $e_{ss} = 0.1$ rad. for unit ramp input. Design a suitable compensator.

8.15. Design a suitable compensator having given that the damping ratio $\xi = 1/\sqrt{2}$, settling time $t_s = 1.0$ sec., and the desired velocity error coefficient $K_v = 20 \text{ sec}^{-1}$. The open-loop transfer function with unity feedback is given as

$$G(s) = \frac{K}{s(s+8)}.$$

CHAPTER 9

Introduction to State Space Analysis of Control Systems

The modern approach, the state space introduced in this chapter deals with the formulation of state model of a control system. The solution of state equation through state transition matrix can be obtained. Similarly transformation transforms state transition matrix into a diagonal form. The diagonalization methods are explained. The transfer function can be obtained for a given control system from the state model using transfer matrix. The chapter concludes with the determination of controllability and observability of a state model.

CONTENTS

- *State Space Representation*
- *The Concept of State*
- *State Space Representation of Systems*
- *Block Diagram for State Equation*
- *Transfer Function Decomposition*
- *Solution of State Equation*
- *Transfer Matrix*
- *Controllability*
- *Observability*
- *Solved Examples*
- *State Variable Feedback*

Introduction. The analysis of control systems carried out in earlier chapters is based on transfer functions and graphical approach such as root locus, Bode and Nyquist plot. The input output relations are in the form of transfer function. In this approach of analysis initial conditions are considered zero meaning that the system is initially at rest and the time solution obtained is in a general form due to input only. However, for multiple-input multiple-output and systems initially not at rest, the transfer function approach for analysis is inadequate and not convenient.

The use of state space approach for the analysis of control systems enables to overcome the shortcomings of transfer function approach. The state space approach to be described hereunder can be conveniently used for analysis of control system with basic knowledge of matrix algebra.

The analysis of a control system using state space approach carried out in time domain by representing a system in the form of first order differential equations by selecting suitable state variables wherein first order derivative terms are arranged on left hand side and right hand side terms are free from derivatives. The variables in such a form of representation are known as state variables. In state space approach multiple input- multiple-output system equations can be arranged in matrix form which facilitates their solution.

9.1 STATE SPACE REPRESENTATION

The representation of a second order differential equation based on state space approach is obtained below :

The differential equation under consideration is

$$a\frac{d^2x(t)}{dt^2} + b\frac{dx(t)}{dt} + cx(t) = u(t) \quad \text{...(9.1)}$$

with initial conditions given as

$$\frac{dx(t)}{dt} = 0 \text{ and } x(t) = 0 \text{ at } t = 0$$

Eq. (9.1) can be rearranged in the following form :

$$\frac{d^2x(t)}{dt^2} = \frac{1}{a}u(t) - \frac{b}{a}\frac{dx(t)}{dt} - \frac{c}{a}x(t) \quad \text{...(9.2)}$$

The block diagram representation indicating the integration process of Eq. (9.2) is shown in Fig. 9.1.1.

Let $x(t) = x_1$ and $\frac{dx(t)}{dt} = x_2$

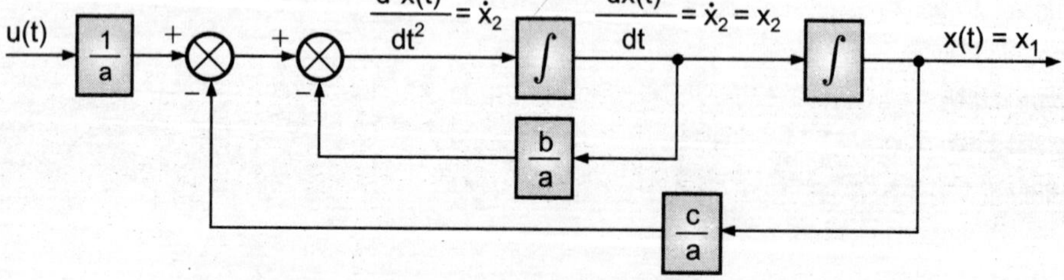

Fig. 9.1.1. Block diagram representation of differential Eq. (9.2).

Therefore, the dynamic equations describing the system equations can be written as

$$\dot{x}_1 = x_2 \quad \text{...(9.3)}$$

and

$$\dot{x}_2 = -\frac{c}{a}x_1 - \frac{b}{a}x_2 + \frac{1}{a}u(t) \quad \text{...(9.4)}$$

In the matrix form Eqs. (9.3) and (9.4) can be written as

$$\begin{bmatrix} \dot{x}_1 \\ \dot{x}_2 \end{bmatrix} = \begin{bmatrix} 0 & 1 \\ -\dfrac{c}{a} & -\dfrac{b}{a} \end{bmatrix} \begin{bmatrix} x_1 \\ x_2 \end{bmatrix} + \begin{bmatrix} 0 \\ \dfrac{1}{a} \end{bmatrix} u(t) \qquad \text{...(9.5)}$$

The representation given by Eq. (9.5) is known as state space representation of the system given by Eq. (9.1).

9.2 THE CONCEPT OF STATE

The dynamic behaviour of the system given by Eq. (9.1) can be determined by two variables $x_1(t)$ and $x_2(t)$. The variables $x_1(t)$ and $x_2(t)$ can be determined by solving Eq. (9.5).

At any instant of time $t > t_0$ the state of the system is given by the values of $x_1(t)$ and $x_2(t)$, therefore, $x_1(t)$ and $x_2(t)$ are called state variables.

Fig. 9.2.1 shows a plot of $x_1(t)$ and $x_2(t)$ on co-ordinate axes. The starting point of the plot is at instant $t = t_0$ and as time increases the plot traces a path.

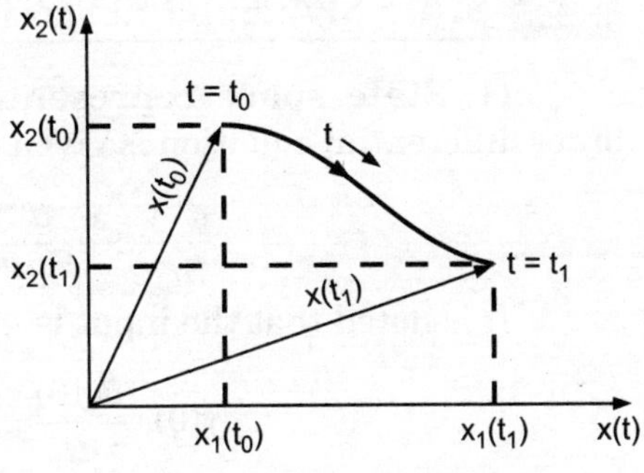

Fig. 9.2.1. Path traced by state vector.

At $t = t_0$ the state of system is given by state vector $x(t_0)$ and $t = t_1$ by state vector $x(t_1)$. At any instant of time $t > t_0$ the state vector is related to state variables as

$$\vec{x(t)} = \vec{x_1(t)} + \vec{x_2(t)}$$

Thus the state of a system at any instant of time $t > t_0$ is determined by a state vector $x(t)$. The state vector $x(t)$ is the sum of state variables. The path traced by a state vector with the increase of time is known as state trajectory.

The analysis given above can be generalized for nth order system where the state of nth order system in matrix form can be written as

$$\mathbf{x} = \begin{bmatrix} x_1 \\ x_2 \\ x_3 \\ \vdots \\ \vdots \\ x_n \end{bmatrix} n \times 1 \text{ ; and the input vector in matrix form can be written as } \mathbf{u} = \begin{bmatrix} u_1 \\ u_2 \\ u_3 \\ \vdots \\ \vdots \\ u_m \end{bmatrix} m \times 1$$

The state vector thus can be represented in n-dimensional space known as state space. For a second order system the state space reduces to a state plane.

As a consequence of the analysis of the above illustrative example the definitions concerning state space approach are given below :

(1) **State Variables.** "The smallest set of variables which determine the state of a dynamic system are called state variables."

It is not necessary that the state variables be physically measurable or observable quantities.

(2) **State.** "The state of a system is the smallest set of state variables such that the knowledge of these variables at $t = t_0$ (initial condition) together with the inputs completely determines the behaviour of the system for any time $t > t_0$."

(3) **State Vector.** "If n state variables are necessary to determine the behaviour of a given system, the variables can be considered as n components of a vector called state vector".

(4) **State Space.** "The n dimensional state variables are elements of n dimensional space called state space".

9.3 STATE SPACE REPRESENTATION OF SYSTEMS

(1) **State space representation of *n*th order differential equation.** The nth order differential equation is given below :

$$\frac{d^n y}{dt^n} + a_1 \frac{d^{n-1} y}{dt^{n-1}} + \dots + a_n y = u(t) \qquad \text{...(9.6)}$$

It is noted that the input is $u(t)$ and the initial conditions

$$y(0), \quad \frac{dy(0)}{dt} \dots \frac{d^{n-1}y(0)}{dt^{n-1}} \quad \text{at} \quad t = 0 \text{ are given.}$$

The dynamic behaviour of differential Eq. (9.6) can be completely determined from the knowledge of $u(t)$, $\dot{y}(t) \dots, \overset{(n-1)}{y}(t)$.

The terms $y(t)$, $\dot{y}(t) \dots y^{n-1}(t)$ can be considered as a set of n state variables.

Let

$$\begin{aligned} y &= x_1 \\ \dot{y} &= x_2 \\ &\vdots \\ y^{n-1} &= x_n \end{aligned} \qquad \text{...(9.6a)}$$

The set of Eqs. (9.6*a*) can be written as

$$\begin{aligned} \dot{x}_1 &= x_2 \\ \dot{x}_2 &= x_3 \\ &\vdots \\ \dot{x}_{n-1} &= x_n \\ \dot{x}_n &= -a_n x_1 - a_{n-1} x_2 \dots - a_1 x_{n-1} + u \end{aligned} \qquad (9.7)$$

A general representation of set of Eqn. (9.7) can be expressed in the form of a state equation as

$$\dot{\mathbf{x}} = \mathbf{A}\mathbf{x} + \mathbf{B}u$$

where

$$\mathbf{x} = \begin{bmatrix} x_1 \\ x_2 \\ \vdots \\ \vdots \\ \vdots \\ \vdots \\ x_n \end{bmatrix}; \mathbf{A} = \begin{bmatrix} 0 & 1 & \dots & 0 & 0 \\ 0 & 0 & \dots & 1 & 0 \\ 0 & & & & 0 \\ 0 & 0 & \dots & 0 & 1 \\ \vdots & \vdots & \dots & \vdots & \vdots \\ \vdots & \vdots & \dots & \vdots & \vdots \\ -a_n & -a_{n-1} & \dots & 0 & a_1 \end{bmatrix}; \mathbf{B} = \begin{bmatrix} 0 \\ 0 \\ 0 \\ \vdots \\ \vdots \\ \vdots \\ 1 \end{bmatrix} \qquad \text{...(9.8)}$$

where $\mathbf{x}$ = state vector ($n \times 1$) u = control (scalar)

$\mathbf{A}$ = matrix ($n \times n$) $\mathbf{B}$ = matrix ($n \times 1$)

The output equation is given by

$$y = [1 \ \ 0 \ \ 0 \ \ .. \ \ 0]\begin{bmatrix} x_1 \\ x_2 \\ : \\ : \\ x_n \end{bmatrix} \quad ...(9.9)$$

or $$y = \mathbf{Cx} \quad ....(9.10)$$

where $\mathbf{C} = [1\ 0\ 0\ ...\ 0]$, y = output (scalar), $\mathbf{C}$ = matrix ($1 \times n$)

The output equation is algebraic and state equation is a set of first order differential equations.

(2) **State space representation for electrical network.** An electrical network is shown in Fig. 9.3.1. If the initial values of the current $i(0)$, the capacitor voltages $v_1(0)$ and $v_2(0)$ are known, then the dynamic behaviour of electrical network can be completely defined for $t > 0$.

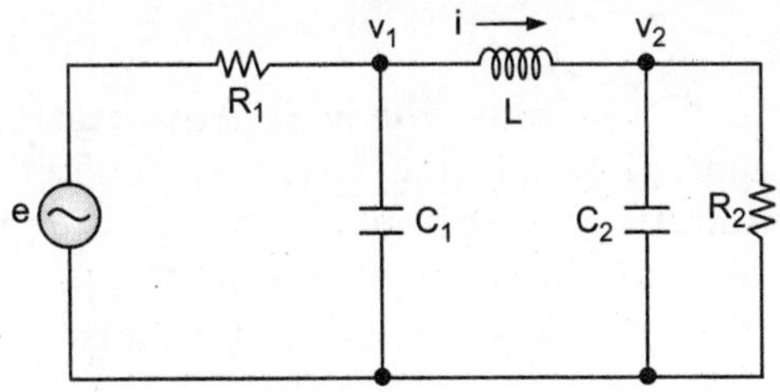

Fig. 9.3.1. Electrical network.

Thus the state of the network for $t > 0$ is completely determined by selecting $v_1(t)$, $v_2(t)$ and $i(t)$ as state variables.

The equations describing dynamic behaviour of the system are

$$\frac{v_1 - e}{R_1} + C_1 \frac{dv_1}{dt} + i = 0 \quad ...(9.11)$$

$$C_1 \frac{dv_2}{dt} + \frac{v_2}{R_2} - i = 0 \quad ...(9.12)$$

$$L\frac{di}{dt} + v_2 - v_1 = 0 \quad ...(9.13)$$

Rearranging and simplifying, following equations are obtained :

$$\frac{dv_1}{dt} = -\frac{v_1}{R_1C_1} - \frac{i}{C_1} + \frac{e}{R_1C_1} \quad ...(9.14)$$

$$\frac{dv_2}{dt} = -\frac{v_2}{R_2C_2} + \frac{i}{C_2} \quad ...(9.15)$$

$$\frac{di}{dt} = \frac{v_1}{L} - \frac{v_2}{L} \quad ...(9.16)$$

As v_1, v_2 and i are selected as state variables, therefore,

$$x_1 = v_1$$

$$x_2 = v_2$$

and $$x_3 = i$$

In view of above set of equations state space representation of the system is obtained below :

$$\dot{x}_1 = -\frac{1}{R_1C_1}x_1 - \frac{1}{C_1}x_3 + \frac{1}{R_1C_1}e$$

$$\dot{x}_2 = -\frac{1}{R_2C_2}x_2 + \frac{1}{C_2}x_3$$

$$\dot{x}_3 = \frac{1}{L}x_1 - \frac{1}{L}x_2$$

In vector-matrix form state space representation is given below :

$$\begin{bmatrix}\dot{x}_1\\ \dot{x}_2\\ \dot{x}_3\end{bmatrix} = \begin{bmatrix}-\frac{1}{R_1C_1} & 0 & -\frac{1}{C_1}\\ 0 & -\frac{1}{R_2C_2} & \frac{1}{C_2}\\ \frac{1}{L} & -\frac{1}{L} & 0\end{bmatrix}\begin{bmatrix}x_1\\ x_2\\ x_3\end{bmatrix} + \begin{bmatrix}\frac{1}{R_1C_1}\\ 0\\ 0\end{bmatrix}e \qquad ...(9.17)$$

(3) **State space representation for transfer function.** The transfer function of a control system can be represented in the form of a state block diagram wherein the variable appearing at the output of each integrator is considered as state variable and state space representation obtained therefrom.

Consider the following transfer function

$$\frac{C(s)}{R(s)} = \frac{1}{s^2 + a_1 s + a_2} \qquad ...(9.18)$$

The corresponding differential equation for the above transfer function is

$$\frac{d^2c}{dt^2} + a_1\frac{dc}{dt} + a_2c = r$$

To draw the state block diagram the equation above is rearranged below :

$$\frac{d^2c}{dt^2} = -a_2c - a_1\frac{dc}{dt} + r$$

The state block diagram is shown in Fig. 9.3.2.

The output of last integrator gives the output c. The output of each integrator is assigned a state variable in ascending order from right to left as shown in Fig. 9.3.2.

As per the block diagram, the state equations for the system are :

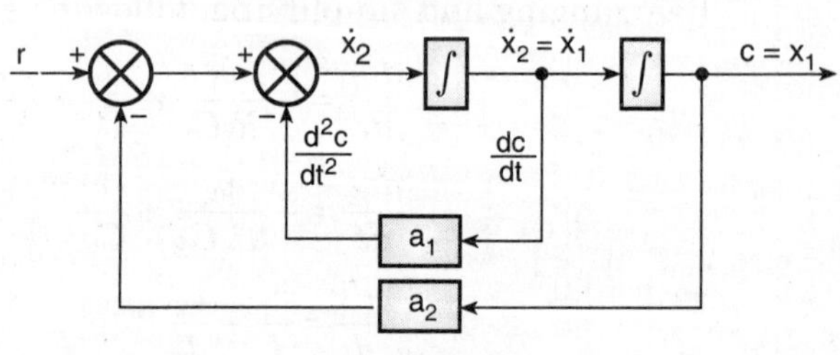

Fig. 9.3.2. State block diagram for $d^2c/dt^2 + a_1(dc/dt) + a_2c = r$.

$$\dot{x}_1 = x_2$$

$$\dot{x}_2 = -a_2x_1 - a_1x_2 + r$$

The output is denoted by y. Therefore,

$$y = x_1$$

As per above equations, the state model in vector-matrix form is given below :

$$\begin{bmatrix} \dot{x}_1 \\ \dot{x}_2 \end{bmatrix} = \begin{bmatrix} 0 & 1 \\ -a_2 & -a_1 \end{bmatrix} \begin{bmatrix} x_1 \\ x_2 \end{bmatrix} + \begin{bmatrix} 0 \\ 1 \end{bmatrix} r \qquad \text{...(1)}$$

$$y = [1 \quad 0] \begin{bmatrix} x_1 \\ x_2 \end{bmatrix}$$

On reversing the order of state variables *i.e.* assigning the integrator output by a state variable in descending order from right to left. Denoting $\bar{x}_2$ in place of x_1 and $\bar{x}_1$ for x_2, the state equations are

$$\bar{x}_1 = -a_1 \bar{x}_1 - a_2 \bar{x}_2 + r$$

$$\bar{x}_2 = \bar{x}_1$$

and $$y = \bar{x}_2$$

The state model in vector-matrix form is given below :

$$\begin{bmatrix} \dot{\bar{x}}_1 \\ \dot{\bar{x}}_2 \end{bmatrix} = \begin{bmatrix} -a_1 & -a_2 \\ 1 & 0 \end{bmatrix} \begin{bmatrix} \bar{x}_1 \\ \bar{x}_2 \end{bmatrix} + \begin{bmatrix} 1 \\ 0 \end{bmatrix} r \qquad \text{...(2)}$$

$$y = [0 \quad 1] \begin{bmatrix} \bar{x}_1 \\ \bar{x}_2 \end{bmatrix}$$

The state model (2) can be obtained from (1) by applying similarity transformation

$$\mathbf{x} = \mathbf{P}\bar{\mathbf{x}}$$

where $\mathbf{P} = \begin{bmatrix} 0 & 1 \\ 1 & 0 \end{bmatrix}; \mathbf{P}^{-1} \begin{bmatrix} 0 & 1 \\ 1 & 0 \end{bmatrix}$

$$\mathbf{x} = \mathbf{A}\mathbf{x} + \mathbf{B}\,u$$

substitute $$\mathbf{x} = \mathbf{P}\bar{\mathbf{x}}$$

$$\therefore \quad \mathbf{P}\dot{\bar{\mathbf{x}}} = \mathbf{A}\,\mathbf{P}\,\bar{\mathbf{x}} + \mathbf{B}\,u$$

on pre multiplying by $\mathbf{P}^{-1}$

$$\dot{\bar{\mathbf{x}}} = \mathbf{P}^{-1}\,\mathbf{A}\,\mathbf{P}\,\bar{\mathbf{x}} + \mathbf{P}^{-1}\,\mathbf{B}\,u$$

$$\mathbf{A} = \begin{bmatrix} 0 & 1 \\ -a_2 & -a_1 \end{bmatrix} \quad \text{and} \quad \mathbf{B} = \begin{bmatrix} 0 \\ 1 \end{bmatrix}$$

$$\therefore \quad \begin{bmatrix} \dot{\bar{x}}_1 \\ \dot{\bar{x}}_2 \end{bmatrix} = \begin{bmatrix} 0 & 1 \\ 1 & 0 \end{bmatrix} \begin{bmatrix} 0 & 1 \\ -a_2 & -a_1 \end{bmatrix} \begin{bmatrix} 0 & 1 \\ 1 & 0 \end{bmatrix} \begin{bmatrix} \bar{x}_1 \\ \bar{x}_2 \end{bmatrix} + \begin{bmatrix} 0 & 1 \\ 1 & 0 \end{bmatrix} \begin{bmatrix} 0 \\ 1 \end{bmatrix} r$$

or $$\begin{bmatrix} \dot{\bar{x}}_1 \\ \dot{\bar{x}}_2 \end{bmatrix} = \begin{bmatrix} -a_1 & -a_2 \\ 1 & 0 \end{bmatrix} \begin{bmatrix} \bar{x}_1 \\ \bar{x}_2 \end{bmatrix} + \begin{bmatrix} 1 \\ 0 \end{bmatrix} r$$

$$y = [1 \quad 0]\,\mathbf{P} \begin{bmatrix} \bar{x}_1 \\ \bar{x}_2 \end{bmatrix}$$

$$= [1 \quad 0]\begin{bmatrix} 0 & 1 \\ 1 & 0 \end{bmatrix}\begin{bmatrix} \overline{x}_1 \\ \overline{x}_2 \end{bmatrix}$$

$$\therefore \qquad y = [0 \quad 1]\begin{bmatrix} \overline{x}_1 \\ \overline{x}_2 \end{bmatrix}$$

The differential equation :

$$\frac{d^2c}{dt^2} + a_1 \frac{dc}{dt} + a_2 c = r \qquad \text{...(9.19)}$$

can be rearranged as

$$\frac{d^2c}{dt^2} + a_1 \frac{dc}{dt} = - a_2 c + r$$

The state block diagram corresponding to above equation is drawn in Fig. 9.3.3.

The output of each integrator is assigned a state variable as shown in Fig. 9.3.3. As per the state block diagram the state equations are :

$$\dot{x}_1 = - a_1 x_1 + x_2$$

$$\dot{x}_2 = - a_2 x_1 + r$$

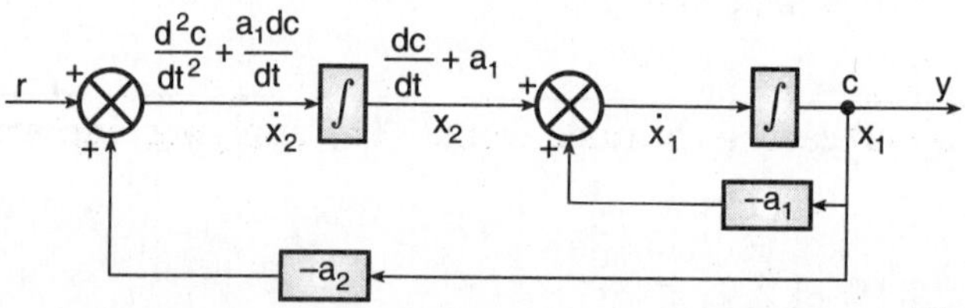

Fig. 9.3.3. State block diagram for $d^2c/dt^2 + a_1(dc/dt) = - a_2c + r$.

The output is denoted as y. Therefore,

$$y = x_1$$

As per above equations, the state model in vector-matrix form is given below :

$$\begin{bmatrix} \dot{x}_1 \\ \dot{x}_2 \end{bmatrix} = \begin{bmatrix} -a_1 & 1 \\ -a_2 & 0 \end{bmatrix}\begin{bmatrix} x_1 \\ x_2 \end{bmatrix} + \begin{bmatrix} 0 \\ 1 \end{bmatrix} r$$

$$y = [1 \quad 0]\begin{bmatrix} x_1 \\ x_2 \end{bmatrix}$$

On reversing the order of state variables in the state block diagram Fig. 9.3.3, the corresponding state model is given below :

$$\begin{bmatrix} \dot{\overline{x}}_1 \\ \dot{\overline{x}}_2 \end{bmatrix} = \begin{bmatrix} 0 & -a_2 \\ 1 & -a_1 \end{bmatrix}\begin{bmatrix} \overline{x}_1 \\ \overline{x}_2 \end{bmatrix} + \begin{bmatrix} 1 \\ 0 \end{bmatrix} r$$

$$y = [0 \quad 1]\begin{bmatrix} \overline{x}_1 \\ \overline{x}_2 \end{bmatrix}$$

The characteristic polynomial for the transfer function

$$\frac{C(s)}{R(s)} = \frac{1}{s^2 + a_1 s + a_2}$$

is $F(s) = s^2 + a_1 s + a_2$ which is of second order.

In general, the nth order polynomial is below :

$$F(s) = s^n + a_1 s^{n-1} + a_2 s^{n-2} + \ldots + a_n$$

and $F(s) = s^4 + a_1 s^3 + a_2 s^2 + a_1$ which has the order $n = 4$ (say).

On obtaining a state model, the system matrix in four different forms can be found to be :

(1) $$\mathbf{A} = \begin{bmatrix} 0 & 1 & 0 & 0 \\ 0 & 0 & 1 & 0 \\ 0 & 0 & 0 & 1 \\ -a_4 & -a_3 & -a_2 & -a_1 \end{bmatrix}$$

The system matrix **A** given above contains upper elements of the diagonal as unity and the last row consists of negative coefficients of the characteristic polynomial. All other elements are zero.

(2) $$\mathbf{A} = \begin{bmatrix} -a_1 & -a_2 & -a_3 & -a_4 \\ 1 & 0 & 0 & 0 \\ 0 & 1 & 0 & 0 \\ 0 & 0 & 1 & 0 \end{bmatrix}$$

The system matrix **A** given above contains lower elements of the diagonal as unity and the first row consists of negative coefficients of the characteristic polynomial. All other elements are zero.

For the case (1) and (2), the system matrix **A** is known to be in Controllable Canonical form. Also known as companion form 1.

(3) $$\mathbf{A} = \begin{bmatrix} -a_1 & 1 & 0 & 0 \\ -a_2 & 0 & 1 & 0 \\ -a_3 & 0 & 0 & 1 \\ -a_4 & 0 & 0 & 0 \end{bmatrix}$$

The system matrix **A** given above contains upper elements of the diagonal as unity and the first column contains negative coefficients of the characteristic polynomial. All other elements are zero.

(4) $$\mathbf{A} = \begin{bmatrix} 0 & 0 & 0 & -a_4 \\ 1 & 0 & 0 & -a_3 \\ 0 & 1 & 0 & -a_2 \\ 0 & 0 & 1 & -a_1 \end{bmatrix}$$

The system matrix **A** given above contains lower elements of the diagonal as unity and the last column consists of negative coefficients of the characteristic polynomial. All other elements are zero.

For the case (3) and (4), the system matrix **A** is known to be in Observable Canonical form. Also known as companion form 2.

On observation of state models mentioned above, it is concluded that these models are not same, though have been derived from the same transfer function. Thus, it is concluded that for a system the state model is not unique.

Application of Signal Flow graph for the determination of state model from a given transfer function.

The overall transfer function $C(s)/R(s)$ can be obtained using signal flow graph technique. In this respect the Mason's gain formula is given below :

$$\frac{C(s)}{R(s)} = \frac{\sum_{k=1}^{k} P_k \Delta_k}{\Delta} \qquad ...(9.20)$$

The term refers to output $C(s)$ and $R(s)$ input. The notations P_k (forward path gain), Δ_k (path factor) and Δ (graph determinant) have already been explained in chapter 3.

The signal flow graph in relation to state variable analysis is also termed as state diagram. The signal flow graph for a given transfer function can be drawn in several ways. However, for convenience the case of all loops touching each other as well as to forward path is considered. This reduces the Mason's gain formula as below :

$$\frac{C(s)}{R(s)} = \frac{\text{Sum of forward path of gains}}{1 - \text{Sum of individual loop gains}}$$

For the transfer function

$$\frac{C(s)}{R(s)} = \frac{1}{s^2 + a_1 s + a_2}$$

The signal flow graph is developed below :

On multiplying the numerator and denominator by s^{-2}, above transfer function is rewritten as

$$\frac{C(s)}{R(s)} = \frac{s^{-2}}{1 - (-a_1 s^{-1} - a_2 s^{-2})}$$

The transfer function takes the form of Mason's gain formula wherein,

$$P = s^{-2} \text{ (forward path gain)}$$
$$L_1 = -a_1 s^{-1} \text{ (loop gain).}$$
$$L_2 = -a_2 s^{-2} \text{ (loop gain).}$$

With the information as above, the signal flow graph is drawn in Fig. 9.3.4.

In the state signal flow graph, the output of each integrator (s^{-1}) is denoted by a state variable in ascending order from right to left. The output is denoted as y and input r.

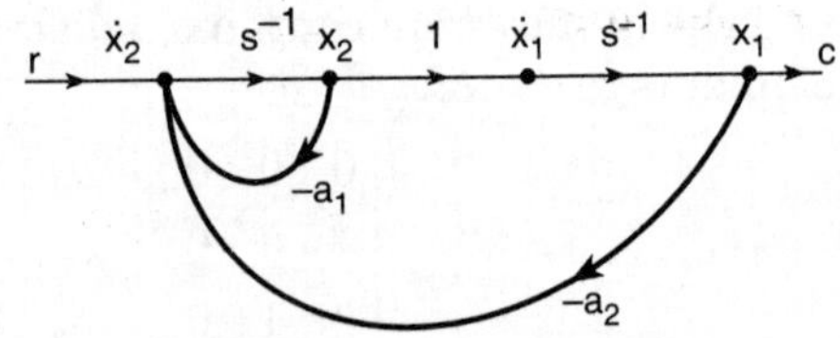

Fig. 9.3.4. State signal flow graph (controllable canonical form).

The state equations are

$$\dot{x}_1 = x_2$$
$$\dot{x}_2 = -a_2 x_1 - a_1 x_2 + r$$

and the output equation

$$y = x_1$$

In vector-matrix form the state model is

$$\begin{bmatrix} \dot{x}_1 \\ \dot{x}_2 \end{bmatrix} = \begin{bmatrix} 0 & 1 \\ -a_2 & -a_1 \end{bmatrix} \begin{bmatrix} x_1 \\ x_2 \end{bmatrix} + \begin{bmatrix} 0 \\ 1 \end{bmatrix} r$$

$$y = [1 \quad 0]\begin{bmatrix} x_1 \\ x_2 \end{bmatrix}$$

On reversing the order of state variables i.e. replacing x_1 and $\bar{x}_2$ and x_2 by $\bar{x}_1$, the state model takes the form given below :

$$\begin{bmatrix} \dot{\bar{x}}_1 \\ \dot{\bar{x}}_2 \end{bmatrix} = \begin{bmatrix} -a_1 & -a_2 \\ 1 & 0 \end{bmatrix}\begin{bmatrix} \bar{x}_1 \\ \bar{x}_2 \end{bmatrix} + \begin{bmatrix} 1 \\ 0 \end{bmatrix} r$$

$$y = [0 \quad 1]\begin{bmatrix} \bar{x}_1 \\ \bar{x}_2 \end{bmatrix}$$

For the transfer function

$$\frac{C(s)}{R(s)} = \frac{s^{-2}}{1-(-a_1 s^{-1} - a_2^{-2})}$$

the state signal flow graph is drawn in a different way as shown in Fig. 9.3.5.

The output of each integrator is assigned a state variable as shown in Fig. 9.3.5. The output and input are y and r respectively. The state equation are

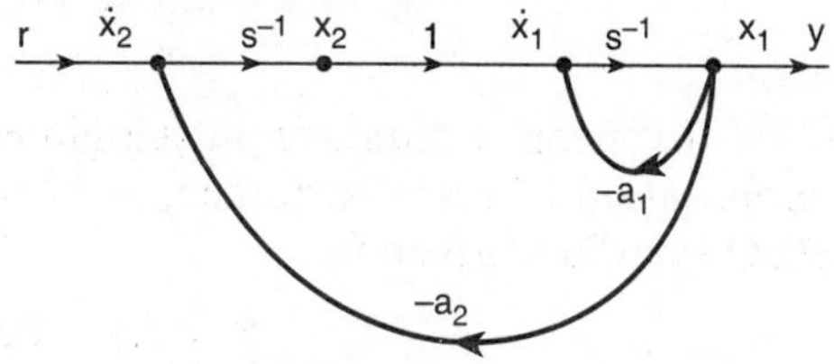

Fig. 9.3.5. State signal flow graph (observable canonical form).

$$\dot{x}_1 = -a_1 x_1 + x_2$$

$$\dot{x}_2 = -a_2 x_2 + r$$

and the output equation

$$y = x_1$$

In vector-matrix form the state model is given below :

$$\begin{bmatrix} \dot{x}_1 \\ \dot{x}_2 \end{bmatrix} = \begin{bmatrix} -a_1 & 1 \\ -a_2 & 0 \end{bmatrix}\begin{bmatrix} x_1 \\ x_2 \end{bmatrix} + \begin{bmatrix} 0 \\ 1 \end{bmatrix} r$$

$$y = [1 \quad 0]\begin{bmatrix} x_1 \\ x_2 \end{bmatrix}$$

On reversing the order of state variables in the state signal flow graph i.e. replacing x_1 by $\bar{x}_2$ and x_2 by $\bar{x}_1$, the state model takes the form :

$$\begin{bmatrix} \dot{\bar{x}}_1 \\ \dot{\bar{x}}_2 \end{bmatrix} = \begin{bmatrix} 0 & -a_2 \\ 1 & -a_1 \end{bmatrix}\begin{bmatrix} \bar{x}_1 \\ \bar{x}_2 \end{bmatrix} + \begin{bmatrix} 1 \\ 0 \end{bmatrix} r$$

$$y = [0 \quad 1]\begin{bmatrix} \bar{x}_1 \\ \bar{x}_2 \end{bmatrix}$$

9.4 BLOCK DIAGRAM FOR STATE EQUATION

The state space representation of multiple input multiple output (MIMO) system is given by the following equations :

$$\dot{\mathbf{x}} = \mathbf{Ax} + \mathbf{Bu} \qquad ...(9.21)$$

$$\mathbf{y} = \mathbf{Cx} + \mathbf{Du} \qquad ...(9.22)$$

where $\mathbf{x}$ = state vector ($n \times 1$), $\mathbf{u}$ = control vector ($m \times 1$)
$\mathbf{y}$ = output vector ($p \times 1$), $\mathbf{A}$ = system matrix ($n \times n$)
$\mathbf{B}$ = Input matrix ($n \times m$), $\mathbf{C}$ = output matrix ($p \times n$)
$\mathbf{D}$ = Input output coupling matrix ($n \times m$)

The block diagram for the system based on Eqs. (9.21) and (9.22) is shown in Fig. 9.4.1, broad arrows represent vector quantities.

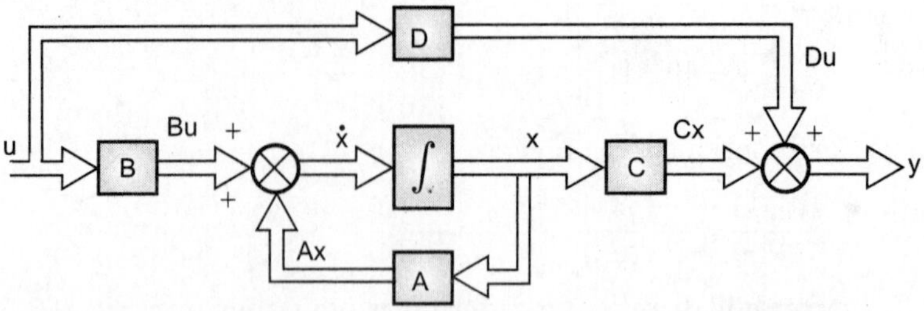

Fig. 9.4.1. Block diagram for state Eqs. (9.21) and (9.22) representing multiple-input multiple-output system.

In most of single-input-single-output (SISO) systems, the output consists of linear combination of state variables, as such $\mathbf{D}$ is a null matrix. Therefore, the state model of a SISO system is given by

$$\dot{\mathbf{x}} = \mathbf{A}\mathbf{x} + \mathbf{B}u \quad \text{...(9.23)}$$

$$y = \mathbf{C}\mathbf{x} \quad \text{...(9.24)}$$

where $\mathbf{x}$ = state vector ($n \times 1$), u = control (scalar)
y = output (scalar), $\mathbf{A}$ = matrix ($n \times n$)
$\mathbf{B}$ = matrix ($n \times 1$), $\mathbf{C}$ = matrix ($1 \times n$)

The block diagram for the SISO system based on Eqs. (9.23) and (9.24) is shown in Fig. 9.4.2.

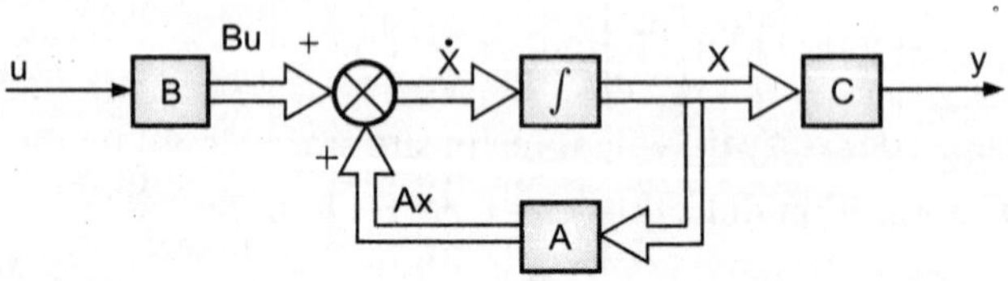

Fig. 9.4.2. Block representation diagram for state Eqs. (9.23) and (9.24) representing SISO system.

9.5 TRANSFER FUNCTION DECOMPOSITION

The process of obtaining state model from the transfer function is called transfer function decomposition.

Let the given transfer function be

$$\frac{Y(s)}{U(s)} = \frac{1}{s^2 + 2s + 5} \quad \text{...(9.25)}$$

On transforming the transfer function given above in time domain, the following differential equation is obtained :

$$\frac{d^2y}{dt^2} + 2\frac{dy}{dt} + 5y = u \qquad ...(9.26)$$

To obtain state model for the system governed by the above differential equation, the block diagram indicating integration process is shown in Fig. 9.5.1.

The output of each integrator is assigned a state variable as shown in Fig. 9.5.1 and the state equations are written below :

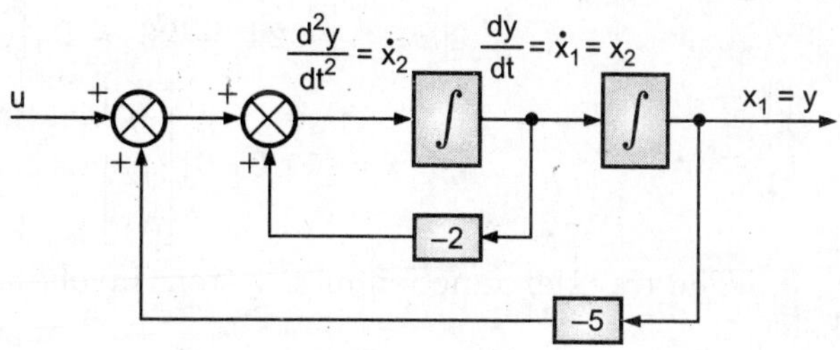

Fig. 9.5.1. Block diagram for Eq. (9.26).

$$\dot{x}_1 = x_2$$

$$\dot{x}_2 = -5x_1 - 2x_2 + u$$

and the output is $y = x_2$

The state model is :

$$\begin{bmatrix}\dot{x}_1\\ \dot{x}_2\end{bmatrix} = \begin{bmatrix}0 & 1\\ -5 & -2\end{bmatrix}\begin{bmatrix}x_1\\ x_2\end{bmatrix} + \begin{bmatrix}0\\ 1\end{bmatrix}u \qquad ...(9.27)$$

$$y = [1 \quad 0]\begin{bmatrix}x_1\\ x_2\end{bmatrix}$$

The transfer function decomposition based on above example is carried out using following methods.

9.5.1 Direct decomposition

The direct decomposition of a transfer function is performed for the system, when the transfer function is given in the following form :

$$\frac{Y(s)}{U(s)} = \frac{1}{s^3 + 9s^2 + 26s + 24} \qquad ...(9.28)$$

The corresponding differential equation for the above transfer function is of third order, therefore, three integrators and three summing junctions are used to draw the block diagram as shown in Fig. 9.5.2.

The output of each integrator is assigned a state variable and the state model obtained therefrom is given below :

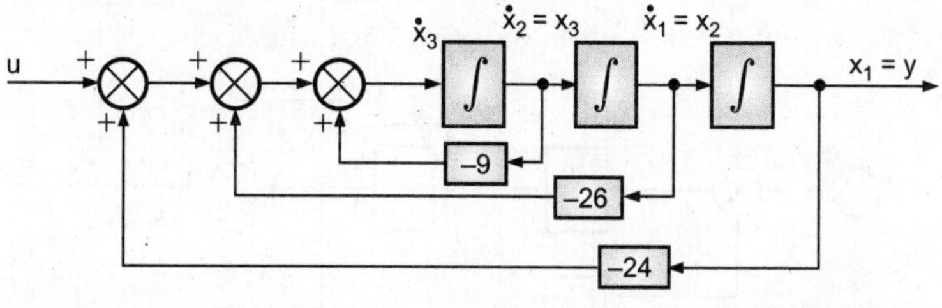

Fig. 9.5.2. Block diagram representation for transfer function (9.28).

$$\dot{x}_1 = x_2$$
$$\dot{x}_2 = x_3$$
$$\dot{x}_3 = -24x_1 - 26x_2 - 9x_3 + u$$

and the output is $y = x_1$

or
$$\begin{bmatrix} \dot{x}_1 \\ \dot{x}_2 \\ \dot{x}_3 \end{bmatrix} = \begin{bmatrix} 0 & 1 & 0 \\ 0 & 0 & 1 \\ -24 & -26 & -9 \end{bmatrix} \begin{bmatrix} x_1 \\ x_2 \\ x_3 \end{bmatrix} + \begin{bmatrix} 0 \\ 0 \\ 1 \end{bmatrix} u \qquad \text{...(9.29)}$$

$$y = [1 \ \ 0 \ \ 0] \begin{bmatrix} x_1 \\ x_2 \\ x_3 \end{bmatrix}$$

The transfer function of a system involving a differential term in its input is given below :

$$\frac{Y(s)}{U(s)} = \frac{2s+1}{s^2+2s+3} \qquad \text{...(9.30)}$$

The transfer function (9.30) is split into two parts as follows :

$$\frac{Y(s)}{U(s)} = \frac{1}{s^2+2s+3} + \frac{2s}{s^2+2s+3} \qquad \text{...(9.31)}$$

The block diagram for the term

$$\frac{Y(s)}{U(s)} = \frac{1}{s^2+2s+3} \qquad \text{...(9.31a)}$$

is drawn as shown in Fig. 9.5.3.

The second part of the transfer function (9.31) *i.e.* $2s/(s^2 + 2s + 3)$ is the twice multiple of the first differential of the first part, which is available as the input to the last integrator as shown in Fig. 9.5.3. Therefore, the block diagram in relation to transfer function (9.30) is drawn as shown in Fig. 9.5.4.

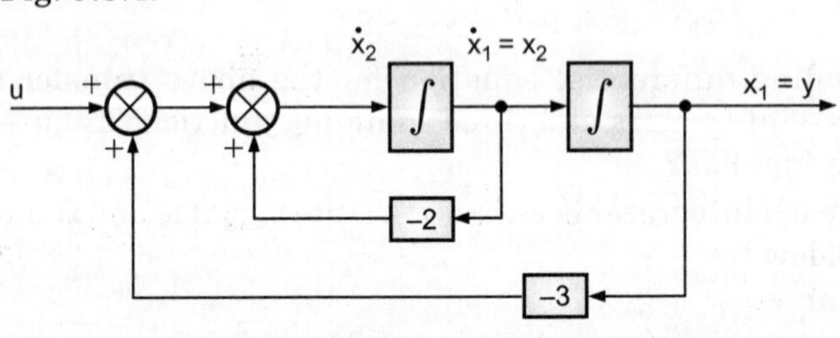

Fig. 9.5.3. Block diagram representation for transfer function (9.31*a*).

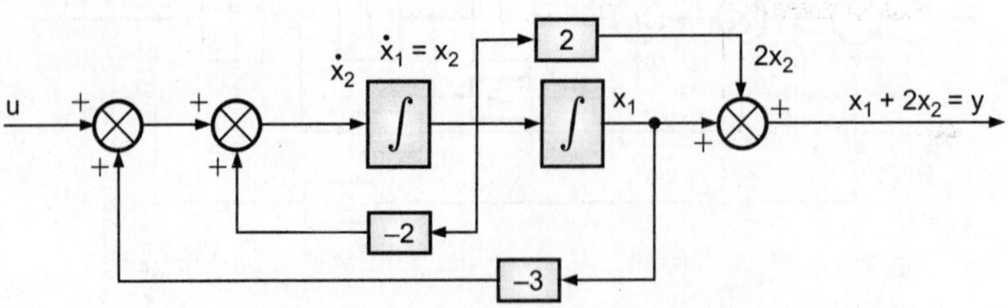

Fig. 9.5.4. Block diagram representation for transfer function (9.30).

The output of each integrator is assigned a state variable as shown in Figs. 9.5.3 and 9.5.4 and the state equations determined therefrom are given below :

$$\dot{x}_1 = x_2$$

$$\dot{x}_2 = -3x_1 - 2x_2 + u$$

The output equation is : $y = x_1 + 2x_2$

The state model : $$\begin{bmatrix} \dot{x}_1 \\ \dot{x}_2 \end{bmatrix} = \begin{bmatrix} 0 & 1 \\ -3 & -2 \end{bmatrix} \begin{bmatrix} x_1 \\ x_2 \end{bmatrix} + \begin{bmatrix} 0 \\ 1 \end{bmatrix} u \qquad ...(9.32)$$

$$y = [1 \quad 2] \begin{bmatrix} x_1 \\ x_2 \end{bmatrix}$$

9.5.2 Cascade decomposition

In the case of cascade decomposition, the transfer function is given in the following form :

$$\frac{Y(s)}{U(s)} = \frac{1}{(s+2)} \cdot \frac{1}{(s+3)} \cdot \frac{1}{(s+4)} \qquad ...(9.33)$$

For decomposition of the transfer function (9.33), the output of the term $1/(s + 2)$ is given as input to $1/(s + 3)$ and the output of $1/(s + 3)$ is given as input to $1/(s + 4)$. The output of the term $1/(s + 4)$ is the system output. The integration process is illustrated in Fig. 9.5.5.

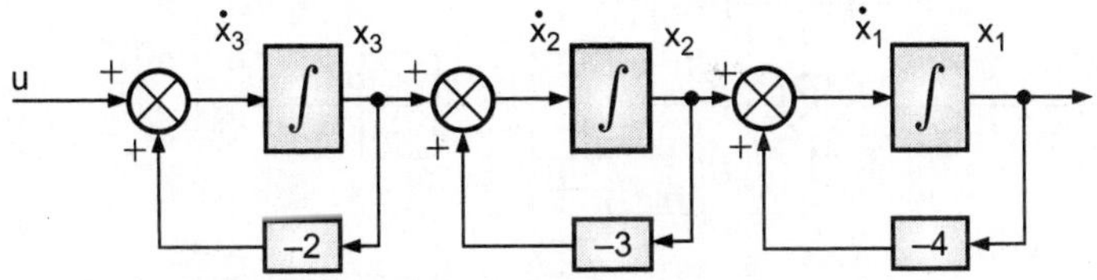

Fig. 9.5.5. Block diagram for the transfer function (9.33).

The output of each integrator is assigned a state variable as shown in Fig. 9.5.5 and the state equations based on Fig. 9.5.5 are given below :

$$\dot{x}_1 = -4x_1 + x_2$$

$$\dot{x}_2 = -3x_2 + x_3$$

$$\dot{x}_3 = -2x_3 + u$$

The output equation is : $y = x_1$

The state model : $$\begin{bmatrix} \dot{x}_1 \\ \dot{x}_2 \\ \dot{x}_3 \end{bmatrix} = \begin{bmatrix} -4 & 1 & 0 \\ 0 & -3 & 1 \\ 0 & 0 & -2 \end{bmatrix} \begin{bmatrix} x_1 \\ x_2 \\ x_3 \end{bmatrix} + \begin{bmatrix} 0 \\ 0 \\ 1 \end{bmatrix} u \qquad ...(9.34)$$

$$y = [1 \quad 0 \quad 0] \begin{bmatrix} x_1 \\ x_2 \\ x_3 \end{bmatrix}$$

The coefficients of matrix **A** of a state equation, obtained as per cascade decomposition of the transfer function give a triangular matrix known as pole-zero form.

9.5.3 Parallel decomposition

The parallel decomposition of a transfer function is carried out by spliting the given transfer function into partial fractions.

Let the transfer function be :

$$\frac{Y(s)}{U(s)} = \frac{1}{(s+2)(s+3)(s+4)} \quad ...(9.35)$$

or
$$\frac{Y(s)}{U(s)} = \frac{1/2}{(s+2)} + \frac{-1}{(s+3)} + \frac{1/2}{(s+4)} \quad ...(9.36)$$

The integration process for each term of the transfer function (9.36) is shown in Fig. 9.5.6.

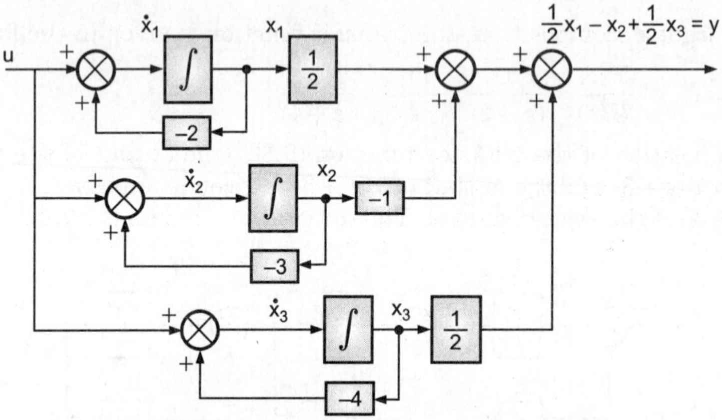

Fig. 9.5.6. Block diagram for transfer function (9.36).

The output of each integrator is assigned a state variable as shown in Fig. 9.5.6 and the state equations are given below :

$$\dot{x}_1 = -2x_1 + u$$
$$\dot{x}_2 = -3x_2 + u$$
$$\dot{x}_3 = -4x_3 + u$$

The output equation is $y = \frac{1}{2}x_1 - x_2 + \frac{1}{2}x_3$

The state model : $$\begin{bmatrix}\dot{x}_1\\ \dot{x}_2\\ \dot{x}_3\end{bmatrix} = \begin{bmatrix}-2 & 0 & 0\\ 0 & -3 & 0\\ 0 & 0 & -4\end{bmatrix}\begin{bmatrix}x_1\\ x_2\\ x_3\end{bmatrix} + \begin{bmatrix}1\\ 1\\ 1\end{bmatrix}u \quad ...(9.37)$$

$$y = \begin{bmatrix}\frac{1}{2} & -1 & \frac{1}{2}\end{bmatrix}\begin{bmatrix}x_1\\ x_2\\ x_3\end{bmatrix}$$

The coefficient matrix **A** obtained as per parallel decomposition of a transfer function is a diagonal one and called simply canonical or diagonal or partial fraction expansion form.

9.5.4 Parallel decomposition of a transfer function having repeated roots of its characteristic equation

The three methods given earlier for obtaining the state model of a transfer function, considered distinct roots of its characteristic equation. For the case of repeated roots the procedure for obtaining the state model of a given transfer function is given below :

Let, the transfer function be

$$\frac{Y(s)}{U(s)} = \frac{2s^2 + 8s + 7}{(s+2)^2 (s+1)} \qquad ...(9.38)$$

On spliting R.H.S. of the transfer function (9.38) into partial fractions, the following form of the transfer function is obtained.

$$\frac{Y(s)}{U(s)} = \frac{1}{(s+1)} + \frac{1}{(s+2)} + \frac{1}{(s+2)^2} \qquad ...(9.39)$$

The block diagram based on integration process similar to that of parallel decomposition case is drawn in Fig. 9.5.7.

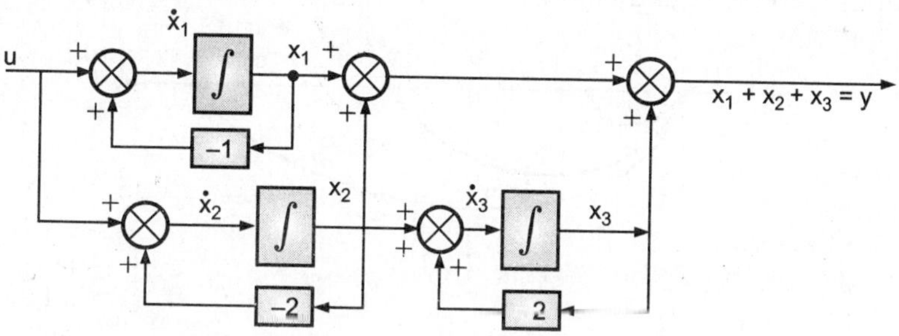

Fig. 9.5.7. Block diagram representation for transfer function (9.39).

The output of each integrator is assigned a state variable as shown in Fig. 9.5.7 and the state equations obtained therefrom are given below :

$$\begin{aligned}\dot{x}_1 &= -x_1 + u\\ \dot{x}_2 &= -2x_2 + u\\ \dot{x}_3 &= +x_2 - 2x_3\end{aligned}$$

The output equation is $y = x_1 + x_2 + x_3$

The state model : $$\begin{bmatrix}\dot{x}_1\\ \dot{x}_2\\ \dot{x}_3\end{bmatrix} = \begin{bmatrix}-1 & 0 & 0\\ 0 & -2 & 0\\ 0 & 1 & -2\end{bmatrix}\begin{bmatrix}x_1\\ x_2\\ x_3\end{bmatrix} + \begin{bmatrix}1\\ 1\\ 0\end{bmatrix} u \qquad ...(9.40)$$

$$y = [1 \;\; 1 \;\; 1]\begin{bmatrix}x_1\\ x_2\\ x_3\end{bmatrix}$$

The matrix **A** of the state equation (9.40) is called the Jordan's canonical form.

It is noted that the transfer functions given by (9.28), (9.33) and (9.36) are same but written in different form. The corresponding state models are given by state equations (9.29), (9.34) and (9.37) respectively.

9.5.5 Transfer function decomposition using State Signal Flow graph

(1) Direct decomposition :

(*a*) The state model for the transfer function

$$\frac{Y(s)}{U(s)} = \frac{1}{s^3 + 9s^2 + 26s + 24}$$

can be obtained using state signal flow graph. On multiplying the numerator and denominator by s^{-3} the transfer function given above is rewritten below :

$$\frac{Y(s)}{U(s)} = \frac{s^{-3}}{1 + 9s^{-1} + 26s^{-2} + 24s^{-3}}$$

$$= \frac{s^{-3}}{1 - (-9s^{-1} - 26s^{-2} - 24s^{-3})}$$

The corresponding signal flow graph is shown in Fig. 9.5.8.

The output of each integrator (s^{-1}) is assigned a state variable as shown in Fig. 9.5.8 and state equations are obtained below :

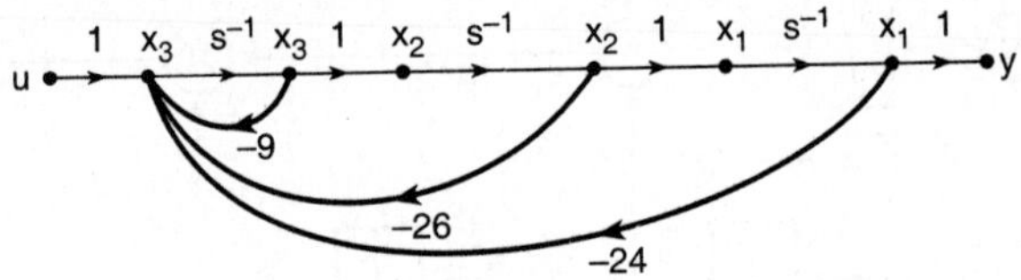

Fig. 9.5.8. State signal flow graph : $Y(s)/U(s) = s^{-3}/1 - (-9s^{-1} - 26s^{-2} - 24s^{-3})$.

$$\dot{x}_1 = x_2$$

$$\dot{x}_2 = x_3$$

$$\dot{x}_3 = -24x_1 - 26x_2 - 9x_3 + u$$

The output equation is $y = x_1$

The equations given above are arranged in vector-matrix form to obtain the state model as below :

$$\begin{bmatrix} \dot{x}_1 \\ \dot{x}_2 \\ \dot{x}_3 \end{bmatrix} = \begin{bmatrix} 0 & 1 & 0 \\ 0 & 0 & 1 \\ -24 & -26 & -9 \end{bmatrix} \begin{bmatrix} x_1 \\ x_2 \\ x_3 \end{bmatrix} + \begin{bmatrix} 0 \\ 0 \\ 1 \end{bmatrix} u$$

$$y = [1 \quad 0 \quad 0] \begin{bmatrix} x_1 \\ x_2 \\ x_3 \end{bmatrix}$$

(*b*) $$\frac{Y(s)}{U(s)} = \frac{2s + 1}{s^2 + 2s + 3}$$

On multiplying the numerator and denominator by s^{-2} the transfer function given above is rewritten below :

$$\frac{Y(s)}{U(s)} = \frac{2s^{-1} + s^{-2}}{1 + 2s^{-1} + 3s^{-2}} = \frac{2s^{-1} + s^{-2}}{1 - (-2s^{-1} - 3s^{-2})}$$

The state signal flow graph for the above transfer function is drawn as shown in Fig. 9.5.9.

The output of each integrator (s^{-1}) is assigned a state variable as shown in Fig. 9.5.9. The state equations are obtained below :

$$\dot{x}_1 = x_2$$

$$\dot{x}_2 = -x_1 - 2x_2 + u$$

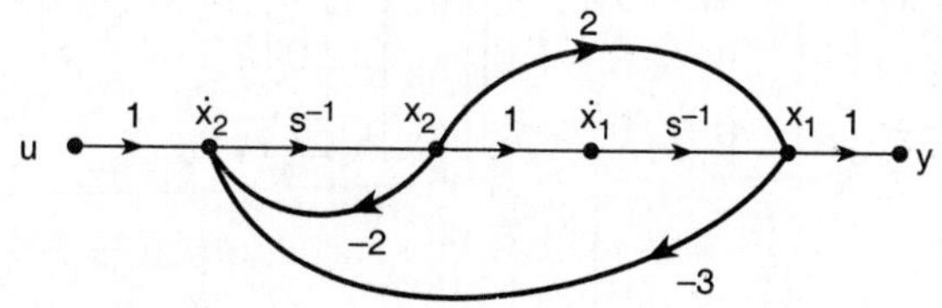

Fig. 9.5.9. State signal flow graph : $Y(s)/U(s) = (2s^{-1} + s^{-2})/[1 - (-2s^{-1} - 3s^{-2})]$.

The output equation is

$$y = x_1 + 2x_2$$

The equations given above are arranged in vector-matrix form to develop state model as below :

$$\begin{bmatrix} \dot{x}_1 \\ \dot{x}_2 \end{bmatrix} = \begin{bmatrix} 0 & 1 \\ -3 & -2 \end{bmatrix} \begin{bmatrix} x_1 \\ x_2 \end{bmatrix} + \begin{bmatrix} 0 \\ 1 \end{bmatrix} u$$

$$y = [1 \quad 2] \begin{bmatrix} x_1 \\ x_2 \end{bmatrix}$$

(2) Cascade Decomposition

The state model for the transfer function

$$\frac{Y(s)}{U(s)} = \frac{1}{(s+2)(s+3)(s+4)}$$

can be obtained in cascade form using state signal flow graph as below :

The transfer function given above is rearranged as

$$\frac{Y(s)}{U(s)} = \frac{s^{-1}}{1 - (-2s^{-1})} \cdot \frac{s^{-1}}{1 - (-3s^{-1})} \cdot \frac{s^{-1}}{1 - (-4s^{-1})}$$

The corresponding state signal flow graph is shown in Fig. 9.5.10.

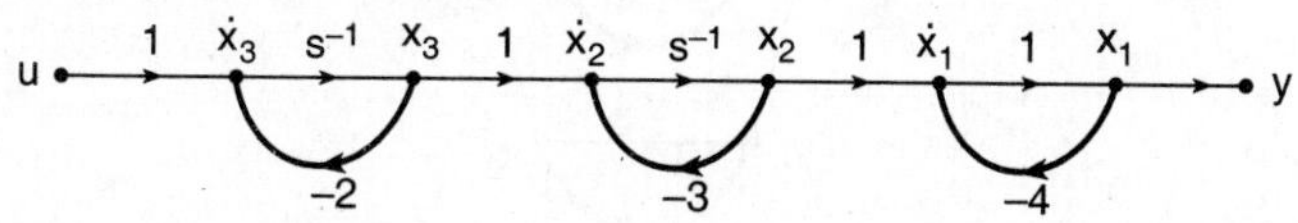

Fig. 9.5.10. State signal flow graph :
$Y(s)/U(s) = s^{-1}/[1 - (-2s^{-1})] \cdot s^{-1}/[1 - (-3s^{-1})] \cdot s^{-1}/[1 - (-4s^{-1})]$.

Assigning a state variable to each integrator (s^{-1}) in Fig. 9.5.10, the state equations are

$$\dot{x}_1 = -4x_1 + x_2$$
$$\dot{x}_2 = -3x_2 + x_3$$
$$\dot{x}_3 = -2x_3 \quad + u$$

The output equation is $y = x_1$

The equations given above are arranged in vector-matrix form to obtain the state model as below :

$$\begin{bmatrix} \dot{x}_1 \\ \dot{x}_2 \\ \dot{x}_3 \end{bmatrix} = \begin{bmatrix} -4 & 1 & 0 \\ 0 & -3 & 1 \\ 0 & 0 & -2 \end{bmatrix} \begin{bmatrix} x_1 \\ x_2 \\ x_3 \end{bmatrix} + \begin{bmatrix} 0 \\ 0 \\ 1 \end{bmatrix} u$$

$$y = [1 \ \ 0 \ \ 0] \begin{bmatrix} x_1 \\ x_2 \\ x_3 \end{bmatrix}$$

(3) Parallel Decomposition

The transfer function $\dfrac{Y(s)}{U(s)} = \dfrac{1}{(s+2)(s+3)(s+4)}$

is factored as

$$\frac{Y(s)}{U(s)} = \frac{1/2}{s+2} + \frac{-1}{s+3} + \frac{1/2}{s+4}$$

The transfer function given above is rewritten as below :

$$\frac{Y(s)}{U(s)} = \frac{\frac{1}{2}s^{-1}}{1-(-2s^{-1})} + \frac{-1s^{-1}}{1-(-3s^{-1})} + \frac{\frac{1}{2}s^{-1}}{1-(-4s^{-1})}$$

The signal flow graph for the transfer function is drawn in Fig. 9.5.11 where in each integrator (s^{-1}) is assigned a state variable.

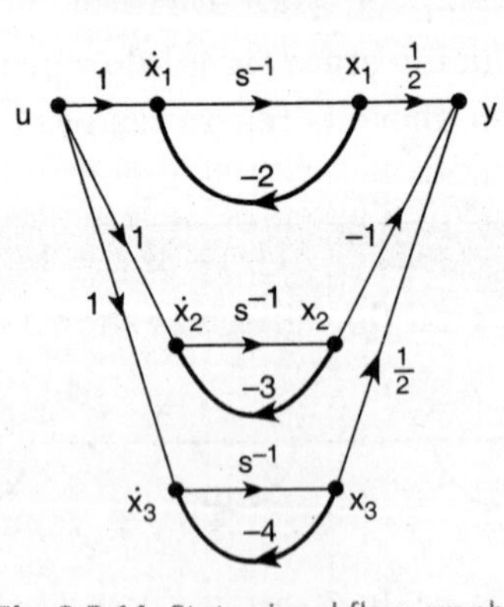

Fig. 9.5.11. State signal flow graph :
$Y(s)/U(s) = (1/2)s^{-1}/[1-(-2s^{-1})] + (-1s^{-1})/[1-(-3s^{-1})] + (1/2)s^{-1}/[1-(-4s^{-1})]$.

As per state signal flow graph, the state equations are

$$\dot{x}_1 = -2x_1 + u$$
$$\dot{x}_2 = -3x_2 + u$$
$$\dot{x}_3 = -4x_3 + u$$

The output equation is $y = \frac{1}{2}x_1 - x_2 + \frac{1}{2}x_3$

The equations given above are arranged in vector-matrix form to obtain the state model in diagonal form as,

$$\begin{bmatrix} \dot{x}_1 \\ \dot{x}_2 \\ \dot{x}_3 \end{bmatrix} = \begin{bmatrix} -2 & 0 & 0 \\ 0 & -3 & 0 \\ 0 & 0 & -4 \end{bmatrix} \begin{bmatrix} x_1 \\ x_2 \\ x_3 \end{bmatrix} + \begin{bmatrix} 1 \\ 1 \\ 1 \end{bmatrix} u$$

$$y = \begin{bmatrix} \frac{1}{2} & -1 & \frac{1}{2} \end{bmatrix} \begin{bmatrix} x_1 \\ x_2 \\ x_3 \end{bmatrix}$$

(4) Parallel Decomposition : Repeated roots of characteristic polynomial

The transfer function

$$\frac{Y(s)}{U(s)} = \frac{2s^2 + 8s + 7}{(s+2)(s+2)^2}$$

is factored in the form given below :

$$\frac{Y(s)}{U(s)} = \frac{1}{(s+1)} + \frac{1}{(s+2)} + \frac{1}{(s+2)^2}$$

Above transfer function is rearranged as

$$\frac{Y(s)}{U(s)} = \frac{s^{-1}}{1-(-s^{-1})} + \frac{s^{-1}}{1 \;\; (\;\; 2s^{-1})} + \frac{s^{-1} \,.\, s^{-1}}{[1-(-2s^{-1})]^2}$$

The state signal flow graph for the above transfer function is drawn as shown in Fig. 9.5.12 wherein each integrator (s^{-1}) is assigned a state variable as indicated.

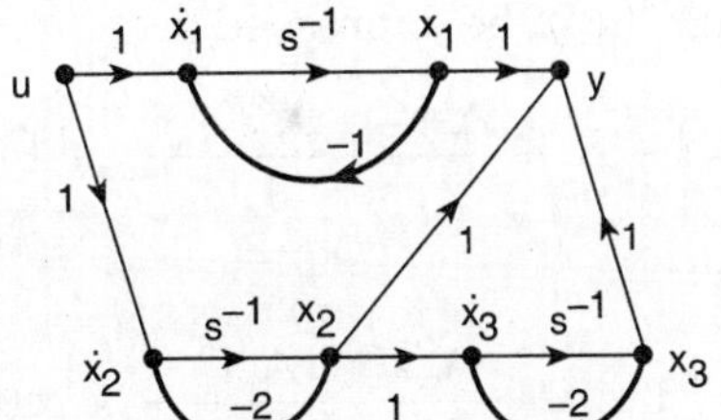

Fig. 9.5.12. State signal flow graph :
$Y(s)/U(s) = s^{-1}/[1-(-s^{-1})] + s^{-1}/[1-(-2s^{-1})] + (s^{-1} . s^{-1})/[1-(-2s^{-1})]^2$.

The state equations are

$$\dot{x}_1 = -x_1 + u$$
$$\dot{x}_2 = -2x_2 + u$$
$$\dot{x}_3 = x_2 - 2x_3$$

The output equation is

$$y = x_1 + x_2 + x_3$$

The equations above are arranged in vector-matrix form as below to form the state model

$$\begin{bmatrix} \dot{x}_1 \\ \dot{x}_2 \\ \dot{x}_3 \end{bmatrix} = \begin{bmatrix} -1 & 0 & 0 \\ 0 & -2 & 0 \\ 0 & 1 & -2 \end{bmatrix} \begin{bmatrix} x_1 \\ x_2 \\ x_3 \end{bmatrix} + \begin{bmatrix} 1 \\ 1 \\ 0 \end{bmatrix} u$$

$$y = [1 \;\; 1 \;\; 1] \begin{bmatrix} x_1 \\ x_2 \\ x_3 \end{bmatrix}$$

9.6 SOLUTION OF STATE EQUATION

The state equation is given below :

$$\dot{\mathbf{x}} = \mathbf{A}\mathbf{x}(t) + \mathbf{B}\mathbf{u}(t) \qquad ...(9.41)$$

or

$$\frac{d\mathbf{x}(t)}{dt} = \mathbf{A}\mathbf{x}(t) + \mathbf{B}\mathbf{u}(t) \qquad ...(9.42)$$

The Eq. (9.42) is linear time invariant and the time solution may be obtained either using Laplace transform or classical method. The solution of a state equation is termed as state transition equation describing the change of state of a system w.r.t. time.

The natural response of a state equation is obtained by considering the effect of initial vector $\mathbf{x}(0)$, while the input vector $\mathbf{u}(t)$ in the equation is deleted from the equation. This makes the state equation as homogeneous which is written below :

$$\frac{d\mathbf{x}(\mathbf{t})}{dt} = A\mathbf{x}(t) \qquad ...(9.43)$$

The solution of Eq. (9.43) is to be obtained for the given initial condition $\mathbf{x}(0)$. The classical method of solving above equation assumes a solution given below :

$$\mathbf{x}(t) = e^{\mathbf{A}t}\,\mathbf{x}(0) \qquad ...(9.44)$$

The matrix exponential $e^{\mathbf{A}t}$ can be expressed as

$$e^{\mathbf{A}t} = \mathbf{I} + \mathbf{A}t + \frac{\mathbf{A}^2t^2}{2!} + \frac{\mathbf{A}^3t^3}{3!} + ..., \; t \geq 0 \qquad ...(9.45)$$

substituting Eq. (9.45) in (9.44)

$$\therefore \qquad \mathbf{x}(t) = \left(\mathbf{I} + \mathbf{A}t + \frac{\mathbf{A}^2t^2}{2!} + \frac{\mathbf{A}^3t^3}{3!} + ...\right)\mathbf{x}(0) \qquad ...(9.46)$$

Differentiate Eq. (9.46)

$$\therefore \qquad \frac{d\mathbf{x}(t)}{dt} = \left(\mathbf{A} + \frac{\mathbf{A}^2 2t}{2!} + \frac{\mathbf{A}^3 3t}{3!} + ...\right)\mathbf{x}(0)$$

or

$$\frac{d\mathbf{x}(t)}{dt} = \mathbf{A}\left(\mathbf{I} + \mathbf{A}t + \frac{\mathbf{A}^2t}{2!} + ...\right)\mathbf{x}(0) \qquad ...(9.47)$$

$$\because \qquad e^{\mathbf{A}t} = \left(\mathbf{I} + \mathbf{A}t + \frac{\mathbf{A}^2 t}{2!} + ...\right) \qquad ...(9.48)$$

substitute Eq. (9.48) in (9.47)

$$\therefore \qquad \frac{d\mathbf{x}(t)}{dt} = \mathbf{A}e^{\mathbf{A}t}\,\mathbf{x}(0) \qquad ...(9.48a)$$

The state Eq. (9.43) is

$$\frac{d\mathbf{x}(t)}{dt} = \mathbf{A}\mathbf{x}(t)$$

Therefore, equating Eqs. (9.48) and (9.43) following relation is obtained,

$$\mathbf{A}\mathbf{x}(t) = \mathbf{A}e^{\mathbf{A}t}\,\mathbf{x}(0)$$

or $\mathbf{x}(t) = e^{\mathbf{A}t} x(0)$ is the solution of state equation $\{d\mathbf{x}(t)/dt\} = \mathbf{A}\mathbf{x}(t)$

The initial state $\mathbf{x}(0)$ at $t = 0$, is shifted to a state $\mathbf{x}(t)$ at time t. This matrix exponential $e^{\mathbf{A}t}$ carries out the transition of the state during time t, $e^{\mathbf{A}t}$ is termed as state transition matrix and denoted as $\phi(t)$.

Solution of non-homogeneous state equation

The non-homogeneous state equation has initial condition $\mathbf{x}(0)$ along with the input $\mathbf{u}(t)$ as given below

$$\frac{d\mathbf{x}(t)}{dt} = \mathbf{A}\mathbf{x}(t) + \mathbf{B}\mathbf{u}(t) \qquad ...(9.49)$$

The solution of Eq. (9.49) is obtained by rearranging the same as below :

$$\frac{d\mathbf{x}(t)}{dt} - \mathbf{A}\mathbf{x}(t) = \mathbf{B}\mathbf{u}(t)$$

On premultiplying both sides by $e^{-\mathbf{A}t}$

$$e^{-\mathbf{A}t}\{\dot{\mathbf{x}}(t) - \mathbf{A}\mathbf{x}(t)\} = e^{-\mathbf{A}t}\,\mathbf{B}\mathbf{u}(t)$$

or

$$e^{-\mathbf{A}t}\,\dot{\mathbf{x}}(t) - \mathbf{A}e^{-\mathbf{A}t}\,\mathbf{x}(t) = e^{-\mathbf{A}t}\,\mathbf{B}\mathbf{u}(t) \qquad ...(9.50)$$

$$\because \qquad \frac{d}{dt}\{e^{-\mathbf{A}t}\,\mathbf{x}(t)\} = e^{-\mathbf{A}t}\,\dot{\mathbf{x}}(t) - \mathbf{A}e^{-\mathbf{A}t}\,\mathbf{x}(t) \qquad ...(9.51)$$

substitute Eq. (9.51) in (9.50) to obtain following relation

$$\frac{d}{dt}\{e^{-\mathbf{A}t}\,\mathbf{x}(t)\} = e^{-\mathbf{A}t}\,\mathbf{B}\mathbf{u}(t) \qquad ...(9.52)$$

On integrating Eq. (9.52) between time limits 0 to t,

$$e^{-\mathbf{A}t}\,\mathbf{x}(t) = x(0) + \int_0^t e^{-A\tau}\,\mathbf{B}\mathbf{u}(\tau)\,d\tau$$

or

$$\mathbf{x}(t) = e^{\mathbf{A}t}\,\mathbf{x}(0) + \int_0^\tau e^{-A(t-\tau)}\,\mathbf{B}\mathbf{u}(\tau)\,d\tau \qquad ...(9.52a)$$

$$\because \qquad \phi(t) = e^{\mathbf{A}t}$$

$$\therefore \qquad \mathbf{x}(t) = \phi(t)\,\mathbf{x}(0) + \int_0^\tau \phi(t-\tau)\,\mathbf{B}\mathbf{u}(\tau)\,d\tau \qquad ...(9.53)$$

The solution of non-homogeneous state Eq. (9.49) is given by Eq. (9.53). The first term on the right hand side of the solution is the natural response (zero input response) where as the second term forced response (zero state response).

The Eq. (9.53) is a state transition equation wherein the initial condition occurs at $t = 0$.

The state equation is given by, $\frac{d\mathbf{x}(t)}{dt} = \mathbf{Ax}(t) + \mathbf{Bu}(t)$ ...(9.54)

The solution of the above linear time invariant homogeneous equation can be obtained by Laplace transform method. Taking Laplace transform of the Eq. (9.54) on both sides,

$$s\,\mathbf{X}(s) - \mathbf{x}(0) = \mathbf{AX}(s) + \mathbf{BU}(s) \qquad ...(9.55)$$

where $\mathbf{x}(0)$ is the initial state at $t = 0$.

On rearranging Eq. (9.55),

$$\mathbf{X}(s) = (s\,\mathbf{I} - \mathbf{A})^{-1}\,\mathbf{x}(0) + (s\,\mathbf{I} - \mathbf{A})^{-1}\,\mathbf{BU}(s)$$

put $\phi(s) = (s\mathbf{I} - \mathbf{A})^{-1}$

$$\therefore \quad \mathbf{X}(s) = \phi(s)\,\mathbf{x}(0) + \phi(s)\,\mathbf{BU}(s) \qquad ...(9.56)$$

where $\phi(t) = \mathcal{L}^{-1}\,\phi(s) = \mathcal{L}^{-1}\,(s\,\mathbf{I} - \mathbf{A})^{-1}$ is state transition matrix.

Taking inverse Laplace transform on both sides of equation (9.56)

$$\mathcal{L}^{-1}\,\mathbf{X}(s) = \mathcal{L}^{-1}\,\phi(s)\,\mathbf{x}(0) + \mathcal{L}^{-1}\,\phi(s)\,\mathbf{BU}(s)$$

or

$$\mathbf{x}(t) = \phi(t)\,\mathbf{x}(0) + \mathcal{L}^{-1}\,\phi(s)\,\mathbf{BU}\,(s)$$

using convolution integral, $\mathbf{x}(t) = \phi(t)\,\mathbf{x}(0) + \int_0^t \phi(t-\tau)\;\mathbf{BU}\,(\tau)\,d\tau.$ ...(9.57)

The equation (9.57) gives the solution of state equation (9.41) using Laplace transform method.

9.6.1 Determination of state transition matrix

The matrix exponential $e^{\mathbf{A}t}$ is termed as state transition matrix and denoted as $\phi(t)$. The methods for determination of state transition matrix are given below,

(1) Infinite series method

$e^{\mathbf{A}t}$ can be expressed in the exponential series form as

$$e^{\mathbf{A}t} = \mathbf{I} + \mathbf{A}t + \frac{\mathbf{A}^2t^2}{2!} + \frac{\mathbf{A}^3t^3}{3!} + \ldots \qquad ...(9.58)$$

The state transition matrix $e^{\mathbf{A}t}$ for $\mathbf{A} = \begin{bmatrix} 1 & 4 \\ -2 & -5 \end{bmatrix}$ is determined on substituting values of $\mathbf{A}, \mathbf{A}^2, \mathbf{A}^3, \ldots$ in the Eq. (9.58)

$$\mathbf{A} = \begin{bmatrix} 1 & 4 \\ -2 & -5 \end{bmatrix}$$

$$\mathbf{A}^2 = \mathbf{AA}$$

$$= \begin{bmatrix} 1 & 4 \\ -2 & -5 \end{bmatrix}\begin{bmatrix} 1 & 4 \\ -2 & -5 \end{bmatrix} = \begin{bmatrix} -7 & -16 \\ 8 & 17 \end{bmatrix}$$

$$\vdots$$

$$\mathbf{A}^n = \mathbf{A\,A\,A} \ldots \text{ up to } n$$

$$\therefore \quad e^{\mathbf{A}t} = \begin{bmatrix} 1 & 0 \\ 0 & 1 \end{bmatrix} + \begin{bmatrix} 1 & 4 \\ -2 & -5 \end{bmatrix}t + \begin{bmatrix} -7 & -16 \\ 8 & 17 \end{bmatrix}\frac{t^2}{2} + \ldots$$

$$= \begin{bmatrix} 1+t-\frac{7}{2}t^2+\ldots & 4t-\frac{16}{2}t^2+\ldots \\ -2t+\frac{8}{2}t^2+\ldots & 1-5t+\frac{17}{2}t^2\ldots \end{bmatrix}$$

$\because$ $$e^{-t} = 1 - t + \frac{t^2}{2!} - \frac{t^3}{3!} + \ldots$$

and $$e^{-3t} = 1 - 3t + \frac{9t^2}{2!} - \frac{27t^3}{3!} + \ldots$$

It can be verified that

$$1 + t - \frac{7}{2}t^2 + \ldots = 2e^{-t} - e^{-3t}$$

$$4t - \frac{16}{2}t^2 + \ldots = 2e^{-t} - 2e^{-3t}$$

$$-2t + \frac{8}{2}t^2 - \ldots = e^{-t} + e^{-3t}$$

$$1 - 5t + \frac{17}{2}t^2 \ldots = -e^{-t} + 2e^{-3t}$$

$\therefore$ $$\phi(t) = e^{\mathbf{A}t} = \begin{bmatrix} 2e^{-t} - e^{-3t} & 2e^{-t} - 2e^{-3t} \\ -e^{-t} + e^{-3t} & -e^{-t} + 2e^{-3t} \end{bmatrix}$$

(2) Laplace transform method

The Laplace transform of the state transition matrix is given by

$$\phi(s) = (s\,\mathbf{I} - \mathbf{A})^{-1} \qquad \ldots (9.59)$$

The state transition matrix $\phi(t)$ is determined by taking inverse Laplace transform on both sides of equation (9.59).

$\therefore$ $$\phi(t) = \mathcal{L}^{-1}(s\,\mathbf{I} - \mathbf{A})^{-1}$$

The state transition matrix $\phi(t)$ for $\mathbf{A} = \begin{bmatrix} 1 & 4 \\ -2 & -5 \end{bmatrix}$ is determined using Laplace transform method as below :

$$(s\,\mathbf{I} - \mathbf{A}) = \begin{bmatrix} s & 0 \\ 0 & s \end{bmatrix} - \begin{bmatrix} 1 & 4 \\ -2 & -5 \end{bmatrix} = \begin{bmatrix} s-1 & -4 \\ 2 & s+5 \end{bmatrix}$$

$$(s\,\mathbf{I} - \mathbf{A})^{-1} = \frac{\text{Adj}\,(s\,\mathbf{I} - \mathbf{A})}{|\,s\,\mathbf{I} - \mathbf{A}\,|}$$

$$= \frac{\begin{bmatrix} s+5 & 4 \\ -2 & s-1 \end{bmatrix}}{\begin{vmatrix} s-1 & -4 \\ 2 & s+5 \end{vmatrix}} = \frac{\begin{bmatrix} s+5 & 4 \\ -2 & s-1 \end{bmatrix}}{s^2 + 4s + 3} = \frac{\begin{bmatrix} s+5 & 4 \\ -2 & s-1 \end{bmatrix}}{(s+1)(s+3)}$$

$$= \begin{bmatrix} \dfrac{s+5}{(s+1)(s+3)} & \dfrac{4}{(s+1)(s+3)} \\ \dfrac{-2}{(s+1)(s+3)} & \dfrac{s-1}{(s+1)(s+3)} \end{bmatrix}$$

$$= \begin{bmatrix} \dfrac{2}{s+1} + \dfrac{-1}{s+3} & \dfrac{2}{s+1} + \dfrac{-2}{s+3} \\ \dfrac{-1}{s+1} + \dfrac{1}{s+3} & \dfrac{-1}{s+1} + \dfrac{2}{s+3} \end{bmatrix}$$

The state transition matrix is

$$\phi(t) = \mathcal{L}^{-1} (s\,\mathbf{I} - \mathbf{A})^{-1}$$

$$= \mathcal{L}^{-1} \begin{bmatrix} \dfrac{2}{s+1} + \dfrac{-1}{s+3} & \dfrac{2}{s+1} + \dfrac{-2}{s+3} \\ \dfrac{-1}{s+1} + \dfrac{1}{s+3} & \dfrac{-1}{s+1} + \dfrac{2}{s+3} \end{bmatrix}$$

or
$$\phi(t) = \begin{bmatrix} 2e^{-t} - e^{-3t} & 2e^{-t} - 2e^{-3t} \\ -e^{-t} + e^{-3t} & -e^{-t} + 2e^{-3t} \end{bmatrix}$$

(3) Cayley-Hamilton Theorem

The Cayley-Hamilton theorem is useful in matrix analysis and applied for the determination of state transition matrix $e^{\mathbf{A}t}$.

"The theorem states that every square matrix satisfies its own characteristic equation."

The characteristic equation of a matrix **A** of order n is given by

$$|\lambda\,\mathbf{I} - \mathbf{A}| = 0 \qquad ...(9.60)$$

which can be expressed in polynomial form of λ as below :

$$f(\lambda) = \lambda^n + a_1 \lambda^{n-1} + a_2 \lambda^{n-2} + ... + a^n = 0 \qquad ...(9.61)$$

As per Cayley-Hamilton theorem matrix **A** satisfies characteristic Eq. (9.61), thus

$$f(\mathbf{A}) = \mathbf{A}^n + a_1\mathbf{A}^{n-1} + a_2\mathbf{A}^{n-2} + ... + a_n\,\mathbf{I} = 0 \qquad ...(9.62)$$

where, **I** is the identity matrix of order n.

The theorem is verified for the matrix **A** as given below :

$$\mathbf{A} = \begin{bmatrix} 1 & 4 \\ -2 & -5 \end{bmatrix}$$

The characteristic equation is

$$|\lambda\,\mathbf{I} - \mathbf{A}| = 0$$

$$\therefore \quad \left| \begin{bmatrix} \lambda & 0 \\ 0 & \lambda \end{bmatrix} - \begin{bmatrix} 1 & 4 \\ -2 & -5 \end{bmatrix} \right| = 0$$

or
$$\begin{vmatrix} \lambda - 1 & -4 \\ 2 & \lambda + 5 \end{vmatrix} = 0$$

$$\therefore \quad (\lambda - 1)(\lambda + 5) - (2)(-4) = 0 \text{ or } \lambda^2 + 4\lambda + 3 = 0$$

$\therefore \quad f(\lambda) = \lambda^2 + 4\lambda + 3$

$\because \quad \lambda^2 + 4\lambda + 3 = 0$

$\therefore \quad f(\lambda) = 0$

$$f(\mathbf{A}) = \mathbf{A}^2 + 4\mathbf{A} + 3\mathbf{I}$$

$$\because \quad \mathbf{A} = \begin{bmatrix} 1 & 4 \\ -2 & -5 \end{bmatrix}$$

$$\therefore \quad f(\mathbf{A}) = \begin{bmatrix} 1 & 4 \\ -2 & -5 \end{bmatrix}\begin{bmatrix} 1 & 4 \\ -2 & -5 \end{bmatrix} + 4\begin{bmatrix} 1 & 4 \\ -2 & -5 \end{bmatrix} + 3\begin{bmatrix} 1 & 0 \\ 0 & 1 \end{bmatrix}$$

$$= \begin{bmatrix} -7 & -16 \\ 8 & 17 \end{bmatrix} + \begin{bmatrix} 4 & 16 \\ -8 & -20 \end{bmatrix} + \begin{bmatrix} 3 & 0 \\ 0 & 3 \end{bmatrix}$$

$$= \begin{bmatrix} -3 & 0 \\ 0 & -3 \end{bmatrix} + \begin{bmatrix} 3 & 0 \\ 0 & 3 \end{bmatrix} = 0$$

or $\quad f(\mathbf{A}) = 0$

Hence, Cayley-Hamilton theorem is verified.

Application of Cayley-Hamilton theorem for determination of state transition matrix

$F(\mathbf{A})$ is function of matrix **A** such that the degree of $F(\mathbf{A})$ is higher than or of the order of **A** and if the characteristic equation of matrix **A** is given as

$$f(\lambda) = \lambda^n + a_1\lambda^{n-1} + a_2\lambda^{n-2} + \ldots + a_n = 0$$

then, according to Cayley-Hamilton theorem

$$f(\mathbf{A}) = 0$$

As already stated the degree of $F(\mathbf{A})$ is higher than the order of **A**, the same can be expressed as

$$F(\mathbf{A}) = D(\mathbf{A})\, f(\mathbf{A}) + R(\mathbf{A})$$

Since, $\quad f(\mathbf{A}) = 0$

$\therefore \quad F(\mathbf{A}) = R(\mathbf{A})$

$R(\mathbf{A})$ can be written in polynomial form as

$$F(\mathbf{A}) = R(\mathbf{A}) = \alpha_0\mathbf{I} + \alpha_1\mathbf{A} + \alpha_2\mathbf{A}^2 + \ldots + \alpha_{n-1}\mathbf{A}^{n-1}$$

$$= \sum_{k=0}^{n-1} \alpha_k \mathbf{A}^k \quad \ldots (9.63)$$

where, the order of matrix **A** is n.

Considering λ as the eigen value of **A**, the function $F(\lambda)$ in relation to $F(\mathbf{A})$ can be written as

$$F(\lambda) = R(\lambda) = \sum_{k=0}^{n-1} \alpha_k \lambda^k \quad \ldots (9.64)$$

The matrix exponential $e^{\mathbf{A}t}$ can be written in the form

$$e^{\mathbf{A}t} = \sum_{k=0}^{k} \alpha_k \mathbf{A}^k \quad \ldots (9.65)$$

and also, $$e^{\lambda t} = \sum_{k=0}^{k} \alpha_k \lambda^k \quad \ldots (9.66)$$

Eqs. (9.65) and (9.66) can be written in expanded form as

$$e^{\mathbf{A}t} = \alpha_0 \mathbf{I} + \alpha_1\mathbf{A} + \alpha_2\mathbf{A}^2 + \ldots + \alpha_n\mathbf{A}^n \quad \ldots(9.67)$$

$$e^{\lambda t} = \alpha_0 + \alpha_1\lambda + \alpha_2\lambda^2 + \ldots + \alpha_n\lambda^n \quad \ldots(9.68)$$

The Eq. (9.68) is a scalar one and values of the coefficients α_0, α_1, α_2 ... can be determined using this equation. Having determined the values of α_0, α_1, α_2 ..., the state transition matrix $e^{\mathbf{A}t}$ can be determined using Eq. (9.67).

The example below gives application of the Cayley-Hamilton theorem to obtain state transition matrix for

$$\mathbf{A} = \begin{bmatrix} 1 & 4 \\ -2 & -5 \end{bmatrix}$$

$$[\lambda\mathbf{I} - \mathbf{A}] = \begin{bmatrix} \lambda & 0 \\ 0 & \lambda \end{bmatrix} - \begin{bmatrix} 1 & 4 \\ -2 & -5 \end{bmatrix} = \begin{bmatrix} \lambda - 1 & -4 \\ 2 & \lambda + 5 \end{bmatrix}$$

$$|\lambda\mathbf{I} - \mathbf{A}| = \begin{vmatrix} \lambda - 1 & -4 \\ 2 & \lambda + 5 \end{vmatrix}$$

$$= (\lambda - 1)(\lambda + 5) - (2)(-4)$$

$$= \lambda^2 + 4\lambda + 3 = (\lambda + 1)(\lambda + 3)$$

The characteristic equation is given by

$$|\lambda\mathbf{I} - \mathbf{A}| = 0$$

$$\therefore \quad (\lambda + 1)(\lambda + 3) = 0$$

The eigen values of matrix **A** are $\lambda_1 = -1$ and $\lambda_2 = -3$ which are distinct.

The scalar equation in terms of eigen values is formed below :

$$e^{\lambda t} = \alpha_0 + \alpha_1\lambda \quad \ldots(9.69)$$

$$\because \quad \lambda_1 = -1 \quad \text{and} \quad \lambda_2 = -3$$

Therefore, the two equations are

$$e^{(-1)t} = \alpha_0 + \alpha_1(-1)$$

and

$$e^{(-3)t} = \alpha_0 + \alpha_1(-3)$$

or

$$e^{-t} = \alpha_0 - \alpha_1$$

and

$$e^{-3t} = \alpha_0 - 3\alpha_1$$

or

$$\begin{bmatrix} e^{-t} \\ e^{-3t} \end{bmatrix} = \begin{bmatrix} 1 & -1 \\ 1 & -3 \end{bmatrix} \begin{bmatrix} \alpha_0 \\ \alpha_1 \end{bmatrix} \quad \ldots(9.70)$$

Solving Eq. (9.70)

$$\alpha_0 = \left(\frac{3}{2}e^{-t} - \frac{1}{2}e^{-3t}\right)$$

$$\alpha_1 = \left(\frac{1}{2}e^{-t} - \frac{1}{2}e^{-3t}\right)$$

$$\because \quad e^{\mathbf{A}t} = \alpha_0 \mathbf{I} + \alpha_1\mathbf{A}$$

$$\therefore \quad e^{\mathbf{A}t} = \left(\frac{3}{2}e^{-t} - \frac{1}{2}e^{-3t}\right)\begin{bmatrix} 1 & 0 \\ 0 & 1 \end{bmatrix} + \left(\frac{1}{2}e^{-t} - \frac{1}{2}e^{-3t}\right)\begin{bmatrix} 1 & 4 \\ -2 & -5 \end{bmatrix}$$

$$= \begin{bmatrix} \frac{3}{2}e^{-t} - \frac{e^{-3t}}{2} & 0 \\ 0 & \frac{3}{2}e^{-t} - \frac{e^{-3t}}{2} \end{bmatrix} \begin{bmatrix} \frac{e^{-t}}{2} - \frac{e^{-3t}}{2} & 2e^{-t} - 2e^{-3t} \\ e^{-t} + e^{-3t} & -\frac{5}{2}e^{-t} - \frac{5}{2}e^{-3t} \end{bmatrix}$$

or
$$e^{\mathbf{A}t} = \begin{bmatrix} 2e^{-t} - e^{-3t} & 2e^{-t} - 2e^{-3t} \\ -e^{-t} + e^{-3t} & -e^{-t} + 2e^{-3t} \end{bmatrix}$$

(4) Determination of state transition matrix using matrix diagonalization

The state transition matrix $e^{\mathbf{A}t}$ is conveniently determined when **A** matrix is given in the form of a diagonal matrix. This is verified as below :

The diagonal form of matrix **A** is given by

$$\mathbf{A} = \begin{bmatrix} \lambda_1 & 0 \\ 0 & \lambda_2 \end{bmatrix} \quad ...(9.71)$$

The state transition matrix $e^{\mathbf{A}t}$ can be expressed in the form of a series as

$$e^{\mathbf{A}t} = \mathbf{I} + \mathbf{A}t + \frac{\mathbf{A}^2t^2}{2!} + \frac{\mathbf{A}^3t^3}{3!} + ... \quad ...(9.72)$$

On substituting the value of **A** in (9.72) from (9.71)

$$e^{\mathbf{A}t} = \begin{bmatrix} 1 & 0 \\ 0 & 1 \end{bmatrix} + \begin{bmatrix} \lambda_1 & 0 \\ 0 & \lambda_2 \end{bmatrix} + \begin{bmatrix} \frac{\lambda_1^2}{2!}t^2 & 0 \\ 0 & \frac{\lambda_1^2}{2!}t^2 \end{bmatrix} + \begin{bmatrix} \frac{\lambda^3t^3}{3!} & 0 \\ 0 & \frac{\lambda^3t^3}{3!} \end{bmatrix} + ...$$

or
$$e^{\mathbf{A}t} = \begin{bmatrix} e^{\lambda_1 t} & 0 \\ 0 & e^{\lambda_2 t} \end{bmatrix}$$

Thus, the state transition matrix of a diagonal matrix having diagonal elements λ_i $(i = 1, 2, 3, ...)$ is also a diagonal one with its diagonal elements as exponential of $\lambda_i t$ $(i = 1, 2, 3, ...)$

In case, the matrix is not in diagonal form, the same may be transformed into diagonal form using a similarity transformation as described in the texts of matrix algebra. A non-diagonal matrix **A** with distinct eigen values is considered for the transformation.

The homogeneous state equation is given below

$$\dot{\mathbf{x}} = \mathbf{Ax} \quad ...(9.73)$$

The variable **x** is transformed using following similarity transformation

$$\mathbf{x} = \mathbf{Mz} \quad ...(9.74)$$

M is known as modal matrix of A

$$\therefore \quad \dot{\mathbf{x}} = \mathbf{M}\dot{\mathbf{z}} \quad ...(9.75)$$

Substituting in Eq. (9.73) from (9.74) and (9.75) the transformed equation Eq. (9.73) is

$$\mathbf{M}\dot{\mathbf{z}} = \mathbf{AMz}$$

Premultiplying on both sides by $\mathbf{M}^{-1}$

$$\dot{\mathbf{z}} = \mathbf{M}^{-1}\mathbf{AMz}$$

The matrix $\mathbf{M}^{-1}\mathbf{AM}$ is a diagonal one and denoted as Λ

$$\therefore \qquad \dot{\mathbf{z}} = \Lambda \mathbf{z} \qquad \text{...(9.76)}$$

The solution of the (9.76) is given by

$$z = e^{\Lambda t}\,\mathbf{z}(0)$$

$$\because \qquad z = \mathbf{M}^{-1}\mathbf{x} \text{ and } \mathbf{z}(0) = \mathbf{M}^{-1}\mathbf{x}(0)$$

$$\therefore \qquad \mathbf{M}^{-1}\mathbf{x} = e^{\Lambda t}\,\mathbf{M}^{-1}\mathbf{x}(0)$$

Premultiplying both sides by **M**

$$\mathbf{x} = \mathbf{M}e^{\Lambda t}\,\mathbf{M}^{-1}\mathbf{x}(0)$$

Thus, the state transition matrix is

$$e^{\mathbf{A}t} = \mathbf{M}e^{\Lambda t}\,\mathbf{M}^{-1}$$

The columns of a Modal matrix **M** are formed by eigen vectors of $m \times n$ matrix **A**, therefore

$$\mathbf{M} = \begin{bmatrix} m_{11} & m_{12} \ldots. & m_{1n} \\ m_{21} & m_{22} & m_{2n} \\ . & . & . \\ . & . & . \\ m_{n1} & m_{n2} & m_{nn} \end{bmatrix}$$

where, $[m_{11}\ m_{21}\ \ldots\ m_{n1}]^T$ is the eigen vector associated with eigen value λ_1

$[m_{12}\ m_{22}\ \ldots\ m_{n2}]^T$ is the eigen vector associated with eigen value λ_2

.... etc.

The procedure given below explains the use of matrix diagonalization method to determine the state transition matrix for a given non-diagonal matrix **A**.

$$\text{Let} \qquad \mathbf{A} = \begin{bmatrix} 1 & 4 \\ -2 & -5 \end{bmatrix}$$

As already calculated, the eigen values of the matrix **A** are $\lambda_1 = -1$ and $\lambda_2 = -3$.

(*i*) The eigen vector $\mathbf{x}_1$ of matrix **A** associated with eigen value $\lambda_1 = -1$ is obtained by solving the following equation,

$$[\lambda_1 \mathbf{I} - \mathbf{A}]\,\mathbf{x}_1 = 0$$

$$\text{where } \mathbf{x}_1 = \begin{bmatrix} m_{11} \\ m_{21} \end{bmatrix}$$

$$\therefore \qquad \left\{(-1)\begin{bmatrix} 1 & 0 \\ 0 & 1 \end{bmatrix} - \begin{bmatrix} 1 & 4 \\ -2 & -5 \end{bmatrix}\right\}\begin{bmatrix} m_{11} \\ m_{21} \end{bmatrix} = 0$$

$$\text{or} \qquad \begin{bmatrix} -2 & -4 \\ 2 & 4 \end{bmatrix}\begin{bmatrix} m_{11} \\ m_{21} \end{bmatrix} = 0$$

$$\therefore \qquad -2m_{11} - 4m_{21} = 0$$

$$\text{and} \qquad 2m_{11} + 4m_{21} = 0$$

$\therefore \qquad m_{11} = -2m_{21}$

select $m_{11} = 1$

$\therefore \qquad m_{21} = -0.5$

Thus,
$$\mathbf{x}_1 = \begin{bmatrix} m_{11} \\ m_{21} \end{bmatrix} = \begin{bmatrix} 1 \\ -0.5 \end{bmatrix}$$

(*ii*) The eigen vector $\mathbf{x}_2$ of matrix **A** associated with eigen value $\lambda_1 = -3$ is obtained by solving the following equation

$$[\lambda_2 \mathbf{I} - \mathbf{A}]\, \mathbf{x}_2 = 0$$

where $\mathbf{x}_2 = \begin{bmatrix} m_{11} \\ m_{22} \end{bmatrix}$

$$\therefore \qquad \left\{(-3)\begin{bmatrix} 1 & 0 \\ 0 & 1 \end{bmatrix} - \begin{bmatrix} 1 & 4 \\ -2 & -5 \end{bmatrix}\right\}\begin{bmatrix} m_{12} \\ m_{22} \end{bmatrix} = 0$$

or
$$\begin{bmatrix} -4 & -4 \\ 2 & 2 \end{bmatrix}\begin{bmatrix} m_{12} \\ m_{22} \end{bmatrix} = 0$$

$\therefore \qquad -4m_{12} - 4m_{22} = 0$

and $\qquad 2m_{12} + 2m_{22} = 0$

$\therefore \qquad m_{12} = -m_{22}$

$\therefore \qquad$ select $m_{12} = 1$

$\qquad m_{22} = -1$

Thus,
$$\mathbf{x}_2 = \begin{bmatrix} m_{12} \\ m_{22} \end{bmatrix} = \begin{bmatrix} 1 \\ -1 \end{bmatrix}$$

The modal matrix **M** is given by

$$\mathbf{M} = [\boldsymbol{x}_1 : \boldsymbol{x}_2]$$

or
$$\mathbf{M} = \begin{bmatrix} 1 & 1 \\ -0.5 & -1 \end{bmatrix}$$

and
$$\mathbf{M}^{-1} = \begin{bmatrix} 2 & 2 \\ -1 & -2 \end{bmatrix}$$

$$\mathbf{\Lambda} = \mathbf{M}^{-1}\ \mathbf{A}\ \mathbf{M}$$

or
$$\mathbf{\Lambda} = \begin{bmatrix} 2 & 2 \\ -1 & -2 \end{bmatrix}\begin{bmatrix} 1 & 4 \\ -2 & -5 \end{bmatrix}\begin{bmatrix} 1 & 1 \\ -0.5 & -1 \end{bmatrix} = \begin{bmatrix} -1 & 0 \\ 0 & -3 \end{bmatrix}$$

$$\therefore \qquad e^{\Lambda t} = \begin{bmatrix} e^{-t} & 0 \\ 0 & e^{-3t} \end{bmatrix}$$

and
$$e^{\mathbf{A}t} = \mathbf{M}e^{\Lambda t}\,\mathbf{M}^{-1}$$

$$= \begin{bmatrix} 1 & 1 \\ -0.5 & -1 \end{bmatrix}\begin{bmatrix} e^{-t} & 0 \\ 0 & e^{-3t} \end{bmatrix}\begin{bmatrix} 2 & 2 \\ -1 & -1 \end{bmatrix}$$

$$= \begin{bmatrix} e^{-t} & e^{-3t} \\ -0.5e^{-t} & -e^{-3t} \end{bmatrix} \begin{bmatrix} 2 & 2 \\ -1 & -2 \end{bmatrix}$$

or
$$e^{\mathbf{A}t} = \begin{bmatrix} 2e^{-t} - e^{-3t} & 2e^{-t} - 2e^{-3t} \\ -e^{-t} + e^{-3t} & -e^{-t} + 2e^{-3t} \end{bmatrix}$$

The matrices **A** and $\Lambda = \mathbf{M}^{-1}\,\mathbf{AM}$ are similar and have the same characteristic equation.

The representation of matrix **A** in phase variable form is given below :

$$\mathbf{A} = \begin{bmatrix} 0 & 1 & 0 & 0 & \dots & 0 \\ 0 & 0 & 1 & 0 & \dots & 0 \\ 0 & 0 & 0 & 1 & \dots & 0 \\ \vdots & \vdots & \vdots & \vdots & & \vdots \\ -a_1 & -a_2 & -a_3 & -a_4 & & -a_n \end{bmatrix}$$

The phase variable form of matrix **A** has a unit element above its diagonal and negative coefficient elements in the last row. All other elements being zero.

The Modal matrix with reference to phase variable form of matrix **A** is denoted as **P** and given by

$$\mathbf{P} = \begin{bmatrix} 1 & 1 & 1 & \dots & 1 \\ \lambda_1 & \lambda_2 & \lambda_3 & \dots & \lambda_n \\ \lambda_1^2 & \lambda_2^2 & \lambda_3^2 & \dots & \lambda_n^2 \\ \vdots & \vdots & \vdots & & \\ \lambda_1^{n-1} & \lambda_2^{n-1} & \lambda_3^{n-1} & \dots & \lambda_n^{n-1} \end{bmatrix}$$

where $\lambda_1, \lambda_2, \lambda_3, \dots \lambda_n$ are distinct eigen values of matrix **A**, which is given in phase variable form.

The matrix $\Lambda = \mathbf{P}^{-1}\,\mathbf{AP}$ is a diagonal one. The matrix **P** is known as Vandermonde's matrix.

The matrices **A** and $\mathbf{P}^{-1}\,\mathbf{AP}$ are similar and have the same characteristic equation.

The procedure given below explains the use of matrix diagonalization method to determine state transition matrix for a given phase variable form of matrix **A**.

Let
$$\mathbf{A} = \begin{bmatrix} 0 & 1 \\ -6 & -5 \end{bmatrix}$$

It is given that matrix **A** is in phase variable form, hence Vandermonde's matrix **P** is used for diagonalization.

The characteristic equation is given by

$$|\lambda \mathbf{I} - \mathbf{A}| = 0$$

$$\therefore \quad \left| \begin{bmatrix} \lambda & 0 \\ 0 & \lambda \end{bmatrix} - \begin{bmatrix} 0 & 1 \\ -6 & -5 \end{bmatrix} \right| = 0$$

or
$$\begin{bmatrix} \lambda & -1 \\ 6 & \lambda + 5 \end{bmatrix} = 0$$

$\therefore \quad \lambda(\lambda + 5) - (6)(-1) = 0$

or $$\lambda^2 + 5\lambda + 6 = 0 \quad \text{or} \quad (\lambda + 2)(\lambda + 3) = 0$$

The eigen values are $\lambda_1 = -2$ and $\lambda_2 = -3$

$\therefore$ Vandermonde's matrix

$$\mathbf{P} = \begin{bmatrix} 1 & 1 \\ \lambda_1 & \lambda_2 \end{bmatrix}$$

or $$\mathbf{P} = \begin{bmatrix} 1 & 1 \\ -2 & -3 \end{bmatrix}$$

and $$\mathbf{P}^{-1} = \begin{bmatrix} 3 & 1 \\ -2 & -1 \end{bmatrix}$$

$\therefore$ $$\Lambda = \mathbf{P}^{-1}\mathbf{AP}$$

$$= \begin{bmatrix} 3 & 1 \\ -2 & -1 \end{bmatrix}\begin{bmatrix} 0 & 1 \\ -6 & -5 \end{bmatrix}\begin{bmatrix} 1 & 1 \\ -2 & -3 \end{bmatrix}$$

$$= \begin{bmatrix} -6 & -2 \\ 6 & 3 \end{bmatrix}\begin{bmatrix} 1 & 1 \\ -2 & -3 \end{bmatrix}$$

$\because$ $$\Lambda = \begin{bmatrix} -2 & 0 \\ 0 & -3 \end{bmatrix}$$

$\therefore$ $$e^{\Lambda t} = \begin{bmatrix} e^{-2t} & 0 \\ 0 & e^{-3t} \end{bmatrix}$$

and $$e^{\mathbf{A}t} = \mathbf{P}\, e^{\Lambda t}\, \mathbf{P}^{-1}$$

or $$e^{\mathbf{A}t} = \begin{bmatrix} 1 & 1 \\ -2 & -3 \end{bmatrix}\begin{bmatrix} e^{-2t} & 0 \\ 0 & e^{-2t} \end{bmatrix}\begin{bmatrix} 3 & 1 \\ -2 & -1 \end{bmatrix}$$

$$= \begin{bmatrix} e^{-2t} & -e^{-3t} \\ -2e^{-2t} & -3e^{-3t} \end{bmatrix}\begin{bmatrix} 3 & 1 \\ -2 & -1 \end{bmatrix}$$

$$= \begin{bmatrix} 3e^{-2t} - 2e^{-3t} & -e^{-2t} - e^{-3t} \\ -6e^{-2t} + 6e^{-3t} & -2e^{-2t} + 3e^{-3t} \end{bmatrix}$$

Invariance of eigen values under similarity transformation

The similarity transformation

$$\mathbf{x} = \mathbf{P}\bar{\mathbf{x}}$$

P is a non-singular matrix used for transforming the system matrix **A** into a desired form. The determinant $|\, s\mathbf{I} - \mathbf{A}\,|$ on such transformation remains unchanged as shown below :

The system matrix on transformation becomes $\mathbf{P}^{-1}\,\mathbf{AP}$ and the determinant is $|\, s\mathbf{I} - \mathbf{P}^{-1}\mathbf{AP}\,|$ which can be written as :

$$= |\, s\mathbf{I} - \mathbf{P}^{-1}\,\mathbf{AP}\,|$$
$$= s\mathbf{P}^{-1}\mathbf{P} - \mathbf{P}^{-1}\,\mathbf{AP}$$

$$= |\mathbf{P}^{-1}(s\mathbf{I} - \mathbf{A})\mathbf{P}|$$
$$= |\mathbf{P}^{-1}||(s\mathbf{I} - \mathbf{A})||\mathbf{P}|$$
$$= |\mathbf{P}^{-1}||\mathbf{P}||(s\mathbf{I} - \mathbf{A})|$$
$$= |\mathbf{P}^{-1}\mathbf{P}||s\mathbf{I} - \mathbf{A}|$$
$$= |s\mathbf{I} - \mathbf{A}|$$

Since the eigen values can be determined by using $|s\mathbf{I} - \mathbf{A}| = 0$, therefore the invariance of eigen values under similarity transformation is established.

9.6.2 Properties of State Transition matrix

(1) $\phi(t) = e^{\mathbf{A}t}$ put $t = 0$

$\therefore$ $\phi(0) = e^{\mathbf{A}\times 0}$ $\quad\therefore\quad \phi(0) = \mathbf{I}$

(2) $\phi(t) = e^{\mathbf{A}t}$

post multiply both sides by $e^{-\mathbf{A}t}$

$\therefore$ $\phi(t)\, e^{-\mathbf{A}t} = e^{\mathbf{A}t}\, e^{-\mathbf{A}t} = \mathbf{I}$

premultiply both sides by $\phi^{-1}(t)$

$\therefore$ $\phi^{-1}(t)\,\phi(t)e^{-\mathbf{A}t} = \phi^{-1}(t) = \mathbf{I} = \phi^{-1}(t)$

$\because$ $e^{-\mathbf{A}t} = \phi(-t)$

$\therefore$ $\phi^{-1}(t)\,\phi(t)\,\phi(-t) = \phi^{-1}(t)$

$\therefore$ $\phi^{-1}(t) = \phi(-t)$

(3) $\phi(t_2 - t_1)\,\phi(t_1 - t_0) = e^{\mathbf{A}(t_2 - t_1)}\, e^{\mathbf{A}(t_1 - t_0)} = e^{\mathbf{A}(t_2 - t_0)}$

$\therefore$ $\phi(t_2 - t_1)\,\phi(t_1 - t_0) = \phi(t_2 - t_0)$

(4) $[\phi(t)]^k = e^{\mathbf{A}t}\, e^{\mathbf{A}t}\, e^{\mathbf{A}t} \ldots = e^{\mathbf{A}kt}$ $\quad$ upto k terms $= e^{\mathbf{A}kt}$

$\therefore$ $[\phi(t)]^k = \phi(kt)$ $\quad$ k = positive integer

(5) $\dot{\phi}(t) = \dfrac{d}{dt} e^{\mathbf{A}t} = \mathbf{A}e^{\mathbf{A}t}$

$\therefore$ $\dot{\phi}(t) = \mathbf{A}\,\phi(t)$

9.7 TRANSFER MATRIX

The matrix relating Laplace transform of output to Laplace transform of input of state space representation of a control system is known as transfer matrix.

The state equations of a system are given below :

$$\dot{\mathbf{x}} = \mathbf{A}\mathbf{x} + \mathbf{B}u \qquad \ldots(9.77)$$
$$y = \mathbf{C}\mathbf{x} + Du \qquad \ldots(9.78)$$

where $\mathbf{x}$ is state vector $(n \times 1)$, u is input, and y is output

$\mathbf{A}$ is matrix $(n \times n)$, $\mathbf{C}$ is matrix $(1 \times n)$ and D is scalar

Taking Laplace transform on both sides of Eqs. (9.77) and (9.78)

$$s\mathbf{X}(s) - \mathbf{x}(0) = \mathbf{A}\mathbf{X}(s) + \mathbf{B}U(s) \qquad \ldots(9.79)$$
$$Y(s) = \mathbf{C}\mathbf{X}(s) + DU(s) \qquad \ldots(9.80)$$

As per definition of transfer function $\mathbf{x}(0) = 0$

$\therefore$ $s\mathbf{X}(s) = \mathbf{A}\mathbf{X}(s) + \mathbf{B}U(s)$

or $\mathbf{X}(s) = (s\mathbf{I} - \mathbf{A})^{-1}\mathbf{B}U(s)$ $\qquad \ldots(9.81)$

Substituting (9.81) in (9.80), the following equation is obtained

$$Y(s) = \mathbf{C}(s\mathbf{I} - \mathbf{A})^{-1}\,\mathbf{B}U(s) + DU(s)$$

or

$$Y(s) = [\mathbf{C}(s\mathbf{I} - \mathbf{A})^{-1}\,\mathbf{B} + D]\,U(s) \quad ...(9.82)$$

The transfer matrix is $G(s) = \dfrac{Y(s)}{U(s)} = \mathbf{C}(s\mathbf{I} - \mathbf{A})^{-1}\,\mathbf{B} + D$ ...(9.83)

If D is null matrix, the transfer matrix takes the following form :

$$G(s) = \frac{Y(s)}{U(s)} = \mathbf{C}(s\mathbf{I} - \mathbf{A})^{-1}\,\mathbf{B} \quad ...(9.84)$$

Transfer matrix (9.84) can also be represented as

$$G(s) = \frac{Q(s)}{|(s\mathbf{I} - \mathbf{A})|} \quad ...(9.85)$$

where $Q(s)$ is a polynomial in s and $|s\mathbf{I} - \mathbf{A}| = 0$ is characteristics polynomial of $G(s)$. Eigen values of $(s\mathbf{I} - \mathbf{A})$ are identical to the poles of $G(s)$.

For a multi-input multi-output system the transfer matrix is given by

$$\mathbf{G}(s) = \frac{\mathbf{Y}(s)}{\mathbf{U}(s)}$$

Therefore, $\mathbf{Y}(s) = \mathbf{G}(s)\,\mathbf{U}(s)$ ...(9.86)

In expanded form Eq. 9.70 can be written as

$$\begin{vmatrix} Y_1(s) \\ Y_2(s) \\ \vdots \\ \vdots \\ \vdots \\ Y_p(s) \end{vmatrix} = \begin{bmatrix} G_{11}(s) & G_{12}(s) & ... G_{1m}(s) \\ G_{21}(s) & G_{22}(s) & .... G_{2m}(s) \\ \vdots & \vdots & \vdots \\ \vdots & \vdots & \vdots \\ \vdots & \vdots & \vdots \\ G_{p1}(s) & G_{p2}(s) & ... G_{pm}(s) \end{bmatrix} \begin{bmatrix} U_1(s) \\ U_2(s) \\ \vdots \\ \vdots \\ \vdots \\ U_m(s) \end{bmatrix} \quad ...(9.87)$$

The transfer function for multiple-input multiple-output system is given below

$$\mathbf{G}(s) = \mathbf{C}(s\mathbf{I} - \mathbf{A})^{-1}\,\mathbf{B} + \mathbf{D} \quad ...(9.88)$$

9.8 CONTROLLABILITY

The controllability is in relation to transfer of a system from one state to another by appropriate input controls in a finite time. Sometimes such a control is not possible and this can be verified by using controllability test matrix. The concept of controllability and derivation of controllability test matrix are given below :

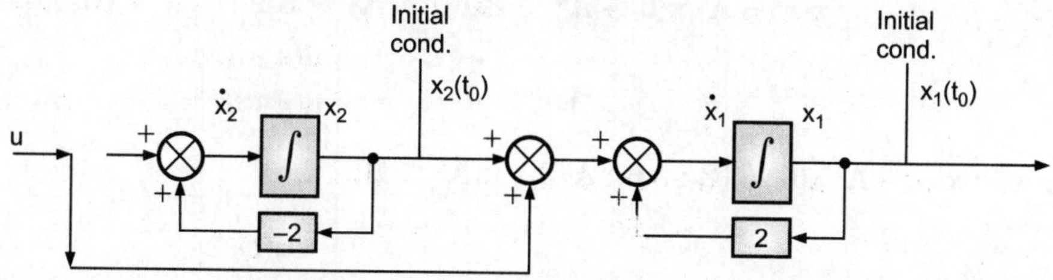

Fig. 9.8.1. Block diagram for Eqs. (9.89) and (9.90).

The block diagram representation of state Eqs. (9.89) and (9.90) is shown in Fig. 9.8.1.

$$\dot{x}_1 = 2x_1 + x_2 \quad + u \qquad \text{...(9.89)}$$

$$\dot{x}_2 = \quad - 2x_2 \qquad \text{...(9.90)}$$

In the block diagram representation (Fig. 9.8.1), it is noted that the state x_1 is controllable whereas the state x_2 is independent of the input control u. Hence, the input has no control over the state x_2 and as such it is not possible to operate the state x_2 from its initial state $x_2(t_0)$ to the final state $x_2\ (t_f)$ in a finite time period $(t_f - t_0)$, using input control u. Thus, the system is state uncontrollable. The term controllability is also referred to as state controllability.

A system is said to controllable, if it is possible to have an input u to transfer the system from any given initial state $x(t_0)$ to any given final state $x(t_f)$ over a specified interval of time $(t_f - t_0)$.

As stated above, the dependence of state variables on the input gives the concept of controllability. Consider the following state equations.

$$\dot{\mathbf{x}} = \mathbf{Ax} + \mathbf{Bu} \qquad \text{...(9.91)}$$

$$\mathbf{y} = \mathbf{Cx} + \mathbf{Du} \qquad \text{...(9.92)}$$

where $\dot{\mathbf{x}}$ = velocity vector $(n \times 1)$
$\mathbf{x}$ = state vector $(n \times 1)$
$\mathbf{u}$ = input vector $(m \times 1)$
$\mathbf{y}$ = output vector $(p \times 1)$
$\mathbf{A}$ = matrix $(n \times n)$
$\mathbf{B}$ = matrix $(n \times m)$
$\mathbf{C}$ = matrix $(p \times n)$
$\mathbf{D}$ = matrix $(p \times m)$

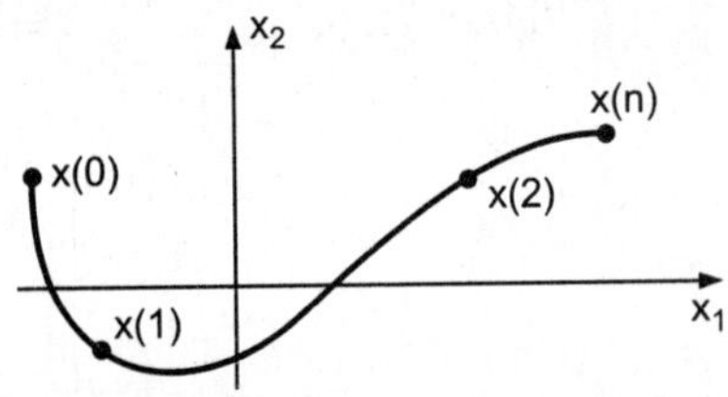

Fig. 9.8.2. Control of state vector.

The initial state of a system having matrix **A** (2 × 2) can be represented as **x**(0) and final as **x**(n) as shown in Fig. 9.8.2. The condition under which an input control can drive the system from initial state **x**(0) to final state **x**(n), concerns controllability. The mathematical relations given below explain the transfer process of a control system from initial state $x(0)$ to final state **x**(n),

$$\mathbf{x}(1) = \mathbf{Ax}(0) + \mathbf{Bu}(0) \qquad \text{...(9.93)}$$

$$\mathbf{x}(2) = \mathbf{Ax}(1) + \mathbf{Bu}(1) = \mathbf{A}\,\{\mathbf{Ax}(0) + \mathbf{Bu}(0)\} + \mathbf{Bu}(1)$$

$$\mathbf{x}(3) = \mathbf{Ax}(2) + \mathbf{Bu}(2) = \mathbf{A}\{\mathbf{Ax}(1) + \mathbf{Bu}(1)\} + \mathbf{Bu}(2)$$

$$\vdots$$

$$\vdots$$

$$\mathbf{x}(n) = \mathbf{A}^n\,\mathbf{x}(0) + \mathbf{A}^{n-1}\,\mathbf{Bu}(0) + \mathbf{A}^{n-2}\,\mathbf{Bu}(1) + \ldots + \mathbf{Bu}(n-1)$$

Thus,
$$\mathbf{x}(n) - \mathbf{A}^n\,\mathbf{x}(0) = [\mathbf{B} : \mathbf{AB} : \mathbf{A}^2\mathbf{B} : \ldots\ \mathbf{A}^{n-1}\,\mathbf{B}] \begin{bmatrix} \mathbf{u}(n-1) \\ \mathbf{u}(n-2) \\ \vdots \\ \vdots \\ \mathbf{u}(1) \\ \mathbf{u}(0) \end{bmatrix} \qquad \text{...(9.94)}$$

For the system described by equations (9.91) and (9.92), L.H.S. of the equation (9.94) gives initial and final state. For the system to be completely state controllable, it is necessary and sufficient that the $n \times nm$ matrix given below has a rank of n.

$$\mathbf{U} = [\mathbf{B} : \mathbf{AB} : \mathbf{A}^2\mathbf{B} : \ldots : \mathbf{A}^{n-1}\mathbf{B}] \quad \ldots(9.95)$$

The matrix **U** is known as controllability test matrix, which is $n \times nm$

The controllability condition of a system depends on the coefficient matrices **A** and **B**, and it is said that the pair **A**, **B** is controllable indicating that the rank of the test matrix is n.

However, if controllability matrix **U** is $n \times n$ *i.e.* square matrix, then the condition for state controllability is $|\mathbf{U}| \neq 0$ *i.e.* matrix be non singular.

Example 9.8.1. *The state equations of a system are given below :*

$$\dot{x}_1 = x_1 + x_2 + u$$
$$\dot{x}_2 = \quad - x_2$$

Check for controllability.

Solution. $$\begin{bmatrix} \dot{x}_1 \\ \dot{x}_2 \end{bmatrix} = \begin{bmatrix} 1 & 1 \\ 0 & -1 \end{bmatrix} \begin{bmatrix} x_1 \\ x_2 \end{bmatrix} + \begin{bmatrix} 1 \\ 0 \end{bmatrix} u$$

$\therefore$ $$\mathbf{B} = \begin{bmatrix} 1 \\ 0 \end{bmatrix} \quad \text{and} \quad \mathbf{A} = \begin{bmatrix} 1 & 1 \\ 0 & -1 \end{bmatrix}$$

$$\mathbf{AB} = \begin{bmatrix} 1 & 1 \\ 0 & -1 \end{bmatrix} \begin{bmatrix} 1 \\ 0 \end{bmatrix} = \begin{bmatrix} 1 \\ 0 \end{bmatrix}$$

$$\mathbf{U} = [\mathbf{B} : \mathbf{AB}] = \begin{bmatrix} 1 & 1 \\ 0 & 0 \end{bmatrix}$$

$$|\mathbf{U}| = (1 \times 0 - 1 \times 0) = 0$$

The test matrix **U** is found to be singular, hence the rank of **U** is not equal to $n(n = 2)$ and the system is uncontrollable. **Ans.**

Example 9.8.2. *Verify whether the following system is controllable :*

$$\begin{bmatrix} \dot{x}_1 \\ \dot{x}_2 \end{bmatrix} = \begin{bmatrix} -2 & 0 \\ 0 & -1 \end{bmatrix} \begin{bmatrix} x_1 \\ x_2 \end{bmatrix} + \begin{bmatrix} 1 \\ 1 \end{bmatrix} u.$$

Solution. $$\mathbf{B} = \begin{bmatrix} 1 \\ 1 \end{bmatrix} \text{ and } \mathbf{A} = \begin{bmatrix} -2 & 0 \\ 0 & -1 \end{bmatrix}$$

$$\mathbf{AB} = \begin{bmatrix} -2 & 0 \\ 0 & -1 \end{bmatrix} \begin{bmatrix} 1 \\ 1 \end{bmatrix} = \begin{bmatrix} -2 \\ -1 \end{bmatrix}$$

$\therefore$ $$\mathbf{U} = [\mathbf{B} : \mathbf{AB}] = \begin{bmatrix} 1 & -2 \\ 1 & -1 \end{bmatrix}$$

$$|\mathbf{U}| = \{1 \times (-1) - 2 \times (-2)\} = 1$$

The test matrix **U** is found to nonsingular, hence the rank of the test matrix **U** is equal to n $(n = 2)$ and the system is controllable.

9.9 OBSERVABILITY

The method of determining the state of a system by observing its output concerns observability. Sometimes it is desirable to get information about the state variables from the measurements of the output and input. In case, it is not possible to obtain any one of the states from the measurements of the output, the state is said to be unobservable and the system as a whole is unobservable. The condition under which the state of a system is measured from the output is explained below :

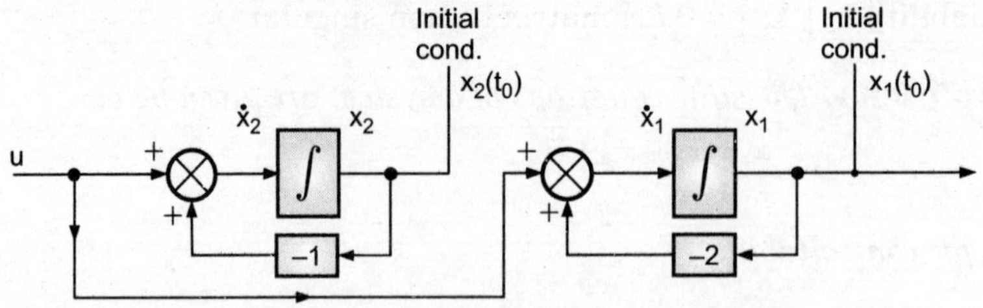

Fig. 9.9.1. Block diagram for Eqs. (9.80) and (9.81).

The block diagram representation of the state Eqs. (9.96) and (9.97) is shown in Fig. 9.9.1.

$$\dot{x}_1 = -2x_1 \qquad + u \qquad ...(9.96)$$

$$\dot{x}_2 = \qquad -2x_2 + u \qquad ...(9.97)$$

In the block diagram (Fig. 9.9.1) it is seen that, the state x_2 is not connected to the output. If the output y is measured, the state x_1 is known, but state x_2 cannot be measured (observed). Thus the system is unobservable.

A system is said to be observable if, every state variable affects the system output.

Consider the system described by the following state equations :

$$\dot{\mathbf{x}} = \mathbf{Ax} + \mathbf{Bu} \qquad ...(9.98)$$

$$\mathbf{y} = \mathbf{Cx} \qquad ...(9.99)$$

where $\dot{\mathbf{x}}$ = velocity vector $(n \times 1)$,
$\mathbf{x}$ = state vector $(n \times 1)$,
$\mathbf{u}$ = input vector $(m \times 1)$,
$\mathbf{y}$ = output vector $(p \times 1)$,
$\mathbf{A}$ = matrix $(n \times n)$,
$\mathbf{B}$ = matrix $(n \times m)$,
$\mathbf{C}$ = matrix $(p \times n)$,

By definition, $\mathbf{y}(0) = \mathbf{Cx}(0)$

Further, at the next instant

$$\mathbf{y}(1) = \mathbf{Cx}(1) = \mathbf{CAx}(0) \qquad \because \mathbf{x}(1) = \mathbf{Ax}(0)$$

$$\mathbf{y}(2) = \mathbf{CA}^2\,\mathbf{x}(0)$$

$$\mathbf{y}(3) = \mathbf{CA}^3\,\mathbf{x}(0)$$

$$\vdots$$

$$\vdots$$

$$\mathbf{y}(n-1) = \mathbf{CA}^{n-1}\,\mathbf{x}(0)$$

or
$$\begin{bmatrix} \mathbf{y}(0) \\ \mathbf{y}(1) \\ \mathbf{y}(2) \\ \vdots \\ \vdots \\ \mathbf{y}(n-1) \end{bmatrix} = \begin{bmatrix} \mathrm{C} \\ \mathrm{CA} \\ \mathrm{CA}^2 \\ \vdots \\ \vdots \\ \mathbf{CA}^{n-1} \end{bmatrix} \mathbf{x}(0)$$

$$\therefore \qquad \begin{bmatrix} \mathbf{C} \\ \mathbf{CA} \\ \mathbf{CA}^2 \\ \vdots \\ \vdots \\ \mathbf{CA}^{n-1} \end{bmatrix}^{-1} \begin{bmatrix} \mathbf{y}(0) \\ \mathbf{y}(1) \\ \mathbf{y}(2) \\ \vdots \\ \vdots \\ \mathbf{y}(n-1) \end{bmatrix} = \mathbf{x}(0) \qquad \text{...(9.100)}$$

For unique determination of state $\mathbf{x}(0)$, the matrix (9.100) must have inverse and therefore

$$\begin{bmatrix} \mathbf{C} \\ \mathbf{CA} \\ \mathbf{CA}^2 \\ \vdots \\ \vdots \\ \mathbf{CA}^{n-1} \end{bmatrix} \qquad \text{...(9.101)}$$

the matrix (9.101) is to have a rank of n. However, if the matrix (9.101) is $n \times n$ *i.e.* square matrix then for establishing observability, it should be nonsingular.

The matrix (9.101) is known as observability test matrix which can be written into an $n \times np$ row matrix form and denoted as **V** being given below:

$$\mathbf{V} = [\mathbf{C}^T : \mathbf{A}^T\mathbf{C}^T : (\mathbf{A}^T)^2\, \mathbf{C}^T : \ldots : (\mathbf{A}^T)^{n-1}\, \mathbf{C}^T] \qquad \text{...(9.102)}$$

If the observability condition is satisfied *i.e.* observability test matrix **V** be of rank n, then it is said that **C**, **A** is observable pair.

Example 9.9.1. *Check observability of the system given below :*

(*a*) $\dot{x}_1 = x_2$

$\dot{x}_2 = -2x_1 - 3x_2 + u$ *and* $y = x_1 + x_2$

(*b*) $\dot{x}_1 = -2x + x_2 + u$

$\dot{x}_2 = \quad -2x_2 + u$ *and* $y = x_1 + x_2$

Solution.
$$\begin{bmatrix} \dot{x}_1 \\ \dot{x}_2 \end{bmatrix} = \begin{bmatrix} 0 & 1 \\ -2 & -3 \end{bmatrix} \begin{bmatrix} x_1 \\ x_2 \end{bmatrix} + \begin{bmatrix} 0 \\ 1 \end{bmatrix} u$$

$$y = [1 \;\; 1] \begin{bmatrix} x_1 \\ x_2 \end{bmatrix}$$

$$\mathbf{A} = \begin{bmatrix} 0 & 1 \\ -2 & -3 \end{bmatrix}, \mathbf{C} = [1 \ \ 1]$$

$$\mathbf{A}^T = \begin{bmatrix} 0 & -2 \\ 1 & -3 \end{bmatrix}, \mathbf{C}^T = \begin{bmatrix} 1 \\ 1 \end{bmatrix}$$

$$\mathbf{A}^T\mathbf{C}^T = \begin{bmatrix} 0 & -2 \\ 1 & -3 \end{bmatrix} \begin{bmatrix} 1 \\ 1 \end{bmatrix} = \begin{bmatrix} -2 \\ -2 \end{bmatrix}$$

The observability test matrix is

$$\mathbf{V} = [\mathbf{C}^T : \mathbf{A}^T\mathbf{C}^T] = \begin{bmatrix} 1 & -2 \\ 1 & -2 \end{bmatrix}$$

$$|\mathbf{V}| = \{1 \times (-2) - 1 \times (-2)\} = 0$$

The observability test matrix V is singular. The rank of matrix V is not equal to $n(n = 2)$. Therefore, the system is unobservable. Ans.

(*b*)
$$\begin{bmatrix} \dot{x}_1 \\ \dot{x}_2 \end{bmatrix} = \begin{bmatrix} -2 & 1 \\ 0 & 1 \end{bmatrix} \begin{bmatrix} x_1 \\ x_2 \end{bmatrix} + \begin{bmatrix} 1 \\ 1 \end{bmatrix} u$$

$$y = [1 \ \ 1] \begin{bmatrix} x_1 \\ x_2 \end{bmatrix}$$

$$\mathbf{A} = \begin{bmatrix} -2 & 1 \\ 0 & -1 \end{bmatrix}, \mathbf{C} = [1 \ \ 1]$$

$$\mathbf{A}^T = \begin{bmatrix} -2 & 0 \\ 1 & -1 \end{bmatrix}, \mathbf{C}^T = \begin{bmatrix} 1 \\ 1 \end{bmatrix}$$

$$\mathbf{A}^T\mathbf{C}^T = \begin{bmatrix} -2 & 0 \\ 1 & -1 \end{bmatrix} \begin{bmatrix} 1 \\ 1 \end{bmatrix} = \begin{bmatrix} -2 \\ 0 \end{bmatrix}$$

The observability test matrix is

$$\mathbf{V} = [\mathbf{C}^T : \mathbf{A}^T\mathbf{C}^T] = \begin{bmatrix} 1 & -2 \\ 1 & 0 \end{bmatrix}$$

$$|\mathbf{V}| = \{1 \times 0 - 1 \times (-2)\} = 2$$

The observability test matrix V is nonsingular. The rank of the matrix V is equal to n ($n = 2$). Therefore, the system is observable. Ans.

9.10 SOLVED EXAMPLES

Example 9.10.1. *For the electrical network shown in Fig. 9.10.1, determine the state model. Consider i_1, i_2 and v_c as state variables. The output variable are i_1 and i_2.*

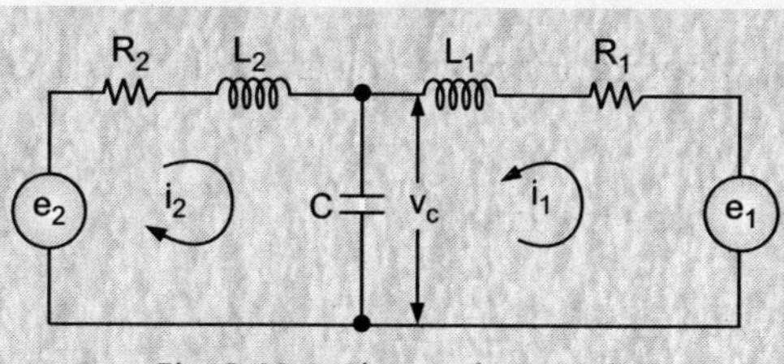

Fig. 9.10.1. Electrical network.

Solution. The differential equations are written below :

$$R_1 i_1 + L_1 \frac{di_1}{dt} + v_C = e_1$$

$$R_2 i_2 + L_2 \frac{di_2}{dt} + v_C = e_2 \qquad \text{and} \qquad c\frac{dv}{dt} = i_1 + i_2$$

on rearranging above equations :

$$\frac{di_1}{dt} = -\frac{R_1}{L_1} i_1 - \frac{1}{L_1} v_C + \frac{1}{L_1} e_1$$

$$\frac{di_2}{dt} = -\frac{R_2}{L_2} - \frac{1}{L_2} v_C + \frac{1}{L_2} e_2 \qquad \text{and} \qquad \frac{dv_C}{dt} = \frac{1}{C} i_1 + \frac{1}{C} i_2$$

The output equation is :

$$y_1 = i_1$$

and

$$y_2 = i_2$$

Put $x_1 = i_1$, $x_2 = i_2$ and $x_3 = v_c$, therefore,

$$\begin{bmatrix} \dot{x}_1 \\ \dot{x}_2 \\ \dot{x}_3 \end{bmatrix} = \begin{bmatrix} -\frac{R_1}{L_1} & 0 & -\frac{1}{L_1} \\ 0 & -\frac{R_2}{L_2} & -\frac{1}{L_2} \\ \frac{1}{C} & \frac{1}{C} & 0 \end{bmatrix} \begin{bmatrix} x_1 \\ x_2 \\ x_3 \end{bmatrix} + \begin{bmatrix} \frac{1}{L_1} & 0 \\ 0 & \frac{1}{L_2} \\ 0 & 0 \end{bmatrix} \begin{bmatrix} e_1 \\ e_2 \\ 0 \end{bmatrix}$$

and

$$[y] = \begin{bmatrix} 1 & 0 \\ 0 & 1 \end{bmatrix} \begin{bmatrix} x_1 \\ x_2 \end{bmatrix} \quad \textbf{Ans.}$$

Example 9.10.2. *Determine the state model for the electrical circuit shown in Fig. 9.10.2. Select state and output variables.*

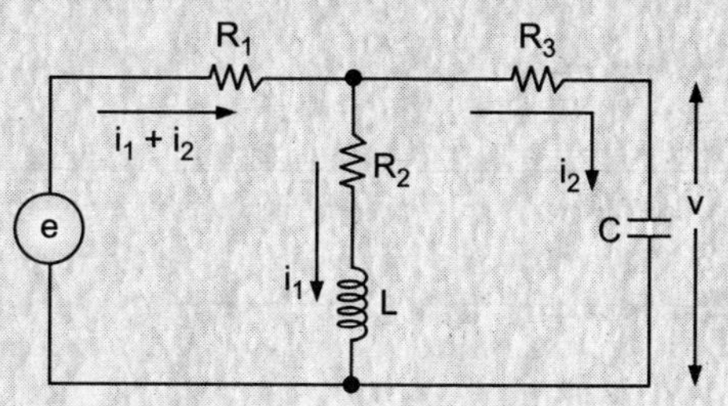

Fig 9.10.2. Electrical circuit.

Solution. The current flowing through the capacitor is $i_2 = c\,\dfrac{dv}{dt}$, where v is the voltage across the capacitor.

The differential equations are given below :

$$e = R_1 i_1 + R_1 C\,\frac{dv}{dt} + R_2 i_1 + L\,\frac{di_1}{dt} \qquad ...(1)$$

and

$$e = R_1 i_1 + R_1 C\,\frac{dv}{dt} + R_3 C\,\frac{dv}{dt} + v \qquad ...(2)$$

on rearranging Eqs. (1) and (2) :

$$L\frac{di_1}{dt} = e - \left(R_1 i_1 + R_1 C\frac{dv}{dt} + R_2 i_1\right) \qquad ...(3)$$

and

$$C\frac{dv}{dt} = \frac{e}{R_1 + R_3} - \frac{R_1 i_1}{R_1 + R_3} - \frac{v}{R_1 + R_3} \qquad ...(4)$$

Substitute for $C\,\dfrac{dv}{dt}$ in Eq. (3) from (2), therefore,

$$L\,\frac{di}{dt} = e - \left\{R_1 i_1 + R_1\left(\frac{e}{R_1 + R_3} - \frac{R_1 i_1}{R_1 + R_3} - \frac{v}{R_1 + R_3}\right) + R_2 i_1\right\}$$

$$= e - \left\{R_1 i_1 + \frac{R_1 e}{R_1 + R_3} - \frac{R_1^2}{R_1 + R_3}\,i_1 - \frac{R_1 v}{R_1 + R_3} + R_2 i_1\right\}$$

$$= e - \left\{\left(R_1 + R_2 - \frac{R_1^2}{R_1 + R_3}\right) i_1 + \frac{R_1 e}{R_1 + R_3} - \frac{R_1 v}{R_1 + R_3}\right\}$$

$$= -\frac{(R_1 R_2 + R_2 R_3 + R_3 R_1)}{R_1 + R_3}\,i_1 + \frac{R_1 v}{R_1 + R_3} + \frac{R_3 e}{R_1 + R_3}$$

or

$$L\,\frac{di_1}{dt} = -\frac{(R_1 R_2 + R_2 R_3 + R_3 R_1)}{R_1 + R_3}\,i_1 + \frac{R_1 v}{R_1 + R_3} + \frac{R_3 e}{R_1 + R_3} \qquad ...(5)$$

On rearranging Eqs. (5) and (4) :

$$\frac{di_1}{dt} = -\frac{(R_1 R_2 + R_2 R_3 + R_3 R_1)\,i_1}{L(R_1 + R_3)} + \frac{R_1 v}{L(R_1 + R_3)} + \frac{R_3 e}{L(R_1 + R_3)} \qquad ...(6)$$

$$\frac{dv}{dt} = -\frac{R_1 i_1}{C(R_1 + R_3)} - \frac{v}{C(R_1 + R_3)} \qquad ...(7)$$

Put $i_1 = x_1$ and $v = x_2$, therefore,

$$\dot{x}_1 = -\frac{(R_1 R_2 + R_2 R_3 + R_3 R_1)x_1}{L(R_1 + R_3)} + \frac{R_1 x_2}{L(R_1 + R_3)} + \frac{R_3 e}{L(R_1 + R_3)} \qquad ...(8)$$

$$\dot{x}_2 = -\frac{R_1 x_1}{C(R_1 + R_3)} - \frac{x_2}{C(R_1 + R_3)} + \frac{e}{C(R_1 + R_3)} \qquad ...(9)$$

Select $(i_1 + i_2)$ as output variable, therefore,

$$y = (i_1 + i_2) \qquad ...(10)$$

The state model is given below :

$$\begin{bmatrix} \dot{x}_1 \\ \dot{x}_2 \end{bmatrix} = \begin{bmatrix} -\dfrac{(R_1R_2 + R_2R_3 + R_3R_1)}{L(R_1 + R_3)} & \dfrac{R_1}{L(R_1 + R_3)} \\ -\dfrac{R_1}{C(R_1 + R_3)} & -\dfrac{1}{C(R_1 + R_3)} \end{bmatrix} \begin{bmatrix} x_1 \\ x_2 \end{bmatrix} + \begin{bmatrix} \dfrac{R_3}{L(R_1 + R_3)} & 0 \\ 0 & \dfrac{1}{C(R_1 + R_3)} \end{bmatrix} \begin{bmatrix} e \\ e \end{bmatrix}$$

and $\quad y = [1\ 1] \begin{bmatrix} x_1 \\ x_2 \end{bmatrix}$ **Ans.**

Example 9.10.3. *Obtain the state equations for the field controlled D.C. motor shown in Fig. 9.10.3*

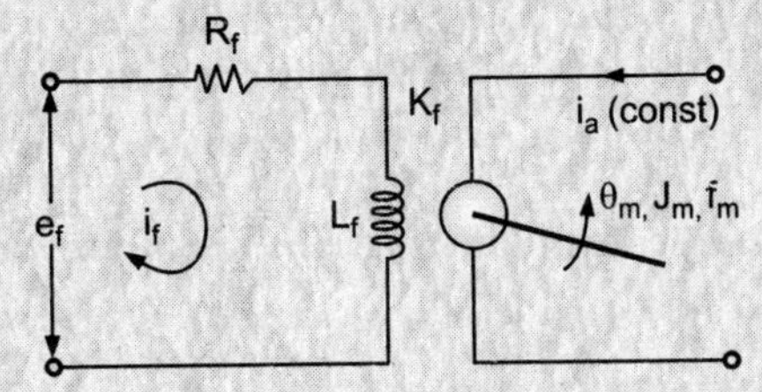

Fig. 9.10.3. Field controlled D.C. motor.

Solution. The differential equations for the field control D.C. motor are given below :

$$e_f = R_f i_f + L_f \frac{di_f}{dt} \qquad ...(1)$$

Torque developed : $T_e = K_f i_f$

Load torque : $\quad T_m = J_m \dfrac{d^2\theta_m}{dt^2} + f_m \dfrac{d\theta_m}{dt}$

Load torque equals developed torque. Therefore,

$$K_f i_f = J_m \frac{d^2\theta_m}{dt^2} + f_m \frac{d\theta_m}{dt} \qquad ...(2)$$

On rearranging Eqs. (1) and (2) following equations are obtained :

$$\frac{di_f}{dt} = -\frac{R_f i_f}{L_f} + \frac{e_f}{L_f} \qquad ...(3)$$

$$\frac{d^2\theta_m}{dt^2} = -\frac{f_m}{J_m} \cdot \frac{d\theta_m}{dt} + \frac{K_f i_f}{J_m} \qquad ...(4)$$

Let, $i_f = x_1$, $\theta_m = x_2$ and $\dfrac{d\theta_m}{dt} = x_3$ be state variables and $y = \theta_m$ be the output variable.

Therefore, state equations are as written below :

$$\dot{x}_1 = -\frac{R_f}{L_f}x_1 + \frac{e_f}{L_f}$$

$$\dot{x}_2 = x_3$$

$$\dot{x}_3 = \frac{K_f}{J_m}x_1 - \frac{f_m}{J_m}x_3 \quad \text{and} \quad y = \theta_m$$

or

$$\begin{bmatrix}\dot{x}_1\\ \dot{x}_2\\ \dot{x}_3\end{bmatrix} = \begin{bmatrix}-\frac{R_f}{L_f} & 0 & 0\\ 0 & 0 & 1\\ \frac{K_f}{J_m} & 0 & -\frac{f_m}{J_m}\end{bmatrix}\begin{bmatrix}x_1\\ x_2\\ x_3\end{bmatrix} + \begin{bmatrix}\frac{1}{L_f}\\ 0\\ 0\end{bmatrix}e_f$$

$$y = [0\ 1\ 0]\begin{bmatrix}x_1\\ x_2\\ x_3\end{bmatrix} \quad \textbf{Ans.}$$

Example 9.10.4. *Draw the state block diagram for the transfer function given below and obtain state equations.*

(*a*) $\frac{C(s)}{R(s)} = \frac{1}{(s+1)(s+3)}$ *and* (*b*) $\frac{C(s)}{R(s)} = \frac{2s+1}{s^2+2}$.

Solution. (*a*) The transfer function is composed of two terms $1/(s + 1)$ and $1/(s + 3)$. The state block diagram for individual terms is drawn and then joined as shown in Fig. 9.10.4.

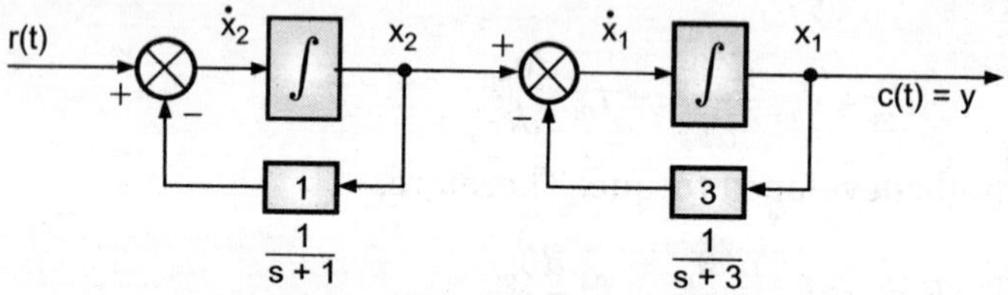

Fig. 9.10.4. State block diagram for $C(s)/R(s) = 1/(s + 1)(s + 3)$.

In Fig. 9.10.4 the output of each integrator is designated by a state variable and thus the state equations are

$$\dot{x}_1 = -3x_1 + x_2$$

$$\dot{x}_2 = -x_2 + r(t) \quad \text{and} \quad y = x_1$$

In matrix form the state equations are

$$\begin{bmatrix}\dot{x}_1\\ \dot{x}_2\end{bmatrix} = \begin{bmatrix}-3 & 1\\ 0 & -1\end{bmatrix}\begin{bmatrix}x_1\\ x_2\end{bmatrix} + \begin{bmatrix}0\\ 1\end{bmatrix}r(t) \quad \text{and} \quad y = [1\ \ 0]\begin{bmatrix}x_1\\ x_2\end{bmatrix} \quad \textbf{Ans.}$$

(*b*) The transfer function can be re-written as

$$\frac{C(s)}{R(s)} = \frac{1}{s^2 + 2} + \frac{2s}{s^2 + 2}$$

The state diagram for the term $1/(s^2 + 2)$ is shown in Fig. 9.10.5.

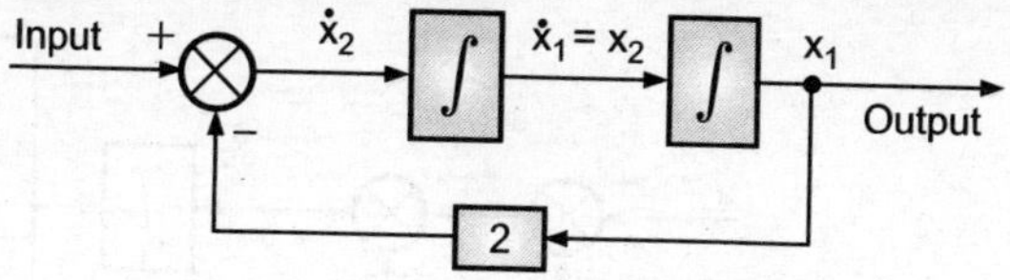

Fig. 9.10.5. State block diagram for term $1/(s^2 + 2)$.

The derivative signal $s/(s^2 + 2)$ is available before the last integrator in Fig. 9.10.5, therefore, the state block diagram for transfer function $(2s + 1)/(s^2 + 1)$ takes the form as shown in Fig. 9.10.6.

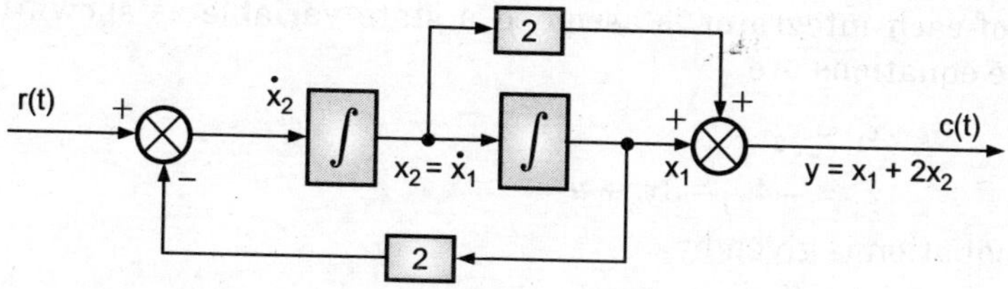

Fig. 9.10.6. State block diagram for $C(s)/R(s) = (2s + 1)/(s^2 + 1)$.

As per Fig. 9.10.6, the state equations are

$$\dot{x}_1 = x_2$$

$$\dot{x}_2 = -2x_1 + r(t)$$

and

$$y = x_1 + 2x_2$$

The matrix form of the state equations are

$$\begin{bmatrix} \dot{x}_1 \\ \dot{x}_2 \end{bmatrix} = \begin{bmatrix} 0 & 1 \\ -2 & 0 \end{bmatrix} \begin{bmatrix} x_1 \\ x_2 \end{bmatrix} + \begin{bmatrix} 0 \\ 1 \end{bmatrix} r(t)$$

and

$$y = [1 \ \ 2] \begin{bmatrix} x_1 \\ x_2 \end{bmatrix}$$ **Ans.**

Example 9.10.5. *Obtain state equations for the differential equation*

$$\frac{d^2y}{dt^2} + \frac{3dy}{dt} + 4y = \frac{du}{dt} + 3u.$$

Solution.

$$\frac{d^2y}{dt^2} + \frac{3dy}{dt} + 4y = \frac{du}{dt} + 3u$$

Assuming zero initial conditions and taking Laplace transform on both sides of the above equation :

$$Y(s)\,(s^2 + 3s + 4) = U(s)\,(s + 3) \quad \text{or} \quad \frac{Y(s)}{U(s)} = \frac{s + 3}{s^2 + 3s + 4}$$

The state block diagram as per above transfer function is shown as drawn in Fig. 9.10.7.

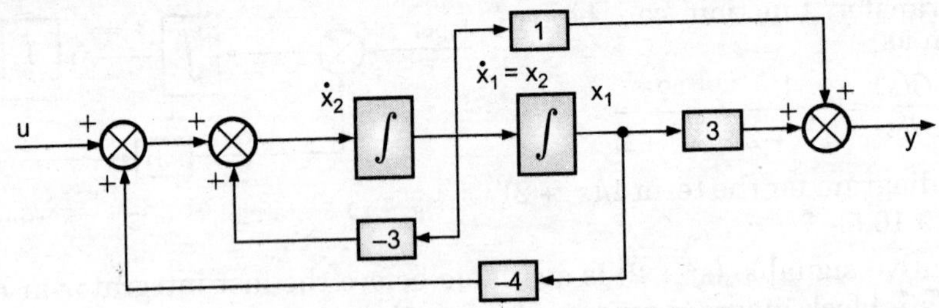

Fig. 9.10.7. State block diagram for $Y(s)/U(s) = (s + 3)/(s^2 + 3s + 4)$.

The output of each integrator is assigned a state variable as shown in Fig. 9.10.7. Therefore, the state equations are

$$\dot{x}_1 = x_2$$

$$\dot{x}_2 = -4x_1 - 3x_2 + u$$

The output equation is given by

$$y = 3x_1 + x_2$$

or
$$\begin{bmatrix} \dot{x}_1 \\ \dot{x}_2 \end{bmatrix} = \begin{bmatrix} 0 & 1 \\ -4 & -3 \end{bmatrix} \begin{bmatrix} x_1 \\ x_2 \end{bmatrix} + \begin{bmatrix} 0 \\ 1 \end{bmatrix} u \quad \text{and} \quad y = [3 \ \ 1] \begin{bmatrix} x_1 \\ x_2 \end{bmatrix}.$$ **Ans.**

Example 9.10.6. *The transfer function of a system is given by*

$$\frac{Y(s)}{U(s)} = \frac{s^2 + 3s + 2}{s^3 + 9s^2 + 26s + 24}$$

Determine the state model. Use direct decomposition method.

Solution.
$$\frac{Y(s)}{U(s)} = \frac{s^2 + 3s + 2}{s^3 + 9s^2 + 26s + 24}$$

or
$$\frac{Y(s)}{U(s)} = \frac{2}{s^3 + 9s^2 + 26s + 24} + \frac{3s}{s^3 + 9s^2 + 26s + 24} + \frac{s^2}{s^3 + 9s^2 + 26s + 24}$$

The state block diagram for the term $\dfrac{2}{s^3 + 9s^2 + 26s + 24}$ is drawn as shown in Fig. 9.10.8 (*a*).

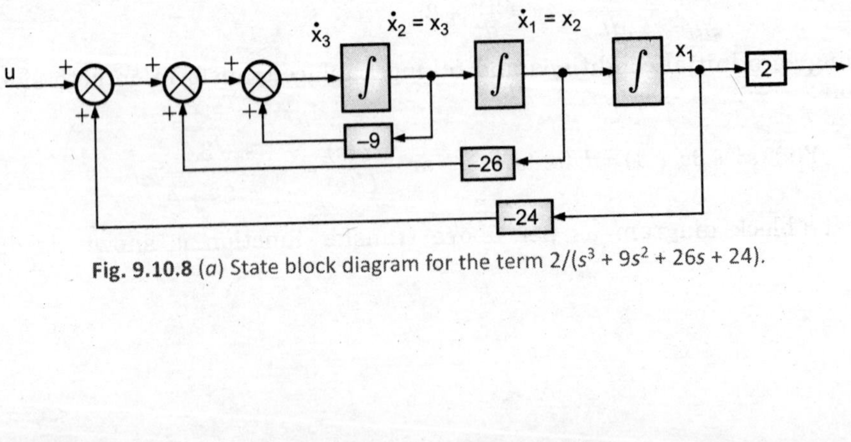

Fig. 9.10.8 (*a*) State block diagram for the term $2/(s^3 + 9s^2 + 26s + 24)$.

The output of each integrator is assigned a state variable as shown in Fig. 9.10.8 (*a*).

1. At the output of integrator no. 1 from RHS, the term $\dfrac{1}{s^3 + 9s^2 + 26s + 24}$ is available.
2. At the output of integrator no. 2 from RHS, the term $\dfrac{s}{s^3 + 9s^2 + 26s + 24}$ is available.
3. At the output of integrator no. 3 from RHS, the term $\dfrac{s^2}{s^3 + 9s^2 + 26s + 24}$ is available.

In view of points (1), (2) and (3) the state block diagram for $\dfrac{Y(s)}{U(s)} = \dfrac{s^2 + 3s + 2}{s^3 + 9s^2 + 26s + 24}$ is drawn as shown in Fig. 9.10.8 (*b*).

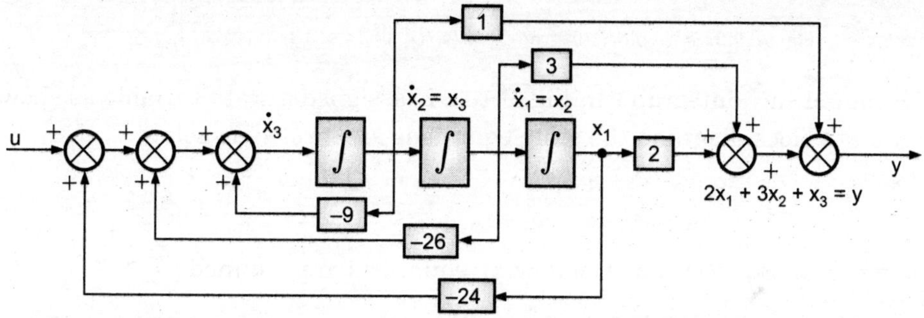

Fig. 9.10.8 (*b*) State block diagram for $Y(s)/U(s) = (s^2 + 3s + 2)/(s^3 + 9s^2 + 26s + 24)$.

As per the state block diagram Fig. [9.10.8 (*b*)], the state equations are written below :

$$\dot{x}_1 = -x_2$$

$$\dot{x}_2 = x_3$$

$$\dot{x}_3 = -24x_1 - 26x_2 - 9x_3 + u$$

and

$$y = 2x_1 + 3x_2 + x_3$$

or

$$\begin{bmatrix} \dot{x}_1 \\ \dot{x}_2 \\ \dot{x}_3 \end{bmatrix} = \begin{bmatrix} 0 & 1 & 0 \\ 0 & 0 & 1 \\ -24 & -26 & -9 \end{bmatrix} \begin{bmatrix} x_1 \\ x_2 \\ x_3 \end{bmatrix} + \begin{bmatrix} 0 \\ 1 \end{bmatrix} u$$

and

$$y = [2 \ \ 3 \ \ 1] \begin{bmatrix} x_1 \\ x_2 \\ x_3 \end{bmatrix}$$ **Ans.**

Example 9.10.7. *Using cascade method decompose the transfer function* $\dfrac{Y(s)}{U(s)} = \dfrac{(s+3)}{(s+1)(s+2)}$ *and obtain the state model.*

Solution. The state block diagram is drawn in two parts

(1) $\dfrac{s+3}{s+1}$ (2) $\dfrac{1}{s+2}$

The output of (1) is connected as input to (2). The input u is given to (1). The state block diagram is drawn as shown in Fig. 9.10.9.

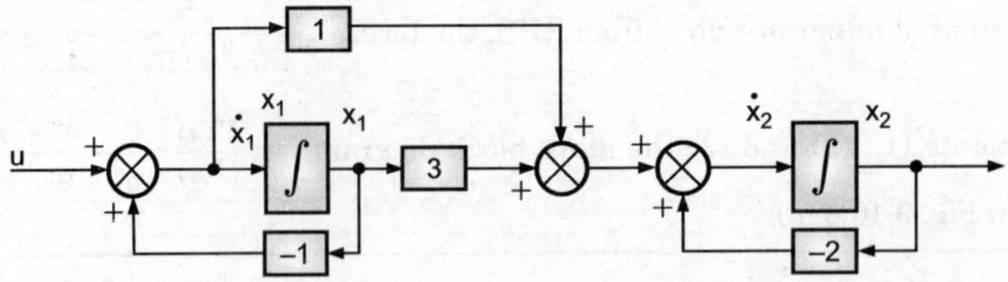

Fig. 9.10.9. State block diagram for $Y(s)/U(s) = (s + 3)/(s + 1)(s + 2)$.

The output of each integrator in Fig. 9.10.9 is assigned a state variable as shown.

As per state block diagram, the state equations are formed as below :

$$\dot{x}_1 = -x_1 + u \quad ...(1)$$

$$\dot{x}_2 = 3x_1 + \dot{x}_1 - 2x_2 \quad ...(2)$$

substituting for $\dot{x}_1$ in Eq. (2) from (1) following equations are obtained

$$\dot{x}_1 = -x_1 + u$$

$$\dot{x}_2 = -2x_1 - 2x_2 + u$$

The output equation is : $y = x_2$

The state model is : $\begin{bmatrix} \dot{x}_1 \\ \dot{x}_2 \end{bmatrix} = \begin{bmatrix} -1 & 0 \\ 2 & -2 \end{bmatrix} \begin{bmatrix} x_1 \\ x_2 \end{bmatrix} + \begin{bmatrix} 1 \\ 1 \end{bmatrix} u$

and $y = [0 \;\; 1] \begin{bmatrix} x_1 \\ x_2 \end{bmatrix}$ **Ans.**

Example 9.10.8. *The transfer function of a system is given by* $\dfrac{Y(s)}{U(s)} = \dfrac{(s+3)}{(s+1)(s+2)}$. *Obtain the state model in canonical form using parallel decomposition method.*

Solution. $\dfrac{Y(s)}{U(s)} = \dfrac{(s+3)}{(s+1)(s+2)}$

$$\frac{(s+3)}{(s+1)(s+2)} = \frac{K_1}{(s+1)} + \frac{K_2}{(s+2)}$$

The coefficients K_1 and K_2 can be determined as : $K_1 = 2$ and $K_2 = -1$.

Therefore, $\dfrac{Y(s)}{U(s)} = \dfrac{2}{(s+1)} + \dfrac{-1}{(s+2)}$

The state block diagram for above transfer function is drawn as shown in Fig. 9.10.10.

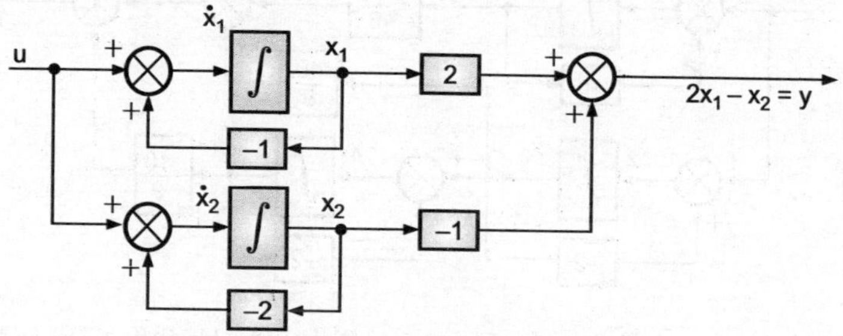

Fig. 9.10.10. State block diagram for parallel decomposition of transfer function $Y(s)/U(s) = (s + 3)/(s + 1)(s + 2)$.

The output of each integrator is assigned a state variable as shown in Fig. 9.10.10 and the state equations are written below :

$$\dot{x}_1 = -x_1 + u$$

$$\dot{x}_2 = -2x_2 + u \quad \text{and} \quad y = 2x_1 - x_2$$

or
$$\begin{bmatrix} \dot{x}_1 \\ \dot{x}_2 \end{bmatrix} = \begin{bmatrix} -1 & 0 \\ 0 & -2 \end{bmatrix} \begin{bmatrix} x_1 \\ x_2 \end{bmatrix} + \begin{bmatrix} 1 \\ 1 \end{bmatrix} u$$

and
$$y = [2 \;\; -1] \begin{bmatrix} x_1 \\ x_2 \end{bmatrix} \quad \textbf{Ans.}$$

Example 9.10.9. *Obtain the state space equations for the transfer function given as,*

$$G(s) = \frac{10(s+1)}{(s+4)(s+2)^2}.$$

Solution. Let,
$$\frac{10(s+1)}{(s+5)(s+2)^2} = \frac{K_1}{(s+5)} + \frac{K_2}{(s+2)} + \frac{K_3}{(s+2)^2}$$

The coefficients can be determined as :

$$K_1 = -\frac{40}{9}, K_2 = \frac{40}{9}, \quad \text{and} \quad K_3 = -\frac{10}{3}$$

Therefore,
$$G(s) = \frac{-\frac{40}{9}}{(s+5)} + \frac{\frac{40}{9}}{(s+2)} + \frac{-\frac{10}{3}}{(s+2)^2}$$

The state block diagram for the transfer function given above is constructed as shown in Fig. 9.10.11. It is to be noted that the term $\frac{1}{(s+2)^2}$ is obtained in the state block diagram by cascading two blocks of the term $\frac{1}{(s+2)}$.

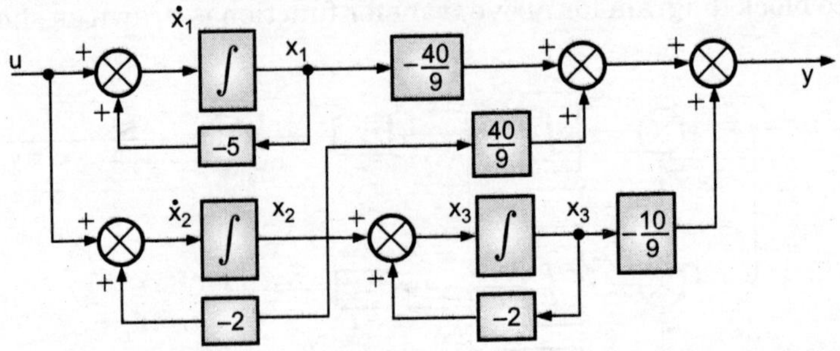

Fig. 9.10.11. State block diagram for $G(s) = (-40/9)/(s + 5) + (40/9)/(s + 2) + (-10/3)/(s + 2)^2$.

The output of each integrator is assigned a state variable and the state equations w.r.t. Fig. 9.10.11 are obtained as below :

$$\dot{x}_1 = -5x_1 + u$$
$$\dot{x}_2 = -2x_2 + u$$
$$\dot{x}_3 = x_2 - 2x_3$$

and
$$y = -\frac{40}{9}x_1 + \frac{40}{9}x_2 - \frac{10}{3}x_3$$

or
$$\begin{bmatrix} \dot{x}_1 \\ \dot{x}_2 \\ \dot{x}_3 \end{bmatrix} = \begin{bmatrix} -5 & 0 & 0 \\ 0 & -2 & 0 \\ 0 & 1 & -2 \end{bmatrix} \begin{bmatrix} x_1 \\ x_2 \\ x_3 \end{bmatrix} + \begin{bmatrix} 1 \\ 1 \\ 0 \end{bmatrix} u$$

and
$$y = \begin{bmatrix} -\frac{40}{9} & \frac{40}{9} & -\frac{10}{3} \end{bmatrix} \begin{bmatrix} x_1 \\ x_2 \\ x_3 \end{bmatrix}$$ **Ans.**

Example 9.10.10. *Apply cascade decomposing method to obtain state space representation of the transfer function given as,* $G(s) = \dfrac{s^2 + 6s + 8}{(s+3)(s^2 + 2s + 5)}$.

Solution. $G(s) = \dfrac{s^2 + 6s + 8}{(s+3)(s^2 + 2s + 5)}$ or $G(s) = \dfrac{(s+2)(s+4)}{(s+3)(s^2 + 2s + 5)}$

For drawing state block diagram, the transfer function given above is split into two parts which are joined in cascade. The two parts are :

(1) $\dfrac{(s+2)}{(s+3)}$ and (2) $\dfrac{(s+4)}{(s^2 + 2s + 5)}$

The state block diagram for the above mentioned two parts is drawn and shown in Fig. 9.10.12. The state equations formed as per Fig. 9.10.12 are given below :

$$\dot{x}_1 = x_2 \quad \text{...(1)}$$
$$\dot{x}_2 = -5x_1 - 2x_2 + \dot{x}_3 + 2x_3 \quad \text{...(2)}$$
$$\dot{x}_3 = -3x_3 + u \quad \text{...(3)}$$

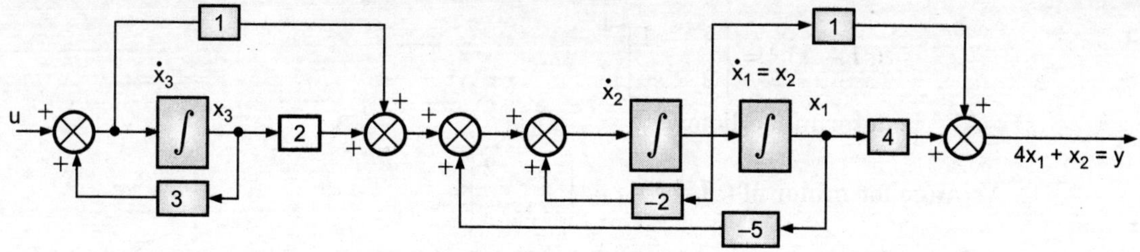

Fig. 9.10.12. State block diagram for $G(s) = (s + 2)(s + 4)/(s + 3)(s^2 + 2s + 5)$.

Substitute for $\dot{x}_3$ from (3) into (2), therefore,

$$\dot{x}_1 = x_2$$
$$\dot{x}_2 = -5x_1 - 2x_2 - 3x_3 + u + 2x_3$$
$$\dot{x}_3 = -3x_3 + u$$

or

$$\dot{x}_1 = x_2$$
$$\dot{x}_2 = -5x_1 - 2x_2 - x_3 + u$$
$$\dot{x}_3 = -3x_3 + u$$

The output equation is : $y = 4x_1 + x_2$

The state space representation is given below :

$$\begin{bmatrix}\dot{x}_1\\ \dot{x}_2\\ \dot{x}_3\end{bmatrix} = \begin{bmatrix}0 & 1 & 0\\ -5 & -2 & -1\\ 0 & 0 & -3\end{bmatrix}\begin{bmatrix}x_1\\ x_2\\ x_3\end{bmatrix} + \begin{bmatrix}0\\ 1\\ 1\end{bmatrix}u$$

and $$y = [4 \;\; 1 \;\; 0]\begin{bmatrix}x_1\\ x_2\\ x_3\end{bmatrix} \quad \textbf{Ans.}$$

Example 9.10.11. *Obtain the time response of the system given below :*

$$\mathbf{x} = \mathbf{A}\mathbf{x}$$

where $$\mathbf{A} = \begin{bmatrix}0 & 1\\ -2 & 0\end{bmatrix}; \textit{given } \mathbf{x}(0) = [1 \;\; 1]^T$$

and $$y = [1 \;\; -1]\begin{bmatrix}x_1\\ x_2\end{bmatrix}.$$

Solution. The time response is given by

$$\mathbf{x}(t) = \boldsymbol{\phi}(t)\,\mathbf{x}(0)$$
$$\boldsymbol{\phi}(t) = \mathcal{L}^{-1}(s\mathbf{I} - \mathbf{A})^{-1}$$

$$(s\mathbf{I} - \mathbf{A}) = \begin{bmatrix}s & 0\\ 0 & s\end{bmatrix} - \begin{bmatrix}0 & 1\\ -2 & 0\end{bmatrix} = \begin{bmatrix}s & -1\\ 2 & s\end{bmatrix}$$

$$(s\mathbf{I}-\mathbf{A})^{-1} = \begin{bmatrix} s & -1 \\ 2 & s \end{bmatrix}^{-1}$$

$(s\mathbf{I}-\mathbf{A})^{-1}$ **is determined below :**

1. **Arrange for minor of** $(s\mathbf{I}-\mathbf{A})$, *i.e.* $\begin{bmatrix} s & 2 \\ -1 & s \end{bmatrix}$

2. **Change to cofactor,** *i.e.* $\begin{bmatrix} s & -2 \\ 1 & s \end{bmatrix}$

3. **Interchange rows and columns,** *i.e.* $\begin{bmatrix} s & 1 \\ -2 & s \end{bmatrix}$

$$\therefore \qquad \text{Adj. } (s\mathbf{I}-\mathbf{A}) = \begin{bmatrix} s & 1 \\ -2 & s \end{bmatrix}$$

4. $$(s\mathbf{I}-\mathbf{A})^{-1} = \frac{\text{Adj } (s\mathbf{I}-\mathbf{A})}{|s\mathbf{I}-\mathbf{A}|} = \frac{\begin{bmatrix} s & 1 \\ -2 & s \end{bmatrix}}{\begin{vmatrix} s & -1 \\ 2 & s \end{vmatrix}} = \frac{\begin{bmatrix} s & 1 \\ -2 & s \end{bmatrix}}{(s^2+2)}$$

$$= \begin{bmatrix} \frac{s}{s^2+2} & \frac{1}{s^2+2} \\ -\frac{2}{s^2+2} & \frac{s}{s^2+2} \end{bmatrix}$$

$$\because \qquad \phi(s) = (s\mathbf{I}-\mathbf{A})^{-1}$$

$$\therefore \qquad \phi(s) = \begin{bmatrix} \frac{s}{s^2+2} & \frac{1}{s^2+2} \\ -\frac{2}{s^2+2} & \frac{s}{s^2+2} \end{bmatrix}$$

$\therefore$ **The state transition matrix** $\phi(t)$ **is**

$$\phi(t) = \mathcal{L}^{-1}\phi(s) = \mathcal{L}^{-1}\begin{bmatrix} \frac{s}{s^2+2} & \frac{1}{s^2+2} \\ -\frac{2}{s^2+2} & \frac{s}{s^2+2} \end{bmatrix}$$

$$= \begin{bmatrix} \cos\sqrt{2}t & \frac{1}{\sqrt{2}}\sin\sqrt{2}t \\ -\frac{2}{\sqrt{2}}\sin\sqrt{2}t & \cos\sqrt{2}t \end{bmatrix} = \begin{bmatrix} \cos\sqrt{2}t & \frac{1}{\sqrt{2}}\sin\sqrt{2}t \\ -\sqrt{2}\sin\sqrt{2}t & \cos\sqrt{2}t \end{bmatrix}$$

$$\mathbf{x}(t) = \phi(t)\,\mathbf{x}(0)$$

$$\therefore \qquad \begin{bmatrix} x_1(t) \\ x_2(t) \end{bmatrix} = \begin{bmatrix} \cos\sqrt{2}t & \frac{1}{\sqrt{2}}\sin\sqrt{2}t \\ -\sqrt{2}\sin\sqrt{2}t & \cos\sqrt{2}t \end{bmatrix}\begin{bmatrix} 1 \\ 1 \end{bmatrix}$$

$$\therefore \qquad x_1(t) = \cos \sqrt{2}t + \frac{1}{\sqrt{2}} \sin \sqrt{2}t$$

$$x_2(t) = -\sqrt{2} \sin \sqrt{2}t + \cos \sqrt{2}t. \quad \textbf{Ans.}$$

$$y = x_1 - x_2$$

$$\therefore \qquad y = \frac{3}{\sqrt{2}} \sin \sqrt{2}t. \quad \textbf{Ans.}$$

Example 9.10.12. *For the system given below :*
Obtain (i) Zero input response (ii) Zero state response (iii) Total response,

$$\dot{\mathbf{x}} = \begin{bmatrix} 0 & 1 \\ -2 & -1 \end{bmatrix} \mathbf{x} + \begin{bmatrix} 0 \\ 1 \end{bmatrix} u(t)$$

where $x_1(0) = 1, x_2(0) = 0$ *and* $u(t) = 1.$

Solution. (*i*) Zero input response is given by

$$\mathbf{x}(t) = \phi(t)\,\mathbf{x}(0)$$

$$\phi(t) = \mathcal{L}^{-1}(s\mathbf{I} - \mathbf{A})^{-1}$$

$$(s\mathbf{I} - \mathbf{A}) = \begin{bmatrix} s & 0 \\ 0 & s \end{bmatrix} - \begin{bmatrix} 0 & 1 \\ -2 & -1 \end{bmatrix} = \begin{bmatrix} s & -1 \\ 2 & s+1 \end{bmatrix}$$

$$\therefore \qquad (s\mathbf{I} - \mathbf{A})^{-1} = \begin{bmatrix} s & -1 \\ 2 & s+1 \end{bmatrix}^{-1}$$

$(s\mathbf{I} - \mathbf{A})^{-1}$ is determined below :

1. Arrange for minor of $(s\mathbf{I} - \mathbf{A})$, *i.e.* $\begin{bmatrix} s+1 & 2 \\ -1 & s \end{bmatrix}$

2. Change to cofactor, *i.e.* $\begin{bmatrix} s+1 & -2 \\ 1 & s \end{bmatrix}$

3. Interchange rows and columns, *i.e.* $\begin{bmatrix} s+1 & 1 \\ -2 & s \end{bmatrix}$

$$\therefore \qquad \text{Adj. } (s\mathbf{I} - \mathbf{A}) = \begin{bmatrix} s+1 & 1 \\ -2 & s \end{bmatrix}$$

4. $$(s\mathbf{I} - \mathbf{A})^{-1} = \frac{\text{Adj } (s\mathbf{I} - \mathbf{A})}{|\,s\mathbf{I} - \mathbf{A}\,|} = \frac{\begin{bmatrix} s+1 & 1 \\ -2 & s \end{bmatrix}}{\begin{vmatrix} s & -1 \\ 2 & s+1 \end{vmatrix}} = \frac{\begin{bmatrix} s+1 & 1 \\ -2 & s \end{bmatrix}}{(s^2 + s + 3)}$$

$$= \begin{bmatrix} \dfrac{s+1}{s^2+s+3} & \dfrac{1}{s^2+s+3} \\ -\dfrac{2}{s^2+s+3} & \dfrac{s}{s^2+s+3} \end{bmatrix}$$

$$\because \quad \phi(s) = (s\mathbf{I} - \mathbf{A})^{-1}$$

$$\therefore \quad \phi(s) = \begin{bmatrix} \dfrac{s+1}{s^2+s+3} & \dfrac{1}{s^2+s+3} \\ -\dfrac{2}{s^2+s+3} & \dfrac{s}{s^2+s+3} \end{bmatrix}$$

The state transition matrix $\phi(t)$ is

$$\phi(t) = \mathcal{L}^{-1}\,\phi(s)$$

$$= \mathcal{L}^{-1} \begin{bmatrix} \dfrac{s+1}{s^2+s+3} & \dfrac{1}{s^2+s+3} \\ -\dfrac{2}{s^2+s+3} & \dfrac{s}{s^2+s+3} \end{bmatrix}$$

$$\mathbf{x}(t) = \mathcal{L}^{-1}\,\phi(s)\,\mathbf{x}(0)$$

$$= \mathcal{L}^{-1} \begin{bmatrix} \dfrac{s+1}{s^2+s+3} & \dfrac{1}{s^2+s+3} \\ -\dfrac{2}{s^2+s+3} & \dfrac{s}{s^2+s+3} \end{bmatrix} \begin{bmatrix} 1 \\ 0 \end{bmatrix} = \mathcal{L}^{-1} \begin{bmatrix} \dfrac{s+1}{s^2+s+3} \\ -\dfrac{2}{s^2+s+3} \end{bmatrix}$$

$$= \mathcal{L}^{-1} \begin{bmatrix} \dfrac{(s+0.5)+0.5}{(s+0.5)^2+(1.65)^2} \\ -\dfrac{2}{1.65}\cdot\dfrac{1.65}{(s+0.5)^2+(1.65)^2} \end{bmatrix}$$

$$= \mathcal{L}^{-1} \begin{bmatrix} \dfrac{(s+0.5)}{(s+0.5)^2+(1.65)^2} + \dfrac{0.5}{1.65}\cdot\dfrac{1.65}{(s+0.5)^2+(1.65)^2} \\ -1.21\cdot\dfrac{1.65}{(s+0.5)^2+(1.65)^2} \end{bmatrix}$$

Taking inverse Laplace transform, the zero input response is

$$\mathbf{x}(t) = \begin{bmatrix} (e^{-0.5t}\,.\cos 1.65\,t + 0.3\,e^{-0.5t}\,.\sin 1.65\,t) \\ -1.21\,e^{-0.5t}\,.\sin 1.65\,t \end{bmatrix}$$

(*ii*) Zero state response is given by

$$\mathbf{x}(t) = \mathcal{L}^{-1}\,\phi(s)\,\mathbf{BU}\,(s)$$

$$= \mathcal{L}^{-1} \begin{bmatrix} \dfrac{s+1}{s^2+s+3} & \dfrac{1}{s^2+s+3} \\ -\dfrac{2}{s^2+s+3} & \dfrac{s}{s^2+s+3} \end{bmatrix} \begin{bmatrix} 0 \\ 1 \end{bmatrix} \frac{1}{s} = \mathcal{L}^{-1} \begin{bmatrix} \dfrac{1}{s(s^2+s+3)} \\ \dfrac{1}{s}\cdot\dfrac{s}{(s^2+s+3)} \end{bmatrix}$$

$$= \mathcal{L}^{-1} \begin{bmatrix} \dfrac{1}{s(s^2+s+3)} \\ \dfrac{1}{(s^2+s+3)} \end{bmatrix} = \mathcal{L}^{-1} \begin{bmatrix} \dfrac{1}{3}\dfrac{1}{s}\,\dfrac{3}{(s^2+s+3)} \\ \dfrac{1}{1.65}\cdot\dfrac{1.65}{(s+0.5)^2+(1.65)^2} \end{bmatrix}$$

Taking inverse Laplace transform, the zero state response is

$$\mathbf{x}(t) = \begin{bmatrix} \frac{1}{3}\left\{1 - \frac{e^{-0.5t}}{\sqrt{1-0.28^2}} . \sin\left(1.65t + \tan^{-1}\frac{\sqrt{1-0.28^2}}{0.28}\right)\right\} \\ \frac{1}{1.65} . e^{-0.5t} . \sin(1.65t) \end{bmatrix}$$

or
$$\mathbf{x}(t) = \begin{bmatrix} 0.33\{1 - 1.04\, e^{-0.5t} . \sin(1.65t + \tan^{-1} 3.42)\} \\ 0.6\, e^{-0.5t} . \sin 1.65t \end{bmatrix}$$

The total response

$$x(t) = \text{Zero input response} + \text{Zero state response}$$

$$\therefore \quad \begin{bmatrix} x_1(t) \\ x_2(t) \end{bmatrix} = \begin{bmatrix} (e^{-0.5t} . \cos 1.65\,t + 0.3\, e^{-0.5\,t} . \sin 1.65\,t) \\ -1.21\, e^{-0.5t} . \sin 1.65t \end{bmatrix} + \begin{bmatrix} 0.33\{1 - 1.04\, e^{-0.5t} . \sin(1.65t + \tan^{-1} 3.42)\} \\ 0.6\, e^{-0.5t} . \sin 1.65t \end{bmatrix}$$

$$\therefore \quad x_1(t) = 0.33 + e^{-0.5\,t}\,[\cos 1.65t + 0.3 \sin 1.65t] - 0.34 \sin(1.65t + \tan^{-1} 3.42)]$$
$$x_2(t) = -0.61\, e^{-0.5t} . \sin 1.65t. \quad \textbf{Ans.}$$

Example 9.10.13. *Obtain the state transition matrix in the form $e^{\mathbf{A}t}$ and determine the time response for the system, $\dot{\mathbf{x}} = \mathbf{Ax}$*

where $\mathbf{A} = \begin{bmatrix} 0 & 1 \\ -2 & 0 \end{bmatrix}$ *and* $x_1(0) = 1,\ x_2(0) = 1.$

Solution. $\phi(t) = e^{\mathbf{A}t} = \mathbf{I} + \mathbf{A}t + \frac{\mathbf{A}t^2}{2!} + \frac{\mathbf{A}t^3}{3!} + \ldots..$

$$\mathbf{A} = \begin{bmatrix} 0 & 1 \\ -2 & 0 \end{bmatrix}$$

$$\mathbf{A}^2 = \begin{bmatrix} 0 & 1 \\ -2 & 0 \end{bmatrix}\begin{bmatrix} 0 & 1 \\ -2 & 0 \end{bmatrix} = \begin{bmatrix} -2 & 0 \\ 0 & -2 \end{bmatrix}$$

$$\mathbf{A}^3 = \begin{bmatrix} -2 & 0 \\ 0 & -2 \end{bmatrix}\begin{bmatrix} 0 & 1 \\ -2 & 0 \end{bmatrix} = \begin{bmatrix} 0 & -2 \\ 4 & 0 \end{bmatrix}$$

$$\phi(t) = \mathbf{I} + \mathbf{A}t + \frac{\mathbf{A}^2t^2}{2!} + \frac{\mathbf{A}^3t^3}{3!} + \ldots$$

$$= \begin{bmatrix} 1 & 0 \\ 0 & 1 \end{bmatrix} + \begin{bmatrix} 0 & t \\ -2t & 0 \end{bmatrix} + \begin{bmatrix} -\frac{2t^2}{1\times 2} & 0 \\ 0 & -\frac{2t^2}{1\times 2} \end{bmatrix} + \begin{bmatrix} 0 & \frac{-2t^3}{1\times 2\times 3} \\ \frac{4t^3}{1\times 2\times 3} & 0 \end{bmatrix} + \ldots$$

$$= \begin{bmatrix} 1 - \frac{2t^2}{2} + ... & t - \frac{2t^3}{2 \times 3} + ... \\ -2t + \frac{4t^3}{2 \times 3} - ... & 1 - \frac{2t^2}{2} + ... \end{bmatrix}$$

$$= \begin{bmatrix} \left(1 - \sqrt{2}^2 \frac{t^2}{2} + ...\right) & \left(\frac{\sqrt{2}}{\sqrt{2}} t - \frac{\sqrt{2}^3}{\sqrt{2}} \frac{t^3}{2 \times 3} + ...\right) \\ -\left(\sqrt{2}\sqrt{2}t - \sqrt{2}\sqrt{2}^3 \frac{t^3}{2 \times 3} + ...\right) & \left(1 - \sqrt{2}^2 \frac{t^2}{2} + ...\right) \end{bmatrix}$$

$$= \begin{bmatrix} \cos \sqrt{2}t & \frac{1}{\sqrt{2}} \sin \sqrt{2}t \\ -\sqrt{2} \sin \sqrt{2}t & \cos \sqrt{2}t \end{bmatrix}$$

$$\mathbf{x}(t) = \boldsymbol{\phi}(t)\, \mathbf{x}(0)$$

$$\therefore \quad \begin{bmatrix} x_1(t) \\ x_2(t) \end{bmatrix} = \begin{bmatrix} \cos \sqrt{2}t & \frac{1}{\sqrt{2}} \sin \sqrt{2}t \\ -\sqrt{2} \sin \sqrt{2}t & \cos \sqrt{2}t \end{bmatrix} \begin{bmatrix} 1 \\ 1 \end{bmatrix}$$

$$x_1(t) = \cos \sqrt{2}t + \frac{1}{\sqrt{2}} \sin \sqrt{2}t$$

$$\therefore \quad x_2(t) = -\sqrt{2} \sin \sqrt{2}t + \cos \sqrt{2}t .$$ **Ans.**

Example 9.10.14. *A control system is described as below :*

$$\dot{\mathbf{x}} = \mathbf{A}\mathbf{x} + \mathbf{B}u \quad \text{and} \quad y = \mathbf{C}\mathbf{x}$$

where,

$$\mathbf{A} = \begin{bmatrix} 1 & 4 \\ -2 & -5 \end{bmatrix}, \mathbf{x}(0) = \begin{bmatrix} 1 \\ 0 \end{bmatrix}, \mathbf{B} = \begin{bmatrix} 0 \\ 1 \end{bmatrix} \text{ and } \mathbf{C} = \begin{bmatrix} 1 \\ 0 \end{bmatrix}^T$$

Determine the time response. Use diagonalization process for matrix **A**.

Solution.

$$\mathbf{A} = \begin{bmatrix} 1 & 4 \\ -2 & -5 \end{bmatrix}$$

The matrix **A** given above is a 2 × 2 general matrix as such eigen vector approach is considered for its diagonalization.

The eigen values of λ_1 and λ_2 matrix of **A** are determined by solving following equation :

$$|\lambda \mathbf{I} - \mathbf{A}| = 0$$

$$|\lambda \mathbf{I} - \mathbf{A}| = \left\{ \begin{bmatrix} \lambda & 0 \\ 0 & \lambda \end{bmatrix} - \begin{bmatrix} 1 & 4 \\ -2 & -5 \end{bmatrix} \right\} = \begin{bmatrix} \lambda - 1 & -4 \\ 2 & \lambda + 5 \end{bmatrix}$$

$$\therefore \quad |\lambda \mathbf{I} - \mathbf{A}| = (\lambda - 1)(\lambda + 5) - (2)(-4)$$

$$= \lambda^2 + 4\lambda + 3 = (\lambda + 1)(\lambda + 3)$$

put $\quad (\lambda + 1)(\lambda + 3) = 0$

Thus, two distinct eigen values are :

$$\lambda_1 = -1 \quad \text{and} \quad \lambda_2 = -3$$

Determination of eigen vectors of matrix **A** is given below :

Eigen vectors of matrix **A** are determined by solving the following equation :

$$\{\lambda_i \mathbf{I} - \mathbf{A}\}\, \mathbf{x}_i = 0$$

where, λ_i and $\mathbf{x}_i$ are respectively the eigen value and corresponding eigen vector of matrix **A**.

(*i*) The eigen vector $\mathbf{x}_1$ associated with eigen values λ_1 is determined by solving the equation given below :

$$\{\lambda_1 \mathbf{I} - \mathbf{A}\}\, \mathbf{x}_1 = 0$$

$$\because \quad \lambda_1 = -1 \quad \text{and} \quad \mathbf{A} = \begin{bmatrix} 1 & 4 \\ -2 & -5 \end{bmatrix}$$

$$\therefore \quad \left\{(-1)\begin{bmatrix} 1 & 0 \\ 0 & 1 \end{bmatrix} - \begin{bmatrix} 1 & 4 \\ -2 & -5 \end{bmatrix}\right\} \mathbf{x}_1 = 0$$

Let, $\mathbf{x}_1 = \begin{bmatrix} m_{11} \\ m_{21} \end{bmatrix}$, $\therefore \begin{bmatrix} -2 & -4 \\ 2 & 4 \end{bmatrix}\begin{bmatrix} m_{11} \\ m_{21} \end{bmatrix} = 0$

$\therefore \quad -2m_{11} - 4m_{21} = 0$

or $\quad 2m_{11} + 4m_{21} = 0$

put $\quad m_{11} = 1 \quad \therefore \quad m_{21} = -0.5$

Therefore, $\mathbf{x}_1 = \begin{bmatrix} m_{11} \\ m_{21} \end{bmatrix} = \begin{bmatrix} 1 \\ -0.5 \end{bmatrix}$

(*ii*) The eigen vector $\mathbf{x}_2$ associated with eigen value λ_2 is determined by solving the equation given below :

$$\{\lambda_2 \mathbf{I} - \mathbf{A}\}\, \mathbf{x}_2 = 0$$

$$\because \quad \lambda_2 = -3 \quad \text{and} \quad \mathbf{A} - \begin{bmatrix} 1 & 4 \\ -2 & -5 \end{bmatrix}$$

$$\therefore \quad \left\{(-3)\begin{bmatrix} 1 & 0 \\ 0 & 1 \end{bmatrix} - \begin{bmatrix} 1 & 4 \\ -2 & -5 \end{bmatrix}\right\} \mathbf{x}_2 = 0$$

Let, $\mathbf{x}_2 = \begin{bmatrix} m_{12} \\ m_{22} \end{bmatrix}$, $\therefore \begin{bmatrix} -4 & -4 \\ 2 & 2 \end{bmatrix}\begin{bmatrix} m_{12} \\ m_{22} \end{bmatrix} = 0$

$\therefore \quad -4m_{12} - 4m_{22} = 0$

or $\quad 2m_{12} + 2m_{22} = 0$

put $\quad m_{12} = 1 \quad \therefore \quad m_{22} = -1$

Therefore, $\mathbf{x}_2 = \begin{bmatrix} m_{12} \\ m_{22} \end{bmatrix} = \begin{bmatrix} 1 \\ -1 \end{bmatrix}$

Thus, the modal matrix is given by :

$$\mathbf{M} = \begin{bmatrix} m_{11} & m_{12} \\ m_{21} & m_{22} \end{bmatrix} = \begin{bmatrix} 1 & 1 \\ -0.5 & -1 \end{bmatrix}$$

The inverse of modal matrix can be determined as

$$\mathbf{M}^{-1} = \frac{\text{Adj}\begin{bmatrix} 1 & 1 \\ -0.5 & -1 \end{bmatrix}}{\begin{vmatrix} 1 & 1 \\ -0.5 & -1 \end{vmatrix}} = \begin{bmatrix} 2 & 2 \\ -1 & -2 \end{bmatrix}$$

The state equation for the system is given as :

$$\dot{\mathbf{x}} = \mathbf{Ax} + \mathbf{Bu}$$

put $\mathbf{x} = \mathbf{Mz}$ in the state equation given above,

$$\therefore \quad \mathbf{M\dot{z}} = \mathbf{AMz} + \mathbf{Bu}$$

or

$$\dot{\mathbf{z}} = \mathbf{M}^{-1}\,\mathbf{AMz} + \mathbf{M}^{-1}\,\mathbf{Bu}$$

Taking Laplace transform on both sides :

$$s\mathbf{Z}(s) - \mathbf{z}(0) = \mathbf{M}^{-1}\,\mathbf{AMZ}(s) + \mathbf{M}^{-1}\,\mathbf{BU}(s)$$

or

$$\{s\,\mathbf{I} - \mathbf{M}^{-1}\,\mathbf{AM}\}\,\mathbf{Z}(s) = \mathbf{z}(0) + \mathbf{M}^{-1}\,\mathbf{BU}\,(s)$$

$$\therefore \quad \mathbf{Z}(s) = \{s\mathbf{I} - \mathbf{M}^{-1}\,\mathbf{AM}\}^{-1}\,\mathbf{z}(0) + \{s\mathbf{I} - \mathbf{M}^{-1}\,\mathbf{AM}\}^{-1}\,\mathbf{M}^{-1}\,\mathbf{BU}(s)$$

$$\mathbf{M}^{-1}\,\mathbf{AM} = \begin{bmatrix} 2 & 2 \\ -1 & -2 \end{bmatrix}\begin{bmatrix} 1 & 4 \\ -2 & -5 \end{bmatrix}\begin{bmatrix} 1 & 1 \\ -0.5 & -1 \end{bmatrix} = \begin{bmatrix} -1 & 0 \\ 0 & -3 \end{bmatrix}$$

$$\because \quad \mathbf{x} = \mathbf{Mz}$$

$$\mathbf{z}(0) = \mathbf{M}^{-1}\,\mathbf{x}(0) = \begin{bmatrix} 2 & 2 \\ -1 & -2 \end{bmatrix}\begin{bmatrix} 1 \\ 0 \end{bmatrix} = \begin{bmatrix} 2 \\ -1 \end{bmatrix}$$

$$\mathbf{M}^{-1}\,\mathbf{B} = \begin{bmatrix} 2 & 2 \\ -1 & -2 \end{bmatrix}\begin{bmatrix} 0 \\ 1 \end{bmatrix} = \begin{bmatrix} 2 \\ -2 \end{bmatrix}$$

substituting the values of $\mathbf{M}^{-1}\,\mathbf{AM}$, $\mathbf{z}(0)$ and $\mathbf{M}^{-1}\,\mathbf{B}$ in equation for $\mathbf{Z}(s)$, the equation given below is obtained :

$$\mathbf{Z}(s) = \left\{\begin{bmatrix} s & 0 \\ 0 & s \end{bmatrix} - \begin{bmatrix} -1 & 0 \\ 0 & -3 \end{bmatrix}\right\}^{-1}\begin{bmatrix} 2 \\ -1 \end{bmatrix} + \left\{\begin{bmatrix} s & 0 \\ 0 & s \end{bmatrix} - \begin{bmatrix} -1 & 0 \\ 0 & -3 \end{bmatrix}\right\}^{-1}\begin{bmatrix} 2 \\ -2 \end{bmatrix}\frac{1}{s}$$

$$= \begin{bmatrix} s+1 & 0 \\ 0 & s+3 \end{bmatrix}^{-1}\begin{bmatrix} 2 \\ -1 \end{bmatrix} + \begin{bmatrix} s+1 & 0 \\ 0 & s+3 \end{bmatrix}^{-1}\begin{bmatrix} \frac{2}{s} \\ -\frac{2}{s} \end{bmatrix}$$

$$= \begin{bmatrix} \frac{1}{s+1} & 0 \\ 0 & \frac{1}{s+3} \end{bmatrix}\begin{bmatrix} 2 \\ -1 \end{bmatrix} + \begin{bmatrix} \frac{1}{s+1} & 0 \\ 0 & \frac{1}{s+3} \end{bmatrix}\begin{bmatrix} \frac{2}{s} \\ -\frac{2}{s} \end{bmatrix}$$

$$= \begin{bmatrix} \frac{2}{s+1} \\ -\frac{1}{s+3} \end{bmatrix} + \begin{bmatrix} \frac{2}{s(s+1)} \\ -\frac{2}{s(s+3)} \end{bmatrix} = \begin{bmatrix} \frac{2}{s+1} \\ -\frac{1}{s+3} \end{bmatrix} + \begin{bmatrix} \frac{2}{s} - \frac{2}{s+1} \\ -\frac{2}{3s} + \frac{2}{s+3} \end{bmatrix}$$

or
$$\mathbf{Z}(s) = \begin{bmatrix} \frac{2}{s} \\ -\frac{2}{3s} - \frac{1}{3(s+3)} \end{bmatrix}$$

$\because$
$$\mathbf{X}(s) = \mathbf{MZ}(s)$$

$\therefore$
$$\mathbf{X}(s) = \begin{bmatrix} 1 & 1 \\ -0.5 & -1 \end{bmatrix} \begin{bmatrix} \frac{2}{s} \\ -\frac{2}{3s} - \frac{1}{3(s+3)} \end{bmatrix} = \begin{bmatrix} \frac{2}{s} - \frac{2}{3s} - \frac{1}{3(s+3)} \\ -\frac{1}{s} + \frac{2}{3s} + \frac{1}{3(s+3)} \end{bmatrix}$$

$\therefore$
$$\mathbf{x} = \mathcal{L}^{-1} \begin{bmatrix} \frac{2}{s} - \frac{2}{3s} - \frac{1}{3(s+3)} \\ -\frac{1}{s} + \frac{2}{3s} + \frac{1}{3(s+3)} \end{bmatrix} = \mathcal{L}^{-1} \begin{bmatrix} \frac{4}{3s} - \frac{1}{3(s+3)} \\ -\frac{1}{3s} + \frac{1}{3(s+3)} \end{bmatrix}$$

$\therefore$
$$\begin{bmatrix} x_1 \\ x_2 \end{bmatrix} = \begin{bmatrix} \frac{4}{3} - \frac{1}{3}e^{-3t} \\ -\frac{1}{3} + \frac{1}{3}e^{-3t} \end{bmatrix} \quad \textbf{Ans.}$$

The output equation is : $y = \begin{bmatrix} 1 \\ 0 \end{bmatrix}^T \begin{bmatrix} x_1 \\ x_2 \end{bmatrix}$

or
$$y = [1 \ \ 0] \begin{bmatrix} x_1 \\ x_2 \end{bmatrix} \quad \therefore \quad y = x_1$$

or
$$y = \frac{4}{3} - \frac{1}{3}e^{-3t} \quad \textbf{Ans.}$$

Alternative solution

As already determined,

$$\mathbf{M} = \begin{bmatrix} 1 & 1 \\ -0.5 & -1 \end{bmatrix}; \mathbf{M}^{-1} = \begin{bmatrix} 2 & 2 \\ -1 & -2 \end{bmatrix}$$

$\therefore$
$$\mathbf{\Lambda} = \mathbf{M}^{-1}\mathbf{AM} = \begin{bmatrix} 2 & 2 \\ -1 & -2 \end{bmatrix} \begin{bmatrix} 1 & 4 \\ -2 & -5 \end{bmatrix} \begin{bmatrix} 1 & 1 \\ -0.5 & -1 \end{bmatrix}$$

$$= \begin{bmatrix} -1 & 0 \\ 0 & -3 \end{bmatrix}$$

$\therefore$
$$e^{\wedge t} = \begin{bmatrix} e^{-t} & 0 \\ 0 & e^{-3t} \end{bmatrix}$$

$$e^{\mathbf{A}t} = \mathbf{M}e^{\wedge t}\mathbf{M}^{-1}$$

$$= \begin{bmatrix} 1 & 1 \\ -0.5 & -1 \end{bmatrix} \begin{bmatrix} e^{-t} & 0 \\ 0 & e^{-3t} \end{bmatrix} \begin{bmatrix} 2 & 2 \\ -1 & -2 \end{bmatrix}$$

$$= \begin{bmatrix} 1 & 1 \\ -0.5 & -1 \end{bmatrix} \begin{bmatrix} 2e^{-t} & 2e^{-t} \\ -e^{-3t} & -2e^{-3t} \end{bmatrix}$$

or
$$e^{\mathbf{A}t} = \begin{bmatrix} 2e^{-t} - e^{-3t} & 2e^{-t} - 2e^{-3t} \\ -e^{-t} + e^{-3t} & -e^{-t} + 2e^{-3t} \end{bmatrix}$$

$$\mathbf{x}(t) = \phi(t)\,\mathbf{x}(0) + \int_0^t \phi(t-\tau)\,\mathbf{B}u(\tau)\,d\tau$$

$\because$
$$\phi(t) = e^{\mathbf{A}t}\,;\ \mathbf{x}(0) = \begin{bmatrix} 1 \\ 0 \end{bmatrix} \text{ and } \mathbf{B} = \begin{bmatrix} 0 \\ 1 \end{bmatrix}$$

$\therefore$
$$\mathbf{x}(t) = \begin{bmatrix} -2e^{-t} - e^{-3t} & 2e^{-t} - 2e^{-3t} \\ -e^{-t} + e^{-3t} & -e^{-t} + 2e^{-3t} \end{bmatrix} \begin{bmatrix} 1 \\ 0 \end{bmatrix}$$

$$+ \int_0^t \begin{bmatrix} 2e^{-(t-\tau)} - e^{-3(t-\tau)} & 2e^{-(t-\tau)} - 2e^{-3(t-\tau)} \\ -e^{-(t-\tau)} + e^{-3(t-\tau)} & -e^{-(t-\tau)} + 2e^{-3(t-\tau)} \end{bmatrix} \begin{bmatrix} 0 \\ 1 \end{bmatrix} d\tau$$

$$= \begin{bmatrix} (2e^{-t} - e^{-3t}) \\ (-e^{-t} - e^{-3t}) \end{bmatrix} + \int_0^t \begin{bmatrix} 2e^{-(t-\tau)} - 2e^{-3(t-\tau)} \\ -e^{-(t-\tau)} + 2e^{-3(t-\tau)} \end{bmatrix} d\tau$$

$\therefore$
$$= \begin{bmatrix} (2e^{-t} - e^{-3t}) \\ (-e^{-t} + e^{-3t}) \end{bmatrix} + \int_0^t \begin{bmatrix} 2e^{-t} \,.\, e^{\tau} - 2e^{-3t} \,.\, e^{3\tau} \\ -e^{-t} \,.\, e^{\tau} + 2e^{-3t} \,.\, e^{3\tau} \end{bmatrix} d\tau$$

$$= \begin{bmatrix} (2e^{-t} - e^{-3t}) \\ (-e^{-t} + e^{-3t}) \end{bmatrix} + \begin{bmatrix} 2e^{-t} \,.\, e^{\tau} - \frac{2}{3}e^{-3t} \,.\, e^{3\tau} \\ -e^{-t} \,.\, e^{\tau} + \frac{2}{3}e^{-3t} \,.\, e^{3\tau} \end{bmatrix}_0^t$$

$$= \begin{bmatrix} 2e^{-t} - e^{-3t} \\ -e^{-t} + e^{-3t} \end{bmatrix} + \left\{ \begin{bmatrix} 2e^{-t} \,.\, e^{t} - \frac{2}{3}e^{-3t} \,.\, e^{3t} \\ -e^{-t} \,.\, e^{t} + \frac{2}{3}e^{-3t} \,.\, e^{3t} \end{bmatrix} - \begin{bmatrix} 2e^{-t} \,.\, e^{0} - \frac{2}{3}e^{-3t} \,.\, e^{0} \\ -e^{-t} \,.\, e^{0} + \frac{2}{3}e^{-3t} \,.\, e^{0} \end{bmatrix} \right\}$$

$$= \begin{bmatrix} 2e^{-t} - e^{-3t} \\ -e^{-t} + e^{-3t} \end{bmatrix} + \left\{ \begin{bmatrix} 2 - \frac{2}{3} \\ -1 + \frac{2}{3} \end{bmatrix} - \begin{bmatrix} 2e^{-t} - \frac{2}{3}e^{-3t} \\ -e^{-t} + \frac{2}{3}e^{-3t} \end{bmatrix} \right\}$$

or
$$\begin{bmatrix} x_1 \\ x_2 \end{bmatrix} = \begin{bmatrix} \frac{4}{3} - \frac{1}{3}e^{-3t} \\ -\frac{1}{3} + \frac{1}{3}e^{-3t} \end{bmatrix}$$ **Ans.**

The output equation is

$$y = [1 \quad 0]\begin{bmatrix} x_1 \\ x_2 \end{bmatrix}$$

$\therefore$ $$y = x_1$$

or $$y = \frac{4}{3} - \frac{1}{3}e^{-3t} \quad \textbf{Ans.}$$

Example 9.10.15. *Use diagonalization of matrix* **A** *to determine the time response of the system :*

$$\begin{bmatrix} \dot{x}_1 \\ \dot{x}_2 \end{bmatrix} = \begin{bmatrix} 0 & 1 \\ -6 & -5 \end{bmatrix}\begin{bmatrix} x_1 \\ x_2 \end{bmatrix} + \begin{bmatrix} 0 \\ 1 \end{bmatrix} u \text{ and } y = [6 \quad 1]\begin{bmatrix} x_1 \\ x_2 \end{bmatrix}; \text{ given that, } x(0) = \begin{bmatrix} 1 \\ 0 \end{bmatrix}.$$

Solution. It is noted that 2 × 2 matrix **A** is given in phase variable form, therefore, Vandermonde's matrix is considered for diagonalization of matrix **A**

The eigen values of matrix **A** are determined by solving following equation :

$$[\lambda \mathbf{I} - \mathbf{A}] = \left\{\begin{bmatrix} \lambda & 0 \\ 0 & \lambda \end{bmatrix} - \begin{bmatrix} 0 & 1 \\ -6 & -5 \end{bmatrix}\right\} = \begin{bmatrix} \lambda & -1 \\ 6 & \lambda+5 \end{bmatrix}$$

$\therefore$ $$|\lambda \mathbf{I} - \mathbf{A}| = \lambda(\lambda+5) - (-1)(6)$$
$$= \lambda^2 + 5\lambda + 6 = (\lambda+2)(\lambda+3)$$

put $$(\lambda+2)(\lambda+3) = 0$$

Thus, the two distinct eigen values of matrix **A** are :

$$\lambda_1 = -2 \text{ and } \lambda_2 = -3$$

The Vandermonde's matrix is given by

$$\mathbf{P} = \begin{bmatrix} 1 & 1 \\ \lambda_1 & \lambda_2 \end{bmatrix} - \begin{bmatrix} 1 & 1 \\ -2 & -3 \end{bmatrix}$$

The inverse of matrix can be determined as :

$$\mathbf{P}^{-1} = \frac{\text{Adj}\begin{bmatrix} 1 & 1 \\ -2 & -3 \end{bmatrix}}{\begin{vmatrix} 1 & 1 \\ -2 & -3 \end{vmatrix}} = \begin{bmatrix} 3 & 1 \\ -2 & -1 \end{bmatrix}$$

The state equation for the system is given by :

$$\dot{\mathbf{x}} = \mathbf{Ax} + \mathbf{Bu}$$

put $\mathbf{x} = \mathbf{Pz}$ in the above equation :

$\therefore$ $$\mathbf{P}\dot{\mathbf{z}} = \mathbf{APz} + \mathbf{Bu}$$

or $$\dot{\mathbf{z}} = \mathbf{P}^{-1}\mathbf{APz} + \mathbf{P}^{-1}\mathbf{Bu}$$

Taking Laplace transform on both sides :

$$s\mathbf{Z}(s) - \mathbf{z}(0) = \mathbf{P}^{-1}\mathbf{APZ}(s) + \mathbf{P}^{-1}\mathbf{BU}(s)$$

or $$\{s\mathbf{I} - \mathbf{P}^{-1}\mathbf{AP}\}\mathbf{Z}(s) = \mathbf{z}(0) + \mathbf{P}^{-1}\mathbf{BU}(s)$$

$\therefore$ $$\mathbf{Z}(s) = \{s\mathbf{I} - \mathbf{P}^{-1}\mathbf{AP}\}^{-1}\mathbf{z}(0) + \{s\mathbf{I} - \mathbf{P}^{-1}\mathbf{AP}\}^{-1}\mathbf{P}^{-1}\mathbf{BU}(s)$$

$$\mathbf{P}^{-1}\mathbf{AP} = \begin{bmatrix} 3 & 1 \\ -2 & -1 \end{bmatrix}\begin{bmatrix} 0 & 1 \\ -6 & -5 \end{bmatrix}\begin{bmatrix} 1 & 1 \\ -2 & -3 \end{bmatrix} = \begin{bmatrix} -2 & 0 \\ 0 & -3 \end{bmatrix}$$

$\because$ $$\mathbf{x} = \mathbf{Pz}$$

$\therefore$ $$\mathbf{z}(0) = \mathbf{P}^{-1}\mathbf{x}(0) = \begin{bmatrix} 3 & 1 \\ -2 & -1 \end{bmatrix}\begin{bmatrix} 0 \\ 1 \end{bmatrix} = \begin{bmatrix} 3 \\ -2 \end{bmatrix}$$

$$\mathbf{P}^{-1}\mathbf{B} = \begin{bmatrix} 3 & 1 \\ 2 & -1 \end{bmatrix}\begin{bmatrix} 0 \\ 1 \end{bmatrix} = \begin{bmatrix} 1 \\ -1 \end{bmatrix}$$

Substituting the values of $\mathbf{P}^{-1}\mathbf{AP}$, $\mathbf{z}(0)$ and $\mathbf{P}^{-1}\mathbf{B}$ in equation for $\mathbf{Z}(s)$, the following equation is obtained :

$$\mathbf{Z}(s) = \left\{\begin{bmatrix} s & 0 \\ 0 & s \end{bmatrix} - \begin{bmatrix} -2 & 0 \\ 0 & -3 \end{bmatrix}\right\}^{-1}\begin{bmatrix} 3 \\ -2 \end{bmatrix} + \left\{\begin{bmatrix} s & 0 \\ 0 & s \end{bmatrix} - \begin{bmatrix} -2 & 0 \\ 0 & -3 \end{bmatrix}\right\}^{-1}\begin{bmatrix} 1 \\ -1 \end{bmatrix}U(s)$$

$$= \begin{bmatrix} s+2 & 0 \\ 0 & s+3 \end{bmatrix}^{-1}\begin{bmatrix} 3 \\ -2 \end{bmatrix} + \begin{bmatrix} s+2 & 0 \\ 0 & s+3 \end{bmatrix}^{-1}\begin{bmatrix} 1 \\ -1 \end{bmatrix}\frac{1}{s}$$

$$= \begin{bmatrix} \dfrac{1}{s+2} & 0 \\ 0 & \dfrac{1}{s+3} \end{bmatrix}\begin{bmatrix} 3 \\ -2 \end{bmatrix} + \begin{bmatrix} \dfrac{1}{s+2} & 0 \\ 0 & \dfrac{1}{s+3} \end{bmatrix}\begin{bmatrix} \dfrac{1}{s} \\ -\dfrac{1}{s} \end{bmatrix}$$

$$= \begin{bmatrix} \dfrac{3}{s+2} \\ -\dfrac{2}{s+3} \end{bmatrix} + \begin{bmatrix} \dfrac{1}{s(s+2)} \\ -\dfrac{1}{s(s+3)} \end{bmatrix} = \begin{bmatrix} \dfrac{3}{s+2} \\ -\dfrac{2}{s+3} \end{bmatrix} + \begin{bmatrix} \dfrac{1}{2s} - \dfrac{1}{2(s+2)} \\ -\dfrac{1}{3s} + \dfrac{1}{3(s+3)} \end{bmatrix}$$

$$= \begin{bmatrix} \dfrac{3}{s+2} + \dfrac{1}{2s} - \dfrac{1}{2(s+2)} \\ -\dfrac{2}{s+3} - \dfrac{1}{3s} + \dfrac{1}{3(s+3)} \end{bmatrix}$$

$\because$ $$\mathbf{X}(s) = \mathbf{PZ}(s)$$

$\therefore$ $$\mathbf{X}(s) = \begin{bmatrix} 1 & 1 \\ -2 & -3 \end{bmatrix}\begin{bmatrix} \dfrac{3}{s+2} + \dfrac{1}{2s} - \dfrac{1}{2(s+2)} \\ -\dfrac{2}{s+3} - \dfrac{1}{3s} + \dfrac{1}{3(s+3)} \end{bmatrix}$$

$$\therefore \quad = \begin{bmatrix} \frac{3}{s+2} + \frac{1}{2s} - \frac{1}{2(s+2)} - \frac{2}{s+3} - \frac{1}{3s} + \frac{1}{3(s+3)} \\ -\frac{6}{s+2} - \frac{1}{s} + \frac{1}{s+2} + \frac{6}{s+3} + \frac{1}{s} - \frac{1}{s+3} \end{bmatrix} = \begin{bmatrix} \frac{1}{6s} + \frac{5}{2(s+2)} - \frac{5}{3(s+3)} \\ -\frac{5}{s+2} + \frac{5}{s+3} \end{bmatrix}$$

$$\mathbf{x} = \mathcal{L}^{-1} \begin{bmatrix} \frac{1}{6s} + \frac{5}{2(s+2)} - \frac{5}{3(s+3)} \\ -\frac{5}{s+2} + \frac{5}{s+3} \end{bmatrix}$$

$$\therefore \quad \begin{bmatrix} x_1 \\ x_2 \end{bmatrix} = \begin{bmatrix} \frac{1}{6} + \frac{5}{2}e^{-2t} - \frac{5}{3}e^{-3t} \\ -5e^{-2t} + 5e^{-3t} \end{bmatrix} \qquad \textbf{Ans.}$$

$$y = [6 \;\; 1] \begin{bmatrix} x_1 \\ x_2 \end{bmatrix} = 6x_1 + x_2$$

$$= 6\left(\frac{1}{6} + \frac{5}{2}e^{-2t} - \frac{5}{3}e^{-3t}\right) + (-5e^{-2t} + 5e^{-3t})$$

$$= (1 + 15e^{-2t} - 10e^{-3t} - 5e^{-2t} + 5e^{-3t})$$

or $$y = (1 + 10e^{-2t} - 5e^{-3t}) \qquad \textbf{Ans.}$$

Alternative solution

As already determined,

$$\mathbf{P} = \begin{bmatrix} 1 & 1 \\ -2 & -3 \end{bmatrix}; \; \mathbf{P}^{-1} = \begin{bmatrix} 3 & 1 \\ -2 & -1 \end{bmatrix}$$

$$\therefore \quad \Lambda = \mathbf{P}^{1}\,\mathbf{AP} = \begin{bmatrix} 3 & 1 \\ -2 & -1 \end{bmatrix}\begin{bmatrix} 0 & 1 \\ -6 & -5 \end{bmatrix}\begin{bmatrix} 1 & 1 \\ -2 & -3 \end{bmatrix}$$

$$= \begin{bmatrix} -2 & 0 \\ 0 & -3 \end{bmatrix}$$

$$\therefore \quad e^{\wedge t} = \begin{bmatrix} e^{-2t} & 0 \\ 0 & e^{-3t} \end{bmatrix}$$

$$e^{\mathbf{A}t} = \mathbf{P}\, e^{\wedge t}\, \mathbf{P}^{-1}$$

$$= \begin{bmatrix} 1 & 1 \\ -2 & -3 \end{bmatrix}\begin{bmatrix} e^{-2t} & 0 \\ 0 & e^{-3t} \end{bmatrix}\begin{bmatrix} 3 & 1 \\ -2 & -1 \end{bmatrix}$$

$$= \begin{bmatrix} 1 & 1 \\ -2 & -3 \end{bmatrix}\begin{bmatrix} 3e^{-2t} & e^{-2t} \\ -2e^{-3t} & -e^{-3t} \end{bmatrix}$$

$$= \begin{bmatrix} 3e^{-2t} - 2e^{-3t} & e^{-2t} - e^{-3t} \\ -6e^{-2t} + 6e^{-3t} & -2e^{-2t} + 3e^{-3t} \end{bmatrix}$$

$$\mathbf{x}(t) = \phi(t)\,\mathbf{x}(0) + \int_0^t \phi(t-\tau)\,\mathbf{B}u(\tau)\,d\tau$$

$\because$ $$\phi(t) = e^{\mathbf{A}t}\,;\ \mathbf{x}(0) = \begin{bmatrix} 1 \\ 0 \end{bmatrix} \text{ and } \mathbf{B} = \begin{bmatrix} 0 \\ 1 \end{bmatrix}$$

$$\mathbf{x}(t) = \begin{bmatrix} -3e^{-2t} - 2e^{-3t} & e^{-2t} - e^{-3t} \\ -6e^{-2t} + 6e^{-3t} & -2e^{-2t} + 3e^{-3t} \end{bmatrix}\begin{bmatrix} 1 \\ 0 \end{bmatrix}$$

$$+ \int_0^t \begin{bmatrix} 3e^{-2(t-\tau)} - 2e^{-3(t-\tau)} & e^{-2(t-\tau)} - e^{-3(t-\tau)} \\ -6e^{-2(t-\tau)} + 6e^{-3(t-\tau)} & -2e^{-2(t-\tau)} + 3e^{-3(t-\tau)} \end{bmatrix}\begin{bmatrix} 0 \\ 1 \end{bmatrix} d\tau$$

$$= \begin{bmatrix} 3e^{-2t} - 2e^{-3t} \\ -6e^{-2t} + 6e^{-3t} \end{bmatrix} + \int_0^t \begin{bmatrix} e^{-2(t-\tau)} - e^{-3(t-\tau)} \\ -2e^{-2(t-\tau)} + 3e^{-3(t-\tau)} \end{bmatrix} d\tau$$

$$= \begin{bmatrix} 3e^{-2t} - 2e^{-3t} \\ -6e^{-2t} + 6e^{-3t} \end{bmatrix} + \int_0^t \begin{bmatrix} e^{-2t}\,.\,e^{2\tau} - e^{-3t}\,.\,e^{3\tau} \\ -2e^{-2t}\,.\,e^{2\tau} + 3e^{-3t}\,.\,e^{3\tau} \end{bmatrix} d\tau$$

$$= \begin{bmatrix} 3e^{-2t} - 2e^{-3t} \\ (-6e^{-2t} + 6e^{-3t}) \end{bmatrix} + \begin{bmatrix} \frac{1}{2}e^{-2t}\,.\,e^{2\tau} - \frac{1}{3}e^{-3t}\,.\,e^{3\tau} \\ -\frac{2}{2}e^{-2t}\,.\,e^{2\tau} + \frac{3}{3}e^{-3t}\,.\,e^{3\tau} \end{bmatrix}_0^t$$

$$= \begin{bmatrix} 3e^{-2t} - 2e^{-3t} \\ -6e^{-2t} + 6e^{-3t} \end{bmatrix} + \begin{bmatrix} \left(\frac{1}{2}e^{0} - \frac{1}{3}e^{0}\right) - \left(\frac{1}{2}e^{-2t} - \frac{1}{3}e^{-3t}\right) \\ (-e^{0} + e^{0}) - (-e^{-2t} + e^{-3t}) \end{bmatrix}$$

$$= \begin{bmatrix} 3e^{-2t} - 2e^{-3t} \\ -6e^{-2t} + 6e^{-3t} \end{bmatrix} + \begin{bmatrix} \left(\frac{1}{2} - \frac{1}{3} - \frac{1}{2}e^{-2t} + \frac{1}{3}e^{-3t}\right) \\ (-1 + 1 + e^{-2t} - e^{-3t}) \end{bmatrix}$$

or $$\begin{bmatrix} x_1 \\ x_2 \end{bmatrix} = \begin{bmatrix} \frac{1}{6} + \frac{5}{2}e^{-2t} - \frac{5}{3}e^{-3t} \\ -5e^{-2t} + 5e^{-3t} \end{bmatrix}$$ **Ans.**

$$y = [6 \quad 1]\begin{bmatrix} x_1 \\ x_2 \end{bmatrix} = 6x_1 + x_2$$

$$= 6\left(\frac{1}{6} + \frac{5}{2}e^{-2t} - \frac{5}{3}e^{-3t}\right) + (-5e^{-2t} + 5e^{-3t})$$

$$= 1 + 10e^{-2t} - 5e^{-3t}$$ **Ans.**

Example 9.10.16. *The state block diagram of a control system is shown below (Fig. 9.10.13). Write state equations and determine transfer matrix.*

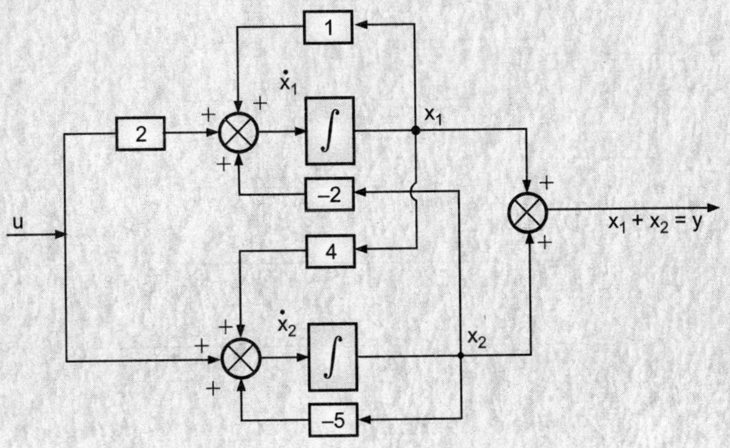

Fig. 9.10.13. State block diagram for example 9.10.16.

Solution. As per state block diagram (Fig. 9.10.13), state equations are :

$$\dot{x}_1 = x_1 - 2x_2 + 2u$$

$$\dot{x}_2 = 4x_1 - 5x_2 + u$$

The output equation is : $y = x_1 + x_2$

$$\therefore \quad \begin{bmatrix} \dot{x}_1 \\ \dot{x}_2 \end{bmatrix} = \begin{bmatrix} 1 & -2 \\ 4 & -5 \end{bmatrix} \begin{bmatrix} x_1 \\ x_2 \end{bmatrix} + \begin{bmatrix} 2 \\ 1 \end{bmatrix} u$$

and
$$y = [1 \quad 1] \begin{bmatrix} x_1 \\ x_2 \end{bmatrix} \quad \textbf{Ans.}$$

The coefficient matrices are :

$$\mathbf{A} = \begin{bmatrix} 1 & -2 \\ 4 & -5 \end{bmatrix}, \mathbf{B} = \begin{bmatrix} 2 \\ 1 \end{bmatrix} \text{ and } \mathbf{C} = [1 \quad 1]$$

$$\{s\mathbf{I} - \mathbf{A}\} = \left\{ \begin{bmatrix} s & 0 \\ 0 & s \end{bmatrix} - \begin{bmatrix} 1 & -2 \\ 4 & -5 \end{bmatrix} \right\} = \begin{bmatrix} (s-1) & 2 \\ -4 & (s+5) \end{bmatrix}$$

$$(s\mathbf{I} - \mathbf{A})^{-1} = \frac{\text{Adj}\,(s\mathbf{I} - \mathbf{A})}{|s\mathbf{I} - \mathbf{A}|} = \frac{\begin{bmatrix} s+5 & -2 \\ 4 & s-1 \end{bmatrix}}{s^2 + 4s + 3}$$

or
$$(s\mathbf{I} - \mathbf{A})^{-1} = \frac{\begin{bmatrix} s+5 & -2 \\ 4 & s-1 \end{bmatrix}}{(s+1)(s+3)} = \begin{bmatrix} \dfrac{s+5}{(s+1)(s+3)} & -\dfrac{2}{(s+1)(s+3)} \\ \dfrac{4}{(s+1)(s+3)} & \dfrac{s-1}{(s+1)(s+3)} \end{bmatrix}$$

The transfer matrix is given by

$$G(s) = \mathbf{C}(s\mathbf{I} - \mathbf{A})^{-1}\,\mathbf{B}$$

$$= [1 \quad 1]\begin{bmatrix} \dfrac{s+5}{(s+1)(s+3)} & \dfrac{-2}{(s+1)(s+3)} \\ \dfrac{4}{(s+1)(s+3)} & \dfrac{s-1}{(s+1)(s+3)} \end{bmatrix}\begin{bmatrix} 2 \\ 1 \end{bmatrix}$$

$$= [1 \quad 1]\begin{bmatrix} \dfrac{2(s+5)}{(s+1)(s+3)} - \dfrac{2}{(s+1)(s+3)} \\ \dfrac{8}{(s+1)(s+3)} + \dfrac{s-1}{(s+1)(s+3)} \end{bmatrix} = [1 \quad 1]\begin{bmatrix} \dfrac{2s+8}{(s+1)(s+3)} \\ \dfrac{s+7}{(s+1)(s+3)} \end{bmatrix}$$

$$= \frac{2s+8}{(s+1)(s+3)} + \frac{s+7}{(s+1)(s+3)} = \frac{3(s+5)}{(s+1)(s+3)} \quad \textbf{Ans.}$$

Example 9.10.17. *Determine the transfer matrix from the data given below :*

$$\mathbf{A} = \begin{bmatrix} -3 & 1 \\ 0 & -1 \end{bmatrix},\ \mathbf{B} = \begin{bmatrix} 1 \\ 1 \end{bmatrix},\ \mathbf{C} = [1 \quad 1] \text{ and } \mathbf{D} = 0.$$

Solution. The transfer matrix is given by :

$$G(s) = \mathbf{C}(s\mathbf{I} - \mathbf{A})^{-1}\,\mathbf{B} + \mathbf{D}$$

$$(s\mathbf{I} - \mathbf{A}) = \left\{\begin{bmatrix} s & 0 \\ 0 & s \end{bmatrix} - \begin{bmatrix} -3 & 1 \\ 0 & -1 \end{bmatrix}\right\} = \begin{bmatrix} s+3 & -1 \\ 0 & s+1 \end{bmatrix}$$

$$(s\mathbf{I} - \mathbf{A})^{-1} = \frac{\text{adj}\,(s\mathbf{I} - \mathbf{A})}{|s\mathbf{I} - \mathbf{A}|} = \frac{\begin{bmatrix} s+1 & 1 \\ 0 & s+3 \end{bmatrix}}{(s+1)(s+3)}$$

$$= \begin{bmatrix} \dfrac{s+1}{(s+1)(s+3)} & \dfrac{1}{(s+1)(s+3)} \\ 0 & \dfrac{s+3}{(s+1)(s+3)} \end{bmatrix} = \begin{bmatrix} \dfrac{1}{s+3} & \dfrac{1}{(s+1)(s+3)} \\ 0 & \dfrac{1}{s+1} \end{bmatrix}$$

$$\therefore \quad G(s) = [1 \quad 1]\begin{bmatrix} \dfrac{1}{s+3} & \dfrac{1}{(s+1)(s+3)} \\ 0 & \dfrac{1}{s+1} \end{bmatrix}\begin{bmatrix} 1 \\ 1 \end{bmatrix},\ \mathbf{D} = 0 \text{ (given)}$$

$$= [1 \quad 1]\begin{bmatrix} \dfrac{1}{s+3} + \dfrac{1}{(s+1)(s+3)} \\ \dfrac{1}{s+1} \end{bmatrix} = \frac{1}{s+3} + \frac{1}{(s+1)(s+3)} + \frac{1}{s+1}$$

$$= \frac{(s+1) + 1 + (s+3)}{(s+1)(s+3)}$$

or $$G(s) = \frac{2s + 5}{(s + 1)(s + 3)} \quad \textbf{Ans.}$$

Example 9.10.18. *Determine the transfer matrix for the system given below and draw the block diagram*

$$\begin{bmatrix} \dot{x}_1 \\ \dot{x}_2 \end{bmatrix} = \begin{bmatrix} 0 & 3 \\ -2 & -5 \end{bmatrix} \begin{bmatrix} x_1 \\ x_2 \end{bmatrix} + \begin{bmatrix} 1 & 1 \\ 1 & 1 \end{bmatrix} u(t) \quad and \; y = \begin{bmatrix} 2 & 1 \\ 1 & 0 \end{bmatrix} \begin{bmatrix} x_1 \\ x_2 \end{bmatrix}.$$

Solution. The transfer matrix is given by

$$\mathbf{G}(s) = \mathbf{C}(s\mathbf{I} - \mathbf{A})^{-1}\,\mathbf{B} + \mathbf{D} \qquad \because \; \mathbf{D} = 0$$

$$\therefore \qquad \mathbf{G}(s) = \mathbf{C}(s\mathbf{I} - \mathbf{A})^{-1}\,\mathbf{B}$$

$$\mathbf{A} = \begin{bmatrix} 0 & 3 \\ -2 & -5 \end{bmatrix} \quad \mathbf{B} = \begin{bmatrix} 1 & 1 \\ 1 & 1 \end{bmatrix}$$

$$\mathbf{C} = \begin{bmatrix} 2 & 1 \\ 1 & 0 \end{bmatrix} \qquad \mathbf{D} = 0$$

$$(s\mathbf{I} - \mathbf{A}) = \begin{bmatrix} s & 0 \\ 0 & s \end{bmatrix} - \begin{bmatrix} 0 & 3 \\ -2 & -5 \end{bmatrix} = \begin{bmatrix} s & -3 \\ 2 & s+5 \end{bmatrix}$$

$$\therefore \qquad (s\mathbf{I} - \mathbf{A})^{-1} = \begin{bmatrix} s & -3 \\ 2 & s+5 \end{bmatrix}^{-1}$$

$(s\mathbf{I} - \mathbf{A})^{-1}$ is determined below :

(1) Arrange for the minor of $(s\mathbf{I} - \mathbf{A})$, *i.e.* $\begin{bmatrix} (s+5) & 2 \\ -3 & s \end{bmatrix}$

(2) Change to cofactor, *i.e.* $\begin{bmatrix} (s+5) & -2 \\ 3 & s \end{bmatrix}$

(3) Interchange rows and columns, to obtain Adj $(s\mathbf{I} - \mathbf{A})$, *i.e.*

$$\text{Adj}\,(s\mathbf{I} - \mathbf{A}) = \begin{bmatrix} (s+5) & 3 \\ -2 & s \end{bmatrix}$$

(4) $$(s\mathbf{I} - \mathbf{A})^{-1} = \frac{\text{Adj}\,(s\mathbf{I} - \mathbf{A})}{|s\mathbf{I} - \mathbf{A}|}$$

$$(s\mathbf{I} - \mathbf{A})^{-1} = \frac{\begin{bmatrix} (s+5) & 3 \\ -2 & s \end{bmatrix}}{\begin{vmatrix} s & -3 \\ 2 & (s+5) \end{vmatrix}} = \frac{\begin{bmatrix} (s+5) & 3 \\ -2 & s \end{bmatrix}}{s^2 + 5s + 6}$$

$$= \frac{\begin{bmatrix} (s+5) & 3 \\ -2 & s \end{bmatrix}}{(s+2)(s+3)} = \begin{bmatrix} \dfrac{(s+5)}{(s+2)(s+3)} & \dfrac{3}{(s+2)(s+3)} \\ -\dfrac{2}{(s+2)(s+3)} & \dfrac{s}{(s+2)(s+3)} \end{bmatrix}$$

$$\therefore \quad \mathbf{G}(s) = \begin{bmatrix} 2 & 1 \\ 1 & 0 \end{bmatrix} \begin{bmatrix} \frac{s+5}{(s+2)(s+3)} & \frac{3}{(s+2)(s+3)} \\ -\frac{2}{(s+2)(s+3)} & \frac{s}{(s+2)(s+3)} \end{bmatrix} \begin{bmatrix} 1 & 1 \\ 1 & 1 \end{bmatrix}$$

$$= \begin{bmatrix} \frac{(2s+8)}{(s+2)(s+3)} & \frac{(s+6)}{(s+2)(s+3)} \\ \frac{(s+5)}{(s+2)(s+3)} & \frac{3}{(s+2)(s+3)} \end{bmatrix} \begin{bmatrix} 1 & 1 \\ 1 & 1 \end{bmatrix}$$

$$= \begin{bmatrix} \frac{(3s+14)}{(s+2)(s+3)} & \frac{(3s+14)}{(s+2)(s+3)} \\ \frac{(s+8)}{(s+2)(s+3)} & \frac{(s+8)}{(s+2)(s+3)} \end{bmatrix} \quad \textbf{Ans.}$$

$$\mathbf{Y}(s) = \mathbf{G}(s)\, \mathbf{U}(s)$$

$$\therefore \quad \begin{bmatrix} Y_1(s) \\ Y_2(s) \end{bmatrix} = \begin{bmatrix} \frac{(3s+14)}{(s+2)(s+3)} & \frac{(3s+14)}{(s+2)(s+3)} \\ \frac{(s+8)}{(s+2)(s+3)} & \frac{(s+8)}{(s+2)(s+3)} \end{bmatrix} \begin{bmatrix} U_1(s) \\ U_2(s) \end{bmatrix}$$

$$\therefore \quad Y_1(s) = \frac{(3s+14)}{(s+2)(s+3)} U_1(s) + \frac{(3s+14)}{(s+2)(s+3)} U_2(s)$$

$$Y_2(s) = \frac{(s+8)}{(s+2)(s+3)} U_1(s) + \frac{(s+8)}{(s+2)(s+3)} U_2(s)$$

The block diagram for the system is shown in Fig. 9.10.14.

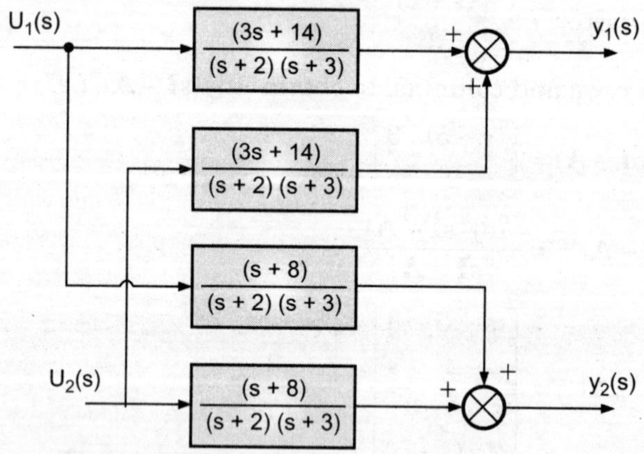

Fig. 9.10.14. Block diagram for example 9.10.18.

Example 9.10.19. *The state equations of a control system are given below : Examine for complete state controllability.*

$$\dot{\mathbf{x}}_1 = -\frac{1}{T_1} x_1 + \frac{1}{T_1} u \quad \text{and} \quad \dot{\mathbf{x}}_2 = -\frac{1}{T_2} x_2 + \frac{1}{T_2} u.$$

Solution.
$$\begin{bmatrix} \dot{x}_1 \\ \dot{x}_2 \end{bmatrix} = \begin{bmatrix} -\frac{1}{T_1} & 0 \\ 0 & -\frac{1}{T_2} \end{bmatrix} \begin{bmatrix} x_1 \\ x_2 \end{bmatrix} + \begin{bmatrix} \frac{1}{T_1} \\ \frac{1}{T_2} \end{bmatrix} u$$

$$\mathbf{B} = \begin{bmatrix} \frac{1}{T_1} \\ \frac{1}{T_2} \end{bmatrix} \text{ and } \mathbf{A} = \begin{bmatrix} -\frac{1}{T_1} & 0 \\ 0 & -\frac{1}{T_2} \end{bmatrix}$$

$$\mathbf{AB} = \begin{bmatrix} -\frac{1}{T_1} & 0 \\ 0 & -\frac{1}{T_2} \end{bmatrix} \begin{bmatrix} \frac{1}{T_1} \\ \frac{1}{T_2} \end{bmatrix} = \begin{bmatrix} -\frac{1}{T_1^2} \\ -\frac{1}{T_2^2} \end{bmatrix}$$

Controllability test matrix is given by

$$\mathbf{U} = [\mathbf{B} : \mathbf{AB}] = \begin{bmatrix} \frac{1}{T_1} & -\frac{1}{T_1^2} \\ \frac{1}{T_2} & -\frac{1}{T_2^2} \end{bmatrix}$$

For complete state controllability, the matrix **U** be of rank 2.

$$\therefore \qquad |\mathbf{U}| \neq 0$$

or
$$\begin{bmatrix} \frac{1}{T_1} & -\frac{1}{T_1^2} \\ \frac{1}{T_2} & -\frac{1}{T_2^2} \end{bmatrix} \neq 0$$

or
$$-\frac{1}{T_1 T_2^2} - \frac{1}{T_2}\left(-\frac{1}{T_1^2}\right) \neq 0 \quad \text{or} \quad \frac{1}{T_2 T_1^2} - \frac{1}{T_1 T_2^2} \neq 0$$

or
$$\frac{1}{T_1 T_2}\left(\frac{1}{T_1} - \frac{1}{T_2}\right) \neq 0 \quad \text{or} \quad \frac{1}{T_1 T_2}\left(\frac{T_2 - T_1}{T_1 T_2}\right) \neq 0$$

or
$$\frac{1}{(T_1 T_2)^2}(T_2 - T_1) \neq 0$$

$$\therefore \qquad T_1 \neq T_2 \quad \textbf{Ans.}$$

Example 9.10.20. *Verify the controllability. Given that :*

$$\dot{x}_1 = x_2 + u_1$$
$$\dot{x}_2 = x_3$$
$$\dot{x}_3 = -2x_2 - 3x_3 + u_1 + u_2.$$

Solution. $$\begin{bmatrix} \dot{x}_1 \\ \dot{x}_2 \\ \dot{x}_3 \end{bmatrix} = \begin{bmatrix} 0 & 1 & 0 \\ 0 & 0 & 1 \\ 0 & -2 & -3 \end{bmatrix}\begin{bmatrix} x_1 \\ x_2 \\ x_3 \end{bmatrix} + \begin{bmatrix} 0 & 1 \\ 0 & 0 \\ 1 & 1 \end{bmatrix}\begin{bmatrix} u_1 \\ u_2 \end{bmatrix}$$

$$\mathbf{B} = \begin{bmatrix} 0 & 1 \\ 0 & 0 \\ 1 & 1 \end{bmatrix} \text{ and } \mathbf{A} = \begin{bmatrix} 0 & 1 & 0 \\ 0 & 0 & 1 \\ 0 & -2 & -3 \end{bmatrix}$$

$$\mathbf{AB} = \begin{bmatrix} 0 & 1 & 0 \\ 0 & 0 & 1 \\ 0 & -2 & -3 \end{bmatrix}\begin{bmatrix} 0 & 1 \\ 0 & 0 \\ 1 & 1 \end{bmatrix} = \begin{bmatrix} 0 & 0 \\ 1 & 1 \\ -3 & -3 \end{bmatrix}$$

$$\mathbf{A^2B} = \begin{bmatrix} 0 & 1 & 0 \\ 0 & 0 & 1 \\ 0 & -2 & -3 \end{bmatrix}\begin{bmatrix} 0 & 0 \\ 1 & 1 \\ -3 & -3 \end{bmatrix} = \begin{bmatrix} 1 & 1 \\ -3 & -3 \\ -7 & 7 \end{bmatrix}$$

Controllability test matrix is given by :

$$\mathbf{U} = [\mathbf{B} : \mathbf{AB} : \mathbf{A^2B}]$$

$$\therefore \qquad \mathbf{U} = \begin{bmatrix} 0 & 1 & 0 & 0 & 1 & 1 \\ 0 & 0 & 1 & 1 & -3 & -3 \\ 1 & 1 & -3 & -3 & 7 & 7 \end{bmatrix}$$

Collect the columns from **U** to form a square 3 × 3 to form non-zero determinant matrix as given below :

$$\therefore \qquad \begin{bmatrix} 0 & 1 & 0 \\ 0 & 0 & 1 \\ 1 & 1 & -3 \end{bmatrix} = 0\begin{vmatrix} 0 & 1 \\ 1 & -3 \end{vmatrix} - (1)\begin{vmatrix} 0 & 1 \\ 1 & -3 \end{vmatrix} + 0\begin{vmatrix} 0 & 0 \\ 1 & 1 \end{vmatrix}$$

$$= -(1)\{0 \times (-3) - (1)(1)\} = 1$$

$$\text{As} \qquad \begin{bmatrix} 0 & 1 & 0 \\ 0 & 0 & 1 \\ 1 & 1 & -3 \end{bmatrix} \neq 0$$

Therefore, the rank of **U** is equal to 3 ($n = 3$), hence the system is completely state controllable. **Ans.**

Example 9.10.21. *Check for controllability and observability of a system having following coefficient matrices.*

$$\mathbf{A} = \begin{bmatrix} 0 & 1 & 0 \\ 0 & 0 & 1 \\ -6 & -11 & -6 \end{bmatrix}, \mathbf{B} = \begin{bmatrix} 1 \\ 0 \\ 1 \end{bmatrix} \text{ and } \mathbf{C}^T = \begin{bmatrix} 10 \\ 5 \\ 1 \end{bmatrix}.$$

Solution. (*a*) Controllability test :

$$\mathbf{B} = \begin{bmatrix} 1 \\ 0 \\ 1 \end{bmatrix}, \mathbf{A} = \begin{bmatrix} 0 & 1 & 0 \\ 0 & 0 & 1 \\ -6 & -11 & -6 \end{bmatrix}$$

$$\mathbf{AB} = \begin{bmatrix} 0 & 1 & 0 \\ 0 & 0 & 1 \\ -6 & -11 & -6 \end{bmatrix} \begin{bmatrix} 1 \\ 0 \\ 1 \end{bmatrix} = \begin{bmatrix} 0 \\ 1 \\ -12 \end{bmatrix}$$

$$\mathbf{A^2B} = \begin{bmatrix} 0 & 1 & 0 \\ 0 & 0 & 1 \\ -6 & -11 & -6 \end{bmatrix} \begin{bmatrix} 0 \\ 1 \\ -12 \end{bmatrix} = \begin{bmatrix} 0 \\ -12 \\ 61 \end{bmatrix}$$

Controllability test matrix is given by :

$$\mathbf{U} = [\mathbf{B} : \mathbf{AB} : \mathbf{A^2B}]$$

$$\therefore \qquad \mathbf{U} = \begin{bmatrix} 1 & 0 & 0 \\ 0 & 1 & -12 \\ 1 & 12 & 61 \end{bmatrix}$$

$$|\mathbf{U}| = 1\begin{vmatrix} 1 & -12 \\ 12 & 61 \end{vmatrix} + 0\begin{vmatrix} 0 & -12 \\ 1 & 61 \end{vmatrix} + 0\begin{vmatrix} 0 & 1 \\ 1 & 12 \end{vmatrix}$$

$$\therefore \qquad |\mathbf{U}| \neq 0$$

Therefore, the rank of **U** is equal to its order *i.e.* $3(n = 3)$ indicating that the system is controllable. **Ans.**

(*b*) Observability test :

$$\mathbf{C}^T = \begin{bmatrix} 10 \\ 5 \\ 1 \end{bmatrix}, \mathbf{A} = \begin{bmatrix} 0 & 1 & 0 \\ 0 & 0 & 1 \\ -6 & -11 & -6 \end{bmatrix}$$

$$\mathbf{A}^T = \begin{bmatrix} 0 & 0 & -6 \\ 1 & 0 & -11 \\ 0 & 1 & -6 \end{bmatrix}$$

$$\mathbf{A}^T\mathbf{C}^T = \begin{bmatrix} 0 & 0 & -6 \\ 1 & 0 & -11 \\ 0 & 1 & -6 \end{bmatrix} \begin{bmatrix} 10 \\ 5 \\ 1 \end{bmatrix} = \begin{bmatrix} -6 \\ -1 \\ -1 \end{bmatrix}$$

$$(\mathbf{A}^T)^2\mathbf{C}^T = \begin{bmatrix} 0 & 0 & -6 \\ 1 & 0 & -11 \\ 0 & 1 & -6 \end{bmatrix} \begin{bmatrix} -6 \\ -1 \\ -1 \end{bmatrix} = \begin{bmatrix} 6 \\ 5 \\ 5 \end{bmatrix}$$

Observability test matrix is given by

$$\mathbf{V} = [\mathbf{C}^T : \mathbf{A}^T\mathbf{C}^T : (\mathbf{A}^T)^2\,\mathbf{C}^T]$$

$$\therefore \qquad \mathbf{V} = \begin{bmatrix} 10 & -6 & 6 \\ 5 & -1 & 5 \\ 1 & -1 & 5 \end{bmatrix}$$

$$|\mathbf{V}| = 10 \begin{vmatrix} -1 & 5 \\ -1 & 5 \end{vmatrix} - (-6) \begin{vmatrix} 5 & 5 \\ 1 & 5 \end{vmatrix} + 6 \begin{vmatrix} 5 & -1 \\ 1 & -1 \end{vmatrix}$$

$$= 10\,[(-1)\,5 - 5\,(-1)] + 6\,(5 \times 5 - 5 \times 1) + 6\,[(5)(-1) - (-1)(1)]$$

$$= 10 \times 0 + 6 \times 20 + 6 \times (-4) = 96$$

Observability matrix $\mathbf{V} \neq 0$, hence its rank *i.e.* 3 equals order *i.e.* 3, therefore system is completely observable.

Example 9.10.22. *The transfer function of a control system is given by,*

$$\frac{Y(s)}{U(s)} = \frac{s+2}{s^3 + 9s^2 + 26s + 24}$$

Check for controllability and observability.

Solution. The state block diagram is drawn as shown in Fig. 9.10.15.

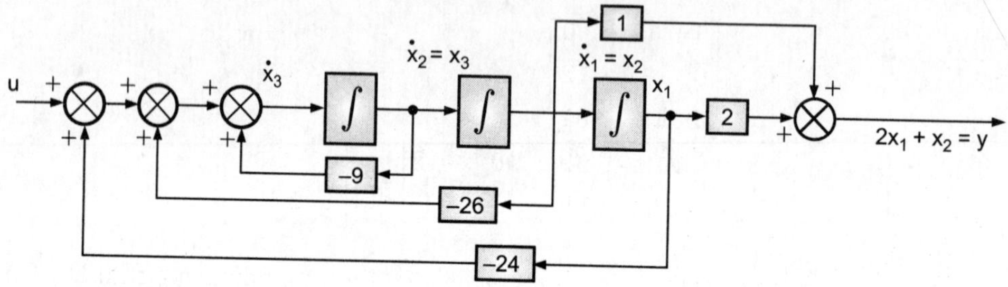

Fig. 9.10.15. State block diagram for $Y(s)/U(s) = (s+2)/(s^3 + 9s^2 + 26s + 24)$.

The state equations are written below :

$$\dot{x}_1 = x_2$$

$$\dot{x}_2 = x_3$$

$$\dot{x}_3 = -24x_1 - 26x_2 - 9x_3 + u$$

and $$y = 2x_1 + x_2$$

or $$\begin{bmatrix} \dot{x}_1 \\ \dot{x}_2 \\ \dot{x}_3 \end{bmatrix} = \begin{bmatrix} 0 & 1 & 0 \\ 0 & 0 & 1 \\ -24 & -26 & -9 \end{bmatrix} \begin{bmatrix} x_1 \\ x_2 \\ x_3 \end{bmatrix} + \begin{bmatrix} 0 \\ 0 \\ 1 \end{bmatrix} u$$

and $$y = [2 \;\; 1 \;\; 0] \begin{bmatrix} x_1 \\ x_2 \\ x_3 \end{bmatrix}$$

The coefficient matrices are :

$$\mathbf{A} = \begin{bmatrix} 0 & 1 & 0 \\ 0 & 0 & 1 \\ -24 & -26 & -9 \end{bmatrix}, \mathbf{B} = \begin{bmatrix} 0 \\ 0 \\ 1 \end{bmatrix}, \mathbf{C} = [2 \;\; 1 \;\; 0]$$

(*a*) Controllability test :

$$\mathbf{B} = \begin{bmatrix} 0 \\ 0 \\ 1 \end{bmatrix}, \mathbf{A} = \begin{bmatrix} 0 & 1 & 0 \\ 0 & 0 & 1 \\ -24 & -26 & -9 \end{bmatrix}$$

$$\mathbf{AB} = \begin{bmatrix} 0 & 1 & 0 \\ 0 & 0 & 1 \\ -24 & -26 & -9 \end{bmatrix} \begin{bmatrix} 0 \\ 0 \\ 1 \end{bmatrix} = \begin{bmatrix} 0 \\ 1 \\ -9 \end{bmatrix}$$

$$\mathbf{A^2B} = \begin{bmatrix} 0 & 1 & 0 \\ 0 & 0 & 1 \\ -24 & -26 & -9 \end{bmatrix} \begin{bmatrix} 0 \\ 1 \\ -9 \end{bmatrix} = \begin{bmatrix} 1 \\ -9 \\ 55 \end{bmatrix}$$

The controllability test matrix is given by :

$$\mathbf{U} = [\mathbf{B} : \mathbf{AB} : \mathbf{A^2B}]$$

$$\therefore \qquad \mathbf{U} = \begin{bmatrix} 0 & 0 & 1 \\ 0 & 1 & -9 \\ 1 & -9 & 55 \end{bmatrix}$$

$$|\mathbf{U}| = 0\begin{vmatrix} 1 & -9 \\ -9 & 55 \end{vmatrix} - 0\begin{vmatrix} 0 & -9 \\ 1 & 55 \end{vmatrix} + 1\begin{vmatrix} 0 & 1 \\ 1 & -9 \end{vmatrix}$$

$$= 0 + 0 + 1\,[0 \times (-9) - (1) \times (1)] = -1$$

$$\therefore \qquad |\mathbf{U}| \neq 0$$

Thus, the rank of **U** is 3 and order is also 3. The system is completely controllable.

Ans.

(*b*) Observability test :

$$\mathbf{A}^T = \begin{bmatrix} 0 & 0 & -24 \\ 1 & 0 & -26 \\ 0 & 1 & -9 \end{bmatrix}, \mathbf{C}^T = \begin{bmatrix} 2 \\ 1 \\ 0 \end{bmatrix}$$

The observability test matrix is given by :

$$\mathbf{V} = [\mathbf{C}^T : \mathbf{A}^T\mathbf{C}^T : (\mathbf{A}^T)^2\,\mathbf{C}^T]$$

$$\therefore \qquad \mathbf{V} = \begin{bmatrix} 2 & 0 & -24 \\ 1 & 2 & -26 \\ 0 & 1 & -7 \end{bmatrix}$$

$$|\mathbf{V}| = 2\begin{vmatrix} 2 & -26 \\ 1 & -7 \end{vmatrix} - 0\begin{vmatrix} 1 & -26 \\ 0 & -7 \end{vmatrix} + (-24)\begin{vmatrix} 1 & 2 \\ 0 & 1 \end{vmatrix}$$

$$= 2[2 \times (-7) - (1) \times (-26)] + 0\,[1 \times (-7) - (-26) \times (0)] - 24\,(1 \times 1 + 2 \times 0)$$

$$= 24 + 0 - 24 = 0$$

$$|\mathbf{V}| = 0,$$

Therefore its rank $\neq 3$, hence, the system is not observable.

9.11 STATE VARIABLE FEEDBACK

State variable feedback for poleplacement : The state model of SISO (single-input-single-output) control system is given by

$$\dot{\mathbf{x}} = \mathbf{Ax} + \mathbf{B}u$$
$$y = \mathbf{Cx}$$

The characteristic polynomial of the system is

$$|\lambda \mathbf{I} - \mathbf{A}| = 0$$

The eigen values of the system matrix **A** are the roots of the characteristic polynomial i.e. closed-loop poles of the system. On application of the state variable feedback, the eigen values of the system matrix **A** can be adjusted to a desired value. For the purpose of applying state variable feedback, the system matrices **A** and **B** should form a controllable pair. The application of state variable feedback is shown in Fig. 9.11.1.

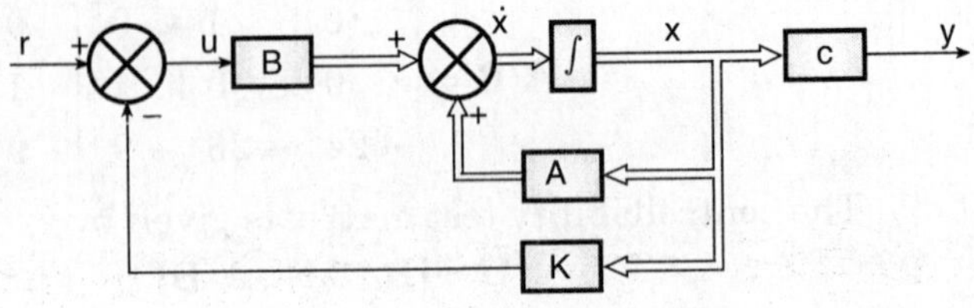

Fig. 9.11.1. State variable feedback.

As per the block diagram (Fig. 9.11.1) following equations are obtained

$$\dot{\mathbf{x}} = \mathbf{Ax} + \mathbf{B}u$$

and $$u = r - \mathbf{Kx}$$

$\therefore$ $$\mathbf{x} = \mathbf{Ax} + \mathbf{B}\,(r - \mathbf{Kx})$$
$$= (\mathbf{A} - \mathbf{BK})\,\mathbf{x} + \mathbf{B}r$$

Due to state variable feedback, the system matrix is (**A** – **BK**). The eigen values of matrix (**A** – **BK**) can be adjusted by selecting the feedback gain matrix **K**.

On application of the state variable feedback the characteristic polynomial is

$$|s\mathbf{I} - (\mathbf{A} - \mathbf{BK})|$$

(**A** – **BK**) is the modified system matrix. The feedback gain matrix **K** is to be determined such that the desired eigen values of matrix (**A** – **BK**) satisfy the characteristic equation $|s\mathbf{I} - (\mathbf{A} - \mathbf{BK})| = 0$.

The feedback gain matrix **K** is determined through following steps :

(1) Determine the characteristic polynomial as per desired closed-loop pole locations.

(2) Determine the characteristic polynomial in terms of feedback gain matrix **K**.

(3) Comparing the coefficients of characteristic polynomials as determined in step (1) and (2) the feedback gain matrix **K** is obtained.

Example 9.11.1. *Given the transfer function,*

$$\frac{Y(s)}{U(s)} = \frac{1}{(s+5)(s+4)}$$

Obtain the state equation using,

(a) Cascade decomposition

(b) Direct decomposition

It is desired that the closed-loop poles are to be placed at s = (– 1 ± j2). Determine the feedback gain matrix **K** *for (a) and (b).*

Solution. (*a*) The transfer function can be written as

$$\frac{Y(s)}{U(s)} = \frac{s^{-1}}{1-(-5s^{-1})} \cdot \frac{s^{-1}}{1-(-4s^{-1})}$$

The state signal flow graph for the above transfer function is drawn as shown in Fig. 9.11.2.

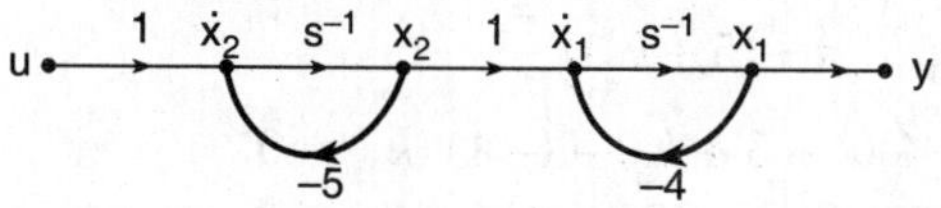

Fig. 9.11.2. State signal flow graph : $Y(s)/U(s) = s^{-1}/[1-(-5s^{-1})] \, . \, s^{-1}/[1-(-4s^{-1})]$.

The state equations are

$$\dot{x}_1 = -4x_1 + x_2$$

$$\dot{x}_2 = -5x_2 + u$$

$$\therefore \quad \begin{bmatrix} \dot{x}_1 \\ \dot{x}_2 \end{bmatrix} = \begin{bmatrix} -4 & 1 \\ 0 & -5 \end{bmatrix} \begin{bmatrix} x_1 \\ x_2 \end{bmatrix} + \begin{bmatrix} 0 \\ 1 \end{bmatrix} u$$

$$\mathbf{A} = \begin{bmatrix} -4 & 1 \\ 0 & -5 \end{bmatrix}; \mathbf{B} = \begin{bmatrix} 0 \\ 1 \end{bmatrix}$$

Test for controllability

$$\mathbf{AB} = \begin{bmatrix} -4 & 1 \\ 0 & -5 \end{bmatrix} \begin{bmatrix} 0 \\ 1 \end{bmatrix} = \begin{bmatrix} 1 \\ -5 \end{bmatrix}$$

Controllability test matrix $\mathbf{U} = [\mathbf{B} \quad \mathbf{AB}] = \begin{bmatrix} 0 & 1 \\ 1 & -5 \end{bmatrix}$

$$\begin{bmatrix} 0 & 1 \\ 1 & -5 \end{bmatrix} \neq 0$$

Thus, the matrix **A** is controllable. Hence, the state variable feedback can be applied.

$$\mathbf{A} - \mathbf{BK} = \begin{bmatrix} -4 & 1 \\ 0 & -5 \end{bmatrix} - \begin{bmatrix} 0 \\ 1 \end{bmatrix} [k_1 \; k_2]$$

$$= \begin{bmatrix} -4 & 1 \\ 0 & -5 \end{bmatrix} - \begin{bmatrix} 0 & 0 \\ k_1 & k_2 \end{bmatrix}$$

$$= \begin{bmatrix} -4 & 1 \\ -k_1 & -(5+k_2) \end{bmatrix}$$

(1) The desired characteristic equation is

$$(s + 1 + j2)\,(s + 1 - j2) = 0$$

or $\quad s^2 + 2s + 5 = 0 \qquad \ldots(1)$

(2) The characteristic equation using state variable feedback is

$$|s\mathbf{I} - (\mathbf{A} - \mathbf{BK})| = 0$$

$$\therefore \quad \left| \begin{bmatrix} s & 0 \\ 0 & s \end{bmatrix} - \begin{bmatrix} -4 & 0 \\ -k_1 & -(5+k_2) \end{bmatrix} \right| = 0$$

or

$$\begin{vmatrix} s+4 & -1 \\ k_1 & s+(5+k_2) \end{vmatrix} = 0$$

$$\therefore \quad (s+4)(s+5+k_2) - (-1)(k_1) = 0$$

or

$$s^2 + (9+k_2)s + (20+4k_2+k_1) = 0 \qquad \text{...(2)}$$

Comparing coefficients in (1) and (2)

$$9 + k_2 = 2$$

and

$$20 + 4k_2 + k_1 = 5$$

$$\therefore \quad k_1 = 13$$

$$k_2 = -7$$

$$\mathbf{K} = [k_1 \quad k_2] = [13 \quad 7] \quad \textbf{Ans.}$$

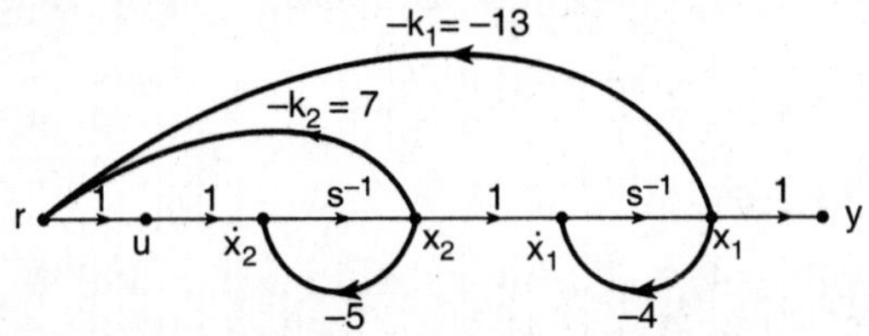

Fig. 9.11.3. State variable feedback for $Y(s)/U(s) = 1/(s+5)(s+4)$.

(*b*) Direct decomposition

$$\frac{Y(s)}{U(s)} = \frac{1}{(s+5)(s+4)}$$

or

$$\frac{Y(s)}{U(s)} = \frac{1}{s^2 + 9s + 20}$$

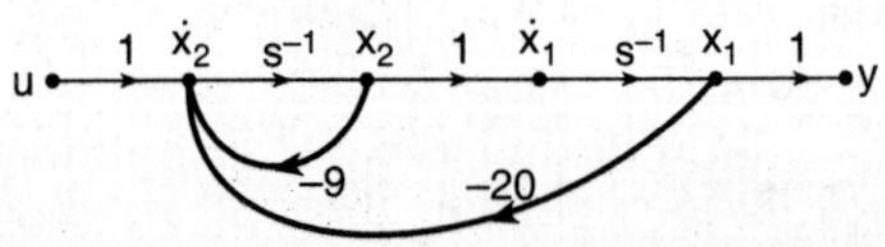

Fig. 9.11.4. State signal flow graph : $Y(s)/U(s) = s^{-2}/[1 - (-9s^{-1} - 20s^{-2})]$.

The transfer function given above is rewritten as below :

$$\frac{Y(s)}{U(s)} = \frac{s^{-2}}{1 - (-9s^{-1} - 20s^{-2})}$$

The signal flow graph for the transfer function is shown in Fig. 9.11.4.

As per the state signal flow graph state equations are

$$\dot{x}_1 = x_2$$

$$\dot{x}_2 = -20x_1 - 9x_2 + u$$

$$\therefore \quad \begin{bmatrix} \dot{x}_1 \\ \dot{x}_2 \end{bmatrix} = \begin{bmatrix} 0 & 1 \\ -20 & -9 \end{bmatrix} \begin{bmatrix} x_1 \\ x_2 \end{bmatrix} + \begin{bmatrix} 0 \\ 1 \end{bmatrix} u$$

The system matrix $\mathbf{A} = \begin{bmatrix} 0 & 1 \\ -20 & -9 \end{bmatrix}$ is in controllable canonical form. Thus, the state variable feedback can be applied. The state equation with state variable feedback is

$$\mathbf{x} = (\mathbf{A} - \mathbf{BK})\,\mathbf{x} + \mathbf{B}r$$

$$(\mathbf{A} - \mathbf{BK}) = \begin{bmatrix} 0 & 1 \\ -20 & -9 \end{bmatrix} - \begin{bmatrix} 0 \\ 1 \end{bmatrix} [k_1 \quad k_2]$$

$$= \begin{bmatrix} 0 & 1 \\ -20 & -9 \end{bmatrix} - \begin{bmatrix} 0 & 0 \\ k_1 & k_2 \end{bmatrix}$$

$$= \begin{bmatrix} 0 & 1 \\ -(20 + k_1) & -(9 + k_2) \end{bmatrix}$$

(1) The desired characteristic equation is

$$s^2 + 2s + 5 = 0 \qquad \text{...(3)}$$

(2) Since, the matrix **(A – BK)** is in controllable canonical form the characteristic equation (with state variable feedback can be written by inspection as

$$s^2 + (9 + k_2)\,s + (20 + k_1) = 0 \qquad \text{...(4)}$$

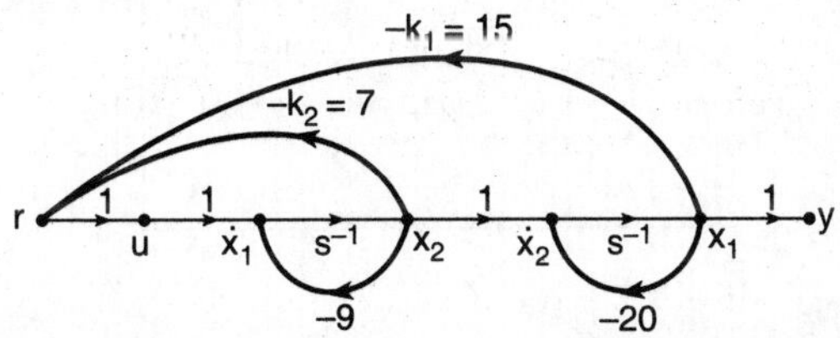

Fig. 9.11.5. State variable feedback for $Y(s)/U(s) = 1/(s^2 + 9s + 20)$.

On comparing coefficients in (3) and (4)

$$(9 + k_2) = 2$$

and

$$(20 + k_1) = 5$$

$$\therefore \quad k_1 = -15$$

$$k_2 = -7$$

$$\mathbf{K} = [k_1 \quad k_2] = [-15 \quad -7] \quad \textbf{Ans.}$$

PROBLEMS

9.1. Obtain state space representation for the electrical network shown in Fig. 9-P-1. Select the current i and voltage across the capacity v_c as state variables.

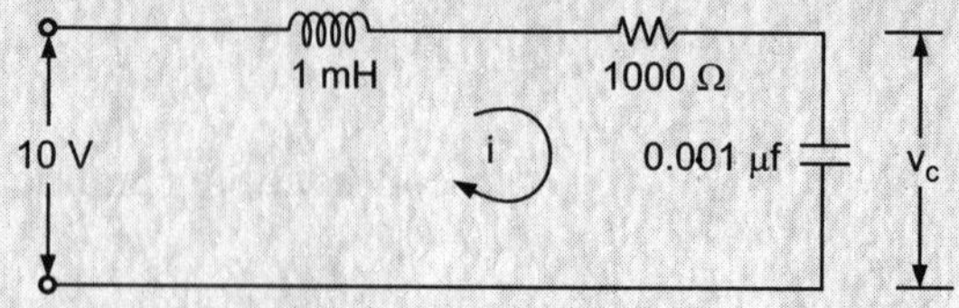

Fig. 9-P-1. Electrical network for problem 9.1.

9.2. Write state equations for the electrical network given below (Fig. 9P-2). Select v_1, v_2 as state variables and i as output variable.

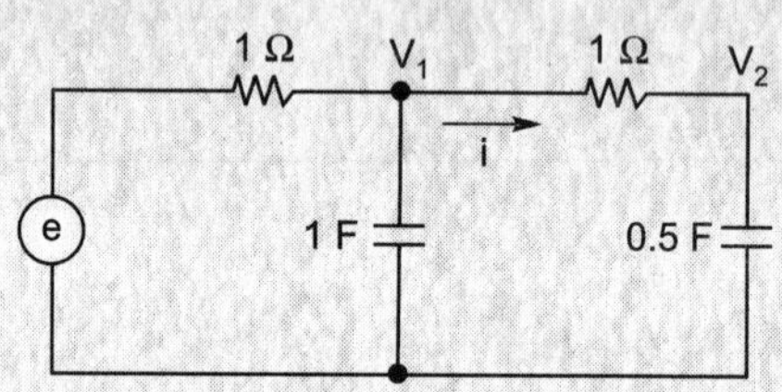

Fig. 9-P-2. Electrical network for problem 9.2.

9.3. In Fig. 9P-3, consider i_1, i_2 and voltage across the capacitor as state variables, and also i_1 and i_2 as output variables. Obtain the state model.

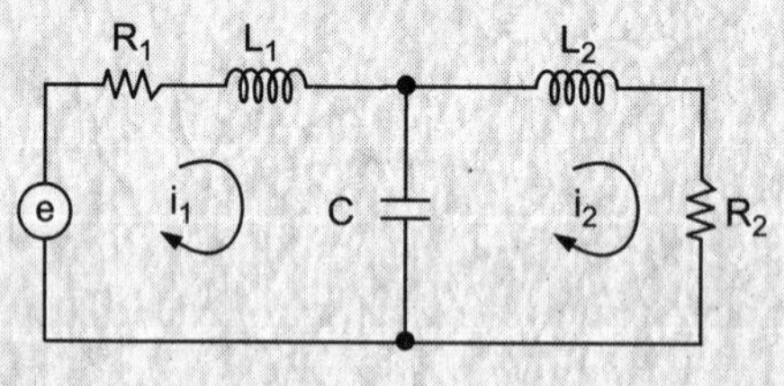

Fig. 9-P-3. Electrical network for problem 9.3.

9.4. Obtain state space representation of the electrical netwrok shown in Fig. 9P-4.

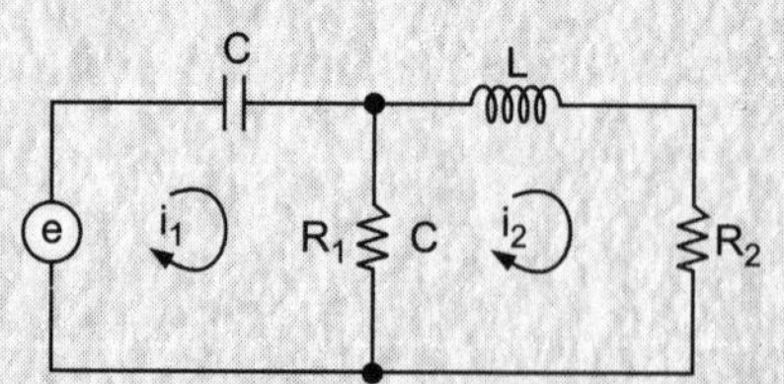

Fig. 9-P-4. Electrical network for problem 9.4.

Select suitable state variables and output variable.

9.5. Represent the equations for the armature controlled D.C. motor (Fig. 9P-5) in state space form. Consider i_a, θ and $\frac{d\theta}{dt}$ as state variables and θ as output variable.

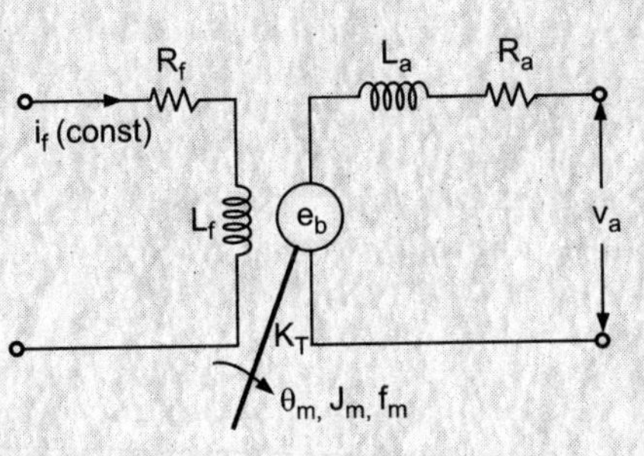

Fig. 9-P-5. Armature controlled d.c. motor.

9.6. A mechanical system is shown in Fig. 9P-6 choose $x(t)$ and $\dot{x}(t)$ as state variables and represent system equations in state space form.

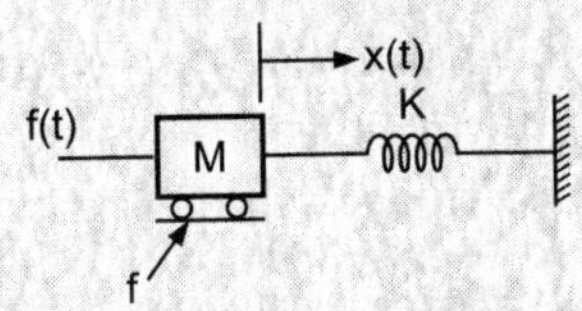

Fig. 9-P-6. Mechanical system for problem 9.6.

9.7. Write state space equations for the system shown in Fig. 9P-7.

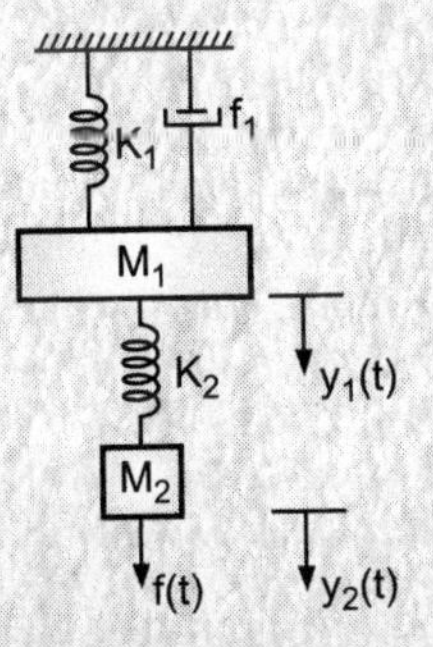

Fig. 9-P-7. System for problem 9.7.

9.8. Obtain state model for the transfer functions given below :

(a) $\frac{C(s)}{R(s)} = \frac{(s+2)}{(s+1)(s+3)}$ (b) $\frac{C(s)}{R(s)} = \frac{1}{(s+1)(s+3)}$

9.9. The transfer function of a system is given below :

$\frac{Y(s)}{U(s)} = \frac{s^2+s+2}{s^3+9s^2+26s+24}$. Determine the state model.

9.10. For the differential equation given below, write state equations.

$$\frac{d^2y}{dt^2} + \frac{4dy}{dt} + 2y = 3\frac{du}{dt} + u$$

9.11. Consider the system given by $\frac{Y(s)}{U(s)} = \frac{6s}{s^3 + 6s^2 + 11s + 6}$ obtain a state variable representation for the system.

9.12. Use parallel decomposition method to determine the state model for the transfer function given below :

$$G(s) = \frac{s^2 + 6s^2 + 8}{(s+3)(s^2 + 2s + 2)}$$

9.13. A system is represented by the equations given below

$$\begin{bmatrix} \dot{x}_1 \\ \dot{x}_2 \end{bmatrix} = \begin{bmatrix} 0 & 1 \\ -4 & -5 \end{bmatrix} \begin{bmatrix} x_1 \\ x_2 \end{bmatrix} + \begin{bmatrix} 0 \\ 1 \end{bmatrix} u. \text{ Given } \quad x(0) = [1 \quad 1]^T$$

Determine (*a*) state transition matrix $\phi(t)$, (*b*) Zero input response, (*c*) Zero state response for $u = 1$ and (*d*) Total response.

9.14. Use Laplace transform method to determine the time response given that :

(*a*) $$\begin{bmatrix} \dot{x}_1 \\ \dot{x}_2 \end{bmatrix} = \begin{bmatrix} -3 & 1 \\ -4 & 0 \end{bmatrix} \begin{bmatrix} x_1 \\ x_2 \end{bmatrix} + \begin{bmatrix} 1 \\ 0 \end{bmatrix} u(t)$$

$$x_1(0) = 0 \,;\, x_2(0) = 0 \quad \text{and} \quad y = [1 \quad 1] \begin{bmatrix} x_1 \\ x_2 \end{bmatrix}$$

(*b*) $$\begin{bmatrix} \dot{x}_1 \\ \dot{x}_2 \end{bmatrix} = \begin{bmatrix} -4 & 1 \\ -3 & 0 \end{bmatrix} \begin{bmatrix} x_1 \\ x_2 \end{bmatrix} + \begin{bmatrix} 2 \\ 0 \end{bmatrix} u(t)$$

$$x_1(0) = 0 \,;\, x_2(0) = 0 \quad \text{and} \quad y = [1 \quad -1] \begin{bmatrix} x_1 \\ x_2 \end{bmatrix}.$$

9.15. Show that the state transition matrix (STM) for the state equations

$$\begin{bmatrix} \dot{x}_1 \\ \dot{x}_2 \end{bmatrix} = \begin{bmatrix} -3 & 1 \\ 0 & -1 \end{bmatrix} \begin{bmatrix} x_1 \\ x_2 \end{bmatrix} \quad \text{is} \quad \begin{bmatrix} e^{-3t} & \dfrac{e^{-t} - e^{-3t}}{2} \\ 0 & -e^{-t} \end{bmatrix}.$$

9.16. Determine the state transition matrix and the output time response, given that

$$\begin{bmatrix} \dot{x}_1 \\ \dot{x}_2 \end{bmatrix} = \begin{bmatrix} 0 & 1 \\ -2 & -3 \end{bmatrix} \begin{bmatrix} x_1 \\ x_2 \end{bmatrix} + \begin{bmatrix} 0 \\ 5 \end{bmatrix} u(t)$$

$$x_1(0) = 1; \, x_2(0) = 0$$

and $$y = \begin{bmatrix} 0 & 1 \\ -2 & -3 \end{bmatrix}\begin{bmatrix} x_1 \\ x_2 \end{bmatrix}.$$

9.17. Represent the following differential equation in state variable form

$$\frac{d^2y}{dt^2} + 3\frac{dy}{dt} + 2y = u(t)$$

and show that the state transition matrix is $\begin{bmatrix} (2e^{-t} - e^{-2t}) & (e^{-t} - e^{-2t}) \\ (-2e^{-t} + 2e^{-2t}) & (-e^{-t} + 2e^{-2t}) \end{bmatrix}$.

9.18. Obtain the output time response for the following state equation

$$\begin{bmatrix} \dot{x}_1 \\ \dot{x}_2 \end{bmatrix} = \begin{bmatrix} 0 & -1 \\ 2 & 0 \end{bmatrix}\begin{bmatrix} x_1 \\ x_2 \end{bmatrix} + \begin{bmatrix} 1 \\ 1 \end{bmatrix} r(t)$$

given that $x_1(0) = 0 : x_2(0) = 0$ and $r(t)$ = unit step function.

9.19. (*a*) A system is described by equations :

$$\dot{\mathbf{x}} = \mathbf{Ax} + \mathbf{B}u$$
$$y = \mathbf{Cx}$$

where, $$\mathbf{A} = \begin{bmatrix} -2 & -1 \\ 5 & 0 \end{bmatrix}, \mathbf{B} = \begin{bmatrix} 1 \\ 0 \end{bmatrix}, \mathbf{C} = [1 \;\; 1]$$

$$u = 1 \text{ and } \mathbf{x}(0) = \begin{bmatrix} 1 \\ -1 \end{bmatrix}$$

Determine the time response $\mathbf{x}(t)$ and $y(t)$.

(*b*) Diagonalise the matrix **A**.

9.20. Obtain the state model for a system whose transfer function is given as,

$\dfrac{C(s)}{R(s)} = \dfrac{1}{s^3 + 9s^2 + 26s + 24}$. Determine the time response for unit step input. Assume initial conditions as nil.

9.21. Show that the transfer function for the system having given

$$\mathbf{A} = \begin{bmatrix} 0 & 1 \\ -6 & -5 \end{bmatrix}, \mathbf{B} = \begin{bmatrix} 1 \\ 0 \end{bmatrix}$$

$$\mathbf{C} = [1 \;\; 0], \mathbf{D} = 0$$

is $$G(s) = \frac{s+5}{(s+2)(s+3)}.$$

9.22. Obtain transfer matrix from the following data

$$\mathbf{A} = \begin{bmatrix} -5 & 1 \\ -6 & 0 \end{bmatrix}, \mathbf{B} = \begin{bmatrix} 1 \\ 2 \end{bmatrix}, \mathbf{C} = [2 \;\; 1] \text{ and } \mathbf{D} = 0.$$

9.23. Write state equations for the state block diagram shown in Fig. 9P-8 and determine the transfer matrix $G(s)$.

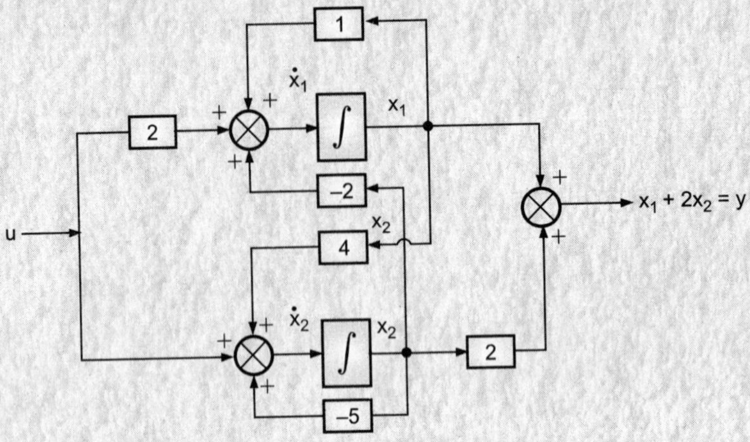

Fig. 9-P-8. State block diagram for problem 9.23.

9.24. Determine the transfer matrix from the data given below:

$$\mathbf{A} = \begin{bmatrix} 0 & 1 \\ -2 & -3 \end{bmatrix}, \mathbf{B} = \begin{bmatrix} 0 \\ 1 \end{bmatrix}, \mathbf{C} = \begin{bmatrix} 0 & 1 \\ 2 & 0 \end{bmatrix} \text{ and } \mathbf{D} = 0.$$

9.25. Check the controllability and observability of the system given below :

$$\begin{bmatrix} \dot{x}_1 \\ \dot{x}_2 \\ \dot{x}_3 \end{bmatrix} = \begin{bmatrix} 1 & 0 & 0 \\ 0 & 2 & 0 \\ 0 & 2 & 1 \end{bmatrix} \begin{bmatrix} x_1 \\ x_2 \\ x_3 \end{bmatrix} + \begin{bmatrix} 0 & 1 \\ 1 & 0 \\ 0 & 1 \end{bmatrix} \begin{bmatrix} u_1 \\ u_2 \end{bmatrix} \text{ and } \begin{bmatrix} y_1 \\ y_2 \end{bmatrix} = \begin{bmatrix} 1 & 0 & 0 \\ 0 & 1 & 0 \end{bmatrix} \begin{bmatrix} x_1 \\ x_2 \\ x_3 \end{bmatrix}.$$

9.26. Comment on the controllability and observability of the system having following coefficient matrices :

$$\mathbf{A} = \begin{bmatrix} -1 & -2 & -1 \\ 0 & -1 & 1 \\ 1 & 0 & -1 \end{bmatrix}, \mathbf{B} = \begin{bmatrix} 2 \\ 1 \\ 1 \end{bmatrix}, \mathbf{C} = [1 \;\; 0 \;\; 1] \text{ and } \mathbf{D} = 0.$$

9.27. Show that the system having following coefficients matrices :

$$\mathbf{A} = \begin{bmatrix} 2 & 0 & 0 \\ 0 & 2 & 0 \\ 0 & 3 & 1 \end{bmatrix} \text{ and } \mathbf{C} = [2 \;\; 2 \;\; 2]$$

is unobservable and if $C = \begin{bmatrix} 1 & 2 & 1 \\ 1 & 1 & 1 \end{bmatrix}$ observable.

CHAPTER 10

Sampled Data Control Systems

The analysis of sampled-data control systems is given in this chapter. The sampled-data control system forms the basis of understanding discrete time control system. The analysis of sampled-data control system is carried out in *z*-domain using *z*-transform. The chapter gives explanation of sampling process and derivation of *z*-transform of sampled functions. The hold-circuit, pulse transfer function, time response and stability, with regards to sampled-data control systems is dealt in this chapter.

CONTENTS

- *Sampler*
- *Sampling Process*
- *Laplace Transform of Sampled Function*
- *z-Transform*
- *z-Transform of Some Useful Functions*
- *Inverse z-Transform*
- *Hold Circuit*
- *Reconstruction of Signal Minimum Sampling Frequency*
- *Pulse Transfer Function (z-Transfer Function)*
- *Stability Analysis of Sampled-Data Control Systems*
- *Solved Examples*

Introduction. The analysis of linear control systems is based on the fact that the signals at various points in the system are continuous with respect to time. However, in some applications it is convenient to use any one or more control signals at discrete time intervals of time, for example in some industrial process control applications the signal is available only in sampled data form as a sequence of pulses. The control systems using one or more signals at discrete time intervals are known as sampled data control systems.

In a sampled-data control system the signal at any one or more places is sampled and appears in the form of a pulse train at periodic intervals. The control signal is thus amplitude sampled train of pulses and the control systems employing such signals are known as sampled data control systems.

Fig. 10.0.1 shows a sampled-data control system. The continuous error signal $e(t)$ is sampled at intervals of time T with the help of a sampler. The output of the controlled process is continuous one with respect to time because a hold circuit precedes the controlled process block.

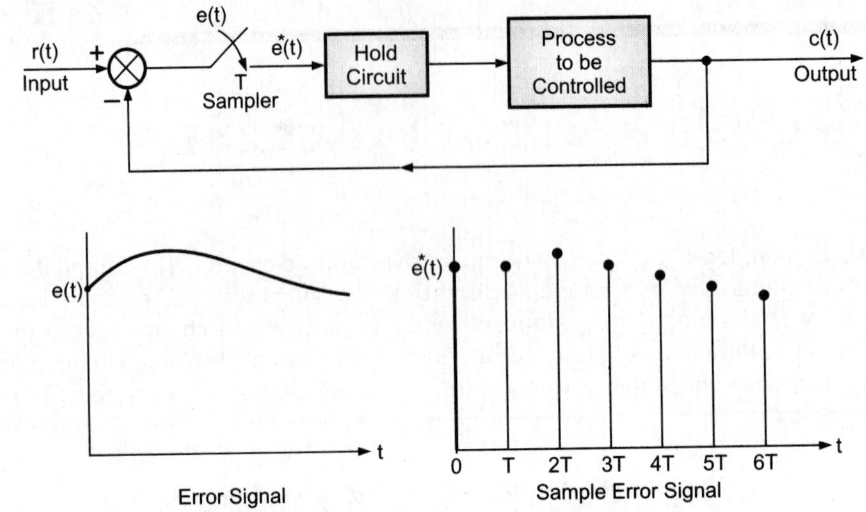

Fig. 10.0.1. Sampled-data control system.

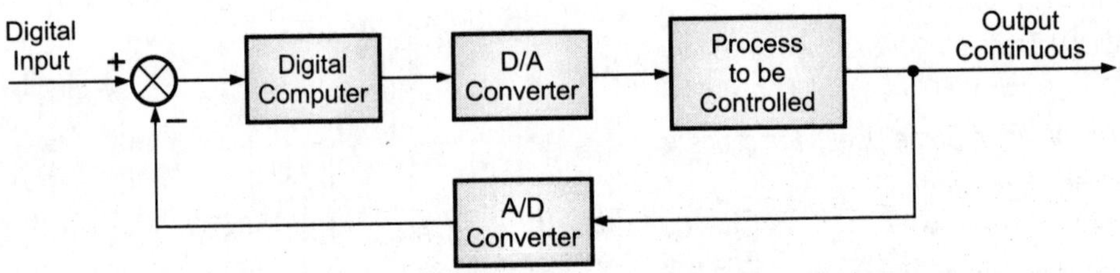

Fig. 10.0.2. Digital control system.

A digital control system makes use of a digital computer and the actuating signal is in the form of a coded digital data at discrete intervals of time. A/D (analog to digital) and D/A (digital to analog) converters are used in digital control systems. Fig. 10.0.2 shows a digital control system.

The sampled-data, digital control systems are placed in the category of discrete-data control systems.

The presence of sampled or digital data in a linear control system as mentioned above does not essentially make the system non-linear. The linear nature of the system being still maintained.

The use of sampled-data in control systems enables time sharing between different input signals using the same control equipment. Different input signals can be sampled periodically by staggering sampling time and thus over the same control equipment number of inputs can be used. This arrangement of control reduces the cost of control equipment. The sampled-data control systems work on the same principle as that of transmitting a number of communication signals on a single transmission channel.

In some control applications the load is to be driven by low power input signal. The use of input signal in sampled form in such a case improves the sensitivity of the system in terms of system gain.

The on-off temperature control is a simplest example of sampled-data control system.

10.1 SAMPLER

The basic element of a sampled-data control system is a sampler which samples the continuous signal into a sequence pulses appearing at regular interval of time.

Fig. 10.1.1 (a) shows a switch being used as a sampler. The input to the switch is a continuous signal denoted $f(t)$ as shown in Fig. 10.1.1 (b). The switch is closed for a very short duration of time say Δt and then remains open for some duration of time. This operation is repeated with a time period T. The time T is known as a sampling time and the switch output appears only for the closing duration (Δt) of the switch. The signal $f(t)$ is thus sampled at a regular interval of time as shown in Fig. 10.1.1 (c).

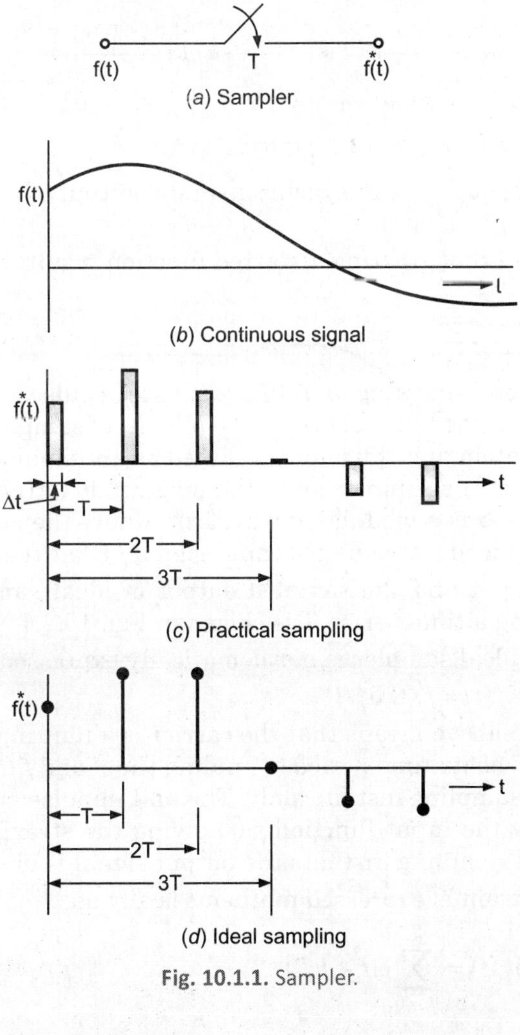

Fig. 10.1.1. Sampler.

The sampled signal is denoted as $f^*(t)$ meaning that $f^*(t)$ is obtained after sampling the input signal $f(t)$ at a regular interval of time.

The actuating signal for a sampled-data system is $f^*(t)$ which is in the form of a modulated pulse train.

Ideal Sampler. The signal modulated pulse train having pulse width Δt is shown in Fig. 10.1.1 (*c*). The pulse width of an ideal sampler approaches to zero and therefore the output $f^*(t)$ of an ideal sampler is the input signal modulated impulse train as shown in Fig. 10.1.1 (*d*).

However, if the pulse width Δt is very small as compared to the dominant time constant of the system as well as to the sampling time T then to a fair approximation the pulse train can be considered as an impulse train. This approximation simplifies the analysis of sampled-data control systems.

The sampled output denoted by $f^*(t)$ is known as starred function of $f(t)$. Fig 10.1.2 shows a function $f(t)$ being sampled twice. It is assumed that the two samplers operate in synchronism *i.e.*, their sampling time and the instants of operation are same.

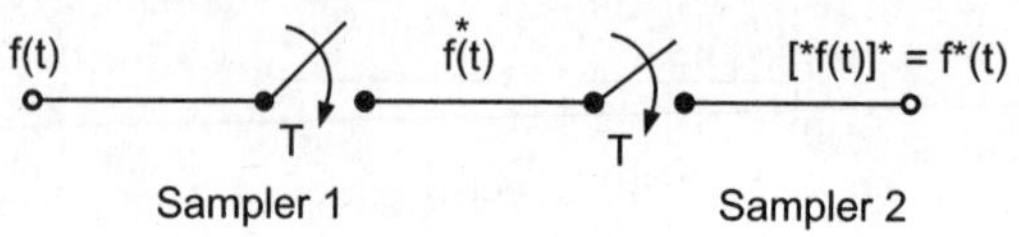

Fig. 10.1.2. Function *f*(*t*) sampled twice.

As the two samplers operate in synchronism the output of the second sampler is also $f^*(t)$ therefore, $[f^*(t)]^* = f^*(t)$

It is thus concluded that starring a starred function results in starred function itself.

10.2 SAMPLING PROCESS

The sampler converts a continuous signal into a sequence of pulses wherein the magnitude of the pulse gives the value of the input signal at the instant of sampling. The pulse considered is of very short time duration and thus approximated as an impulse. Therefore, at the sampling instant the output of the sampler gives the magnitude (strength) of the input signal. The output of ideal sampler is a modulated waveform where the carrier being unit impulse train $\delta_T(t)$ which is being modulated by the input signal $f(t)$ at the sampling instants.

The continuous input and the sampled output of ideal sampler along with carrier impulse train $\delta_T(t)$ having a time period T is shown in Fig. 10.2.1

The output of the ideal sampler is mathematically expressed as

$$f^*(t) = f(t)\,\delta_T(t) \qquad ...(10.1)$$

The notation given above means that the carrier is a unit impluse train $\delta_T(t)$ appearing periodically having T as its time period (sampling time) and $f(t)$ being appearing at the output of the sampler at sampling instants only. The unit impulse appearing at the sampling instant is multiplied by the input function $f(t)$ giving the strength of output signal and between two consecutive sampling instants the output signal is absent.

The unit impulse train is expressed mathematically as

$$\delta_T(t) = \sum_{k=0}^{\infty} \delta(t - kT) \qquad ...(10.2)$$

$\delta(t - kT)^{\dagger}$ is the unit impulse appearing at time $t = kT$, where $k = 0, 1, 2, ...$

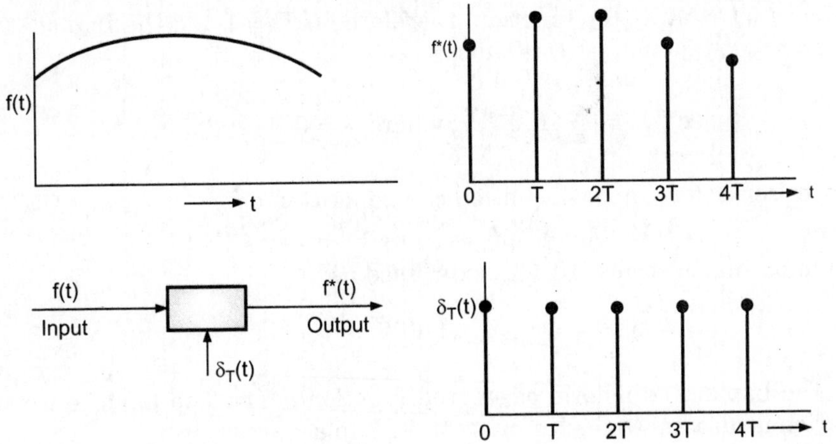

Fig. 10.2.1. Sampling process.

As the function $f(t)$ appears only at sampling instants in the output the same is denoted as $f(kT)$ and the starred function given by (10.1) can thus be expressed as

$$f^*(t) = \sum_{k=0}^{\infty} f(kT)\,\delta(t - kT) \qquad ...(10.3)$$

The above equation gives the sampler output which is a weighted train of pulses. At the sampling instants ($t = kT$) the sampler output $f^*(t)$ appears in the form of pulses as shown in Fig. 10.2.1.

10.3 LAPLACE TRANSFORM OF SAMPLED FUNCTION

The sampled output denoted by $f^*(t)$ is a starred function which gives the values of input function $f(t)$ at sampling instants only *i.e.*, at $t = kT$ where $k = 0, 1, 2, 3 ...$

Since
$$f^*(t) = \sum_{k=0}^{\infty} f(kT)\,\delta(t - kT) \qquad ...(10.4)$$

and the Laplace transform of a unit impulse train is e^{-skT}. Therefore the Laplace transform of Eq. (10.4) is

$$F^*(s) = \sum_{k=0}^{\infty} f(kT)\,e^{-skT} \qquad ...(10.5)$$

$^{\dagger}\delta_T(t) = \delta(t - kT)$ is a unit impulse train wherein the unit impulse occurs at sampling instants $t = kT$: where $k = 0, 1, 2, 3 ...$ The unit impulse train is the first time derivative of sampled unit step function $u(t - kT)$ occurring at $t = kT$. Therefore,

$$\mathcal{L}\delta_T(t) = \mathcal{L}\delta(t - kT) = \mathcal{L}\frac{d}{dt}[u(t - kT)] = s\mathcal{L}u(t - kT) = s\frac{1}{s}e^{-skT}$$

or
$$\mathcal{L}\delta_T(t) = e^{-skT}$$

$f^*(t)$ is starred function and $F^*(s)$ is known as Laplace transform of starred function or starred transform.

Consider $f(t)$ as unit step function, therefore $f(kT) = 1$ and the Laplace transform of the starred unit step function $f^*(t)$ is thus

$$F^*(1) = \sum_{k=0}^{\infty} 1 \,.\, e^{-skT} \text{ ; where, } k = 0, 1, 2, 3...$$

In power series form above equation can be written as

$$F^*(1) = 1 + e^{-sT} + e^{-2sT} + e^{-3sT} + ... + e^{-ksT} \qquad ...(10.6)$$

The summation of series (10.6) is expressed as

$$F^*(1) = \frac{1}{1 - e^{-sT}}$$

which gives the Laplace transform of starred function $f^*(1)$. The Laplace tansform of the starred function is also known as "impulse train Laplace transform".

10.4 z-TRANSFORM

The Laplace transform of starred $f^*(t)$ known as impulse train Laplace transform is given by

$$F^*(s) = \sum_{k=0}^{\infty} f(kT) e^{-skT} \qquad ...(10.7)$$

Substituting $z = e^{sT}$

i.e. $s = \frac{1}{T} \log_e z$ in Eq. (10.7)

and considering $z > 1$ following relation is obtained

$$F^*(s) \Big|_{s = \frac{1}{T} . \log_e z} = \sum_{k=0}^{\infty} f(kT)\, z^{-k} \text{ ; where } k = 0, 1, 2, 3...$$

The relation given above is known as z-transform of the function $f(t)$ and denoted as $Z[f(t)]$ or $F(z)$, therefore

$$Z[f(t)] = F(z) = \sum_{k=0}^{\infty} f(kT)\, z^{-k} \qquad ...(10.8)$$

where $k = 0, 1, 2, 3...$

In series form $F(z)$ given by Eq. (10.8) can be written as

$$F(z) = f(0)z^0 + f(T)\, z^{-1} + f(2T)\, z^{-2} + f(3T)\, z^{-3} + ... + f(kT)\, z^{-k} \qquad ...(10.9)$$

On comparing (10.9) and (10.6) it is noted that the values of $f(t)$ at sampling instants kT are given by the coefficients of z^{-k} terms in (10.9). $f(0)$, $f(T)$, $f(2T)$, ... are the values of function $f(t)$ at instants $k = 0, 1, 2...$ respectively.

Properties of z-Transform. $Z[f(t)]$ or $F(z)$ denotes z-transform of $f(t)$. Some properties of z-transform are listed below :

1. Addition : $Z[f_1(t) + f_2(t)] = Z[f_1(t)] + Z[f_2(t)] = F_1(z) + F_2(z)$
2. Substraction : $Z[f_1(t) - f_2(t)] = Z[f_1(t)] - Z[f_2(t)] = F_1(z) - F_2(z)$
3 Multiplication by constant : $Z[a\, f(t)] = a\, Z[f(t)] = a\, F(z)$
4. Delayed function: $Z[f(t - kT)] = z^{-k}\, Z[f(t)] = z^{-k}\, F(z)$
5. Multiplication by t : $Z[t\, f(t)] = -\, Tz \dfrac{d}{dz} Z[f(t)] = -\, Tz \dfrac{d}{dz} F(z)$;

where T is sampling time.

6. Division by t : $Z\left[\frac{1}{t} f(t)\right] = -\frac{1}{T}\int \frac{Z[f(t)]}{z}\, dz = -\frac{1}{T}\int \frac{F(z)}{z}\, dz$

where T is sampling time.

7. $Z[e^{\pm at} f(t)] = F(ze^{\pm aT})$.
8. Initial value theorem.

The strength of the first sample $f(0)$ is given by

$$f(0) = \lim_{z \to \infty} F(z) \qquad \text{if the limit exists}$$

9. Final value theorem.

The strength of $f(kT)$ as $k \to \infty$ is given by

$$f(\infty) = \lim_{z \to 1} (1 - z^{-1})\, F(z)$$

Provided that $(1 - z^{-1})\, F(z)$ has no pole on or outside the unit circle in z-plane having origin as its centre.

10.5 z-TRANSFORM OF SOME USEFUL FUNCTIONS

(1) z-transform of unit step function $f(t) = u(t) = 1$:

$$f(t) = 1$$
$$t \geq 0$$

put $t = kT$ $\quad\therefore\quad f(kT) = 1$

$k = 0, 1, 2, 3...$

$$\because \qquad F^*(s) = \sum_{k=0}^{\infty} f(kT)\, e^{-skT}$$

put $f(kT) = 1$ and $z = e^{sT}$

$$\therefore \qquad F(z) = F^*(s)\Big|_{z = e^{sT}} = \sum_{k=0}^{\infty} 1 \,.\, z^{-k}$$

or

$$Z[1] = \sum_{k=0}^{\infty} 1 \,.\, z^{-k}$$

$$k = 0, 1, 2, 3...$$

In series form above expression can be written as

$$Z[1] = 1 + z^{-1} + z^{-2} + z^{-3}...$$

In summation form above series can be written as

$$Z[1] = \frac{1}{1-z^{-1}} = \frac{z}{z-1}$$

Fig. 10.5.1 (*a*) shows continuous unit step function $f(t) = u(t)$ and Fig. 10.5.1 (*b*) gives $u(t)$ being sampled periodically with time period T.

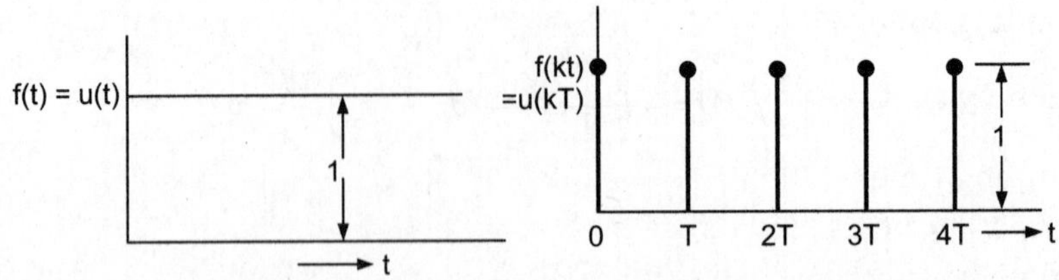

Fig. 10.5.1. Unit-step function and its sampled form.

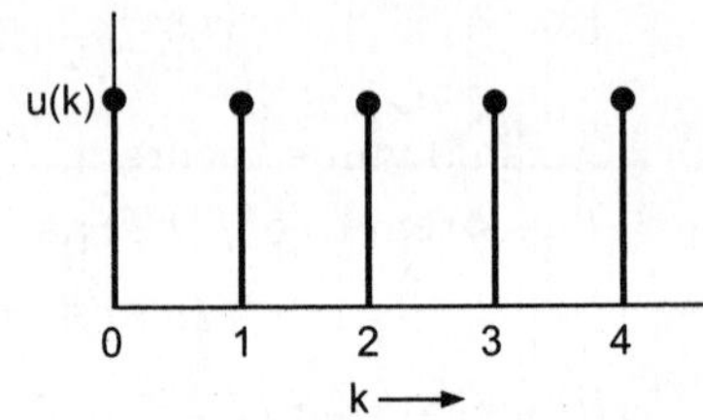

Fig. 10.5.2. Unit step sequence.

Fig. 10.5.1 (*b*) can also be expressed in number sequence form $f(k) = u(k)$ where k being the sampling number. The unit step sequence is shown in Fig. 10.5.2 which gives the value of the sampled unit step function at the sampling sequence number $k = 0, 1, 2, 3...$

If the sampling time is T, the corresponding sampling time at the sampling number $k = 0, 1, 2, 3...$ is given by $t = 0, T, 2T, 3T...$ respectively.

The unit step sequence can also be written as, $u(k) = (1)^k$ meaning that the unit step function $u(t)$ being sampled periodically with period T at the sampling number $k = 0, 1, 2, 3...$

The coefficients of the series

$$Z[1] = 1 \,.\, z^0 + 1 \,.\, z^{-1} + 1 \,.\, z^{-2} + 1 \,.\, z^{-3} + ... \; 1 \,.\, z^{-k}$$

give the values of the unit step function at the sampling number $k = 0, 1, 2, 3...$ respectively as shown in Fig. 10.5.2.

As the z-transform of $u(t)$ is $(z/z - 1)$ the inverse z-transform of $(z/z - 1)$ yields $u(t)$ at the sampling instants only. The inverse z-transform* is denoted as $Z^{-1}[F(z)]$.

The z-transform of unit step function $u(t)$ is $(z/z - 1)$, therefore, in number sequence form

$$Z^{-1}\left[\frac{z}{z-1}\right] = (1)^k$$

For which the corresponding continuous time function $u(t) = 1$.

(2) z-transform of unit ramp function $f(t) = r(t) = t$:

$$f(t) = t$$
$$t \geq 0$$

Put $t = kT$ $\quad \therefore \quad f(kT) = kT$

$$k = 0, 1, 2, 3...$$

*The inverse z-transform is not unique as two functions may have different values between two sampling instants but same value at the sampling instants.

$\because \quad F^*(s) = \sum_{k=0}^{\infty} f(kT)\, e^{-skT}$

put $\quad f(kT) = kT$ and $z = e^{sT}$

$$\therefore \quad F(z) = F^*(s)\Big|_{z=e^{sT}} = \sum_{k=0}^{\infty} (kT)\, z^{-k} \quad \text{or} \quad Z(t) = \sum_{k=0}^{\infty} f(kT)\, z^{-k}$$

In series form above expression can be written as :

$$Z[t] = 0Tz^0 + Tz^{-1} + 2Tz^{-2} + 3Tz^{-3} + \ldots$$

or $\quad Z[t] = Tz^{-1} + 2Tz^{-2} + 3Tz^{-3} + \ldots$

Multiplying both sides by $(z-1)$

$$(z-1)\, Z[t] = (z-1)(Tz^{-1} + 2Tz^{-2} + 3Tz^{-3} + \ldots)$$
$$= T(1 + z^{-1} + z^{-2} + z^{-3} + \ldots)$$

In summation form above expression can be written as

$$(z-1)\, Z[t] = \frac{T}{1-z^{-1}} = \frac{zT}{(z-1)}$$

$$\therefore \quad Z[t] = \frac{zT}{(z-1)^2}$$

and $\quad Z^{-1}\left[\frac{zT}{(z-1)^2}\right] = t$

The coefficients of the series, $Z[t] = 0Tz^0 + Tz^{-1} + 2Tz^{-2} + 3Tz^{-3} + \ldots$ give the values of the unit ramp function $r(t) = t$ at the sampling number $k = 0, 1, 2, 3, \ldots$ respectively as shown in Fig. 10.5.3.

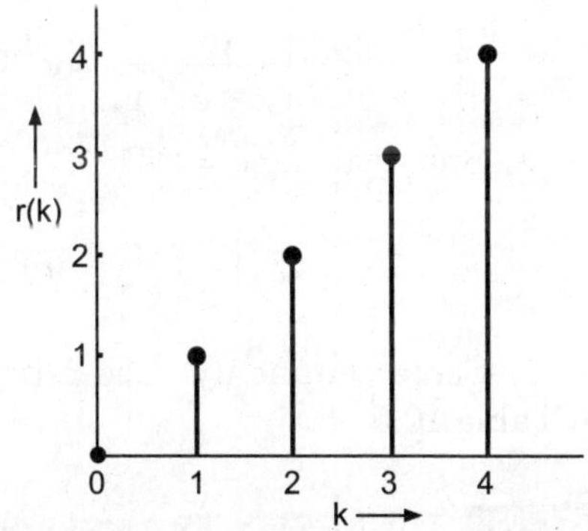

Fig. 10.5.3. Sampled sequence of unit ramp function $r(t) = t$.

As the z-transform of the unit ramp function $r(t) = t$ is $\left[\frac{zT}{(z-1)^2}\right]$ therefore, in number sequence form the corresponding inverse z-transform is $Z^{-1}\left[\frac{zT}{(z-1)^2}\right] = k$ for which the corresponding time function is $f(t) = t$.

(3) z-transform of $f(t) = e^{-at}$

$$f(t) = e^{-at} \quad t \geq 0$$

put $\quad t = kT \quad \therefore \quad f(kT) = e^{-akT}$

$$k = 0, 1, 2, 3, \ldots$$

$$\because \quad F^*(s) = \sum_{k=0}^{\infty} f(kT)\, e^{-skT}$$

put $\quad f(kT) = e^{-akT}$ and $z = e^{sT}$,

$$\therefore \qquad F(z) = F^*(s) \Big|_{z = e^{sT}} = \sum_{k=0}^{\infty} e^{-akT} z^{-k}$$

or

$$Z[e^{-at}] = \sum_{k=0}^{\infty} e^{-akT} z^{-k}$$

In series from above expression can be written as

$$Z[e^{-at}] = 1 + e^{-aT} z^{-1} + e^{-2aT} z^{-2} + e^{-3aT} z^{-3} + ...$$

In summation form above series can be written as

$$Z[e^{-at}] = \frac{z}{z - e^{-aT}}$$

$$\therefore \qquad Z^{-1}\left[\frac{z}{z - e^{-aT}}\right] = e^{-at}$$

The coefficients of the series

$$Z[e^{-aT}] = 1 + e^{-aT} z^{-1} + e^{-2aT} z^{-2} + e^{-3aT} z^{-3} + ...$$

give the value of the function e^{-at} at the sampling number $k = 0, 1, 2, 3, ...$ respectively as shown in Fig. 10.5.4.

As the z-transform of function e^{-at} is $\left(\frac{z}{z - e^{-aT}}\right)$ therefore, in number sequence form the corresponding inverse z-transform is

$$Z^{-1}\left[\frac{z}{z - e^{-aT}}\right] = e^{-akT} = (e^{-aT})^k$$

Substituting $c = e^{-aT}$

$$Z^{-1}\left[\frac{z}{z - c}\right] = (c)^k$$

Some commonly used z-transforms are given in Table 10.5

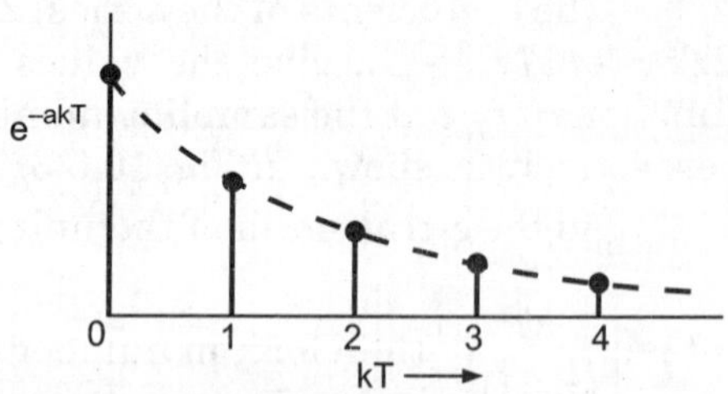

Fig. 10.5.4. Sample sequence of $f(t) = e^{-at}$.

10.6 INVERSE z-TRANSFORM

Inverse z-transform is used to obtain starred function $f^*(t)$ or number sequence $f(k)$ from the corresponding z-transform $F(z)$. The inverse z-transform is denoted as $Z^{-1}[F(z)] = f(kT)$ or in number sequence form $Z^{-1}[F(z)] = f(k)$. The inverse z-transform gives the value of the continuous time function $f(t)$ at sampling instants only *i.e.* at kT, where $k = 0, 1, 2, 3...$ and T being the sampling period or $f(k)$ gives the value at the sampling number k. The inverse z-transform does not explain the behaviour of continuous time function between consecutive sampling instants. The inverse z-transform is obtained by partial fraction and power series method described below :

Partial Fraction Expansion Method. z-transforms $F(z)$ can be expanded into partial fractions and using z-transform tables corresponding inverse z-transform is obtained. As indicated in z-transform table "z" appears in the numerator of z-transform and as such it is convenient to expand $F(z)/z$ into partial fractions instead of $F(z)$.

Table 10.5

z-transform F(z)	*Number sequence form*	*Laplace transform F(s)*	*Continuous time function f(t)*
(1)	*(2)*	*(3)*	*(4)*
1	$\delta(k)$ discrete unit impulse	1,	
$\dfrac{z}{z-1}$	$(1)^k$	$\dfrac{1}{s}$	$u(t) = 1$
$\dfrac{zT}{(z-1)^2}$	k	$\dfrac{1}{s^2}$	$r(t) = t$
$\dfrac{z(z+1)T}{(z-1)^2}$	$(k)^2$	$\dfrac{1}{s^3}$	$f(t) = \dfrac{t^2}{2}$
$\dfrac{z}{z-e^{-aT}} = \dfrac{z}{z-c}$	$(c)^k$	$\dfrac{1}{s+a}$	$f(t) = e^{-at}$
$\dfrac{zTe^{-aT}}{(z-e^{-aT})^2} = \dfrac{zcT}{(z-c)^2}$	$k(c)^k$	$\dfrac{1}{(s+a)^2}$	$f(t) = te^{-at}$
$\dfrac{z \sin \omega T}{z^2 - 2z \cos \omega T + 1}$	$\sin k\omega T$	$\dfrac{\omega}{s^2+\omega^2}$	$f(t) = \sin \omega t$
$\dfrac{z(z - \cos \omega T)}{z^2 - 2z \cos \omega T + 1}$	$\cos k\omega T$	$\dfrac{s}{s^2+\omega^2}$	$f(t) = \cos \omega t$
$\dfrac{z(e^{-at} \sin \omega T)}{[z - e^{(-a+j\omega)T}][z - e^{(-a-j\omega)T}]}$	$(e^{-aT})^k \sin k\omega T$		
$= \dfrac{zc \sin \omega T}{z^2 - z(2c \cos \omega T) + c^2}$	$= (c)^k \sin k\omega T$	$\dfrac{\omega}{(s+a)^2+\omega^2}$	$f(t) = e^{-at} \sin \omega t$
$\dfrac{z(z - e^{-aT} \cos \omega T)}{[z - e^{(-a+j\omega)T}][z - e^{(-a-j\omega)T}]}$	$(e^{-aT})^k \cos k\omega T$		
$= \dfrac{z(z - c \cos \omega T)}{z^2 - z(2c \cos \omega T) + c^2}$	$= (c)^k \cos k\omega T$	$\dfrac{s+a}{(s+a)^2+\omega^2}$	$f(t) = e^{-at} \cos \omega t$

Power Series Method. Function $F(z)$ can be expressed as a ratio of two polynomials in "z" as follows :

$$F(z) = \frac{a_0 z^n + a_1 z^{n-1} + a_2 z^{n-2} + \ldots + a_n}{b_0 z^m + b_1 z^{m-1} + b_2 z^{m-2} + \ldots + b_m} \qquad \ldots(10.10)$$

If $n \le m$ then, dividing the numerator by denominator the above function $F(z)$ can be written as :

$$F(z) = f_0 z^0 + f_1 z^{-1} + f_2 z^{-2} + f_3 z^{-3} + \ldots \qquad \ldots(10.11)$$

On comparing the coefficients of the power series Eq. (10.11) and z-transform expression Eq. (10.9), it is concluded that the values of $f^*(t)$ at the sampling instants are given by the coefficients of the power series (10.11).

However, the power series method gives the value of the corresponding function of $f^*(t)$ at the sampling instants *i.e.* $f(0)$, $f(T)$, $f(2T)$, $f(3t)$, ... whereas the partial fraction expansion method gives an expression for $f^*(t)$ in a closed form.

Relation between z-transform and Impulse Train Laplace Transform

As the z-transform is obtained by putting $z = e^{sT}$ in the impulse train Laplace transform $F^*(s)$. Therefore, $F^*(s)$ is obtained by substituting e^{sT} for "z" in z-transform expression.

For example the z-transform, $F(z) = \dfrac{z}{z - e^{-sT}}$

has the following impulse train Laplace transform

$$F^*(s) = \frac{e^{sT}}{e^{sT} - e^{-aT}} \qquad \text{or} \qquad F^*(s) = \frac{1}{1 - e^{-aT}\, e^{-sT}}.$$

Illustrative Example 10.6.1. *Find the inverse z-transform of the function given below, given that the sampling time T = 1 sec.*

$$F(z) = \frac{0.632z}{z^2 - 1.368z + 0.368}.$$

Solution. 1. Partial Fraction Expansion Method

$$F(z) = \frac{0.632z}{z^2 - 1.368z + 0.368}$$

$$\therefore \qquad \frac{F(z)}{z} = \frac{0.632}{z^2 - 1.368z + 0.368}$$

The denominator can be factored as

$$(z - 1)(z - 0.368)$$

$$\therefore \qquad \frac{F(z)}{z} = \frac{0.632}{(z-1)(z-0.368)}$$

Expanding R.H.S. into partial fractions

$$\frac{F(z)}{z} = \frac{K_1}{z-1} + \frac{K_2}{z-0.368}$$

The coefficients can be determined as

$$K_1 = 1 \quad \text{and} \quad K_2 = -1$$

$$\therefore \quad \frac{F(z)}{z} = \frac{1}{z-1} - \frac{1}{z-0.368}$$

Since $e^{-1} = 0.368$ and $T = 1$ sec (given)

$$\therefore \quad \frac{F(z)}{z} = \frac{1}{z-1} - \frac{1}{z-e^{-T}} \quad \text{or} \quad F(z) = \frac{z}{z-1} - \frac{z}{z-e^{-T}}$$

From z-transform table

$$Z^{-1}\left[\frac{z}{z-1}\right] = 1 \quad \text{and} \quad Z^{-1}\left[\frac{z}{z-e^{-T}}\right] = (e^{-t})^*$$

$$\therefore \quad f^*(t) = (1 - e^{-t})^*$$

Substituting $t = kT$; $k = 0, 1, 2, 3...$

$f^*(t)$ can be written as : $f(kT) = \sum_{k=0}^{\infty} (1 - e^{-kt})\ \delta(t - kT)$ **Ans.**

From z-transform tables the inverse z-transform in number sequence form is

$$Z^{-1}\left[\frac{z}{z-1}\right] = (1)^k \quad \text{and} \quad Z^{-1}\left[\frac{z}{z-0.368}\right] = (0.368)^k$$

Therefore, in the number sequence form

$Z^{-1}\left[\frac{z}{z-1} - \frac{z}{z-0.368}\right]$ is written as : $f(k) = (1)^k - (0.368)^k$ **Ans.**

2. Power Series Method :

$$F(z) = \frac{0.632z}{z^2 - 1.368z + 0.368}$$

$F(z)$ can be expressed in the form of power series by long division as follows :

$$\begin{array}{r|l}
 & 0.632z^{-1} + 0.864z^{-2} + 0.949z^{-3} + 0.981z^{-4} + 0.993z^{-5} + ... \\
z^2 - 1.368z + 0.368 & 0.632z \\
 & -0.632z \mp 0.864 \pm 0.232z^{-1} \\
\hline
 & 0.864 - 0.232z^{-1} \\
 & -0.864 \mp 1.181z^{-1} \pm 0.317z^{-2} \\
\hline
 & 0.949z^{-1} - 0.317z^{-2} \\
 & -0.949z^{-1} \mp 1.298z^{-2} \pm 0.349z^{-3} \\
\hline
 & 0.981z^{-2} - 0.349z^{-3} \\
 & -0.981z^{-2} \mp 1.342z^{-3} \pm 0.361z^{-4} \\
\hline
 & 0.993z^{-3} - 0.361z^{-4} \\
 & \vdots \qquad \vdots \\
\hline
 & \vdots \qquad \vdots
\end{array}$$

Hence $$F(z) = \frac{0.632z}{z^2 - 1.368z + 0.368}$$

$$= 0z^0 + 0.632z^{-1} + 0.864z^{-2} + 0.949z^{-3} + 0.981z^{-4} + 0.993z^{-5} + \ldots$$

The first few terms of the sampled function $f^*(t)$ at the sampling instants $k = 0, 1, 2, 3, 4, 5$ are given by the $F(z)$ series therefore,

$$f(0) = 0, f(1) = 0.632, f(2) = 0.864, f(3) = 0.949, f(4) = 0.981,$$
$$f(5) = 0.993$$

It is noted that the sampled sequence *i.e.* $f^*(t)$ can be obtained by two methods mentioned above. However, the power series method requires long division to get more terms whereas number sequence approach using z-transform table appears to be more convenient. Moreover, if it is not convenient to factorize the denominator of function $F(z)$ then the power series can be obtained by long division to get the value of $f^*(t)$ at the sampling instants.

The graphical representation of the sampled sequence $f(k) = (1)^k - (0.368)^k$ is shown in Fig. 10.6.1.

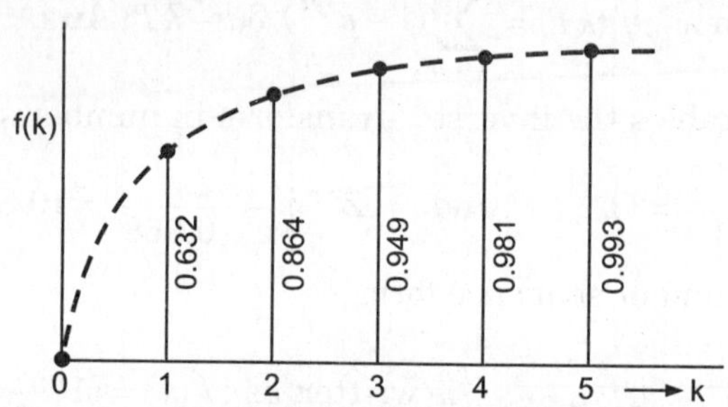

Fig. 10.6.1. $f(k) = (1)^k - (0.368)^k$.

The value of $f(k)$ as k approaches higher values is obtained by using final value theorem as follows:

$$f(\infty) = \lim_{z \to 1} (1 - z^{-1}) F(z) = \lim_{z \to 1} (1 - z^{-1}) \frac{0.632z}{(z-1)(z-0.368)}$$

$$= \lim_{z \to 1} \left(\frac{z-1}{z}\right) \left[\frac{0.632z}{(z-1)(z-0.368)}\right] = \lim_{z \to 1} \frac{0.632}{(z-0.368)} = \frac{0.632}{0.632}$$

or $$f(\infty) = 1$$

Difference equations

The analysis of sampled-data control system can be carried out in terms of difference equations because the input to the system is in the form of a signal occurring at regular intervals (sampling instants) of time. The analytical results obtained by this method are equivalent to those obtained by z-transform analysis, because in the latter case, the values of the function only at the sampling instants are considered.

In terms of z-transform, the output function $C(z)$ is given by

$$C(z) = \frac{G(z)}{1 + G(z)H(z)} R(z)$$

The pulse transfer function $G(z)/[1 + G(z) H(z)]$ can be expressed as a power series in inverse powers of z.

$$\therefore \qquad C(z) = \left(a_0 + \frac{a_1}{z} + \frac{a_2}{z^2} + ... \right) R(z)$$

or
$$C(z) = (a_0 + a_1 z^{-1} + a_2 z^{-2} + ...)\, R(z)$$

The time function at the sampling instants of the above equation can be written as follows :

$$c(kT) = r(kT) + a_1 r(kT - 1) + r(kT - 2) + ...$$

In number sequence form

$$c(k) = a_0 r(k) + a_1 r(k - 1) + a_2 r(k - 2) + ...$$

As per above equation the output of a second order sampled-data control system at any given sampling instant is given by the input at the same instant, at one sampling instant earlier, at two sampling instant earlier.

$$\therefore \qquad c(k + 2) + a_1 c(k + 1) + a_0 c(k) = b_0 r(k) + b_1 r(k + 1)$$

is the difference equation representation of a second order sampled-data control system.

The difference equation representation of a third order sampled-data control system is given by

$$c(k + 3) + a_2 c(k + 2) + a_1 c(k + 1) + a_0 c(k)$$
$$= b_0 r(k) + b_1 r(k + 1) + b_2 r(k + 2)$$

The z-transform is related to difference equation in a similar manner as the Laplace transform to differential equation. The Laplace transform converts linear differential equation with constant coefficients into algebraic equation in terms of "s". Similarly, z-transform converts difference equation with constant coefficients into algebraic equation in terms of "z".

z-transform of $x(k)$ is denoted as $X(z)$.

z-transform of $x(k + 1)$ is determined below :

$$Z[x(k + 1)] = \sum_{k=0}^{\infty} x(k + 1)\, z^{-k} = \sum_{k=1}^{\infty} x(k)\, z^{-k+1}$$

$$= z \left[\sum_{0}^{\infty} x(k) z^{-1} - x(0) \right]$$

$$\because \qquad \sum_{0}^{\infty} x(k) z^{-1} = x(z)$$

$$\therefore \qquad Z[x(k + 1)] = z[X(z) - x(0)] \quad \text{or} \quad Z[x(k + 1)] = z\, X(z) - z\, x(0)$$

Similarly,

$$Z\, [x\, (k + 2)] = zZ\, [x\, (k + 1)] - zx\, (1) = z\, [zX\, (z) - zx\, (0)] - zx\, (1)$$

$$\therefore \qquad Z\, [x\, (k + 2)] = z^2\, X(z) - z^2 x\, (0) - zx\, (1)$$

In general, $Z\, [x\, (k + n)] = z^n\, X\, (z) - z^n\, x\, (0) - z^{n-1}\, x\, (1) ... - zx\, (n - 1),\ k \geq n$

Where n being a positive integer. It is noted that, when a difference equation is transformed into an algebraic one, the initial data is already included in the algebraic representation.

Illustrative Example 10.6.2. *Solve the difference equation given as,*

$$c\,(k+2) - 5c\,(k+1) + 6c\,(k) = u\,(k)$$

Given that, c (0) = 0 and c (1) = 1.

Solution.

$$c(k+2) = 5c(k+1) + 6c(k) + u(k)$$

put $k = 0$

$$c(0+2) = 5c(0+1) - 6c(0) + u(0)$$

or $$c(2) = 5c(1) - 6\,c(0) + u(0) = 5 \times 1 - 6 \times 0 + 1 = 6$$

put $k = 1$

$$c(1+2) = 5c(1+1) - 6c(1) + u(1)$$

or $$c(3) = 5c(2) - 6c(1) + u(1) = 5 \times 6 - 6 \times 1 + 1 = 25$$

put $k = 2$

$$c(2+2) = 5c(2+1) - 6c(2) + u(2)$$

or $$c(4) = 5c(3) - 6c(2) + u(2) = 5 \times 25 - 6 \times 6 + 1 = 90$$

$k = 0, 1, 2, 3, 4 \ldots$

$\therefore$ $c(k) = 0, 1, 6, 25, 90, \ldots$ **Ans.**

The difference equation given above is solved below by z-transform method.

$$c(k+2) - 5c\,(k+1) + 6\,c(k) = u(k)$$

Taking z-transform on both sides,

$$Z[c\,(k+2) - 5Z\,[c\,(k+1)] + 6Z\,[c\,(k)] = Z\,[u\,(k)]$$

$$\therefore \quad [z^2\,C(z) - z^2c(0) - zc(1)] - 5\,[zC(z) - zc\,(0)] + 6C(z) = \frac{z}{z-1}$$

given that $c(0) = 0$ and $c\,(1) = 1$

$$\therefore \quad [z^2\,C\,(z) - z^2 \times 0 - z \times 1] - 5\,[zC(z) - z \times 0] + 6C(z) = \frac{z}{z-1}$$

Rearranging, $$C(z) = \frac{z^2}{(z-1)\,(z^2 - 5z + 6)}$$

or $$\frac{C(z)}{z} = \frac{z}{(z-1)\,(z-2)\,(z-3)} = \frac{K_1}{z-1} + \frac{K_2}{z-2} + \frac{K_3}{z-3}$$

The coefficients K_1, K_2 and K_3 can be determined as $K_1 = \frac{1}{2}$, $K_2 = -2$ and $K_3 = \frac{3}{2}$

$$\therefore \quad \frac{C\,(z)}{z} = \frac{1}{2}\cdot\frac{1}{z-1} - 2.\,\frac{1}{z-2} + \frac{3}{2}.\,\frac{1}{z-3}$$

or $$C\,(z) = \frac{z}{2\,(z-1)} - \frac{2z}{(z-2)} + \frac{3z}{2\,(z-3)}$$

Taking inverse z-transform on both sides

$$c(k) = \frac{1}{2}(1)^k - 2\,(2)^k + \frac{3}{2}\,(3)^k$$

$\therefore$ for $k = 0, 1, 2, 3, 4$ ...

$$c(k) = 0, 1, 6, 25, 90 \text{ ... } \textbf{Ans.}$$

It is noted that the initial values given in the problem automatically appear in the generated sequence.

10.7 HOLD CIRCUIT

The sampled signal provided by the sampling process is a weighted impulse train $f^*(t)$ which being the input signal to the system transfer function. The system output is also an impulse train, the envelop of which gives the output $c^*(t)$ at the sampling instants only as shown in Fig. 10.7.1.

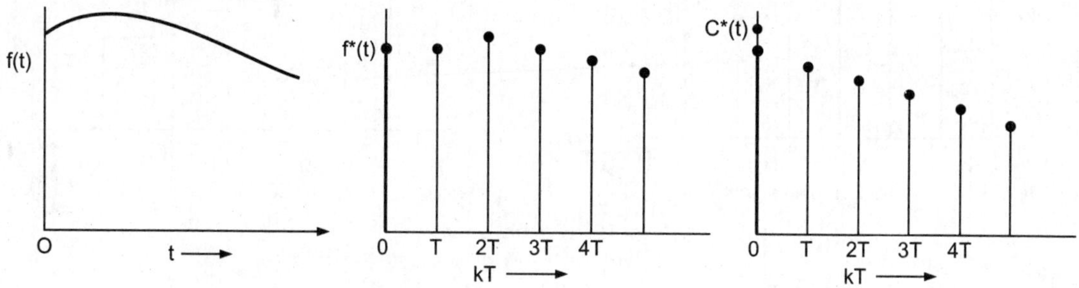

Fig. 10.7.1. Sampled-data control system : Input and output.

The original signal is reconstructed from the sampled signal by using a hold circuit. The hold circuit brings smoothness in the sampled output. At the sampling instants the hold signal and the original signal have the same value. The use of hold circuit enables to hold the signal between two consecutive sampling instants at the preceded value till the next sampling instant is reached.

Zero-order Hold. The hold circuit mentioned above holds the output signal at a fixed level between two consecutive sampling instants such that the slope of the hold circuit output signal is zero or in other words hold signal is the zeroth derivative of an impulse. Such a holding device is called zero-order hold (ZOH). The ZOH device is also known as "box-car" generator. The input and output of a zero-order hold are shown in Fig. 10.7.2.

Transfer Function of Zero-order Hold Circuit. Fig. 10.7.3 shows a unit impulse input given to a zero-order hold circuit which holds the input signal for a duration T and therefore, the output appears to be a unit step function till duration T.

As the output of ZOH is a unit step function appearing upto time T, the same can be written as

$$h(t) = u(t) - u(t - T) \qquad \because \qquad \mathcal{L}u(t) = \frac{1}{s}$$

and
$$\mathcal{L}u(t - T) = \frac{1}{s}\cdot e^{-sT} \qquad \therefore \qquad \mathcal{L}h(t) = \frac{1}{s} - \frac{1}{s}\cdot e^{-sT}$$

therefore, the Laplace transform of the output of ZOH is

$$H(s) = \frac{1}{s} - \frac{1}{s}\cdot e^{-sT}$$

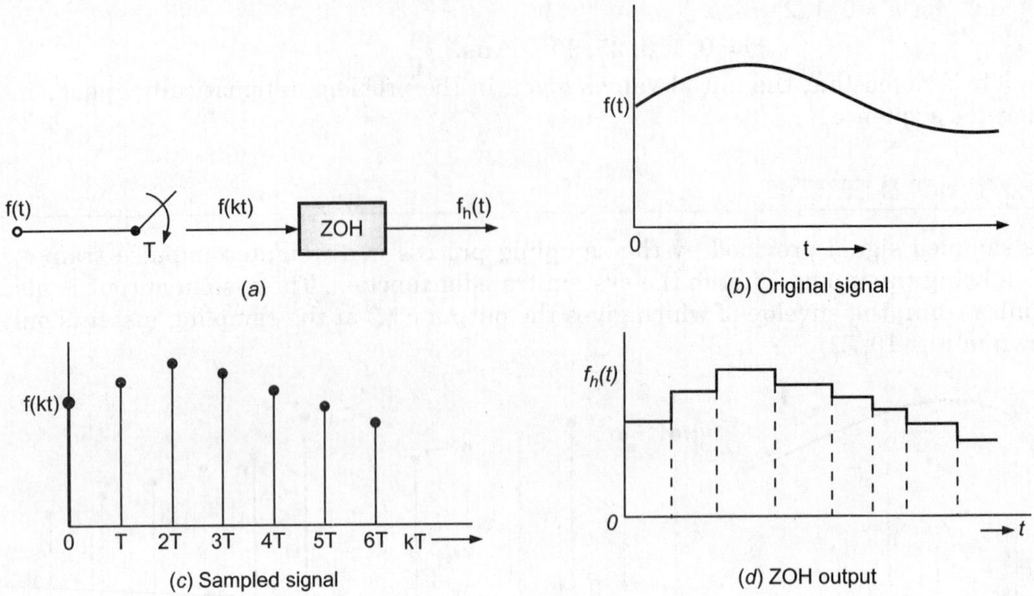

Fig. 10.7.2. Hold circuit.

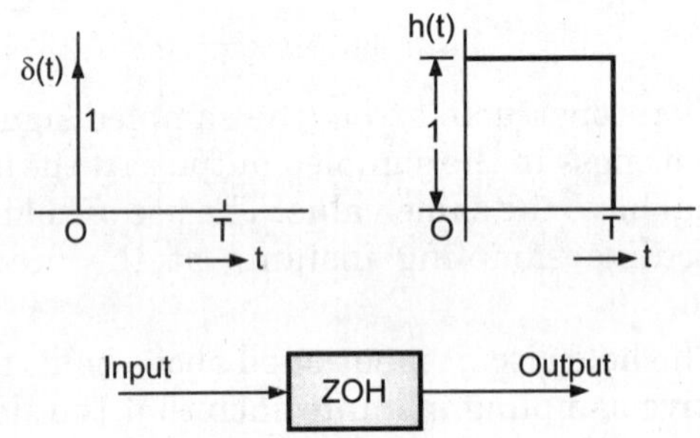

Fig. 10.7.3. Unit impulse input to ZOH.

As the input to ZOH is $\delta(t)$ *i.e.*, unit impulse function, the Laplace transform of the input is $\mathcal{L}\delta(t) = 1$.

The transfer function of the ZOH is denoted by $G_{h0}(s)$, therefore,

$$G_{h0}(s) = \frac{\mathcal{L}\,(\text{output of ZOH})}{\mathcal{L}\,(\text{input to ZOH})} = \frac{\frac{1}{s} - \frac{1}{s} \cdot e^{-sT}}{1}$$

or
$$G_{h0}(s) = \frac{1 - e^{-sT}}{s}$$

Gain frequency plot of zero-order hold circuit. The transfer function of zero-order hold is given by

$$G_{h0}(s) = \frac{1 - e^{-sT}}{s}$$

The sinusoidal form of above transfer function is obtained by putting $s = j\omega$

$$\therefore \quad G_{h0}(j\omega) = \frac{1 - e^{-j\omega T}}{j\omega}$$

which can also be written as :

$$G_{h0}(j\omega) = 2e^{j\omega T/2}.e^{-j\omega T/2}\frac{(1 - e^{-j\omega T})}{j2\,\omega} = \frac{2e^{-j\omega T/2}}{\omega}\cdot\frac{(e^{j\omega T/2} - e^{-j\omega T/2})}{j2}$$

$$= \frac{2e^{-j\omega T/2}}{\omega}\sin\left(\frac{\omega T}{2}\right) = \frac{Te^{-j\omega T/2}}{\frac{\omega T}{2}}\sin\left(\frac{\omega T}{2}\right)$$

or
$$G_{h0}(j\omega) = \frac{T\sin(\omega T/2)}{\omega T/2}\cdot e^{-j\omega T/2}$$

If the sampling frequency is ω_s, then the sampling time T is given by

$$T = \frac{2\pi}{\omega_s}$$

$$\therefore \quad G_{h0}(j\omega) = \frac{T\sin\left(\frac{\omega\times 2\pi}{2\omega_s}\right)}{\frac{\omega\times 2\pi}{2\,\omega_s}}\cdot e^{-j\left(\frac{\omega\times 2\pi}{2\,\omega_s}\right)}$$

or
$$G_{h0}(j\omega) = \frac{T\sin\left(\frac{\pi\omega}{\omega_s}\right)}{\frac{\pi\omega}{\omega_s}}\cdot e^{-j\left(\frac{\pi\omega}{\omega_s}\right)} \quad \text{which gives } |G_{h0}(j\omega)| = T\left|\frac{\sin\left(\frac{\pi\omega}{\omega_s}\right)}{\frac{\pi\omega}{\omega_s}}\right|$$

and
$$\angle G_{h0}(j\omega) = -\left(\frac{\pi\omega}{\omega_s}\right) + \theta$$

where
$$\theta = 0, \text{ for } \sin\left(\frac{\pi\omega}{\omega_s}\right) > 0, \text{ and } \theta = \pi, \text{ for } \sin\left(\frac{\pi\omega}{\omega_s}\right) < 0$$

In view of above derivations gain-frequency and phase frequency graphs for ZOH circuit are shown in Fig. 10.7.4.

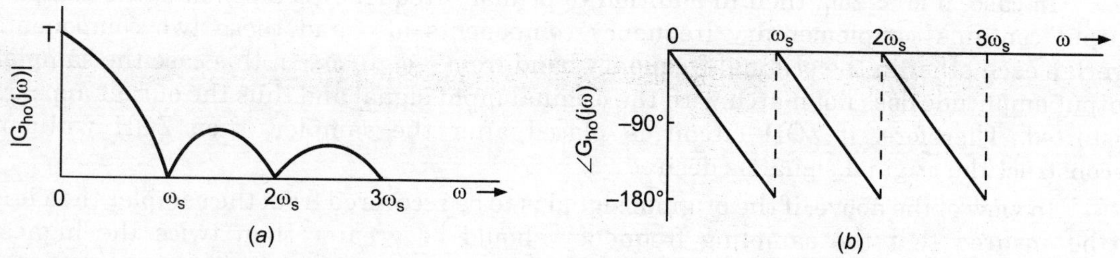

Fig. 10.7.4 (*a*) Gain-frequency graph for ZOH (*b*) Phase-frequency graph for ZOH.

10.8 RECONSTRUCTION OF SIGNAL : MINIMUM SAMPLING FREQUENCY

The use of sampler in control system generates high frequency components in the sampled output. Fig. 10.8.1 shows the frequency spectrum of the input signal given to the sampler, the highest frequency contained in the input signal being ω_1.

The sampler output contains primary (original) components of the input frequency ω_1 and in addition complementary high frequency components as shown in Fig. 10.8.2 where it is assumed that the sampling frequency is greater than $2\omega_1$ *i.e.,* $\omega_s > 2\omega_1$.

However, if the sampling frequency is reduced *i.e.* sampling time increased such that $\omega_s < 2\omega_1$, the sampler output frequency spectrum appears to be as shown in Fig. 10.8.3.

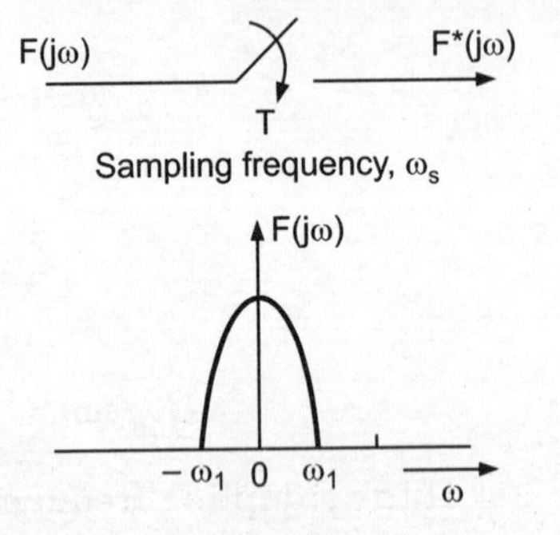

Fig. 10.8.1. Sampler input frequency.

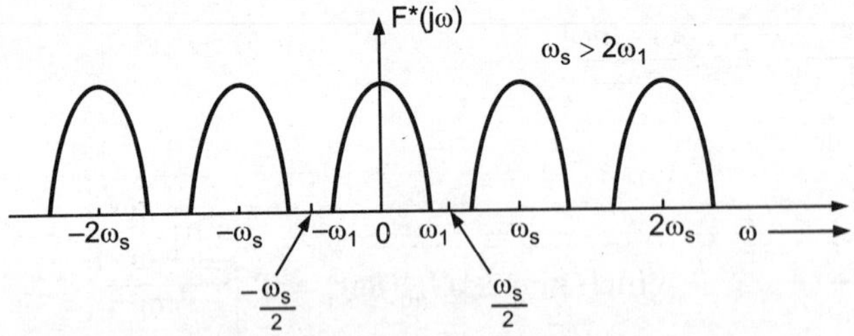

Fig. 10.8.2. Sampler output frequency spectrum : $\omega_s > 2\omega_1$.

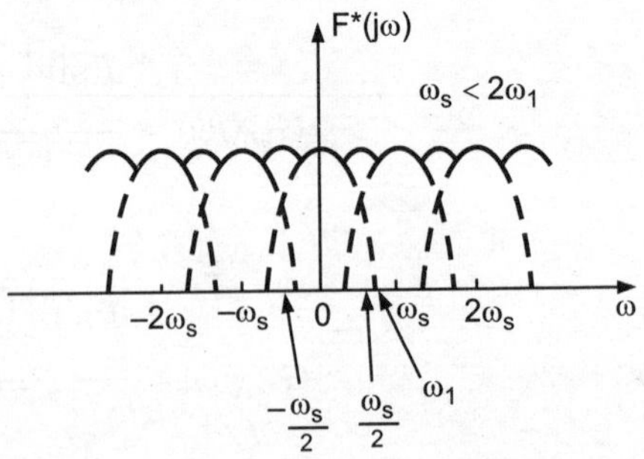

Fig. 10.8.3. Sampler output frequency spectrum : $\omega_s < 2\omega_1$.

A comparison of Figs. 10.8.2 and 10.8.3 shows that for the case $\omega_s > 2\omega_1$, the primary (original) frequency components of the input signal are isolated from the complementary frequency components in the sampler output and thus the original signal is reconstructed. From Fig. 10.7.4, it is noted that the zero-order hold circuit has a characteristic of low-pass filter and therefore, if a ZOH circuit is placed after the sampler, the high frequency (complementary) components are blocked and at the output of ZOH the original signal appears.

In case, if $\omega_s < 2\omega_1$, then in addition to primary frequency components, the sampler output contains complementary frequency components also and these two components overlap each other in the original frequency band from $-\omega_1$ to ω_1. In this case the sampler output amplitude does not match with the original input signal and thus the output appears distorted. Therefore, if ZOH circuit is placed after the sampler, then ZOH will not reconstruct the original signal as desired.

In view of the above, if the original signal is to be recovered from the sampler then it is to be ensured that the sampling frequency should be greater than twice the highest frequency contained in the input signal *i.e.* $\omega_s \geq 2\omega_1$.

The above result is known as sampling theorem.

10.9 PULSE TRANSFER FUNCTION (z-TRANSFER FUNCTION)

The block diagram or a unity feedback sampled-data control system is shown in Fig. 10.9.1. The error signal $e(t)$ is sampled and fed to the forward path transfer function.

Ordinarily, a sampled-data control system contains a hold circuit placed after the sampler. The hold circuit furnishes certain amount of smoothing in the output, therefore, in s-domain the output can be denoted as $C(s)$ which is considered to be continuous and the block diagram in s-domain is shown in Fig. 10.9.2.

The Laplace transform and z-transform of the output $c(t)$ being $C(s)$ and $C(z)$ respectively. The z-transform of the output describes the behaviour of the output signal at the sampling instants only.

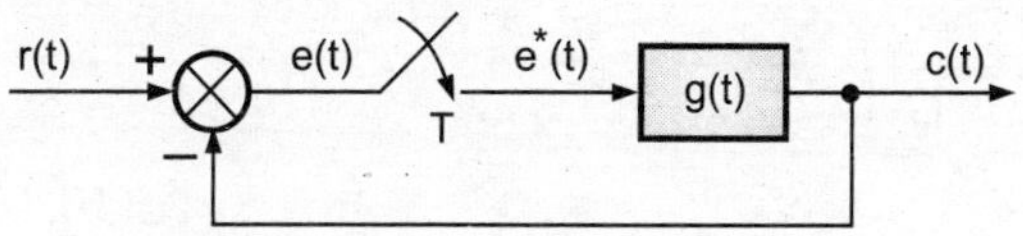

Fig. 10.9.1. Sampled-data control system : Block diagram in time domain.

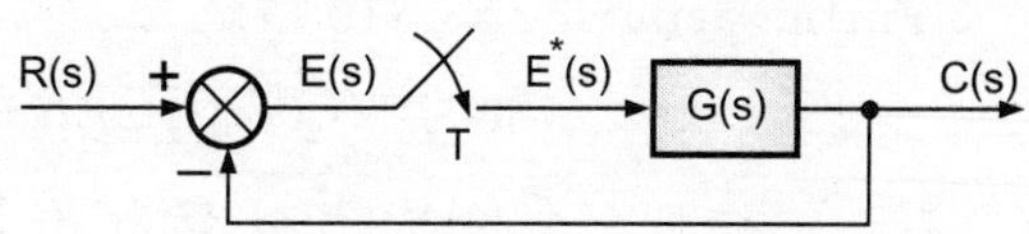

Fig. 10.9.2. Sampled-data control system : Block diagram in s-domain.

Fig. 10.9.3 shows a sampled-data control system wherein the input $R(s)$ is sampled and fed to the system transfer function $G(s)$ and the output is $C(s)$.

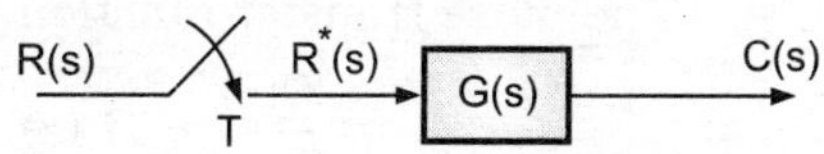

Fig. 10.9.3. Input sampling.

From Fig. 10.9.3 it is noted that, the input to the transfer function $G(s)$ is sampled *i.e.* $R^*(s)$. Therefore, the output is considered sampled one and thus

$$[C(s)]^* = [G(s)\, R^*(s)]^*$$

or $$C^*(s) = G^*(s)\, [R^*(s)]^* \qquad ...(10.12)$$

As starring a starred function results in starred function itself, therefore

$$C^*(s) = G^*(s)\, R^*(s) \qquad ...(10.13)$$

In z-transform form the equation (10.13) can be written as

$$C(z) = G(z)\, R(z)$$

or $$\frac{C(z)}{R(z)} = G(z) \qquad ...(10.14)$$

The pulse transfer function (z-transfer function) of the sampled-data control system shown in Fig. 10.9.3 is $G(z)$.

The z-domain block diagram is shown in Fig. 10.9.4.

PULSE TRANSFER FUNCTION FROM BLOCK DIAGRAMS

(1) Blocks in Cascade

(*a*) Two blocks in cascade each using a sampler in the input

Fig. 10.9.5 shows two blocks having transfer functions $G_1(s)$ and $G_2(s)$ connected in cascade such that the input to each block is sampled and the sampler S_1 and S_2 operate in synchronism.

From Fig. 10.9.5 it is noted that

$$C_1(s) = G_1(s)\, R^*(s) \qquad ...(10.15)$$

and $$C(s) = G_2(s)\, C_1^*(s) \qquad ...(10.16)$$

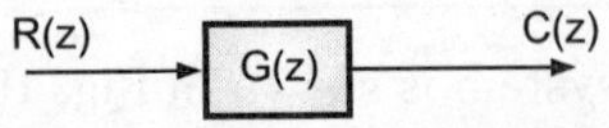

Fig. 10.9.4. Sampled-data control system : Block diagram in z-domain.

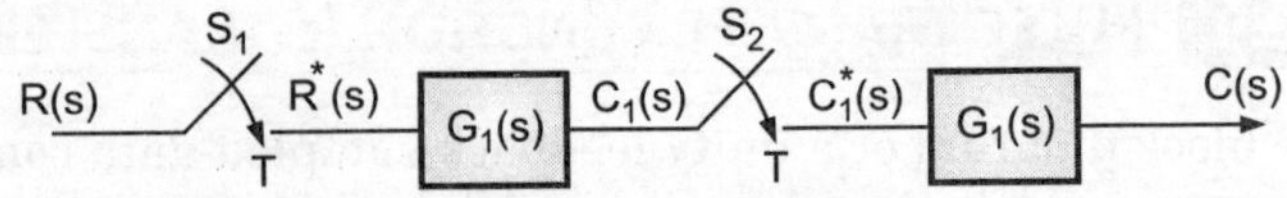

Fig. 10.9.5. Two blocks in cascade each using input sampler.

Since $\quad C_1^*(s) = [G_1(s)\,R^*(s)]^*$

$$C(s) = G_2(s)\,[G_1(s)\,R^*(s)]^* = G_2(s)\;G_1^*(s)\;[R^*(s)]^*$$

noting that starring a starred function result in the starred function itself.

$$\therefore \qquad C(s) = G_2(s)\;G_1^*(s)\;R^*(s) \qquad ...(10.17)$$

Further, starring Eq. (10.17)

$$C^*(s) = [G_2(s)\;G_1^*(s)\;R^*(s)]^* = G_2^*(s)\;G_1^*(s)\;R^*(s)$$

or
$$C^*(s) = G_1^*(s)\;G_2^*(s)\;R^*(s) \qquad ..(10.18)$$

In z-transform form Eq. (10.18) can be written as :

$$C(z) = G_1(z)\,G_2(z)\,R(z)$$

The pulse transfer function is

$$\frac{C(z)}{R(z)} = G_1(z)\;G_2(z) \qquad ...(10.19)$$

The block diagram representation of pulse transfer function (10.19) in z-domain is shown in Fig. 10.9.6.

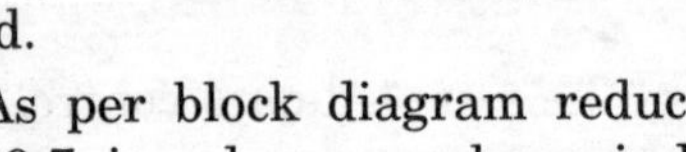

Fig. 10.9.6. Block diagram representation of pulse transfer function (10.19).

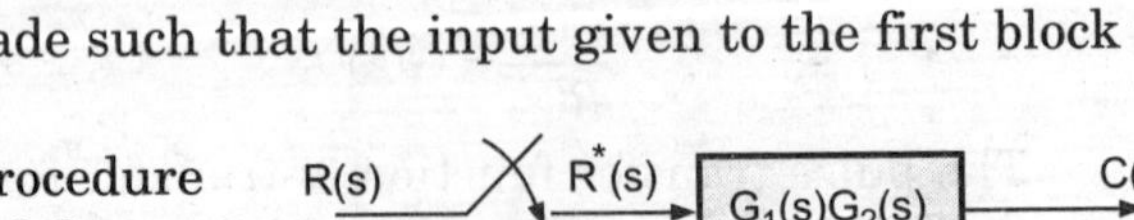

Fig. 10.9.7. Two blocks in cascade using sampler before first block.

(*b*) **The blocks in cascade using sampler in the input of first block :**

Fig. 10.9.7 shows two blocks in cascade such that the input given to the first block is sampled.

As per block diagram reduction procedure Fig. 10.9.7. is redrawn as shown in Fig. 10.9.8.

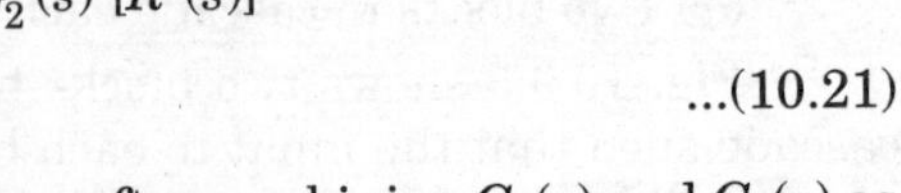

Fig. 10.9.8. Fig. 10.9.7 redrawn.

It is noted from Fig. 10.9.8 that

$$C(s) = G_1(s)\,G_2(s)\,R^*(s) \qquad ...(10.20)$$

starring Eq. (10.20), $C^*(s) = [G_1(s)\,G_2(s)\,R^*(s)]^* = G_1\;G_2^*(s)\;[R^*(s)]^*$

or
$$C^*(s) = G_1 G_2^*(s)\,R^*(s) \qquad ...(10.21)$$

The terms $G_1 G_2^*(s)$ means that the starring is done after combining $G_1(s)$ and $G_2(s)$ as per block diagram reduction rules.

In terms of z-transform Eq. (10.21) can be written as

$$C(z) = G_1G_2(z)\,R(z) \qquad ...(10.22)$$

The pulse transfer function is

$$\frac{C(z)}{R(z)} = G_1G_2(z) \qquad \text{...(10.23)}$$

The block diagram representation of Eq. (10.23) is shown in Fig. 10.9.9.

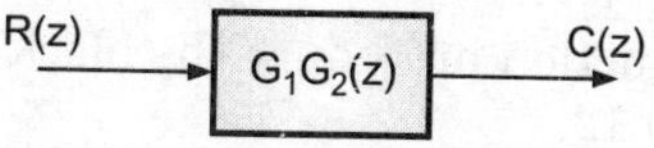

Fig. 10.9.9. Block diagram representation of pulse transfer function (10.23).

It should be noted that $G_1G_2(z) \neq G_1(z)\, G_2(z)$, this is illustrated in example given below :

Illustrative Example. 10.9.1 *Determine the pulse transfer function for the system represented by the block diagram shown in Figs. 10.9.10 (i) and 10.9.10 (ii).*

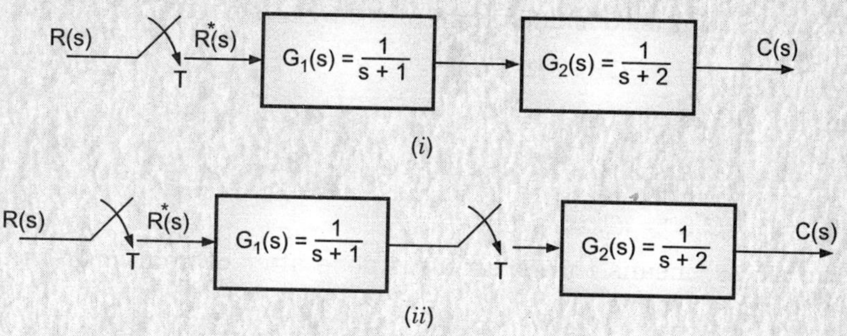

Fig. 10.9.10. Illustrative Example.

Solution. (*i*) $G_1G_2(s) = \dfrac{1}{s+1} \cdot \dfrac{1}{s+2} = \dfrac{1}{s+1} - \dfrac{1}{s+2}$

In terms of z-transform

$$G_1G_2(z) = \frac{z}{z - e^{-T}} - \frac{z}{z - e^{-2T}} = \frac{z(e^{-T} - e^{-2T})}{(z - e^{-T})(z - e^{-2T})}$$

The pulse transfer function is

$$\frac{C(z)}{R(z)} = G_1G_2(z)$$

or
$$\frac{C(z)}{R(z)} = \frac{z(e^{-T} - e^{-2T})}{(z - e^{-T})(z - e^{-2T})} \qquad \text{...(10.24)}$$

(*ii*) In terms of z-transform

$$G_1(z) = \frac{z}{z - e^{-T}} \quad \text{and} \quad G_2(z) = \frac{z}{z - e^{-2T}}$$

The pulse transfer function is given by

$$\frac{C(z)}{R(z)} = G_1(z)\, G_2(z) = \frac{z^2}{(z - e^{-T})(z - e^{-2T})}$$

or
$$\frac{C(z)}{R(z)} = \frac{z^2}{(z - e^{-T})(z - e^{-2T})} \qquad \text{...(10.25)}$$

On comparing pulse transfer functions (10.24) and (10.25) it is noted that

$$G_1G_2(z) \neq G_1(z)\, G_2(z)$$

(c) Two blocks in cascade using sampler in the input to second block :

Fig. 10.9.11 shows two blocks in cascade such that the input to second block is sampled.

As per block diagram reduction procedure the block diagram shown in Fig. 10.9.11 is redrawn and shown in Fig. 10.9.12.

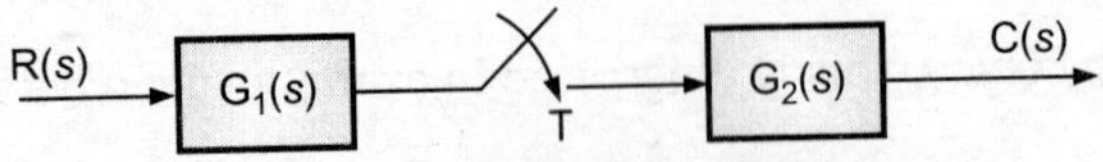

Fig. 10.9.11. Two blocks in cascade : Sampler before second block.

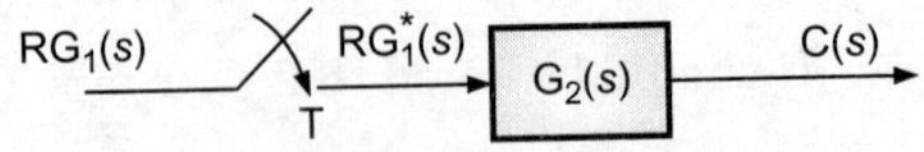

Fig. 10.9.12. Fig. 10.9.11 is redrawn.

From Fig. 10.9.12 it is concluded that

$$C(s) = G_2(s)\ RG_1^*(s) \qquad ...(10.26)$$

Starring Eq. (10.26)

$$C^*(s) = [G_2(s)\ RG_1^*(s)]^* = G_2^*(s)\ [RG_1^*(s)]^*$$

or

$$C^*(s) = G_2^*(s)\ RG_1^*(s) \qquad ...(10.27)$$

The term $RG_1(s)$ means that starring is done after combining $R(s)$ and $G(s)$ as per block diagram reduction rules.

The Eq. (10.27) giving output at the sampling instants can be written in z-transform as

$$C(z) = G_2(z)\, RG_1(z) \qquad ...(10.28)$$

Illustrative Example 10.9.2. *Determine the z-transform of the output for the sampled-data system shown in Fig. 10.9.13 consider input function to be a unit step.*

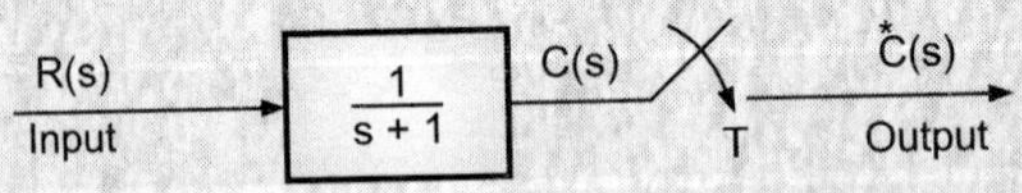

Fig. 10.9.13. Sampled-data system : Illustrative example 10.9.2.

Solution.
$$C(s) = R(s)\,.\,\frac{1}{s+1} = \frac{1}{s}\cdot\frac{1}{s+1}$$

or

$$C(s) = \frac{1}{s} - \frac{1}{s+1}$$

Taking z-transform on both sides

$$C(z) = \frac{z}{z-1} - \frac{z}{z-e^{-T}} = \frac{z(z-e^{-T}) - z(z-1)}{(z-1)(z-e^{-T})}$$

or

$$C(z) = \frac{z(z-e^{-T})}{(z-1)(z-e^{-T})} \quad \textbf{Ans.}$$

2. Closed-Loop Control systems

(a) Error signal sampling. Fig. 10.9.14 shows a closed-loop control system wherein the error signal is sampled.

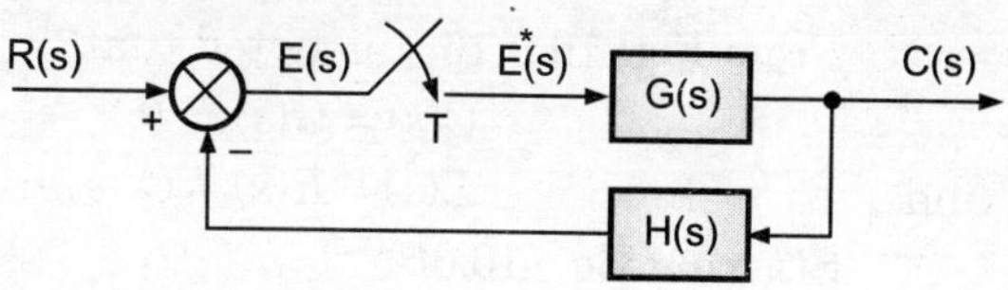

Fig. 10.9.14. Error Sampling.

From Fig. 10.9.14 it is noted that

$$C(s) = G(s)\, E^*(s) \qquad ...(10.29)$$

and

$$E(s) = R(s) - C(s)\, H(s) \qquad ...(10.30)$$

Substituting for $C(s)$ in Eq. (10.30) from (10.29)

$$E(s) = R(s) - G(s)\, E^*(s)\, H(s) \qquad ...(10.31)$$

Starring Eq. (10.31)

$$E^*(s) = [R(s) - G(s)\, E^*(s)\, H(s)]^*$$

$$= R^*(s) - E^*(s)\, [G(s)\, H(s)]^* = R^*(s) - E^*(s)\, GH^*(s)$$

or

$$E^*(s)\, [1 + GH^*(s)] = R^*(s)$$

$\therefore$

$$E^*(s) = \frac{R^*(s)}{1 + GH^*(s)} \qquad ...(10.32)$$

Starring Eq. (10.29)

$$C^*(s) = [G(s)\, E^*(s)]^* = G^*(s)\, [E^*(s)]^*$$

or

$$C^*(s) = G^*(s)\, E^*(s) \qquad ...(10.33)$$

Substituting Eq. (10.32) in (10.33)

$$C^*(s) = G^*(s)\,.\, \frac{R^*(s)}{1 + GH^*(s)}$$

or

$$\frac{C^*(s)}{R^*(s)} = \frac{G^*(s)}{1 + GH^*(s)} \qquad ...(10.34)$$

In terms of z-transform relation (10.34) can be written as the required pulse transfer function

$$\frac{C(z)}{R(z)} = \frac{G(z)}{1 + GH(z)} \qquad ...(10.35)$$

The block diagram representation of the pulse transfer function (10.35) is shown in Fig. 10.9.15.

(b) Error and feedback sampling. Fig. 10.9.16 shows a closed-loop control system wherein the error signal as well as feedback signal being sampled such that the samplers are synchronized.

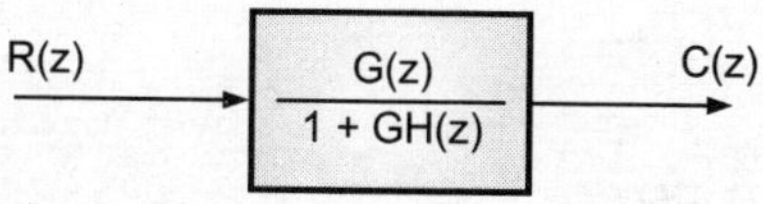

Fig. 10.9.15. Block diagram in z-domain : Error sampling.

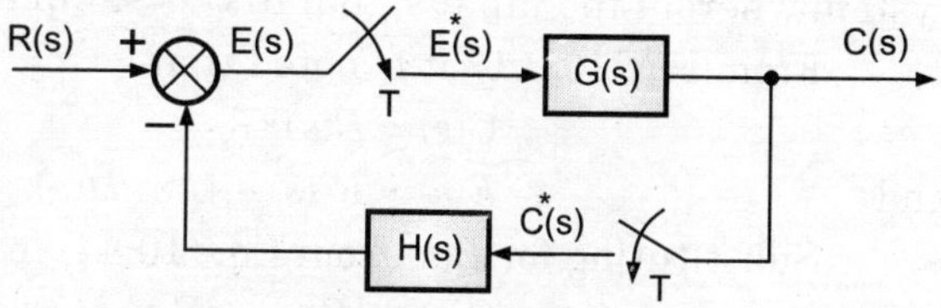

Fig. 10.9.16. Error and feedback sampling.

From Fig. 10.9.16 it is noted that

$$C(s) = G(s)\, E^*(s) \quad ...(10.36)$$

and

$$E(s) = R(s) - C^*(s)\, H(s) \quad ...(10.37)$$

Starring Eq. (10.36)

$$C^*(s) = [G(s)\, E^*(s)]^* = G^*(s)\, E^*(s)$$

or

$$C^*(s) = G^*(s)\, E^*(s) \quad ...(10.38)$$

Substituting for $C^*(s)$ in Eq. (10.37) from (10.38)

$$E(s) = R(s) - G^*(s)\, E^*(s)\, H(s) \quad ...(10.39)$$

Starring Eq. (10.39)

$$E^*(s) = [R(s) - G^*(s)\, E^*(s)^*\, H(s)]^* = R^*(s) - G^*(s)\, E^*(s)\, H^*(s)$$

or

$$E^*(s)\, [1 + G^*(s)\, H^*(s)] = R^*(s)$$

$\therefore$

$$E^*(s) = \frac{R^*(s)}{1 + G^*(s)\, H^*(s)} \quad ...(10.40)$$

Substituting Eq. (10.40) in (10.38)

$$C^*(s) = G^*(s) \cdot \frac{R^*(s)}{1 + G^*(s)\, H^*(s)}$$

or

$$\frac{C^*(s)}{R^*(s)} = \frac{G^*(s)}{1 + G^*(s)\, H^*(s)} \quad ...(10.41)$$

In terms of z-transform relation (10.41) can be expressed as the required pulse transfer function

$\therefore$

$$\frac{C(z)}{R(z)} = \frac{G(z)}{1 + G(z)\, H(z)} \quad ...(10.42)$$

The block diagram representation of the pulse transfer function (10.42) is shown in Fig. 10.9.17.

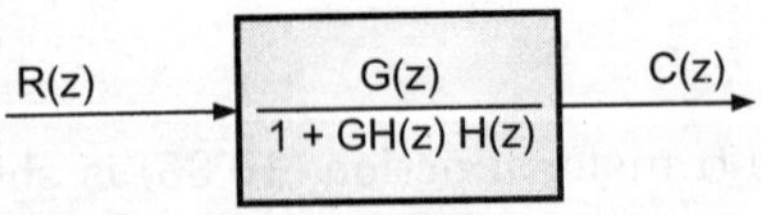

Fig. 10.9.17. Block diagram in z-domain : Error and feedback sampling.

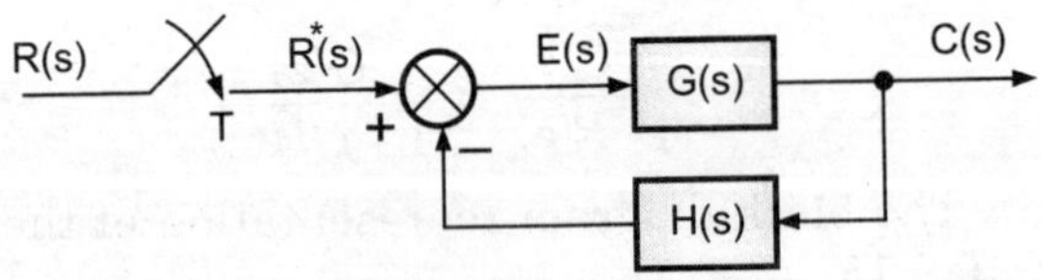

Fig. 10.9.18. Closed-loop system : Input sampling.

(c) *Closed-loop system with input sampling*. Fig. 10.9.18 shows a closed-loop control system wherein the input signal $R(s)$ is sampled.

From Fig. 10.9.18 it is noted that

$$C(s) = E(s)\, G(s) \quad ...(10.43)$$

and

$$E(s) = R^*(s) - C(s)\, H(s) \quad ...(10.44)$$

Substituting for $E(s)$ from Eq. (10.44) in Eq. (10.43)

$$C(s) = [R^*(s) - C(s)\, H(s)]\, G(s) = R^*(s)\, G(s) - C(s)\, H(s)\, G(s)$$

or

$$C(s)\, [1 + G(s)\, H(s)] = G(s)\, R^*(s)$$

$$\therefore \qquad C(s) = \frac{G(s)}{1 + G(s)\,H(s)} R^*(s) \qquad \text{...(10.45)}$$

Starring Eq. (10.45)

$$C^*(s) = \left[\frac{G(s)}{1 + G(s)\,H(s)} R^*(s)\right]^* = \left[\frac{G}{1 + GH}(s)^*\right] [R^*(s)]^*$$

$$= \left[\frac{G}{1 + GH}(s)^*\right] R^*(s) \qquad \text{...(10.46)}$$

The term $\left[\frac{G}{1 + GH}(s)^*\right]$ means that the starring is done after combining $G(s)$ and $H(s)$ as per block diagram reduction rules.

In terms of z-transform the relation (10.46) can be written as

$$C(z) = \left[\frac{G}{1 + GH}(z)\right] R(z)$$

and thus the required pulse transfer function is

$$\frac{C(z)}{R(z)} = \frac{G}{1 + GH}(z) \qquad \text{...(10.47)}$$

The block diagram representation of the pulse transfer function is shown in Fig. 10.9.19.

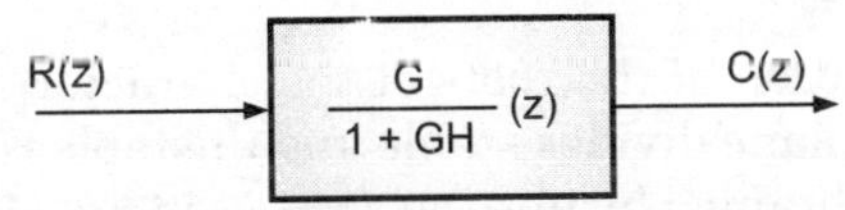

Fig. 10.9.19. Block diagram representation in z-domain : Closed-loop system with input signal sampling.

10.10 STABILITY ANALYSIS OF SAMPLED-DATA CONTROL SYSTEMS

The overall transfer function of a sampled-data feedback control system shown in Fig. 10.10.1 is given by

$$\frac{C(s)}{R(s)} = \frac{G(s)}{1 + GH^*(s)} \qquad \text{...(10.48)}$$

The characteristic equation of the system is

$$1 + GH^*(s) = 0 \qquad \text{...(10.49)}$$

The stability of a sampled-data system is determined by the location of the roots (poles of overall transfer function) characteristic equation

$$1 + GH^*(s) = 0$$

The roots being plotted in s-plane. For stability the roots of this characteristic equation should lie in the left-half of s-plane. The region of stability in s-plane is shown in Fig. 10.10.2.

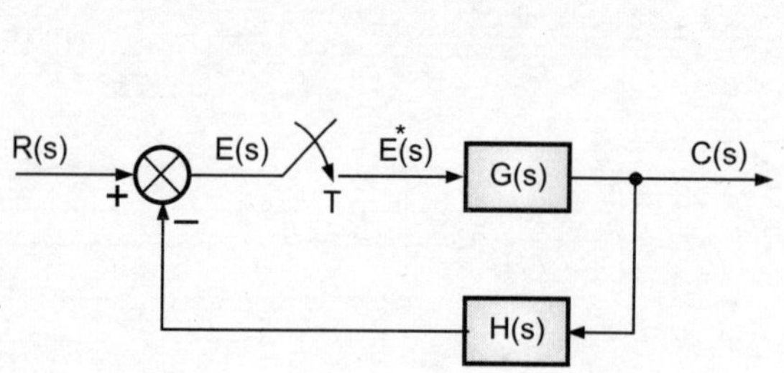

Fig. 10.10.1. Sampled data feedback control system with error sampling.

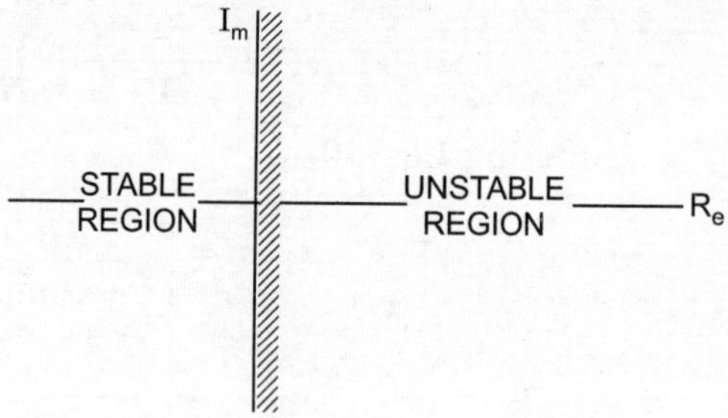

Fig. 10.10.2. Stability region in s-plane.

The overall transfer function given by Eq. (10.48) can be expressed in z-transform as

$$\frac{C(z)}{R(z)} = \frac{G(z)}{1+GH(z)} \quad ...(10.50)$$

The characteristic equation of z-transfer function (10.50) is

$$1 + GH(z) = 0 \quad ...(10.51)$$

In terms of z-transform form of the characteristic equation *i.e.* $1 + GH(z) = 0$, the stability region in z-plane corresponding to s-plane is located by mapping from s-plane to z-plane.

Mapping from s-plane to z-plane

As the z-transform is obtained by substituting $z = e^{sT}$ in the corresponding impulse train Laplace transform, therefore the relation between the variables 's' and 'z' is

$$z = e^{sT}$$

With reference to location of the roots of the characteristic equation is s-domain the imaginary ($j\omega$) axis in the s-plane divides stable and unstable regions and the corresponding regions in z-domain can be obtained by putting $s = \pm j\omega$ in Eq. (10.52) and plotting the values of 'z' thus obtained in another complex plane called z-plane. Therefore,

$$z = e^{\pm j\omega T} \quad ...(10.53)$$

or

$$z = (\cos \omega T \pm j \sin \omega T) \quad ...(10.54)$$

From the Eq. (10.54) the magnitude

$$|z| = 1 \quad ...(10.55)$$

and phase angle

$$\angle z = \pm \omega T \quad ...(10.56)$$

In view of the Eqs. (10.55) and (10.56) the corresponding variation of the dependent variable 'z' in the z-plane as the independent variable 'ω' varied along the imaginary axis in s-plane is given by a circle of unit radius centred at the origin of z-plane shown in Fig. 10.10.3.

The L.H.S. of s-plane is mapped as the inside and R.H.S. outside of the unit circle in z-plane which is varified below :

(*a*) Let $s = -\alpha \pm j\omega$; a point in the L.H.S. of s-plane. The corresponding point in z-plane is given by

$$z = e^{(-\alpha \pm j\omega)T} = e^{-\alpha T} (\cos \omega T \pm j \sin \omega T)$$

$$\therefore \quad |z| = e^{-\alpha T}$$

and

$$\angle z = \pm \omega T$$

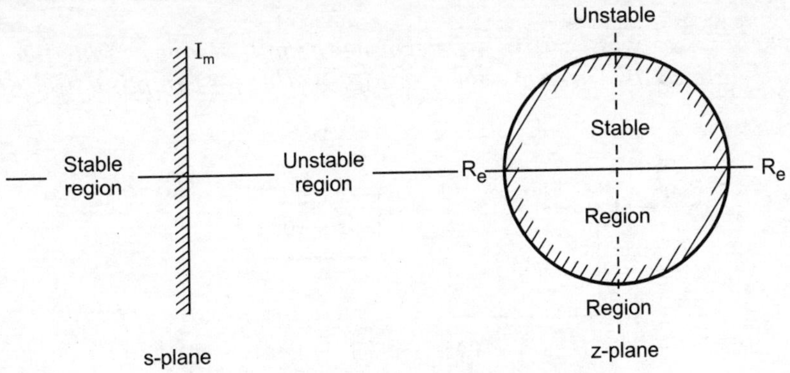

Fig. 10.10.3. Mapping from s-plane to z-plane.

As α the real part of the point under consideration lies in the L.H.S. s-plane and T being positive

$$\therefore \quad |z| < 1$$

Hence, the point $(-\alpha \pm j\omega)$ with negative real part located in s-plane lies inside the unit circle when mapped into z-plane.

(b) Let $s = \alpha \pm j\omega$; a point in the R.H.S. of s-plane.

The corresponding point in z-plane is given by

$$z = e^{(\alpha \pm j\omega)T} = e^{\alpha T} (\cos \omega T \pm j \sin \omega T)$$

$$\therefore \quad |z| = e^{\alpha T}$$

and

$$\angle z = \pm \omega T$$

As α the real part of the point under consideration lies in the R.H.S. of s-plane and T being positive

$$\therefore \quad |z| > 1$$

Hence, the point $(\alpha \pm j\omega)$ with positive real part located in s-plane lies outside the unit circle mapped into z-plane.

In view of the above analysis it is concluded that for the sampled data system to be stable all the roots of the z-transformed characteristic equation $1 + GH(z) = 0$ should be located inside the unit circle centred at the origin in z-plane and in case, any root is located outside the unit circle centred at the origin in z-plane makes the system unstable.

If the roots of the characteristic equation $1 + GH(z) = 0$ are located inside the unit circle centred at the origin in z-plane then the time response of a sampled data system decays with the sampling number $k = 0, 1, 2, 3, \ldots$ indicating stability while any root located outside the unit circle centred at the origin in z-plane leads to growing response with $k = 0, 1, 2, 3 \ldots$ and thus the system is unstable.

If the characteristic equation is of higher degree then it is not simple to determine the roots. However, the roots of the characteristic equation $1 + GH(z) = 0$ which are located inside the unit circle centered at the origin in z-plane can be identified without factoring the characteristic equation. This is done by changing the variable from 'z' to 'w'. This change of variable maps the interior of the z-plane unit circle into left-half of complex w-plane while the exterior of the z-plane unit circle is mapped into right-half of w-plane and therefore the Routh-Hurwitz criterion can be applied to determine the stability. The change of variable from 'z' to 'w' is known as bilinear transformation.

The example given below illustrates the determination of the stability by modifying Routh-Hurwitz test to suit sampled data control system.

Illustrative Example 10.10.1. *Determine the pulse transfer function and stability of the sampled-data control system shown in Fig. 10.10.4 for sampling time (a) T = 0.5 sec. (b) T = 1 sec.*

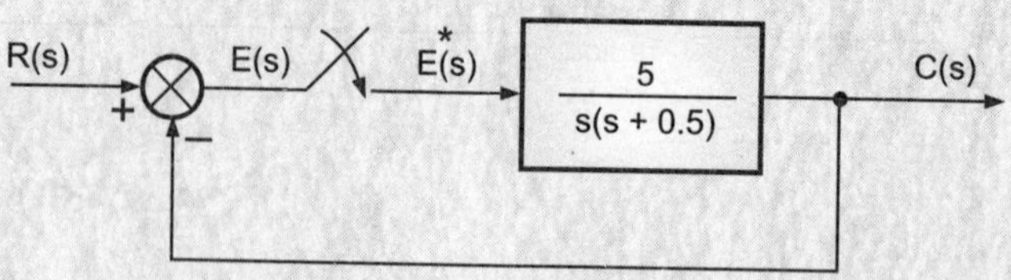

Fig. 10.10.4. Sampled-data system for illustrative example 10.10.1.

Solution. $$G(s) = \frac{5}{s\,(s+0.5)} = \frac{K_1}{s} + \frac{K_2}{s+0.5}$$

The coefficients can be determined as

$$K_1 = 10 \text{ and } K_2 = -10$$

$$\therefore \qquad G(s) = \frac{10}{s} - \frac{10}{s+0.5}$$

Taking z-transform, $$G(z) = 10\left[\frac{z}{(z-1)} - \frac{z}{(z-e^{-0.5T})}\right] = \frac{10z\,(1-e^{-0.5T})}{z^2 - z\,(1-e^{-0.5T}) + e^{-0.5T}}$$

Since, $$G(s) = \frac{5}{s(s+0.5)} \text{ and } H(s) = 1$$

$$\therefore \qquad G(s)\,H(s) = \frac{5}{s(s+0.5)} \times 1 = \frac{5}{s(s+0.5)} = \frac{10}{s} - \frac{10}{s+0.5}$$

Taking z-transform, $$GH(z) = 10\left[\frac{z}{(z-1)} - \frac{z}{(z-e^{-0.5T})}\right] = \frac{10z\,(1-e^{-0.5T})}{z^2 - z\,(1+e^{-0.5T}) + e^{-0.5T}}$$

The pulse transfer function for the sampled-data control system shown in Fig. 10.10.4 is given by

$$\frac{C(z)}{R(z)} = \frac{G(z)}{1+GH(z)}$$

Substituting for the values of $G(z)$ and $GH(z)$ determined above, the pulse transfer function is

$$\frac{C(z)}{R(z)} = \frac{\dfrac{10z\,(1-e^{-0.5T})}{z^2 - z\,(1+e^{-0.5T}) + e^{-0.5T}}}{1 + \dfrac{10z\,(1-e^{-0.5T})}{z^2 - z\,(1+e^{-0.5T}) + e^{-0.5T}}}$$

or $$\frac{C(z)}{R(z)} = \frac{10z\,(1-e^{-0.5T})}{z^2 - z\,(11\,e^{-0.5T} - 9) + e^{-0.5T}} \qquad \textbf{Ans.}$$

The characteristic equation is, $z^2 - z\,(11\,e^{-0.5T} - 9) + e^{-0.5T} = 0$

(*a*) As the sampling time is $T = 0.5$ sec the characteristic equation becomes

$$z^2 - z\,(11\,e^{-0.5 \times 0.5} - 9) + e^{-0.5 \times 0.5} = 0$$

or
$$z^2 + 0.42z + 0.78 = 0$$

Applying bilinear transformation from *z*-plane to *w*-plane *i.e.* putting $z = (1 + w)/(1 - w)$ gives

$$\left(\frac{1+w}{1-w}\right)^2 + 0.42\left(\frac{1+w}{1-w}\right) + 0.78 = 0$$

or
$$1.36w^2 + 0.44w + 2.2 = 0$$

The Routh array for the above equation is formed below

w^2	1.36	2.2
w^1	0.44	0
w^0	2.2	

There is no change of sign in the first column of the Routh array, hence, all the roots of the characteristic equation lie within the unit circle centred at the origin of *z*-plane, therefore, the system is stable. **Ans.**

(*b*) As the sampling time $T = 1$ sec the characteristic equation is

$$z^2 - z\,(11\,e^{-0.5 \times 1} - 9) + e^{-0.5 \times 1} = 0$$

or
$$z^2 + 2.33z + 0.606 = 0$$

Applying bilinear transformation from *z*-plane *w*-plane *i.e.* putting $z = (1 + w)/(1 - w)$ gives

$$\left(\frac{1+w}{1-w}\right) + 2.33\left(\frac{1+w}{1-w}\right) + 0.606 = 0$$

or
$$-\,0.724w^2 + 0.788w + 3.93 = 0$$

The Routh array for the above equation is formed below

w^2	− 0.724	3.93
w^1	0.788	0
w^0	3.93	

There is one change of sign in the first column of the Routh array, hence, one root of the characteristic equation is located outside the unit circle in *z*-plane, therefore, the system is unstable. **Ans.**

The corresponding continuous-time system given by the above illustative example is of second order and as such absolutely stable. Whereas, if the system is made sampled data, it becomes unstable for the sampling time $T = 1$ sec. However, decreasing the sampling time to $T = 0.5$ sec, the system is table.

Thus, it is concluded that a sampled-data control system behaves more like a continuous time system when the sampling time is reduced (sampling frequency increased).

Jury's stability test :

The identification of the roots of a characteristic polynomial $F(z) = 1 + G(z)\,H(z)$ within the unit circle for determining system stability of a sampled-data control system is done by Jury's stability criterion. This criterion is an algebraic one. Consider the following *n*th order polynomial in *z*

$$F(z) = a_n z^n + a_{n-1} z^{n-1} + a_{n-2} z^{n-2} + \dots + a_0;\ a_n > 0$$

(1) Necessary conditions for the roots of the above polynomial $F(z)$ to lie within a unit circle centred at the origin of z-plane are :

$$F(1) > 0$$

and $$(-1)^n F(-1) > 0$$

$$|a_0| < a_n$$

(2) Sufficient conditions for the roots of the above polynomial $F(z)$ to lie within a unit circle in z-plane are determined as follows :

The application of Jury's stability test through steps, requires the construction of array of coefficients as described below.

This array consists of $(2n - 3)$ rows. The elements of the first two rows are the coefficients of the characteristic polynomial and the elements of subsequent rows are defined in terms of the elements of preceding rows. The array is given in table below :

Row	z^0	z^1	z^2	...	z^{n-k}...	z^{n-2}	z^{n-1}	z^n
1	a_0	a_1	a_2	...	a_{n-k} ...	a_{n-2}	a_{n-1}	a_n
2	a_n	a_{n-1}	a_{n-2}	...	a_k ...	a_2	a_1	a_0
3	b_0	b_1	b_2	...	b_{n-k} ...	b_{n-2}	b_{n-1}	
4	b_{n-1}	b_{n-2}	b_{n-3}	...	b_{k-1} ...	b_1	b_0	
5	c_0	c_1	c_2	...	c_{n-k} ...	c_{n-2}		
6	c_{n-2}	c_{n-3}	b_{n-4}	...	c_{k-2} ...	c_0		
	⋮	⋮	⋮	⋮				
$2n-6$	p_0	p_1	p_2	p_3				
$2n-5$	p_3	p_2	p_1	p_0				
$2n-4$	q_0	q_1	q_2					
$2n-3$	q_2	q_1	q_0					

In the array mentioned above, the elements of 2nd, 4th, 6th ...etc. upto $(2n - 4)$ rows are respectively the elements of 1st, 3rd, 5th...etc. upto $(2n - 5)$ rows in reverse order. The elements of 3rd, 5th ... etc. upto $(2n - 3)$ rows are determined by following second order determinants.

$$b_k = \begin{vmatrix} a_0 & a_{n-k} \\ a_n & a_k \end{vmatrix} \quad \text{for 3rd row, } k = 0, 1, \dots (n-1)$$

$$c_k = \begin{vmatrix} b_0 & b_{n-1-k} \\ b_{n-1} & b_k \end{vmatrix} \quad \text{for 5th row, } k = 0, 1, \dots (n-2)$$

$$d_k = \begin{vmatrix} c_0 & c_{n-2-k} \\ c_{n-2} & c_k \end{vmatrix} \quad \text{for 7th row, } k = 0, 1, \dots (n-3)$$

$$q_2 = \begin{vmatrix} p_3 & p_0 \\ p_0 & p_3 \end{vmatrix}$$

$$q_1 = \begin{vmatrix} p_3 & p_1 \\ p_0 & p_2 \end{vmatrix} \text{ for } (2n-3) \text{ row}$$

$$q_0 = \begin{vmatrix} p_3 & p_2 \\ p_0 & p_1 \end{vmatrix}$$

Sufficient conditions for the roots of the characteristic polynomial $F(z)$ to lie within a unit circle in z-plane are given by

$$| a_0 | < | a_n |$$
$$| b_0 | > | b_{n-1} |$$
$$| c_0 | > | c_{n-2} |$$
$$| p_3 | > | p_0 |$$
$$| p_2 | > | q_0 | \text{ with } (n-1) \text{ constraints}$$

Illustrative Example 10.10.2. *Determine the stability of a sampled-data control system having following characteristic polynomial.*

$$2z^4 + 8z^3 + 12z^2 + 5z + 1 = 0.$$

Solution. $F(1) = [2(1)^4 + 8(1)^3 + 12(1)^2 + 5(1) + 1] = 28$

$\therefore$ $F(1) > 0$

$$(-1)^n F(-1) = (-1)^4 [2(-1)^4 + 8(-1)^3 + 12(-1)^2 + 5(-1) + 1] = 2$$

$\therefore$ $(-1)^n F(-1) > 0$

Row	z^0	z^1	z^2	z^3	z^4
1	1	5	12	8	2
2	2	8	12	5	1
3	b_0	b_1	b_2	b_3	
4	b_3	b_2	b_1	b_0	
5	c_0	c_1	c_2		

$$n = 4,\ a_0 = 1,\ a_1 = 2,\ a_2 = 12,\ a_3 = 8 \text{ and } a_4 = 2$$

$$b_k = \begin{vmatrix} a_0 & a_{n-k} \\ a_n & a_k \end{vmatrix}$$

$k = 0;$ $$b_0 = \begin{vmatrix} a_0 & a_4 \\ a_4 & a_0 \end{vmatrix} = \begin{vmatrix} 1 & 2 \\ 2 & 1 \end{vmatrix} = 1 \times 1 - 2 \times 2 = -3$$

$k = 1;$ $$b_1 = \begin{vmatrix} a_0 & a_3 \\ a_4 & a_1 \end{vmatrix} = \begin{vmatrix} 1 & 8 \\ 2 & 5 \end{vmatrix} = 1 \times 5 - 2 \times 8 = -11$$

$$k = 2; \qquad b_2 = \begin{vmatrix} a_0 & a_2 \\ a_4 & a_2 \end{vmatrix} = \begin{vmatrix} 1 & 8 \\ 2 & 8 \end{vmatrix} = 1 \times 8 - 2 \times 8 = -8$$

$$k = 3; \qquad b_3 = \begin{vmatrix} a_0 & a_1 \\ a_4 & a_3 \end{vmatrix} = \begin{vmatrix} 1 & 5 \\ 2 & 8 \end{vmatrix} = 1 \times 8 - 2 \times 5 = -2$$

$$c_k = \begin{vmatrix} b_0 & b_{n-1-k} \\ b_{n-1} & b_k \end{vmatrix}$$

$$k = 0; \qquad c_0 = \begin{vmatrix} b_0 & b_{4-1-0} \\ b_{4-1} & b_0 \end{vmatrix} = \begin{vmatrix} b_0 & b_3 \\ b_3 & b_0 \end{vmatrix} = \begin{vmatrix} -3 & -2 \\ -2 & -3 \end{vmatrix}$$

$$= -3 \times -3 - (-2 \times -2) = 5$$

$$k = 1; \qquad c_1 = \begin{vmatrix} b_0 & b_{4-1-1} \\ b_{4-1} & b_1 \end{vmatrix} = \begin{vmatrix} b_0 & b_2 \\ b_3 & b_1 \end{vmatrix} = \begin{vmatrix} -3 & -8 \\ -2 & -11 \end{vmatrix}$$

$$= (-3 \times -11) - (-2 \times -8) = 17$$

$$k = 2; \qquad c_2 = \begin{vmatrix} b_0 & b_{4-1-2} \\ b_{4-1} & b_2 \end{vmatrix} = \begin{vmatrix} b_0 & b_1 \\ b_3 & b_2 \end{vmatrix} = \begin{vmatrix} -3 & -11 \\ -2 & -8 \end{vmatrix}$$

$$= (-3 \times -8) - (-2 \times -11) = 2$$

Sufficient conditions for stability are

$| a_0 | < | a_4 |$ or $| 1 | < | 2 |$ satisfied

$| b_0 | > | b_3 |$ or $| 3 | > | 2 |$ satisfied

$| c_0 | > | c_2 |$ or $| 5 | > | 2 |$ satisfied

The required necessary and sufficient conditions are satisfied, hence, the system is stable. **Ans.**

10.11 SOLVED EXAMPLES

Example 10.11.1. *Find the impulse train Laplace transform (starred transform) $F^*(s)$ for the following :*

(a) $f(t) = u(t)$ *(b)* $f(t) = e^{-at}$ *(c)* $F(z) = \dfrac{z}{z - e^{-aT}}$

Solution. *(a)* $f^*(t) = f(kt)$

Because $f(t) = 1 \quad \therefore \quad f(kT) = 1$

The function $f^*(t) = 1$ is shown in Fig. 10.11.1.

The impulse train Laplace transform (starred transform) is given by

$$F^*(s) = \sum_{k=0}^{\infty} (f(kT)\, e^{-skT} \; ; k = 0, 1, 2, 3...$$

$$= \sum_{k=0}^{\infty} 1 . e^{-skT}$$

$$= 1 + e^{-sT} + e^{-2sT} + e^{-3sT} + \ldots$$

The summation of the above series is

$$F^*(s) = \frac{1}{1 - e^{-sT}} \quad \textbf{Ans.}$$

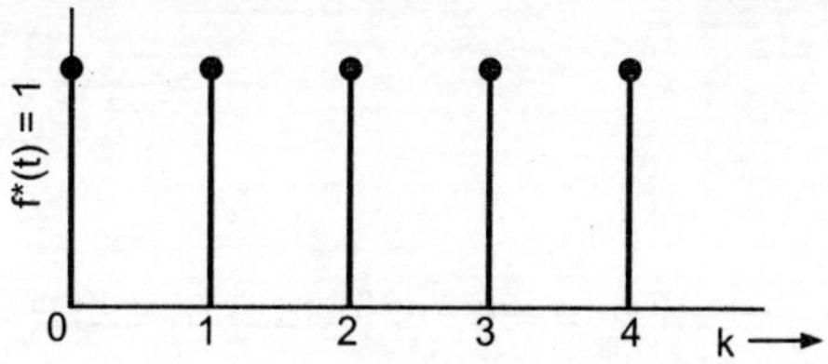

Fig. 10.11.1. Function $f^*(t) = 1$.

(b) $f^*(t) = f(kT)$

$\because \quad f(t) = e^{-at} \quad \therefore \quad f(kT) = e^{-akT}$

The function $f(kT) = e^{-akT}$ is shown in Fig. 10.11.2.

The impulse train Laplace transform is given by

$$F^*(s) = \sum_{k=0}^{\infty} f(kT) e^{-skT} \quad ; k = 0, 1, 2, 3 \ldots$$

$$= \sum_{k=0}^{\infty} e^{-akT} e^{-skT}$$

$$= 1 + e^{-(a+s)T} + e^{-2(a+s)T} + e^{-3(a+s)T} + \ldots$$

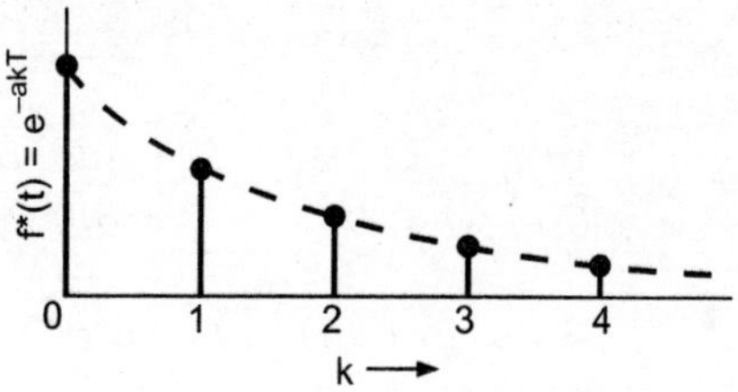

Fig. 10.11.2. Function $f^*(t) = e^{-akt}$.

The summation of the above series is

$$F^*(s) = \frac{1}{1 - e^{-(a+s)T}}$$

or

$$F^*(s) = \frac{1}{1 - e^{-aT} e^{-sT}} \quad \textbf{Ans.}$$

(c)

$$F(z) = \frac{z}{z - e^{-aT}}$$

The impulse train Laplace transform is obtained from the corresponding z-transform by substituting

$$z = e^{sT}$$

$$\therefore \quad F^*(s) = \frac{e^{sT}}{e^{sT} - e^{-\alpha T}} \quad \text{or} \quad F^*(s) = \frac{1}{1 - e^{-\alpha T} e^{-sT}} \quad \textbf{Ans.}$$

Example 10.11.2. *Find z-transform of the following functions :*
(a) $f(t) = u(t) - e^{-2t}$ *(b) $f(t) = (1 - e^{-5t})$ sampling time $T = 0.2$ sec.*

Solution. (a) $f(t) = u(t) - e^{-2t}$

From z-transform table $Z[u(t)] = \dfrac{z}{z-1}$

and

$$Z[e^{-2t}] = \frac{z}{z - e^{-2T}}$$

$$\therefore \quad Z[u(t) - e^{-2t}] = \frac{z}{z-1} - \frac{z}{z - e^{-2T}} = \frac{z^2 - ze^{-2T} - z^2 + z}{(z-1)(z - e^{-2T})}$$

$$= \frac{z(1-e^{-2T})}{(z-1)(z-e^{-2T})} \quad \textbf{Ans.}$$

(*b*) $f(t) = (1 - e^{-2T})$

From z-transform table, $Z[1] = \dfrac{z}{z-1}$ and $Z[e^{-5t}] = \dfrac{z}{z-e^{-5T}}$

$$\therefore \quad Z[1-e^{-5t}] = \frac{z}{z-1} - \frac{z}{z-e^{-5T}}$$

as the sampling time $T = 0.2$ sec.

$$\therefore \quad Z[1-e^{-5t}] = \frac{z}{z-1} - \frac{z}{z-e^{-5\times 0.2}} = \frac{z}{z-1} - \frac{z}{z-0.368}$$

$$= \frac{z(z-0.368) - z(z-1)}{(z-1)(z-0.368)} = \frac{0.632z}{z^2 - 1.368z + 0.368} \quad \textbf{Ans.}$$

Example 10.11.3. *Find z-transform of the following :*
(a) $f(t) = e^{at}$, (b) $f(t) = \sin \omega t$ and (c) $f(t) = e^{-\alpha t} \cos \omega t$.

Solution. (*a*) $f(t) = e^{at}$ $\therefore$ $f(kT) = e^{akT}$

$$\because \quad Z[f(t)] = \sum_{k=0}^{\infty} f(kT) z^{-k}; \quad k = 0, 1, 2, 3, \ldots$$

$$\therefore \quad Z[e^{at}] = \sum_{k=0}^{\infty} e^{akT} z^{-k} = 1 + e^{+aT} z^{-1} + e^{+2aT} z^{-2} + e^{+3aT} z^{-3} + \ldots$$

The summation of the above series is

$$Z[e^{at}] = \frac{1}{1 - e^{aT} z^{-1}} = \frac{z}{z - e^{aT}} \quad \textbf{Ans.}$$

(*b*) $f(t) = \sin \omega t$

$\therefore$ $f(kT) = \sin k\omega T; \quad k = 0, 2, 3...$

$$\because \quad Z[f(t)] = \sum_{k=0}^{\infty} f(kT) z^{-k}$$

$$\therefore \quad Z[f(t)] = \sum_{k=0}^{\infty} \sin k\omega T . z^{-k}$$

$$\because \quad \sin k\omega T = \frac{e^{j\omega kT} - e^{-j\omega kT}}{2j}$$

$$\therefore \quad Z[f(t)] = Z[\sin \omega t]$$

$$= \frac{1}{2j}\left[\sum_{k=0}^{\infty} e^{j\omega kT} z^{-k} - \sum_{k=0}^{\infty} e^{-j\omega kT} z^{-k}\right]$$

The summation of the above series is

$$Z[\sin \omega t] = \frac{1}{2j}\left[\frac{1}{1-e^{j\omega T}\, z^{-1}} - \frac{1}{1-e^{-j\omega T}\, z^{-1}}\right] = \frac{1}{2j}\left[\frac{z}{z-e^{j\omega T}} - \frac{z}{z-e^{-j\omega T}}\right]$$

$$= \frac{z}{2j}\left[\frac{(z-e^{-j\omega T})-(z-e^{j\omega T})}{(z-e^{j\omega T})(z-e^{-j\omega T})}\right] = \frac{z}{2j}\left[\frac{e^{j\omega T}-e^{-j\omega T}}{(z-e^{j\omega T})(z-e^{-j\omega T})}\right]$$

$$= z\,.\,\frac{e^{j\omega T}-e^{-j\omega T}}{2j}\cdot\frac{1}{z^2-2z\left(\dfrac{e^{j\omega T}+e^{-j\omega T}}{2}\right)+1}$$

$$= \frac{z\sin \omega T}{z^2-2z\cos \omega T+1} \quad \textbf{Ans.}$$

(c) $f(t) = e^{-\alpha t}\cos \omega T \qquad \therefore\ f(kT) = e^{-\alpha kT}\cos \omega kT$

$$\because \quad Z[f(t)] = \sum_{k=0}^{\infty} f(kT) z^{-k}\,; \quad k = 0, 1, 2, 3...$$

$$\therefore \quad Z[e^{-\alpha kT}\cos \omega t] = \sum_{k=0}^{\infty} e^{-\alpha kT}\cos k\omega T.\, z^{-k}$$

$$\because \quad \cos k\omega T = \frac{e^{j\omega kT}+e^{-j\omega kT}}{2}$$

$$\therefore \quad Z[e^{-\alpha kT}\cos \omega kT] = \frac{1}{2}\left[\sum_{k=0}^{\infty} e^{-\alpha kT}\, e^{j\omega kT} z^{-k} + \sum_{k=0}^{\infty} e^{-\alpha kT}\, e^{-j\omega kT} z^{-k}\right]$$

The summation of the above series is

$$Z[e^{-\alpha t}\cos \omega t] = \frac{1}{2}\left[\frac{1}{1-e^{-\alpha T}e^{j\omega T}z^{-1}} + \frac{1}{1-e^{-\alpha T}e^{-j\omega T}z^{-1}}\right]$$

$$= \frac{1}{2}\left[\frac{z}{z-e^{-\alpha T}e^{j\omega T}} + \frac{z}{z-e^{-\alpha T}e^{-j\omega T}}\right] = \frac{z}{2}\left[\frac{(z-e^{-\alpha T}e^{j\omega T})+(z-e^{-\alpha T}e^{-j\omega T})}{(z-e^{-\alpha T}e^{j\omega T})(z-e^{-\alpha T}e^{-j\omega T})}\right]$$

$$= \frac{z}{2}\left[\frac{2z-e^{-\alpha T}e^{j\omega T}-e^{-\alpha T}e^{-j\omega T}}{z^2-ze^{-\alpha T}(e^{j\omega T}+e^{-j\omega T})+e^{-2\alpha T}}\right]$$

$$= z\left[z-\frac{e^{-\alpha T}(e^{j\omega T}+e^{-j\omega T})}{2}\right]\times\frac{1}{z^2-2ze^{-\alpha T}\left(\dfrac{e^{j\omega T}+e^{-j\omega T}}{2}\right)+e^{-2\alpha T}}$$

$$= \frac{z(z-e^{-\alpha T}\cos \omega T)}{z^2-2ze^{-\alpha T}\cos \omega T+e^{-2\alpha T}} \quad \textbf{Ans.}$$

Example 10.11.4. *Determine z-transform of the function* $F(s) = \frac{1}{s^2 + 2s + 2}$ *and assume sampling time T = 1 sec.*

Solution.
$$F(s) = \frac{1}{s^2 + 2s + 2} = \frac{1}{[s + (1 - j1)][s + (1 + j1)]}$$
$$= \frac{K_1}{s + (1 - j1)} + \frac{K_2}{s + (1 + j1)}$$

The coefficients can be determined as
$$K_1 = \frac{1}{j2} \text{ and } K_2 = -\frac{1}{j2}$$

$$\therefore \quad F(s) = \frac{1}{j2}\left[\frac{1}{s + (1 - j1)} - \frac{1}{s + (1 + j1)}\right]$$

Taking z-transform
$$F(z) = \frac{1}{j2}\left[\frac{z}{z - e^{-(1-j1)T}} - \frac{1}{z - e^{-(1+j1)T}}\right]$$

put $T = 1$ sec (given)
$$= \frac{1}{j2} \cdot \frac{z\,[z - e^{-(1+j1)}] - z\,[z - e^{-(1-j1)}]}{[z - e^{-(1-j1)}]\,[z - e^{-(1+j1)}]} = \frac{1}{j2} \cdot \frac{z^2 - ze^{-(1+j1)} - z^2 + ze^{-(1-j1)}}{[z - e^{-(1+j1)}]\,[z - e^{-(1+j1)}]}$$
$$= \frac{1}{j2} \cdot \frac{z\,[e^{-(1-j1)} - e^{-(1+j1)}]}{[z - e^{-(1-j1)}]\,[z - e^{-(1+j1)}]} = \frac{1}{j2} \cdot \frac{z\,[e^{-(1-j1)} - e^{-(1+j1)}]}{z^2 - z\,[e^{-(1-j1)} + e^{-(1+j1)}] + e^{-2}}$$
$$= \frac{1}{j2} \cdot \frac{z(e^{-1}e^{j1} - e^{-1}\,e^{-j1})}{z^2 - z\,[e^{-(1-j1)} + e^{-(1+j1)}] + e^{-2}}$$
$$= ze^{-1} . \frac{(e^{j1} - e^{-j1})}{j2} \times \frac{1}{z^2 - z\,[e^{-1}(\cos 1 + j\sin 1) + e^{-1}\,(\cos 1 - j\sin 1) + e^{-2}]}$$
$$= ze^{-1} . \frac{(e^{j1} - e^{-j1})}{j2} \cdot \frac{1}{z^2 - 2ze^{-1}\cos 1 + e^{-2}} = \frac{ze^{-1}\sin 1}{z^2 - 2ze^{-1}\cos 1 + e^{-2}} \quad \textbf{Ans.}$$

Example 10.11.5. *Find z-transform of the function whose Laplace transform is*
$$F(s) = \frac{3}{s^2 + 9} + \frac{-2\,e^{0.2s}}{s\,(s + 2)}$$
given that sampling time T = 0.05 sec.

Solution.
$$F(s) = \frac{3}{s^2 + 9} + \frac{2\,e^{-0.2s}}{s\,(s + 2)} = \frac{3}{s^2 + (3)^2} + 2e^{-0.2s}\left(\frac{K_1}{s} + \frac{K_2}{s + 2}\right)$$

The coefficients K_1 and K_2 can be determined as
$$K_1 = \frac{1}{2} \quad \text{and} \quad K_2 = -\frac{1}{2}$$

$$\therefore \quad F(s) = \frac{3}{s^2 + (3)^2} + 2e^{-0.2s}\left[\frac{1}{2}\cdot\frac{1}{s} - \frac{1}{2}\cdot\frac{1}{(s+2)}\right]$$

The term $e^{-0.2s}$ can be written as

$e^{-0.2s} = e^{-sT(0.2/T)}$, where T is the sampling period.

put $z = e^{sT}$ $\therefore$ $Z[e^{-0.2s}] = z^{-0.2/T}$

From z-transform table

$$Z\left[\frac{3}{s^2 + (3)^2}\right] = \frac{z \sin 3T}{z^2 - 2z\cos 3T + 1}$$

$$Z\left[\frac{1}{s}\right] = \frac{z}{z-1} \quad \text{and} \quad Z\left[\frac{1}{s+2}\right] = \frac{z}{z - e^{-2T}}$$

$$\therefore \quad F(z) = Z\left[\frac{3}{s^2 + (3)^2} + 2e^{-0.2s}\left(\frac{1}{2}\cdot\frac{1}{s} - \frac{1}{2}\cdot\frac{1}{s+2}\right)\right]$$

$$= \frac{z \sin 3T}{z^2 - 2z\cos 3T + 1} + 2z^{-0.2/T}\left[\frac{1}{2}\cdot\frac{z}{(z-1)} - \frac{1}{2}\cdot\frac{z}{(z - e^{-2T})}\right]$$

since $T = 0.05$ sec

$$\therefore \quad F(z) = \frac{z \sin(3 \times 0.05)}{z^2 - 2z\cos(3 \times 0.05) + 1} + 2z^{-0.2/0.5}\left[\frac{1}{2}\cdot\frac{z}{(z-1)} - \frac{1}{2}\cdot\frac{z}{(z - e^{-2\times 0.05})}\right]$$

$$= \frac{z \sin 0.15}{z^2 - 2z\cos 0.15 + 1} + 2z^{-4}\left[\frac{1}{2}\cdot\frac{z}{(z-1)} - \frac{1}{2}\cdot\frac{z}{(z - e^{-0.1})}\right]$$

$$= \frac{0.149z}{z^2 - 2\times 0.989z + 1} + z^{-4}\cdot z\left[\frac{1}{(z-1)} - \frac{z}{(z - 0.905)}\right]$$

$$= \frac{0.149z}{z^2 - 1.978z + 1} + \frac{z^{-3}(z - 0.905 - z + 1)}{(z-1)(z-0.905)}$$

$$= \frac{0.149z}{z^2 - 1.978z + 1} + \frac{0.095z^{-3}}{z^2 - 1.905z + 0.905}$$

$$\text{Simplifying } F(z) = \frac{0.149z^6 - 0.283z^5 + 0.135z^3 + 0.095z^2 - 0.188z + 0.095}{z^3(z^4 - 3.89z^3 + 5.67z^2 - 3.69z + 0.905)} \quad \textbf{Ans.}$$

Example 10.11.6. *Find inverse z-transform of the function* $F(z) = \dfrac{0.5z}{(z-1)^2}$.

Solution. The z-transform of the unit ramp function

$f(t) = t$ is given by $zT/(z-1)^2$

where T is the sampling time. Therefore, if

$$F(z) = \frac{0.5z}{(z-1)^2}$$

the sampling time is $T = 0.5$ sec.

Hence, $$f(t) = Z^{-1}[F(z)] = Z^{-1}\left[\frac{0.5z}{(z-1)^2}\right] = 0.5t$$

The function $f(t) = 0.5t$ is sampled and $T = 0.5$ sec. Therefore in terms of sample number sequence

$$f(kT) = kT = 0.5k$$

The sampled unit ramp function mentioned above with, $T = 0.5$ sec is shown in Fig. 10.11.3.

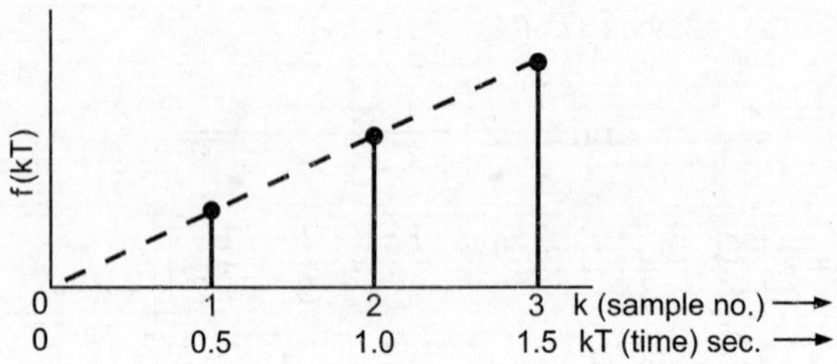

Fig. 10.11.3. Function $f(t) = t$ sampled with $T = 0.5$ sec.

Example 10.11.7. *Find the inverse z-transform and number sequence for the function*

$$F(z) = \frac{0.632z}{z^2 - 1.368z + 0.368}$$

given that sampling time (a) T = 1 sec (b) T = 0.2 sec.

Solution. (*a*) $$F(z) = \frac{0.632z}{z^2 - 1.368z + 0.368}$$

The denominator can be factorized as $(z-1)(z-0.368)$

$$\therefore \quad F(z) = \frac{0.632z}{(z-1)(z-0.368)}$$

or $$\frac{F(z)}{z} = \frac{0.632}{(z-1)(z-0.368)}$$

Let $$\frac{0.632}{(z-1)(z-0.368)} = \frac{K_1}{z-1} + \frac{K_2}{z-0.368}$$

The coefficients can be determined as

$$K_1 = 1, \quad K_2 = -1$$

$$\therefore \quad \frac{F(z)}{z} = \frac{1}{z-1} - \frac{1}{z-0.368}$$

or $$F(z) = \frac{z}{z-1} - \frac{z}{z-0.368}$$

It is noted that $e^{-1} = 0.368$

Sampling time $T = 1$ sec (given)

$\therefore \qquad F(z) = \frac{z}{z-1} - \frac{z}{z-e^{-T}}$

$\therefore \qquad f(t) = Z^{-1}[F(z)] = (1 - e^{-t})$

$\therefore \qquad f(kT) = (1 - e^{-kT})$

or $\qquad f(k) = (1 - e^{-k})$ **Ans.**

In terms of number sequence form $Z^{-1} F(z)$ can be written as

$$f(k) = (1)^k - (0.368)^k \; ; T = 1 \text{ sec.} \quad \textbf{Ans.}$$

The function $f(k)$ is plotted as shown in Fig. 10.11.4.

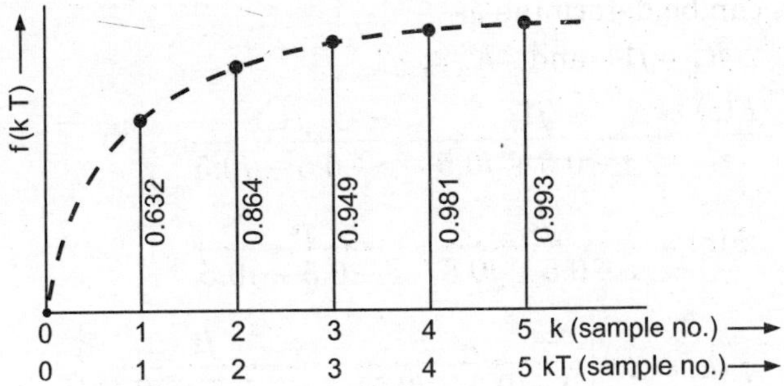

Fig. 10.11.4. $f(k) = (1)^k - (0.368)^k$; $T = 1$ sec.

(b) $\qquad F(z) = \frac{z}{z-1} - \frac{z}{z-0.368}$

It is noted that $e^{-1} = 0.368$ and sampling time $T = 0.2$ sec. (given)

$\therefore \qquad e^{-5T} = 0.368$

Hence $\qquad F(z) = \frac{z}{z-1} - \frac{z}{z-e^{-5T}}$

$\therefore \qquad f(t) = Z^{-1}[F(z)] = (1 - e^{-5T})$

$\therefore \qquad f(kT) = (1 - e^{-5kT})$

In number sequence form

$f(k) = (1)^k - (0.368)^k \; ; T = 0.2$ sec.

$f(k) = (1)^k - (0.368)^k$ **Ans.**

with sampling time $T = 0.2$ sec is plotted in Fig. 10.11.5.

In case (a) $T = 1$ sec and that in (b) $T = 0.2$ sec. Therefore the time corresponding to sampling instants $k = 0, 1, 2, 3, \ldots$ is

for case (a) $t = 0, 1, 2, 3, \ldots$ sec.

for case (b) $t = 0, 0.2, 0.4, 0.6, \ldots$ sec.

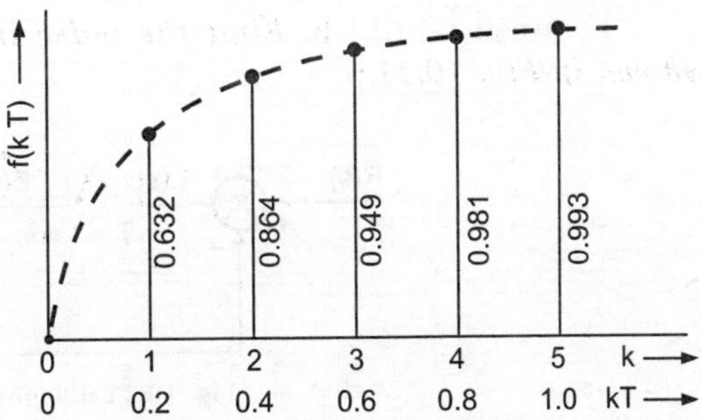

Fig. 10.11.5. $f(k) = (1)^k - (0.368)^k$; $T = 0.2$ sec.

Example 10.11.8. *Find inverse z-transform in number sequence form for the function*

$$F(z) = \frac{z}{z^2 - z + 0.5}.$$

Solution. $$F(z) = \frac{z}{z^2 - z + 0.5}$$

The denominator $(z^2 - z + 0.5)$ can be factorized as

$$(z^2 - z + 0.5) = (z - 0.5 + j0.5)\,(z - 0.5 - j0.5)$$

$$\therefore \quad \frac{F(z)}{z} = \frac{1}{(z - 0.5 + j0.5)\,(z - 0.5 - j0.5)}$$

Let $$\frac{1}{(z - 0.5 + j0.5)\,(z - 0.5 - j0.5)} = \frac{K_1}{(z - 0.5 + j0.5)} + \frac{K_2}{(z - 0.5 - j0.5)}$$

The coefficients can be determine as

$$K_1 = j1 \quad \text{and} \quad K_2 = -j1$$

$$\therefore \quad \frac{F(z)}{z} = \frac{j1}{z - 0.5 + j0.5} - \frac{j1}{z - 0.5 - j0.5}$$

or $$F(z) = \frac{jz}{z - 0.5 + j0.5} - \frac{jz}{z - 0.5 - j0.5}$$

$$\therefore \quad f(k) = Z^{-1}\left[\frac{jz}{j - (0.5 - j0.5)} - \frac{jz}{j - (0.5 + j0.5)}\right]$$

$$= j(0.5 - j0.5)^k - (0.5 + j0.5)^k]$$

$$= j(0.5\sqrt{2}\, e^{-j(\pi/4)})^k - (0.5\sqrt{2}\, e^{j(\pi/4)})^k] = j(0.5\sqrt{2}\,)^k\, [e^{-j(\pi/4)k} - e^{j(\pi/4)k}]$$

$$= j(0.5\sqrt{2}\,)^k \left[\left(\cos\frac{\pi}{4}k - j\sin\frac{\pi}{4}k\right) - \left(\cos\frac{\pi}{4}k + j\sin\frac{\pi}{4}k\right)\right]$$

$$= j(0.5\sqrt{2}\,)^k \left(-j2\sin\frac{\pi}{4}k\right)$$

or $$f(k) = 2(0.5\sqrt{2}\,)^k \sin\frac{\pi}{4}\,k \quad \textbf{Ans.}$$

Example 10.11.9. *Find the pulse transfer function for the error sampled system shown in Fig. 10.11.6.*

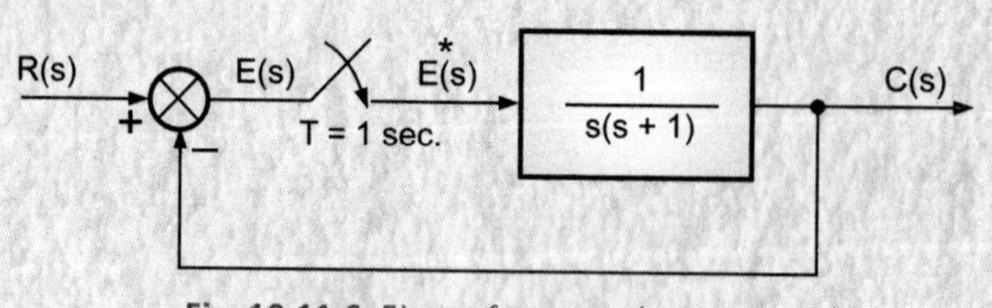

Fig. 10.11.6. Figure for example 10.11.9.

Solution. (1) The forward path transfer function is

$$G(s) = \frac{1}{s(s+1)}$$

Let $$\frac{1}{s(s+1)} = \frac{K_1}{s} + \frac{K_2}{s+1}$$

The coefficients can be determined as

$$K_1 = 1 \quad K_2 = -1$$

$$\therefore \quad G(s) = \frac{1}{s} - \frac{1}{s+1}$$

Taking z-transform, $G(z) = \frac{z}{z-1} - \frac{z}{z-e^{-T}}$

Since, $T = 1$ sec.

$$\therefore \quad G(z) = \frac{z}{z-1} - \frac{z}{z-e^{-1}} = \frac{z}{z-1} - \frac{z}{z-0.368} = \frac{0.632z}{(z-1)(z-0.368)}$$

(2) Since, $H(s) = 1$

$$\therefore \quad GH(z) = \frac{1}{s(s+1)} \times 1 \quad \text{or} \quad GH(s) = \frac{1}{s} - \frac{1}{s+1}$$

Taking z-transform

$$GH(z) = \frac{z}{z-1} - \frac{z}{z-e^{-T}}$$

Since, $T = 1$ sec

$$\therefore \quad GH(s) = \frac{z}{z-1} - \frac{z}{z-e^{-1}} = \frac{z}{z-1} - \frac{z}{z-0.368}$$

or

$$GH(z) = \frac{0.632z}{(z-1)(z-0.368)}$$

(3) The pulse transfer function for the error sampling is given by

$$\frac{C(z)}{R(z)} = \frac{G(z)}{1+GH(z)} = \frac{\frac{0.632z}{(z-1)(z-0.368)}}{1+\frac{0.632z}{(z-1)(z-0.368)}}$$

$$= \frac{0.632z}{(z-1)(z-0.368)+0.632z} = \frac{0.632z}{z^2-0.736z+0.368} \quad \textbf{Ans.}$$

Example 10.11.10. *Obtain the pulse transfer function for the system shown in Fig. 10.11.7.*

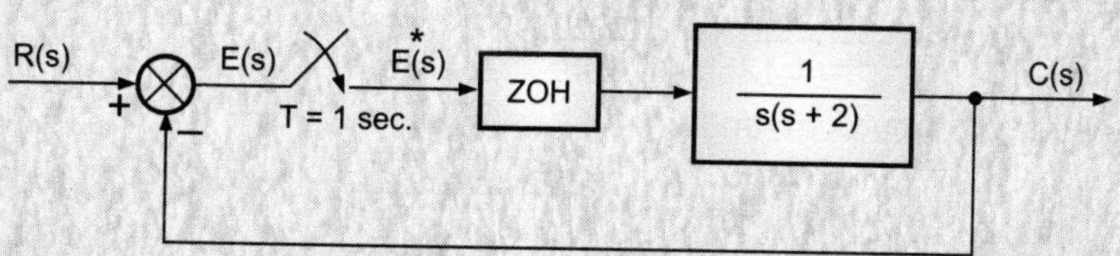

Fig. 10.11.7. Figure for example 10.11.10.

Solution. The transfer function of ZOH is given by

$$G_{h0}(s) = \frac{(1-e^{-sT})}{s}$$

The forward path transfer function is

$$G(s) = G_{h0}(s) \cdot \frac{1}{s(s+2)} = \frac{1-e^{-sT}}{s} \cdot \frac{1}{s(s+2)} = (1-e^{-sT}) \cdot \frac{1}{s^2(s+2)}$$

Let $\quad \dfrac{1}{s^2(s+2)} = \dfrac{K_1}{s^2} + \dfrac{K_2}{s} + \dfrac{K_3}{s+2}$

The coefficients can be determined as :

$$K_1 = \frac{1}{2}\ ; \quad K_2 = -\frac{1}{4} \quad \text{and} \quad K_3 = \frac{1}{4}$$

$$\therefore \qquad G(s) = (1-e^{-sT}) \left[\frac{1}{2} \cdot \frac{1}{s^2} - \frac{1}{4} \cdot \frac{1}{s} + \frac{1}{4} \cdot \frac{1}{(s+2)}\right]$$

Taking z-transform

$$G(z) = (1-z^{-1}) \left[\frac{1}{2} \cdot \frac{zT}{(z-1)^2} - \frac{1}{4} \cdot \frac{z}{(z-1)} + \frac{1}{4} \cdot \frac{z}{(z-e^{-2T})}\right]$$

put $\quad T = 1$ (given)

$$\therefore \qquad G(s) = (1-z^{-1}) \left[\frac{1}{2} \cdot \frac{z \times 1}{(z-1)^2} - \frac{1}{4} \cdot \frac{z}{(z-1)} + \frac{1}{4} \cdot \frac{z}{(z-e^{-2\times 1})}\right]$$

$$= \frac{z-1}{z} \left[\frac{1}{2} \cdot \frac{z}{(z-1)^2} - \frac{1}{4} \cdot \frac{z}{(z-1)} + \frac{1}{4} \cdot \frac{z}{(z-0.135)}\right]$$

$$= (z-1) \left[\frac{2(z-0.135) - (z-1)(z-0.135) + (z-1)^2}{4(z-1)^2 (z-0.135)}\right]$$

$$= \left[\frac{2(z-0.135) - (z-1)(z-0.135) + (z-1)^2}{4(z-1)(z-0.135)}\right] = \frac{0.283z + 0.148}{z^2 - 1.135z + 0.135}$$

Since $H(s) = 1$

and $$G(s) = \frac{1-e^{-sT}}{s} \cdot \frac{1}{s(s+2)}$$

$$\therefore \qquad GH(s) = \frac{(1-e^{-sT})}{s} \cdot \frac{1}{s(s+2)} \times 1 = (1-e^{-sT}) \cdot \frac{1}{s^2(s+2)}$$

Taking z-transform (as determined above)

$$GH(z) = \frac{0.283z + 0.148}{z^2 - 1.135z + 0.135}$$

Pulse transfer function is given by

$$\frac{C(z)}{R(z)} = \frac{G(z)}{1 + GH(z)} = \frac{\dfrac{0.283z + 0.148}{z^2 - 1.135z + 0.135}}{1 + \dfrac{0.283z + 0.148}{z^2 - 1.135z + 0.135}}$$

or $$\frac{C(z)}{R(z)} = \frac{0.282z + 0.148}{z^2 - 0.852z + 0.283} \quad \textbf{Ans.}$$

Example 10.11.11. *Determine the unit step time response for the pulse transfer function*

$$\frac{C(z)}{R(z)} = \frac{z}{z^2 - z + 0.5}.$$

Solution. $$\frac{C(z)}{R(z)} = \frac{z}{z^2 - z + 0.5}$$

The term $(z^2 - z + 0.5)$ can be factorized as follows :

Let $(z^2 - z + 0.5) = 0$

the two roots of the above equation are

$$z_1 = (0.5 - j0.5) \text{ and } z_2 = (0.5 + j0.5)$$

$$\therefore \quad (z^2 - z + 0.5) = (z - 0.5 + j0.5)(z - 0.5 - j0.5)$$

$$\therefore \quad \frac{C(z)}{R(z)} = \frac{z}{(z - 0.5 + j0.5)(z - 0.5 - j0.5)}$$

Since the input is unit step function *i.e.* $r(t) = 1$, therefore,

$$R(z) = \frac{z}{z-1}$$

$$\therefore \quad C(z) = \frac{z}{z-1} \cdot \frac{z}{(z - 0.5 + j\,0.5)(z - 0.5 - j0.5)}$$

or $$\frac{C(z)}{z} = \frac{z}{(z-1)(z - 0.5 + j\,0.5)(z - 0.5 - j0.5)}$$

$$= \frac{K_1}{(z-1)} + \frac{K_2}{(z - 0.5 + j0.5)} + \frac{K_3}{(z - 0.5 - j0.5)}$$

The coefficients can be determined as

$$K_1 = 2, \quad K_2 = -1 \quad \text{and} \quad K_3 = -1$$

$$\therefore \quad \frac{C(z)}{z} = \frac{2}{z-1} - \frac{1}{z - 0.5 + j0.5} - \frac{1}{z - 0.5 - j0.5}$$

or $$C(z) = \frac{2z}{z-1} - \frac{z}{z - 0.5 + j0.5} - \frac{z}{z - 0.5 - j0.5}$$

$$\therefore \quad c(k) = Z^{-1}\left[\frac{2z}{z-1} - \frac{z}{z - 0.5 + j0.5} - \frac{z}{z - 0.5 - j0.5}\right]$$

$$= 2(1)^k - (0.5 - j0.5)^k - (0.5 + j0.5)^k$$

$$= 2(1)^k - (0.5\sqrt{2}\,e^{-j(\pi/4)})^k - (0.5\sqrt{2}\,e^{j(\pi/4)})^k$$

$$= 2(1)^k - (0.5\sqrt{2}\,)^k\left(\cos\frac{\pi}{4}k - j\sin\frac{\pi}{4}k\right) - (0.5\sqrt{2})^k\left(\cos\frac{\pi}{4}k - j\sin\frac{\pi}{4}k\right)$$

or $$c(k) = 2(1)^k - 2(0.5\sqrt{2}\,)^k \cos\frac{\pi}{4}k \quad \textbf{Ans.}$$

PROBLEMS

10.1. Find z-transform

(a) $f(t) = 2u(t) + t$, (b) $\cos \omega t$

(c) $e^{-\alpha t} \sin \omega t$ (d) $F(s) = \dfrac{s^2 + 2s + 2}{s^3 + 3s^2 + 2s}$.

10.2. Find inverse z-transform

(a) $F(z) = \dfrac{z}{(z+1)(z+2)}$ (b) $F(z) = \dfrac{z}{(z-0.5)(z-0.3)}$.

10.3. Find inverse z-transform, $F(z) = \dfrac{z^2}{(z-1)(z^2 - 1.6z + 1.64)}$.

10.4. (a) Determine pulse transfer function for the sampled-data control system shown in Fig. 10-P-1.

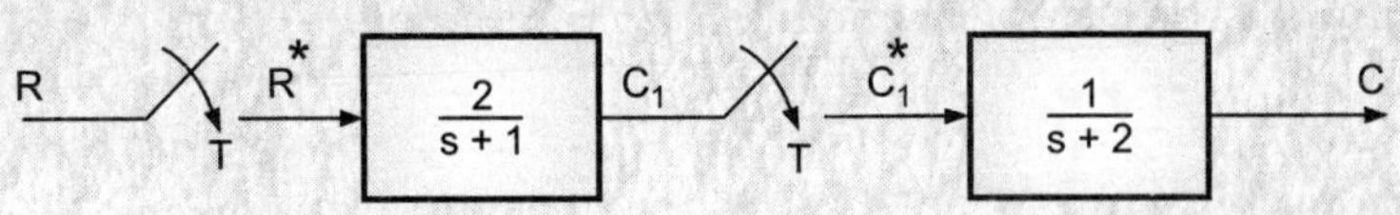

Fig. 10.P-1. Sampled-data control system for problem 10.4 (a).

(b) Determine output z-transform for the sampled-data control system shown in Fig. 10-P-2. Given that the input is unit step function.

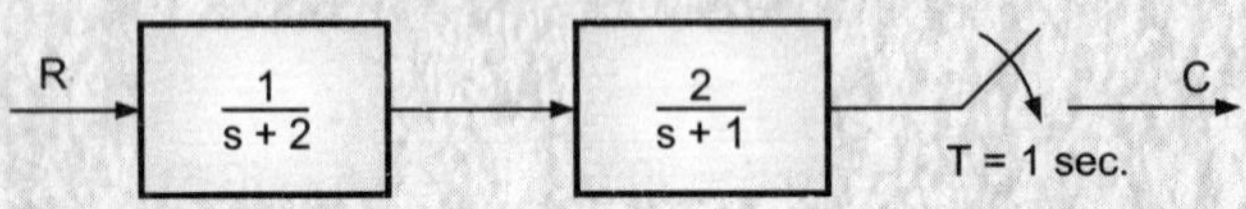

Fig. 10.P-2. Sampled-data control system for problem 10.4 (b).

10.5. Determine pulse transfer function (z-transfer function) for the sampled-data control system shown in Fig. 10-P-3.

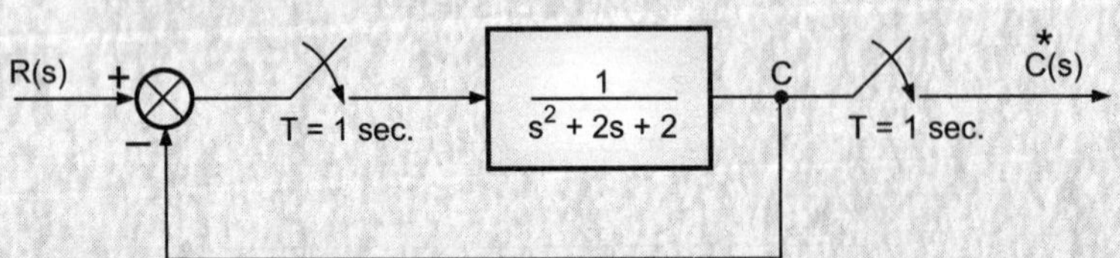

Fig. 10.P-3. Sampled-data control system for problem 10.5.

10.6. Obtain z-transfer function for the block diagram shown in Fig. 10-P-4.

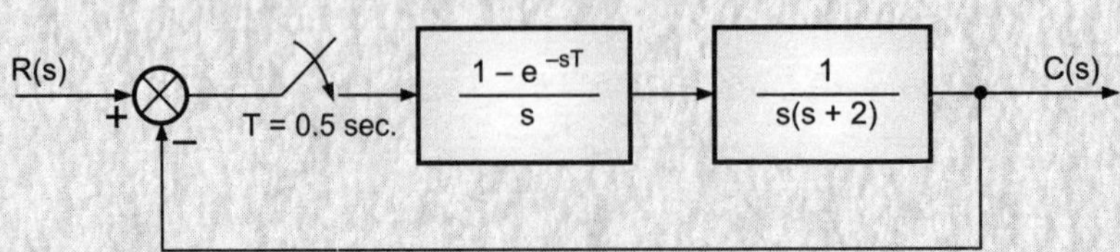

Fig. 10.P-4. Block diagram for problem 10.6.

10.7. Obtain z-transfer function for the error sampled system shown in Fig. 10-P-5 and deter-mine time response if the input is unit step. Given that $T = 1$ sec and $G(s) = 1/[(s + 1)(s + 2)]$.

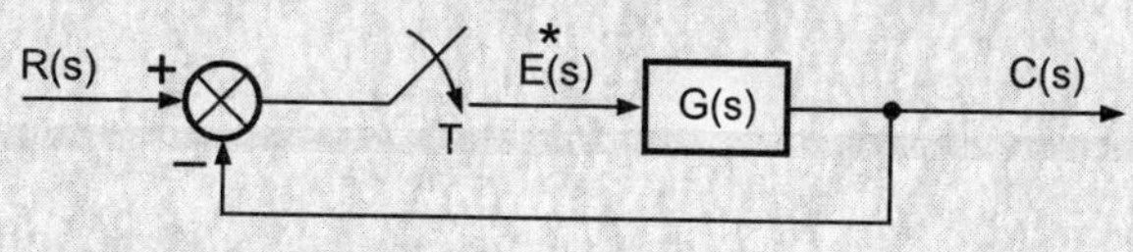

Fig. 10.P-5. Error sampled system for problem 10.7.

10.8. Determine the unit step time response for the sampled-data control system shown in Fig. 10-P-6. Given that $G(s) = 1/[s(s + 1)]$; $H(s) = (s + 4)$ and $T = 1$ sec.

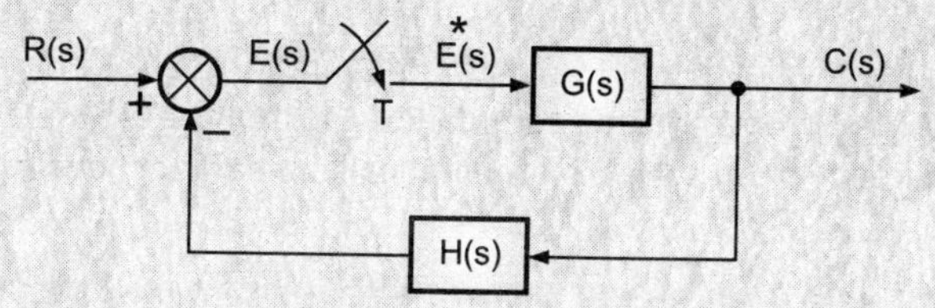

Fig. 10.P-6. Sampled-data control system for problem 10.8.

10.9. Determine the stability of sampled-data control system shown in Fig. 10-P-7. Use bilinear transformation.

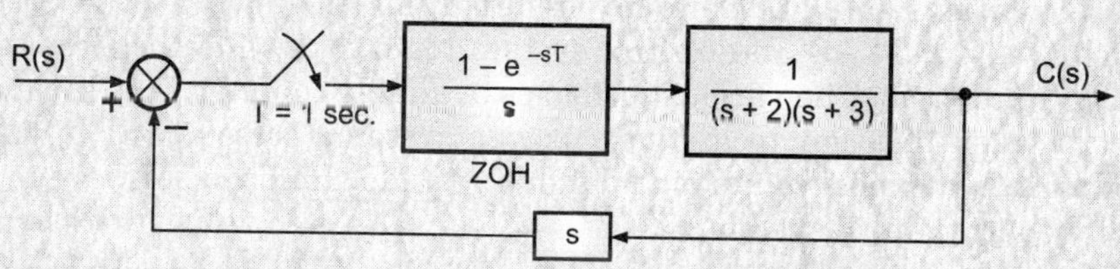

Fig. 10.P-7. Sampled-data control system for problem 10.9.

10.10. A sampled-data control system is shown in Fig. 10-P-8. Determine the value of K such that the system is stable. Sampling time $T = 0.5$sec.

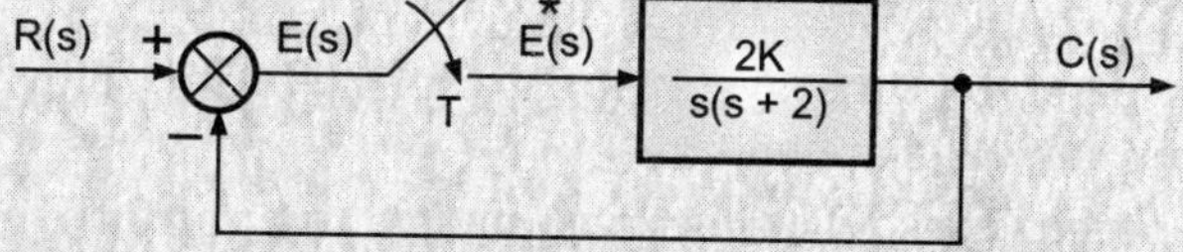

Fig. 10.P-8. Sampled-data control system for problem 10.10.

CHAPTER 11

Solution of Problems Using Computer

This chapter deals with the application of MATLAB* software for the solution of examples. A brief note explains about the MATLAB procedure and commands. Most of the example solutions obtained using MATLAB approach have already been solved mathematically in earlier chapters and thus the results are varified using a computer.

Introduction. MATLAB is a high level technical computing environment. It is suitable for solving scientific and engineering problems. MATLAB and control system toolboxes are commercial software products. MATLAB is an interactive program for scientific and engineering calculations. MATLAB family of programs includes a base program and a variety of toolboxes. MATLAB and control system toolbox provide the user to design and analyse the control system. The various toolboxes available commercially are *e.g.*, fuzzy and neural network, Identification techniques, power system toolbox, DSP toolbox etc. The MATLAB software is the most common and industry wide standard software in the analysis and design of modern control system.

MATLAB has a rich collection of functions which are useful to control engineers. The control system toolbox is built on the MATLAB platform. Control system platform has functions designed for control engineering problem solutions using computer. The toolbox is generally written as M files which implements control system design, analysis and modelling techniques. Graphical User Interface (GUI) simplify the control system problems. A control system can be modelled as a transfer function, pole-zero-gain or state-space form. Continuous time and discrete time systems can be modelled. Time response, frequency response and root locus can be graphed.

The text given herein uses LTI viewer which is a Graphical User Interface (GUI) and this further simplifies the analysis of Linear time invariant systems. LTI viewer is used for displaying and manipulating the response plots of linear models. The LTI viewer is used to find

1. Step and impulse response plots.
2. Bode and Nyquist plots.
3. Pole-zero plots.
4. Nichol's plots.

---

* MATLAB is a trade mark of Mathwork, Inc.

At the MATLAB icon user is required to double click the left button mouse. At the command window. MATLAB > >, the transfer function can be created and it is used extensively in the solution of problems given in the following pages. Once the transfer function is created then at MATLAB prompt > > type load `lti` examples and press enter, this loads several LTI models. When the `lti` window opens, then appropriate name is to be high lighted and OK button is clicked. The next picture shows the `lti` viewer and by default the step response is seen in the `lti` window. Alternatively, `lti` viewer is opened and the appropriate transfer function assigned by a variable name is imported from the MATLAB prompt.

```
>> lti viewer ('step', sys)
```

Alternatively in the MATLAB command window type `sisotool (sys)`, then next window is opened by default with root locus editor and Bode plot editor. The root lacus plot and Bode plot are simultaneously displayed on the computer screen. The plots can be modified as per the constraints given in the problem and by left clicking the mouse on the given plot either gain can be changed by dragging the plot up and down or the poles/zeros simple or complex can be added/deleted.

MATLAB functions used for the solution of examples.

Function name	*Function description*
bode	Generates bode frequency plot
cloop	Computes the closed-loop system with unity feedback
conv	Multiplies two polynomials (convolution)
czdm	Converts a continuous time state variable system representation to a discrete-time system representation
ctrb	Computes the controllability matrix
d2cm	Converts discrete-time system representation to a continuous-time state variable system representation
dbode	Generates a bode frequency response of digital system
dstep	Generates a unit step response of a discrete time system
dnyquist	Generates a nyquist plot of discrete time system
eig	Computes the eigen values and eigen vectors
exp	Computes the exponential with base e
expm	Computes the matrix exponential with base e
feedback	Calculates the feedback connection of two systems
for	Generates a loop
grid	Adds grid to current graph
help	Prints a list of help topics
impulse	Computes the unit impulse response of a system

load	Loads variables saved in file
lsim	Computes the time response of a system of an arbitrary input and initial condition
ltiview	Accesses the ltiviewer in the MATLAB command window
margin	Calculates gain margin, phase margin, gain crossover frequency and phase crossover frequency
nyquist	Computes nyquist frequency response
nichols	Computes the Nichols plot
parallel	Adds the transfer function
plot	Generates a linear plot
poly	Computes a polynomial from roots
polyval	Evaluates a polynomial
printsys	Prints the transfer function
pzmap	Plots a pole zero plot of a linear system
rank	Calculates the rank of a matrix
residue	Computes a partial fraction expansion
roots	Determine the roots of a polynomial
rootsl	Same as roots function but gives more accurate answers when there are repeated roots
rlocus	Computes the root locus
series	Finds transfer function when blocks are in cascade
ss2tf	Converts state variable form to transfer function form
step	Computes unit step response of a system
tf2ss	Converts transfer function form to state variable form
tf	Computes the transfer function in s variable

Command Summary for Symbolic Math Commands

ilaplace(X)	Find inverse laplace transform of $X(s)$
int(s, v, a, b)	Integrate s.w.r. to v from lower limit a to upper limit b *i.e.*, $\int_a^b sdv$
iztrans(F, k)	Find inverse Z transform. Find $f(KT)$ given $F(Z)$ *i.e.* find $f(nT)$
laplace(X)	Finds $L[x(t)]$
pretty(x)	Pretty prints x
sym(v)	Convert v to symbolic object
syms x y z	Declare x, y and z to be symbolic objects
ztrans(f)	Find Z transform of $f(nT)$

The comment % gives more readibility for the programme. The sentence followed by the comment % is non executable statement and no code is generated in the MATLAB.

The notation semicolon (;) after the MATLAB command suppresses output on the terminal as well as print out.

If the semicolon (;) is not used the output is displayed either on the terminal or print out.

It facilitates the beginner not to use semicolon (;) after the command because it gives the visual display of the data.

Running MATLAB

M files or script files store a sequence of commands. MATLAB is capable of executing the M file. A program can be written and saved in ASCII format in a file. The file has m extension in directory. Another type of M file is a function file. The function file has a name and it follows the work function.

Example 11.1. *The transfer function of a system is given below :*

$$G(s) = \frac{8(s+3)(s+4)}{s(s+2)(s^2+2s+5)}$$

Determine the poles and zeros and show the pole-zero configuration in s-plane.

Solution. The numerator and denominator polynomials is s for the given transfer function are as

$$p1 = 8s^2 + 56s + 96$$
$$q1 = s^4 + 4s^3 + 9s^2 + 10s$$

The numerator is assigned a variable name num and the coefficients of s are arranged in descending order, same procedure is followed for denominator polynomial. The transfer function given variable name q1 and further commands are followed sequentially as below :

```
> > p1 = [8 56 96] ; q1 = [1 4 9 10 0]
q1 =
      1      4      9      10      0
> >    sys4 = tf (p1, q1)
Transfer function :
        8 s^2 + 56 s + 96
  ---------------------------
  s^4 + 4 s^3 + 9 s^2 + 10 s
> > pzmap (sys4)
```

The pole-zero map is obtained as shown in Fig. 11.1. From the pole-zero map of the transfer function, poles and zeros are indicated as below :

poles : $s = 0$; $s = -2$; $s = -1 + j1$; $s = -1 - j1$

zeros : $s = -3$; $s = -4$.

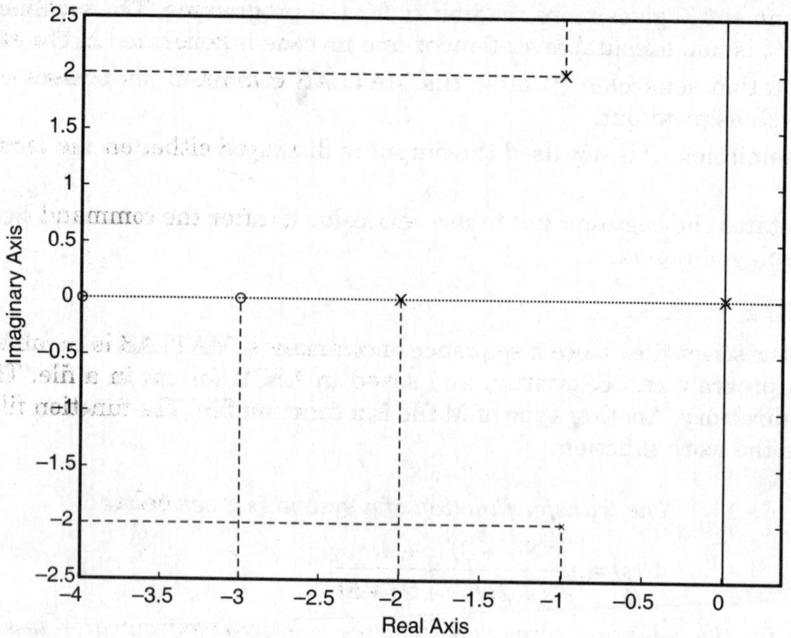

Fig. 11.1

Example 11.2. *A control system is shown in Fig. 11.2. Determine the transfer function.*

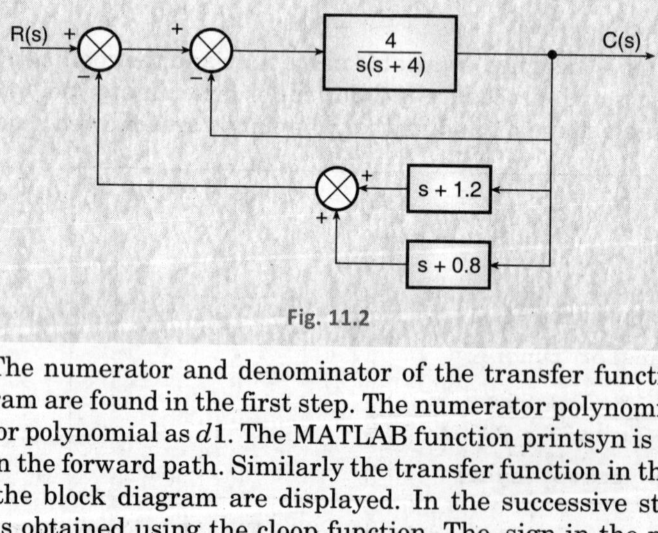

Fig. 11.2

Solution. The numerator and denominator of the transfer function in the forward path of block diagram are found in the first step. The numerator polynomial is considered as n1 and denominator polynomial as $d1$. The MATLAB function printsyn is used to display the transfer function in the forward path. Similarly the transfer function in the second and third feedback path of the block diagram are displayed. In the successive step the closed-loop transfer function is obtained using the cloop function. The–sign in the paranthesis for the cloop indicates the –ive feedback. Next the parallel function is used to obtain the resultant transfer function in the feedback path. Finally feedback function is used to obtain the transfer function as

$$\frac{C(s)}{R(s)} = \frac{4}{s^2 + 12s + 12}$$

Following commands are used to obtain, the transfer function

```
> > n1 = [4]
n1 =
        4
> > d1 = [1   4   0]
d1 =
        1      4      0
> > printsys (n1, d1)
num/den =
         4
     ---------
     s^2 + 4 s
> > n2 = [1   1.2]
n2 =
     1.0000     1.2000
> > d2 = [0   1]
> > d2 =
        0   1
> > printsys (n2, d2)
num/den =
       s + 1.2
       -------
          1
> > n3 = [1   0.8]
> > d3 = [0   1]
d3 =
        0   1
> > printsys (n3, d3)
num/den =
       s + 0.8
       -------
          1
> > [n4, d4] = cloop(n1, d1, -1)
n4 =
     0   0   4
d4 =
     1   4   4
> > printsys (n4, d4)
num/den =
             4
       -----------
       s^2 + 4s + 4
> > [n5, d5] = parallel (n2, d2, n3, d3)
n5 =
      0      2      2
d5 =
      0      0      1
> > printsys (n5, d5)
```

```
num/den =

        2s + 2
        ------
          1
> > [n6, d6] = feedback (n4, d4, n5, d5, - 1)
n6 =
         0         0         0         0         4
d6 =
         0         0         1        12        12
> > printsys (n6, d6)
num/den =
           4
    --------------
    s^2 + 12s + 12
```

Example 11.3. *The figure 11.3 represents a control system, obtain closed-loop transfer function.*

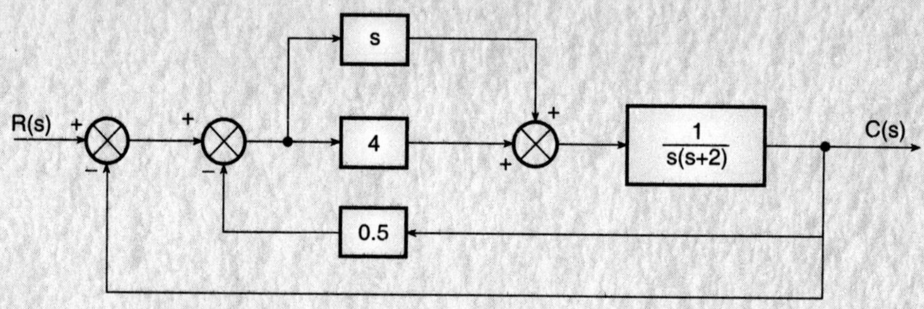

Fig. 11.3

Solution. The MATLAB script for obtaining the closed-loop transfer function is as follows :

```
> > n = [1]: d1 = [1 2 0]:
> > printsys (n1, d1)
num/den =
       1
    -------
    s^2 + 2s
> > n2 = [4]; d2 = [0 1];
> > printsys (n2, d2)
num/den =
    4
    -
    1
> > n3 = [1  0] ; d3 = [0  1];
> > [n4, d4] = parallel (n2, d2, n3, d3);
> > printsys (n4, d4)
num/den =
    s + 4
    -----
      1
```

```
> > [n5, d5] = series (n1, d1, n4, d4);
> > printsys (n5, d5)
num/den =

     s + 4
    -------
    s^2 + 2s

> > n6 = [0.5]; d6 = [0.1];
> > [n7, d7] = cloop (n5, d5, n6, d6 -1);
> > [n7, d7] = feedback (n5, d5, n6, d6, -1);
> > printsys (n7, d7)
num/den =

       s + 4
    ---------------
    s^2 + 2.5s + 2

> > [n8, d8] = cloop (n7, d7, -1);
> > printsys (n8, d8)
num/den =

       s + 4
    ---------------
    s^2 + 3.5s + 6
```

The transfer function for the blocks in the toward path are obtained. The function parallel is used to get the transfer function $\frac{s+4}{1}$. As two blocks are in cascade then function series is used to get the resultant transfer function $\frac{s+4}{s^2+2s}$. The inner loop involving the above two transfer functions with a feedback path gain of 0.5 is reduced to one block with function cloop. The final answer is obtained after reducing, the inner loop with forward path block having a unity feedback. Then answer is

$$\frac{C(s)}{R(s)} = \frac{s+4}{s^2+3.5s+6}.$$

Example 11.4. *Determine the damping ratio, damped frequency of oscillatory roots and % overshoot for a unit step response given that*

$$\frac{C(s)}{E(s)} = \frac{1}{s(1+0.5s)(1+0.2s)}$$

The system is unity feedback type.

The closed-loop transfer function for the given condition is determined as below :

$$\frac{C(s)}{E(s)} = \frac{\frac{1}{s(1+0.5s)(1+0.2s)}}{1+\frac{1}{s(1+0.5s)(1+0.2s)}}$$

$$\text{or} \qquad = \frac{10}{(s+5.5)(s^2+1.5s+1.8)}$$

$$\text{or} \qquad \frac{C(s)}{R(s)} = \frac{10}{s^3+7s^2+10s+10}.$$

Solution. The commands are given below :

```
> > num = 10
num =
    10
> >
> > den = [1      7      10      10]
den =
   1      7      10      10
> > sysl = tf (num, den)
Transfer function :
        10
-----------------------
s^3 + 7 s^2 + 10s + 10
> > load ltiexamples
> > ltiview
```

The time response is shown in Fig. 11.4 and following results are noted.

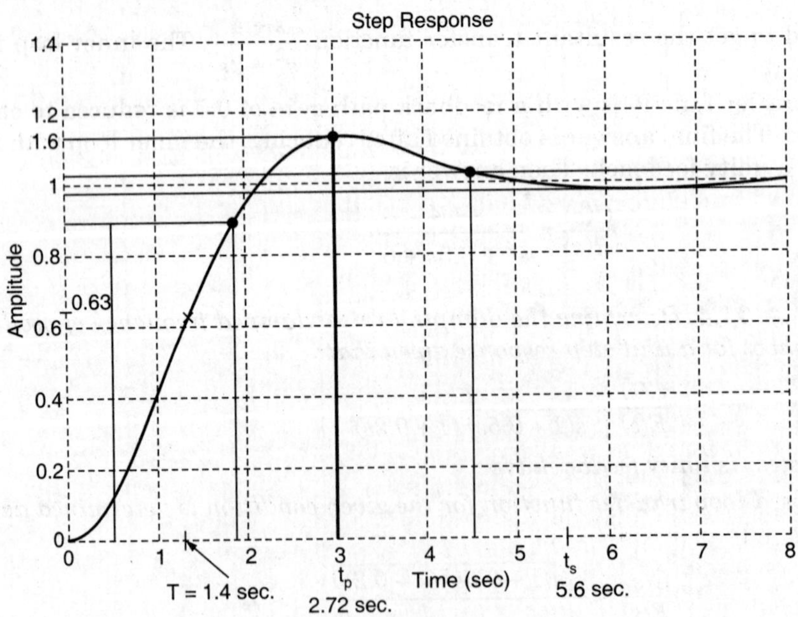

Fig. 11.4

$\% M_p = 16\%, \xi = 0.5$ (as per calculation)

$$\omega_d = \frac{1/2}{2.72} \times 2\pi = 1.15 \text{ rad/sec}$$

$$\omega_n = \frac{\omega_d}{\sqrt{1-\xi^2}} = \frac{1.15}{\sqrt{1-0.5^2}} = 1.32 \text{ rad/sec}$$

$$t_s = 5.6 \text{ sec.}$$

$$T = \frac{1}{\xi\omega_n} = 1.4 \text{ sec.}$$

Example 11.5. *The open-loop transfer function of a unity feedback control system is given by*

$$G(s) = \frac{K(s+2)}{s^3 + \beta s^2 + 4s + 1}$$

Determine the value of K and β such that the closed-loop response to unit step function has $\omega_d = 4$ rad / sec.

Solution. The commands are as below :

```
> > num = 6.81* [1 2]

num =

      6.8100          13.6200

> > den = [1     2.71     4     1]

den =

      1.0000     2.7100    4.0000    1.0000

> > sys2 = tf (num, den)

Transfer function :

      6.81s + 13.62
--------------------------
s^3 + 2.71 s^2 + 4s + 1

> > load ltiexamples

> > sisotool
```

The value of K and β are arbitrarily chosen and the step response for the transfer function is displayed. Then by trial and error method new values are selected and same procedure is repeated till the desired step response having $\omega_d = 4$ rad/sec. is obtained.

The step response obtained as per above procedure is shown in Fig. 11.5.

The time for half cycle is noted as 0.78 sec. Hence, the frequency of oscillation is determined as

$$\omega_d = 2\pi \times \frac{(1/2)}{0.78} = 4 \text{ rad/sec.}$$

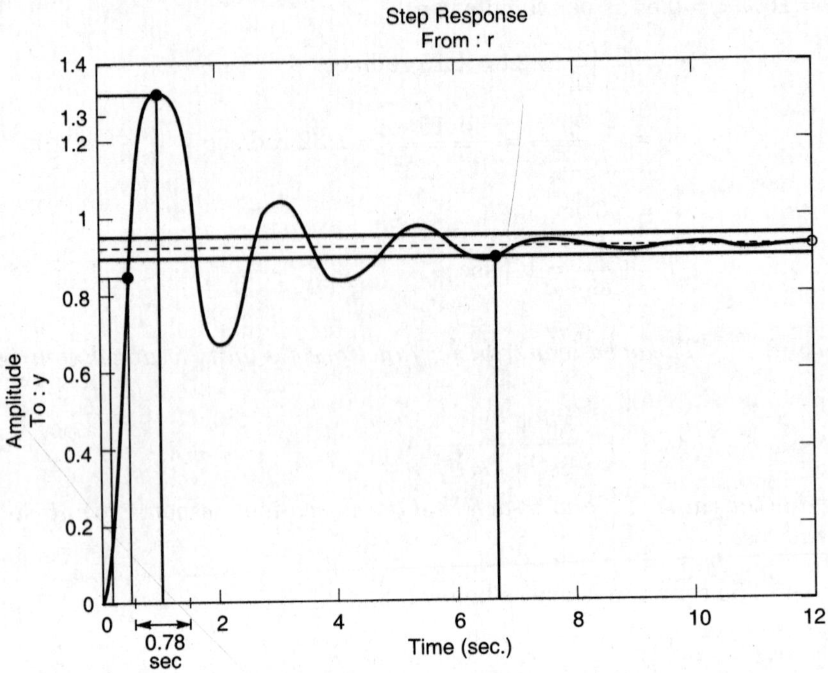

Fig. 11.5. Time response for example 11.5.

Example 11.6. *A unity feedback control system has forward path transfer function given by*

$$G(s) = \frac{s + 2}{s(s + 1)}$$

Determine (a) the time response for unit step response. (b) the rise time (c) the maximum overshoot.

Solution. Initially, in the given transfer function, the coefficients of numerator polynomial and denominator polynomial in s are arranged and variable names n3, d3 are assigned. The MATLAB scripts are used to get the transfer function $G(s)$. The MATLAB function cloop computes the closed-loop transfer function. The printsys function prints the transfer function representation of linear systems in a readable form. Subsequently the use of step function gives the transient response for a given $G(s)$

Following commands are given

```
> > [n3] = [1 2]; [d3] = [1 1 0];
> > printsys (n3, d3)
num/den =
   s + 2
  -------
  s^2 + s
> > [n4, d4] = cloop (n3, d3, - 1)
n4 =
     0     1     2
```

```
d4 =
    1     2     2
> > printsys (n4, d4)
num/den =
          s + 2
      -------------
      s^2 + 2 s + 2
> > step (n4, d4); grid
```

The rise time is 1.12 sec. and overshoot (%) is 6.7 as shown in Fig. 11.6.

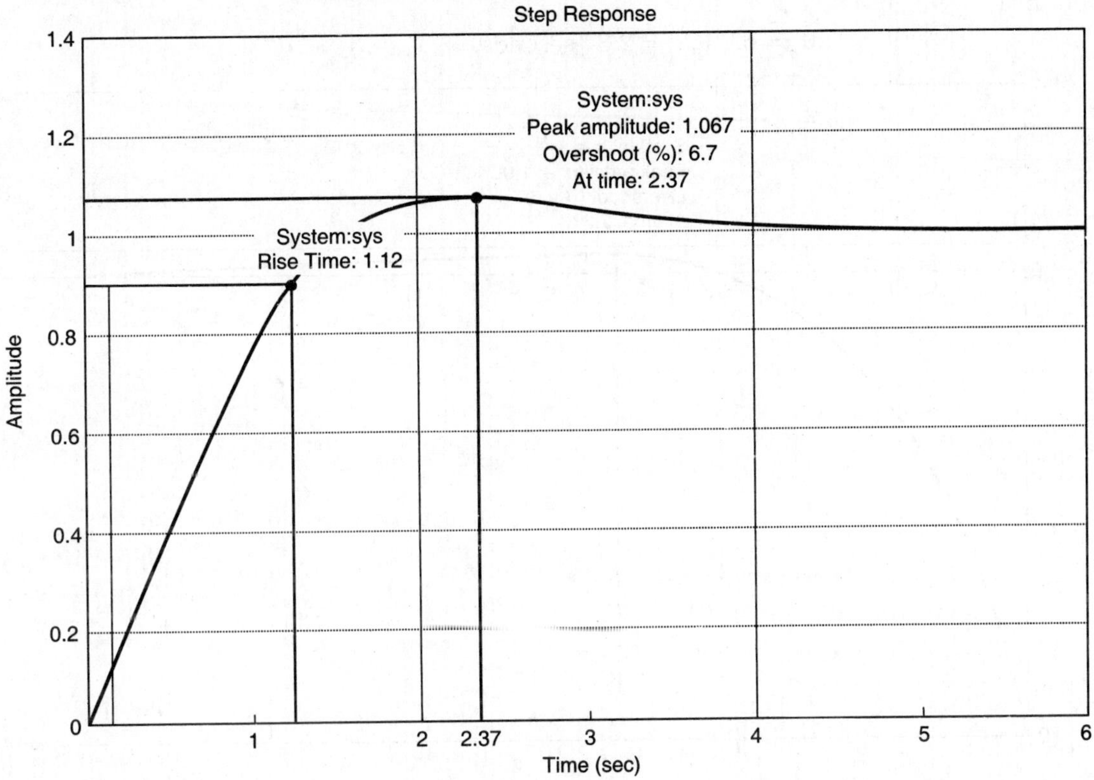

Fig. 11.6

Example 11.7. *Determine the time response for the following transfer function*

$$\frac{C(s)}{R(s)} = \frac{s+2}{s^2+2s+2}.$$

Solution. Following commands are given

```
> > num = [1     2]
num =
    1     2
> > den = [1     2     2]
```

```
den =
      1      2      2
> > sys5 = tf (num, den)
Transfer function :
          s + 2
     ---------------
     s^2 + 2 s + 2
> > load ltiexamples
> > ltiview
```

The unit step response is given in Fig. 11.7. From Fig. 11.7, the following results are obtained

$$M_p = 6.7\%,\ t_p = 2.37 \text{ seconds.}$$

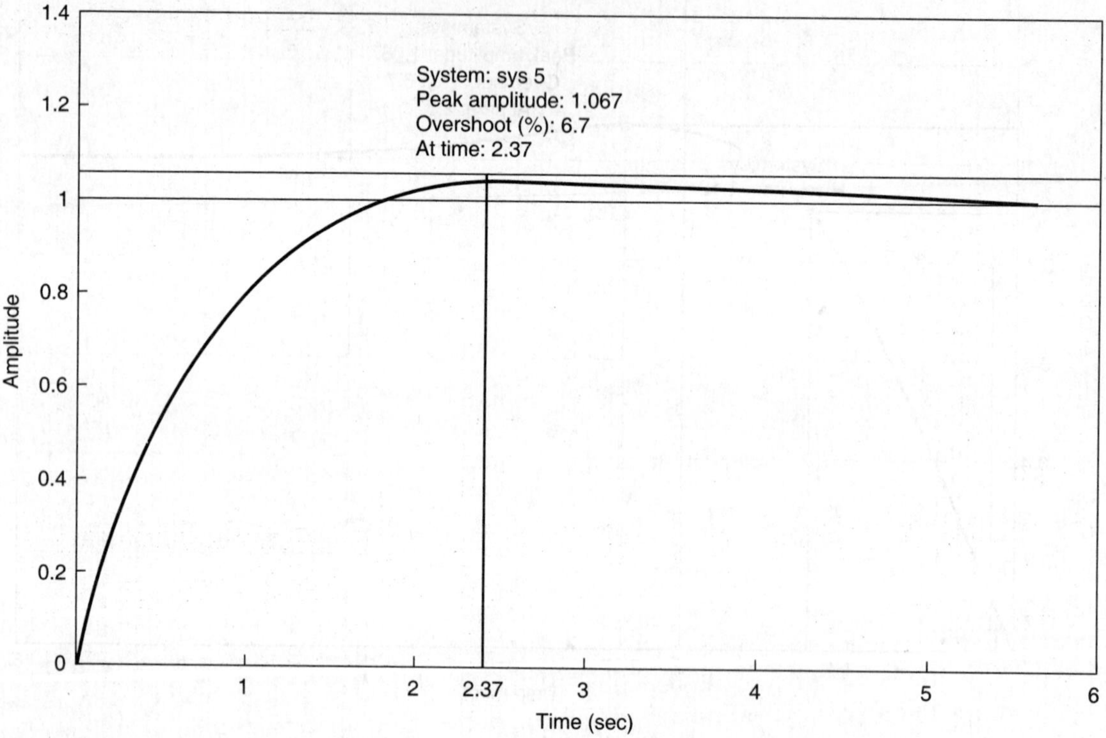

Fig. 11.7

Example 11.8. *The block diagram of a unity feedback control system is shown in figure 11.8.*

Find M_p, t_p the time period of oscillations and the number of cycles completed before reaching the steady state.

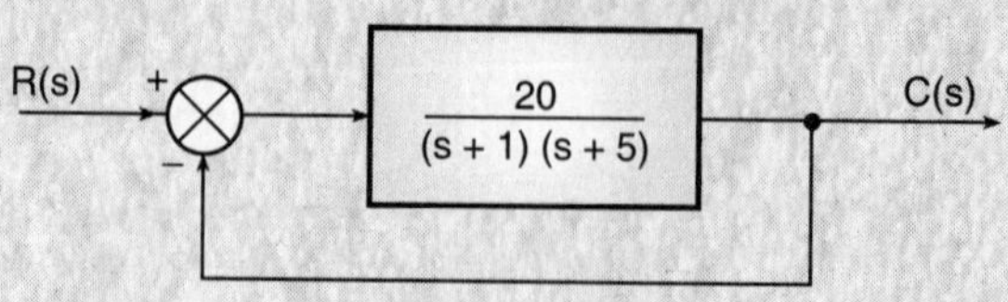

Fig. 11.8

Solution. Commands are given below :

```
> > n1 = [0  20], d1 = [1   6   5];
n1 =
     0    20
> > [n2, d2] = cloop (n1, d1, -1)
n2 =
     0     0       20
d2 =
     1     6       25
> > sys9 = tf (n2, d2)
Transfer function :
          20
    --------------
    s^2 + 6 s + 25
> > load ltiexamples
> > ltiview
```

From the step response as shown in Fig. 11.9 t_p = 0.78 seconds. The period of oscillation is 2 seconds, t_s = 1.19 seconds

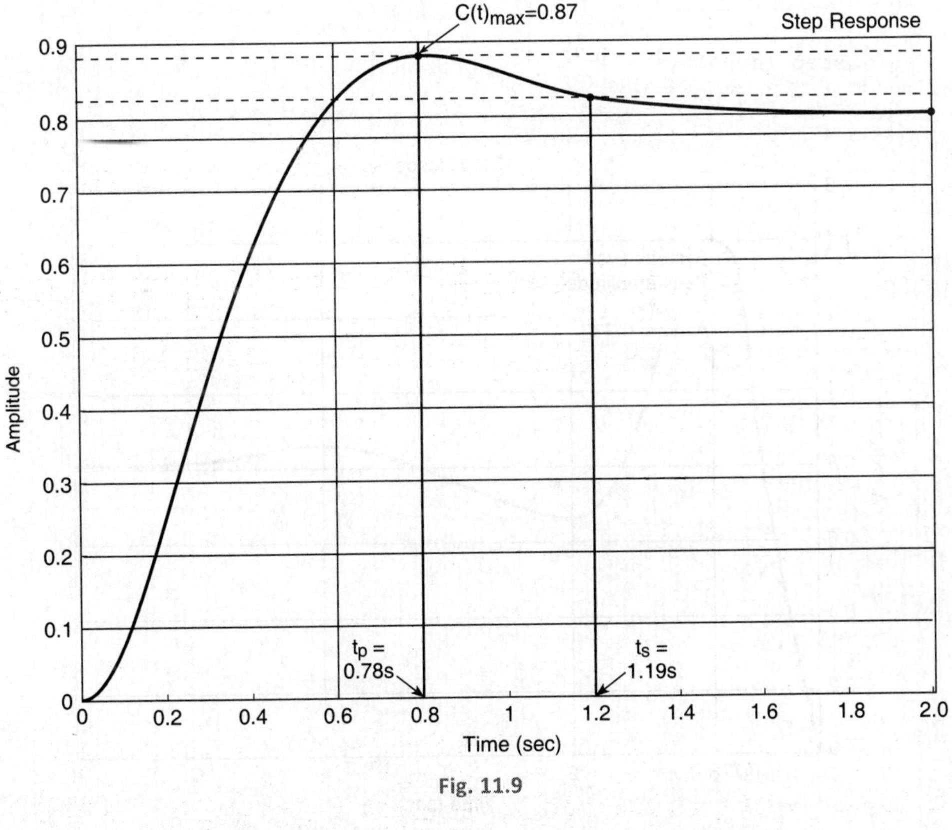

Fig. 11.9

$$\% M_p = \frac{c(t)_{\max} - C_{ss}}{C_{ss}} = \frac{0.87 - 0.8}{0.8} \times 100 = 8.75\%.$$

Example 11.9. *Determine the time response for a given transfer function*

$$\frac{C(s)}{R(s)} = \frac{4s + 4}{s^2 + 2s + 5}$$

for a step input and find $\%M_p$ *and* t_p.

Solution. The commands are given as follows

```
> > num = [4   4]
num =
     4      4
> > den = [1   2  5]
den =
      1  2  5
> > sys4 = tf(num, den)
Transfer function :
        4s + 4
    ---------------
    s^2 + 2 s + 5
> > step (sys4)
```

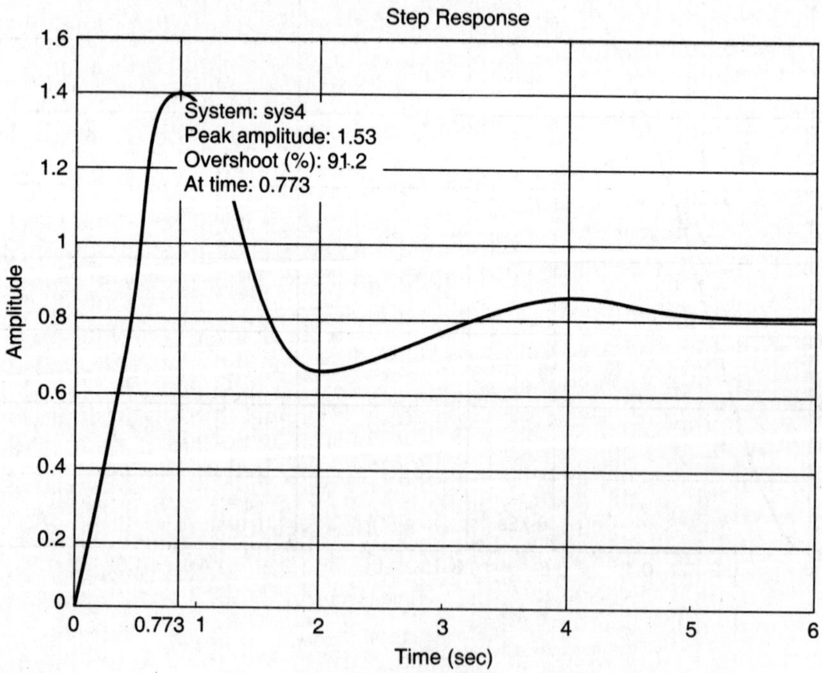

Fig. 11.10

The transfer function is created using the `tf` function and variable name `sys4` is assigned. Using the step function, the step response is displayed in Fig. 11.10. The mouse is right clicked in one corner of window in Fig. 11.10 then peak response is clicked with left button of mouse. The peak response point is clicked, % overshoot is 91.2 and t_p = 0.773 second.

Example 11.10. *A unity feedback control system is shown in Fig. 11.11 given below. By using derivative control the damping ratio is to be made 0.8. Determine the value of T_d and compare, the rise time, peak time and maximum overshoot for the system*

(a) without derivative control

(b) with derivative control

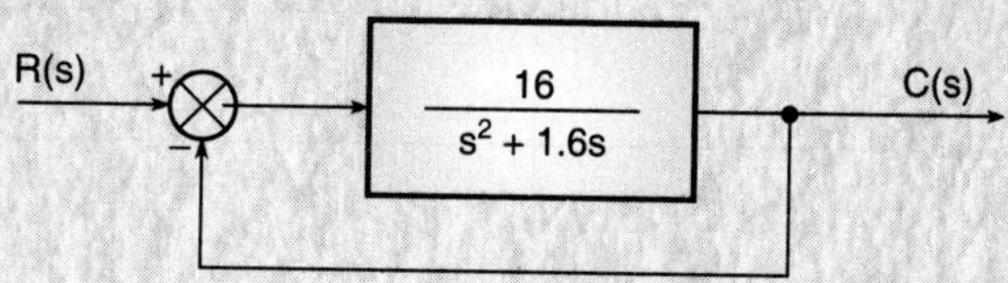

Fig. 11.11

Solution. To, initiate

(*a*) transfer function w/o derivative feedback is obtained as below :

$$\frac{C(s)}{R(s)} = \frac{16}{s^2 + 1.6s + 16}$$

Following commands are given,

```
> > num = 16
num =
     16
> > den = [1   1.6   16]
den =
     1.0000     1.6000    16.0000
> > sys = tf (num, den)
Transfer function :
           16
     ---------------
     s^2 + 1.6s + 16
> > load ltiexamples
> > ltiview
```

In Fig. 11.12, the unit step time response is displayed.

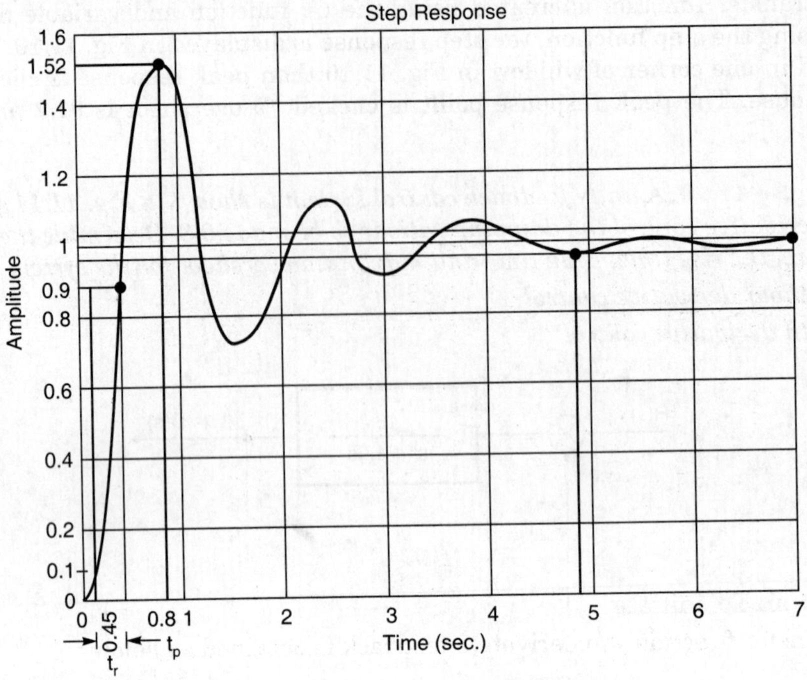

Fig. 11.12. Time response for example 11.10 (without derivative control).

From Fig. 11.12 following results are obtained :

t_r = 0.45 sec. (time needed to reach output response from 10% to 90%)

t_p = 0.8 sec.

M_p = 52.6%

(*b*) The derivative control is shown in following block diagram.

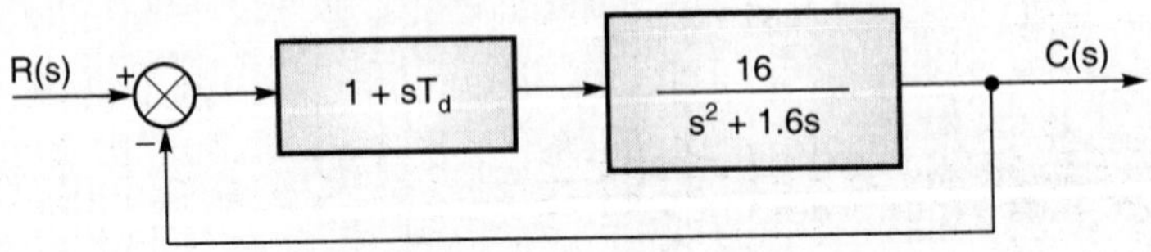

Fig. 11.13

The transfer function with derivative control is determined as below :

$$\frac{C(s)}{R(s)} = \frac{16(1+sT_d)}{s^2+(1.6+16T_d)s+16}$$

for damping ratio to be equal to 0.8

$$2\xi\omega_n = 1.6 + 16T_d$$

$$\because \quad \omega_n = \sqrt{16} = 4$$

$$\therefore \quad 2 \times 0.8 \times 4 = 1.6 + 16T_d$$

$$\therefore \quad T_d = 0.3$$

The transfer function takes the form as below :

$$\frac{C(s)}{R(s)} = \frac{16 + 4.8s}{s^2 + 6.4s + 6}$$

Following commands are given

```
> > num = (4.8 16]
To get started, select ''MATLAB Help'' from the Help menu.
num =
     4.8000        16.0000
> > den = [1 6.4 16]
den =
     1.0000        6.4000        16.0000
> > sys1 = tf (num, den)
Transfer function :
      4.8s + 16
   ---------------
   s^2 + 6.4 s + 16
>> load ltiexamples
>> ltiview
```

The time response for unit step input is shown in Fig. 11.14.

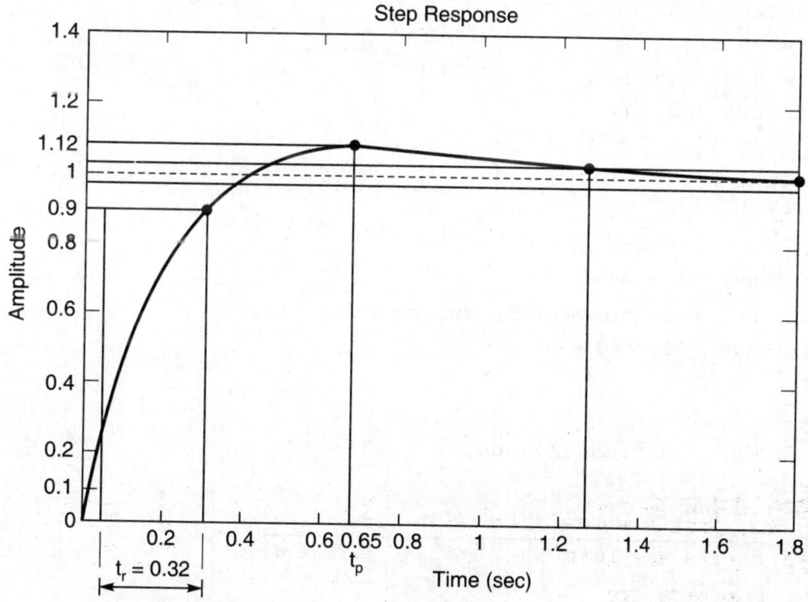

Fig. 11.14. Time response for example 11.10 (with derivative control).

From Fig. 11.14. following results are obtained :

t_r = 0.32 sec (time needed to reach output response from 10% to 90%)

t_p = 0.65 sec

M_p = 12%.

Example 11.11. *Determine an expression for a unit step response for the control system shown in figure given below :*

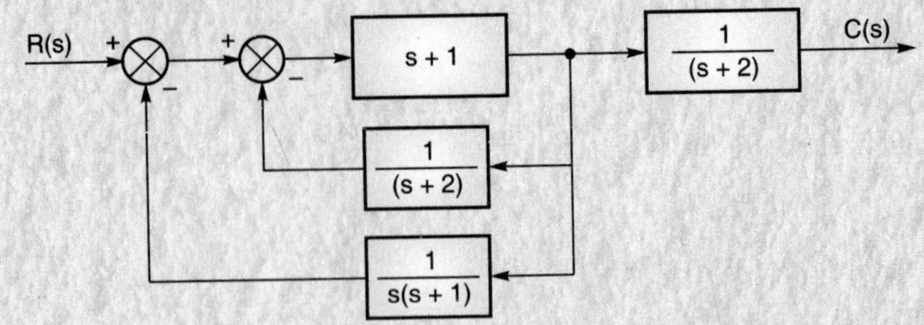

Fig.11.15

Solution. The MATLAB script to get the unit step response is as follows :

```
>> syms s t
n1 = [1 1];
d1 = [0 1];
printsys (n1, d1)
n2 = [0 1];
d2 = [1 2];
printsys (n2, d2)
n3 = [0 1];
d3 = [1 1 0];
printsys (n3, d3)
n4 = [0 1];
d4 = [1 2];
printsys (n4, d4)
[n5, d5] = feedback (n1, d1, n2, d2, -1);
printsys (n5, d5)
[n6, d6] = feedback (n5, d5, n3, d3, -1);
printsys (n6, d6)
n7 = [0 1];
d7 = [1 2];
[n8, d8] = series (n6, d6, n7, d7);
printsys (n8, d8)
                 s^4 + 4 *s^3 + 5*s^2 + 2*s
C =  -------------------------------------------
     s*(2*s^4 + 10*s^3 + 18*s^2 + 14*S + 4)
c = ilaplace (c)
c =
1/2*exp (-t)
>>
```

The transfer functions in the forward and feedback path are obtained by feeding the coefficients in the row vectors corresponding to numerator and denominator polynomial in each case. The feedback function for inner and outer loop is non unity *e.g.* $1/(s + 2)$ and

$1/s(s + 1)$ respectively. Hence, MATLAB function feedback is used to obtain the closed-loop transfer function for both the loops. The MATLAB function series is subsequently used for obtaining the resultant transfer function as the two blocks are connected in cascade.

The answer for the unit step response is

$$c(t) = 0.5e^{-t}.$$

Example 11.12. *A unity feedback control system has its open-loop transfer function given by $G(s) = (4s + 1)/4s^2$. Determine an expression for the time response when the system is subjected to*

(a) unit impulse input function.

(b) unit step input function.

Solution. The MATLAB is as follows :

```
>> % (a) time response to unit impulse input
syms s t
n1 = [4 1]
d1 = [4 0 0]
sys1 = tf(n1, d1)
[n2, d2] = cloop (n1, d1, -1)
F = (4*s + 1)/(4*s^2 + 4*s + 1)
f = ilaplace (F)
% (b) time response to unit step input
F = (4*s + 1)/(s*(4*s^2 + 4*s + 1))
f = ilaplace (F)
n1 =
     4     1
d1 =
     4     0     0
Transfer function :
    4s + 1
    ------
     4s^2
n2 =
     0     4     1
d2 =
     4     4     1
F =
(4*s + 1)/(4*s^2 + 4*s + 1)
f =
-1/4*t*exp (-1/2*t) + exp(-1/2*t)
F =
(4*s + 1)/(s*(4*s^2 + 4*s + 1))
f =
1 + 1/2*t*exp(-1/2*t)-exp(-1/2*t)
```

The open-loop transfer function is assigned a variable name sys1. Next the closed-loop transfer function of a –ive feedback is obtained. The closed-loop transfer function is given by the variable name F. For a unit impulse input $R(s) = 1$, ilaplace command of the command summary table is used to obtain the function f or $c(t)$. The answer is as follows :

$$c(t) = -\frac{1}{4} t\, e^{-0.5t} + e^{-0.5t}$$

The same procedure is used further but the input is unit step *i.e.*, $R(s) = \frac{1}{s}$. Hence the function of which the inverse Laplace transform is to be found is modified. The function $c(t)$ in time domain is obtained as

$$c(t) = 1 + 0.5\, te^{-0.5t} - e^{-0.5t}.$$

Example 11.13. *The impulse response of a system is given by $g(t) = e^{-t}(1 - \cos 2t)$. Determine the transfer function of the system.*

Solution. The MATLAB script is as follows :

```
>> syms t
>> f = exp (-t) * (1 - cos (2*t))
f =
exp (-t) * (1 - cos (2*t))
>> F = laplace (f)
F =
1/(1 + s) - (1 + s)/((1 + s)^2 + 4)
>>
```

The first command is symst constructs a symbolic object for time variable t. The Laplace transform of f [timefunction $f(t)$] is thus obtained using command Laplace (f).

The transfer function obtained is

$$G(s) = \frac{1}{s+1} - \frac{s+1}{(s+1)^2 + 4} \quad \textbf{Ans.}$$

Example 11.14. *The block diagram of a unit feedback control system is shown in Fig. 11.16.*

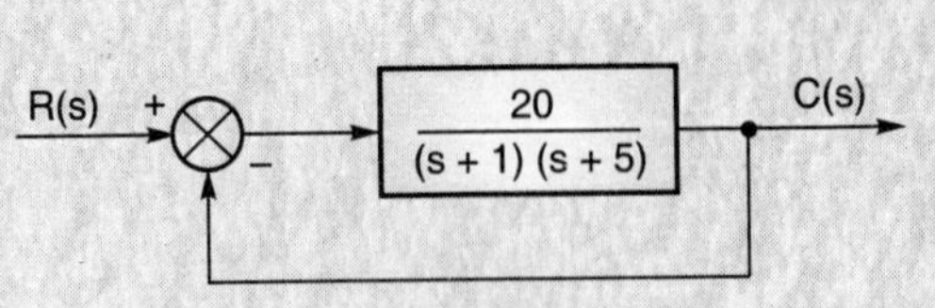

Fig. 11.16

Solution. Determine the characteristic equation of the system, ω_n, ξ, ω_d, t_p, $\%M_p$, the time at which the first understood occurs and the time period of oscillations and the number of cycles completed before reaching the steady state.

```
>> num = 20 ; den = [1 6 5] ;
>> [n1, d1] = cloop (num, den, -1)
n1 =
     0      0      20
d1 =
     1      6      25
>> sys = t1(n1, d1)
Transfer function :
      20
-------------
s^2 + 2s + 25
>> wn = sqrt (25)
wn =
5
>> ζ = 6/(2*wn)
zeta =
0.6000
>> wd = wn* sqrt (1 - zeta^2)
wd =
4
>> tp = pi/wd
tp =
     0.7854
>> mp = exp (- zeta*wn*tp)* 100
mp =
     9.4780
>> tfirst- 2* pi/wd       % time for first undershoot
tfirst =
     1.5708
>> period = 2* pi/wd     % period of oscillation
period-
     1.5708
>> oscillations = (wd/2* pi))* (4/(zeta*wn)) % number of
oscillation before steady state
oscillations-
     0.8488
>>
```

The results obtained are as follows, ω_n = 5 rad./sec., ξ = 0.6, ω_d = 4 rad./sec., t_p = 0.7854 sec., M_p = 9.478%, time to reach first undershoot = 1.5708 sec., the time period of oscillations = 1.5708 sec., the number of oscillations before reaching the steady state = 0.8488 cycle.

Example 11.15. *The open-loop transfer function of unity feedback control system is given by*

$$G(s) = \frac{25}{s(s + 5)}.$$

(a) Calculate (i) the natural frequency of oscillations, damped frequency of oscillations, damping factor, damping ratio, the maximum overshoot for a unit step input. (ii) e_{ss} for a unit ramp input.

(b) If the damping ratio is to be made 0.75 using a tachometer feedback, calculate the tachometer constant and determine the maximum overshoot.

Solution. The MATLAB script is as follows,

```
n1 = 25 ; d1 = [1    5     0] ;
[n2, d2] = cloop (n1, d1, -1) ;
sys2 = tf(n2, d2)
wn = √(25)
zeta = 5/(2* wn)
wd = wn* sqrt (1 - zeta^2)
mp = exp ((- zeta* pi)/sqrt (1 - zeta^2))* 100
ess = (2* zeta)/wn
% increase of damping ratio from 0.5 to 0.75
% with tacho feedback
zetan = 0.75
wn = sqrt (25)
kt = (2* zetan* wn - 5)/25
mp = exp ((- zetan* pi)/sqrt(1 - zetan^2))* 100
Transfer function :

      25
---------------
s^2 + 5s + 25

wn =
     5
zeta =
     0.5000
wd =
     4.3301
mp =
     16.3034
ess =
     0.2000
zetan =
     0.7500
wn =
     5
kt =
     0.1000
mp =
     2.8375
>>
```

For the given open-loop transfer function, from the results,
ω_n = 5 rad/sec., ω_d = 4.3301 rad./sec., damping ratio = 0.5, % M_p = 16.3034, e_{ss} = 0.2
With damping ratio = 0.75 and tachometer feedback K_t = 0.1, % M_p = 2.8375.

Example 11.16. *The open-loop transfer function of a unity feedback control system is given by*

$$G(s) = \frac{10}{s^2 + 6s + 10}$$

Find (a) K_p *(b)* K_v *(c)* K_a.

Solution. The MATLAB script of obtain the values of K_p, K_v and K_a is written below :

```
>> % calculation of error coefficients
numg  =  10
deng=  [1 6 10]
% step input
% calculation of error coefficient kp
G=tf (numg, deng)
kp=dcgain (G)
Ess=1/(1+kp)
% ramp input
numsg=conv ([1 0], numg)
densg=[1  6  10]
sG=tf (numsg, densg)
kv=dcgain (sG)
ess=1/kv
% parabolic input
nums2g=conv ([1 0 0], numg)
dens2g= [1 6 10]
s2G=tf (nums2g, dens2g)
ka=dcgain (s2G)
ess=1/ka
pause

numg=
    10
deng=
    1       6       10
Transfer function:
      10
--------------
s^2 + 6 s + 10
kp  =
    1
Ess=
    0.5000
```

```
numsg=
     10     0
densg=
     1      6      10
Transfer function :

     10 s
---------------
s^2 + 6 s + 10

kv =
     0
Warning: Divide by zero.
ess=
     Inf
nums2g =
     10     0      0
dens2g =
     1      6      10
Transfer function:

     10 s^2
---------------
s^2 + 6 s + 10

ka =
     0
Warning: Divide by zero.
ess =
     Inf
>>
```

The open-loop transfer function is assigned a variable G. The command d.c. gain (G) finds the dc gain for $G(s)$ *i.e.*, $s = 0$ and as per the formula $K_p = \lim_{s \to 0} G(s)$, it is the value of K_p, $K_p = 1$ is obtained. For finding K_v, as per theory $K_v = \lim_{s \to 0} sG(s)$, hence $sG(s)$ is obtained and d.c. gain (sG) given $K_v = 0$ and using $\lim_{s \to 0} s^2G(s)$, $K_a = 0$ is calculated. This solution uses same commands to obtain steady error in each case for step, ramp and parabolic inputs accordingly.

Example 11.17. *Determine the position, velocity and acceleration error coefficients for a system given by*

$$G(s)H(s) = \frac{100(s + 2)(s + 40)}{s^3(s^2 + 4s + 200)}.$$

Solution. The MATLAB script for finding the position, velocity and acceleration error coefficients is as follows.

```
num = 100* (poly ([-2, -40]))
den = conv ([1   0   0   0], [1   4   200])
GH = tf (num, den)
Kp = dcgain (GH)
num1 = conv ([1 0], num)
sGH = tf (num1, den)
kv = dcgain (sGH)
num2 = conv ([1 0], num1)
s2GH = tf (num2, den)
ka = dcgain (s2GH)
num =
    100          4200         8000
den =
    1     4     200     0     0     0
Transfer function:
 100s^2+4200s+8000
 -----------------
  s^5+4s^4+200s^3
Kp =
    Inf
num1 =
    100   4200   8000   0
Transfer function:
 100s^3+4200s^2+8000s
 --------------------
   s^5+4s^4+200s^3
kv =
    Inf
num2 =
    100   4200   8000   0     0
Transfer function:
 100s^4+4200s^3+8000s^2
 ----------------------
    s^5+4s^4+200s^3
ka =
    Inf
>>
```

$K_p = \infty$, $K_v = \infty$ and $K_a = \infty$.

Example 11.18. *The open-loop transfer function of a control system is given below*

$$G(s)H(s) = \frac{2(s^2 + 3s + 20)}{s(s + 2)(s^2 + 4s + 10)}$$

Determine the static error coefficients and steady state error for the input given as :

(a) 5 *(b) 4t* *(c) $4t^2/2$.*

Solution. The MATLAB script for obtaining the static error coefficients and a steady state error is as follows :

```
>> % STATIC ERROR COEFFICIENT
num = 2* [1 3 20]
den = conv ([1 2 0], [1 4 10])
GH = tf(num, den)
% calculation of error coeff for R = 4, 4t, 4t^2
kp = dcgain(GH)
ess = 1/(1 + kp)
numg = conv ([1 0], num)
sGH = tf (numg, den)
kv = dcgain (sGH)
ess = (4*1)/kv
num1 = conv ([1 0], numg)
sys = tf (num1, den)
ka = dcgain (sys)
ess = (4*1)/ka
num =
          2      6      40
den =
          1      6      18      20      0
Transfer function:
    2s^2+6 s+40
-------------------
s^4+6s^3+18 s^2+20s
kp =
       Inf
ess =
       0
numg =
       2      6      40      0
Transfer function:
   2s^3+6 s^2+40s
-------------------
s^4+6s^3+18 s^2+20s
kv =
       2
ess =
       2
num1 =
       2     6     40     0     0
Transfer function:
  2s^4+6 s^3+40s^2
-------------------
s^4+6s^3+18 s^2+20s
```

```
ka =
    0
Warning : Divide by zero.
ess =
    Inf
>>
```

Example 11.19. *A closed-loop control system when subjected to a unit step input has an expression for time response given by*

$$c(t) = 0.5 + 1.25e^{-t} - 1.75e^{-12t}$$

Determine the overall transfer function.

Solution. The MATLAB script is written below :

```
>> syms s t
>> c = 0.5+1.25*exp (-3*t)-1.75*exp (12*t)
c =
1/2+5/4*exp (-3*t)-7/4*exp (-12*t)
>> C = laplace (c)
C =
1/2/s+5/4/(s+3)-7/4/(s+12)
>>
```

The first command after execution constructs the symbolic objects for s and t. Next, the output $c(t)$ is fed. The MATLAB function laplace finds the Laplace transform of $c(t)$. The output $c(t)$ of the closed-loop transfer function is for a specified reference step input $r(t)$ and $R(s) = \frac{1}{s}$.

The Laplace transform of the output $c(t)$ as obtained from MATLAB application is

$$C(s) = \left(\frac{1/2}{s} + \frac{5/4}{s+3} - \frac{7/14}{s+12}\right)$$

$$\because \quad R(s) = \frac{1}{s}$$

$$\therefore \quad \frac{C(s)}{R(s)} = s\left(\frac{1/2}{s} + \frac{5/4}{s+3} - \frac{7/14}{s+12}\right)$$

or

$$\frac{C(s)}{R(s)} = \frac{17.25(s+1.043)}{(s+3)(s+12)} \quad \textbf{Ans.}$$

Example 11.20. *A unit step input is applied to a unity feedback control system whose open-loop transfer function is given by*

$$G(s) = \frac{K}{s(sT+1)}$$

Determine the values of K and T given that maximum overshoot is 25%. Calculate the gain crossover frequency and phase margin.

Solution. In the given transfer function the numerical values of K and T are to be chosen and the corresponding time response plot is obtained in order to get the M_p = 25%. The same procedure is to be followed number of times by trial and error and ultimately the final value of M_p = 25% is obtained from the displayed transient response plot. The commands to be followed sequentially are as below :

```
>> num = 11.9
num =
    11.9000
>> den = [.13    1    0]
den =
    0.1300      1.0000
>> sys 10 = tf(num, den)
Transfer function :
    11.9
 ------------
 0.13 s^2+s
>> sisotool (sys10)
```

After execution of this command the corresponding time response plot is selected from the menu of loop response after clicking the loop (step) response. The time response is highlighted with the mouse and in the window, time response plot is displayed as shown in Fig. 11.17.

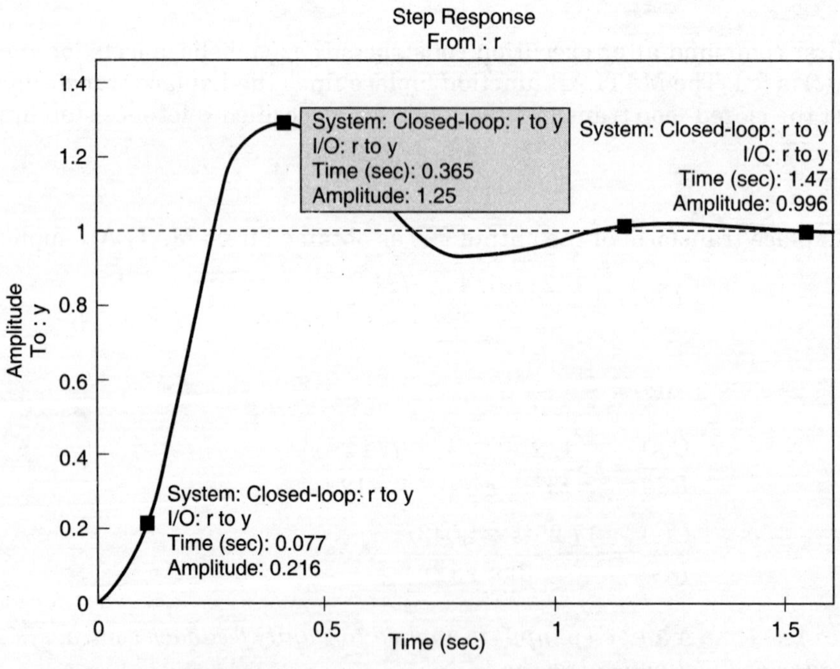

Fig. 11.17. Unit step response for example 11.20.

The value of K = 11.9 and T = 0.13 gives 20% maximum overshoot.

The same command *i.e.* >> `sisotool (sys10)` is to be executed and instead of step response the Bode plot is highlighted and OK button is clicked for the display box. The Bode plot is shown in Fig. 11.18 and phase margin of 43.3° at a given crossover frequency ω = 8.16 rad/sec is noted.

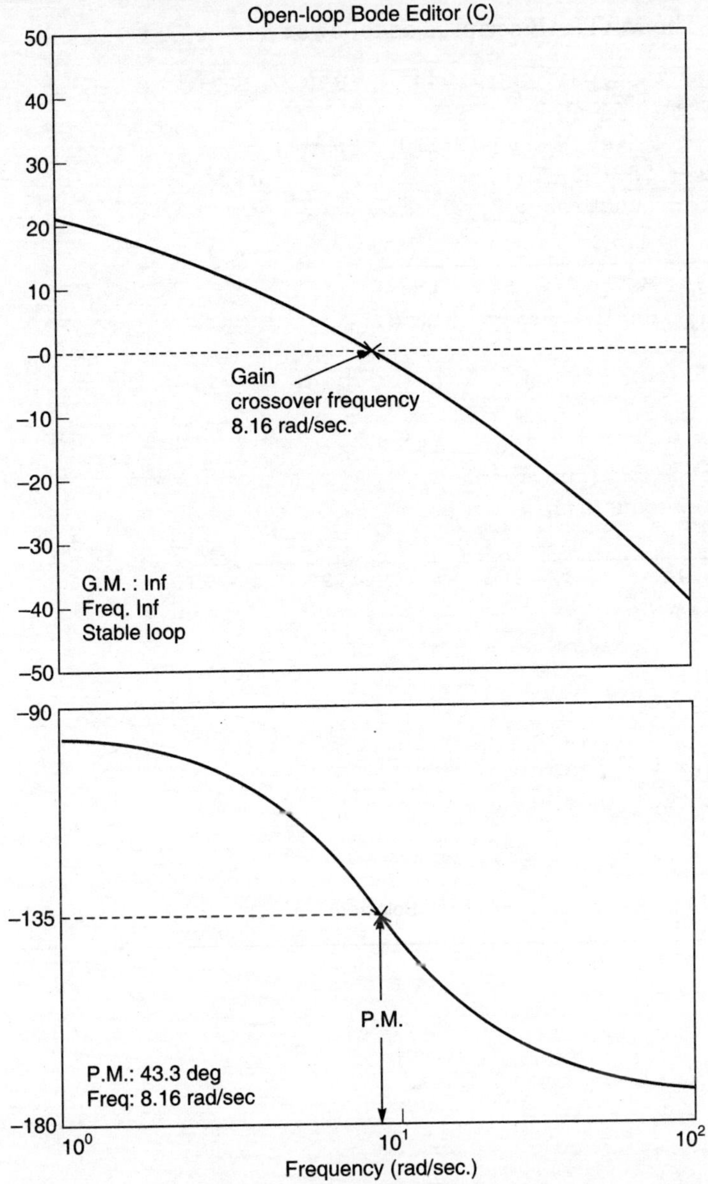

Fig. 11.18. Bode plot for example 11.20.

Example 11.21. *A unity feedback control system has its forward path transfer function as*

$$G(s) = \frac{210}{s(s+2)(s^2+12s+6)}$$

Give a MATLAB programme to determine

(a) The roots of the characteristic equation

(b) The Bode plot for open-loop transfer function.

Solution. The MATLAB script is as follows :

```
>> n1 = 210 ; d1 = [1    44      390     192      0]
d1 =
     1      44      390      192      0
>> sys = tf (n1, d1)
Transfer function :
                  210
-----------------------------------
s^4+44  s^3+390  s^2+192s
>> [num, den] = cloop (n1, d1)
num =
     0       0       0       0       210
den =
     1       44      390     192     210
>> sys1 = tf (num, den)
Transfer function :
                          210
------------------------------------------------------
s ^ 4 + 44    s ^ 3 + 390    s ^ 2 + 192    s + 210
>> roots (den)
ans =
- 31.9898
- 11.5578
- 0.2262 + 0.7189i
- 0.2262 - 0.7189i
>> sisotool (sys)
>>
```

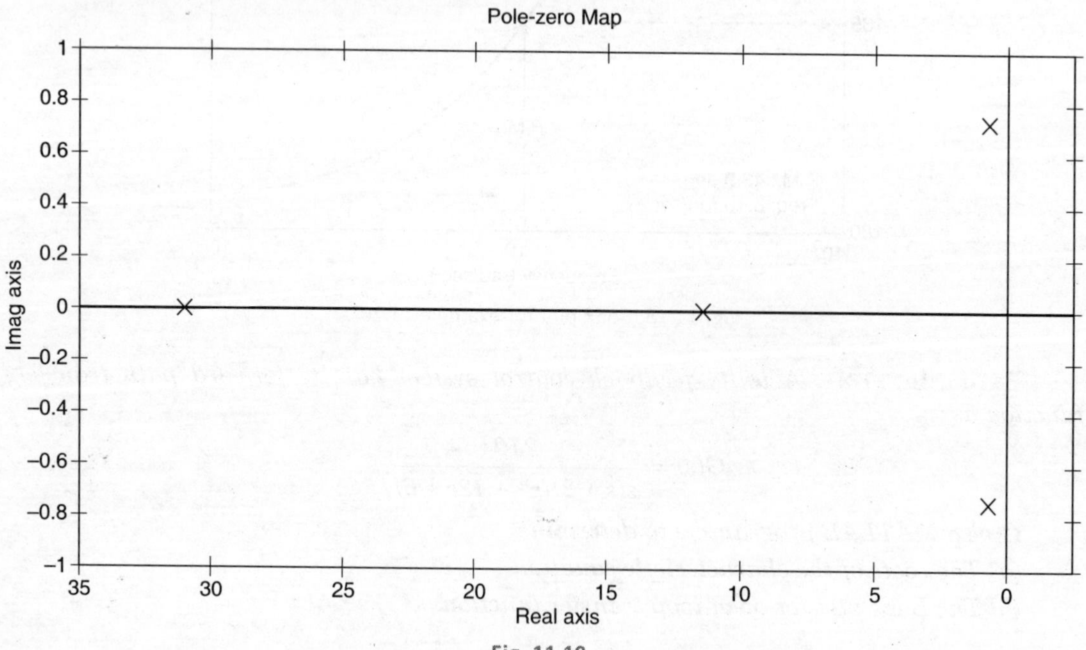

Fig. 11.19

For the given open-loop transfer function $G(s)$, the closed-loop transfer function is calculated first. The characteristic equation is

$$s^4 + 44s^3 + 390s^2 + 192s + 210 = 0$$

The roots of the characteristic equation are obtained using MATLAB function roots and the roots are

$$-31.9898, -11.5578, -0.2262 + 0.7181i \text{ and } -0.2262 - 0.7181i$$

The roots are shown in Fig. 11.19.

The Bode plot is obtained using function `sisotool` and shown in Fig. 11.20.

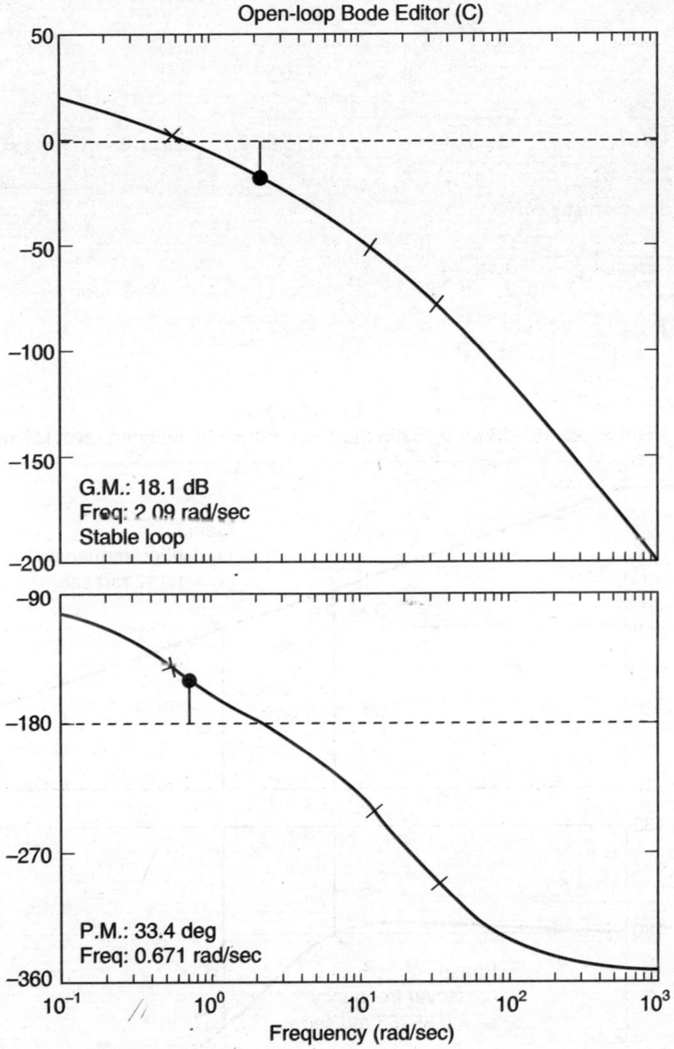

Fig. 11.20

Example 11.22. *Sketch the Bode plot for the transfer function given below :*

$$G(s)H(s) = \frac{2(s + 0.25)}{s^2 (s + 1)(s + 0.5)}$$

From the Bode plot determine :
(a) The phase crossover frequency *(b) The gain crossover frequency*
(c) The gain margin *(d) The phase margin.*

Solution. The given transfer function $G(s)H(s)$ is modelled as `sys1` using following commands.

```
>> num = [2 0.5]
num =
     2.0000          0.5000
>> den = [1          1.5     0.5     0 0]
den =
     1.0000          1.5000          0.5000          0          0
>> sys1 = tf (num, den)
Transfer function :
          2s + 0.5
    ----------------------
    s^4 + 1.5 s^3 + 0.5 s^2
>> margin (sys1)
```

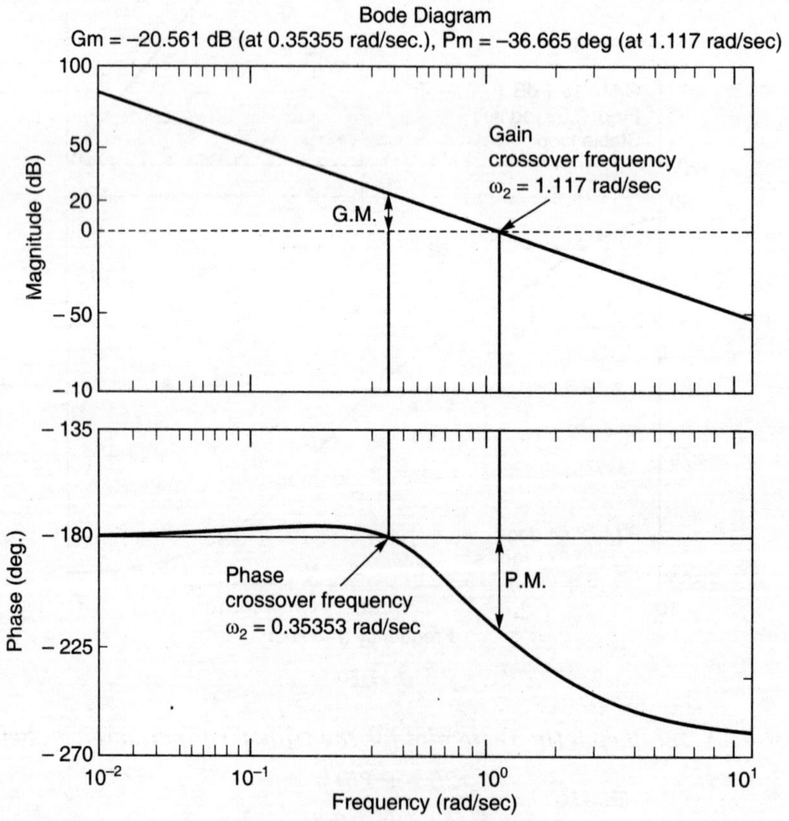

Fig. 11.21. Bode plot for example 11.22.

The function margin sys1 computes and plots the gain margin and phase margin directly as displayed in Fig. 11.21

The phase crossover frequency = 0.35355 rad/sec

The gain crossover frequency = 1.117 rad/sec

The gain margin = – 20.561 db

The phase margin = – 36.665 deg.

As both margins are negative, hence the system is unstable.

Example 11.23. *The open-loop transfer function of a system is given by*

$$G(s)H(s) = \frac{4}{s(1+0.5s)(1+0.08s)}$$

Determine (a) gain margin(b) phase margin (c) closed-loop stability.

The open-loop transfer function can be rewritten as

$$G(s)H(s) = \frac{4}{0.04s^3 + 0.58s^2 + s}.$$

Solution. The given transfer function is modelled using following commands

```
>> num = 4
num =
    4
>> den = [.04    .58    1      0]
den =
    0.0400       0.5800       1.0000       0
>> sys = tf (num, den)
Transfer function :
              4
 ---------------------------
 0.04s ^ 3 + 0.58s ^ 2 + s
>> load ltiexamples
>> ltiview
margin (sys)
```

The gain margin and phase margin are positive, as shown in Fig. 11.22 the system is stable.

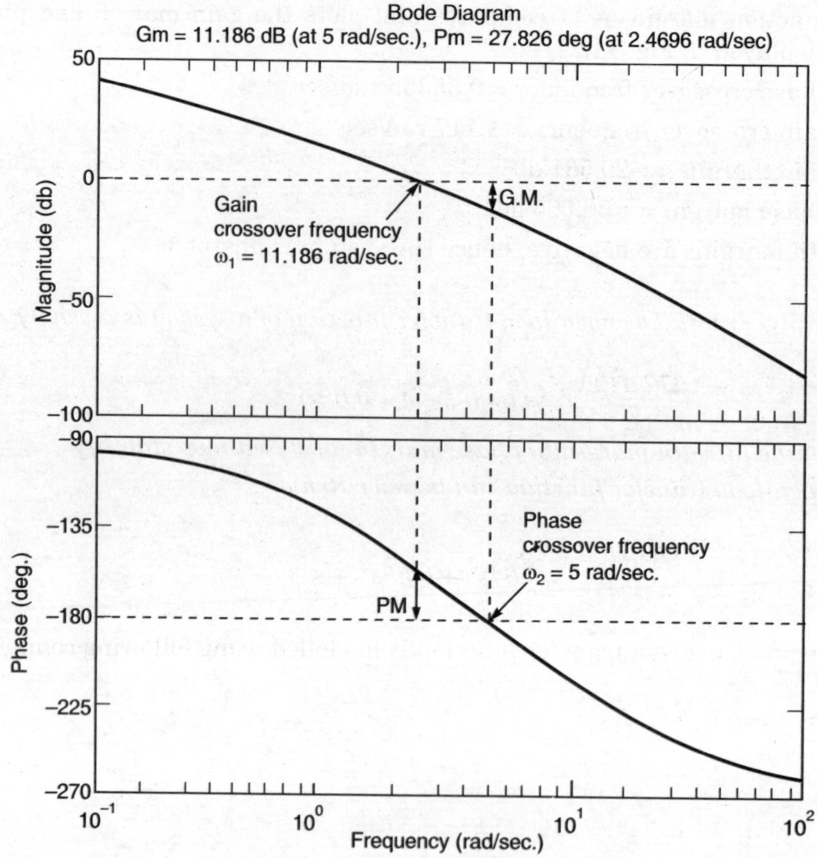

Fig. 11.22. Bode plot for example 11.23.

Example 11.24. *Examine the closed-loop stability of a system whose open-loop transfer function is given by*

$$G(s)H(s) = \frac{50}{(s+1)(s+2)}.$$

Solution. The commands are to be used sequentially as mentioned below :

```
>> num = 50
num =
    50
>> den = [1      3      2]
den =
    1      3      2
>> sys 3 = tf (num, den)
Transfer function :
    50
-----------
s^2+3s+2
>> load ltiexamples
>> ltiview
```

The transfer function model sys3 is obtained. The lti examples are loaded. The lti view browser is opened next. Under file menu, sys3 is imported. Nyquist plot is chosen (Fig. 11.23). The closed-loop stability for the given open-loop transfer function as indicated is stable.

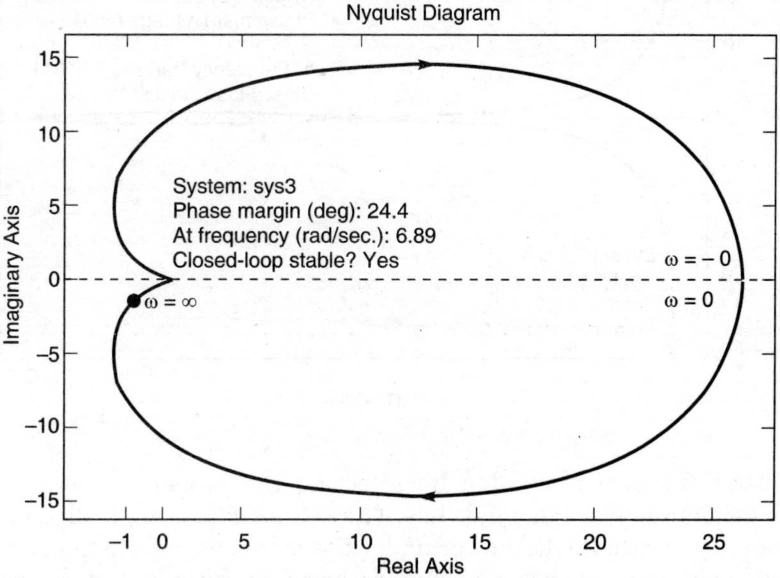

Fig. 11.23. Nyquist plot example 11.24.

Example 11.25. *Draw the Nyquist plot for the open-loop transfer function given below and comment on closed-loop stability*

$$G(s)H(s) = \frac{2.2}{s(s^2 + 2s + 2)(s + 1)}.$$

Solution. The commands are to be used sequentially as mentioned below :

```
>> num = 2.2
num =
    2.2000
>> den = [1      3      4      2      0]
den =
    1      3      4      2      0
>> sys 4 = tf (num, den)
Transfer function :
           2.2
-------------------------
s^4+3  s^3+4  s^2+2  s
>> load ltiexamples
>> ltiview
```

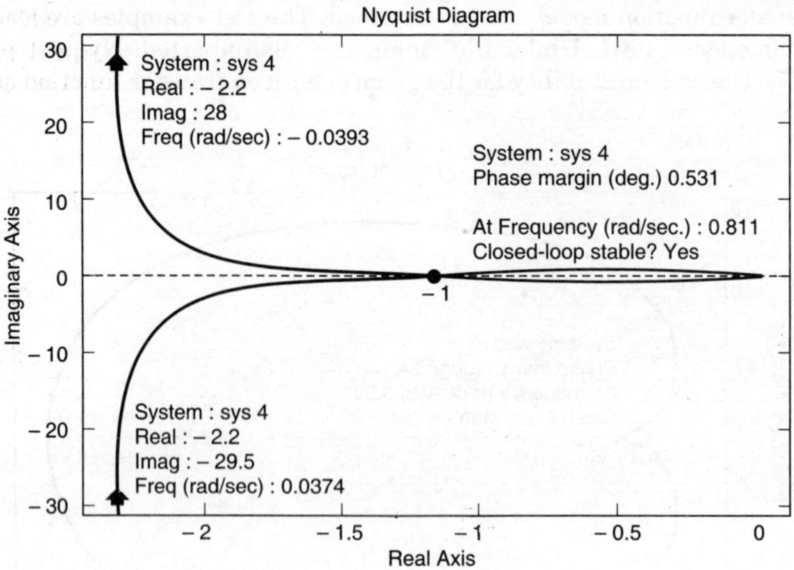

Fig. 11.24. Nyquist plot for example 11.25.

The transfer function is created by selecting the proper values of coefficients of numerator and denominator polynomials in *s*. The `lti` examples are loaded in the next step. To load the model `sys4` into the lti viewer under file menu, as a first step import is selected and the desired model `sys4` is chosen in the `lti` browser. After importing the model in the `lti` viewer, the plot type *i.e.* Nyquist is highlighted. The Nyquist plot is obtained as shown in Fig. 11.24.

The phase margin obtained is 0.531° which is very small ; as such the system can be considered as marginally stable.

Example 11.26. *The open-loop transfer function of a control system is given by*

$$G(s)H(s) = \frac{k(s-1)}{(s+2)(s+3)}$$

using Nyquist plot, determine the stability for (i) K = 10 (ii) K = 4.

Solution.

(*i*) Following MATLAB Commands are given

```
>> n1 = [10 - 10]
n1 =
    10 - 10
>> d1 = [1  5  6]
d1 =
    1  5  6
>> sys 5 = tf (n1, d1)
Transfer function:
   10s - 10
  -------------
  s ^ 2 + 5s + 6
>> load ltiexamples
>> ltiview
```

The given transfer function is obtained using `tf` command. After execution of `ltiview`, the `litviewer` window is opened. In the `litviewer`, from the file menu click the button on import. In the `ltibrowser` window select the system to import *e.g.* `sys5` and click the OK button. In the `litviewer` by default the step response is displayed. Then in the right hand corner right click the mouse and select plot type. Then select Nyquist from the given menu. The Nyquist diagram is obtained as shown in Fig. 11.25.

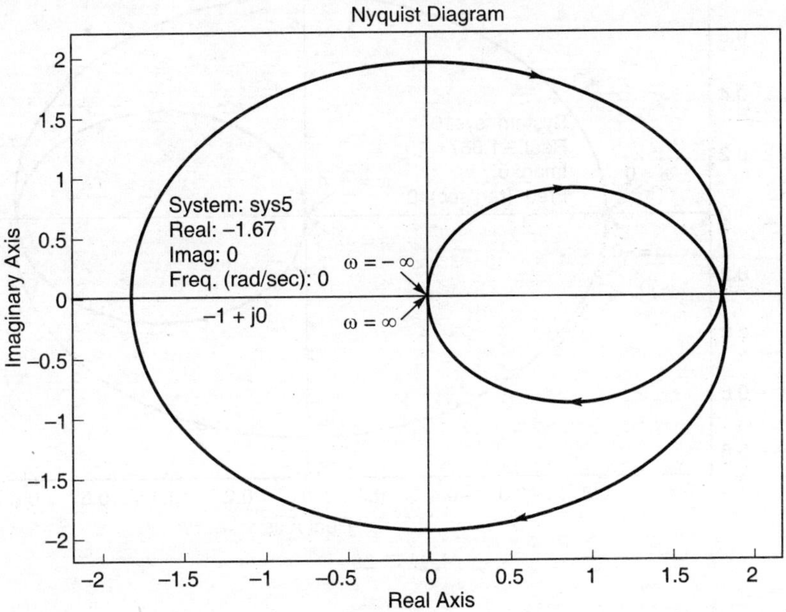

Fig. 11.25

At point *a*, the plot crosses the –ive real axis (ω = 0). The critical point (– 1 + *j*0) is encircled once in clockwise direction

$$\therefore \qquad N = -1$$

The number of open-loop poles is nil

$$\therefore \qquad P_+ = 0$$

$$\therefore \qquad N = P_+ - Z_+$$

$$\text{or} \qquad -1 = 0 - Z_+$$

$$\therefore \qquad Z_+ = 1$$

The closed-loop system is unstable.

(*ii*) Following MATLAB commands are given

```
>> n1 = [4   -4]
n1 =
     4      - 4
>> sys10 = tf (n1, d1)
Transfer function:
   4s - 4
-----------
s^2 + 5s + 6
>> load ltiexamples
>> ltiview
```

As per procedure given in (*ii*) The Nyquist plot is obtained as shown in Fig. 11.26.

The critical point $(-1 + j0)$ is not encircled by the Nyquist plot the closed-loop system is stable.

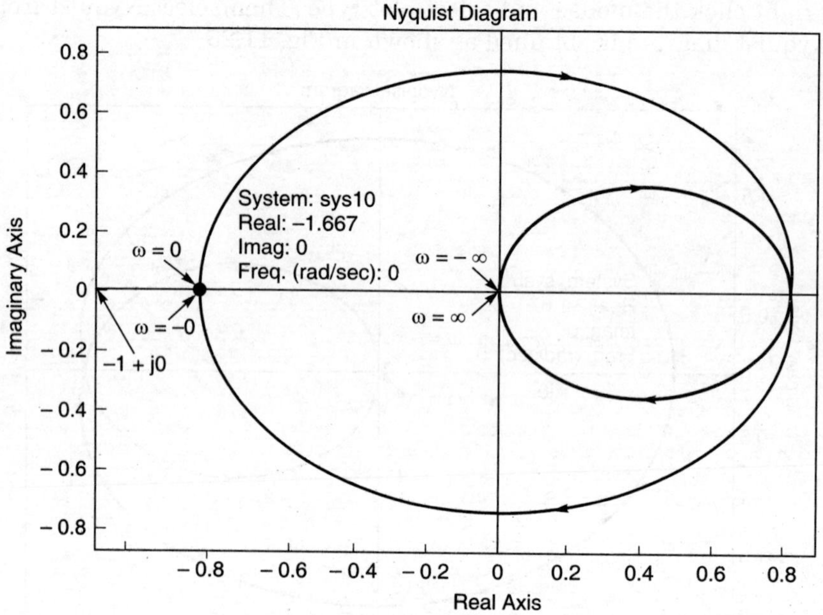

Fig. 11.26

Example 11.27. *Sketch the root locus plot for the open-loop transfer function given below :*

$$G(s)H(s) = \frac{K(s^2 + 4)}{s(s + 2)}$$

Calculate the value of K at (a) breakaway point and (b) s = – 0.69 + j0.88.

Solution.

The following commands are given

```
>> p = [1   0   4]; q = [1   2   0];
>> sys 9 = tf (p, q)
Transfer function:
 s^2 + 4
---------
s^2 + 2s
>> rlocus (p, q)
>> rlocfind (p, q)
Select a point in the graphics window
selected point = - 0.6940 + 0.8885i.
ans = 0.4575
```

```
>> rlocfind (p, q)
Select a point in the graphics window
selected point = - 0.8184.
ans =
0.2072
```

The transfer function is obtained with $K = 1$ using the `tf` function with p and q variables. The p and q variables correspond to the coefficients in the numerator and denominator polynomials respectively. After execution of rlocus function, the root locus is displayed. The `rlocfind` command when executed open, the graphic window displaying the root locus and the cursor. The cursor is placed on the breakaway point $s = -0.81$ and then $K = 0.2072$ is displayed.

After execution of `rlocfind` command and placing the cursor at $s = -0.69 + 0.88i$ $K = 0.4575$, is obtained as shown in Fig. 11.27.

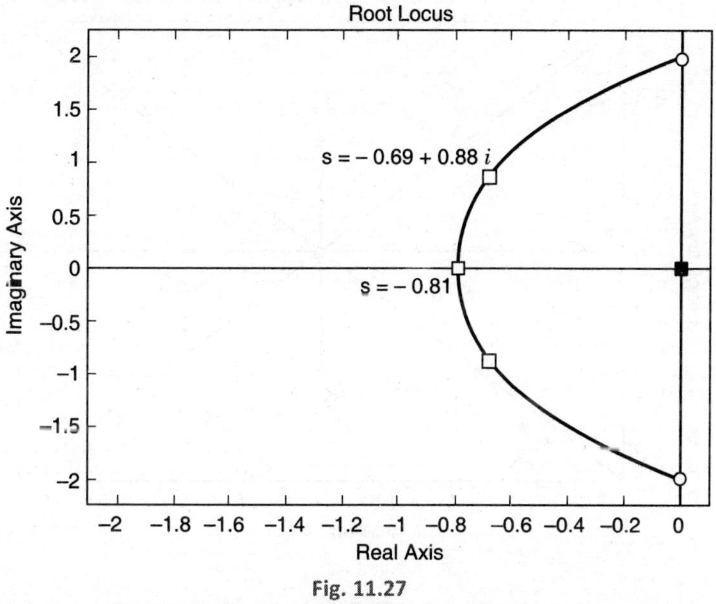

Fig. 11.27

Example 11.28. *The open-loop transfer function of a unity feedback system is given below :*

$$G(s) = \frac{K}{s(s+2)(s^2+2s+2)}$$

Plot the root locus to determine the value of gain K (i) at the breakaway point (ii) for marginal stability.

Solution. The given open-loop transfer function can be written as

$$G(s) = \frac{K}{s^4 + 4s^3 + 6s^2 + 4s}$$

Following commands are given to obtain the root locus plot as shown in Fig. 11.28.

```
>> p = 1
p = 1
>> q = [1   4   6   4   0]
q =
    1    4    6    4    0
>> sys 5 = tf (p, q)
Transfer function:
        1
-------------------------
s^4 + 4s^3 + 6s^2 + 4s
>> rlocus (sys 5)
```

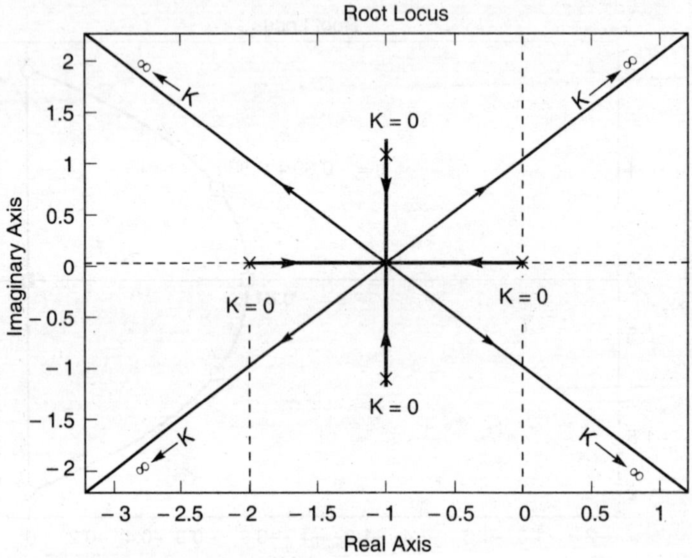

Fig. 11.28

As per root locus magnitude condition

$$|G(s)| = 1$$

$\because$ $$G(s) = \frac{K}{s(s+2)(s^2+2s+2)}$$

(*i*) at the breakaway point $s = -1$

$$\left|\frac{K}{(-1)(-1+2)[(-1)^2+2(-1)+2]}\right| = 1$$

or $$\left|\frac{K}{-1(1)(1)}\right| = 1 \quad \therefore \quad K = 1$$

(*ii*) The root locus plot crosses the imaginary axis at $s = j1$, for marginal stability at $s = j1$

$$\left|\frac{K}{(j1)(j1+2)\,[(-j1)^2 + 2(-j1) + 2]}\right| = 1$$

or
$$\left|\frac{K}{(j1)(j1+2)(j2+1)}\right| = 1$$

or
$$\frac{K}{1\sqrt{1^2+2^2}\,\sqrt{2^2+1^2}} = 1 \qquad \therefore \quad K = 5.$$

Example 11.29. *Sketch the root locus plot for the open-loop transfer function*

$$G(s)H(s) = \frac{K}{s(s+1)(s+3)}$$

Determine (i) the value of K at s = – 4

(ii) the frequency of sustained oscillations.

Solution. The given open-loop transfer function can be written as

$$G(s)H(s) = \frac{K}{s^3 + 4s^2 + 3s}$$

Following commands are given to get the transfer function assigned by variable `sys 3`.

```
>> p = 1
p = 1
>> q = [1     4     3     0]
q =
    1     4     3     0
>> sys 3 = tf (p, q)
Transfer function:
       1
--------------------
s^3 + 4s^2 + 3s
>> rlocus (p, q)
```

The root locus plot is obtained as shown in Fig. 11.29.

$$G(s)H(s) = \frac{K}{s(s+1)\,(s+3)}$$

Applying root locus magnitude condition *i.e.*

$$|G(s)H(s)| = 1$$

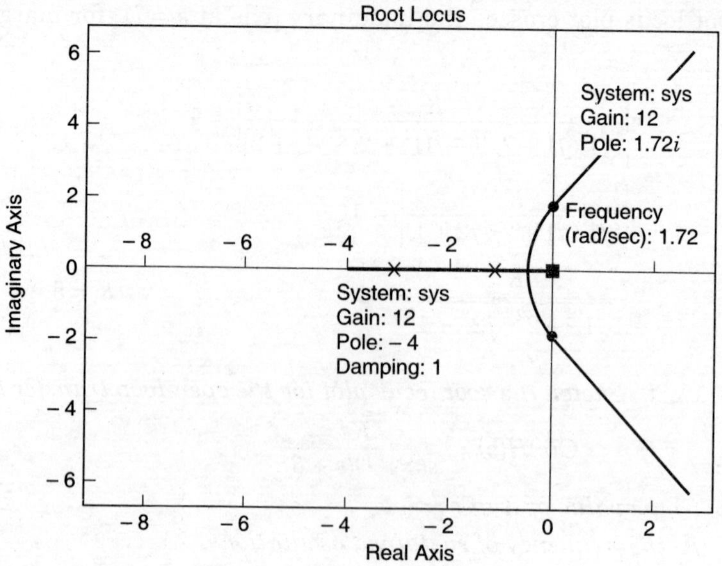

Fig. 11.29

(*i*) The value of gain K at $s = -4$ is obtained below

$$\left|\frac{K}{-4(-4+1)(-4+3)}\right| = 1$$

or

$$\left|\frac{K}{-4(-3)(-1)}\right| = 1 \quad \therefore \quad K = 12 \text{ at } s = -4$$

(*ii*) The root locus plot crosses the imaginary axis at $s = j\,1.72$, hence the frequency of sustained oscillations is $\omega = 1.72$ rad./sec.

Example 11.30. *Sketch the root locus plot for the system when the open-loop transfer function is given by*

$$G(s) = \frac{K}{s(s+4)(s^2+4s+13)}.$$

Solution. The following commands are given

```
>> p = 1
p = 1
>> q = [1    8    29    52    0]
q =
     1    8    29    52    0
>> sys4 = tf (p, q)
Transfer function:
          1
-----------------------
s^4 + 8s^3 + 29s^2 + 52s
>> rlocus (p, q)
```

The root locus plot is shown in Fig. 11.30.

Setting the cursor at the selected location on the root locus plot, the desired information with reference to system performance can be obtained.

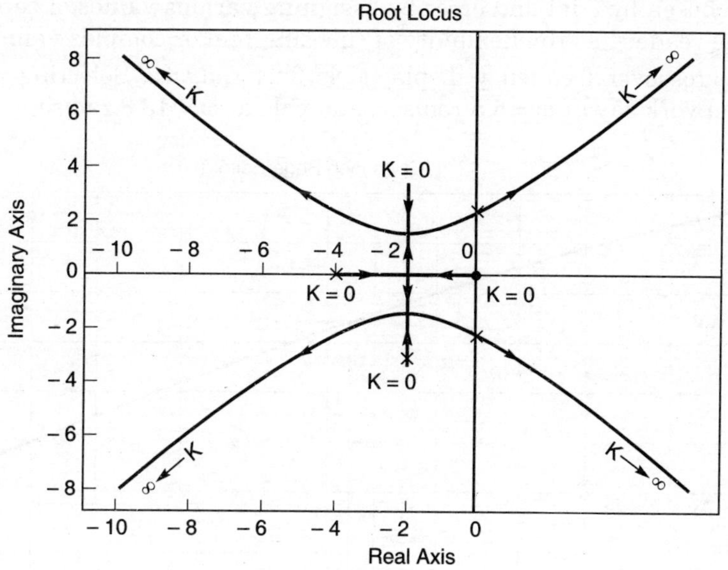

Fig. 11.30

Example 11.31. *The open-loop transfer function of a unity feedback control system is given by*

$$G(s) = \frac{K}{s(1+0.2s)}$$

Design a lead compensator such that the system will have $K_v = 10$ and P.M. = 57°.

Solution. $$K_v = \lim_{s \to 0} s\ G(s) = \lim_{s \to 0} s \cdot \frac{K}{s(1+0.2s)}$$

$\therefore$ $$10 = \lim_{s \to 0} s \cdot \frac{K}{s(1+0.2s)}$$

or $$K = 10$$

and in order to satisfy steady state error requirement

$$G(s) = \frac{10}{s(1+0.2s)}$$

The following MATLAB commands are given to plot the Bode plot for uncompensated system.

```
>> n = 10; d = [0.2   1   0];
 sys 6 = tf (n, d)
Transfer function:
   10
---------
0.2s^2+s
>> sisotool (sys 6)
```

In the graphical window of `sisotool`, the Bode plot for the open-loop transfer function is shown as in Fig. 11.31. The phase margin of 38.7° is displayed.

The phase margin is less than the desired value. The transfer function of the lead compensator is chosen by trial and error by assigning various values of zeros and poles for the network. The reader has the flexibility of choosing real or complex values.

The gain crossover frequency displayed is 6.24 rad./sec. Selecting zero of the lead compensating network say at $\omega = 5.5$ rad/sec. and pole at $\omega = 13.8$ rad/sec.

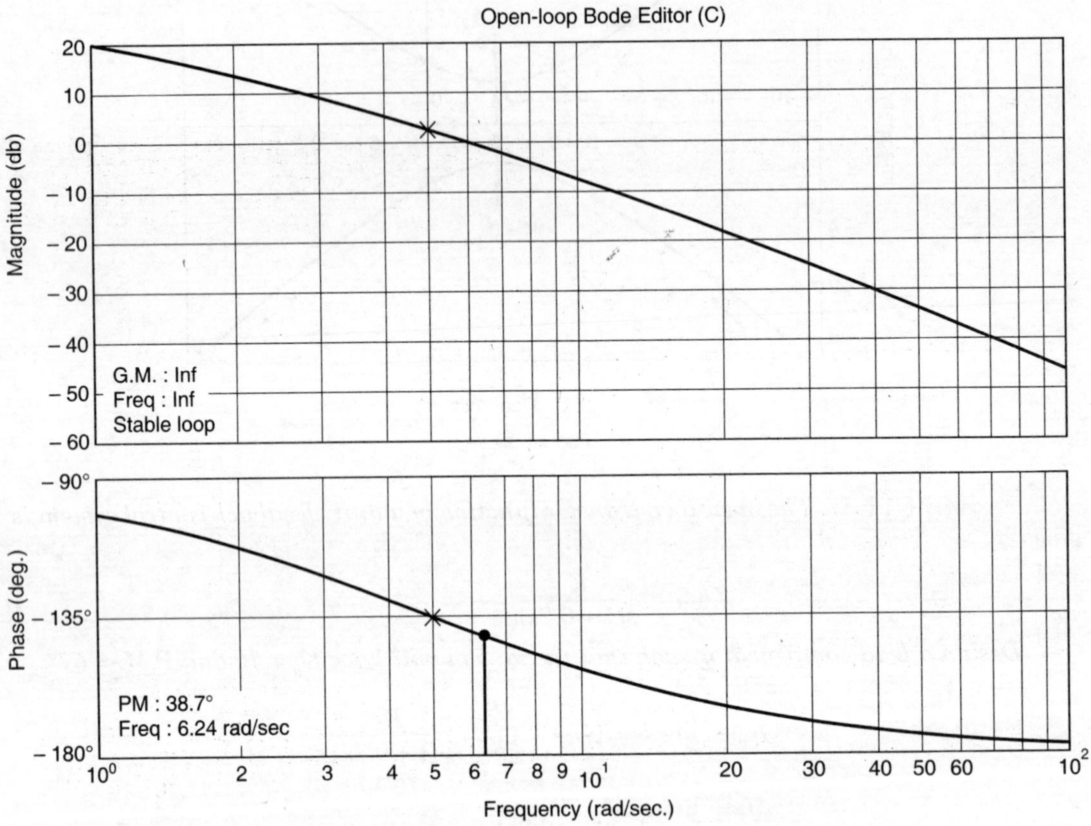

Fig. 11.31. Bode plot for uncompensated system.

The transfer function of the lead compensator is

$$G_c(j\omega) = \frac{j\omega + 5.5}{j\omega + 13.8}$$

or

$$G_c(j\omega) = \frac{5.5(j0.018\omega + 1)}{13.8(j0.072\omega + 1)}$$

or

$$G_c(j\omega) = \frac{0.398(j0.18\omega + 1)}{(j0.072\omega + 1)}$$

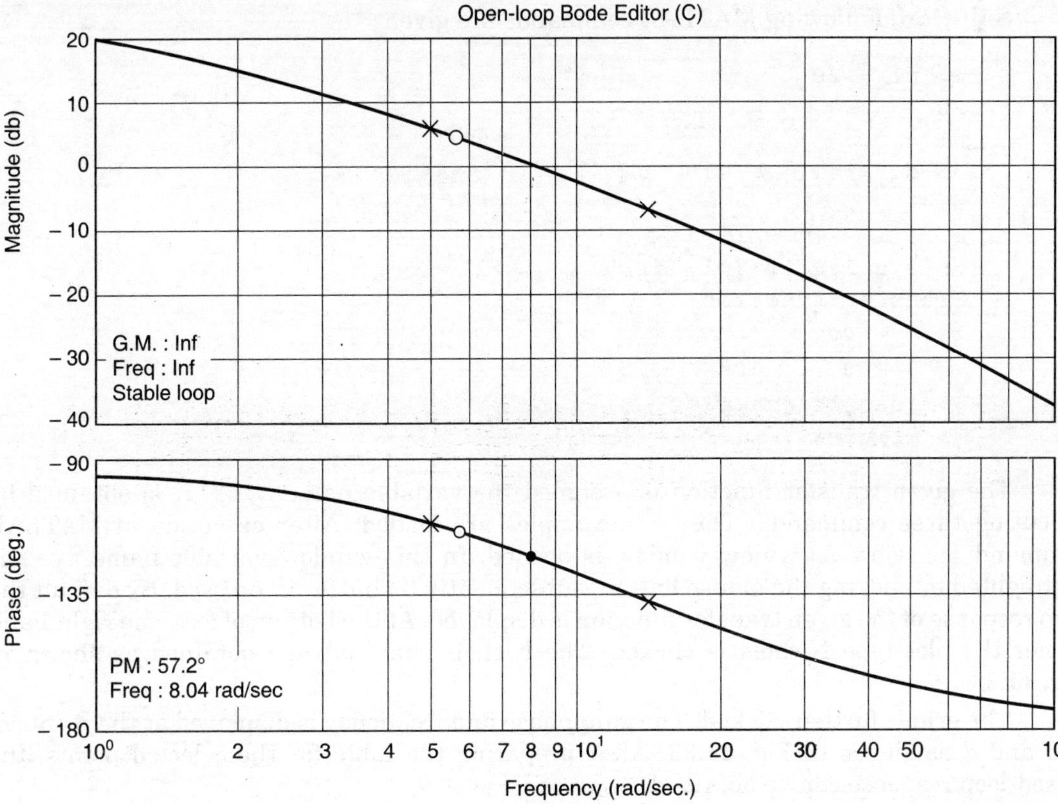

Fig. 11.32

Accounting for attenuation factor of 0.398 by increasing gain by 1/0.398. The transfer function of lead compensator can be modified as

$$G_c(j\omega) = \frac{(j0.18\omega + 1)}{(j0.072\omega + 1)}$$

The transfer function of the compensated system is

$$G_c(j\omega)\, G(j\omega) = \frac{(j0.18\,\omega + 1) \times 10}{(j0.072\,\omega + 1)\; j\omega(1 + j0.2\,\omega)}$$

or

$$G_c(s)\, G(s) = \frac{(0.18s + 1) \times 10}{(j0.072s + 1)\; s(1 + 0.2s)}$$

The Bode plot for the above compensated transfer function is displayed in Fig. 11.32 and the phase margin in noted as 57.2° **Ans.**

Example 11.32. *Use Nichols chart to obtain the closed-loop frequency response $C(j\omega)/R(j\omega)$ for a unity feedback control system having open-loop transfer function given below :*

$$G(s) = \frac{20}{s(s + 2)(s + 5)}$$

Solution. Following MATLAB commands are given

```
>> n1 = 20
20
n1 =
>> d1 = [1   7   10   0]
d1 =
     1    7    10    0
>> sys 3 = tf (n1, d1)
Transfer function:
         20
--------------------
s ^ 3 + 7s ^ 2 + 10s
>> load ltiexamples
>> ltiview
```

The given transfer function is assigned the variable name sys3. It is obtained by executing three commands. The lti examples are loaded. After execution of MATLAB command lti view, lti view window is opened. In this window variable name sys3 is highlighted by clicking the mouse button, subsequently OK buttor in pressed. By default the step response of the given transfer function is displayed. At the bottom of extreme right hand corner the plot type Nichols is clicked. The Nichols chart is then obtained as shown in Fig. 11.33.

The grid is further clicked. The gain phase and frequency is displayed at the points *a*, *b*, *c* and *d* as shown in Fig. 11.33. After preparing the table for the selected points, the closed-loop response can be obtained.

ω rad./sec	0.374	0.723	3.07	12.5
$\lvert C(j\omega)/R(j\omega) \rvert$ db	14.4	8.25	– 10.3	– 40.5
$\angle C(j\omega)/R(j\omega)°$	– 105	– 118	– 178	– 239

Ans.

Example 11.33. *Obtain the state model for the transfer function given below :*

$$\frac{C(s)}{R(s)} = \frac{(s+2)}{(s+1)(s+3)}.$$

Solution. The MATLAB script is as follows :

```
%State model for the transfer function
%(s+2)/(s+1)(s+3)
>> num=[1 2];
num =
     1     2
>> den=[1   4   3];
den =
     1    4    3
>> [A, B, C, D] = tf2ss (num, den);
```

```
p = [0  1;  1  0]
% Phase variable form
Ap=inv(P)*A*P
Bp=inv(P)*B
Cp=C*P
Dp=D
```

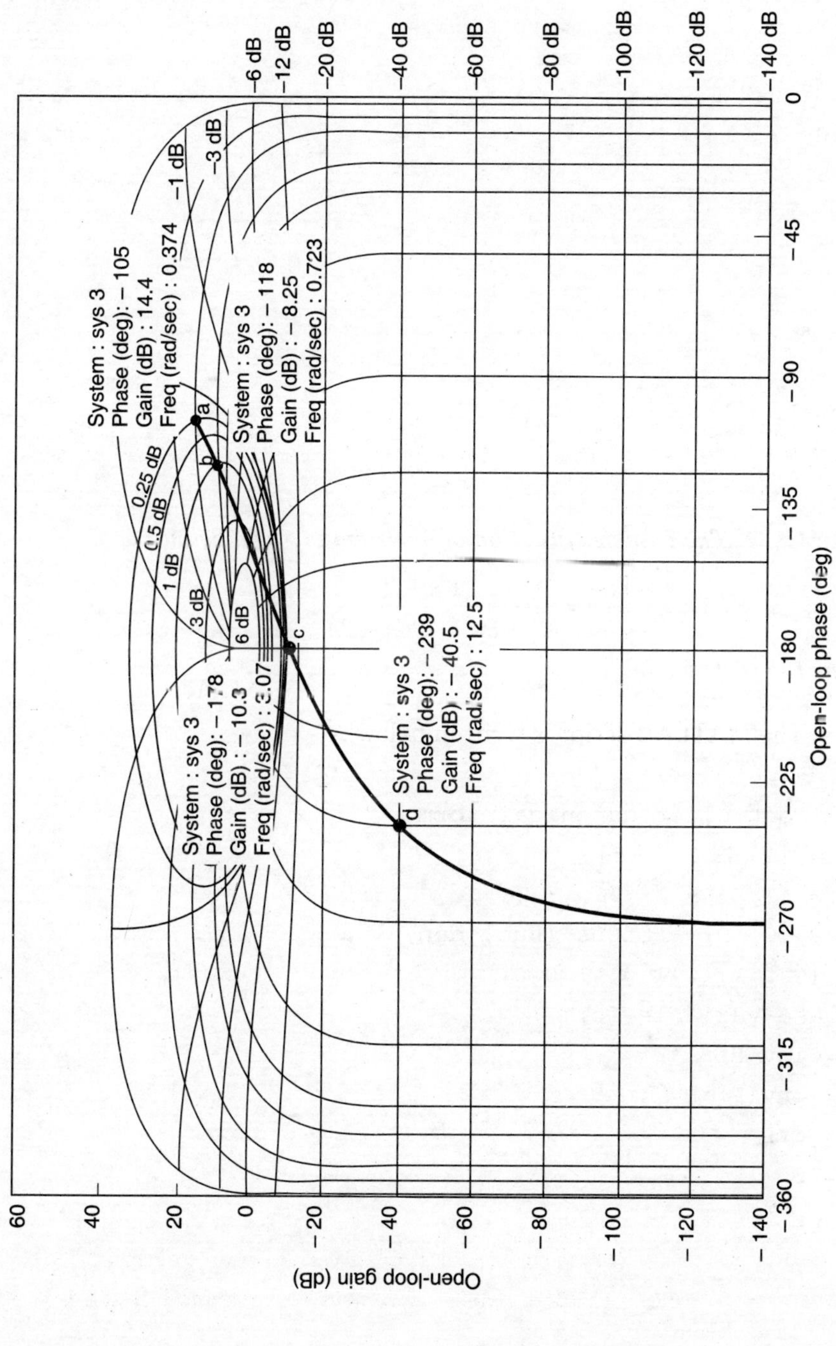

Fig. 11.33. Nichols chart.

The row vectors corresponding to numerator and denominator polynomials of a given transfer function are assigned the variables num and den respectively. The state model corresponding to controllable canonical form is obtained using `tf2ss` command. This model is converted to phase variable form using the transformation. P is the transformation matrix. The matrices `Ap, Bp, Cp, Dp` in phase variable are as under :

```
P =
     0     1
     1     0
Ap =
     0     1
    -3    -4
Bp =
     0
     1
Cp =
     2     1
Dp =
     0
```

Example 11.34. *The transfer function of a system is given below :*

$$\frac{Y(s)}{U(s)} = \frac{s^2 + s + 2}{s^3 + 9s^2 + 26s + 24}$$

Determine the state model.

Solution. The MATLAB script is written below :

```
>> % controller canonical form
num = [1    1    2];
den = [1    9    26    24];
[A, B, C, D] = tf2ss (num, den)
P = [0 0 1; 0 1 0; 1 0 0]
% Phase variable form
Ap = inv(P)*A*P
Bp = inv(P)*B
Cp = C*P
Dp = D
```

The row vectors for the numerator and denominator of the given transfer function are assigned by the variable names `num` and `den` respectively. The transfer function is converted to controller canonical form using `tf2ss` command. The state model is thus obtained. The obtained state model is converted to phase variable form using similarity transformation wherein similarity matrix P is used. The answer is as follows :

(1)

```
A =
     -9    -26    -24
      1      0      0
      0      1      0
B =
      1
      0
      0
C =
      1      1      2
D =
      0
```

(2)

```
P =
      0      0      1
      0      1      0
      1      0      0
Ap =
      0      1      0
      0      0      1
    -24    -26     -9
Bp =
      0
      0
      1
Cp =
      2      1      1
Dp =
      0
```

Example 11.35. *Obtain the state model for the following transfer function by direct decomposition.*

$$\frac{Y(s)}{U(s)} = \frac{1}{s^2 + 2s + 5}.$$

Solution. The state equations can be written as

$$\dot{x}_1 = x_2$$

$$\dot{x}_2 = -5x_1 - 2x_2 + u$$

and the output is

$$y = x_2$$

The state model is

$$\begin{bmatrix} \dot{x}_1 \\ \dot{x}_2 \end{bmatrix} = \begin{bmatrix} 0 & 1 \\ -5 & -2 \end{bmatrix} \begin{bmatrix} x_1 \\ x_2 \end{bmatrix} + \begin{bmatrix} 0 \\ 1 \end{bmatrix} u$$

$$y = [1 \quad 0] \begin{bmatrix} x_1 \\ x_2 \end{bmatrix}$$

The MATLAB script for obtaining the state model is as follows :

```
>> % controller canonical form
num = [1];
den = [1   2   5];
[A, B, C, D] = tf2ss (num, den)
```

```
P = [0 1; 1 0]
% Phase variable form
Ap = inv(P)*A*P
Bp = inv(P)*B
Cp = C*P
Dp = D
```

The row vectors corresponding to numerator and denominator polynomials of a given transfer function are assigned the variables `num` and `den` respectively. The state model corresponding to controllable canonical form is obtained using `tf2ss` command. This model is converted to phase variable form using the transformation. P is the transformation matrix. The matrices `Ap, Bp, Cp, Dp` in phase variable form are as under :

```
A =
     - 2   - 5
       1     0
B =
       1
       0
C =
       0     1
D =
       0
```

```
P =
       0     1
       1     0
Ap =
       0     1
     - 5   - 2
Bp =
       0
       1
Cp =
       1     0
Dp =
       0
>>
```

Example 11.36. *Write MATLAB script to determine the state transition matrix for*

$$A = \begin{bmatrix} 1 & 4 \\ -2 & -5 \end{bmatrix}.$$

Solution. The MATLAB script for determination of the state transition matrix is as follows :

```
>> % calculation of state transition matrix using
   inverse technique sys t
a = [1  4;  -2  -5]
phi = exmp (a*t)
a =
     1     4
    -2    -5
phi =
[2*exp(-t) -exp (-3*t), -2*exp (-3*t)+2*exp(-t)]
[exp(-3*t)-exp(-t), -exp(-t)+2*exp(-3*t)]
```

The first line begins with % and it is a comment and ignored by MATLAB. The first command in the MATLAB script constructs a symbolic object for the time variable 't'. The state transition matrix is obtained and assigned a name phi and the elements of the state transition matrix are in exponential form.

The state transition matrix is

$$\phi(t) = \begin{bmatrix} 2e^{-t} - e^{-3t} & -2e^{-3t} + 2e^{-t} \\ e^{-3t} - e^{-t} & -e^{-t} + 2e^{-3t} \end{bmatrix}.$$

Example 11.37. *The state equations of a system are given below :*

$$\dot{x}_1 = x_1 + x_2 + u$$

$$\dot{x}_2 = -x_2$$

Check for the controllability.

Solution.

$$\begin{bmatrix} \dot{x}_1 \\ \dot{x}_2 \end{bmatrix} = \begin{bmatrix} 1 & 1 \\ 0 & -1 \end{bmatrix}\begin{bmatrix} x_1 \\ x_2 \end{bmatrix} + \begin{bmatrix} 0 \\ 1 \end{bmatrix} u$$

Hence, the system matrix $\mathbf{A} = \begin{bmatrix} 1 & 1 \\ 0 & -1 \end{bmatrix}$

$$\mathbf{B} = \begin{bmatrix} 0 \\ 1 \end{bmatrix}$$

The MATLAB script is as follows :

```
>> A = [1 1; 0 -1]
A =
      1      1
      0     -1
>> B = [1; 0]
B =
      1
      0
>> P = ctrb(A, B)
P =
      1      1
      0      0
>> RANK(P)
ans =
      1
>>
```

The controllability matrix P is a square matrix (2 × 2) and it is a singular matrix, hence the rank of P matrix is 1, not equal to $n = 2$. The system is therefore uncontrollable.

Example 11.38. *Obtain the transfer function from the following data*

$$\mathbf{A} = \begin{bmatrix} -5 & 1 \\ -6 & 0 \end{bmatrix}, \mathbf{B} = \begin{bmatrix} 1 \\ 2 \end{bmatrix}, \mathbf{C} = [2 \quad 1], \mathbf{D} = 0.$$

Solution. The MATLAB script is given below :

```
>> % obtain transfer matrix
A = [-5   1 ;   -6   0]
B = [1 ;   2]
C = [2     1]
D = [0]
[num, den] = ss2tf (A, B, C, D)
printsys (num,den)
A =
      -5     1
      -6     0
B =
       1
       2
C =
       2     1
D =
       0
num =
       0     4     8
den =
       1     5     6
num/den =
   4s+8
 ---------
 s^2+5s+6
>>
```

The ss2tf command converts state variable form to transfer function. The transfer function is

$$\frac{Y(s)}{U(s)} = \frac{4s + 8}{s^2 + 5s + 6}.$$

Example 11.39. *Determine the transfer matrix having given*

$$\mathbf{A} = \begin{bmatrix} 0 & 1 \\ -6 & -5 \end{bmatrix}, \mathbf{B} = \begin{bmatrix} 0 \\ 1 \end{bmatrix}, \mathbf{C} = [2 \quad 1].$$

Solution. The MATLAB script is as follows :

```
>> A = [0   1 ;   -6   -5]
A =
       0      1
      -6     -5
>> B = [0 ;   1]
B =
       0
       1
>> C = [2 1]
C =
       2     1
>> D = [0]
D =
       0
>> [num, den] = ss2tf (A,B,C,D)
num =
           0        1.0000        2.0000
den =
      1.0000        5.0000        6.0000
>> printsys (num, den)
num/den =
     s+2
  ----------
  s^2+5s+6
```

The matrices **A, B** and **C** are arranged first, the first command displays the matrix **A**.

The `ss2tf` converts the state variable form to transfer function form. The transfer function obtained is

$$\frac{Y(s)}{U(s)} = \frac{s+2}{s^2+5s+6}. \quad \textbf{Ans.}$$

Example 11.40. *Obtain the pulse transfer function for the system shown in figure below :*

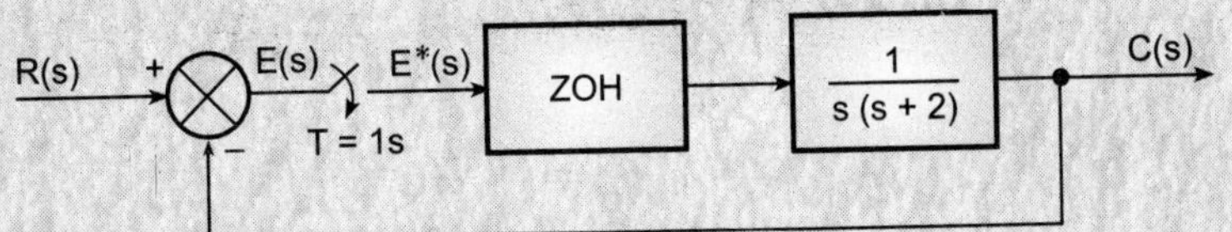

Fig. 11.34

Solution. The transfer function of ZOH is given by

$$G_{ho}(s) = \frac{1-e^{-st}}{s}$$

The feedback transfer function : $H(s) = 1$

Tha MATLAB script is as follows :

```
>> num=1; den=[1  2  0]; T=1;
>> [numz,denz]=c2dm(num,den,T,'ZOH')
numz=
        0          0.2838        0.1485
denz =
    1.0000      -1.1353        0.1353
>> printsys(numz, denz, 'Z']
num/den=
    0.28383 Z+0.1485
 ----------------------
 Z^2-1.1353 Z+0.13534
```

The pulse transfer function obtained is

$$\frac{C(z)}{R(z)} = \frac{0.28383z + 0.1485}{z^2 - 1.1353z + 0.13534} \qquad \textbf{Ans.}$$

CHAPTER 12

Classified Solved Examples

Example 12.1. *Determine the transfer function C/R from the block diagram shown in Fig. 12.1.*

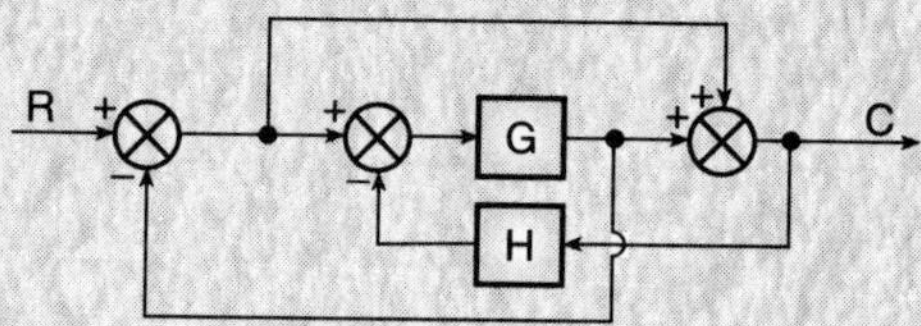

Fig. 12.1

Solution. Shifting take off point after block G to a position before block G, (Fig. 12.2). Shifting takeoff point after summing point to a position before summing point (Fig. 12.3).

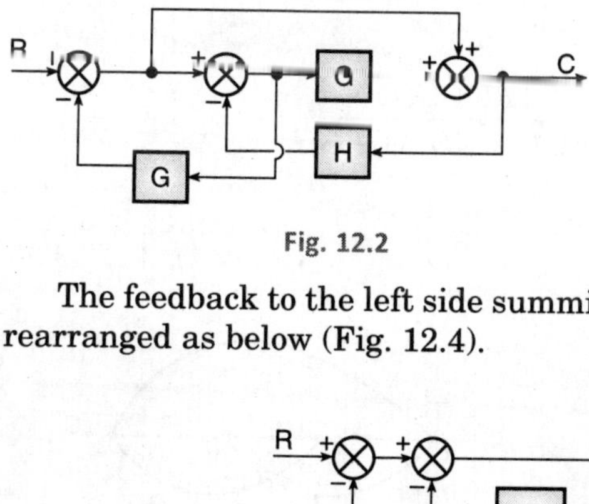

Fig. 12.2

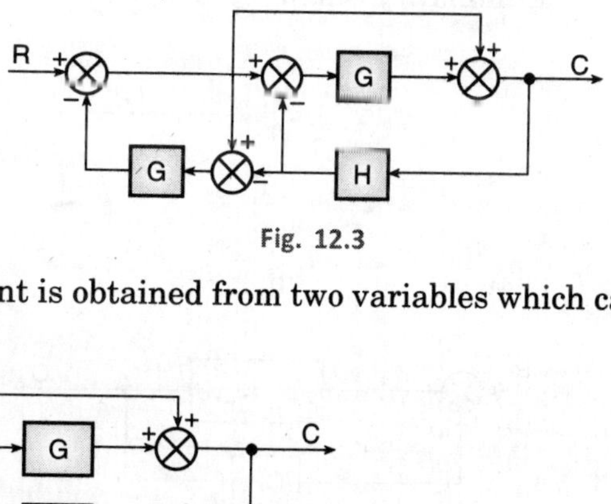

Fig. 12.3

The feedback to the left side summing point is obtained from two variables which can be rearranged as below (Fig. 12.4).

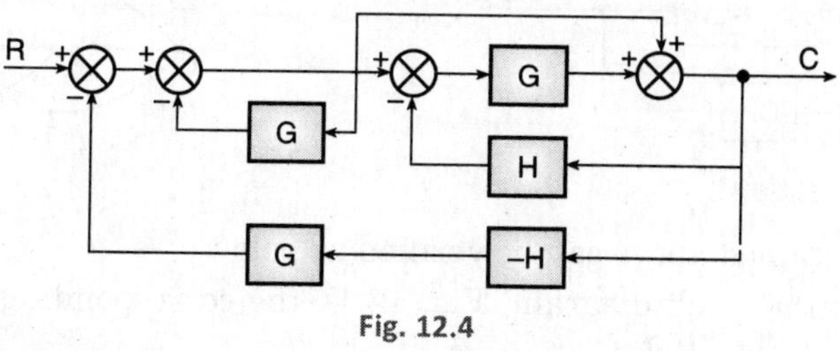

Fig. 12.4

Shifting take off point in the forward path before block G to a position after block G (Fig. 12.5).

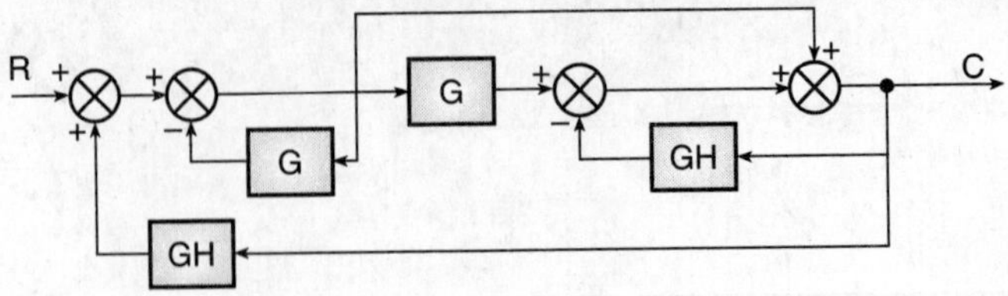

Fig. 12.5

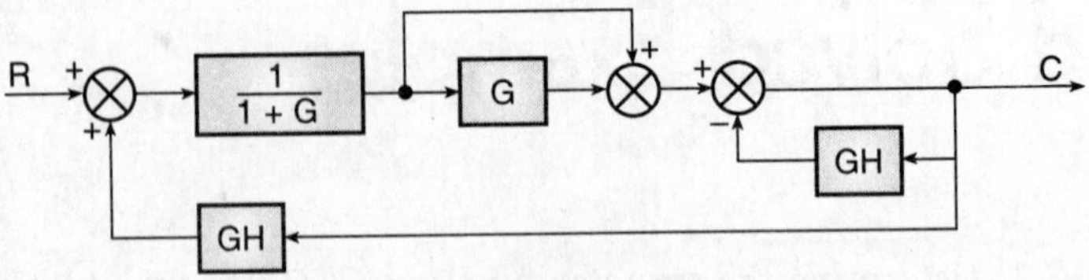

Fig. 12.6

Interchanging consecutive summing points after block G (Fig. 12.6)
Eliminating summing points after block G (Fig. 12.7).

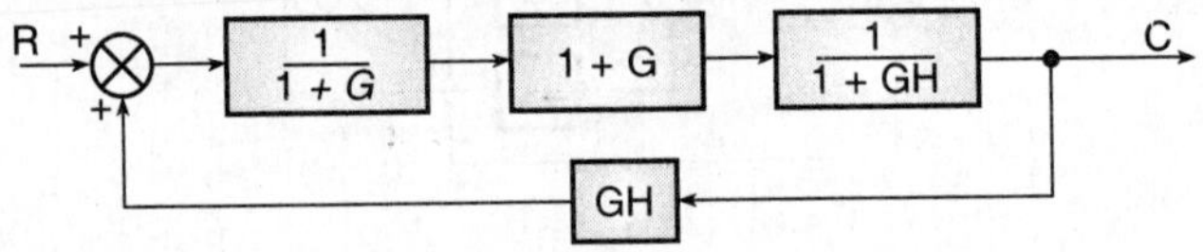

Fig. 12.7

Cascading forward path blocks (Fig. 12.8).
Eliminating summing point

$$\frac{C}{R} = \frac{\dfrac{1}{1+GH}}{1 - \dfrac{1}{1+GH} \cdot GH} = \frac{1/(1+GH)}{(1+GH-GH)\,(1+GH)}$$

or $\quad \dfrac{C}{R} = 1$ **Ans.**

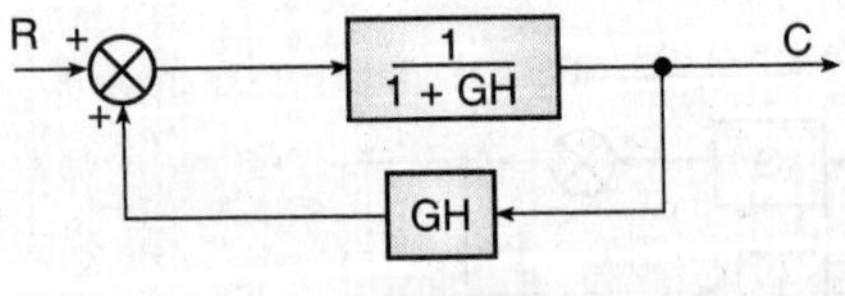

Fig. 12.8

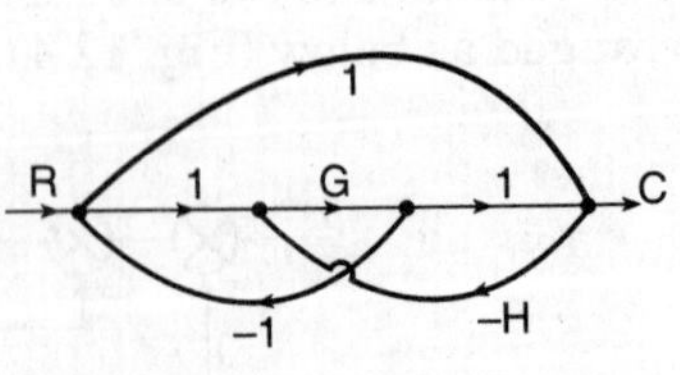

Fig.12.9

The answer obtained above can be verified as below.

On inspecting the block diagram (Fig. 12.1), the corresponding signal flow graph is developed as shown in Fig. 12.9.

The forward paths are

$$P_1 = (1)\,(G)\,(1) = G\;; \quad \Delta_1 = 1$$
$$P_2 = 1 \qquad ; \quad \Delta_2 = 1$$

The loops are

$$L_1 = (1)(G)(-1) = -G$$
$$L_2 = (G)(1)(-H) = -GH$$
$$L_3 = (1)(-H)(G)(-1) = GH$$

Applying Mason's gain formula

$$\frac{C}{R} = \frac{P_1\Delta_1 + P_2\Delta_2}{1-(L_1+L_2+L_3)}$$

$$= \frac{G\times1+1\times1}{1-(-G-GH+GH)}$$

$$= \frac{G+1}{G+1}$$

or $$\frac{C}{R} = 1 \text{ Verified.}$$

Example 12.2. *Determine the overall transfer function for the system whose block diagram representation is given below in Fig. 12.10.*

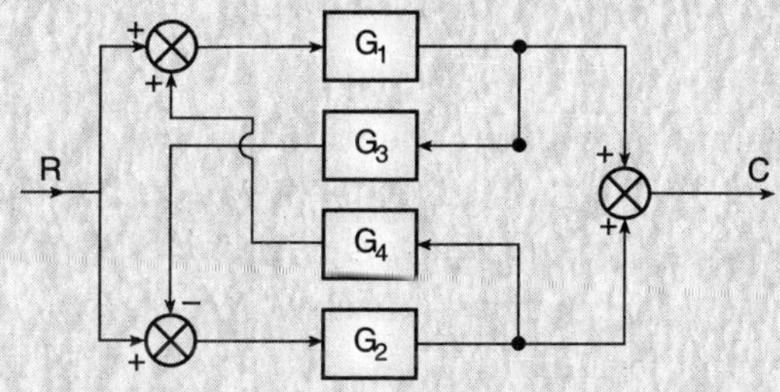

Fig 12.10

Solution. The signal flow graph for the given block diagram is shown in Fig. 12.11. The takeoff point before right side summing point corresponds to a transmittance of 1 between such points.

As per the signal flow graph, the forward paths and loops are identified as below :

$$P_1 = (1)(G_1)(1) = G_1;\ \Delta_1 = 1,\ P_2 = (1)(G_2)(1) = G_2;\ \Delta_2 = 1$$
$$P_3 = (1)(G_1)(-G_3)G_2(1) = -G_1G_2G_3;\ \Delta_3 = 1,$$
$$P_4 = (1)(G_2)(G_4)(G_1)(1) = G_1G_2G_4;\ \Delta_4 = 1$$
$$L_1 = (G_1)(-G_3)(G_2)(G_4) = -G_1G_2G_3G_4$$

Applying Mason's gain formula

$$\frac{C}{R} = \frac{P_1\Delta_1 + P_2\Delta_2 + P_3\Delta_3 + P_4\Delta_4}{1-L_1}$$

$$= \frac{G_1\cdot1 + G_2\cdot1 - G_1G_2G_3\cdot1 + G_1G_2G_4\cdot1}{1-(-G_1G_2G_3G_4)}$$

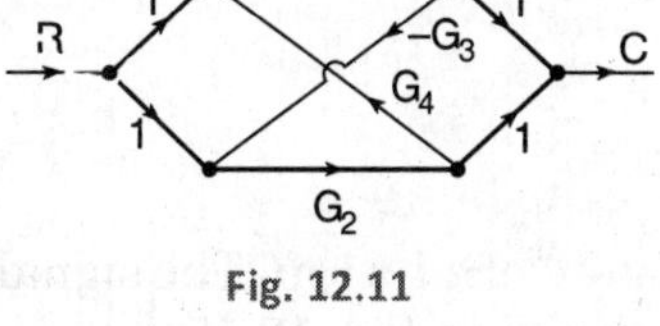

Fig. 12.11

$$\therefore \quad \frac{C}{R} = \frac{G_1+G_2+G_1G_2(G_4-G_3)}{1+G_1G_2G_3G_4} \textbf{ Ans.}$$

Example 12.3. *Develop signal flow graph for the block diagram shown in Fig. 12.12 and determine the ratio C/R.*

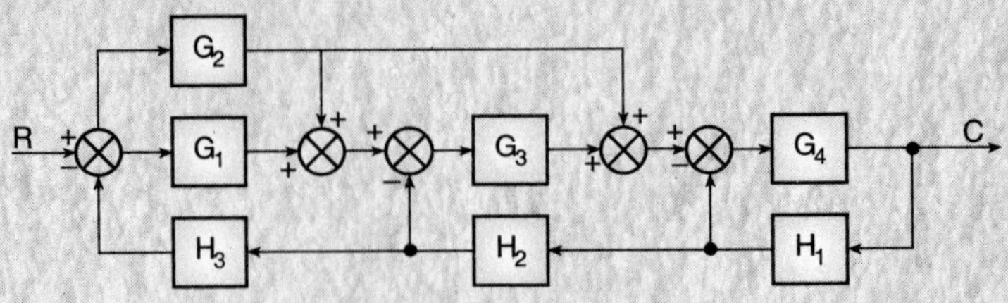

Fig. 12.12

Solution. The signal flow graph is drawn as shown in Fig. 12.13. The take off point before a summing point corresponds to transmittance of 1 between such points.

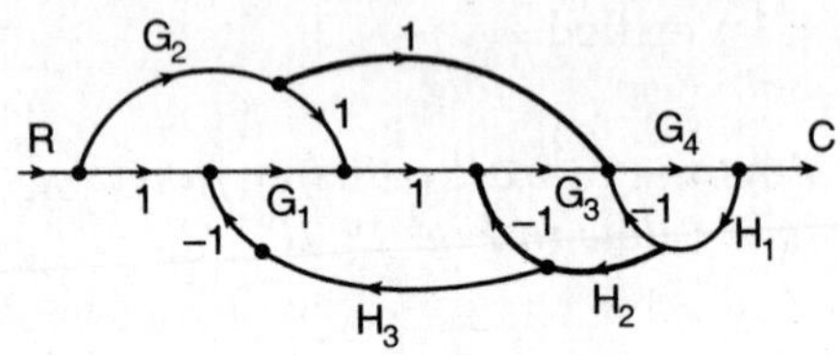

Fig. 12.13

The forward paths and the loops are identified below :

$$P_1 = G_2 G_4 \,;\, \Delta_1 = 1, P_2 = G_2 G_3 G_4 \,;\, \Delta_2 = 1, P_3 = G_1 G_3 G_4 \,;\, \Delta_3 = 1$$

$$L_1 = G_4 H_1 (-1) = -G_4 H_1, L_2 = G_3 G_4 H_1 H_2 (-1) = -G_3 G_4 H_1 H_2$$

$$L_3 = G_1 G_3 G_4 H_1 H_2 H_3 (-1) = -G_1 G_3 G_4 H_1 H_2 H_3$$

Applying Manson's gain formula

$$\frac{C}{R} = \frac{P_1 \Delta_1 + P_2 \Delta_2 + P_3 \Delta_3}{1-(L_1 + L_2 + L_3)} = \frac{G_2 G_4 \cdot 1 + G_2 G_3 G_4 \cdot 1 + G_1 G_3 G_4 \cdot 1}{1-(-G_4 H_1 - G_3 G_4 H_1 H_2 - G_1 G_3 G_4 H_1 H_2 H_3)}$$

or
$$\frac{C}{R} = \frac{G_2 G_4 + G_2 G_3 G_4 + G_1 G_3 G_4}{1 + G_4 H_1 + G_3 G_4 H_1 H_2 + G_1 G_3 G_4 H_1 H_2 H_3}$$ **Ans.**

Example 12.4. *Determine the transfer function C/R for the block diagram shown in Fig. 12.14.*

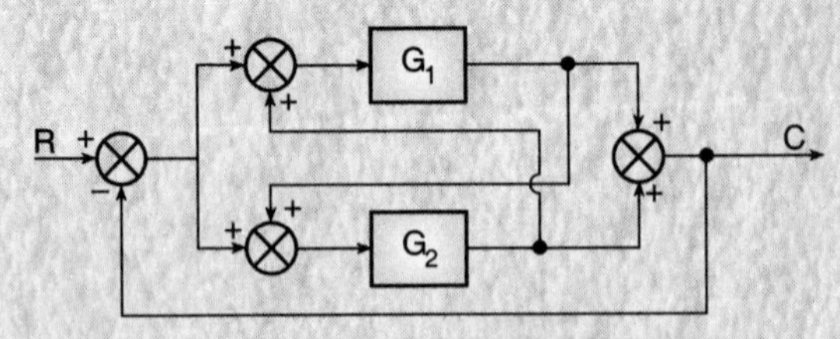

Fig. 12.14

Solution. The signal flow graph corresponding to the given block diagram is drawn as shown in Fig. 12.15.

As per the signal flow graph, the forward paths, path factors and closed-loops are identified as below :

$P_1 = (1)(G_1)(1) = G_1, \Delta_1 = 1$

$P_2 = (1)(G_2)(1) = G_2, \Delta_2 = 1$

$P_3 = (1)(G_1)(1)(G_2)(1) = G_1 G_2, \ \Delta_3 = 1$

$P_4 = (1)(G_2)(1)(G_1)(1) = G_1 G_2, \ \Delta_4 = 1$

$L_1 = (1)(G_1)(1)(-1) = -G_1$

$L_2 = (1)(G_2)(1)(-1) = -G_2$

$L_3 = (1)(G_1)(1)(G_2) = G_1 G_2$

$L_4 = (1)(G_1)(1)(G_2)(-1) = -G_1 G_2$

$L_5 = (1)(G_2)(G_1)(1)(-1) = -G_1 G_2$

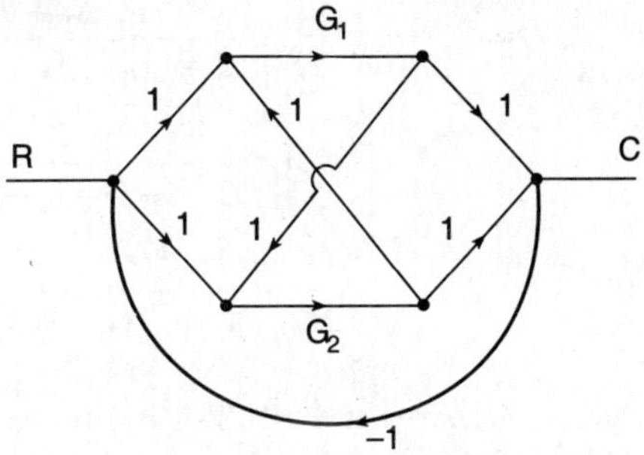

Fig. 12.15

Applying Mason's gain formula,

$$\frac{C}{R} = \frac{P_1\Delta_1 + P_2\Delta_2 + P_3\Delta_3 + P_4\Delta_4}{1-(L_1+L_2+L_3+L_4+L_5)} = \frac{G_1 \,.\, 1 + G_2 \,.\, 1 + G_1 G_2 \,.\, 1 + G_1 G_2 \,.\, 1}{1-(-G_1 - G_2 + G_1 G_2 - G_1 G_2 - G_1 G_2)}$$

or
$$\frac{C}{R} = \frac{G_1 + G_2 + 2G_1 G_2}{1 + G_1 + G_2 + G_1 G_2} \quad \textbf{Ans.}$$

Example 12.5. *Apply Mason's gain formula to determine the overall transfer function of a control system having the block diagram given below in Fig. 12.16.*

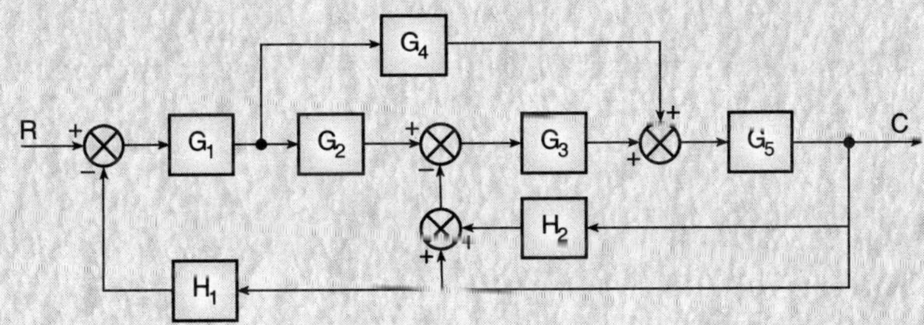

Fig. 12.16

Solution. The signal flow graph for the given below diagram is developed as shown in Fig. 12.17.

The forward paths and the loops are given below :

$P_1 = G_1 G_4 G_5\,;\ \Delta_1 = 1,\ P_2 = G_1 G_2 G_3 G_5\,;\ \Delta_2 = 1$

$L_1 = G_3 G_5(-1),\ L_2 = G_3 G_5(-H_2),\ L_3 = G_1 G_4 G_5(-H_1),$

$L_4 = G_1 G_2 G_3 G_5(-H_1)$

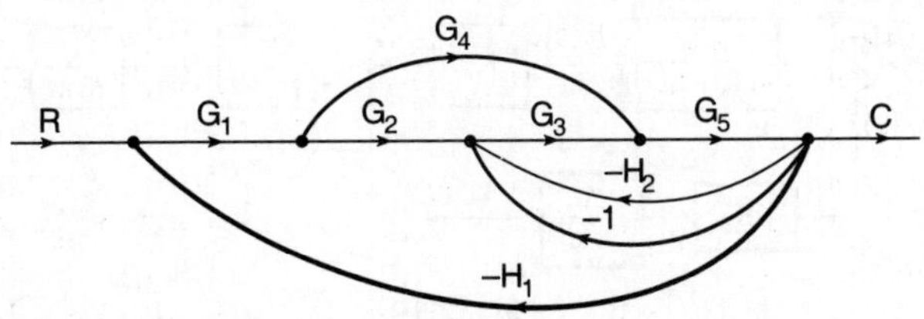

Fig. 12.17

Applying Mason's gain formula

$$\frac{C}{R} = \frac{P_1 \Delta_1 + P_2 \Delta_2}{1-(L_1 + L_2 + L_3 + L_4)}$$

$$= \frac{G_1 G_4 G_5.1 + G_1 G_2 G_3 G_5.1}{1-(-G_3G_5 - G_3 G_5 H_2 - G_1 G_4 G_5 H_1 - G_1 G_2 G_3 G_5 H_1)}$$

or $$\frac{C}{R} = \frac{G_1 G_4 G_5 + G_1 G_2 G_3 G_5}{1+G_3G_5 + G_3 G_5 H_2 + G_1 G_4 G_5 H_1 + G_1 G_2 G_3 G_5 H_1}$$ **Ans.**

Example 12.6. *An armature controlled d.c. motor is used for the lift as shown in Fig. 12.18.*

Determine the transfer function $X(s)/V_a(s)$.

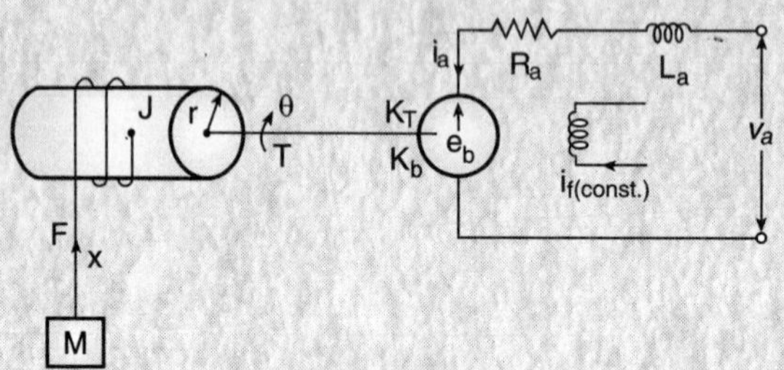

Fig. 12.18

Solution. The equations are as follows — Laplace transformed equations

$$v_a - e_b = R_a i_a + L_a \frac{di_a}{dt} \qquad V_a(s) - E_b(s) = R_a I_a(s) + sL_a I_a(s)$$

$$e_b = K_b \frac{d\theta}{dt} \qquad E_b(s) = s K_b \theta(s)$$

$$T = K_T i_a \qquad T(s) = K_T I_a(s)$$

$$T = \frac{J d^2\theta}{dt^2} \qquad T(s) = s^2 J \theta(s)$$

$$T = Fr \qquad F(s) = \frac{T(s)}{r}$$

$$F = M\frac{d^2x}{dt^2} \qquad F(s) = s^2 MX(s)$$

Using Laplace transformed equations, the block diagram (Fig. 12.19) is developed below :

Fig. 12.19

From the block diagram the transfer function is determined as

$$\frac{X(s)}{V_a(s)} = \frac{JK_T}{s\,Mr\,\{J(R_a + sL_a)\,s + K_T K_b\}} \quad \textbf{Ans.}$$

Example 12.7. *A unity feedback control system having open-loop transfer function*

$$G(s) = \frac{s+2}{s(s+1)}$$

is subjected to a unit step input. Determine the peak time t_p and % maximum overshoot.

Solution. The overall transfer functions of the system is given by

$$\frac{C(s)}{R(s)} = \frac{G(s)}{1+G(s)} = \frac{(s+2)/s\,(s+1)}{1+[(s+2)/s\,(s+1)]} = \frac{s+2}{s^2+2s+2}$$

$$\because \quad R(s) = \frac{1}{s}$$

$$\therefore \quad C(s) = \frac{1}{s}\cdot\frac{(s+2)}{s^2+2s+2} = \frac{K_1}{s} + \frac{K_2 s + K_3}{s^2+2s+2}$$

The coefficients can be determined as, $K_1 = 1$ and $K_2 = K_3 = -1$

$$\therefore \quad C(s) = \frac{1}{s} - \frac{s+1}{s^2+2s+2} = \frac{1}{s} - \frac{s+1}{(s+1)^2+(1)^2}$$

Taking inverse Laplace transform on both sides

$$c(t) = (1 - e^{-t}\cos t)$$

To determine peak time t_p, put $\dfrac{dc\,(t)}{dt} = 0$

$$\therefore \quad \frac{dc\,(t)}{dt} - 0 - [e^{-t}\,(-\sin t) + \cos t\,(-1)\,e^{-t}] = 0$$

or $\quad -[e^{-t}\,(-\sin t) - e^{-t}\cos t] = 0$

or $$\frac{\sin t}{\cos t} = -1$$

$$\therefore \quad \tan t = -1$$

or $$t_p = \tan^{-1}(-1)$$

$$\therefore \quad t_p = \pi - \tan^{-1} 1 = \pi - \frac{\pi}{4} = \frac{3\pi}{4} \text{ sec.}$$

$$c(t)_{max} = 1 - e^{-t_p}\cos t_p = 1 - e^{\frac{-3\pi}{4}}\cos\frac{3\pi}{4}$$

$$= 1 - 0.0947\,(-0.707) = 1 + 0.067 = 1.067$$

$$c_{ss} = \lim_{s\to 0} s\,C(s) = \lim_{s\to 0} s\,.\,\frac{1}{s}\cdot\frac{s+2}{s^2+2s+2} = 1$$

$$\%M_p = \frac{c(t)_{max} - c_{ss}}{c_{ss}} \times 100 = \frac{1.067-1}{1} \times 100 = 6.7\% \quad \textbf{Ans.}$$

Example 12.8. *Determine c(t) and % overshoot having given r(t) = 1 u(t) and*

$$\frac{C(s)}{R(s)} = \frac{8(s+1)}{s^2+2s+2}.$$

Solution. $\dfrac{C(s)}{R(s)} = \dfrac{8(s+1)}{s^2+2s+2}$; $r(t) = 1$, $\therefore$ $R(s) = \dfrac{1}{s}$

$$\therefore \quad \frac{C(s)}{R(s)} = \frac{1}{s}\cdot\frac{8\,(s+1)}{s^2+2s+2} = 8\left[\frac{1}{s}\cdot\frac{(s+1)}{(s^2+2s+2)}\right]$$

$$\frac{(s+1)}{s(s^2+2s+2)} = \frac{K_1}{s} + \frac{K_2 s + K_3}{s^2+2s+2}$$

or $$(s+1) = K_1\,(s^2+2s+2) + (K_2\, s + K_3\,)s$$

put $s = 0$

$$\therefore \quad 1 = K_1 \times 2 \quad \text{and} \quad K_1 = \frac{1}{2}$$

$$(s+1) = K_1\,[(s+1)^2+1] + (K_2 s + K_3)\,s$$

put $s = -1 + j1$

$$(-1+j1+1) = K_1\,[(-1+j1+1)^2+1] + [K_2\,(-1+j1) + K_3]\,(-1+j1)$$

$$\therefore \quad j\,1 = \text{K}_1 \times 0 + K_2\,(-1+j\,1)^2 + K_3\,(-1+j\,1)$$

or $$j\,1 = -j\,2K_2 - K_3 + j\,K_3$$

Equating real and imaginary part on both sides

$$-2\,K_2 + K_3 = 1$$

and $$K_3 = 0 \qquad \therefore \quad K_2 = -\frac{1}{2}$$

$$\therefore \quad C(s) = 8\left\{\frac{1}{2s} - \frac{1}{2}\cdot\frac{s}{(s^2+2s+2)}\right\} = 4\left\{\frac{1}{s} - \frac{s}{(s+1)^2+(1)^2}\right\}$$

$$= 4\left\{\frac{1}{s} - \frac{s+1}{(s+1)^2+(1)^2} + \frac{1}{(s+1)^2+(1)^2}\right\}$$

Taking inverse Laplace transform on both sides

$$c(t) = 4\,\{1 - (e^{-t}\cos t - e^{-t}\sin t)\}$$

$$= 4\left\{1 - \sqrt{2}\,e^{-t}\left(\frac{1}{\sqrt{2}}\cos t - \frac{1}{\sqrt{2}}\sin t\right)\right\}$$

$$= 4\left\{1 - \sqrt{2}\,e^{-t}\left(\sin\frac{\pi}{4}\cos t - \cos\frac{\pi}{4}\sin t\right)\right\}$$

$$= 4\left\{1 - \sqrt{2}\,e^{-t}\sin\left(\frac{\pi}{4} - t\right)\right\}$$

or $$c(t) = 4\left\{1 + \sqrt{2}\,e^{-t}\sin\left(t - \frac{\pi}{4}\right)\right\}$$

To determine peak time t_p, put $\frac{dc(t)}{dt} = 0$

$$\frac{dc(t)}{dt} = 0 + \sqrt{2}\left\{e^{-t}\cos\left(t-\frac{\pi}{4}\right) + \sin\left(t-\frac{\pi}{4}\right)(-1)\,e^{-t}\right\}$$

$\therefore$ $$e^{-t}\cos\left(t-\frac{\pi}{4}\right) - e^{-t}\sin\left(t-\frac{\pi}{4}\right) = 0$$

$\therefore$ $$\cos\left(t-\frac{\pi}{4}\right) = \sin\left(t-\frac{\pi}{4}\right) \qquad \therefore \quad \frac{\sin\left(t-\frac{\pi}{4}\right)}{\cos\left(t-\frac{\pi}{4}\right)} = 1$$

$\therefore$ $$\tan\left(t-\frac{\pi}{4}\right) = 1 \qquad \therefore \quad t_p - \frac{\pi}{4} = \tan^{-1} 1$$

or $$t_p = \frac{\pi}{4} + \frac{\pi}{4} = \frac{\pi}{2} \text{ sec.}$$

$$c(t)_{\max} = 4\left\{1+\sqrt{2}\,e^{-t_p}\sin\left(t_p-\frac{\pi}{4}\right)\right\} = 4\left\{1+\sqrt{2}\,e^{-\frac{\pi}{2}}\sin\left(\frac{\pi}{2}-\frac{\pi}{4}\right)\right\}$$

$$= 4\left\{1+\sqrt{2}\,e^{-\frac{\pi}{2}}\sin\frac{\pi}{2}\right\} = 4\left\{1+\sqrt{2}\,e^{-\frac{\pi}{2}}\times\frac{1}{\sqrt{2}}\right\}$$

$$= 4\left\{1+e^{-\frac{\pi}{2}}\right\} = 4\,\{1 + 0.2\} - 4.8$$

$$c_{ss} = \lim_{s\to 0} s\,C(s) = \lim_{s\to 0} s\cdot\frac{1}{s}\cdot\frac{8\,(s+1)}{s^2+2s+2} = 4$$

$$\%M_p = \frac{c(t)_{\max} - c_{ss}}{c_{ss}}\times 100 = \frac{4.8-4}{4}\times 100 = 20\% \quad \textbf{Ans.}$$

Example 12.9. *A second order servo system has poles (– 1 ± j2) and zero at (–1 + j 0). Its steady state output for a unit step is 0.8. Determine the transfer function. What is the peak overshoot for a unit step input ?*

Solution. $$\frac{C(s)}{R(s)} = \frac{K(s+1)}{(s+1-j2)\,(s+1+j2)} = \frac{K\,(s+1)}{s^2+2s+5}$$

$\therefore$ $$C(s) = R(s)\,.\,\frac{K\,(s+1)}{s^2+2s+5}$$

$\because$ $$R(s) = \frac{1}{s}$$

$\therefore$ $$C(s) = \frac{1}{s}\cdot\frac{K(s+1)}{s^2+2s+5}$$

$$C_{ss} = \lim_{s \to 0} sC(s)$$

$$= \lim_{s \to 0} s \cdot \frac{1}{s} \cdot \frac{K(s+1)}{s^2 + 2s + 5} = \lim_{s \to 0} \frac{K(s+1)}{s^2 + 2s + 5}$$

$$= \frac{K}{5} \qquad \because c_{ss} = 0.8$$

$$\therefore \quad 0.8 = \frac{K}{5} \qquad \therefore \quad K = 4$$

Hence, the transfer function is

$$\frac{C(s)}{R(s)} = \frac{4(s+1)}{s^2 + 2s + 5} \quad \textbf{Ans.}$$

$$C(s) = R(s) \cdot \frac{4(s+1)}{s^2 + 2s + 5}$$

$$\because \quad R(s) = \frac{1}{s}$$

$$C(s) = \frac{1}{s} \cdot \frac{4\,(s+1)}{s^2 + 2s + 5} = 4\left\{\frac{(s+1)}{s\,(s^2 + 2s + 5)}\right\}$$

$$= 4\left\{\frac{K_1}{s} + \frac{K_2 s + K_3}{(s^2 + 2s + 5)}\right\}$$

The coefficients can be determined as

$$K_1 = \frac{1}{5}\ ;\ K_2 = -\frac{1}{5} \text{ and } K_3 = \frac{3}{5}$$

$$\therefore \quad C(s) = 4\left\{\frac{1}{5s} - \frac{1}{5} \cdot \frac{(s-3)}{(s^2 + 2s + 5)}\right\} = \frac{4}{5}\left\{\frac{1}{s} - \frac{s-3}{(s^2 + 2s + 5)}\right\}$$

$$= \frac{4}{5}\left\{\frac{1}{s} - \frac{s+1}{(s^2 + 2s + 5)} + \frac{4}{(s^2 + 2s + 5)}\right\}$$

$$= \frac{4}{5}\left\{\frac{1}{s} - \frac{s+1}{(s+1)^2 + (2)^2} + 2 \cdot \frac{2}{(s+1)^2 + (2)^2}\right\}$$

Taking inverse Laplace transform on both sides

$$c(t) = \frac{4}{5}\,(1 - e^{-t}\cos 2t + 2\,e^{-t}\sin 2t)$$

or

$$c(t) = \frac{4}{5}\,\{1 - e^{-t}\,(2\sin 2t - \cos 2t)\}$$

$$\frac{dc(t)}{dt} = -\frac{4}{5}\,[e^{-t}\,\{2 \times 2\cos 2t - 2(-\sin t)\} + (2\sin t - 2\cos 2t)\,(-1)\}\,e^{-t}]$$

$$= -\frac{4}{5}\,e^{-t}\,(4\cos 2t + 2\sin 2t - 2\sin 2t + 2\cos 2t)$$

$$= -\frac{4}{5}\,e^{-t}\,5\cos 2t$$

To determine peak time t_p, put $\frac{dc(t)}{dt} = 0$

$$\therefore \quad -e^{-t}(5\cos 2t) = 0$$

or $$5\cos 2t = 0$$

$$\therefore \quad 2t = \cos^{-1}(0)$$

or $$2t_p = \cos^{-1}(0) = \frac{\pi}{2}$$

$$\therefore \quad t_p = \frac{\pi}{2\times 2} = 0.785 \text{ sec.}$$

$$c(t)_{\max} = \frac{4}{5}\,(1 - e^{-t_p}\cos 2t_p + e^{-t_p}\times 2\sin 2t_p)$$

$$= \frac{4}{5}\left(1 - e^{-0.785}\cos\frac{2\pi}{4} + e^{-0.785}\times 2\sin\frac{2\pi}{4}\right)$$

$$= \frac{4}{5}\left(1 - 0.456\cos\frac{\pi}{2} + 0.456\times 2\sin\frac{\pi}{2}\right)$$

$$= \frac{4}{5}(1 - 0 + 0.456\times 2\times 1) = 1.53$$

$\because \quad c_{ss} = 0.8$ (given)

$$\%M_p = \frac{c(t)_{\max} - c_{ss}}{c_{ss}}\times 100 = \frac{1.53 - 0.8}{0.8} = 91.2\% \quad \textbf{Ans.}$$

Example 12.10. *Determine the unit step response for a system having overall transfer function given as*

$$\frac{C(s)}{R(s)} = \frac{1}{(s+1)(s^2+1)}.$$

Solution. $$\frac{C(s)}{R(s)} = \frac{1}{(s+1)(s^2+1)} \quad ; \because \quad R(s) = \frac{1}{s}$$

$$\therefore \quad C(s) = \frac{1}{s\,(s+1)(s^2+1)}$$

$$\frac{1}{s\,(s+1)(s^2+1)} = \frac{K_1}{s} + \frac{K_2}{s+1} + \frac{K_3 s + K_4}{(s^2+1)}$$

$$\therefore \quad 1 = K_1(s+1)(s^2+1) + K_2(s^2+1)s + (K_3 s + K_4)\,s\,(s+1)$$

put $s = 0$ $$1 = K_1\times 1\times 1 + 0 + 0\,;\ K_1 = 1$$

put $s = -1$ $$1 = 0 + K_2(1+1)(-1) + 0\,;\ K_2 = -1/2$$

put $s = j1$ $$1 = 0 + 0 + (j1\,K_3 + K_4)(j1)(j1+1)$$

$$\therefore \quad -K_3 + jK_4 = \frac{1}{j+1} \quad \text{or} \quad -K_3 + jK_4 = \frac{-j+1}{2}$$

Equating real and imaginary part on both sides

$$K_3 = -\frac{1}{2} \quad \text{and} \quad K_4 = -\frac{1}{2}$$

$$\therefore \qquad C(s) = \frac{1}{s} - \frac{1}{2}\cdot\frac{1}{s+1} - \frac{1}{2}\frac{s+1}{s^2+1}$$

$$= \frac{1}{s} - \frac{1}{2}\cdot\frac{1}{s+1} - \frac{1}{2}\cdot\frac{s}{s^2+1} - \frac{1}{2}\cdot\frac{1}{s^2+1}$$

Taking inverse Laplace transform on both sides

$$c(t) = 1 - \frac{1}{2}e^{-t} - \frac{1}{2}\cos t - \frac{1}{2}\sin t$$

$$= 1 - \frac{1}{2}e^{-t} - \frac{1}{2}(\cos t + \sin t)$$

$$= 1 - \frac{1}{2}e^{-t} - \frac{\sqrt{2}}{2}\left(\frac{1}{\sqrt{2}}\cos t + \frac{1}{\sqrt{2}}\sin t\right)$$

$$= 1 - \frac{1}{2}e^{-t} - \frac{1}{\sqrt{2}}\left(\sin\frac{\pi}{4}\cos t + \cos\frac{\pi}{4}\sin t\right)$$

or $$c(t) = 1 - \frac{1}{2}e^{-t} - \frac{1}{\sqrt{2}}\sin\left(t + \frac{\pi}{4}\right) \quad \textbf{Ans.}$$

Example 12.11. *Determine the time response to (i) unit step input (ii) unit impulse input for the closed-loop control system shown in Fig. 12.20*

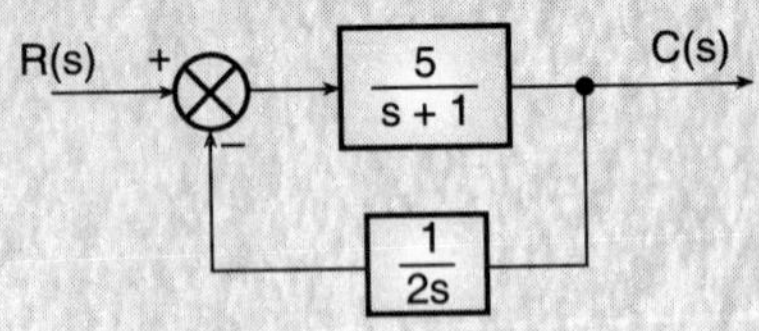

Fig. 12.20

Solution. $$\frac{C(s)}{R(s)} = \frac{G(s)}{1 + G(s)\,H(s)} = \frac{\dfrac{5}{s+1}}{1 + \dfrac{5}{(s+1)} \times \dfrac{1}{2s}} = \frac{10s}{2s^2 + 2s + 5}$$

or $$\frac{C(s)}{R(s)} = \frac{5s}{s^2 + s + 2.5}$$

(*i*) Unit step response : $R(s) = \dfrac{1}{s}$

$$\therefore \qquad C(s) = \frac{1}{s}\cdot\frac{5s}{s^2 + s + 2.5} = \frac{5}{s^2 + s + 0.25 + 2.5 - 0.25}$$

$$= \frac{5}{(s+0.5)^2 + (1.5)^2} = \frac{5}{1.5}\left\{\frac{1.5}{(s+0.5)^2 + (1.5)^2}\right\}$$

Taking inverse Laplace transform on both sides

$$c(t) = \frac{10}{3} \mathcal{L}^{-1} \frac{1.5}{(s+0.5)^2 + (1.5)^2}$$

or $$c(t) = \frac{10}{3} e^{-0.5t} \sin 1.5t \textbf{ Ans.}$$

(*ii*) Unit impulse response : $R(s) = 1$

$$\therefore \quad C(s) = \frac{5s}{s^2 + s + 5} = \frac{5s}{s^2 + s + 0.25 + 2.5 - 0.25}$$

$$= \frac{5s}{(s+0.5)^2 + (1.5)^2}$$

$$= 5\left\{\frac{(s+0.5)}{(s+0.5)^2 + (1.5)^2} - \frac{0.5}{(s+0.5)^2 + (1.5)^2}\right\}$$

$$= 5\left\{\frac{(s+0.5)}{(s+0.5)^2 + (1.5)^2} - \frac{0.5}{1.5} \cdot \frac{1.5}{(s+0.5)^2 + (1.5)^2}\right\}$$

Taking inverse Laplace transform on both sides

$$c(t) = 5\left\{e^{-0.5t} \cos 1.5t - \frac{0.5}{1.5} e^{-0.5t} \sin 1.5t\right\}$$

$$= \frac{5}{1.5} e^{-0.5t} (1.5 \cos 1.5t - 0.5 \sin 1.5t)$$

$$= \frac{5}{1.5}\sqrt{1.5^2 + 0.5^2}\; e^{-0.5t}$$

$$\left(\frac{1.5}{\sqrt{1.5^2 + 0.5^2}} \cos 1.5t - \frac{0.5}{\sqrt{1.5^2 + 0.5^2}} \sin 1.5t\right)$$

$$= 5.27\, e^{-0.5t} (\sin 71.56° \cos 1.5t - \cos 71.56° \sin 1.5t)$$

or $$c(t) = 5.27\, e^{-0.5t} \sin (71.56° - 1.5t) \quad \textbf{Ans.}$$

Example 12.12. *Determine* ω_d, ω_n, $\zeta\omega_n$, $\% M_p$, t_s, *the time interval between maximum and minimum values of c(t). The block diagram is shown in Fig. 12.21 and the input being a unit step.*

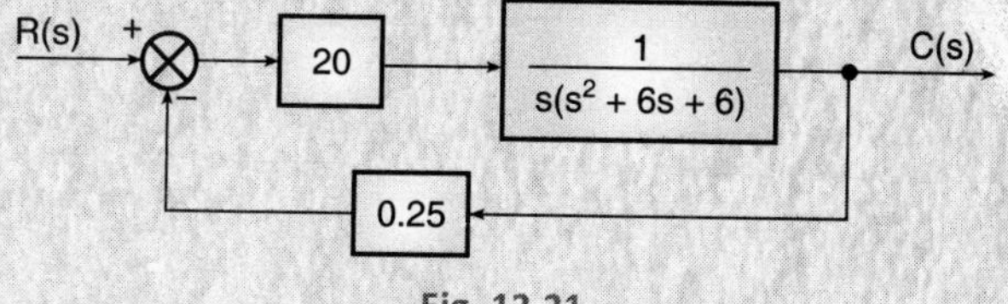

Fig. 12.21

Solution. $$G(s) = \frac{20}{s\,(s^2 + 6s + 6)}\ ; H(s) = 0.25$$

The characteristic equation is given by $1 + G(s) H(s) = 0$

$$\therefore \quad 1 + \frac{20}{s(s^2 + 6s + 6)} \times 0.25 = 0$$

or

$$s^3 + 6s^2 + 6s + 5 = 0$$

Above equation is factored as below

$$(s + 5)(s^2 + s + 1) = 0$$

The time constant due to first order term : $T_1 = \frac{1}{5} = 0.2$ sec

The roots of the quadratic term are :

$$s = \frac{-1 \pm \sqrt{1 - 4 \times 1}}{2} = -0.5 \pm j\,0.89$$

The time constant due to quadratic term : $T_2 = \frac{1}{0.5} = 2$ sec

Neglect T_1 in comparison to T_2, then the characteristic equation is approximated as below

$$s^2 + s + 1 = 0$$

As per the second order characteristic equation

$$\omega_n = \sqrt{1} = 1 \text{ rad/sec.} \quad \textbf{Ans.}$$

$$2\zeta\omega_n = 1$$

$$\zeta = \frac{1}{2\,\omega_n} \qquad \therefore \quad \zeta = \frac{1}{2 \times 1} = 0.5$$

$$\omega_d = \omega_n \sqrt{1 - \xi^2} = 1\sqrt{1 - 0.5^2} = 0.86 \text{ rad./sec.} \quad \textbf{Ans.}$$

$$\% M_p = \frac{-\zeta\pi}{e^{\sqrt{1-\zeta^2}}} \times 100 = \frac{-0.5\pi}{e^{\sqrt{1-0.5^2}}} \times 100 = 16.3\% \quad \textbf{Ans.}$$

Settling time $$t_s = \frac{4}{\zeta\omega_n} = \frac{4}{0.5 \times 1} = 8.0 \text{ sec.} \quad \textbf{Ans.}$$

Peak time $$t_p = \frac{\pi}{\omega_n \sqrt{1 - \zeta^2}} = \frac{\pi}{1\sqrt{1 - 0.5^2}} = 3.65 \text{ sec.}$$

$$\underset{\text{1st under shoot}}{t_u} = \frac{2\pi}{\omega_n \sqrt{1 - \zeta^2}} = \frac{\pi}{1\sqrt{1 - 0.5^2}} = 7.3 \text{ sec.}$$

time interval between max. and min. values of $c(t)$

$$= t_u - t_p = 7.3 - 3.65 = 3.65 \text{ sec.} \quad \textbf{Ans.}$$

Example 12.13. *A second order control system incorporating proportional-plus-derivative control is shown in Fig. 12.22.*

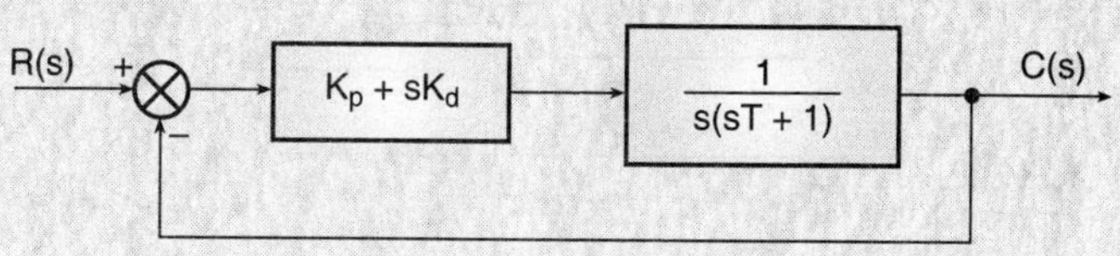

Fig. 12.22

Derive an expression for the steady state error when the system is subjected to a unit ramp input.

Solution.
$$\frac{E(s)}{R(s)} = \frac{1}{1+G(s)\,H(s)}$$

$\because$
$$G(s) = \frac{(K_p + s\,K_d)K}{s(sT+1)} \; ; H(s) = 1$$

$\therefore$
$$\frac{E(s)}{R(s)} = \frac{1}{1+\dfrac{(K_p+sK_d)K\times 1}{s(sT+1)}} = \frac{s^2\,T+s}{s^2T+s+K_pK+s\,K_dK}$$

or
$$E(s) = R(s)\,\frac{s^2\,T+s}{s^2T+s+K_pK+s\,K_dK}$$

$$e_{ss} = \lim_{s\to 0} s\,E\,(s) = \lim_{s\to 0} s\,R\,(s)\,\frac{s^2T+s}{s^2T+s+K_pK+sK_dK}$$

$\because$
$$R(s) = \frac{1}{s^2}$$

$\therefore$
$$e_{ss} = \lim_{s\to 0} s\cdot\frac{1}{s^2}\,\frac{s(sT+1)}{s^2T+s+K_pK+s\,K_dK}$$

or
$$e_{ss} = \frac{1}{K_pK} \quad \textbf{Ans.}$$

Example 12.14. *A closed-loop control system shown in Fig. 12.23 uses a PID controller.*

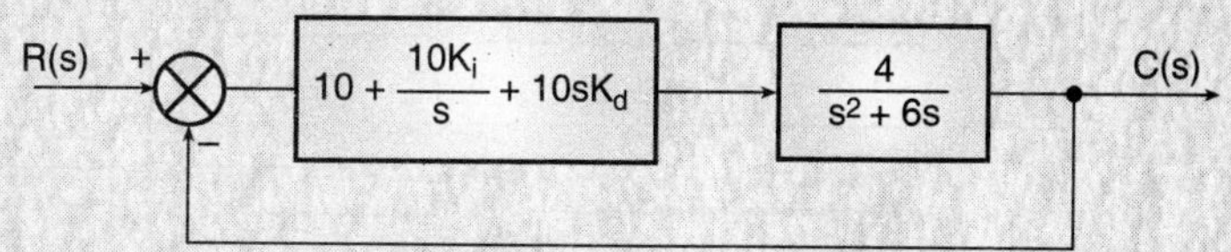

Fig. 12.23

(i) If $K_i = 0$, determine the value of K_d for critical damping.

(ii) With K_d as determined in (i) determine the limiting value of K_i such that stability is maintained.

Solution. (*i*) $K_i = 0$, the block diagram is shown in Fig. 12.24

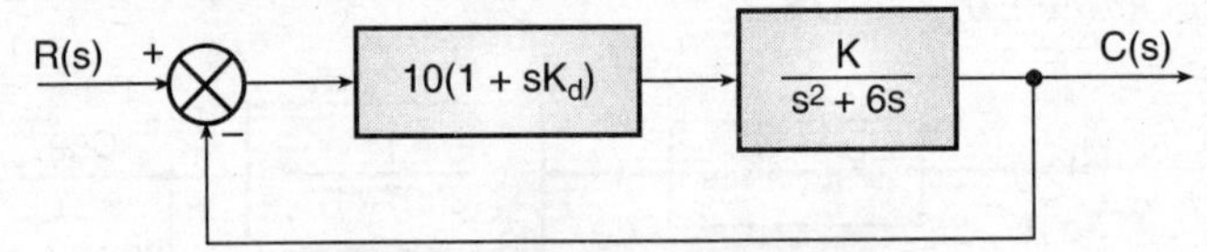

Fig 12.24

The characteristic equation is given by $1 + G(s)\,H(s) = 0$.

$$G(s) = \frac{10\,(1 + sK_d) \times 4}{s^2 + 6\,s} \;;\; H(s) = 1$$

$$\therefore \qquad 1 + \frac{10\,(1 + sK_d) \times 4 \times 1}{s^2 + 6\,s} = 0$$

or $\quad s^2 + 6s + 40 + 40\,sK_d = 0 \quad$ or $\quad s^2 + (6 + 40K_d)s + 40 = 0$

As per second order characteristic equation

$$\omega_n = \sqrt{40} \text{ rad/sec.}$$

$$2\zeta\omega_n = (6 + 40\,K_d)$$

For critical damping $\zeta = 1$

$$\therefore \qquad 2 \times 1 \times \sqrt{40} = (6 + 40K_d)$$

$$\therefore \qquad K_d = \frac{2\sqrt{40} - 6}{40} = 0.166 \quad \textbf{Ans.}$$

$$G(s) = \frac{10s + 10K_i + 1.66s^2}{s} \times \frac{4}{s^2 + 6s}$$

or

$$G(s) = \frac{40s + 40K_i + 6.64s^2}{s^3 + 6s^2}$$

$$H(s) = 1$$

The characteristic equation is given by

$$1 + \frac{(40s + 40K_i + 6.64s^2) \times 1}{s^3 + 6s^2} = 0$$

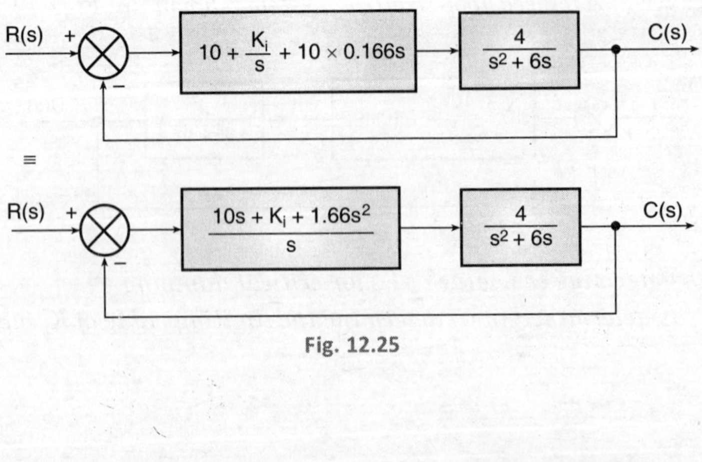

Fig. 12.25

or $$s^3 + 6s^2 + 40s + 40K_i + 6.64s^2 = 0$$

or $$s^3 + 12.64s^2 + 40s + 40K_i = 0$$

The Routh array is formed below :

s^3	1	40
s^2	12.64	$40K_i$
s^1	$\dfrac{12.64 \times 40 - 40K_i \times 1}{12.64}$	
s^0	$40K_i$	

For stability, $K_i > 0$ and $(12.64 \times 40 - 40K_i \times 1) \geq 0$

$\therefore$ limiting value of K_i for stability is

$$K_i = \frac{12.64 \times 40}{40} = 12.64 \quad \textbf{Ans.}$$

Example 12.15. *A closed-loop control system is shown in Fig. 12.26. Determine the range of K for stability.*

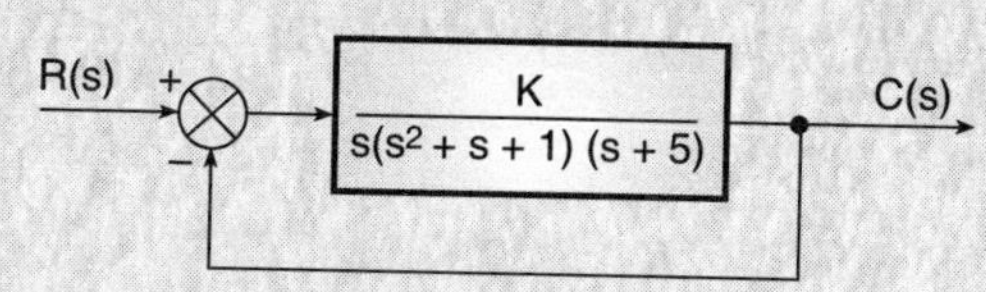

Fig. 12.26

Solution. $$G(s) = \frac{K}{s(s^2 + s + 1)(s + 5)}$$

$$H(s) = 1$$

The characteristic equation is

$$1 + G(s)\,H(s) = 0$$

$$\therefore \quad 1 + \frac{K}{s(s^2 + s + 1)(s + 5)} = 0$$

or $$s(s^2 + s + 1)(s + 5) + K = 0$$

or $$s^4 + 6s^3 + 6s^2 + 5s + K = 0$$

Applying Routh Hurwitz criterion

s^4	1	6	K
s^3	6	5	
s^2	5.1	K	
s^1	$\dfrac{5.1 \times 5 - 6\,K}{5.1}$		
s^0	K		

For stability,

(*i*) $$\frac{5.1 \times 5 - 6K}{5.1} < 0$$

or $$6K < 5.1 \times 5$$

$$K < \frac{5.1 \times 5}{6}$$

$$K < 4.25$$

(*ii*) $K > 0$ **Ans.**

Example 12.16. *Determine the stability range of K for a system having characteristic equation,*

$$s^4 + 2s^3 + 10s^2 + (K - 10)s + K = 0.$$

Solution. The Routh array is formed below :

s^4	1	10	K
s^3	2	$(K-10)$	
s^2	$\frac{2 \times 10 - 1(K-10)}{2} = \frac{(30-K)}{2}$	K	
s^1	$\frac{\frac{(30-K)}{2}(K-10) - 2K}{(30-K)/2} = \frac{-(K-22.9)(K-13.1)}{30-K}$		
s^0	K		

For no change of sign in the first column elements, following conditions are to be satisfied,

(*i*) $K < 30$

(*ii*) $K < 22.9$ and $K > 13.1$

(*iii*) $K > 0$

Thus, for stability, the range of K is determined as below,

$$13.1 < K < 22.9$$ **Ans.**

Example 12.17. *The forward transfer function of a unity feedback control system is given below :*

$$G(s) = \frac{K}{(s^2 + 3s + 2)(s^2 + 6s + 24)}$$

Determine the limiting value of K for stability.

Solution. The characteristic equation is

$$1 + G(s) = 0$$

$$\therefore \quad 1 + \frac{K}{(s^2 + 3s + 2)(s^2 + 6s + 24)} = 0$$

or $$(s^2 + 3s + 2)(s^2 + 6s + 24) + K = 0$$

or $$s^4 + 9s^3 + 44s^2 + 84s + K + 48 = 0$$

The Routh array is formed below

s^4	1	44	$K + 48$
s^3	9	84	
s^2	$\frac{9 \times 44 - 84 \times 1}{9} = 34.66$	$K + 48$	
s^1	$\frac{34.66 \times 84 - 9(K + 48)}{34.66} = 74.84 - 0.259K$		
s^0	$K + 48$		

For stability, $74.84 - 0.259K > 0$

$$\therefore \quad K < \frac{74.84}{0.259} \quad \therefore \quad K < 288.9$$

The limiting value for stability is $K = 288.9$ **Ans.**

Example 12.18. *The characteristic equation of a feedback control system is given below :*

$$s^5 + s^4 + 4s^3 + 4s^2 + s + 1 = 0. \text{ Determine the stability.}$$

Solution. The Routh table is formed below :

s^5	1	4	1	
s^4	1	4	1	A(s)
s^3	0	0		
s^3	*4	8		*coeffs. of $\frac{dA(s)}{ds}$
s^2	2	1		
s^1	6	0		
s^0	1			

The elements of the s^3-row are zero. The auxiliary equation is formed using elements of the s^4-row

$$\therefore \quad A(s) = s^4 + 4s^2 + 1 \quad \text{and} \quad \frac{dA(s)}{ds} = 4s^3 + 8s$$

The coefficients of $\frac{dA(s)}{ds}$ are substituted in the s^3-row of the Routh table and the Routh array is completed. There is no sign change in the first column of the Routh array.

The four roots of the auxiliary equation are also the four roots of the characteristic equation.

The roots of the auxiliary equation are determined below

$$s^4 + 4s^2 + 1 = 0, \text{ put } s^2 = x$$

$$\therefore \quad x^2 + 2x + 1 = 0 \; ; x = -0.265, -1.935 \therefore s_1, s_2 = \pm j\sqrt{0.265} \; ; s_3, s_4 = \pm \sqrt{1.935}$$

The roots are non-repeating on the s-plane imaginary axis. Hence, the system is marginally stable. **Ans.**

Example 12.19. *Determine the stability of a closed-loop control system having characteristic equation as*

$$s^5 + s^4 + 4s^3 + 4s^2 + 4s + 4 = 0$$

Solution. The Routh table is formed below :

s^5	1	4	4	
s^4	*1	4	4	$*A(s) = s^4 + 4s^2 + 4$
s^3	0	0		$\frac{dA(s)}{ds} = 4s^3 + 8s^2$
s^3	**4	8		*coeffs. of $\frac{dA(s)}{ds}$
s^2	† 2	4		† $A(s) = 2s^2 + 4$
s^1	0	0		$\frac{dA(s)}{ds} = 4s$
s^1	†† 4	0		†† coeff. of $\frac{dA(s)}{ds}$

There is no sign change in the first column of the Routh table formed, but two rows s^3 and s^1 have zero elements. The stability is examined as below :

The roots of the auxiliary equation for the s^4-row are also the roots of the characteristic equation and determined below

$$s^4 + 4s^2 + 4 = 0$$

put $\quad s^2 = x \quad \therefore \quad x^2 + 4x + 4 = 0$

or $\quad (x + 2)^2 = 0 \quad \therefore \quad (s^2 + 2)^2 = 0$

The roots are $s_1, s_2 = \pm j\sqrt{2}$;

$s_3, s_4 = \pm J\sqrt{2}$

Thus the characteristic equation has multiple roots on the s-plane imaginary axis. Hence the system is unstable. **Ans.**

Example 12.20. *The presence of roots of the characteristic equation on the imaginary axis $s = \pm j\omega$, such that the number of roots in RHS of s-plane being nil make the system to have sustained oscillations of frequency of ω.*

The open-loop transfer function of a unity feedback control system is given below,

$$G(s) = \frac{K(s+2)}{s^3 + \beta s^2 + 4s + 1}$$

The system has to oscillate at a frequency $\omega = 3.0$ rad./sec.

Determine the value of the parameter K and β.

Solution. The characteristic equation is

$$1 + \frac{K(s+2)}{s^3 + \beta s^2 + 4s + 1} = 0$$

or $$s^3 + \beta s^2 + 4s + 1 + K(s + 2) = 0$$

or $$s^3 + \beta s^2 + (K + 4)s + (2K + 1) = 0$$

The Routh table is tabulated below

$$\begin{array}{lll} s^3 & 1 & K+4 \\ s^2 & \beta & 2K+1 \\ s^1 & \dfrac{\beta(K+4) - 1 \times (2K+1)}{\beta} = (K+4) - \dfrac{(2K+1)}{\beta} & \\ s^0 & 2K+1, & \end{array}$$

1. For sustained oscillations

$$(K + 4) - \frac{(2K+1)}{\beta} = 0 \quad \text{or} \quad \beta = \frac{(2K+1)}{(K+4)} \qquad ...(1)$$

2. The auxiliary equation is formed by s^2-row as below

$$\beta s^2 = (2K + 1) = 0 \quad \text{or} \quad s^2 = -\frac{(2K+1)}{\beta}$$

As the roots are to lie on imaginary axis for sustained oscillations, put $s = j\omega$

$$\therefore \quad (j\omega)^2 = -\frac{(2K+1)}{\beta}$$

$$-\omega^2 = -\frac{(2K+1)}{\beta}; \qquad \therefore \beta = \frac{(2K+1)}{\omega^2} \qquad ...(2)$$

Equating (1) and (2) $\dfrac{2K+1}{K+4} = \dfrac{2K+1}{\omega^2}$

$$\therefore \quad \omega^2 = K + 4 \qquad \because \omega = 3.0 \text{ rad./sec. (given)}$$

$$\therefore \quad 3^2 = K + 4 \qquad \therefore K = 5 \quad \textbf{Ans.}$$

and $$\beta = \frac{2K+1}{\omega^2} = \frac{2 \times 5 + 1}{3^2} = \frac{11}{9} \quad \textbf{Ans.}$$

Example 12.21. *Apply Routh-Hurwitz criterion to determine the limiting value of K such that the feed back control system given below exhibits sustained oscillations. Determine the frequency of oscillations.*

Solution.

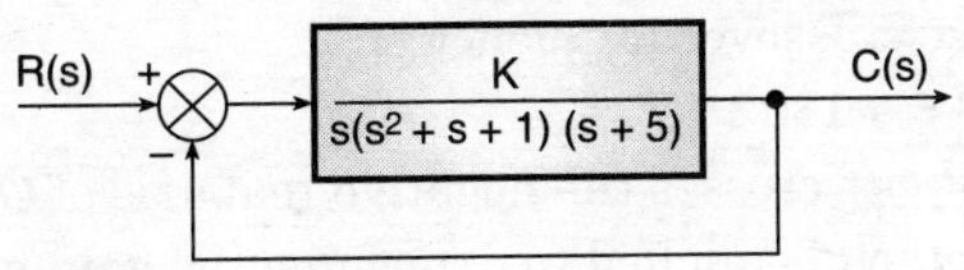

Fig. 12.27

The characteristic equation is given by

$$1 + \frac{Ks(2s+1)}{s^3 + 3s + 20} = 0$$

$$\therefore \quad s^3 + 3s + 20 + Ks(2s + 1) = 0 \quad \text{or} \quad s^3 + 2K s^2 + (K + 3)s + 20 = 0$$

The Routh table is formed below

$$\begin{array}{lll} s^3 & 1 & K+3 \\ s^2 & 2K & 20 \\ s^1 & \dfrac{2K(K+3) - 1 \times 20}{2K} = \dfrac{2K^2 + 6K - 20}{2K} & \\ s^0 & 20 & \end{array}$$

For sustained oscillations

$$2K^2 + 6K - 20 = 0 \quad \text{or} \quad K^2 + 3K - 10 = 20$$

$$\therefore \qquad K = \frac{-3 \pm \sqrt{3^2 + 4 \times 10}}{2} = \frac{-3 \pm 7}{2} = 2, -5$$

Considering positive value, $K = 2$ **Ans.**

The frequency of oscillations is calculated by solving equation formed by s^2-row of the Routh table.

$$\therefore \qquad 2Ks^2 + 20 = 0 \quad \text{or} \quad 2 \times 2s^2 + 20 = 0$$

$$\therefore \qquad s = \pm j\sqrt{5} \quad \text{or} \quad \omega = \sqrt{5} \text{ rad./sec.} \quad \textbf{Ans.}$$

Example 12.22. *The open-loop transfer function of a unity feedback control system is given below :*

$$G(s) = \frac{K}{s(s\,T_1 + 1)(s\,T_2 + 1)}$$

K_1, T_1 and T_2 have positive values. Determine the gain margin.

Solution. $$G(s) = \frac{K}{s(s\,T_1 + 1)(s\,T_2 + 1)}$$

put $s = j\omega$

$$\therefore \qquad G(j\omega) = \frac{K}{j\omega(j\omega T_1 + 1)(j\omega T_2 + 1)}$$

$$|G(j\omega)| = \frac{K}{\omega\sqrt{\omega^2 T_1^2 + 1}\,\sqrt{\omega^2 T_2^2 + 1}}$$

and $$\angle G(j\omega) = -90° - \tan^{-1} \omega T_1 - \tan^{-1} \omega T_2$$

As ω is varied from 0 to ∞, the corresponding $\angle G(j\omega)$ changes from $-90°$ to $-270°$, indicating that at the phase crossover frequency ω_2

$$\angle G(j\omega_2) = -180°$$

This means that $|G(j\omega)|$ crosses the negative real axis of $G(j\omega)$ – plane

The Nyquist plot method is applied to determine the gain margin

$$\angle G(j\omega) = -90° - (\tan^{-1} \omega T_1 + \tan^{-1} \omega T_2)$$

$$= -90° - \tan^{-1} \frac{\omega T_1 + \omega T_2}{1 - \omega T_1 \cdot \omega T_2}$$

$$= -90° - \tan^{-1} \frac{\omega(T_1 + T_2)}{1 - \omega^2 T_1 T_2}$$

At the phase crossover frequency ω_2,

$$\angle G(j\omega_2) = -180°$$

$$\therefore \tan^{-1}\frac{\omega_2(T_1+T_2)}{1-\omega_2^2T_1T_2} = 90° \quad \text{or} \quad \frac{\omega_2(T_1+T_2)}{1-\omega_2^2T_1T_2} = \tan 90°$$

$$\therefore \frac{\omega_2(T_1+T_2)}{1-\omega_2^2T_1T_2} \to \infty$$

This requires that, $1-\omega_2^2\,T_1T_2 = 0$

$$\therefore \omega_2 = \frac{\pm 1}{\sqrt{T_1T_2}}$$

$$\therefore |G(j\omega_2)| = \frac{K}{\frac{1}{\sqrt{T_1T_2}}\sqrt{(1/T_1T_2)T_1^2+1}\,\sqrt{(1/T_1T_2)T_2^2+1}}$$

$$= \frac{K\sqrt{T_1T_2}}{\sqrt{(T_1/T_2)+1}\,\sqrt{(T_2/T_1)+1}} = \frac{K\sqrt{T_1T_2}\,\sqrt{T_1T_2}}{\sqrt{T_1+T_2}\,\sqrt{(T_1+T_2)}}$$

or
$$|G(j\omega_2)| = \frac{K\,T_1T_2}{T_1+T_2}$$

The shape of the Nyquist plot is shown in Fig. 12.28

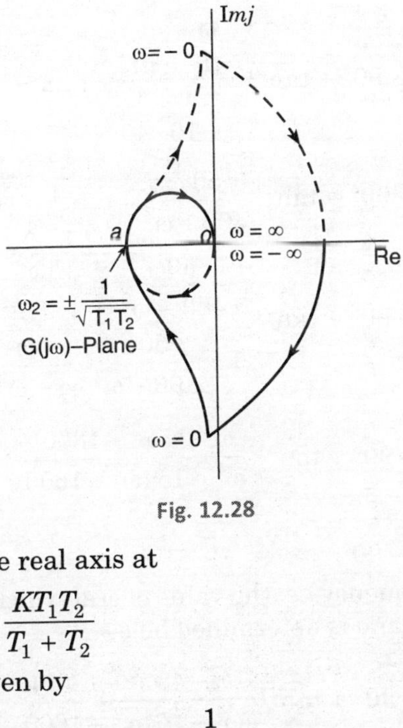

Fig. 12.28

The plot intersects –ive real axis at

$$|oa| = \frac{KT_1T_2}{T_1+T_2}$$

The gain margin is given by

$$\text{G.M.} = 20\ \log_{10}\frac{1}{KT_1T_2/(T_1+T_2)}$$

$$= 20\log_{10}\left(\frac{T_1+T_2}{KT_1T_2}\right) \quad \textbf{Ans.}$$

Example 12.23. *The open-loop transfer function of a unity feedback control system is given below :*

$$G(s) = \frac{K\,(s+10)\,(s+40)}{s\,(s+1)\,(s+4)}$$

Apply Nyquist plot method to examine the closed-loop stability for (i) K = 0.1 (ii) K = 1 (iii) K = 10.

Solution.
$$G(s) = \frac{K\,(s+10)\,(s+40)}{s(s+1)\,(s+4)}$$

put $s = j\omega$

$$\therefore \quad G(j\omega) = \frac{K(j\omega+10)\,(j\omega+40)}{j\omega(j\omega+1)\,(j\omega+4)}$$

$$|G(j\omega)| = \frac{K\sqrt{\omega^2+10^2}\sqrt{\omega^2+40^2}}{\omega\sqrt{\omega^2+1}\sqrt{\omega^2+4^2}}$$

and
$$\angle G(j\omega) = -90° + \tan^{-1}\frac{\omega}{10} + \tan^{-1}\frac{\omega}{40} - \tan^{-1}\frac{\omega}{1} - \tan^{-1}\frac{\omega}{4}$$

$$= -90° + \left(\tan^{-1}\frac{\omega}{10} + \tan^{-1}\frac{\omega}{40}\right) - \left(\tan^{-1}\frac{\omega}{1} + \tan^{-1}\frac{\omega}{4}\right)$$

$$= -90° + \tan^{-1}\frac{\frac{\omega}{10}+\frac{\omega}{40}}{1-\frac{\omega}{10}\cdot\frac{\omega}{40}} - \tan^{-1}\frac{\frac{\omega}{1}+\frac{\omega}{4}}{1-\frac{\omega}{1}\cdot\frac{\omega}{4}}$$

$$= -90° + \tan^{-1}\frac{50\,\omega}{400-\omega^2} - \tan^{-1}\frac{5\,\omega}{4-\omega^2}$$

$$= -90° + \tan^{-1}\frac{\frac{50\omega}{400-\omega^2} - \frac{5\omega}{4-\omega^2}}{1+\frac{50\omega}{400-\omega^2}\cdot\frac{5\omega}{4-\omega^2}}$$

$$= -90° + \tan^{-1}\frac{\omega(-45\omega^2-1800)}{\omega^4-154\omega^2+1600}$$

As $\omega \to 0$ $|G(j\omega)| \to \infty$

and $\angle G(j\omega) \to -90°$

The phase crossover frequency *i.e.* the value of ω at which the Nyquist plot crosses the negative real axis of $G(j\omega)$ – plane is determined below :

$$\angle G(j\omega) = -90° + \tan^{-1}\frac{\omega(-45\omega^2-1800)}{\omega^4-154\omega^2+1600}$$

or
$$\angle G(j\omega) = -90° - \tan^{-1}\frac{\omega(+45\omega^2+1800)}{\omega^4-154\omega^2+1600}$$

Intersection of the negative real axis requires that at the phase crossover frequency ω_2

$$\angle G(j\omega_2) = -180°$$

This is possible if, $\tan^{-1} \dfrac{\omega_2(45\omega_2^2 + 1800)}{\omega_2^4 - 154\omega_2^2 + 1600} = +90°$

i.e.
$$\frac{\omega_2(45\omega_2^2 + 1800)}{\omega_2^4 - 154\omega_2^2 + 1600} \rightarrow \infty$$

$\therefore$
$$\omega_2^4 - 154\omega_2^2 + 1600 = 0$$

Solving for ω_2^2

$$\omega_2^2 = 11.2,\ 142.8$$

$\therefore$
$$\omega_2 = \pm 3.35,\ \pm 11.95 \text{ rad./sec.}$$

$$|G(j3.35)| = \frac{K\sqrt{3.35^2 + 10^2}\ \sqrt{3.35^2 + 40^2}}{3.35\ \sqrt{3.35^2 + 1^2}\ \sqrt{3.35^2 + 4^2}} = 6.9\text{ K}$$

$$|G(j11.95)| = \frac{K\sqrt{11.95^2 + 100}\ \sqrt{11.95^2 + 1600}}{11.95\ \sqrt{11.95^2 + 1^2}\ \sqrt{11.95^2 + 4^2}} = 0.36\text{ K}$$

As $\omega \rightarrow \infty$ $\quad |G(j\,\infty)| \rightarrow 0$

and $\angle G\,(j\omega) \rightarrow -90°$

As per above calculations the shape of the Nyquist form takes the form as shown in Fig. 12.29.

The transfer function $G(j\omega)$ is of type 1, hence from $\omega = -0$ to $\omega = +0$, the plot is completed through an angle of $-180°$ with a radius $\rightarrow \infty$.

(*a*) At $\omega = 3.35$ rad./sec., the plot intersects –ive real axis such that $|G\,(j\,3.35)| = 6.9$ K

(*b*) At $\omega = 11.95$ rad./sec., the plot intersects –ive real axis such that $|G\,(j\,11.95)| =$ 0.36 K

(*i*) For $K = 0.1$

$$|G(j3.35)| = 6.9 \times 0.1 = 0.69$$

and
$$|G(j11.95)| = 0.36 \times 0.1 = 0.036$$

The critical point $(-1 + j0)$ is outside the Nyquist plot (Fig. 12.30)

$\therefore$ $\quad N = 0$

$\because$ $\quad P_+ = 0$

$\quad N = P_+ - Z_+$

$\therefore$ $\quad 0 = 0 - Z_+$

$\therefore$ $\quad Z_+ = \text{Nil}$

Indicating stable system **Ans.**

(*ii*) $K = 1$

$$|G(j3.35)| = 6.9 \times 1 = 6.9$$

and
$$|G(j11.95)| = 0.36 \times 1 = 0.36$$

The critical point $(-1 + j0)$ Fig. 12.31 is encircled twice in the clockwise direction

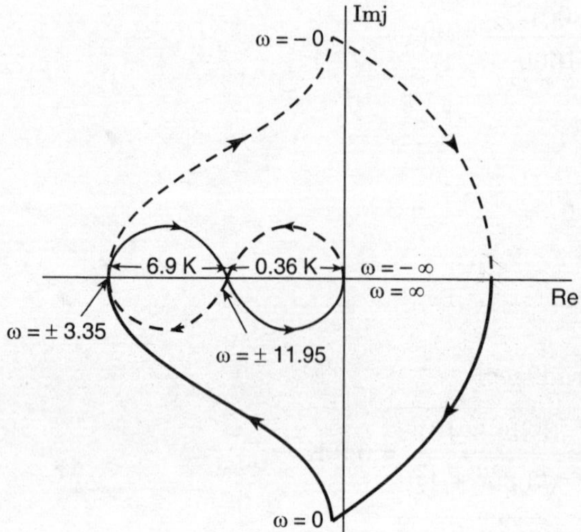

Fig. 12.29

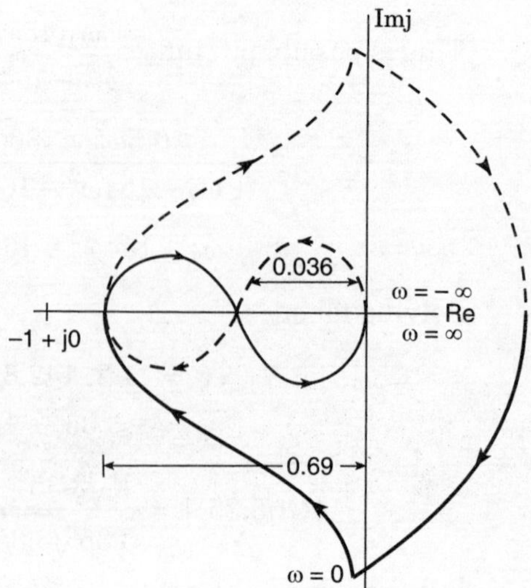

Fig. 12.30

$$\therefore \quad N = -2$$

$$N = P_+ - Z_+$$

$$\therefore \quad -2 = 0 - Z_+$$

$$\therefore \quad Z_+ = 2$$

Indicating unstable system. **Ans.**

(*iii*) $K = 10$

$$|G(j3.35)| = 6.9 \times 10 = 69$$

$$|G(j11.95)| = 0.36 \times 10 = 3.6$$

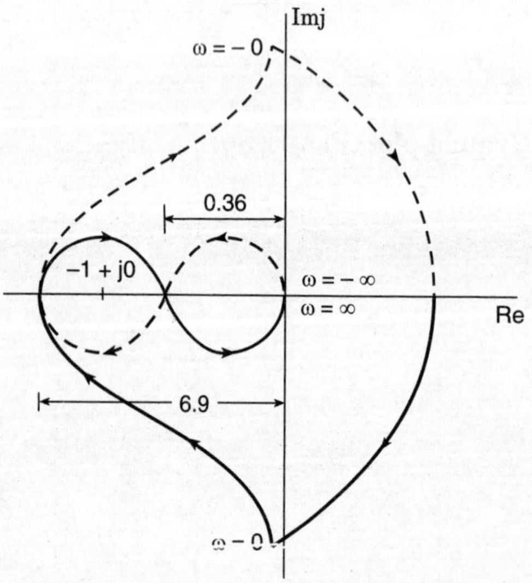

Fig. 12.31

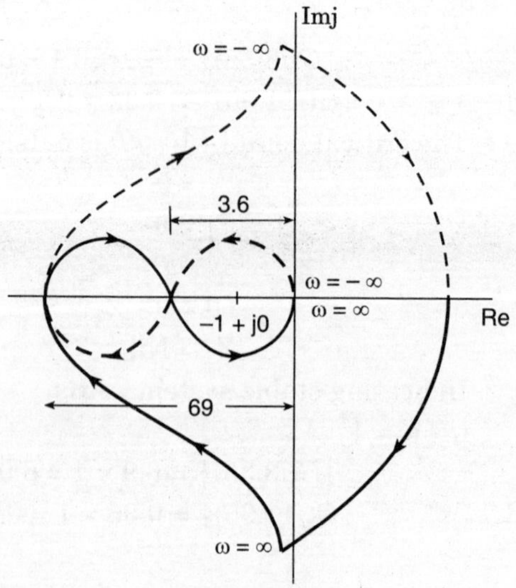

Fig. 12.32

The critical point $(-1 + j0)$ (Fig. 12.32) is encircled once in the clockwise direction and again once in the counter clockwise direction.

$$\therefore \quad N = 0$$

$$N = P_+ - Z_+ \qquad \therefore \qquad 0 = P_+ - Z_+$$

$\therefore$ Z_+ = Nil. Indicating stable system **Ans.**

Example 12.24. *Use Nyquist plot approach to determine the gain margin for a system having its open-loop transfer function as*

$$G(s)\,H(s) = \frac{2\,(s+0.2)}{s^2(s+1)\,(s+2)}.$$

Solution. $G(s)\,H(s) = \dfrac{2\,(s+0.2)}{s^2(s+1)\,(s+2)}$

put $s = j\omega$

$$\therefore \quad G(j\omega)\,H(j\omega) = \frac{2\,(j\omega+0.2)}{-\,\omega^2(j\omega+1)\,(j\omega+2)}$$

$$|G(j\omega)\,H(j\omega)| = \frac{2\sqrt{\omega^2+0.2^2}}{\omega^2\sqrt{\omega^2+1^2}\,\sqrt{\omega^2+2^2}}$$

and

$$\angle G(j\omega)\,H(j\omega) = -\,180° + \tan^{-1}\frac{\omega}{0.2} - \tan^{-1}\frac{\omega}{1} - \tan^{-1}\frac{\omega}{2}$$

$$= -\,180° + \tan^{-1}5\omega - \tan^{-1}\omega - \tan^{-1}0.5\,\omega$$

$$= -\,180° + \tan^{-1}\frac{5\omega-\omega}{1+5\,\omega.\omega} - \tan^{-1}0.5\omega$$

$$= -\,180° + \tan^{-1}\frac{\dfrac{4\omega}{1+5\omega^2} - 0.5\omega}{1+\dfrac{4\omega}{1+5\omega^2}\,.\,0.5\omega}$$

$$= -\,180° + \tan^{-1}\frac{\omega\,(3.5-2.5\omega^2)}{1+7\omega^2}$$

As $\omega \to 0$

$$|G(j\omega)\,H(j\omega)| \to \infty$$

and

$$\angle G(j\omega)\,H(j\omega) \to -\,180°$$

For the intersection of – ive real axis of $G(j\omega)\,H(j\omega)$ – plane, at $\omega = \omega_2$, it is required that

$$\angle G(j\omega)\,H(j\omega) = -\,180°$$

$$\because \quad \angle G(j\omega)\,H(j\omega) = -\,180° + \tan^{-1}\frac{\omega(3.5-2.5\omega^2)}{1+7\omega^2}$$

Therefore, the condition for intersection of the –ive real axis is given by

$$\angle G(j\omega_2)\,H(j\omega_2) = -\,180°$$

$$\because \quad \angle G(j\omega)\,H(j\omega) = -\,180° + \tan^{-1}\frac{\omega(3.5-2.5\omega^2)}{1+7\omega^2}$$

$$\therefore \quad \tan^{-1} \frac{\omega_2(3.5 - 2.5\omega_2^2)}{1 + 7\omega_2^2} = 0 \qquad \therefore \quad \omega_2(3.5 - 2.5\,\omega_2^2) = 0$$

Hence, $\omega_2 = 0,\ \omega_2 = \sqrt{\dfrac{3.5}{2.5}} = \pm\ 1.18$ rad./sec.

At $\omega = 1.18$ rad./sec., the magnitude is calculated below :

$$|G(j1.18)\,H(1.18)| = \frac{2\sqrt{1.18^2 + 0.2^2}}{1.18^2\sqrt{1.18^2 + 1^2}\sqrt{1.18^2 + 2^2}} = 0.48$$

As $\omega \to \infty$ $|G(j\,\infty)\,H(j\,\infty)| \to 0$ and $\angle G(j\,\infty)\,H(j\,\infty) \to -\,270°$

As per above calculations the shape of the Nyquist plot is shown in Fig. 12.33

The given transfer function is of type 2, hence the Nyquist plot is completed from $\omega = -\,0$ to $\omega = 0$ through $-\,360°$ with radius $\to \infty$.

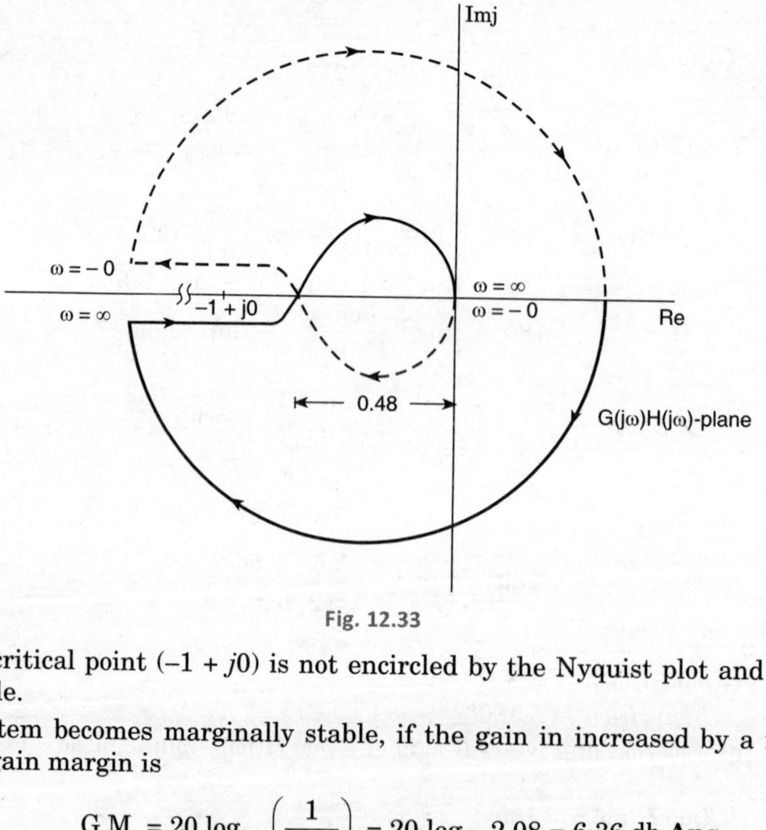

Fig. 12.33

As the critical point $(-1 + j0)$ is not encircled by the Nyquist plot and P_+ = nil, the system is stable.

The system becomes marginally stable, if the gain in increased by a factor 1/0.48, therefore the gain margin is

$$\text{G.M.} = 20\log_{10}\left(\frac{1}{0.48}\right) = 20\log_{10} 2.08 = 6.36 \text{ db } \textbf{Ans.}$$

Example 12.25. *Apply Nyquist plot method to determine the value of K such that the gain margin is 10 db. The open-loop transfer function is given below :*

$$G(s)\,H(s) = \frac{K}{s\,(s+2)\,(s+4)}$$

Calculate the phase margin for the value of K as determined above.

Solution. $G(s)\,H(s) = \dfrac{K}{s\,(s+2)\,(s+4)}$ put $s = j\omega$

$$\therefore \quad G(j\omega)\,H(j\omega) = \frac{K}{j\omega\,(j\omega+2)\,(j\omega+4)}$$

$$|G(j\omega)\,H(j\omega)| = \frac{K}{\omega\sqrt{\omega^2+4}\sqrt{\omega^2+16}}$$

$$\angle G(j\omega)\,H(j\omega) = -90° - \tan^{-1}\frac{\omega}{2} - \tan^{-1}\frac{\omega}{4}$$

$$= -90° - \tan^{-1}\left[\frac{\frac{\omega}{2}+\frac{\omega}{4}}{1-\frac{\omega}{2}\cdot\frac{\omega}{4}}\right]$$

$$= -90° - \tan^{-1}\left(6\frac{\omega}{8-\omega^2}\right)$$

$$\omega \to 0\,;\ \angle G(j0)\,H(j0) \to -90°$$

$$\omega \to \infty\,;\ \angle G(j\infty)\,H(j\infty) \to -270°$$

The general shape of the plot from $\omega = 0$ to $\omega = \infty$ is shown in Fig. 12.34.

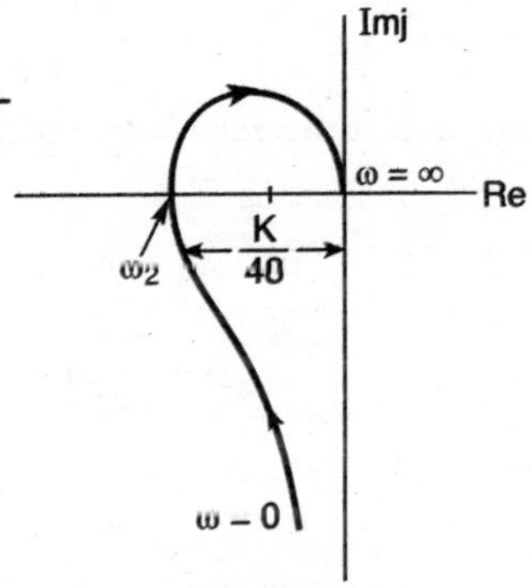

Fig. 12.34

The plot crosses the –ive real axis at phase crossover frequency ω_2 and $\angle G(j\omega_2)\,H(j\omega_2) = -180°$

$$\therefore \quad -90 - \tan^{-1}\left(\frac{6\omega_2}{8-\omega_2^2}\right) = -180°$$

or $$\tan^{-1}\left(\frac{6\omega_2}{8-\omega_2^2}\right) = 90°$$

or $$\frac{6\omega_2}{8-\omega_2^2} = \tan 90°$$

or $$\frac{6\omega_2}{8-\omega_2^2} = \infty$$

$$\therefore \quad 8 - \omega_2^2 = 0$$

Therefore, phase crossover frequency $\omega_2 = \sqrt{8}$ rad./sec.

The magnitude at ω_2 is given by

$$|G(j\omega_2)\,H(j\omega_2)| = \frac{K}{\omega_2\sqrt{\omega_2^2+4}\sqrt{\omega_2^2+16}} = \frac{K}{\sqrt{8}\,\sqrt{8+4}\sqrt{8+16}} = \frac{K}{48}$$

The gain margin is given as 10 db. The intersection of the Nyquist plot with the –ive real axis occurs at a magnitude $\dfrac{K}{48}$.

$$\therefore \quad 20 \log_{10}\left(\frac{1}{K/48}\right) = 10 \quad \text{or} \quad \log_{10}\left(\frac{48}{K}\right) = 0.5$$

$$\therefore \quad \left(\frac{48}{K}\right) = \text{Antilog}_{10}(0.5) = 3.163$$

$$\therefore \quad K = \frac{48}{3.162} = 15.17 \quad \textbf{Ans.}$$

To have a gain margin of 10 db, $K = 15.18$, therefore the transfer function is given by

$$G(j\omega)\, H(j\omega) = \frac{15.17}{j\omega\,(j\omega+2)\,(j\omega+3)}$$

At gain crossover frequency ω_1

$$|G(j\omega_1)\, H(j\omega_1)| = 1$$

$$\therefore \quad \left|\frac{15.17}{j\omega_1\,(j\omega_1+2)\,(j\omega_1+4)}\right| = 1$$

or

$$\frac{15.17}{\omega_1\sqrt{\omega_1^2+4}\,\sqrt{\omega_1^2+16}} = 1$$

$$\therefore \quad \omega_1^2(\omega_1^2+4)(\omega_1^2+16) = (15.17)^2 \qquad \text{put } \omega_1^2 = x$$

$$\therefore \quad x\,(x+4)\,(x+16) = 230.12$$

by trial and error the solution of the above equation is

$$x = 2.09$$

$$\therefore \quad \omega_1 = \sqrt{2.09} = 1.446 \text{ rad./sec.}$$

$$\angle G(j\omega_1)\, H(j\omega_1) = -\,90° - \tan^{-1}\frac{\omega_1}{2} - \tan^{-1}\frac{\omega_1}{4}$$

$$= -\,90° - \tan^{-1}\frac{1.446}{2} - \tan^{-1}\frac{1.446}{2}$$

$$= -\,90° - 35.86° - 19.87° = -\,145.73°$$

Phase margin $= 180° + \angle G(j\omega_1)\, H(j\omega_1)$

$= 180° - 145.73° = 34.27°$ **Ans.**

Example 12.26. *The open-loop transfer function of a unity feedback control system is given by*

$$G(s) = \frac{K}{s\,(s+2)\,(s+4)}$$

Determine the value of K such that the phase margin in 30°. Calculate the gain margin for the value of k determine as above.

Solution. $G(s) = \dfrac{K}{s\,(s+2)\,(s+4)}$ put $s = j\omega$

$$\therefore \quad G(j\omega) = \frac{K}{j\omega\,(j\omega+2)\,(j\omega+4)}$$

$$\angle G(j\omega) = -90° - \tan^{-1}\frac{\omega}{2} - \tan^{-1}\frac{\omega}{4}$$

The phase margin given by, P.M. = 180° + $\angle G(j\omega_1)\, H(j\omega_1)$
where ω_1 is the gain crossover frequency

$$\therefore \qquad 30° = 180° + \left(-90° - \tan^{-1}\frac{\omega_1}{2} - \tan^{-1}\frac{\omega_1}{4}\right)$$

$$= 90° - \tan^{-1}\left(\frac{\frac{\omega_1}{2}+\frac{\omega_1}{4}}{1-\frac{\omega_1}{2}\cdot\frac{\omega_1}{4}}\right)$$

$$= 90° - \tan^{-1}\left(\frac{6\,\omega_1}{8-\omega_1^2}\right)$$

or $$\tan^{-1}\left(\frac{6\omega_1}{8-\omega_1^2}\right) = 60°$$

$$\therefore \qquad \frac{6\omega_1}{8-\omega_1^2} = \tan 60°$$

or $$\frac{6\omega_1}{8-\omega_1^2} = \sqrt{3}$$

or $$8\sqrt{3} - \omega_1^2\sqrt{3} = 6\omega_1$$

$$\omega_1^2\sqrt{3} + 6\omega_2 - 13.856 = 0$$

Solving above equation ω_1 = 1.588 rad./sec.
For ω_1 = 1.588 rad./sec.

$$|G(j\omega_1)| = \frac{K}{\omega_1\sqrt{\omega_1^2+4}\sqrt{\omega_1^2+16}}$$

At ω_1, $|G(j\omega_1)| = 1$

$$\therefore \qquad 1 = \frac{K}{1.588\sqrt{1.588^2+4}\sqrt{1.588^2+16}}$$

or $$K = 1.588\sqrt{1.588^4+4}\sqrt{1.588^2+16}$$

or $K = 17.4$ **Ans.**

$$\angle G(j\omega) = -90° - \tan^{-1}\frac{6\omega}{8-\omega_2^2}$$

At phase crossover frequency ω_2, $\angle G(j\omega_2) = -180°$

$$\therefore \qquad -180° = -90° - \tan^{-1}\frac{6\omega_2}{8-\omega_2^2}$$

or $$\tan^{-1}\frac{6\omega_2}{8-\omega_2^2} = 90° \quad \text{or} \quad \frac{6\omega_2}{8-\omega_2^2} = \tan 90°$$

or $$\frac{6\omega_2}{8-\omega_2^2} = \infty$$

Hence, $$8-\omega_2^2 = 0$$

or $$\omega_2 = \sqrt{8} \text{ rad/sec}$$

$$|G(j\omega_2)| = \left|\frac{17.4}{j\omega_2\sqrt{j\omega_2+2}\sqrt{j\omega_2+4}}\right| = \frac{17.4}{\omega_2\sqrt{\omega_2^2+4}\sqrt{\omega_2^2+16}}$$

$$= \frac{17.4}{\sqrt{8}\sqrt{\sqrt{8}^2+4}\sqrt{\sqrt{8}^2+16}} = 0.3625$$

$\therefore$ Gain Margin $$= 20\log_{10}\left|\frac{1}{G(j\omega_2)}\right| = 20\log_{10}\frac{1}{0.3625}$$

$$= 20\log_{10} 2.758 = 8.82 \text{ db}$$ **Ans.**

Example 12.27. *Sketch the Nyquist plot having given the open-loop transfer function of a unity feedback control system as*

$$G(s) = \frac{K}{s(s+1)^2}$$

and (i) determine the limiting value of K for closed-loop stability

(ii) determine the gain and phase margin for K = 0.5.

Solution. $$G(s) = \frac{K}{s(s+1)^2}$$

put $s = j\omega$

$$\therefore \quad G(j\omega) = \frac{K}{j\omega(j\omega+1)^2}$$

$$= -\left\{\frac{-jK}{\omega(j\omega+1)^2} \times \frac{(-j\omega+1)^2}{(-j\omega+1)^2}\right\} = -\left\{\frac{jK\{(1-\omega^2)-j2\omega\}}{\omega(\omega^2+1)^2}\right\}$$

or $$G(j\omega) = -\frac{2K}{(\omega^2+1)^2} + \frac{jK(1-\omega^2)}{\omega(\omega^2+1)^2}$$

As $$\omega \to 0 \quad \left.\begin{matrix}\text{Re } G(j0) \to -2K \\ \text{Imj } G(j0) \to -j\infty\end{matrix}\right\}$$

$$\omega \to \infty \quad \left.\begin{matrix}\text{Re } G(j\infty) \to -0 \\ \text{Imj } G(j\infty) \to +j0\end{matrix}\right\}$$

The plot crosses the negative real axis at a frequency determined by equating imaginary part to zero

$$\therefore \quad (1-\omega^2) = 0 \quad \text{or} \quad \omega = \pm 1.0 \text{ rad/sec.}$$

This is the phase crossover frequency ω_2

$$\therefore \quad \omega_2 = \pm 1.0 \text{ rad/sec.}$$

As the transfer function is type 1 the Nyquist plot from $\omega = -0$ to $\omega = 0$ is completed by an arc of infinite radius through an angle $-\pi$ rad. The shape of the Nyquist plot is shown in Fig. 12.35.

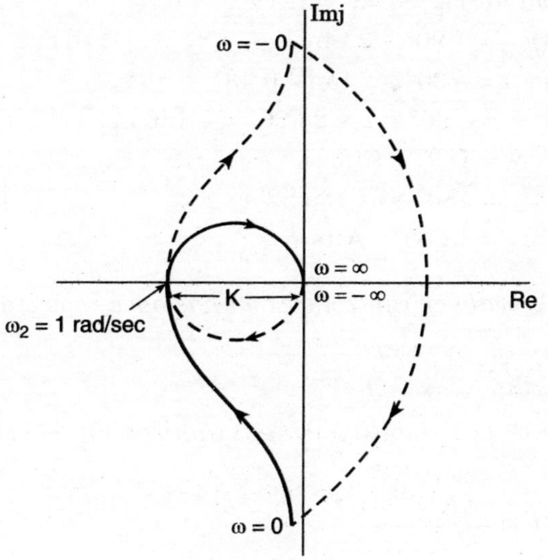

Fig. 12.35

$$G(j\omega) = -\frac{2K}{\omega^2 + 1^2} + \frac{jK(1-\omega^2)}{\omega(\omega^2+1)}$$

At phase crossover frequency ω_2 – 1.0 rad/sec.

$$G(j1) = -\frac{2K}{(1^2+1^2)} + \frac{jK(1-1)}{1(1^2+1^2)} = -K$$

For stability : $K < 1$

For marginal stability : $K = 1$ **Ans.**

Given value of $K = 0.5$

$$\therefore \text{ Gain margin } = 20\log_{10}\frac{1}{0.5} = 20\log_{10} 2$$

$$= 6 \text{ db} \quad \textbf{Ans.}$$

At the gain crossover frequency ω_1

$$|G(j\omega_1)| = 1 \quad \because \quad K = 0.5$$

The transfer function for $K = 0.5$ is given by,

$$G(j\omega) = \frac{0.5}{j\omega(j\omega+1)^2}$$

$$\therefore \quad |1| = \left|\frac{0.5}{j\omega_1(j\omega_1+1)^2}\right|$$

$$\therefore \quad 1 = \frac{0.5}{\omega_1\sqrt{\omega_1^2+1}\sqrt{\omega_1^2+1}}$$

or $\omega_1(\omega_1^2 + 1) = 0.5$

or $\omega_1^3 + \omega_1 = 0.5$

above equation is satisfied for $\omega_1 = 0.43$ rad./sec.

$$\angle G(j\omega) = -90° - 2\tan^{-1}\omega$$

$$\therefore \quad \angle G(j\omega_1) = -90° - 2\tan^{-1} 0.43$$

$$= -90° - 2 \times 23.26° = -136.52°$$

$$\text{Phase Margin} = 180° + G(j\omega_1)$$

$$= 180° + (-136.52°)$$

$$= 43.48° \quad \textbf{Ans.}$$

Example 12.28. *A feedback control system has open-loop transfer function*

$$G(s)\,H(s) = \frac{K(1-s)}{s(s+2)}$$

Determine the closed-loop stability using Nyquist stability criterion.

Solution. $G(s)\,H(s) = \dfrac{K(1-s)}{s(s+2)}$

put $s = j\omega$

$$\therefore \quad G(j\omega)\,H(j\omega) = \frac{K(1-j\omega)}{j\omega(j\omega+2)}$$

$$= -jK\left\{\frac{(2-\omega^2) - j\,3\omega}{\omega(\omega^2+4)}\right\}$$

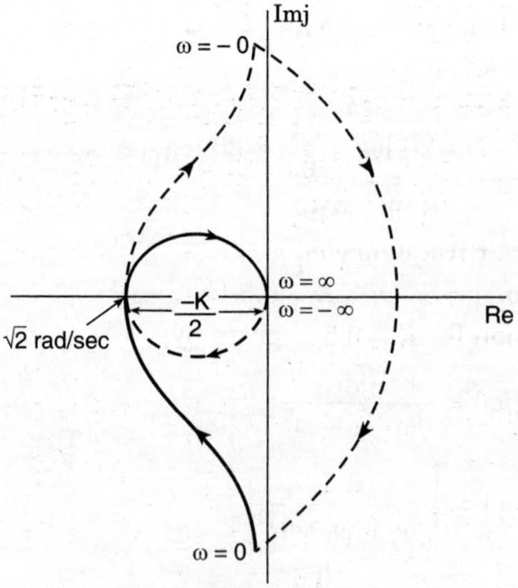

Fig. 12.36

$$= -\frac{3K}{\omega^2 + 4} - \frac{jK(2-\omega^2)}{\omega(\omega^2+4)}$$

As $\omega \to 0$ $\left.\begin{array}{l} \text{Re } G(j0)\,H(j0) \to 0 \\ \text{Imj } G(j0)\,H(j0) \to -\infty \end{array}\right\}$

Intersection with the – ive real axis is obtained by equating Imj $G(j\omega)\,H(j\omega)$ to zero

i.e. $$\frac{jK(2-\omega^2)}{\omega(\omega^2+4)} = 0$$

$\therefore$ $$2 - \omega^2 = 0$$

or $$\omega = \pm 2 \text{ rad./sec.}$$

Thus the phase crossover frequency is

$$\omega_2 = \pm \sqrt{2} \text{ rad./sec.}$$

$$G(j\omega_2) = -\frac{3K}{(\sqrt{2})^2 + 4} = -\frac{K}{2}$$

As per above results, shape of the Nyquist plot is shown in Fig. 12.36

The type of the transfer function is 1, hence the Nyquist plot from $\omega = -0$ to $\omega = 0$ is closed through $-\pi$ rad. with a semi circle of infinite radius.

For closed-loop stability the critical point $(-1 + j0)$ lies outside the Nyquist plot if

$$\frac{K}{2} < 1 \quad \text{or} \quad K < 2 \quad \textbf{Ans.}$$

Example 12.29. *Examine the stability of a closed-loop control system having its open-loop transfer function*

$$G(s)\,H(s) = \frac{K(s+4)}{s(s-2)}$$

Determine the phase crossover frequency and the value of the intercept on the negative real axis of G (jω) H(jω)-plane

Solution. $$G(s)\,H(s) = \frac{K(s+4)}{s(s-2)}$$

put $s = j\omega$

$\therefore$ $$G(j\omega)\,H(j\omega) = \frac{K(j\omega+4)}{j\omega(j\omega-2)}$$

Separating R.H.S. into real and imaginary parts

$$G(j\omega)\,H(j\omega) = -\frac{jK}{\omega}\left\{\frac{(j\omega+4)(-j\omega-2)}{(j\omega-2)(-j\omega-2)}\right\}$$

$$= -\frac{jK}{\omega}\left\{\frac{(\omega^2-8)-j6\omega}{\omega^2+4}\right\}$$

$$= -\frac{6K}{\omega^2+4} - \frac{jK(\omega^2-8)}{\omega^2+4}$$

(*i*) $\omega \to 0$, Re $G(j0)\,H(j0) \to -\dfrac{6K}{4}$

Imj $G(j0)\,H(j0) \to +j\,\infty$

(*ii*) $\omega \to \infty$, Re $G(j\infty)\,H(j\infty) \to -0$

Imj $G(j\infty)\,H(j\infty) \to -j0$

(*iii*) The frequency at which the Nyquist plot crosses – ive real axis is obtained by equating imaginary part of $G(j\omega)\,H(j\omega)$ to zero.

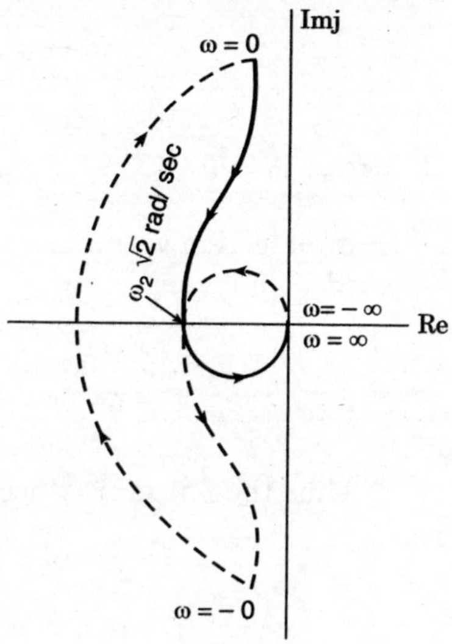

Fig. 12.37

i.e. $$\frac{K(\omega^2-8)}{\omega^2+4} = 0$$

∴ $$\omega = \pm\sqrt{8} \text{ rad/sec.}$$

Thus, the phase crossover frequency is $\omega_2 = \sqrt{8}$ rad/ sec.

The real axis intercept is determined by substituting $\omega = \omega_2 = \sqrt{8}$ rad./sec. in the expression for $G(j\omega)\,H(j\omega)$

∴ $$G(j\omega_2)\,H(j\omega_2) = -\frac{6K}{\omega_2^2+4} - \frac{jK(\omega_2^2-8)}{\omega_2(\omega^2+4)}$$

or $$G(j\omega_2)\,H(j\omega_2) = \frac{-6K}{(\sqrt{2})^2+4} - 0$$

or $$G(j\omega_2)\,H(j\omega_2) = \frac{-6K}{12} = -\frac{K}{2}$$

The Nyquist plot is shown in Fig. 12.37

As the transfer function is type 1. The Nyquist plot is completed from $\omega = -0$ to $\omega = 0$ through – 180° with infinite radius.

If the critical point $(-1+j0)$ is located inside the small circle (Fig. 12.37), then

$N = 1$; $P_+ = 1$ (given)

∵ $N = P_+ - Z_+$

∴ $1 = 1 - Z_+$, thus $Z_+ = 0$ and system is stable.

For stability $\dfrac{K}{2} > 1$ or $K > 2$ **Ans.**

Example 12.30. *Determine the transfer function for the Bode plot shown in Fig. 12.38*

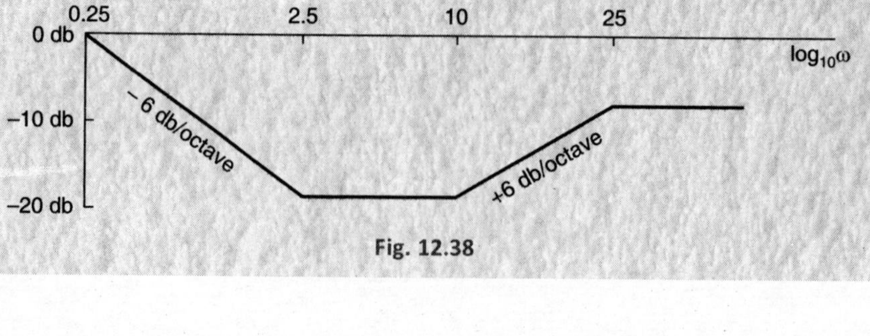

Fig. 12.38

Solution. 6 db/octave = 20 db/decade

The initial slope is – 6 db/octave (– 20 db/decade)

∴ The type of system is 1

1. The initial slope intersects 0 db axis at $\omega = 0.25$ rad/s.

$$\therefore \quad K = 0.25$$

2. Beyond $\omega = 2.5$ rad/s, the slope changes by (+6 db/octave) = (+ 20 db/decade) which corresponds to the term

$$(j\omega T_1 + 1) = \left(j\omega\frac{1}{2.5} + 1\right) = (j0.4\omega + 1)$$

3. Beyond $\omega = 10$ rad/s, the slope changes by +6 db/octave (+20 db/decade) which corresponds to the term

$$(j\omega T_2 + 1) = \left(j\omega\frac{1}{10} + 1\right) = (j0.1\omega + 1)$$

4. Beyond $\omega = 25$ rad/s, the slope changes by – 6 db/octave (–20 db/decade) which corresponds to the term

$$\frac{1}{(j\omega T_3 + 1)} = \left(\frac{1}{j\omega\frac{1}{25} + 1}\right) = \frac{1}{(j\,0.04\omega + 1)}$$

$$\therefore \quad G(j\omega)\,H(j\omega) = \frac{0.25\,(j\,0.4\omega + 1)\,(j0.1\omega + 1)}{j\omega\,(j\,0.04\omega + 1)}$$

$$\therefore \quad G(s)\,H(s) = \frac{0.25\,(0.4s + 1)\,(0.1s + 1)}{s\,(0.04s + 1)} \quad \textbf{Ans.}$$

Example 12.31. *The open-loop transfer function of a feedback control system is given by*

$$G(s)\,H(s) = \frac{K}{(s + 1)\,(2s + 1)\,(3s + 1)}$$

Find the value of K such that the gain margin is 20 db.

Solution. Plot the Bode considering $K = 1$

$$\therefore \quad G(s)\,H(s) = \frac{1}{(s + 1)\,(2s + 1)\,(3s + 1)}$$

put $s = j\omega$

$$\therefore \quad G(j\omega)\,H(j\omega) = \frac{1}{(j\omega + 1)\,(j2\omega + 1)\,(j3\omega + 1)}$$

1. Type 0 system : Intercept with the 0 db axis is

$$20 \log_{10} K = 20 \log_{10} 1 = 0 \text{ db}$$

2. Corner frequencies : $\omega = 1$ rad/sec

$$\omega = 1/3$$
$$= 0.33 \text{ rad/sec}$$

$$\omega = 1/2$$
$$= 0.5 \text{ rad/sec}$$

As the corner frequencies are contributed due to poles of the transfer function, the slope of the Bode plot changes by – 20 db/decade successively beyond each corner frequency

3. $\angle G(j\omega)\, H(j\omega) = -\tan^{-1}\omega - \tan^{-1} 2\omega - \tan^{-1} 3\omega$

The $\angle G(j\omega)\, H(j\omega)$ for some frequencies is calculated below :

ω rad/s :	0.4	0.8	1	5
$\angle G(j\omega)\, H(j\omega)$:	– 110.6°	– 164°	– 180°	– 249.1°

The Bode plot for $K = 1$ is drawn in Fig. 12.39.

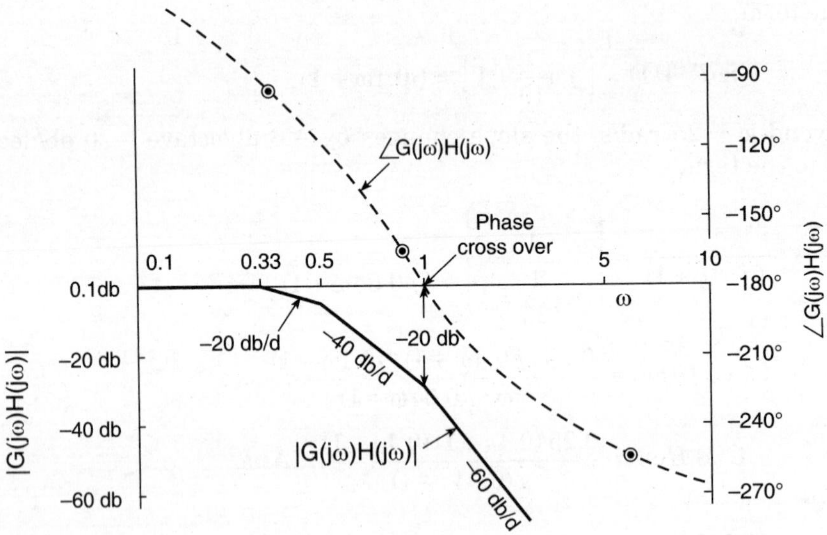

Fig. 12.39

As the phase crossover frequency gain is –20 db.

Hence the G.M. = 20 db.

Thus, $K = 1$ **Ans.**

Example 12.32. *The open-loop transfer function of a unity feed-unity feedback control system is given below :*

$$G(s) = \frac{K}{s\,(s+1)\,(s+10)}$$

Plot asymptotic Bode plot and determine the value of K such that the gain margin is 10 db. For the value of K thus obtained determine the new gain crossover frequency and phase margin.

Solution.
$$G(s) = \frac{K}{s\,(s+1)\,(s+10)}$$

or
$$G(s) = \frac{10K}{s\,(0.1s+1)\,(s+1)} = \frac{K'}{s\,(0.1s+1)\,(s+1)} \; ; K' = 10K$$

put $s = j\omega$

$$\therefore \qquad G(j\omega) = \frac{K'}{j\omega\,(j0.1\omega + 1)\,(j\omega + 1)}$$

The type of the transfer function is 1, therefore, the initial plot of the Bode plot is – 20 db/decade, intersecting 0 db axis at $\omega = K' = 1.0$ ($K' = 1.0$ is considered)

The two corner frequencies due to open-loop poles are $\omega = 1.0$ rad./sec. and $\omega = \frac{1}{0.1}$ = 10 rad./sec.

The slope changes by – 20 db/decade beyond $\omega = 1.0$ rad./sec and this slope (– 20 – 20) = – 40 db/decade

The slope again changes by – 20 db/decade beyond $\omega = 10$ rad/sec and this slope is (– 20 – 40) = – 60 db/decade

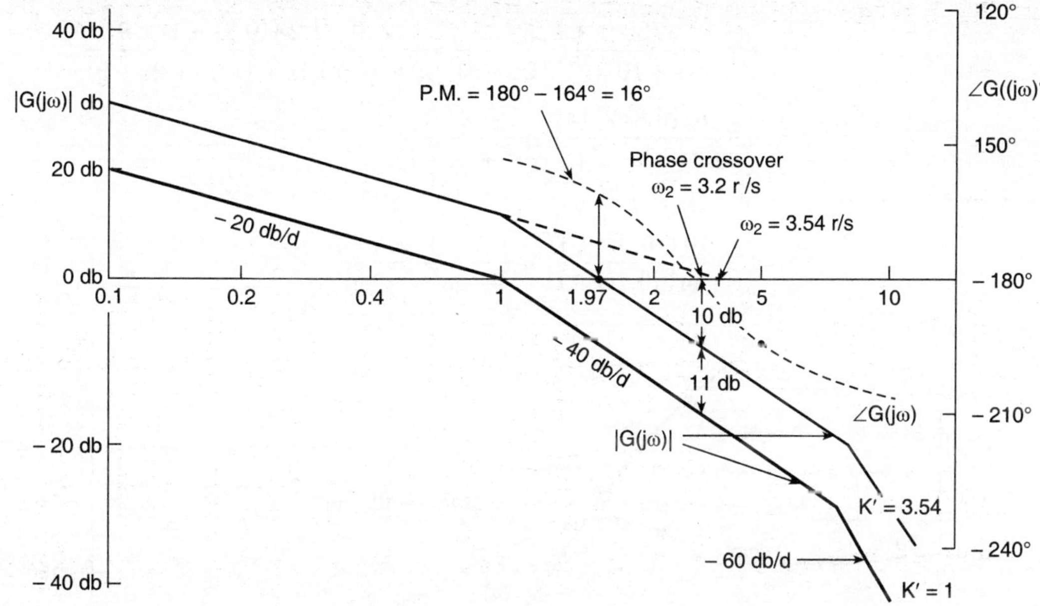

Fig. 12.40

The frequency range selected being $\omega = 0.1$ rad./sec. to $\omega = 10$ rad./sec.

$$G(j\omega) = -90° - \tan^{-1} 0.1\omega - \tan^{-1} \omega$$

$G(j\omega)$ for some values of ω is calculated and tabulated below

ω rad./sec.	1	2	3	5	10
$G(j\omega)°$	– 140.7°	– 165°	– 168.8°	– 195.2°	– 219.2°

As per above data the Bode plot is drawn as shown in Fig. 12.40.

At the phase crossover point $\omega = 3.2$ rad./sec., the gain is – 21 db. To make the gain marin 10 db, the magnitude (db) plot is to be lifted upwards by (21 – 10) db = 11 db. On lifting up the magnitude (db) plot by 11 db, the initial part of the magnitude (db) plot intersects 0 db axis at $\omega = 3.54$ rad/sec.

Since the transfer function is type 1,

Therefore, $K' = 3.54$

and $$K = \frac{3.54}{10} = 0.354 \quad \textbf{Ans.}$$

Example 12.33. *The transfer function of a unity feedback control system is given below :*

$$G(s) = \frac{720\,(s + 1.25)}{s\,(s + 10)\,(s^2 + 2s + 9)}$$

Determine the closed-loop stability applying Bode plot method.

Solution. Transforming the transfer function in time constant form and expressing quadratic term in the standard second order form, the given transfer function is obtained as below :

$$G(s) = \frac{720\,(s + 1.25)}{s\,(s + 10)\,(s^2 + 2s + 9)} = \frac{720 \times 1.25\,(0.8s + 1) \times 9}{9 \times 10s(0.1s + 1)\,(s^2 + 2s + 9)}$$

$$= \frac{10\,(0.8s + 1)}{s\,(0.1s + 1)} \times \frac{9}{(s^2 + 2s + 9)}$$

put $s = j\omega$

$$\therefore \quad G(j\omega) = \frac{10\,(j0.8\omega + 1)}{j\omega\,(j0.1\omega + 1)} \times \frac{9}{(9 - \omega^2 + j2\omega)}$$

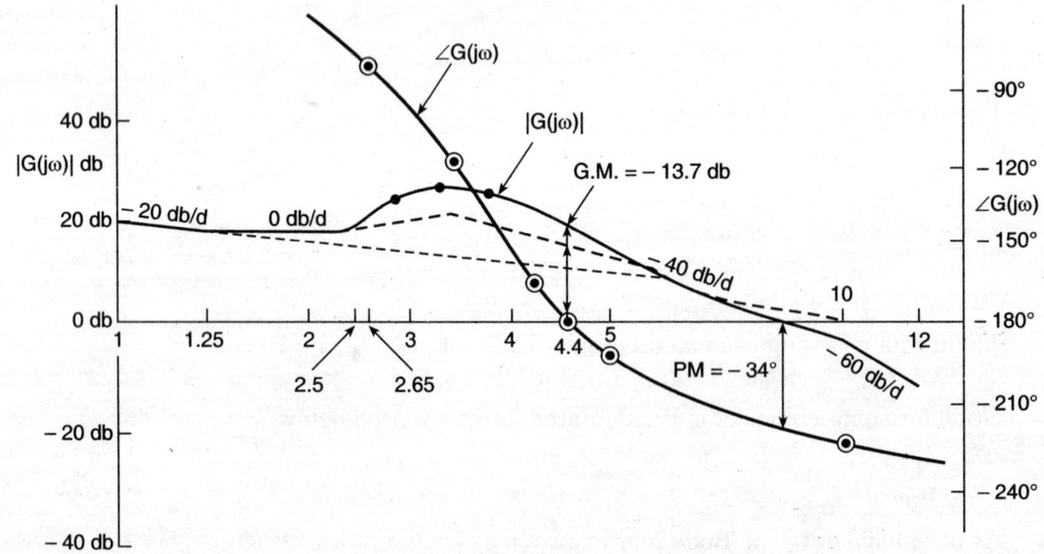

Fig. 12.41

The type of the transfer function is 1 and $K = 10$, hence, the initial slope is – 20 db/decade intersecting 0 db. axis at $\omega = K = 10$. The corner frequencies are,

$$\omega = \frac{1}{0.8} = 1.25 \text{ rad./sec. due to open-loop zero.}$$

$$\omega = \frac{1}{0.1} = 10 \text{ rad./sec due to open-loop pole.}$$

$$\omega = \omega_n = \sqrt{9} = 3 \text{ rad./sec. due to complex open-loop poles (quadratic term)}$$

At

(*i*) $\omega = 1.25$ rad./sec. slope of the $|G\ |(j\omega)|_{db}$ plot changes by + 20 db/decade.

(*ii*) $\omega = \omega_n = 3$, slope of the $|G(j\omega)|_{db}$ plot changes by – 40 db/decade. However, as the second order term $(s^2 + 2s + 9)$, presents underdamped response because

$$2\zeta\omega_n = 2$$

or
$$\zeta = \frac{1}{\omega_n} = \frac{1}{3} = 0.33\ (\zeta < 1)$$

As $\zeta < 1$ correction to $|G\ (j\omega)|_{db}$ asymptotic plot has to be applied for frequencies near about ω_n. The maximum value of the magnitude occurs at resonant frequency given by

$$\omega_r = \omega_n\sqrt{1 - 2\zeta^2} = 3\sqrt{1 - 2 \times 0.33^2} = 2.65 \text{ rad./sec.}$$

The exact magnitude at $\omega_r = 2.65$ rad./sec. is calculated below

$$|(j\ 2.65)| = \frac{10\sqrt{(2.65 \times 0.8)^2 + 1}}{2.65\sqrt{(2.65 \times 0.1)^2 + 1}} \times \frac{9}{\sqrt{(9 - 2.65)^2 + (2 \times 2.65)^2}}$$

$$= 14.02$$

$$|G(j\ 2.265)|_{db} = 20 \log_{10} 14.02 = 22.96 \text{ db}$$

Also,
$$|G(j3)|_{db} = 21.9 \text{ db}$$
$$|G(j3.5)|_{db} = 19.3 \text{ db}$$
$$|G(j4.4)|_{db} = 13.7 \text{ db}$$

(*iii*) $\omega = 10$, slope of the $|G(j\omega)|_{db}$ plot changes by – 20 db/decade.

The phase angle of $G\ (j\omega)$ is given by

(*i*) $\angle G(j\omega) = -90° - \tan^{-1} 0.1\omega + \tan^{-1} 0.8\omega - \tan^{-1} \frac{2\omega}{9 - \omega^2}$ ... for $\omega < 3$

(*ii*) $\angle G(j\omega) = -90° - \tan^{-1} 0.1\omega + \tan^{-1} 0.8\omega - 180° - \tan^{-1} \frac{2\omega}{9 - \omega^2}$ ... for $\omega > 3$

The frequency range for Bode plot is selected from $\omega = 1$ rad./sec. to $\omega = 12$ rad./sec.

The phase angle $G(j\omega)$ for some values of ω in the above range is calculated and tabulated below,

ω rad/sec.	2.5	3	4	5	10
G(jω)	– 102°	– 129.3°	– 170.3°	– 188°	– 219°

As per above calculations the Bode plot is drawn as shown in Fig. 12.41

It is noted from the Bode plot that,

Gain Margin = – 13.7 db and

Phase Margin = – 34°

As the gain margin and phase margin being negative, the closed-loop system is unstable. **Ans.**

Example 12.34. *The open-loop transfer function of a unity feedback control system incorporating a time delay is given below :*

$$G(s) = \frac{Ke^{-0.5s}}{(0.5s+1)}$$

(i) Plot the asymptotic Bode plot for K = 1.

(ii) Determine the limiting value of K for stability.

Solution. $G(s) = \frac{Ke^{-0.5s}}{(0.5s+1)}$

put $s = j\omega$

$$\therefore \quad G(j\omega) = \frac{Ke^{-j0.5\omega}}{(j\,0.5\omega+1)}$$

For $K = 1$; $\quad G(j\omega) = \frac{e^{-j0.5\omega}}{(j0.5\omega+1)}$

$|e^{-j0.5\omega}| = 1$, this term does not contribute to $|G(j\omega)|_{db}$ plot.

$\angle e^{-j0.5\omega} = -0.5\omega$ rad.

$= -57.3 \times 0.5\omega°$, this term contributes to $\angle G(j\omega)$ plot

(*i*) The given open-loop transfer function is type 1 and $K = 1$, hence, intercept with the $|G(j\omega)|_{db}$ axis is given by

$$20 \log_{10} K = 20 \log_{10} 1 = 0 \text{ db}$$

i.e. the plot $|G(j\omega)|_{db}$ for low frequencies lies on the 0 db axis.

(*ii*) At corner frequency $\omega = \frac{1}{0.5} = 2$ rad./sec. (due to open-loop pole) the slope of $|G(j\omega)|_{db}$ plot changes by – 20 db/decade

The frequency range for plotting the Bode plot is selected as $\omega = 0.1$ rad./sec. to $\omega = 10$ rad./sec.

(*iii*) $\angle G(j\omega) = -\tan^{-1} 0.5\omega - 57.3 \times 0.5\omega$

The phase angle $\angle G(j\omega)$ for some values of ω without the selected frequency range is calculated and tabulated below :

ω rad./sec.	2	3	4	5
$\angle G(j\omega)$	– 102°	– 142°	– 177°	– 211°

As per above calculations, the Bode plot is drawn as shown in Fig. 12.42

The phase crossover frequency $\omega = 4.08$ rad./sec. At $\omega = 4.08$ rad./sec., the gain is – 7.1 db *i.e.* G.M. = 7.1 db.

To make the closed-loop system marginally stable (G.M. = 0 db), $|G(j\omega)|_{db}$ plot is to be lifted up by 7.1 db so that the gain crossover and phase crossover frequencies coincide.

$$\therefore \quad 20 \log_{10} K = 7.1$$

or $$K = \text{anti} \log_{10} \frac{7.1}{20}$$

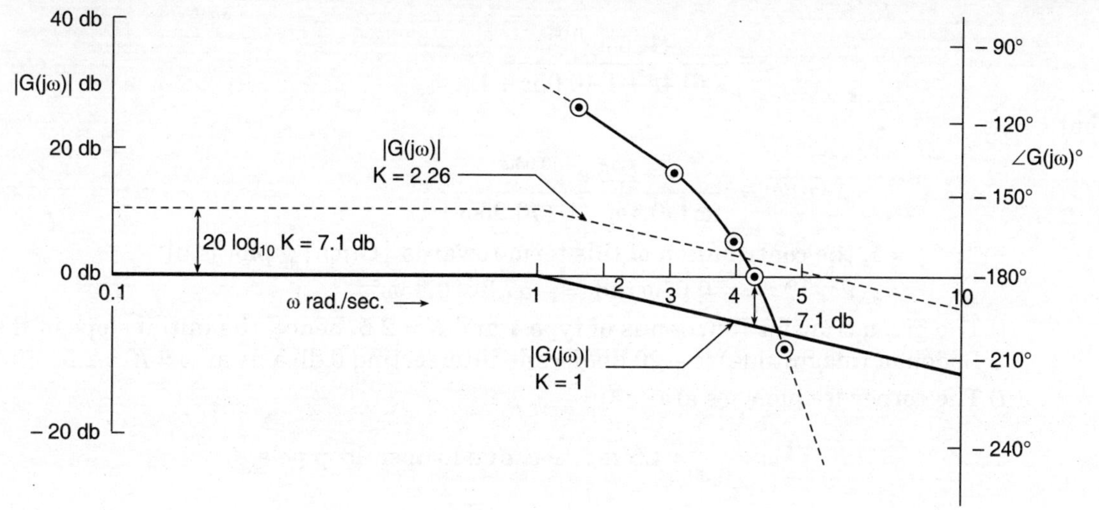

Fig. 12.42

or $$K = \text{antilog}_{10}\, 0.355$$

or $$K = 2.26 \quad \textbf{Ans.}$$

Example 12.35. *A unity feedback control system has open-loop transfer function given below*

$$G(s) = \frac{125\, e^{-0.05s}}{s\,(s + 2.5)\,(s + 20)}$$

Sketch the Bode plot and determine the gain margin and phase margin.

Solution. $$G(s) = \frac{125\, e^{-0.05s}}{s\,(s + 2.5)\,(s + 20)} = \frac{125\, e^{-0.05s}}{2.5 \times 20s(0.4s + 1)\,(0.05s + 1)}$$

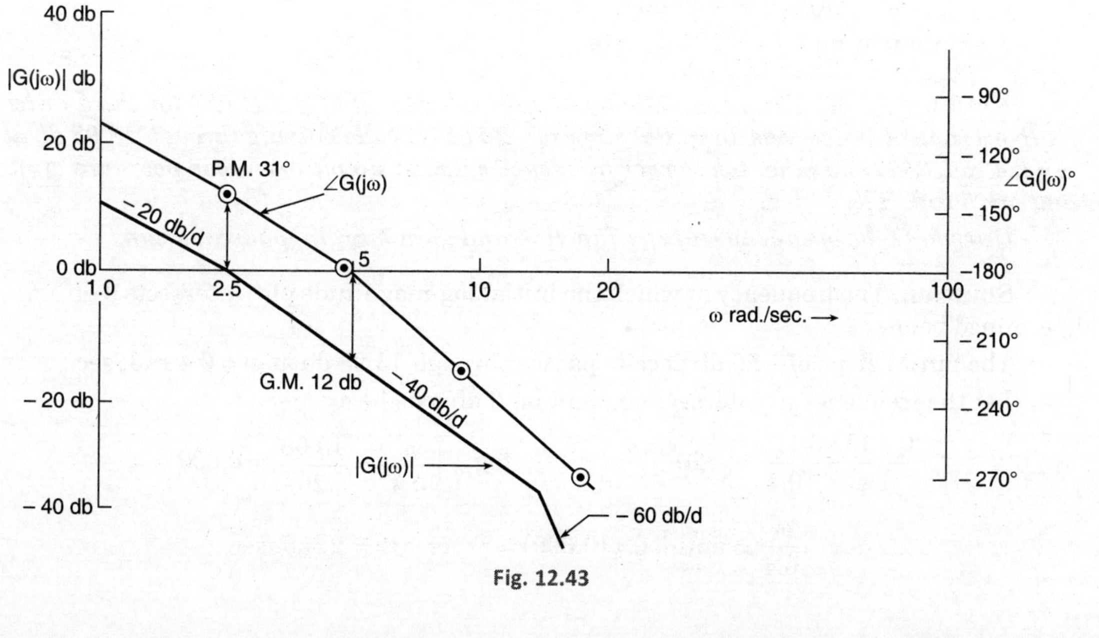

Fig. 12.43

$$= \frac{125\, e^{-0.05s}}{s\,(0.4s+1)\,(0.05s+1)}$$

put $s = j\omega$

$$\therefore \qquad G(j\omega) = \frac{125\, e^{-j0.05\omega}}{j\omega\,(j0.4\omega+1)\,(j0.05\omega+1)}$$

$|e^{-j0.05\omega}| = 1$, the contribution of this term towards $|G(j\omega)|_{db}$ plot is nil

$$\angle e^{-j0.05\omega} = -\,0.05\omega \text{ rad.} = -\,57.3 \times 0.05\omega°$$

(*i*) The given transfer function is of type 1 and $K = 2.5$, hence, the initial slope of the Bodeplot (magnitude) is – 20 db/decade intersecting 0 db axis at $\omega = K = 2.5$.

(*ii*) The corner frequencies are

$$\omega = \frac{1}{0.4} = 25 \text{ rad./sec. due to open-loop pole.}$$

$$\omega = \frac{1}{0.05} = 20 \text{ rad./sec. due to open-loop pole.}$$

At, $\omega = 2.5$ rad./sec. the slope of $|G(j\omega)|_{db}$ changes by – 20 db/decade

$\omega = 20$ rad./sec. the slope of $|G(j\omega)|_{db}$ changes by 20 db/decade

(*iii*) $\angle G(j\omega) = -\,90° - \tan^{-1} 0.4\omega - \tan^{-1} 0.05\omega - 57.3 \times 0.05\omega$

$\angle G(j\omega)$ for some values of ω is calculated and tabulated below :

ω rad./sec.	1	2.5	5	10	20
$G(j\omega)$	– 114°	– 149°	– 180°	– 221°	– 274°

As per above data, the Bode plot is drawn as shown in Fig. 12.43

The gain crossover frequency : $\omega_1 = 2.5$ rad./sec.

$$G(j\omega_1) = -\,149°$$

$\therefore$ Phase margin $= 180° + (-\,149°) = 31°$ **Ans.**

The phase crossover frequency $\omega_2 = 5$ rad./sec.

$$|G(j\omega_2)| = -\,12 \text{ db}$$

$\therefore$ Gain margin $= 12$ db **Ans.**

Example 12.36. *The asymptotic log-magnitude plot of G (jω) H (jω) for third order feedback control system has an initial slope of – 20 db / decade passing through 13.98 db at ω = 0.4 rad. / sec. and other two corner frequencies due to simple open-loop poles are ω = 2 and ω = 5 rad. / sec.*

Determine the open-loop transfer function and gain margin, phase margin.

Solution. The frequency at which the initial log-magnitude plot intersects 0 db axis is determined below :

The initial slope of – 20 db/decade passes through 13.98 db at $\omega = 0.4$ rad./sec.

Let the frequency at intersection point on 0 db axis be ω_i

$$\therefore \quad \frac{0 - 13.98}{\log_{10} \frac{\omega_i}{0.4}} = -\,20 \qquad \therefore \quad \log_{10}\left(\frac{\omega_i}{0.4}\right) = \frac{13.98}{20} = 0.699$$

$$\therefore \quad \frac{\omega_i}{0.4} = \text{anti}\log_{10}(0.699) = 5 \quad \text{or} \quad \omega_i = 2 \text{ rad./sec.}$$

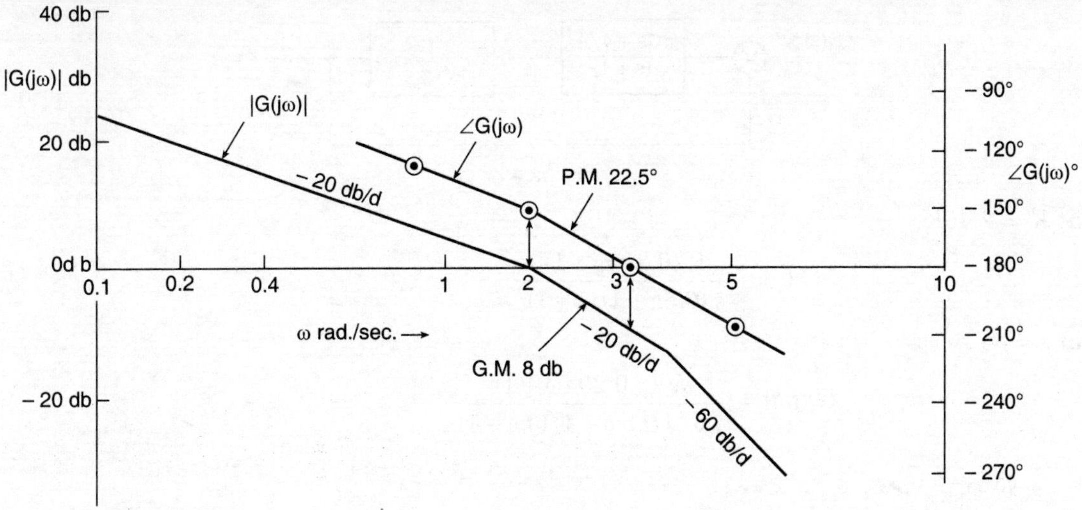

Fig. 12.44

Initial slope – 20 db/decade intersects 0 db at $\omega = 2 = K$, hence type of the open-loop transfer function is 1

The two simple corner frequencies due to open-loop poles are $\omega = 2$ and $\omega = 5$ rad./sec.

Thus, the open-loop transfer function is

$$G(j\omega)\,H(j\omega) = \frac{2}{j\omega\,(j0.2\omega + 1)\,(j\,0.5\omega + 1)}$$

or

$$G(s)\,H(s) = \frac{2}{s\,(0.2s + 1)\,(0.5s + 1)} \quad \textbf{Ans.}$$

$$\angle G(j\omega) = -90° - \tan^{-1} 0.2\omega - \tan^{-1} 0.5\omega$$

ω rad./sec.	1	2	3	5
∠G(jω)	– 127.8°	– 156.8°	– 177°	– 203°

As per above calculations $G(j\omega)$ plot is shown in Fig. 12.44 and it is noted therefrom, P.M. = 22.5° and G.M. = 8 db **Ans.**

Indicating stable system.

Example 12.37. *The block diagram of a control system is shown in Fig. 12.45. Sketch the asymptotic Bode plot and determine the value of K for the system to have a phase margin of 30°*

Solution.
$$G(s) = \frac{K\,(s + 4.9)}{s\,(s+1)\,(s+2)} = \frac{K \times 4.9\left(\frac{1}{4.9}s + 1\right)}{2s\left(\frac{1}{2}s + 1\right)(s+1)}$$

$$= \frac{2.45K\,(0.204s + 1)}{s\,(0.5s + 1)\,(s + 1)}$$

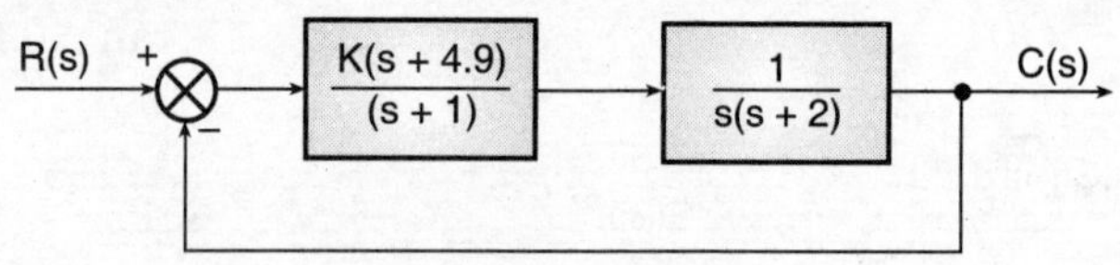

Fig. 12.45

put $K' = 2.45K$

$$\therefore \qquad G(s) = \frac{K'(0.204s + 1)}{s(0.5s + 1)(s + 1)}$$

put $s = j\omega$

$$\therefore \qquad G(j\omega) = \frac{K'(j\,0.204\omega + 1)}{j\omega(j\,0.5\omega + 1)(j\omega + 1)}$$

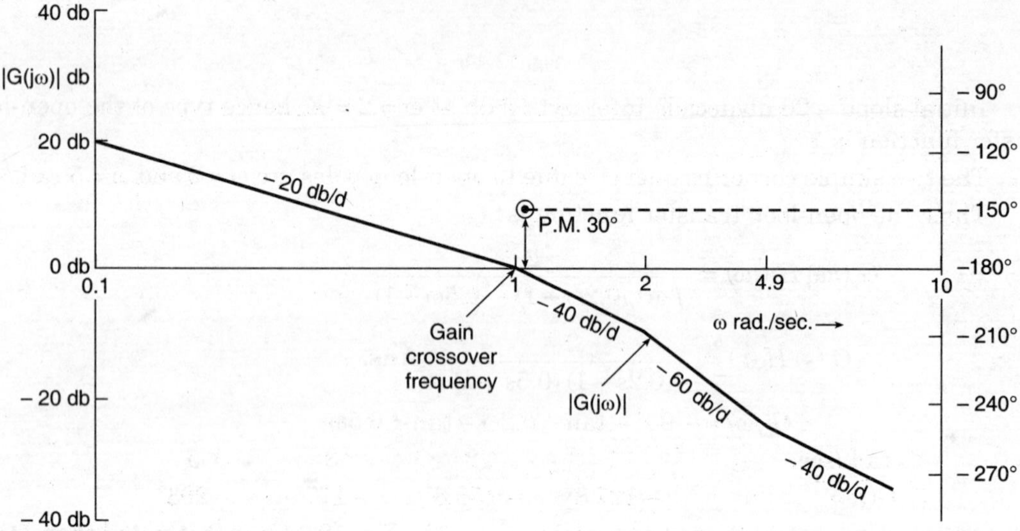

Fig. 12.45 (*a*)

Consider $K' = 1$

(*i*) Type 1 system, initial slope – 20 db/decade intersecting 0 db axis at $\omega = K' = 1$.

(*ii*) Corner frequencies :

At,

$\omega = 1$ rad./sec. due to open-loop pole, the magnitude slope changes by – 20 db/decade.

$\omega = \dfrac{1}{0.5} = 2$ rad./sec. due to open-loop pole, the magnitude slope changes by – 20 db/decade.

$\omega = \dfrac{1}{0.204} = 4.49$ rad./sec. due to open-loop zero, the magnitude slope changes by + 20 db/decade. The Bode plot is drawn as shown in Fig. 12.45 (*a*)

(*iii*) $\angle G(j\omega) = -90° - \tan^{-1}\omega - \tan^{-1} 0.5\omega + \tan^{-1} 0.204\omega$

The gain crossover frequency is $\omega = 1$ rad./sec.

$$\angle G(j1) = -90° - \tan^{-1} 1 - \tan^{-1} 0.5 \times 1 + \tan^{-1} 0.0204 \times 1$$
$$= -90° - 45° - 26.56° + 11.53°$$
$$\approx -150°$$

Phase margin $= 180° + (-150°) = 30°$

As the Bode (magnitude) plot is drawn for $K' = 1$

$\therefore$ $K' = 1$; For phase margin = 30°

$$2.45\, K = K'$$

$$\therefore \quad K = \frac{K'}{2.45} = \frac{1}{2.45} = 0.408 \quad \textbf{Ans.}$$

Example 12.38. *Having given the open-loop transfer function of a feedback control system as*

$$G(s)\, H(s) = \frac{K}{s\,(s^2 + 2s + 5)}$$

(i) Determine the value of gain K such that the phase margin is 45°

(ii) Determine the gain margin for the value of K obtained in (i)

Solution.

(i) $\quad$ P.M. $= 180° + G(j\omega_1)\, H(j\omega_1)$

where ω_1 = gain crossover frequency

$$\therefore \quad 45° = 180° + G(j\omega_1)\, H(j\omega_1) \;\therefore\; \angle G(j\omega_1)\, H(j\omega_1) = 45° - 180° = -135°$$

$$G(j\omega)\, H(j\omega) = \frac{K}{j\omega\,(5 - \omega^2 + j2\omega)} \quad \therefore \quad \angle G(j\omega)\, H(j\omega) = -90° - \tan^{-1} \frac{2\omega}{5 - \omega^2}$$

at $\omega = \omega_1$

$$\angle G(j\omega_1)\, H(j\omega_1) = -90° - \tan^{-1} \frac{2\omega_1}{5 - \omega_1^2}$$

$$\therefore \quad -90 - \tan^{-1} \frac{2\omega_1}{5 - \omega_1^2} = -135°$$

$$\text{or} \quad \tan^{-1} \frac{2\omega_1}{5 - \omega_1^2} = 45° \quad \text{or} \quad \frac{2\omega_1}{5 - \omega_1^2} = \tan 45°$$

$$\text{or} \quad \frac{2\omega_1}{5 - \omega_1^2} = 1 \quad \text{or} \quad \omega_1^2 + 2\omega_1 - 5 = 0$$

Solving above equation gain crossover frequency $\omega_1 = 1.445$ rad./sec.

At gain crossover frequency, $|G(j\omega_1)\, H(j\omega_1)| = 1$

$$\therefore \quad 1 = \left| \frac{K}{j\omega_1\,[(5 - \omega_1^2) + j2\omega_1]} \right|$$

or $\quad \omega_1 = 1.445$ rad./sec.

$$1 = \left| \frac{K}{j1.445\,[(5 - 1.445)^2 - j2 \times 1.445]} \right|$$

$$= \frac{K}{1.445\sqrt{(-2.91)^2 + (2.89)^2}} = \frac{K}{1.445\sqrt{8.47 + 8.35}} = \frac{K}{5.92}$$

$\therefore \quad K = 1 \times 5.92 = 5.92$ **Ans.**

(*ii*) $K = 5.92$

$$\therefore \quad G(j\omega)\, H(j\omega) = \frac{5.92}{j\omega\,(5 - \omega^2 + j2\omega)}$$

At the phase crossover frequency ω_2,

$$\angle G(j\omega_2)\, H(j\omega_2) = -180° \qquad \therefore \qquad -90° - \tan^{-1}\frac{2\omega_2}{5 - \omega^2} = -180°$$

or $$\tan^{-1}\frac{2\omega_2}{5 - \omega_2^2} = 90° \qquad \text{or} \qquad \frac{2\omega_2}{5 - \omega_2^2} = \tan 90° = \infty$$

$\therefore \quad 5 - \omega_2^2 = 0$, hence $\omega_2 = \sqrt{5}$ rad./sec.

The magnitude $|G(j\omega_2)\, H(j\omega_2)|$ is determined below :

$$|G(j\omega_2)\, H(j\omega_2)| = \left|\frac{5.9^2}{j\omega_2\,(5 - \omega_2^2 + j2\omega_2)}\right|$$

$\omega_2 = \sqrt{5}$

$$\therefore \quad |G(j\sqrt{2})\, H(j\sqrt{2})| = \left|\frac{5.92}{j\sqrt{5}\,(5 - \sqrt{5}^2 + j2\sqrt{5})}\right| = \left|\frac{5.92}{j\sqrt{5} \times 2\sqrt{5}}\right| = \frac{5.92}{10} = 0.592$$

Gain margin $$= 20 \log_{10}\left(\frac{1}{0.592}\right) = 4.55 \text{ db. Ans.}$$

Example 12.39. *(i) Determine the gain K such that the G.M. = 6 db. The feedback control system transfer function is given below :*

$$G(s)\, H(s) = \frac{K}{s\,(s^2 + 2s + 5)}$$

(ii) Determine the P.M. for the value of K obtained as above.

Solution.

(*i*) $$G(s)\, H(s) = \frac{K}{s\,(s^2 + 2s + 3)}$$

put $s = j\omega$

$$\therefore \quad G(j\omega)\, H(j\omega) = \frac{K}{j\omega\,(5 - \omega^2 + j2\omega)}$$

$$\angle G(j\omega)\, H(j\omega) = -90° - \tan^{-1}\frac{2\omega}{5 - \omega^2}$$

At the phase crossover frequency ω_2, $\angle G(j\omega_2)\, H(j\omega_2) = -180°$

$$\therefore \quad -90° - \tan^{-1}\frac{2\omega_2}{5 - \omega_2^2} = -180° \quad \text{or} \quad \tan^{-1}\frac{2\omega_2}{5 - \omega_2^2} = 90°$$

$$\therefore \quad \frac{2\omega_2}{5-\omega_2^2} = \tan 90° \quad \text{or} \quad \frac{2\omega_2}{5-\omega_2^2} = \infty$$

hence, $5 - \omega_2^2 = 0$

or $\omega_2 = \sqrt{5}$ rad./sec.

As the gain margin is 6 db, $|G(j\omega_2)\,H(j\omega_2)|$ is calculated as below :

$$20 \log_{10}\left(\frac{1}{x}\right) = 6$$

where $x = |G(j\omega_2)\,H(j\omega_2)|$

or $$\log_{10}\left(\frac{1}{x}\right) = \frac{6}{20}$$

$$\therefore \quad \frac{1}{x} = \text{antilog}_{10}\left(\frac{6}{20}\right) = 2 \quad \therefore \quad x = 0.5$$

and $$0.5 = \left|\frac{K}{j\omega_2\,(5-\omega_2^2 + j2\omega_2)}\right| \quad \because \quad \omega_2 = \sqrt{5} \text{ rad./sec.}$$

$$0.5 = \left|\frac{K}{j\sqrt{5}\,(5-5+j\,2\times\sqrt{5})}\right| = \left|\frac{K}{j\sqrt{5}\times j\,2\sqrt{5}}\right|$$

$$= \frac{K}{\{\sqrt{5^2}\times 2\sqrt{5}^2\}^{\frac{1}{2}}} = \frac{K}{10}$$

$\therefore$ $K = 0.5 \times 10 - 5$ **Ans.**

(*ii*) $$G(j\omega)\,H(j\omega) = \frac{5}{j\omega\,(5-\omega^2 + j\,2\omega)}$$

At the gain crossover frequency ω_1, $|G\,(j\omega_1)\,H(j\omega_1)| = 1$

$$\therefore \quad 1 = \left|\frac{5}{j\omega_1\,(5-\omega_1^2 + j2\omega_1)}\right|$$

$$= \left|\frac{5}{\omega_1\sqrt{(5-\omega_1^2)^2 + (2\omega_1)^2}}\right|$$

or $\omega_1^2\,(5-\omega_1^2)^2 + (2\omega_1)^2 = 5^2$

By trial and error, the solution of the above equation is $\omega_1 = 1.153$ rad./sec.

$$G(j\omega_1)\,H(j\omega_1) = -90° - \tan^{-1}\frac{2\omega_1}{5-\omega_1^2}$$

$$= -90° - \tan^{-1}\frac{2\times 1.153}{5-1.153^2} = -90° - \tan^{-1}\frac{2.306}{3.67}$$

$$= -90° - \tan^{-1} 0.628 = -90° - 32.12° = -122.12°$$

$$\text{P.M.} = 180° + G(j\omega_1)\,H(j\omega_1) = 180° + (-122.12°)$$

$$= 57.88°$$ **Ans.**

Example 12.40. *For the given open-loop transfer function of a unity feedback control system*

$$G\,(j\omega) = \frac{K}{j\omega\,(1 + j0.25\omega)\,(1 + j\omega)}.\ \textit{Plot the gain-phase plot.}$$

Determine (i) gain margin for K = 1 ; (ii) gain K′ for gain margin = 10 db ; (iii) gain K″ for phase margin = 25°.

Solution. $\quad \angle G(j\omega) = -\,90° - \tan^{-1} 0.25\omega - \tan^{-1}\omega$

For $K = 1$, $\quad |G(j\omega)|_{db} = 20\log_{10}\left|\dfrac{1}{\omega\sqrt{1+(0.25\omega)^2}\,\sqrt{1+\omega^2}}\right|$

As per above relations values of $\angle G(j\omega)$ and $|G(j\omega)|_{db}$ are calculated and tabulated below :

ω rad./sec.	0.5	0.78	1	2	2.5	3
$\|G(j\omega)\|_{db}$	4.96	0	– 3.35	7.1	– 18	– 21.5
$\angle G(j\omega)$	– 123.6°	– 139.2°	– 149°	– 180°	– 190°	– 198°

The gain-phase plot for $K = 1$ is drawn as shown in Fig. 12.46

(*i*) From the gain-phase plot it is noted that G.M. = – 7.1 db (for $K = 1$)

To have a G.M. = 10 db, the value of K is to be reduced the plot is shifted down wards by {– 10 – (– 7.1)} = – 2.9 db

$$\therefore \qquad 20\log_{10} K' = -\,2.9$$

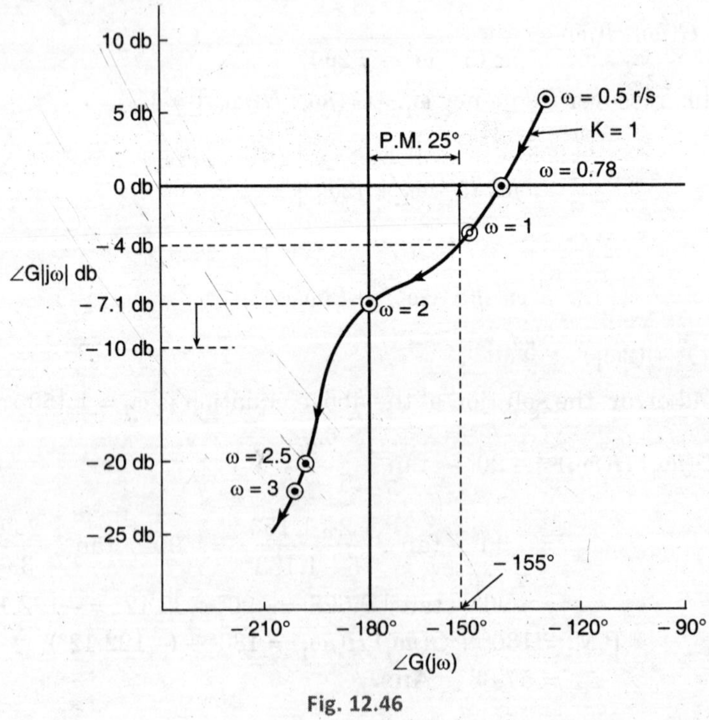

Fig. 12.46

or $$K' = \text{antilog}_{10}\left(\frac{-2.9}{20}\right) = 0.716 \quad \textbf{Ans.}$$

(*ii*) P.M. = 25°

$$G(j\omega)\, H(j\omega) = -180° + 25° = -155°$$

The corresponding gain at – 155°, is – 4 db, whereas at – 155°, the gain should be 0 db. This is achieved by shifting the gain-phase plot by {0 – (– 4)} = 4 db

$$\therefore \quad 20 \log_{10} K'' = 4 \quad \text{or} \quad K''' \text{ antilog}_{10}\left(\frac{4}{20}\right) = 1.58 \quad \textbf{Ans.}$$

Example 12.41. *The open-loop transfer function of a unity feedback control system having a time delay element is given below*

$$G(s) = \frac{Ke^{-0.5s}}{(0.5s + 1)}$$

Draw the gain-phase plot and comment on the stability.

Solution. $$G(s) = \frac{Ke^{-0.5s}}{(0.5s + 1)}$$

put $s = j\omega$

$$\therefore \quad G(j\omega) = \frac{Ke^{-j0.5\omega}}{(j0.5\omega + 1)}$$

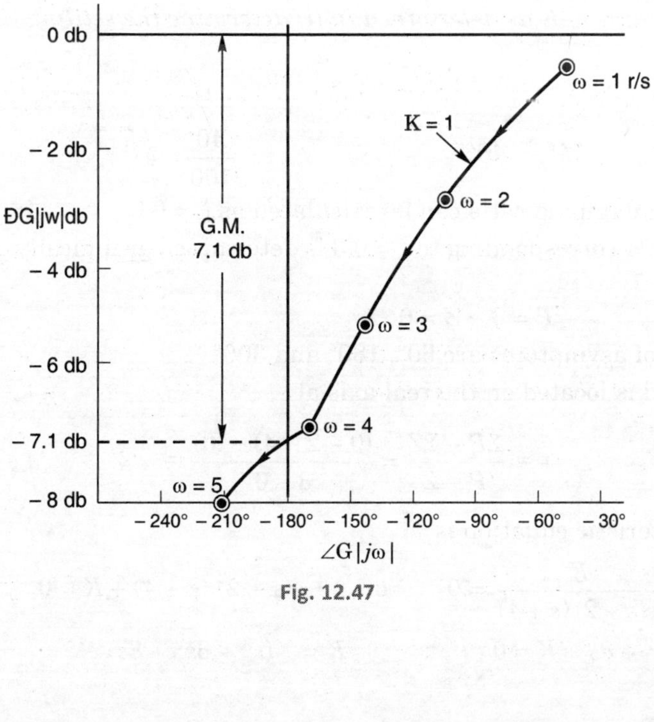

Fig. 12.47

The magnitude of the term $e^{-0.5\omega}$ is unity for all values of ω and phase angle is – 57.3 × 0.5ω°. Therefore,

$$|G(j\omega)| = \left|\frac{1}{\sqrt{(0.5\omega)^2 + 1}}\right| \text{ ; for } K = 1$$

and
$$\angle G(j\omega) = -\tan^{-1} 0.5\omega - 57.3 \times 0.5\omega$$

The magnitude (db) and phase angle for some values of ω are calculated and tabulated below :

ω rad./sec.	1	2	3	4	5
$\|G(j\omega)\|$ db	– 0.91	– 3	– 5.19	– 7	– 8.46
$G(j\omega)°$	– 55.2	– 102.3	– 142.5	– 177	– 211

The gain-phase plot for $K = 1$ is shown in Fig. 12.47 and noted that G.M. = 7.1 db.

The plot can be lifted upwards by increasing the value of gain K such that the gain-phase plot passes through (0 db, – 180°) point. This makes the system marginally stable. Thus the limiting value of gain for stability is given by

$$20 \log_{10} K = 7.1$$

or
$$K = \text{antilog}_{10}\left(\frac{7.1}{20}\right) = 2.26 \quad \textbf{Ans.}$$

Example 12.42. *The open-loop transfer function of a feedback control is given as*

$$G(s)\, H(s) = \frac{K}{s\,(s+2)\,(s+4)}$$

(i) It is desired to have M_p = 40% for a unit step input, determine the value of K
(ii) For the value of K as determined in (i) determine the settling time and K_v.

Solution.

As $\quad M_p = 40\% \quad \therefore \quad \dfrac{40}{100} = e^{\frac{-\zeta\pi}{\sqrt{1-\zeta^2}}}$

The value of damping ratio can be calculated as $\zeta = 0.3$

The value of K corresponding to $\zeta = 0.3$ is determined graphically using root locus plot as below :

$\because \quad P = 3, \; Z = 0$

The angles of asymptotes are 60°, 180° and 300°

The centroid is located on the real axis at

$$x = \frac{\Sigma P - \Sigma Z}{P - Z} = \frac{(0 - 2 - 4) - (0)}{3 - 0} = -2$$

The characteristic equation is

$$1 + \frac{K}{s(s+2)(s+4)} = 0 \quad \text{or} \quad s(s+2)(s+4) + K = 0$$

or
$$s^3 + 6s^2 + 8s + K = 0 \quad \therefore \quad K = -(s^3 + 6s^2 + 8s)$$

$$\frac{dK}{ds} = -(3s^2 + 12s + 8)$$

put $\frac{dK}{ds} = 0$ for determining the breakaway point

$\therefore \quad 3s^2 + 12s + 8 = 0 \quad$ or $\quad s^2 + 4s + 2.66 = 0$

The solution of the above equation is

$$s_1 s_2 = \frac{-4 \pm \sqrt{4^2 - 4 \times 1 \times 2.66}}{2} = -0.842, -3.15$$

$s_1 = -0.842$ lies on the root locus, hence breakaway point

$s_2 = -3.15$ is not a breakaway point as it does not lie on the root locus

The characteristic equation is given by

$s^3 + 6s^2 + 8s + K = 0$

Arranging Routh array for the above equation

s^3	1	8
s^2	6	K
s^1	$\frac{48 - K}{6}$	
s^0	K	

The value of K for marginal stability is given by

$$48 - K = 0 \quad \therefore \quad K = 48$$

The intersection of root locus with the s-plane imaginary axis is determined by solving the equation

$$6s^2 + K = 0 \quad \therefore \quad 6s^2 + 48 = 0$$

or
$$s = \sqrt{\frac{-48}{6}} = \pm j2.82$$

The root locus plot is shown in Fig. 12.48

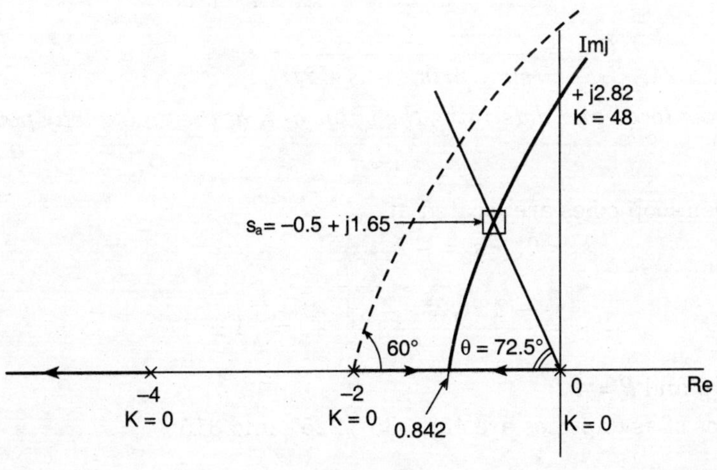

Fig. 12.48

The line $\theta = \cos^{-1} \zeta = 72.5°$ intersects root locus at $s_a = (-0.5 + j\,1.65)$

The value of gain K at $s = s_a$ is determined as per the root locus magnitude condition *i.e.*

$$|1| = \left|\frac{K}{s_a\,(s_a + 2)\,(s_a + 4)}\right|$$

$$\therefore \quad 1 = \left|\frac{K}{(-0.5 + j\,1.65)\,(-0.5 + j\,1.65 + 2)\,(-0.5 + j\,1.65 + 4)}\right|$$

$$= \left|\frac{K}{(-0.5 + j\,1.65)\,(1.5 + j\,1.65)\,(3.5 + j\,1.65)}\right|$$

$$= \frac{K}{\sqrt{0.5^2 + 1.65^2}\,\sqrt{1.5^2 + 1.65^2}\,\sqrt{3.5^2 + 1.65^2}} = \frac{K}{14.87}$$

(*i*) $\therefore$ $K = 14.87$ for $M_p = 40\%$ **Ans.**

(*ii*) $\zeta\,\omega_n = 0.5$

$$\therefore \quad \text{Settling time } t_s = \frac{4}{\zeta\,\omega_n} = \frac{4}{0.5} = 8 \text{ sec.} \quad \textbf{Ans.}$$

The effect of the third root located on the –ive real axis of the s-plane can be neglected as the same is far away from the dominant root s_a

$$K_v = \lim_{s \to 0} s\,G(s)\,H(s)$$

$$= \lim_{s \to 0} s.\frac{K}{s(s+2)\,(s+4)} = \frac{K}{8}$$

$$\therefore \quad K_v = \frac{14.87}{8} = 1.85 \quad \textbf{Ans.}$$

Example 12.43. *The open-loop transfer function of a unity feedback control system is given below :*

$$G(s) = \frac{K}{s(s+2)\,(s^2 + 2s + 2)}$$

Plot the root locus and determine the value of K at the breakaway point.

Solution.

1. The open-loop poles are located at

$$s_1 = 0, \quad s_2 = -2$$

and
$$s_3, s_4 = \frac{-2 \pm \sqrt{2^2 - 4 \times 1 \times 2}}{2} = -1 \pm j1$$

2. As $Z = 0$ and $P = 4$

The angles of asymptotes are 45°, 135°, 225°, and 315°.

The asymptotes intersect on real axis at

$$x = \frac{\Sigma P - \Sigma Z}{P - Z} = \frac{(0 - 2 - 1 + j1 - 1 - j1) - 0}{4 - 0} = -1$$

3. The characteristic equation is given by

$$1 + \frac{K}{s(s+2)(s^2+2s+2)}$$

or $$s(s+2)(s^2+2s+2) + K = 0$$

or $$s^4 + 4s^3 + 6s^2 + 4s + K = 0$$

$\therefore$ $$K = -(s^4 + 4s^3 + 6s^2 + 4s)$$

$$\frac{dK}{ds} = -(4s^3 + 12s^2 + 12s + 4)$$

The breakaway points are determined by solving $(dK/ds) = 0$

$\therefore$ $$4s^3 + 12s^2 + 12s + 4 = 0$$

or $$(s+1)^3 = 0$$

The solution of the above equation is $s = -1$

Thus, the breakaway point is located at $s = -1$

4. The characteristic equation is $s^4 + 4s^3 + 6s^2 + 4s + K = 0$

Applying Routh-Hurwitz criterion to the characteristic equation.

s^4	1	6	K
s^3	4	4	
s^2	5	K	
s^1	$\frac{20-4K}{5}$		
s^0	K		

The value of K for marginal stability is determined below

$$\frac{20 - 4K}{5} = 0$$

$\therefore$ $K = 5$ for marginal stability

The auxiliary equation is

$$5s^2 + K = 0$$

$\therefore$ $$5s^2 + 5 = 0$$

or $$s^2 = -1$$

$\therefore$ $s = \pm j1$ are the intersection points of the root locus with the imaginary axis.

The angles of departure from the complex poles are determined below

$$\underset{(-1+j1)}{\phi_d} = 180 - (90° + 45° + 135°) = -90°$$

Similarly

$$\underset{(-1-j1)}{\phi_d} = 180 - (-90° + 45° + 135°) = +90°$$

From the data calculated above the root locus plot is drawn as shown in Fig. 12.49.

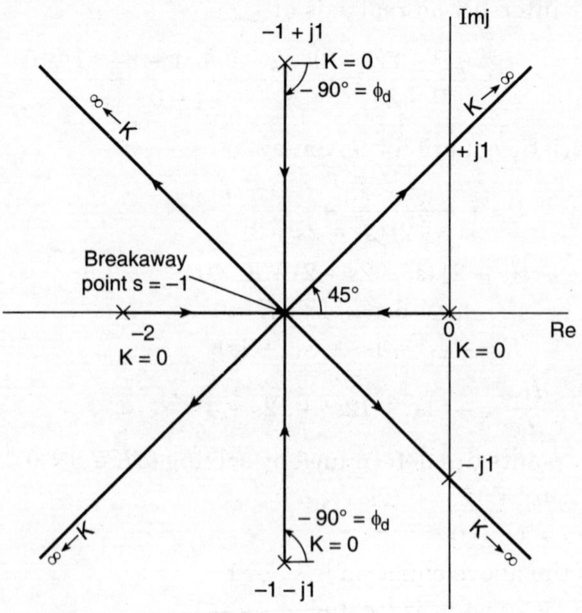

Fig. 12.49

The breakaway point and the imaginary axis intersection point of the root locus are located at $s = -1$ and $s = \pm\, j1$ respectively.

The root locus passing through these points makes an angle of 45°, with the real axis. The angle of asymptotes are 45°, 135° (– 45°) etc. Thus, the root locus branches break from the breakaway point forming straight lines as indicated on the root locus plot.

The value of K at the breakaway point *i.e.* $s = -1$ is determined by applying root locus magnitude condition to the open-loop transfer function, $G(s) = K/[s(s + 2)(s^2 + 2s + 2)]$

i.e.
$$1 = \left|\frac{K}{(-1)(-1+2)[(-1)^2 + 2(-1) + 2]}\right|$$

$$= \left|\frac{K}{(-1)(1)(1)}\right| = \left|\frac{K}{-1}\right|$$

∴ K = 1 at the breakaway point **Ans.**

Example 12.44. *Sketch the root locus plot and determine the value of gain K for marginal stability given that*

$$G(s)\,H(s) = \frac{K}{(s^2 + 2s + 2)(s^2 + 2s + 5)}.$$

Solution. The open-loop poles are determined below

1. (*i*) $s^2 + 2s + 2 = 0$

$$s_1, s_2 = \frac{-2 \pm \sqrt{2^2 - 4 \times 1 \times 2}}{2} = -1 \pm j1$$

(*ii*) $s^2 + 2s + 5 = 0$

$$s_3, s_4 = \frac{-2 \pm \sqrt{2^2 - 4 \times 1 \times 5}}{2} = -1 \pm j2$$

2. $P = 4, Z = 0$

The angles of asymptotes are :

45°, 135°, 225° (– 135°) and 315° (– 45°)

The asymptotes intersect on the real axis at

$$x = \frac{\Sigma P - \Sigma Z}{P - Z} = \frac{\{(-1 + j1) + (-1 - j1) + (-1 + j2) + (-1 - j2) - (0)\}}{4 - 0}$$

$$= -1$$

3. The characteristic equation is given by

$$1 + \frac{K}{(s^2 + 2s + 2)(s^2 + 2s + 5)} = 0$$

or $$(s^2 + 2s + 2)(s^2 + 2s + 5) + K = 0$$

or $$s^4 + 4s^3 + 11s^2 + 14s + (K + 10) = 0$$

$\therefore$ $$K = -(s^4 + 4s^3 + 11s^2 + 14s + 10)$$

$\therefore$ $$\frac{dK}{ds} = -(4s^3 + 12s^2 + 22s + 14)$$

The breakaway point is determined by solving $(dK/ds) = 0$

$\therefore$ $$4s^3 + 12s^2 + 22s + 14 = 0$$

or $$2s^3 + 6s^2 + 11s + 7 = 0$$

or $$(s + 1)(2s^2 + 4s + 7) = 0$$

or $$(s + 1)(s + 1 + j\,1.59)(s + 1 - j1.59) = 0$$

The solution of the above equation is

$$s = -1$$
$$s = -1 + j\,1.59$$
$$s = -1 - j\,1.59$$

As $s = -1$ does not lie on the root locus, the same cannot be considered as a breakaway point.

The points $s = (-1 \pm j1.59)$ are located on the root locus, hence, considered as breakaway points.

4. The angle of departure from the complex open-loop poles is calculated below :

$$\underset{(-1 \pm j2)}{\phi_d} = 180° - (\theta_{p_1} + \theta_{p_2} + \theta_{p_3})$$

$$= 180° - (90° \mp 90° \mp 90°) = \mp 90°$$

$$\underset{(-1 \pm j1)}{\phi_d} = 180° - (90° \pm 90° \pm 90°) = \pm 90°$$

The intersection of the root locus with the imaginary axis is determined by applying Routh-Hurwitz criterion to the characteristic equation.

$$s^4 + 4s^3 + 11s^2 + 14s + (K + 10) = 0$$

Therefore,

s^4	1	11	$K + 10$
s^3	4	14	
s^2	$\frac{4\times11-14\times1}{4} = 7.5$	$K + 10$	
s^1	$\frac{7.5\times14-4(K+10)}{7.5}$		
s^0	$K + 10$		

The value of K for marginal stability is obtained by solving the equation

$$\frac{7.5\times14-4(K+10)}{7.5} = 0$$

or $$7.5 \times 14 - 4(K + 10) = 0 \quad \text{or} \quad 105 - 4K - 40 = 0$$

$\therefore$ $$K = \frac{65}{4} = 16.25 \text{ (for marginal stability)}$$ **Ans.**

The auxiliary equation is

$$7.5s^2 + (K + 10) = 0$$

or $$7.5s^2 + (16.25 + 10) = 0$$

$\therefore$ $$s^2 = \frac{-26.25}{7.5} = -3.5$$

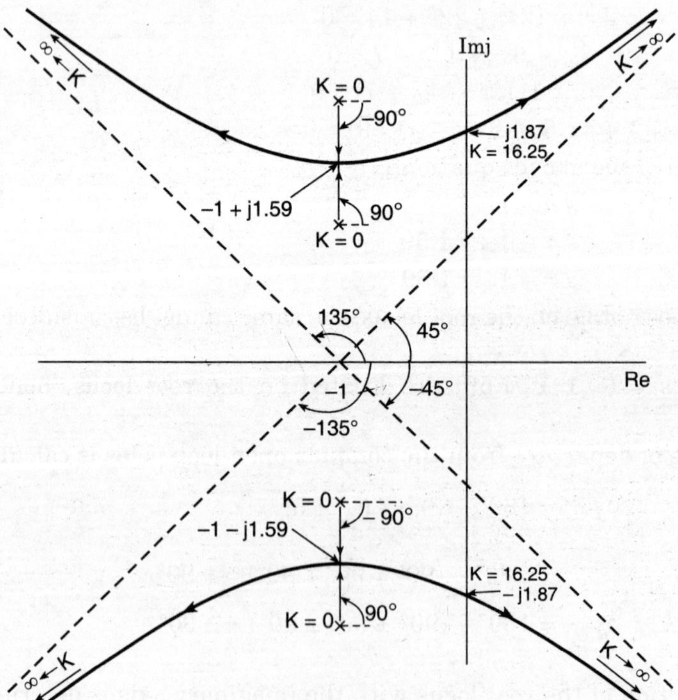

Fig. 12.50

$\therefore \qquad s = \pm\sqrt{3.5} = \pm j\,1.87$

which gives the intersection of the root locus with imaginary axis at $s = \pm j\,1.87$

As per the above calculations, the root locus is plotted as shown in Fig. 12.50. **Ans.**

Example 12.45. *The open-loop transfer function of a feedback system is given below*

$$G(s)\,H(s) = \frac{K\,(s+1)}{s\,(s+2)\,(s^2+4s+8)}$$

Sketch the root locus for $K \geq 0$ and determine the value of K for marginal stability.

Solution. The open-loop poles are

$$s = 0,\ s = -2,\ s = \frac{-4 \pm \sqrt{16 - 4 \times 8}}{2} = -2 \pm j2$$

The open-loop zero is $s = -1$

The root locus starts ($K = 0$) from open-loop poles and one branch terminates ($K = \infty$) at zero and other two at ∞

Number of root locus branches $N = P = 4$ ($P > Z$).

Angle of asymptotes for higher values of K

$$= \frac{(2k+1)}{P-Z} \times 180°, \quad k = 0, 1, 2$$

(*i*) $$= \frac{(2 \times 0 + 1)}{4-1} \times 180° = 60°$$

(*ii*) $$= \frac{(2 \times 1 + 1)}{4-1} \times 180° = 180°$$

(*iii*) $$= \frac{(2 \times 2 + 1)}{4-1} \times 180° = 300°\ (-60°)$$

Asymptotes intersect on – ive real axis at

$$x = \frac{\Sigma P - \Sigma Z}{P-Z} = \frac{(10 - 2 - 2 + j2 - 2 - j2) - (-1)}{4-1} = -1.66$$

Angle of departure from complex pole (Fig. 12.51)

$$\theta_d = 180° - (\theta_{p_1} + \theta_{p_2} + \theta_{p_3} + \theta_z)$$

$$\theta_{p_1} = 90°$$

$$\theta_{p_2} = 90°$$

$$\theta_{p_3} = 90° + \tan^{-1}\frac{2}{2} = 90° + 45° = 135°$$

$$\theta_z = 90° + \tan^{-1}\frac{1}{2} = 90° + 26.56° = 116.56°$$

$$\therefore \qquad \theta_{d\,(-2+j2)} = 180° - (90° + 90° + 135° - 116.56°)$$

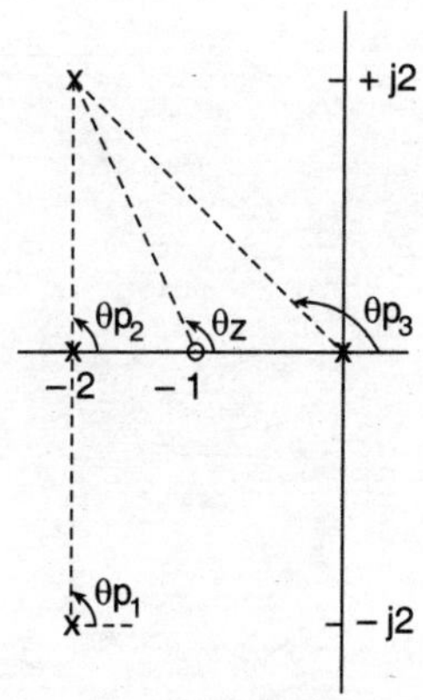

Fig. 12.51

$$= -18.44°$$

Similarly, $\underset{(-2-j2)}{\theta_d} = +18.44°$

Intersection of root loci with imaginary axis :

The characteristic equation is given by $1 + G(s)\,H(s) = 0$

$$\therefore \quad 1 + \frac{K(s+1)}{s(s+2)(s^2+4s+8)} = 0$$

or $\quad s(s+2)(s^2+4s+8) + K(s+1) = 0$ or $s^4 + 6s^3 + 16s^2 + (K+16)s + K = 0$

The Routh table is formed below :

s^4	1	16	K
s^3	6	$K+16$	
s^2	$\dfrac{80-K}{6}$	K	
s^1	$\dfrac{\frac{(80-K)}{6}(K+16)-6K}{(80-K)/6} = -K^2 + 28K + 1280$		
s^0	K		

The limiting value of K for stability is determined by solving the following equation

$$-K^2 + 28K + 1280 = 0$$

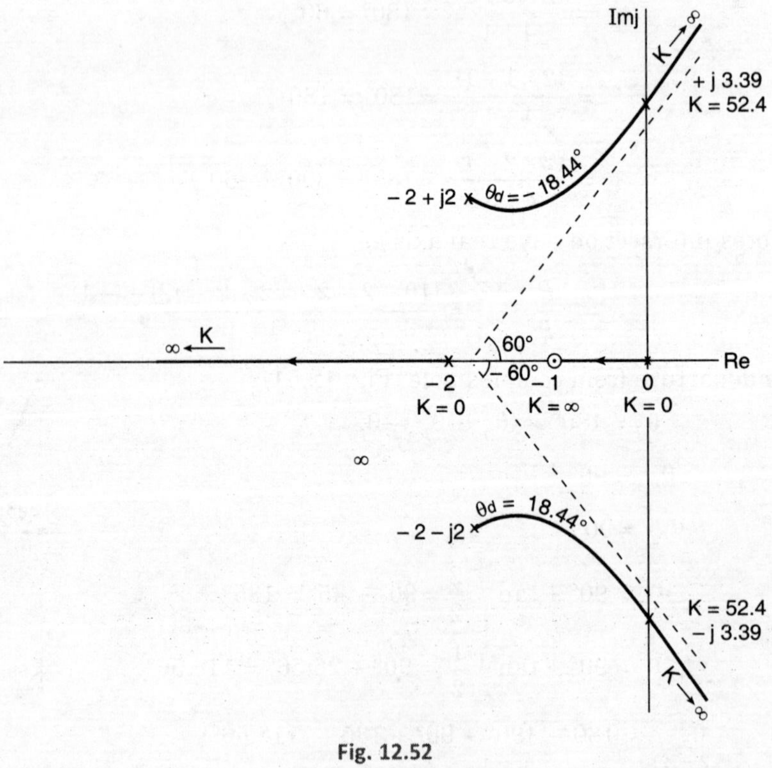

Fig. 12.52

Solving above equation $K = 52.4, -24.42$

Considering + ive value, the root locus intersects imaginary axis for $K = 52.4$

The intersecting point is obtained by solving following equation

$$\frac{80 - K}{6} s^2 + K = 0$$

$$\therefore \qquad \frac{80 - 52.4}{6} s^2 + 52.4 = 0$$

or $$s = \pm j\,3.39$$

The root locus branches cross the imaginary axis at $s = \pm j\,3.39$

The root locus plot is drawn as shown in Fig. 12.52 and for marginal stability, $K = 52.4$

Ans.

Example 12.46. *Sketch the root locus for the open-loop transfer function of a unity feedback control system having given*

$$G(s) = \frac{K}{s\,(s + 6)^2}$$

Determine (i) the value of centroid and the frequency at which the root locus branches cross the imaginary axis

(ii) the value of K such that the response to unit step input is critically damped.

Solution. The open-loop poles are

$$s = 0$$

$$s = -6, \quad 6 \text{ (multiple)}$$

The number of root locus branches $N = P = 3$

The root locus branches start ($K = 0$) from open-loop poles and terminate ($K = \infty$) at infinity

Three angles of asymptotes are given by

$$\frac{(2k + 1)}{P - Z} \times 180° \;; \quad k = 0, 1, 2$$

(i) $$\frac{(2 \times 0 + 1)}{3 - 0} \times 180° = 60°$$

(ii) $$\frac{(2 \times 1 + 1)}{3 - 0} \times 180° = 180°$$

(iii) $$\frac{(2 \times 2 + 1)}{3 - 0} \times 180° = 300° \; (-60°)$$

The asymptotes intersect on the real axis at

$$x = \frac{\Sigma P - \Sigma Z}{P - Z} = \frac{(0 - 6 - 6) - (0)}{3 - 0} = -4 \quad \textbf{Ans.}$$

The characteristic equation is $1 + G(s) = 0$

$$\therefore \qquad 1 + \frac{K}{s(s + 6)^2} = 0$$

or $$s^3 + 12s^2 + 36s + K = 0 \quad \therefore \quad K = -(s^3 + 12s^2 + 36s)$$

Breakaway point is given by solving the equation $\frac{dK}{ds} = 0$ for s

$$\frac{dK}{ds} = -(3s^2 + 24s + 36)$$

put $dK/ds = 0$

$$\therefore \quad 3s^2 + 24s + 36 = 0 \quad \text{or} \quad s^3 + 8s + 12 = 0$$

$$\therefore \quad s_1, s_2 = -6, -2$$

The point $s_2 = -2$ is selected as breakaway point as it is located between two open-loop poles.

The characteristic equation is $s^3 + 12s^2 + 36s + K = 0$

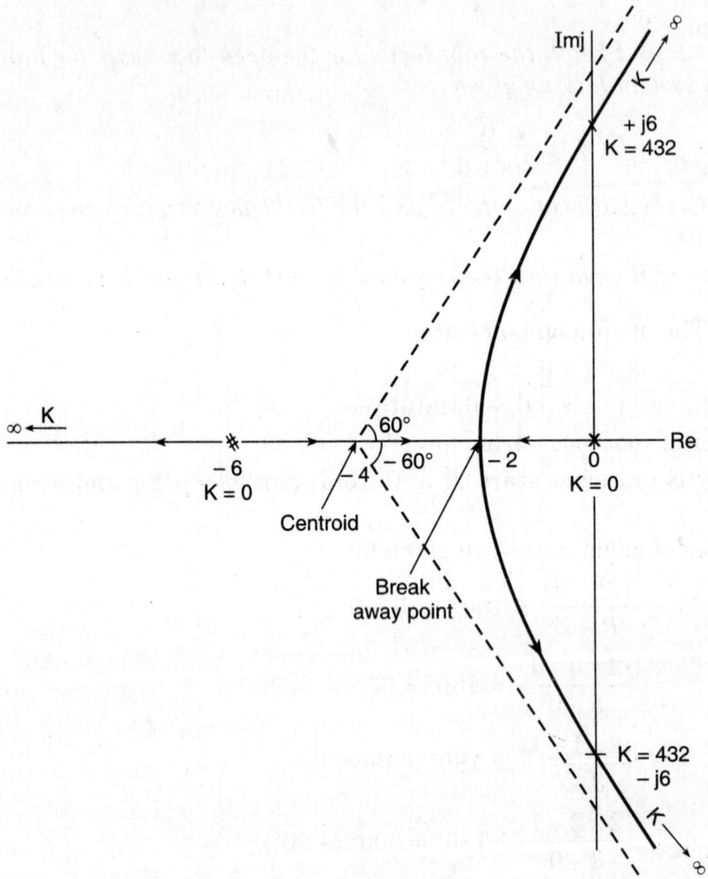

Fig. 12.53

The Routh table is formed below :

s^3	1	36
s^2	12	K
s^1	$\frac{12 \times 36 - K \times 1}{12}$	
s^0	K	

The value of K for marginal, stability is determined by equating s^1 term to zero,

$$\therefore \quad \frac{12\times 36 - K\times 1}{12} = 0$$

$\therefore$ $K = 432$ for marginal stability.

The crossing point of the root locus branches on the imaginary axis is determined below :

$$12s^2 + K = 0 \quad \because \quad K = 432$$

$$\therefore \quad 12s^2 + 432 = 0$$

or $$s = \sqrt{\frac{-432}{12}} = \pm j6 \quad \textbf{Ans.}$$

The root locus plot is drawn as shown in Fig. 12.53

At $s = -2$ (breakaway point), the two roots of the characteristic equation are coincident with no imaginary part, hence the system is critically damped

The value of K at $s = -2$ is calculated below,

As the point $s = -2$ lies on the root locus

$$\therefore \quad |G(-2)| = 1$$

$$\therefore \quad 1 = \left|\frac{K}{(-2)[-2+6]^2}\right| = \left|\frac{K}{-2\times 4^2}\right|$$

or $$K = 32 \quad \textbf{Ans.}$$

Example 12.47. *The open-loop transfer function of a feedback control system is given below :*

$$G(s)\,H(s) = \frac{K\,(s+2)}{s\,(s+1)}$$

(i) Plot the root locus and show that the complex roots lie on a circle.

(ii) Determine the value of gain factor K corresponding to minimum damping ratio.

Solution. $G(s)\,H(s) = \dfrac{K\,(s+2)}{s\,(s+1)}$

Let any point $s = (\sigma + j\omega)$ lie on the root locus.

$$\therefore \quad G(\sigma + j\omega) = \frac{K\,(\sigma + j\omega + 2)}{(\sigma + j\omega)\,(\sigma + j\omega + 1)} = \frac{K\,(\sigma + 2 + j\omega)}{(\sigma + j\omega)\,(\sigma + 1 + j\omega)}$$

$$G(\sigma + j\omega) = \tan^{-1}\frac{\omega}{\sigma+2} - \tan^{-1}\frac{\omega}{\sigma} - \tan^{-1}\frac{\omega}{\sigma+1}$$

For the point $s = (\sigma + j\omega)$ to lie on the root locus, then as per phase angle condition,

$$\angle G(\sigma + j\omega) = \pm\,(2k+1)\,180° \qquad k = 0, 1, 2...$$

Consider $k = 0$

$$\therefore \quad G(\sigma + j\omega) = 180°$$

$$\therefore \quad \tan^{-1}\frac{\omega}{\sigma+2} - \tan^{-1}\frac{\omega}{\sigma} - \tan^{-1}\frac{\omega}{\sigma+1} = -\,180°$$

or $$\tan^{-1}\frac{\omega}{\sigma+2} - \left\{\tan^{-1}\frac{\left(\frac{\omega}{\sigma}+\frac{\omega}{\sigma+1}\right)}{\left(1-\frac{\omega}{\sigma}\cdot\frac{\omega}{\sigma+1}\right)}\right\} = -180°$$

or $$\tan^{-1}\frac{\omega}{\sigma+2} - \tan^{-1}\frac{\omega(\sigma+1)+\omega\sigma}{\sigma(\sigma+1)-\omega^2} = -180°$$

or $$-\tan^{-1}\frac{\omega(\sigma+1)+\omega\sigma}{\sigma(\sigma+1)-\omega^2} = -180° - \tan^{-1}\frac{\omega}{\sigma+2}$$

$\because$ $$\tan^{-1} 0 = -180°$$

$\therefore$ $$-\tan^{-1}\frac{\omega(\sigma+1)+\omega\sigma}{\sigma(\sigma+1)-\omega^2} = \tan^{-1} 0 - \tan^{-1}\frac{\omega}{\sigma+2}$$

or $$-\tan^{-1}\frac{\omega(\sigma+1)+\omega\sigma}{\sigma(\sigma+1)-\omega^2} = \tan^{-1}\left(\frac{0-\frac{\omega}{\sigma+2}}{1+0\cdot\frac{\omega}{\sigma+2}}\right)$$

or $$-\tan^{-1}\frac{\omega(\sigma+1)+\omega\sigma}{\sigma(\sigma+1)-\omega^2} = \tan^{-1}\left(\frac{-\omega}{\sigma+2}\right)$$

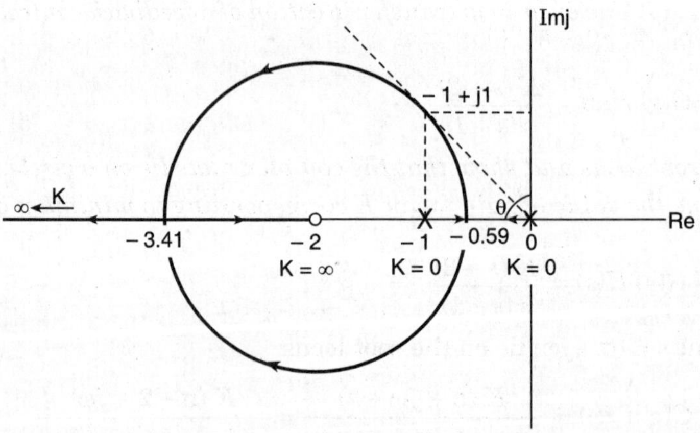

Fig. 12.54

Taking tangent on both sides,

$$\frac{\omega(\sigma+1)+\omega\sigma}{\sigma(\sigma+1)-\omega^2} = \frac{\omega}{\sigma+2}$$

or $$\frac{2\sigma+1}{\sigma^2+\sigma-\omega^2} = \frac{1}{\sigma+2}$$

or $$(2\sigma+1)(\sigma+2) = \sigma^2+\sigma-\omega^2$$

or $$2\sigma^2+5\sigma+2 = \sigma^2+\sigma-\omega^2$$

Simplifying $\sigma^2 + 4\sigma + 2 + \omega^2 = 0$

or $\sigma^2 + 4\sigma + 4 - 4 + 2 + \omega^2 = 0$

or $(\sigma + 2)^2 + (\omega - 0)^2 = (\sqrt{2})^2$

The equation derived above is that of a circle with centre $(-2 + j0)$ and radius $\sqrt{2}$.

The root locus plot is drawn as shown in Fig. 12.54

(ii) Tangential point on the root locus from the origin of the *s*-plane determines the value of minimum damping ratio as below :

$$\theta = \cos^{-1}\zeta$$

$$\therefore \quad \zeta = \cos\theta$$

$$= \frac{1}{\sqrt{1^2 + 1^2}} = 0.707 \quad \textbf{Ans.}$$

Example 12.48. *Sketch the root locus for the open-loop transfer function of a unity feedback control system*

$$G(s) = \frac{K}{s(s+2)(s+4)}\ ; K \geq 0$$

From root locus plot

(i) Obtain the value of gain factor K such that the closed-loop system operates with a damping ratio $\zeta = 0.277$.

(ii) The roots of the characteristic equation for the value of K as determined in (i).

(iii) The overall transfer function.

Solution. The open-loop poles are $s = 0$, $s = -2$ and $s = -3$

The root locus branches start ($K = 0$) from poles

The root locus branches terminate ($K = \infty$) at ∞

The number of root locus branches $N = P = 3$ $(Z = 0)$

The angle of asymptotes for higher values of K are given by

$$\frac{(2k+1)}{P-Z} \times 180°, \quad k = 0, 1, 2$$

(i) $$\frac{(2\times 0+1)}{3-0} \times 180° = 60°$$

(ii) $$\frac{(2\times 1+1)}{3-0} \times 180° = 180°$$

(iii) $$\frac{(2\times 2+1)}{3-0} \times 180° = 300° \ (-60°)$$

The asymptotes intersect on real axis at a point

$$x = \frac{\Sigma P - \Sigma Z}{P - Z}$$

$$= \frac{(0-2-4)-0}{3} = -2$$

Between poles $s = 0$ and $s = -2$ the root locus breaks. The breakaway point is determined by solving $\dfrac{dK}{ds} = 0$

The characteristic equation is

$$1 + G(s) = 0$$

$$1 + \frac{K}{s(s+2)(s+4)} = 0$$

$$\therefore \quad s(s+2)(s+4) + K = 0$$

or

$$s^3 + 6s^2 + 8s + K = 0$$

$$\therefore \quad K = -(s^3 + 6s^2 + 8s)$$

$$\frac{dK}{ds} = -(3s^2 + 12s + 8)$$

put $dK/ds = 0$

$$\therefore \quad 3s^2 + 12s + 8 = 0$$

The solution of above equation is $s = -0.85, -3.15$. The point $s = -0.85$ lies on the root locus, hence, breakaway point.

The characteristic equation : $s^3 + 6s^2 + 8s + K = 0$

The Routh table is formed below :

s^3	1	8
s^2	6	K
s^1	$\dfrac{6 \times 8 - K}{6}$	
s^0	K	

The root locus intersects the imaginary axis at the value of K given by

$$\frac{6 \times 8 - K}{6} = 0 \quad \therefore \quad K = 48$$

The point of crossing imaginary is calculated by solving,

$$6s^2 + K = 0$$

$$\therefore \quad 6s^2 + 48 = 0 \quad \text{or} \quad s^2 = -8$$

$$\therefore \quad s = \pm j\,2.82$$

As per the data calculated above the root locus is plotted as shown in Fig. 12.55

The damping ratio line ($\zeta = 0.277$) is drawn at an angle $\theta = \cos^{-1}\zeta = \cos^{-1} 0.277 = 73.9°$ as shown in Fig. 12.55. This line intersects the root locus at $s = -0.5 + j\,1.66$. The value of K at this point is determined below :

The point $s = -0.5 + j\,1.66$ lies on the root locus,

$$\therefore \quad 1 = \left| \frac{K}{s(s+2)(s+4)} \right|_{s = -0.5 + j\,1.66}$$

$$= \left| \frac{K}{(-0.5 + j1.66)(-0.5 + j1.66 + 2)(-0.5 + j1.66 + 4)} \right|$$

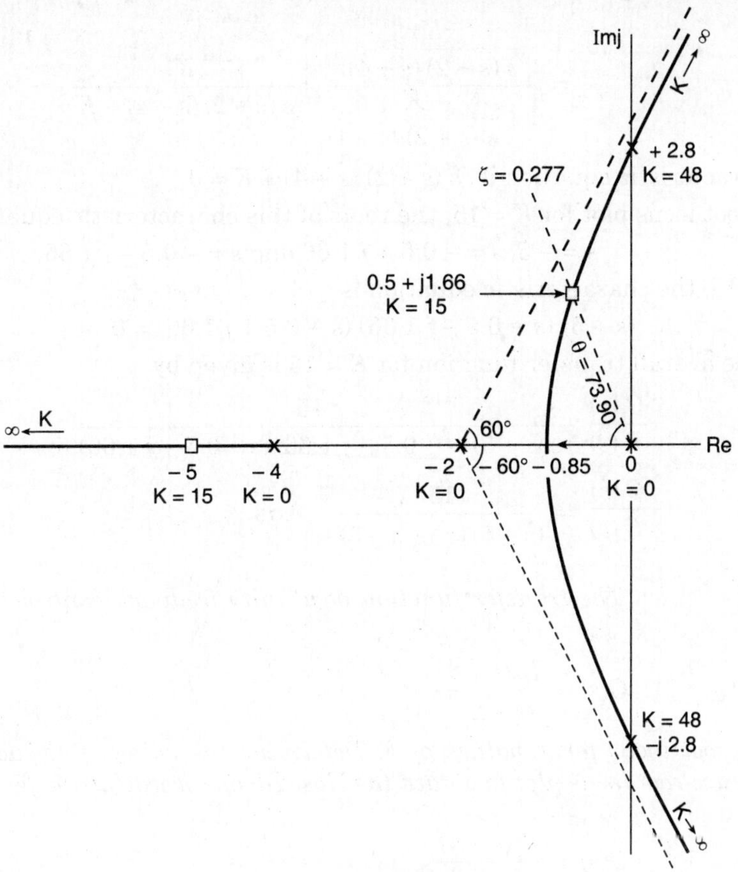

Fig. 12.55

$$= \left|\frac{K}{(-0.5 + j1.66)(1.5 + j1.66)(3.5 + j1.66)}\right|$$

$$= \frac{K}{\sqrt{0.5^2 + 1.66^2}\,\sqrt{1.5^2 + 1.66^2}\,\sqrt{3.5^2 + 1.66^2}} = \frac{K}{15}$$

$\therefore \qquad K = 15$ **Ans.**

At $K = 15$, the two complex roots of the characteristic equation are $s = -0.5 \pm j\,1.66$

By trial and error third real root is at $s = -5$ as verified below

$$1 = \left|\frac{K}{s(s+2)(s+4)}\right|_{s=-5} = \left|\frac{K}{-5(-5+2)(-5+4)}\right| = \frac{K}{15}$$

$\therefore \qquad K = 15$

Determination of overall transfer function :

The overall transfer function is given by

$$\frac{C(s)}{R(s)} = \frac{G(s)}{1 + G(s)\,H(s)}$$

$$= \frac{\dfrac{K}{s\,(s+2)\,(s+4)}}{1+\dfrac{K}{s\,(s+2)\,(s+4)}} = \frac{K}{s\,(s+2)\,(s+4)+K}$$

The characteristic equation is, $s\ (s + 2)\ (s + 4) + K = 0$

As per root locus plot for $K = 15$, the roots of this characteristic equation are

$$s = -5,\ s = -0.5 + j\,1.66 \text{ and } s = -0.5 - j\,1.66,$$

hence, for $K = 15$, the characteristic equation is

$$(s + 5)\ (s + 0.5 - j\,1.66)\ (s + 0.5 + j\,1.66) = 0$$

Thus, the overall transfer function for $K = 15$ is given by

$$\frac{C\,(s)}{R\,(s)} = \frac{15}{(s+5)\,(s+0.5-j\,1.66)\,(s+0.5-j\,1.66)}$$

or
$$\frac{C\,(s)}{R\,(s)} = \frac{15}{(s+5)\,(s^2+s+3)} \quad \textbf{Ans.}$$

Example 12.49. *The transfer function of a unity feedback control system is given below :*

$$G(s) = \frac{K\,(s+4)}{(s+2)^2}$$

Plot the root locus for + values of K. Determine the value of K such that $\omega_d = 2$ rad./sec. and use root locus plot to obtain the closed-loop (overall) transfer function.

Solution. $G(s) = \dfrac{K\,(s+4)}{(s+2)^2}$

The open-loop poles are $s = -2, -2$ (coincident)

The open-loop zero is $s = -4$

Number of branches $N = P = 2\ (P > Z)$

The root locus branches start ($K = 0$) from open-loop poles ($s = -2, -2$).

One root locus branch terminates

($K = \infty$) at zero ($s = -4$) while the other terminates at ∞.

The angle of asymptotes is given by $\dfrac{(2k+1)}{P-Z} \times 180°$

$\because$ $P - Z = 2 - 1 = 1$, only one asymptotes exists and its angle is

$$\frac{(2\times 0+1)}{2-1} \times 180° = 180°$$

The breakaway points are determined below

The characteristic equation is

$$(s + 2)^2 + K\,(s + 4) = 0$$

$$\therefore \qquad K = \frac{(s+2)^2}{s+4}$$

$$\therefore \qquad \frac{dK}{ds} = \frac{(s+4)\,2\,(s+2)\times 1 - (s+2)^2 \times 1}{(s+4)^2} = \frac{(s+2)\,(s+6)}{(s+4)^2}$$

put $\frac{dK}{ds} = 0$

$$\therefore \qquad (s+2)\,(s+6) = 0$$

Hence, breakaway points are $s = -2$ and $s = -6$. The root locus plot is drawn as shown in Fig. 12.56

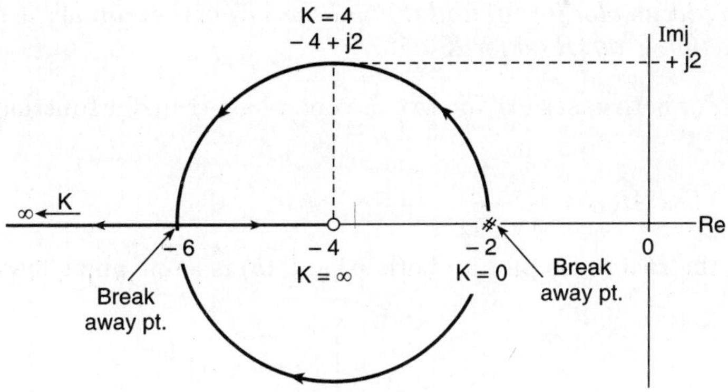

Fig. 12.56

For, $\omega_d = 2$ rad./sec., the corresponding point on the root locus is $s = -4 + j\,2$. The value of K at $s = -4 + j\,2$ is determined below

$$1 = \left|\frac{K\,(s+4)}{(s+2)}\right|_{s=-4+j\,2}$$

$$\therefore \qquad 1 = \left|\frac{K\,(-4+j2+4)}{(-4+j2+2)^2}\right| = \left|\frac{K \times j2}{(-2+j2)^2}\right| = \frac{K \times 2}{(\sqrt{2^2+2^2}\,)^2} = \frac{K}{4}$$

$$\therefore \qquad K = 4$$

The closed-loop transfer function is given by

$$\frac{C(s)}{R(s)} = \frac{G(s)}{1+G(s)} = \frac{K\,(s+4)\,/\,(s+2)^2}{1+\dfrac{K\,(s+4)}{(s+2)^2}} = \frac{K\,(s+4)}{(s+2)^2 + K\,(s+4)}$$

As per root locus plot, at $K = 4$, the two complex roots are

$s = -4 + j2$ and $s = -4 - j\,2$, hence

the characteristic equation is $(s + 4 - j2)\,(s + 4 + j\,2) = 0$

$$\therefore \qquad \frac{C(s)}{R(s)} = \frac{4\,(s+4)}{(s+4-j2)\,(s+4+j2)} = \frac{4\,(s+4)}{s^2+8s+20}$$ **Ans.**

Example 12.50. *Consider the following transfer functions of a feedback control system.*

(a) Forward path transfer function : $G(s) = \dfrac{K}{s(s+4)}$

Feedback transfer function : $H(s) = 1$

(b) Forward path transfer function : $G(s) = \dfrac{K}{s+4}$

Feedback path transfer function : $H(s) = 1/s$

Plot the root locus plot for (a) and (b) and therefrom determine the closed-loop (overall) transfer function C (s)/R (s) for K = 8.

Solution. For both cases (*a*) and (*b*), the open-loop transfer function $G(s)\,H(s)$ is same *i.e.*

$$G(s)\,H(s) = \frac{K}{s\,(s+4)}$$

Therefore, the root locus plot for both (*a*) and (*b*) is same and shown in Fig. 12.57.

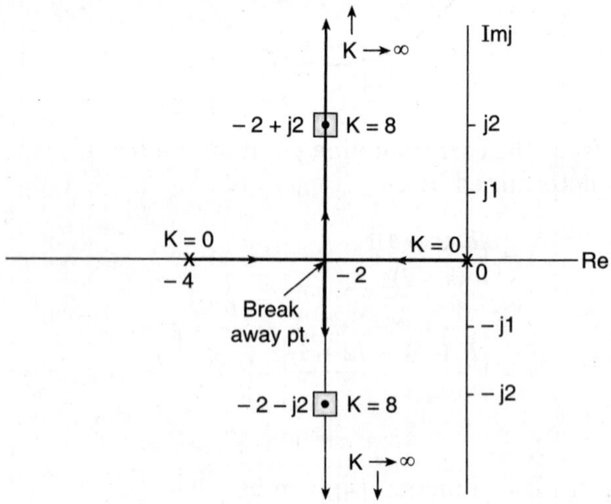

Fig. 12.57

Applying magnitude condition of root locus, the corresponding point on the root locus for $K = 8$ is located as below :

$$1 = |G\,(s)\,H\,(s)|$$

$$1 = \left|\frac{K}{s\,(s+4)}\right| \text{ at any point on the root locus}$$

Applying trial-and-error procedure,

(*i*) select $s = -2$

$$\therefore \qquad 1 = \left|\frac{K}{s\,(s+4)}\right|_{s=-2} = \left|\frac{K}{-2\,(-2+4)}\right| = \frac{K}{4}\;;\; K = 4$$

(*ii*) select $s = -2 \pm j\,1$

$$1 = \left|\frac{K}{s\,(s+4)}\right|_{s=-2\pm j1} = \left|\frac{K}{(-2\pm j\,1)(-2\pm j1+4)}\right|$$

$$= \frac{K}{\sqrt{2^2+1^2}\,\sqrt{2^2+1^2}} = \frac{K}{\sqrt{5}\,\sqrt{5}}\,;\, K = 5$$

(*iii*) select $s = -2 \pm j\,2$

$$1 = \left|\frac{K}{s\,(s+4)}\right|_{s=-2\pm j2} = \left|\frac{K}{(-2\pm j\,2)(-2\pm j\,2+4)}\right|$$

$$= \frac{K}{\sqrt{2^2+2^2}\,\sqrt{2^2+2^2}} = \frac{K}{\sqrt{8}\,\sqrt{8}}\,;\, K = 8$$

Thus, for $K = 8$, the two roots of the characteristic equation are $s = -2 + j\,2$ and $s = -2 - j\,2$

(*a*)
$$\frac{C(s)}{R(s)} = \frac{G(s)}{1+G(s)\,H(s)} = \frac{K/s\,(s+4)}{1+\dfrac{K}{s\,(s+4)}\times 1} = \frac{K}{s\,(s+4)+K}$$

For $K = 8$
$$\frac{C(s)}{R(s)} = \frac{8}{s\,(s+4)+8}$$

The characteristic equation is $s(s + 4) + 8 = 0$.

A point on the root locus plot is the root of the characteristic equation corresponding to a given value of K. Thus, the characteristic equation in the factored form can be obtained from the root locus plot.

For $K = 8$, the two roots are $s = (-2 - j2)$ and $s = (-2 + j2)$, hence, the characteristic equation is $(s + 2 + j2)\,(s + 2 - j2) = 0$

$$\therefore \quad \frac{C(s)}{R(s)} = \frac{8}{(s+2+j2)\,(s+2-j2)} = \frac{8}{s^2+4s+8}$$ **Ans.**

(*b*)
$$\frac{C(s)}{R(s)} = \frac{G(s)}{1+G(s)\,H(s)} = \frac{K/(s+4)}{1+\dfrac{K}{(s+4)}\times\dfrac{1}{s}} = \frac{Ks}{s\,(s+4)+K}$$

For $K = 8$
$$\frac{C(s)}{R(s)} = \frac{8s}{s\,(s+4)+8}$$

The characteristic equation is $s\,(s + 4) + 8 = 0$

In the factored form the characteristic equation is $(s + 2 + j2)\,(s + 2 - j2) = 0$

$$\therefore \quad \frac{C(s)}{R(s)} = \frac{8s}{(s+2+j2)\,(s+2-j2)}$$

$$= \frac{8s}{s^2+4s+8}$$ **Ans.**

Example 12.51. *Plot the root locus and determine the gain margin and phase margin for K = 3.46. The system is shown in Fig. 12.58.*

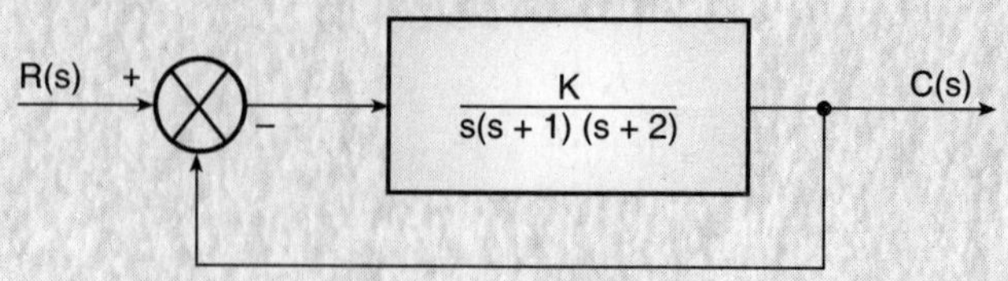

Fig. 12.58

Solution. $G(s) = \dfrac{K}{s\,(s+1)\,(s+2)}$

The open-loop poles are : $s = 0$, $s = -1$, and $s = -2$.

The number of root locus branches $N = P = 3$; $(Z = 0)$

As per the rules for drawing root locus plot, following data can be calculated,

Breakaway point : $s = -0.424$

Angle of asymptotes : 60°, 180° and – 60°

Asymptotes intersect on real axis at $x = -2$

The value of K for marginal stability = 6

Intersection in the imaginary axis at $s = \pm j\,1.41$

As per above data part of the root locus is drawn as shown in Fig. 12.59

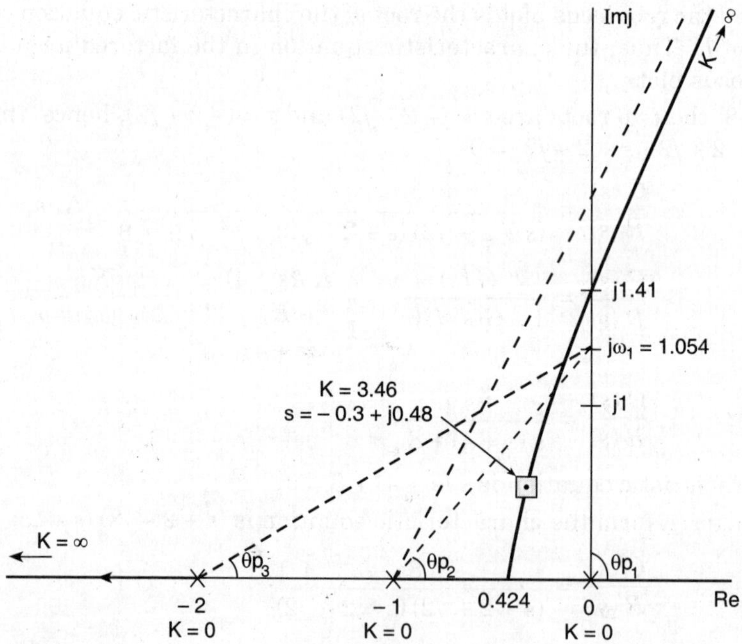

Fig. 12.59

1. Gain margin :

The value of gain factor for marginal stability : $K = 6$

The desired value of gain factor : $K_1 = 3.46$

$$\therefore \qquad \text{Gain margin} = \frac{3}{3.46} = 1.734$$

or $$\text{G.M.}\,|_{db} = 20 \log_{10} 1.734 = 20 \times 0.239 = 4.78 \text{ db} \quad \textbf{Ans.}$$

2. At the gain crossover frequency ω_1, $|G(j\omega_1)| = 1$, therefore corresponding to $K = 3.46$, the gain crossover frequency ω_1 is calculated below :

$$1 = \left|\frac{3.46}{j\omega_1(j\omega_1 + 1)(j\omega_1 + 2)}\right|$$

$$1 = \frac{3.46}{\omega_1\sqrt{\omega_1^2 + 1}\sqrt{\omega_1^2 + 4}}$$

or $$\omega_1^2(\omega_1^2 + 1)(\omega_1^2 + 4) = (3.46)^2$$

By trial-and-error, it is noted that above equation is satisfied to a reasonable approximation at $\omega_1 = 1.054$ rad./sec.

$$\text{Phase margin} = 180° + (\theta_{p_1} + \theta_{p_2} + \theta_{p_3})$$

$$= 180° + \left(-90° - \tan^{-1}\frac{1.054}{1} - \tan^{-1}\frac{1.054}{2}\right)$$

$$= 180° + (-90° - 46.50° - 27.78°)$$

$$= 180° - 164.28° = 15.72° \quad \textbf{Ans.}$$

Example 12.52. *Obtain the state model for the network shown in Fig. 12.32. Consider $v_1 = x_1$, $v_2 = x_2$ and $y = i$.*

Solution. Apply Kirchhoff's current law at node 1 and 2

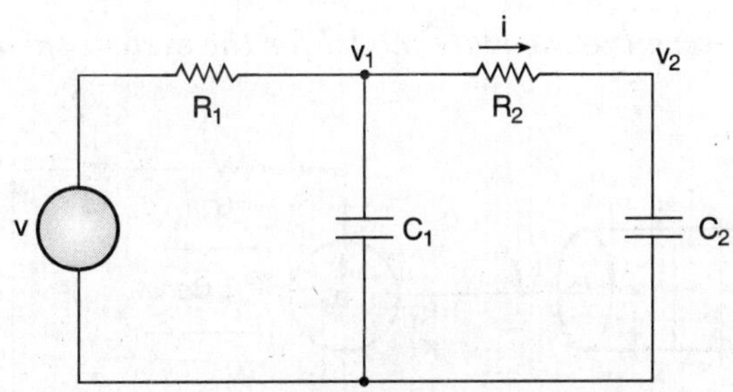

Fig. 12.60

$$\frac{v_1 - v}{R_1} + C_1\frac{dv_1}{dt} + \frac{v_1 - v_2}{R_2} = 0 \qquad ...(1)$$

$$\frac{v_2 - v_1}{R_2} + C_2\frac{dv_2}{dt} = 0 \qquad ...(2)$$

$$i = \frac{v_1 - v_2}{R_2} \quad ...(3)$$

Rearranging (1), (2) and (3)

$$\frac{dv_1}{dt} = \frac{1}{C_1(R_1 + R_2)} v_1 - \frac{1}{C_1 R_2} v_2 + \frac{1}{C_1 R_1} v \quad ...(4)$$

$$\frac{dv_2}{dt} = -\frac{1}{C_1 R_2} v_1 + \frac{1}{C_2 R_2} v_2 \quad ...(5)$$

$$i = \frac{1}{R_1} v_1 - \frac{1}{R_2} v_2 \quad ...(6)$$

put $v_1 = x_1$, $v_2 = x_2$ and $y = i$

Therefore,
$$\dot{x}_1 = + \frac{1}{C_1(R_1 + R_2)} x_1 - \frac{1}{C_1 R_2} x_2 + \frac{1}{R_1} v_1$$

$$\dot{x}_2 = -\frac{1}{C_1 R_2} x_1 + \frac{1}{C_1 R_2} x_2$$

$$y = \frac{1}{R_2} x_1 - \frac{1}{R_2} x_2$$

The state model is obtained as below :

$$\begin{bmatrix} \dot{x}_1 \\ \dot{x}_2 \end{bmatrix} = \begin{bmatrix} \frac{1}{C_1(R_1 + R_2)} & -\frac{1}{C_1 R_2} \\ -\frac{1}{C_2 R_2} & +\frac{1}{C_2 R_2} \end{bmatrix} \begin{bmatrix} x_1 \\ x_2 \end{bmatrix} + \begin{bmatrix} \frac{1}{R_1} \\ 0 \end{bmatrix} v$$

$$y = \begin{bmatrix} \frac{1}{R_1} & -\frac{1}{R_2} \end{bmatrix} \begin{bmatrix} x_1 \\ x_2 \end{bmatrix} \quad \textbf{Ans.}$$

Example 12.53. *Determine the state model for the system shown in Fig. 12.33 given below.*

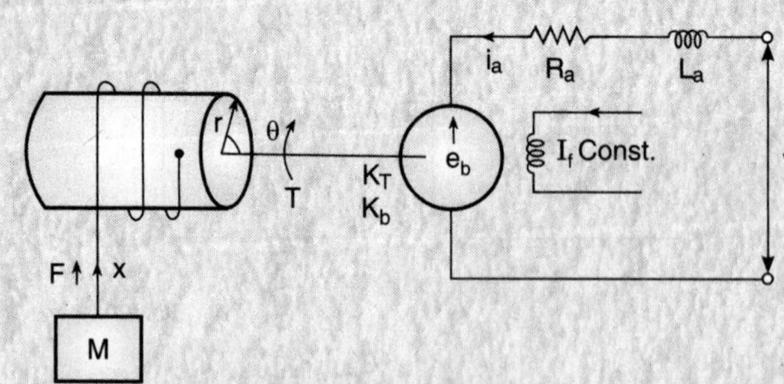

Fig. 12.61

Solution. As per Fig. 12.33 equations are given below

$$v_a - e_b = R_a i_a + L_a \frac{di_a}{dt} \quad ...(1)$$

$$e_b = K_b \frac{d\theta}{dt} \quad \text{...(2)}$$

$$T = K_T i_a \quad \text{...(3)}$$

$$T = J\frac{d^2\theta}{dt^2} \quad \text{...(4)}$$

$$F = M \frac{d^2x}{dt^2} \quad \text{...(5)}$$

$$T = Fr \quad \text{...(6)}$$

From (3) and (4),

$$K_T\, i_a = J\frac{d^2\theta}{dt^2} \quad \text{...(7)}$$

From (5) and (6),

$$T = Mr\frac{d^2x}{dt^2} \quad \text{...(8)}$$

From (3) and (8),

$$K_T i_a = Mr\frac{d^2x}{dt^2} \quad \text{...(9)}$$

Rearranging (1), (2), (7) and (9)

$$\frac{di_a}{dt} = -\frac{R_a}{L_a}i_a - \frac{K_b}{L_a}\frac{d\theta}{dt} + \frac{v_a}{L_a}$$

$$\frac{d^2\theta}{dt^2} = \frac{K_T}{J}i_a\,;\; \frac{d^2x}{dt^2} = \frac{K_T}{Mr}i_a$$

Select state variables as

$$i_a = x_1\,;\; \theta = x_2$$

$$\frac{d\theta}{dt} = x_3\,;\; x = x_4$$

$$\frac{dx}{dt} = x_5$$

and put x as the output variable. Therefore,

$$\dot{x}_1 = -\frac{R_a}{L_a}x_1 - \frac{K_b}{L_a}x_3 + \frac{v_a}{L_a}$$

$$\dot{x}_2 = x_3\,;\; \dot{x}_3 = \frac{K_T}{J}x_1$$

$$\dot{x}_4 = x_5\,;\; \dot{x}_5 = \frac{K_T}{Mr}x_1$$

and $$y = x_4$$

$$\therefore \quad \begin{bmatrix} \dot{x}_1 \\ \dot{x}_2 \\ \dot{x}_3 \\ \dot{x}_4 \\ \dot{x}_5 \end{bmatrix} = \begin{bmatrix} -\frac{R_a}{L_a} & 0 & -\frac{K_b}{L_a} & 0 & 0 \\ 0 & 0 & 0 & 1 & 0 & 0 \\ \frac{K_T}{J} & 0 & 0 & 0 & 0 & 0 \\ 0 & 0 & 0 & 0 & 0 & 1 \\ \frac{K_T}{Mr} & 0 & 0 & 0 & 0 & 0 \end{bmatrix} \begin{bmatrix} x_1 \\ x_2 \\ x_3 \\ x_4 \\ x_5 \end{bmatrix} + \begin{bmatrix} \frac{1}{L_a} \\ 0 \\ 0 \\ 0 \\ 0 \end{bmatrix} v_a$$

$$y = [0 \quad 0 \quad 0 \; 1 \quad 0] \begin{bmatrix} x_1 \\ x_2 \\ x_3 \\ x_4 \\ x_5 \end{bmatrix} \quad \textbf{Ans.}$$

Example 12.54. *The closed-loop transfer function of a feedback control system is given below :*

$$\frac{C(s)}{R(s)} = \frac{5(s+2)}{s(s+1)(s+5)}$$

Derive the state model in (i) Phase variable form (ii) Canonical variable form.

Solution. (*i*) $\quad \dfrac{C(s)}{R(s)} = \dfrac{5s + 10}{s^3 + 6s^2 + 5s}$

The state diagram for the above transfer function is shown in Fig. 12.62. The output of each integrator is assigned a state variable

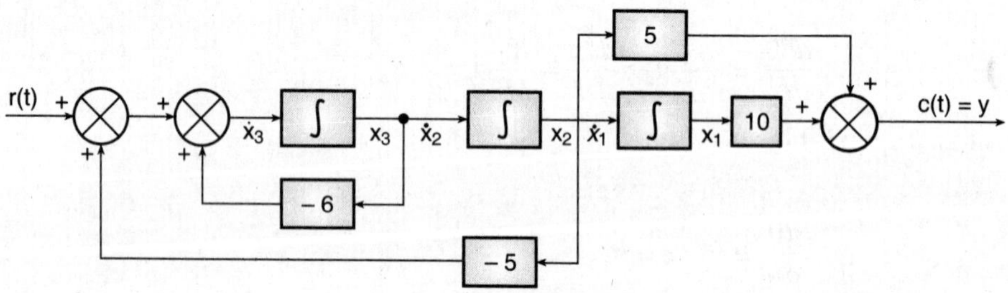

Fig. 12.62

The state model as per Fig. 12.62 is given below :

$$\dot{x}_1 = x_2$$

$$\dot{x}_2 = x_3$$

$$\dot{x}_3 = -5x_2 - 6x_3 + r(t)$$

and $\quad y = 10x_1 + 5x_2$

or $\quad \begin{bmatrix} \dot{x}_1 \\ \dot{x}_2 \\ \dot{x}_3 \end{bmatrix} = \begin{bmatrix} 0 & 1 & 0 \\ 0 & 0 & 1 \\ 0 & -5 & -6 \end{bmatrix} \begin{bmatrix} x_1 \\ x_2 \\ x_3 \end{bmatrix} + \begin{bmatrix} 0 \\ 0 \\ 1 \end{bmatrix} r(t)$

and $$y = [10 \quad 5 \quad 0]\begin{bmatrix} x_1 \\ x_2 \\ x_3 \end{bmatrix} \quad \textbf{Ans.}$$

(*ii*) $$\frac{C(s)}{R(s)} = \frac{5s + 10}{s(s+1)(s+5)}$$

Breaking R.H.S. into partial fractions

$$\frac{C(s)}{R(s)} = \frac{2}{s} - \frac{1.25}{s+1} - \frac{0.75}{s+5}$$

The state diagram for the above transfer function is shown in Fig. 12.63

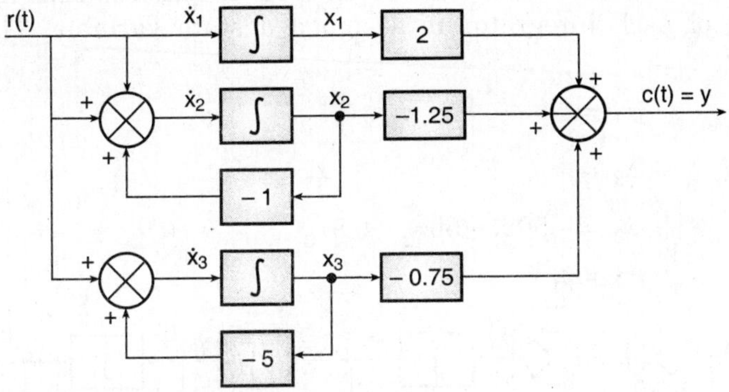

Fig. 12.63

Assigning the output of each integrator with a state variable, the state model is obtained below,

$$\dot{x}_1 = \qquad\qquad r(t)$$
$$\dot{x}_2 = -x_2 \qquad\qquad + r(t)$$
$$\dot{x}_3 = \qquad -5x_3 \qquad + r(t)$$
$$y = 2x_1 - 1.25x_2 - 0.75x_3$$

or $$\begin{bmatrix} \dot{x}_1 \\ \dot{x}_2 \\ \dot{x}_3 \end{bmatrix} = \begin{bmatrix} 0 & 0 & 0 \\ 0 & -1 & 1 \\ 0 & 0 & -5 \end{bmatrix}\begin{bmatrix} x_1 \\ x_2 \\ x_3 \end{bmatrix} + \begin{bmatrix} 1 \\ 1 \\ 1 \end{bmatrix} r(t)$$

and $$y = [2 \quad -1.25 \quad -0.75]\begin{bmatrix} x_1 \\ x_2 \\ x_3 \end{bmatrix} \quad \textbf{Ans.}$$

Example 12.55. *The open-loop transfer function of a unity feedback control system is given below,*

$$G(s) = \frac{50}{s(s^2 + 0.5s + 50)}$$

Develop a state model for the closed-loop system.

Solution. The closed-loop transfer function is given by

$$\frac{C(s)}{R(s)} = \frac{G(s)}{1+G(s)}$$

$$\therefore \quad \frac{C(s)}{R(s)} = \frac{\dfrac{50}{s(s^2+0.5s+50)}}{1+\dfrac{50}{s(s^2+0.5s+50)}}$$

or $$\frac{C(s)}{R(s)} = \frac{50}{s^3+0.5s^2+50s+50}$$

The state block diagram for the above transfer function is drawn as shown in Fig. 12.64. The output of each integrator is assigned a state variable. The state model is developed as below,

$$\dot{x}_1 = x_2$$
$$\dot{x}_2 = x_3$$
$$\dot{x}_3 = -50x_1 - 50x_2 - 0.5x_3 + r(t)$$
$$y = x_1$$

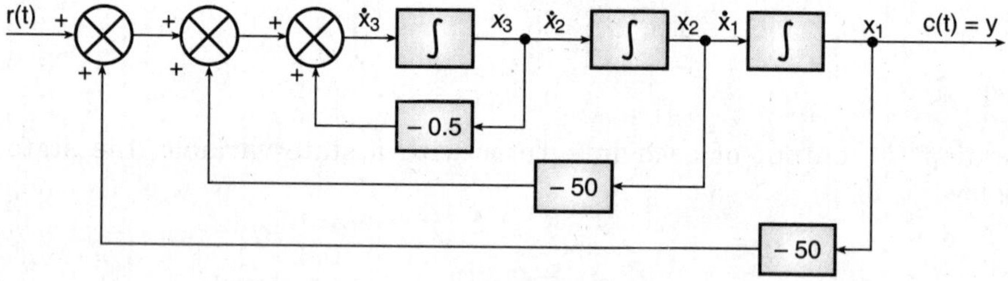

Fig. 12.64

or $$\begin{bmatrix}\dot{x}_1\\ \dot{x}_2\\ \dot{x}_3\end{bmatrix} = \begin{bmatrix}0 & 1 & 0\\ 0 & 0 & 1\\ -50 & -50 & -0.5\end{bmatrix}\begin{bmatrix}x_1\\ x_2\\ x_3\end{bmatrix} + \begin{bmatrix}0\\ 0\\ 1\end{bmatrix} r(t)$$

$$y = [1 \quad 0 \quad 0]\begin{bmatrix}x_1\\ x_2\\ x_3\end{bmatrix}$$ **Ans.**

Example 12.56. *Determine the state transition matrix given that*

$$\begin{bmatrix}\dot{x}_1\\ \dot{x}_2\end{bmatrix} = \begin{bmatrix}0 & 1\\ 0 & -3\end{bmatrix}\begin{bmatrix}x_1\\ x_2\end{bmatrix}.$$

Solution. $$\mathbf{A} = \begin{bmatrix}0 & 1\\ 0 & -3\end{bmatrix}$$

$$(s\,\mathbf{I}-\mathbf{A}) = \begin{bmatrix} s & 0 \\ 0 & s \end{bmatrix} - \begin{bmatrix} 0 & 1 \\ 0 & -3 \end{bmatrix} = \begin{bmatrix} s & -1 \\ 0 & s+3 \end{bmatrix}$$

$$(s\,\mathbf{I}-\mathbf{A})^{-1} = \frac{\text{Adj}\,(s\,\mathbf{I}-\mathbf{A})}{|s\,\mathbf{I}-\mathbf{A}|}$$

$$= \frac{\begin{bmatrix} s+3 & 1 \\ 0 & s \end{bmatrix}}{\begin{vmatrix} s & -1 \\ 0 & s+3 \end{vmatrix}} = \frac{\begin{bmatrix} s+3 & 1 \\ 0 & s \end{bmatrix}}{s\,(s+3)}$$

$$= \begin{bmatrix} \dfrac{s+3}{s(s+3)} & \dfrac{1}{s(s+3)} \\ 0 & \dfrac{s}{s(s+3)} \end{bmatrix} = \begin{bmatrix} \dfrac{1}{s} & \dfrac{1}{3s} - \dfrac{1}{3(s+3)} \\ 0 & \dfrac{1}{s+3} \end{bmatrix}$$

$$\phi(t) = \mathcal{L}^{-1}\,(s\,\mathbf{I}-\mathbf{A})^{-1} = \mathcal{L}^{-1} \begin{bmatrix} \dfrac{1}{s} & \dfrac{1}{3s} - \dfrac{1}{3(s+3)} \\ 0 & \dfrac{1}{s+3} \end{bmatrix}$$

Taking inverse Laplace transform

or $$\phi(t) = \begin{bmatrix} 1 & \dfrac{1}{3} - \dfrac{1}{3}e^{-3t} \\ 0 & e^{-3t} \end{bmatrix} \quad \textbf{Ans.}$$

Example 12.57. *Use Laplace transform method to determine time response of a system having state model as*

$$\dot{x}_1 = x_2 + u(t)$$

$$\dot{x}_2 = -2x_1 - 3x_2$$

where, $$u(t) = e^{-t} \text{ for } t \geq 0$$
$$= 0 \text{ for } t < 0$$

and $$x_1(0) = x_2(0) = 0.$$

Solution. In matrix form the state model is

$$\begin{bmatrix} \dot{x}_1 \\ \dot{x}_2 \end{bmatrix} = \begin{bmatrix} 0 & 1 \\ -2 & -3 \end{bmatrix} \begin{bmatrix} x_1 \\ x_2 \end{bmatrix} + \begin{bmatrix} 1 \\ 0 \end{bmatrix} u(t)$$

The time response is given by

$$\boldsymbol{x}(t) = \mathcal{L}^{-1}\,\phi(s)\,\boldsymbol{x}(0) + \mathcal{L}^{-1}\,\phi(s)\,\mathbf{BU}(s)$$

$\because$ $$\phi(s) = (s\,\mathbf{I}-\mathbf{A})^{-1}$$

$\therefore$ $$\boldsymbol{x}(t) = \mathcal{L}^{-1}\,(s\,\mathbf{I}-\mathbf{A})^{-1}\,\boldsymbol{x}(0) + \mathcal{L}^{-1}\,(s\,\mathbf{I}-\mathbf{A})^{-1}\,\mathbf{BU}\,(s)$$

$\because$ $$\boldsymbol{x}(0) = 0$$

$\therefore$ $$\boldsymbol{x}(t) = \mathcal{L}^{-1}\,(s\,\mathbf{I}-\mathbf{A})^{-1}\,\mathbf{BU}\,(s)$$

$$(s\,\mathbf{I}-\mathbf{A}) = \begin{bmatrix} s & 0 \\ 0 & s \end{bmatrix} - \begin{bmatrix} 0 & 1 \\ -2 & -3 \end{bmatrix} = \begin{bmatrix} s & -1 \\ 2 & s+3 \end{bmatrix}$$

$$(s\,\mathbf{I}-\mathbf{A})^{-1} = \frac{Adj\,(s\,\mathbf{I}-\mathbf{A})}{|\,s\,\mathbf{I}-\mathbf{A}\,|} = \frac{\begin{bmatrix} s+3 & 1 \\ -2 & s \end{bmatrix}}{\begin{vmatrix} s & 1 \\ 2 & s+3 \end{vmatrix}}$$

$$= \begin{bmatrix} \dfrac{s+3}{(s+1)(s+2)} & \dfrac{1}{(s+1)(s+2)} \\ -\dfrac{2}{(s+1)(s+2)} & \dfrac{s}{(s+1)(s+2)} \end{bmatrix}$$

$\because$ $u(t) = e^{-t}$, $\therefore$ $\mathbf{u}(s) = \dfrac{1}{s+1}$ and $\mathbf{B} = \begin{bmatrix} 1 \\ 0 \end{bmatrix}$

$\therefore$
$$\mathbf{x}(t) = \mathcal{L}^{-1} \begin{bmatrix} \dfrac{s+3}{(s+1)(s+2)} & \dfrac{1}{(s+1)(s+2)} \\ -\dfrac{2}{(s+1)(s+2)} & \dfrac{s}{(s+1)(s+2)} \end{bmatrix} \begin{bmatrix} 1 \\ 0 \end{bmatrix} \frac{1}{s+1}$$

$$= \mathcal{L}^{-1} \begin{bmatrix} \dfrac{s+3}{(s+1)(s+2)} & \dfrac{1}{(s+1)(s+2)} \\ \dfrac{-2}{(s+1)(s+2)} & \dfrac{s}{(s+1)(s+2)} \end{bmatrix} \begin{bmatrix} \dfrac{1}{s+1} \\ 0 \end{bmatrix}$$

$$= \mathcal{L}^{-1} \begin{bmatrix} \dfrac{(s+3)}{(s+1)^2\,(s+2)} \\ \dfrac{-2}{(s+1)^2\,(s+2)} \end{bmatrix}$$

$$\frac{(s+3)}{(s+1)^2\,(s+2)} = \frac{K_1}{(s+1)} + \frac{K_2}{(s+1)^2} + \frac{K_3}{(s+2)}$$

$$\frac{s+3}{s+2} = K_1\,(s+1) + K_2 + K_3\,\frac{(s+1)^2}{(s+2)}$$

put $s = -1$

$\therefore$
$$\frac{-1+3}{-1+2} = K_2,\; K_2 = 2$$

$$(s+3) = K_1\,(s+1)\,(s+2) + K_2\,(s+2) + K_3\,(s+1)^2$$

put $s = -2$

$\therefore$
$$K_3 = 1$$

$$\frac{d}{ds}\left[\frac{s+3}{s+2}\right] = K_1 + 0 + K_3\left[\frac{(s+2)\,(s+1) - (s+1)^2\,(1)}{(s+2)^2}\right]$$

or $$\frac{(s+2)(1)-(s+3)(1)}{(s+2)^2} = K_1 + K_3\left[\frac{(s+2)(s+1)-(s+1)^2(1)}{(s+2)^2}\right]$$

put $s = -1$

$$\therefore \quad K_1 = \frac{1-2}{1} = -1$$

Hence, $$\frac{s+3}{(s+1)^2(s+2)} = \frac{-1}{(s+1)} + \frac{2}{(s+1)^2} + \frac{1}{(s+2)}$$

Similarly the term, $\dfrac{-2}{(s+1)^2(s+2)}$ can be expressed as

$$\frac{-2}{(s+1)^2(s+2)} = \frac{2}{s+1} + \frac{-2}{(s+1)^2} + \frac{-2}{(s+2)}$$

$$\therefore \quad \boldsymbol{x}(t) = \mathcal{L}^{-1}\begin{bmatrix} -\dfrac{1}{s+1} + \dfrac{2}{(s+1)^2} + \dfrac{1}{(s+2)} \\ \dfrac{2}{s+1} + \dfrac{-2}{(s+1)^2} + \dfrac{-2}{(s+2)} \end{bmatrix}$$

Taking inverse Laplace transform

$$\boldsymbol{x}(t) = \begin{bmatrix} -e^{-t} + 2\,te^{-t} + e^{-2t} \\ 2\,e^{-t} - 2\,te^{-t} - 2\,e^{-2t} \end{bmatrix} \quad \textbf{Ans.}$$

Example 12.58. *The state model of a system is given below :*

$$\begin{bmatrix} \dot{x}_1 \\ \dot{x}_2 \end{bmatrix} = \begin{bmatrix} 0 & 1 \\ -2 & -3 \end{bmatrix}\begin{bmatrix} x_1 \\ x_2 \end{bmatrix} + \begin{bmatrix} 1 \\ 0 \end{bmatrix} u(t)$$

Given that $u(t) = e^{-t}$ *for* $t \geq 0$

$= 0$ *for* $t < 0$

and $x_1(0) = x_2(0) = 0$

Obtain the time response using canonical transformation of **A** *matrix.*

Solution. As the matrix $\mathbf{A} = \begin{bmatrix} 0 & 1 \\ -2 & -3 \end{bmatrix}$ is in phase variable form, therefore, Vandermonde matrix is used for transformation.

Tha eigen values of matrix **A** are determined below

$$(s\,\mathbf{I} - \mathbf{A}) = \begin{bmatrix} s & 0 \\ 0 & s \end{bmatrix} - \begin{bmatrix} 0 & 1 \\ -2 & -3 \end{bmatrix} = \begin{bmatrix} s & -1 \\ 2 & s+3 \end{bmatrix}$$

$$|s\,\mathbf{I} - \mathbf{A}| = \begin{bmatrix} s & -1 \\ 2 & s+3 \end{bmatrix} = s(s+3) - (-1)(-2) = s^2 + 3s + 2$$

put $s^2 + 3s + 2 = 0$

or $(s+1)(s+2) = 0$

$\therefore$ Eigen values are $\lambda_1 = -1$ and $\lambda_2 = -2$

The Vandermonde matrix is $\mathbf{P} = \begin{bmatrix} 1 & 1 \\ \lambda_1 & \lambda_2 \end{bmatrix} = \begin{bmatrix} 1 & 1 \\ -1 & -2 \end{bmatrix}$

The inverse of matrix **P** can be determined as

$$\mathbf{P}^{-1} = \begin{bmatrix} 1 & 1 \\ -1 & -2 \end{bmatrix}^{-1} = \begin{bmatrix} 2 & 1 \\ -1 & -1 \end{bmatrix}$$

The diagonal matrix $\wedge = \mathbf{P}^{-1}\mathbf{AP}$ is determined below

$$\wedge = \mathbf{P}^{-1}\mathbf{AP} = \begin{bmatrix} 2 & 1 \\ -1 & -1 \end{bmatrix}\begin{bmatrix} 0 & 1 \\ -2 & -3 \end{bmatrix}\begin{bmatrix} 1 & 1 \\ -1 & -2 \end{bmatrix} = \begin{bmatrix} -1 & 0 \\ 0 & -2 \end{bmatrix}$$

$$\therefore \qquad e^{\wedge t} = \begin{bmatrix} e^{-t} & 0 \\ 0 & e^{-2t} \end{bmatrix}$$

The state transition matrix $\phi(t) = \mathbf{P}\, e^{\wedge t}\, \mathbf{P}^{-1}$

$$\therefore \qquad \phi(t) = \begin{bmatrix} 1 & 1 \\ -1 & -2 \end{bmatrix}\begin{bmatrix} e^{-t} & 0 \\ 0 & e^{-2t} \end{bmatrix}\begin{bmatrix} 2 & 1 \\ -1 & -1 \end{bmatrix}$$

$$= \begin{bmatrix} 2e^{-t} - 2e^{-2t} & e^{-t} - e^{-2t} \\ -2e^{-t} + 2e^{-2t} & -e^{-t} + 2e^{-2t} \end{bmatrix}$$

The time response is given by

$$\boldsymbol{x}(t) = \mathcal{L}^{-1}\phi(s)\, x(0) + \mathcal{L}^{-1}\phi(s)\, \mathbf{BU}(s)$$

Applying convolution integral to second term of the R.H.S.

$$\boldsymbol{x}(t) = \phi(t)\,\boldsymbol{x}(0) + \int_0^t \phi(t-\tau)\, \mathbf{B}u(\tau)\, d\tau$$

$\because \quad \boldsymbol{x}(0) = 0,\ \mathbf{B} = \begin{bmatrix} 1 \\ 0 \end{bmatrix}$ and $u(t) = e^{-t}$

$$\therefore \qquad \boldsymbol{x}(t) = \int_0^t \phi(t-\tau)\, \mathbf{B}u(\tau)\, d\tau$$

$$= \int_0^t \begin{bmatrix} 2e^{-(t-\tau)} - 2e^{-2(t-\tau)} & e^{-(t-\tau)} - e^{-2(t-\tau)} \\ -2e^{-(t-\tau)} + 2e^{-2(t-\tau)} & -e^{-(t-\tau)} + 2e^{-2(t-\tau)} \end{bmatrix}\begin{bmatrix} 1 \\ 0 \end{bmatrix} e^{-\tau}\, d\tau.$$

$$= \int_0^t \begin{bmatrix} 2e^{-(t-\tau)} - 2e^{-2(t-\tau)} \\ -2e^{-(t-\tau)} + 2e^{-2(t-\tau)} \end{bmatrix} e^{-\tau}\, d\tau = \int_0^t \begin{bmatrix} (2e^{-(t-\tau)} - e^{-2(t-\tau)})e^{-\tau} \\ (-2e^{-(t-\tau)} + 2e^{-2(t-\tau)})\, e^{-\tau} \end{bmatrix} d\tau$$

$$= \int_0^t \begin{bmatrix} 2e^{-t} - e^{-2t}e^{\tau} \\ -2e^{-t} + 2e^{-2t}e^{\tau} \end{bmatrix} d\tau = \begin{bmatrix} 2\tau e^{-t} - e^{-2t}e^{\tau} \\ -2\,\tau\, e^{-t} + 2e^{-2t}\, e^{\tau} \end{bmatrix}_{\tau=0}^{\tau=t}$$

$$= \begin{bmatrix} (2t\, e^{-t} - e^{-2t}\, e^{t}) - (2 \times 0 \times e^{-t} - e^{-2t} e^{0}) \\ (-2t\, e^{-t} + 2\, e^{-2t}\, e^{t}) - (-2 \times 0 \times e^{-t} - 2\, e^{-2t} e^{0}) \end{bmatrix}.$$

or
$$\boldsymbol{x}(t) = \begin{bmatrix} 2te^{-t} - e^{-t} + e^{-2t} \\ -2te^{-t} + 2e^{-t} - 2e^{-2t} \end{bmatrix} \quad \textbf{Ans.}$$

Example 12.59. *Determine the time response for a system given below,*

$$\dot{x}_1 = -x_1$$
$$\dot{x}_2 = x_1 - x_2 + u(t)$$

given that $\boldsymbol{x}(0) = \begin{bmatrix} 1 \\ 0 \end{bmatrix}$.

Solution.
$$\begin{bmatrix} \dot{x}_1 \\ \dot{x}_2 \end{bmatrix} = \begin{bmatrix} -1 & 0 \\ 1 & -1 \end{bmatrix} \begin{bmatrix} x_1 \\ x_2 \end{bmatrix} + \begin{bmatrix} 0 \\ 1 \end{bmatrix} u(t)$$

$$\mathbf{A} = \begin{bmatrix} -1 & 0 \\ 1 & -1 \end{bmatrix}, \mathbf{B} = \begin{bmatrix} 0 \\ 1 \end{bmatrix}$$

$$(s\mathbf{I} - \mathbf{A}) = \begin{bmatrix} s & 0 \\ 0 & s \end{bmatrix} - \begin{bmatrix} -1 & 0 \\ 0 & -1 \end{bmatrix} = \begin{bmatrix} s+1 & 0 \\ -1 & s+1 \end{bmatrix}$$

$$\phi(s) = (s\mathbf{I} - \mathbf{A})^{-1} = \frac{\text{Adj}\,(s\mathbf{I} \quad \Lambda)}{|\,s\mathbf{I} - \mathbf{A}\,|} + \frac{\begin{bmatrix} s+1 & 0 \\ 1 & s+1 \end{bmatrix}}{(s+1)^2} = \begin{bmatrix} \frac{1}{s+1} & 0 \\ \frac{1}{(s+1)^2} & \frac{1}{(s+1)} \end{bmatrix}$$

$$\boldsymbol{x}(t) = L^{-1}\{\phi(s)\,\boldsymbol{x}(0) + \phi(s)\,\mathbf{B}\,U(s)\}\ ;\ U(o) = \frac{1}{s}$$

$$\therefore \qquad \boldsymbol{x}(t) = L^{-1}\left\{\begin{bmatrix} \frac{1}{(s+1)} & 0 \\ \frac{1}{(s+1)^2} & \frac{1}{(s+1)} \end{bmatrix}\begin{bmatrix} 1 \\ 0 \end{bmatrix} + \begin{bmatrix} \frac{1}{(s+1)} & 0 \\ \frac{1}{(s+1)^2} & \frac{1}{(s+1)} \end{bmatrix}\begin{bmatrix} 0 \\ 1 \end{bmatrix}\frac{1}{s}\right\}$$

$$= L^{-1}\left\{\begin{bmatrix} \frac{1}{(s+1)} & 0 \\ \frac{1}{(s+1)^2} & \frac{1}{(s+1)} \end{bmatrix}\begin{bmatrix} 1 \\ 0 \end{bmatrix} + \begin{bmatrix} \frac{1}{(s+1)} & 0 \\ \frac{1}{(s+1)^2} & \frac{1}{(s+1)} \end{bmatrix}\begin{bmatrix} 0 \\ \frac{1}{s} \end{bmatrix}\right\}$$

$$= L^{-1}\left\{\begin{bmatrix} \frac{1}{(s+1)} \\ \frac{1}{(s+1)^2} \end{bmatrix} + \begin{bmatrix} 0 \\ \frac{1}{s\,(s+1)} \end{bmatrix}\right\}$$

$$= \mathcal{L}^{-1}\left\{\begin{bmatrix} \dfrac{1}{(s+1)} \\ \dfrac{1}{(s+1)^2} + \dfrac{1}{s\,(s+1)} \end{bmatrix}\right\}$$

$$= \mathcal{L}^{-1}\begin{bmatrix} \dfrac{1}{(s+1)} \\ \dfrac{1}{(s+1)^2} + \dfrac{1}{s} - \dfrac{1}{(s+1)} \end{bmatrix}$$

or $$\boldsymbol{x}(t) = \begin{bmatrix} e^{-t} \\ te^{-t} + 1 - e^{-t} \end{bmatrix}$$

$\therefore$ $$x_1(t) = e^{-t}$$

$$x_2\,(t) = t\,e^{-t} + 1 - e^{-t} \textbf{ Ans.}$$

Example 12.60. *Determine the state transition matrix having given the time response as below :*

$$\boldsymbol{x}(t) = \begin{bmatrix} e^{-t} \\ -e^{-t} \end{bmatrix} \text{ for } \boldsymbol{x}(0) = \begin{bmatrix} 1 \\ -1 \end{bmatrix}$$

and $$\boldsymbol{x}(t) = \begin{bmatrix} e^{-3t} \\ -e^{-3t} \end{bmatrix} \text{ for } \boldsymbol{x}(0) = \begin{bmatrix} 1 \\ -3 \end{bmatrix}$$

Solution. Let the system matrix **A** be in phase variable form. Therefore,

$$\mathbf{A} = \begin{bmatrix} 0 & 1 \\ -p & -q \end{bmatrix}$$

$$(s\,\mathbf{I} - \mathbf{A}) = \begin{bmatrix} s & 0 \\ 0 & s \end{bmatrix} - \begin{bmatrix} 0 & 1 \\ -p & -q \end{bmatrix} = \begin{bmatrix} s & -1 \\ p & s+q \end{bmatrix}$$

$$\mathbf{X}\,(s) = (s\,\mathbf{I} - \mathbf{A})^{-1}\,\boldsymbol{x}(0)$$

$\therefore$ $$\boldsymbol{x}(0) = (s\,\mathbf{I} - \mathbf{A})\,\mathbf{X}\,(s)$$

$\because$ $$\boldsymbol{x}(t) = \begin{bmatrix} e^{-t} \\ -e^{-t} \end{bmatrix} \text{ for } \boldsymbol{x}(0) = \begin{bmatrix} 1 \\ -1 \end{bmatrix}$$

$\therefore$ $$\mathbf{X}(s) = \begin{bmatrix} \dfrac{1}{s+1} \\ -\dfrac{1}{s+1} \end{bmatrix}$$

$\therefore$ $$\begin{bmatrix} 1 \\ -1 \end{bmatrix} = \begin{bmatrix} s & -1 \\ p & s+q \end{bmatrix} \begin{bmatrix} \dfrac{1}{s+1} \\ -\dfrac{1}{s+1} \end{bmatrix} \qquad \text{...(1)}$$

Similarly $\begin{bmatrix} 1 \\ -3 \end{bmatrix} = \begin{bmatrix} s & -1 \\ p & s+q \end{bmatrix} \begin{bmatrix} \dfrac{1}{s+3} \\ \dfrac{-1}{s+3} \end{bmatrix}$ ...(2)

From (1) and (2) $-1 = \dfrac{p}{s+1} - \dfrac{s+q}{s+1}$ ...(3)

and $-3 = \dfrac{p}{s+3} - \dfrac{s+q}{s+3}$ ...(4)

Solving (3) and (4), $p = 3$ and $q = 4$

$$\therefore \quad \mathbf{A} = \begin{bmatrix} 0 & 1 \\ -3 & -4 \end{bmatrix}$$

and
$$(s\,\mathbf{I} - \mathbf{A}) = \begin{bmatrix} s & 0 \\ 0 & s \end{bmatrix} - \begin{bmatrix} 0 & 1 \\ -3 & -4 \end{bmatrix}$$

$$= \begin{bmatrix} s & -1 \\ 3 & s+4 \end{bmatrix}$$

$$\text{Adj}\,(s\,\mathbf{I} - \mathbf{A}) = \begin{bmatrix} s+4 & 1 \\ -3 & s \end{bmatrix}$$

$$\phi(s) = (s\,\mathbf{I} - \mathbf{A})^{-1} = \frac{\text{Adj.}\,(s\,\mathbf{I} - \mathbf{A})}{|\,s\,\mathbf{I} - \mathbf{A}\,|}$$

$$= \frac{\begin{bmatrix} s+4 & 1 \\ -3 & s \end{bmatrix}}{s\,(s+4) - (-1)\,(3)} = \frac{\begin{bmatrix} s+4 & 1 \\ -3 & s \end{bmatrix}}{(s+1)\,(s+3)}$$

$$= \begin{bmatrix} \dfrac{s+4}{(s+1)\,(s+3)} & \dfrac{1}{(s+1)\,(s+3)} \\ \dfrac{-3}{(s+1)\,(s+3)} & \dfrac{s}{(s+1)\,(s+3)} \end{bmatrix}$$

Taking inverse Laplace transform on both sides

$$\phi(t) = \mathcal{L}^{-1} \begin{bmatrix} \dfrac{s+4}{(s+1)\,(s+3)} & \dfrac{1}{(s+1)\,(s+3)} \\ \dfrac{-3}{(s+1)\,(s+3)} & \dfrac{s}{(s+1)\,(s+3)} \end{bmatrix}$$

$$= \mathcal{L}^{-1} \begin{bmatrix} \dfrac{3/2}{(s+1)} + \dfrac{-1/2}{(s+3)} & \dfrac{1/2}{(s+1)} + \dfrac{-1/2}{(s+3)} \\ \dfrac{-3/2}{(s+1)} + \dfrac{3/2}{(s+3)} & \dfrac{-1/2}{(s+1)} + \dfrac{3/2}{(s+3)} \end{bmatrix}$$

or $$\phi(t) = \begin{bmatrix} \frac{3}{2}e^{-t} - \frac{1}{2}e^{-3t} & \frac{1}{2}e^{-t} - \frac{1}{2}e^{-3t} \\ -\frac{3}{2}e^{-t} + \frac{3}{2}e^{-3t} & -\frac{1}{2}e^{-t} + \frac{3}{2}e^{-3t} \end{bmatrix}$$ **Ans.**

Example 12.61. *Determine the transfer matrix for the state model given below :*

$$\begin{bmatrix} \dot{x}_1 \\ \dot{x}_2 \end{bmatrix} = \begin{bmatrix} -1 & -1 \\ 3 & -5 \end{bmatrix} \begin{bmatrix} x_1 \\ x_2 \end{bmatrix} + \begin{bmatrix} 1 \\ 1 \end{bmatrix} u$$

$$y = [1 \;\; 2] \begin{bmatrix} x_1 \\ x_2 \end{bmatrix}.$$

Solution. $$\mathbf{A} = \begin{bmatrix} -1 & -1 \\ 3 & -5 \end{bmatrix}$$

$$(s\,\mathbf{I} - \mathbf{A}) = \begin{bmatrix} s & 0 \\ 0 & s \end{bmatrix} - \begin{bmatrix} -1 & -1 \\ 3 & -5 \end{bmatrix} = \begin{bmatrix} s+1 & 1 \\ -3 & s+5 \end{bmatrix}$$

$$(s\,\mathbf{I} - \mathbf{A})^{-1} = \frac{Adj\,(s\,\mathbf{I} - \mathbf{A})}{|\,s\,\mathbf{I} - \mathbf{A}\,|}$$

$$= \frac{\begin{bmatrix} s+5 & -1 \\ 3 & s+1 \end{bmatrix}}{(s+1)(s+5) - (-3)(1)} = \frac{\begin{bmatrix} s+5 & -1 \\ 3 & s+1 \end{bmatrix}}{s^2 + 6s + 8} = \frac{\begin{bmatrix} s+5 & -1 \\ 3 & s+1 \end{bmatrix}}{(s+2)(s+4)}$$

or $$(s\,\mathbf{I} - \mathbf{A})^{-1} = \begin{bmatrix} \frac{s+5}{(s+2)(s+4)} & \frac{-1}{(s+2)(s+4)} \\ \frac{3}{(s+2)(s+4)} & \frac{s+1}{(s+2)(s+4)} \end{bmatrix}$$

$$\mathbf{B} = \begin{bmatrix} 1 \\ 1 \end{bmatrix}, \mathbf{C} = [1 \;\; 2]$$

$$G(s) = \mathbf{C}\,(s\,\mathbf{I} - \mathbf{A})^{-1}\,\mathbf{B}$$

$$= [1 \;\; 2] \begin{bmatrix} \frac{s+5}{(s+2)(s+4)} & \frac{-1}{(s+2)(s+4)} \\ \frac{3}{(s+2)(s+4)} & \frac{s+1}{(s+2)(s+4)} \end{bmatrix} \begin{bmatrix} 1 \\ 1 \end{bmatrix}$$

$$= [1 \;\; 2] \begin{bmatrix} \frac{s+5}{(s+2)(s+4)} - \frac{-1}{(s+2)(s+4)} \\ \frac{3}{(s+2)(s+4)} - \frac{s+1}{(s+2)(s+4)} \end{bmatrix}$$

$$= [1 \quad 2]\begin{bmatrix} \dfrac{1}{s+2} \\ \dfrac{2s+1}{(s+2)(s+4)} \end{bmatrix} = \frac{1}{s+2} + \frac{2(2s+1)}{(s+2)(s+4)}$$

or $$G(s) = \frac{5s+6}{(s+2)(s+4)} \quad \textbf{Ans.}$$

Example 12.62. *Show that the pair* **AB** *given below is controllable for all values of α, β and γ*

$$\mathbf{A} = \begin{bmatrix} 0 & 1 & 0 \\ 0 & 0 & 1 \\ -\alpha & -\beta & -\gamma \end{bmatrix}; \mathbf{B} = \begin{bmatrix} 0 \\ 0 \\ 2 \end{bmatrix}.$$

Solution. $$\mathbf{A} = \begin{bmatrix} 0 & 1 & 0 \\ 0 & 0 & 1 \\ -\alpha & -\beta & -\gamma \end{bmatrix}; \mathbf{B} = \begin{bmatrix} 0 \\ 0 \\ 2 \end{bmatrix}$$

$$\mathbf{AB} = \begin{bmatrix} 0 & 1 & 0 \\ 0 & 0 & 1 \\ -\alpha & -\beta & -\gamma \end{bmatrix}\begin{bmatrix} 0 \\ 0 \\ 2 \end{bmatrix} = \begin{bmatrix} 0 \\ 2 \\ -2\gamma \end{bmatrix}$$

$$\mathbf{A^2B} = \mathbf{AAB} = \begin{bmatrix} 0 & 1 & 0 \\ 0 & 0 & 1 \\ -\alpha & -\beta & -\gamma \end{bmatrix}\begin{bmatrix} 0 \\ 0 \\ -2\gamma \end{bmatrix} = \begin{bmatrix} 2 \\ -\gamma \\ -2\beta + 2\gamma^2 \end{bmatrix}$$

$$\mathbf{U}_c = [\mathbf{B\ AB\ A^2\ B}]$$

$$= \begin{bmatrix} 0 & 0 & 2 \\ 0 & 2 & -\gamma \\ 2 & -2\gamma & -2\beta + 2\gamma^2 \end{bmatrix}$$

$$|\mathbf{U}_c| = 2\begin{vmatrix} 0 & 2 \\ 2 & -2\gamma \end{vmatrix} = -8$$

The det $|\mathbf{U}_c| \neq 0$ and independent of any values of α, β and γ, therefore, the pair **AB** is controllable for all values of α, β and γ.

Example 12.63. *Determine the state controllability of the system shown in Fig. 12.65*

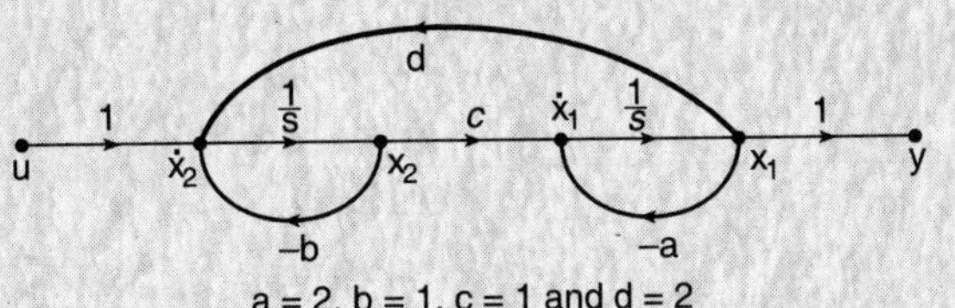

a = 2, b = 1, c = 1 and d = 2

Fig. 12.65

Solution. $\begin{bmatrix} \dot{x}_1 \\ \dot{x}_2 \end{bmatrix} = \begin{bmatrix} -a & c \\ d & -b \end{bmatrix} \begin{bmatrix} x_1 \\ x_2 \end{bmatrix} + \begin{bmatrix} 0 \\ 1 \end{bmatrix} u$

$$\mathbf{A} = \begin{bmatrix} -a & c \\ d & b \end{bmatrix} \quad \mathbf{B} = \begin{bmatrix} 0 \\ 1 \end{bmatrix}$$

$$\mathbf{AB} = \begin{bmatrix} -a & c \\ d & -b \end{bmatrix} \begin{bmatrix} 0 \\ 1 \end{bmatrix} = \begin{bmatrix} c \\ -b \end{bmatrix}$$

$$\mathbf{U}_c = [\mathbf{B} \quad \mathbf{AB}] = \begin{bmatrix} 0 & c \\ 1 & -b \end{bmatrix}$$

$$|\mathbf{U}_c| = \{0 \times (-b) - (1) \times c\} = -c$$

For state controllability $c \neq 0$ **Ans.**

Example 12.64. *Determine the pulse transfer function for the sampled-data control system shown in Fig. 12.66*

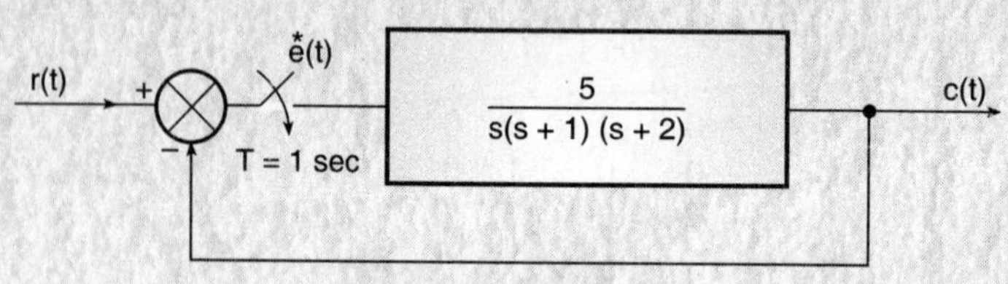

Fig. 12.66

Solution. $G(s) = \dfrac{5}{s(s+1)(s+2)} = \dfrac{K_1}{s} + \dfrac{K_2}{s+1} + \dfrac{K_3}{s+2}$

The coefficients can be determined as

$$K_1 = 2.5, K_2 = -5 \text{ and } K_3 = 2.5$$

$$\therefore \qquad G(s) = \frac{2.5}{s} - \frac{5}{s+1} + \frac{2.5}{s+2}$$

Taking z-transform on both sides,

$$G(z) = 2.5\left\{\frac{z}{z-1} - \frac{2z}{z-e^{T}} + \frac{z}{z-e^{-2T}}\right\}$$

$$T = 1 \text{ sec}$$

$$\therefore \qquad G(z) = 2.5\left\{\frac{z}{z-1} - \frac{2z}{z-e^{-1}} + \frac{z}{z-e^{-2}}\right\}$$

$$= 2.5\left\{\frac{z}{z-1} - \frac{2z}{z-0.368} + \frac{z}{z-0.135}\right\}$$

$$= 2.5\left\{\frac{z(z-0.368)(z-0.135) - 2z(z-1)(z-0.135) + z(z-1)(z-0.368)}{(z-1)(z-0.368)(z-0.135)}\right\}$$

$$= 2.5 \left\{ \frac{z(z^2 - 0.5z + 0.05) - 2z\,(z^2 - 1.135z + 0.135) + z(z^2 - 1.368z + 0.368)}{(z-1)\,(z-0.368)\,(z-0.135)} \right\}$$

$$= 2.5 \left\{ \frac{z^3 - 0.5z^2 + 0.05z - 2z^3 + 2.27z^2 - 0.135z + z^3 - 1.368z + 0.368z}{(z-1)\,(z-0.368)\,(z-0.135)} \right\}$$

$$= 2.5 \left\{ \frac{0.402z^2 + 0.283z}{z^3 - 1.503z^2 + 0.532z - 0.05} \right\}$$

or $$G(z) = \frac{1.005z^2 + 0.7075z}{z^3 - 1.50z^2 + 0.532z - 0.05}$$

$\because \quad H(s) = 1$

$\therefore \quad GH(z) = G(z)$

The pulse transfer function is given by

$$\frac{C(z)}{R(z)} = \frac{G(z)}{1 + G(z)} = \frac{\dfrac{1.005z^2 + 0.7075z}{z^3 - 1.505z^2 + 0.532z - 0.05}}{1 + \dfrac{1.005z^2 + 0.7075z}{z^3 - 1.505z^2 + 0.532z - 0.05}}$$

$$= \frac{1.005z^2 + 0.7075z}{z^3 - 1.505z^2 + 0.532z - 0.05 + 1.005z^2 + 0.7075z}$$

or $$\frac{C(z)}{R(z)} = \frac{z(1.005z + 0.7075)}{z^3 - 0.498z^2 + 1.2395z - 0.05}$$ **Ans.**

APPENDIX I

Answers to Typical Problems

Chapter 1

1.1. (i) $f(t) = \frac{1}{3}e^{-t} - \frac{1}{3}e^{-4t}$ (ii) $f(t) = e^{-3t}\cos 2t - \frac{3}{2}e^{-3t}\sin 2t$

(iii) $f(t) = \cos t - \cos 2t$ (iv) $f(t) = e^{-0.5t}(1 - 0.25t)$

(v) $f(t) = \frac{2}{3} - e^{-t} + \frac{1}{3}e^{-3t}$

1.3. $\frac{C(s)}{R(s)} = \frac{1}{s^2 + 4s + 1}$; $e(t) = \frac{2+\sqrt{3}}{2\sqrt{3}}[e^{-(2+\sqrt{3})t} + e^{-(2-\sqrt{3})t}]$

1.5. (i) $F(\infty) = 0$ (ii) $F(\infty) = 0.4$.

Chapter 2

2.1. Poles : $s_a = 0, s_b = s_c = -2, s_d, s_s = (-1 \pm j2)$

2.3. $G(s) = \frac{4}{(s+1)(s^2 + 2s + 5)}$

2.5. $\frac{I(s)}{E(s)} = \frac{1}{sL + R}$

2.7. $\frac{E_o(s)}{E_i(s)} = \frac{K}{(sR_1C_1 + 1)(sR_2C_2 + 1)}$.

Chapter 3

3.1. $\frac{C}{R} = \frac{G_1G_2 + G_3}{1 + G_2H_1}$

3.3. Overall transfer function $= \frac{G_1G_2 + G_1G_3}{1 + G_1G_2H_1 + G_2H_2 + G_3H_2}$

3.5.

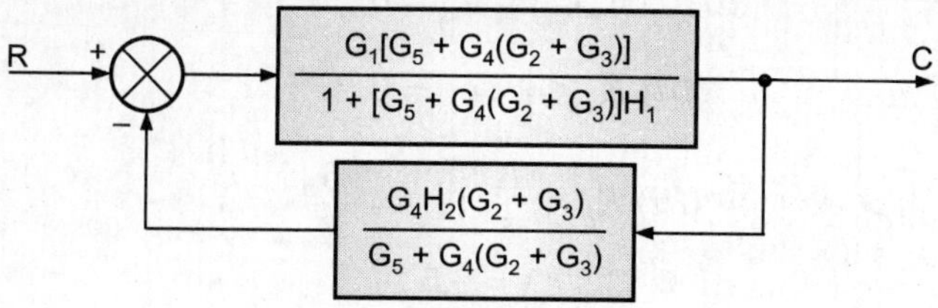

3.7. $\dfrac{C}{R} = \dfrac{G_1G_4 + G_1G_2G_3}{1 + G_1G_2 + G_2G_3H_1 - G_1G_2G_4H_1}$

3.9. $\dfrac{C}{R_1} = \dfrac{G_3(G_1 + G_2)}{1 + G_3H_1}, \dfrac{C}{R_2} = \dfrac{G_3}{1 + G_3H_1}$

3.11. $C = \dfrac{(G_1 + G_2)G_3G_4}{1 + G_3H_1 + G_1G_3G_4H_2} . R + \dfrac{G_4}{1 + G_3H_1 + G_1G_3G_4H_2}$

3.12. $\dfrac{C}{R} = \dfrac{G_1G_2 + G_2G_3}{1 + G_2G_3 + G_3H_1 + G_1G_2H_2 + G_1G_3H_2}$

3.13. $\dfrac{C}{R} = \dfrac{G_1G_2 + G_3(1 + G_1G_2H_1)}{1 + G_2H_2 + G_1G_2H_1}$.

Chapter 4

Refer to the answer of problems of Chapter 3.

Chapter 5

5.1. $\dfrac{Y(s)}{X(s)} = \dfrac{\dfrac{f_2}{K}s}{\dfrac{(f_1 + f_2)s}{K} + 1}$

5.3. $M_1\ddot{x}_1 + f_1(\dot{x}_1 - \dot{x}_2) + K_2x_1 = f(t)$

$M_2\ddot{x}_2 + f_1(\dot{x}_2 - \dot{x}_1) + K_1x_2 = 0$

5.5. $K_A = 3.96$ A/V

5.7. $\dfrac{\omega(s)}{I_f(s)} = \dfrac{2000}{1 + 0.4s}$

5.9. $\dfrac{\omega(s)}{V_f(s)} = \dfrac{1}{(1 + 0.02s)(1 + 0.0025s)}$

5.11. (*a*) 20 V (*b*) 951 RPM.

Chapter 6

6.1. $c(t) = 0.5e^{-t}$

6.3. $\dfrac{C(s)}{R(s)} = \dfrac{4}{s^2 + 2s + 3}$ $c(t)$ 1.33 $[1 - 1.22e^{-0.998t} \sin(1.41t + \tan^{-1} 1.41)]$

6.5. $\zeta = 0.59, \omega_n = 1.69$ rad/sec

6.7. $M_p = 27.5\%, t_s = 2$ sec.

6.9. $\dfrac{C(s)}{R(s)} = \dfrac{17.25\,(s + 1.043)}{(s + 3)\,(s + 12)}$

6.11. $c(t) = 1 - \dfrac{e^{-2t}}{2} - \dfrac{1}{\sqrt{2}} \sin(t + 45°)$

6.13. (*a*) (*i*) $K_p = \infty$; $e_{ss} = 0$

(*ii*) $K_v = \dfrac{K}{100}$; $e_{ss} = \dfrac{100}{K}$

(*iii*) $K_a = 0$; $e_{ss} = \infty$

(*b*) (*i*) $K_p = \infty$; $e_{ss} = 0$

(*ii*) $K_v = \infty$; $e_{ss} = 0$

(*iii*) $K_a = \dfrac{K}{5}$; $e_{ss} = \dfrac{5}{K}$

6.15. $\omega_n = 48$ rad/sec

6.17. $e_{ss} = 0.5$

6.19. Feedback factor $b = 0.16$, $e_{i2} = 5e_{i1}$

6.21. $K = 20$.

Chapter 7

7.1. (*a*) stable, (*b*) stable, (*c*) unstable, (*d*) unstable and (*e*) limitedly stable

7.3. Unstable, 2 roots with positive real parts

7.5. $K < 69$

7.7. (*a*) stable, (*b*) unstable, (*c*) unstable (*d*) stable ; G.M. 6.84 db and (*e*) Marginally stable.

7.9. (*a*) stable, if $K < 6$, (*b*) $K > 1$

7.11. (*a*) $G(s) = \dfrac{9}{s(0.5s+1)(sT+1)}$ and (*b*) $T < 0.125$

7.13. (*a*) G.M. = 0, P.M. = 0 (marginally stable)
(*b*) 0.316 (*c*) 0.421

7.15. Unstable

7.17. (*a*) $K = 2.63$ (*b*) $K = 10$

7.19. (*a*) (*i*) $N = 2$ (*ii*) Angle of asymptotes 90°, 270°
(*iii*) Intersection of asymptotes on real axis – 1.25
(*b*) (*i*) $N = 3$ (*ii*) Angle of asymptotes 60°, 180°, 300°
(*iii*) Intersection of asymptotes on real axis – 1.33
(*e*) (*i*) $N = 3$ (*ii*) Angle of asymptotes 90°, 270°
(*iii*) Intersection of asymptotes on real axis – 1.5
(*d*) (*i*) $N = 3$ (*ii*) Angle of asymptotes 90°, 270°
(*iii*) Intersection of asymptotes on real axis + 1.5

7.21. (*a*) $K = 109$ (*b*) $K = 25.5$

7.23. $K = 21$

7.25. (*i*) Root locus is asymptotic to, and never crosses the imaginary axes, therefore, the system is inherently stable.
(*ii*) Breakaway point is located at $s = -0.465$ and at this point the value of $K = 0.15$ makes the system critically damped.

7.27. $K = 2.6$, G.M. = 17 db, P.M. = 53°.

Chapter 8

8.1. $K = 2$

8.3. $\beta = 2.99$, $\dfrac{1}{T} = 2$ rad/sec, $\dfrac{1}{\beta T} = 0.668$ rad/sec.

8.5. $K_t = 0.4$.

Chapter 9

9.1. $x_1 = i, x_2 = v_c$

$$\begin{bmatrix} \dot{x}_1 \\ \dot{x}_2 \end{bmatrix} = \begin{bmatrix} 10^6 & 10^3 \\ 10^9 & 0 \end{bmatrix} \begin{bmatrix} x_1 \\ x_2 \end{bmatrix} + \begin{bmatrix} 10^3 \\ 0 \end{bmatrix} 10$$

9.3. $$\begin{bmatrix} \dot{x}_1 \\ \dot{x}_2 \\ \dot{x}_3 \end{bmatrix} = \begin{bmatrix} -\dfrac{R_1}{L_1} & 0 & -\dfrac{1}{L_1} \\ 0 & -\dfrac{R_2}{L_2} & \dfrac{1}{L_2} \\ \dfrac{1}{C} & \dfrac{1}{C} & 0 \end{bmatrix} \begin{bmatrix} x_1 \\ x_2 \\ x_3 \end{bmatrix} + \begin{bmatrix} \dfrac{1}{L_1} \\ 0 \\ 0 \end{bmatrix} e, \begin{bmatrix} y_1 \\ y_2 \end{bmatrix} = \begin{bmatrix} 1 & 0 \\ 0 & 1 \end{bmatrix} \begin{bmatrix} x_1 \\ x_2 \end{bmatrix}$$

9.5. $$\begin{bmatrix} \dot{x}_1 \\ \dot{x}_2 \\ \dot{x}_3 \end{bmatrix} = \begin{bmatrix} -\dfrac{R_a}{L_a} & 0 & -\dfrac{R_b}{L_a} \\ 0 & 0 & 1 \\ \dfrac{K_b}{J} & 0 & -\dfrac{f}{J} \end{bmatrix} \begin{bmatrix} x_1 \\ x_2 \\ x_3 \end{bmatrix} + \begin{bmatrix} \dfrac{1}{L_a} \\ 0 \end{bmatrix} e_a, y = [0 \quad 1 \quad 0] \begin{bmatrix} x_1 \\ x_2 \\ x_3 \end{bmatrix}$$

9.7. $x_1 = y_1, x_3 = y_2 \quad x_2 = \dot{y}_1, x_4 = \dot{y}_2$

$$\begin{bmatrix} \dot{x}_1 \\ \dot{x}_2 \\ \dot{x}_3 \\ \dot{x}_4 \end{bmatrix} = \begin{bmatrix} 0 & 1 & 0 & 0 \\ \left(-\dfrac{K_1}{M_1} - \dfrac{K_2}{M_2}\right) & -\dfrac{f}{M_1} & \dfrac{K_2}{M_2} & 0 \\ 0 & 0 & 0 & 1 \\ \dfrac{K_2}{M_2} & 0 & -\dfrac{K_2}{M_2} & 0 \end{bmatrix} \begin{bmatrix} x_1 \\ x_2 \\ x_3 \\ x_4 \end{bmatrix} + \begin{bmatrix} 0 \\ 0 \\ 0 \\ \dfrac{1}{M_2} \end{bmatrix} f(t)$$

9.9. $$\mathbf{A} = \begin{bmatrix} 0 & 1 & 0 \\ 0 & 0 & 1 \\ -24 & -26 & -9 \end{bmatrix}, \mathbf{B} = \begin{bmatrix} 0 \\ 0 \\ 1 \end{bmatrix}, \mathbf{C} = [2 \quad 1 \quad 1]$$

9.11. $$\mathbf{A} = \begin{bmatrix} 0 & 1 & 0 \\ 0 & 0 & 1 \\ -6 & -11 & -6 \end{bmatrix}, \mathbf{B} = \begin{bmatrix} 0 \\ 0 \\ 1 \end{bmatrix}, \mathbf{C} = [1 \quad 6 \quad 0]$$

9.13. $$\phi(t) = \begin{bmatrix} \left(\dfrac{4}{3}e^{-t} - \dfrac{1}{3}e^{-4t}\right) & \left(\dfrac{1}{3}e^{-t} - \dfrac{1}{3}e^{-4t}\right) \\ \left(-\dfrac{4}{3}e^{-t} + \dfrac{1}{3}e^{-4t}\right) & \left(-\dfrac{1}{3}e^{-t} + \dfrac{4}{3}e^{-4t}\right) \end{bmatrix}$$

$$\underset{(ZIR)}{\begin{bmatrix} x_1(t) \\ x_2(t) \end{bmatrix}} = \begin{bmatrix} \dfrac{5}{3}e^{-t} - \dfrac{2}{3}e^{-4t} \\ -\dfrac{5}{3}e^{-t} + \dfrac{8}{3}e^{-4t} \end{bmatrix}, \underset{(ZSR)}{\begin{bmatrix} x_1(t) \\ x_2(t) \end{bmatrix}} = \begin{bmatrix} 1 - \dfrac{1}{3}e^{-t} + \dfrac{1}{12}e^{-4t} \\ \dfrac{1}{3}e^{-t} - \dfrac{1}{3}e^{-4t} \end{bmatrix}$$

$$\underset{(ZIR + ZSR)}{\text{Total responce}} \begin{bmatrix} x_1(t) \\ x_2(t) \end{bmatrix} = \begin{bmatrix} 1 + \dfrac{4}{3}e^{-t} - \dfrac{7}{12}e^{-4t} \\ -\dfrac{4}{3}e^{-t} + \dfrac{7}{3}e^{-4t} \end{bmatrix}$$

9.19. (*a*) $\begin{bmatrix} x_1(t) \\ x_2(t) \end{bmatrix} = \begin{bmatrix} e^{-t}\cos 2t + 0.5e^{-t}\sin 2t \\ 1 - 2e^{-t}\cos 2t + 1.5e^{-t}\sin 2t \end{bmatrix}$

$$y = 1 - e^{-t}\cos 2t + 2e^{-t}\sin 2t$$

(*b*) $\mathbf{M} = \begin{bmatrix} 1 & 1 \\ (-1-j2) & (-1+j2) \end{bmatrix}$, $\mathbf{M}^{-1}\,\mathbf{AM} = \begin{bmatrix} (-1+j2) & 0 \\ 0 & (-1-j2) \end{bmatrix}$

9.21. $\dfrac{Y(s)}{U(s)} = \dfrac{s+5}{(s+2)(s+3)}$

9.23. $\dfrac{Y(s)}{U(s)} = \dfrac{3}{(s+1)}$

9.25. Rank of $\mathbf{U} = 3$: controllable
Rank of $\mathbf{V} = 2$: unobservable.

Chapter 10

10.1. (*a*) $\dfrac{z(z-2+T)}{(z-1)^2}$ (*b*) $\dfrac{z(z-\cos\omega T)}{z^2 - 2z\cos\omega T + 1}$

(*c*) $\dfrac{ze^{-\alpha T}\sin\omega T}{z^2 - 2ze^{-\alpha T}\cos\omega T + e^{-2\alpha T}}$

(*d*) $\dfrac{z\,[z^2 - e^{-T}(2z + e^{-T} - e^{-2T} - 1)]}{(z-1)(z-e^{-T})(z-e^{-2T})}$

10.3. $f(kT) = 0.96(1)^k + (1.29)^k\,(0.96\cos 51.3k° - 0.8\sin 51.3k°)$

10.5. $\dfrac{C(z)}{R(z)} = \dfrac{0.309z}{z^2 - 0.88z + 0.135}$

10.7. $\dfrac{C(z)}{R(z)} = \dfrac{0.233z}{z^2 - 0.272z + 0.0497}$

$c(k) = 0.3(1)^k - 0.3\,(0.22)^k\,(\cos 52.3k° + 0.49\sin 52.3k°)$

10.9. Stable.

APPENDIX II

Objective Type Questions

(✓) **Tick mark the correct answer :**

1. The Laplace transform of e^{-2t} sin $2t$ is

(a) $\frac{4}{(s+2)^2+4}$ (b) $\frac{4}{s^2+4}$

(c) $\frac{2}{s^2+4s+8}$ (d) $\frac{2}{s^2+4}$

2. The value of function $F(s) = \frac{2}{s^2+3}$ at $t = 0+$ is

(a) 3 (b) 2

(c) 3/2 (d) zero

3. A system having transfer function $G(s) = \frac{1}{2(s+0.5)}$ is subjected to a unit step input, the steady value of the output is

(a) 1 (b) 2

(c) 1/2 (d) 1/10

4. The solution of the differential equation $\frac{d^2x}{dt^2} + 2\frac{dx}{dt} + 2x = 1$ is

(a) Critically damped (b) Underdamped

(c) Overdamped (d) Steady

5. The natural frequency of oscillations of the output for the equation $\frac{d^2x}{dt^2} + 1.5\frac{dx}{dt} + 4x = 1$ is

(a) 0 rad/sec (b) 1.5 rad/sec

(c) 2 rad/sec (d) 4 rad/sec

6. The output of a system is given by $C(s) = \frac{1}{s}\cdot\frac{2}{(s-3)}$. The output will

(a) increase w.r.t. time (b) decrease w.r.t. time

(c) remain steady w.r.t. time (d) become zero w.r.t. time

7. For input $u(t) = 2$ the steady output for the transfer function $G(s) = \dfrac{1}{s^2 + 1.5s + 4}$ is

(*a*) 2 (*b*) 0.5

(*c*) 1.5 (*d*) 1

8. The forward-path gain of a control system is 2.5 and the pole-zero configuration of the overall transfer function is shown in Fig. Q. 8. The overall transfer function is

(*a*) $\dfrac{2.5(s+1)}{s(s+2)(s+3)}$

(*b*) $\dfrac{2.5(s+2)}{s(s+1)(s+3)}$

(*c*) $\dfrac{2.5(s+3)}{s(s+1)(s+2)}$

(*d*) $\dfrac{s+3}{2.5(s+1)(s+2)}$

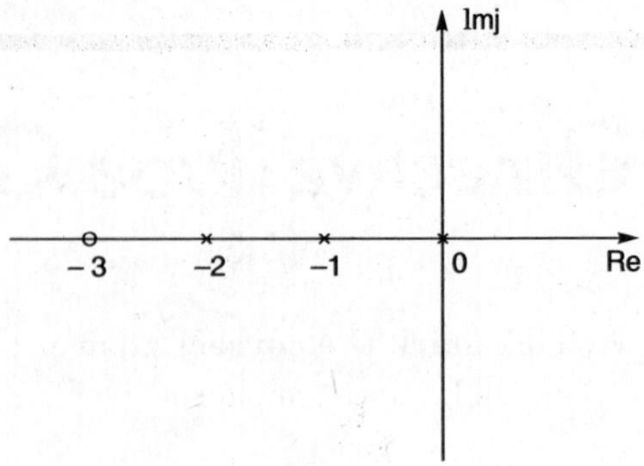

Fig. Q. 8

9. The inverse Laplace transform of $1.5/s\,(s + 1)$ is

(*a*) $(1 - e^{-1.5t})$ (*b*) $1.5\,(1 - e^{-1.5t})$

(*c*) $1.5\,(1 - e^{-t})$ (*d*) $(1.5 + e^{-1.5t})$

10. The transfer function is defined as

(*a*) **the ratio of Laplace transform of output to Laplace transform of input considering initial condition as zero**

(*b*) **the ratio of Laplace transform of input to Laplace transform of output considering initial conditions as zero**

(*c*) **the ratio of input to output**

(*d*) **the ratio of output to input**

11. The error detector element in a control system gives

(*a*) **the sum of the reference signal and feedback signal**

(*b*) **the sum of the reference signal and error signal**

(*c*) **the difference of the reference signal and feedback signal**

(*d*) **the difference of the reference signal and output signal**

12. Unit impulse response of a system in Laplace transform form gives

(*a*) **transfer function** (*b*) **system gain**

(*c*) **unit step function** (*d*) **unit ramp function**

13. A system is represented by the differential equation

$$M\frac{d^2x}{dt^2} + F\frac{dx}{dt} + Kx = u(t)$$

The transfer function relating $X(s)$ and $U(s)$ is

(a) $\dfrac{M}{Ms^2 + Fs + K}$ (b) $\dfrac{F}{Ms^2 + Fs + K}$

(c) $\dfrac{K}{Ms^2 + Fs + K}$ (d) $\dfrac{1}{Ms^2 + Fs + K}$

14. The equation governing a control system is given by $\dfrac{d^2c(t)}{dt^2} + \dfrac{5dc(t)}{dt} + 4c(t) = 3r(t)$

The transfer function for the system is

(a) $\dfrac{C(s)}{R(s)} = \dfrac{3}{(s+1)(s+4)}$ (b) $\dfrac{C(s)}{R(s)} = \dfrac{5}{(s+1)(s+4)}$

(c) $\dfrac{C(s)}{R(s)} = \dfrac{1}{s^2 + 3s + 4}$ (d) $\dfrac{C(s)}{R(s)} = \dfrac{4}{s^2 + 3s + 5}$

15. The transfer function relating output $I(s)$.... and input $E(s)$ in a series RLC circuit is given by

(a) $\dfrac{Cs}{(s^2LC + RCs + LC)}$ (b) $\dfrac{Cs}{(s^2LC + RCs + 1)}$

(c) $\dfrac{1}{C} \cdot \dfrac{s}{(s^2LC + RCs + 1)}$ (d) $\dfrac{Cs}{(s^2 + LCs + R)}$

16. The transfer function relating $\theta(s)$ and $E(s)$ for the figure given below

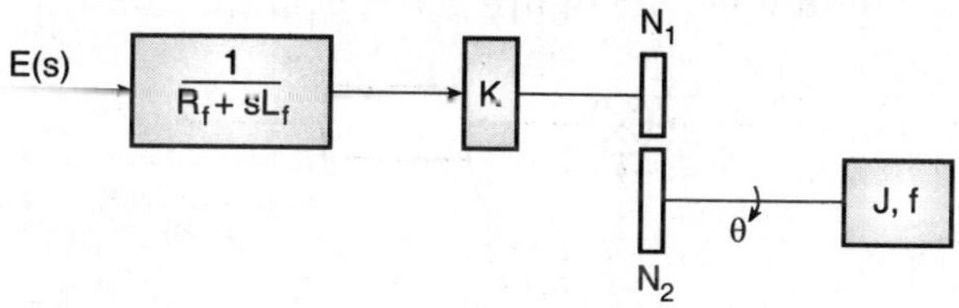

is

(a) $\dfrac{K(N_1)}{s(R_f + sL_f)(Js + f)(N_2)}$ (b) $\dfrac{K}{s(R_f + sL_f)[J(N_1/N_2)^2 s + f(N_1/N_2)^2]}$

(c) $\dfrac{K(N_2)}{s(sR_f + sL_f)(Js^2 + f)(N_1)}$ (d) $\dfrac{K}{s(R_f + sL_f)[J(N_1/N_2)s + f(N_1/N_2)]}$

17. A servomotor is connected through a gear ratio of 10 to a load having moment of inertia J and coefficient of friction f. The equivalent parameters referred to motor shaft side are

(a) $J_{eq} = 0.01J, f_{eq} = 0.01f$ (b) $J_{eq} = 10J, f_{eq} = 10f$

(c) $J_{eq} = 0.1J, f_{eq} = 0.1f$ (d) $J_{eq} = 100J, f_{eq} = 100f$

18. A speed control system is expressed by the transfer function $\dfrac{\omega(s)}{V(s)} = \dfrac{100}{2 + 10s}$, 1 volt of the input corresponds to an output of

(a) 100 rad/sec (b) 10 rad/sec

(c) $\dfrac{100}{12}$ rad/sec (d) 50 rad/sec

19. The generator constant K_g for a generator driven at 1000 RPM is 50 V/A; if the generator is made to run at 1500 RPM the new value of the generator constant will be

(a) 75 V/A (b) $33\frac{1}{2}$ V/A

(c) 150 V/A (d) 100 V/A

20. The transfer function relating E and R for the block diagram given below :

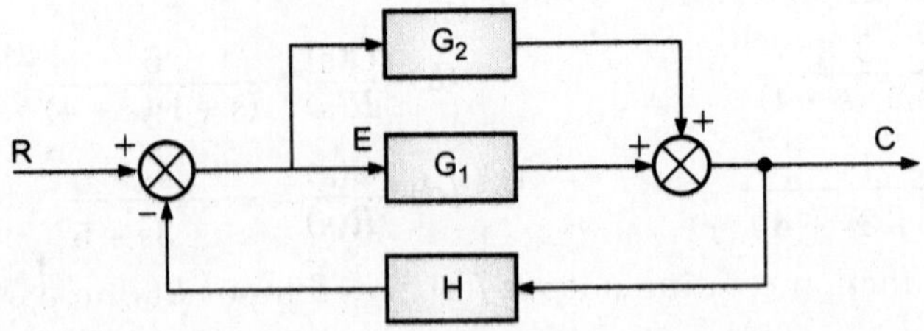

is

(a) $\dfrac{1}{1+(G_1+G_2)H}$ (b) $\dfrac{1}{1+G_1G_2H}$

(c) $\dfrac{G_1G_2}{1+(G_1+G_2)H}$ (d) $\dfrac{G_1G_2}{1+G_1G_2H}$

21. Equivalent block diagram for the figure given below :

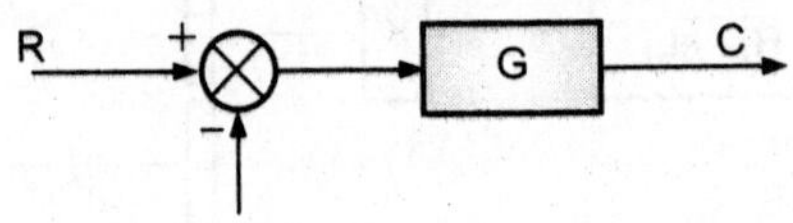

is

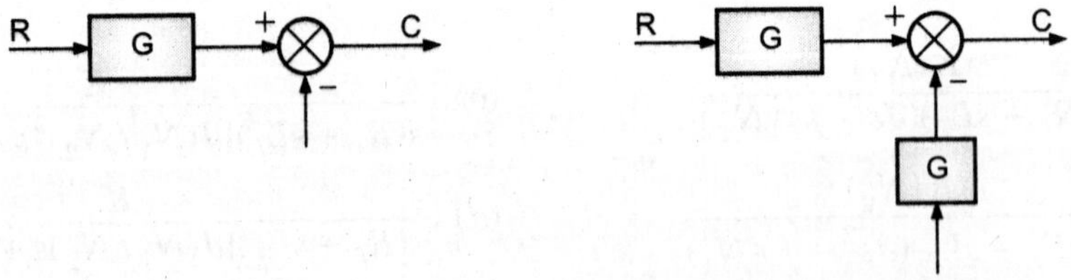

(a) (b)

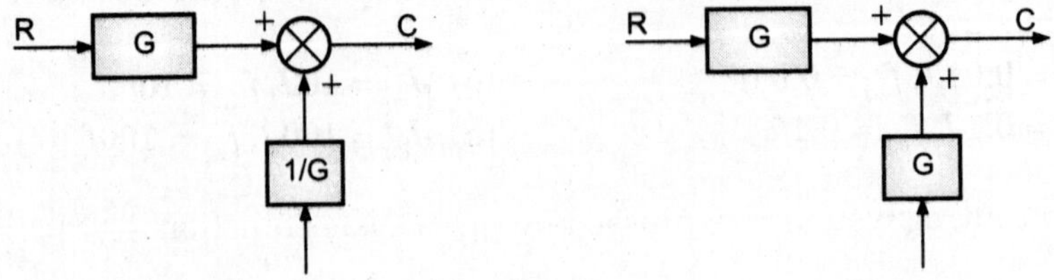

(c) (d)

22. Indicate the correct equivalent of the block diagram given below :

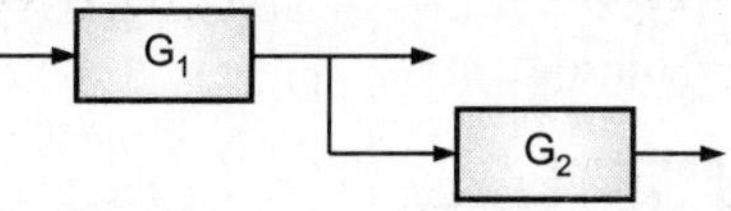

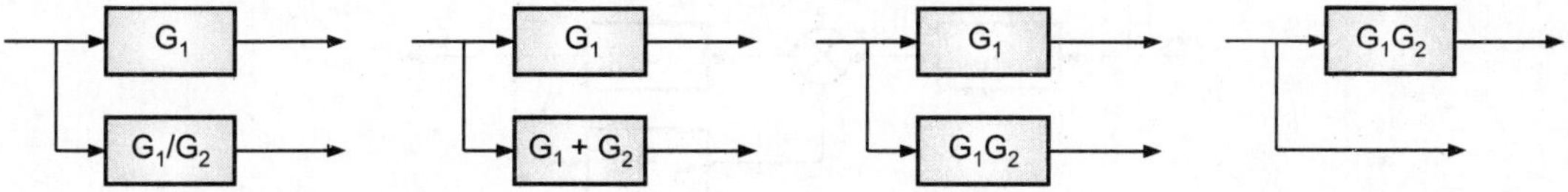

(*a*) (*b*) (*c*) (*d*)

23. The overall gain of the system given below :

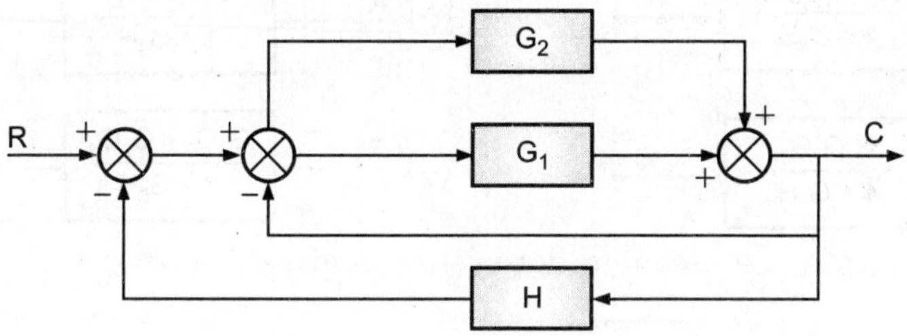

is

(*a*) $\dfrac{G_1 + G_2}{1 + G_1G_2 + G_1H + G_2H}$ (*b*) $\dfrac{G_1G_2}{1 + G_1 + G_2 + G_1H + G_2H}$

(*c*) $\dfrac{G_1G_2}{1 + G_1G_2 + G_1H + G_2H}$ (*d*) $\dfrac{G_1 + G_2}{1 + G_1 + G_2 + G_1H + G_2H}$

24. The transfer function for the network given below :

is

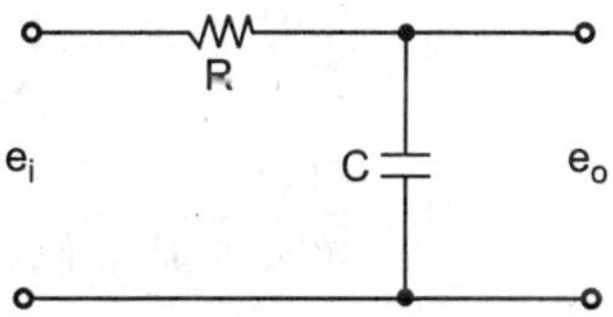

(*a*) $\dfrac{1}{RCs + 1}$ (*b*) $RCs + 1$

(*c*) $\dfrac{1}{\frac{R}{C}s + 1}$ (*d*) $\dfrac{R}{C}s + 1$

25. *C*/*R* for the block diagram shown below :

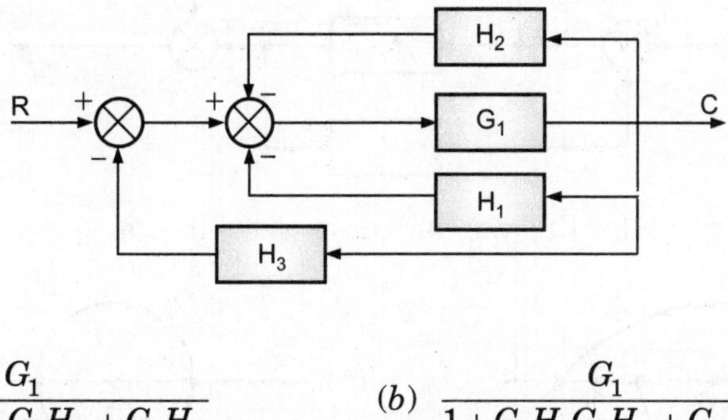

is

(*a*) $\dfrac{G_1}{1 + G_1H_1 + G_1H_2 + G_1H_3}$ (*b*) $\dfrac{G_1}{1 + G_1H_1G_1H_2 + G_1H_3}$

(c) $\dfrac{1}{G_1(G_1H_1 + G_1H_2 + G_1H_3)}$ (d) $\dfrac{1}{G_1(G_1H_1G_1H_2 + G_1H_3)}$

26. The block diagram given below :

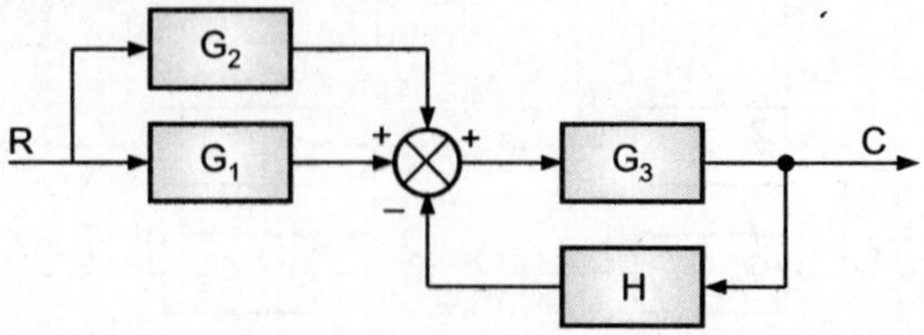

is equivalent to

(a) 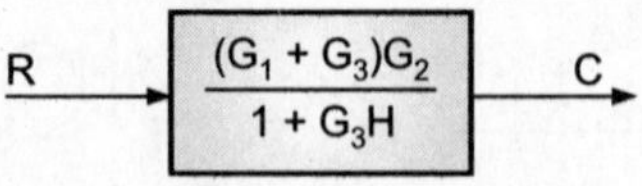

(b) 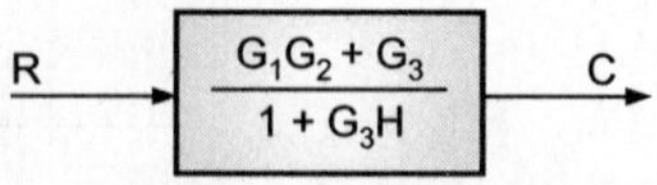

(c) 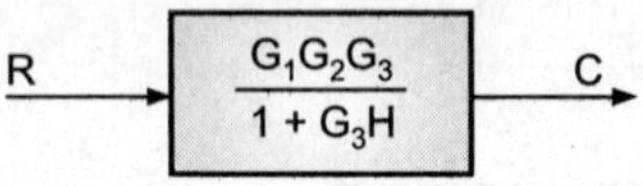

(d) 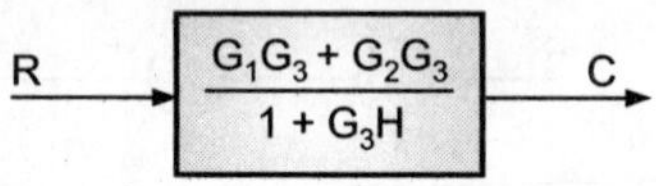

27. The block diagram of a second order system is given below :

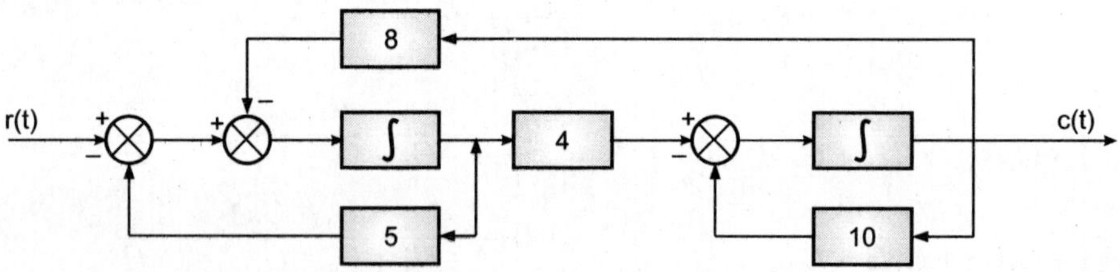

The transfer function of the system is

(a) $\dfrac{C(s)}{R(s)} = \dfrac{4}{s^2 + 10s + 15}$ (b) $\dfrac{C(s)}{R(s)} = \dfrac{4}{s^2 + 8s + 20}$

(c) $\dfrac{C(s)}{R(s)} = \dfrac{4}{s^2 + 15s + 80}$ (d) $\dfrac{C(s)}{R(s)} = \dfrac{4}{s^2 + 15s + 82}$

28.

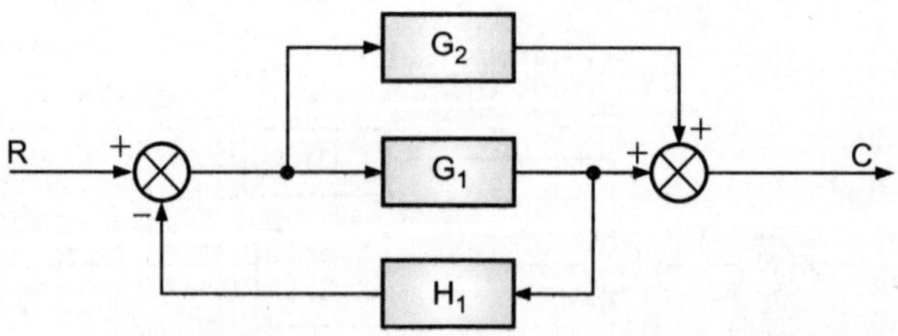

The signal flow graph for the system whose block diagram shown above will be

(a) 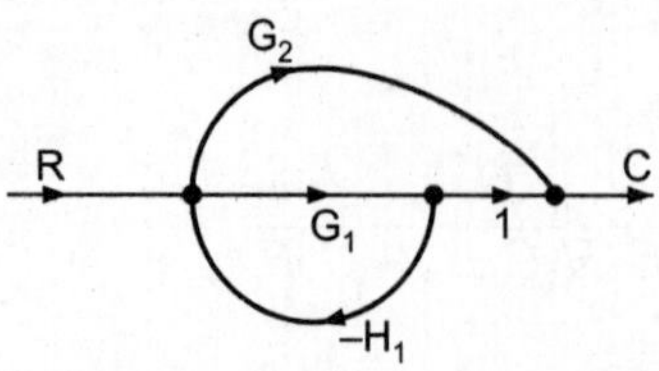

(b) 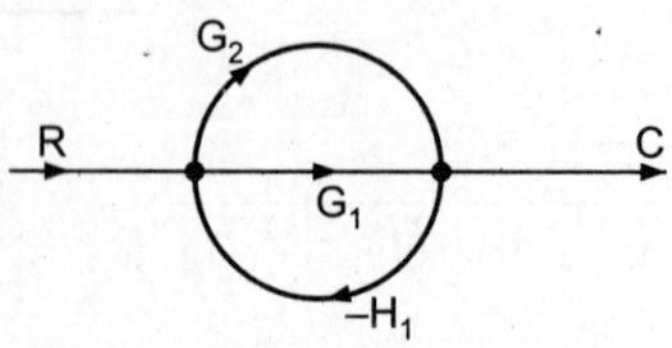

(c) 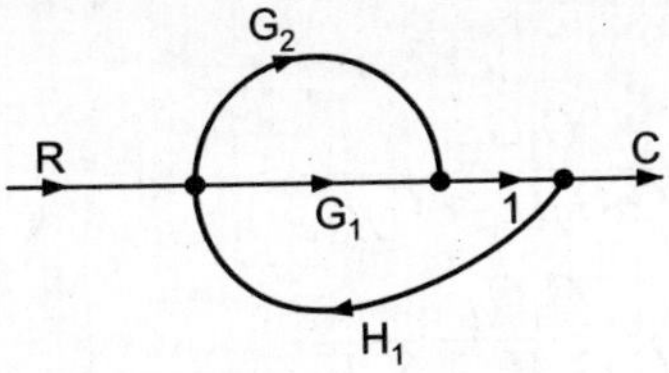

(d) 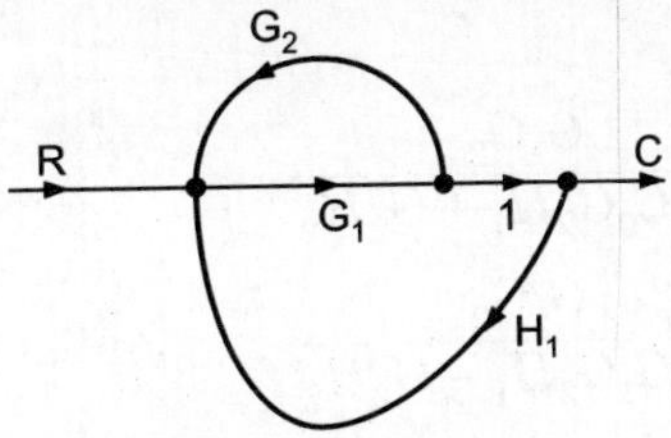

29. The number of closed-loops in the signal flow graph shown below :

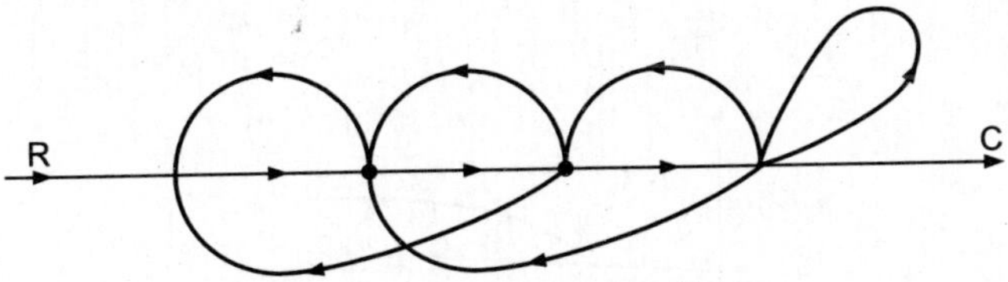

is

(a) 6 (b) 8

(c) 7 (d) 5

30. The number of forward paths in the signal flow graph shown below :

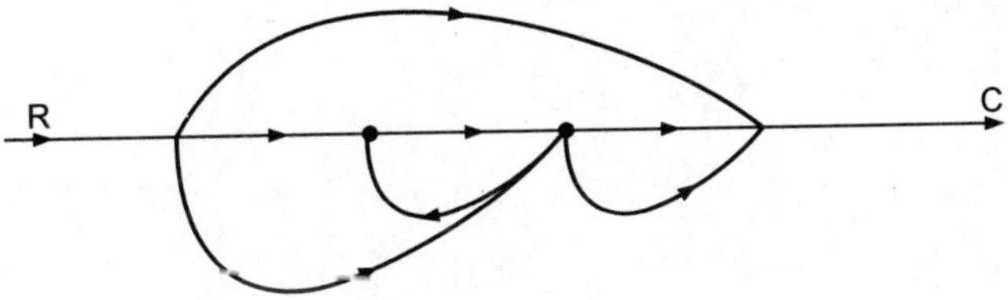

is

(a) 1 (b) 2

(c) 3 (d) 5

31. The graph determinant for the signal flow graph shown below :

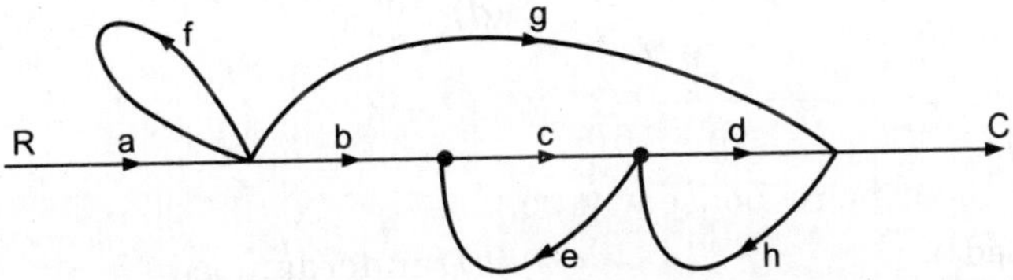

is

(a) $\Delta = -f - ce - dh$ (b) $\Delta = 1 + f + ce + dh + fce + fdh$

(c) $\Delta = 1 - f - ce - dh$ (c) $\Delta = 1 - f - ce - dh + fce + fdh$

32. The overall transfer function for the signal flow graph given below :

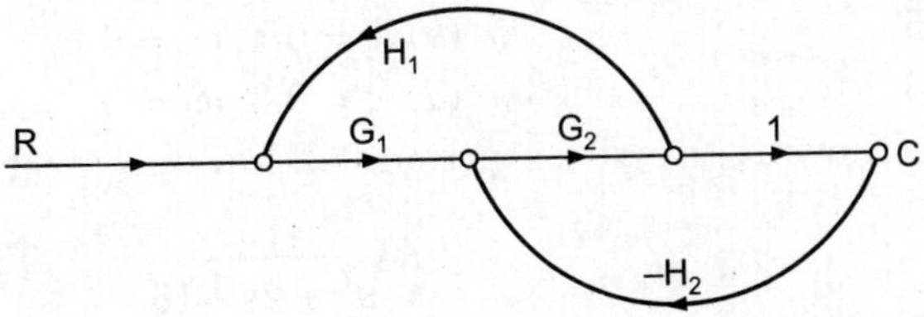

is

(a) $\dfrac{G_1G_2}{1-G_1G_2H_1+G_2H_2}$ (b) $\dfrac{G_1G_2}{1+G_1G_2H_1+G_2H_2}$

(c) $\dfrac{G_1G_2}{1-G_1G_2H_1-G_2H_2}$ (d) $\dfrac{G_1G_2}{1+G_1G_2H_1+G_2H_1}$

33. Figure given below represents unit step response; the time constant of the system is

(a) 0.005 sec (b) 2 sec

(c) 0.02 sec (d) 0.5 sec

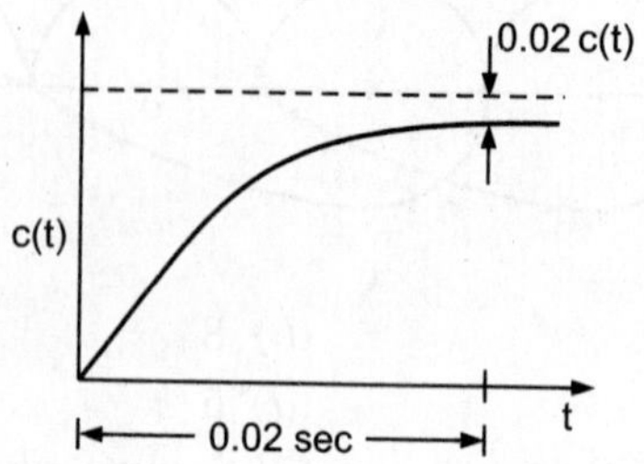

34. The steady value of the output on application of unit step input to transfer function

$$\frac{C(s)}{R(s)} = \frac{2.5}{s^2+2.5s+10}, \text{ will be}$$

(a) $\dfrac{1}{2.5}$ (b) 1

(c) 0.25 (d) 2.5

35. A closed-loop control system has $G(s) = \dfrac{100}{s^2}$ and $H(s) = (s + 1)$

The steady state output for a unit step input is

(a) 1.0 (b) 1/100

(c) 1/10 (d) 10

36. For the system $\dfrac{C(s)}{R(s)} = \dfrac{16}{s^2+8s+16}$

The nature of the time response will be

(a) overdamped (b) underdamped

(c) critically damped (d) none of these

37. For the system $\dfrac{C(s)}{R(s)} = \dfrac{25}{s^2+6s+25}$ the damping factor ζ and damped frequency of oscillations ω_d will be

(a) $\zeta = 0.6, \omega_d = 5$ (b) $\zeta = 0.4, \omega_d = 6$

(c) $\zeta = 0.5, \omega_d = 3$ (d) $\zeta = 0.3, \omega_d = 5$

38. Which one of the following transfer function will have greatest maximum overshoot

(a) $\dfrac{2}{s^2+2s+9}$ (b) $\dfrac{16}{s^2+2s+16}$

(c) $\frac{25}{s^2 + 2s + 25}$ (d) $\frac{36}{s^2 + 2s + 36}$

39. The location of the roots of a second order system is given in figure below, the system has an actual damping

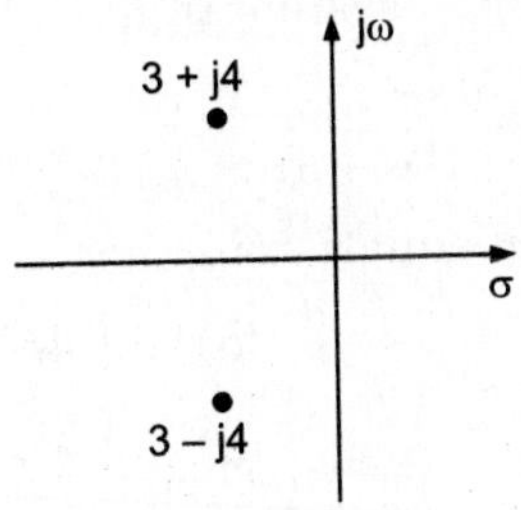

(a) 3 (b) 5
(c) 4 (d) 2

40. The expression given below is the impulse response of a feedback control system $c(t) = e^{-0.6t} \sin 0.8t$
The damping ratio and natural frequency of oscillations are
(a) 0.6, 1 rad/sec (b) 1, 0.6 rad/sec
(c) 0.8, 0.6 rad/sec (d) 1, 0.8 rad/sec

41. The transfer function of a control system is given by $\frac{C(s)}{R(s)} = \frac{800}{s(s+1)(s+20)(s+25)}$ can be approximated as

(a) $\frac{C(s)}{R(s)} = \frac{40}{s(s+1)(s+25)}$ (b) $\frac{C(s)}{R(s)} = \frac{3.2}{s(s+1)(s+20)}$

(c) $\frac{C(s)}{R(s)} = \frac{1.6}{s(s+1)}$ (d) $\frac{C(s)}{R(s)} = \frac{80}{(s+20)(s+25)}$

42. The closed-loop poles for a system having open-loop transfer function $G(s)\,H(s) = K/s\,(s+2)$ at $K = 2$ are
(a) $1 \pm j1$ (b) $-1 \pm j1$
(c) $-2 \pm j1$ (d) $2 \pm j1$

43. To a fair approximation the unit step time response for the system

$$\frac{C(s)}{R(s)} = \frac{4}{(s+100)(s^2 + 2s + 4)}$$

is

(a) $e^{-0.5t} \sin\left(2\sqrt{0.75}t + \tan^{-1}\frac{\sqrt{0.75}}{0.5}\right)$

(b) $1 - \frac{e^{-2t}}{\sqrt{0.75}} \sin\left(2\sqrt{0.75}t + \tan^{-1}\frac{\sqrt{0.75}}{0.5}\right)$

(c) $e^{-2t} \sin\left(2\sqrt{0.75}t + \tan^{-1}\frac{\sqrt{0.75}}{0.5}\right)$

(d) $1 - \frac{e^{0.5t}}{\sqrt{0.75}} \sin\left(2\sqrt{0.75}t + \tan^{-1}\frac{\sqrt{0.75}}{0.5}\right)$

44. The overall transfer function of a system is given below :

$$\frac{C(s)}{R(s)} = \frac{2.5(s+2)}{5(s+1)(s+3)}$$

The corresponding differential equation is

(a) $\ddot{c} + 3\dot{c} + 4c = 0$
(b) $\ddot{c} + 4\dot{c} + 2c = \dot{r} + r$
(c) $\ddot{c} + 4\dot{c} + 3c = 0.5\dot{r} + r$
(d) $\ddot{c} + \dot{c} + c = 0.5\dot{r} + r$

45. The open-loop transfer function of a feedback control system is given below :

$$G(s)\,H(s) = \frac{K}{s(s+10)}$$

The value of gain factor K for critical damping is

(a) 10
(b) 15
(c) 20
(d) 25

46. A linear time invariant system initially at rest, when subjected to unit step input gives a response $y(t) = te^{-t}$, $t > 0$. The corresponding transfer function is

(a) $\frac{1}{(s+1)^2}$
(b) $\frac{1}{s\,(s+1)^2}$
(c) $\frac{s}{(s+1)^2}$
(d) $\frac{1}{s\,(s+1)}$

47. A control system is defined by the mathematical relationship $\frac{d^2x}{dt^2} + 6\frac{dx}{dt} + 5x = 12u(t)$

The response of the system as $t \to \infty$ is

(a) $x = 6$
(b) $x = 2$
(c) $x = 2.4$
(d) $x = 0.2$

48. A unity feedback control system having $G(s) = \frac{4}{s\,(s+2.4)}$ is subjected to a unit step input. The expression for $e(t)$ is given by

(a) $e(t) = \left(\frac{1}{0.8}\right)e^{-1.2t} \sin\left(1.6t + \tan^{-1}\frac{4}{3}\right)$

(b) $e(t) = \left(-\frac{1}{0.8}\right)e^{-1.2t} \sin\left(1.6t + \tan^{-1}\frac{4}{3}\right)$

(c) $e(t) = \left(\frac{1}{0.6}\right)e^{-1.6t} \sin\left(1.2t + \tan^{-1}\frac{4}{3}\right)$

(d) $e(t) = \left(-\frac{1}{0.6}\right)e^{-1.6t} \sin\left(1.6t + \tan^{-1}\frac{4}{3}\right)$

49. The impulse response of the system

$$\frac{C(s)}{R(s)} = \frac{8}{s(s+2)(s+4)} \text{ is}$$

(*a*) $c(t) = 2 - e^{-2t} + e^{-4t}$ (*b*) $c(t) = 1 + 2e^{-2t} - 4e^{-4t}$

(*c*) $c(t) = 1 - 2e^{-2t} + e^{-4t}$ (*d*) $c(t) = 2 + e^{-2t} - 4e^{-4t}$

50. The damping ratio of a second order control system is increased from 0.3 to 0.6, the corresponding change in the % overshoot is

(*a*) 37.08% to 9.47% (*b*) 9.47% to 37.08%

(*c*) 60% to 30% (*d*) 20% to 50%

51. For a second order system given that $\omega_n = 4$ rad/sec and $\zeta = 0.8$. The settling and peak times are

(*a*) 1.3 s, 1.25 s (*b*) 1.25 s, 1.3 s

(*c*) 1.3 s, 2.5 s (*c*) 2.5 s, 3.1 s

52. The unit step response of a feedback control system is given by $c(t) = 1 - 1.25\, e^{-3t} \sin (4t + 53.1°)$. The closed-loop transfer function is

(*a*) $\frac{C(s)}{R(s)} = \frac{2.5}{s^2 + 6s + 2.5}$ (*b*) $\frac{C(s)}{R(s)} = \frac{2.5}{s^2 + 5s + 25}$

(*c*) $\frac{C(s)}{R(s)} = \frac{25}{s^2 + 10s + 25}$ (*d*) $\frac{C(s)}{R(s)} = \frac{25}{s^2 + 6s + 25}$

53. Insertion of negative feedback in a control system affects :

(*a*) the transient response to vanish uniformly

(*b*) the transient response to decay very fast

(*c*) no change in transient response

(*d*) the transient response decays at a slow rate.

54. The damped frequency of oscillations for a system whose roots are located as shown in adjoining figure will be

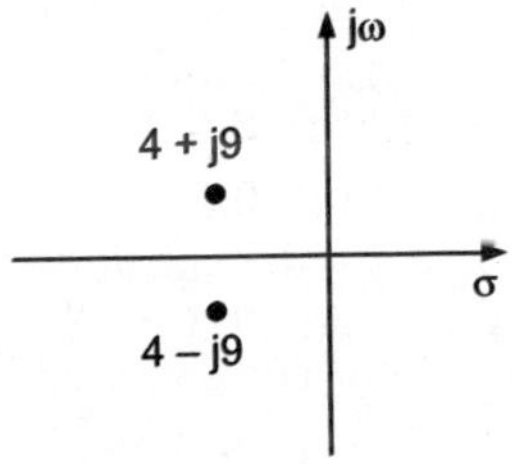

(*a*) 4 rad/s

(*b*) 9 rad/s

(*c*) 2 rad/s

(*d*) 3 rad/s

55. The root locations for a second order system is given in adjoining figure, the damping ratio is

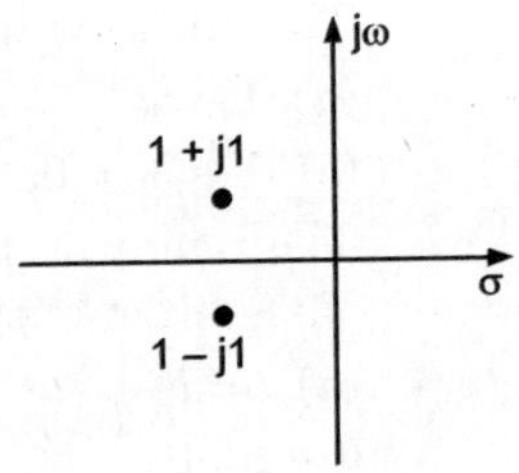

(*a*) 1

(*b*) 1.41

(*c*) 0.707

(*d*) 0.141

56. The natural frequency of oscillations due to dominant roots of the system transfer function

$$\frac{C(s)}{R(s)} = \frac{3}{(s+5)(s^2+2s+3)} \text{ will be}$$

(*a*) 1.73 rad/sec (*b*) 2 rad/sec
(*c*) 1.41 rad/sec (*d*) 3 rad/sec

57. The unit step response is given below, the forward path gain is

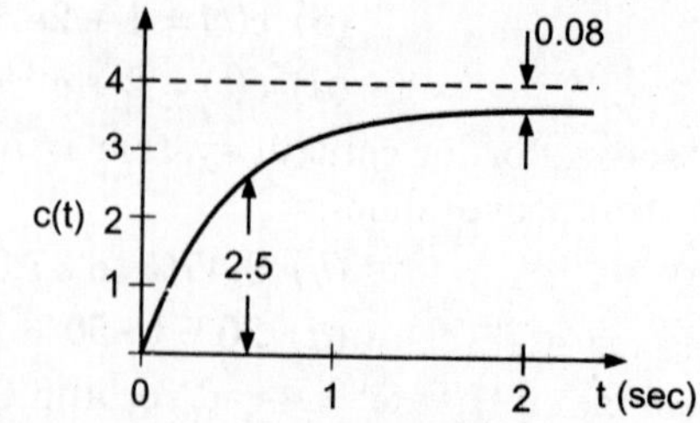

(*a*) 2.0 (*b*) 4.0
(*c*) 5.0 (*d*) 50

58. For 0% overshoot the damping ratio for a second order system is
(*a*) < 1 (*b*) > 1
(*c*) zero (*d*) 1

59. The forward path transfer function of a unit feedback control system is given by $G(s) = \dfrac{20}{s^2 + 3s + 60}$. The steady state error to unit step input is
(*a*) 0 (*b*) 1
(*c*) 0.75 (*d*) ∞

60. A control system is described by the equation $\dfrac{d^2c(t)}{dt^2} + \dfrac{5\,dc(t)}{dt} = 100\,\{r(t) - 0.25c(t)\}$
where, $r(t)$ and $c(t)$ being reference input and controlled variable respectively, the damping ratio is
(*a*) 2.5 (*b*) 1
(*c*) 0.5 (*d*) 0.25

61. For the system given below

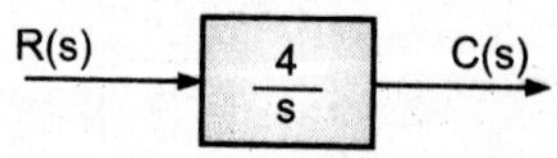

The initial value of the output is $c(0) = 1$, the output will be
(*a*) $1 - 4t$ (*b*) $1 + 4t$
(*c*) $1 + 3t = 0$ (*d*) $1 - 3t$

62. The steady state error is determined as the difference between the reference input and the system output at
(*a*) $t = t_p$ (*b*) $t = \infty$
(*c*) $t = 0$ (*d*) t = time constant

63. The steady state error for a unity feedback system having open-loop transfer function as, $G(s) = \dfrac{9}{s(0.2s + 1)}$, when subjected to a unit step input will be
(*a*) 0.1 (*b*) 1/9
(*c*) 0.2 (*d*) 0

64. The steady state error for an input $2u(t)$ applied to a type 0 system is

(a) $\dfrac{1}{2(1+K_p)}$ (b) $\dfrac{2}{1+K_p}$

(c) $\dfrac{1}{K_p}$ (d) 0

65. The velocity error coefficient is given by

(a) $\lim_{s \to 0} sG(s)\,H(s)$ (b) $\lim_{s \to 0} G(s)\,H(s)$

(c) $\lim_{s \to 0} \dfrac{1}{G(s)\,H(s)}$ (d) $\lim_{s \to 0} \dfrac{s}{G(s)\,H(s)}$

66. The steady state error for a type 2 system subjected to unit ramp input is

(a) 1 (b) ∞

(c) $\dfrac{1}{K}$ (d) 0

67. Finite steady state error

(a) varies inversely with K (b) varies directly with K

(c) is independent of K (d) none of the above

68. The type of the transfer function, $G(s)\,H(s) = \dfrac{K}{s^3 + 2s^2 + 3s}$ is

(a) 1 (b) 2

(c) 3 (d) 4

69. A unity feedback system having transfer function, $G(s) = \dfrac{100}{s(0.1s+1)}$ is subjected to unit ramp input. The steady state error will be

(a) 1 (b) 0

(c) 0.01 (d) 0.05

70. With the increase of type of a system the steady state error for a particular input function

(a) increases (b) decreases

(c) remains changed (d) first increases then decreases

71. A type 1 system is subjected to acceleration input signal leads to steady state error as

(a) 1 (b) K

(c) $\dfrac{1}{K}$ (d) ∞

72. Derivative feedback control

(a) increases rise time (b) increases overshoot

(c) decreases steady state error (d) does not affect the steady error

73. Which one of the following transfer function exhibit least steady state error for ramp input

(a) $\dfrac{9}{s^2 + 2s + 9}$ (b) $\dfrac{16}{s^2 + 2s + 16}$

(c) $\dfrac{25}{s^2 + 2s + 25}$ (d) $\dfrac{36}{s^2 + 2s + 36}$

74. The time response of a system is given by

$$c(t) = 1 - \frac{e^{-1.2t}}{0.8}\left[\sin\left(1.6t + \tan^{-1}\frac{4}{3}\right)\right]$$

characteristic equation of the system is

(a) $s^2 + s + 0.8 = 0$ (b) $s^2 + 2.4s + 4 = 0$

(c) $s^2 + 4s + 2.4 = 0$ (d) $s^2 + s + 4 = 0$

75. Which of the following complex roots relate to greatest value of ω_n

(a) $s_1, s_2 = -1 \pm j1$ (b) $s_1, s_2 = -1 \pm j2$

(c) $s_1, s_2 = -4 \pm j3$ (d) $s_1, s_2 = -4 \pm j2$

76. Which one of the following complex roots relate to least value of ω_d

(a) $s_1, s_2 = -2 \pm j2$ (b) $s_1, s_2 = -4 \pm j3$

(c) $s_1, s_2 = -1 \pm j4$ (d) $s_1, s_2 = -0.5 \pm j2.5$

77. ω_d and settling time are related as

(a) $\omega_d = t_s \cdot \dfrac{4\sqrt{1-\zeta^2}}{\zeta}$ (b) $\omega_d = \dfrac{\zeta}{t_s 4\sqrt{1-\zeta^2}}$

(c) $\omega_d = \dfrac{1}{t_s} \cdot \dfrac{4\sqrt{1-\zeta^2}}{\zeta}$ (d) $\omega_d = \dfrac{t_s \zeta}{4\sqrt{1-\zeta^2}}$

78. The roots of a closed-loop characteristic equation of the system shown below :

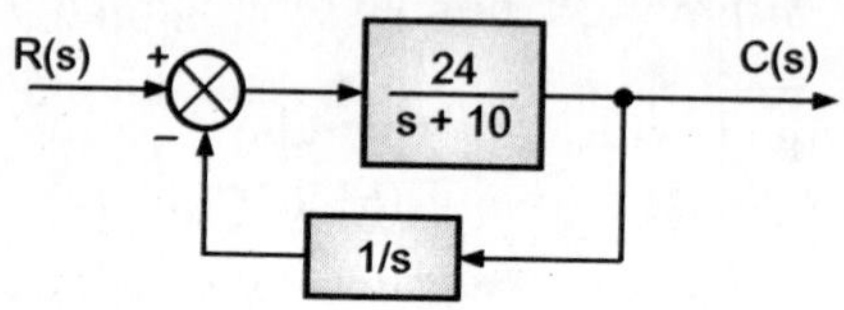

are :

(a) −4, −10 (b) −4, −6

(c) −4, +6 (d) −4, +10

79. For a unity gain open-loop pole zero configuration is shown below :

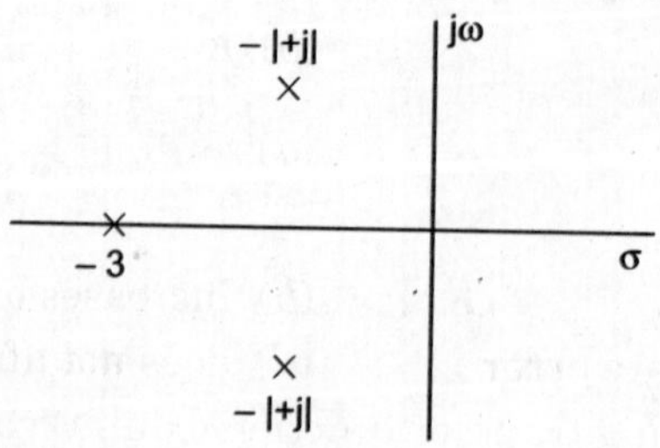

The characteristic equation is

(a) $s^3 + 5s^2 + 8s + 6 = 0$ (b) $s^3 + 4s^2 + 8s + 6 = 0$

(c) $s^3 + 5s^2 + 9s + 6 = 0$ (d) $2s^3 + 5s^2 + 9s + 6 = 0$

80. In the case of overdamped system rise time is considered as the time required to reach the unit step response

(*a*) 10% to 90% of final value (*b*) 0 to 100% of final value
(*c*) 0 to 50% of final value (*d*) 10% to 50% of final value

81. The time needed for the response to reach 50% of the final value is termed as

(*a*) rise time (*b*) nothing
(*c*) settling time (*d*) delay time

82. The unit impulse response of a system is given by $y(t) = -2e^{-t} + 3e^{-2t}$. The response to $2u(t)$ is

(*a*) $4e^{-t} + 3e^{-2t}$ (*b*) $5e^{-t} + e^{-2t}$
(*c*) $1 + 4e^{-t} - 3e^{-2t}$ (*d*) $-1 + 4e^{-t} - 3e^{-2t}$

83. A feedback control system is shown below in figure: for $\xi = 0.5$ and $\omega_n = 3$

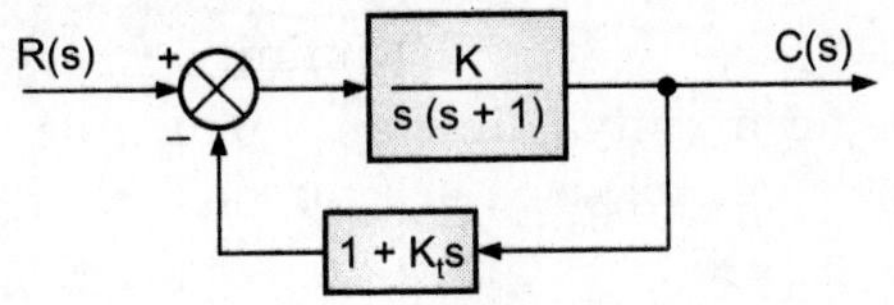

(*a*) $K = 1, K_t = 0.75$ (*b*) $K = 9, K_t = 0.88$
(*c*) $K = 10, K_t = 0.5$ (*d*) $K = 3, K_t = 0.5$

84. The response $f(t) = e^{-\xi t} \cos \omega t$; $\xi < 0$ is

(*a*) exponentially decreasing oscillatory
(*b*) exponentially increasing oscillatory
(*c*) exponentially decaying
(*d*) oscillatory steady

85. For non-oscillatory response the acceptable value of damping ratio is

(*a*) 1 (*b*) 0
(*c*) 0.5 to 0.8 (*d*) > 1

86. ω_n and peak time are related as

(*a*) $\omega_n = \dfrac{t_p\sqrt{1-\zeta^2}}{\pi}$ (*b*) $\omega_n = \dfrac{\pi}{t_p\sqrt{1-\zeta^2}}$
(*c*) $\omega_n = \dfrac{t_p}{\pi\sqrt{1-\zeta^2}}$ (*d*) $\omega_n = \dfrac{1}{\pi t_p\sqrt{1-\zeta^2}}$

87. The settling time of a second order system on 2% basis is given by

(*a*) $t_s = \dfrac{4}{\zeta\omega_n}$ (*b*) $t_s = \dfrac{\zeta\omega_n}{4}$
(*c*) $t_s = \dfrac{4\zeta}{\omega_n}$ (*d*) $t_s = 4\zeta\omega_n$

88. If the dominant complex poles of a higher order control system are nearer to the imaginary axis in *s*-plane then the nature of the time response will be

(*a*) oscillatory type (*b*) non-oscillatory type
(*c*) sustained oscillation type (*d*) steady type

89. In a critically damped system

(*a*) oscillations are prominent (*b*) no oscillations are observed

(*c*) oscillations just disappear (*d*) oscillations are sustained

90. An increase in damping ratio

(*a*) increases rise time (*b*) decreases rise time

(*c*) does not affect rise time (*d*) keeps the time within limits

91. Without affecting steady state error the maximum overshoot can be decreased by incorporating

(*a*) derivative error control (*b*) integral error control

(*c*) gain adjustment (*d*) proportional error control

92. If M_p = 100%, the damping ratio is

(*a*) 1 (*b*) 0

(*c*) 0.5 (*d*) infinity

93. The unit step response for a unity gain transfer function having pole zero configuration as given in figure is

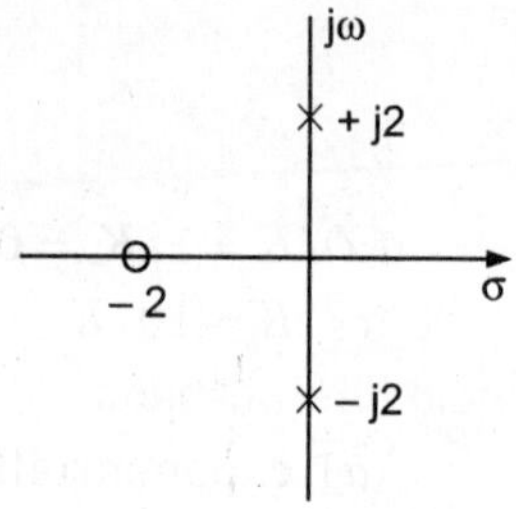

(*a*) $(\cos 2t + 2 \sin 2t)$

(*b*) $\frac{1}{2}(1 + \sin 2t - \cos 2t)$

(*c*) $(\cos 2t - \sin 2t)$

(*d*) $\frac{1}{2}(1 - \sin 2t + \cos 2t)$

94. Out of the figures given below which one represents the impulse response of a second order underdamped system

(*a*)

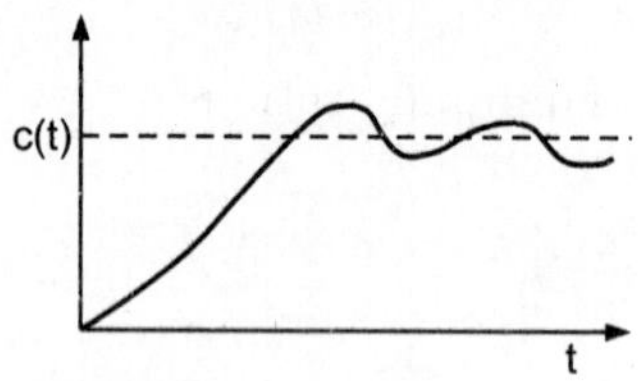

(*b*)

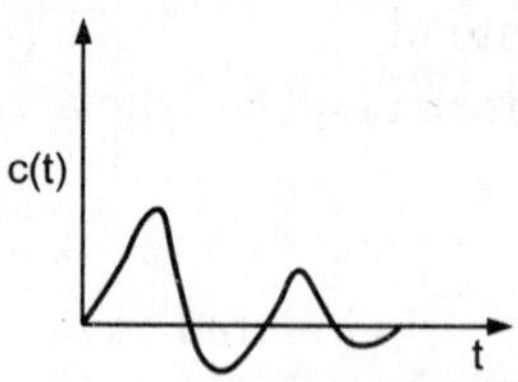

(*c*)

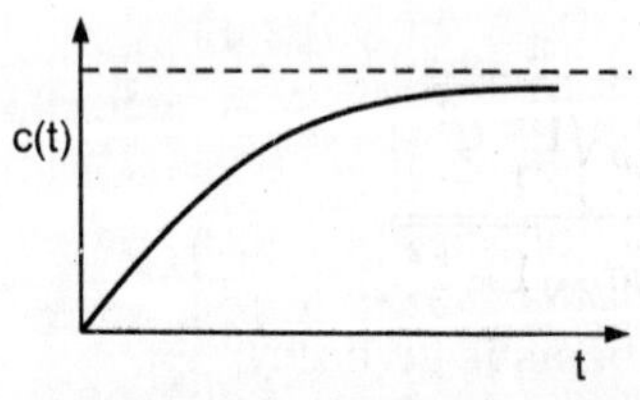

(*d*)

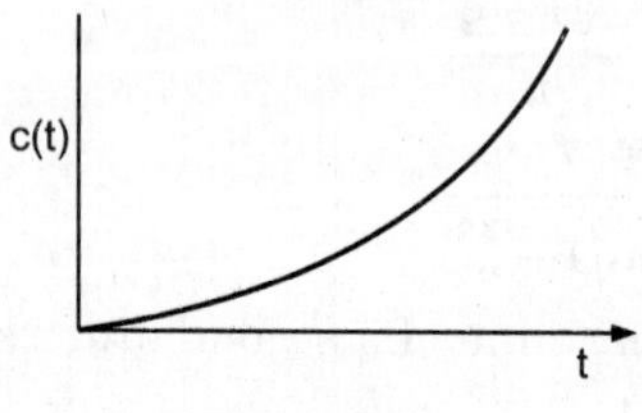

95. The stability of a system

(*a*) decreases as the type of system increases

(*b*) increases as the type of system increases

(*c*) does not change as the type of system increases

(*d*) none of the above

96. A second order characteristic equation having all the coefficients of same sign represents a system which is

(*a*) inherently stable (*b*) marginally stable

(*c*) inherently unstable (*d*) conditionally stable

97. The Routh-Hurwitz criterion gives

(*a*) relative stability (*b*) absolute stability

(*c*) gain margin (*d*) phase margin

98. The Routh array for the characteristic equation of a system is given below

$$\begin{array}{c|cc} s^3 & T_1T_2 & 1 \\ s^2 & (T_1+T_2) & K \\ s^1 & \dfrac{(T_1+T_2)-KT_1T_2}{(T_1+T_2)} & \\ s^0 & K & \end{array}$$

the system is stable if

(*a*) $K < \frac{1}{T_1} + \frac{1}{T_2}$ (*b*) $K = \frac{1}{T_1} + \frac{1}{T_2}$

(*c*) $K > \frac{1}{T_1} + \frac{1}{T_2}$ (*d*) $K = T_1 + T_2$

99. For the Routh array given below :

$$\begin{array}{c|cc} s^3 & 1 & 3 \\ s^2 & 4 & K \\ s^1 & \dfrac{(12-K)}{4} & \\ s^0 & K & \end{array}$$

the system is unstable if

(*a*) $K > 12$ (*b*) $K = 12$

(*c*) $K > 39$ (*d*) $K < 3$

100. The location of the closed-loop conjugate pair of pole on $j\omega$ axis indicates that the system is

(*a*) stable (*b*) unstable

(*c*) marginally stable (*d*) critically stable

101. A closed-loop control system has a pair of complex pole of multiplicity 2 located on the imaginary axis of the *s*-plane. This indicates

(*a*) stability (*b*) unstability

(*c*) absolute stability (*d*) relative stability

102. As the type of a open-loop transfer function increases the stability

(*a*) increases (*b*) decreases

(*c*) does not change (*d*) none of (*a*), (*b*) and (*c*)

103. A second order control system with $\xi < 0$ is always

(*a*) stable (*b*) marginally stable

(*c*) unstable (*d*) inherently stable

104. A second order control system with $\xi = 0$ is always

(*a*) marginally stable (*b*) stable

(*c*) unstable (*d*) none of these

105. The roots of the characteristic equation are same as

(*a*) closed-loop zeros (*b*) closed-loop poles

(*c*) open-loop zeros (*d*) open-loop poles

106. The loop gain $G(s)\,H(s)$ of a closed-loop system is given by $K/s\,(s + 2)\,(s + 4)$. The value of K for which the system just becomes unstable is

(*a*) $K = 6$ (*b*) $K = 4$

(*c*) $K = 24$ (*d*) $K = 48$

107. In Routh-Hurwitz table if first column element in any row is zero to be denoted as ε and the sign of coefficients of element above ε is same as below, then it is concluded that the system has

(*a*) unstability

(*b*) pair of imaginary conjugate roots

(*c*) two real roots with equal magnitude

(*d*) two real roots with opposite sign

108. Integral error control

(*a*) increase the order of the system (*b*) decreases the order of the system

(*c*) increases steady state error (*d*) does not affect the steady state error

109. For the system having open-loop transfer function

$$G(s)\,H(s) = \frac{70}{s(s+2)(s+5)}$$

the gain margin is

(*a*) 2 db (*b*) 5 db

(*c*) 7 db (*d*) 0 db

110. The number of encirclement of the point $(-1 + j0)$ by the Nyquist plot as s is varied along the path shown in figure : is given by

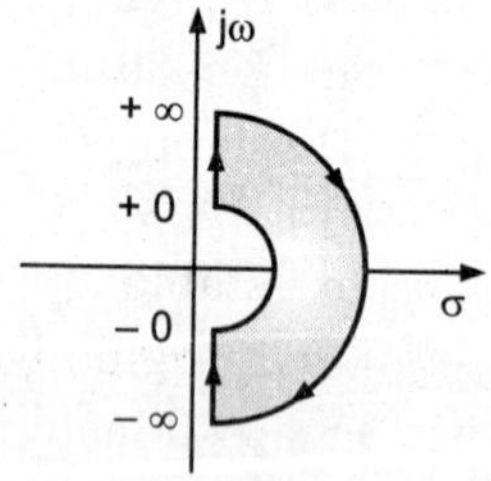

(*a*) $N = (Z - P)$

(*b*) $N = (P - Z)$

(*c*) $N = 2\,(P - Z)$

(*d*) $N = 2\,(Z - P)$

111. The number of open-loop poles having positive real part is 1 and the Nyquist plot is drawn in figure : The value of K for the closed-loop system to be stable is

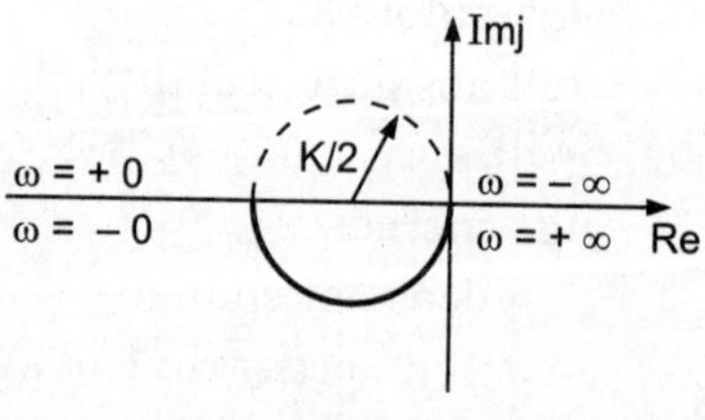

(*a*) $K > 1$ (*b*) $K < 1$

(*c*) $K < 0.5$ (*d*) $K = 0.5$

112. In $G(s)\,H(s)$-plane the Nyquist plot from $\omega = -0$ to $\omega = +0$ is closed through an angle

(*a*) $n\pi$ (*b*) $2n\pi$

(*c*) $\frac{n\pi}{2}$ (*d*) $-n\pi$

where n is the number of open-loop poles at the origin.

113. The Nyquist plot for a control system is shown in figure : the type of the system is

(a) 0 (b) 1

(c) 2 (d) 3

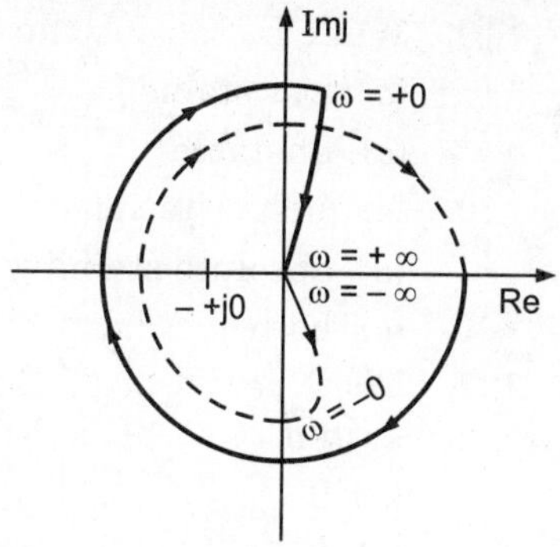

114. In problem no. 100, if the number of poles having positive real parts is nil, then the number is encirclements of the critical point is

(a) 2 (b) – 2

(c) 2 (d) – 3

115. The origin for the investigation of closed-loop stability in relation to Nyquist criterion is

(a) $-1+j0$ (b) $+1+j0$

(c) E0^ + ^j1 (d) $0-j1$

116. The Nyquist plot for a system is having the number of open-loop positive poles as 1 is shown below :

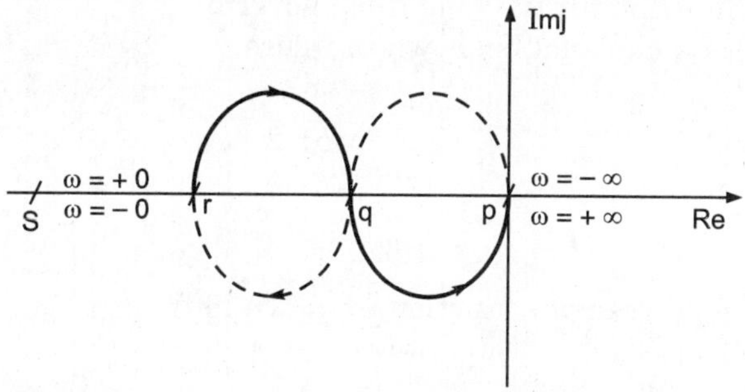

For closed-loop stability the critical point should be located

(a) between p and q (b) between q and r

(c) between r and s (d) beyond s

117. The number of poles at the origin s-plane in relation to Nyquist plot shown below :

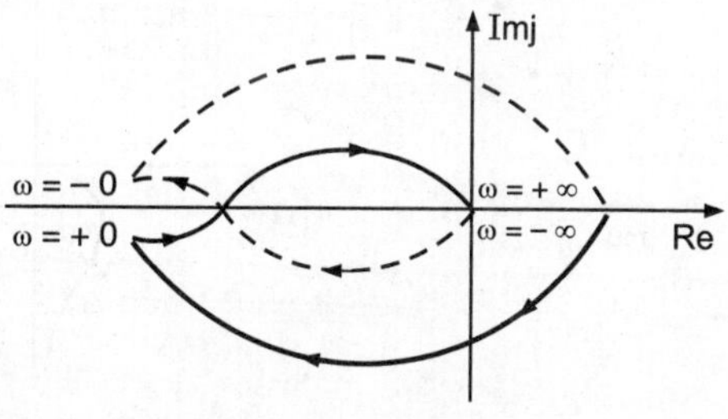

is

(a) 0 (b) 1

(c) 2 (d) 4

118. The open-loop transfer function of a unity feedback control system is given as,

$G(s) = \dfrac{1000}{s(s+10)^2}$ the phase crossover frequency ω_2 is

(a) 1000 rad/sec (b) 500 rad/sec

(c) 10 rad/sec (d) 5 rad/sec

119. While increasing the value of gain factor K the system becomes :

(*a*) less stable (*b*) more stable

(*c*) unstable (*d*) absolutely stable

120. Nyquist criterion enables to determine

(*a*) absolute stability (*b*) relative stability

(*c*) both (*d*) either (*a*) or (*b*)

121. The Nyquist plot is shown in figure, what will be the type of the system

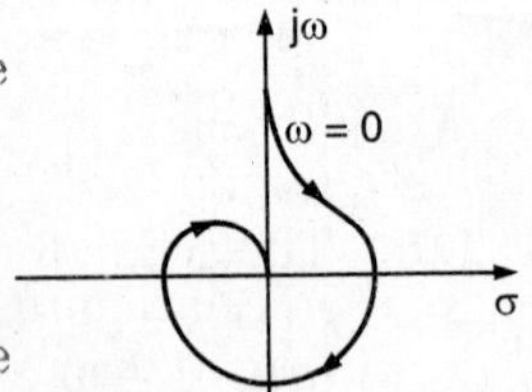

(*a*) 0 (*b*) 1

(*c*) 2 (*d*) 3

122. At gain factor $K = 0$, the roots of the characteristic equation are placed at

(*a*) open-loop poles (*b*) open-loop zeros

(*c*) both (*a*) and (*b*) (*d*) none of these position

123. The order of the differential equation governing a system having its polar plot is shown in figure :

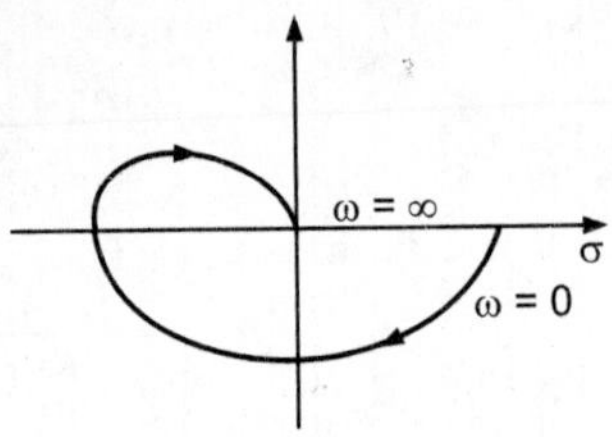

is

(*a*) 3 (*b*) 2

(*c*) 1 (*d*) 0

124. The polar plot for a transfer function is shown in figure :

An addition of a zero will change the plot as

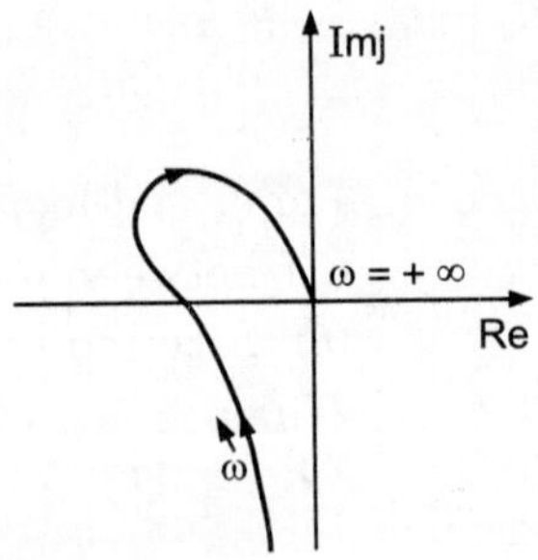

(*a*)

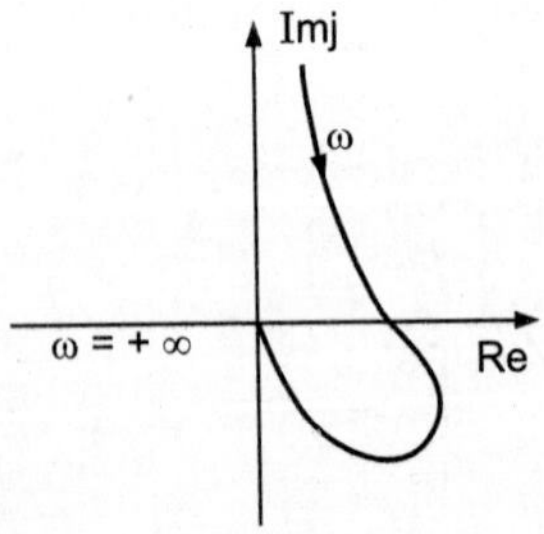

(*b*)

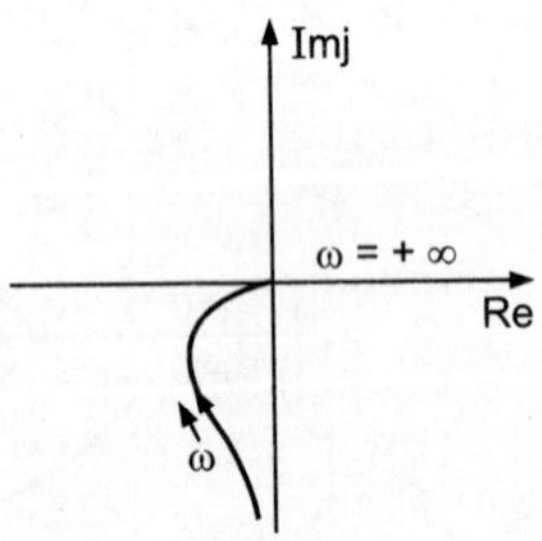

(*c*)

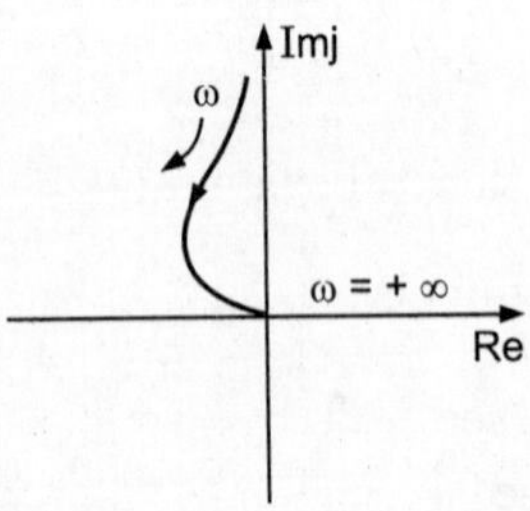

(*d*)

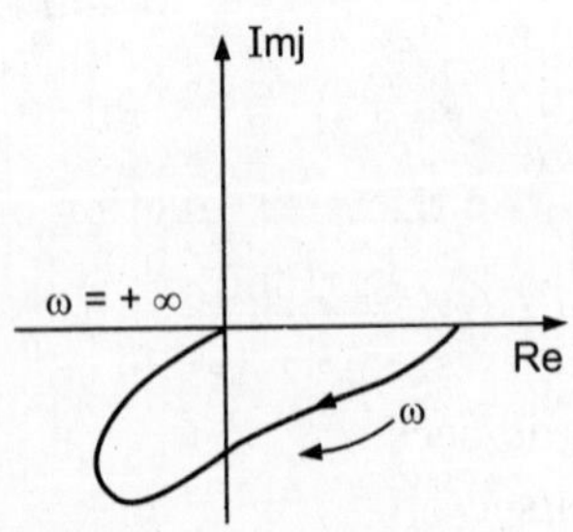

125. The gain margin for a system whose Nyquist plot drawn below :

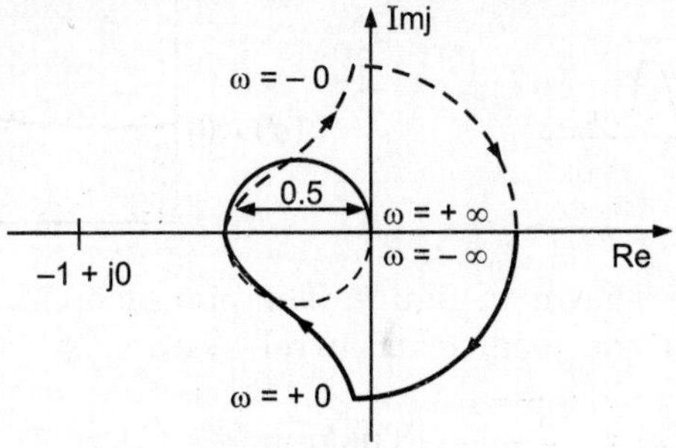

is

(*a*) 0.5 db (*b*) – 1.5 db

(*c*) 5 db (*d*) 6 db

126. The phase margin in relation to following polar plot is

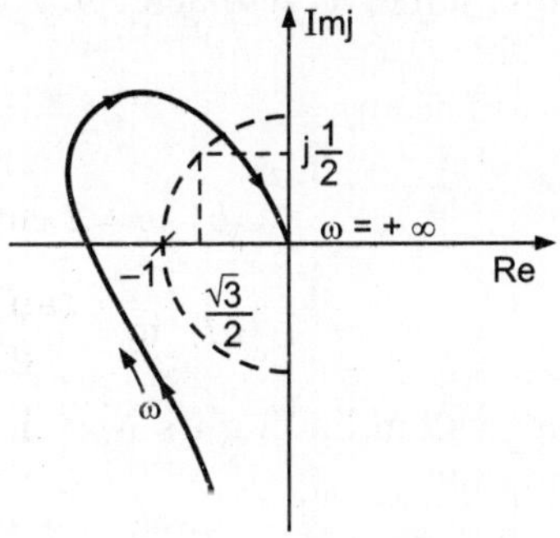

(*a*) 30° (*b*) – 30°

(*c*) 60° (*d*) – 60°

127. The nature of time response corresponding to polar plot shown below will be :

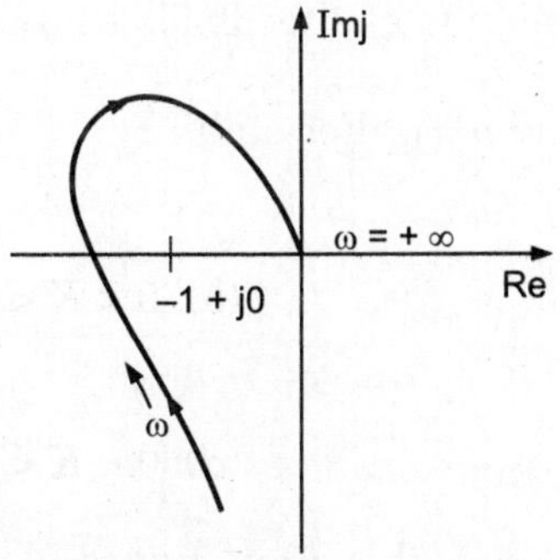

(*a*) 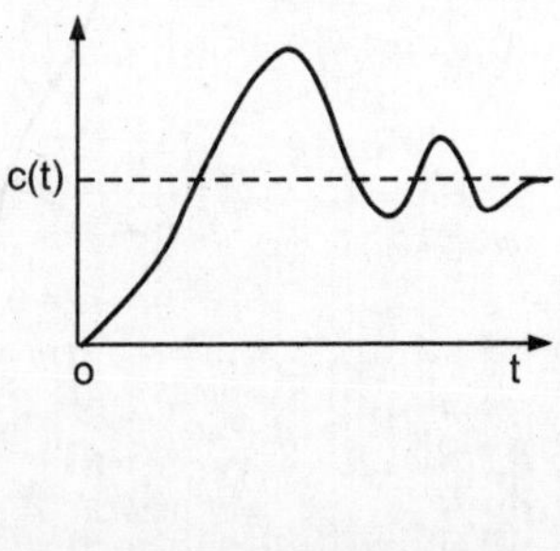

(*b*) 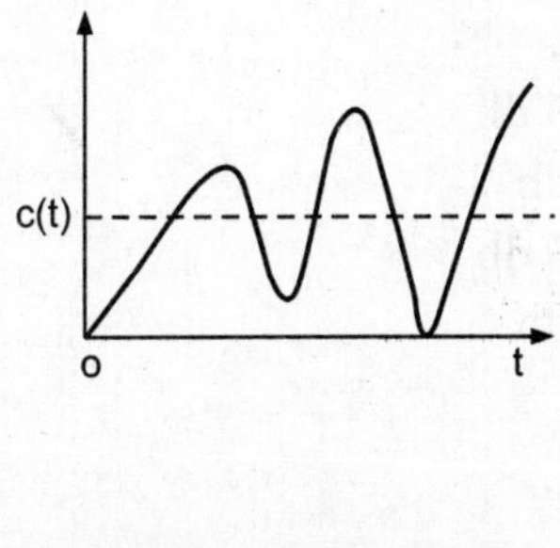

(c) 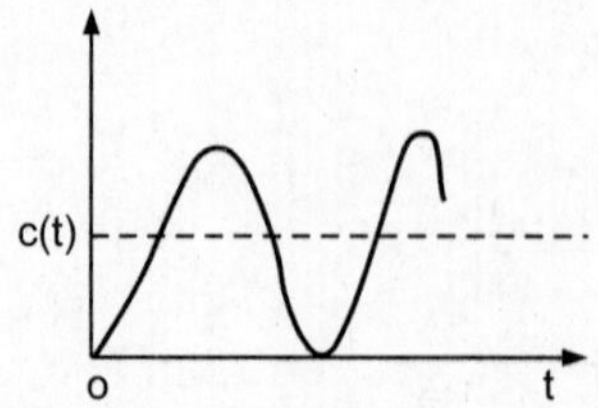

(d) 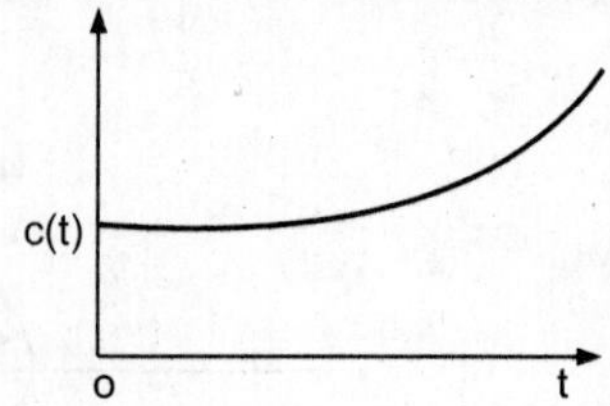

128. The Nyquist plot is shown in figure. The open-loop transfer function of the feedback control system is given below, $G(s)H(s) = \dfrac{s+2}{s^2(s-4)}$. The number of + ve real part roots of the characteristic equation is

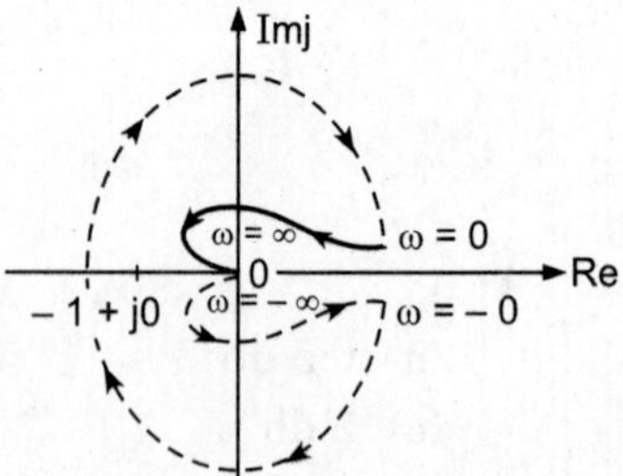

(a) nil (b) 3
(c) 1 (d) 2

129. The open-loop transfer function of a feedback control system is given below, $G(s)H(s) = \dfrac{e^{-Ts}}{s(s+2)}$. $\angle G(j\omega)H(j\omega) = 0$ occurs at

(a) $\omega = 2 \tan \omega T$ (b) $\omega = 2 \cot \omega T$

(c) $\omega = \dfrac{\cot \omega}{2T}$ (d) $\omega = \dfrac{\tan \omega T}{2T}$

130. The order and type of the system having its Nyquist plot shown in figure are

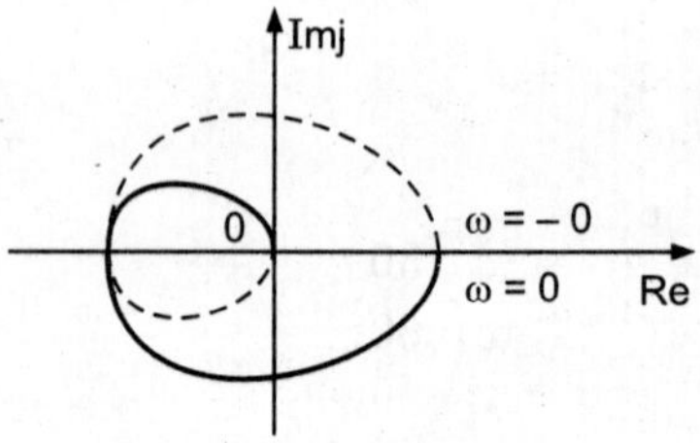

(a) 3, 2
(b) 3, 0
(c) 2, 1
(d) 2, 0

131. The open-loop transfer function of a feedback control system is given below, $G(s)H(s) = \dfrac{K}{(s+p)(s+q)}$; $\begin{matrix} p > 0 \\ q > 0 \end{matrix}$ and $p \neq q$. For stability

(a) $0 < K < \dfrac{pq}{p+q}$ (b) $0 < K < pq(p+q)$

(c) $0 < K < \dfrac{p}{q}(p+q)$ (d) $0 < K < \dfrac{p+q}{pq}$

132. With reference to polar plot shown in figure. The gain margin is

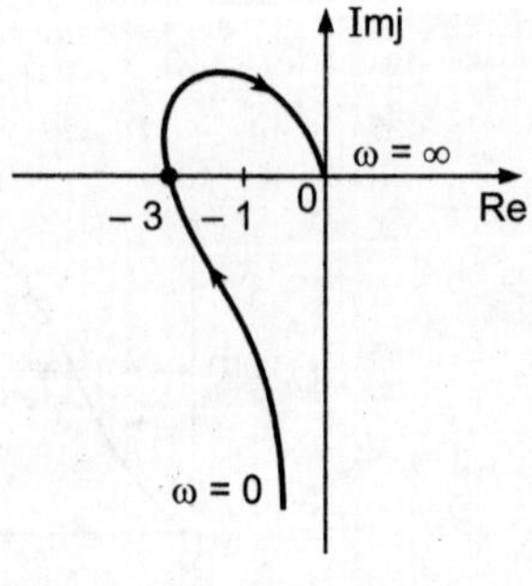

(a) – 9.55 db
(b) 9.55 db
(c) – 5.95 db
(d) 5.95 db

133. The part of Nyquist plot for the open-loop transfer function of a feedback control system is shown in figure. The gain and phase margins are,

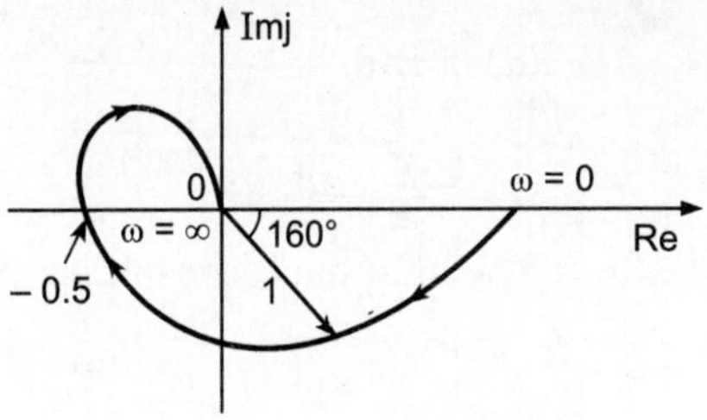

(*a*) 6 db, 120°
(*b*) 2 db, 60°
(*c*) 6 db, 30°
(*d*) 6 db, 90°

134. The shape of the Nyquist plot is shown in figure. It is known that the number of open-loop poles having + ve real part is two. For the system to be stable, the range of positive values of *K* being

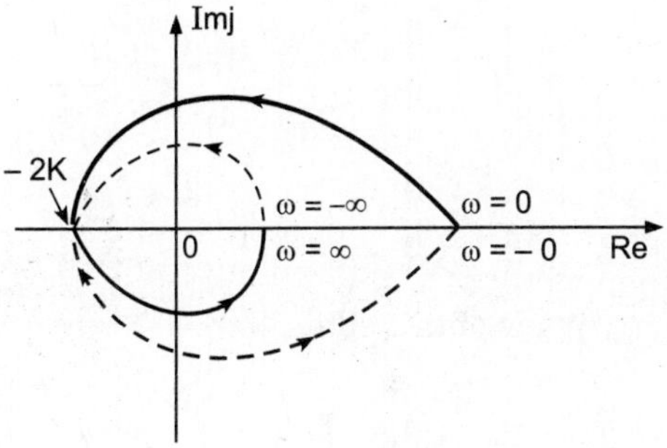

(*a*) $K > 0.5$
(*b*) $K < 0.5$
(*c*) $K = 5$
(*d*) $K > 2$

135. A part of the Nyquist plot is shown in figure, the corresponding transfer function is

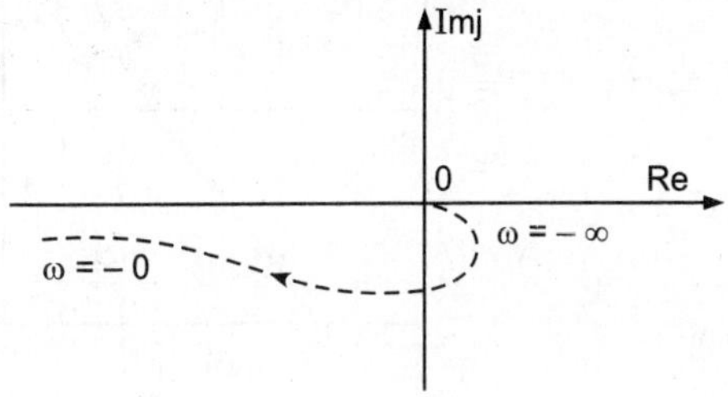

(*a*) $\dfrac{K}{s^2(sT_1+1)}$

(*b*) $\dfrac{K}{s(sT_1+1)}$

(*c*) $\dfrac{K}{s(sT_1+1)(sT_2+1)}$

(*d*) $\dfrac{K}{s^3(sT_1+1)}$

136. Consider the Nyquist plot shown in figure, which one of the following transfer functions represent this plot

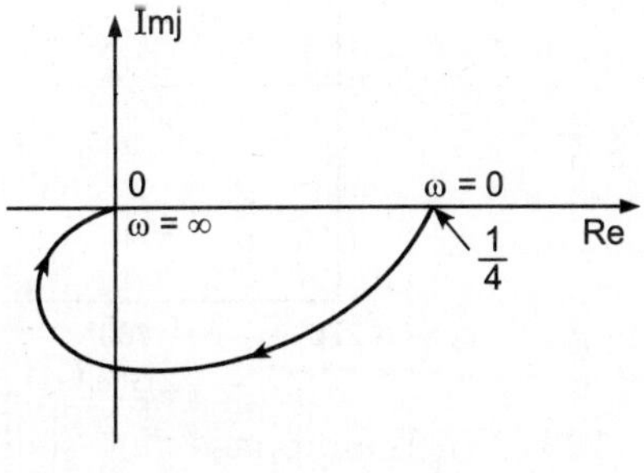

(*a*) $\dfrac{1}{(s+2)^2}$

(*b*) $\dfrac{1}{s^2+3s+2}$

(*c*) $\dfrac{1}{(s+2)^3}$

(*d*) $\dfrac{1}{(s+2)}$

137. A unit feedback control system has the following open-loop transfer function

$G(s) = \dfrac{1}{(s+2)^2}$. The phase margin is

(*a*) π rad. (*b*) $\frac{\pi}{3}$ rad.

(*c*) $-\frac{\pi}{2}$ rad. (*d*) $\frac{\pi}{2}$ rad.

138. The shape of gain-phase plot corresponding to Nyquist plot given below :

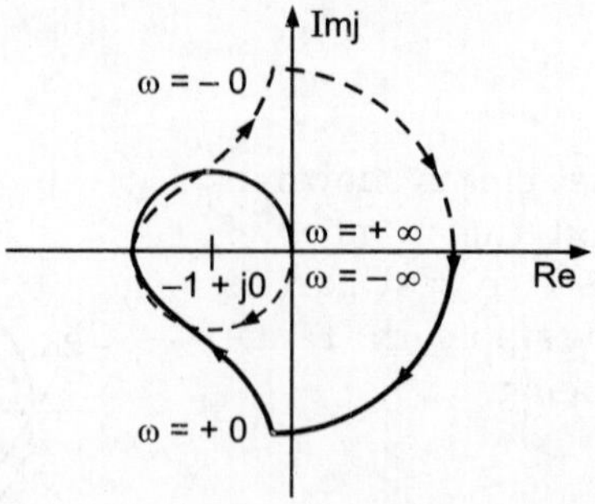

is obtained as

(*a*)

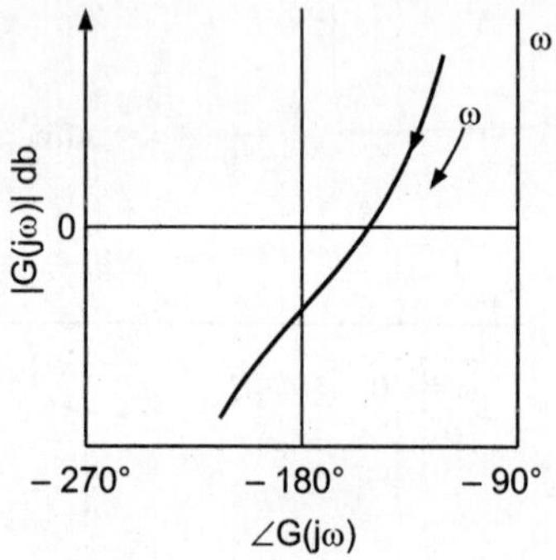

(*b*)

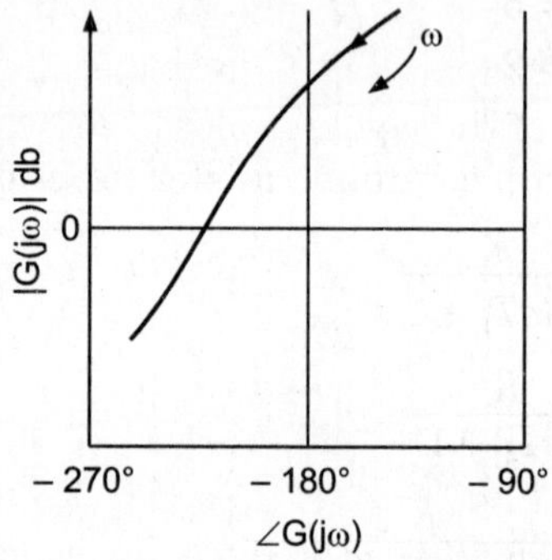

(*c*)

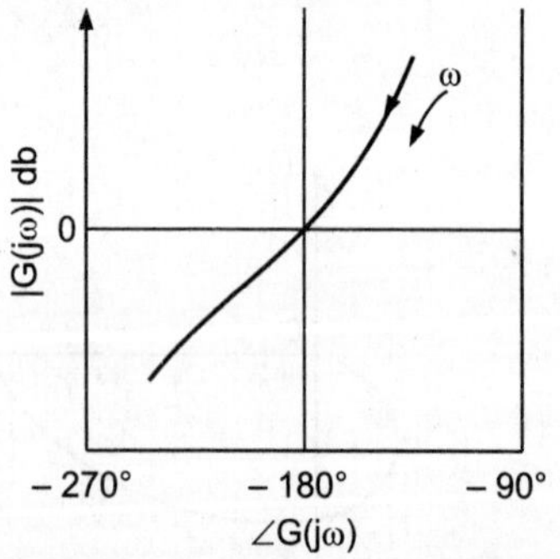

(*d*)

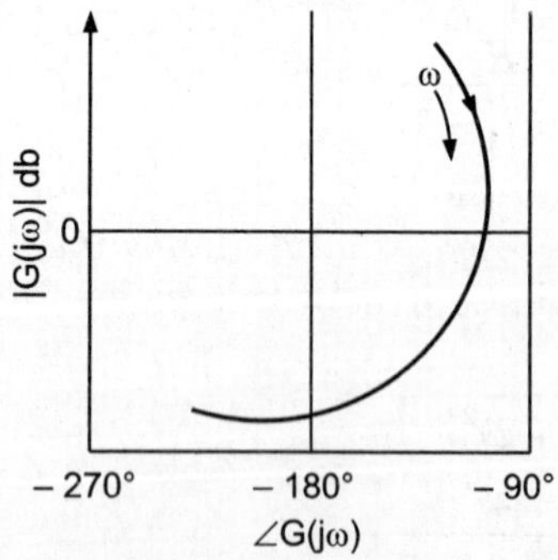

139. The gain phase plot is superimposed on Nichol's chart to obtain

(*a*) open-loop frequency response (*b*) closed-loop frequency response

(*c*) time response (*d*) settling time

140. The table gain below gives the open-loop frequency response

ω	4	5	6	8	10	12.5
$\lvert G(j\omega) \rvert$	3.2	2.3	1.7	1.0	0.5	0.65
$\angle G(j\omega)$	− 145°	− 153°	− 158°	− 168°	− 180°	− 200°

(*a*) 12°, 6 db (*b*) – 12°, – 6 db

(*c*) 168°, 0 db (*d*) 10°, 6 db

141. The gain-phase plot is shown in figure, the gain margin, phase margin, gain crossover frequency, phase crossover frequency are

(*a*) – 4 db, – 30°, 100 rad/sec, 10 rad/sec

(*b*) 4 db, – 30°, 10, rad/sec, 100 rad/sec

(*c*) – 2 db, – 30°, $\sqrt{10}$ rad/sec, 10 rad/sec

(*d*) 2 db, – 30°, 10 rad/sec, $\sqrt{10}$ rad/sec

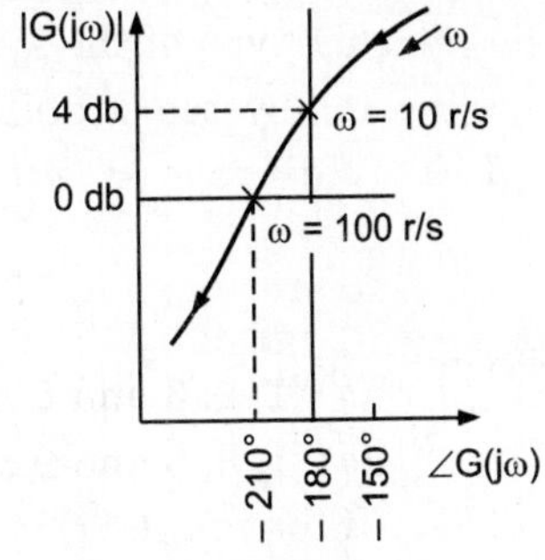

142. As observed from the polar plot is shown in figure

(*a*) G.M. is + ve, P.M. is + ve

(*b*) G.M. is – ve, P.M. is – ve

(*c*) G.M. is – ve, P.M. is + ve

(*d*) G.M. is + ve, P.M. is – ve

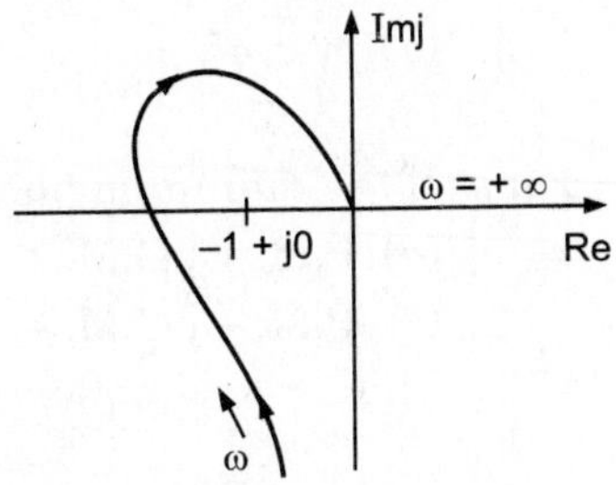

143. The initial slope of the Bode plot for a transfer function having no poles at the origin is

(*a*) – 10 db/decade (*b*) 0 db/decade

(*c*) 10 db/decade (*d*) 24 db/decade

144. The initial slope of the Bode plot for a type 2 system intersects 0 db axis at

(*a*) $\omega = 0$ (*b*) $\omega = K$

(*c*) $\omega = \sqrt{K}$ (*d*) $\omega = K^2$

145. The error at the corner frequency due to the term $(1 + j\omega T)^{\pm N}$ is

(*a*) ± 5 N db (*b*) ± 3 db

(*c*) ± 6 db (*d*) ± 3 db

146. For a stable system

(*a*) G.M. and P.M. both are positive

(*b*) G.M. and P.M. both are negative

(*c*) G.M. is positive and P.M. is negative

(*d*) G.M. is negative and P.M. is positive

147. The gain for constructing the Bode plot in relation to transfer function

$$G(s)\,H(s) = \frac{20}{s(s+2)} \text{ is}$$

(*a*) 20 (*b*) 10

(*c*) 2 (*d*) 40

148. The gain for drawing the Bode plot for the transfer function

$$G(s) = \frac{320(s+2)}{s(s+1)(s^2+8s+64)}$$

(*a*) 10 (*b*) 5

(*c*) 16 (*d*) 160

149. The initial slope of the Bode plot gives an indication of

(*a*) type of the system (*b*) nature of the system time response

(*c*) system stability (*d*) gain margin

150. With reference to Bode plot the corner frequencies for the transfer function

$$G(s)\,H(s) = \frac{20(1+0.5s)}{s(1+0.2s)(1+s)(s^2+2s+9)}$$

(*a*) 1, 2, 3 and 5 rad/sec (*b*) 1, 2, 3 and 4 rad/sec

(*c*) 1, 3, 5 and 9 rad/sec (*d*) 1, 2, 5 and 9 rad/sec

151. If the input $r(t) = X \sin \omega t$ is applied to a system whose phase angle is $\pm 45°$, the output will have a phase shift of

(*a*) $\omega t \pm 45°$ (*b*) $\omega t \mp 45°$

(*c*) $-45°$ (*d*) $+45°$

152. If the minimum phase system is to be stable

(*a*) P.M. + ; G.M. + (*b*) P.M. – ; G.M. –

(*c*) P.M. + ; G.M. – (*d*) P.M. – ; G.M. +

153. The corner frequencies for the open-loop transfer function

$$G(j\omega)\,H(j\omega) = \frac{10(j\omega+4)}{(j\omega)^2 + j2\omega + 3}$$

are

(*a*) 2 rad/sec and 3 rad/sec (*b*) 4 rad/sec and $\sqrt{3}$ rad/sec

(*c*) $\sqrt{2}$ rad/sec and $\sqrt{3}$ rad/sec, (*d*) 0.25 rad/sec and 0.866 rad/sec

154. The term $j\omega$ on log-magnitude plot has a slope of

(*a*) – 20 db/decade (*b*) – 6 db/decade

(*c*) 6 b/octave (*d*) 20 db/octave

155. The slope of asymptotic Bode plot at low frequencies for a type 3 transfer function is

(*a*) – 30 db/decade (*b*) – 60 db/decade

(*c*) – 20 db/decade (*d*) – 40 db/decade

156. The slope of the Bode plot for the region $4 < \omega < 25$ rad/sec for the open-loop transfer function $G(s)\,H(s) = \dfrac{100}{s(s+4)(s+25)}$, is

(*a*) 12 db/octave (*b*) 6 db/octave

(*c*) – 6 db/octave (*d*) – 12 db/octave

157. The asymptotic magnitude plot for the open-loop transfer function is shown in figure. If the input is unit step, then

(*a*) $e_{ss} = \dfrac{1}{101}$

(*b*) $e_{ss} = \dfrac{1}{1001}$

(*c*) $e_{ss} = \dfrac{1}{110}$

(*d*) $e_{ss} = \dfrac{1}{1010}$

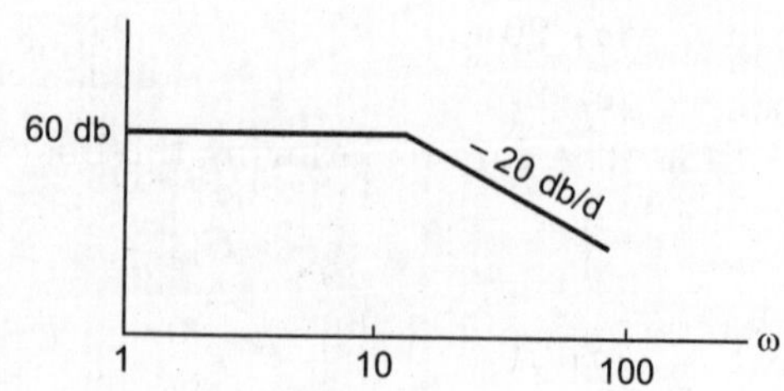

158. The overall transfer function of a control system is given below

$\dfrac{C(s)}{R(s)} = \dfrac{25}{s^2 + 5\sqrt{2}s + 25}$. The resonant peak M_r is

(*a*) 0.5 (*b*) $\sqrt{2}$

(*c*) 2 (*d*) 1.0

159. The open-loop transfer function is $G(s)\,H(s) = \dfrac{50(s+0.25)}{s^2(s+1)(s+0.5)}$ at $\omega = 0.25$ rad/sec the magnitude (db) versus frequency (log) slope changes by

(*a*) + 20 db/decade (*b*) + 40 db/decade

(*c*) – 40 db/decade (*d*) – 20 db/ decade

160. The magnitude (db) versus $\log_{10}\omega$ asymptotic plot is shown in figure. The transfer function is

(*a*) $\dfrac{2s(0.08s+1)}{(0.4s+1)}$ (*b*) $\dfrac{s(0.08s+1)}{(0.4s+1)}$

(*c*) $\dfrac{4(0.4s+1)}{s(0.08s+1)}$ (*d*) $\dfrac{5(0.4s+1)}{s(0.08s+1)}$.

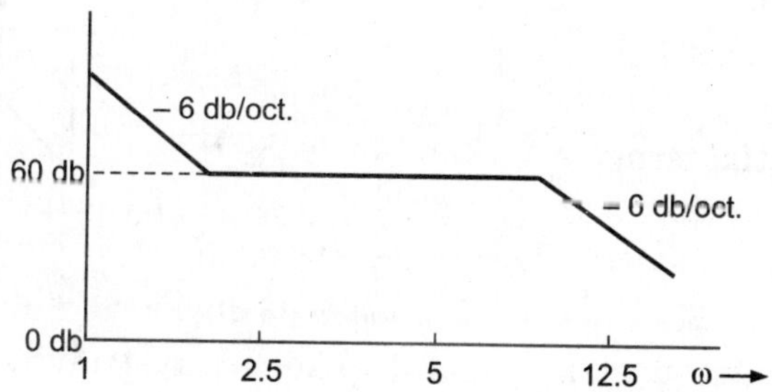

161. The phase crossover frequency for the open-loop transfer function

$$G(s)\,H(s) = \frac{20}{(s+1)(s+\frac{1}{2})(s+\frac{1}{3})}, \text{ is}$$

(*a*) 1.0 rad/sec (*b*) 2.0 rad/sec

(*c*) $\sqrt{2}$ rad/sec (*d*) $\sqrt{3}$ rad/sec

162. The Bode plot of a feedback control system has a constant slope of – 20 db/decade for all frequencies and a fixed phase angle – 90°. The phase margin is

(*a*) 90° (*b*) 0°

(*c*) – 90° (*d*) 180°

163. The Bode plot shown in figure is for

(*a*) pure integration

(*b*) pure differentiation

(*c*) constant term

(*d*) integro-differential term

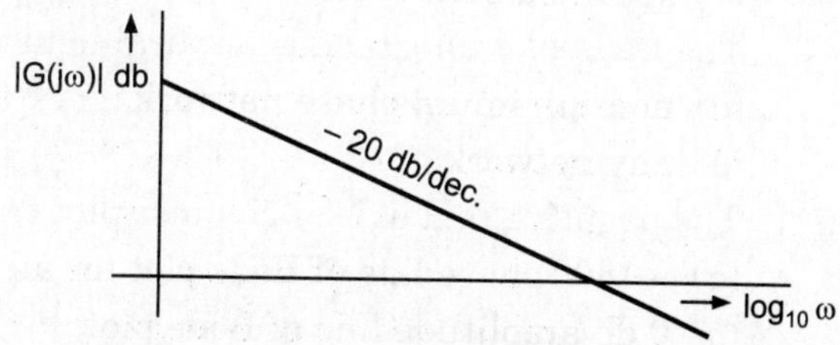

164. For the transfer, $G(s)\,H(s) = \dfrac{K}{\left(\dfrac{1}{b}s+1\right)}$. Asymptotic Bode plot is shown below :

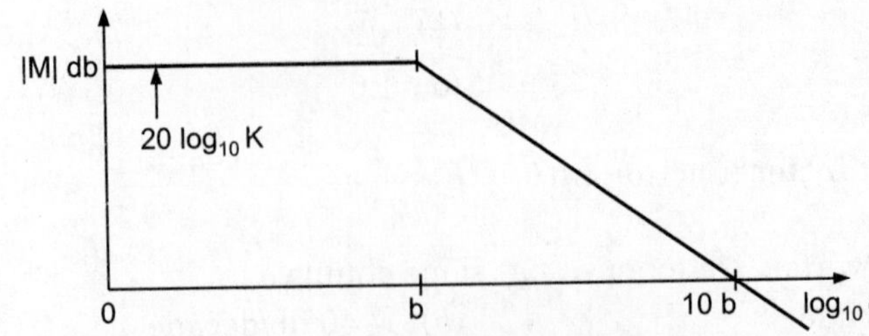

the error in the diagram at $\omega = 5b$ is determined as

(*a*) 0.16 db (*b*) 0.52 db

(*c*) 0.17 db (*d*) 0.2 db

165. At gain crossover frequency $\omega = 150$ rad./sec. $G(j\omega)\,H(j\omega) = -200°$, the phase margin is

(*a*) 30° (*b*) – 40°

(*c*) 20° (*d*) – 20°

166. The Bode plot in given figure is for

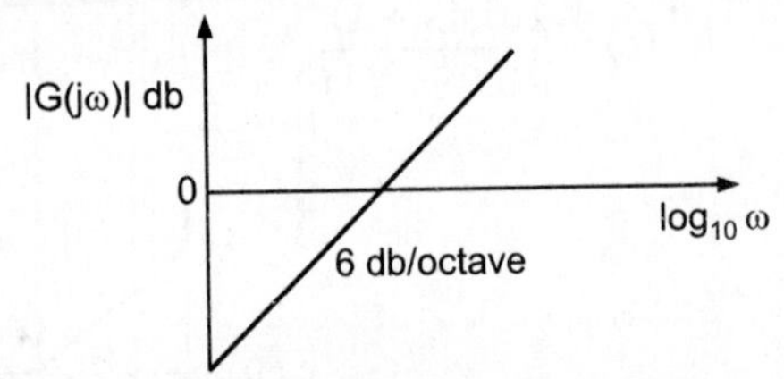

(*a*) pure integration

(*b*) pure differentiation

(*c*) constant term

(*d*) integrodifferential term

167. $|G(j\omega_2)\,H(j\omega_2)| = 0.125$ where ω_2 is phase crossover frequency the gain margin is

(*a*) 12.04 db (*b*) 18.06 db

(*c*) 6.02 db (*d*) 9.03 db

168. The open-loop transfer function of a unity feedback system is given below :

$$G(js) = \frac{sK}{(0.5s+1)}$$

The initial slope of Bode plot intersects 0 db axis at

(*a*) $\omega = K$ (*b*) $\omega = \dfrac{1}{K}$

(*c*) $\omega = \sqrt{K}$ (*d*) $\omega = 1/\sqrt{K}$

169. The unit circle of the Nyquist plot transforms into 0 db amplitude line of the Bode plot for

(*a*) low frequencies (*b*) all frequencies

(*c*) specified bandwidth (*d*) high frequencies

170. The Bode plot approach is applicable to

(*a*) non-minimum phase network (*b*) minimum phase network

(*c*) any network (*d*) none of the above

171. The negative real axis of Nyquist plot transforms into

(*a*) – 180° phase line of Bode plot for all frequencies

(*b*) 0 db amplitude line of Bode plot

(*c*) 0 db amplitude line of Bode plot at low frequencies
(*d*) 0 db amplitude line of Bode plot at high frequencies

172. For a type "1" system the intersection of the initial slope of the Bode plot with 0 db axis gives
(*a*) steady state error (*b*) error constant
(*c*) phase margin (*d*) crossover frequency

173. The frequency at which the magnitude of the Bode plot crosses 0 db axis is termed as
(*a*) natural frequency (*b*) phase crossover frequency
(*c*) gain crossover frequency (*d*) corner frequency

174. The frequency at which the phase curve of a Bode plot crosses – 180° line is called
(*a*) natural frequency (*b*) phase crossover frequency
(*c*) gain crossover frequency (*d*) corner frequency

175. For a stable system
(*a*) the gain crossover occurs earlier than phase crossover
(*b*) the phase crossover occurs earlier than gain crossover
(*c*) the gain crossover and phase crossover frequencies are very near to each other
(*d*) the gain crossover and phase crossover frequencies are coincident

176. The phase angle at $\omega = \infty$ for the transfer function

$$G(j\omega) = \frac{90(1 + j0.5\omega)}{j\omega(1 + j\omega)(1 + j2\omega)}$$

is
(*a*) 90° (*b*) – 90°
(*c*) 180° (*d*) – 180°

177. The phase angle for the transfer function, $G(j\omega) = \dfrac{1}{(1 + j\omega T)^2}$ at corner frequency is
(*a*) 45° (*b*) – 45°
(*c*) – 90° (*d*) 90°

178. The Bode plot given below

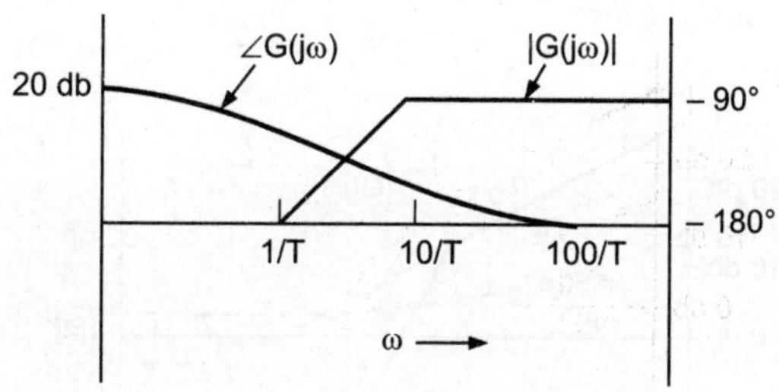

is for a
(*a*) non-minimum phase transfer function
(*b*) minimum phase transfer function
(*c*) all pass transfer function
(*d*) type "2" transfer function

179. The magnitude vs frequency plot is shown below :

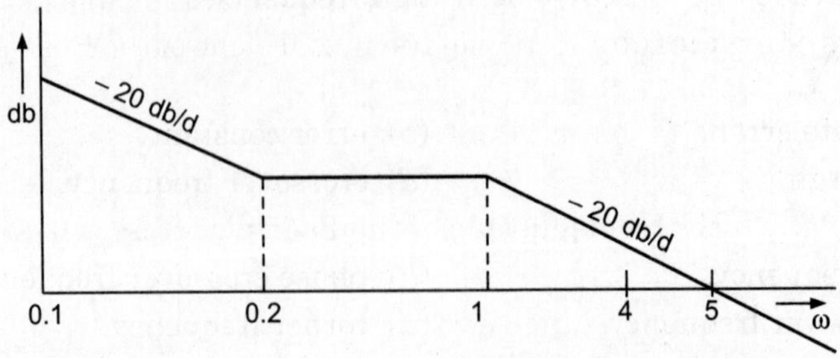

the time constant are

(*a*) 0.2, 1 and 5 sec (*b*) 0.1, 0.2 and 5 sec

(*c*) 0.2, 0.25 and 1 sec (*d*) 0.4, 0.5 and 0.8 sec

180. The transfer function in relation to Bode plot given below

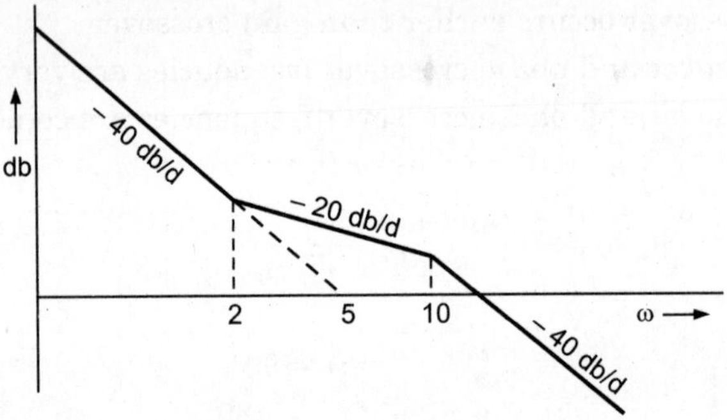

is

(*a*) $\dfrac{25(1+0.5s)}{s^2(1+0.1s)}$ (*b*) $\dfrac{5(1+0.5s)}{s^2(1+0.1s)}$

(*c*) $\dfrac{25(1+0.s)}{s(1+0.1s)}$ (*d*) $\dfrac{25(1+2s)}{s^2(1+10s)}$

181. For the Bode plot shown below

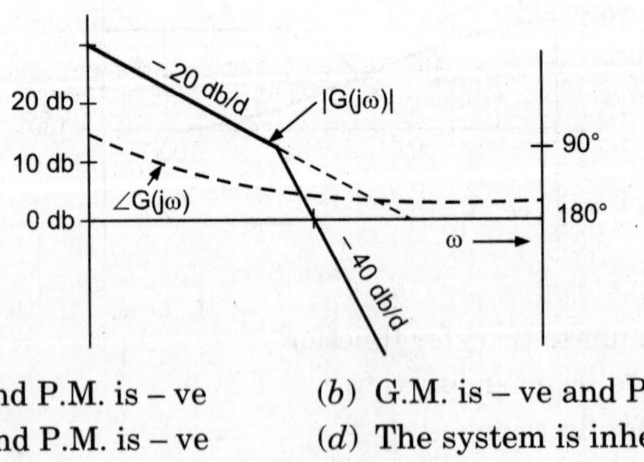

(*a*) G.M. is – ve and P.M. is – ve (*b*) G.M. is – ve and P.M. is + ve

(*c*) G.M. is + ve and P.M. is – ve (*d*) The system is inherently stable

182. If the horizontal distance of logarithmic scale between $\omega = 1$ to $\omega = 10$ is 1 unit, the distance between $\omega = 1$ to $\omega = 20$ is

(a) 2 unit (b) 1.301 unit
(c) 0.602 unit (d) 1.2 unit

183. For the gain term K, the log-magnitude (db) sketch in Bode plot is

(a) zero db
(b) negative
(c) negative or positive depending on the value of K
(d) positive

184. With reference to frequency response resonance peak occurs only if the value of damping ratio

(a) equals to 1 (b) 0.707
(c) 0.707 (d) equals to 1.41

185. The Bode plot is obtained using

(a) characteristic equation (b) open-loop transfer function
(c) closed-loop transfer function (d) overall transfer function

186. As K approaches zero the root loci approach towards

(a) open-loop poles (b) open-loop zeros
(c) breakaway points (d) break in points

187. The break away point is determined by solving

(a) $\frac{dK}{ds} = 0$ (b) characteristic equation
(c) $\int \frac{dK}{ds} = 0$ (d) $G(s)\,H(s) = 0$

188. As K approaches infinity the root loci are asymptotic to lines which make an angle with real axis given by

(a) $\frac{2k\pi}{P - Z}$ (b) $\frac{(2k + 1)\pi}{P - Z}$
(c) $\frac{(k + 1)\pi}{P - Z}$ (d) $\frac{(2k + 1)}{P - Z} \cdot \frac{\pi}{2}$

189. With reference to following characteristic equation of a feedback control system. The centroid of the root locus plot

$$s^3 + 2s^2 + Ks + K = 0$$

(a) 0.5 (b) – 0.5
(c) – 1 (d) 1

190. The intersection of the asymptotes of root locus w.r.t. following open-loop transfer function $G(s)\,H(s) = \frac{K(s + 1)}{s(s - 1)(s^2 + 5s + 6)}$ is given by

(a) $s = + 1$ (b) $s = - 1$
(c) $s = - 5$ (d) $s = - 6$

191. For $G(s)\,H(s) = \dfrac{1}{s^2}$ the root locus plot will be

(a) 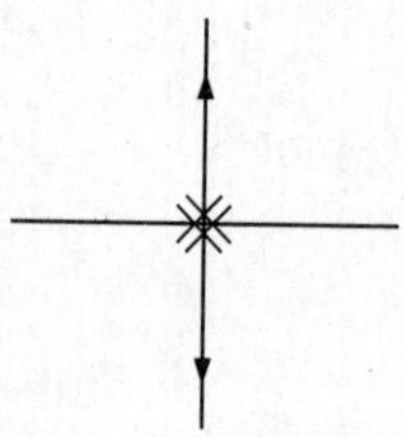

(b) 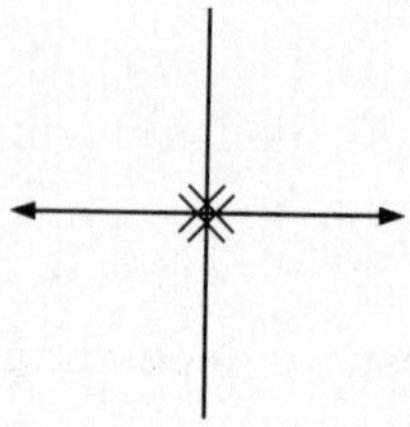

(c) 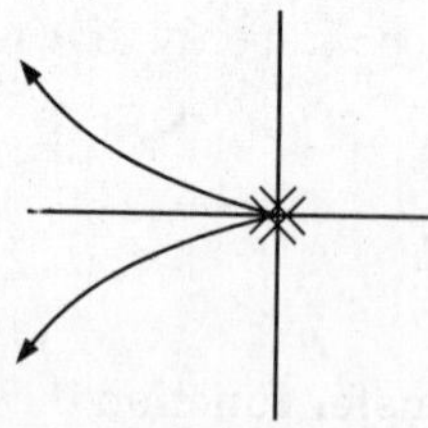

(d) 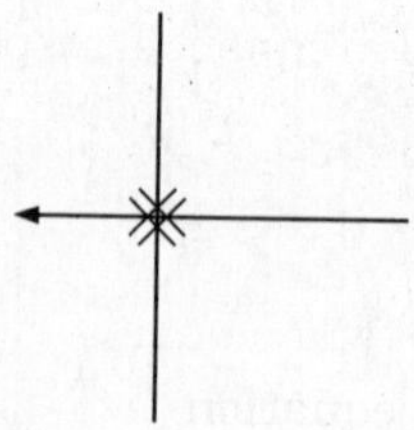

192. Which of the following RL plot indicates a stable system for all values of K.

(a) 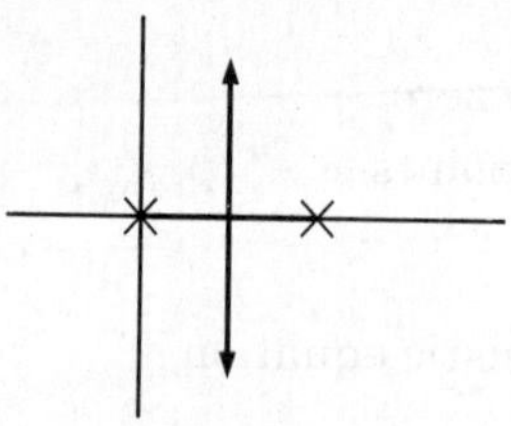

(b) 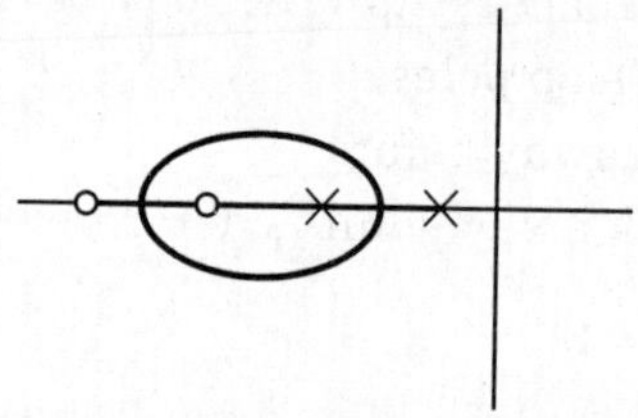

(c) 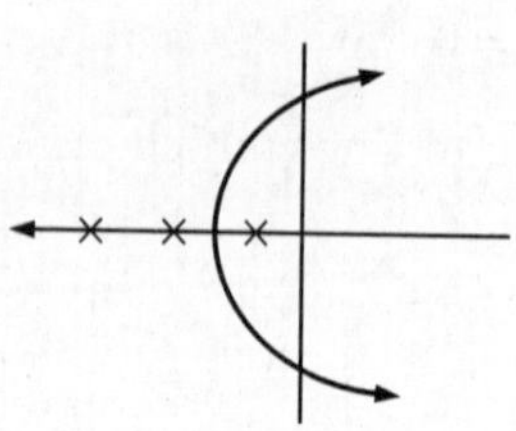

(d) 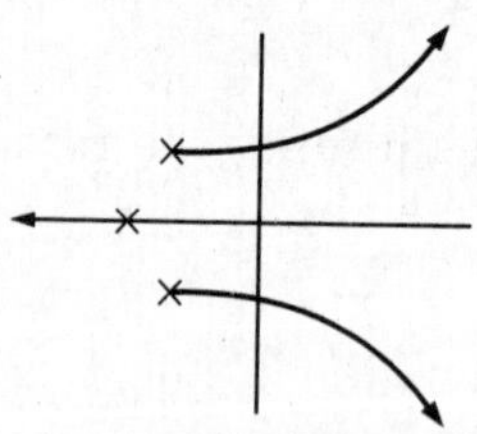

193. The intersection point of RL branches with s-phase imaginary axis is determined using

(a) Bode plot

(b) Routh-Hurwitz array

(c) gain-phase plot

(d) $\dfrac{dK}{ds} = 0$

194. The root locus plot of a feedback control system is shown below :
The open-loop transfer function is given by

(a) $\dfrac{K(s+3)}{s(s+1)}$

(b) $\dfrac{K(s-3)}{s(s+1)}$

(c) $\dfrac{K(s+1)}{s(s+3)}$

(d) $\dfrac{K(s-1)}{s(s+3)}$

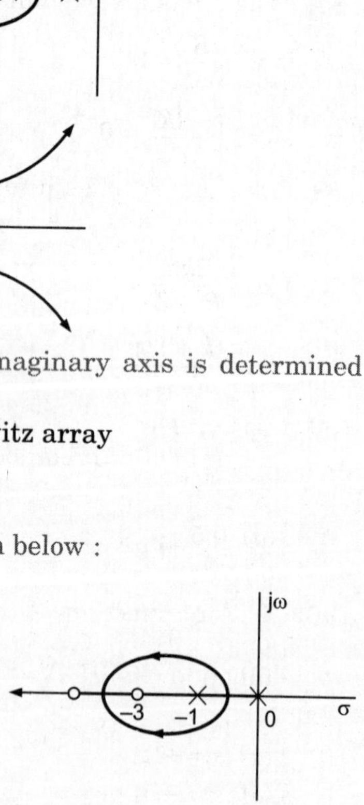

195. The characteristic equation of a UFB control system is given by

$$s^3 + 3s^2 + 2s + Ks + K = 0$$

The center of gravity of RL asymptotes is located at

(*a*) $s = -2$ (*b*) $s = 1$

(*c*) $s = -3 + j0$ (*d*) $s = 3 + j0$

196. The root locus branches for the open-loop transfer function

$$G(s) = \frac{K}{s(s+1)(s+2)}$$

of a unity feedback control system intersects *s*-plane imaginary axis at

(*a*) $s = \pm j1.41$ (*b*) $s = \pm j0.866$

(*c*) $s = \pm j1.732$ (*d*) $\pm j0.707$

197. The angle of arrival of root locus branches at zeros of the transfer function

$$G(s)\,H(s) = \frac{K(s^2+4)}{s(s+2)}$$

is

(*a*) $\pm 135°$ (*b*) $\pm 180°$

(*c*) $\mp 145°$ (*d*) $\pm 60°$

198. With refrence to following open-loop transfer function

$$G(s)\,H(s) = \frac{K}{s(s^2+1)}$$

The angles of departure of root locus branches from open-loop poles located on imaginary axis are givon by

(*a*) $\pm 90°$ (*b*) $\pm 60°$

(*c*) $\pm 45°$ (*d*) $\pm 180°$

199. The open-loop transfer function of a feedback control system is given below :

$$G(s)\,H(s) = \frac{K(s^2+4)}{(s^2+1)}$$

The angles of departure from the poles located on $j\omega$ axis are given by

(*a*) $+90°$ (*b*) $180°, 180°$

(*c*) $\pm 30°$ (*d*) $-180°, -180°$

200. The angle of asymptotes w.r.t. root locus plot for a control system having

$$G(s)\,H(s) = \frac{s^2+1}{s(s+1)}$$

are

(*a*) $\pm 90°$ (*b*) nil

(*c*) $\pm 270°$ (*d*) $\pm 180°$

201. The root locus plot is shown in figure :
The damping ratio for the value $K = K_1$ is

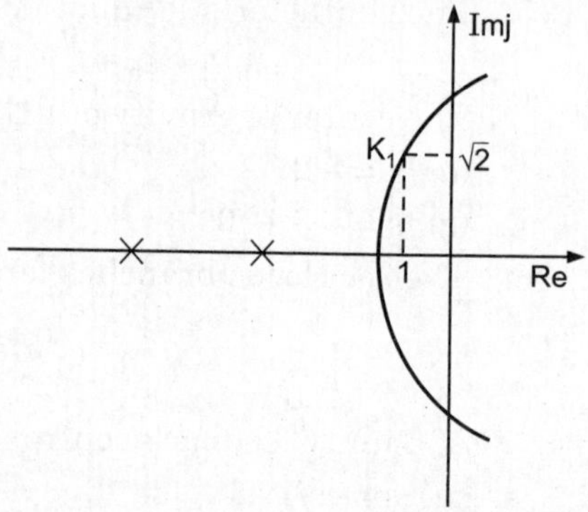

(*a*) 1.41 (*b*) 1.73
(*c*) 0.577 (*d*) 0.707

202. The root locus separates at a point between two open-loop poles, the point is called
(*a*) Critical point
(*b*) Crossover point
(*c*) Shift point
(*d*) Breakaway point

203. A point on the root locus satisfies
(*a*) $|G(s)\,H(s)| = 1$ and $\angle G(s)\,H(s) = (2k + 1)\,\pi$
(*b*) $|1 + G(s)\,H(s)| = 1$ and $\angle G(s)\,H(s) = (2k + 1)\,\pi$
(*c*) $\left|\dfrac{1}{G(s)\,H(s)}\right| = 1$ and $\angle\dfrac{1}{G(s)\,H(s)} = (2k + 1)\,\pi$
(*d*) $\left|\dfrac{G(s)}{1 + G(s)\,H(s)}\right| = 1$ and $\angle\dfrac{G(s)}{1 + G(s)\,H(s)} = (2k + 1)\,\pi$

204. If the root locus branches cross the imaginary axis the system becomes
(*a*) overdamped (*b*) critically damped
(*c*) stable (*d*) unstable

205. If the root locus lies only on the negative real axis then the time response is
(*a*) overdamed (*b*) underdamped
(*c*) oscillatory (*d*) sustained oscillations

206. The intersection of the root locus branch with the imaginary axis is determined by using equation
(*a*) $G(s)\,H(s) = 0$ (*b*) $1 - G(s)\,H(s) = 0$
(*c*) $1 + G(s)\,H(s) = 0$ (*d*) $[1 + G(s)\,H(s)]^{-1} = 0$

207. The asymptotes corresponding to root locus plot for the open-loop transfer function of a unity feedback control system $G(s) = \dfrac{K}{s(s+5)}$. Cross the imaginary axis at
(*a*) $\omega = \pm 2$ (*b*) $\omega = 0$
(*c*) $\omega = \pm\sqrt{5}$ (*d*) never cross.

208. The root locus plot is shown below in the figure.

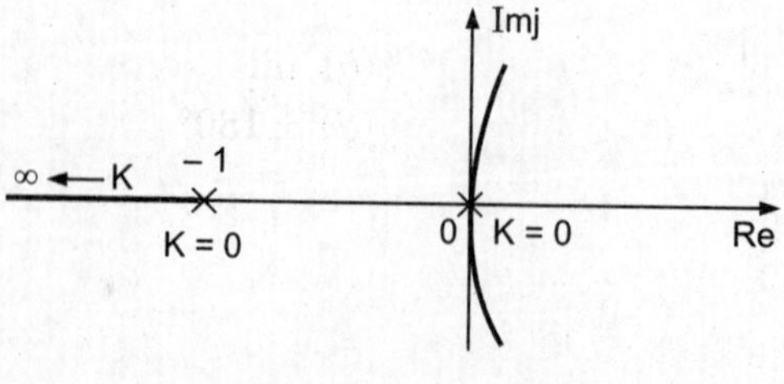

The corresponding open-loop transfer function is

(*a*) $\dfrac{K}{s^2(s+1)}$ (*b*) $\dfrac{K}{(s+1)(s+2)}$

(*c*) $\dfrac{K}{(s+1)^3}$ (*d*) $\dfrac{K}{(s+1)(s+2)^2}$

209. The open-loop transfer function of a unity feedback control system is given by $G(s) = \dfrac{K(s+2)}{s(s+4)(s+5)}$. The root locus lies on the real axis between

(*a*) $s = -2$ and $s = -4$ (*b*) $s = -4$ and $s = -5$
$s = -5$ and $s \to \infty$ $s = 0$ and $s = -2$

(*c*) $s = 0$ and $s = 2$ (*d*) $s = 4$ and $s = 5$

210. The open-loop transfer function of a feedback control system is given by $G(s)\,H(s) = \dfrac{K}{s(s+4)(s^2+4s+13)}$; $0 < K < \infty$. The number of breakaway points on the root locus plot is /are

(*a*) Nil (*b*) one

(*c*) two (*d*) three

211. The angle of departure of the root locus from the complex open-loop pole $(-1 - j1)$, having given the open-loop transfer function as $G(s)\,H(s) = \dfrac{K}{s(s+2)(s^2+2s+2)}$

(*a*) 90° (*b*) 270°

(*c*) 180° (*d*) – 60°

212. For the open-loop transfer function $G(s)\,H(s) = \dfrac{K}{s(s^2+4s+5)}$, the number of breakaway points is/are

(*a*) Nil (*b*) two

(*c*) one (*d*) three

213. With reference to root locus plot the complex conjugate roots of the characteristic equation for $G(s)\,H(s) = \dfrac{K(s+3)}{(s+1)^2}$, lie on

(*a*) straight line (*b*) parabola

(*c*) circle (*d*) semi circle

214. The asymptotes to root locus branches intersect on the real axis at a point given by

(*a*) $\dfrac{\Sigma P - \Sigma Z}{(P - Z)}$ (*b*) $\dfrac{\Sigma P + \Sigma Z}{(P - Z)}$

(*c*) $\dfrac{\Sigma P + \Sigma Z}{(P - Z)}$ (*d*) $\dfrac{\Sigma Z - \Sigma P}{(P + Z)}$

215. If the root locus branches do not cross the imaginary axes the system is

(*a*) conditionally stable (*b*) inherently stable

(*c*) maginally stable (*d*) likely to be unstable

216. The value of K at which the root locus crosses the imaginary axis makes the system

(*a*) stable (*b*) underdamped

(*c*) marginally stable (*d*) unstable

217. The number of root loci branches for the following open-loop transfer function

$$G(s)\,H(s) = \frac{K(s+2)(s+4)}{s(s+1)(s+3)(s^2+5s+6)}, \text{ is}$$

(*a*) 5 (*b*) 4

(*c*) 3 (*d*) 2

218. The transfer function of an integral compensator is given by

(*a*) $\frac{1}{s}$ (*b*) $\frac{1}{s^2}$

(*c*) $\frac{K}{s}$ (*d*) Ks

219. While incorporating lead-compensation network the gain factor of the amplifier is

(*a*) α (*b*) $\frac{1}{\alpha}$

(*c*) α^2 (*d*) $\frac{1}{\alpha^2}$

220. The lag-compensation has a

(*a*) zero nearer to the origin (*b*) pole nearer to the origin

(*c*) pole at the origin (*d*) zero at the origin

221. The lag-compensator

(*a*) decreases the bandwidth (*b*) increases the bandwidth

(*c*) does not affect the bandwidth (*d*) increases steady state error

222. The open-loop transfer function of a system is

$$G(s)\,H(s) = \frac{10(1+0.25\,\text{sec})}{s(1+0.6s)}$$

the phase shift as $\omega \to 0$ and $\omega \to \infty$ will be

(*a*) 0° and – 90° (*b*) – 180° and + 180°

(*c*) + 90° and – 90° (*d*) 0° and 0°

223. The transfer function given below

$$G_c(s) = \frac{1+R_2C_2s}{1+(R_1+R_2)C_2s}, \text{ represents}$$

(*a*) lead-network (*b*) lag-network

(*c*) lag-lead network (*d*) feedback network

224. The phase lag-lead network shifts the phase of a control signal in order that the output phase

(*a*) lags at low frequencies and leads at high frequencies
(*b*) leads at low frequencies and lags at high frequencies
(*c*) becomes independent of frequencies
(*d*) leads at all frequencies

225. A lead compensation needed for a closed-loop system has the following transfer function

$$G_c(s) = \frac{K\left(1+\frac{1}{a}s\right)}{\left(1+\frac{1}{b}s\right)}, \text{ for such a network}$$

(*a*) $a < b$ (*b*) $b < a$
(*c*) $a > Kb$ (*d*) $b < Ka$

226. The lead compensation network is considered to be
(*a*) high pass filter (*b*) low pass filter
(*c*) equalizer (*d*) none of the above

227. The transfer function of a lead compensation network is given below :

$$G_c(s) = \frac{\alpha(1+sT)}{(1+s\alpha T)}$$

The mid-corner frequency is given by

(*a*) $\omega_m = \frac{1}{\alpha T}$ (*b*) $\omega_m = \frac{1}{T\sqrt{a}}$
(*c*) $\omega_m = \alpha T$ (*d*) $\omega_m = \frac{1}{\alpha\sqrt{T}}$

228. The maximum phase lead is contributed by a phase lead network at
(*a*) mid corner frequency (*b*) lower corner frequency
(*c*) upper corner frequency (*d*) phase corner frequency

229. Which one of the following gives the transfer function of a phase-lag compensation network?

(*a*) $\frac{1+s\alpha T}{1+sT}$; $\alpha < 1$ (*b*) $\frac{1+s\alpha T}{1+sT}$; $\alpha > 1$
(*c*) $\frac{1+s\alpha T}{1+sT}$; $\alpha = 1$ (*d*) $\frac{1-s\alpha T}{1+sT}$; $\alpha < 1$

230. The analysis of multiple-input multiple-output system is conveniently studied by
(*a*) state-space approach (*b*) root locus approach
(*c*) characteristic equation approach (*d*) Nichols chart

231. The state transition matrix $\phi(t)$ is given by
(*a*) $[sI] - [A]$ (*b*) $\{[sI] - [A]\}^{-1}$
(*c*) $L^{-1}\{[sI] - [A]\}^{-1}$ (*d*) $L^{-1}\{[sI] - [A]\}$

232. In exponential series form the state transition matrix is
(*a*) $e^{-[A]t}$ (*b*) $e^{[A]t}$
(*c*) $e^{[A]}$ (*d*) $e^{-[A]}$

233. For the state model $[\dot{x}] = [A][x] + [B]u, y = [c]\,[x]$ the transfer matrix is

(a) $[C]\{[sI] - [A]\}^{-1}[B]$
(b) $[C]\,\{[sI] - [A]\}\,[B]$
(c) $[C]\,[B]\,\{[sI] - [A]\}^{-1}$
(d) $[C]\,[B]\,\{[sI] - [A]\}$.

234. For second order system with an initial condition of $\begin{bmatrix} 1 \\ 2 \end{bmatrix}$ without any external input, the state transition matrix for the system is given below :

$$\phi(t) = \begin{bmatrix} e^{-2t} & 0 \\ 0 & e^{-t} \end{bmatrix}$$

The state of the system at the end of one second is

(a) $\begin{bmatrix} 0.135 \\ 0.735 \end{bmatrix}$
(b) $\begin{bmatrix} 0.135 \\ 0.368 \end{bmatrix}$
(c) $\begin{bmatrix} 0.271 \\ 0.735 \end{bmatrix}$
(d) $\begin{bmatrix} 0.135 \\ 1.47 \end{bmatrix}$

235. The differential equation governing the dynamics of a separately excited d.c. motor is given below :

$$J\frac{d^2\theta}{dt^2} + f\frac{d\theta}{dt} + \frac{K^2}{L}\theta = \frac{K}{L}V_a$$

The above equation is arranged in the following form

$$\begin{bmatrix} \dfrac{d^2\theta}{dt^2} \\ \dfrac{d\theta}{dt} \end{bmatrix} = [A]\begin{bmatrix} \dfrac{d\theta}{dt} \\ \theta \end{bmatrix} + [B]V_a$$

The value of coefficient matrix $[A]$ is given by

(a) $\begin{bmatrix} \dfrac{f}{J} & -\dfrac{K^2}{J} \\ 1 & 0 \end{bmatrix}$
(b) $\begin{bmatrix} \dfrac{K^2}{LJ} & -\dfrac{f}{J} \\ 0 & 1 \end{bmatrix}$
(c) $\begin{bmatrix} 0 & 1 \\ -\dfrac{K^2}{LJ} & -\dfrac{f}{J} \end{bmatrix}$
(d) $\begin{bmatrix} 1 & 0 \\ -\dfrac{f}{J} & -\dfrac{K^2}{LJ} \end{bmatrix}$

236. With reference to state space representation of the equation,

$\dfrac{d^2y(t)}{dt^2} + \dfrac{2dy(t)}{dt} + 3y(t) = u(t)$. The matrices **A** and **B** are,

(a) $\mathbf{A} = \begin{bmatrix} 0 & 1 \\ -3 & -2 \end{bmatrix}; \mathbf{B} = \begin{bmatrix} 0 \\ 1 \end{bmatrix}$
(b) $\mathbf{A} = \begin{bmatrix} -2 & -3 \\ 0 & 1 \end{bmatrix}; \mathbf{B} = \begin{bmatrix} 0 \\ 1 \end{bmatrix}$
(c) $\mathbf{A} = \begin{bmatrix} 0 & -2 \\ 1 & -3 \end{bmatrix}; \mathbf{B} = \begin{bmatrix} 1 \\ 0 \end{bmatrix}$
(d) $\mathbf{A} = \begin{bmatrix} -2 & 0 \\ 1 & -3 \end{bmatrix}; \mathbf{B} = \begin{bmatrix} 1 \\ 0 \end{bmatrix}$

237. A control system is represented by $\begin{bmatrix} \dot{x}_1 \\ \dot{x}_2 \end{bmatrix} = \begin{bmatrix} -1 & 0 \\ 0 & -3 \end{bmatrix} \begin{bmatrix} x_1 \\ x_2 \end{bmatrix} + \begin{bmatrix} 0 \\ 1 \end{bmatrix} u$ and $y = [1\ 1] \begin{bmatrix} x_1 \\ x_2 \end{bmatrix}$

The system transfer function is

(*a*) $\dfrac{s}{s+1}$ (*b*) $\dfrac{1}{s+3}$

(*c*) $\dfrac{1}{s+1}$ (*d*) $\dfrac{s}{s^2+4s+3}$

238. The state model is given below, $\begin{bmatrix} \dot{x}_1 \\ \dot{x}_2 \end{bmatrix} = \begin{bmatrix} -1 & 0 \\ 0 & -2 \end{bmatrix} \begin{bmatrix} x_1 \\ x_2 \end{bmatrix} + \begin{bmatrix} 0 \\ 1 \end{bmatrix} u$. If the input is unit step, the time response is

(*a*) undamped (*b*) underdamped

(*c*) critically damped (*d*) overdamped

249. An atonomous state euaqtion is $[\dot{x}] = [A]\ [x]$ where, $[A] = \begin{bmatrix} 0 & 2 \\ 8 & 0 \end{bmatrix}$. The polers are located at

(*a*) $j4, -j4$ (*b*) $-4, 2$

(*c*) $j2, -j2$ (*d*) $-8, 4$

240. The characteristic equation for a system having state equation

$$\begin{bmatrix} \dot{x}_1 \\ \dot{x}_2 \end{bmatrix} = \begin{bmatrix} 1 & 4 \\ -2 & -5 \end{bmatrix} \begin{bmatrix} x_1 \\ x_2 \end{bmatrix} + \begin{bmatrix} 0 \\ 1 \end{bmatrix} u \text{ is}$$

(*a*) $s^2 + 5s + 8 = 0$ (*b*) $s^2 + 4s + 3 = 0$

(*c*) $s^2 + 4s + 5 = 0$ (*d*) $s^2 + 4s + 2 = 0$

241. For the system $\begin{bmatrix} \dot{x}_1 \\ \dot{x}_2 \end{bmatrix} = \begin{bmatrix} 2 & 0 \\ 0 & 4 \end{bmatrix} \begin{bmatrix} x_1 \\ x_2 \end{bmatrix} + \begin{bmatrix} 1 \\ 1 \end{bmatrix} r(t)$

$$y = [4\quad 0] \begin{bmatrix} x_1 \\ x_2 \end{bmatrix}$$

where $r(t)$ is a unit impulse function. Having initial conditions as nil, the output y is determined as

(*a*) $y(t) = 2e^{2t}$ (*b*) $y(t) = 4e^{2t}$

(*c*) $y(t) = 2e^{4t}$ (*d*) $y(t) = 4e^{4t}$

242. For the homogenous state equation

$$\begin{bmatrix} \dot{x}_1 \\ \dot{x}_2 \end{bmatrix} = \begin{bmatrix} -3 & 1 \\ 0 & -2 \end{bmatrix} \begin{bmatrix} x_1 \\ x_2 \end{bmatrix}$$

with $\begin{bmatrix} x_1 & (0) \\ x_2 & (0) \end{bmatrix} = \begin{bmatrix} 2 \\ -1 \end{bmatrix}$

The steady state value of x is

(a) $\begin{bmatrix} 0 \\ 0 \end{bmatrix}$ (b) $\begin{bmatrix} -2 \\ -1 \end{bmatrix}$

(c) $\begin{bmatrix} -1 \\ 3 \end{bmatrix}$ (d) $\begin{bmatrix} 0 \\ 1 \end{bmatrix}$

243. For the matrix

$$[A] = \begin{bmatrix} 1 & 4 \\ -2 & -5 \end{bmatrix}$$

The eigen values are

(a) $\lambda_1 = -1 ; \lambda_2 = -3$ (b) $\lambda_1 = -1 ; \lambda_2 = -4$

(c) $\lambda_1 = -2 ; \lambda_2 = -4$ (d) $\lambda_1 = -1 ; \lambda_2 = -2$

244. $[A] = \begin{bmatrix} 0 & 1 & 0 \\ 0 & 0 & 1 \\ -6 & -11 & -6 \end{bmatrix}$

(a) $[P] = \begin{bmatrix} 0 & 1 & 0 \\ 1 & 2 & 3 \\ 1 & 4 & 9 \end{bmatrix}$ (b) $[P] = \begin{bmatrix} 1 & 1 & 1 \\ -1 & -2 & -3 \\ 1 & 4 & 9 \end{bmatrix}$

(c) $[P] = \begin{bmatrix} 1 & 4 & 9 \\ -1 & -2 & -3 \\ 1 & 4 & 9 \end{bmatrix}$ (d) $[P] = \begin{bmatrix} -1 & 0 & 0 \\ 0 & -2 & 0 \\ 0 & 0 & -3 \end{bmatrix}$

245. For a given state space representation $\mathbf{A} = \begin{bmatrix} 1 & 0 \\ 0 & 4 \end{bmatrix}$. The state transition matrix is

(a) $\phi(t) = \begin{bmatrix} e^{-t} & 0 \\ 0 & e^{-4t} \end{bmatrix}$ (b) $\phi(t) = \begin{bmatrix} 0 & e^{-t} \\ e^{-4t} & 0 \end{bmatrix}$

(c) $\phi(t) = \begin{bmatrix} e^{-4t} & 0 \\ 0 & e^{-t} \end{bmatrix}$ (d) $\phi(t) = \begin{bmatrix} 0 & e^{-4t} \\ e^{-t} & 0 \end{bmatrix}$

246. A system is described as, $\boldsymbol{x} = \mathbf{A}\boldsymbol{x} + \mathbf{B}\boldsymbol{u}$. The controllability test matrix is given by

(a) ... : $\mathbf{A^2B : AB : B}$ (b) $\mathbf{B : AB : A^2B}$: ...

(c) $\mathbf{B : BA : BA^2}$: ... (d) ... $\mathbf{BA^2 : BA : B}$

247. In z-plane the unit circle corresponds to

(a) imaginary axis of s-plane (b) + real axis of s-plane

(c) – real axis of s-plane (d) origin of s-plane

248. On transformation using $z = e^{sT}$, a point $(-1 + j2)$ in the s-plane is placed in z-plane

(a) inside the unit circle (b) outside the unit circle

(c) on the unit circle (d) cannot be placed in the z-plane

249. The faithful reconstruction of a signal on account of sampling is obtained if

(a) $\omega_s = \omega_1$ (b) $\omega_s \geq 2\omega_1$

(c) $\omega_s \leq \omega_1$ (d) $\omega_s \leq 2\omega_1$

Answers

1. (c)	2. (d)	3. (a)	4. (b)	5. (c)	6. (a)
7. (b)	8. (c)	9. (c)	10. (a)	11. (c)	12. (a)
13. (d)	14. (a)	15. (b)	16. (b)	17. (a)	18. (d)
19. (a)	20. (a)	21. (b)	22. (c)	23. (c)	24. (a)
25. (a)	26. (d)	27. (d)	28. (b)	29. (a)	30. (d)
31. (d)	32. (a)	33. (a)	34. (c)	35. (a)	36. (c)
37. (a)	38. (d)	39. (a)	40. (a)	41. (c)	42. (b)
43. (d)	44. (c)	45. (d)	46. (c)	47. (c)	48. (a)
49. (c)	50. (a)	51. (b)	52. (d)	53. (b)	54. (b)
55. (c)	56. (a)	57. (b)	58. (d)	59. (c)	60. (c)
61. (b)	62. (b)	63. (d)	64. (b)	65. (a)	66. (d)
67. (a)	68. (a)	69. (c)	70. (b)	71. (d)	72. (a)
73. (d)	74. (b)	75. (c)	76. (a)	77. (c)	78. (b)
79. (a)	80. (a)	81. (d)	82. (a)	83. (b)	84. (b)
85. (a)	86. (b)	87. (a)	88. (a)	89. (c)	90. (b)
91. (a)	92. (b)	93. (b)	94. (b)	95. (a)	96. (a)
97. (b)	98. (a)	99. (a)	100. (c)	101. (b)	102. (b)
103. (c)	104. (a)	105. (b)	106. (d)	107. (b)	108. (a)
109. (d)	110. (b)	111. (a)	112. (d)	113. (d)	114. (b)
115. (a)	116. (a)	117. (c)	118. (c)	119. (a)	120. (d)
121. (b)	122. (a)	123. (a)	124. (b)	125. (d)	126. (b)
127. (b)	128. (d)	129. (b)	130. (b)	131. (b)	132. (a)
133. (a)	134. (b)	135. (c)	136. (a)	137. (d)	138. (b)
139. (b)	140. (a)	141. (a)	142. (a)	143. (b)	144. (c)
145. (d)	146. (a)	147. (c)	148. (a)	149. (a)	150. (a)
151. (a)	152. (a)	153. (b)	154. (c)	155. (b)	156. (d)
157. (b)	158. (d)	159. (a)	160. (d)	161. (a)	162. (c)
163. (a)	164. (c)	165. (d)	166. (b)	167. (b)	168. (b)
169. (b)	170. (b)	171. (a)	172. (b)	173. (c)	174. (b)
175. (a)	176. (d)	177. (c)	178. (a)	179. (a)	180. (a)
181. (d)	182. (b)	183. (c)	184. (b)	185. (b)	186. (a)
187. (a)	188. (b)	189. (b)	190. (b)	191. (a)	192. (b)
193. (b)	194. (a)	195. (c)	196. (a)	197. (a)	198. (c)
199. (a)	200. (b)	201. (c)	202. (d)	203. (a)	204. (d)

205. (*a*)	**206.** (*c*)	**207.** (*d*)	**208.** (*a*)	**209.** (*b*)	**210.** (*d*)
211. (*a*)	**212.** (*a*)	**213.** (*c*)	**214.** (*a*)	**215.** (*b*)	**216.** (*c*)
217. (*a*)	**218.** (*c*)	**219.** (*b*)	**220.** (*b*)	**221.** (*a*)	**222.** (*a*)
223. (*b*)	**224.** (*a*)	**225.** (*a*)	**226.** (*a*)	**227.** (*b*)	**228.** (*a*)
229. (*a*)	**230.** (*a*)	**231.** (*c*)	**232.** (*a*)	**233.** (*a*)	**234.** (*a*)
235. (*c*)	**236.** (*a*)	**237.** (*c*)	**238.** (*d*)	**239.** (*a*)	**240.** (*b*)
241. (*b*)	**242.** (*a*)	**243.** (*a*)	**244.** (*b*)	**245.** (*a*)	**246.** (*b*)
247. (*a*)	**248.** (*a*)	**249.** (*b*)			

APPENDIX III

Short Answer Questions

One compulsory question having 10 or 20 parts usually appear in question papers. The said compulsory question is invariably short answer type. An effort is made to make the readers aware of this fact and as such short answer type questions are framed hereunder. These type of questions will be of help towards viva-voce examinations and job related interviews.

Q. 1. Explain open-loop control system.

Ans. In an open-loop control system the output is neither measured nor feedback for comparison with the input. The accuracy of an open-loop control system depends on the input calibration.

Q. 2. Explain closed-loop control system.

Ans. The control system wherein the output is feedback and compared with the reference input in order to maintain the desired output value is called a closed-loop control system.

Q. 3. Mention the advantages of closed-loop control system.

Ans. (1) The error is continuously measured through output feedback and as such the closed-loop control system works more accurately.

(2) The closed-loop control system adjusts to the effects of non-linearity present in its elements.

Q. 4. What are the components of a feedback control system ?

Ans. The feedback control system has the following components :

(*i*) Forward path elements (*ii*) Feedback path elements (*iii*) Reference input.

The feedback control system is shown below.

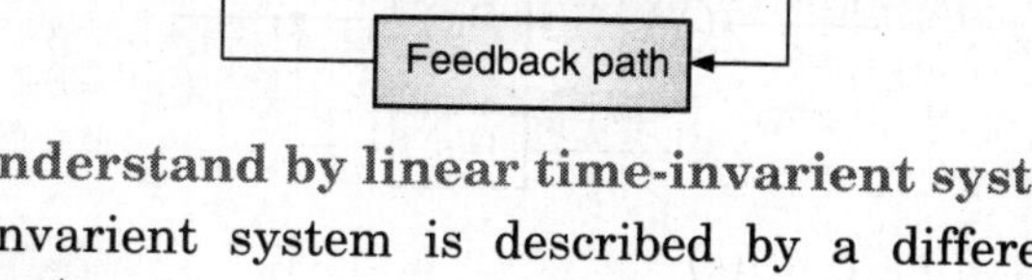

Q. 5. What do you understand by linear time-invarient system ?

Ans. A linear time-invarient system is described by a differential equation having constant coefficients.

Q. 6. Determine the time domain function corresponding to pole-location s = 2 ± j2.

Ans. $$G(s) = \frac{1}{(s-2) \mp j2}$$

$$= \frac{1}{(s-2-j2)(s-2+j2)} = \frac{1}{(s+2)^2+(2)^2} = \frac{1}{2} \cdot \frac{2}{(s+2)^2+2^2}$$

Taking inverse Laplace transform on both sides

$$g(t) = \frac{1}{2} e^{-2t} \sin t.$$

Q. 7. The transfer function of a control system is given below

$$\frac{C(s)}{R(s)} = \frac{2(s+1)}{(s+2)(s+3)}$$

Write the corresponding differential equation.

Ans. $$\frac{C(s)}{R(s)} = \frac{2(s+1)}{s^2+5s+6}$$

The corresponding differential equation is

$$\frac{dc^2}{dt^2} + 5\frac{dc}{dt} + 6c = 2r + 2\frac{dr}{dt}.$$

Q. 8. Write the transfer function for the system described by the following differential equation

$$a_0 \frac{d^2y}{dt^2} + a_1 \frac{dy}{dt} + a_2 y = b_0 x + b_1 \frac{dx}{dt}.$$

Ans. Taking Laplace transform on both sides are rearranging

$$\frac{Y(s)}{X(s)} = \frac{b_0 + b_1 s}{a_0 s^2 + a_1 s + a_2}.$$

Q. 9. Give two advantages of block diagram representation of a control system.

Ans. (1) The block diagram gives the interconnection of components used in a control system.

(2) The block diagram gives the relation between the components of a control system.

Q. 10. Explain what is meant by open-loop transfer function in relation to a closed-loop control system.

Ans. The block diagram of a closed-loop control system is shown below

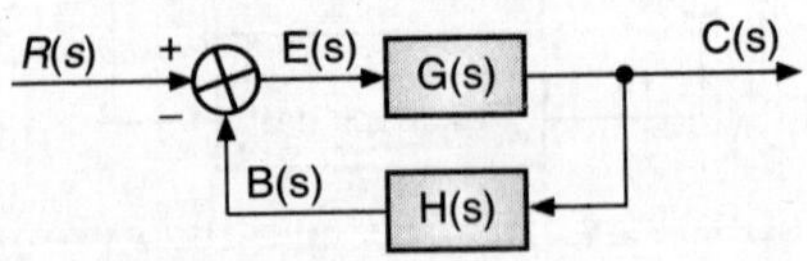

$$B(s) = C(s)\,H(s)$$

$$\because \quad C(s) = E(s)\,G(s)$$

$$\therefore \quad B(s) = G(s)\,H(s)\,E(s)$$

The ratio $\frac{B(s)}{E(s)} = G(s)\,H(s)$ is known as open-loop transfer function.

Q. 11. What is a block diagram ?

Ans. The symbolic representation of a transfer function is represented by a block diagram. The complete control system can be represented by interconnecting block diagrams of components as per the requirements.

Q. 12. Define non-touching loops.

Ans. The non-touching loops in a signal flow graph are those which do not share a common node.

Q. 13. Write Mason's gain formula.

Ans. Overall gain (transmittance) $= \dfrac{\sum\limits_{k=1}^{k} P_k \Delta_k}{\Delta}$

where P = Number of forward paths

P_k = Forward path transmittance of the kth path.

Graph determinant

Δ = 1 – sum loop gains (transmittances)
+ sum of products of all possible pairs of non-touching loop gains
– sum of products all possible triplets of non-touching loop gains
+ ....

Δ_k = Path factor associated with the kth path is equal to the graph determinant of the signal flow graph which remains after erasing the kth path.

Q. 14. What is the use of gear train in a control system ?

Ans. In a control system the load demands high torque at low speed while the motor supplies low torque at high speed. In such a case a gear train is used having a gear ratio $(N_2/N_1) > 1$. N_1 is the gear on the motor side and N_2 on the load side.

Q. 15. Show that in mks system $K_T = K_b$.

Ans. The mechanical power output of the motor $= T_M\omega_m$

The electrical power input to the motor $= e_b i_a$

Neglecting armature and rotational losses

$$T_M\omega_m = e_b i_a$$

$\because$ $T_M = K_T i_a$ and $e_b = K_b\omega_m$

$\therefore$ $K_T\, i_a\omega_m = K_b\omega_m i_a$, or $K_T = K_b$.

Q. 16. Show how the transfer function of a motor is modified if a gear train is attached to it.

Ans.

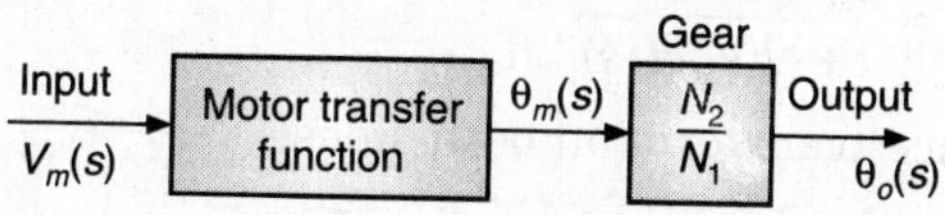

$$\frac{N_2}{N_1} = \text{Gear ratio.}$$

Q. 17. State the assumption made while obtaining the transfer function of an armature controlled d.c. motor.

Ans. The armature inductance L_a is neglected.

Q. 18. Write the transfer function of an armature controlled d.c. motor relating angular displacement with applied voltage V_a.

Ans.
$$\frac{\theta_m(s)}{V_a(s)} = \frac{K_T}{(R_a + sL_a)(s^2J + sf) + sK_TK_b}$$

Neglect L_a in comparison with R_a

$$\therefore \quad \frac{\theta_m(s)}{V_a(s)} = \frac{K_T}{sR_a(sJ + f) + sK_TK_b}.$$

Q. 19. Write equivalent moment of inertia and coefficient of friction with reference motor side. The ratio is 20.

Ans.
$$J_{eq} = J\left(\frac{N_1}{N_2}\right)^2 = J\left(\frac{1}{20}\right)^2$$

$$f_{eq} = f\left(\frac{N_1}{N_2}\right)^2 = f\left(\frac{1}{20}\right)^2.$$

Q. 20. What is a techogenerator ?

Ans. The electromechanical device which generates the output voltage proportional to shaft speed is called a tachogenerator.

Q. 21. Name the test signals used in time response analysis.

Ans. The commonly used test signals are unit (1) impulse (2) unit step (3) unit ramp (4) unit acceleration.

Q. 22. What is meant by time response of a control system ?

Ans. The time response of a control means as to how the output of a system varies in accordance with time when a specified test signal is applied.

Q. 23. Distinguish between the static response and dynamic response.

Ans. The static response corresponds to steady state part and dynamic response refers to transient part of the time response of control system.

Q. 24. How the response of control system is determined ?

Ans. The time response denoted as $c(t)$ gives the variation of output of a control system as a function time on application of a specified input signal.

The closed-loop transfer function is given by

$$\frac{C(s)}{R(s)} = \frac{G(s)}{1 + G(s)\,H(s)}$$

$$\therefore \quad C(s) = \frac{R(s)\,G(s)}{1 + G(s)\,H(s)}$$

Taking inverse Laplace transform on both sides,

$$C(t) = \mathcal{L}^{-1}\frac{R(s)\,G(s)}{1 + G(s)\,H(s)}.$$

Q. 25. What is meant by characteristic equation ?

Ans. The characteristic equation is obtained by equating the denominator polynomial of the overall transfer function $C(s)/R(s)$ to zero.

For a control system,

$$\frac{C(s)}{R(s)} = \frac{G(s)}{1+G(s)\,H(s)}$$

The characteristic equation is

$$1 + G(s)\,H(s) = 0.$$

Q. 26. The open-loop transfer function is given below :

$$\mathbf{G(s)\,H(s) = \frac{K}{s(s+2)}}$$

Write the characteristic equation.

Ans. The characteristic equation is given by

$$1 + G(s)\,H(s) = 0$$

$$\therefore \qquad 1 + \frac{K}{s(s+2)} = 0$$

$$\text{or} \qquad s^2 + 2s + K = 0.$$

Q. 27. Draw the typical unit step time response of underdamped second order control system.

Ans.

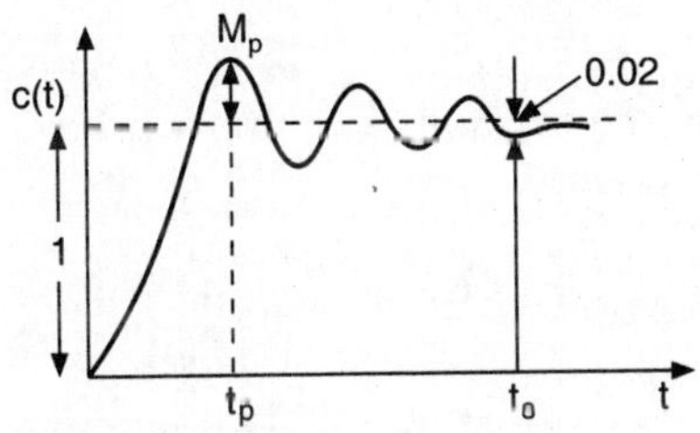

Q. 28. Explain the term "time constant" of a first order system.

Ans. The time response of a first system to a unit step input is given by,

$$c(t) = (1 - e^{-at})$$

The exponent "a" is +ive real. At $t = 1/a$ the time response reaches 63.2% of the steady state value. The time constant denoted as $T = 1/a$.

Q. 29. Determine the closed-loop poles of G(s) H(s) = $\frac{K}{s(s+4)}$ for K = 6.

Ans. The closed-loop poles are roots of the characteristics equation.

The characteristic equation

$$1 + G(s)\,H(s) = 0$$

$$\therefore \qquad 1 + \frac{K}{s(s+4)} = 0$$

$$\text{or} \qquad s^2 + 4s + K = 0$$

$$\text{put} \qquad K = 6$$

$$\therefore \qquad s^2 + 4s + 6 = 0$$

$$s_1, s_2 = \frac{-4 \pm \sqrt{4^2 - 4 \times 1 \times 6}}{2} = -2 \pm j\sqrt{2}.$$

Q. 30. Define settling time in relation to time response.

Ans. The settling time is the time required to settle the time response within a tolerance band of steady state value. Usually the tollerance band is either 2% or 5%.

$$t_s = 4T \text{ on } 2\% \text{ basis}$$
$$t_s = 3T \text{ on } 5\% \text{ basis}$$

where T = time constant.

Q. 31. Define damping ratio.

Ans. The damping ratio ζ is defined as the ratio actual damping ($\zeta\omega_n$) to critical damping ($1 \times \omega_n$).

Q. 32. Write the relationship between the % maximum overshoot (M_p) and damping ratio ζ.

Ans. $\% M_p = e^{\frac{-\zeta\pi}{\sqrt{1-\zeta^2}}} \times 100, \quad \zeta < 1$

The decrease in the value of damping ratio increases the maximum overshoot. For $\zeta = 0$, $\% M_p = 100\%$.

Q. 33. Write the expression for peak time (t_p) in terms of ω_n and ζ.

Ans. $t_p = \dfrac{\pi}{\omega_n\sqrt{1-\zeta^2}}$.

Q. 34. Sketch the impulse response corresponding to closed-loop transfer function poles shown below.

Ans.

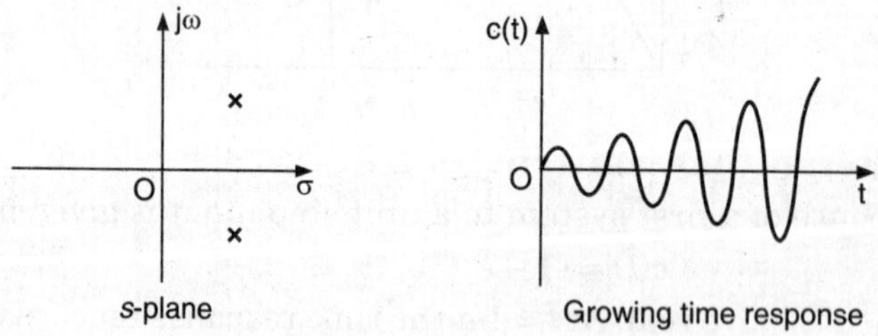

Q. 35. A second order control system has a damping ratio ζ = 0.8 and natural frequency of oscillation ω_n = 12 rad./sec. Determine the damped frequency of oscillation.

Ans. $\omega_d = \omega_n\sqrt{1-\zeta^2} = 12\sqrt{1-0.8^2} = 7.2$ rad./sec.

Q. 36. The open-loop transfer function of a unity feedback control system is $G(s) = \dfrac{16}{s(s+6)}$. What is the nature of closed-loop time response ? The input is unit step function.

Ans.
$$\frac{C(s)}{R(s)} = \frac{G(s)}{1+G(s)}$$
$$= \frac{16}{s^2 + 6s + 16}$$
$$\omega_n = \sqrt{16} = 4 \text{ rad./sec.}$$
$$2\zeta\omega_n = 6$$
$$\therefore \quad 2\zeta \times 4 = 6, \quad \zeta = 0.75$$

$\zeta < 1$, the response is underdamped oscillatory.

Q. 37. Mention the nature of transient response of second order control system for different types of damping.

Ans. (1) Oscillatory (sustained oscillations) for $\zeta = 0$

(2) Damped oscillatory (underdamped) for $\zeta < 1$

(3) Exponentially rising (critical damping) for $\zeta = 1$

(4) Exponentially rising (overdamped) for $\zeta > 1$. The rise time is more as compared to critical damping.

Q. 38. The closed-loop transfer function is given below :

$$\frac{C(s)}{R(s)} = \frac{25}{s^2 + 4s + 25}$$

Determine ω_n and ζ.

Ans. The characteristic equation is

$$s^2 + 4s + 25 = 0$$
$$\omega_n = \sqrt{25} = 5 \text{ rad./sec.}$$
$$2\zeta\omega_n = 4$$
$$\therefore \quad 2\zeta \times 5 = 4$$
$$\zeta = 0.4.$$

Q. 39. If the damping ratio is increased, how will it affect the settling time ?

Ans. The settling time $t_s = \dfrac{4}{\zeta\omega_n}$.

The settling decreases as the damping ratio is increased (keeping constant ω_n).

Q. 40. What is the value of damping ratio for an overdamped response?

Ans. $\zeta > 1$.

Q. 41. What is the frequency of oscillation for underdamped system ?

Ans.
$$\omega_d = \omega_n\sqrt{1-\zeta^2}, \quad \zeta < 1.$$

Q. 42. The characteristic equation of a system is given below :

$$(s + 2)(s + 3)(s^2 + 2s + 2) = 0$$

Determine the dominant roots.

Ans. The roots of the characteristic equation are

$$s_1 = -2,\ s_2 = -3,\ s_3, s_4 = \frac{-2 \pm \sqrt{2^2 - 4 \times 1 \times 2}}{2} = -1 \pm j1$$

The roots $-1 \pm j1$ being near to s-plane imaginary axis are dominant.

Q. 43. What will be the nature of time response if the roots of the characteristic equation are located on s-plane imaginary axis.

Ans. (1) Complex conjugate (non-multiple) : oscillatory (sustained oscillations)

(2) Complex conjugate (multiple) : unstable (growing oscillations).

Q. 44. What is understood by impulse response ?

Ans. The response to unit impulse input is known as impulse response. The impulse response is given by inverse Laplace transform of the system transfer function.

Q. 45. The natural frequency of oscillations and the damping ratio of a system are 0.6 and 10 rad./sec. respectively. Determine the peak time t_p and % overshoot M_p.

Ans.

$$t_p = \frac{\pi}{\omega_n\sqrt{1-\zeta^2}} = \frac{\pi}{10\sqrt{1-0.6^2}} = 0.39 \text{ sec}$$

$$\%M_p = e^{\frac{-\zeta\pi}{\sqrt{1-\zeta^2}}} \times 100 = e^{\frac{-0.6\pi}{\sqrt{1-0.6^2}}} = 9.8\%.$$

Q. 46. The damping ratio and natural frequency of oscillations are $\zeta = 0.8$ and $\omega_n = 12$ rad./sec.

Detemine the rise time.

Ans. Rise time :

$$t_r = \frac{\pi - \tan^{-1}\dfrac{\sqrt{1-\zeta^2}}{\zeta}}{\omega_n\sqrt{1-\zeta^2}}$$

$$= \frac{\pi - \tan^{-1}\dfrac{\sqrt{1-0.8^2}}{0.8}}{12\sqrt{1-0.8^2}} = \frac{\pi - \tan^{-1}0.75}{12 \times 0.6} = \frac{\pi - 0.643}{7.2} = 0.346 \text{ sec.}$$

Q. 47. What is a dominant pole pair ?

Ans. The closed-loop L.H.S. complex pole pair nearer to s-plane imaginary axis is known as dominant pole pair. These poles dominate the time response while the other L.H.S. poles far away from the s-plane imaginary axis have less effect on the time response.

Q. 48. What is the order of the system whose transfer function is $G(s) = \dfrac{k}{s^2(s+2)}$?

Ans. The total number of poles of the transfer function is 3, hence the order of the system is 3.

Q. 49. Distinguish between the type and order of a system transfer function.

Ans. (*i*) The number of poles at the origin gives the type of a transfer function.

(*ii*) The total number of poles give the order of a transfer function.

Q. 50. What is static positional error coefficient for a type 0 system ?

Ans. The static positional error coefficient is associated with unit step input and given by

$$K_p = K$$

where K is gain of the system.

The steady state error is $e_{ss} = \frac{1}{1+K_p}$.

Q. 51. What is the steady state error of a type 1 system subjected to a parabolic input ?

Ans. The transfer function of a type 1 system is

$$G(s)\,H(s) = \frac{K(1+sT_a)\ldots}{s(1+sT_1)\ldots}$$

The acceleration error coefficient is

$$K_a = \lim_{s\to 0} s^2\, G(s)\, H(s)$$

$$= \lim_{s\to 0} s^2 \cdot \frac{K(1+sT_a)\ldots}{s(1+sT_1)\ldots} = 0$$

$$l_{ss} = \frac{1}{K_a} = \frac{1}{0} = \infty.$$

Q. 52. A second order control system has damping ratio ζ = 0.6 and natural frequency of oscillations ω_n = 10 rad./sec. Determine the resonant peak and resonant frequency.

Ans. Resonant peak : $M_r = \frac{1}{2\zeta\sqrt{1-\zeta^2}} = \frac{1}{2\times 0.6\sqrt{1-0.6^2}} = 1.041$

Resonant frequency : $\omega_r = \omega_n\sqrt{1-2\zeta^2} = 10\sqrt{1-2\times 0.6^2} = 5.3$ rad./sec.

Q. 53. A second order control system is subjected to a unit ramp input. What is the steady state error and how to reduce it ?

Ans. $e_{ss} = \frac{2\zeta}{\omega_n}$.

The steady state error can be reduced by increasing the gain K.

Q. 54. Why negative feedback is preferred in a closed-loop control system ?

Ans. The error is continuously measured through negative feedback as such the closed-loop system works more accurately.

The closed-loop system with negative feedback is less sensitive to effects of parameter variations of the elements placed in the forward path.

Q. 55. Name the controllers used in control system.

Ans. The controllers used in a control system are

(1) On-off controller

(2) Proportional controller

(3) Proportional-plus-integral controller

(4) Proportional + derivative controller

(5) Proportional-plus-derivative-plus integral controller.

Q. 56. What is integral control action ?

Ans. The control action where the control signed consists of integral of the error signal is known as integral control action.

Q. 57. Obtain the transfer function for PI controller.

Ans. The block diagram for the PI controller is shown below :

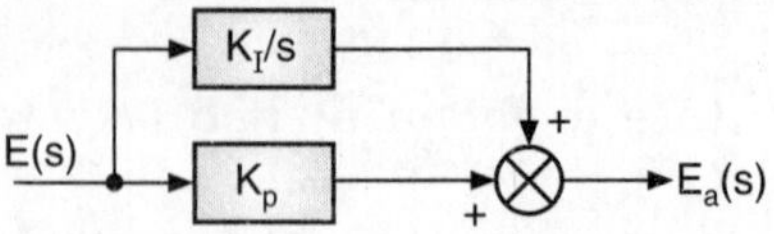

$$\frac{E_a(s)}{E(s)} = K_p + \frac{K_I}{s}.$$

Q. 58. What is the effect of PD controller on the time response ?

Ans. The PD controller reduces the peak overshoot by increasing the damping ratio.

Q. 59. What is the effect of PI controller on system performance ?

Ans. The PI controller introduces a pole in the open-loop transfer function of a system. This increases the order of the system which reduces the steady state error. The system becomes less stable on increases its order.

Q. 60. What is derivative control and its effects ?

Ans. The control which produces a control action using derivative of the error signal is known as derivative controller. The controller has no effect under steady, therefore, works only under transient state where the error is a function of time as such the derivative control is not used alone.

Q. 61. Discuss the effect of adding a pole to open-loop transfer function.

Ans. The steady state error is reduced due to addition of a pole (having –ive real part). This addition increases the order of the transfer function which makes the system less stable compared to the original system.

Q. 62. Define BIBO stability.

Ans. If every bounded (finite) input results in bounded (finite) output, then the system is said to have BIBO stability.

Q. 63. How are the *s*-plane locations of the roots of the characteristics equation are related to stability ?

Ans.
1. Stable system : All roots of lie in the L.H.S. of *s*-plane.
2. Marginally stable system : Non-repeated complex conjugate roots lie on the *s*-plane imaginary axis.
3. Unstable system : Presence of any one of the root (complex/simple) in R.H.S. of *s*-plane.

Q. 64. What is limitedly stable system ?

Ans. If the output of a control system has sustained oscillations for a bounded input, the control system is limitedly (marginally) stable.

Q. 65. What is conditional stability ?

Ans. A system is said to be conditionally stable with respect to a system parameter if it is stable only within a certain range of the values of the said parameter.

Q. 66. Explain absolute stability.

Ans. The term absolute stability is used in relation to qualitative analysis of stability. The absolute stability can be determined by identifying the presence of the roots of the characteristic equations in L.H.S. of s-plane.

Q. 67. Explain the term relative stability.

Ans. The relative stability gives the stability of any system in comparison to another system. It is measured in terms of gain margin, phase margin and location of the closed-loop poles (the roots of the characteristic equation) in s-plane.

Q. 68. What conclusion is drawn w.r.t. stability when multiple roots of the characteristic equation lie on the s-plane imaginary axis?

Ans. Ustable (growing response).

Q. 69. What is the necessary condition for stability ?

Ans. The necessary condition for the stability: All the coefficients of characteristic polynomial be positive and none of the coefficients be missing except either all even or odd coefficients.

Q. 70. A system has the characteristic equation as

$$(s + 1)\ (s + 2)\ (s^2 - 2s + 2) = 0.$$

State whether the system is stable/unstable.

Ans. Unstable because, the factor $(s^2 - 2s + 2)$ has roots $(+1 \mp j1)$ with +ive real part.

Q. 71. In a Routh array if all elements in the intermediate row become zero. What is the damping ratio ?

Ans. Complex conjugate roots lies on the s-plane imaginary axis. The response will be oscillatory and $\zeta = 0$.

Q. 72. How to assess relative stability from RH criterion ?

Ans. Apply RH criterion with reference to line say $s = -s_1$

1. For system (A) the roots of the characteristic equation lie right side of line $s = -s_1$.
2. For the system (B) the roots of the characteristic equation lie to the left side of line $s = -s_1$. Then, the system (B) is more stable than (A).

Q. 73. What is polar plot ?

Ans. $|G(j\omega)|$ and corresponding $\angle G(j\omega)$ plotted in a complex $G(j\omega)$-plane with variation of frequency from $\omega = 0$ to $\omega = \infty$ is known as polar plot. Thus, the plot is the locus of phasor $|G(j\omega)|\ \angle G(j\omega)$ as the frequency is varied from $\omega = 0$ to $\omega = \infty$.

Q. 74. Sketch the shape of the polar plot for the transfer function $G(s) = \dfrac{k}{sT + 1}$.

Ans.

$$G(j\omega) = \frac{K}{j\omega T + 1}$$

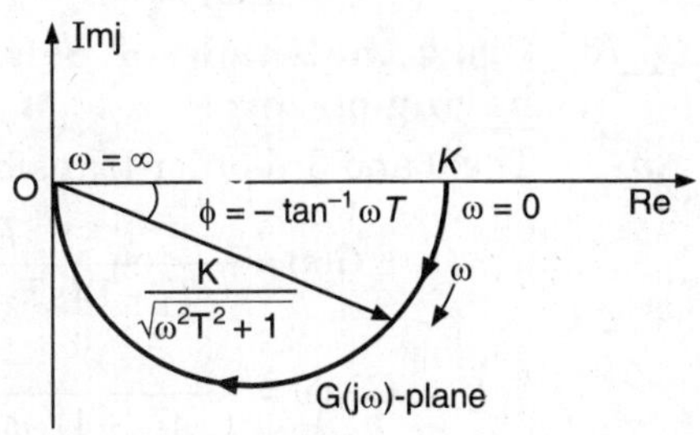

In polar form

$$G(j\omega) = \frac{K}{\sqrt{\omega^2 T^2 + 1}} \angle -\tan^{-1}\frac{\omega T}{1}$$

At $\omega = 0$, $G(j0)| = K$

$\angle G(j0) = -0°$

At $\omega = \infty$, $|G(j\infty)| = 0$

$\angle G(j\infty) = -90°$.

Q. 75. Show the shape of the polar plot for

$$G(s)\,H(s) = \frac{1}{s(sT_1+1)(sT_2+1)}.$$

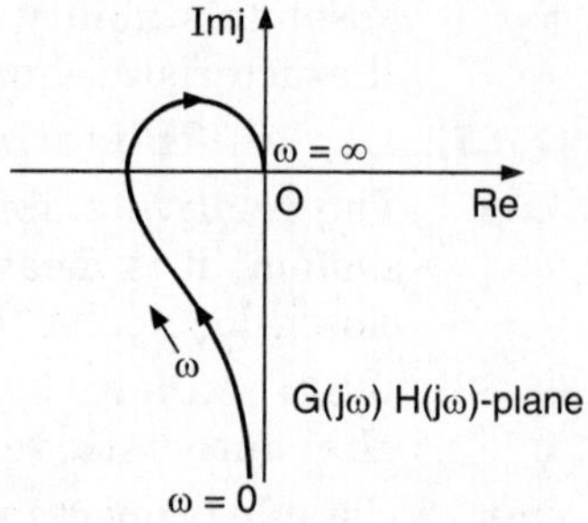

Ans. $G(j\omega)\,H(j\omega) = \dfrac{1}{j\omega(j\omega T_1+1)(j\omega T_2+1)}$

As $\omega \to 0$, $|G(j0)\,H(j0)| \to \infty$

$\angle G(j0)\,H(j0) \to -90°$

As $\omega \to \infty$, $|G(j\infty)\,H(j\infty)| \to 0$

$\angle G(j\infty)\,H(j\infty) \to 270°$.

Q. 76. Draw the polar plot for the transfer function

$$G(j\omega) = \frac{K}{j\omega}.$$

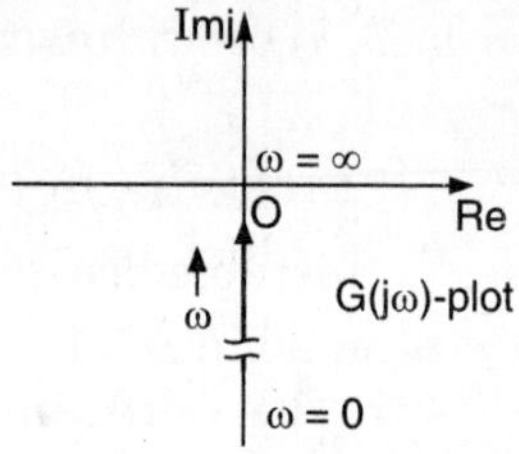

Ans. $|G(j\omega)| = \dfrac{K}{\omega}$

$\angle G(j\omega) = -90°$

As $\omega \to 0$, $|G(j0)| \to \infty$

$\angle G(j0) \to -90°$

As $\omega \to \infty$, $|G(j\infty)| \to 0$

$\angle G(j\infty) \to -90$.

Q. 77. Sketch the polar plot of $G(s) = \dfrac{K}{s^2}$.

Ans. $G(j\omega) = \dfrac{5}{(j\omega)^2}$

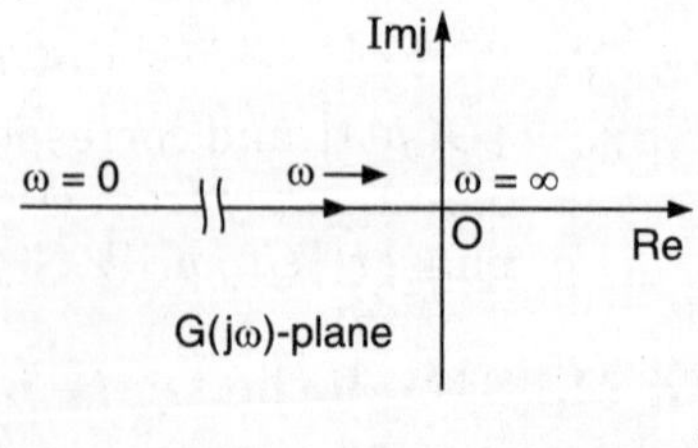

$|G(j\omega)| = \dfrac{5}{\omega^2}$

$\angle G(j\omega) = -180°$

As $\omega \to 0$, $|G(j0)| \to \infty$

$\angle G(j0) \to -180°$

As $\omega \to \infty$, $|G(j\infty)| \to 0$

$\angle G(j\infty) \to -180°$.

Q. 78. Draw the shape of polar plot of type 0 and 3rd order transfer function having no zeros.

Ans. The 0 and 3rd order transfer function is given below :

$$G(s) = \frac{K}{(sT_1+1)(sT_2+1)(sT_3+1)}$$

$$G(j\omega) = \frac{K}{(j\omega T_1+1)(j\omega T_2+1)(j\omega T_3+1)}$$

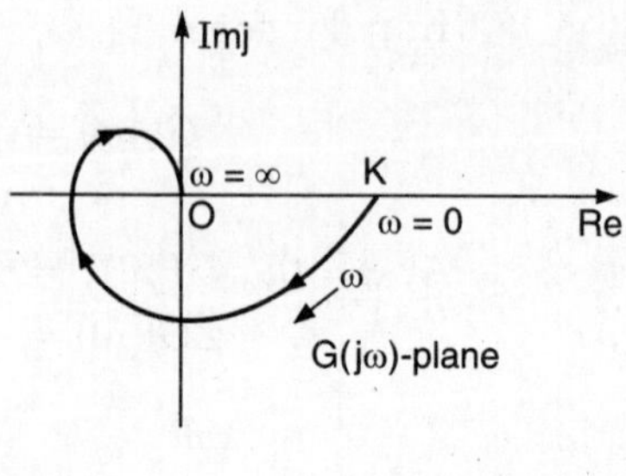

$$|G(j\omega)| = \frac{K}{\sqrt{\omega^2 T_1^2 + 1}\sqrt{\omega^2 T_2^2 + 1}\sqrt{\omega^2 T_3^2 + 1}}$$

$$\angle G(j\omega) = -\tan^{-1}\frac{\omega T_1}{1} - \tan^{-1}\frac{\omega T_2}{1} - \tan^{-1}\frac{\omega T_3}{1}$$

As $\omega \to 0$, $|G(j0)| \to K$

$\angle G(j0) \to -0°$

As $\omega \to \infty$, $|G(j\infty)| \to 0$

$\angle G(j\infty) \to -270°$.

Q. 79. Sketch the polar plot for $G(s) = \frac{1}{s(sT+1)}$.

Ans.

$$G(j\omega) = \frac{1}{j\omega(j\omega T + 1)}$$

$$|G(j\omega)| = \frac{1}{\omega\sqrt{\omega^2 T^2 + 1}}$$

$$\angle G(j\omega) = -90° - \tan^{-1}\frac{\omega T}{1}$$

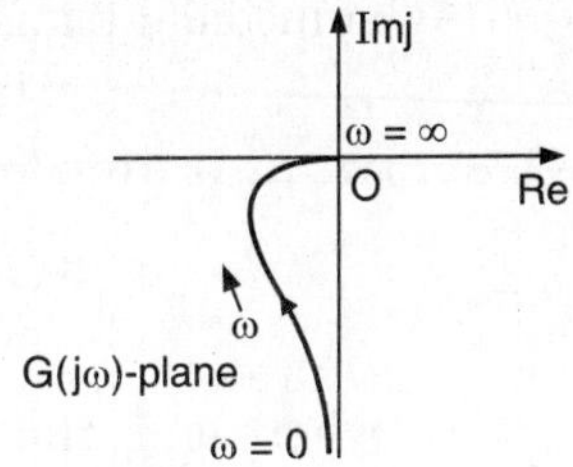

As $\omega \to 0$, $|G(j0)| \to \infty$

$\angle G(j0) \to -90°$

As $\omega \to \infty$, $|G(j\infty)| \to 0$

$\angle G(j\infty) \to -180°$.

Q. 80. What is meant by frequency response of a control system ?

Ans. The steady state response of a system to sinusoidal input is known as frequency response. $|G(j\omega)|$ and $\angle G(j\omega)$ as a function of frequency ω give the plot of frequency response.

Q. 81. State Nyquist criterion.

Ans. The expression for the Nyquist criterion is given below :

$$N = P - Z$$

where N = Number of counter clockwise encirclements of the point $(-1 + j0)$ by the Nyquist plot.

Z = Number of zeroes of the characteristics polynomial $[1 + G(s)H(s)]$ or the roots of the characteristic equation $1 + G(s)H(s) = 0$ having +ive real part.

P = Number of poles of the open-loop transfer function $G(s)H(s)$ having + real part *i.e.*, poles located in R.H.S. of s-plane.

For stability $Z = 0$

$\therefore$ $N = P$.

Q. 82. What is Nyquist plot ?

Ans. The plot of open-loop sinusoidal form of transfer function $G(s)H(s)$ in a complex plane corresponding to Nyquist contour traced in clockwise direction in s-plane is called Nyquist plot.

Q. 83. What is the principle of argument ?

Ans. The principle of argument refers to Nyquist plot and written as $\Delta \text{Aug} = 2\pi (P - Z)$

where ΔAug = is the change in argument of Nyquist plot w.r.t. $(-1 + j0)$ point in $G(s) H(s)$-plane

P = Number of open-loop poles having +ive real part

Z = Number of zeroes of polynomial $[1 + G(s) H(s)]$ having + real part.

Q. 84. Define phase crossover frequency.

Ans. The phase crossover frequency ω_2 is the frequency at which the phase angle of sinusoidal transfer function $G(j\omega) H(j\omega)$ is 180°.

$$\therefore \quad \angle G(j\omega_2) H(j\omega_2) = 180°.$$

Q. 85. Define gain crossover frequency.

Ans. The gain crossover frequency ω_1 is the frequency at which the magnitude of sinusoidal transfer $G(j\omega) H(j\omega)$ is unity.

$$\therefore \quad |G(j\omega_1) H(j\omega_1)| = 1.$$

Q. 86. What is gain margin ?

Ans.
$$\text{G.M.} = \left| \frac{1}{G(j\omega) H(j\omega)} \right|_{\omega = \omega_2}$$

where ω_2 is the phase crossover frequency.

$$\text{G.M.} \Big|_{\text{db}} = 20 \log_{10} \left| \frac{1}{G(j\omega) H(j\omega)} \right|_{\omega = \omega_2}$$

Q. 87. If G(jω) = 0.5 ∠180°, determine the gain margin.

Ans.
$$\text{G.M.} = \frac{1}{0.5} = 2$$

$$\text{G.M.} \Big|_{\text{db}} = 2 \log_{10} 2 = 6.02 \text{ db.}$$

Q. 88. The gain at the phase crossover frequency is $\frac{KT_1 T_2}{T_1 + T_2}$, determine the gain margin.

Ans.
$$\text{G.M.} = 20 \log_{10} \left| \frac{1}{G(j\omega_2) H(j\omega_2)} \right| \text{db}$$

$$= 20 \log_{10} \left| \frac{T_1 + T_2}{KT_1 T_2} \right| \text{db.}$$

Q. 89. Define phase margin.

Ans. The phase margin (ϕ_m) is defined as the additional phase lag required at the gain crossover frequency ω_1 to make the system marginally stable. Thus,

$$\phi_m = 180° + \angle G(j\omega_1) H(j\omega_1).$$

Q. 90. What is the advantage of frequency response analysis ?

Ans. The frequency response plot of an open-loop transfer function gives the assessment of relative stability.

Q. 91. Write the advantages of Bode plot.

Ans. 1. The method of sketching the Bode plot is simple.

2. The multiplication of magnitudes of terms is converted to addition.

3. From the experimental Bode plot the transfer function can be determined.

4. The low frequency and high frequency behaviour of the transfer function can be determined.

Q. 92. What is Bode plot ?

Ans. The Bode plot is a frequency response plot of the open-loop transfer function expressed in sinusoidal form. It consists of two parts as below :

1. Transfer function magnitude in db versus $\log_{10} \omega$.

2. Phase angle of the transfer function versus $\log_{10} \omega$.

Q. 93. Show that 20 db/decade = 6 db/octave.

Ans.

$$1 \text{ decade} = \log_{10} \frac{10}{1} = 1$$

$$1 \text{ octave} = \log_{10} \frac{2}{1} = 0.301$$

$$\therefore \quad 20 \text{ db/decade} = 20 \times 0.301 \text{ db/octave} = 6 \text{ db/octave}.$$

Q. 94. What is the maximum error due to asymptotic magnitude plot for the term (1 + jωT) ?

Ans. The maximum error is 3 db occuring at the corner frequency $\omega = \frac{1}{T}$.

Q. 95. Find the corner frequencies for the transfer function

$$\mathbf{G(s)\,H(s)} = \frac{\mathbf{K}}{\mathbf{s\,(s+1)(s+4)}}.$$

Ans. The corner frequencies due to zeroes are

(*i*) $\omega = 1$ rad./sec.

(*ii*) $\omega = 4$ rad./sec.

Q. 96. Determine the corner frequencies and initial slope of the Bode plot for the open-loop transfer function given below :

$$\mathbf{G(s)\,H(s)} = \frac{\mathbf{20\,(s+0.25)}}{\mathbf{(s+0.5)\,(s+10)}}.$$

Ans. The corner frequencies are

(*i*) $\omega = 0.25$ rad./sec due to zero

(*ii*) $\omega = 0.5$ rad./sec due to pole

(*iii*) $\omega = 10$ rad./sec due to pole

The type of the given transfer function is 0, therefore, the initial slope is 0 db. At the frequencies lower than the lowest corner frequency the Bode plot magnitude starts at $20 \log_{10} K$ and

$$K = \frac{20 \times 0.25}{0.5 \times 10} = 1.$$

Q. 97. How G.M. and P.M. can be determined from the Bode plot ?

Ans. G.M. $= - \mid G(j\omega_2)\, H(j\omega_2) \mid$ db, where ω_2 is the phase crossover frequency.

P.M. $= 180° + \angle G(j\omega_1)\, H(j\omega_1)$, where ω_1 is the gain crossover frequency.

Q. 98. Draw the Body plot for $G(s) = \frac{8}{s}$.

Ans. (1) Type 1 transfer function, the initial slope of the magnitude plot is – 20 db/decade intersecting 0 db axis at $\omega = K = 8$ rad/sec.

(2) $\angle G(j\omega) = -90°$.

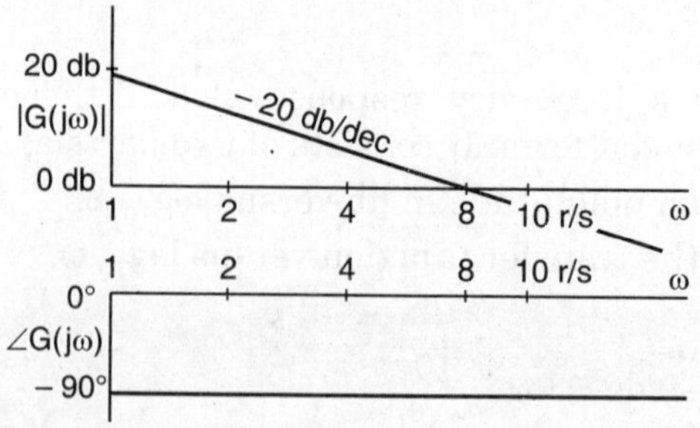

Q. 99. Sketch the magnitude (db) plot for $G(s) = \frac{9}{s^2}$.

Ans. Type 2 transfer function, the initial slope of the magnitude plot is – 40 db/decade intersecting 0 db axis at $\omega = \sqrt{K} = \sqrt{9} = 3$ rad./sec.

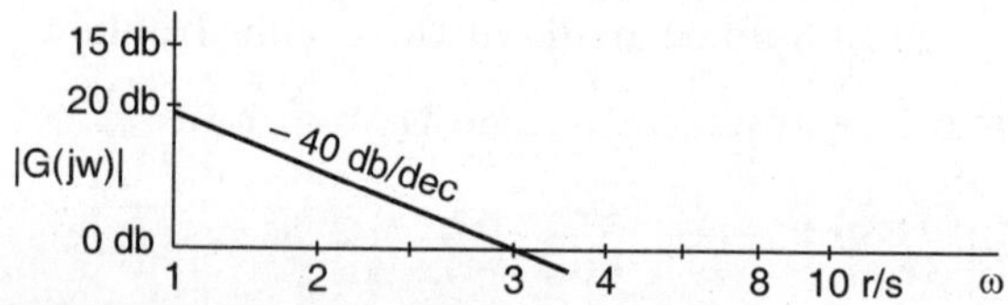

Q. 100. What is minimum phase transfer function ?

Ans. The transfer function having poles and zeroes located to the left hand side of *s*-plane imaginary axis is called minimum phase transfer function.

Q. 101. What is all pass transfer function ?

Ans. The transfer function having symmetrical pole-zero locations w.r.t. *s*-plane imaginary axis is called all pass transfer function. The magnitude is always unity.

Q. 102. What is non-minimum phase transfer function ?

Ans. The transfer function having one or more zeroes in the right hand side of *s*-plane is called non-minimum phase transfer function.

Q. 103. Define corner frequency.

Ans. The corner frequency is defined as the meeting point of low frequency and high frequency asymptotes on 0 db axis of the Bode plot.

Q. 104. What is asymptotic Bode plot ?

Ans. The logarithmic magnitude plot of the frequency response can be approximated as below :

(1) For the term $(1 + j\omega T)^{\pm N}$, by two straight line asymptotes, one for the frequencies less than the corner frequency $\omega = \frac{1}{T}$. This line is 0 db axis. The other line has a slope of ± 20 N db/sec. starting at the corner frequency $\omega = \frac{1}{T}$.

The maximum error due to this approximation is ± 3 N db occuring at the corner frequency $\omega = \frac{1}{T}$.

(2) For the term $(s^2 + 2\zeta\omega_n s + \omega_n^2)^{\pm 1}$, by two straight line asymptotes, one for the frequencies less than the corner frequency ω_n. This line is 0 db axis.
The other line has a slope of ± 40 db/sec. starting at the corner frequency ω_n.
The exact value near about the corner frequency ω_n or resonant frequency ω_r depends on the value of damping ratio ζ.

Q. 105. Define resonant frequency.

Ans. The frequency at which the resonance peak occurs is known as resonant frequency ω_r. The input being sinusoidal.
The resonant frequency is related to natural frequency of oscillations due to step input response as below :

$$\omega_r = \omega_n \sqrt{1 - 2\zeta^2}\,, \zeta < 0.707$$

where ω_n = natural frequency of oscillations.
and ζ = damping ratio.

Q. 106. Distinguish between the resonant frequency and natural frequency of oscillations.

Ans. At resonant frequency ω_r the resonance occurs due sinusoidal input. The resonant frequency ray is given by

$$\omega_r = \omega_n \sqrt{1 - 2\zeta^2}\,, \zeta < 0.707$$

The natural frequency of oscillations ω_n is the undamped frequency oscillations due to step input.

Q. 107. Determine the resonant peak of a second order control system having damping ratio ζ = 0.6.

Ans.

$$M_r = \frac{1}{2\zeta\sqrt{1-\zeta^2}}, \zeta < 0.707$$

$$= \frac{1}{2 \times 0.6\sqrt{1 - 0.6^2}} = 1.041.$$

Q. 108. M-circle centre is located at.

Ans.

$$x = -\frac{M^2}{M^2 - 1}, y = 0.$$

Q. 109. What is the equation of a line on which the centre of constant N-circles is located ?

Ans.

$$x = -\frac{1}{2}.$$

Q. 110. What are M-circles ?

Ans. The magnitude "*M*" of a unity feedback control system's transfer function forms a circle in a complex plane for each constant value of "*M*". The family of such circles are called *M*-circles.

Q. 111. Mention the use of Nichol's chart.

Ans. The gain phase plot of a unity feedback control system is superimposed on the Nichol's chart to obtain the closed loop frequency of a control system.

Q. 112. What are N-circles ?

Ans. For a unity feedback control system, $N = \tan \phi$, where ϕ is the phase angle of the open-loop transfer function. For every value of "N" a circle can be drawn in a complex plane. The family of such circles are called N-circles.

Q. 113. Explain the assessment of stability using Bode plot.

Ans. Using Bode plot the stability can be assessed as below :

(1) If G.M. and P.M. are +ive : stable system

(2) If G.M. and P.M. are –ive : Unstable system

(3) If G.M. and P.M. are zero : Marginally stable system.

Q. 114. What is meant by the term magnitude criterion with reference to root locus technique ?

Ans. The characteristic equation is

$$1 + G(s)\,H(s) = 0$$

or
$$G(s)\,H(s) = -1$$

The point s_1 located on the root locus plot satisfies the following magnitude condition,

$$|\,G(s_1)\,H(s_1)\,| = 1.$$

Q. 115. What is phase angle criterion with reference to root locus technique ?

Ans. The characteristic equation is

$$1 + G(s)\,H(s) = 0$$

$$\therefore \qquad G(s)\,H(s) = -1$$

The point s_1 located on the root locus plot satisfies the phase angle condition.

$$\angle\, G(s_1)\,H(s_1) = (2k + 1)\,180°$$

where, $k = 0, 1, 2, \ldots\ldots$

Q. 116. State the rule for obtaining the breakaway point on the root locus plot.

Ans. Breakaway point on the root locus between two open-loop poles can be determined by arranging the characteristic equation as below :

$$K = F(s)$$

and solving $\dfrac{dK}{ds} = 0$ for "s". K is the gain factor of the open-loop transfer function.

Q. 117. What is the centroid of root locus plot ?

Ans. The intersection point of the root locus asymptotes on the real axis is known as the centroid and determined using the following relation

$$x = \frac{\Sigma \text{ Poles} - \Sigma \text{ zeros}}{P - Z}.$$

Q. 118. What are asymptotes in relation to root locus ?

Ans. The straight lines parallel to the root locus branches as the gain factor K approaches ∞ are called asymptotes. The angle of asymptotes is given by

$$\angle\text{Asymptotes} = \frac{(2k + 1) \times 180°}{P - Z} \qquad k = 0, 1, 2, \ldots\ (P - Z)$$

Q. 119. State rule for the existence of root locus on the s-plane real axis.

Ans. The existence of root locus on a section of s-plane real axis is confirmed if the sum of the number of poles and zeroes to the right of the section is odd.

Q. 120. The open-loop transfer function is given below :

$$G(s) = \frac{K}{s\,(s+6)^2}$$

determine the breakaway point on the root locus.

Ans. The characteristic equation is

$$1 + G(s) = 0$$

$$\therefore \qquad 1 + \frac{K}{s\,(s+6)^2} = 0$$

Rearranging $s(s+6)^2 + K = 0$

$$\therefore \qquad K = -\,s\,(s+6)^2$$

$$= -\,(s^3 + 12s^2 + 36s)$$

$$\frac{dK}{ds} = -\,(3s^2 + 24s + 36) = -\,3\,(s+2)\,(s+6)$$

Put $\frac{dK}{ds} = 0$

$$\therefore \qquad -\,3(s+2)\,(s+6) = 0$$

$$s_1 = -\,2,\ s_2 = -\,6$$

$s = -\,2$ is the breakaway point.

Q. 121. How to determine intersection of the root locus branches with the s-plane imaginary axis ?

Ans. Apply Routh-Hurwitz criterion to the polynomial $F(s) = 1 + G(s)\,H(s)$ and determine the value of gain factor K for marginal stability, put $s = j\omega$ in the s^2 row and equate the row equation to zero. Solve for ω. This complex value of ω gives the crossing point of the root locus with s-plane imaginary axis.

Q. 122. State relation for the determination of angle of departure of root locus from a complex open-loop pole.

Ans. The angle of departure from a complex open-loop pole is given by

$$\phi_d = 180° - (\phi_p - \phi_z)$$

ϕ_p = the sum of all angles subtended by remaining poles at the pole under consideration.

ϕ_z = the sum of all angles subtended by the zeroes at the pole under consideration.

Q. 123. Determine the angle of asymptotes for the root locus.

$$G(s)\,H(s) = \frac{K}{s(s+4)}.$$

Ans.

$$\angle\text{Asymptotes} = \frac{(2k+1)}{P-Z} \times 180°, \quad k = 0, 1.$$

(*i*) $\frac{(2\times 0+1)\times 180°}{2-1} = 90°$

(*ii*) $\frac{(2\times 1+1)\times 190°}{2-0} = 270°.$

Q. 124. What is control system compensation ?

Ans. The method of incorporating a compensation network in a control system to improve its performance is known as compensation.

Q. 125. What are the effects of Lead/Lag/Lag-lead compensation network ?

Ans. **Lead network :** Produces phase lead, shifts gain crossover frequency to a higher value reduced settling time, improvement in response speed, not much improvement in steady state error.

Lag network : Provides phase lag, reduction in steady state error and increase in steady state error.

Lag-Lead network : Reduction in steady state error and fast response.

Q. 126. Write an expression for ϕ_m (maximum lead) and the corresponding frequency ω_m for lead compensation network.

Ans.

$$\phi_m = \tan^{-1} \frac{1-\alpha}{2\sqrt{\alpha}}$$

$$\omega_m = \frac{1}{T\sqrt{\alpha}}.$$

Q. 127. What is the purpose of a compensating network in a feedback control system ?

Ans. The compensation for a feedback control system is necessary in the situation given below :

(1) For unstable system, the compensation is used to make the system stable with desired performance specifications.

(2) For stable systems compensation is used to meet the desired performance specifications.

Q. 128. Sketch the approximate (asymptotic) Bode plot for a lead compensator.

Ans. The sinusoidal form of the lead compensator is given by

$$G_c(j\omega) = \frac{\alpha(1 + j\omega T)}{(1 + j\omega\alpha T)}, \ \alpha < 1, T > 0$$

lower corner frequency : $\omega = \dfrac{1}{T}$

upper corner frequency : $\omega = \dfrac{1}{\alpha T}$

The transfer function is type 0, intercept on 0 db axis for frequencies lower than the lowest corner frequency $\omega = \dfrac{1}{T}$ is $20 \log_{10} \alpha$.

$$\angle G_c(j\omega) = \tan^{-1} \frac{\omega T}{1} - \tan^{-1} \frac{\omega\alpha T}{1}$$

The approximate Bode plot is shown in adjoining figure.

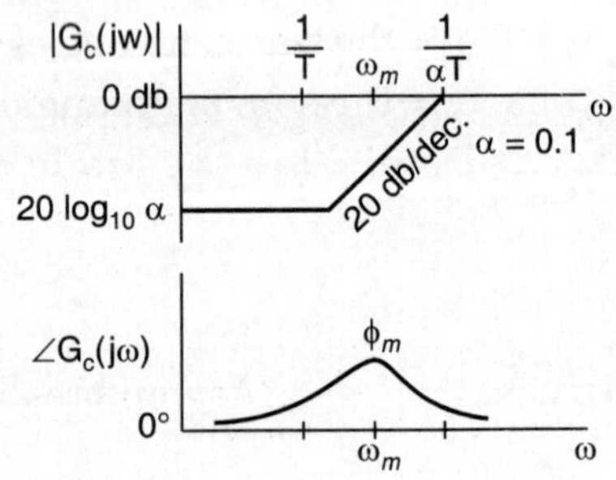

Q. 129. Write the transfer function of a lead compensation network and draw pole-zero configuration.

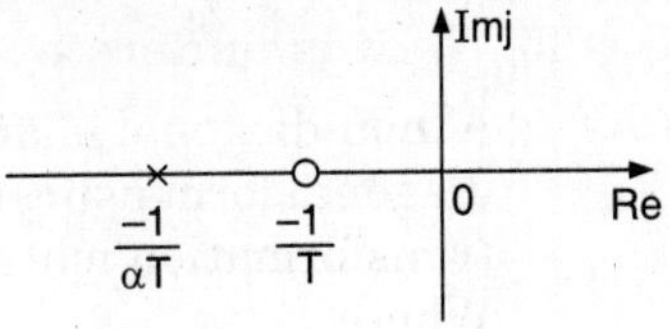

Ans. $$\underset{\text{(lead)}}{G_c(s)} = \frac{\alpha(1+1+sT)}{(1+s\alpha T)}$$

$\alpha < 0$, $T > 0$

Q. 130. Write the transfer function of Lag compensator and draw pole-zero configura-tion.

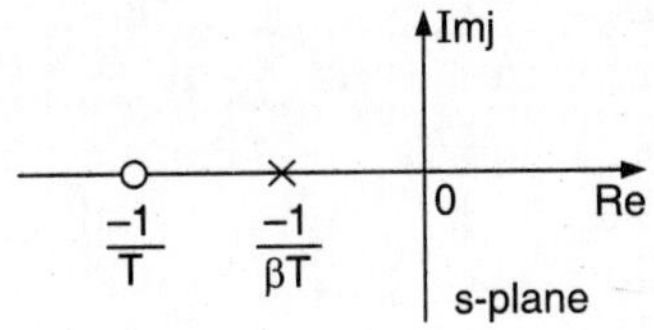

Ans. $$\underset{\text{(lead)}}{G_c(s)} = \frac{1+sT}{1+s\beta T},$$

$\beta > 1$; $T > 0$

Q. 131. Give the state equation for a LTI system.

Ans. $\dot{\mathbf{x}} = \mathbf{Ax} + \mathbf{Bu}$ MIMO system.

$\dot{\mathbf{x}} = \mathbf{Ax} + \mathbf{Bu}$ SISO system.

Q. 132. What is understood by the term "state transition matrix" ?

Ans. The solution of a homogeneous state equation is given by $\mathbf{x}(t) = e^{\mathbf{A}t}\,\mathbf{x}(0)$

The matrix exponential $e^{\mathbf{A}t}$ carrier out the transition of the state from one instant to another during time period t. The matrix exponential $e^{\mathbf{A}t}$ is known as state transition matrix denoted as $\phi(t)$.

Q. 133. What are phase variables ?

Ans. $$\mathbf{A} = \begin{bmatrix} 0 & 1 & 0 & \cdots & \cdots & 0 \\ 0 & 0 & 1 & \cdots & \cdots & 0 \\ \vdots & \vdots & \vdots & & & \vdots \\ -a_n & -a_{n-1} & \cdots & \cdots & \cdots & -a_1 \end{bmatrix}$$

The system matrix **A** given above is known to be as Bush's form/controllable canonical form/phase variable form where in the upper elements off the are 1 and the last row consists of negated coefficients of the characteristic polynomial, all other elements are zero. The state variables forming the above matrix **A** are called phase variables.

Q. 134. Write the state model for the system having transfer function given below :

$$\frac{\mathrm{C(s)}}{\mathrm{R(s)}} = \frac{2}{\mathrm{s^3 + 4s^2 + 6s + 8}}.$$

Ans. $$\begin{bmatrix} \dot{x}_1 \\ \dot{x}_2 \\ \dot{x}_3 \end{bmatrix} = \begin{bmatrix} 0 & 1 & 0 \\ 0 & 0 & 1 \\ -8 & -6 & -4 \end{bmatrix} \begin{bmatrix} x_1 \\ x_2 \\ x_3 \end{bmatrix} + \begin{bmatrix} 0 \\ 0 \\ 2 \end{bmatrix} r(t)$$

$$y = [1 \;\; 0 \;\; 0] \begin{bmatrix} x_1 \\ x_2 \\ x_3 \end{bmatrix}.$$

Q. 135. What is understood by diagonalization of a matrix ?

Ans. A non-diagonal matrix **A** with distinct eigen values can be transformed into a diagonal form using similarity transformation. For the given matrix **A** the similarity transformation matrix **M** (modal matrix) is used to get $\mathbf{M}^{-1}\mathbf{AM}$ as diagonal matrix. Where,

$$\mathbf{M} = \begin{bmatrix} m_{11} & m_{12} & \cdots & \cdots \\ m_{21} & m_{22} & \cdots & \cdots \\ \vdots & \vdots & & \end{bmatrix}$$

$m_{11}, m_{12}, \ldots\ldots$

$m_{21}, m_{22}, \ldots\ldots$

are the eigen vectors of the given non-diagonal matrix.

Q. 136. $\dot{\mathbf{x}} = \mathbf{Ax} + \mathbf{Bu}$

$\mathbf{y} = \mathbf{Cx} + \mathbf{Du}$

The equations given above represent a state model. Write expression for the transfer matrix.

Ans. $$\frac{\mathbf{Y}(s)}{\mathbf{U}(s)} = \mathbf{C}(s\mathbf{I} - \mathbf{A})^{-1}\,\mathbf{B} + \mathbf{D}.$$

Q. 137. What is Vander Monde's matrix ?

Ans. Vander Monde's matrix **P** is used to transform **A** matrix (having distinct eigen values) given in phase variable/controllable canonical form into a diagonal form. The transformed matrix $\mathbf{P}^{-1}\mathbf{AP}$ is diagonal.

$$\mathbf{P} = \begin{bmatrix} 1 & 1 & 1 & \ldots\ldots \\ \lambda_1 & \lambda_2 & \lambda_3 & \ldots\ldots \\ -\lambda_1^2 & -\lambda_2^2 & -\lambda_3^2 & \ldots\ldots \\ \vdots & \vdots & \vdots & \end{bmatrix}$$

where, $\lambda_1, \lambda_2, \lambda_3, \ldots\ldots$ are eigen values of the given matrix **A**.

Q. 138. What is understood by pulse transfer function ?

Ans. The function $G(z) = \dfrac{C(z)}{R(z)}$ is known as pulse transfer function.

$C(z)$ is the z-transform of the output function and $R(z)$ is the z-transform of the input function.

Q. 139. Give relation between s-domain and z-domain.

Ans. $z = e^{sT}$.

Q. 140. Write z-transform of unit step function.

Ans. $$Z[1] = \frac{z}{z-1}.$$

Q. 141. What is hold circuits ?

Ans. The hold circuit enables to hold the signal between two consecutive sampling instants at the preceding value till the next sampling instant is reached.

Q. 142. Sketch the stable region in z-plane.

Ans. The inside region of a unit circle centred at the origin of z-plane is the stable region for discrete-data (sampled data) control systems.

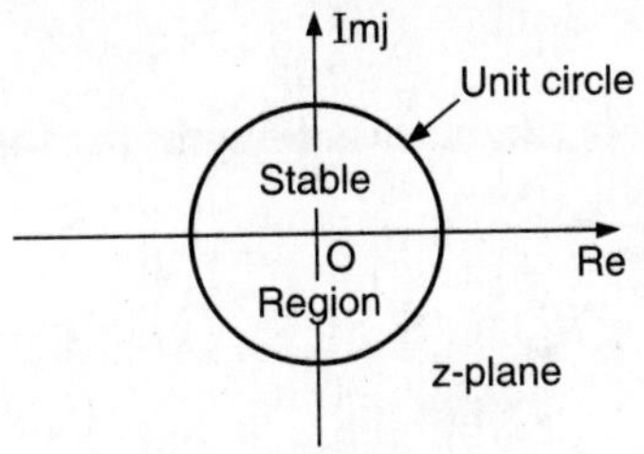

Q. 143. Write the transfer function of a zero order hold.

Ans. $G_{ho}(s) = \dfrac{1-e^{-sT}}{s}$ where, T is sampling period.

Q. 144. What is bilinear transformation in relation to characteristic polynomial in z-domain ?

Ans. The change of variable from z to w is known as bilinear transformation, where

$$z = \frac{1-w}{1+w}$$

Routh-Hurwitz's criterion can be applied to characteristic polynomial 1 + $GH(z)$ for the stability determination of a discrete-data (sampled-data) control system.

APPENDIX IV

Key Formulae Tables, Charts

Feedback control system

Closed-loop (overall transfer function) :

$$\frac{C(s)}{R(s)} = \frac{G(s)}{1 + G(s)\,H(s)}$$

Characteristic equation : $1 + G(s)\,H(s) = 0$

Second order control system

Closed-loop transfer function : $$\frac{C(s)}{R(s)} = \frac{\omega_n^2}{s^2 + 2\xi\omega_n s + \omega_n^2}$$

Damping ratio : ξ

Natural frequency of oscillation : ω_n

Damped frequency of oscillations : $\omega_d = \omega_n\sqrt{1-\xi^2}$

Time response to unit step input

$\xi < 1$ (underdamped)

$$c(t) = 1 - \frac{e^{-\xi\omega_n t}}{\sqrt{1-\xi^2}} \sin\left\{(\omega_n\sqrt{1-\xi^2})t + \tan^{-1}\frac{\sqrt{1-\xi^2}}{\xi}\right\}$$

Time constant : $$\tau = \frac{1}{\xi\omega_n}$$

Rise time : $$t_r = \frac{\pi - \phi}{\omega_n\sqrt{1-\xi^2}}\,;\ \phi = \tan^{-1}\frac{\sqrt{1-\xi^2}}{\xi}$$

Peak time : $$t_p = \frac{\pi}{\omega_n\sqrt{1-\xi^2}}$$

Settling time : $$t_s = \frac{4}{\xi\omega_n}\ \text{(2\% basis)}$$

$$t_s = \frac{3}{\xi\omega_n}\ \text{(5\% basis)}$$

% Max. overshoot : $M_p = e^{\frac{-\xi\pi}{\sqrt{1-\xi^2}}} \times 100\%$

$\xi = 0$ (no damping)

$$c(t) = (1 - \cos \omega_n t)$$

$\xi = 1$ (critical damping)

$$c(t) = \{1 - e^{-\omega_n t}\ (1 + \omega_n t)\}$$

$\xi < 1$ (over damped)

$$c(t) = \frac{e^{-(\xi - \sqrt{\xi^2 - 1})\,\omega_n t}}{2\sqrt{\xi^2 - 1}\,(\xi - \sqrt{\xi^2 - 1})} + \frac{e^{-(\xi - \sqrt{\xi^2 - 1})\,\omega_n t}}{2\sqrt{\xi^2 - 1}\,(\xi + \sqrt{\xi^2 - 1})}$$

Characteristic equation : $s^2 + 2\xi\omega_n s + \omega_n^2 = 0$

Root of the characteristic equation : $s = -\xi\omega_n \pm j\omega_n\sqrt{1-\xi^2}$

Response to impulse input

$\xi < 1$

$$c(t) = \frac{\omega_n}{\sqrt{1-\xi^2}}\, e^{-\xi\omega_n t} \sin\left\{(\omega_n\sqrt{1-\xi^2})t\right\}$$

$\xi = 1$

$$c(t) = t\,(\omega_n^2 e^{-\omega_n t})$$

$\xi > 1$

$$c(t) = \frac{\omega_n}{2\sqrt{\xi^2 - 1}}\left[e^{-(\xi - \sqrt{\xi^2 - 1})\,\omega_n t} - e^{-(\xi + \sqrt{\xi^2 - 1})\,\omega_n t}\right]$$

Steady state error $e_{ss} = \dfrac{1}{1 + K_p}$

$$l_{ss} = \frac{e}{K_v}$$

$$l_{ss} = \frac{e}{K_a}$$

Error transfer function $\dfrac{E(s)}{R(s)} = \dfrac{1}{1 + G(s)\,H(s)}$

Second order system

Transfer function with proportional control : $\dfrac{C(s)}{R(s)} = \dfrac{\omega_n^2}{s^2 + 2\xi\omega_n s + \omega_n^2}$

Transfer function with derivative control : $\dfrac{C(s)}{R(s)} = \dfrac{(1 + T\alpha)\omega_n^2}{s^2 + (2\xi\omega_n + \omega_n^2 T_\alpha)s + \omega_n^2}$

Transfer function with integral control : $\dfrac{C(s)}{R(s)} = \dfrac{(s + K_i)\omega_n^2}{s^3 + 2\xi\omega_n s^2 + \omega_n^2 s + K_i\omega_n^2}$

Transfer function with derivative feedback control : $\dfrac{C(s)}{R(s)} = \dfrac{\omega_n^2}{s^2 + (2\xi\omega_n + \omega_n^2 K_t)s + \omega_n^2}$

Frequency response analysis

Gain crossover frequency : The frequency ω_1 is the point on the $G(j\omega)\,H(j\omega)$ plot such that $|G(j\omega_1)\,H(j\omega_1) = 1$ *i.e.* $|G(j\omega_1)\,H(j\omega_1)|_{db} = 0$

Phase crossover frequency : The frequency ω_2 is the point on the $G(j\omega)\,H(j\omega)$ plot such that $\angle G(j2)\,H(j2) = 180°$ *i.e.* $\angle G(j\omega)\,H(j\omega) = 180°$

The gain crossover and phase crossover frequencies can be determined on the three plots mentioned below :

Nyquist/Polar plot

Locate ω_1 and ω_2 such that

For gain crossover ; $|G(j\omega_1)| = 1$

For phase crossover ; $\angle G(j\omega_2) = 180°$

Bode (Asymptotic) plot

For gain crossover : The magnitude-db plot intersects 0 db axis at ω_1

For phase crossover : The phase curve intersects – 180° axis at ω_2

Gain phase plot

For gain crossover : The $G(j\omega)$ at $H(j\omega)$ curve crosses 0 db axis at ω_1

For phase crossover : The $G(j\omega)\,H(j\omega)$ curve crosses – 180° axis

Gain Margin : $$\text{G.M.}_{db} = 20 \log_{10} \left| \frac{1}{G(j\omega_2)\,H(j\omega_2)} \right|$$

Phase Margin : $$\text{P.M.} = 180° + \angle G(j\omega_1)\,H(j\omega_1)$$

Resonant frequency : $$\omega_r = \omega_n \sqrt{1 - 2\xi^2}\ ;\ \xi < 0.707$$

Resonant peak : $$M_r = \frac{1}{2\xi\sqrt{1-\xi^2}}$$

The initial slope of the Bode plot depends on the type of open-loop transfer function $G(j\omega)\,H(j\omega)$. The table below give the slope and corresponding type.

Transfer function type	*Initial slope of Bode plot*	*Remark*
0	0	A line $\lvert G(j\omega)\,H(j\omega)\rvert_{db} = 20 \log_{10} K$ Parallel to ω axis
1	– 20 db/decade	Intersect 0 db axis at $\omega = K$
2	– 40 db/decade	Intersect 0 db axis at $\omega = K^{1/2}$
3	– 60 db/decade	Intersect 0 db axis at $\omega = K^{1/3}$
N	– 20 N db/decade	Intersect 0 db axis at $\omega = K^{1/N}$

Bode plot : $(1 + j\omega T)^{\pm 1}$ magnitude error table

The actual bode plot for the term $(1 + j\omega T)^{\pm 1}$ deviates from the asymptotic (straight line) plot. The error due to deviation is tabulated in the table given below :

ωT	ω	$\lvert(1 + j\omega T)^{\pm 1}\rvert\, db$	*Straight line approximation* $\lvert(1 + j\omega T)^{\pm 1}\rvert\, db$	*Error-db*
0.01	$\frac{0.01}{T}$	± 0.00043	0	± 0.00043
0.1	$\frac{0.1}{T}$	± 0.043	0	± 0.043
0.5	$\frac{0.5}{T}$	± 1.0	0	± 1.0
0.76	$\frac{0.76}{T}$	± 2.0	0	± 2.0
1.0	$\frac{1}{T}$	± 3.0	0	± 3.0
1.31	$\frac{1.31}{T}$	+ 4.3	± 2.3	± 2.0
2.0	$\frac{2}{T}$	+ 7.0	± 6.0	± 1.0
10	$\frac{10}{T}$	± 20.043	± 20	± 0.043
100	$\frac{100}{T}$	± 40.00043	± 40	± 0.00043

Semilog scale

The Bode plot consists of $\lvert G(j\omega) H(j\omega)\rvert$ in decible versus $\log_{10} \omega$. A unit change in $\log_{10} \omega$ corresponds to one decade (ratio of 101) variation of ω *i.e.* ..., $\omega = 1$ rad/sec to $\omega = 10$ rad/sec, $\omega = 10$ rad/sec to $\omega = 100$ rad/sec, ...

This change, as independent variable and $\lvert G(j\omega) H(j\omega)\rvert_{db}$ as dependent variable.

The semilog scale can be plotted on plain paper. Select some scale say 1 decade = 6 cms. The selection of scale is arbitrary and to be adjusted according to the size of the paper. The calibration is given below :

Frequency		*Distance from initial point semilog scale*
ω	Initial frequency point on semilog scale	0
2ω	$\log_{10}\left(\frac{2\omega}{\omega}\right) = 0.301$	$0.301 \times 6 = 1.8$ cm
3ω	$\log_{10}\left(\frac{3\omega}{\omega}\right) = 0.4771$	$0.4771 \times 6 = 2.86$ cm

(Contd...)

Frequency		*Distance from initial point semilog scale*
4ω	$\log_{10}\left(\frac{4\omega}{\omega}\right) = 0.602$	0.602 × 6 = 3.6 cm
5ω	$\log_{10}\left(\frac{5\omega}{\omega}\right) = 0.699$	0.699 × 6 = 4.19 cm
6ω	$\log_{10}\left(\frac{6\omega}{\omega}\right) = 0.7782$	0.7782 × 6 = 4.66 cm
7ω	$\log_{10}\left(\frac{7\omega}{\omega}\right) = 0.8451$	0.8451 × 6 = 5.07 cm
8ω	$\log_{10}\left(\frac{8\omega}{\omega}\right) = 0.9031$	0.9031 × 6 = 35.72 cm
9ω	$\log_{10}\left(\frac{9\omega}{\omega}\right) = 0.9542$	0.9542 × 6 = 5.41 cm
10 ω	$\log_{10}\left(\frac{10\omega}{\omega}\right) = 1.0$	1 × 6 = 6 cm

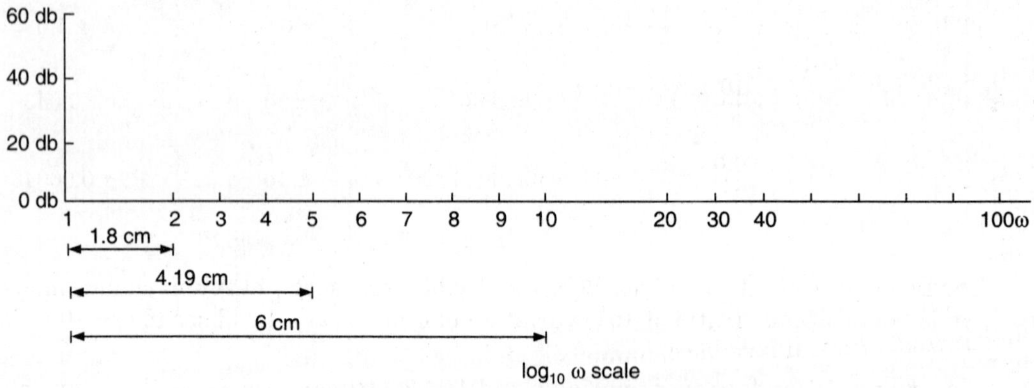

M loci for Polar plot

M	*Centre x coordinate y = 0*	*Radius*	*ψ = sin⁻¹ (1/M)°*
0.5	0.333	0.67	
0.6	0.562	0.94	
0.7	0.96	1.37	
0.8	1.777	2.22	
0.9	4.26	4.70	
1.0	∞	∞	90

(Contd...)

M	*Centre x coordinate* $y = 0$	*Radius*	$\psi = \sin^{-1}(1/M)°$
1.1	– 5.77	5.24	65.2
1.2	– 3.27	2.73	56.5
1.3	– 2.45	1.88	50.2
1.4	– 2.04	1.46	45.6
1.5	– 1.8	1.2	41.8
1.6	– 1.64	1.03	38.7
1.7	– 1.53	0.9	36.0
1.8	– 1.47	0.84	33.7
1.9	– 1.38	0.73	31.7
2.0	– 1.33	0.666	30.0
2.2	– 1.26	0.573	27.0
2.4	– 1.21	0.505	24.6
2.5	– 1.19	0.476	23.6
2.6	– 1.17	0.451	22.6
2.7	– 1.15	0.41	20.9
3.0	– 1.12	0.375	19.5
3.5	– 1.10	0.34	16.6
4.0	– 1.07	0.266	14.8
4.5	– 1.05	0.234	12.8
5.0	– 1.04	0.208	11.5

N Loci for Polar plot

$\phi°$	*Centre y co-ordinate* $x = -0.5$	*Radius*
0	∞	∞
– 5	– 5.73	5.75
– 10	– 2.84	2.88
– 15	– 1.84	1.90
– 20	– 1.37	1.46
– 25	– 1.07	1.17
– 30	– 1.86	1.0
– 35	– 0.714	0.87

(Contd...)

φ°	*Centre y co-ordinate* $x = -0.5$	*Radius*
– 40	– 0.59	0.77
– 45	– 0.5	0.707
– 50	– 0.42	0.65
– 55	– 0.35	0.61
– 60	– 0.29	0.57
– 65	– 0.22	0.54
– 70	– 0.18	0.53
– 75	– 0.13	0.518
– 80	– 0.087	0.506
– 90	– 0.00	0.50

Conversion from Ratio to Decibel

Ratio	5	4.5	4	3.5	3.0	2.5	2.0	1.9	1.8
Decibel	14	13	12	10.85	9.51	7.39	6.0	5.57	5.10

Ratio	1.7	1.6	1.5	1.4	1.3	1.2	1.0	0.9	0.8
Decibel	4.6	4.08	3.52	2.91	2.28	1.58	0	– 0.906	– 1.938

Ratio	0.7	0.6	0.5	0.4	0.3	0.2	0.1	0.05
Decibel	– 3.10	– 4.44	– 6.0	– 7.39	– 10.44	– 14	– 20	– 28

Conversion from Decibel to Ratio

Decibel	20	18	15	12	10	9	8	7	6	5
Ratio	10	7.94	5.62	3.99	3.16	2.82	2.51	2.25	2.0	1.78

Decibel	4	3	2	1	0.5	0	– 0.5	– 1.0	– 2.0	– 3.0	– 4.0
Ratio	1.59	1.41	1.26	1.12	1.06	1.0	0.945	0.894	0.795	0.707	0.63

Decibel	– 5	– 6	– 7	– 8	– 9	– 10	– 12	– 15	– 18	– 20
Ratio	0.562	0.5	0.447	0.398	0.354	0.316	0.251	0.178	0.125	0.10

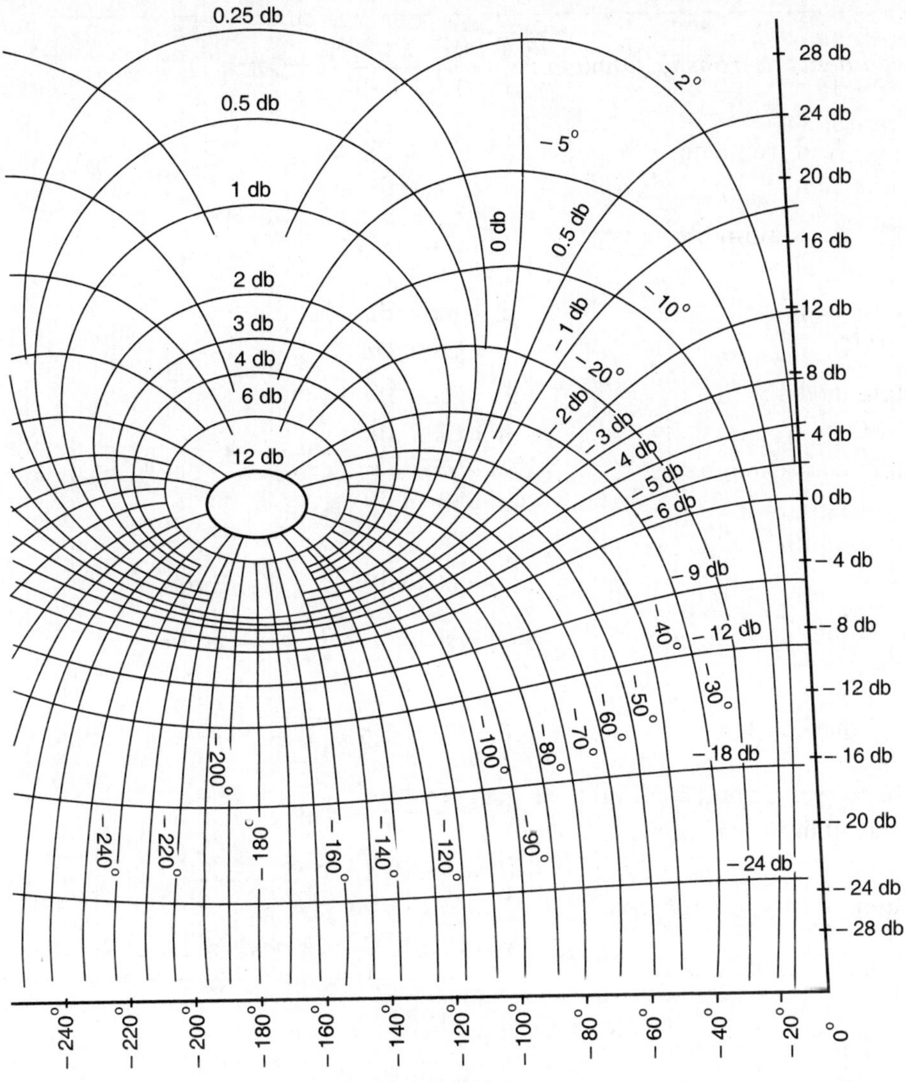

Nichols Chart

1. *Lead network transfer function* : $\dfrac{E_o(s)}{E_i(s)} = \dfrac{\alpha(1+sT)}{(1+s\alpha T)}$; $\alpha < 1$

Mid-frequency : $\omega_m = \dfrac{1}{(\sqrt{\alpha})T}$

Maximum phase load : $\phi_m = \sin^{-1}\dfrac{(1-\alpha)}{(1+\alpha)}$

2. *Lead network transfer function :* $\dfrac{E_o(s)}{E_i(s)} = \dfrac{1+sT}{1+s\beta T}$; $\beta > 1$

Mid-frequency : $\omega_m = \dfrac{1}{(\sqrt{\beta})T}$

Maximum phase lag : $\phi_m = \sin^{-1} \dfrac{(1-\beta)}{(1+\beta)}$

State model : $\dot{\boldsymbol{x}} = \mathbf{A}\boldsymbol{x} + \mathbf{B}\boldsymbol{u}$
(MIMO) $\boldsymbol{y} = \mathbf{C}\boldsymbol{x} + \mathbf{D}\boldsymbol{u}$

State model : $\dot{\boldsymbol{x}} = \mathbf{A}\boldsymbol{x} + \mathbf{B}u$
(SISO) $\boldsymbol{y} = \mathbf{C}\boldsymbol{x} + \mathbf{D}u$

State transition matrix : $\boldsymbol{\phi}(t) = \mathcal{L}^{-1}(s\,\mathbf{I} - \mathbf{A})^{-1}$
$\boldsymbol{\phi}(t) = e^{\mathbf{A}t}$

Time solution :

$$\boldsymbol{x}(t) = \boldsymbol{\phi}(t)\,\boldsymbol{x}(0) + \mathcal{L}^{-1}\,\boldsymbol{\phi}(s)\,\mathbf{B}\boldsymbol{u}(s)$$

or $$\boldsymbol{x}(t) = \boldsymbol{\phi}(t)\,\boldsymbol{x}(0) + \int_0^t \boldsymbol{\phi}(t-\tau)\,\mathbf{B}\boldsymbol{u}(\tau)\,d\tau$$

Transfer matrix : $\dfrac{\mathbf{Y}(s)}{\mathbf{U}(s)} = \mathbf{C}\,(s\mathbf{I} - \mathbf{A})^{-1}\,\mathbf{B} + \mathbf{D}$

Characteristic polynominal : $|s\mathbf{I} - \mathbf{A}| = 0$

Controllability test matrix :

$$\mathbf{U} = \mathbf{B} : \mathbf{AB} : \mathbf{A}^2\mathbf{B} : ...... : \mathbf{A}^{n-1}\,\mathbf{B}$$

Observability test matrix :

$$\mathbf{V} = \mathbf{C}^T : \mathbf{A}^T\mathbf{C}^T : (\mathbf{A}^T)^2\,\mathbf{C}^T : ...... (\mathbf{A}^T)^{n-1}\,\mathbf{C}^T$$

References

1. *Aggarwal, K.K.,* 'Control System Analysis and Design', Khanna Publishers, Delhi, 1981.
2. *Buckley, R.V.,* 'Control Engineering : Worked Examples', Macmillan.
3. *Chen, C.F., and Hass, I.J.,* 'Elements of Control System Analysis', Prentice-Hall, 1969.
4. *Chenge, David K.,* 'Analysis of Linear Systems', Addison Wesley, 1959.
5. *Das Gupta S.,* 'Control Systems Theory', Khanna Publishers, Delhi, 1973.
6. *D' Azzo, J.J. and Houpis, C.H.,* 'Feedback Control Systems Analysis and Synthesis', McGraw-Hill, 1966.
7. *Del Toro, V.,* 'Principles of Electrical Engineering', Prentice-Hall, 1964.
8. *Distefano, J.J., Stubberud, A.R. and Williams, I.J.,* 'Theory and Problems of Feedback and Centrol Systems, Schaum's Outline Series, McGraw-Hill, 1967.
9. *Dorf, R.C., Bishop, R.H.* Modern Control Systems, 8th edition, Addison-Wesley.
10. *Fitzgerald, R.P. and C. Kingsley (Jr),* 'Electric Machinery', McGraw-Hill, 1961.
11. *Kuo, B.C.,* 'Automatic Control Systems' Prentice-Hall, 1962.
12. *Gopal, M.,* 'Modern Control System Theory', Wiley Eastern Ltd., 1984.
13. *Nagrath, I.J., and Gopal, M.,* 'Control Systems Engineering', Wiley-Eastern Ltd. 1975.
14. *Nise, Norman,* Control Systems Engineering, John, Wiley.
15. *Ogata, K.,* 'Modern Control Engineering', Prentice Hall, 1970.
16. *Subba Rao, A., Desai, P.R., Kulkarni, S.K., and Tandon, I.K.,* 'A Course in Control Engineering', Dhanpat Rai and Sons, Delhi, 1984.
17. *Thaler, G.J., and Brown, R.G.,* 'Analysis and Design of Feedback Control Systems', McGraw-Hill, 1960.
18. *Trauxal, J.G.,* 'Automatic Feedback Control Synthesis', McGraw-Hill, 1955.
19. *Van Valkenburg, M.E.,* 'Network Analysis', Prentice-Hall, 1964.

Index

A

Absolute stability, 218, 712
A.C. servomotor, 95
A.C. tachogenerator, 101
Acceleration error coeff, 156
Acceleration function, 129
Actual damping, 135, 136
Actuating signal, 173
All pass transfer function, 306
Amplidyne generator, 99
Angle of arrival from complex zero, 324
Angle of departure from complex pole, 324
Argument principle, 750
Armature controlled D.C. motor, 93
Asymptotes : Bode plot, 294
Asymptotes : Root locus, 321, 322
Asymptotic bode plot, 752

B

Back e.m.f. constant, 92
Bandwidth, 258
Bilinear transformation, 527
Block diagram, 25
— closed-loop, 30
— of control system, 17
— of transfer system, 16
— reduction, 25
Bode plot, 289, 294, 751
— initial slope, 295
— magnitude error table, 763
— procedure for drawing, 301
— solved examples, 308
— time delay, 304
— transfer functions, 289
Breakaway point, 323

C

Cascade compensation, 364
Cayley-Hamilton theorem, 442
— application, 443
Characteristic equation, 140, 741
Classification of transfer function, 154
Closed-loop control system, 2, 169, 488
Closed-loop transfer function : equivalence, 30
Close-loop frequency response, 250
Comparator, 2
Compensation network, 364–368
Compensation network root locous approach 380
Compensation, 363
— phase lag, 366
— phase lead, 364
— phase-lag-lead, 368
Conditionally stable system, 245
Constant phase angle loci, 287
Control actions, 173
— derivative, 173
— example of, 2
— proportional, 173
Control system, 1, 734
— block diagram of, 17
— controllers in, 745
— determination, 740
— determining transfer function, 19
Controllability, 451
Controllable canonical form, 426
Corner frequency, 293, 752
Correlation between transient and frequency response, 315
Critical damping, 138, 139
Critically damped system, 137
Cut-off frequency, 257

D

Damped frequency, 136
Damping factor, 136
Damping ratio, 135, 742
Decade of frequency, 291
Decaying oscillations, 151
Decibel, 766
D.C. technogenerator, 101
Derivative control, 173, 176
Derivative feed-back control, 181
Diagonalization of matrix, 410, 758
Difference equations, 512
Digital control system, 500
Displacement function, 129
Dominant pole pair, 744
Dominant poles, 151

E

Eigen value, 444, 445
Eigen vector, 446, 447
Electrical network, 80
Equivalent moment of inertia, 23
Equivalent viscous friction, 23
Error coeff. acceleration, 155
Error coeff. position, 155
Error coeff. velocity, 155
Error detector, 102
— potentiometer, 102
— synchro (selsyn), 103
Error signal, 2
— sampling, 523
Error transfer function, 31
Excitation (Cause), 16

F

Feedback compensation, 369
Feedback path transfer function, 30
Feedback, 2
Field controlled D.C. motor, 90
Final value theorem, 7
Force current analogy, 82
Force voltage analogy, 82
Forward path transfunction, 31
Forward path transmittance, 63
Frequency response, 315
— analysis, 762

G

Gain adj : 1/M circle, 288
Gain adj : M-circle, 289
Gain cross-over frequency, 243
Gain margin from Bode plot, 301
Gain margin, 242, 244, 715
Gain phase plot, 245
Gain term K,
— graphs, 290
Gear, 22
Generalized error coeffs, 159
Generators, 97
— excited d.c., 97
Generator constant, 97
Graph determinant, 66

H

Heat transfer system, 89
Higher order system, 22
Hold circuits, 515
Hurwitz determinant, 220
Hydraulic actuator, 86
Hydraulic system, 86

I

Impulse function, 129
Impulse response, 146
Impulse train, 504
Infinite series, 440
Initial value theorem, 6
Input node, 63
Input test signals, 128
Integral absolute error, 162
Integral control, 178
Integral-of-time-multiplied-absolute error, 162
Integral-of-time-multiplied-square error, 162
Inverse Laplace transform, 3
Inverse Nyquist plot, 284
Inverse *z*-transform, 508
— partial expansion, 509
— power series, 510

J

Jordan's canonical form, 433
Jury's stability test, 529

L

Lag lead compensator design 404
Laplace transform pairs, 5
Laplace transform theorems, 6
Laplace transform, 3, 441
— derivation of, 4
— impulse train, 504
— sampled function, 503
— use in control systems, 3
Linear control system, 1
Loop transmittance, 63, 64

M

1/M circles, 287
Mapping from *s*-plane to $G(s)H(s)$-plane, 229
Mapping from *s*-plane to *z*-plane, 526
Mason's gain formula, 67, 704
MATLAB, 500
Maximum overshoot, 142
Maximum phase transfer function, 279
M-Circles, 247, 718
Mechanical circuit diagram, 83
Mechanical systems, 81
— rotational, 82
— translational, 81
Minimum phase function, 306
Minimum sampling frequency, 518
Modal matrix, 445, 447
Motor gain constant, 94
Motor time constant, 94
Multiple poles, 18

N

Natural frequency of oscillations, 135
N-circles, 249, 719
Nichol's chart, 254, 732
Non-minimum phase transfer function, 306
Number sequence, 511
Nyquist criterion application, 231
Nyquist criterion, 229
Nyquist plot e^{-st}, 241
— relative stability, 244
— solved examples, 259

O

Observability, 454
Observable canonical form, 427
Octave of frequency, 291
Open-loop control system, 2, 169
Open-loop transfer function, 154
Oscillatory (overdamped) response, 135
Output node, 63
Output signal, 2
Overall transfer function, 30
Overdamped response, 138
Overshoot, 142

P

Parabolic (acceleration) function, 129
Pass transfer function, 306
Path factor, 67
Peak time, 142
Performance indices, 162
Phase crossover frequency, 272
Phase margin bode plot, 301
Phase margin, 244, 715
Phase variables, 757
Phase-lag compensation, 80, 366
Phase lead compensator design 381
Phase lag compensator design 394
Phase-lag network, 79
Phase-lead network, 364
PI control, 178
PID control, 181

Pneumatic actuating valve, 88
Pneumatic system, 88
Polar plot, 747
Poles : complex, multiple, simple, 17
— in s-plane, 18
— of transfer function, 140
Positional error coeff., 155
Proportional control, 173
Pulse transfer function, 519, 758
— from block diagrams, 519

R

Ramp (velocity) function, 129
Relative stability, 218, 244, 747
Resonant frequency, 299, 753
Resonant peak, 300
Response (effect), 16
Rise time, 141
Root contours, 351
Root locus, 319
— angle of arrival, 324
— angle of asymptote, 323
— angle of departure, 324
— asymptote intersection, 324
— breakaway point, 323
— imaginary axis intersection, 324
— plotting procedure, 322
— salient features, 323
— solved examples, 324
— starting point, 323
— terminating points, 323
Rotational mechanical system, 84
Routh-Hurwitz criterion, 221
— solved examples, 222
Rules for drawing signal flow graph, 74

S

Sampled-data control system, 500
Sampler, 501
— ideal, 502
Sampling process, 502
Sampling theorem, 518
Sensitivity, 168
Separately excited D.C. generator, 97
Servomotors, 90
— two phase a.c., 95
Settling time, 144
— first order system, 131
— second order system, 144
Signal flow graph, 62
Stability definition, 218
Stability: sampled-data, 490
Stable system, 217, 244
Starred function, 503
Starred transform, 504
State equation: solution, 438
— of non-homogeneous, 439
State signal flow graph, 391
State space representation, 420, 421, 422
State transition matrix, 438
— determination of, 441
— Laplace transform method, 441
— properties, 450
State variables, 429
— feedback, 490
State, 420
— concept, 419
— space, 420
— vector, 420
Static error coeffs., 155
Steady state error, 153, 158
Steady state response, 128
Step (displacement) function, 128
Summing point, 27
Synchro error detector, 17

T

Take-off point, 26
Techogenerator constant, 101
Thermal system, 89
Time delay element
— Bode plot, 304
— Nyquist plot, 240
Time response :
— first order system (ramp input), 132
— first order system (step input), 130
— higher order system, 152
— second order system (ramp input), 144
— second order system (step input), 134
— third order system, 147
Transfer function decomposition, 428
— cascade, 431
— direct, 429
— parallel, 432
Transfer function, 16
— block diagram of, 16
— poles and zeros of, 17
— impulse response, 19
Transfer matrix, 450
Transient response, 315
— specifications, 128
Translational spring-mass damper system, 81
Transmittance, 63, 64
Two phase A.C. servomotor, 95

U

Undamped (sustained oscillation) response, 136
Unit parabolic function, 129
Unity feedback, 31
Unstable system, 217

V

Vander Monde's matrix, 449, 758
Velocity error coeff., 155
Velocity function, 129

Z

Zero input response, 439
Zero order hold, 515
— transfer function of, 515
Zeros of transfer function, 17
Zero state response, 439
z-transform, 508
— inverse, 508
— of useful functions, 505
— properties, 504
z-transfer function, 519